Hoppe-Seyler's Zeitschrift für Physiologische Chemie

Band 359 – 1. Jahreshälfte

1978

Hoppe-Seyler's Zeitschrift für Physiologische Chemie

Band 359 – 1. Jahreshälfte

Fortgeführt von A. Kossel, F. Knoop und K. Thomas · Herausgegeben von

A. Butenandt, F. Lynen, G. Weitzel

unter Mitwirkung von K. Bernhard, K. Decker, J. Engel, H. Fritz, B. Hamprecht, H. Hanson, E. Helmreich, H. Herken, B. Hess, N. Hilschmann, H. Hilz, P. W. Jungblut, P. Karlson, H. L. Kornberg, K. Kühn, F. Leuthardt, D. Oesterhelt, K. Rajewsky, J. Seelig, G. Siebert, H. Simon, Hj. Staudinger, W. Stoffel, H. Tuppy, H. Wiegandt, H. G. Wittmann, H. G. Zachau, H. Zahn

Redaktion A. Dillmann, G. Peters

W DE G

1978

WALTER DE GRUYTER · BERLIN · NEW YORK

Anzeigenverwaltung: Merkur-Werbung, Postfach 12 45, D-5210 Troisdorf 1, Tel. (0 22 41) 4 20 51. Für den Anzeigenteil verantwortlich: Wolfgang Frieben, Troisdorf. – Printed in Germany – Satz: R. & J. Blank, München; Druck: Grafik + Druck, München; Bindearbeiten: F. Steinberger & Sohn, München.

Autoren-Verzeichnis der ersten Jahreshälfte 1978

d 359 Heft 1 Januar 1978

N 0018-4888 A 20679 E

oppe-Seyler's Zeitschrift für hysiologische Chemie

Herausgegeben von

A. Butenandt, F. Lynen, G. Weitzel

WALTER DE GRUYTER · BERLIN · NEW YORK

Hoppe-Seyler's Z. Physiol. Chem. Bd. 359 Heft 1 S. 1 - 136 Januar 1978
HSZPAZ 359 (1) 1 - 136 (1978)

W DE G

Walter de Gruyter
Berlin · New York

W. Bandlow
R. J. Schweyen
D. Y. Thomas
K. Wolf
F. Kaudewitz
(Editors)

Genetics, Biogenesis and Bioenergetics of Mitochondria

Proceedings of a Symposium held at the Genetisches Institut der Universität München,
Munich, September 11–13, 1975

1976. Large-octavo. XV, 418 pages. With many illustrations.
Bound DM 140,–; $63.70 ISBN 3 11 006865 6

W. Bandlow
R. J. Schweyen
K. Wolf
F. Kaudewitz
(Editors)

Mitochondria 1977

Genetics and Biogenesis of Mitochondria.

Proceedings of a colloquium, held at Schliersee, Germany, August 1977.

1977. Large-octavo. XX, 669 pages. Bound DM 190,–; $86.40
ISBN 3 11 007321 8

A comprehensive view of the present state of research on genetics and biogenesis of mitochondria.

This book is an essential source of information for all workers in the field of mitochondrial genetics, genetics of chloroplasts and other extrakaryotic genetic systems and of general molecular biology.

Prices are subject to change without notice

Hoppe-Seyler's Zeitschrift für Physiologische Chemie
Begründet 1877 · Fortgeführt von A. Kossel, F. Knoop und K. Thomas

Bd. 359 Januar 1978 Heft 1

Inhaltsverzeichnis / Contents

Fortsetzung nächste Seite *Continued*

Redaktionelle Hinweise siehe gegenüber Seite 136

Indexed in Current Contents

Autoren-Verzeichnis

Titel Nr. 3 109 002 688

 – Printed in Germany – Satz: R. & J. Blank, München; Druck: Grafik + Druck München; Bindearbeiten: F. Steinberger & Sohn, München.

Hoppe-Seyler's Z. Physiol. Chem.
Bd. 359, S. 1–11, Januar 1978

Trihexosyldiacylglycerol and Acyltrihexosyldiacylglycerol as Lipid Anchors of the Lipoteichoic Acid of *Lactobacillus casei* DSM 20021

Masuo NAKANO and Werner FISCHER

Institut für Physiologische Chemie der Universität Erlangen–Nürnberg

(Received 16 November 1977)

Summary: The lipoteichoic acid of *Lactobacillus casei* DSM 20021 was isolated. It consists of glycolipid and a poly(glycerophosphate), 37 - 44 units in length. The lipid portions can be released by treatment with hydrofluoric acid or acetic acid. They were determined to be 1,2-di-*O*-acyl-3-*O*-[*O*-β-D-glucopyranosyl-(1→6)-*O*-α-D-galactopyranosyl-(1→2)-α-D-glucopyranosyl]glycerol and 1,2-di-*O*-acyl-3-*O*-[*O*-β-D-glucopyranosyl-(1→6)-*O*-α-D-galactopyranosyl-(1→2)-(6-*O*-acyl-α-D-glucopyranosyl)]glycerol.
Lipoteichoic acid with a third carbohydrate-linked fatty acid in the glycolipid portion represents a novel subclass.

Trihexosyldiacylglycerin und Acyltrihexosyldiacylglycerin als Lipid-Anker der Lipoteichonsäure von Lactobacillus casei DSM 20021

Die Lipoteichonsäure von *Lactobacillus casei* DSM 20021 wurde isoliert. Sie besteht aus Glykolipid und einem Poly(glycerophosphat) mit einer Kettenlänge von 37 - 44 Einheiten. Die Lipidanteile sind freisetzbar durch Fluorwasserstoffsäure oder Essigsäure. Sie wurden identifiziert als 1,2-Di-*O*-acyl-3-*O*-[*O*-β-D-glucopyranosyl-(1→6)-*O*-α-D-galaktopyranosyl-(1→2)-α-D-glucopyranosyl]glycerin und 1,2-Di-*O*-acyl-3-*O*-[*O*-β-D-glucopyranosyl-(1→6)-*O*-α-D-galaktopyranosyl-(1→2)-(6-*O*-acyl-α-D-glucopyranosyl)]glycerin.
Lipoteichonsäure mit einer dritten an den Kohlenhydratanteil des Glykolipids gebundenen Fettsäure ist ein neuer Lipoteichonsäuretyp.

Key words: Lipoteichoic acid, trihexosyldiacylglycerol, acyltrihexosyldiacylglycerol.

Lipoteichoic or membrane teichoic acids are widespread among Gram-positive bacteria[1,2]. They are thought to be uniformly made up of a poly(glycerophosphate) backbone which is anchored at the outer layer of the cytoplasmic membrane by a phosphodiester bond to a glycolipid. The

Enzymes:
Alkaline phosphatase, orthophosphoric-monoester phosphohydrolase (alkaline optimum) (EC 3.1.3.1);
Deoxyribonuclease I, deoxyribonucleate 5′-oligonucleotidohydrolase (EC 3.1.4.5);
α-Galactosidase, α-D-galactoside galactohydrolase (EC 3.2.1.22);
α-Glucosidase, α-D-glucoside glucohydrolase (EC 3.2.1.20);
Ribonuclease I, ribonucleate 3′-pyrimidino-oligonucleotidohydrolase (EC 3.1.4.22).

glycolipid portions, so far identified, were identical with the respective membrane dihexosyldiacylglycerols[1-4] or a phosphatidyl derivative thereof[5,6]. Recently we found in *Lactobacillus casei sn*-glycero-1-phospho-glycolipids which were exclusively derived from the membrane trihexosyldiacylglycerol and its acylated derivative[6a]. Since according to our working hypothesis[7], *sn*-glycero-1-phospho-glycolipids are enzymic breakdown products or biosynthetic intermediates of lipoteichoic acids, the lipid anchors of *L. casei* lipoteichoic acid should also consist of acyltrihexosyldiacylglycerol and trihexosyldiacylglycerol. The present communication substantiates this prediction and shows for the first time that lipoteichoic acids can be anchored in the membrane by glycoplipids containing three fatty acids.

Methods

Materials.
Enzymes and cosubstrates were purchased from Boehringer Mannheim GmbH. Reference glycolipids isolated from *L. casei* DSM 20021 have been described recently[8]. Partially methylated alditol acetates were from previous work[8].

Analytical procedures.
Protein[9], carbohydrate[10], D-glucose[11], D-galactose[12], hexosamine[13], formaldehyde[14], and phosphorus[15] were measured by established procedures. For enzymic determination of glycerol the method of Hohorst[16] was adapted. Rhamnose was quantified by gas-liquid chromatography using the accompanying glucose and galactose as internal standards (see acid hydrolysis). Acyl groups of glycolipids were measured by a hydroxamate method[17], those of lipoteichoic acid as described under alkaline hydrolysis. Fatty acid methyl esters and alditol acetates were analyzed by gas-liquid chromatography as in previous work[8].

Thin-layer chromatography.
The following solvent systems were used: A, chloroform/methanol/water 65 : 25 : 4 (v/v); B, petroleum ether/diethyl ether/acetic acid 50 : 50 : 4 (v/v); C, 1-propanol/pyridine/water 7 : 4 : 2 (v/v); D, 1-propanol/25% ammonia/water 6 : 4 : 3 (v/v); E, phenol/water/ethanol/acetic acid 75 : 33 : 4.5 : 4.5 (w/v/v/v). Solvents A - C were used on precoated silica gel plates (Merck 60), D and E on cellulose plates (Merck). The reagents used for the detection of lipids, glycolipids, glycosides, 1,2-glycol groups, phosphates and phospholipids were the same as in a previous work[18]. Preparative thin-layer chromatography and elution of lipids from silica gel was carried out as before[19].

Growth of bacteria.
Lactobacillus casei ssp. rhamnosus DSM 20021 was grown and harvested as described[8]. One batch was harvested in the middle of the growth phase (A_{578} = 2.67; 5.6 g wet cells/*l* medium), and another at the end of the growth phase (A_{578} = 4.13; 11.6 g wet cells/*l* medium).

Preparation of crude lipoteichoic acid
*Preparation 1.** Immediately after harvesting, the wet cells (114 g) of the first batch were suspended in chloroform/methanol 1 : 2 (v/v; 300 m*l*) and stirred for 30 min at room temperature. After centrifugation the supernatant was withdrawn and the residue was extracted twice more with stirring for 30 min in chloroform/methanol/water 1 : 2 : 0.8 (v/v; 380 m*l*) and in chloroform/methanol 1 : 2 (v/v; 150 m*l*), respectively. The defatted material was dried in vacuo (27.6 g).

Chloroform (250 m*l*) and water (290 m*l*) were added to the combined lipid extracts for phase separation. The organic phase was evaporated to dryness under reduced pressure. The extracted lipids (850 mg) were the same as recently described[8]. The aqueous layer was concentrated by evaporation (200 m*l*), dialyzed against water and freeze-dried (124 mg, 157 μmol phosphorus). It contained no lipoteichoic acid as evidenced by chromatography on Sepharose 6B (Fig. 1a).

Defatted cells were suspended in water (400 m*l*) in a Potter Elvejhem homogenizer, and extracted with 80% phenol (400 m*l*) at 68 °C as described by Osborn[20]. The combined water layers were dialyzed against running water for 48 h and then concentrated (100 m*l*) under reduced pressure. The same volume of 0.1 M Tris/HCl buffer, pH 7.7, containing 0.02M $MgCl_2$, 0.02% NaN_3, and 0.005% chloramphenicol was added. The mixture was treated with ribonuclease (2 mg) and deoxyribonuclease (3 mg) at 37 °C for 72 h. Then it was concentrated by evaporation (100 m*l*) and stirred with 80% phenol (100 m*l*) at room temperature for 40 min. After phase separation by centrifugation, the aqueous layer was dialyzed against running water for 48 h and freeze-dried (605 mg, 1.78 mmol phosphorus). No phosphorus was extractable from defatted cells by stirring them in water at room temperature for 24 h or at 80 °C for 1 h.

Preparation 2. The lipoteichoic acid of the cells harvested at the end of the growth phase was prepared according to the method of Coley et al.[21], which starts with membrane fragments of mechanically disrupted

* The procedure described with preparation 1 was worked out by *H. U. Koch* in this laboratory.

cells. 924 mg crude lipoteichoic acid (1.56 mmol phosphorus) was obtained from 232 g wet cells. The contamination with polysaccharide (see Results) was larger than in the first preparation (75 and 28% of the total carbohydrate, respectively).

Chromatography on Sepharose 6B

All separations were performed on columns (45 cm × 2.5 cm) of Sepharose 6B that had been equilibrated with 0.2M ammonium acetate, pH 6.9, containing 0.01% NaN_3[21]. Lipoteichoic acid was dissolved in the same buffer, insoluble material was removed by centrifugation at 16 000 × *g*. The load was 500 - 600 μmol phosphorus (3 m*l*) in crude fractionations,

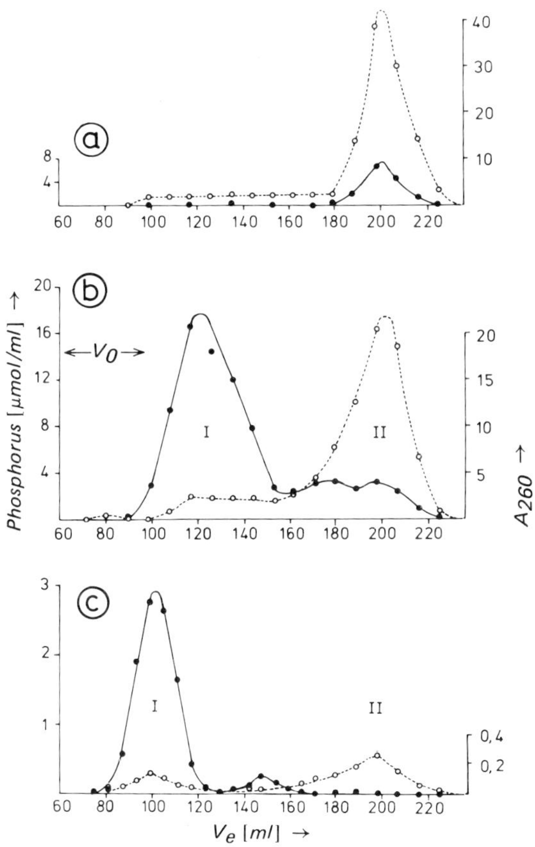

Fig. 1. Chromatography on Sepharose 6B.

For details see Methods. a, Water-phase of the lipid extract; b, crude lipoteichoic acid preparation; c, rechromatography of the material corresponding to peak I in b. The exclusion volume (V_0) was similar in all three experiments. •——•, Phosphorus (μmol/m*l* effluent); o- - -o, A_{260} of undiluted samples.

25 - 50 μmol (2 m*l*) for further purification. The column was eluted with the same buffer by upward flow at a rate of 10 m*l*/h. Fractions (3 m*l*) were collected and every third one was analyzed for phosphorus and nucleic acids (A_{260}). Appropriate fractions were pooled, dialyzed and then freeze-dried. The void volume of the columns was determined by the elution volume of Blue Dextran.

Purification of lipoteichoic acid on DE 32-cellulose

Final purification was achieved by chromatography on DE 32-cellulose following the method of Fiedler and Glaser[22]. Polymer (200 μmol phosphorus) obtained on chromatography on Sepharose 6B was dissolved in 2 m*l* 0.01M Tris/HCl, pH 8.0, containing 0.1% Triton X-100, and was equilibrated by dialysis against this buffer overnight. Then it was applied to a column (44 cm × 0.9 cm) of DE 32-cellulose that was preequilibrated in and eluted with the same buffer by downward flow at a rate of 9 m*l*/h. Elution was continued with a linear gradient of NaCl (0 - 0.25 M in 24 h) in the aforementioned buffer. The gradient was established by a LKB Ultrograd. Fractions (4 m*l*) were collected and assayed by determination of carbohydrate and phosphorus. Pooled fractions were concentrated under reduced pressure and dialyzed with constant stirring at 4 °C, during 72 h against twelve changes of water. Thin-layer chromatography in solvent A revealed only a little residual Triton X-100.

Acid hydrolysis.

Lipoteichoic acid (10 μmol phosphorus/m*l*) was hydrolyzed in 2M HCl at 100 °C for 2.5 h in sealed tubes. Fatty acids were extracted with petroleum ether. Samples for enzymic determinations of monosaccharides were neutralized with 2M NaOH. A sample was passed through a small column of anion exchange resin (Merck M 5080), acetate form, the effluent was evaporated to dryness under reduced pressure, and the monosaccharides were converted into polyol acetates[8] for analysis by gas-liquid chromatography. Glycerol/phosphorus ratios were determined after continued hydrolysis at 125 °C for 48 h[18]. Polysaccharides, glycolipids and glycosides were hydrolyzed in 2M HCl at 100 °C for 2.5 h; for hexosamine determination samples were hydrolyzed overnight.

Alkaline hydrolysis.

Samples of polymer (10 - 15 μmol phosphorus/m*l*) were hydrolyzed in 1M NaOH at 100 °C for 5 h in screw-capped teflon vessels. $Na^{\oplus}$ was removed by passage through small columns of cation exchange resin (Merck S 1080), $H^{\oplus}$ form. After extraction of fatty acids with petroleum ether, the hydrolysate was neutralized with ammonia. For quantitation of fatty acids, methyl heptadecanoate (30 nmol/μmol phosphorus) was added, prior to hydrolysis, as internal standard; the extracted fatty acids were methylated[23] and examined by gas-liquid chromatography[8]. Water-soluble products were analyzed after separation on Sephadex G-25 or De 32-cellulose as described under Results.

Degradation with 48% HF.

Lipoteichoic acid (100 μmol phosphorus) was dissolved in 48% (w/w) HF (1 m*l*), previously cooled at – 20 °C, in polyethylene beakers fitted with caps. The mixture was kept at 2 °C for 36 h. In earlier experiments HF was neutralized with vigorous stirring at 0 °C; 95% of HF was neutralized with saturated LiOH[5], then the pH was adjusted to 7 by addition of 0.15 M $LiHCO_3$. The precipitate was washed successively with water and chloroform/methanol 2 : 1 (v/v); the washings were combined and subjected to Folch partition. Later on the degradation mixtures were frozen at – 70 °C and taken to dryness in vacuo over KOH at 0 °C overnight[24]. Lipid and water-soluble products were separated in chloroform/-methanol/water 1 : 1 : 0.9 (v/v). The resulting water phase was analyzed for inorganic phosphate and glycerol and then extracted with water-saturated 1-butanol to avoid loss of polar lipids such as lyso-glycolipids. The organic extracts were combined for analysis.

Hydrolysis with 98% acetic acid.

After suspending lipoteichoic acid (15 μmol phosphorus) in 98% (v/v) acetic acid (1 m*l*) by sonication for 5 min in a screw-capped tube with a Teflon-lined cap, it was hydrolyzed at 100 °C for 1 h. Acetic acid was removed by repeated evaporation with addition of CCl_4/methanol 1 : 1 (v/v). Lipid and water-soluble products were separated by addition of chloroform/methanol/water 1:1:0.9 (v/v). The resulting water phase was extracted with butanol as above.

Enzymic dephosphorylation.

Glycerophosphates (1 μmol) were dissolved in 0.1 M glycine buffer, pH 9.5 (0.3 m*l*) and digested with alkaline phosphatase (30 U/m*l* reaction mixture) at 37 °C overnight. Enzyme was precipitated with 4% trichloroacetic acid (1.5 m*l*) and removed by centrifugation. Samples were analyzed for total phosphorus and inorganic phosphate; glycerol was measured after neutralization with 0.3 M NaOH.

CrO_3 oxidation.

CrO_3 oxidation followed by successive ester cleavage and treatment with glycosidases was performed as described[8].

Methylation procedures.

Methylation and analysis of the partially methylated alditol acetates by gas-liquid chromatography and gas chromatography-mass spectrometry were the same as in previous work[8].

Results

Final purification and composition of lipoteichoic acid

Preparations 1 and 2 of crude lipoteichoic acid were from cells harvested in the middle and at the end of the growth phase, respectively, and had been prepared in different ways (see Methods). Chromatographed on Sepharose 6B, the preparations yielded similar patterns (Fig. 1b). Lipoteichoic acid (peak I) eluted near the void volume of the column and was separated from most of the breakdown products of nucleic acids (peak II). After rechromatography of the material corresponding to peak I (Fig. 1c), the resulting lipoteichoic acid contained only negligible amounts of of nucleic acids and less than 1% protein. The composition of the purified preparations is given in Table 1. Both preparations released fatty acids on alkaline hydrolysis and contained glycerol and phosphorus in equimolar ratios. They differed significantly, however, in the content of carbohydrate and contained rhamnose, suggesting a contamination with variable amounts of polysaccharide[25,26].

To probe that, samples of both preparations were subjected to alkaline hydrolysis (1 M NaOH, 100 °C, 5 h), whereby the non-glycosylated poly-(glycerophosphate) chain of *L. casei* lipoteichoic acid is hydrolyzed[27]. The neutralized hydrolysates (see Methods) were passed through columns (38 cm × 1.5 cm) of Sephadex G-25 fine which were eluted with water. With both samples, part of the carbohydrate emerged at the void volume (30 - 42 m*l*), the rest near the inner volume (48 - 68 m*l*) and most of the phosphorus eluted at the inner volume (55 - 72 m*l*). Fractions containing material of high and low molecular weight (30 - 42 m*l*, 45 - 72 m*l*) were combined and analyzed (Table 1). The fractions of high molecular weight were composed of D-glucose, D-galactose and rhamnose in similar proportions as the type-specific polysaccharide of *L. casei ssp. rhamnosus**[25]. The low-molecular-weight fractions from both preparations, which contain the water-soluble

* The anionic character of the polysaccharide (Fig. 2) and the alkali lability of the hexosamine residues (Table 1) are other properties of the type-specific polysaccharide[26].

Table 1. Composition of preparation 1 and 2 (see Methods) of lipoteichoic acid of *L. casei.*

Products obtained on alkaline hydrolysis were fractionated on Sephadex G-25 (see text). For separation on DE 32-cellulose see Fig. 2. Values are given as molar ratios related to phosphorus. Values of polysaccharide fractions are related to the total phosphorus eluted on the respective separation.

n.d., not determined; –, not detectable.

	Phosphorus	D-Glucose	D-Galactose	Rhamnose	Hexosamine	Glycerol	Fatty acid ester
Preparation 1							
Purified on Sepharose 6B	1.00	0.083	0.038	0.029	0.015	1.02	0.063
Products of alkaline hydrolysis							
i) Polysaccharide fraction*	–	0.010	0.005	0.011	0.002	–	
ii) Low-molecular-weight fraction	1.00	0.053	0.025	–	0.001	0.99	
Separation on DE 32-cellulose							
i) Polysaccharide	–	0.023	0.010	0.026	0.004	–	
ii) Lipoteichoic acid	1.00	0.055	0.028	–	0.002	1.00	
Preparation 2							
Purified on Sepharose 6B	1.00	0.194	0.095	0.233	0.025	1.04	0.055
Products of alkaline hydrolysis							
i) Polysaccharide fraction*	–	0.108	0.050	0.149	0.005	–	
ii) Low-molecular-weight fraction	1.00	0.047	0.021	–	n.d.	1.00	

* The recovery of polysaccharide was less than calculated, probably owing to partial alkaline degradation from the reducing end.

breakdown products of lipoteichoic acid, had a quite similar composition, in contrast to the parent material (Table 1).

Separation of polysaccharide and native lipoteichoic acid was achieved by chromatography on DE 32-cellulose (Fig. 2). The composition of the pooled peak fractions is given in Table 1. The lipoteichoic acid was free of polysaccharide as judged by the absence of rhamnose and the content of glucose and galactose, which was virtually identical with that of the breakdown products obtained on alkaline degradation. On periodate oxidation the purified lipoteichoic acid released 0.03 mol formaldehyde per mol phosphorus.

The fatty acid composition, given in Table 2, showed no significant differences before and after chromatography on DE 32-cellulose and was similar to that of preparation 2. The same fatty acids, in similar proportions, were found in the membrane glycolipids of *L. casei* DSM20021[8].

Table 2. Fatty acid composition of preparation 1 and 2 of lipoteichoic acid of *L. casei.*

(a), after chromatography on Sepharose 6B; (b), after further purification on DE 32-cellulose (see text). Values are given as percent. 19cy, C_{19} cyclopropanecarboxylic acid (it co-chromatographed in gas-liquid chromatography with authentic methyl 11,12-methyleneoctadecanoate).

Preparation	14:0	16:0	16:1	18:0	18:1	19cy	20:1
1 (a)	1.7	28.2	3.3	6.4	52.4	5.4	2.6
1 (b)	1.8	27.4	5.7	6.6	49.7	6.1	2.6
2 (a)	1.7	25.4	6.0	5.7	49.7	7.1	3.7

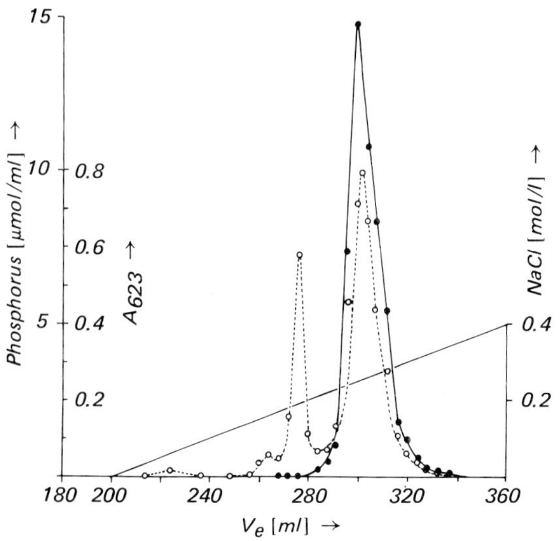

Fig. 2. Separation of polysaccharide and lipoteichoic acid on DE 32-cellulose.

For details see Methods. The column was eluted with 0.01 M Tris/HCl buffer, pH 8.0, containing 0.1% Triton X-100 (200 m*l*), and then with a linear gradient of NaCl in the same buffer.

●——●, Phosphorus (μmol/m*l* effluent); ○---○, A_{623}, carbohydrate (0.2 m*l* effluent, 3 m*l* anthrone reagent); ——, NaCl gradient.

Products of alkaline hydrolysis

Lipoteichoic acid purified on DE 32-cellulose was subjected to alkaline hydrolysis (1 M NaOH, 100°C, 5 h). The neutralized hydrolysate was applied to a column (38 cm × 1.5 cm) of DE 32-cellulose, $HCO_3^{\ominus}$ form, which was eluted with water and a stepwise concentration gradient of triethylamine carbonate buffer, pH 8.5. Free glycerol (18% of total glycerol) and the total carbohydrate eluted with water. On thin-layer chromatography in solvent C, a single glycoside was detected which had the same mobility as Glc(β1-6)Gal(α1-2)Glc(α1-3)-glycerol and stained rapidly with the periodate Schiff reagent. 69% and 31% of the phosphorus eluted with 0.06 M and 0.25 M buffer. Enzymic dephosphorylation gave glycerol and phosphorus in the ratios of 1 : 1 and 0.44 : 1, respectively. Accordingly, a mixture of α- and β-glycerophosphate and glycerobisphosphate were detectable on thin-layer chromatography in solvent D and E. Glycerobisphosphate was accompanied by a compound of lower mobility, possibly di(phosphoglycerol)phosphate, which was earlier characterized among the alkali hydrolysis products of intracellular teichoic acid of *L. casei*[27].

The non-ionic products were passed through a small column of charcoal/celite 1 : 1 (w/w) from which glycerol and the glycoside were eluted with water and 50% (v/v) ethanol, respectively. The glycoside showed on analysis D-glucose, D-galactose, glycerol and 1,2-glycol groups in the proportions 2.0 : 0.9 : 1.0 : 1.2.

Degradation of lipoteichoic acid with 48% HF and 98% acetic acid

In order to release the glycolipid portion, lipoteichoic acid was treated with 48% (w/w) HF (2 °C, 36 h). Thereby organic phosphorus is released as inorganic phosphate, whereas glycosidic bonds are not attacked and fatty acid esters are fairly stable[5,18,28]. After removal of HF by precipitation as LiF or by evaporation in vacuo at 0 °C, lipid and water-soluble breakdown products were separated (see Methods). In the water phase 97% of the initial phosphorus was recovered; total phosphorus, inorganic phosphate and free glycerol were found in molar proportions of 1.00 : 0.91 : 0.95. On thin-layer chromatography in solvent A, the organic phase showed two glycolipids with the mobilities of Glc(β1-6)Gal(α1-2)-acyl→6Glc(α1-3)diacylglycerol and Glc(β1-6)Gal-(α1-2)Glc(α1-3)diacylglycerol, respectively (Fig. 3). Lyso-glycolipids were detectable in small amounts, their proportion being higher when HF had been neutralized. The organic phase contained neither di- nor monoacylglycerol, as evidenced by thin-layer chromatography in solvent B. The released glycolipids were separated by preparative thin-layer chromatography (solvent A) and were quantitated (Table 3). Calculated from the mean carbohydrate content (Table 1) and the ratio of glycolipid : phosphorus (Table 3), the recovery of glycolipid was 78% and 84%. In preparations 1 and 2, the molar ratios of acyltrihexosyldiacylglycerol and trihexosyldiacylglycerol were 0.27 : 1.00 and 0.49 : 1.00, respectively.

Lipoteichoic acid was heated in 98% (v/v) acetic acid at 100 °C for 1 h. On this treatment, phosphodiester linkages with a neighbouring free hydroxyl group are hydrolyzed, while glycosidic and fatty acid ester bonds remain intact[18,29,30], but some acetylation of glycolipids is unavoidable[30]. After evaporation, lipid and water-soluble products were separated. The water phase contained 96% of the initial phosphorus, 92 ± 2% of which was susceptible to phosphomonoesterase, indicating an extensive cleavage of the poly(glycerophosphate) chain. Besides glycerol, glycerophosphates and glycerobisphosphate had been formed in the molar ratio of approx. 2 : 1. The organic phase contained the same two glycolipids as found after HF degradation (Fig. 3), accompanied by faster moving partially acetylated derivatives (cf. ref.[30]). Neither phosphatidylglycolipids (cf. ref.[7]) nor lyso-phosphatidic acid were detected. Deacylation gave a single glycoside of the expected chromatographic mobility.

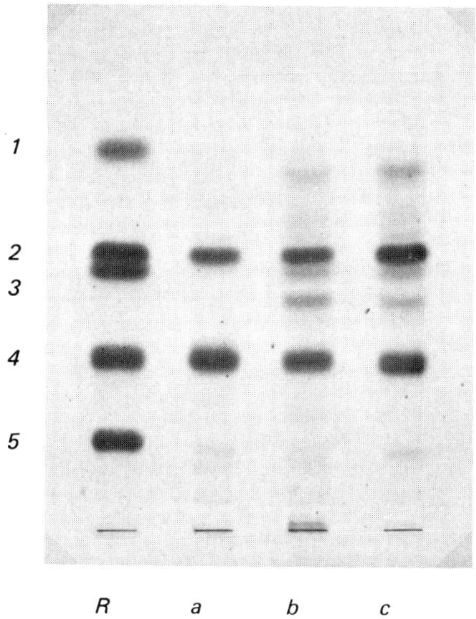

Fig. 3. Chromatographic identification of the lipid portions of lipoteichoic acid.
R, reference glycolipids of *L. casei*[8]: 1, Glc(α1-3)diacylglycerol; 2, Glc(β1-6)Gal(α1-2)acyl→6Glc(α1-3)diacylglycerol; 3, Gal(α1-2)Glc(α1-3)diacylglycerol; 4, Glc-(β1-6)Gal(α1-2)Glc(α1-3)diacylglycerol; 5, Glc(β1-6)-Glc(β1-6)Gal(α1-2)Glc(α1-3)diacylglycerol; a, lipid products after degradation of lipoteichoic acid with HF; b, lipid products after degradation of lipoteichoic acid with acetic acid; c, reference glycolipid 2 and 4 treated with acetic acid under the same conditions as sample b. Thin-layer chromatography on precoated silica gel plate in solvent A, visualization with α-naphthol/H_2SO_4.

The structure of the glycolipid portion

The purified glycolipids obtained from preparation 1 and 2 by HF degradation were combined for structural studies which were performed as recently described for the membrane glycolipids of *L. casei*[8]. The two glycolipids gave on analysis D-glucose, D-galactose, glycerol and acyl groups in the molar ratio of 1.9 : 0.9 : 1.0 : 2.9 and 1.8 : 0.9 : 1.0 : 2.0, respectively.

Both glycolipids were acetylated and subjected to CrO_3 oxidation followed by ester cleavage[8]. On thin-layer chromatography in solvent C, in both cases the same glycoside was detected. It had the mobility of authentic Gal(α1-2)Glc(α1-3)glycerol and was hydrolyzed by α-galactosidase and α-glucosidase in this order. Thus, in both lipids a β-D-glucopyranosyl residue is joined to an αGal→α-Glc→glycerol unit.

Trihexosyldiacylglycerol was permethylated according to the method of Hakomori[31] in the previous version[8]. Acyltrihexosyldiacylglycerol was subjected to the recently described two-step methylation procedure in which originally esterified and free hydroxyl groups are methylated with C^2H_3I and CH_3I, respectively[8]. The results

Table 3. Glycolipids released on degradation with HF.
The glycolipids were isolated by preparative thin-layer chromatography and quantitated by an anthrone method, using glucose : galactose 2 : 1 as standard mixture. Proportions refer to total carbohydrate recovered, total glycolipid is related to the initial phosphorus of the lipoteichoic acid degraded.

Lipoteichoic acid	Proportions			Total glycolipid / Phosphorus
	Acyltrihexosyl-diacylglycerol	Trihexosyl-diacylglycerol	Lysoglycolipid	
Preparation 1	0.19	0.72	0.08	0.021
Preparation 2	0.29	0.59	0.11	0.019

Table 4. Gas chromatography-mass spectral analysis of the partially methylated alditol acetates derived from permethylated trihexosyldiacylglycerol and permethylated acyltrihexosyldiacylglycerol.

t_R, Retention time relative to authentic 1,5-di-*O*-acetyl-2,3,4,6-tetra-*O*-methylglucitol; the t_R values of authentic 1,2,5-tri-*O*-acetyl-3,4,6-tri-*O*-methylglucitol and 1,5,6-tri-*O*-acetyl-2,3,4-tri-*O*-methylgalactitol were 1.88 and 3.07, respectively; x, signal at the mass number given; ↑, signal shifted upward by three mass units.

Parent compound and interpretation of the substitution of the monosaccharide units	t_R	Peak area ratios	Mass number of primary and some secondary fragments								
			45	101	117	129	145	161	189	205	233
Trihexosyldiacylglycerol											
1,5-di-*O*-acetyl-2,3,4,6-tetra-*O*-methylglucitol	1.00	1.00	x	x	x	x	x	x		x	
1,2,5-tri-*O*-acetyl-3,4,6-tri-*O*-methylglucitol	1.87	1.05	x	x		x	x	x	x	x	x
1,5,6-tri-*O*-acetyl-2,3,4-tri-*O*-methylgalactitol	3.06	0.91		x	x	x		x	x		x
*Acyltrihexosyldiacylglycerol**											
1,5-di-*O*-acetyl-2,3,4,6-tetra-*O*-methylglucitol	1.00	1.00	x	x	x	x	x	x		x	
1,2,5-tri-*O*-acetyl-3,4,6-tri-*O*-methylglucitol	1.88	0.99	↑	↑		x	↑	↑	x	↑	x
1,5,6-tri-*O*-acetyl-2,3,4-tri-*O*-methylgalactitiol	3.07	0.93		x	x	x		x	x		x

* The originally esterified and free hydroxyl groups of acyltrihexosyldiacylglycerol had been methylated with C^2H_3I and CH_3I, respectively.

of gas chromatography-mass spectral analysis of the resulting partially methylated alditol acetates are summarized in Table 4. These data, together with the findings of the sequential degradation above, prove the basic glycoside of both glycolipids to be Glc(β1-6)Gal(α1-2)Glc(α1-3)glycerol. The shifted fragments of partially methylated alditol acetate C derived from acyltrihexosyldiacylglycerol (Table 4) locate the third fatty acid at position 6 of the inner glycerol-linked glucosyl moiety (cf. ref.[8]). A 1,2-di-*O*-acylglycerol portion can be assumed for both glycolipids, since they co-chromatographed with the respective membrane glycolipids of *L. casei* (Fig. 3) and gave no reaction for 1,2-glycol groups, in contrast to their deacylation product.

Discussion

The hydrophilic portion of the intracellular or lipoteichoic acid of *Lactobacillus casei* ATCC 7469* has earlier been shown to be a non-glycosylated 1,3-poly(glycerophosphate) bearing alanyl residues[27]. Accordingly, the alkaline hydrolysis products and the water-soluble components obtained on degradation with HF or acetic acid in the present study indicate a non-glycosylated poly-(glycerophosphate), the alanyl residues of which had apparently been lost on purification[21]. An estimation of chain length became possible after separation of native lipoteichoic acid or its alkaline hydrolysis products from contaminating type-specific polysaccharide. Analysis of the neutral fraction of the alkaline hydrolysate showed that all the glucose and galactose of lipoteichoic acid was present in the form of a glycoside which had the composition and chromatographic properties of Glc(β1-6)Gal(α1-2)Glc(α1-3)glycerol. On the assumption that each poly(glycerophosphate) is linked to this trihexosylglycerol, from the values of glucose and galactose given in Table 1, a mean chain-length of 37 and 44 glycerophosphate units is calculated for preparation 1 und 2*. Preparation 1 released 0.03 mol formaldehyde per mol phosphorus on periodate oxidation. Therefore, as

* *L. casei* DSM 20021 is a derivative of this strain.

expected from biosynthetic considerations, each chain seems to have one glycerol terminus.

For the calculated chain-lengths, the ratios of fatty acids to phosphorus, which were 0.063 and 0.055 (Table 1), indicate on the average 2.3 and 2.4 acyl groups per chain in preparation 1 and 2*, respectively. In agreement with these values, acyltrihexosyldiacylglycerol was detected in addition to trihexosyldiacylglycerol after degradation of the lipoteichoic acid with HF. From the proportions of these glycolipids (Table 3), mean values of 2.2 and 2.3 acyl groups per chain are computed. These are minimum values, since partial deacylation by HF lowers the ratio acyltrihexosyldiacylglycerol: trihexosyldiacylglycerol.

Phosphatidylglycolipids as lipid anchors can be precluded; they would have survived degradation with acetic acid[30] and would have given diacylglycerol on degradation with HF[5,30,32]. Also another possibility, suggested by some authors [2,33,34], can be cancelled, namely a third fatty acid joined to position 2 of the innermost unit of the poly(glycerophosphate) chain. In this case acetic acid would have released 2-*O*-acyl-glycerophosphoglycoplipid and/or lysophosphatidic acid after subsequent acyl migration, whereas monoacylglycerol would have been liberated by HF.

The absence of native and deacylated dihexosyldiacylglycerol from the breakdown products of lipoteichoic acid is noteworthy, since dihexosyldiacylglycerol is one of the major membrane glycolipids of *L. casei*[8]. Its absence indicates, therefore, that the lipoteichoic acid preparations were not contaminated with membrane glycolipids**. Moreover, that only trihexosyldiacylglycerol and its acylated derivative were detected, points to a high glycolipid acceptor specificity, selecting two out of six membrane glycolipids[8] in the biosynthesis of this lipoteichoic acid.

The release of the glycolipids or their deacylation products by degradation with HF, acetic acid and alkaline hydrolysis leave no doubt that the poly(glycerophosphate) is linked to the glycolipids through a phosphodiester bond. As point of attachment, C-6 of the terminal glucosyl residue can be suggested, since the glycerophosphoglycolipids of *L. casei* have been established to be *sn*-glycero-1-phospho→6Glc(β1-6)Gal(α1-2)acyl→6Glc(α1-3)diacylglycerol and *sn*-glycero-1-phospho→6Glc(β1-6)Gal(α1-2)Glc(α1-3)diacylglycerol[6a].

Having Glc(β1-6)Gal(α1-2)acyl→6Glc(α1-3)diacylglycerol and Glc(β1-6)Gal(α1-2)Glc(α1-3)diacylglycerol as lipid anchors, the lipoteichoic acid of *L. casei* is so far unique, since each of the previously studied lipoteichoic acids contained the respective membrane dihexosyldiacylglycerol[1-4] or, as in the case of *Streptococcus faecalis*, phosphatidyldihexosyldiacylglycerol[5,6]. Whereas the species with trihexosyldiacylglycerol may be viewed as an interesting modification of a more common principle, the species with acyltrihexosyldiacylglycerol represents a novel subclass. The finding of an analogous species in *Streptococcus lactis* (Fischer, W. and Koch, H.U., unpublished) and the recently discussed[8] occurence of acylated diacylglyceroglycolipids in other Gram-positive bacteria suggest that lipoteichoic acids with three fatty acids at the glycolipid portion are more widely spread.

The space-filling model depicted in Fig. 4 shows that three fatty acids in a spatial arrangement as found in the lipoteichoic acid of *L. casei* readily join to generate one hydrophobic region. Thus, one functional aspect seems to be a more firm anchoring of this species in the cytoplasmic membrane.

Whether there is a relation to the lipoteichoic acid carrier which serves as acceptor in the biosynthesis of wall teichoic acid[27] needs further investigation. Recently the lipoteichoic acid carrier and lipoteichoic acid of *Staphylococcus aureus* H were separated and shown to differ mainly in the average content of three and two acyl groups, respectively. So far, however, the third fatty acid of the carrier has not been localized[34]. On the other hand lipoteichoic acid carrier that was active with the ribitol phosphate polymerase of *S. aureus* H was earlier detected also in *L. casei*[35]. Interestingly, the assembly point of the poly(ribitol phosphate) seems to be neither the free glycerol terminus nor any other

---

* Whether the differences between preparation 1 and 2 can be related to the growth phases in which the bacteria had been harvested, needs further investigation. Changes in the pattern of membrane lipids have been observed[8].

** Contaminating phospholipids can also be precluded because they would have released diacylglycerol on degradation with HF and acetic acid.

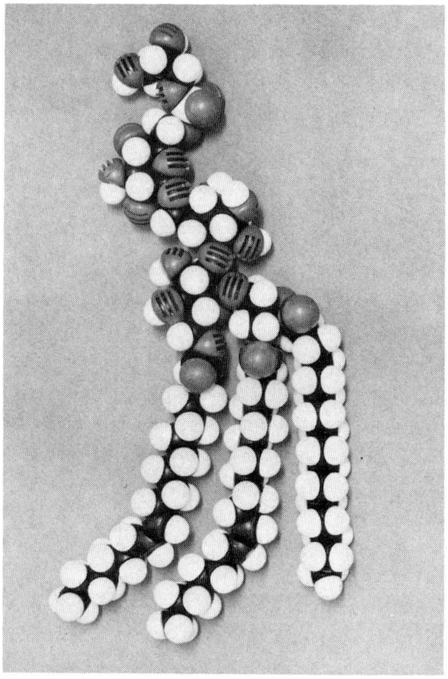

Fig. 4. Space-filling model of *sn*-glycero-1-phospho→6Glc(β1-6)Gal(α1-2)acyl→6Glc(α1-3)diacylglycerol. Diacylglycerol portion, right; carbohydrate-linked fatty acid, left; glycerophosphate residue, top. Fatty acid composition as in the main molecular species of acyltrihexosyldiacylglycerol of *L. casei* (cf. ref.[8]).

glycerophosphate of the lipoteichoic acid carrier of *S. aureus* H but an otherwise substituted glycerol which was tentatively localized at the hydrophobic end[24]. If, therefore, the previously discussed[3] possibility were true, that the acceptor glycerol is the diacylglycerol portion from which one fatty acid would first have to be removed, then a carbohydrate-linked fatty acid, as found in *L. casei* lipoteichoic acid (Fig. 4), together with the remaining fatty acid at the glycerol could keep the loaded carrier still anchored in the membrane.

We thank *Dr. H.J. Aschoff* for advice in column chromatography and *Dr. K. Wartenberg* for the facility of gas chromatography – mass spectral analysis. M.N. thanks the *Alexander von Humboldt Stiftung* for a research fellowship. This work was supported by the *Deutsche Forschungsgemeinschaft.*

Literature

1 Coley, J., Duckworth, M. & Baddiley, J. (1972) *J. Gen. Microbiol.* **73**, 587 - 591.
2 Knox, K.W. & Wicken, A.J. (1973) *Bacteriol. Rev.* **37**, 215 - 257.
3 Duckworth, M., Archibald, A.R. & Baddiley, J. (1975) *FEBS Lett.* **53**, 176 - 179.
4 Button, D. & Hemmings, N.L. (1976) *Biochemistry* **15**, 989 - 995.
5 Toon, P., Brown, P.E. & Baddiley, J. (1972) *Biochem. J.* **127**, 399 - 409.
6 Ganfield, M.-C.W. & Pieringer, R.A. (1975) *J. Biol. Chem.* **250**, 702 - 709.
6a Fischer, W., Laine, R.A. & Nakano, M. (1978) *Biochim. Biophys. Acta*, in press.
7 Fischer, W. (1976) in Lipids, (Paoletti, R., Porcellati, G. & Jacini, G., eds.) vol. I, pp. 255 - 266, Raven Press, New York.
8 Nakano, M. & Fischer, W. (1977) *Hoppe-Seyler's Z. Physiol. Chem.* **358**, 1439 - 1453.
9 Lowry, O.H., Rosebrough, N.J., Farr, A.L. & Randall, R.J. (1951) *J. Biol. Chem.* **193**, 265 - 275.
10 Ough, L.D. (1964) in Methods in Carbohydrate Chemistry (Whistler, R., ed.) vol. IV, pp. 91 - 98, Academic Press, New York, London.
11 Bergmeyer, H.U., Bernt, E., Schmidt, F. & Stork, H. (1970) in Methoden d. Enzymat. Analyse (Bergmeyer, H.U., ed.) 2nd ed., vol. II, pp. 1163 - 1165, Verlag Chemie, Weinheim/Bergstraße.
12 Kurz, G. & Wallenfels, K. (1970) in Methoden d. Enzymat. Analyse (Bergmeyer, H.U., ed.) 2nd ed., vol. II, pp. 1240 - 1244, Verlag Chemie, Weinheim/Bergstraße.
13 Gatt, R. & Berman, E.R. (1966) *Anal. Biochem.* **15**, 167 - 171.
14 Mc Fadayen, D.A. (1945) *J. Biol. Chem.* **158**, 107 - 135.
15 Schnittger, H., Papenberg, K., Gause, E., Czok, R., Bücher, Th. & Adam, H. (1959) *Biochem. Z.* **332**, 167 - 185.
16 Hohorst, H.J. (1970) in Methoden d. Enzymat. Analyse (Bergmeyer, H.U., ed.) 2nd ed., vol. II, pp. 1379 - 1383, Verlag Chemie, Weinheim/Bergstraße.
17 Snyder, F. & Stephens, N. (1959) *Biochim. Biophys. Acta* **34**, 244 - 245.
18 Fischer, W., Ishizuka, I., Landgraf, H.R. & Herrmann, J. (1973) *Biochim. Biophys. Acta* **296**, 527 - 547.
19 Fischer, W. (1977) *Biochim. Biophys. Acta* **487**, 74 - 88.
20 Osborn, M.J. (1966) in Methods of Enzymology (Neufeld, E.F. & Ginsburg, V., eds.) vol. VIII, pp. 161 - 164, Academic Press, New York, London.
21 Coley, J., Duckworth, M. & Baddiley, J. (1975) *Carbohydr. Res.* **40**, 41 - 52.
22 Fiedler, F. & Glaser, L. (1974) *J. Biol. Chem.* **249**, 2684 - 2689.
23 James, A.T. (1960) *Methods Biochem. Anal.* **8**, 1 - 59.

24 Fiedler, F. & Glaser, L. (1974) *Carbohydr. Res.* **37**, 37 - 46.
25 Knox, K.W. (1963) *J. Gen. Microbiol.* **31**, 59 - 72.
26 Hall, E.A. & Knox, K.W. (1965) *Biochem. J.* **96**, 310 - 318.
27 Kelemen, M.V. & Baddiley, J. (1961) *Biochem. J.* **80**, 246 - 254.
28 Shaw, N. & Stead, A. (1974) *Biochem. J.* **143**, 461 - 464.
29 Coulon-Morelec, M.J., Faure, M. & Marechal, J. (1960) *Bull. Soc. Chim. Biol.* **42**, 867 - 876.
30 Fischer, W. & Landgraf, H.R. (1975) *Biochim. Biophys. Acta* **380**, 227 - 244.
31 Hakomori, S.-I. (1964) *J. Biochem.* **55**, 205 - 208.
32 Fischer, W., Landgraf, H.R. & Herrmann, J. (1973) *Biochim. Biophys. Acta* **306**, 353 - 367.
33 Wicken, A.J. & Knox, K.W. (1975) *Science* **187**, 1161 - 1167.
34 Lambert, P.A., Coley, J. & Baddiley, J. (1977) *FEBS Lett.* **79**, 327 - 330.
35 Fiedler, F. & Glaser, L. (1974) *J. Biol. Chem.* **249**, 2690 - 2695.

Prof. Dr. W. Fischer, Institut für Physiologische Chemie der Universität Erlangen–Nürnberg; Wasserturmstraße 5, D-8520 Erlangen

Dr. M. Nakano, Obihiro University, Obihiro, Hokkaido, Japan

Hoppe-Seyler's Z. Physiol. Chem.
Bd. 359, S. 13 - 17, Januar 1978

Synthesis of Collagen by Chick Calvaria in the Presence of 2,2′-Bipyridyl Lack of the Aminoterminal Extension Peptides

Peter K. MÜLLER, Kurt RAISCH, Helene KADNER and Hanna WIEDEMANN

Max-Planck-Institut für Biochemie Abteilung für Bindegewebsforschung, Martinsried b. München

(Received 14 October 1977)

Summary: A precursor of type I collagen was accumulated when cranial bones from embryonic chicken were cultured in vitro in the presence of 2,2′-bipyridyl. While biochemical studies demonstrate that the precursor was present as a covalently linked trimer, evaluation of electron micrographs of segment-long-spacing aggregates show that the molecules of this precursor have an extension at the carboxy terminus only. These data provide further evidence that the inter-chain disulfide bonds in procollagen are located at the carboxy-terminal extension, and secondly, that 2,2′-bipyridyl does not interfere with a normal cleavage of procollagen peptides at the amino terminus, but inhibits the cleavage of the carboxy-terminal procollagen peptides.

Kollagen-Synthese durch Hühnchen-Kalvarien in Gegenwart von 2,2′-Bipyridyl. Fehlen aminoterminaler Extensionspeptide

Zusammenfassung: Eine Vorstufe von Typ-I-Kollagen reichert sich an, wenn die Kalvarien embryonaler Hühnchen unter In-vitro-Bedingungen in Anwesenheit von 2,2′-Bipyridyl kultiviert werden. Während biochemische Untersuchungen zeigen, daß die Vorstufe in Form eines kovalent gebundenen Trimers vorliegt, ist an Segment-long-spacing-Kristalliten im Elektronenmikroskop eine Extension von Prokollagenpeptiden nur am carboxyterminalen Ende zu erkennen. Diese Untersuchungen beweisen erstens, daß sich die interkatenaren Disulfidbrücken des Prokollagens innerhalb der carboxyterminalen Extension befinden, und zweitens, daß 2,2′-Bipyridyl nur die Peptidase hemmt, die die Prokollagenpeptide vom carboxyterminalen Ende abspaltet.

Key words: Procollagen, carboxy terminus, electron micrographs, 2,2′-bipyridyl, calvaria.

Collagen is synthesized in a precursor form[1-4] (see ref.[4] for review). While earlier structural studies of collagen precursors suggested that there was a non-collagenous region at the amino-terminal end of each precursor chain[5-9] more recent investigations also demonstrated the existence of such an extension at the carboxy terminus[10-13]. Using tadpole collagenase to cleave the procol-

---

Abbreviations:
EDTA, ethylendiaminetetracetic acid; ATP, adenosine triphosphate.

lagen molecules into two triple-stranded fragments, it was possible to demonstrate that the sites for interchain disulfide bonds are located at the carboxy-terminal extension. This conclusion is based on the fact that type I collagen does not contain any disulfide bridges within the triple helical part of the collagen molecule.

In our present study, we used 2,2′-bipyridyl, a chelating agent, in the culture medium to accumulate the precursor of type I collagen in the tissue of cranial bones of embryonic chicken [2,14]. Chromatography of the newly synthesized collagen on Agarose A 1.5 showed a collagen trimer which contains cysteine and tryptophan residues. Reduction of this trimer by 2-mercaptoethanol resulted in the respective precursor α chains. Examination of segment-long-spacing aggregates by electron microscopy showed molecules with a carboxy-terminal extension only, giving additional evidence for a disulfide linkage region at this end of the procollagen molecule.

Materials and Methods

Incubation of chick calvaria

Calvaria from 16 day-old chick embryos were incubated for 3 h at 37 °C in Dulbecco's modified Eagle's medium lacking cysteine and tryptophan. The medium for pulse labeling of the procollagen was supplemented with 2-aminopropionitrile (100 μg/m*l*), sodium ascorbate (50 μg/m*l*), penicillin (400 U/m*l*) and 2,2′-bipyridyl (1.4mM). 1.4mM 2,2′-bipyridyl was used because lower concentrations of this drug do not completely inhibit hydroxylation of collagen α chains in chick calvaria. Generally, 50 calvaria were labeled with 12.5 μCi/m*l* L-[2,3-^{3}H]proline (20 - 40 Ci/mmol) and 7.5 μCi/m*l* [^{35}S]cysteine (20 - 60 Ci/mmol) New England Nuclear Corp.). In some experiments L-[G-^{3}H]tryptophan (1 - 5 Ci/mmol) (New England Nuclear Corp.) was used instead of L-[2,3-^{3}H]proline.

Preparation of radioactively labeled precursor collagen

After incubation, the cranial bones were extracted for 24 h at 4 °C with 1M NaCl, 50mM Tris/HCl, pH 7.5, to which 10mM EDTA, 50mM iodoacetamide, 0.2μM phenylmethylsulfonyl fluoride and 0.1μM 4-(chloro)mercuribenzoate had been added to inhibit proteolysis. The precursor collagen in the medium was precipitated in the presence of inhibitors by adding ammonium sulfate to a final concentration of 40% saturation. Both calvaria extract and medium precursor collagen were dissolved in 1M NaCl, 0.05M Tris, pH 7.4 plus inhibitors and further purified by precipitation with 20% NaCl. The precipitate was collected by centrifugation and redissolved in 0.2M NaCl, 50mM Tris, pH 7.4.

Procedures used to characterize the newly synthesized collagen

Samples were dialysed against appropriate buffers for molecular sieve chromatography on Agarose A 1.5, or polyacrylamide gel electrophoresis with dodecylsulfate, respectively[15,16]. To samples used for either of the procedures, lathyritic type I collagen from rat skin was added to mark the positions of collagen constituents. In some experiments, limited digestion with pepsin was used to convert the precursor of collagen to the respective collagen[17]. This is possible because the originally randomly coiled trimers form triple-helical and thus pepsin-resistant molecules during extraction at 4 °C. The precursor preparation for electron microscopic investigations was collected from 150 chickens and did not con-

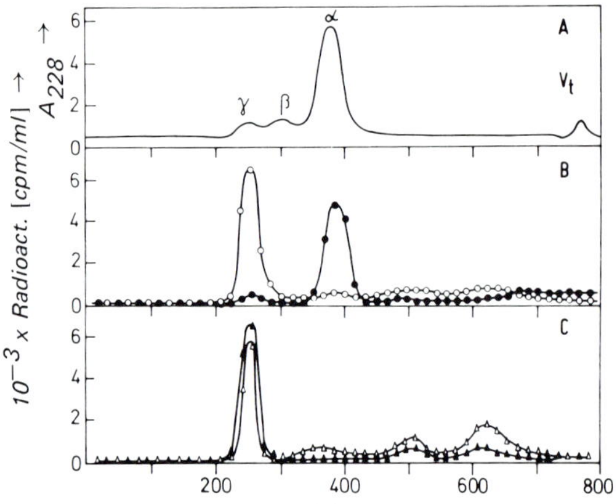

Fig. 1. Chromatography on Agarose A 1.5 in 1M $CaCl_2$, 0.05M Tris, pH 7.4 under denaturing conditions.

0.5 m*l* portions of the 5-m*l* fractions were mixed with 10 m*l* Aquasol and radioactivity was determined in a Beckman LS250 liquid scintillation spectrometer.

A: Elution profile of lathyritic type I collagen used as an internal marker for collagen constituents.

B: Pattern of radioactivity of L-[^{3}H]proline-labeled protein, synthesized in the presence of 1.4mM 2,2′-bipyridyl before (○–○–○) and after (●–●–●) treatment with pepsin.

C: Pattern of radioactivity of [^{35}S]cysteine (△–△–△) and L-[^{3}H]tryptophan (▲–▲–▲) labeled protein synthesized in the presence of 1.4mM 2,2′-bipyridyl. The [^{3}H]Trp counts were multiplied by 10^{-4}, not 10^{-3}.

tain any additional carrier collagen. In order to form segment-long-spacing crystallites, the sample was briefly dialysed against dilute acetic acid and subsequently against 0.4% disodium ATP. The segment-long-spacing crystallites were positively stained with phosphotungstic acid and uranyl acetate and examined in a Siemens Elmiskop I.

Results

The cranial bones of embryonic chickens are an excellent system to study synthesis of collagen in vitro. For this reason, they have been widely used to isolate and characterize the biosynthetic precursor of collagen, procollagen, which has been shown to be a disulfide-linked trimer of pro-α chains of collagen. The precursor of unhydroxylated collagen, which is synthesized in the presence of the chelator 2,2′-bipyridyl, can also be isolated as a trimer when acidic conditions are prevented and protease inhibitors are added to the extraction buffer[14]. Fig. 1 depicts the elution profile of unhydroxylated type I collagen labeled with either L-[2,3-^{3}H]proline (Fig. 1, B) or L-[^{35}S]cysteine and L-[^{3}H]tryptophan (Fig. 1, C). Essentially all collagenous material is

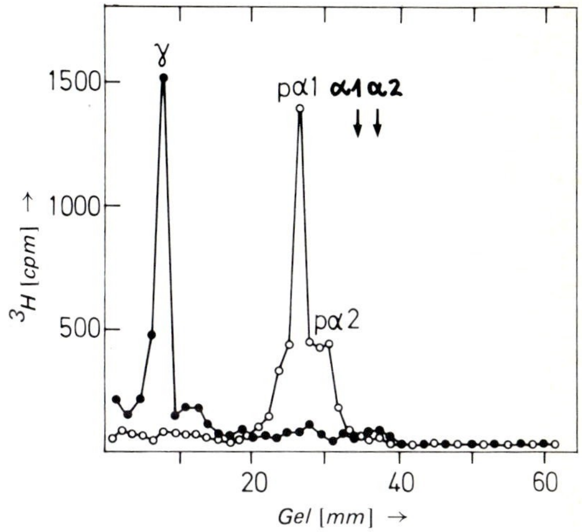

Fig. 2. Polyacrylamide gel electrophoresis, in the presence of dodecylsulfate, of procollagen synthesized in the presence of 1.4mM 2,2′-bipyridyl.

(Without reduction •–•–•, and with reduction with mercaptoethanol ○–○–○.) The positions of collagen constituents used as internal markers are denoted by arrows. The gels were sliced at 1-mm intervals and digested with 30% H_2O_2. The radioactivity was measured by standard methods using Aquasol.

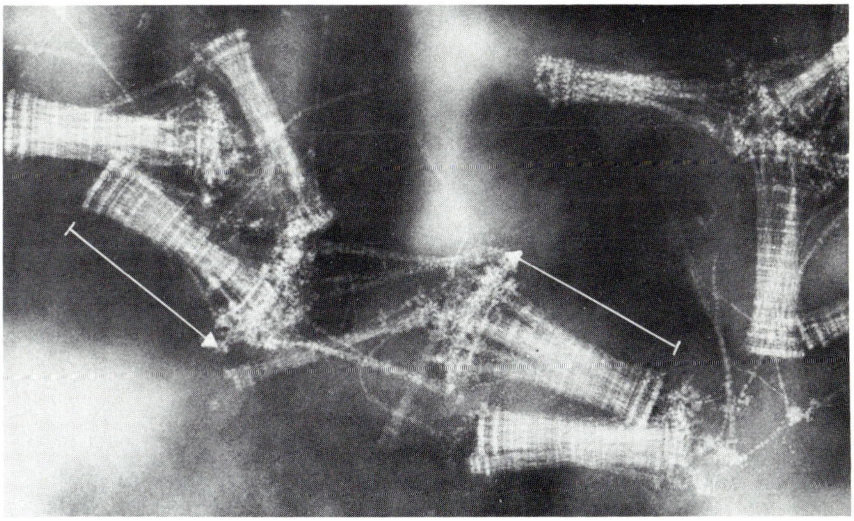

Fig. 3. Segment-long-spacing aggregates prepared from procollagen synthesized by cranial bones in the presence of 1.4mM 2,2′-bipyridyl.

The sample was prepared by dialyzing the tissue extract against 0.4% ATP. Positive staining was achieved with phosphotungstic acid and uranyl acetate. Almost all the aggregates show an extension at the carboxy-terminal end. Magnification × 95 000 present as a trimer, and this trimer contains cysteine and tryptophan residues (Fig. 1, B, C). It is also shown that the trimer can be cleaved to collagen α chains by treatment with pepsin (Fig. 1, B, filled circles). The trimeric state of the precursor is further demonstrated by dodecylsulfate gel electrophoresis (Fig. 2). In the unreduced state, most of the ^{3}H-labelled protein migrated in a single peak with molecules the size of precursor components (closed circles). When the ^{3}H-labeled protein was reduced with 2-mercaptoethanol before application to the gels, most of the radioactivity was recovered in two peaks identified as the respective α chains of pro-type I collagen (open circles). Similar results were obtained when the unhydroxylated procollagen from the medium was analysed by identical procedures.

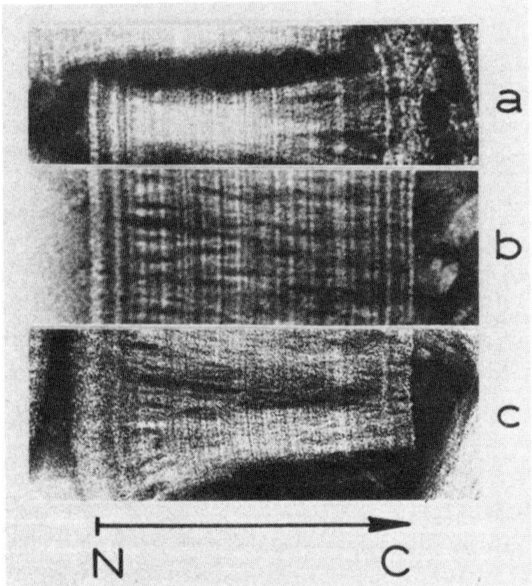

Fig. 4. Segment-long-spacing aggregates of procollagen and collagen.

a) Synthesized in the presence of 1.4mM 2,2′-bipyridyl,

b) isolated from normal calf skin,

c) isolated from dermatosparactic calf skin[6].

Positive staining with phosphotungstic acid and uranyl acetate. N: amino terminus; C: carboxy terminus. While b represents the normal collagen molecule, a has an extension at the carboxy terminus and c has an extension at the amino terminus.
Magnification × 180 000.

Another preparation of unhydroxylated type I procollagen synthesized and secreted by some 150 cranial bones in the presence of 1.4mM 2,2′-bipyridyl was used to precipitate segment-long-spacing crystallites by dialysis against 0.4% ATP. Even a simple survey of the precipitated crystallites showed a substantial number of precursor molecules which were longer than normal collagen molecules (Fig. 3). A more detailed analysis, using higher magnification and comparing segment-long-spacing crystallites from normal calf skin and dermatosparactic calf skin collagen with these precursor molecules, clearly indicates that the majority of the molecules synthesized and processed in the presence of 2,2′-bipyridyl have a peptide extension at the carboxy-terminal end, while an extension is not seen at the amino terminus (Fig. 4a). Only a small fraction of molecules is seen to have extensions at both ends of the molecule. When segment-long-spacing crystallites were investigated from collagen which was synthesized by calvaria in the absence of 1.4mM 2,2′-bipyridyl, no extension could be found[18], either at the amino-terminal or at the carboxy-terminal end of the collagen molecule. On the other hand, segment-long-spacing crystallites from collagen of dermatosparactic calf have an extension at the amino terminus (Fig. 4c)[6].

Discussion

2,2′-Bipyridyl has been used in a number of studies to interfere with normal collagen synthesis by preventing hydroxylation of proline and lysine residues, disturbing chain selection in type I collagen, and by inhibiting the conversion of precursor collagen to collagen[2,9,14,19-21] (for review see ref.[21]). Taking into account recent studies on the nature of the precursor of type I collagen, which suggest that it has peptide extensions at both the amino-terminal and the carboxy-terminal end of the molecule, we have reexamined the influence of 2,2′-bipyridyl on the conversion of procollagen to collagen. Our data bring further evidence that the unhydroxylated procollagen trimer is linked by disulfide bonds which are located in the carboxy-terminal extension peptide, since only this extension is apparent in the segment-long-spacing crystallites prepared from the precursor synthesized in the presence of 2,2′-bipyridyl.

Our data also indicate that at least two enzymes are involved in the cleavage of the procollagen peptide extensions. 2,2′-Bipyridyl apparently does not interfere with cleavage of the amino-terminal extension when present in the culture medium of embryonic cranial bones. Under the same conditions, however, the action of a second enzyme normally cleaving off the carboxy-terminal extension seems to be inhibited. However it should be emphasized that the inhibiting effect of 2,2′-bipyridyl is not necessarily a direct one, but arises from the fact that 2,2′-bipyridyl alters the substrate (e.g. the collagen molecule is underhydroxylated and consists largely of $\alpha 1(I)$-trimers[20].) In addition, no evidence was obtained whether the conversions occur intracellularly or extracellularly.

On the other hand, the observation that freshly isolated tendon fibroblasts from chicken accumulate procollagen with an amino-terminal extension only[9], may be explained in the following way. Firstly, the carboxy-terminal extension was probably lost due to the acidic conditions used for the extraction of procollagen. Secondly, as shown by the same authors, conversion of procollagen to collagen is impaired in freshly isolated tendon fibroblasts even in the absence of 2,2′-bipyridyl[8,13]. By contrast, cells confined in cranial bones obviously retained the full enzymatic capacity to cleave amino-terminal and carboxy-terminal extensions. The cleavage of the latter is inhibited when 2,2′-bipyridyl is present during synthesis of collagen. It is obvious, however, that collagen molecules with a carboxy-terminal extension only may also be found in the absence of 2,2′-bipyridyl as transient forms, which are rapidly processed to collagen molecules.

The authors wish to thank *Prof. Dr. K. Kühn* for helpful discussions and for encouraging this work. This study was supported by a grant from the *Deutsche Forschungsgemeinschaft* (Mu 376/4) with in the project group study "Biopolymere und Biomechanik von Bindegewebssystemen".

Literature

1 Bellamy, G. & Bornstein, P. (1971) *Proc. Natl. Acad. Sci. U.S.A.* **68**, 1138 - 1142.
2 Müller, P.K., McGoodwin, E.B. & Martin, G.R. (1971) *Biochem. Biophys. Res. Commun.* **44**, 110 - 117.
3 Lenaers, A., Ansay, M., Nusgens, B.V. & Lapiere, C.M. (1971) *Hoppe-Seyler's Z. Physiol. Chem.* **352**, 14 - 18.
4 Martin, G.R., Byers, P.H. & Piez, K.A. (1975) in Advances in Enzymology (Kaplan, N. & Colowick, S.P., eds.) vol. 42, pp. 167 - 191, Academic Press, New York.
5 Jimenez, S.A., Dehm, P. & Prockop, D.J. (1971) *FEBS Lett.* **17**, 245 - 248.
6 Stark, M., Lenaers, A., Lapiere, C.M. and Kühn, K. (1971) *FEBS Lett.* **18**, 225 - 227.
7 Lenaers, A., Ansay, M., Nusgens, B.V. & Lapiere, C.M. (1971) *Eur. J. Biochem.* **23**, 533 - 543.
8 Dehm, P., Jimenez, S.A., Olsen, B.R. & Prockop, D.J. (1972) *Proc. Natl. Acad. Sci. U.S.A.* **69**, 60 - 64.
9 Jimenez, S.A., Dehm, P., Olsen, B.R. & Prockop, D.J. (1973) *J. Biol. Chem.* **248**, 720 - 729.
10 Tanzer, M.L., Church, R.L., Yaeger, J.A., Wampler, D.E., & Park, E.D. (1974) *Proc. Natl. Acad. Sci. U.S.A.* **71**, 3009 - 3013.
11 Byers, P.H., Click, E.M., Harper, E. & Bornstein, P. (1975) *Proc. Natl. Acad. Sci. U.S.A.* **72**, 3009 - 3013.
12 Fessler, L.I., Morris, N.P. & Fessler, J.H. (1975) *Proc. Natl. Acad. Sci. U.S.A.* **72**, 4905 - 4909.
13 Olsen, B.R., Hoffmann, H.P. & Prockop, J.D. (1976) *Arch. Biochem. Biophys.* **175**, 341 - 350.
14 Fessler, L.J. & Fessler, J.H. (1974) *J. Biol. Chem.* **23**, 7637 - 7646.
15 Piez, K.A. (1968) *Anal. Biochem.* **26**, 205 - 212.
16 Furthmayr, H. & Timpl, R. (1971) *Anal. Biochem.* **41**, 510 - 516.
17 Pontz, B., Muller, P.K. & Meigel, W.N. (1973) *J. Biol. Chem.* **248**, 7558 - 7564.
18 Raisch, K. (1976) Ph.D. Thesis, Universität München.
19 Hurych, J. & Chvapil, M. (1965) *Biochim. Biophys. Acta* **97**, 361 - 363.
20 Müller, P.K., Meigel, W.N., Pontz, B.F. & Raisch, K. (1974) *Hoppe-Seyler's Z. Physiol. Chem.* **355**, 985 - 996.
21 Prockop, D.J., Berg, R.A., Kivirikko, K.J. & Uitto, J. (1976) in Biochemistry of Collagen (Ramachandran, G.N. & Reddi, A.H., eds.) 163 - 273, Plenum, New York.

Dr. P.K. Müller, Dr. K. Raisch, H. Kadner and H. Wiedemann, Max-Planck-Institut für Biochemie, Abteilung für Bindegewebsforschung,
D-8033 Martinsried b. München.

Hoppe-Seyler's Z. Physiol. Chem.
Bd. 359, S. 19 - 27, Januar 1978

Properties of Two *Clostridia* Strains Acting as Catalysts for the Preparative Stereospecific Hydrogenation of 2-Enoic Acids and 2-Alken-1-ols with Hydrogen Gas

Johann BADER, Helmut GÜNTHER, Bernhard RAMBECK und Helmut SIMON

Institut für Organische Chemie der Technischen Universität München

(Received 16 August 1977)

Summary: A *Clostridium* strain growing on crotonate/hydrogencarbonate, which is able to hydrogenate stereospecifically 2-enoates as well as other unsaturated compounds with hydrogen gas, has been isolated and methods for its propagation elucidated. For this strain and for *Clostridium kluyveri* DSM 555 grown on ethanol/acetate/hydrogencarbonate, samples of 200-*l* batches were assayed to determine the hydrogenation activity for (*E*)-2-methyl-2-butenoate and (*E*)-2-buten-1-ol as a function of time during the exponential and stationary growth phases. For the strain growing on crotonate/hydrogencarbonate, the hydrogenation rate as a function of substrate concentration, pH and temperature has been measured. Storage conditions for both strains are given.

Eigenschaften von zwei Clostridien-Stämmen als Katalysatoren zur präparativen stereospezifischen Hydrierung von 2-Encarbonsäuren und 2-Alken-1-olen

Zusammenfassung: Es wurde ein *Clostridium*-Stamm isoliert, der auf Crotonat/Hydrogencarbonat wächst und die Fähigkeit hat, unter Wasserstoff neben anderen ungesättigten Verbindungen insbesondere 2-Encarbonsäuren stereospezifisch zu hydrieren. Methoden für dessen Anzucht wurden ausgearbeitet. Für diesen Stamm und für *Clostridium kluyveri* DSM 555, gewachsen auf Ethanol/Acetat/Hydrogencarbonat wurde die Hydrieraktivität gegenüber (*E*)-2-Methyl-2-butenoat und (*E*)-2-Buten-1-ol in Abhängigkeit von der Wachstums- und stationären Phase verfolgt. Dazu wurden aus einer 200-*l*-Anzucht aliquote Teile gezogen. Für den auf Crotonat/Hydrogencarbonat wachsenden Stamm wurde die Hydriergeschwindigkeit in Abhängigkeit von Substratkonzentration, pH und Temperatur bestimmt. Für beide Stämme werden Lagerbedingungen angegeben.

Key words: Clostridium kluyveri, biohydrogenation.

Enzyme:
Butyryl-CoA dehydrogenase, butyryl-CoA:(acceptor) oxidoreductase (EC 1.3.99.2).

Recently we observed that it is possible to use microorganisms for stereospecific hydrogenation on a preparative scale according to the general scheme[1-4]

$$>C=X \xrightarrow[\text{Microorganisms}]{H_2} H-\overset{|}{\underset{|}{C}}-X-H.$$

In the following we want to report defined culture conditions for two different clostridial strains in 200-*l* batches and some other practical aspects, such as the dependence of hydrogenation activity on growth time, storage and hydrogenation conditions, etc. The two clostridial strains hydrogenate many different 2-enoates[1] and allyl alcohols in the following fashion.

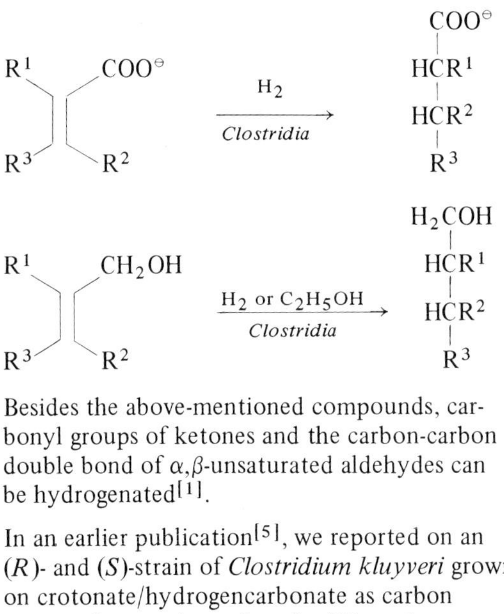

Besides the above-mentioned compounds, carbonyl groups of ketones and the carbon-carbon double bond of α,β-unsaturated aldehydes can be hydrogenated[1].

In an earlier publication[5], we reported on an (*R*)- and (*S*)-strain of *Clostridium kluyveri* grown on crotonate/hydrogencarbonate as carbon sources. It turned out that the (*R*)-strain possesses, besides the butyryl-CoA dehydrogenase, an enzyme which seemed to be unknown and which we call 2-enoate reductase. It hydrogenates the nonactivated carboxylate anions in the above formulated stereochemical manner[5-7]. (*E*)-2-Butenoate and (*E*)-2-methyl-2-butenoate are hydrogenated by whole cells or lysates by this 2-enoate reductase, as well as by butyryl-CoA dehydrogenase after activation to the corresponding CoA esters. However, the stereochemical course is different[5,7]. Cells which are not able to achieve this activation reduce these substrates only via 2-enoate reductase. 2-Enoates such as cinnamate and others[1,2] which are less similar to crotonate are substrates for the 2-enoate reductase only.

The enzyme system which hydrogenates allyl alcohols is different from that of the α,β-unsaturated acids[8]. The formulated stereochemistry of this reaction was determined by the reduction of (*E*)-2-methyl-2-buten-1-ol, (*E*)-2-methyl-3-phenyl-2-propen-1-ol and (*E*)-3-methyl-2-penten-1-ol to the corresponding saturated alcohols of complete stereochemical purity*. Details will be published elsewhere.

During our studies we observed typical differences between cells growing on the ethanol/acetate/hydrogencarbonate medium and on crotonate/hydrogencarbonate, respectively. Such differences and the question whether the organism growing on crotonate/hydrogencarbonate, as first described by Bergmeyer et al.[9] is *Clostridium kluyveri* as defined by Bornstein and Barker[10,11] will be discussed in another paper. In the following we describe some experiments with *Clostridium kluyveri* DSM 555 and another organism which we call *Clostridium* strain La 1.

Materials and Methods

Chemicals

Methylviologen and resazurin were purchased from Serva, D-6900 Heidelberg, sodium thioglycolate from Oxoid Deutschland GmbH, D-4230 Wesel, yeast extract from Difco, Detroit and crotonic acid [(*E*)-2-butenoic acid] from Baker Chemikalien, D-6080 Groß-Gerau. All other chemicals were products of E. Merck AG, D-6100 Darmstadt. If available, the purity was analytical grade. The ethanol was denatured with 0.5% 2-butanone.

Organisms

C. kluyveri DSM 555 and 562 were obtained from Deutsche Sammlung von Mikroorganismen, D-3400 Göttingen. Crotonate strain La 1 was isolated from freeze-dried cells, which descended from a culture of *C. kluyveri* supplied by Boehringer, D-8132 Tutzing, by repeated streaking on agar plates.

Media

The following media were used:
Medium I, similar to that of Bornstein and Barker[10,11]: 16 *ml* ethanol, 7.5 g sodium acetate, 2.5 *ml* glacial

* Simon, H. & Angermaier, L., unpublished.

acetic acid, 150 mg $(NH_4)_2HPO_4$, 100 mg K_2HPO_4, 33 mg $MgCl_2 \cdot 6\,H_2O$, 0.6 mg $MgSO_4 \cdot 7\,H_2O$, 40 mg $CaCl_2 \cdot 2\,H_2O$, 0.4 mg $MnSO_4 \cdot 2\,H_2O$, 0.4 mg $FeSO_4 \cdot 7\,H_2O$, 50 mg $(NH_4)Cl$, 10 mg $(NH_4)_6Mo_7O_{24} \cdot 4\,H_2O$, 0.04 mg biotin, 0.8 mg *p*-aminobenzoic acid, 1 mg resazurin, 6 m*l* 50% solution of K_2CO_3, 1 *l* deionized water. To 200 *l* batches, 1.0 g $FeCl_2 \cdot 4\,H_2O$ was added.

Medium II, similar to that of Bergmeyer et al.[9]:
The ethanol, sodium acetate and acetic acid of medium I were replaced by 6 g crotonic acid and 0.3 g sodium hydroxide, and 8 m*l* of 50% K_2CO_3 solution instead of 6 m*l*. In most cases, medium II was supplemented with 0.1% tryptic digested casein-peptone and 0.1% yeast extract.

For preparation of surface colonies, the media were supplemented with 0.6% peptone and 0.6% yeast extract. An agar addition of 2% for roll tubes and of 1.2% for agar plates was used.

Anaerobic conditions, roll tubes, agar plates and flasks

With the exception of the preparation of liquid cultures of more than 50 m*l*, the media were purified of oxygen by boiling under a stream of nitrogen for 10 min. After cooling, sodium thioglycolate was added to an end concentration of 0.05% and the ethanol required in medium I was admixed. The pH was adjusted to 6.4 with 50% K_2CO_3 solution. The anaerobic culture tubes (Bellco Glass Inc. Vineland, U.S.A.) were filled with highly purified nitrogen and the medium was transferred into them in amounts of 10 m*l*. The tubes were closed with butyl rubber stoppers (Fisher Scientific Co., Pittsburg, U.S.A.) and locked with wire clamps for autoclaving. In order to solidify the medium, 100 mg agar powder was added. After autoclaving, the pH value was 6.8 - 7.0. For preparation of roll tubes[12], 10 m*l* oxygen-free medium was filled into isolation roll streak tubes (Bellco Glass Inc.) containing 200 mg agar powder. The tubes were closed with rubber stoppers.

For preparation of agar plates, the oxygen-free agar medium (pH 6.4) was filled into a pressure-proof flask for autoclaving. The Petri dishes were equipped with spacers. The plates were kept in an anaerobic jar containing catalyst pellets of the BBL Gaspak anaerobic system (Becton, Dickinson GmbH, D-6900 Heidelberg) and silica blue gel beads in an atmosphere of one volume of carbon dioxide and two volumes of hydrogen gas.

Liquid cultures were mixed in 500 m*l* or 2000 m*l* measuring flasks or in 20 *l* glass vessels with a magnetic bar. The medium without potassium carbonate was autoclaved according to the size for 20 to 60 min at 121 °C. After autoclaving, the flasks were closed immediately with rubber stoppers equipped with glass tubes filled with some cotton. Screw clamps regulated the gas exchange. During cooling an overpressure of nitrogen was applied. Before inoculation, the pH of the media was adjusted to 7.0 by addition of the separately autoclaved oxygen-free solution of 50% K_2CO_3. In the case of medium I, filter sterilized ethanol (cellulose acetate filters, Sartorius, D-3400 Göttingen) was added. If necessary, traces of oxygen, indicated by the colour of oxidized resazurin, were eliminated by dropping in a 1% filter-sterilized solution of sodium dithionite until the resazurin was decolourized, indicating a redox potential of < -0.042 V.

For a large scale culture the bacteria were grown in a 200 *l* glass bulb (Schott u. Gen., D-6500 Mainz) with a special flat cover with lead throughs for the stirrer and other equipment. The ingredients for 200 *l* medium were dissolved in 140 *l* deionized water and heated to 100 °C by passing steam into the medium for 6 h. After cooling to 37 °C under a stream of nitrogen, about 180 *l* fluid was in the bulb. The pH was adjusted to 7.0 by adding 50% K_2CO_3 solution through a rubber tubing.

General culture conditions

All cultures were incubated at 37 °C and preserved at room temperature. Roll tubes were inoculated under a carbon dioxide stream as described by Holdeman and Moore[13]. Liquid cultures were transferred to the bottom of the tubes with a Pasteur pipette bent at the tip. They were streaked on the surface with the reverse side of the bend. Tubes were incubated for up to two weeks. During inoculation of agar plates the time of oxygen admission was minimized. The plates were incubated for one week. Liquid cultures were magnetically stirred. In the 200-*l* bulb, the temperature was kept at 35 °C by an infrared heater. This volume was stirred by an overhead stirrer while a stream of filter-sterilized nitrogen was passed in. The 20-*l* inoculum was transferred anaerobically through glass tubing.

Harvest

Cells of 200-*l* batches were harvested simultaneously with two Sorvall RC2-B centrifuges equipped with the Sorvall KSB tube-type continuous flow system (Sorvall Inc., Newton, U.S.A.) at 39000 × *g* at 0 - 5 °C. A maximal throughput of 25 *l* per hour was obtained with the centrifuge by nitrogen overpressure. If washing of cells was necessary, the pellets were suspended at 0 °C with a magnetic stirrer in two volumes of oxygen-free 0.05M potassium phosphate buffer, pH 7.0, containing 0.14M NaCl and 1 mg resazurin per litre. The washing fluid was reduced to decolorization of the resazurin by the addition of about 10 mg sodium dithionite/*l*. Cells are washed twice in this manner. The pellet was stored frozen at − 18 °C under nitrogen. Freeze-dried cell material was produced by freezing the cell paste in a bulb under a stream of nitrogen. The powder was kept at – 18 °C.

Other general procedures

Oxygen was excluded from all solutions and during all operations. The density of the cultures was estimated at 578 nm. Cuvettes were filled with carbon dioxide and closed with a rubber stopper to avoid disturbance by oxidized resazurin present in the media.

Dry weight was determined by filtration of 20 m*l* cell suspension on cellulose acetate filters SM 11106 (Sartorius GmbH, D-3400 Göttingen). The filters were washed twice with 5 m*l* distilled water and dried in an evacuated desiccator over silica blue gel.

Standard hydrogenation tests were performed in Warburg vessels containing in a total volume of 3.0 m*l* 0.1M phosphate buffer pH 7.0, 0.4 g wet packed cells, 300 μmol sodium (*E*)-2-methyl-2-butenoate dissolved in 0.1 m*l* distilled water, or 295 μmol (*E*)-2-butenol, 75 μg tetracycline · HCl (Boehringer Mannheim) in 0.05 m*l* distilled water. Hydrogen at a pressure of 1 atm was applied for 15 min at 35 °C before the reaction was started by adding the substrate. The vessels were shaken with a frequency of about 130 oscillations/min.

For gas chromatography, samples of 0.2 m*l* were removed. For the determination of carboxylic acids, the 0.2 m*l* samples were acidified with 1N HCl and centrifuged. The suitably diluted supernatant was analyzed in an Aerograph, Series 1400 (Varian, D-6100 Darmstadt) or in a Fractovap Modell 2400 (Carlo Erba, D-6238 Hofheim/Taunus), on a stainless steel column (0.3 × 180 cm) packed with 20% neopentenylglycol succinate and 2% phosphoric acid on chromosorb (Günther, D-4000 Düsseldorf) at 140 °C.

For the determination of alcohols, the neutral supernatant was analyzed on a stainless steel column (0.3 × 180 cm) packed with 10% polyethyleneglycol 400 on chromosorb at 90 °C (ethanol) or 100 °C (C_4-alcohols). Samples were applied to the columns by an automatic sampler A 7176 (Hewlett Packard, D-6000 Frankfurt/M.).

Results

The media suggested in the literature[9-11] were modified and optimized. The propagation method using surface colonies from roll tubes or agar plates as described in Materials and Methods and mentioned already in ref.[5] turned out to be reliable for obtaining cells with the capacity to hydrogenate 2-enoates, 2-alken-1-ols, ketones, nitro compounds etc. The starting material from which the strain La 1 has been isolated contained different strains which were less suitable for hydrogenation experiments or for propagation, and was probably even a mixture of different clostridial species*. Also strain DSM 562 was less efficient than strain DSM 555 for the hydrogenation of (*E*)-2-buten-1-ol. In general, roll tubes were used for the more oxygen-sensitive *C. kluyveri* DSM 555 and agar plates for *Clostridium* La 1. The surface colonies of *Clostridium* La 1, once prepared, can be preserved at room temperature for more than 18 months without degeneration of reviveability and hydrogenation activity. The recovery time for DSM 555 was 3 - 15 days and for La 1 three days.

The cell mass was increased in steps of 10 m*l*, 500 m*l*, 20 and 200 *l*. The average yield is about 1.5 g wet packed cells per litre for *Clostridium* La 1 and 0.75 g for *C. kluyveri* DSM 555, respectively. Cultures of 5 and more litres were stirred in order to increase the cell yield[14]. For *Clostridium* La 1, doubling times of 3 - 4.5 h, and for *C. kluyveri*, 8 - 10 h have been observed.

For the determination and comparison of the hydrogenation rate of different cultures of *Clostridia* we used (*E*)-2-methyl-2-butenoate and (*E*)-2-buten-1-ol. The products and substrates can easily be detected by gas liquid chromatography. In our experience, the hydrogenation rate of (*E*)-2-methyl-2-butenoate is representative for that of many other α,β-unsaturated carboxylic acids[1,2]. So far this has not been proved for (*E*)-2-buten-1-ol.

Table 1 reveals the initial hydrogenation rates of the two different clostridial strains for (*E*)-2-methyl-2-butenoate and (*E*)-2-buten-1-ol. On the average, the α,β-unsaturated acid was hydrogenated about twice as fast as the unsaturated alcohol by strain La 1. The opposite is true for the strain DSM 555. The former organism showed less deviation from the average in the hydrogenation rate of α,β-unsaturated acids and the latter for the unsaturated alcohols, respectively. Usually a certain amount of the substrate was hydrogenated by reducing equivalents deriving from endogenous material**.

* Bader, J., Günther, H., Weiss, N., Schleicher, E. & Simon, H., in preparation.

** Günther, H., Sato, N. and Simon, H., unpublished.

Table 1. Rate of hydrogen uptake and product formation catalyzed by cells obtained from nine 200-*l* cultures of strain DSM 555 and 5 cultures of strain La 1, respectively.

In a Warburg vessel 400 mg wet packed cells and 300 µmol substrate were incubated in 3.0 m*l* 0.1M phosphate buffer, pH 7.0, at 35 °C.

Strain	(*E*)-2-Methyl-2-butenoate as substrate				(*E*)-2-Buten-1-ol as substrate			
	Initial H_2-uptake [µmol/h]		2-Methylbutyrate formation [µmol/h]		Initial H_2-uptake [µmol/h]		n-Butanol formation [µmol/h]	
	Average	Extreme values	Average	Extreme values	Average	Extreme values	Average	Extreme values
DSM 555	37	7 - 95	41	5 - 106	91	53 - 167	115	25 - 190
La 1	88	50 - 142	95	55 - 168	27	12 - 50	48	21 - 114

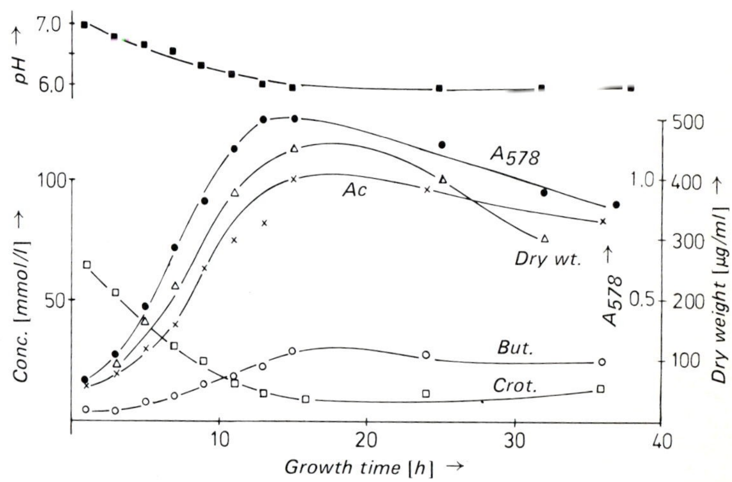

Fig. 1a). Concentrations of crotonate □ □ □, acetate x x x, and butyrate ○ ○ ○, pH ■ ■ ■, dry weight △ △ △ and optical density at 578 nm ● ● ● during the growth of a typical 200-*l* batch of *Clostridium* La 1 on crotonate, peptone and yeast extract.

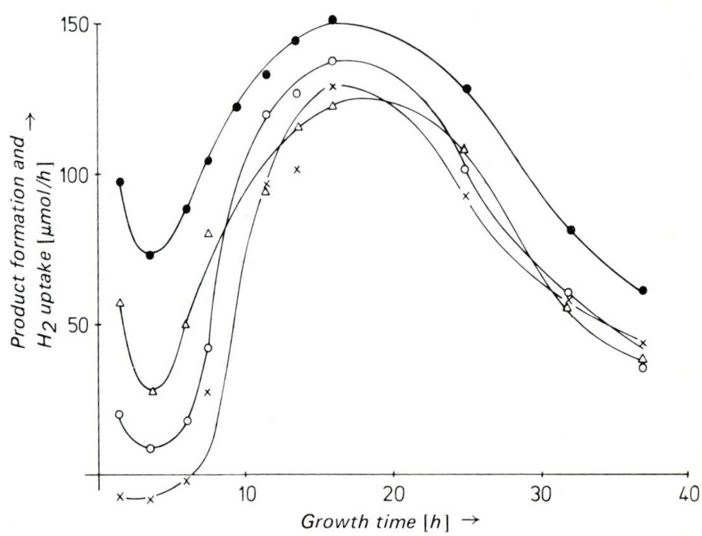

Fig. 1. b) Rate of hydrogen uptake and product formation after 1 h catalyzed by 400 mg wet packed cells harvested as portions of the 200-*l* batch depicted in Fig. 1a.

△ △ △ Hydrogen uptake in the presence of (*E*)-2-methyl-2-butenoate, ● ● ● amount of 2-methylbutanoate, ○ ○ ○ and x x x are the symbols for hydrogen uptake and butan-1-ol formation in the presence of (*E*)-2-buten-1-ol. Negative hydrogen uptake means gas evolution.

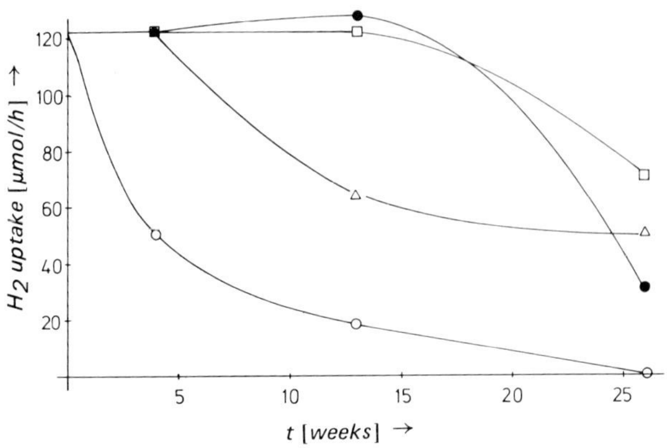

Fig. 2. Rate of hydrogen uptake of lyophilized cells of *Clostridium* strain La 1 after different storage conditions and times.
● ● ● Temperature – 30 °C, under nitrogen;
□ □ □ Temperature – 30 °C, under air;
△ △ △ Temperature – 8 °C, under air;
○ ○ ○ Temperature 20 °C, under air.
In 3.0 m*l* phosphate buffer, pH 7.0, 100 mg lyophilized cells were incubated with (*E*)-2-methyl-2-butenoate (100mM) and methylviologen (1.0mM).

Fig. 1 reveals the kinetics of crotonate consumption and formation of butyrate and acetate during a growth experiment. The hydrogenation rate under standard conditions depends heavily on the growth time of the culture. It decreases right after the start, increases with the logarithmic phase and decreases again in the stationary phase.
Fig. 1b demonstrates that the hydrogen consumption and product formation run rather parallel. Differences are mainly due to the fact that the cells contain endogenous material(s) from which reducing equivalents derive. One of the main reasons for the differences in the hydrogenation rates shown in Table 1 is very probably differences in the time of harvesting the cells. Washing of cells with 0.05M phosphate buffer, pH 7.0, or 0.9% sodium chloride solution can reduce the hydrogenation activity to less than 50%.

Experiments to enhance the rate of hydrogenation for (*E*)-2-methyl-2-butenoate, cinnamate of (*E*)-2-buten-1-ol by growing the cells in the presence of one of these substrates were never successful. Even after three passages, the amount of cells and their hydrogenation capacity was always much lower than of such cells which were grown in the absence of these unsaturated compounds.

The cells harvested after reaching the stationary phase can be stored under different conditions. Fig. 2 shows the hydrogenation rate of lyophilized cells of *Clostridium* La 1 which were kept under different storage conditions for different times. At – 30 °C the cells are stable for at least 3 months. Sometimes during storage or lyophilization of cells, the hydrogenation capacity decreases. Addition of methylviologen (1mM) usually restores the hydrogenation rate for 2-enoates.

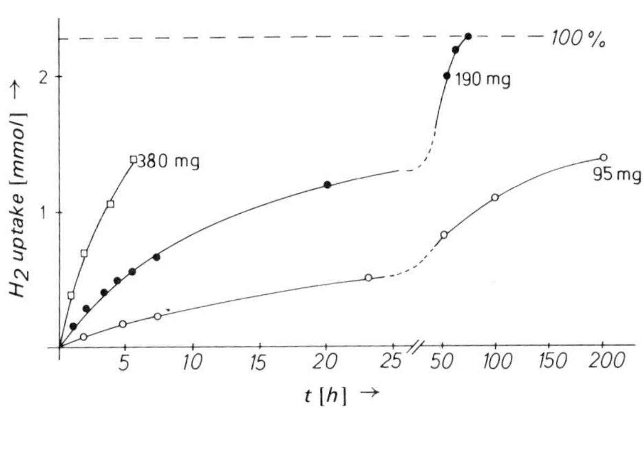

Fig. 3. Rate of hydrogen uptake catalyzed by the indicated amounts of lyophilized cells of *Clostridium* strain La 1 in the presence of 100mM cinnamate and 1.0mM methylviologen in 23 m*l* 0.1M phosphate buffer, pH 7.

The rate of hydrogenation of cinnamate by freeze-dried cells of strain La 1 as function of the amount of cells is shown by Fig. 3. This also reveals that the cells in the presence of methylviologen are active even after 100 h at 35 °C. Normally methylviologen has no effect on the hydrogenation rate for the first 20 - 30 h. After this time, it shows a considerable stabilization effect on the hydrogenation rate for up to 200 h.

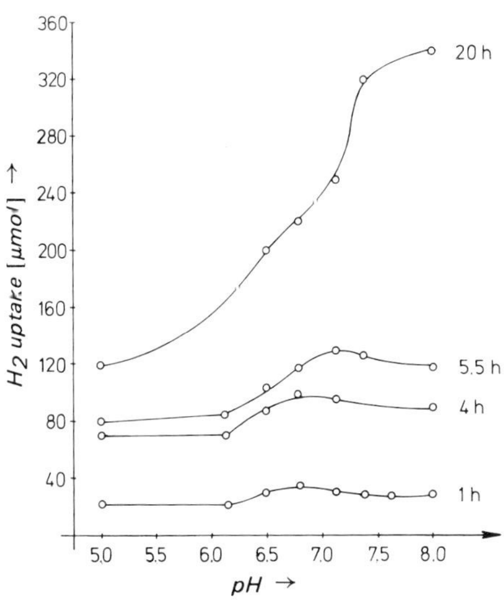

Fig. 4. Total hydrogen uptake during the times indicated, depending on the pH.
In 3.0 m*l* 0.1M phosphate buffer, 200 mg wet packed cells *Clostridium* strain La 1 were shaken at 35 °C with 500 μmol (*E*)-2-methyl-2-butenoate.

The dependence of hydrogen uptake on the pH value is rather complex (Fig. 4). For the first hour there is only a slight rate maximum between pH 5.0 and 8.0. After longer incubation times, the rate enhancement with pH increasing from pH 5.0 to 7.1 is rather pronounced, and after 20 h the rate of hydrogen uptake is the highest at pH 8.0. It seems that the enzymes involved, or at least that which is rate limiting, are more stable at higher pH values. According to Fig. 3, at pH 8 the cells may be active even longer.

As can be seen from Fig. 5 the hydrogenation rate is also a function of the concentration of (*E*)-2-methyl-2-butenoate. The rate increases up

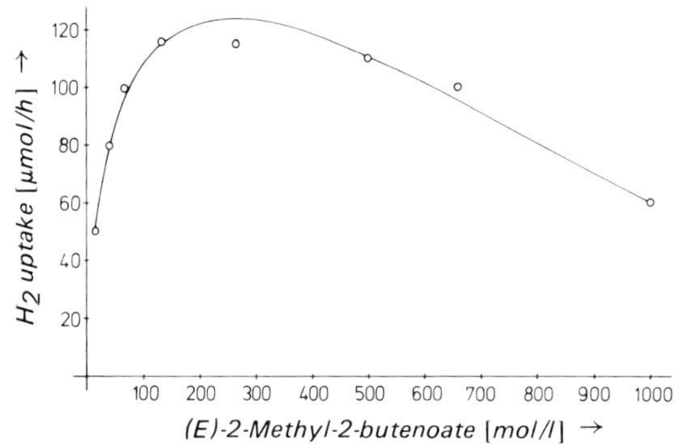

Fig. 5. Rate of hydrogen uptake depending on the concentration of (*E*)-2-methyl-2-butenoate.
Conditions as given for Fig. 4.

to a concentration of about 300mM and decreases at a concentration of 1000mM to about half of the maximum.

Fig. 6 shows an experiment with *C. kluyveri* DSM 555 which corresponds to that revealed by Fig. 1 for *Clostridium* La 1. The growth time for *C. kluyveri* is about three times longer than that of *Clostridium* La 1. In contrast to *Clostridium* La 1, *Clostridium kluyveri* forms a lot of hexanoate. The maximum of the hydrogenation rate of (*E*)-2-buten-1-ol is about 25 h after the batch has reached its stationary phase. The hydrogenation capacity of this batch is about 7 times higher for (*E*)-2-buten-1-ol than for (*E*)-2-methyl-2-butenoate.

Table 2 indicates that wet packed as well as lyophilized cells of *C. kluyveri* DSM 555 can be stored for a rather long time without a tremendous loss of activity. *C. kluyveri* DSM 555 also shows losses of hydrogenation activity by washing the cells with 0.05M phosphate buffer pH 7.0 or 0.9% sodium chloride solution at 0 °C. The reason is presumably the breaking open of cells and the solubility of enzymes. The hydrogenation rate for 2-enoates can decrease in one washing to less than half, and that for (*E*)-2-buten-1-ol to about 80%. This is in agreement with the fact that the reductase for allyl alcohols is located in the membrane fraction, while that for 2-enoates is in the supernatant of broken cells[8]. Presence of air can lead to an additional loss of more than 50% of the hydrogenation rate.

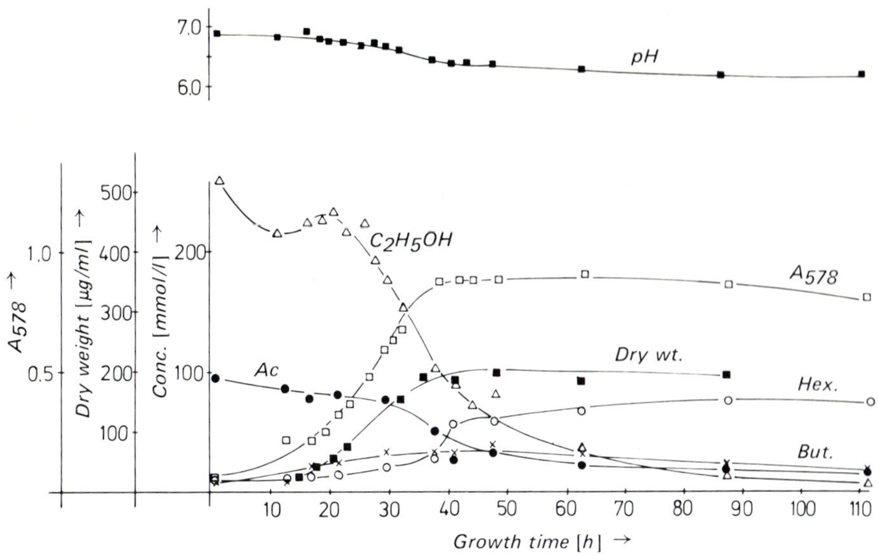

Fig. 6a). Concentration of ethanol △ △ △, acetate ● ● ●, butyrate x x x, and hexanoate ○ ○ ○, pH ■ ■ ■, dry weight ■ ■ ■ and optical density at 578 nm □ □ □ during the growth of a typical 200-*l* batch of *C. kluyveri* DSM 555 on ethanol and acetate.

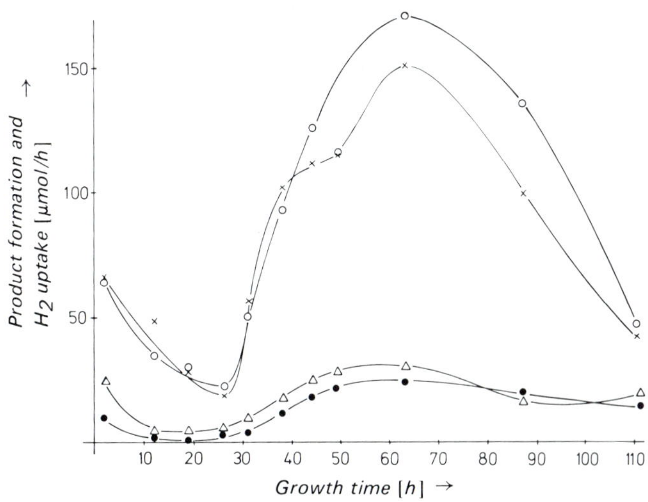

Fig. 6b). Rate of hydrogen uptake and product formation after 1 h catalyzed by 400 mg wet packed cells harvested as portions of the 200-*l* batch as depicted in Fig. 6a.
x x x Hydrogen uptake in the presence of (*E*)-buten-1-ol, ○ ○ ○ amount of butanol, △ △ △ and ● ● ● are the symbols for hydrogen uptake and formation of 2-methylbutyrate in the presence of (*E*)-2-methyl-2-butenoate.

Table 2. Rate of hydrogen uptake of *Clostridia kluyveri* DSM 555 stored under different conditions.

Time	Storage Conditions			
	Temperature	Atmosphere	Container	Hydrogenation rate (*E*)-2-buten-1-ol[a]
[Days]	[°C]			[μmol/h]
45	−18	N_2	Stainless steel tube	94
45	−18	N_2	Glass tube	86
45	−18	N_2	Polypropylene tube	129
335	−18	N_2	Polypropylene tube	82
45	−18	O_2	Polyethylene tube	74
10	−18	N_2	Lyophilized cells	76
273	−18	N_2	Lyophilized cells	84

[a] Hydrogen uptake [μmol/h] catalyzed by 400 mg wet packed cells or 100 mg lyophilized cells under standard conditions.

The preceding procedures allow the reliable preparation of highly selective and stereospecific hydrogenation catalysts. They are especially useful for the stereospecific deuteration of methylene groups. Such compounds can be applied for mechanistic and stereochemical studies in organic chemistry and biochemistry[15].

We gratefully acknowledge financial support by *Bundesministerium für Forschung und Technologie* and *Fonds der Chemischen Industrie*. We thank *Dr. N. Weiß* for valuable advice, *Dr. H. U. Holz*, Boehringer, for cell material and Miss *H. Pflaumer* as well as Mr. *H. Leichmann* for their patient and skilled technical assistance.

Literature

1 Simon, H. (1975) *Nachr. Chem. Tech.* **23**, 66 - 68.
2 Günther, H., Krezdorn, E. & Simon, H. (1976) in Fifth Int. Fermentation Symposium (Dellweg, H., ed.) p. 317, Verlag Versuchs- und Lehranstalt für Spiritusfabrikation und Fermentationstechnologie im Institut für Gärungsgewerbe und Biotechnologie, Berlin.
3 Krezdorn, E., Höcherl, S. & Simon, H. (1977) *Hoppe-Seyler's Z. Physiol. Chem.* **358**, 945 - 948.
4 Bartl, K., Cavalar, C., Krebs, T., Ripp, E., Rétey, J., Hull, W.E., Günther, H. & Simon, H. (1977) *Eur. J. Biochem.* **72**, 247 - 250.
5 Hashimoto, H., Rambeck, B., Günther, H., Mannschreck, A. & Simon, H. (1975), *Hoppe-Seyler's Z. Physiol. Chem.* **356**, 1203 - 1208.
6 Rambeck, W. & Simon, H. (1975) *FEBS Lett.* **56**, 356 - 358.
7 Rambeck, W., Günther, H., Schmidtchen, F.P. & Simon, H. (1976) *Hoppe-Seyler's Z. Physiol. Chem.* **357**, 275 - 276.
8 Kim, M.A., Günther, H., Rambeck, B. & Simon, H. (1976), *Hoppe-Seyler's Z. Physiol. Chem.* **357**, 268.
9 Bergmeyer, H.U., Holz, G., Klotzsch, H. & Lang, G. (1963) *Biochem. Z.* **338**, 114 - 121.
10 Bornstein, B.T. & Barker, H.A. (1948) *J. Biol. Chem.* **172**, 659 - 669.
11 Bornstein, B.T. & Barker, H.A. (1948) *J. Bacteriol.* **55**, 223 - 230.
12 Hungate, R.E. (1969) *Methods Microbiol.* **3B**, 117 - 132.
13 Holdeman, L.V. & Moore, W.E.C. (eds.) (1975) in Anaerobe laboratory manual, 2nd edn., Virginia Polytechnic Institute and State University, Blacksburg.
14 Schoberth, S. (1970) Diss. Göttingen.
15 Simon, H. & Kraus, A. (1976) in Isotopes in Organic Chemistry (Buncel, E. & Lee, C.C., eds.), vol. II, pp. 153 - 229, Elsevier, Amsterdam.

Dr. J. Bader, Dr. H. Günther, Prof. Dr. H. Simon, Lehrstuhl für Organische Chemie und Biochemie der TU München, Lichtenbergstr. 4, D-8046 Garching,

Dr. B. Rambeck, Gesellschaft für Epilepsie-Forschung, Biochemische Laboratorien, Maraweg 13, Bethel, D-4800 Bielefeld.

Hoppe-Seyler's Z. Physiol. Chem.
Bd. 359, S. 29 - 36, Januar 1978

Carbon and Hydrogen Metabolism of Xylitol and Various Sugars in Human Erythrocytes*

Karl-Hugo QUADFLIEG and Karl BRAND

Physiologisch-Chemisches Institut der Universität Erlangen–Nürnberg

(Received 16 August/14 October 1977)

Summary:

1) Erythrocytes are able to metabolize D-ribose, D-xylitol, D-xylulose, D-fructose and D-glucose; the rates of metabolism increase in that order from 2430 to 26200 ng atom C/m*l* packed cells per 120 min of incubation.

2) The utilization of the carbon of these substrates and its recovery in the products were found to be in balance, when the change in the 2,3-bisphosphoglycerate concentration was taken into account.

3) The metabolic rates strongly affected the 2,3-bisphosphoglycerate level. Without addition of substrate the decomposition rate of this intermediate was found to be 1030 nmol/m*l* packed cells per 120 min.

4) The net decrease of the 2,3-bisphosphoglycerate concentration and the conversion of this compound into lactate provides a $NAD^{\oplus}$ regeneration system which enables the red blood cells to utilize xylitol.

5) The rate of carbon metabolism via the pentose phosphate cycle is determined by the NADPH requirement of the erythrocytes which was found to be 160 nmol/m*l* packed cells per 120 min under the experimental conditions employed.

Stoffwechsel des Kohlenstoffs und Wasserstoffs von Xylit und verschiedenen Zuckern in menschlichen Erythrozyten

Zusammenfassung:

1) Erythrozyten sind in der Lage, die Substrate D-Ribose, D-Xylit, D-Xylulose, D-Fructose und D-Glucose mit steigenden Umsatzgeschwindigkeiten von 2430 bis zu 26200 ng-Atome C/m*l* Zellsediment während zweistündiger Inkubation umzusetzen.

2) Wurde die Änderung der 2,3-Bisphosphoglycerat-Konzentration einbezogen, so konnte in allen Fällen eine ausgeglichene Bilanz zwischen Aufnahme der Kohlenstoffatome von diesen Substraten und Wiederfindung in verschiedenen Produkten gefunden werden.

3) Die unterschiedlichen Stoffwechselgeschwindigkeiten beeinflußten den 2,3-Bisphosphoglycerat-Spiegel erheblich. Ohne Zusatz von Substraten war die Zerfallgeschwindigkeit dieses Intermediates 1030 nmol/m*l* Erythrozytensediment in 120 min.

4) Der Abbau des 2,3-Bisphosphoglycerats zu Lactat führt zur Regenerierung von $NAD^{\oplus}$, wo-

---

* A preliminary report of this work has been presented at the VIIIth International Berlin Symposium on Structure and Function of Erythrocytes, August 1st - 5th, 1976

durch den roten Blutzellen die Verwertung von Xylit ermöglicht wird.

5) Die Umsatzgeschwindigkeit über den Pentosephosphat-Weg wird durch den NADPH-Bedarf der Erythrozyten bestimmt, der unter den experimentellen Bedingungen mit 160 nmol/m*l* Erythroyztensediment in 120 min ermittelt wurde.

Key words: Xylitol, sugar, metabolism, human erythrocytes.

Mature human erythrocytes are restricted in their metabolic activities to the glycolytic and the pentose phosphate pathway. The metabolic rates of the two pathways and the carbon balances of glucose metabolism in erythrocytes under different conditions are well established[1-9].

Studies on the utilization of xylitol by different organs of the rat revealed that only liver [10-12], and to a smaller extent kidney, are able to utilize xylitol directly, while other organs like brain, adipocytes, lung and muscle cannot take up xylitol[13].

Erythrocytes, on the other hand, have been shown to utilize a variety of sugars and they are also capable of metabolizing xylitol[14-16]. In contrast to glucose metabolism, no extensive studies on the rates of fructose and xylitol metabolism via the different pathways in red cells and their relationship to the 2,3-bisphosphoglycerate bypass have been reported in the literature. Asakura et al.[15] studied the effect of xylitol metabolism in human erythrocytes on the level of several glycolytic intermediates and the reduction of methemoglobin. They reported a decrease of ATP and 2,3-bisphosphoglycerate when erythrocytes were incubated with xylitol, whereas the levels of these intermediates were not affected by xylitol in the presence of glucose. These authors, however, did not measure the amount of xylitol uptake and $^{14}CO_2$ production. Therefore, their data neither permit the construction of a balance of xylitol metabolism nor the calculation of metabolic rates.

Since glucose, fructose and xylitol, separately or in mixtures, are widely used as an energy source in parenteral nutrition, knowledge about the complete intracellular metabolism of these compounds in human erythrocytes is of special interest.

In this paper we report carbon balance studies with xylitol and various other sugars, which provide information about their complete intracellular metabolism. Our results indicate that xylitol uptake in erythrocytes is coupled with a decrease of 2,3-bisphosphoglycerate. This is due to the degradation of 2,3-biphosphoglycerate to lactate coupled to the oxidation of NADH to $NAD^{\oplus}$. Thus the extra NADH formed in the metabolism of xylitol is reoxidized, and a new balance is established which enables the red blood cell to metabolize xylitol.

Materials and Methods

Chemicals

The Krebs-Ringer phosphate buffer pH 7.4 or 7.6 (free from calcium ions and glucose) contained 123mM NaCl; 4.9mM KCl; 1.24mM $MgSO_4$; 16mM Na_2HPO_4.

Enzymes, coenzymes, nucleotides and biochemical substrates were obtained from Boehringer Mannheim GmbH.

[U-^{14}C]D-Xylitol, [U-^{14}C]D-fructose and [U-^{14}C]D-glucose were from Amersham Buchler GmbH & Co., K.G., Braunschweig. D-Xylulose was obtained from Koch-Light Laboratories Ltd., Colnbrook, U.K. All other chemicals were of reagent grade from E. Merck, Darmstadt.

Incubation

For the incubations fresh blood was taken from a cubital vein of healthy donors and heparin added to avoid coagulation. After centrifugation at 4 °C plasma and leucocytes were removed and the red cells washed four times with buffer at 4 °C. 2.5 m*l* of packed cells were suspended in 10 m*l* of Krebs-Ringer phosphate buffer pH 7.4 or 7.6 and incubated at 37 °C for 120 min in stoppered 100-m*l* Erlenmeyer flasks with a center well containing a vial for CO_2 collection. The final volume of 12.5 m*l* resulted in high metabolic rates and guaranteed a constant pH. The incubation was started by the addition of the appropriate substrate(s) and terminated after 120 min by injecting 2 m*l* of 2.2 M perchloric acid.

After acidification agitation was continued for a further 45 min for the collection of $^{14}CO_2$. With the exception of D-xylulose all substrates used were uniformly labelled with the following spec. act.: D-ribose, 244000 dpm/μmol; D-xylitol 220000 dpm/μmol; D-fructose 444000 dpm/μmol; D-glucose 328000 dpm/μmol.

Analytical Procedures

The acidified incubation mixtures were centrifuged and the pellets washed twice with 0.2 M perchloric acid. Acid supernatants plus washings were combined and neutralized with 2 M KOH. In the neutralized perchloric acid extracts, metabolic intermediates, products and non-utilized substrate(s) were determined enzymatically or in the case of [^{14}C]D-ribose by isotopic techniques described by Katz et al.[17]. From these measurements at time 0 and after 120 min of incubation, the quantity of substrate taken up, the change in the concentrations of intermediates and the quantity of products formed was calculated, and balances of the carbon metabolism of the different substrates was constructed.

Xylitol was determined with L-iditol dehydrogenase (EC 1.1.1.14)* in the following assay: 0.2 m*l* 120 mM NAD$^{\oplus}$ solution, 0.01 m*l* of L-iditol dehydrogenase (5 mg/m*l*) and 1 m*l* of glycine-semicarbazide buffer (1 M + 0.2 M) pH 9.0. Portions of sample were used to give a change of optical density in the range of 0.15 to 0.3. This method gives a 90% recovery. For correction to 100%, xylitol standards were used.

The uptake of D-xylitol was determined from the difference between the values at time 0 and after 120 min.

D-Xylulose was determined enzymatically at pH 7.4 using L-iditol dehydrogenase and NADH.

2,3-Bisphosphoglycerate was determined according to Schröter and Winter[18] and in later experiments according to Ericson and de Verdier[19].

The $^{14}CO_2$ liberated was trapped by 0.5 m*l* of 2-phenylethylamine in the center well and radioactivity determined in a Packard liquid scintillation counter using the solvent of Bray.

Results

Fig. 1 shows that xylitol uptake and lactate formation in erythrocytes proceed linearly with time up to four hours. For all further experiments an incubation period of two hours was chosen.

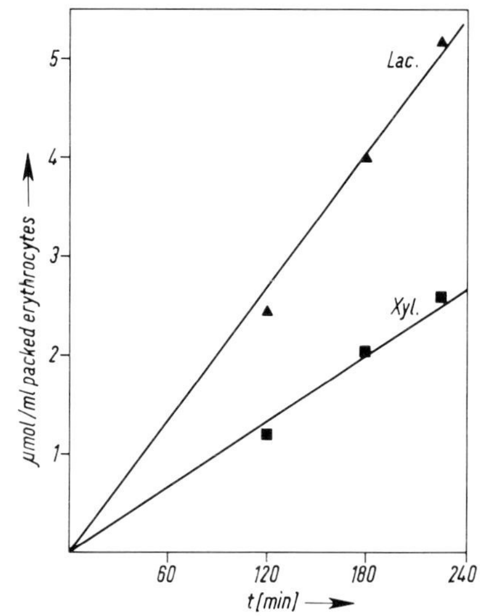

Fig. 1: Time course of xylitol utilization and lactate formation in human erythrocytes.

For experimental conditions and details see Materials and Methods. ▲——▲ = Lactate; ■——■ = xylitol.

In order to get quantitative information about the complete metabolism of xylitol in erythrocytes, balance studies of carbon utilization from substrate(s) and formation of products and metabolites were carried out.

Table 1 shows the results of balance studies without and with a variety of substrates under different experimental conditions.

Since the 2,3-bisphosphoglycerate level decreases during incubation with lower rates of substrate utilization, this compound must also be considered, together with the substrates, as a carbon donor in red cells.

At time 0 the concentration of 2,3-bisphosphoglycerate was in the range 3200 - 3600 nmol/m*l* packed cells. When the erythrocytes were incubated without substrate 1040 nmol of 2,3-bisphosphoglycerate per m*l* packed cells disappeared during 120 min of incubation and was found to be converted into lactate and pyruvate.

In the presence of substrates the utilization rate of carbon atoms varied between 2430 and 26200 ng atom C per m*l* of cells, depending on the incubation conditions (pH) and the identity of the

* Also named sorbitol dehydrogenase.

Table 1. Balances of carbon metabolism from various substrates in human erythrocytes.

Mean values of 2 - 6 experiments are expressed as ng atom C/m*l* packed erythrocytes per 120 min of incubation at pH 7.4 or 7.6 with the ranges in brackets. n.d. = not determined. For details see Materials and Methods.

Substrates	mM	pH	No. of exp.	Utilization of C-atoms from Substrate	Utilization of C-atoms from 2,3-P_2-Glycerate	ΣC	Recovery of C-atoms as Lactate	Recovery of C-atoms as Pyruvate	Recovery of C-atoms as $^{14}CO_2$	Others	ΣC	Recovery [%]
—	—	7.4	8	—	3 130 (3 003 - 3 228)	3 130	494 (399 - 654	2 950 (2 766 - 3 192)	—	—	3 444	110
	—	7.6	6	—	3 070 (2 955 - 3 147)	3 070	483 (339 - 582)	2 630 (2 544 - 2 724)	—	—	3 113	101
D-Ribose	2	7.4	4	2 430 (2 260 - 2 608)	2 650 (2 447 - 2 856)	5 080	2 590 (2 436 - 2 748)	2 600 (2 561 - 2 637)	28 (25 - 30)	—	5 118	103
Xylitol	4	7.4	6	5 440 (3 960 - 6 693)	2 420 (2 223 - 2 757)	7 860	6 730 (5 682 - 7 906)	119 (0 - 357)	35 (33 - 37)	994[a] (590 - 1 595)	7 878	100
D-Xylulose	4	7.4	2	6 940	1 540	8 480	7 460	1 780	n.d.	—	9 240	109
D-Fructose	2	7.4	4	6 500 (5 919 - 7 074)	2 100 ((1 530 - 2 667)	8 600	6 930 (6 648 - 7 218)	1 540 (953 - 2 115)	88 (86 - 89)	—	8 558	100
	1.6	7.6	2	5 980	1 770	7 750	5 970	1 580	57	—	7 607	98
D-Glucose	2	7.4	6	9 430 (8 229 - 10 776)	1 120 (465 - 1 539)	10 550	9 060 (8 352 - 1 038)	932 (741 - 1 038)	105 (101 - 109)	—	10 097	96
	1.6	7.6	2	26 200	—	26 200	20 600	—	n.d.	4 860[b]	25 460	97

[a] Formation of D-xylulose. [b] Formation of dihydroxyacetone phosphate, D-glyceraldehyde 3-phosphate and D-fructose 1,6-bisphosphate.

substrate. The lowest rate of uptake was obtained with ribose as substrate followed by xylitol, D-fructose, D-xylulose and D-glucose. Depending on the conditions, most of the carbon atoms taken up were found in lactate and/or pyruvate, while only small amounts of CO_2 were released. The accumulation of pyruvate is negatively correlated to the quantity of carbon taken up. In the case of xylitol as substrate, however, very little pyruvate was formed, although the rate of uptake was very low. This is due to the fact that excess NADH formed in the D-xylulose reductase reaction was reoxidized by converting pyruvate to lactate. D-xylulose accumulated when xylitol was applied, while with glucose as substrate at pH 7.6, glyceraldehyde 3-phosphate, dihydroxyacetone phosphate and fructose 1,6-bisphosphate accumulated. The accumulation of D-xylulose observed with xylitol as substrate indicates that the D-xylulose-kinase reaction is the rate-limiting step in xylitol uptake. As can be seen from the last column of Table 1, carbon utilization and product formation are in balance for all substrates studied. This, however, is only the case when the decrease in the level of 2,3-bisphosphoglycerate is taken into account. Xylitol utilization in mature erythrocytes depends on the ability of the cells to remove "extra NADH" formed during the D-xylulose-reductase reaction.

Degradation of one mol of 2,3-bisphosphoglycerate via pyruvate to lactate is accompanied by the regeneration of one mol of $NAD^{\oplus}$. As shown on Table 2, uptake of xylitol and decrease of the 2,3-bisphosphoglycerate concentration was almost stoichiometrical. From this result one might deduce that the "extra NADH" formed upon xylitol utilization causes the decomposition of equimolar amounts of 2,3-bisphosphoglycerate. In the presence of the oxidant phenazine methosulphate, or with D-xylulose as substrate, however, a similar decrease of this intermediate did occur. Therefore, a different or an additional reason other than reoxidation of "extra NADH" has to be assumed for the observed decrease of the 2,3-bisphosphoglycerate level.

When a mixture of glucose and xylitol was offered as substrate, no decrease in the 2,3-bisphosphoglycerate content was observed; the xylitol was not utilized and the glucose alone served as substrate. In the presence of electron acceptors,

Table 2. Dependence of xylitol utilization on NADH reoxidation.
Mean values of 2 - 6 experiments are expressed as nmol/m*l* packed cells per 120 min of incubation at pH 7.6. For details see Materials and Methods.
2,3-P_2-glycerate = 2,3-bisphosphoglycerate; Fru-1,6-P_2 = fructose 1,6-bisphosphate; Grn-3-*P* glycerone 3-phosphate; Gra-3-*P* = glyceral 3-phosphate; PMS = phenazine methosulfate.

Substrate	Conc. [mM]	ΔXylitol	ΔD-Xylulose	Δ[2,3-P_2-Glycerate]	Δ[Fru-1,6-P_2 + Grn-3-*P* + Gra-3-*P*]
Xylitol	1.6	− 828	+ 466	− 819	−
Xylitol + PMS	1.6 0.02	− 679	+ 466	− 722	−
D-Xylulose	0.4	+ 78	− 1229	− 709	−
Glucose + xylitol	1.6 1.6	− 7	−	+ 153	+ 1212
Glucose + xylitol + PMS	1.6 1.6 0.02	− 1067	+ 311	+ 1194	−
Glucose + xylutol + pyruvate	1.6 0.4 0.4	− 266	+ 19	+ 818	−
Glucose + D-xylulose	1.6 0.4	+ 536	− 1318	+ 521	+ 443

such as phenazine methosulphate or pyruvate, however, xylitol is metabolized in the presence of glucose, in spite of the fact that the 2,3-bisphosphoglycerate concentration increases extensively. D-Xylulose, on the other hand, can be metabolized together with glucose, without the addition of oxidants. These results indicate the necessity of reoxidation of NADH for xylitol metabolism either by degradation of 2,3-bisphosphoglycerate to lactate or by oxidants. As Table 2 shows, whenever xylitol was metabolized, D-xylulose accumulated regardless of whether oxidants were present or not. As pointed out above, this again indicates that the D-xylulose-kinase reaction is the rate-limiting step in xylitol metabolism.

When the red cells were incubated with mixtures of substrates in the presence of oxidants a pronounced increase of the 2,3-bisphosphoglycerate concentration was detected. With glucose and xylitol as substrates the increase of the 2,3-bisphosphoglycerate amounted to only 152 nmol/m*l* packed cells in 120 min in the absence of oxidants. Under these conditions, however, this accumulation gives rise to a corresponding formation of NADH and consequently to an inhibition of D-glyceraldehyde-3-phosphate dehydrogenase as indicated by the accumulation of fructose 2,6-bisphosphate and triosephosphate.

Fig. 2 shows the production of $^{14}CO_2$ from various uniformly labelled substrates against the rate of carbon utilization. Since in mature erythrocytes CO_2 is liberated only in the oxidative route of the pentose cycle, the amount of CO_2 formed during 120 min of incubation corresponds to the rate of carbon metabolism via this pathway. As can be seen in this figure, $^{14}CO_2$ production increases with increasing carbon utilization. At a metabolic rate of about 7 μg atom C/m*l* packed cells in 120 min, CO_2 production amounts to about 80 nmol and remains within this range, even at higher rates of carbon utilization.

Discussion

The main point of this paper was to compare xylitol metabolism in human erythrocytes with that of pentoses and hexoses, which is of special

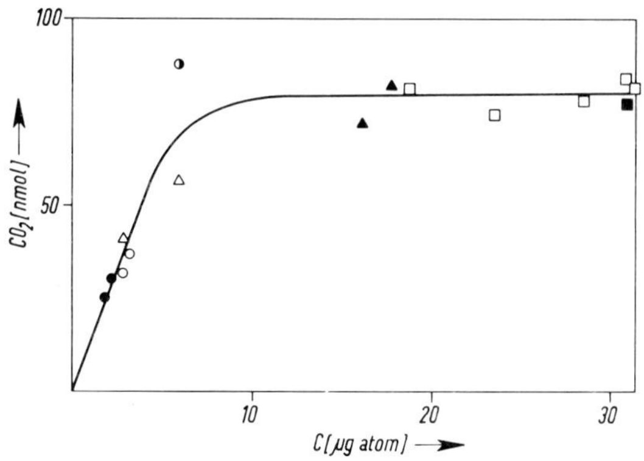

Fig. 2: Pentose cycle activity in human erythrocytes in relation to carbon utilization rates from various substrates.

Values of individual experiments (n = 15) performed in duplicate are given as µg atom C taken up from substrate and nmol CO_2 formed per m*l* packed cells during 120 min of incubation at pH 7.4 or 7.6. For details see Materials and Methods section. Symbols for substrates and incubation conditions:

- ●: D-ribose 2mM, pH 7.4;
- ◑: D-fructose 2mM, pH 7.4;
- ○: xylitol 4mM, pH 7.4.
- △: D-fructose 1.6mM or 2mM, pH 7.6;
- ▲: D-glucose 0.8mM or 1.6mM pH 7.4;
- □: D-glucose 0.8mM or 1.6mM, pH 7.6;
- ■: D-glucose 1.6mM and xylitol 0.4mM, pH 7.6;

interest in connection with the clinical use of these compounds as a source of energy in parenteral nutrition. In order to obtain quantitative and complete information about the fate of all carbon atoms of the various substrates taken up by the red cells, carbon balances have been constructed. From the balance studies it can be concluded that erythrocytes are able to metabolize a variety of substrates as a source of energy but with significantly different rates, which strongly affect the 2,3-bisphosphoglycerate level. There are no data available in the literature which allow a comparison of D-ribose, xylitol, D-xylulose and D-fructose uptake by erythrocytes. Willgerodt et al.[20] concluded from their measurements of xylitol concentration in human blood after infusion with 0.76 g xylitol/kg body weight and hour that this polyalcohol does not penetrate the erythrocytes, since they did not find a change in the xylitol concentration.

In experiments under comparable conditions, in which glucose and xylitol were applied as substrates (Table 2), we were also unable to observe an uptake of xylitol. Our direct in vitro measurements, however, clearly show that erythrocytes are able to metabolize xylitol. This is consistent with results described by Baessler and Reimold[14] and Asakura et al.[15] who were able to determine lactate formation from xylitol in red blood cells.

In contrast to fructose, the uptake of glucose strongly depends on pH. Our results obtained with glucose are in good agreement with data already reported[4,7-9]. When the red cells were incubated without any substrate a decomposition rate of 2,3-bisphosphoglycerate of 500 nmol/m*l* packed cells in 60 min was determined, which is in excellent agreement with that reported by Duhm et al.[21] and I. Rapoport et al.[22]. With all compounds listed in Table 1 the carbon atoms utilized from the substrates applied and from the endogenous 2,3-bisphosphoglycerate pool were completely recovered in products. Since the carbon metabolism of the various substrates has been found to be in balance, the formation of products other than lactate, pyruvate and CO_2 can be excluded. As can be seen from Table 1 the extent of the decrease of the 2,3-bisphosphoglycerate concentration strongly depends on the rate of utilization of carbon atoms from substrate. At a carbon-utilization rate of 14 µg atom C/m*l* packed cells during 120 min of incubation it has been shown[23] that the level of 2,3-bisphosphoglycerate remains constant. The large deviation in the decrease of this compound shown in this table with glucose as substrate at pH 7.4 is due to different carbon utilization rates in the individual experiments (8230 to 10800 ng atom C) which cause a different decrease in the level of 2,3-bisphosphoglycerate.

Our results indicate that xylitol metabolism in red blood cells is coupled with the decrease of 2,3-bisphosphoglycerate which in turn is converted into lactate, thus regenerating NAD$^{\oplus}$. The observation that uptake of xylitol and decrease of the 2,3-bisphosphoglycerate concentration were almost equal, and the fact that with a mixture of glucose and xylitol no decrease in the 2,3-bisphosphoglycerate content was observed and no xylitol but only glucose was found to be utilized (Table 2), clearly indicate the necessity for the reoxidation of NADH by converting 2,3-bisphosphoglycerate via pyruvate to lactate. This concept is further supported by the finding that the addition of electron acceptors such as pyruvate or phenazine methosulphate led to the utilization of xylitol even in the presence of glucose.

Since the 2,3-bisphosphoglycerate concentration, however, also decreased in the presence of xylitol and phenazine methosulphate or with D-xylulose as substrate, it can be concluded that net decomposition of 2,3-bisphosphoglycerate during the incubation period primarily occured in order to keep the ATP concentration constant, as has been shown by Quadflieg and Brand[23]. When small amounts of substrate were metabolized, 2,3-bisphosphoglycerate therefore obviously acts as an "energy source", yielding ATP in the pyruvate-kinase step. Breakdown of 2,3-bisphosphoglycerate to lactate, however, concomitantly provides the reoxidation of the "extra NADH" thus facilitating uptake and metabolism of polyalcohols like xylitol in erythrocytes.

The rate limiting step for xylitol metabolism in red blood cells was shown to be the D-xylulose-kinase reaction. This is proven by the accumulation of D-xylulose, observed with xylitol as a substrate, even when regeneration of $NAD^{\oplus}$ was sufficient, either by 2,3-bisphosphoglycerate breakdown or by the addition of electron acceptors (Table 2).

Our results show that xylitol per se causes a decrease in the 2,3-bisphosphoglycerate content and a marked increase of the lactate-pyruvate ratio in erythrocytes (Table 1). These effects would restrict the use of xylitol in parenteral nutrition. Under physiological conditions, however, glucose is always present and under these conditions, erythrocytes do not utilize xylitol, as can be seen from Table 2.

Our results obtained from $^{14}CO_2$ measurements (Fig. 2) suggest that the quantity of carbon atoms from the various substrates entering the oxidative route of the pentose phosphate pathway is determined by the demand of the red cells for NADPH. An increase in the $^{14}CO_2$ production was evident only up to a carbon uptake of 7 μg atom C/m*l* cells in 120 min. Further increase in carbon uptake did not result in a corresponding production of $^{14}CO_2$. If the demand of NADPH is covered – in our experiments 160 nmol/m*l* cells in 120 min – no further metabolism via the pentose cycle occurs, and all carbon atoms are degraded in the Embden-Meyerhof pathway. The same conclusion could also be drawn from experiments in which erythrocytes were incubated at different pH ranges[9].

This work was in part supported by the *Jacques-Pfrimmer-Gedächtnisstiftung*, Erlangen.
The authors wish to thank Miss *Ch. Ostermann* and Mrs. *R. Hinrichs* for the excellent technical assistance and Mrs. *Racano* for the preparation of the figures.

Literature

1 Brin, M. & Yonemoto, R.H. (1958) *J. Biol. Chem.* **230**, 307 - 317.
2 Murphy, J.R. (1960) *J. Lab. Clin. Med.* **55**, 286 - 302.
3 Blum, K.O. & Glogner, F. (1962) *Z. Exp. Med.* **136**, 52 - 58.
4 Jahrmärker, H. (1964) *Int. Un. Biochem.* **32**, 514 - 518.
5 Minakami, S., Suzuki, C., Saito, T. & Yoshikawa, H. (1965) *J. Biochem.* **58**, 543 - 550.
6 Minakami, S. & Yoshikawa, H. (1966) *J. Biochem.* **59**, 145 - 150.
7 Rapoport, S. & Jacobasch, G. (1968) in Stoffwechsel und Membranpermeabilität von Erythrozyten und Thrombozyten (Deutsch, E., Gerlach, E. & Moser, K., eds.) p. 1, G. Thieme-Verlag, Stuttgart.
8 Roigas, H., Zoellner, E., Jacobasch, G., Schultze, M. & Rapoport, S. (1970) *Eur. J. Biochem.* **12**, 24 - 30.
9 Brand, K., Arese, P. & Rivera, M. (1970) *Hoppe-Seyler's Z. Physiol. Chem.* **351**, 501 508.
10 Müller, F., Strack, E., Kuhfahl, E. & Dettmer, D. (1967) *Z. Gesamte Exp. Med.* **142**, 338 - 350.
11 Wang, M.C. & Meng, H.C. (1971) *Z. Ernährungswiss. Suppl.* **11**, 8 - 16.
12 Woods, H.F. & Krebs, H.A. (1973) *Biochem. J.* **134**, 437 - 443.
13 Quadflieg, K.-H. & Brand, K. (1976) *Z. Ernährungswiss.* **15**, 345 - 354.
14 Bässler, K.H. & Reimold, W.V. (1965) *Klin. Wochenschr.* **43**, 169 - 171.
15 Asakura, T., Adachi, K., Minakami, S. & Yoshikawa, H. (1967) *J. Biochem.* **62**, 184 - 193.
16 Quadflieg, K.-H. & Brand, K. (1975) *Z. Ernährungswiss.* **14**, 268 - 271.
17 Katz, J., Brand, K., Golden, S. & Rubinstein, D. (1974) *Cancer Res.* **34**, 872 - 877
18 Schröter, W. & Winter, P. (1967) *Klin. Wochenschr.* **45**, 255 - 261.
19 Ericson, A. & de Verdier, C.H. (1972) *Scand. J. Clin. Lab.* **29**, 85 - 90.
20 Willgerodt, H., Beyreiss, K. & Thiele, H. (1972) *Acta Biol. Med. Ger.* **28**, 651 - 665.

21 Duhm, J., Deuticke, B. & Gerlach, E. (1969) *Pfluegers Arch.* **306**, 329 - 340.
22 Rapoport, I., Berger, H., Elsner, R. & Rapoport, S.M. (1976) in Abstracts of the VIII International Berlin Symposium on Structure and Function of Erythrocytes, pp. 98 - 99.
23 Quadflieg, K.-H. & Brand, K. (1977) *Eur. J. Biochem.* in press.

K.H. Quadflieg u. Prof. Dr. K. Brand, Inst. f. Physiol. Chem. d. Univ., Lehrstuhl II, Egerlandstr. 7, D-8520 Erlangen.

Hoppe-Seyler's Z. Physiol. Chem.
Bd. 359, S. 37 - 45, Januar 1978

The Functional Tryptophan and Tyrosine Residues of F_1 ATPase of *Micrococcus sp.* ATCC 398

Sergio RISI, Christina SCHRÖDER, Hans-Joachim KAISER, José CARREIRA* and Klaus DOSE

Institut für Biochemie der Universität Mainz and
* Instituto de Biología Celular, C.S.I.C., Madrid

(Received 19 August 1977/14 November 1977)

Dedicated to Theodor Wieland on the occasion of his 65th birthday

Summary: Bacterial F_1 ATPase from *Micrococcus sp.* ATCC 398 contains a total of 16 tryptophan residues per molecule.

In the native enzyme, six tryptophan residues per molecule are readily accessible to oxidation with *N*-bromosuccinimide.

The correlation between inactivation and modification of the six accessible tryptophan residues is linear on a semilogarithmic plot.

The present data indicate that the destruction of only about one tryptophan residue per molecule completely inactivates the enzyme.

All six accessible tryptophan residues are probably part of nucleotide binding sites, inasmuch as these residues are fully protected against oxidation in the presence of adenosine 5′-[β,γ-imido]-triphosphate (Ado*PP*NH*P*).

A cooperativity between the subunits carrying nucleotide binding sites is suggested.

The activity of F_1 ATPase is reduced to 37% after about 15 tyrosine residues have become nitrated with tetranitromethane.

The substrate analogue Mg · Ado*PP*NH*P* protects two of the tyrosine residues from nitration without, however, completely protecting the enzyme from inactivation.

Only 30% of the enzymic activity is lost after 32 (out of 60) tyrosine residues are acetylated by *N*-acetylimidazole.

Mg · Ado*PP*NH*P* also protects two tyrosine residues from acetylation, without complete protection from inactivation.

It is shown that these two tyrosine residues are required to keep the pH optimum of the enzyme at pH 8.2.

Die funktionellen Tryptophan- und Tyrosinreste der F_1-ATPase aus Micrococcus sp. ATCC 398

Zusammenfassung: Die F_1-ATPase aus *Micrococcus sp.* ATCC 398 enthält 16 Tryptophanreste pro Molekül, sechs davon werden – in der nativen ATPase – durch *N*-Bromsuccinimid bei neutralen pH-Werten leicht oxidiert. Die Beziehung zwischen Inaktivierung des Enzyms und Modifizie-

---

Enzymes:
Adenosinetriphosphatase (ATPase, F_1), ATP phosphohydrolase (EC 3.6.1.3);
Deoxyribonuclease I (DNase I), deoxyribonucleate 5′-oligonucleotidohydrolase (EC 3.1.4.5);
Ribonuclease I (RNase I), ribonucleate 3′-pyrimidino-oligonucleotidohydrolase (EC 3.1.4.22).

Abbreviation:
Ado*PP*NH*P* Adenosine 5′-[β,γ-imido]triphosphate.

rung der sechs zugänglichen Tryptophanreste ist in einer semilogarithmischen Auftragung linear. Die Auswertung dieser Daten erlaubt die Aussage, daß die Zerstörung eines einzigen Tryptophanrestes die vollständige Inaktivierung des Enzyms verursacht.

Alle sechs zugänglichen Tryptophanreste sind mit großer Wahrscheinlichkeit Bestandteile von Nucleotidbindungsstellen, denn diese Tryptophanreste werden durch das Substratanalogon Adenosin-5'-[β,γ-imido]triphosphat vollständig gegen Oxidation geschützt. Eine Kooperativität unter den Untereinheiten, welche Nucleotidbindungsstellen enthalten, wird vorgeschlagen.

Auch Tyrosinreste sind an der Funktion der ATPase beteiligt. Die Aktivität des Enzyms wird auf 37% reduziert, nachdem 15 Tyrosinreste durch Tetranitromethan nitriert worden sind.

Adenosin-5'-[β,γ-imido]triphosphat schützt zwei Tyrosinreste vor Nitrierung, ohne jedoch das Enzym vollständig gegen Inaktivierung zu schützen; dasselbe gilt, wenn die ATPase mit *N*-Acetylimidazol acetyliert wird. In diesem Fall wird die ATPase um etwa 30% inaktiviert, wenn 32 (von 60) Tyrosinreste acetyliert wurden. Es konnte gezeigt werden, daß die zwei schützbaren Tyrosinreste für die Aufrechterhaltung des pH-Optimums (8.2) der ATPase notwendig sind.

Key words: Adenosinetriphosphatase, tryptophan, tyrosine.

The membrane-bound ATPase from *Micrococcus sp.* ATCC 398 is thought to be a coupling factor of the respiratory chain[1]. In vivo it supposedly catalyzes the final steps in the synthesis of ATP. The solubilized coupling factor F_1, however, only catalyzes the hydrolysis of ATP to ADP and P_i.

The bacterial F_1 ATPases are related to the corresponding enzymes from mitochondria or chloroplasts. The subunit structure of bacterial F_1 ATPases appears to be similarly complex, $\alpha_3 \beta_3 \gamma\delta\epsilon$ [1-4], although also simpler, yet fully active molecular assemblies such as $\alpha_3 \beta_3 \gamma$ or even $\alpha_3 \beta_3$ or $\alpha'_3 \beta_3$ have been prepared[5,6].

The role of tryptophan in the function of F_1 ATPase has not been investigated in detail, although tryptophan appears suitable for noncovalent binding of the adenine nucleus by forming a charge transfer complex[7,8]. Moreover, the tryptophan residue seldom appears in the tables on the amino acid composition of ATPases reported in the literature, and when it is listed, there are significant discrepancies[9-11].

The significance of tyrosine residues for the ATPase function has found more consideration.

Evidence has been accumulated which indicates that a special tyrosine residue of mitochondrial ATPase can be selectively modified by the agent 4-chloro-7-nitro-benzofurazan[12].

The ATPase so modified is inactive. After incubation at pH 9.0, the nitrobenzofurazan group undergoes an intramolecular transfer to an amino group[13]. The ATPase of *Micrococcus denitrificans* can also be inactivated by modification of tyrosine residues[14]. Less well established, so far, is the mode of inactivation of the solubilised membrane ATPase of *Escherichia coli* by the same agent[15,16].

In preliminary experiments we have failed to modify tyrosine residues of the solubilised F_1 ATPase of *Micrococcus sp.* ATCC 398 with 4-chloro-7-nitrobenzofurazan. Although this enzyme is readily inactivated by the agent at pH 7.5, one or two amino groups per molecule become labeled instead of tyrosine residues[17].

In order to establish whether tyrosine residues are essential for the function of the solubilised ATPase of *Micrococcus sp.* ATCC 398, we have therefore applied other tyrosine-modifiying agents, *N*-acetylimidazole according to Simpson et al.[18], and tetranitromethane according to Sokolowsky et al.[19]. These results and our data on the oxidation of functional tryptophan residues with *N*-bromosuccinimide according to Spande and Witkop[20] are described below.

Materials and Methods

Materials

Reagents were purchased as follows:
Tetranitromethane from Fluka, Buchs; RNase I, DNase I (purity grade I), ATP and adenosine 5'-[β,γ-imido]triphosphate (Ado*PP*NH*P*) from Boehringer Mannheim GmbH, and 2-mercaptoethanol from Serva, Heidelberg.

N-Bromosuccinimide was purchased from Serva (Heidelberg) and was twice recrystallized from water. Methanesulfonic acid and tryptamine (purissimum) were from Roth (Karlsruhe). 5,5′-Dithiobis(2-nitrobenzoic acid) was purchased from Calbiochem (Los Angeles).

All other chemicals were of analytical grade and purchased from Merck (Darmstadt).

Methods

Soluble bacterial F_1 ATPase from the membranes of *Micrococcus sp.* ATCC 398 was prepared and purified according to Risi et al.[21]. This ATPase preparation largely consists of a molecular assembly with the subunit composition $\alpha_3 \beta_3 \gamma \delta \epsilon$. γ, δ and ϵ account for only 10% of the total protein. The amino acid compositions of the subunits α and β appear to be closely related (unpublished results). The ultraviolet absorption spectrum of this preparation is that of a pure protein and agrees with the presence of 16 tryptophan residues and 60 tyrosine residues[22]. Thus the UV absorption spectrum would not be compatible with the presence of more than two tightly bound nucleotides per molecule. The activity of the ATPase samples was assayed semiautomatically (system Technicon) by determination of inorganic phosphate released during hydrolysis of ATP [23-25].

The amount of protein was determined according to Lowry et al.[26].

The amino acid composition of the protein was determined after hydrolysis in either 6N HCl for 24 h or 4M methanesulfonic acid containing 0.2% tryptamine for 16 h[27] at 110 °C in vacuo with a Biotronik amino acid analyzer.

The modification of the tryptophan residues with *N*-bromosuccinimide was carried out by the method of Spande and Witkop[20]. The decrease in optical density at 280 nm was determined with a Cary 118C spectrophotometer. The oxidation was performed either in 8M urea at pH 4 (adjusted by addition of acetic acid) or in the absence of urea in 100mM Tris/HCl buffer, pH 7.5. For the calculation, a molecular mass of 350000 daltons was assumed.

The exceptional protection of the tryptophan residues against oxidation with *N*-bromosuccinimide after preincubation with Ado*PP*NH*P* could indicate that Ado*PP*NH*P* was converted into a more strongly binding nucleotide analogue by *N*-bromosuccinimide. Therefore, *N*-bromosuccinimide-treated samples of Ado*PP*NH*P* were analysed by thin-layer chromatography on cellulose plates MN 300 (Machery & Nagel, Düren) with the following solvents: isobutyric acid/conc. NH_3/water 60/1/33 (by vol.) and amyl alcohol/formic acid/water 3/2/1 (by vol.). The spots were detected by the UV absorption of the purine nucleus.

The following experiments were carried out in order to reveal whether or not the amino acids methionine or cysteine are essential:

Modification of the enzyme with 2-bromo-4-nitroacetophenone according to Sigman and Blout[28]; with *p*-chloromercuribenzoate according to Boyer[29]; with iodoacetic acid and iodoacetamide according to Gurd[30]; with 5,5′-dithiobis(2-nitrobenzoate) according to Ellman[31]; and photooxidation in the presence of crystal violet according to Jori et al.[32].

Histidine residues were alkylated with bromoacetone according to Beeley and Neurath[33].

The acetylation of the ATPase was carried out essentially according to Simpson et al.[18] using *N*-acetylimidazole. The agent was incubated with the enzyme in 10mM Tris/HCl at pH 7.5[34] for periods up to 2 h at 20 °C. It was found that a 20-fold excess of *N*-acetylimidazole was optimal in order to study the kinetics of the inactivation within about 90 min. The number of acetylated tyrosine residues was determined spectroscopically at 278 nm[18]. The reaction was stopped by dilution with a 500-fold volume of buffer or by gel filtration on Sephadex G-50.

The nitration was carried out with tetranitromethane according to Sokolowsky et al.[19]. In order to avoid crosslinking[35,36], and to control the reaction in a way which allowed kinetic studies, the following modification of the above method was applied.

The reaction was carried out with 1 to 2 mg protein/m*l* in 50mM Tris/HCl, pH 8.0, at 4 °C. In order to avoid an undue excess of the reagent at the beginning of the reaction, every 30 min about a three-fold excess of tetranitromethane (compared with the number of tyrosine residues) was added. The reagent was dissolved in ethanol (16.8mM). Because of the photochemical activity of 3-nitrotyrosine, the reaction was carried out in the dark. Termination of the reaction and removal of the reagents and other low molecular materials was again achieved by gel filtration on a Sephadex G-50 column (9 × 120 mm).

When the kinetics of the reaction between tetranitromethane and the ATPase was studied, the reaction was stopped by adding an excess of mercaptoethanol (0.06M solution).

Results and Discussion

The total content of tryptophan

The titration of the bacterial ATPase in 8M urea at pH 4 with *N*-bromosuccinimide according to Spande and Witkop[20] indicated the presence of 16.2 ± 0.6 mol tryptophan per mol F_1 ATPase. The amino acid analysis of the hydrolysate obtained by hydrolysis in 4N methanesulfonic acid yielded a tryptophan content of 0.5 ± 0.1 mol-%. This value corresponds with 16 ± 3.2 tryptophan residues for an F_1 ATPase consisting of 3 200 amino acid residues (M_r 350 000). The spectrophotometric analysis according to Goodwin and Morton[37] yielded 17.2 ± 1.4 tryptophan residues per molecule. The similar results obtained by three different methods thus focus on a value of 16 to 17 tryptophan residues per molecule. The amino acid composition of the α- and β-subunits, which are essential for ATPase function, is very similar (in preparation); this result also applies to their tryptophan content. Nieto et al.[11] have found 23 tryptophan residues per molecule ATPase from *Micrococcus lysodeicticus* by application of the method of Goodwin and Morton[37]. These values are significantly higher than those reported here. Tightly bound nucleotides may interfere with the method of Goodwin and Morton. The ATPase preparation used in the present study contained less than two, if any, tightly bound nucleotides per molecule. Some enzyme samples were treated according to Garrett and Penefsky[38] in order to remove tightly bound nucleotides, but the reactivity of the tryptophan residues was the same in these samples.

Exposed and essential tryptophan residues

In order to determine the number of exposed tryptophan residues the titration with *N*-bromosuccinimide was carried out under various conditions, as shown in Table 1.

Unfortunately, the enzyme is irreversibly inactivated both by 8M urea and by acidic solutions (pH 4). In order to correlate the number of modified tryptophan residues with the loss of enzymic activity, the titration with *N*-bromosuccinimide therefore had to be carried out at pH values at which the enzyme is stable. For this purpose, we chose a pH of 7.5, at which value the reactivity of *N*-bromosuccinimide is only slightly decreased in comparison with pH 4.0. In the presence of 8M urea, 15 tryptophan residues are readily oxidized (Table 1, line 2). In the absence of 8M urea, however, only six tryptophan residues are oxidized (Table 1, line 3).

The oxidation with *N*-bromosuccinimide depends on the presence of urea[20,39]. In the presence of urea, between 6 and 7 mol *N*-bromosuccinimide is consumed per mol tryptophan modified. In the absence of urea, the corresponding value is about 20. Even in the absence of urea, *N*-bromosuccinimide leads to no detectable degradation of other amino acids within the pH-range 4.0 - 7.5. $Mg^{2\oplus}$ ions (Table 1, line 4) have no protective effect, whereas 25μM Mg · Ado*PP*NH*P* completely protects all enzyme tryptophan residues against attack by *N*-bromosuccinimide. Control experiments with Mg · Ado*PP*NH*P* in the presence of free tryptophan showed no influence of Mg · Ado*PP*NH*P* on the modification of tryptophan by *N*-bromosuccinimide.

Table 1: Results of *N*-bromosuccinimide titration of ATPase in different media.

Line	Conditions	Tryptophan residues/ ATPase molecule
1	8M urea, pH 4.0 (acetic acid)	16.2 ± 0.6
2	8M urea, pH 7.5 (Tris buffer 100mM)	15 ± 1
3	pH 7.5 (Tris buffer 100mM)	6 ± 1
4	2mM $MgSO_4$, pH 7.5 (Tris buffer 100mM)	6 ± 1
5	25μM.₅ Mg · Ado*PP*NH*P*, pH 7.5 (Tris buffer 100mM)	0

The enzyme could not be protected against *N*-bromosuccinimide by 25μM ADP under the present conditions. There could be at least two reasons for this: firstly, Mg-ADP and Mg-Ado*PP*NH*P* react in different ways with the present F_1 ATPase*, and secondly, the ADP-enzyme complex ($K_i = 3.0 \times 10^{-5}$M) is not as tight as the Ado*PP*NH*P*-enzyme complex ($K_i = 1.2 \times 10^{-6}$M).

Fig. 1 shows the correlation between the number of tryptophan residues destroyed and the residual activity of the enzyme. The oxidation was carried out under the conditions described in Table 1, line 3. The enzyme is almost completely inactivated after five to six of the six to seven accessible

* Höckel, M., Hulla, F.W., Risi, S. & Dose, K. (1978) *J. Biol. Chem.*, submitted.

tryptophan residues are oxidized. After destruction of an average of three tryptophan residues per molecule, the residual activity is less than 20%. As can be deduced from Fig. 1, about 60% inactivation is observed after an average destruction of a single tryptophan residue per molecule.

Control experiments

Particular attention has been paid to the possible oxidation of methionine, cysteine, histidine, and tyrosine residues by *N*-bromosuccinimide. Additional analytical experiments with agents modifying methionine and/or cysteine, such as 2-bromo-4-nitroacetophenone[28], *p*-chloromercuribenzoate[29], iodoacetic acid[30], iodoacetamide[30], 5,5'-dithiobis(2-nitrobenzoate)[31] and photooxidation in the presence of crystal violet[32] have shown that these two amino acids are not essential for the function of the enzyme. No decrease in the number of tyrosine and histidine residues was observed by amino acid analysis after all six reactive tryptophan residues had been oxidized by *N*-bromosuccinimide. Also, experiments with bromoacetone, which were designed to alkylate exposed histidine residues according to Beeley and Neurath[33], have not shown the presence of functional histidine residues. Our results thus indicate that the inactivation obtained by oxidation with *N*-bromosuccinimide must be very largely, if not exclusively, attributed to the specific destruction of up to six tryptophan residues.

The number of easily accessible tryptophan residues had not been determined earlier. Inasmuch as all six accessible tryptophan residues are completely protected against *N*-bromosuccinimide by 25µM Mg · Ado*PP*NH*P*, we suggest that all six tryptophan residues are involved in nucleotide binding. Ado*PP*NH*P* is a competitive inhibitor of this ATPase[17]. It is not yet clear, however, whether it binds preferentially to the catalytic sites or protects them via conformational changes induced by binding to regulatory sites, or both. Photoaffinity labelling experiments with 8-azido-purine nucleotides are under way in this laboratory to resolve this question. Mg · Ado*PP*NH*P* is a more powerful protective agent than Mg · ADP, because the K_i for the Mg · Ado*PP*NH*P* is 1.2µM [21], whereas it is 30µM for Mg · ADP. Mg · Ado*PP*NH*P* has no effect on the oxidation of free tryptophan by *N*-bromosuccinimide, nor does Mg · Ado*PP*NH*P* appear to be altered by treatment with *N*-bromosuccinimide, as revealed by thin-layer chromatography with two different solvents and UV spectroscopy.

The presence of up to six functional tryptophan residues is compatible with the finding of about five tight binding sites for adenine nucleotides [38,40-42] as established for mitochondrial ATPase. The inference that each of the six major subunits (α- and β-subunits) of the bacterial F_1 ATPase contains one binding site is so far inconclusive.

If all six functional tryptophan residues were equally sensitive to *N*-bromosuccinimide and operated independently during the enzymatic reaction, 50% of the activity would remain after destruction of an average of three tryptophan

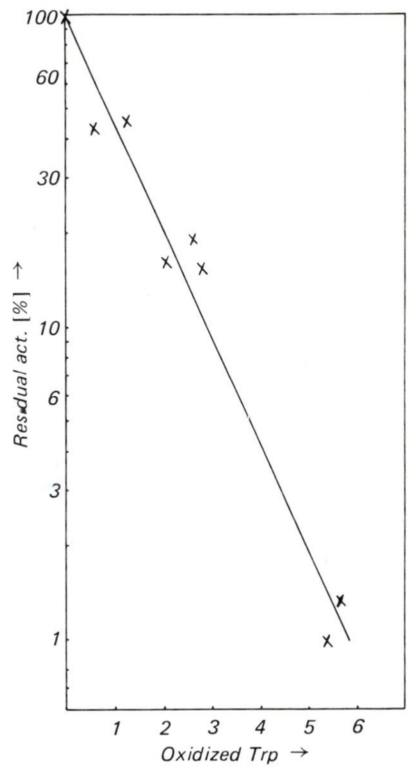

Fig. 1. Correlation between ATPase activity and the number of tryptophan residues oxidized with *N*-bromosuccinimide.

(Solutions of ATPase in 100mM Tris buffer, pH 7.5).

residues per molecule. Only about 40% of the activity remains, however, after destruction of a single tryptophan residue. This result may be interpreted to indicate a cooperativity between the subunits carrying tryptophan at the nucleotide binding sites, and/or a preferential oxidation of a single essential tryptophan residue.

The assumption of cooperativity between the subunits required for nucleotide binding and ATPase function is also in agreement with earlier findings concerning the cooperativity between the α- and β-subunits:

Trypsin degradation of the bacterial F_1 ATPase primarily yields an active $\alpha'_3\beta_3$ complex which becomes successively converted into an inactive α'_2 (x, y)β_3-complex[6].

Reaction of tyrosine residues with N-acetylimidazole

Under the experimental conditions described here, not only the OH-groups of tyrosine, but also the ϵ-NH_2 groups of lysine could become acetylated, to a minor degree. The number of acetylated tyrosine residues was therefore calculated from the change in the optical density at 278 nm[18,34].

Table 2 shows that after acetylation of no less than about 32 out of 60 tyrosine residues, a residual ATPase activity of about 69% still remains. The present data thus suggest that only a limited number of functional tyrosine residues, if any, are acetylated at the time of the ATPase assay. If the acetylation is carried out in the presence of 0.1mM Mg · Ado*PP*NH*P*, about two fewer tyrosine residues (30 instead of 32) become acetylated. Under these conditions, a residual ATPase activity of about 85% is observed. The data thus indicate that the selective protection of two tyrosine residues by Mg · Ado*PP*NH*P* yields only a partial protection against inactivation.

The corresponding pH-activity diagrams shown in Fig. 2 suggest that the two tyrosine residues which are protected in the presence of Mg · Ado*PP*NH*P* may be required to maintain the pH optimum at pH 8.2. The pH optimum completely disappears

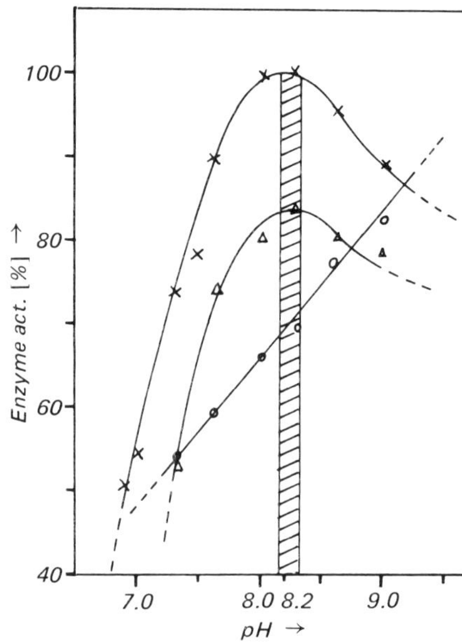

Fig. 2. pH-activity diagram of F_1 ATPase after acetylation with *N*-acetylimidazole.

x—x—x: native enzyme; ○—○—○: acetylation without protection; △—△—△: acetylation in the presence of 0.1mM Mg · Ado*PP*NH*P*.

See also Table 2.

Table 2: Protection of tyrosine residues of F_1 ATPase by 0.1mM Mg · Ado*PP*NH*P* against modification with *N*-acetylimidazole and tetranitromethane.

	Nitrated residues per molecule	Residual activity [%]	Acetylated residues per molecule	Residual activity [%]
ATPase unprotected	14.6	37	31.9	69
ATPase protected	12.1	52	30.0	85
Protected tyrosine residues	2.5 ± 0.4		1.9 ± 0.4	

if the ATPase is acetylated in the absence of Mg · Ado*PP*NH*P*. On the other hand, if about 12 tyrosine residues per molecule are nitrated in the presence of Mg · Ado*PP*NH*P*, only a slight shift in the pH optimum is observed (Fig. 3). In Fig. 4 the kinetics of inactivation by acetylation with *N*-acetylimidazole in the absence and in the presence of Mg · Ado*PP*NH*P* is shown. The data reveal the presence of at least two types of inactivation mechanism. The fast inactivation reaction is suppressed in the presence of Mg · Ado*PP*NH*P*.

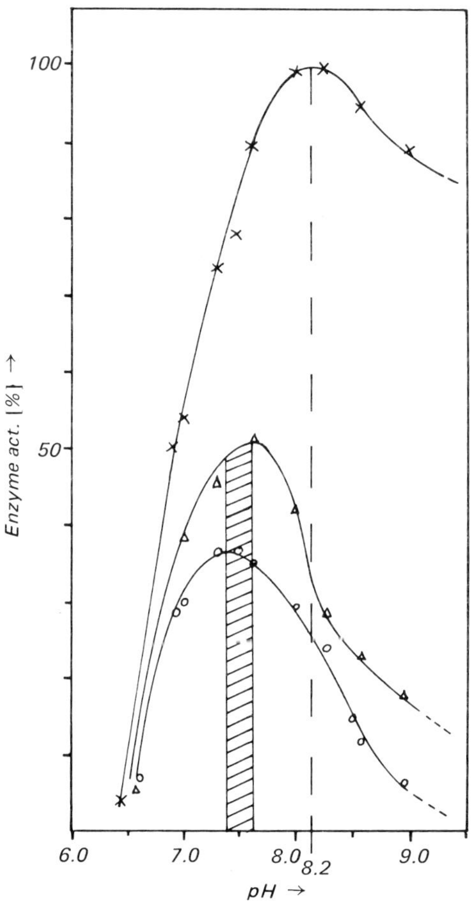

Fig. 3. Shift of pH optimum of F_1 ATPase by nitration with tetranitromethane.

x–x–x: native enzyme; ○–○–○: nitration without protection; △–△–△: nitration in the presence of 0.1mM Mg · Ado*PP*NH*P*.
See also Table 2.

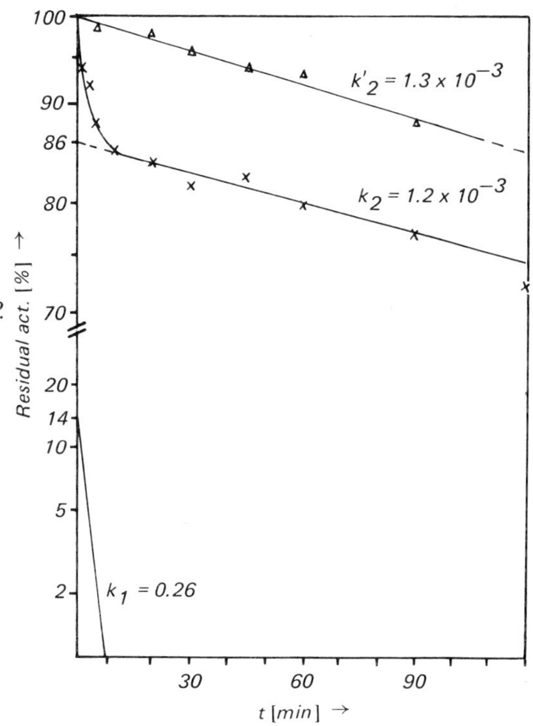

Fig. 4. Inactivation of F_1 ATPase by acetylation with *N*-acetylimidazole in the presence of 0.1mM Mg · Ado*PP*NH*P* (△–△–△) and without protection (x–x–x).

The inference that the reactive species ($k_1 = 0.26$) which is protected in the presence of 0.1mM Mg · Ado*PP*NH*P* is identical with the two protected tyrosine residues shown in Table 2 is tempting, but not yet conclusive.

The present data appear to be in disagreement with the results obtained by Ferguson et al.[12], who found that mitochondrial ATPase is almost completely inactivated after the OH group of a single tyrosine residue becomes blocked by the nitrobenzofurazan group.

Our preliminary data on the reaction of the present bacterial ATPase with chloronitrobenzofurazan show that this reagent ultimately blocks functional amino groups of this enzyme[17].

The same effect was found by Ferguson et al.[13], but only after incubation at pH 9.0[13]. No simple explanation for these discrepancies is presently available.

Reaction with tetranitromethane

Since we have failed substantially to inactivate this bacterial ATPase at pH 7.5 by blocking the OH groups of tyrosine residues with acetylimidazole or chloronitrobenzofurazan, we have tried to modify the tyrosine residues irreversibly by nitration with tetranitromethane.

Table 2 shows that the residual activity of the enzyme is 37% after about 15 tyrosine residues are nitrated.

In the presence of 0.1mM Mg · Ado*PP*NH*P*, the residual activity is above 50% and two tyrosines less are nitrated under these conditions.

In Fig. 3, the pH activity diagram of the nitrated enzyme is compared with that of the control enzyme. The shift in the pH optimum, however, can be only qualitatively related to the difference between the pK_3' of tyrosine (pK_3' = 10) and 3-nitrotyrosine (pK_3' = 7.2).

Fig. 5 shows that the kinetics of inactivation by nitration is also significantly changed in the presence of Mg · Ado*PP*NH*P*. A substantial fraction of the more reactive species is protected in the presence of the substrate analogue.

The interpretation of the present data is complicated by at least two facts. Firstly, it is unlikely that an essential tyrosine residue will completely lose its function after being nitrated. Secondly, tetranitromethane also modifies histidine, tryptophan, and cysteine residues[43]. The latter restriction is probably negligible, inasmuch as the data of the amino acid analysis indicate that amino acids other than tyrosine are not significantly modified by tetranitromethane.

Related nitration experiments have been carried out by Senior[44] with soluble beef-heart mitochondrial ATPase. A total of 9 to 10 tyrosine residues had been nitrated by the time inactivation by tetranitromethane was almost complete. In the presence of ATP, the amount of inactivation was reduced to 50% and the amount of tyrosine residues nitrated to 6 - 7. However, this author did not base his conclusions on amino acid analysis, but only on measuring the increase in absorption at 428 nm, which is assumed to be due to formation of nitrotyrosine. Observations regarding a shift of the pH-optimum of the nitrated enzyme have not been reported.

The present data suggest that tyrosine residues play an important but not crucial role in the function of the ATPase of *Micrococcus sp.* ATCC 398. The two exposed tyrosine residues per molecule which are masked by Mg · Ado*PP*NH*P* are probably only required to maintain the pH optimum at 8.2. It is unlikely that the failure to protect completely the funtional tyrosine residues by Mg · Ado*PP*NH*P* is due to insufficient binding of the analogue, because the functional tryptophan residues are efficiently protected from oxidation with *N*-bromosuccinimide in related experiments.

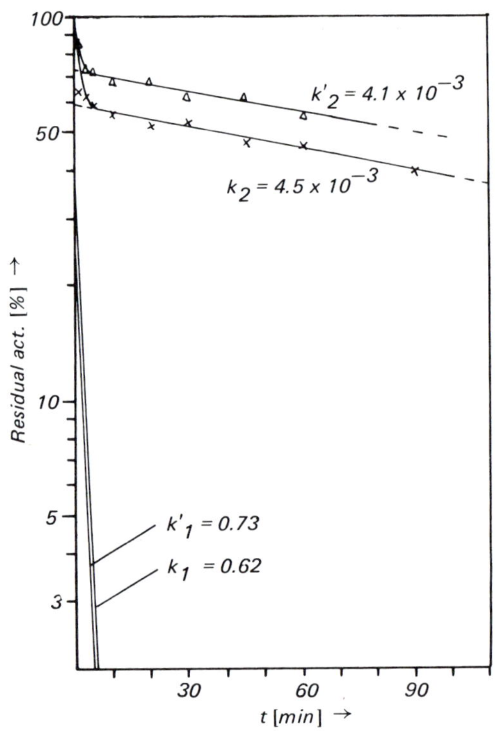

Fig. 5. Inactivation of F_1 ATPase by nitration with tetranitromethane in the presence of 0.1mM Mg·Ado*PP*NH*P* (△–△–△) and without protection (x–x–x).

This work was supported by grants from *Deutsche Forschungsgemeinschaft* (Ri 268/3; Do 192/1).
The authors wish to thank Miss *Christel Brand* for carrying out the amino acid analysis and for revising the manuscript.

Literature

1 Salton, M. R. J. & Schor, M. T. (1972) *Biochem. Biophys. Res. Commun.* **49**, 350 - 357.
2 West, I. C. (1974) in Membrane Adenosine Triphosphatases and Transport Processes (Bronk, J. R., ed.) pp. 27 - 38, *Biochem. Soc. Spec. Publ.* **4**, London.
3 Beechey, R. B. (1974) in Membrane Adenosine Triphosphatases and Transport Processes (Bronk, J. R., ed.) pp. 41 - 62, *Biochem. Soc. Spec. Publ.* **4**, London.
4 Bragg, P. D. & Hou, C. (1973) *Biochem. Biophys. Res. Commun.* **50**, 729 - 736.
5 Hanson. R. L. & Kennedy, E. P. (1973) *J. Bacteriol.* **114**, 772 - 781.
6 Höckel, M., Hulla, F. W., Risi, S. & Dose, K. (1976) *Biochim. Biophys. Acta* **429**, 1020 - 1028.
7 Morita, F. (1967) *J. Biol. Chem.* **242**, 4501 - 4506.
8 Morita, F. (1974) *Biochim. Biophys. Acta* **343**, 674 - 681.
9 Andreu, J. M., Albendea, J. A. & Muñoz, E. (1973) *Eur. J. Biochem.* **37**, 505 - 515.
10 Schor, M. T., Heincz, M. C., Salton, M. R. J. & Zaboretzky, F. (1974) *Microbios* **10** A, 145 - 150.
11 Nieto, M., Muñoz, E., Carreira, J. & Andreu, J. M. (1975) *Biochim. Biophys. Acta* **413**, 394 - 414.
12 Ferguson, S. J., Lloyd, W. J., Lyons, M. H. & Radda, G. K. (1975) *Eur. J. Biochem.* **54**, 117 - 126.
13 Ferguson, S. J., Lloyd, W. J. & Radda, G. K (1975) *Eur. J. Biochem.* **54**, 127 - 133.
14 Ferguson, S. J., John, P., Lloyd, W. J., Radda, G. K. & Whatley, F. R. (1974) *Biochim. Biophys. Acta* **357**, 457 - 461
15 Nelson, N., Kanner, B. I. & Gutnick, D. L. (1974) *Proc. Natl. Acad. Sci. U.S.A.* **71**, 2720 - 2724.
16 Futai, M., Sternweis, P. C. & Heppel, L. A. (1974) *Proc. Natl. Acad. Sci. U.S.A.* **71**, 2725 - 2729.
17 Reitershan, G. (1976) Diplomarbeit, Mainz, FRG.
18 Simpson, R. T., Riordan, J. F. & Vallee, B. L. (1963) *Biochemistry* **2**, 616 - 622.
19 Sokolowsky, M., Riordan, J. F. & Vallee, B. L. (1966) *Biochemistry* **4**, 3582 - 3589.
20 Spande, T. F. & Witkop, B. (1967) *Methods Enzymol.* **11**, 498 - 506 and 528 - 532.
21 Risi, S., Höckel, M., Hulla, F. W. & Dose, K. (1977) *Eur. J. Biochem.* **81**, 103 - 109.
22 Schröder, C. (1976) Diplomarbeit, Mainz, FRG.
23 Fiske, C. H. & Subbarow, Y. (1925) *J. Biol. Chem.* **66**, 375 - 400.
24 Arnold, A., Wolf, H. U., Ackermann, B. P. & Bader, H. (1976) *Anal. Biochem.* **71**, 209 - 213.
25 Schmitt, M. (1975) Diplomarbeit, Mainz, FRG.
26 Lowry, O. H., Rosebrough, N. J., Farr, A. L. & Randall, R. J. (1951) *J. Biol. Chem.* **193**, 256 - 275.
27 Glazer, A. N., Delange, R. J. & Sigman, D. S. (1975) in Chemical Modification of Proteins. Selected Methods and Analytical Procedures p. 31, American Elsevier Publishing Co., Inc., New York.
28 Sigman, D. S. & Blout, E. R. (1967) *J. Am. Chem. Soc.* **89**, 1747 - 1748.
29 Boyer, P. D. (1954) *J. Am. Chem. Soc.* **76**, 4331 - 4337.
30 Gurd, F. R. N. (1967) *Methods Enzymol* **11**, 532 - 541.
31 Ellman, G. L. (1959) *Arch. Biochem. Biophys.* **82**, 70 - 77.
32 Jori, G., Galiazzo, G. & Scoffone, E. (1969) *Int. J. Protein Res.* **1**, 289 - 298.
33 Beeley, J. G. & Neurath, H. (1968) *Biochemistry* **7**, 1239 - 1251.
34 Riordan, J. F., Wacker, W. E. C. & Vallee, B. L. (1965) *Biochemistry* **4**, 1758 - 1765.
35 Bruice, T. C., Gregory, J. J. & Walters, S. L. (1968) *J. Am. Chem. Soc.* **90**, 1612 - 1619.
36 Teuwissen, B., Masson, P. L., Osinski, P. & Heremans, J. F. (1973) *Eur. J. Biochem.* **35**, 366 - 371.
37 Goodwin, T. W. & Morton, R. A. (1946) *Biochem. J.* **40**, 628 - 632.
38 Garrett, N. E. & Penefsky, H. S. (1975) *J. Biol. Chem.* **250**, 6640 - 6647.
39 Funatsu, M., Green, N. M. & Witkop, B. (1964) *J. Am. Chem. Soc.* **86**, 1846 - 1848.
40 Harris, D. A., Rosing, J., van der Stadt, R. J. & Slater, E. C. (1973) *Biochim. Biophys. Acta* **314**, 149 - 153.
41 Garrett, N. E. & Penefsky, H. S. (1975) *J. Supramol. Struct.* **3**, 469 - 478.
42 Pedersen, P. L., Cintrón, N. & Soper, J. W. (1976) in 10th Int. Congress of Biochem., Hamburg, Abstr. 06 - 8 - LO12.
43 Glazer, A. N., Delange, R. J. & Sigman, D. S. (1975) in Chemical Modification of Proteins (Work, T. S. & Work, E.) pp. 95 - 99, North-Holland.
44 Senior, A. E. (1973) *Biochemistry* **12**, 3622 - 3626.

Dr. S. Risi, Dipl.-Chem. Ch. Schröder, Dipl.-Chem. H.-J. Kaiser and Prof. Dr. K. Dose, Institut für Biochemie, Johannes-Gutenberg-Universität Mainz,
Joh.-Joachim-Becher-Weg 30, D-6500 Mainz.

Dr. J. Carreira, Instituto de Biología Celular, C.S.I.C.,
Velázquez, 144, Madrid-6, Spain.

Hoppe-Seyler's Z. Physiol. Chem.
Bd. 359, S. 47 - 61, Januar 1978

Yeast Dipeptidase: Active Site Mapping by Kinetic Studies with Substrates and Substrate Analogs

Volker HERRMANN, Ingo KNACK and Klaus-Heinrich RÖHM

Physiologisch-Chemisches Institut II der Universität Marburg

(Received 15 June/14 November 1977)

Summary: In order to characterize the active site of yeast dipeptidase in more detail, kinetic studies with a variety of dipeptide substrates and substrate analogs were performed. To analyze kinetic data, computer programs were developed which first calculate initial velocities from progress curves and then evaluate the kinetic parameters by nonlinear regression analysis.

A free carboxyl group is a prerequisite for binding of dipeptidase substrates; its position relative to the peptide bond must not deviate from the normal L-dipeptide conformation. The spatial arrangement of the terminal ammonium ion seems to be less crucial.

The enzyme's substrate specificity clearly reflects the interactions of the substrate amino acid side chains with complementary dipeptidase subsites. The domain of the enzyme in contact with the *C*-terminal substrate side chain seems to be an open structure of moderately hydrophobic character. In contrast, the binding site for the amino-terminal side chain is a more strongly hydrophobic "pocket" of limited dimensions.

The kinetics of inhibition by free amino acids points to an ordered release of products from the enzyme.

Kinetische Untersuchungen zur Struktur des aktiven Zentrums der Hefe-Dipeptidase

Zusammenfassung: Zur näheren Charakterisierung des aktiven Zentrums der Hefe-Dipeptidase wurden kinetische Untersuchungen mit einer Reihe von Dipeptid-Substraten und Substratanalogen durchgeführt. Bei der Verarbeitung kinetischer Daten wurden dabei Computerprogramme eingesetzt, die zunächst aus Zeit-Umsatzkurven Anfangsgeschwindigkeiten berechnen und dann in einem zweiten Schritt durch nichtlineare Ausgleichsrechnung die kinetischen Konstanten bestimmen.

Für die Bindung von Substraten an das Enzym ist das Vorliegen einer freien Carboxylatgruppe unerläßlich. Ihre Lage relativ zur Peptidbindung darf nicht von der in Dipeptiden mit L-Konfiguration abweichen. Dagegen scheint die räumliche Orientierung der terminalen Ammoniumgruppe weniger wesentlich zu sein.

Enzyme:
Dipeptidase, L-aminoacyl-L-amino-acid hydrolase (EC 3.4.13.?)
Abbreviations:
Ac-, Acetyl-; Boc-, *t*-butoxycarbonyl-; DCCI, dicyclohexyl carbodiimide; -OBu^t, *t*-butyl ester; -OMe, methyl ester; *Z*-, benzyloxylcarbonyl.

Die Spezifität der Dipeptidase gegenüber nativen Dipeptiden spiegelt in erster Linie die Wechselwirkungen der Substrat-Seitenketten mit komplementären Bindungsstellen am Enzym wieder. Derjenige Bereich des aktiven Zentrums, der mit der *C*-terminalen Seitenkette der Substrate in Kontakt steht, scheint eine offene Struktur von mäßig hydrophobem Charakter zu sein. Dagegen ist die Bindungsstelle für den aminoterminalen Seitenrest eine stärker hydrophobe „Tasche" von begrenzten Abmessungen.

Die Kinetik der Produkthemmung durch freie Aminosäuren weist auf eine obligatorische Reihenfolge bei der Freisetzung der Reaktionsprodukte hin.

Key words: Yeast dipeptidase, kinetics, substrate specificity.

The final steps in intracellular protein breakdown are catalyzed by dipeptidases (L-amino-acyl-L-amino-acid hydrolases) the substrate specificity of which restricts them to the hydrolysis of dipeptides derived from L-amino acids. Although it has been claimed that there are dipeptidases with a narrow specificity towards only one or a few types of dipeptide substrates (such as "Gly-Gly dipeptidases" or "Gly-Leu dipeptidases"[1]) it is now well documented that most dipeptidases are able to attack a rather broad spectrum of dipeptides. In some cases, specificity of dipeptidases for a single type of dipeptides was shown to be only an apparent one, brought about by the inclusion of metal ion modifiers in the assay mixture, which specifically activated the hydrolysis of certain peptides while concomitantly inhibiting the splitting of others[2,3]. Nevertheless, the catalytic efficiency of dipeptidases varies greatly with the type of substrate under attack. Since the L-dipeptide structure is the same for all substrates, these differences must be due mainly to the properties of the amino acid side chains and their position in the dipeptide.

The present study deals mainly with the effect of the substrate side chains on the catalytic efficiency of yeast dipeptidase, an enzyme which has been purified and characterized in some detail in our laboratory[3,4]. The kinetics of hydrolysis of a variety of dipeptides and substrate analogs was studied in the absence of metal activators in Tris buffers at pH 8.0. Under these conditions, the kinetic anomalies which characterize the behaviour of the enzyme at lower pH[4] are of minor importance. From these experiments some general features of the enzyme's active site are deduced. In particular, the properties of the dipeptidase subsites which accept the side chains of the substrates are discussed.

Materials and Methods

Enzyme

Yeast dipeptidase was purified from autolysates of stationary brewer's yeast as described previously[3]. Samples kept frozen in 10mM dimethylglutarate buffer, pH 6.2, containing 25% glycerol, were used for all experiments.

Substrates

If not specified otherwise, substrates and dipeptide analogs were purchased from Fluka, Buchs, or Serva, Heidelberg. In general, they were chromatographically uniform and used without further purification. 5-Amino-laevulinic acid was from Fluka, Buchs. Formylglycylglycine was kindly provided by *Dr. G. Metz*. Glycyl-dehydroalanine was prepared according to ref.[5].

The synthetic routes leading to the stereoisomers of Ala-Lys and Ala-Lys(Ac) (see below) were designed in view of the possible use of these compounds as ligands for affinity chromatography. Therefore protecting groups were chosen which could be removed by mild acidic treatment after coupling to carboxyl derivatives of cellulose or agarose via the ϵ-amino group of lysine. In fact, in this way it was possible to obtain affinity adsorbents containing the Ala-Lys moiety in relatively high amounts*.

YLys(Z)-OBu^t·HCl (I; Y = D or L) was prepared according to ref.[6].

Boc-XAla-YLys(Z)-OBu^t (II; X, Y = D or L). These intermediates were obtained with 80 - 95% yields by coupling of I and Boc-XAla with DCCI in ethyl acetate. The products were oils still containing traces of Boc-Ala.

* Herrmann, V., Röhm, K.-H. & Schneider, F., unpublished results.

Boc-XAla-YLys-OBu^t (III) The *Z* protecting group was removed by hydrogenation in the presence of Pd/charcoal. The resulting compounds are strongly hydrophobic and were not characterized in detail.

XAla-YLys (IV) The free dipeptides were obtained by treatment of III with 4N HCl in dioxane. Since they could not be crystallized by the usual techniques, chromatography of the products on alumina (Aluminiumoxid S, elution with ammonium acetate buffer, pH 5) and Sephadex G-10 (elution with water) was used for the final purification of the products which all are hygroscopic. Yields: 40 - 60%.

	$[\alpha]_D^{22}$
LAla-LLys	− 7.1° (*c* = 1, 0.5N HCl)
LAla-DLys	+ 19.3° (*c* = 1, 0.5N HCl)
DAla-LLys	− 27.1° (*c* = 1, 0.5N HCl)

The compounds were uniform in thin layer chromatography; amino acids analyses showed an Ala/Lys ratio of 1.0.

XAla-YLys(Ac)-OBu^t (V) was obtained by acetylation of III with acetic acid anhydride in acetone/water 1 : 1. The resulting oils could not be crystallized before chromatography on alumina columns (elution with dioxane). The main peak was collected and the product remaining after evaporation of the solvent dissolved in a small volume of ether. Petroleum ether was added; crystals appeared after some time. The compounds were uniform in thin-layer chromatography and had the expected elemental composition.

XAla-YLys(Ac) Compound V could be converted to the free acetyl dipeptides by treatment with trifluoroacetic acid, whereas HCl/dioxane led to the formation of the corresponding diketopiperazines in large amounts. The products remaining after removal of the acid were dissolved in dry ethanol and crystallization was induced by addition of ether. After several weeks of standing in the cold, precipitation of the products was complete. Recrystallization from methanol/ethanol. Yields: 80 - 96%.

Analyses and properties: $C_{11}H_{21}N_3O_4$ (259.31)

	Analysis [%]			m.p.	$[\alpha]_D^{22*}$
	C	H	N	[°C]	
Found L-L	50.55	7.78	15.98	225 - 230	+ 9.1°
L-D	50.85	8.01	16.16	219 - 221	+ 16.7°
D-L	50.86	7.79	15.94	219 - 221	− 17.6°
Calcd.	50.95	8.16	16.20		

* *c* = 1 in water.

N-Glycyl-N'-acetylhydrazine · HCl was synthesized by reaction of Boc-Gly-OMe[7] with hydrazine hydrate in methanol, followed by acetylation of the product with acetyl chloride and removal of the protecting Boc residue with HCl/dioxane. The product was uniform in thin-layer chromatography. m.p. 207 - 210 °C (decomp.).

$C_4H_9N_3O_2 \cdot HCl$ (167.59)

Calcd.:	C 28.67	H 6.01	N 25.07
Found:	C 29.04	H 6.31	N 24.72

Kinetic experiments

Dipeptidase activities were determined spectrophotometrically from the absorption change due to the splitting of the peptide bond as described previously[4]. If not specified otherwise, the enzyme was preincubated with 0.1M Tris/HCl buffer, pH 8.0, for 5 min at 40 °C, then the reaction was initiated by addition of prewarmed substrate solution in the same buffer. During the runs, the absorption was sampled at regular intervals and stored in a digital form by a DASAR data acquisition system (American Instruments Co., Silver Springs, Md.). After visual inspection the data were transferred to punched paper tape and later processed by computer (see below).

With most substrates, hydrolysis was followed at 230 nm. In contrast, the hydrolysis of Gly-LTrp was monitored at 300 - 320 nm where the indole absorption decreases as the peptide bond is split.

The turnover of dipeptides reacting very slowly or having too high an ultraviolet absorption was measured by incubation with enzyme as usual, removing samples after appropriate times and determining the free amino acids on the amino acid analyzer. Usually, analysis times of about 30 min on a 0.9 x 10 cm column of Beckman M 82 resin were sufficient to separate the compounds of interest.

Evaluation of kinetic experiments

For the acquisition and evaluation of steady-state kinetic data, a procedure has been developed which avoids the subjective bias necessarily connected with the conventional graphical methods and, in addition, leads to more accurate results in a considerably shorter time. The method comprises automatic collection of data and processing by computer. During the kinetic experiments absorption values (or any other analog signal which may serve as a measure of substrate or product concentration) are sampled and digitalized at regular time intervals by the DASAR data acquistition system. These data together with the information necessary for processing are then transferred to punched paper tape. All of the subsequent calculations are performed by computer.

In a first step, a polynomial (usually of second degree) is fitted to the progress curves of $A = f(t)$ by means of a FORTRAN computer program*. The first derivative of this polynomial provides values for the reaction rate at

* The programs are available from the authors on request.

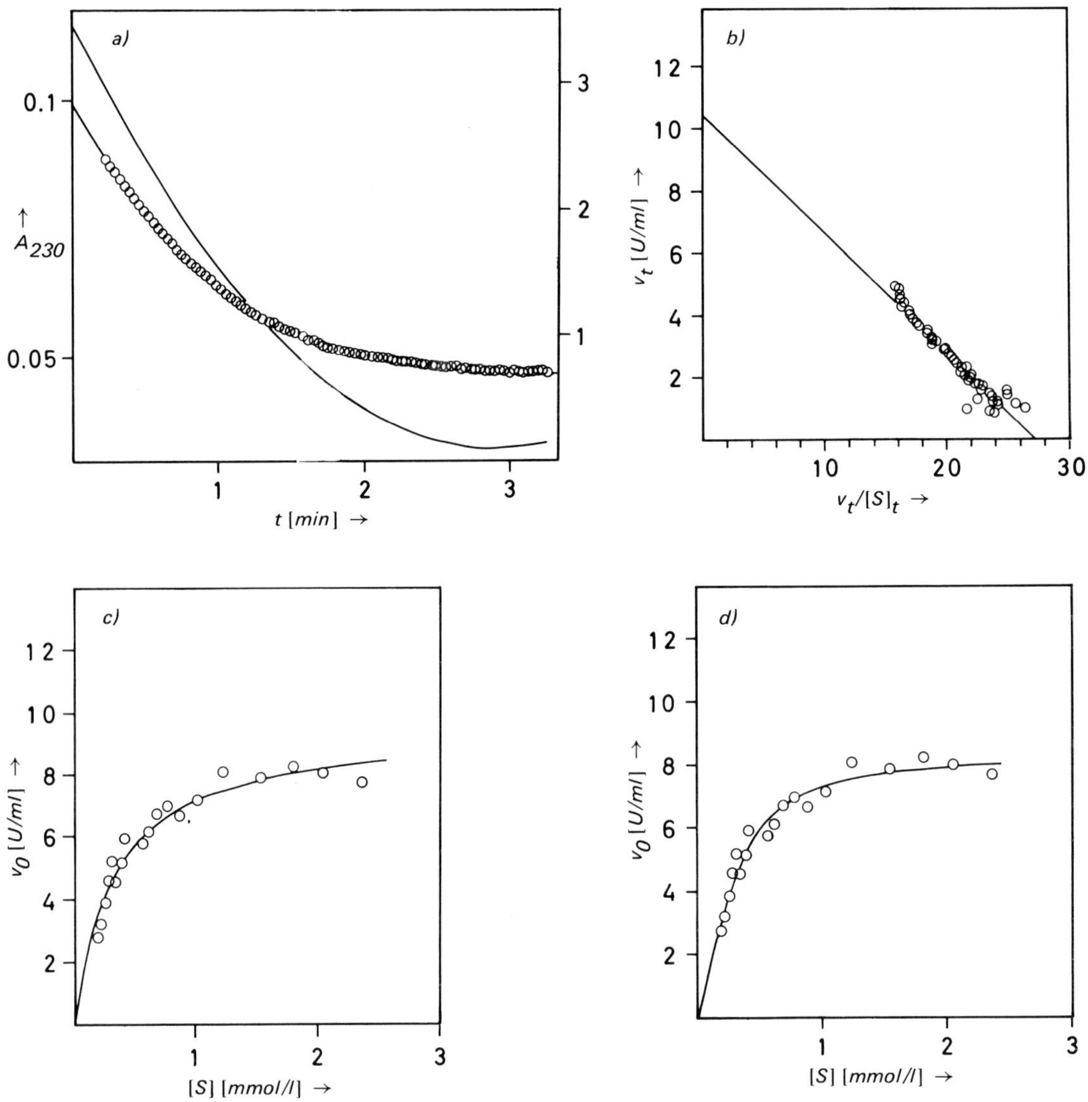

any point of the curve. Extrapolation to zero time yields true initial velocities v_0. In contrast to methods based on integrated rate equations, the results of our procedure are not dependent on any special reaction mechanism. Therefore the quantitative analysis of lag phases and other transient features is feasible.

If entire progress curves are recorded, apart from the temporary velocities v_t, the substrate concentrations at any point may be calculated from the relative absorption changes. Thus, in principle, the kinetic parameters may be estimated from a single absorption vs. time curve. It is well known, however, that the calculation of kinetic constants from progress curves is strongly subject to systematic error (product inhibition, partial inactivation of the enzyme during the run, drifts of the photometric baseline etc.). Therefore, we limit the use of this method to the calculation of estimates for V and K_{app} which are subsequently employed as starting values for the fitting procedure described below.

Routinely, with each dipeptide substrate the absorption change with time is recorded with 10 - 20 substrate concentrations. 30 - 40 data points at 2-s intervals are sufficient to obtain good values for the initial velocity v_0. A few of the reactions are allowed to go to completion.

◁

Fig. 1. Computer evaluation of kinetic experiments.

The graphs of this figure are to visualize the individual procedures applied in the evaluation of steady-state kinetic experiments. As an example, data for the hydrolysis of LLeu-LAla in 0.1M Tris/HCl buffer, pH 8.0 were chosen.

Fig. 1a shows the absorption change with time observed upon incubation of a 0.51mM Leu-Ala solution with enzyme. Absorption values (○) were sampled and stored digitally at 2-s intervals. The solid line intersecting the ○ symbols represents the 3rd degree polynomial fitted to the progress curve by the initial-velocity program (see "methods"). The first derivative of this polynomial (solid line; right-hand scale) provides values for the velocities v_t. Absorption values recorded during the first ten seconds of the reaction, usually being less reliable, have been omitted. Nevertheless, a true initial velocity v_0 is obtained from the value of $v_t = f(t)$ at $t = 0$.

Fig. 1b illustrates the calculation of initial estimates of V and K_{app} from the progress curve shown in Fig. 1a. Local substrate concentrations $[S]_t$ were computed for each point from the relative absorption change. Together with the reaction rates v_t an Eadie-Hofstee plot of v_t vs. $v_t/[S]_t$ was generated. A linear regression on these points (solid line) yields values of V = 10.5 U/ml and K_{app} = 0.38mM.

Fig. 1c Initial velocities v_0 determined as depicted in Fig. 1a with 19 different substrate concentrations are plotted vs. the respective values of $[S]$. The solid line represents the best fit of the Michaelis-Menten equation to these data. The initial estimates of V and K_{app} calculated as shown in Fig. 1b were used. The final values of V and K_{app} as derived from the fitting program (V = 9.5 U/ml, K_{app} = 0.33mM) are in a reasonable agreement with these provisional values. However, Fig. 1c shows that the fit is unsatisfactory, especially at high and low values of $[S]$. The mean relative error (i.e. the deviation of the experimental points from the curve) was 6.7%.

Fig. 1d shows the fit of the Hill equation with a constant Hill coefficient to the same set of data. Clearly, a considerably improved fit is achieved, quantitatively reflected by a lower relative error (4.8%). The set of parameters obtained was V = 8.3 U/ml, K_{app} = 0.14mM and n_H = 1.53.

From these data the program calculates the respective extinction coefficient and determines provisional values of V and K_{app} by a linear regression performed on $v_t = f(v_t/[S]_t)$, i.e. using the Eadie-Hofstee linearization. An example illustrating this procedure is shown in Fig. 1.

After inspection of the initial velocities, obtained in this way and – if necessary – omission of grossly deviating values, the v_0, $[S]$ data pairs and the starting estimates for V and K_{app} are used as input data for a second program* which calculates final values of the kinetic parameters by nonlinear regression analysis. Depending on the type of saturation curve under examination, the experimental points may be fitted alternatively by

a) the conventional Michaelis-Menten equation

$v = V \times [S]/(K_{app} + [S])$

b) by the Hill equation with a constant Hill coefficient

$v = V \times [S]^{n_H}/(K_{app} + [S]^{n_H})$ or

c) by the Hill equation with a variable Hill coefficient.

Method c requires extensive and rather precise data, and is not a routine procedure for normal kinetic studies[8]. Most of the experiments performed in the present study were evaluated according the method b in order to detect possible deviations from Michaelis-Menten kinetics. Our program uses the method of least squares, i.e. it determines a set of values for V, K_{app} and n_H for which the sum of squares of the residuals ($v_{calcd.} - v_{obs.}$) reaches a minimum. Since the Hill equation yields a nonlinear set of normal equations which cannot be solved in a straightforward way, the program approaches the minimum of the sum of squares by an iterative search, in the course of which all three of the parameters are alternately varied until optimum values are attained. Method a proceeds analogously but with n_H fixed at unity.

Since the rate equations are fitted directly to saturation curves rather than to some linearized form, the problem of introducing statistical weights to the data is considerably simplified. With our program, weights may be introduced or automatically calculated in several different ways.

With parameters determined by method b it has to be kept in mind that, if n_H differs from unity, the coefficient K_{app} of the Hill equation is no longer a Michaelis constant but an empirical parameter related to the substrate concentration at half saturation by

$[S]_{0.5} = \sqrt[n_H]{K_{app}}$.

Results

Substrate analogs

In order to establish the minimal structural requirements for the binding of a peptide to yeast dipeptidase we have studied the behaviour of the enzyme towards a series of compounds containing the dipeptide backbone in a slightly modified form. The results are compiled in Table 1 and may be summarized as follows:

1) Blocking of the carboxylate moiety of the basic Gly-Gly structure or shifting its position relative to the peptide bond greatly impairs the interaction of the respective compounds with the enzyme.

Thus dipeptide amides, even if they contain a favourable *C*-terminal amino acid (cf. Gly-Leu-NH_2), are not attacked. The rather small shift of the carboxylate ion in Leu-βAla as compared to Leu-Gly (about 2 Å away from the peptide bond) greatly reduces the turnover rate. Tripeptides are not split at all. The failure of the enzyme to bind peptides with D-amino acids at the COOH-termi-

Table 1. Response of dipeptidase to Gly-Gly derivatives with a modified dipeptide backbone.

50mM solutions of the compounds in 0.1M Tris/HCl buffer, pH 8.0, were incubated with enzyme at 40 °C. Samples were analyzed for Gly (or Leu) on the amino acid analyzer. Relative activities refer to the rate of Gly-Gly hydrolysis taken as 100. Rates less than 0.1 are denoted by 0.

Compound	Structure	v_{rel}	Comments
LPro-Gly	$^{\oplus}H_2N$—CH—CO-NH—CH_2—$COO^{\ominus}$ (ring: H_2N and CH bridged by $(CH_2)_3$)	28	v_{rel} = 230 with 0.1mM $Mn^{2\oplus}$
Formyl-Gly-Gly	HCO—NH—CH_2—CO-NH—CH_2—$COO^{\ominus}$	2	
Gly-Gly	$^{\oplus}H_3N$—CH_2—CO-NH—CH_2—$COO^{\ominus}$	100	v_{rel} = 430 with 0.1mM $Co^{2\oplus}$
5-Aminolaevulinic acid	$^{\oplus}H_3N$—CH_2—CO-CH_2—CH_2—$COO^{\ominus}$	0	Inhibitor, see Fig. 4
Glycyl-glycolic acid	$^{\oplus}H_3N$—CH_2—CO-O—CH_2—$COO^{\ominus}$	0	Inhibitor, see ref.[9]
Gly-NH-NH-Ac	$^{\oplus}H_3N$—CH_2—CO-NH—NH—$COCH_3$	0	Activator, see Fig. 4
Gly-Gly amide	$^{\oplus}H_3N$—CH_2—CO-NH—CH_2—$CONH_2$	0	
Gly-LLeu[a]	$^{\oplus}H_3N$—CH_2—CO-NH—CH(R)-$COO^{\ominus}$	1280	
Gly-LLeu amide[a]	$^{\oplus}H_3N$—CH_2—CO-NH—CH(R)-$CONH_2$	0	
LLeu-Gly[a]	$^{\oplus}H_3N$—CH(R)-CO-NH—CH_2—$COO^{\ominus}$	210	
LLeu-β-alanine[a,b]	$^{\oplus}H_3N$—CH(R)-CO-NH—CH_2—CH_2-$COO^{\ominus}$	4	
Gly-LAla	$^{\oplus}H_3N$—CH_2—CO-NH—CH(CH_3)—$COO^{\ominus}$	250	
Gly-DAla	$^{\oplus}H_3N$—CH_2—CO-NH—CH($COO^{\ominus}$)—CH_3	0	
Gly-dehydro-Ala	$^{\oplus}H_3N$—CH_2—CO-NH—C(=CH_2)—$COO^{\ominus}$	105	
Gly-sarcosine	$^{\oplus}H_3N$—CH_2—CO-N(CH_3)-CH_2—$COO^{\ominus}$	0	
Gly-LPro	$^{\oplus}H_3N$—CH_2—CO-N—CH($COO^{\ominus}$) (ring: N and CH bridged by $(CH_2)_3$)	0	

[a] R = $-CH_2-CH(CH_3)_2$ [b] (S) = 8mM

nus is most probably due to the same effect. It is obvious from Table 2 that the LD-derivatives of Ala-Lys and Ala-Lys(Ac) do not even inhibit the hydrolysis of the respective LL-dipeptides. The activation observed with LAla-D-Lys(Ac) is probably mediated by the binding of this compound to a site different from the substrate binding domain (see below). In contrast, the structural distortion caused by the double bond in Gly-dehydro-Ala is much less severe.

2) Although the positively charged terminal ammonium ion is of great importance for substrate binding, its interaction with the enzyme seems to be less critical than that of the carboxyl end. Thus, formyl-Gly-Gly, which lacks the positive charge, is still hydrolyzed; Pro-Gly is a rather good substrate, at least in the presence of $Mn^{2\oplus}$, although the orientation of its imino group considerably deviates from the position of the Gly-Gly ammonium nitrogen. Likewise, the DL-

Table 2a: Kinetic constants for the hydrolysis of dipeptides by yeast dipeptidase. All experiments were performed at 40 °C in Tris/HCI buffer, pH 8.0. Maximal velocities V are given in relation to the maximal rate of LLeu-Gly hydrolysis, taken as 1.0. The apparent Michaelis constants K_{app} are defined as $K_{app} = [S]^{n_H}\ 0.5$, n_H being the mean Hill coefficient (see methods).

Dipeptide	V	K_{app} [mmol/l]	n_H	V/K_{app}
LAsp-Gly	< 0.001			
LLys-Gly[a]	0.002			
Gly-Gly	0.8	90	MM[b]	
LThr-Gly	2.8	6.0	MM	0.5
LAla-Gly	2.6	2.1	1.0	1.2
LVal-Gly	1.2	0.56	1.1	2.2
LPhe-Gly	1.1	0.36	1.6	3.1
LLeu-Gly	–1–	0.17	1.6	5.9
LLeu-LGlu	0.7	0.73	1.5	1.0
LTrp-Gly	< 0.001			K_i = 0.14mM[c]
LLys(Ac)-Gly	< 0.001			K_i > 10mM[c]
Gly-LAsp[a]	0.02	90	MM	0.001
Gly-LSer	0.8	18	MM	0.04
Gly-LAla	1.6	11	1.0	0.13
Gly-LVal	4.3	4.2	1.1	1.0
Gly-LLeu	6.0	4.1	1.2	1.5
LGlu-LLeu	0.6	4.3	1.0	0.14
Gly-LPhe	4.8	2.4	1.2	2.0
Gly-LTrp	1.3	0.8	1.2	1.6
LAla-LAla	4.2	2.0	1.0	2.1
LAla-LVal	5.8	1.7	0.9	3.4
LAla-LIle	6.5	1.4	1.0	4.6
LAla-LLeu	2.3	0.46	1.0	5.0
LLeu-LAla	2.2	0.18	1.5	12
LLeu-LVal	2.3	0.15	1.5	15
LLeu-LLeu	2.5	0.11	1.7	23
LSer-LAla	2.9	2.5	1.1	1.2
LVal-LAla	2.0	0.43	1.1	4.7

[a] Measured with the amino acid analyzer;
[b] MM = computer fit based on the Michaelis-Menten equation;
[c] Determined from the competitive inhibition of Ala-Gly hydrolysis.

Table 2b: Action of yeast dipeptidase on stereoisomers of Ala-Lys and Ala-Lys(Ac). The kinetic constants of the LL dipeptides (without and in the presence of the respective LD and DL isomers) were determined photometrically in phosphate/borate buffer at 40 °C. The velocities given refer to $V_{Leu\text{-}Gly} = 1$. Inhibition and activation effects are expressed as the percentage change in V. The turnover of dipeptides containing D-amino acids was checked by measuring Ala on the amino acid analyzer.

Dipeptide	Conc. [mmol/l]	V_{rel}	K_{app} [mmol/l]	Effect on LL hydrolysis [% inhibition]
LAla-LLys	0.5 - 5	2.9[a]	2.8	–
DAla-LLys	4	0.002		9
DAla-LLys	8			19
LAla-DLys	4	0		none
LAla-LLys(Ac)	0.5 - 5	3.4[a]	2.6	–
DAla-LLys(Ac)	3.5	0		15
DAla-LLys(Ac)	7			33
LAla-DLys(Ac)	4			– 12 (activation)
LAla-DLys(Ac)	8	0		– 11 (activation)

[a] Maximal velocities

derivative of Ala-Lys is slowly hydrolysed, while the corresponding Ala-Lys(Ac) isomer effectively inhibits the enzyme (see Table 2b). With these compounds either the amino group or the side chain of alanine, but not both, may be brought near the position they occupy in the LL-peptides.

3) Gly-Sar, a Gly-Gly analog with a methylated peptide nitrogen, is not hydrolyzed by yeast dipeptidase. Likewise Gly-Pro, which also contains a substituted peptide bond, is not a dipeptidase substrate. In the latter case, however, molecular models show that the carboxyl group of Gly-Pro is fixed in an orientation so different from the position it occupies in normal dipeptides that this effect alone would account for the absence of turnover. The participation of the peptide NH-group in substrate binding is not a probable explanation for the inertness of Gly-Sar, since 5-aminolaevulinic acid and glycyl-glycolic acid[9], both lacking a peptide nitrogen, are bound to the enzyme, as indicated by their inhibitory effects (see Fig. 4). Possibly the methyl group of Gly-Sar prevents binding by steric hindrance or, alternatively, the peptide NH-group is involved in the catalytic step. This would also account for the failure of the enzyme to split glycyl-glycolic acid.

Kinetics of dipeptide hydrolysis

The role of the substrate side chains in dipeptidase catalysis is reflected by the data of Table 2a, which contains the kinetic parameters characterizing the hydrolysis of a selected series of native LL-dipeptides. Clearly, the catalytic efficiency of the enzyme, as indicated by the ratio V/K_{app} varies within wide limits. With a few exceptions the replacement of glycine by amino acids with larger side chains leads to an increase in the maximal velocity V and a concomitant decrease in the apparent Michaelis constant K_{app}, and thus to a much higher value of V/K_{app}. Inspection of Table 2 reveals a close correlation between the magnitude of the kinetic parameters and the nature and positions of side chains. Without as yet entering into the question as to the physical significance of V and K_{app}, one can make the following points (as indicated in Fig. 2, the enzyme site interacting with the side chain R_n of the amino terminal amino acid A_n will be denoted by S_n; the binding site for the side chain of the carboxyl terminal residue A_c will be abbreviated by S_c):

1) Michaelis constants: Generally, dipeptides containing acidic residues A_n or A_c and glycine in the other position are very poor substrates,

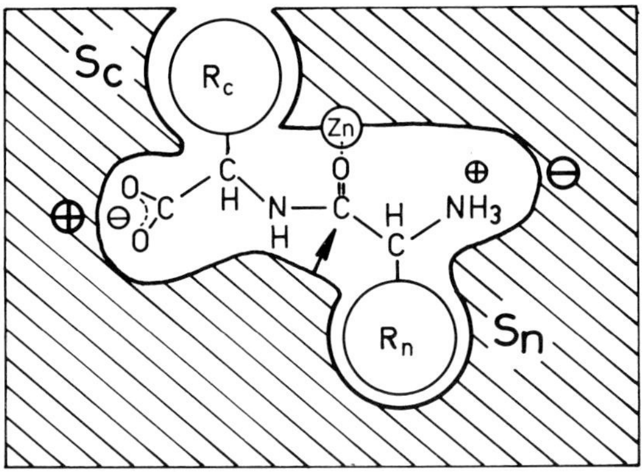

Fig. 2. Schematic representation of the presumed yeast dipeptidase active site.

Binding of both the carboxyl and the amino group of the substrate via ionic interactions is depicted. The sites interacting with the dipeptide side chains are denoted by S_n and S_c, respectively. The position of the essential zinc atom has been drawn in analogy to the situation established with other peptidases, but is as yet purely hypothetical for yeast dipeptidase.

due to a high K_{app} (Asp-Gly, Gly-Asp). Likewise, S_n does not accept the strongly basic lysine side chain (Lys-Gly). Ala-Lys, on the other hand, is a rather good substrate. If leucine is present in a dipeptide, it is accepted as a substrate even if it, in addition, contains an acidic residue (Leu-Glu, Glu-Leu). If both A_n and A_c are uncharged, the magnitude of K_{app} primarily depends on the side chain in the amino terminal position. If A_n is distinctly hydrophobic (Ala, Leu), K_M will be low and not very dependent on the properties of A_c. If A_n is glycine which does not contribute to substrate binding via side chain interactions, the magnitude of K_{app} steadily decreases with increasing size and hydrophobicity of A_c.

2) Maximal velocities: The highest turnover rates are found with substrates containing a small A_n (Gly, Ala) and an A_c side chain of moderate size (Val, Leu, Ile). As a rule, substrates with a low K_{app} have also a relatively low V (cf. substrates of the type Leu-A_c). In spite of their relatively high K_{app} values dipeptides with hydrophilic but uncharged A_n are rather good substrates. Under conditions of saturating substrate concentrations, Thr-Gly is split much faster than the sterically similar Val-Gly. Likewise, Ser-Ala has a higher V than Leu-Ala.
The presence of the strongly hydrophobic Phe and Trp side chains greatly reduces the turnover rates of the corresponding dipeptides, regardless of whether these residues interact with S_n or S_c. Surprisingly, Trp-Gly is not noticeably split by the enzyme, although this peptide is a strong competitive inhibitor with an inhibition constant K_i = 0.14mM. L Lys(Ac)-Gly, on the other hand, another dipeptide with a large bulky side chain in the amino terminal position, does not measureably inhibit the hydrolysis of other dipeptides.

3) Hill coefficients: The only substrates whose Hill coefficients significantly differ from unity (i.e. which do not show Michaelis-Menten kinetics under the conditions of our experiments) are those containing a strongly hydrophobic A_n (Leu, Phe).

Inhibition by amino acids

As shown earlier by Nishi[10] with a crude yeast dipeptidase preparation, neutral and hydrophobic amino acids are potent inhibitors of the enzyme.

Our experiments demonstrate that the type of inhibition encountered is not the same in all cases, but depends on the particular combination of substrate and amino acid inhibitor. Generally, amino acids which are not components of the dipeptide, employed as a substrate are competitive inhibitors. If one of the reaction products is chosen as an inhibitor, the observed type of inhibition depends on the position this amino acid occupies in the substrate. This is illustrated in Fig. 3, which shows the inhibition of Val-Gly and Gly-Val hydrolysis, respectively, by valine. Obviously, in the first case the inhibition is competitive, whereas valine affects both V and K_{app} of Gly-Val hydrolysis. Similarily, glycine greatly reduces the maximal velocity of Ala-Gly hydrolysis, whereas alanine does not.

Some representative inhibition constants measured with our enzyme are summarized in Table 3. Data on amino acids with charged side chains are not included, since these compounds are very poor inhibitors. In agreement with the report of Nishi[10], the table shows that derivatives with modified α-COOH groups inhibit less effectively than native amino acids.

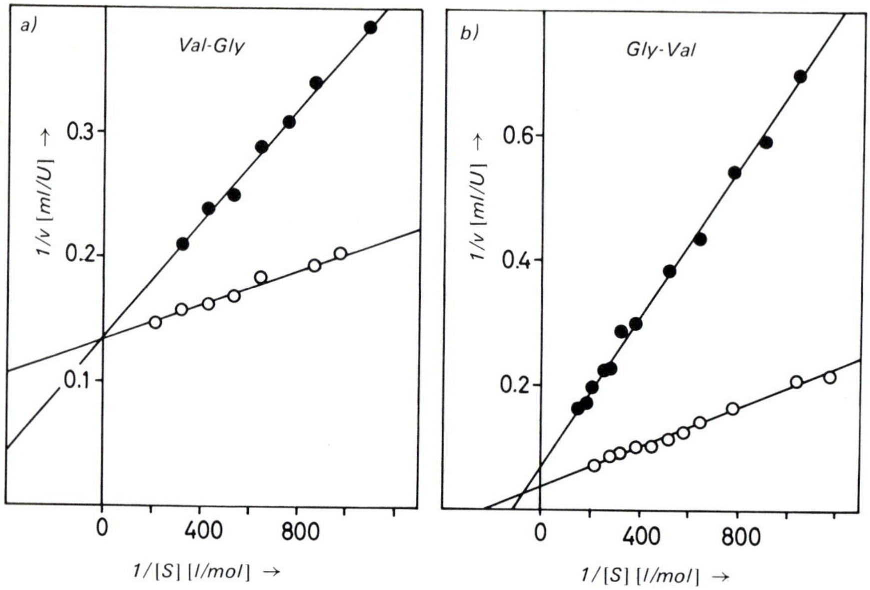

Fig. 3. Inhibition of Val-Gly and Gly-Val hydrolysis by valine.

Kinetic experiments were carried out and evaluated as described in "Methods". Fig. 3a shows double-reciprocal plots of Val-Gly hydrolysis without (○) and in the presence of 15mM valine (●). In Fig. 3b), similar plots for the splitting of Gly-Val without (○) and with 15mM Val (●) are given. The solid lines correspond to computer-generated best fits with the following kinetic constants:

a) Val-Gly, no Val: V = 7.6 U/ml, K_{app} = 0.56mM, n_H = 1.1
with Val: V = 7.4 U/ml, K_{app} = 1.67mM, n_H = 1.1
b) Gly-Val, no Val: V = 27.0 U/ml, K_{app} = 4.3mM, n_H = 1.0
with Val: V = 16.0 U/ml, K_{app} = 9.6mM, n_H = 1.0.

Modifier effects

It has been shown previously[4] that yeast dipeptidase is very sensitive to modifier effects. This is indicated not only by the frequent occurence of initial lag phases, but also by the observation of activation effects exerted by substrate analogs not too similar to the native dipeptide substrates.

In Fig. 4, relative dipeptidase activities towards Ala-Gly or Ala-Ala are plotted vs. the logarithms of the concentration of some modifiers.

As mentioned above, 5-aminolaevulinic acid, which closely resembles the dipeptide backbone of Gly-Gly, acts as a pure inhibitor. *N*-Glycyl-*N'*-acetylhydrazine, on the other hand, which lacks a carboxylate function, activates the enzyme. Leucine has both effects, activating at low and inhibiting at higher concentrations. A similar

Table 3. Inhibition constants of free amino acids.

The constants were calculated from the effect on K_{app} of the substrate indicated in each case (competitive inhibition).

Amino acid	K_i [mM]	Substrate
Gly	> 100	Ala-Ala
LAla	33	Gly-Gly
LVal	7.7	Ala-Gly
LLeu	3.7	Ala-Gly
LLeu-NH_2	> 50	Ala-Gly
Boc-LLeu	3.9	Ala-Gly
LPhe	4.1	Ala-Gly
DPhe	16	Ala-Gly
LTrp	3.0	Ala-Gly

activation is found with the LD isomer of Ala-Lys-(Ac) which, due to the D configuration of the COOH terminal residue, is not accepted as substrate. These findings suggest that the activation phenomena are brought about by the binding of modifiers to an allosteric site different from the substrate binding domain. This is supported by the data of Fig. 4, which show that the activation occurs at modifier concentrations far below the Michaelis constants of typical dipeptidase substrates.

The systematic study of modifier effects is complicated by the fact that the response of the enzyme to such compounds markedly differs with different preparations, even if the purification procedure is carefully controlled. Similarly, the extent to which initial lag phases occur is different from one enzyme preparation to another. At the moment we do not understand the structural basis of this observation. It is important to note, however, that the results compiled in Tables 1 and 2 were all obtained with an enzyme preparation showing only a weak tendency to display lag phases.

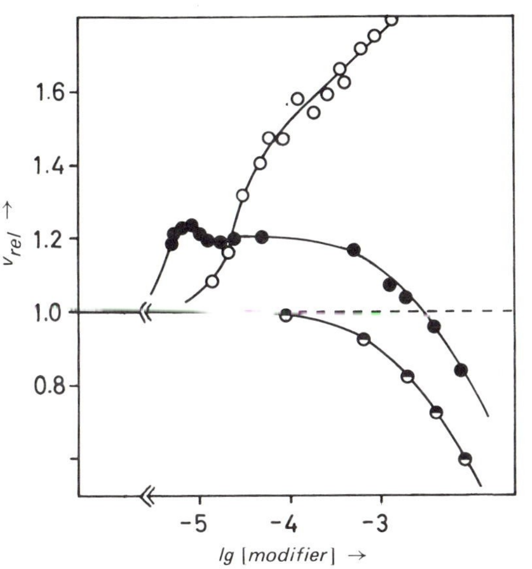

Fig. 4. Effects of modifiers on yeast dipeptidase. Relative velocities v_{rel} (v in the absence of modifiers = 1.0) are plotted vs. the logarithm of modifier concentration. The modifiers are: *N*-glycyl-*N'*-acetylhydrazine (○) and 5-aminolaevulinic acid (◒), the effects of which were determined in 0.1M phosphate/borate buffer, pH 7.8 with Ala-Ala as a substrate; and L-leucine (●), tested with Ala-Gly as a substrate in Tris buffer, pH 8.0.

Discussion

In a general way, the enzyme-catalyzed splitting of a substrate S to give two products P and Q (i.e. a Uni-Bi mechanism) may be represented as follows:

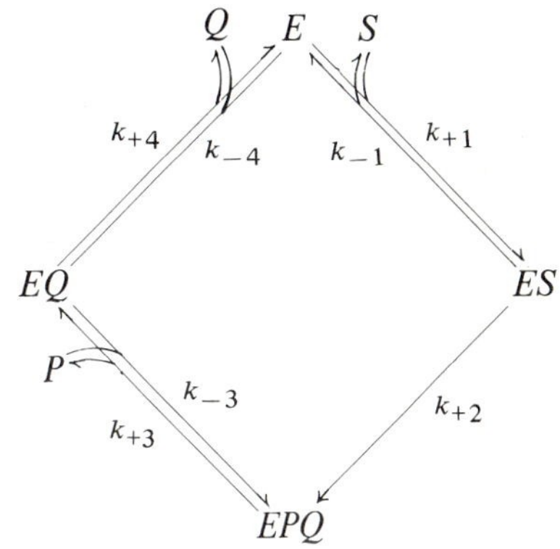

The scheme implies that the products are liberated in a compulsory order rather than in a random fashion. These situations may be distinguished according to the product inhibition patterns observed with P and Q[11]. In the case of a random release, both P and Q should exert the same type of inhibition, either competitive (if the liberation of products is fast as compared to the bond-breaking step) or mixed. With ordered Uni-Bi mechanisms the product Q will be a competitive inhibitor, whereas P shows a mixed-type inhibition. Thus, the data of Fig. 3 suggest that the release of products from yeast dipeptidase proceeds in an ordered way, the *C*-terminal amino acid being the first one to leave the enzyme-product complex ($A_c = P$, $A_n = Q$).

Under initial-velocity conditions, i.e. if the concentrations of P and Q are still negligibly small, one gets the following expressions for the macroscopic kinetic constants.

$$V = \frac{k_{+2} \times \left(\frac{k_{+3} \times k_{+4}}{k_{+3} + k_{+4}}\right)}{k_{+2} + \left(\frac{k_{+3} \times k_{+4}}{k_{+3} + k_{+4}}\right)} \times [E]_0$$

$$K_{app} = \frac{\left(\frac{k_{-1} + k_{+2}}{k_{+1}}\right) \times \left(\frac{k_{+3} \times k_{+4}}{k_{+3} + k_{+4}}\right)}{k_{+2} + \left(\frac{k_{+3} \times k_{+4}}{k_{+3} + k_{+4}}\right)}$$

and

$$V/K_{app} = \frac{k_{+1} \times k_{+2}}{k_{-1} + k_{+2}} \times [E]_0$$

This means that the catalytic constant $k_{cat} = V/[E]_0$ must not be taken as a measure of the rate of step $ES \rightarrow EPQ$ unless k_{+2} is negligibly small as compared to k_{+3} and k_{+4}. Likewise, the apparent Michaelis constant equals the substrate dissociation constant $K_s = k_{-1}/k_{+1}$ only if k_{-1}, k_{+3} and k_{+4} are much greater than k_{+2}. With yeast dipeptidase one cannot exclude the possibility that k_{+3} and/or k_{+4} may be low enough to limit the magnitude of V and K_{app}, since with hydrophobic dipeptides the reaction products are strongly bound to the enzyme. In fact, we believe that the observed variation of the kinetic parameters with the hydrophobicity of the substrate side chains is indicative of such a kinetic situation. Thus, with dipeptides of the type A_n-Gly, k_{+3} should be high and constant for all substrates of the series, since Gly is a poor ligand for site S_c. The value of k_{+4}, on the other hand, should decrease as A_n becomes more and more hydrophobic. If $k_{+3} \gg k_{+4}$ we have

$$V = \frac{k_{+2} \times k_{+4}}{k_{+3} + k_{+4}} \times [E]_0$$

and

$$K_{app} = \frac{\dfrac{k_{-1} + k_{+2}}{k_{+1}}}{1 + \dfrac{k_{+2}}{k_{+4}}}$$

If, in addition, k_{+4} is comparable to or lower than k_{+2}, any further decrease in k_{+4} will result in a concomitant decrease in *both* V and K_{app}. Inspection of Table 2a shows that with A_n-Gly type dipeptides, V is highest with Thr-Gly and then steadily decreases with increasing hydrophobicity of A_n. Concomitantly, K_{app} falls by more than two orders of magnitude. The assumption that k_{+4} limits the kinetic constants with these substrates is supported by the behaviour of L-Trp-Gly, which strongly inhibits the enzyme without being hydrolyzed. A very low value of k_{+4} due to the strongly hydrophobic Trp side chain would account for this finding.

With substrates of the type Gly-A_c, the situation is somewhat different. Again K_{app} decreases with increasing hydrophobicity of A_c. However, the maximal velocity is highest with Gly-Leu. Thus, it seems that k_{+3} does not become limiting before S_c is occupied by the strongly hydrophobic Phe and Trp residues.

In contrast to V and K_{app}, the ratio of the two constants does not depend on k_{+3} and k_{+4}. Therefore with yeast dipeptidase, V/K_{app} seems to be the most suitable parameter if the effects of enzyme-substrate interactions are to be quantitatively characterized.

In Fig. 5, V/K_{app} values for the substrate series A_n-Gly and Gly-A_c are plotted vs. ΔF_t, a parameter introduced by Tanford[12] as a measure of amino acid side chain hydrophobicity. ΔF_t is equivalent to the free energy change upon transfer of a side chain from ethanol to water. For both types of dipeptides, V/K_{app} steadily increases with increasing hydrophobicity of the variable amino acid, levelling off at higher values of ΔF_t.

Similar graphs may be constructed for other substrate series such as A_n-Ala, Ala-A_c, Leu-A_c, and so on. In each case, V/K_{app} increases with the hydrophobicity of the variable residue. Thus, as judged the value of V/K_{app}, the hydrophobicity rather than steric features of its side chains seems to be the main factor in properly aligning the dipeptide within the active site. From the data of Fig. 5, it is obvious that the A_n-S_n interaction is much more effective in this respect than the A_c-S_n contact.

So, for instance, the replacement of the *N*-terminal glycine of Gly-Gly by an alanine residue leads to almost the same enhancement of V/K_{app} as does the introduction of the much more hydrophobic leucine side chain into position A_c. On the other hand, the release of Ala from S_n is fast enough not to reduce the overall turnover due to a low value of k_{+4}. Therefore dipeptides of the class Ala-A_c have the highest maximal velocities of all dipeptides tested as yet. As mentioned above, a leucine side chain in contact with S_n results in particularly high values of V/K_{app} even if A_c is an unfavourable ligand.

Unfortunately, there is no straightforward method to separate V/K_{app} into the component

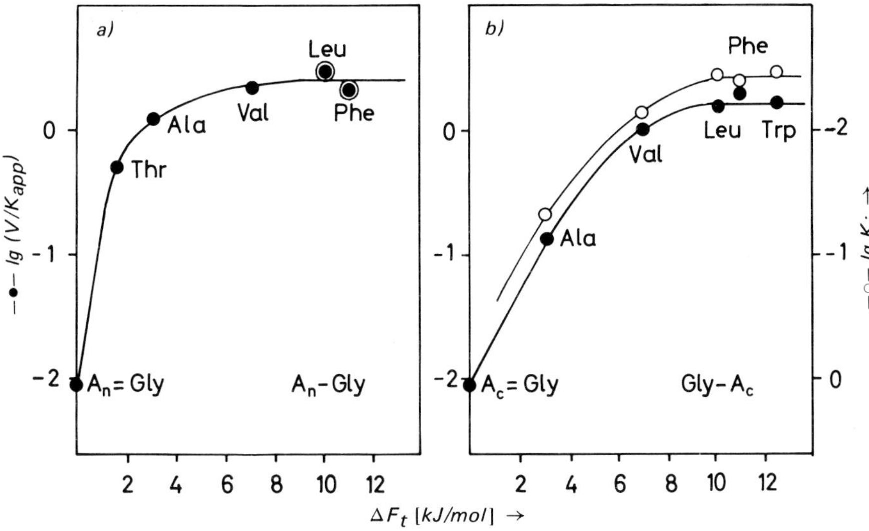

Fig. 5. Dependence of V/K_{app} for selected dipeptides on the hydrophobicity of their side chains.
The figure summarizes data on a) dipeptides of the general structure A_n-Gly with various N-terminal residues and b) on Gly-A_c-type substrates. Logarithms of V/K_{app} are plotted vs. ΔF_t of the variable residues. The parameter ΔF_t according to Tanford[12], quantitatively characterizes the hydrophobicity of amino acid side chains. With Leu-Gly and Phe-Gly (5a), the Hill coefficients of which deviate markedly from unity, the substrate concentrations at half saturation $[S]$ 0.5 were used instead of K_{app}. (○).
In addition, Fig. 5b contains the logarithms of inhibition constants K_i (right hand scale) measured with some L-amino acids (cf. Table 3) as a function of ΔF_t for the inhibitor.

constants K_s and k_{+2}. It seems reasonable, however, to assume that the enhancement of V/K_{app} with increasing ΔF_t comes from a decrease of K_s due to improved substrate binding. Evidence in favour of this view is provided by the observed inhibition constants of free amino acids, which are true thermodynamic binding constants. As shown in Fig. 5b, the dependence of K_i on ΔF_t closely resembles the plot of V/K_{app} vs. ΔF_t for Gly-A_c type substrates, whereas the similarity with the corresponding graph for dipeptides of class A_n-Gly is less conspicuous. Possibly, free amino acids are preferentially bound to site S_c. This would account for the markedly higher inhibition constants of amino acid amides or D isomers as compared to L-amino acids (cf. Table 3).

Although both S_n and S_c readily interact with the bulky Phe and Trp residues, it seems that S_n is spatially more restricted than S_c. So S_c, in spite of its distinctly hydrophobic character, tolerates the charged lysine side chain (but not the much shorter aspartic acid) and also accommodates the still longer N^{ϵ}-acetyllysine. Most probably – in contrast to the Asp side chain – both residues extend beyond the limits of S_c, which therefore should be "open" at the end. S_n, on the other hand, does not cope with Lys(Ac)-Gly, which does not even inhibit the enzyme (see Table 2). The strong interaction between S_n and hydrophobic amino acids could provide a common explanation for the initial lag phases (see ref.[4]) and the deviations from Michaelis-Menten kinetics encountered with substrates like Leu-Gly or Phe-Gly. Both effects could be the direct consequence of a conformational change in the enzyme triggered by the S_n-R_n interaction and leading to a more active conformation. It is not surprising that with Gly in amino terminal position of a substrate, such an event does not occur. If this interpretation is correct, the observed positive cooperativity is not an expression of subunit interactions but rather the reflection of an "induced fit" phenomenon. As was shown by Ainslie et al.[13], both transients and cooperative steady state kinetics are symptomatic of such a mechanism.

A similar conformational adaptation in the S_n region of the enzyme could be the reason for the rather selective activation effects exerted by metal ions like $Co^{2\oplus}$ and $Mn^{2\oplus}$. It was shown previously with yeast dipeptidase[3] and also for other enzymes[14] that $Co^{2\oplus}$ specifically activates the hydrolysis of dipeptides of the type Gly-A_c but inhibits the turnover of others. Similarily, $Mn^{2\oplus}$ is most effective with Pro-Gly, and to a minor degree, with Leu-A_c type substrates. An exchange of the essential zinc atom for $Co^{2\oplus}$ modifying the structural features of the S_n domain and thus facilitating the accomodation of certain substrates, would account for the observed specific activation.

Relatively few kinetic studies on dipeptidases have been published so far. Data are, however, available for two enzymes from mouse ascites tumor cells and *Escherichia coli*, respectively, which have been thoroughly studied by Patterson and coworkers[15].

A comparison of substrate specificities reveals a striking similarity between yeast dipeptidase and the tumor enzyme. Not only their general binding properties but also the numerical values of the kinetic parameters for a variety of substrates are almost identical. (See Table 4.) The same close relationship appears in the effects of metal ions on both enzymes. Other mammalian dipeptidases[16,17] have many substrate specificity characteristics in common with yeast and ascites tumor dipeptidases. The *E. coli* enzyme, on the other hand, behaves differently not only with

Table 4: Comparison of kinetic constants measured with some purified dipeptidases. Michaelis constants are given as mmol/*l*. Velocities were calculated in relation to the rate of LLeu-Gly hydrolysis taken as 1.

Dipeptide	Enzyme isolated from						
	Yeast		Ascites tumor[a]		Human milk[b]	*E. coli*[c]	
	V	K_{app}	V	K_{app}	V_{rel}	V	K_{app}
Gly-Gly	0.8	90	0.2	24	0	7.5	15
Gly-LAla	1.6	11	1.9	22	1.3	5.8	0.88
Gly-LVal	4.3	4.2	4.6	5.1	3.3	0	
Gly-LLeu	6.0	4.1	3.9	2.5	3.8	1.7	2.4
Gly-LPhe	4.8	2.4	2.0	1.7	2.3	1.5	0.5
Gly-LTrp	1.3	0.8	nd	nd	2.0	nd	nd
LAla-Gly	2.6	2.8	3.3	3.4	3.6	20	3.3
LVal-Gly	1.2	0.6	1.3	1.4	1.0	6.5	3.6
LLeu-Gly	–1–	0.17	–1–	0.33	–1–	–1–	2.0
LPhe-Gly	1.1	0.36	0.13	4.4	0.1	20	1.6
LAla-LLys	2.9	2.8	0.9	1,4	nd	27	1.2
LPro-Gly	0.2	nd	0.11	4.1	nd	nd	nd
Pro-Gly($Mn^{2\oplus}$)[d]	1.8	nd	1.1	21	0.6		
Gly-LAsp	0.02	80	0.02	nd	0.06	2.4	2.4
LAsp-Gly	0		0		0	0	
LLys-Gly	0		0		nd	0.02	2

[a] Dipeptidase from mouse ascites tumor cells, data from Patterson et al.[14] Conditions: phosphate/borate buffer pH 8.3 at 30 °C
[b] Dipeptidase from human milk, data from Gründig and Hanson[15]. Conditions: pH = 8.0 at 25 °C. Since Michaelis constants were not determined, relative velocities at $[S] = 10$ mmol/*l* are given.
[c] Dipeptidase from *E. coli*, data from Patterson et al.[14]. Same conditions as with the tumor enzyme.
[d] Measured in the presence of 0.1 mM $Mn^{2\oplus}$. Data for the ascites tumor and *E. coli* enzymes from ref.[13].

respect to its preferred substrates but also in its response to metal ions.

Why organisms so diverse in evolutionary terms as yeast and mammals should have preserved so strikingly similar dipeptidase active sites is a question open to speculation.

We would like to thank Mr. *Henseling* for competent technical assistance. The present study was supported by the *Deutsche Forschungsgemeinschaft.*

Literature

1 Smith, E.L. (1960) in The Enzymes (Boyer, P.D., Lardy, H. & Myrbäck, K., eds.) 2nd ed., vol. 4, p. 1 - 10, Academic Press, New York.
2 Hayman, S. & Patterson, E.K. (1970) *J. Biol. Chem.* **246**, 660 - 669.
3 Röhm, K.-H. (1974) *Hoppe-Seyler's Z. Physiol. Chem.* **355**, 675 - 686.
4 Röhm, K.-H. (1975) *Hoppe-Seyler's Z. Physiol. Chem.* **356**, 903 - 914.
5 Greenstein, J.P. & Winitz, M. (1961) Chemistry of the Amino Acids, vol. 2, p. 851 f., J. Wiley, New York.
6 Roeske, R. (1963) *J. Org. Chem.* **28**, 1251 - 1253.
7 Schwyzer, R., Sieber, P. & Kappeler, H. (1959) *Helv. Chim. Acta* **42**, 2622 - 2624.
8 Knack, I. & Röhm, K.-H. (1977) *Hoppe-Seyler's Z. Physiol. Chem.* **358**, 262.
9 Baum, F. & Schneider, F. (1970) *Naturwissenschaften* **57**, 498 - 499.
10 Nishi, A. (1960) *J. Biochem. (Tokyo)* **47**, 47 - 53.
11 Fromm, H.J. (1975) Initial Rate Enzyme Kinetics, p. 123 f., Springer, Berlin.
12 Tanford, C. (1962) *J. Am. Chem. Soc.* **84**, 4240 - 4247.
13 Ainslie, G.R., Skill, J.P. & Neet, K.E. (1972) *J. Biol. Chem.* **247**, 7088 - 7096.
14 Patterson, E.K., Gatmaitan, J.S. & Hayman, S. (1975) *Biochemistry* **14**, 4261 - 4266.
15 Patterson, E.K., Gatmaitan, J.S. & Hayman, S. (1973) *Biochemistry* **12**, 3701 - 3709.
16 Gründig, C.A. & Hanson, H. (1973) *Hoppe-Seyler's Z. Physiol. Chem.* **354**, 487 - 500.
17 Das, M. & Radhakrishnan, A.N. (1973) *Biochem. J.* **135**, 609 - 615.

V. Herrmann, I. Knack and K.-H. Röhm, Physiologische Chemie II, Lahnberge, D-3550 Marburg/L.

Hoppe-Seyler's Z. Physiol. Chem.
Bd. 359, S. 63 - 67, Januar 1978

Lysosomal Phospholipids from Rat Liver after Treatment with Different Drugs

Hong Boe TJIONG*, Jochen LEPTHIN and Hildegard DEBUCH

Institut für Physiologische Chemie der Universität zu Köln, Lehrstuhl II

(Received 5 July 1977)

Summary: Rats were treated with 5 different drugs *p*-ethoxyacetanilide (I), indometacin (II) and nor-amidopyrine-methanesulfonate (III), *O,O'*-bis(diethylaminoethyl)hexestrol (IV) and chloroquine (V) for 3 - 4 weeks.

Liver cell fractions were isolated by discontinuous gradient centrifugation and the specific activity of acid phosphatase was determined in each. Lysosomal fractions contained widely varying amounts of this marker enzyme, indicating that the concentration of lysosomes within these fractions differed.

The amounts and patterns of phospholipids reflected this fact. Since we assumed bis(monoacylglycero)phosphate [$(MAG)_2P$; synonym:lysobisphosphatidic acid] is a marker lipid for secondary lysosomes, we expected and found significant quantities of this acidic phospholipid only in those lysosomal fractions which were also rich in acid phosphatase activity. 12% of the lysosomal phospholipids from animals receiving the hexestrol derivative (IV), and 19% of those from the chloroquine (V) experiment were present as $(MAG)_2P$.

The fatty acid compositions of this lysosomal phospholipid were not the same in all lysosome fractions. The more $(MAG)_2P$ present in the lysosomes, the more unsaturated are the fatty acids. Thus, after treatment with chloroquine, more than 90% of the fatty acids from $(MAG)_2P$ are unsaturated; $C_{22:6}$ represents about 70% of the total.

Lysomale Phospholipide von Rattenlebern nach Behandlung mit verschiedenen Medikamenten

Zusammenfassung: Nach 3 - 4wöchiger Behandlung mit 5 verschiedenen Medikamenten: *p*-Äthoxyacetanilid (I), Indometacin (II), Nor-amidopyrinmethansulfonat (III), *O,O'*-Bis(diethylaminoethyl)-hexöstrol (IV) und Chloroquin (V) wurden aus Rattenlebern die Zellfraktionen durch Zentrifugation mit diskontinuierlichen Gradienten dargestellt und die spezifische Aktivität der sauren Phosphatase darin bestimmt. Die lysosomalen Fraktionen wiesen sehr unterschiedlich starke Anreicherungen dieses Leitenzyms auf. Die Menge und Zusammensetzung der Phospholipide deutete

---

Enzyme:
Acid phosphatase, orthophosphoric-monoester phosphohydrolase (acid optimum) (EC 3.1.3.2).
Abbreviation:
$(MAG)_2P$ = Bis(monoacylglycero)phosphate, formerly named lysobisphosphatidic acid.
* Part of the thesis of H. B. Tjiong is included.

ebenfalls in diese Richtung. Da wir Bis(monoacylglycero)phosphat [$(MAG)_2P$; Synonym: Lysobisphosphatidsäure] als „Markerlipid" sekundärer Lysosomen ansehen, erwarteten wir dieses saure Phospholipid in beachtlichen Mengen nur in solchen lysosomalen Fraktionen, die reich an saurer Phosphatase-Aktivität waren. Die Analyse bestätigte dies: Nur in den lysosomalen Fraktionen derjenigen Tiere, die das Hexöstrolderivat (IV) oder Chloroquin (V) erhalten hatten, waren 12 bzw. 19% der Gesamtphospholipide als $(MAG)_2P$ vorhanden.

Die Fettsäuregemische dieses sauren Phospholipids waren in den einzelnen Präparaten verschieden. Je größer der Anteil an $(MAG)_2P$ im Gesamtlipid war, umso größer war der Anteil der ungesättigten Fettsäuren. Nach Behandlung mit Chloroquin bestand das Fettsäuregemisch von $(MAG)_2P$ zu mehr als 90% aus ungesättigten Verbindungen, zu etwa 70% aus Docosahexaensäure.

Key words: Lysosomes, phospholipidosis, bis(monoacylglycero)phosphate, lysobisphosphatidic acid, liver lipids.

Recently we[1] showed that not only tritosomes but also liver lysosomes from rats treated with dextran, silver or iron contain an acidic phospholipid in fairly high amounts (6 - 28% of the total phospholipids). We concluded from its typical location in two dimensional thin-layer chromatographs that it is the same compound as that, isolated earlier[2] from liver and spleen of a child who suffered from a special type of Niemann-Pick's disease. This compound was detected for the first time in pig lung by Body and Gray[3], in tritosomes from rat liver by Wherrett and Huterer[4,5]. It was named by the authors[3] lysobisphosphatidic acid. Since, however, the ratio of glycerol : phosphorus : fatty acid in this lipid is 2:1:2 we proposed the name bis(monoacylglycero)phosphate [abbreviated $(MAG)_2P$] for it[1,2].

Since Yamamoto et al.[6,7] reported an increase in human liver phospholipids after administration of *O,O'*-Bis(diethylaminoethyl)hexestrol, it has been shown by several authors (see ref.[8]) that quite a number of drugs induce a phospholipidosis. Usually, lamellar bodies ("myeloid bodies"[8]) can be seen in the tissue by electron microscopy.

We wanted to test some drugs for their ability to induce phospholipid synthesis, to see whether we could obtain "lamellar bodies" (myeloid bodies) and whether these are secondary lysosomes. In this case, we expected to find acid phosphatase activity as well as $(MAG)_2P$ in the isolated organelles, as we did before[1].

Experimental

1. Analytical methods

1.1 Lipid phosphorus, protein and acid phosphatase were estimated as published earlier[1], for cholesterol see ref.[2].

Thin-layer chromatography

1.2 After two-dimensional separation on silica gel MNHR (Fa. Macherey, Nagel & Co., Düren) according to Debuch et al.[9], the various lipids were demonstrated with special stains (see ref.[1]). For quantitative analysis of the phospholipid mixture, the P content of the individual spots was estimated[9].

1.3 Unidimensional separation: on silica gel H (Merck, Darmstadt) with solvent system chloroform/methanol/ammonia = 60 : 35 : 8 (v/v/v).

Gas-liquid chromatography

1.4 The apparatus and the performances for the analyses of fatty acids were the same as described before[1].

2. Preparation of the lysosomal fractions

2.1 Application of the drugs and preparation of the tissue

Five different drugs were administered daily in different ways and in different amounts to adult rats (300 - 350 g) of either sex.

Phenacetin (I) (*p*-ethoxyacetanilide, DAB 7, from Merck, Darmstadt) per os: 1 µg/g rat (Wistar/Han) for 21 days.

Indometacin (II) (1-(*p*-chlorobenzoyl)-5-methoxy-2-methyl-3-indolylacetic acid, Sharp & Dohme GmbH, München) per os: 3 µg/g rat (Wistar/Han) for 21 days.

Nor-amidopyrin-methanesulfonate (III) (sodium [*N*-(2,3-dimethyl-5-oxo-1-phenyl-3-pyrazolin-4-yl)-*N*-methylamino]methanesulfonate, Novalgin from Hoechst, Frankfurt/Main) intramuscular: 1.2 mg/g rat (Wistar/Han) for 21 days.

O,O'-Bis(diethylaminoethyl)hexestrol (IV) (2,2'-[(1,2-diethyl-1,2-ethanediyl)bis(4,1-phenylenoxy)]bis[*N,N*-diethylethanamine], Trimanyl from Maggioni & C., Milano) intraperitoneal: 10 μg/g rat (Sprague-Dawley) for 4 weeks.

Chloroquine (V) (7-chloro-4-[4-(diethylamino)-1-methylbutylamino]quinoline, diphosphate, Resochin from Bayer, Leverkusen) intraperitoneal: 60 μg/g rat (Wistar/Han) for 21 days.

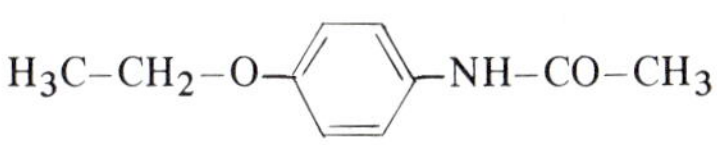
I

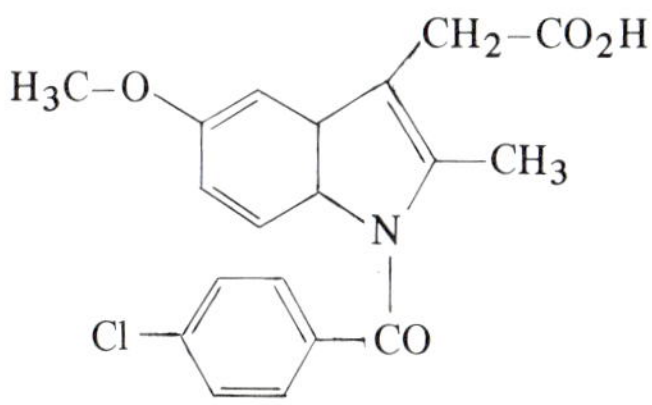
II

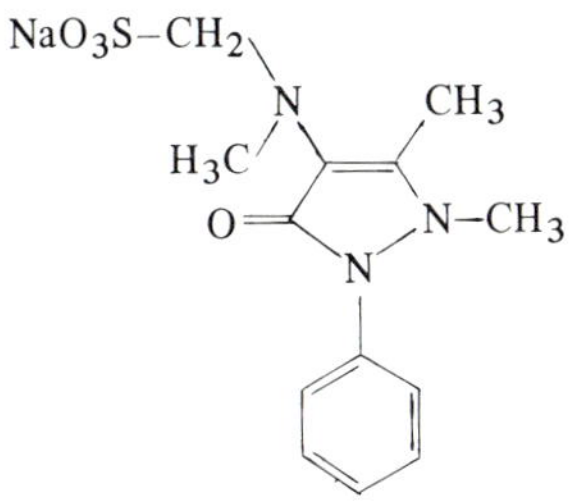
III

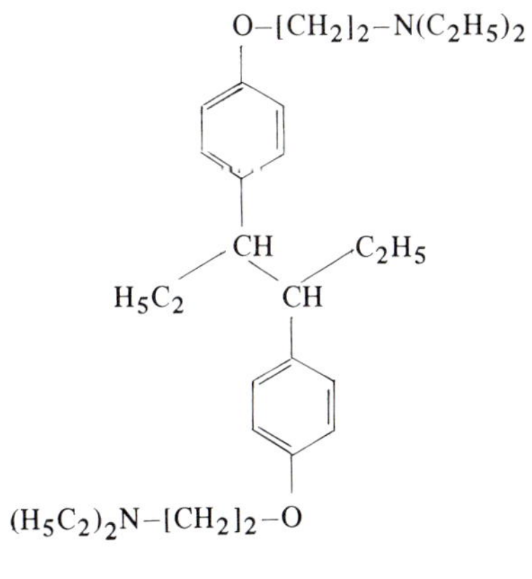
IV

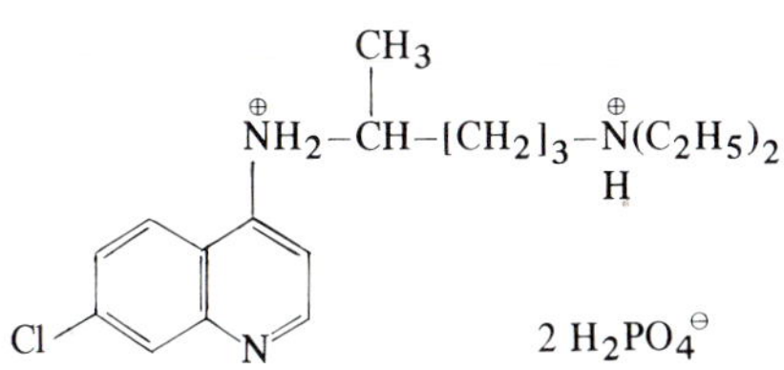
V

Before decapitation the animals were fasted for 24 h. The abdomen, the thoracic cavity and the rigth ventricle of the heart were opened immediately after death. (Every manipulation was performed at 4 °C.) The livers were perfused with 0.25M sucrose (4 °C) through the vena portae and – in a retrograde way – through the vena cava inf.

The tissue was chopped into 0.4 mm^2 pieces (McIlwain tissue chopper, Mickle, England) and homogenized in 9 vol. of 0.25M sucrose using a homogenizer according to Aldridge et al.[10] (1000 rpm and 25 strokes/min.).

2.2 Differential centrifugation

This method is based on that of Trouet[11] and Aronson and Davidson[12], but was modified in the following way: After a pellet had been spun down from the homogenate (750 x *g* for 10 min), the supernatant (S_1) was separated into a crude mitochondrial-lysosomal fraction (pellet: P_2) and a supernatant (S_2) by centrifugation at 12 000 x *g* for 10 min. P_2 was resuspended in 0.25M sucrose, centrifuged as before, and the P_3 obtained was washed in the same way. The resulting mitochondrial-lysosomal fraction (P_4) was resuspended in 1.6M sucrose (1 m*l* 1.6M sucrose per g of tissue fresh weight). The next centrifugation step (100 000 x *g* for 120 min) was carried out in a discontinuous gradient consisting of 5 vol. of the above mentioned suspension in 1.6M sucrose, 4 vol. 1.0M and finally 3 vol. 0.25M sucrose. We obtained a pellet (P_5) and 5 fractions differing in density, designated from the top of the tube: 1, 2 (lysosomal), 3, 4 (mitochondrial) and 5, which were separated and analysed for enzyme activity.

3. Lipid examination

3.1 Extraction

The liver *homogenate* was prepared as described under section 2.1, but using 0.9% NaCl solution for perfusion instead of sucrose. The lipids were then extracted according to Folch[13].

The *lysosomal* fractions obtained after density gradient centrifugation (section 2.2) were transferred into 19 vol. of chloroform/methanol (2:1 v/v) and the lipids also extracted by Folch's procedure[13].

3.2 Analysis

Lipid extracts of liver homogenates (from controls and from experiments with I and V) were analyzed for total cholesterol and phospholipids (section 1.), those from lysosomes for phospholipid composition only.

The phospholipid pattern was determined after thin-layer chromatography[9].

Methyl esters of fatty acids from $(MAG)_2P$ were analyzed by gas-liquid chromatography. Therefore we scraped off the corresponding spots from 5 - 10 plates

Table 1. Specific activity of acid phosphatase [U/mg protein] in different cell fractions.
All values are averages of 2 - 4 experiments. Structures of compounds I - V, see Scheme.

Cell fraction	Treatment for 21 days				Control
	I	II	IV[a]	V	
Homogenate	0.033	0.033	0.038	0.037	0.033
1	0.133	0.203	0.196	0.231	0.119
2 (lysos. fract.)	0.226	0.446	0.336	1.043	0.245
3	0.143	0.302	0.415	0.574	0.160
4 (mitoch. fract.)	0.102	0.235	0.165	0.361	0.140
5	0.094	0.232	0.164	0.286	0.140
Ratio $\frac{\text{Homogenate}}{\text{Fract. 2}}$	1:6.9	1:13.5	1:8.9	1:28.2	1:7.4

[a] Treatment for 28 days.

and transesterified the fatty acids by treatment with methanolic HCl (see ref.[11]).

Results and Discussion

In most of the experiments, we found that administration of the drugs did not affect the specific activity of acid phosphatase in the liver homogenates. The hexestrol derivative (IV) and chloroquine (V) (see Table 1) induced an increase of about 10%. Therefore, we tested the lipid composition of liver homogenates only after treatment with one drug [phenacetin (I)] which did not alter the specific activity of acid phosphatase, and after treatment with one drug (V) which produced an increase in enzyme activity (see Table 2). Of all the cell fractions, the lysosomal always contained the highest acid phosphatase activity (with one exception: the experiment with IV), even with control animals (Table 1). However, the enrichment of this enzyme was usually quite poor, but large in the experiment with V (nearly 30-fold) compared with that of homogenate.

The phospholipid content of liver homogenate is slightly (about 10%) increased only in the experiment with V. The phospholipid patterns re-

Table 2. Lipids from liver homogenates after treatment with different drugs.
Treatment with I was for 21 days; treatment with V for 14 days.

	I	V	Control
Phospholipid [mg/10 g liver]	310	335	294
Total cholesterol [mg/10 g liver]	57	55	51
Phospholipid composition [% of Lipid P]			
Phosphatidylcholine	50.8	44.8	52.1
Lysophosphatidylcholine	0.9	1.2	1.1
Phosphatidylethanolamine	28.7	24.0	26.6
Phosphatidylserine	3.1	4.3	3.1
Phosphatidylinositol	7.8	8.2	8.3
Sphingomyelin	3.7	4.8	3.9
Cardiolipin	4.4	6.3	3.6
$(MAG)_2P$	not detectable	4.7	not detectable
Start and front	0.8	2.0	1.4

semble each other, but we again find some differences in liver lipids after treatment with V: choline and ethanolamine phosphoglycerides seem to be somewhat decreased. Astonishingly enough, we could detect $(MAG)_2P$ in the homogenate lipids, where it represents about 5% of the lipid phosphorus. The phospholipid patterns of the lysosomal fractions after different drug treatments (Table 3), however, show slight qualitative, but some very large absolute differences.

Lysophosphatidylcholine, for example, is present in every lipid extract, but mostly in small amounts. In the experiments with IV and V, however, it accounts for about 8% of the lipid phosphorus, the same % as phosphatidylethanolamine – the latter being much decreased. It should be pointed out that the amount of phosphatidylinositol is doubled in the experiment with V.

We found relatively large amounts of lysophosphatidylcholine, indicating that phospholipase A_2 was probably quite active. Therefore, we tested the phosphatidylinositol for contamination with lysophosphatidylglycerol in the following way; since phosphatidylinositol and lysophosphatidylglycerol [which is identical with the lyso-compound of $(MAG)_2P$] do not separate in our thin-layer chromatography system, we scraped off the corresponding lipid spots after two-dimensional chromatography, eluted the phospholipid and tried to identify it with reference substances in unidimensional thin-layer chromatography in an alkaline solvent system (see section 1 and Figure). We conclude from this that the sample was indeed phosphatidylinositol.

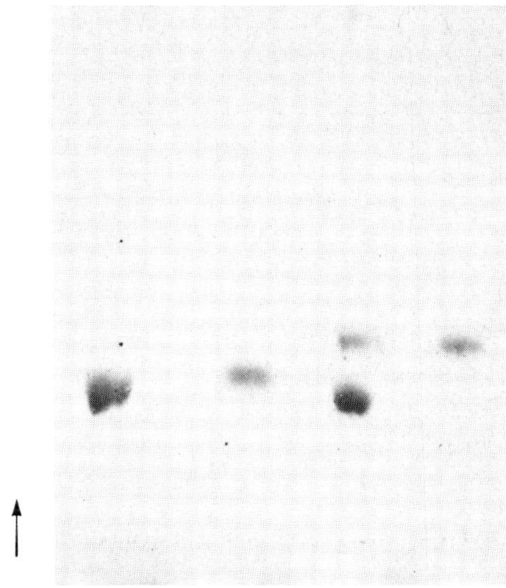

Figure. Thin-layer chromatography of phosphatidylinositol from lysosomal fraction after treatment of the rats with chloroquine (V). 1) phosphatidylinositol (reference); 2) phosphatidylinositol sample of the experiment with V (separated from the total lipids by two-dimensional chromatography); 4) lysophosphatidylglycerol; 3) mixture of 1 and 4. (Method see section 1.3.)

The absolute amounts of $(MAG)_2P$ (see Table 3) are most remarkable. In the lysosomal lipids from untreated animals, $(MAG)_2P$ could only be de-

Table 3. Phospholipids from lysosomal fractions [% of total lipid phosphorus].

<table>
<tr><th rowspan="2">Phospholipid</th><th colspan="5">Treatment for 21 days</th><th rowspan="2">Control</th></tr>
<tr><th>I</th><th>II</th><th>III</th><th>IV[a]</th><th>V</th></tr>
<tr><td>Phosphatidylcholine</td><td>50.3</td><td>47.7</td><td>43.4</td><td>33.3</td><td>37.0</td><td>45.3</td></tr>
<tr><td>Lysophosphatidylcholine</td><td>1.7</td><td>1.9</td><td>1.3</td><td>7.5</td><td>7.5</td><td>4.0</td></tr>
<tr><td>Phosphatidylethanolamine</td><td>26.8</td><td>27.9</td><td>32.3</td><td>5.9</td><td>9.1</td><td>22.8</td></tr>
<tr><td>Phosphatidylserine</td><td>2.0</td><td>5.0</td><td>2.6</td><td>3.9</td><td>4.8</td><td>2.9</td></tr>
<tr><td>Phosphatidylinositol</td><td>10.7</td><td>7.8</td><td>9.5</td><td rowspan="2">25.6</td><td>18.0</td><td>8.9</td></tr>
<tr><td>Sphingomyelin</td><td>3.1</td><td>4.4</td><td>3.2</td><td>3.2</td><td>8.5</td></tr>
<tr><td>Cardiolipin</td><td>2.0</td><td>3.8</td><td>4.8</td><td>1.2</td><td>(1.7)[b]</td><td>4.6</td></tr>
<tr><td>Start and front</td><td>1.8</td><td>0.7</td><td>1.3</td><td>8.8</td><td>2.1</td><td>4.3</td></tr>
<tr><td>$(MAG)_2P$</td><td>1.8</td><td>0.9</td><td>1.6</td><td>12.0</td><td>18.6</td><td>trace</td></tr>
</table>

[a] Also lysophosphatidylserine: 2.1%. Treatment for 28 days.
[b] The value in parantheses was found to represent acyl$(MAG)_2P$ (see following paper).

Table 4. Total phospholipids and $(MAG)_2P$ of lysosomal fractions (µmol/100 g liver wet weight).

	I	II	III	IV	V	Control
Phospholipids[a]	15	35	20	40	81	7.5
$(MAG)_2P$[b]	0.26	0.33	0.31	4.50	14.20	trace

[a] Mean M_r: 800. [b] Mean: M_r: 850.

tected by overloading the thin-layer plate. It is present in small amounts after treatment with I, II or III, but is highly accumulated in the experiments with IV and V. This fact seems to indicate, that the drugs IV and V are very slowly metabolized in the liver. The other drugs in contrast are less accumulated or probably more easily „detoxified" and therefore do not cause a significant multiplication of lysosomes. A relation between duration of drug treatment (V) and the protein content of the lysosomal fraction of liver could be demonstrated[14]. This seems to indicate that accumulation of lysosomes occurs under such conditions.

The special effects of IV and V as representatives of a variety of cationic amphiphilic drugs have been discussed by Michell et al.[15]. The authors presume that the inhibition of phosphatidate phosphatase (EC 3.1.3.4) by those drugs (which could be shown to occur in a cell-free system from liver[16]) would be responsible for the lipid changes in lysosomes. Taking the results of Tables 1 and 3 together, we see that most of the lysosomal fractions low in marker enzyme activity are also low in $(MAG)_2P$ content.

In contrast, the lysosomal fractions from IV-treated livers contain a large amount of $(MAG)_2P$ and a relatively low specific activity of acid phosphatase, probably because these animals were treated not 3, but 4 weeks with the drug. The enzyme activity could be lost during this time. The phospholipids of fraction 3, just below the lysosomal fraction, also contained about 10 - 12% of $(MAG)_2P$, as we could expect, since it was rich

Table 5. Fatty acids of $(MAG)_2P$ from lysosomes [% of total fatty acids].

Fatty acid	Treatment for 21 days			
	I[a]	II	IV[b]	V[c]
14:0	5.8	6.6	1.4	0.8
15:0	2.7	2.3	1.4	–
16:0	21.8	24.7	11.9	3.3
16:1	5.7	7.4	2.5	–
18:0	10.4	14.4	5.4	2.4
18:1	11.6	13.5	14.3	6.8
18:2	5.2	3.8	8.1	6.2
20:0	2.6	3.5	–	–
20:2	3.4	3.2	trace	1.0
20:3	3.6	3.6	1.5	0.7
20:4	7.3	7.2	4.9	4.9
20:5	–	–	2.4	1.4
22:3	4.8	5.4	2.4	1.1
22:4	–	–	4.8	2.2
22:6	14.4	8.9	36.8	67.9
saturated	43.3	51.5	20.1	6.5
polyenoic	39.4	27.6	63.1	86.7

[a] plus C 14:1 3.4%; [b] plus C 22:2 2.2%, treatment for 28 days; [c] plus C 22:5 2.3%.

in marker enzyme activity. One has to consider this fact with regard to the results shown in Table 4. We conclude that the content of total phospholipids in the lysosomal fractions is related to that of $(MAG)_2P$.

We have no explanation yet for the different fatty acid compositions of $(MAG)_2P$ derived from the lysosomal lipids after different drug treatments (see Table 5). The same type of phospholipid within the same cell organelle contains about 50% of docosahexaenoic acid after application of triton[1,5], between 20 and 30% after administration of dextran, iron or silver[1], and somewhat less after treatment with I or II. However, $(MAG)_2P$ from the experiments with V seems to be one of the most highly unsaturated phospholipids ever observed in lipids of mammalian origin, since it contains nearly 70% $C_{22:6}$ acid and only 6.5% saturated fatty acids.

Therefore the question arises: what are the mechanisms which stimulate the biosynthesis and concentration of docosahexaenoic acid within liver lysosomal phospholipids?

We thank Mrs. *Chr. Geyer* and Mrs. *A. Blachetta* for their valuable technical help, *Maggioni & Co.*, Milano, for the gift of the hexestrol derivative (IV) and *Bayer, Leverkusen*, for the chloroquine.
We also thank the *Stiftung Volkswagenwerk* for supporting this work.

Literature

1 Stremmel, W. & Debuch, H. (1976) *Hoppe-Seyler's Z. Physiol. Chem.* **357**, 803 - 810.
2 Seng, P.N., Debuch, H., Witter, B. & Wiedemann, H.R. (1971) *Hoppe-Seyler's Z. Physiol. Chem.* **352**, 280 - 288.
3 Body, D.R. & Gray, G.M. (1967) *Chem. Phys. Lipids* **1**, 424 - 428.
4 Wherrett, J.R. & Huterer, S. (1972) *J. Biol. Chem.* **247**, 4114 - 4120.
5 Wherrett, J.R. & Huterer, S. (1973) *Lipids* **8**, 531 - 533.
6 Yamamoto, A., Adachi, S., Kitani, T., Shinji, Y., Seki, K., Nasu, T. & Nishikawa, M. (1971) *J. Biochem.* **69**, 613 - 615.
7 Adachi, S., Matsuzawa, Y., Yokomura, T., Ishikawa, K., Uhara, S., Yamamoto, A. & Nishikawa, M. (1972) *Lipids* **7**, 1 - 7.
8 Lüllmann, H., Lüllmann-Rauch, R. & Wassermann, O. (1973) *Dtsch. Med. Wochenschr.* **98**, 1616 - 1625.
9 Debuch, H., Mertens, W. & Winterfeld, M. (1968) *Hoppe-Seyler's Z. Physiol. Chem.* **349**, 896 - 902.
10 Aldridge, W.N., Emery, R.C. & Street, B.C. (1960) *Biochem. J.* **77**, 326 - 327.
11 Trouet, A. (1964) *Arch. Int. Physiol. Biochem.* **72**, 698 - 700.
12 Aronson, U.N., Jr. & Davidson, E.A. (1968) *J. Biol. Chem.* **243**, 4494 - 4499.
13 Folch, J., Lees, M. & Sloane-Stanley, G.H. (1957) *J. Biol. Chem.* **226**, 497 - 509.
14 Tjiong, H.B. & Debuch, H. (1978) *Hoppe-Seyler's Z. Physiol. Chem.* **359**, 71 - 79, succeeding paper.
15 Michell, H., Allan, D., Bowley, M. & Brindley, D.N. (1976) *J. Pharm. Pharmacol.* **28**, 331 - 332.
16 Brindley, D.N. & Bowley, M. (1975) *Biochem. J.* **148**, 461 - 469.

H. B. Tjiong and Prof. Dr. H. Debuch, Institut für Physiologische Chemie der Universität zu Köln, Lehrstuhl II, Postfach 41 05 80, D-5000 Köln 41.

Dr. J. Lepthin, I. Medizin. Universitätsklinik, Schittenhelmstr. 12, D-2300 Kiel.

Hoppe-Seyler's Z. Physiol. Chem.
Bd. 359, S. 71 - 79, Januar 1978

Lysosomal Bis(monoacylglycero)phosphate of Rat Liver, its Induction by Chloroquine and its Structure

Hong Boe TJIONG* and Hildegard DEBUCH

Institut für Physiologische Chemie der Universität zu Köln, Lehrstuhl II

(Received 5 July 1977)

Summary: Since chloroquine is able to induce phospholipidosis, and in particular formation of bis(monoacylglycero)phosphate [$(MAG)_2P$] in rat liver lysosomes, we tested the influence of the drug on this acidic phospholipid

1) after application of the drug for different periods, 3, 7, 14, 21 and 28 days, and
2) after application of different doses of the drug for 7 days.

In both cases, the amounts of the total phospholipids and of $(MAG)_2P$ within the lysosomal lipids showed a direct relationship to the quantity of the drug. Lysosomal fractions were identified by acid phosphatase activity and by electron microscopy.

After 14 days of treatment, the lysosomal fractions showed an increase in the marker enzyme to more than 30 times the level in the homogenate, and consisted of almost pure "lamellar bodies". $(MAG)_2P$ was isolated from those lysosomal lipid fractions by column chromatography.

We also isolated this phospholipid from "tritosomal" lipids and tested the two preparations in different ways. They both showed glycerophosphoglycerol as the only *P*-containing product after mild alkaline hydrolysis, and molar ratios for phosphorus: fatty acid: glycerol close to 1:2:2; and yielded diacetylacyl glycerol after acetolysis.

Thus, there is no doubt that $(MAG)_2P$ occurs in these secondary lysosomes – as we described in others – and that its production is a consequence of accumulation of lysosomes.

There is some indication for the occurrence of an acylated $(MAG)_2P$ (which is the same as acylphosphatidylglycerol) within the same cell organelles.

Lysosomales Bis(monoacylglycero)phosphat aus Rattenleber, seine Induktion durch Chloroquin und seine Struktur

Zusammenfassung: Chloroquin induziert eine Phospholipidose und besonders die Bildung von Bis(monoacylglycero)phosphat [$(MAG)_2P$] in Leberlysosomen der Ratte. Um dies näher zu prüfen, untersuchten wir den Einfluß des Medikamentes auf folgende Weise:

1) Nach Verabreichung der gleichen Chloroquindosis über verschiedene Zeiten: nach 3, 7, 14, 21 und 28 Tagen;
2) nach siebentägiger Verabreichung verschiedener Mengen des Medikamentes.

Enzyme:
Acid phosphatase, orthophosphoric – monoester phosphohydrolase (acid optimum) (EC 3.1.3.2).
Abbreviation:
$(MAG)_2P$ = Bis(monoacylglycero)phosphate, formerly named lysobisphosphatidic acid.

* Part of H. B. T.'s thesis is included.

In beiden Fällen wurde eine – fast – direkte Beziehung zwischen der Dosis und der Menge sowohl an Phospholipiden als auch an $(MAG)_2P$ in den lysosomalen Leberlipiden aufgefunden.

Die lysosomalen Fraktionen wurden durch die Bestimmung der spezifischen Aktivität der sauren Phosphatase und mit Hilfe der Elektronenmikroskopie untersucht, wobei sich ergab, daß bereits nach 14tägiger Medikamentengabe eine Anreicherung des Leitenzyms auf das 30fache (verglichen mit dem Leberhomogenat) auftrat und die Fraktion fast ausschließlich aus Vesikeln mit lamellärer Struktur bestand.

$(MAG)_2P$ wurde säulenchromatographisch aus den lysosomalen Lipidfraktionen isoliert. – Auf die gleiche Weise stellten wir dieses saure Phospholipid aus Rattenlebertritosomen dar und verglichen die entsprechenden Substanzen miteinander.

Nach milder alkalischer Hydrolyse erhielten wir als einzigen Phosphorsäureester Glycerophosphoglycerin. Die molaren Verhältnisse von Phosphor: Fettsäuren: Glycerin entsprachen bei allen etwa 1:2:2, nach der Acetolyse fanden wir nur Diacetylacylglycerin. Demnach kann als gesichert angenommen werden, daß auch in diesen sekundären Lysosomen – so, wie wir früher von anderen berichteten – $(MAG)_2P$ als charakteristisches Phospholipid vorkommt und sein Auftreten als Folge der Lysosomenvermehrung angesehen werden kann.

Es ergab sich außerdem ein Hinweis auf das Vorkommen eines Acyl$(MAG)_2P$ (Acylphosphatidylglycerol) in diesen Zellorganellen.

Key words: Lysosomes, chloroquine phospholipidosis, liver lipids, bis(monoacylglycero)phosphate, lysobisphosphatidic acid.

We found[1] in a case of Niemann-Pick's disease, that the liver lipids were not only rich in sphingomyelin and cholesterol – as expected – but also in an acidic phospholipid, which we showed to be $(MAG)_2P$. Before publishing our results, we found a paper by Rouser et al.[2], who also described the occurrence of this lipid in the same type of "lysosomal disease" as this kind of "inborn error of metabolism" is sometimes called.

It seems to be impossible to isolate lysosomes from human organs after death. However, since Wherrett and Huterer[3,4] described this phospholipid within rat liver tritosomes, we tried to find out whether it is also present in lysosomal fractions after application of some more or less "inert" substances like dextran, iron or silver[5], and recently, after treatment of the animals with different drugs[6]. We found this substance to be present in different quantities in liver lysosomes. Because of the difficulty in preparing enough material, we used thin-layer chromatographic identication only[6].

One drug which could be given in different concentrations to induce phospholipidosis in the animals was chloroquine[6]. Adachi et al.[7] first described a "drug-induced phospholipidosis" using *O,O'*-Bis(diethylaminoethyl)hexestrol.

On the other hand, some histological or ultrastructural studies showed that chloroquine is concentrated in rat liver lysosomes[8] and that "abnormal granules" can be seen, e.g. in various white blood cells of patients after chloroquine therapy[9].

Therefore we wanted to test whether or not there is a relationship between the amount of drug and the quantity of acidic phospholipid within the lysosomes. Are these cell organelles identical with the lamellar bodies which are seen in different organs by many authors (for ref. see[10])?

Furthermore, is this phospholipid really identical with $(MAG)_2P$, which, in the meantime, has also been found in lysosomes from cultured hamster fibroblasts[11,12]?

Experimental

1. Materials

The animals used were adult Wistar rats (200 - 300 g, either sex). All animals were starved for 24 h before decapitation. The abdomen and the right ventricle of the heart were opened immediately and the livers perfused through the vena portae and retrogradely through the vena cava inf. Different solutions were used, depending from the purpose:

a) 0.9 % NaCl solution 4 °C (when the lipids of the homogenate were to be analyzed);
b) 0.25M sucrose 4 °C, for preparing the subcellular fractions;
c) peristone solution, 37 °C, followed by 3 - 5 % glutaraldehyde in 0.1M phosphate buffer (pH 7.6) for 5 min, for preparing liver slices for electron microscopy.

1.1. Liver homogenates were obtained after perfusion with a) or b) as described before[6].

1.2. Liver slices for electron microscopy were prepared in the following way: after several washings (to remove the phosphate buffer) further fixation of tissue samples was performed with 2 % OsO_4 in cacodylate buffer, (pH 7.6) followed by 0.5 % uranyl acetate in maleate buffer (pH 5). After dehydration with a graded series of ethanol, the slices were embedded in Araldite. Thin sections were treated with uranyl acetate and lead citrate.

The electron microscope used was the Siemens type 101 A.

1.3. Lysosomal fractions were obtained from liver homogenates after perfusion of the organ with solution b) and subsequent cell fractionation as described before[6]. These fractions were used for protein and enzyme estimations and the lipids were extracted (Folch et al.[13]).

For electron microscopic control, the isolated lysosomal fractions were suspended in the same vol. of 0.32M NaCl, sedimented (40 000 x *g*, 5 min) and fixed in glutaraldehyde. (Further manipulations see ref.[5]).

Most of the methods used in these experiments have been described in the preceding paper[6] or earlier[5]. In addition, the following techniques were used.

2. Analytical methods

2.1. Quantitative estimations: glycerol[14] (modification according to Renkonen[15]). We used 1-glycerophosphate (Koch Light Lab. Engl.) as a standard, calculating the glycerol content according to the *P*-content of the standard solution.

2.2. Qualitative determinations

2.2.1. Glycerophosphoric esters, by paper chromatography. A sample of lipid (about 6 mg) was hydrolyzed according to Dawon's[16] procedure (modification by Debuch and Winterfeld[17]), in mild alkali, and the upper phase obtained was brought to a vol. of 5 m*l* with methanol. Samples containing about 50 µg phosphorus were chromatographed on paper (descending; Schleicher & Schüll 2043 b, acid washed) with solvent: 50 m*l* phenol (saturated with water), 5 m*l* acetic acid, 6 m*l* ethanol for 20 - 22 h at 18 - 20 °C. After the paper had dried, it was washed with ether. The spots were detected after treatment with different reagents[17].

2.2.2. Acetolysis products, by thin-layer chromatography. Samples of about 3 mg were acetolyzed with acetic acid anhydride for 16 h at 145 °C according to Renkonen[18]. The dried products were dissolved in 12 m*l* of chloroform/methanol (1 : 1 v/v) and washed with 3 m*l* water. 4 m*l* of chloroform was added to the lower phase, which was brought to dryness after several washings. The lipophilic hydrolysis products were chromatographed on thin-layer plates (silica gel H; [Merck]) and developed with light petroleum/ether (4 : 1 v/v).

2.2.3. For identification of acidic, nitrogen-free glycerophospholipids we used unidimensional thin-layer chromatography on silica gel H (Merck, Darmstadt); solvent system: chloroform/methanol/acetic acid (65 : 30 : 8 v/v/v).

3. Preparative methods

3.1. Separation of lipids on silicic acid columns. Silicic acid (Roth, Karlsruhe) (0.125 - 0.250 mm) was washed 3 times with distilled water, dried and activated for 16 h at 120 °C. About 100 mg of total lipid extract from lysosomal fractions were separated on 20 g of silicic acid columns (∅ 1 cm) Eluants: I, chloroform, 400 m*l*; II, chloroform/methanol (95 : 5 v/v), 450 m*l*; III, chloroform/methanol (9 : 1 v/v), 500 m*l*.

3.2. Separation of lipids on florisil columns. This method is based on that of Harzer et al.[19], slightly modified by Etzrodt and Debuch[20] and was performed as described earlier[20,21]. However, here we used 20 g florisil (Serva, Heidelberg) per 120 mg lipid extract.

4. Influence of drug on lysosomal phospholipids

4.1. Different duration of treatment with chloroquine: Each rat received 60 µg chloroquine/g body weight daily by intraperitoneal injection. Groups of animals were decapitated after 3, 7, 14, 21 or 28 days of treatment. Preparation of the homogenate and cell fractions, which were analyzed for acid phosphatase activity, was performed as described before[6]. The lipids were extracted from the lysosomal fractions for analyses.

4.2. Application of different doses of chloroquine for 7 days. Each rat was injected intraperitoneally with chloroquine for 7 days. One group of animals received 20 µg, a second 40 µg, a third 60 µg and another 80 µg of the drug per g body weight. From these animals too, the lysosomal fractions were obtained in the usual way and the contents of $(MAG)_2P$ in lipid extracts were determined.

Results and discussion

The lysosomal fractions (as described in the preceding paper[6]) from the livers of rats which

Table 1. Acid phosphatase activity [U/mg protein] of different liver cell fractions after treatment with chloroquine.

Cell fraction	Length of treatment [days]					Control
	3	7	14	21	28	
Homogenate	0.035	0.037	0.037	0.037	0.036	0.033
1	0.126	0.202	0.201	0.231	0.214	0.119
2 (lysos. fract.)	0.502	0.801	1.207	1.043	1.086	0.245
3	0.340	0.429	0.613	0.574	0.418	0.160
4 (mitoch. fract.)	0.190	0.206	0.414	0.361	0.290	0.140
5	0.190	0.188	0.242	0.286	0.153	0.138
Ratio $\frac{\text{Homogenate}}{\text{Fract. 2}}$	1:14.3	1:21.5	1:32.6	1:28.2	1:30.2	1:7.4
	Protein content of lysosomal fractions					
[mg/100 g liver] (wet wt.)	25	53	52	46	60	9

had received the same dose of chloroquine for different time periods contained not only more protein than the controls, but also a higher specific activity of the marker enzyme (Table 1). The latter can hardly be detected in the homogenate. But although the protein content of the lysosomal fractions has already reached a maximum after 7 days of treatment, the peak of acid phosphatase activity is reached after another 7 days treatment and represents 33-fold enrichment (see Table 1). Total lipid as well as phospholipid and $(MAG)_2P$ concentrations also reach a plateau after daily application of 60 μg chloroquine/g body weight (see Fig. 1). We found a similar effect by giving different doses (20 - 80 μg/g body weight) for 1 week. Here also, the more drug administered, the more lipid, phospholipid or $(MAG)_2P$ was found in the lysosomal fractions (see Fig. 2).

Abraham et al.[8] already described structural changes in liver lysosomes after chloroquine treatment using electron microscopic cytochemical methods. They obtained "myelin bodies" as well as autophagic vacuoles.

In our experiments, hepatocytes from rats treated for two weeks with chloroquine contain lysosomes with complex structures, the number of peroxisomes is markedly increased, and the mitochondrial cristae seem to be diminished (see Fig. 3). Lysosomes often have a lamellar structure and are situated mainly around the bile capillaries. After cell fractionation, the lysosomal fractions consist mostly of these lamellar bodies

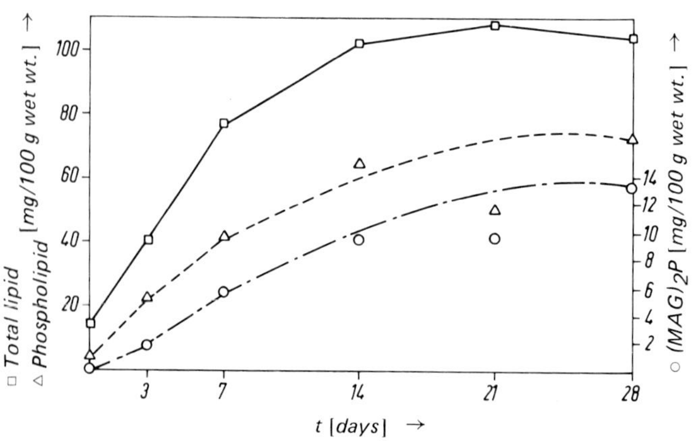

Fig. 1. Total lipids, phospholipids and $(MAG)_2P$ [mg/100 g liver wet weight] of rat liver lysosomes after administration of chloroquine (daily, 60 μg/g body weight) for different times.

(Fig. 4a and b). Since $(MAG)_2P$ is accumulated in these fractions, it is obvious that the lamellar bodies contain $(MAG)_2P$.

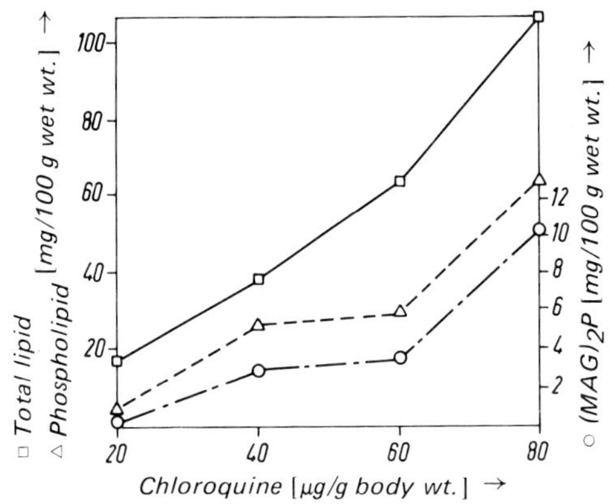

Fig. 2. Total lipids, phospholipids and $(MAG)_2P$ [mg/100 g liver wet weight] of rat liver lysosomes after administration of different quantities of chloroquine (daily, for 7 days).

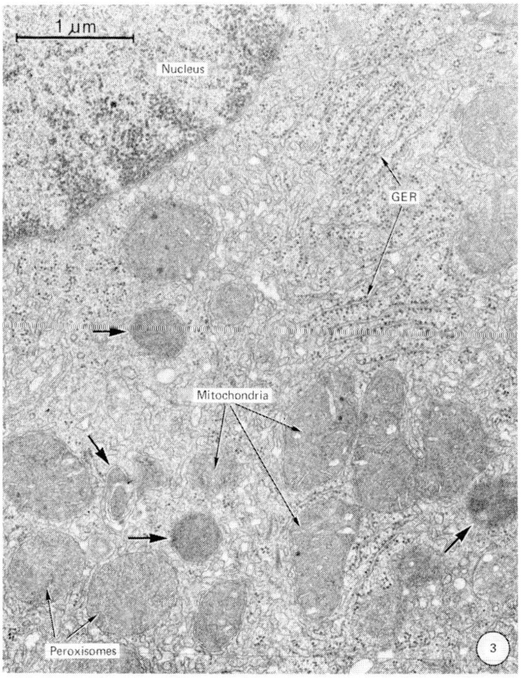

Fig. 3. Hepatocytes of rat liver after 14 days treatment with chloroquine (60 µg/g body weight).

GER: Endoplasmic reticulum (rough). The short arrows point to lysosomes.

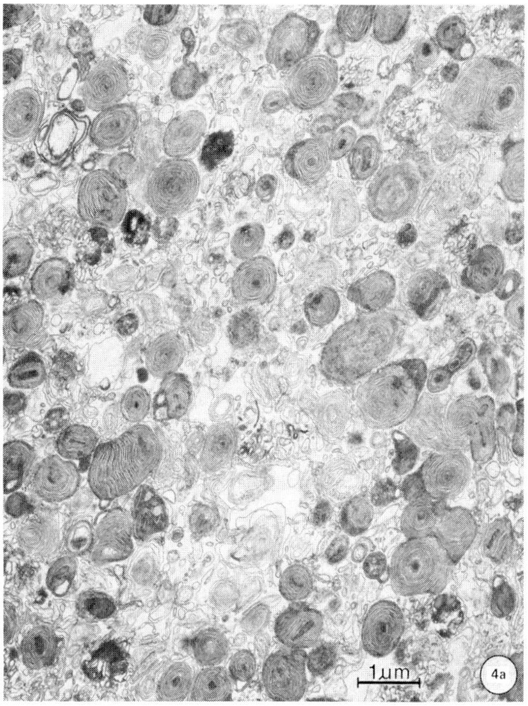

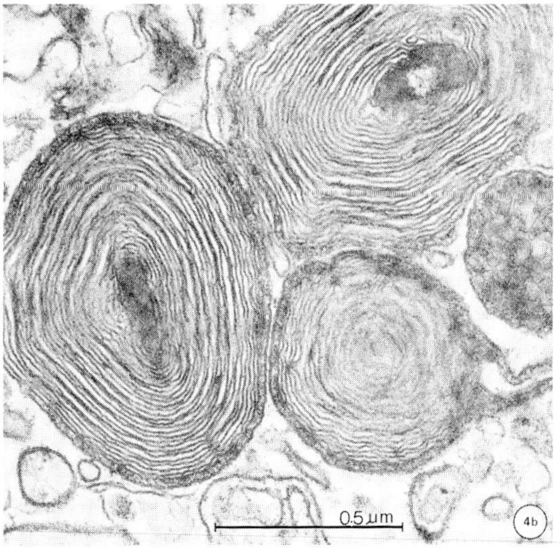

Fig. 4. a) Lysosomal fraction isolated from rat liver after 14 days treatment with chloroquine (60 µg/g body weight).

b) Lamellar bodies from 4a at higher magnification.

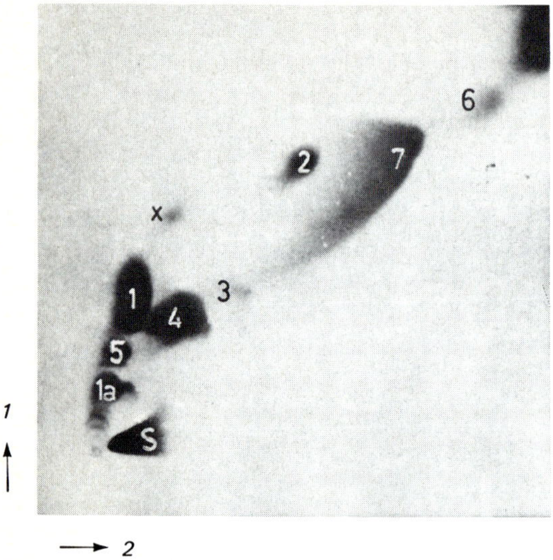

Fig. 5. Thin-layer chromatography of total lipids from lysosomal fraction of rat liver (after treatment with chloroquine, 60 μg/g body weight, for 14 days).

1 Phosphatidylcholine
1a Lysophosphatidylcholine
2 Phosphatidylethanolamine
3 Phosphatidylserine
4 Phosphatidylinositol
5 Sphingomyelin
6 Acyl$(MAG)_2P$
7 $(MAG)_2P$
x Unidentified
S Sucrose

(Method see ref.[6], charring with sulfuric acid).

Therefore we prepared the total lipid extract (see Fig. 5) from a great number of lysosomal fractions from chloroquine-treated animals and isolated two acidic, nitrogen-free phospholipid samples in a pure state by column chromatography (see Fig. 6). We determined their chemical nature in different ways and compared them with the phospholipid from tritosomes.

$(MAG)_2P$ from tritosomes

The acidic phospholipid was isolated from rat liver lysosomal lipid fractions after administration of triton to the animals. The "tritosomes" were obtained in the same way as described before[5].

We carried out many different experiments, obtaining about 8 mg lysosomal lipids per animal on the average (more than 100 rats). However, since at least part of the trition dissolves in the organic solvent phase, the P content of the dried lipid extract was only 0.6 - 0.7%.

We tried to separate triton from the phospholipids in different ways but with poor success. Since we observed that purified $(MAG)_2P$ was very easily decomposed, at least when it was isolated by column chromatography, we analyzed freshly prepared samples only. Thus, we omitted drying, weighing etc. and estimated the different molar ratios within one preparation.

Using silicic acid chromatography [see section 3.1: 60 g adsorbent for 150 mg lipids; eluants: I, 700 m*l*; chloroform/methanol (99:1 v/v), 270 m*l*; II, 590 m*l*; III, 350 m*l*] we obtained *substance A*, which was eluted from the column between solvent II and III and which was not contaminated with any other phospholipid (proved by thin-layer chromatography).

Compound A had a *P*/glycerol ratio of 1:2.17, since a solution of it containing 31.2 (30.9) μg phosphorus contained 197 (205) μg glycerol.

Another sample of A served for estimation of the fatty acid composition by gas chromatography of the methyl esters.

Using a florisil column (60 g for 360 mg of lipids) with gradient elution[20,21], $(MAG)_2P$ was eluted between 600 - 900 m*l* of eluant. This preparation was somewhat contaminated with phosphatidylethanolamine. Therefore it was further purified by unidimensional prepara-

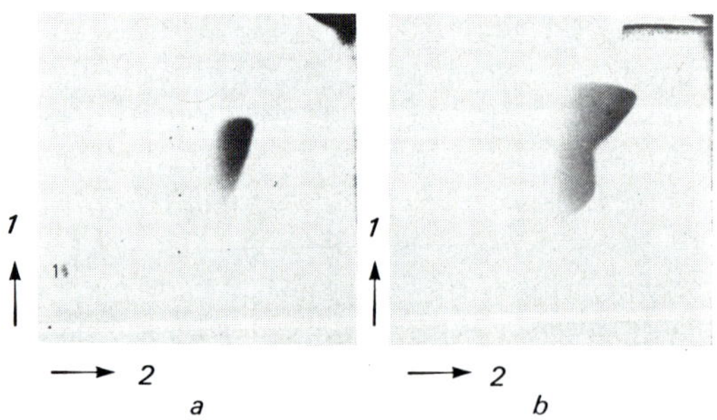

Fig. 6. Thin-layer chromatogram (two dimensional method[6]) of

a) Isolated $(MAG)_2P$: preparation E (see text) (supplemented with phosphatidylcholine (1) for orientation) from lysosomal lipids of rat liver after 14 days treatment with chloroquine.
b) $(MAG)_2P$ and acyl$(MAG)_2P$: preparation D (see text); same origin as a).

tive thin-layer chromatography on silica gel MNHR, (Fa. Macherey, Nagel & Co., Düren) developed in chloroform/methanol/glacial acetic acid 65 : 25 : 4 (v/v/v). After exposure to iodine vapor, the lipids were marked and scraped off the plates after evaporation of the iodine. They were eluted from the adsorbent with chloroform/methanol 2:1 (v/v) to yield *preparation B*, which again, was pure $(MAG)_2P$.

After mild alkaline hydrolysis[16] of sample B, the water-soluble phosphoric ester was identified by paper chromatography, using the corresponding product obtained from phosphatidylglycerol as reference.

Finally, we used the same column chromatographic separation with florisil and obtained a phospholipid sample which we hydrogenated in the same way as described earlier for ethanolamine phospholipids[22]. This compound could be stored for several months in solution (chloroform: methanol 2:1) without decomposition. Before estimation of the *P*/fatty acid ratio it was purified as sample B was by thin-layer chromatography, yielding *substance C*, which again was chromatographically uniform.

A sample of C contained 38.6 μg phosphorus and 815.0 μg of fatty acids. The latter were estimated as methyl esters by gas chromatography. Heptadecanoic acid was added as internal standard before transesterification. Since the mean M_r of fatty acids was 314, ratio of *P*/ fatty acids was 1 : 2.08.

Since, according to the thin-layer chromatographs (two-dimensional), substances A, B and C were identical, we can summarize: the molar ratios of *P*/ fatty acid/glycerol were 1 : 2.1 : 2.2, the deacylation product was glycerophosphoglycerol.

The fatty acid pattern of $(MAG)_2P$ from tritosomes resembled very much the one published by us earlier[5], except that the amount of docosahexaenoic acid was higher than before, about 60%.

$(MAG)_2P$ from lysosomes after treatment with chloroquine

Isolation of $(MAG)_2P$ from the lysosomal lipid fractions after treatment of the animals with chloroquine was much easier than from tritosomes and resulted in chromatographically pure samples (D and E; Table 2, Fig. 6).

A large number of animals received intraperitoneal injections of *chloroquine* (60 μg/g body weight) for 14 days.

The total lipids of the lysosomal fractions (420 mg; *P*: 2.25%) extracted according to Folch's method[13] were chromatographed on 3 silicic acid columns (see section 3.1). $(MAG)_2P$ was eluted between solvent sys-

Table 2. Analysis of $(MAG)_2P$ preparations from rat liver lysosomes.

Preparation	*P* [%]	*P*/fatty acid	*P*/glycerol	Product after Deacylation	Product after Acetolysis
A, B, or C[a]	n.d.	1 : 2.1	1 : 2.2	Glycerophosphoglycerol	n.d.
D[b]	2.80	1 : 2.4	1 : 2.0	Glycerophosphoglycerol	n.d.
E[b]	3.14	1 : 2.0	1 : 2.1	Not determined	Diacetylacylglycerol
Calculated[c]					
Acyl$(MAG)_2P$ or acylphosphatidylglycerol	2.73	1 : 3.0	1 : 2.0	Glycerophosphoglycerol	Diacetylacylglycerol and acetyldiacylglycerol
$(MAG)_2P$	3.66	1 : 2.0	1 : 2.0	Glycerophosphoglycerol	Diacetylacylglycerol

[a] Lipid preparation from lysosomes after application of triton to the animals.
[b] From lysosomal fractions after 2 weeks of chloroquine treatment of the animals.
[c] According to M_r of 1137 for acylphosphatidylglycerol and 846 for $(MAG)_2P$, resp. (based on the fatty acid pattern found).

tems II and III, followed by phosphatidylethanolamine. The fractions were identified at least by unidimensional thin-layer chromatography in chloroform/methanol/water 65 : 25 : 4 (v/v/v). Although we found some small differences in the R_F values of the samples leaving the column at the beginning and at the end of the elution of the nitrogen-free phospholipid, we combined them accordingly and finally we obtained two substances: D: 12.10 mg and E: 8.30 mg.

Sample D was analyzed: *P:* 2.79 (2.82); glycerol: 15.9 (15.0)%; The water-soluble phosphoric ester obtained after mild alkaline hydrolysis[16] was identified by paper chromatography (see Fig. 9). 4.40 mg of fatty acid methyl esters was isolated from 7.71 mg D after transesterification. Gas-liquid chromatography showed, that the mean molecular weight of the fatty acids was 315.0. Analysis of *product E* resulted in *P:* 3.15 (3.12)%; glycerol: 20.4 (20.0)%; 1.45 mg fatty acid methyl esters was isolated from 2.20 mg of sample E after transesterification. They were analyzed by gas-liquid chromatography. The mean molecular weight of fatty acids was 318. The products after acetolysis of E were chromatographed in one dimension on thin layers (see section 2.2.2 and Fig. 8. The corresponding products after acetolysis of phosphatidylglycerol, glycerol and monoglyceride served as reference substances.)

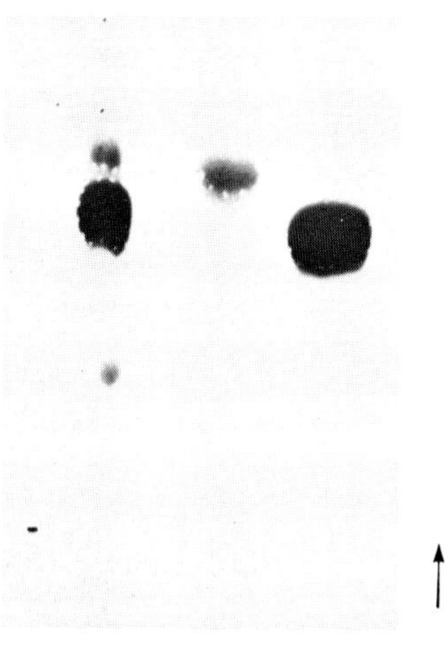

Fig. 7. Thin-layer chromatography of acidic glycerophospholipids.

a) Mixture of b) and c) plus phosphatidic acid; b) $(MAG)_2P$; c) phosphatidylglycerol.

Furthermore, sample E was identified by unidimensional thin-layer chromatography in a basic solvent system (see section 2.2.2) for the following reasons. The separation of phosphatidyl glycerol, phosphatidic acid and $(MAG)_2P$ in the two-dimensional thin-layer chromatographic system varies with the amount of acidic phospholipid. To clarify this, we took those lipids thought to be $(MAG)_2P$ from the plates, eluted them with chloroform/methanol/water (65 : 25 : 2 v/v/v) and compared them with reference substances (see Fig. 7).

Substance E was pure $(MAG)_2P$, showing nearly theoretical values for molar ratios and only diacetylacylglycerol after acetolysis (see Fig. 8). This fact is the only one, apart from the thin-layer chromatographic pattern in the alkaline solvent, mentioned above, which rules out the presence of phosphatidylglycerol. Substance D however, was eluted from the column prior to E and showed little difference on thin-layer chromatography (see Fig. 6). Its *P*/fatty acid ratio seems to indicate that both $(MAG)_2P$ and its acylated form may occur. After alkaline hydrolysis, only one phosphoric ester was found, namely glycerophosphoglycerol (see Fig. 9). That excluded the presence of cardiolipin.

We cannot decide from our experiments, whether the acylated compound is a precursor or a product of $(MAG)_2P$.

Poorthuis and Hostetler[23], who worked on the biosynthesis of those acidic phospholipids, de-

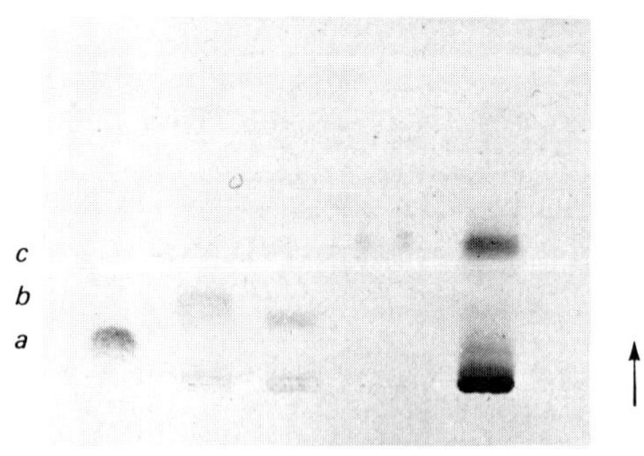

Fig. 8. Thin-layer chromatography of the products after acetolysis of 1) glycerol; 2) monoglyceride; 3) $(MAG)_2P$ (sample E); 4) diglyceride; 5) phosphaditylcholine; a) triacetylglycerol; b) diacetylacylglycerol; c) acetyldiacylglycerol.

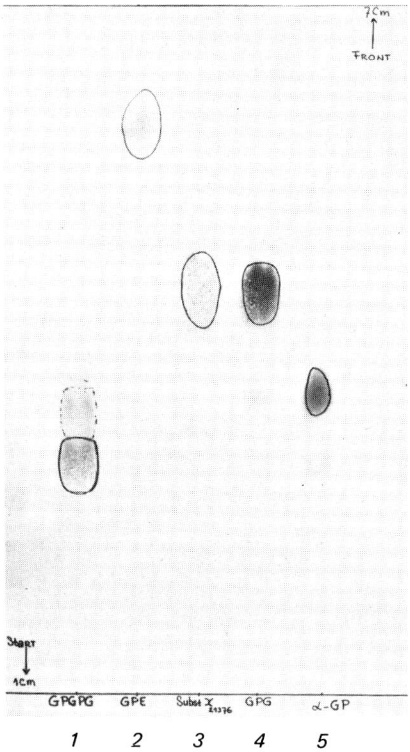

Fig. 9. Paper chromatogram of phosphoric ester of sample D ($(MAG)_2P$ plus acyl$(MAG)_2P$).
Products after alkaline hydrolysis of 1) cardiolipin; 2) phosphatidylethanolamine; 3) sample D, see Fig. 6b; 4) phosphatidylglycerol; 5) α-glycerophosphate as reference.

scribed enzyme systems within the crude mitochondrial fractions from normal rat liver which were able to synthesize $(MAG)_2P$ and acylphosphatidylglycerol from cytidine diphosphate diglyceride and *sn*-glycerol 3-phosphate. However, phosphatidylglycerol was the main product. We found in our experiments no evidence for the presence of phosphatidylglycerol.

We are very grateful to *Priv. Doz. Dr. K. Gorgas* (Anatomisches Institut, Universität Köln) for preparation and interpretation of the electron microscope photographs. Phosphatidylglycerol was a gift, for which we thank *Prof. Dr. E. Heinz* (Botanisches Institut, Universität Köln).

We thank also Miss *U. Nauen* for her skilful technical help, Mrs. *Chr. Geyer* and Mrs. *A. Blachetta* for the extensive analyses and Mr. *K. Braun* for further technical assistance.
The financial support of the *Stiftung Volkswagenwerk* is gratefully acknowledged.

Literature

1 Seng, P.N., Debuch, H., Witter, B. & Wiedemann, H.R. (1971) *Hoppe-Seyler's Z. Physiol. Chem.* **352**, 280 - 288.
2 Rouser, G., Kritchevsky, G., Yamamoto, A., Knudson Jr., A.G. & Simon, G. (1968) *Lipids* **3**, 287 - 290.
3 Wherrett, J.R. & Huterer, S. (1972) *J. Biol. Chem.* **247**, 4114 - 4120.
4 Wherrett, J.R. & Huterer, S. (1973) *Lipids* **8**, 531 - 533.
5 Stremmel, W. & Debuch, H. (1976) *Hoppe-Seyler's Z. Physiol. Chem.* **357**, 803 - 810.
6 Tjiong, H.B., Lepthin, J. & Debuch, H. (1978) *Hoppe-Seyler's Z. Physiol. Chem.* **359**, 63 - 69, preceding paper.
7 Adachi, S., Matsuzawa, Y., Yokomura, T., Ishikawa, K., Uhara, S., Yamamoto, A. & Nishikawa, M. (1972) *Lipids* **7**, 1 - 7.
8 Abraham, R., Hendy, R. & Grasso, P. (1968) *Exp. Mol. Pathol.* **9**, 212 - 229.
9 Fedorko, M. (1967) *J. Clin. Invest.* **46**, 1932 - 1942.
10 Lüllmann, H., Lüllmann-Rauch, R. & Wassermann, O. (1973) *Dtsch. Med. Wochenschr.* **98**, 1616 - 1625.
11 Brotherus, J., Renkonen, O., Herrmann, J. & Fischer, W. (1974) *Chem. Phys. Lipids* **13**, 178 - 182.
12 Brotherus, J. & Renkonen, O. (1977) *J. Lipid Res.* **18**, 191 - 202.
13 Folch, J., Lees, M. & Sloane-Stanley, G.H. (1957) *J. Biol. Chem.* **226**, 497 - 509.
14 Hanahan, D.J. & Olley, J.N. (1958) *J. Biol. Chem.* **231**, 813 - 828.
15 Renkonen, O. (1962) *Biochem. Biophys. Acta* **56**, 367 - 369.
16 Dawson, R.M.C. (1960) *Biochem. J.* **75**, 45 - 53.
17 Debuch, H. & Winterfeld, M. (1966) *Z. Klin. Chem.* **4**, 149 - 153.
18 Renkonen, O. (1964) *Acta Chem. Scand.* **18**, 271 - 272.
19 Harzer, K., Sandhoff, K. & Jatzkewitz, H. (1967) *Hoppe-Seyler's Z. Physiol. Chem.* **348**, 119 - 120.
20 Etzrodt, A. & Debuch, H. (1970) *Hoppe-Seyler's Z. Physiol. Chem.* **351**, 603 - 612.
21 Fürniss, H. & Debuch, H. (1972) *Hoppe-Seyler's Z. Physiol. Chem.* **353**, 1377 - 1384.
22 Debuch, H., Friedemann, H. & Müller, J. (1970) *Hoppe-Seyler's Z. Physiol. Chem.* **351**, 613 - 621.
23 Poorthuis, B.J.H.M. & Hostetler, K.Y. (1975) *J. Biol. Chem.* **250**, 3297 - 3302.

H. B. Tjiong and Prof. Dr. H. Debuch, Institut für Physiologische Chemie der Universität zu Köln, Lehrstuhl II, Postfach 410580, D-5000 Köln 41.

Hoppe-Seyler's Z. Physiol. Chem.
Bd. 359, S. 81 - 88, Januar 1978

2α-Hydroxylierung von Testosteron und Dihydrotestosteron in Ratten in vivo und in vitro

Ursula LEMM und Martin WENZEL

Institut für Pharmazie der Freien Universität Berlin, Biologisch-Chemische Abteilung

(Der Schriftleitung zugegangen am 5. August 1977)

Zusammenfassung: In Ratten wurde die 2α-Hydroxylierung von Testosteron und 5α-Dihydrotestosteron in vivo und in vitro erstmals quantitativ mit Hilfe „Stoffwechsel-labil" Tritium-markierter Substrate bestimmt.

Die Untersuchungen ergaben eine geschlechts- und altersabhängige Entwicklung der Hydroxylase-Aktivität. Dabei zeigte sich zwischen In-vivo- und In-vitro-Befunden eine gute Übereinstimmung. Während Testosteron ausschließlich in der Leber hydroxyliert wurde, ließen sich für 5α-Dihydrotestosteron auch in der Nebenniere, Niere, Lunge und Prostata nennenswerte Hydroxylierungsaktivitäten nachweisen. In der Nebenniere und Lunge war die Hydroxylierung nicht geschlechtsspezifisch.

2α-Hydroxylation of Testosterone and Dihydrotestosterone in Rats in vivo and in vitro

Summary: The hydroxylation of testosterone and 5α-dihydrotestosterone at C-2 was measured quantitatively in vivo and in vitro in rats by using "metabolically labile" tritium labelled substrates.

The enzyme activity was dependent on the sex and age of the animals. There was a good correlation between in vivo and in vitro experiments. Whereas testosterone was hydroxylated only in liver, 5α-dihydrotestosterone was hydroxylated in adrenals, kidney, lung and prostate as well. Hydroxylation was not sex-dependent in adrenals and lungs.

Key words: Hydroxylation, testosterone, 5α-dihydrostestosterone, HTO, sex and organ specificity.

Über den Stoffwechsel der Androgene Testosteron und 5α-Dihydrotestosteron durch Hydroxylierung ist bisher relativ wenig publiziert worden, verglichen mit der Vielzahl an Arbeiten über die Steroid-Umwandlung durch Redoxprozesse (Lit. s.l.c.[1]). Die meisten Ergebnisse über die Androgen-Hydroxylierung wurden in vitro mit Lebermikrosomen von Ratten gewonnen[2-4]. Dagegen sind bisher kaum Aussagen über die Androgen-Hydroxylierung in vivo bekannt geworden. In vivo Untersuchungen über den Stoffwechsel von Steroiden bereiten erhebliche methodische Schwierigkeiten, da aus dem Gewebe und den Ausscheidungen eine Vielzahl von Metaboliten abgetrennt werden müs-

Trivialnamen: Testosteron = 17β-Hydroxy-4-androsten-3-on; Dihydrotestosteron = 17β-Hydroxy-5α-androstan-3-on.

sen. Mit Hilfe der „stoffwechsellabilen Tritium-Markierung“[5,6] können manche dieser Probleme umgangen werden.

Nach der Injektion eines in geeigneter Position mit Tritium markierten Steroids wird das durch Hydroxylierung abgespaltene Tritium homogen im Körperwasser verteilt. Es kann als HTO im Atemwasser bestimmt werden und ist ein Maß für die „Mindest“-Hydroxylierung[6]. Das Verfahren hat den Vorteil, spezifisch für die Position zu sein, gibt aber keine Auskunft über weitere Umwandlungen im Steroid-Molekül.

Da wir beweisen konnten, daß auch die 2α-Hydroxylierung von Androgenen mit Hilfe stoffwechsellabil markierter Substrate zuverlässig bestimmt werden konnte[7], wollten wir mit dieser Methode folgende Fragen klären:

1) Gibt es bei der Androgen-Hydroxylierung eine Parallelität der In-vitro- und In-vivo-Ergebnisse?
2) Läßt sich für die 2α-Hydroxylierung von Testosteron und 5α-Dihydrotestosteron eine Sexualspezifität und eine Organspezifität nachweisen?

Material und Methodik

Versuchstiere

Als Versuchstiere dienten männliche und weibliche Wistar-Ratten der Fa. Hagemann, Bösingfeld/Hannover. Die erwachsenen Ratten hatten ein mittleres Gewicht von 200 g, die infantilen Ratten wogen 30 - 40 g und waren 18 - 21 Tage alt.

Substanzen

Nichtmarkierte Steroide stammten von den Firmen E. Merck, Darmstadt oder Schering AG, Berlin. Die radioaktiven Substanzen wurden von der Fa. Buchler, Amersham, bezogen und hatten folgende spezifische Radioaktivitäten: [1α,2α-T]Testosteron, 49 Ci/mmol; [1α,2α-T]5α-Dihydrotestosteron, 47 Ci/mmol.

Vor der Verwendung der radioaktiven Steroide erfolgte eine chromatographische Reinheitskontrolle[8]. Außerdem wurde die spezifische Radioaktivität durch Zugabe von nichtradioaktivem Testosteron um den Faktor 10^2 gesenkt.

Inkubationsansätze

Von jedem Organ wurden mehrere Inkubationsansätze (n = 4 - 5) durchgeführt. Die dafür verwendete Tierzahl richtete sich nach der benötigten Gewebemenge (pro Inkubationsansatz ca. 50 mg). Da die verfügbare Gewebemenge abhängig war vom Alter der Tiere und von dem Gewicht der Organe, war die Zahl der verwendeten Tiere unterschiedlich.

Ein Inkubationsansatz bestand aus 1.0 m*l* Krebs-Ringer-Phosphat-Puffer, pH 7.4, 50 mg Gewebeschnitten als Enzymquelle und dem Steroid (0.01 - 0.1mM). Die Steroide wurden – gelöst in 90proz. Äthanol – der für den Versuch insgesamt bereiteten Puffermenge zugemischt.

Die Inkubationen wurden durch Einbringen der Gefäße in Eis gestoppt. Anschließend wurde das gebildete HTO durch Gefrierdestillation[9] abgetrennt. Weitere Angaben siehe l.c.[9,10].

Bei allen Versuchen wurden mindestens 4 Ansätze, die außer Gewebe alle Komponenten enthielten, zur Kontrolle mitinkubiert und aufgearbeitet.

Radiospirometrie

Die entsprechende Menge an Tritium-markiertem Steroid wurde den Tieren – gelöst in 90proz. Äthanol/0.9proz. NaCl 1 : 1 (v/v) – i.p. injiziert. Das Gewinnen der Körperwasserproben und die Bestimmung des HTO-Gehaltes im Atemwasser sind an anderer Stelle ausführlich beschrieben[6,11,12]. Die Berechnung der Tritium-Labilisierung aus dem injizierten Steroid erfolgte nach der Formel

% Mindesthydroxylierung =

$$= \frac{\mu\text{Ci HTO}/\text{m}l\ \text{Körperwasserdestillat}}{\mu\text{Ci inj. Steroid}/\text{m}l\ \text{Körperwasser}} \times 100$$

Bei den in dieser Arbeit verwendeten Substraten ist zu berücksichtigen, daß an der HTO-Bildung nur eines der beiden im [1α,2α-T]Testosteron bzw. [1α,2α-T]5α-Dihydrotestosteron befindlichen T-Atome beteiligt ist. Deshalb gibt das entstandene HTO (bezogen auf die eingesetzte Tritium-Aktivität) nur die Hälfte des tatsächlich entstandenen Hydroxylierungsproduktes an, und die Auswertung muß erfolgen nach der Formel: % Hydroxylierung = % HTO × 2. Dies gilt auch für die Berechnung der entstandenen Mole Hydroxylierungsprodukt bei In-vitro-Experimenten.

Ergebnisse

Sexualspezifische Hydroxylierung von Testosteron und 5α-Dihydrotestosteron in Ratten in vivo

In der Abbildung ist die In-vivo-Hydroxylierung von Testosteron und 5α-Dihydrotestosteron in männlichen und weiblichen Ratten in Abhängigkeit von der Zeit aufgetragen. Als Substrat dienten die entsprechenden Androgene, die Tritium in 1α- und 2α-Position enthielten, bei deren Hydroxylierung Tritium als HTO frei wird und damit im Körperwasser nachgewiesen werden kann. Die Se-

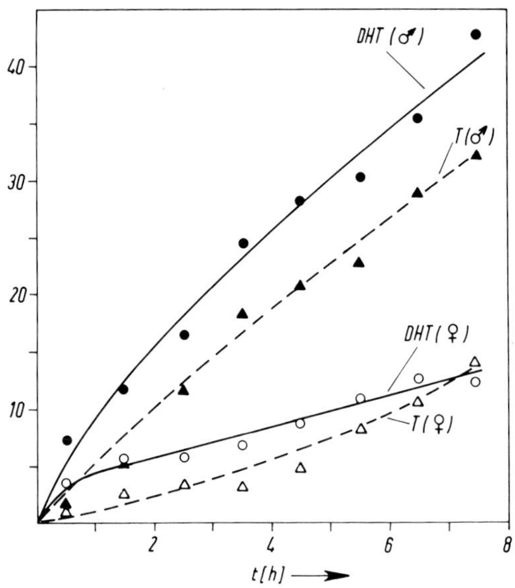

Abbildung. 2α-Hydroxylierung von Testosteron und 5α-Dihydrotestosteron in vivo (Radiospirometrie).

[1α,2α-T]Testosteron bzw. [1α,2α-T]5α-Dihydrotestosteron wurde den Ratten i.p. injiziert und anschließend das abgespaltene Tritium als HTO im Körperwasser bestimmt (vgl. Tab. 1 und Methodik). Ordinate: % der injizierten Aktivität als HTO im Körperwasser (100% ≙ 1 μCi inj. Steroid/m*l* Körperwasser); T = Testosteron; DHT = 5α-Dihydrotestosteron.

xualspezifität der Androgen-Hydroxylierung zeigt die Abbildung sehr deutlich.

In männlichen Ratten ist nach 8 h die injizierte Menge an Testosteron zu 64% und an 5α-Dihydrotestosteron zu 84% an C-2 hydroxyliert. In weiblichen Tieren dagegen werden in der gleichen Zeitdauer bei Testosteron weniger als die Hälfte und bei 5α-Dihydrotestosteron sogar nur ein Drittel dieser Werte erreicht.

Die 2α-Hydroxylierung von Testosteron und 5α-Dihydrotestosteron in adulten und infantilen Ratten

Für die Hydroxylierung von Testosteron wiesen Leybold und Staudinger[2] als erste geschlechtsspezifische Unterschiede nach. Ähnliche Sexualcharakteristika stellten Albaum und Staib[3] für das Metabolitmuster von Androstendion fest. Nach Schriefers et al.[13] zeigen viele Enzyme des Steroidhormon-Stoffwechsels in der Rattenleber einen sexuell indifferenten Aktivitätsanstieg während der ersten 30 Lebenstage mit einem darauf folgenden sexualspezifischen Aktivitätsplateau.

Da die 2α-Hydroxylierung von Testosteron und 5α-Dihydrotestosteron ebenfalls geschlechtsspezifisch ist (s. oben), sollte bei infantilen und erwachsenen Ratten ein entsprechender Vergleich in vivo und in vitro durchgeführt werden. Wir bestimmten deshalb in vivo durch Radiospirometrie und in vitro mit Leberschnitten die Hydroxylierung dieser Androgene bei 20 Tage alten männlichen und weiblichen Ratten. Die entsprechenden Ergebnisse und Vergleiche zeigt Tab. 1. Sowohl in vivo als auch in vitro ist bei infantilen männlichen Tieren die Hydroxylase-Aktivität geringer als bei erwachsenen Männchen.

In weiblichen Tieren wird dagegen in vivo und in vitro Testosteron in beiden Altersstufen in fast demselben Maße hydroxyliert. Erstaunlich sind die Befunde bei 5α-Dihydrotestosteron: Sowohl in vivo als auch in vitro ist in *infantilen* Tieren die Hydroxylierung höher als bei erwachsenen Weibchen.

Vergleicht man die Hydroxylase-Aktivität der Leber [nmol Hydroxylierungsprodukt/(min × g Leber]*, so wird deutlich, daß eine sexuelle Differenzierung bei 20 Tage alten Tieren noch nicht vorhanden ist. Diese Differenzierung setzt erst in späteren Altersstufen ein und führt dann in adulten männlichen Ratten zu 3 - 10fach höheren Hydroxylierungsaktivitäten als bei adulten Weibchen.

Organspezifität der 2-Hydroxylierung von Testosteron und 5α-Dihydrotestosteron in Ratten

Ziel der Untersuchungen dieses Abschnittes ist die Beantwortung der Frage: Ist ausschließlich die Leber für die Hydroxylierung der Hormone verantwortlich oder findet diese Reaktion auch in anderen Organen statt?

Die Ergebnisse der entsprechenden Experimente gibt Tab. 2 wieder. Bei *männlichen Ratten* ist die Leber das einzige Organ, in welchem aus beiden Steroiden eine gleich große Menge Tritium abgespalten wird.

* Berechnung vgl. Methodik

Tab. 1. Vergleich der 2α-Hydroxylierung von Testosteron und 5α-Dihydrotestosteron in vivo und in vitro.
in vivo: pro Tier wurden folgende Mengen an [1α,2α-T]Testosteron bzw. [1α,2α-T]5α-Dihydrotestosteron i.p. injiziert: 5 μCi ≙ 12 μg/kg (adult) und 1 μCi ≙ 18 μg/kg (infantil). Die in der Tabelle angegebenen Werte (% der eingesetzten Aktivität, als HTO im Körperwasser) wurden 7 h post injectionem gemessen.
in vitro: 50 mg Leberschnitte wurden mit [1α,2α-T]Testosteron (10^{-5}M) bzw. [1α,2α-T]5α-Dihydrotestosteron (10^{-5}M) bei 37 °C in 1.0 m*l* Krebs-Ringer-Phosphat-Puffer (pH 7.4) 10 min inkubiert (*n* = 4 Inkubationsansätze, vgl. Methoden). HTO in % der eingesetzten Tritium-Aktivität.

Substrat	Testosteron				5α-Dihydrotestosteron			
Reaktionsprodukt	in vivo		in vitro		in vivo		in vitro	
	HTO		2α-OH-Testosteron		HTO		2α-OH-5α-Dihydrotestosteron	
	[%]	Faktor	$\left[\frac{\text{nmol}}{\text{g} \times \text{min}}\right]$	Faktor	[%]	Faktor	$\left[\frac{\text{nmol}}{\text{g} \times \text{min}}\right]$	Faktor
adult ♂	28.8	1.00	2.66 ± 0.057	1.00	35.2	1.00	2.64 ± 0.058	1.00
infantil ♂	17.0	0.59	0.31 ± 0.008	0.12	25.2	0.72	0.83 ± 0.068	0.32
adult ♀	10.6	0.37 (1.00)	0.34 ± 0.014	0.13 (1.00)	12.5	0.36 (1.00)	0.38 ± 0.013	0.15 (1.00)
infantil ♀	13.7	0.48 (1.30)	0.34 ± 0.008	0.13 (1.00)	18.3	0.52 (1.46)	0.60 ± 0.04	0.23 (1.57)

Im Gegensatz zu Testosteron wird 5α-Dihydrotestosteron auch in verschiedenen anderen Organen hydroxyliert: In der Niere und Prostata macht die Hydroxylase-Aktivität immerhin noch die Hälfte, in Samenblase und Testis ein Fünftel des Leberwertes aus.

Noch mehr erstaunt, daß die Dihydrotestosteron-Hydroxylierung in der Nebenniere so groß ist wie in der Leber. Erhöht man die Substrat-Konzentration auf 10^{-4} mol/*l* (Tab. 2, Exp. II), so ist die Hydroxylierung in der Nebenniere sogar höher als in der Leber. Auch die Lunge enthält nicht unwesentliche Mengen an 5α-Dihydrotestosteron-Hydroxylase.

In der Leber *weiblicher Ratten* (Tab. 2) beträgt die Hydroxylierungsaktivität nur 20 - 30% dessen, was bei männlichen Tieren gemessen wurde. Die Hydroxylierungsaktivitäten für Testosteron und 5α-Dihydrotestosteron sind fast identisch, was gut mit den in vivo gewonnenen Ergebnissen übereinstimmt.

Überraschend ist auch hier wieder der Umsatz von 5α-Dihydrotestosteron in Nebenniere und Lunge: Im Vergleich zur Leber ist die Enzymaktivität in der Nebenniere 6mal, in der Lunge 1.5mal so hoch. Auffälligerweise gibt es in diesen beiden Organen keine geschlechtsspezifischen Unterschiede.

Zwar konnte auch in Niere, Ovar und Uterus für beide Androgene eine geringfügige HTO-Bildung nachgewiesen werden, doch liegen die Werte nur wenig über dem Nullwert; auf ihre Darstellung wurde daher verzichtet.

Diskussion

Auch diese Arbeit zeigt die Eignung der „Stoffwechsel-labilen Tritium-Markierung“ für In-vivo-Bestimmungen von Steroid-Metabolisierungen[5,6]. Besonders interessant erscheint diese Technik für klinisch-pharmakologische Fragestellungen, wo sie auch in bezug auf den Östrogen-Stoffwechsel angewendet wurde[14-16].

Sexualspezifität

Nach Untersuchungen verschiedener Autoren[2,3,13] verläuft die Metabolisierung von

Tab. 2. Organspezifität der 2α-Hydroxylierung von Testosteron und 5α-Dihydrotestosteron.

Gewebeschnitte wurden mit [1α,2α-T]Testosteron bzw. [1α,2α-T]5α-Dihydrotestosteron bei 37 °C in 1.0 m*l* Krebs-Ringer-Phosphat-Puffer (pH 7.4) 10 min inkubiert (*n* = 4 Inkubationsansätze, vgl. Methoden). HTO in % der eingesetzten Tritium-Aktivität. ♂ (Exp. I): 50 mg Gewebe, Steroid-Konz. 10^{-5} mol/*l*; ♂ (Exp. II): 50 mg Gewebe, Steroid-Konz. 10^{-4} mol/*l*; ♀: 50 mg Gewebe, Steroid-Konz. 10^{-4} mol/*l*. 1 nmol Hydroxylierungsprodukt/(g × min) entsprach in Experiment I 2.5% HTO, in Experiment II und bei den Weibchen 0.25% HTO.

Substrat		Testosteron		5α-Dihydrotestosteron	
Reaktionsprodukt		2α-OH-Testosteron		2α-OH-5α-Dihydrotestosteron	
		$\left[\frac{\text{nmol}}{\text{g} \times \text{min}}\right]$	Faktor (Leber = 1)	$\left[\frac{\text{nmol}}{\text{g} \times \text{min}}\right]$	Faktor (Leber = 1)
♂ Exp. I	Leber	2.72 ± 0.08	1.00	2.57 ± 0.01	1.00
	Niere	0.10 ± 0.02	0.04	1.20 ± 0.03	0.48
	Nebenniere	0.15 ± 0.01	0.06	2.37 ± 0.63	0.92
	Samenblase	0.01 ± 0.001	0.004	0.44 ± 0.01	0.18
	Testis	0.06 ± 0.004	0.02	0.65 ± 0.01	0.25
	Prostata	0.06 ± 0.01	0.02	1.22 ± 0.03	0.48
♂ Exp. II	Leber	25.1 ± 3.1	1.00	34.1 ± 3.9	1.00
	Nebenniere	1.07 ± 0.04	0.04	46.6 ± 1.2	1.37
	Lunge	0.17 ± 0.06	0.006	13.4 ± 1.1	0.39
♀	Leber	6.52 ± 0.99	1.00	8.16 ± 0.1	1.00
	Nebenniere	0.84 ± 0.16	0.13	50.52 ± 1.0	6.19
	Lunge	0.34 ± 0.09	0.05	11.5 ± 2.1	1.41

Steroid-Hormonen in Ratten sexualspezifisch. Auch bei der von uns untersuchten 2α-Hydroxylierung von Testosteron und 5α-Dihydrotestosteron findet man eine Sexualspezifität. Beide Androgene werden in männlichen Ratten stärker hydroxyliert als in Weibchen. Diese Unterschiede sind in vitro (bei der Leber) wesentlich ausgeprägter als in vivo (Tab. 1). Außer vom Geschlecht ist die Aktivität der 2α-Hydroxylase abhängig vom Alter der Tiere. Nach Schriefers et al.[17,18] erfolgt eine sexuelle Differenzierung der Hydroxylase-Aktivität nicht vor dem 30. Lebenstag. Damit übereinstimmend fanden wir in der Leber *infantiler* Tiere für beide Geschlechter nahezu identische Hydroxylierungsaktivitäten. Auch in vivo zeigte sich diese Tendenz.

Im Gegensatz zu den männlichen nimmt bei weiblichen Tieren die Hydroxylase-Aktivität mit dem Alter weder in vivo noch in vitro zu (Tab. 1). Entweder sie bleibt gleich (Testosteron) oder verringert sich um etwa ein Drittel (5α-Dihydrotestosteron). Für die altersbedingte Entwicklung der 2α-Hydroxylase von 5α-Dihydrotestosteron in weiblichen Ratten kann demnach ein ähnlicher Verlauf diskutiert werden wie für die 11β-Steroid-Hydroxylase (l.c.[17], Abb. 1b). Diese unterschiedliche Entwicklung der Hydroxylase-Aktivität kann – im Zusammenhang mit den früher ermittelten biochemischen Eigenschaften[7] – ein Hinweis dafür sein, daß es sich bei der 2α-Hydroxylase von Testosteron und 5α-Dihydrotestosteron um verschiedene Enzymsysteme handelt[19].

Die Angaben von Schriefers (l.c.[17], Tab. 1), nach denen in der Leber erwachsener Männchen doppelt soviel Hydroxylierungsprodukt entsteht wie in der infantiler, konnten wir für die 2α-Hydroxylase in vivo bestätigen, in vitro fanden wir in der Leber sogar den Faktor 8. Generell war die altersbedingte Differenzierung in vitro ausgeprägter als in vivo (vgl. Tab. 1). Ein quantitativer Vergleich der Androgen-Hydroxylierung in vivo und in vitro ist nicht möglich, u.a. wegen der unterschiedlichen Reaktionszeiten. Dennoch läßt sich anhand der Tab. 1 aus den Faktoren eine Parallelität in den Relationen zwischen den In-vivo- und In-vitro-Befunden erkennen.

Organspezifität

In männlichen Ratten wird Testosteron an C-2 ausschließlich in der Leber hydroxyliert (Tab. 2). Da in den Sexualorganen – außer der 7α-Hydroxylierung in der Testis[20,21] – keine weiteren Hydroxylierungen nachweisbar sind, steht Testosteron in diesen Organen uneingeschränkt zur Umwandlung in das wirksamere Androgen 5α-Dihydrotestosteron zur Verfügung.

Im Gegensatz dazu wird 5α-Dihydrotestosteron nicht nur in der Leber, sondern auch in den Erfolgsorganen insbesondere in der Prostata hydroxyliert (Tab. 2, Exp. I bezogen auf 1 g Organ: Leber/Prostata = 2.1). Dies ist ein Beitrag zu der oft diskutierten Frage nach der Inaktivierung der im Sexualorgan grundsätzlich nur wenige Stunden wirksamen Steroid-Hormone.

In der Nebenniere von Ratten ist die Hydroxylierung von 5α-Dihydrotestosteron größer als in der Leber (bezogen auf 1 g Organ: Nebenniere/Leber = 1.4 (♂ in Exp. II) bzw. 6.2 (♀)).

Eine quantitative Auswirkung auf den Androgenstoffwechsel ist unwahrscheinlich, da die Nebenniere aufgrund ihres geringen Gewichts – trotz hoher Hydroxylasekonzentration – weniger Steroid umsetzt als die ca. 130mal schwerere Leber. Erstaunlich ist, daß nur 5α-Dihydrotestosteron, nicht aber Testosteron in der Nebenniere hydroxyliert wird. Dafür haben wir bisher keine Erklärung.

Für den Abbau der endogen gebildeten Hormone scheint die Hydroxylierung von 5α-Dihydrotestosteron in der Lunge fast so wichtig zu sein wie in der Leber, da durch die Leber nur 30% des aus dem Herzen kommenden Blutstroms fließt, durch die Lunge aber 100%[24]. In der Lunge gebildete Metaboliten wirken sich – wegen der fehlenden Exkretion – auf den Hormonspiegel im Blut unmittelbar aus. Auch in der Lunge ist eine sexuelle Differenzierung der 2α-Hydroxylase-Aktivität nicht nachweisbar.

Wir danken der *Stiftung Volkswagenwerk* für materielle Unterstützung.

Literatur

1 Schriefers, H. (1967) *Vitamins and Hormones* **25**, 271 - 314.
2 Leybold, K. & Staudinger, H.J. (1959) *Biochem. Z.* **331**, 389 - 398.
3 Albaum, G. & Staib, W. (1965) *Biochem. Z.* **342**, 120 - 128.
4 Abraham, R. & Staudinger, H. (1966) *Hoppe-Seyler's Z. Physiol. Chem.* **346**, 198 - 207.
5 Wenzel, M., Kleucker, H. & Schulze, P.E. (1966) *Z. Naturforsch.* **21b**, 1178 - 1183.
6 Wenzel, M., Köster, G. & Hönicke, J. (1970) *Z. Klin. Chem. Klin. Biochem.* **8**, 231 - 234.
7 Wenzel, M. & Lemm, U. (1976) *Hoppe-Seyler's Z. Physiol. Chem.* **357**, 1145 - 1151.
8 Simon, H. (Hrsg.) (1974) Messung von radioaktiven und stabilen Isotopen, Springer Verlag Berlin.
9 Lemm, U. (1976) Dissertation, Freie Universität Berlin, Fachbereich Pharmazie.
10 Wenzel, M. & Mützel, W. (1970) *Hoppe-Seyler's Z. Physiol. Chem.* **351**, 1221 - 1231.
11 Wenzel, M. (1971) *Biochem. Pharmacol.* **20**, 2531 - 2534.
12 Mützel, W. & Wenzel, M. (1971) *Hoppe-Seyler's Z. Physiol. Chem.* **352**, 304 - 308.
13 Schriefers, H., Ghraf, R., Hoff, H.-G. & Ockenfels, H. (1972) *Acta Endocrinol.* **69**, 789 - 800.
14 Wenzel, M. & Stahl, J. (1970) *Hoppe-Seyler's Z. Physiol. Chem.* **351**, 761 - 762.
15 Wenzel, M., Blum, K. & Kraas, E. (1968) *Z. Klin. Chem. Klin. Biochem.* **6**, 265 - 268.
16 Bolt, H.M. & Bolt, W.H. (1974) *Eur. J. Clin. Pharmacol.* **7**, 295 - 305.
17 Schriefers, H., Hoff, H.-G., Ghraf, R. & Ockenfels, H. (1971) *Hoppe-Seyler's Z. Physiol. Chem.* **352**, 1363 - 1371.
18 Ghraf, R., Lax, E.R. & Schriefers, H. (1972) *Acta Endocrinol.* **71**, 781 - 791.

19 Hoff, H.-G., Ghraf, R., Raible, M. & Schriefers, H. (1973) *Hoppe-Seyler's Z. Physiol. Chem.* **354**, 306-311.
20 Inano, H. & Tamaoki, B. (1971) *Biochemistry* **10**, 1503 - 1509.
21 Inano, H., Tsuno, K. & Tamaoki, B. (1970) *Biochemistry* **9**, 2253 - 2259.
22 Jensen, E.V. & Jacobson, H.J. (1960) Biol. Activ. of Steroids in Relation to Cancer, Academic Press 16.
23 Jensen, E.V. & Jacobson, H.J. (1962) *Recent Progr. Horm. Res.* **18**, 387 - 414.
24 Nicholas, T.E. & Kim, P.A. (1975) *Steroids* **25**, 387 - 402.

Dr. U. Lemm u. Prof. M. Wenzel, Freie Univ. Berlin, FB 22, Königin-Luisenstr. 2/4, D-1000 Berlin 33.

Hoppe-Seyler's Z. Physiol. Chem.
Bd. 359, S. 89 - 101, Januar 1978

tRNA Isopentenyltransferase from *Zea mays* L.

Characterization of the Isopentenylation Reaction of tRNA, Oligo (A) and Other Nucleic Acids

Jürgen Holtz* and Dieter Klämbt

Botanisches Institut der Universität Bonn

(Received 8 June 1977)

Summary: The extraction and purification of the tRNA isopentenyltransferase from maize root tips and kernels are reported. The relative amounts of this enzyme in different organs of maize have been determined. Root tips have the highest enzyme activities, followed by kernels and young leaves. Old leaves exhibit very low activity.

The molecular mass of the monomeric enzyme was determined to be 57000 - 63000 dalton. pH and $Mg^{2⊕}$ optima are in full agreement with the data reported for the enzymes from yeast and *Escherichia coli*. Spermine enhances the isopentenylation of tRNA from kernels The "K_m" values of $KMnO_4$-treated or untreated bulk tRNA from yeast and maize root tips and kernels were in the range of 10 to 34μM while the K_m values of of single species like $tRNA_4^{Ser}$ from rat liver and $tRNA_{KMnO_4}^{Tyr}$ from *E. coli* were 1.1 and 6.6μM, respectively.

The tRNA isopentenyltransferase from maize root tips and kernels catalyzes the incorporation of 2-isopentenyl groups into $(Ap)_{3-7}A$, endogenous bulk oligonucleotes from maize root tips and kernels, and into poly(A), RNA from MS-2 phages and, to a very low extent, into adenosine. The K_m values of $(Ap)_{3-7}A$ varied in the range of 250 to 750μM. Although oligonucleotides have less affinity to the enzyme, the formation of i^6A within oligonucleotides may occur in vivo, due to their higher concentrations in cells.

---

tRNA-Isopentenyltransferase aus Zea mays L. – Charakterisierung der Isopentenylierungsreaktion von tRNA, oligo(A) und anderen Nucleinsäuren

Zusammenfassung: Aus Wurzelspitzen und Karyopsen von *Zea mays* wurde tRNA Isopentenyltransferase hochgereinigt und charakterisiert. Außerdem wurden die relativen Mengen dieses Enzyms in verschiedenen Organen der Mais-Pflanze bestimmt. Die höchsten Enzymaktivitäten be-

---

Enzymes:
Alkaline phosphatase, orthophosphoric-monoester phosphohydrolase (alkaline optimum) (EC 3.1.3.1);
Isopentenyldiphosphate △-isomerase, isopentenyldiphosphate Δ^3-Δ^2-isomerase (EC 5.3.3.2);
tRNA isopentenyltransferase, 2-isopentenyldiphosphate: tRNA 2-isopentenyltransferase (EC 2.5.1.8).

Abbreviations:
$tRNA_{KMnO_4}$ = $KMnO_4$-treated tRNA; isopentenyl group = 2-isopentenyl group; oligo (A) = $(Ap)_{1-9}A$ = [adenylyl-(3′-5′)]$_{1-9}$adenosine; MS-2 RNA = RNA from MS-2 phages; i^6A = N^6-(2-isopentenyl)adenosine.

* Part of the doctoral thesis, Bonn 1976

sitzen Wurzelspitzen; wachsende Karyopsen und junge Blätter zeigen kleinere und alte Blätter nur sehr geringe Enzymaktivität.

Das monomere Enzym besitzt eine Molekülmasse von 57000 - 63000 dalton; pH- und $Mg^{2\oplus}$-Optima der Enzymaktivität stimmen mit den Angaben der entsprechenden Enzyme aus Hefe und *Escherichia coli* überein. Für die Isopentylierung von tRNA aus Maiskaryopsen konnte ein positiver Spermin-Effekt nachgewiesen werden. Die „K_m"-Werte $KMnO_4$-behandelter und unbehandelter tRNA aus Hefe, Wurzelspitzen und Karyopsen aus Mais lagen zwischen 10 und 34μM. Die K_m Werte für die einzelnen tRNA Species $tRNA_4^{Ser}$ aus Rattenleber und $tRNA^{Tyr}_{KMnO_4}$ aus *E. coli* konnten mit 1.1μM bzw. 6.6μM bestimmt werden.

Die tRNA-Isopentenyltransferase aus Mais-Wurzelspitzen und -Karyopsen katalysiert außerdem die 2-Isopentenylierung von $(Ap)_{3-7}A$, von Oligonucleotidfraktionen aus Mais-Wurzeln und -Karyopsen, sowie poly(A), RNA aus MS-2-Phagen und in sehr geringem Ausmaß Adenosin. Die K_m-Werte für $(Ap)_{3-7}A$ liegen zwischen 250 und 750μM. Obwohl Oligonucleotide eine geringere Affinität zum Enzym besitzen, ist wegen ihrer höheren molaren Konzentration gegenüber dem tRNA-Gehalt eine Isopentenylierung in vivo denkbar.

Key words: tRNA isopentenyltransferase, tRNA, oligo(A), N^6-(2-isopentenyl)adenosine, spermine.

The hypermodified i^6A nucleotide in tRNA is unique in one respect: the sequence $Ai^6AA\Psi$ or AAi^6AAC in the anticodon loop is common to all tRNAs possessing an i^6A nucleotide[1,2]. Only the tRNAs coding for Ser, Cys, Trp, Phe, Tyr, and Leu with code triplets UXY are involved in this modification.

While it has been suggested that the i^6A nucleotide in this position in tRNA serves as a reinforcing element for the codon-anticodon interaction[1], the significance of the common sequence for the isopentenylation reaction is not known. Possibly an oligonucleotide as short as $AAA\Psi$ or AAAAC fulfills the minimum condition of a specific substrate[3]. However all published investigations of this possibility have had negative results[3,4*], i.e. the tRNA-isopentenyltransferase seemed to accept only intact tRNA molecules[3-7]. In contrast, our results indicate that the substrate specificity of this enzyme is not restricted to tRNA but includes oligonucleotides, poly(A), MS-2 RNA and adenosine (very weakly).

It is assumed that i^6A nucleotides and their derivatives serve as precursors of the potent plant hormones known as cytokinins, which are the related bases and nucleosides. However, the biosynthesis of this class of substances is still unknown. The hypothetical pathway via isopentenylation at the polymer level (tRNA) followed by degradation to cytokinins is supported by determinations of the half-life of tRNA in various organisms[8,9]. These data do not agree with those of other laboratories[10,11]. Furthermore, isopentenylation of an artificially modified adenosine which cannot be incorporated into tRNA is reported[12]. The pathway of cytokinin synthesis still remains to be resolved. Aside from the special significance for cytokinin production, the properties of the tRNA isopentenyltransferase are another facet in understanding nucleic acid-protein interactions. Since the isopentenylation rates of oligo(A) are comparable with those of tRNA, it seemed to be necessary to measure the concentration of oligonucleotides in vivo.

* Vickers, J. D., personal communication.

Experimental

MATERIAL

Alkaline phosphatase *(E.coli)*, 20 U/mg protein, 1 mg/m*l* Boehringer Mannheim GmbH, $tRNA^{Tyr}$ from *E. coli* MRE 600, 1000 pmol Tyr/A_{260} unit, Boehringer; bulk tRNA from *Saccharomyces spec.*, degree of purity 75%, Boehringer; $tRNA_4^{Ser}$, 260 pmol Ser/A_{260} unit, a gift of *Dr. D. Söll*; A oligonucleotides, adenylyl-(3' → 5')$_{1-9}$-adenosine [$(Ap)_{1-9}A$], Boehringer; poly(A), 3.5 × 10^5 dalton, 20 A_{260} units/mg, Boehringer; RNA from MS-2 phages, 3.6 × 10^6 dalton, 20 A_{260} units/mg, Boehringer; ^{14}C-labelled bulk tRNA from *L. acidophilus*, 2 × 10^5 cpm/-A_{260} unit, labelled by [^{14}C]orotic acid, a gift of *Dr. M. Leineweber*; polyethyleneglycol type 6000, Serva, Heidelberg; DEAE-cellulose (DE 52), Whatman, Maidstone, England; Sephadex G-25 fine and G-100 superfine, Pharmacia, Uppsala, Sweden; Dowex Type 50 W-X 4($H^{\oplus}$), cap. 1.7 meq/m*l* bed volume, 50 - 100 mesh; silica gel 60, grain size 0.063 - 0.20 mm, 70 - 220 mesh, Merck, Darmstadt; silica gel plates with fluorescence indicator, 10 x 20 cm, Merck; acrylamide, *N,N'*-methylenbis(acrylamide), *N,N,N',N'*-tetramethylenethylendiamine, Serva.

ORGANISMS

Lactobacillus acidophilus were cultured by the method described[7]. *Zea mays* L. cv. Perdux: The caryopses were sterilized with Ceresan (mercury product of Bayer, Leverkusen) and soaked overnight at 25 °C under sterile conditions. Within a further 48 h incubation primary roots grew on sterile wet filter paper. Kernels were harvested during the milky stage from open field plants. The diameter was 3 - 4 mm measured at the base of the young fruit. Young leaves, the youngest 3 apical leaves (light green, 18 - 30 cm long) were used from young open field plants (1 m high) and old leaves, 40 - 60 cm long dark green leaves of the same plants.

METHODS

For the preparation of tRNA from maize roots or kernels, the plant material was homogenized 3 x 10 s in 10mM Tris/HCl, pH 7.5. The extraction was performed as described[8]. The preparations (A_{280}/A_{260} = 0.5; A_{230}/A_{260} = 0.5) contained less than 10% 5S RNA and accepted 11 pmol [^{14}C]leucine/A_{260} unit. tRNA was treated as described with KMnO[6]. Pretreatment of tRNA with $KMnO_4$ did not change the amino acylation reaction.

The preparation of tRNA isopentenyltransferase from *L. acidophilus* has been reported[7], likewise the prepara tion and assay of isopentenyldiphosphate-Δ-isomerase[13] with modifications[7]. The enzyme preparation did not contain prenyltransferase or alkaline phosphate activity[7]. 15% of the incubated 3-isopentenyldiphosphate was converted to the Δ^2-isomer.

Purification of the tRNA-isopentenyltransferase from roots of Zea mays

All purification steps were carried out at 4 °C. 8 - 12 g roots was homogenized in 20 - 30 m*l* extraction buffer (10mM Tris/HCl, pH 7.8, 5mM $MgCl_2$, 10mM 2-mercaptoethanol, 10% glycerol (v/w); 3 x 10 s), filtered through 4 layers of cheesecloth and centrifuged (3000 x *g*, 35 min). This supernatant (F-I) was applied to a Sephadex G-100 column (2.5 x 40 cm) equilibrated with the extraction buffer, and eluted with the same buffer (2 m*l*/h flow rate; 5.5 m*l*/fraction). Fractions containing proteins were dialysed and concentrated against 25% polyethylene glycol in extraction buffer and tested for tRNA isopentenyltransferase activity. Proteins of a molecular mass of 60000 to 80000 dalton eluted in fractions 8 - 10, which contain all the tRNA isopentenyltransferase activity. They were collected and referred to as F-II. F-II was chromatographed on a DEAE-cellulose column (1 x 20 cm). An active sample of tRNA isopentenyltransferase (F-III) was collected by front elution with 0.2M NaCl ectraction buffer. F-III was dialysed and concentrated in extraction buffer. containing 25% polyethylene glycol, and stored in the presence of 50% glycerol at – 18 °C. It was subjected to preparative discontinuous polyacrylamide gel electrophoresis according to Nees and Richter[14] as modified by Leineweber[15]. The concentration of the separation gel was 5.7%. F-III was adjusted to 50mM Tris/HCl, pH 7.5, 10mM $MgAc_2$ before application to the gel. After the elution of bromphenolblue, fractions of 1.76 m*l* were collected and tested for tRNA isopentenyltransferase activity. The most active fraction, F-IV, was stored in a solution of 50% glycerol and bovine serum albumin (0.59 mg/m*l*) at – 18 °C. Analytical and dodecylsulfate gel electrophoresis of this F-IV fraction was performed as described[16].

Preparation of tRNA isopentenyltransferase from caryopses and leaves

The procedures were similar to the one described for the isolation from roots. tRNA isopentenyltransferase from caryopses (the F-III fraction) was further analyzed by analytical gel electrophoresis[16]. 0.1 mg of F-III protein in 50% glycerol was applied to the gel and developed by concentrating and separating at 6 V/cm and 12 V/cm, respectively. After electrophoresis the gel was cut into 0.5-cm pieces, ground and tested for isopentenylation activity. Parallel gels of the same separations were stained with Coomassie Brilliant Blue G 250[16].

Assay for tRNA isopentenyltransferase activity

Except where indicated, the incubation mixture contained 15mM Tris/HCl, pH 7.8, 4mM $MgCl_2$, 8mM 2-mercaptoethanol, 8μM [^{14}C]2-isopentenyldiphosphate. The amount of [1-^{14}C]2-isopentenyldiphosphate was calculated from samples of the preincubated isopentenyldiphosphate isomerase reaction mixture. The amount of RNA substrate varied. The test volume was 100 μl.

tRNA isolation: After incubation at 37 °C for 10 min, the tRNA was isolated by chromatography on a DEAE-cellulose column from the incubation mixture. The incorporation of [^{14}C]isopentenyl groups in tRNA was measured by a scintillation counter[7]. The recovery of tRNA was 40 - 60%; the pmol values given in this paper are corrected for the loss of tRNA. Controls were carried out with enzyme preparations boiled for 1 min.

Oligonucleotide isolation: The procedure for extraction of oligonucleotides from the incubation mixture was similar to that for tRNA, except that the washing buffer after application of the test mixture to the DEAE-cellulose column contained 0.25M NaCl in 10mM Tris/-HCl, pH 7.5. Oligonucleotides $(Ap)_{4-9}A$ were eluted with 1M NaCl, 10mM Tris/HCl, pH 7.5. The recovery of $(Ap)_5A$ was 30 - 40% and of $(Ap)_{6-9}A$, 40 - 80%. $(Ap)_{1-4}A$ cannot be separated from proteins by DEAE-cellulose chromatography. Therefore the extraction of these small nucleotides was performed on a silica gel column (0.4 × 3 cm). The $(Ap)_{1-4}A$ oligonucleotides were separated from proteins by eluting the small column with 4 - 6 m*l* 10mM Tris/HCl, pH 7.5. The oligonucleotides were precipitated with 5% trichloroacetic acid and tested for incorporation of radioactive isopentenyl groups. The nucleotide recovery after separation on silica gels was 60 - 90%. Another method based on paper chromatography: after elution from the silica gel column, the eluate was chromatographed on Whatman no. 1 paper (solvent: isobutyric acid/0.5M ammonium hydroxide (5 : 3)), the UV-spots were cut out and their radioactivity measured in the scintillation counter*.

The incorporation of isopentenyl groups in $(Ap)_{6-7}A$ measured after separation on silica gel columns and paper chromatography coincided with the results after DEAE-cellulose separation, although the background was higher after silica gel separation and paper chromatography.

Poly A and MS 2-RNA isolation: Both polymers were extracted by addition of 0.9 m*l* H_2O and 1 m*l* water-saturated phenol (2% sodium dodecylsulfate, 0.1% hydroxychinoline). The polynucleotides in the water phase were precipitated after addition of 0.5 mg bovine serum albumin with 5% trichloroacetic acid, collected on membrane filters and measured in the scintillation counter.

AMP, ADP and ATP isolation: After the incubation and adjustment to pH 8.6, the probes were treated with alkaline phosphatase. After three different thin-layer chromatography runs (solvents: chloroform/acetic acid (8 : 2), 2-propanol/NH_4OH/H_2O (8 : 1 : 1), chloroform/methanol/(9 : 1)), the radioactivity of the spot corresponding to i^6A was determined by scintillation counting.

Adenosine and adenine isolation: After the incubation, the riboside and the free base were separated by thin-layer chromatography (solvents as mentioned above). Spots corresponding to i^6A and N^6 (2-isopentenyl)-adenine were further analyzed by high-pressure liquid chromatography[17]. The column was developed with 40% methanol at 800 psi. UV absorption was continuously recorded, and 0.25-m*l* fractions were collected and poured onto Whatman G/C filters, dried and measured in the scintillation counter.

Chracterization of the product of the tRNA isopentenyltransferase reaction

After incubation and isolation of the nucleic acid substrates, tRNA, poly(A) and the oligonucleotides were hydrolysed by KOH, chromatographed on Dowex WX 50 ($H^{\oplus}$), digested by alkaline phosphatase and separated by thin-layer chromatography[7] (solvents as mentioned above).

Extraction of oligonucleotides from roots and kernels of Zea mays

10 g roots or kernels was homogenized in 50 m*l* buffer (10mM Tris/HCl, pH 7.5; 3 × 10 s) and the homogenate extracted twice with phenol (2% sodium dodecylsulfate, 0.1% hydroxyquinoline). The residual phenol in the water phase was extracted with ether (3 times) and the remaining ether finally removed by a stream of N_2. This extract was applied to a DEAE-cellulose column (1 × 25 cm) equilibrated with 10mM Tris/HCl, pH 7.5, and washed extensively with the same buffer. The oligonucleotides were eluted with 0.35M NaCl, 10mM Tris/HCl, pH 7.5. UV-absorbing fractions were pooled and desalted by Sephadex G-25 chromatography, yielding the bulk oligonucleotides. The chain lengths of oligonucleotides were determined by chromatography on Sephadex G-25 (buffer, 0.5M $(NH_4)_2CO_3$, 6M urea pH 8.6; column size, 1 × 50 cm). The base composition of the oligonucleotides was determined as described[18]. Protein was measured by the method of Lowry.
RNA was measured by UV-absorption:
tRNA: 1 A_{260}unit = 1.88 nmol.
A-oligonucleotides ($(Ap)_{1-9}A$): 25 A_{260}units = 1 mg.

Results

Preparation of tRNA isopentenyltransferase from maize root tips

The homogenate was centrifuged at 3000 × *g*. Centrifugation at 10000 × *g* resulted in hardly detectable activities of tRNA isopentenyltransferase within the supernatant and only minor enzymatic activity within the pellet. Apparently this enzyme is associated with membrane particles found in pellets after 10000 × *g* centrifugation. Sephadex G-100 chromatography was employed as the next step of purification. Since the enzyme activity elutes in one peak at a marker region of 60000 - 80000 dalton, its binding to membranes should be very weak.

* Söll, D., personal communication.

Table 1: Purification of tRNA isopentenyltransferase from root tips of maize.

Fraction	Vol. [m*l*]	Prot. conc. [mg/m*l*][a]	Total prot. [mg]	Total act. [units][b]	Spec. act. [units/mg]	Purif.
F-I: 3000 × *g* supernatant	16	10.2	163.0	8150	50	1
F-II: after Sephadex G-100 chromatography	1.5	2.1	3.15	2800	890	17.8
F-III: after DEAE-cellulose chromatography	4.0	0.074	0.30	2110	6250	125
F-IV: after preparative discont. electrophoresis	3.4	0.006[c]	0.02	5710	286000	5700

[a] Protein determination according to Lowry.
[b] 1 unit is the amount of enzyme catalyzing the incorporation of 1 pmol 2-isopentenyl groups into 5 A_{260} units $tRNA_{KMnO_4}$ from yeast in 10 min incubation at 37 °C.
[c] An estimation by comparison of colour intensity of Coomassie Brilliant Blue stained dodecylsulfate gels with bovine serum albumin standards.

Table 1 gives the purification data. Taking the total tRNA isopentenyltransferase activity in F-I as 100%, 70% of the enzyme activity was recovered in F-IV. The lower total activities for F-II and F-III may be due to inhibitory substances still present in these fractions. The method of preparative gel electrophoresis seems optimal with respect to parameters like pH, buffer solution, ionic strength and acrylamide concentration (Fig. 1). The acrylamide concentration turned out to be crucial, with 5.7% the optimal concentration.

The total purification factor was 5700. To test the homogeneity of the F-IV, 5 µg protein of this fraction was applied to analytical and dodecylsulfate gel electrophoresis (Fig. 2a). Analytical gel electrophoresis separated F-IV into 2 protein bands, a minor and major one, while dodecylsulfate gel electrophoresis resulted in 3 bands. The middle one was clearly prominent. The plot yielded a molecular mass of 60000 ± 3000 dalton for the prominent protein band (Fig. 2b) and 20000 ± 1000 and 120000 ± 6000 for the minor bands, not shown. Since chromatography on Sephadex G-100 indicated a molecular mass of 60000 - 80000, the 60000-dalton band on the dodecylsulfate gel corresponds to tRNA isopentenyltransferase. Because the 60000 dalton protein is prominent, F-IV represents an almost homogeneous enzyme.

No interfering enzymatic activities such as isopentenyldiphosphate Δ-isomerase, phosphatase or RNase were detected in F-III and F-IV. F-III fractions from several maize organs lost no enzyme activity during storage for 3 months in 50% glycerol at – 18 °C.

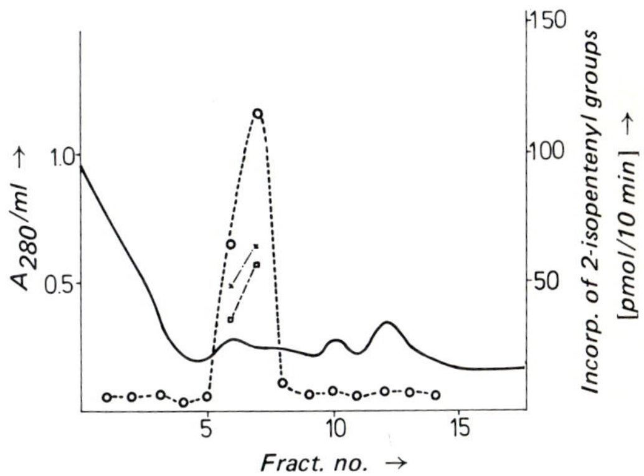

Fig. 1. Preparative discontinuous polyacrylamide gel electrophoresis of tRNA isopentenyltransferase from root tips of maize.

0.25 mg F-III protein was applied to a polyacrylamide gel (6.5 × 14 × 0.7 cm; concentration 5.7%). The assay of the tRNA isopentenyltransferase activity is described in methods.

——— , A_{280}/m*l*; activity of tRNA isopentenyltransferase in the presence of 5 A_{260}units $tRNA_{KMnO_4}$ from yeast (○– – –○), in the presence of 0.7mM $(Ap)_6A$ (x–·–·x), in the presence of 0.9mM $(Ap)_6A$ (□– – –□).

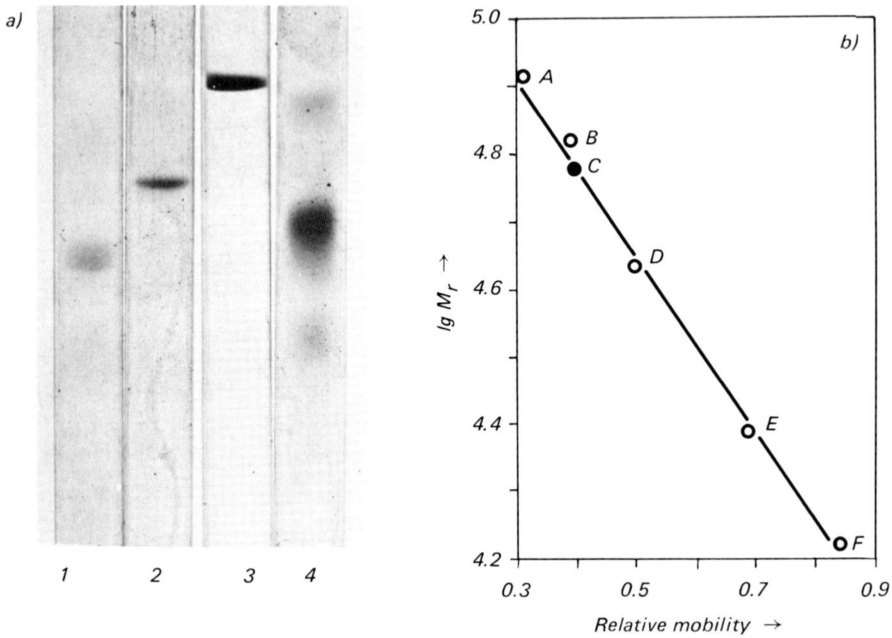

Fig. 2. a) Sodium dodecylsulfate electropherogram of the tRNA isopentenyltransferase from root tips of maize.
1, ovalbumin, 45 000 dalton;
2, bovine serum albumin, 67 000 dalton;
3, β-galactosidase 130 000 dalton;
4, 5 μg F-IV protein from root tips of maize.
b) Plot for determination of the molecular weight of the prominent band on the dodecylsulfate gel.
A, ribonuclease II;
B, bovine serum albumin (monomer);
C, tRNA isopentenyltransferase;
D, ovalbumin (monomer);
E, chymotrypsinogen;
F, myoglobin.

F-IV from root tips was unstable even on storage in 50% gylcerol and 0.59 mg bovine serum albumin/m*l* at −18 °C. After 14 days only 20% of the original activity was left.

Preparation of tRNA isopentenyltransferase from maize kernels and leaves

The extracts of kernels and leaves were purified by the same procedure as the enzyme from root tips, without the last step. The enrichment of tRNA isopentenyltransferase (F-III) from kernels (76-fold) and young leaves (73-fold) was equal and comparable with the 125-fold purification of the enzyme from roots. The purification of the enzyme from old leaves resulted in only a 31-fold enrichment. The elution profiles from Sephadex G-100 columns were similar to that of roots.

Comparison of tRNA isopentenyltransferase activities from different maize organs

A comparison is hindered by a lack of well defined reference data. The protein content (mg protein/g fresh weight) in kernels is not much higher than in young leaves, but root tips contain 2 - 3 times as much. tRNA isopentenyltransferase activity/mg protein is a more reliable value than activity/g fresh weight, since the protein content reflects the biochemical composition and the metabolic activity much better.

The enzyme activity of root tips varied by a factor of 4 in various extractions, but the lowest value still represents a 7-fold higher activity than was extracted from kernels and young leaves. Old leaves contained one tenth of the activity of young leaves.

Isopentenylation of tRNA and oligo (A) by the same enzyme

Fig. 1 demonstrates the coincidence of isopentenylation of tRNA and of oligo (A). The analytical and dodecylsulfate gel electrophoresis, in connection with Sephadex G-100 chromatography, indicate only one monomeric protein responsible for the tRNA isopentenyltransferase activity. The protein bands at 20000 and 120000 dalton (Fig. 2a) probably represent impurities.

Analytical gel electrophoresis of F-III from kernels shows one protein band which is responsible for the activity for isopentenyl group incorporation into tRNA and $(Ap)_6A$ as well (Fig. 3). These results confirm the data of Fig. 1.

Characteristics of tRNA isopentenyltransferase reaction

Regardless of the tRNA used as substrate, the reaction rate is linear for 10 min, and after that slowly levels off.

The pH optimum of the tRNA isopentenyltransferase was 7.8 for all preparations tested. The temperature optimum is 37 °C. The optimal $Mg^{2\oplus}$ concentration comprises the narrow interval of 3 - 5mM. 5mM EDTA completely inhibits the enzyme activity.

The activity of all preparations was directly proportional to the protein concentration (7 - 35 μg protein of F-III from kernels per assay; 0.06 - 0.4 μg protein of F-IV from root tips per assay). Boiled enzyme was inactive. Only 2-isopentenyldiphosphate, not 3-isopentenyldiphosphate, was a suitable substrate. The transferase reaction is completely inhibited by pyrophosphate (1.4mM). N^6-(2-Isopentenyl)adenine and i^6A in concentrations of 10^{-5}M or 10^{-6}M did not influence the isoprenylation reaction at all.

Spermine effect on tRNA isopentenyltransferase reaction

Methylation and also amino acylation of tRNA is stimulated by polyamines[19]. The isopentenylation reaction showed similar characteristics: direct addition of spermine was ineffective[7], while a preincubation of tRNA with spermine produced a remarkable enhancement (Table 2).

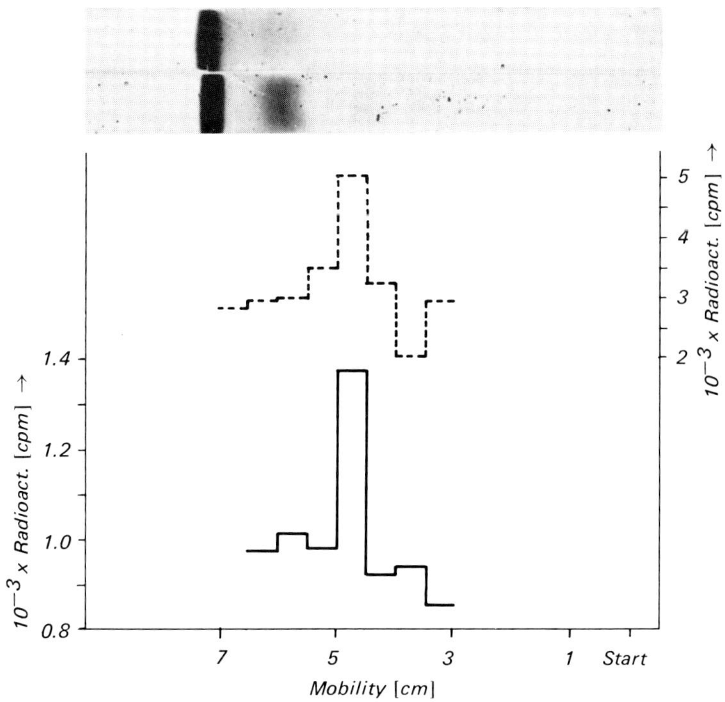

Fig. 3. Analytical electropherogram of tRNA isopentenyltransferase from maize kernels.

The lower gel was loaded with 100 μg F-III protein and 50 μg bovine serum albumin. The upper gel was loaded with 50 μg bovine serum albumin only. Protein was stained by Coomassie blue.
(Lower part: Assays of the corresponding gel slides for tRNA isopentenyltransferase activity with tRNA from *E. coli* (left ordinate) and $(Ap)_6A$ (right ordinate). 5.0 A_{260} units of tRNA was used, and 4.3 A_{260} units $(Ap)_6A$.

Table 2. Spermine effect on the isopentenylation reaction.

The amount of isopentenyl incorporation into 2 A_{260} units of $tRNA_{KMnO_4}$ was measured after preincubation of the latter with spermine. The preincubation mixture contained 2 A_{260} units $tRNA_{KMnO_4}$ from caryopses and the amount of spermine indicated in 33mM Tris/HCl, pH 7.8. Preincubation in 15 μl, 30 min at 37 °C.

Spermine	Incorporation[a] of isopentenyl groups	
	F-III enzyme from caryopses[b]	F-III enzyme from root tips[c]
[mM]	[pmol/10 min]	[pmol/10 min]
0	17	5.4
0.07		5.0
0.7	27.1	25.0
3.4	29.0	18.4
6.7	48.2	8.6
14.0	26.0	4.2

[a] Data varied up to ± 10%.
[b] 14 μg protein per assay
[c] 2 μg protein per assay

However, the stimulation was only observed with tRNA from kernels. The isopentenylation of tRNA from roots – regardless of the origin of the enzyme – was not stimulated by spermine at all. This is difficult to explain. The concentration of $Mg^{2\oplus}$ ions, which influence the isopentenylation reaction, was kept constant for all tRNA stock solutions.

Spermidine concentrations of 0.08 - 1.5mM never caused an enhancement of the reaction.

K_m values and maximum isopentenylation rates of various tRNAs

The K_m values for 2-isopentenyldiphosphate were nearly identical for the tRNA isopentenyltransferases from kernels (1.3μM) and from roots (2.0μM). The determinations of K_m values for tRNAs were performed at 8μM 2-isopentenyldiphosphate.

The "K_m" values of bulk tRNA preparations should be interpreted with some care:
1) Not all the tRNA molecules are suitable substrates.
2) Product molecules are already present.
3) The molecular weights of tRNA species differ.

The "K_m" values of all investigated bulk tRNA in the reactions with the different enzyme preparations were estimated in the range of 10 - 34μM (0.5 - 2.0 A_{260} units/assay, Table 3). Pretreatment of tRNA with $KMnO_4$ reduced the dissociation constants only insignificantly, while the maximum extent of isopentenylation (mol incorporated isopentenyl groups per mol tRNA) increased by a factor of 5. Since the "K_m" values of $tRNA_{KMnO_4}$

Table 3: K_m and maximum isopentenylation extent of various tRNA.
The K_m values varied up to ± 15%. Experiments in which the activity could not be estimated are marked by "–". The maximum isopentenylation extent = mol incorporated isopentenyl groups per mol tRNA. The conversion of A_{260} units to molarities is given in Methods.

tRNA source	Enzyme from root tips		Enzyme from caryopses	
	K_m [μM]	Isopentenylation [%]	K_m [μM]	Isopentenylation [%]
Maize root tips $KMnO_4$-treated	17	4.5	24	0.4
Maize root tips	34	0.9	–	–
Maize caryopses $KMnO_4$-treated	16	1.5	10	3.3
Maize caryopses	–	–	29	0.6
Tyrosine specific from *E. coli* $KMnO_4$-treated	6.6	4.2	–	–
Serine-specific from rat liver	1.4	3.4	1.1	1.5
Yeast $KMnO_4$-treated	28	1.4	19	0.3
E. coli	–	0.2	–	0.2

from yeast differed only slightly, it can be concluded that the tRNA isopentenyltransferases from roots and kernels have equal affinity to this tRNA. The "K_m" values of $tRNA_{KMnO_4}$ from root tips and kernels with the enzymes from root tips and kernels are quite similar. Therefore the tRNA populations of the two organs seemed comparable.

Competitive inhibition by tRNA molecules which are not substrates cannot be expected[4]. $tRNA_4^{Ser}$ from rat liver has the same sequence in the anticodon loop and stem as $tRNA_3^{Ser}$, but lacks the isopentenyl group[3]. $tRNA_{KMnO_4}^{Tyr}$ from *E. coli* is de-isoprenylated to an extent of 25%[6]. Both are suitable substrates for the enzyme. The K_m values of $tRNA_4^{Ser}$ and $tRNA_{KMnO_4}^{Tyr}$ lay in the range of 1.1 - 2.6µM (Table 3).

Isopentenylation of oligo(A)

$(Ap)_{1-9}A$ were used to investigate possible substrate functions for the tRNA isopentenyltransferase. Since AAAΨ and AAAAC were not available[20], $(Ap)_{1-9}A$ were tested.

Different methods of determination gave comparable results on the incorporation of isopentenyl groups into $(Ap)_{3-7}A$. The identity of i^6A was confirmed by KOH hydrolysis, Dowex ($H^{\oplus}$) chromatography, phosphatase digestion and thin-layer chromatography. The radioactivity comigrated with a standard of i^6A. Since i^6A could be isolated in small amounts after KOH hydrolysis without phosphatase digestion, some 3′-terminal nucleotides are modified. The incorporation of isopentenyl groups was proportional to the enzyme concentration (up to 2 µg protein of F-III) and to the incubation time. Only 2-isopentenyldiphosphate, not 3-isopentenyldiphosphate, served as substrate in this reaction. 2.0mM pyrophosphate inhibited the reaction. Preincubation with spermine increased the isopentenyl group incorporation into $(Ap)_5A$ by 20%. The K_m of 2-isopentenyldiphosphate in the presence of $(Ap)_5A$ was 5.0µM, comparable to the K_m of 1.3 to 2.0µM with tRNA as the second substrate.

The K_m of $(Ap)_{3-7}A$ was in the range of 250 to 720µM. The minor differences in the K_m do not show any relation of chain length to the affinity of the enzyme. The maximum extent of isopentenylation varied from 1/700 to 1/2200 (Table 4). Endogenous bulk oligonucleotides (1.2 A_{260}-units/assay) were isopentenylated by tRNA isopentenyltransferase from root tips to an extent of about 1/700 and by the enzyme from kernels to an extent of about 1/1500.

ApA, $(Ap)_2A$ and $(Ap)_9A$, at concentrations up to 5mM, 6mM and 0.4mM respectively, showed no isoprenylation. The same is true for $(Ap)_3A$ and $(Ap)_7A$ up to 0.5mM with the enzyme from *L. acidophilus.*

Isoprenylation of poly(A) and MS-2 RNA

poly(A) and f2 RNA are weak substrates for the tRNA isopentenyltransferase of *E. coli*[3]. In our systems, poly(A) and MS-2 RNA were suitable substrates. The K_m of poly(A) with the enzyme

Table 4: K_m values and maximum isopentenylation extent of oligo(A) and poly(A). K_m values varied up to ± 15%. For definition of maximum isopentenylation, see Table 3.

Enzyme from	$10^4 \times K_m$ [M]					
	$(Ap)_3A$	$(Ap)_4A$	$(Ap_5)A$	$(Ap_6)A$	$(Ap)_7A$	poly A
L. acidophilus	0	5.5	6.0	3.9	0	0.006
Maize root tips	3.1	2.5	2.8	7.0	7.2	0.034
Maize caryopses	+[a]	4.1	5.0	4.1	2.5	0
	Max. isopentenylation [%]					
L. acidophilus	0	0.1	0.04	0.07	0	100.00
Maize root tips	0.5	0.07	0.15	0.07	0.05	4.2
Maize caryopses	0	0.03	0.04	0.09	0.08	0

[a] Isopentenylation was measured, but K_m determination was not possible.

from *L. acidophilus* was 0.66μM, and 3.4μM with the enzyme from maize roots (Table 4). The maximum extents of isopentenylation were 1 : 1 (enzyme from *L. acidophilus*) and 1 : 24 (enzyme from maize root tips). In MS-2 RNA (0.5 A_{260}-units/assay), two isopentenyl groups per RNA molecule were incorporated by the enzyme from roots. rRNA, $KMnO_4$-treated or not, from roots or kernels – was not isopentenylated.

Isopentenylation of adenosine

After incubation of 18mM adenosine with tRNA isopentenyltransferase from roots (F-III and F-IV) under the above conditions, i^6A could be isolated by thin-layer chromatography (Fig. 4). A maximum extent of isopentenylation of 1 : 360000 was determined. Lower concentrations of adenosine, or AMP, ADP and ATP up to 5mM and adenine up to 7mM, were not isopentenylated.

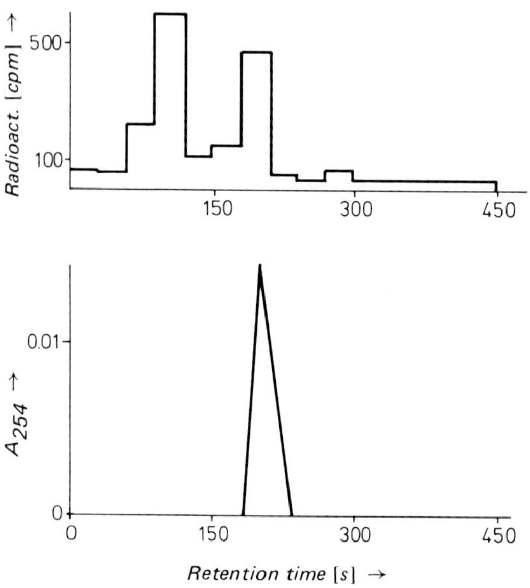

Fig. 4. Products from the incubation mixture with an R_F value similar to i^6A were isolated by 3-fold thin-layer chromatography runs, mixed with authentic i^6A and applied to the high-pressure liquid chromatography column[17].

Upper graph: ^{14}C radioactivity profile of the reaction products.

Lower graph: UV absorption profile of i^6A.

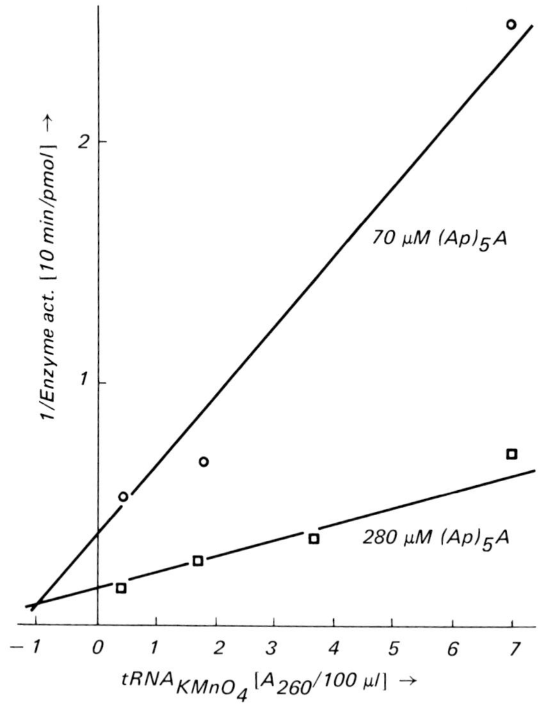

Fig. 5. Competitive inhibition of isopentenylation of $(Ap)_5A$ by $tRNA_{KMnO_4}$ from root tips of maize (Dixon plot).

Invariable parameters: 3 μg protein from root tips of maize (F-III) per assay; 8μM [^{14}C]2-isopentenyldiphosphate; 10 min incubation at 37 °C.

Competition of $(Ap)_5A$ isopentenylation by $tRNA_{KMnO_4}$ from roots

The Dixon plot (Fig. 5) is another indication that tRNA and oligo(A) are isopentenylated by the same enzyme. $(Ap)_5A$ and tRNA compete for the same site at the tRNA isopentenyltransferase. The measured K_i of 17μM for $tRNA_{KMnO_4}$ from roots could be expected, since their respective K_m values differ by a factor of 10. In the reversed assay – $(Ap)_5A$ as inhibitor up to 600μM there was no decrease in the isopentenylation rate of the tRNA.

Extraction of oligonucleotides from root tips and kernels of maize

Since endogenous oligonucleotides are isopentenylated, the concentrations of this class of mole-

Table 5: Content of tRNA and oligonucleotides in root tips and caryopses of maize per g fresh weight. The data varied up to ± 15%.

Source	tRNA		Oligonucleotides[a]				mol oligonucl. / mol tRNA
	A_{260}	[nmol]	A_{260}	$\frac{A_{280}}{A_{260}}$	A-nucleotides [%][b]	[nmol][c]	
Root tips	3.5	6.6	14.0	0.55	25	200	30
Caryopses	5.0	9.5	6.2	0.40	26	88	9

[a] The oligonucleotides were eluted from DEAE-cellulose column with 0.35M NaCl.
[b] Determined as described[18].
[c] An average molecular weight of 2800 determined by Sephadex G-25 chromatography.

cules in maize root tips and kernels were determined (Table 5). While the amounts of tRNA extracted from root tips and kernels were almost equal (3.5 and 5.0 A_{260}units/g fresh weight, respectively) the bulk oligonucleotide concentrations differed by a factor of 3 (14 and 5 A_{260}-units/g fresh weight).

Discussion

Highly purified maize tRNA isopentenyltransferase with new properties is described. The estimated molecular mass of 57000 - 63000 is in agreement with the mass reported for the enzyme from *E. coli*. The 20000 and 120000 dalton protein bands shown by dodecylsulfate gel electrophoresis can be considered impurities. Our data support the view that tRNA isopentenyltransferase is a monomeric protein and that only one enzyme is responsible for the isopentenylation of all tRNA species.

Since isopentenylation activity is detected in all investigated organs of maize, the cytokinin production of these tissues via the tRNA can be considered. A measurable tRNA turnover in *Zea mays* roots has been described[8]. Our data are consistent with the hypothesis that roots are the principal source of cytokinins, but they are not the only one.

The kinetics, $Mg^{2\oplus}$ and pH optima of the maize tRNA isopentenyltransferase have the same shape as those for enzymes from other sources [3,4,6,7]. It has been suggested that the stimulation of tRNA reactions by polyamines is due to a specific association of polyamines with tRNA[21]. Preincubation of tRNA from maize kernels with spermine stimulates the isopentenylation reaction by a factor of 3.5 to 4.5, while spermine added directly to the assay mixtures has no effect. This indicates an influence of spermine on the tRNA and not on the enzyme molecule. The isoprenylation of tRNA from maize root tips is not stimulated by spermine. This could be explained by the presence of substances occupying the spermine binding sites on this tRNA. Since maturation processing of tRNA is enhanced by polyamines[19], the observed stimulation of the isoprenylation reaction by spermine further supports the crucial role of polyamines in macromolecular metabolism[22]. The K_m values of 2-isopentenyldiphosphate and of bulk tRNA from maize are in accord with the values reported for other systems[6,7]. $KMnO_4$-treated tRNA has about the same K_m values as untreated tRNA, but the extent of isopentenylation is increased by a factor of 5 by the treatment.

Isopentenylation of oligonucleotides was reported to be unsuccessful[3,4,*]. Furthermore, a dodecanucleotide representing the anticodon loop and a part of the anticodon stem of suppressor $tRNA_{III}^{Tyr}$ was not isopentenylated[4]. These experiments may have failed due to an inadequate procedure for isolation of the oligonucleotides from the incubation mixture* or due to inadequate concentrations of 2-isopentenyldiphosphate and oligonucleotides[3] or due to insufficient concentrations or unacceptable conformations of the oligomers tested.

* Vickers, J. D., personal communication.

Apparently the successful incorporation of isopentenyl groups into $(Ap)_{3-7}A$ by tRNA isopentenyltransferase from maize is due to the higher concentrations used in our experiments. The enzymatic character of the reaction is confirmed by the substrate specificity. Only 2-isopentenyldiphosphate and oligonucleotides with a chain length from 4 to 8 nucleotides serve as substrates. Further indications that the reaction is enzymatically catalyzed are its kinetics, its inhibition by PP_i, its linear proportion to the protein concentration and the K_m values of 2-isopentenyldiphosphate in the presence of $(Ap)_5A$ as the second substrate. The identity of tRNA isopentenyltransferase as the modifying enzyme of both tRNA and oligo(A) is confirmed by discontinuous preparative gel electrophoresis with the enzyme from root tips and by analytical gel electrophoresis with the enzyme from kernels.

$(Ap)_{3-7}A$ reflect in some respects the nucleotide sequences in anticodon loops of tRNA which can be isopentenylated. The K_m values of the individual A-oligomers are similar. Oligomers shorter than $(Ap)_3A$ and longer than $(Ap)_8A$ are not isopentenylated. It should be worthwhile to examine whether oligomers exhibiting one stable loop with A sequences are substrates with an intermediate affinity to the enzyme. $(Ap)_{3-7}A$ may represent the minimal structural condition for a recognizable substrate being isopentenylated.

The secondary and tertiary structure of tRNA may offer further recognition sites for the tRNA isopentenyltransferase. In analogous experiments with a tRNA methyltransferase, it was shown that a 3/4 fragment of the tRNA molecule has lost essential information for the methylation reaction[23].

rRNA is not isopentenylated, although oligo(A) sequences occur in this class of RNA[24]. The same negative results were found with *E. coli* rRNA[3]. Unlike rRNA, poly(A) and MS-2 RNA were substrates.

Adenosine was the only monomer which could be isopentenylated, although only at high concentrations. Therefore it seems unlikely that cytokinins are produced in vivo via this reaction. However, it may be possible that the optimal conditions for this reaction could not be achieved in vitro, but that membrane-associated tRNA isopentenyltransferase exhibits much higher affinity for adenosine.

Our data on oligonucleotides in maize root tips and kernels and their isopentenylation in vitro can be considered a first tentative indication of an additional pathway of cytokinin biogenesis.

The molar concentration of oligonucleotides is 30 times the tRNA molarity in the root tips and is 9 times higher in the kernels. The relatively high concentration of oligonucleotides may compensate their lower affinity to the enzyme. Thus isopentenylation could be comparable for endogenous oligonucleotides and tRNA.

This work has been supported by the *Deutsche Forschungsgemeinschaft*. We thank Mrs. *E. Hertweck* for technical assistance and Mr. *G. Golden* for reading the manuscript.

Literature

1 Nishimura, S. (1972) *Prog. Nucleic. Acid Res. Mol. Biol.* **12**, 49 - 71.
2 Holmquist, R., Jukes, T.H. & Pangburn, S. (1973) *J. Mol. Biol.* **78**, 91 - 116.
3 Bartz, J.K. & Söll, D. (1972) *Biochimie* **54**, 31 - 39.
4 Rosenbaum, N. & Gefter, M.L. (1972) *J. Biol. Chem.* **247**, 5675 - 5680.
5 Vickers, J.D. & Logan, D.M. (1970) *Biophys. Soc. Annu. Meet. Abstr.* **10**, 166a.
6 Kline, L.K., Fittler, F. & Hall, R.H. (1969) *Biochemistry* **8**, 4361 - 4371.
7 Holtz, J. & Klämbt, D. (1975) *Hoppe Seyler's Z. Physiol. Chem.* **356**, 1459 - 1464.
8 Klemen, F. & Klämbt, D. (1974) *Physiol. Plant.* **31**, 186 - 188.
9 Hahn, H., Heitmann, I. & Blumbach, M. (1976) *Z. Pflanzenphysiol.* **79**, 143 - 153.
10 Hall, R.H. (1973) *Annu. Rev. Plant Physiol.* **56**, 415 - 444.
11 Teyssendier de la Serve, B. & Péaud Leonël, C. (1976) The 9th Intern. Conf. Plant Growth Subst., Collected Abstr., 395 - 396.
12 Chen, C., Eckert, R.L. & McChesney, J.D. (1976) *FEBS Lett.* **64**, 429 - 434.
13 Holloway, P.W. & Popjak, G. (1968) *Biochem. J.* **106**, 836 - 840.
14 Nees, S. & Richter, K. (1972) Gerätebeschreibung und Bedienungsanleitung des Ultraphor, Colora.
15 Leineweber, M. (1976) Thesis, Bonn.
16 Weber, K. & Osborne, M. (1969) *J. Biol. Chem.* **244**, 4406 - 4412.
17 Hahn, H. (1976) *Plant Cell Physiol.* **17**, 1053 - 1058.
18 Katz, S. & Comb, D.G. (1963) *J. Biol. Chem.* **238**, 3065 - 3067.
19 Nau, F. (1975) *Biochimie* **58**, 629 - 645.

20 Schulte, H. (1975) Thesis, Bonn.
21 Bagni, N. & Serafini-Fracassini, D. (1974) Plant Growth Subst., Tokyo.
22 Leboy, P. S. (1970) *Biochemistry* **9**, 1577 - 1583.
23 Kuchino, Y. & Nishimura, S. (1971) *Biochem. Biophys. Res. Commun.* **40**, 306 - 313.
24 Birnboim, H. C. & Coakley, B. V. (1972) *Biochem. Biophys. Res. Commun.* **42**, 1169 - 1175.

Dr. J. Holtz and Prof. Dr. D. Klämbt, Botanisches Institut der Universität Bonn, Meckenheimer Allee 170, D-5300 Bonn 1.

Hoppe-Seyler's Z. Physiol. Chem.
Bd. 359, S. 103 - 111, Januar 1978

Preparation and Application of N^{α}-B1,N^{ϵ}-B29-Bis(tert.-butyloxycarbonyl)insulin

Heinz-Jürgen FRIESEN, Vinod K. NAITHANI and Hans-Gregor GATTNER

Deutsches Wollforschungsinstitut an der Technischen Hochschule Aachen

(Received 17 November 1977)

Summary: Two methods are described for the preparation of $N^{\alpha B1}$,$N^{\epsilon B29}$-Boc$_2$-insulin from N^{α}A1-trifluoroacetyl-insulin and N^{α}A1-citraconyl-insulin in 80 - 90% and 65% yields, respectively. Removal of the Boc protections afforded the fully active insulin. Application of this derivative was demonstrated by the preparations of des-Gly^{A1}-insulin and [A1-guanidinoacetyl]insulin. The former compound exhibited 2% activity in the in vitro free fat cell assay and the latter 88 ± 5% while $N^{\alpha B1}$,-$N^{\epsilon B29}$-Boc$_2$-insulin showed 45 ± 3% activity only.

Darstellung von N^{α}-B1,N^{ϵ}-B29-Bis(tert.-butyloxycarbonyl)-insulin und seine Verwendung zur Synthese von Des-Gly^{A1}-insulin und [A1-Guanidinoacetyl]insulin

Zusammenfassung: Die Darstellung des $N^{\alpha B1}$,$N^{\epsilon B29}$-Boc$_2$-Insulins aus N^{α}A1-Trifluoracetylinsulin und N^{α}A1-Citraconyl-insulin mit Ausbeuten von 80 - 90% bzw. 65% wird beschrieben. Nach Abspaltung der Boc-Gruppen wurde aus dem Derivat wieder voll aktives Insulin erhalten. Die Verwendbarkeit des $N^{\alpha B1}$,$N^{\epsilon B29}$-Boc$_2$-Insulins für Synthesen wird anhand der Darstellungen des Des-Gly^{A1}-insulins und des [A1-Guanidinoacetyl]-Insulins demonstriert. Das erstere zeigte im Fettzelltest 2% Aktivität, während das letztere Derivat eine Aktivität von 88 ± 5% zeigte. $N^{\alpha B1}$,$N^{\epsilon B29}$-Boc$_2$-Insulin besaß eine Aktivität von 45 ± 3%.

Key words: $N^{\alpha B1}$,$N^{\epsilon B29}$-Boc$_2$-Insulin, des-Gly^{A1}-insulin, [A1-guanidinoacetyl]insulin, insulin semisynthesis.

Insulin contains three amino groups, two from the *N*-terminal Gly^{A1} and Phe^{B1} residues, and one from Lys^{B29}. These have been the primary sites of chemical modification in structure-function studies (For reviews see ref.[1] and[2]). In general modification of insulin at the *N* terminus of the A-chain affects significantly the activity in vitro. The removal of the *N*-terminal Gly^{A1}

Abbreviations:
Boc = tert.-butyloxycarbonyl; Msc = methylsulfonylethyloxycarbonyl; Mtc = methylthioethyloxycarbonyl.
Modifications described in this paper are carried out at N^{α} or N^{ϵ} amino groups only. Thus, for the sake of simplicity, N^{α}A1-trifluoroacetyl-insulin is referred to as A1-trifluoroacetyl-insulin, for example. $N^{\alpha B1}$,$N^{\epsilon B29}$-Boc$_2$-insulin, which is mentioned several times in the text, is referred to as Boc$_2$-insulin.

causes the most severe drop in activity. This was first unequivocally shown by the preparations of des-Gly^{A1},des-Phe^{B1}-insulin[3] and des-Phe^{B1}-insulin[4].

Thus far, four different routes for the preparation of Gly^{A1}-modified insulins have been described. The first method is the modification of separated A-chains followed by recombination with separated B-chains to yield insulin derivatives e.g. acetyl-insulin[5], arginyl-insulin[6], and des-Gly^{A1}-insulin[7] etc.[8,9]. The yields, however, are poor (less than 3% starting from insulin in general). In the second procedure, native insulin is partially modified with various reagents and Gly^{A1}-modified insulins are isolated by separation of the reaction products. This method has proven suitable for preparing carbamoylated-[10], acylated and aminoacylated[5,8,11-19], or arylated[20] insulins. In the third procedure, N^{B1}-Msc-insulin[21] is selectively reacted at Gly^{A1} with phenylisothiocyanate, and the remaining free B29 amino function then protected with the Msc group. Edman degradation yields des-Gly^{A1}-, N^{B1},N^{B29}-Msc$_2$-insulin. This has been reacted with several amino acids to obtain, for example, D-Ala^{A1}-[21], L-Ala^{A1}-[21] or β-Ala^{A1}-[22] insulins. Selective introduction of carbamoyl or Mtc residues into N^{B1}-Msc-insulin could be used to obtain the corresponding A1-acylated insulins after removal of the Msc protection group and purification[23]. In the fourth procedure, N-protected methionines[24] or the Msc groups[19] were introduced at the B1 and B29 amino groups in N^{A1}-Boc-insulin to yield the corresponding B1,-B29 di-protected insulins after removal of the protection at A1. These derivatives were used to prepare insulins with elongations at A1 or replacements in position A1.

We wish to describe an alternative method of preparation of B1,B29 di-protected insulins which allows selective modification at A1. The preparations of A1-trifluoroacetyl-insulin[17,19],* and A1-citraconyl-insulins* from native insulin has been investigated. The yields for the A1-trifluoroacetyl-insulin preparation exceed 50%[19], thus giving the best yield for the preparation of an A1-protected insulin so far described. The trifluoroacetyl- and the citraconyl-insulin can be further selectively protected at the free B1 and B29 amino functions with Boc groups. Upon removal of the A1 amino protection, N^{B1},N^{B29}-Boc$_2$-insulin can be obtained. This derivative has only one free amino group, which is available for modification.

The aim of this work has been to develop an improved method of preparation of insulin derivatives with modified N termini in the A-chain. Such derivatives will facilitate systematic structure-function studies at A1.

The present paper describes the preparation of N^{B1},N^{B29}-Boc$_2$-insulin from A1-trifluoroacetyl- and A1-citraconyl-insulin. The application of this derivative is demonstrated by the preparation of des-Gly^{A1}-insulin and [A1-guanidinoacetyl]insulin.

* Detailed publication in preparation.

Materials and Methods

Crystalline *bovine insulin* (25 U/mg) obtained from Brunnengräber, Lübeck, or as a gift from Hoechst AG, Frankfurt, was used throughout.

A1-Trifluoroacetyl-insulin was prepared from native insulin as described[19],* and contained less than 1% carbamoylated and deblocked products.

A1-Citraconyl-insulin was prepared from native insulin by reaction with citraconic anhydride (1.2 - 1.5 equivalents) at pH 8.6 and separation of the reaction products by ion exchange chromatography on DEAE-Sephadex C-25 at pH 7.3*.

Boc-Azide was prepared according to Carpino[25].

Phenyltrifluoroacetate was prepared according to Weygand and Röpsch[26] or according to Benoiton[27].

Phenylisothiocyanate was distilled in vacuo in a nitrogen atmosphere before use.

3,5-Dimethyl-1-pyrazolformamidinium nitrate ("Zur Synthese") was obtained from Merck, Darmstadt.

Dimethylsulfoxide ("getrocknet, Zur Analyse"; 0.03% maximal water content) was obtained from Merck, Darmstadt.

Sephadex gels were obtained from Deutsche Pharmacia, Frankfurt.

Sulfoethylcellulose (SE-23) was obtained from Serva, Heidelberg.

Dialysis tubing (Wisking 18/32) was purchased from Fischer, Freiburg.

* Detailed publication in preparation.

Ampholytes for isoelectric focussing were purchased from LKB Producter AB, Bromma.

The following *buffers* were *used for analytical electrophoreses.*
pH 2.2: 4M urea containing 2.4 mol HCOOH/*l*[28].
pH 8.6: 4M urea containing 25 mmol barbital-Na, 25 mmol sodium acetate and 5 g EDTA/*l*.

Cellulose acetate electrophoreses were performed on 180 x 25 mm strips obtained from Schleicher & Schüll, Dassel. Solutions (2 - 5 µ*l*) of the protein in the electrophoresis buffer (ca. 20 mg/m*l*) were applied. The electrophoreses were run for 3 - 5 h at 15 V/cm. Pherograms were stained in a 0.2% Ponceau S/3% trichloroacetic acid solution.

Analytical sulfitolysis was carried out essentially according to Bailey and Cole[29]. 1 mg insulin, 1.5 mg Na_2SO_3 and 1.5 mg $Na_2S_4O_6 \cdot 2\,H_2O$ were dissolved in 50 µ*l* 0.1M phosphate buffer (pH 7) containing 7 mol urea/*l*. After 3 - 5 h reaction time the solutions were applied directly to paper strips (2043b Mgl from Schleicher & Schüll, Dassel) and the electrophoresis run for 7 h at 8 V/cm or for 15 h at 4 V/cm in pH 2.2 buffer. The strips were dried for at least 10 h at room temperature and then stained with Pauly's reagent.

Reaction time curves[19],* were obtained on the samples containing 1 mg insulin taken from the reaction mixture. The product was precipitated with 1.5 vol. of ether/-ethanol/acetic acid 8 : 1 : 1 (v/v), washed with ether and dried in vacuo for 1 h. The product was dissolved in 50 µ*l* pH 2.2 buffer and the cellulose acetate electrophoresis was performed as described above. The stained pherograms were evaluated at 435 nm with a Zeiss PMQ II spectrophotometer equipped with a chromatogram scanner. Staining factors used are published elsewhere[5],*.

Amino acid analyses were performed according to Spackman et al.[30] with a Biochrom analyzer. Hydrolyses were done in evacuated tubes at 105 °C for 24 h using constant boiling HCl containing 1% phenol.

End group determinations were performed according to Hartley[31]. Dansyl amino acids obtained after hydrolysis of dansylated protein with constant boiling HCl at 105 °C were identified by chromatography on 5 x 5 cm polyamide sheets (BDH, Poole) according to Woods and Wang[32].

Chromatography was performed at room temperature, using an Uvicord I detector and Ultrorac fraction collector attached to a recorder (LKB Producter, Bromma).

Conductivity was measured continuously in the eluates with a CDM 2 conductolyzer (Radiometer, Copenhagen) equipped with a micro flowthrough cell.

In vitro biological activities were determined by the free fat cell assay of Moody et al.[33]. Monocomponent porcine insulin (27 U/mg) used as standard was kindly supplied by Novo, Copenhagen.

Preparation and purification of N^{B1},N^{B29} Boc$_2$-insulin

A) From A1-trifluoroacetyl-insulin

A1-Trifluoroacetyl-insulin (1 g; 0.174 mmol) was dried in vacuo over P_2O_5/KOH for 2 days before use. It was dissolved in a mixture of dimethylsulfoxide (20 m*l*) and triethylamine (0.75 m*l*; 5.4 mmol). *Boc-azide* (freshly prepared; 4.5 m*l*) was added and the reaction flask placed in a water bath at 37 °C for 2.5 h. The flask was then placed in an ice-bath and two vol. of ice-cold 1M acetic acid were added with stirring. The turbid solution was adjusted to pH 4.4 and dialysed against deionized water (three 10-*l* portions in 5 h). The resulting slurry was cautiously evaporated in vacuo to half the starting volume. To this, enough solid NH_4HCO_3 was added to give a 0.2M solution, and the pH was adjusted to 9.0 with 25% ammonia solution. After 16 - 20 h at room temperature the solution was chromatographed on Sephadex G-25f (5 x 40 cm) equilibrated with 0.05M NH_4HCO_3. The protein fraction was pooled and lyophilized. The yield was 80 - 90%.

The above product (450 mg) was dissolved in 50 m*l* pH 3 urea buffer (9 mg/m*l*) and applied to a SE-cellulose column (3.8 x 40 cm) equilibrated with the same buffer. Elution was performed at a flow-rate of 10 m*l*/(cm^2 x h) using a linear salt gradient generated by mixing 1 *l* pH 3 urea buffer containing 0.17M NaCl into 1 *l* pH 3 urea buffer. The main peak fractions (Fig. 1) were pooled (150 m*l*) immediately after elution and adjusted to pH 8 with 25% ammonia solution. The yield, after desalting on a Sephadex G-25f column (4.4 x 60 cm; 0.05M NH_4HCO_3) and freeze drying, was 200 mg (44%).

B) From A1-citraconyl-insulin

A1-Citraconyl-insulin (99 mg; 16.6 µmol) was dissolved in dimethylformamide (6 m*l*)/triethylamine (35 µ*l*; 275 µmol). After reaction with *Boc-azide* (0.5 m*l*) for 24 h at room temperature the solvent was concentrated in vacuo to about 2 m*l*. The product was precipitated with 5 vol. of ether, washed with ether by decantation and dried. The residue was taken up in 20 m*l* of 0.1M sodium formate/4M urea buffer (pH 3.5) and stored at 37 °C for 48 h. The solution was desalted on a Sephadex G-25f column (3 x 35 cm) in 1M acetic acid and lyophilized.

* Friesen, H.-J., Weimann, J., Nowak, J. & Brandenburg, D. "Acylation and Deacylation of Insulins: Analytical Investigations and Consequences for the Preparation of Derivatives" in Proceedings of the *Meeting on Semisynthetic Peptides and Proteins* at Brixen/Bressanone (Italy), Academic Press, in prep.

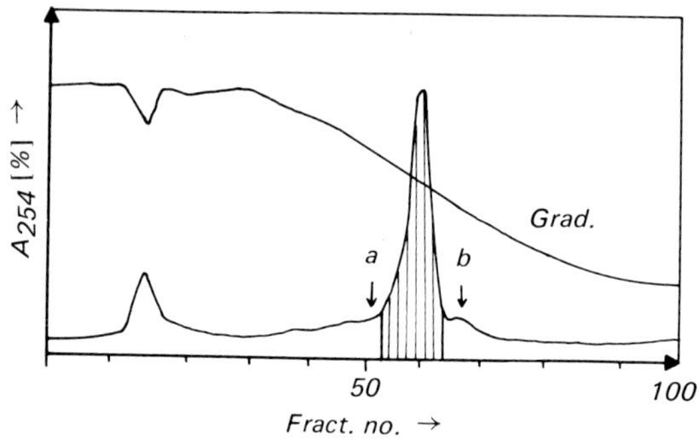

Fig. 1. Purification of Boc_2-insulin on SE-cellulose (3.8 × 40 cm) in 7M urea/1.5M acetic acid buffer (pH 3) with 0 - 0.17M NaCl gradient.

25-m*l* fractions were collected. Expected elution positions relative to the main product for tri- (a) and mono-substituted (b) insulin are indicated by arrows.

Ion exchange chromatography was performed on a CM-cellulose column (2 × 25 cm) equilibrated with 0.1M acetic acid/7M urea solution. The product was dissolved in the same solution and applied to the column which was developed with a linear salt gradient by mixing 300 m*l* of 0.1M acetic acid/0.1M NaCl/7M urea solution into the starting solution. The flow rate was 10 m*l*/(cm^2 × h) and fractions of 6 m*l* were collected. The main fraction was desalted on Sephadex G-25f. The yield was 65 mg (65% overall).

Preparation and purification of des-Gly^{A1}-insulin

Crude *Boc_2-insulin* (210 mg; 36 μmol) was dissolved in 7 m*l* pyridine/water 95 : 5 (v/v) and the solution was flushed with nitrogen for half an hour. *Phenylisothiocyanate* (50 μ*l*; 0.42 mmol) was added and the reaction mixture was kept for 16 h under nitrogen. The solution was evaporated to dryness in vacuo, and the residue was washed with ether before being dissolved in 4 m*l* of absolute dimethylformamide. *Triethylamine* (50 μ*l*, 0.36 mmol) and *phenyltrifluoroacetate* (50 μ*l*; 0.33 mmol) were added and the reaction mixture was stirred for 3 h. The product was precipitated with 15 m*l* ether, collected by centrifugation, washed three times with ether and dried over KOH in vacuo overnight. It was then treated with 3 m*l* of anhydrous trifluoroacetic acid for 1 h. The product was precipitated with 35 m*l* of absolute ether, washed twice with 300 m*l* of ether and dried over KOH overnight. The yield of crude des-Gly^{A1}-insulin was 150 mg (71%).

This material was dissolved in a mixture of 2 m*l* of 6M urea solution and 100 μ*l* of pH 3 - 10 ampholyte (40% solution). The solution was pipetted into a trough of the Sephadex G-75sf gel bed (18 × 18 × 0.2 cm) swollen in 6M urea containing 1.5% ampholyte. The trough was then filled with gel and current was transferred by carbon rods and Whatman 3 MM paper strips soaked in 2N NaOH and 2N H_2SO_4, respectively. Isoelectric focussing was performed by applying 150 V, 200 V and 300 V for 0.5, 1.5 and 14 h, respectively, at 15 °C. Focussed protein appeared as one major and two minor opaque bands. The bands were isolated and suspended in 5 vol. of water. The pH values of the suspensions were found to be 6.05, 5.22 and 4.8. The major (pH 6.05) fraction was adjusted to pH 2 with 1N HCl, chromatographed on Sephadex G-50f (2 × 50 cm) equilibrated with 1M acetic acid, and lyophilized. The yield was 70 mg (33% overall).

For the removal of the ampholyte adsorbed to the protein, the product was subjected to paper electrophoresis in 33% acetic acid (3 h at 8 V/cm). The yield, after elution from the strips and lyophilization, was 30 mg (14% starting from Boc_2-insulin).

Preparation and purification of [A1-guanidinoacetyl]-insulin

The amidination procedure used was essentially as described by Bodanszky et al.[33]. *Boc_2-Insulin* (100 mg; 17 μmol) was dissolved in a mixture of dimethylsulfoxide

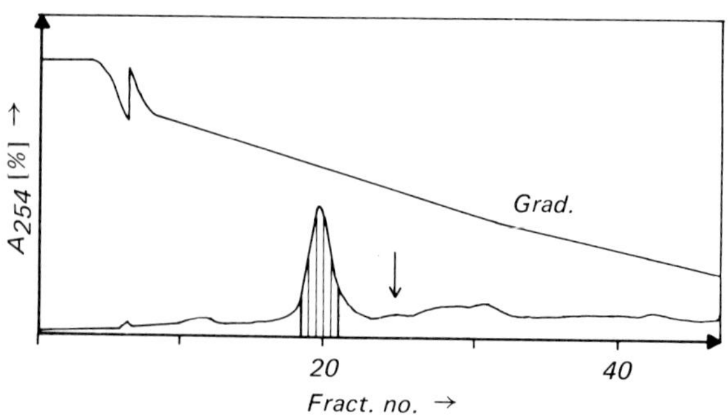

Fig. 2. Purification of [A1-guanidoacetyl]-insulin on DEAE-cellulose (2.5 × 40 cm) in 0.05M Tris/HCl/7M urea buffer (pH 7.3) at 4 °C with 0 - 0.17M NaCl gradient.

25-m*l* fractions were collected. The expected elution position of unmodified insulin is indicated by an arrow.

(10 m*l*) and triethylamine (40 μl; 0.29 mmol). *3,5-Dimethyl-1-pyrazolformamidinum nitrate* (0.8 g; 4 mmol) and triethylamine (350 μl; 2.5 mmol) were added to give a 'pH' 8.6 solution as measured by the glass electrode. After stirring for 18 h at room temperature, the reaction mixture was cooled in an ice-bath and ice-cold 0.05M NH_4HCO_3 (30 m*l*) added with stirring. The resulting clear solution was immediately desalted on a Sephadex G-25f column (5 × 40 cm) equilibrated with 0.05M NH_4HCO_3 and lyophilized. The yield was 80 mg (80%).

The product was treated with 2 m*l* of trifluoroacetic acid for 1 h and then precipitated with 25 m*l* of ether, washed with ether and dried in vacuo over KOH pellets. The product was taken up in 0.05M Tris/HCl/7M urea buffer (pH 7.3), and applied to a DEAE-cellulose column (2.5 × 40 cm) equilibrated with the same buffer at 5 °C. The elution was performed by applying a salt gradient formed by mixing 1 *l* of starting buffer containing 0.14M NaCl into 1 *l* of starting buffer. The main fractions (Fig. 2) were pooled, adjusted to pH 5.5 and dialyzed. They were finally desalted on a Sephadex G-25f column (3.5 × 40 cm) equilibrated with 0.05M NH_4HCO_3, and lyophilized. The yield was 43 mg (43% overall).

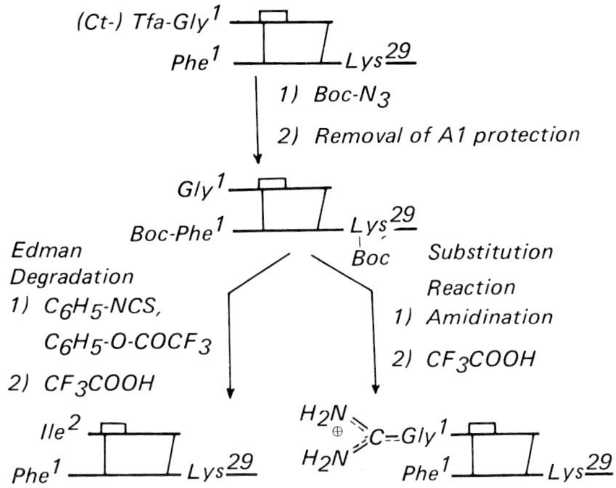

Fig. 3. Diagram showing the routes for the preparations of Boc_2-insulin, des-Gly^{A1}-insulin and [A1-guanidinoacetyl]insulin.

Ct = citraconyl; Tfa = trifluoroacetyl.

Results

The preparation of Boc_2-insulin from A1-trifluoroacetyl- or A1-citraconyl-insulin and its application to the preparation of des-Gly^{A1}-insulin and [A1-guanidinoacetyl]insulin are illustrated in Fig. 3.

Thoroughly dried A1-trifluoroacetyl-insulin was reacted with Boc-azide to yield the tri-substituted derivative (A1-trifluoroacetyl-,N^{B1},N^{B29}-Boc_2-insulin). The kinetics of this reaction were followed (Fig. 4). The formation of more than 97% of tri-substituted derivative after 120 min of reaction was observed. In practice the reaction was carried out for an additional 30 min to ensure more than 99% formation of the tri-substituted derivative.

The A1-trifluoroacetyl group was selectively cleaved from the tri-substituted derivative by overnight exposure to 0.2M NH_4HCO_3/NH_3 (pH 9) solution at room temperature to yield Boc_2-insulin. Low-molecular weight impurities and salt were removed by gel chromatography. The yield of Boc_2-insulin starting from A1-trifluoroacetyl-insulin was 80 - 90% in 6 different experiments. Ion-exchange chromatography of this Boc_2-insulin on SE-cellulose (Fig. 1) showed Boc_2-insulin as the major product (ca. 99%) with traces of mono-substituted material. Boc_2-Insulin was obtained in a yield of 44% after desalting of the product.

The reaction of A1-citraconyl-insulin with Boc-azide was carried out at room temperature overnight. The A1-citraconyl group was subsequently removed by overnight incubation with pH 3.5 buffer at 37 °C. The recovered product was a mixture which was purified on CM-cellulose. Boc_2-Insulin was obtained in 24% yield.

The Boc_2 insulins obtained by both methods were analyzed by electrophoretic methods.

Paper electrophoresis at pH 2.2 after sulfitolysis showed only the band of the unsubstituted A-chain (charge, − 3). The protected B-chain remained at the origin when electrophoresis was carried out at 4 V/cm. At 8 V/cm some of the Boc_2-B-chain moved to the position corresponding to its charge (+ 1). This effect is due to warming of the strips at 8 V/cm which increases the solubility of the protected B-chain.

The Boc_2-insulin obtained via A1-trifluoroacetyl-insulin after chromatography on SE-cellulose was homogeneous (less than 1% minor bands) as shown by pH 2.2 cellulose acetate electrophoresis, while the unpurified product showed traces

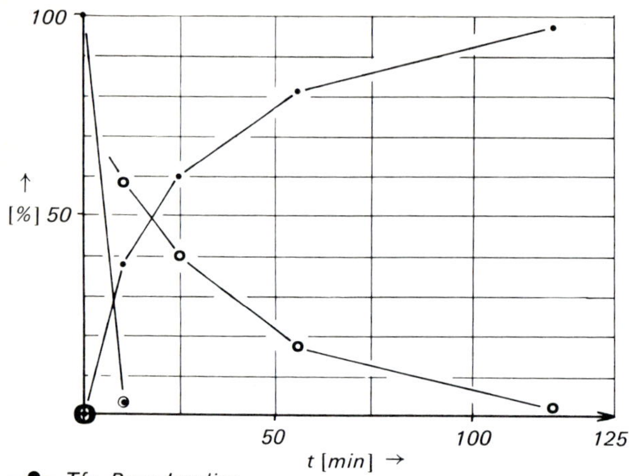

Fig. 4. Time course of the acylation of A1-trifluoroacetyl-insulin with Boc-azide.
The Tfa, Boc-insulin is mainly the A1, B29 derivative[35]. Tfa = trifluoroacetyl.

(ca. 1 - 2%) of monosubstituted insulin. Pherograms of the product at pH 8.6 and after sulfitolysis at pH 2.2 were indistinguishable from those before ion-exchange chromatography. The Boc_2-insulin prepared from A1-citraconyl-insulin contained 10 - 15% mono-substituted derivative before ion-exchange purification as shown by pH 2.2 electrophoresis.

Boc_2-Insulin on treatment with trifluoroacetic acid yielded native insulin which was analyzed by pH 2.2 and 8.6 electrophoresis. The pH 2.2 pherograms indicated traces of *N*-substituted products, which might result from acylation during treatment with anhydrous trifluoroacetic acid.

Des-Gly^{A1}-insulin was obtained from Boc_2-insulin (prepared via A1-trifluoroacetyl-insulin but not purified by ion-exchange chromatography) by Edman degradation. The reaction with phenyl isothiocyanate was carried out as described by Geiger and Langner[36] for N^{A1},N^{B29}-Boc_2-insulin. Any unreacted A1 amino groups were trifluoroacetylated with phenyltrifluoroacetate. The resulting product, upon treatment with trifluoroacetic acid, yielded crude des-Gly^{A1}-insulin, which was purified by preparative isoelectric focussing in a granulated gel[37]. The main band (p*I* 6.05) contained the desired product, while minor bands corresponded to A1-trifluoroacetyl-insulin and desamido-insulins. The des-Gly^{A1}-insulin was separated by chromatography on Sephadex G-50f in 1M acetic acid and lyophilized (yield 33%). However this material still contained some ampholyte, as was shown by pH 2.2 electrophoresis where a brownish stained band migrating twice as fast as the insulin was observed. The ampholyte was finally removed by paper electrophoresis in 33% acetic acid. The pure des-Gly^{A1}-insulin was eluted and lyophilized to give a yield, starting from Boc_2-insulin, of 14%.

End group determination on the des-Gly^{A1}-insulin by the dansyl technique exhibit the spots of dansyl-Phe, dansyl-Ile, dansyl-Ile-Val, *O*-dansyl-Tyr, ε-dansyl-Lys and an extremely faint spot of dansyl-Gly. Pherograms (pH 2.2 and pH 8.6) indicated that the product was homogeneous within the limits of these methods (less than 1% and 3% minor bands, respectively).

Amidination of the Boc_2insulin was carried out in dimethylsulfoxide with Habeeb's reagent[38] 3,5-Dimethyl-1-pyrazolformamidinium nitrate. From the evaluation of the electrophoresis at pH 8.6 after 12 h reaction time, a half-life time of 5.3 h was calculated, assuming the reaction to be first order. The reaction was stopped after 18 h when more than 90% reaction was expected. After removal of the Boc groups, the product was purified on DEAE-cellulose, as shown in Fig. 2. The derivative was pure by all criteria described above for the Boc_2-insulin.

Crystals of the [A1-guanidinoacetyl]insulin suitable for X-ray crystallography were obtained under conditions for growing rhombohedral 2 Zn insulin crystals[39].

The in vitro activities of the derivatives and of insulin regenerated from the Boc_2-insulin are shown in Table 1.

Table 1. The in vitro biological activities of the insulin derivatives. NOVO's monocomponent porcine insulin (27 U/mg) was used as the standard.

	Insulin derivative	Activity [%]
1	$N^{\alpha B1}$,$N^{\epsilon B29}$-Boc_2-Insulin	45 ± 3
2	Des-Gly^{A1}-insulin	2
3	[A1-Guanidinoacetyl]Insulin	88 ± 5
4	Insulin (obtained from 1 by treatment with trifluoroacetic acid)	100

Results of amino acid analysis are shown in Table 2.

Table 2. The diagnostic Gly/Ala ratio obtained after the amino acid analysis of the acid hydrolysate of the insulin derivatives.

Amino acids	Theoretical	Found	
		Des-Gly^{A1}-insulin	[A1-Guanidinoacetyl]Insulin
Gly	3	3.02	3.09[a]
Ala	3	3.00	3.00

[a] Guanidinoacetic acid is partially destroyed during hydrolysis[40]; but does not seem to be converted to glycine.

Discussion

Reversible protecting groups have been used in the preparations of various modified insulins (see Ref.[2] for examples). In these studies trifluoroacetyl and citraconyl groups were selected for the temporary protection of A1 amino groups and Boc protection for B1 and B29 amino groups. This combination allows selective removal of A1 protection in the presence of the Boc groups.

The preparations of A1-trifluoroacetyl- and A1-citraconyl-insulin are one-step reactions which proceed in yields of 45 - 55%[19] and 40% respectively. The introduction of Boc groups at B1 and B29 proceeded in excellent yields. In the case of A1-trifluoroacetyl-insulin the presence of moisture, however, causes partial cleavage of the trifluoroacetyl groups (5 - 15% over a period of 2.5 h) which leads to the formation of the side-product, Boc_3-insulin. This was indicated by cellulose acetate electrophoresis at pH 2.2 after removal of either the trifluoroacetyl or the Boc groups. Therefore this reaction was carried out under strictly anhydrous conditions for only 2.5 h to ensure the stability of the trifluoroacetyl groups. They were removed under mild conditions without disturbing the Boc groups to give pure Boc_2-insulin.

On the other hand Boc_2-insulin obtained via A1-citraconyl-insulin contained 10 - 15% monosubstituted derivative. This monoderivative was formed during pH 3.5 buffer treatment required for the removal of the A1-citraconyl group. In peptide chemistry several examples are known which clearly indicate that the N^{ϵ}-Boc or benzyloxycarbonyl protections are in general more sensitive to acidolytic cleavage than the corresponding N^{α} groups (see Ref.[41] for an example). Due to the requisite purification steps Boc_2-insulin was obtained only in 24% yield, starting from A1-citraconyl-insulin.

Obviously the preparation of Boc_2-insulin from A1-trifluoroacetyl-insulin, which itself can be prepared from native insulin in gram quantities and pure form[19], is the method of choice for large-scale work. The additional purification of this derivative is not necessary in the preparation of those derivatives where the introduced group will affect the net charge of the molecule. This difference will facilitate the separation of the desired product either by ion-exchange chromatography or by isoelectric focussing. Thus, [A1-guanidinoacetyl]insulin was formed by converting the A1-amino group into a guanidino group. The difference in the pK values of these groups resulted in the easy purification of the product. It was homogeneous by electrophoresis, and amino acid analysis agreed with theoretically expected values (Table 2). These results, in combination with the results of ion-exchange chromatography, reveal the chosen amidination procedure to be very specific with no significant side-reactions detectable. So far, only mixtures of products have been isolated after amidination of native insulin[40,42]. The guanidino-insulin reported here is the first that can be considered to be pure within the limits of several now generally accepted criteria.

In other cases, where the modification does not change the charge, the purification of the derivative would be cumbersome if the reaction were not quantitative. In such cases the unreacted derivative should either be separated at this step, or blocked with an acylating reagent. The poor solubility of the di- and tri-substituted derivatives hampers the first possibility. We selected trifluoroacetylation for the second approach. This group remains intact during the acidolytic removal of the Boc groups[43]. The mixture of the deprotected insulin can be separated to remove A1-trifluoroacetyl-insulin from the desired derivative. As an example we have prepared des Gly^{A1}-insulin by this procedure. The isoelectric point found for the des-Gly^{A1}-insulin correlates well with the isoelectric point of native insulin determined in a urea column by Welinder[44]. In addition, amino acid analysis (Table 2), end group determination and electrophoretic data meet expected values, within the limits of the methods. The biological/ activity of the des-Gly^{A1}-insulin preparation (2%) in vitro is in good agreement with values found for similar preparations made via the isolated natural A-chain (1 - 2%)[7], and via synthetic des-Gly^{A1}-A-chain (0.4 - 0.6%)[45].

Although Levy and Carpenter[46] had earlier shown that Boc_3-insulin, upon treatment with trifluoroacetic acid, is quantitatively converted to insulin, we carried out studies under the same conditions. The product, analysed by pH 2.2 and pH 8.6 electrophoresis after the trifluoroacetic acid treatment, showed traces of *N*-substituted material. This might result from acylation during treatment with anhydrous trifluoroacetic acid. Boc_2-Insulin was checked for side products containing *O*-Boc at Tyr (positions A14, A19, B16, B26). Such groups would probably not be removed by trifluoroacetic acid treatment (compare Ref.[47] and[48]. The UV spectra of the trifluoroacetic acid-treated and untreated Boc_2-insulin showed no evidence for such side products.

These studies prove the value of Boc_2-insulin for making specific modifications at A1. We have prepared several additional A1-modified insulins.

Their preparation, circular dichroism spectra, results of crystallization attempts, X-ray data (including the A1-guanidinoacetyl derivative) and biological properties will be presented in separate papers.

This work was supported by the *Bundesministerium für Forschung und Technologie* (Projekt 01 093-B13 SN 0015, formerly SN 1054), the *Minister für Wissenschaft und Forschung des Landes Nordrhein Westfalen*.
Our sincere thanks are due to Mrs. *C. Diaconescu* and Dr. *A. J. Moody* for performing the bioassays and to Dr. *D. Brandenburg* and Prof. *H. Zahn* for their continuous interest in the progress of the experimental work. We thank them and Dr. *D. J. Saunders* for helpful and critical remarks on the manuscript.

Literature

1 Blundell, T. L., Dodson, G. G., Hodgkin, D. C. & Mercola, D. A. (1972) *Adv. Protein Chem.* **26**, 279 - 402.
2 Geiger, R. (1976) *Chem. Ztg.* **100**, 111 - 129.
3 Brandenburg, D. & Ooms, H. A. (1968) in Protein and Polypeptide Hormones (Margoulies, M., ed.) Excerpta Medica Found. Int. Congress Series, vol. 161, pp. 482 - 484.
4 Brandenburg, D. (1969) *Hoppe-Seyler's Z. Physiol. Chem.* **350**, 741 - 750.
5 Brandenburg, D., Gattner, H.-G. & Wollmer, A. (1972) *Hoppe-Seyler's Z. Physiol. Chem.* **353**, 599 - 617.
6 Weinert, M., Kircher, K., Brandenburg, D. & Zahn, H. (1971) *Hoppe-Seyler's Z. Physiol. Chem.* **352**, 719 - 724.
7 Brandenburg, D., Biela, M., Herbertz, L. & Zahn, H. (1975) *Hoppe-Seyler's Z. Physiol. Chem.* **356**, 961 - 979.
8 Krail, G. (1973) Thesis, Techn. Hochschule Aachen.
9 Krail, G., Brandenburg, D. & Zahn, H. (1975) *Hoppe-Seyler's Z. Physiol. Chem.* **356**, 981 - 996.
10 Lindsay, D. G., Loge, D., Losert, W. & Shall, S. (1972) *Biochim. Biophys. Acta* **263**, 658 - 665.
11 Lindsay, D. G. & Shall, S. (1969) *Biochem. J.* **115**, 587 - 595.
12 Lindsay, D. G. & Shall, S. (1970) *Eur. J. Biochem.* **15**, 547 - 554.
13 Lindsay, D.G. & Shall, S. (1971) *Biochem. J.* **121**, 737 - 745.
14 Levy, D. (1973) *Biochim. Biophys. Acta* **317**, 473 - 481.
15 Levy, D. (1973) *Biochim. Biophys. Acta* **322**, 329 - 336.
16 Levy, D. (1973) *Biochim. Biophys. Acta* **328**, 107 - 113.

17 Paselk, R. A. & Levy, D. (1974) *Biochim. Biophys. Acta* **359**, 215 - 221.
18 Nowak, J. (1976) diploma thesis, Techn. Hochsch. Aachen.
19 Friesen, H.-J. (1976) Dr. rer. nat. thesis, Techn. Hochsch. Aachen.
20 Sutton, D. A., Easter, B.R.D. & Drewes, S. E. (1976) *Hoppe-Seyler's Z. Physiol. Chem.* **357**, 1261 - 1266.
21 Geiger, R., Geisen, K., Summ, H.-D. & Langner, D. (1975) *Hoppe-Seyler's Z. Physiol. Chem.* **356**, 1635 - 1649.
22 Geiger, R., Geisen, K., Regitz, G. & Summ, H.-D. (1976) *Hoppe-Seyler's Z. Physiol. Chem.* **357**, 1267 - 1270.
23 Friesen, H.-J., Brandenburg, D., Diaconescu, C., Gattner, H.-G., Naithani, V. K., Nowak, J., Zahn, H., Dockerill, S., Wood, S.P. & Blundell, T.L. (1977) in Proceedings of the 5th American Peptide Symposium (Goodman, M. & Meienhofer, J., eds.), in press.
24 Saunders, D. J. & Offord, R. E. (1977) *Hoppe-Seyler's Z. Physiol. Chem.* **358**, 1469 - 1474.
25 Carpino, L. A. (1957) *J. Am. Chem. Soc.* **79**, 98-101.
26 Weygand, F. & Röpsch, H. (1959) *Chem. Ber.* **92**, 2095 - 2099.
27 Benoiton, L., Rydon, H. N. & Willett, J. E. (1960) *Chem. Ind.*, p. 1060.
28 Du, Y.C., Zhang, Y. S., Lu, Z. X. & Tsou, C. L. (1961) *Sci. Sin.* **10**, 84 - 104.
29 Bailey, J. L. & Cole, R. D. (1959) *J. Biol. Chem.* **234**, 1733 - 1739.
30 Spackman, D. H., Stein, W. H. & Moore, S. (1958) *Anal. Chem.* **30**, 1190 - 1206.
31 Hartley, B. S. (1970) *Biochem. J.* **119**, 805 - 822.
32 Woods, K. R. & Wang, K. T. (1967) *Biochim. Biophys. Acta* **133**, 369 - 370.
33 Moody, A. J., Stan, M. A., Stan, M. & Gliemann, J. (1974) *Horm. Metab. Res.* **6**, 12 - 16.
34 Bodanszky, M., Ondetti, M. A., Birkhimer, C. A. & Thomas, P. L. (1964) *J. Am. Chem. Soc.* **86**, 4452 - 4459.
35 Friesen, H.-J. (1975) diploma thesis, Techn. Hochsch. Aachen.
36 Geiger, R. & Langner, D. (1973) *Hoppe-Seyler's Z. Physiol. Chem.* **354**, 1285 - 1290.
37 Radola, B. J. (1974) *Biochim. Biophys. Acta* **356**, 181 - 195.
38 Habeeb, A. F. S. A. (1972) *Methods Enzymol.* **XXV**, 558 - 566.
39 Schlichtkrull, J. (1956) *Acta Chem. Scand.* **10**, 1455 - 1458.
40 Evans, R. L. & Saroff, H. A. (1957) *J. Biol. Chem.* **228**, 295 - 304.
41 Klostermeyer, H. & Schwertner, E. (1973) *Z. Naturforsch.* **28b**, 334 - 338.
42 Geschwind, J. J. & Li, C. H. (1957) *Biochim. Biophys. Acta* **25**, 171 - 178.
43 Borrás, F. & Offord, R. E. (1970) *Nature (London)* **227**, 716 - 718.
44 Welinder, B. S. (1971) *Acta Chem. Scand.* **25**, 3737 - 3742.
45 Berndt, H., Gattner, H.-G. & Zahn, H. (1975) *Hoppe-Seyler's Z. Physiol. Chem.* **356**, 1469 - 1472.
46 Levy, D. & Carpenter, F. H. (1967) *Biochemistry (Washington)* **6**, 3559 - 3568.
47 Schnabel, E., Herzog, H., Hoffmann, P., Klauke, E. & Ugi, I. (1968) *Liebigs Ann. Chem.* **716**, 175 - 185.
48 Geiger, R., Jäger, G., Volk, A. & Siedel, W. (1968) *Chem. Ber.* **101**, 2189 - 2196.

Dr. H.-J. Friesen, Dr. V. K. Naithani u. Dr. H.-G. Gattner, Deutsches Wollforschungsinstitut, Veltmannplatz 8, D-5100 Aachen.

Hoppe-Seyler's Z. Physiol. Chem.
Bd. 359, S. 113 - 123, Januar 1978

Synthesis and Biological Activity of Two Disulphide Bond Isomers of Human Insulin: [A7-A11,A6-B7-Cystine]- and [A6-A7,A11-B7-Cystine]insulin (Human)

Peter SIEBER, Karel EISLER, Bruno KAMBER, Bernhard RINIKER, Werner RITTEL, Fritz MÄRKI* and Marc DE GASPARO*

Chemische Forschungslaboratorien und
* Biologische Forschungslaboratorien, Ciba-Geigy AG, Basel

(Received 28 October 1977)

Summary: Two unnatural disulphide bond isomers of insulin, [A7-A11,A6-B7-cystine]insulin (human) (II) and [A6-A7,A11-B7-cystine]insulin (human) (III), have been synthesized utilizing a novel strategy developed recently for the total synthesis of human insulin. The synthetic procedure provides an unequivocal structural assignment for II and III.

The two isomers differ distinctly from each other as well as from human insulin in some physico-chemical properties, e.g. behaviour in thin-layer chromatography, electrophoresis and counter-current distribution. Chemical stability at acid and neutral pH is comparable to that of insulin. In 0.1 N ammonia (pH > 10), especially in the presence of mercaptoethanol, II and III are markedly less stable than insulin. Under these unphysiological conditions, varying amounts of insulin are formed in addition to mainly polymeric products.

The biological activity of the isomers is qualitatively indistinguishable from insulin in several test systems. In isolated fat cells in vitro, II and III stimulate glucose oxidation with potencies of 0.14 and 0.17, respectively (insulin = 1), and inhibit lipolysis with potencies of 0.28 and 0.26. In the rat hypoglycaemia test, potencies of II and III are 0.37 and 0.21, respectively. In the mouse convulsion assay, II exhibits a potency of 0.29. Both isomers displace labelled insulin from binding to anti-insulin serum in vitro with potencies of 0.20 and 0.13.

The different physico-chemical behaviour of the two isomers in comparison with insulin suggests alterations of the three-dimensional structure of the molecule. On the other hand, the absence of any qualitative difference in the biological activity between the isomers and insulin, together with the narrow range of potencies observed in the various tests, indicates that essential regions of the tertiary structure of the isomers may be similar to corresponding regions of the insulin molecule.

Key words: Insulin isomers, human insulin, total synthesis, biological activity, anti-insulin serum binding.

Abbreviations:
Acm = acetamidomethyl; Boc = tert. butyloxycarbonyl; Bpoc = 2(biphenyl-4-yl)isopropyloxycarbonyl; Bu^t = tert. butyl; OBu^t = tert. butyloxy; OMe = methoxy; ONp = *p*-nitrophenoxy; ONSu = succinimido-oxy; Hy = hydrazide; Trt = trityl (= triphenylmethyl); DCCI = dicyclohexylcarbodiimide; HOBt = 1-hydroxybenzotriazol; TLC = thin-layer chromatography.

Synthese und biologische Aktivität zweier Disulfid-Isomerer von Humaninsulin: [A7-A11,A6-B7-Cystin]- und [A6-A7,A11-B7-Cystin]Humaninsulin

Zusammenfassung: Unter Verwendung des kürzlich zur Synthese von Humaninsulin ausgearbeiteten Prinzips der stufenweisen und selektiven Bildung der Cystin-Brücken wurden zwei unnatürliche Disulfid-Isomere von Insulin hergestellt, nämlich [A7-A11,A6-B7-Cystin]Humaninsulin (II) und [A6-A7,A11-B7-Cystin]Humaninsulin (III).

In einigen physikalisch-chemischen Eigenschaften, wie z.B. im Verhalten in der Dünnschichtchromatographie, Elektrophorese und Gegenstromverteilung, unterscheiden sich die beiden Isomeren deutlich, sowohl voneinander als auch von Humaninsulin. Die chemische Stabilität in saurer und neutraler Lösung ist vergleichbar mit derjenigen von Insulin. In 0.1 N Ammoniak (pH > 10), besonders in Gegenwart von Mercaptoäthanol, sind II und III deutlich weniger stabil als Insulin. Unter diesen unphysiologischen Bedingungen bilden sich neben vorwiegend polymeren Produkten auch wechselnde Mengen Insulin.

Die biologische Aktivität der Isomeren ist in verschiedenen Testsystemen qualitativ nicht von Insulin unterscheidbar. Die Aktivitäten von II und III bei der Glucoseoxidation an isolierten Fettzellen in vitro betragen 0.14 und 0.17 (Insulin = 1), bei der Hemmung der Lipolyse 0.28 und 0.26. Im Hypoglykämie-Test an der Ratte wurden Aktivitäten von 0.37 und 0.21 gemessen. II zeigt im Mauskrampftest eine Aktivität von 0.29. Bei der Bindung an Anti-Insulinserum – bestimmt durch Verdrängung von markiertem Insulin – besitzen II und III eine Aktivität von 0.20 respektive 0.13.

Das unterschiedliche physikalisch-chemische Verhalten der beiden Isomeren im Vergleich zu Insulin deutet auf eine teilweise veränderte Raumstruktur. Anderseits gibt es keinen qualitativen Unterschied in der biologischen Aktivität zwischen den Isomeren und Insulin, und die Aktivitäten in verschiedenen Tests liegen nahe beisammen. Für die biologische Aktivität essentielle Bereiche der Tertiärstruktur der Isomeren und von Insulin dürften daher weitgehend ähnlich sein.

The insulin molecule contains two polypeptide chains and six half-cystine residues which form three disulphide bridges (Fig. 1). In addition to the natural hormone, eleven further isomers can be constructed by changing the connections between the half-cystines.

In most syntheses of insulin, that have been described so far, the last step is the oxidative combination of the two separate chains (for review see[1]). Large amounts of by-products are formed in this reaction and we may assume that, in addition to di- and oligomers, unnatural monomeric disulphide isomers are also obtained to a certain extent. However, none of the eleven possible isomers has as yet been isolated and characterized chemically or biologically.

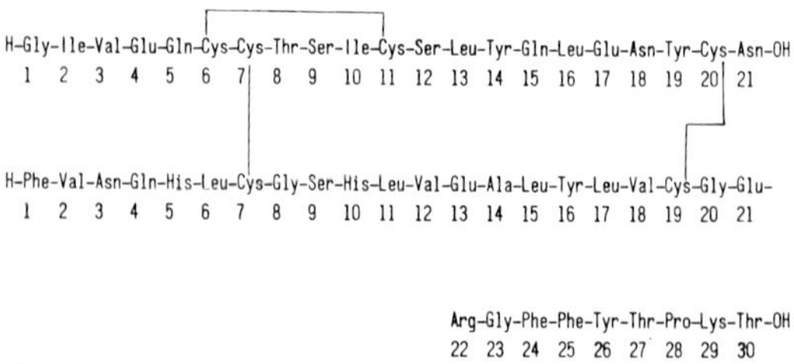

Fig. 1. Human insulin.

Some time ago we published[2] a total synthesis of human insulin (formula, see Fig. 1) whose specific feature is the formation of the three disulphide bonds in separate steps. This synthesis made it possible for the first time to prepare analogues of insulin in which the disulphide bonds are arranged differently as compared with the natural hormone.

We describe here the synthesis and biological activity of two such disulphide isomers of human insulin, i.e. [A7-A11,A6-B7-cystine]insulin (II) and [A6-A7,A11-B7-cystine]insulin (III). They both differ from natural human insulin (I) in the position of two out of three disulphide bridges as represented in Fig. 2.

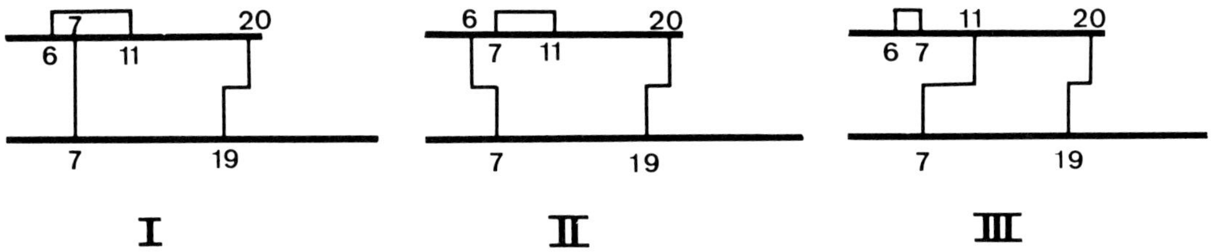

Fig. 2. Schematic structural formulas of insulin (I), [A7-A11,A6-B7-cystine]insulin (II) and [A6-A7,A11-B7-cystine]-insulin (III), showing the arrangement of the three disulphide bonds.

The model of the insulin monomer, as determined by X-ray analysis[3], shows that in the region of the modified disulphide rings the three-dimensional structure of these isomers must differ considerably from natural insulin. Therefore, we considered it of interest to compare the physico-chemical and biological properties of the isomers with those of insulin.

Materials and Methods

Materials

Highly purified bovine serum albumin (ORHD) was obtained from Behringwerke, Marburg, and guinea pig anti-bovine insulin serum (Lot 21) from Pentex, Kankakee, Ill., U.S.A. [1-^{14}C]D-Glucose, spec. act. 2 - 4 mCi/mmol, code no. CFA. 72, was purchased from the Radiochemical Centre, Amersham, Great Britain, and iodinated [^{125}I]-insulin, spec. act. 80 - 110 μ Ci/μg, from Sorin, Saluggia, Italy. Reagents for glucose and glycerol assays were obtained from Boehringer, Mannheim. Synacthen* is a product of Ciba-Geigy. Stock solutions of insulins were prepared in saline (0.154M) acidified to pH 4 and stored frozen at − 20 °C in small aliquots. Dilutions for in vitro assays were made up in the appropriate buffer immediately before use.

Animals

Female rats (Tif: RaIf, SPF) and male mice (Tif: MaGf, SPF) from Tierfarm Sisseln AG, were used at body weights of 170 - 200 g and 25 - 35 g, respectively.

Analytical methods

Thin-layer chromatography and electrophoresis were performed as described earlier[2]. R_F values and migration distances listed in Tables 1 and 2 were obtained by analyzing the various compounds on the same plate in order to produce reliable relative values. The absolute values may vary considerably in different chromatograms or electrophoreses.

Syntheses

The isomer II was prepared as shown in Fig. 3 and 5 using the methods worked out for the synthesis of human insulin[2].

The synthesis of III was planned in such a manner as to allow wherever possible the use of intermediate fragments prepared earlier. For this reason the disulphide bonds A6-A7 and A11-B7 were formed from peptides containing one *S*-trityl and one *S*-acetamidomethyl group with the aid of iodine (s. Fig. 4 and 6). This method will be published in detail later. In order to prevent a partial cleavage of the *S*-trityl group from cysteine A11 by trifluoroacetic acid during the conversion of intermediate **5** to III, the two steps were carried out in reversed order as compared with II: The disulphide ring was first closed by means of iodine and the protecting groups were then removed by trifluoroacetic acid.

Table 1. Properties of new intermediates.

Intermediate	TLC 100[a]	TLC 52A[a]	*K* values
2	0.70	–	0.39[b]
4[c]	–	–	–
3	0.40	0.35	1.4[d]
5	0.57	0.40	1.78[d]

[a] R_F values obtained by thin-layer chromatography on silica gel in solvent systems 100 and 52A[2].
[b] *K* value obtained by countercurrent distribution in the solvent system methanol/buffer (0.05M ammonium acetate, 0.05M acetic acid)/chloroform/carbon tetrachloride 10 : 3 : 8 : 4 (v/v).
[c] Very poorly soluble, purified by trituration in 90% trifluoroethanol. R_F 0.78 obtained by chromatography in solvent system chloroform/trifluoroethanol/conc. aqueous ammonia 50 : 60 : 5 (v/v).
[d] *K* value obtained by countercurrent distribution in the solvent system methanol/20% acetic acid/1,2-dichloroethane/chloroform 9 : 3 : 11 : 3 (v/v).

* Corticotrophin-(1-24)-tetracosapeptide.

Table 2. Properties of II and III in comparison with human insulin (I).

	TLC 112E[a]	TLC 155[a]	EP 1.9[b]	EP 8.6[b]	*K* values[c]
I	0.23	0.12	– 5.3	+ 5.4	0.93
II	0.34	0.20	– 4.5	+ 4.9	0.78
III[d]	0.37	0.24	– 4.6	+ 4.8	1.24

[a] R_F values obtained by thin-layer chromatography on cellulose in solvent system 112E and 155[2].
[b] Mobilities in cm during electrophoresis on cellulose acetate foils, pH 1.9 (90 min) and pH 8.6 (120 min), 20 V/cm.
[c] *K* values obtained by countercurrent distribution in the solvent system: 1-butanol/pyridine/0.1% acetic acid 5:3:11 (v/v).
[d] Aqueous solutions of III at pH 3 have a distinct tendency to form fibrils.

The starting material for the syntheses was the asymmetrical disulphide peptide **1**. It was prepared as published earlier[2] and therefore also contained the D-tyrosine B16 stereoisomer. Although this by-product could not be eliminated from the intermediates **3** or **5**, the separation by countercurrent distribution from the end-products II and III presented no problems. The intermediates **2** and **4** with the preformed unnatural disulphide rings A7-A11 and A6-A7, respectively, were each coupled with **1** using the HOBt/DCCI method[4] to produce the bis-disulphides **3** and **5**. The syntheses of **2** and **4** were carried out from the intermediates described in[5] as shown in Fig. 5 and 6. The properties of the new intermediates are given in Table 1, those of the end-products in Table 2.

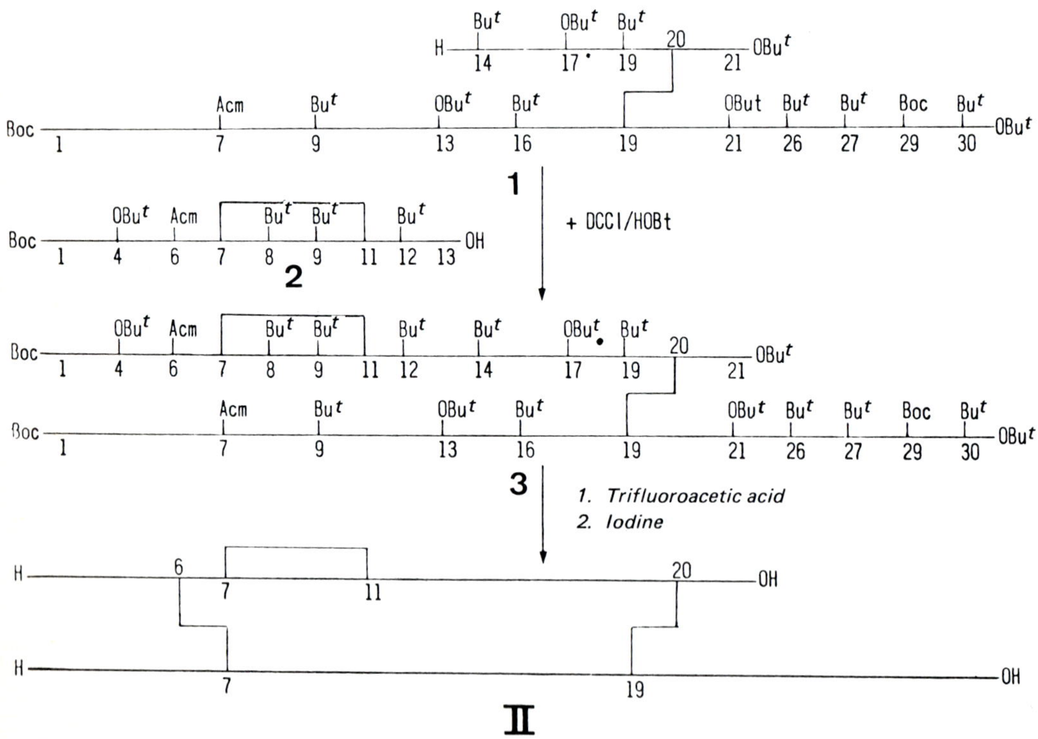

Fig. 3. Scheme of the synthesis of human [A7-A11,A6-B7-cystine]insulin (II).

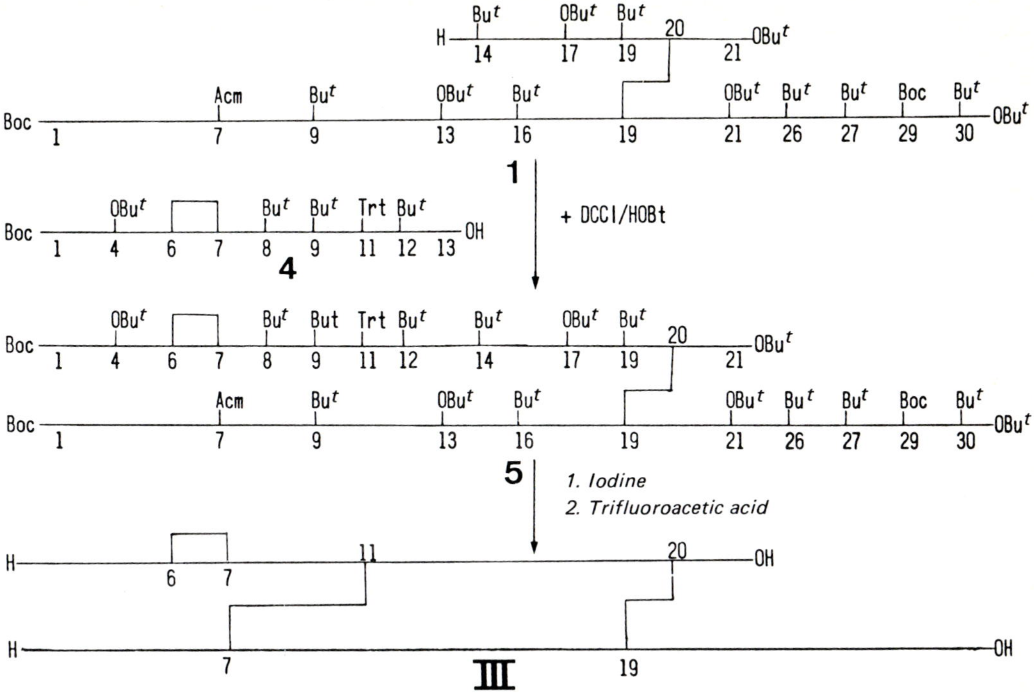

Fig. 4. Scheme of the synthesis of human [A6-A7,A11-B7-cystine]insulin (III).

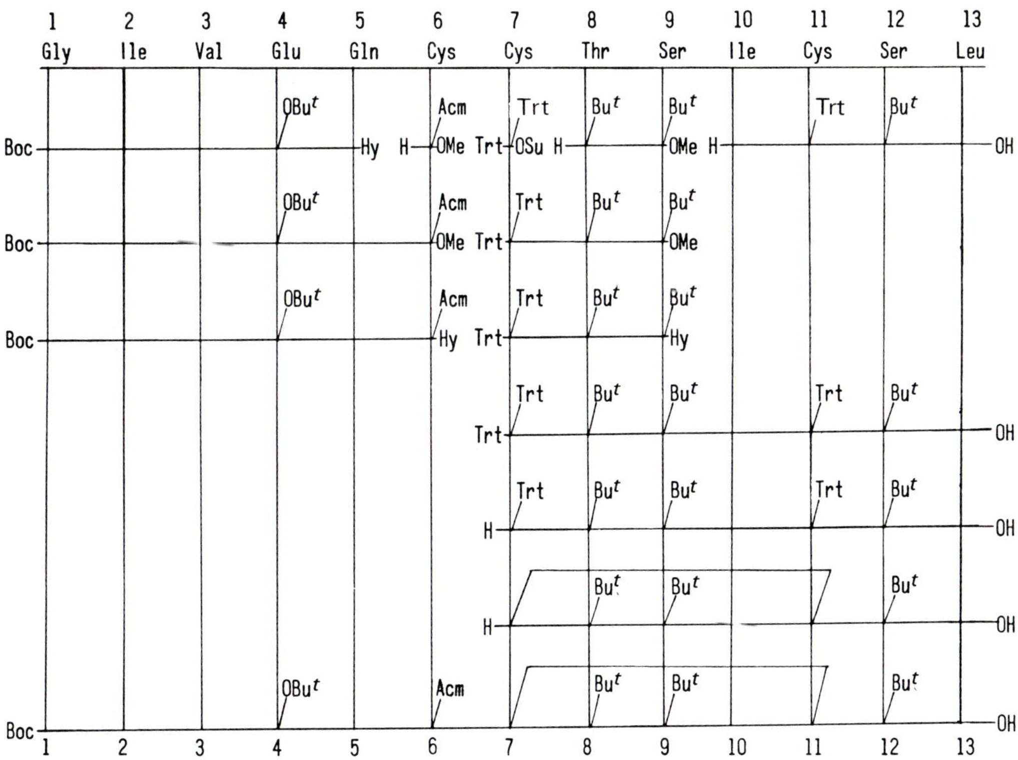

Fig. 5. Scheme of the synthesis of the protected tridecapeptide A1-13 (**2**).

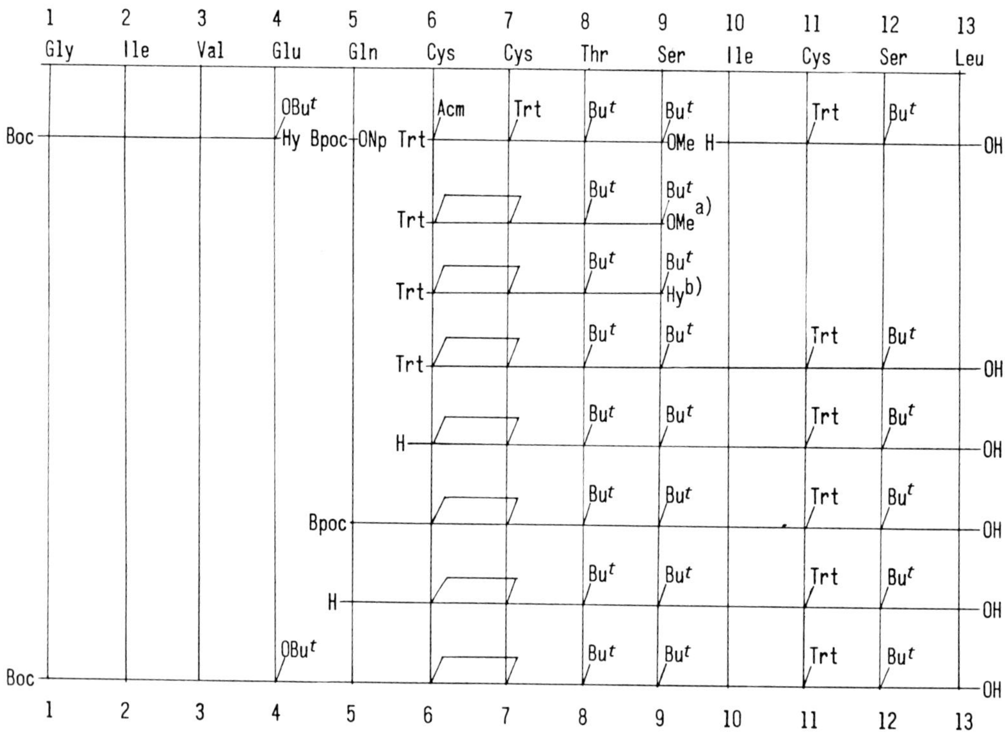

Fig. 6. Scheme of the synthesis of the protected tridecapeptide A1-13 (4).

a The iodine oxidation was carried out at pH 3.5 to prevent cleavage of the *N*-trityl group.

b Partial disulphide exchange to dimeric material occurred during the hydrazinolysis.

Stability tests. Partial conversion of II and III to insulin (I)

a) In acidic and neutral solution: 50 μg samples were dissolved in 10 μl of 0.5N acetic acid, 95% trifluoroacetic acid or aqueous pyridine acetate (pH 7.0) and left at room temperature for the indicated time (cf. Table 3). The samples were evaporated and treated for 15 min with trifluoroacetic acid in order to reconvert the fibrils into soluble material. Especially III has a tendency to form fibrils. After evaporation of the trifluoroacetic acid the samples were dissolved in 10 μl of 0.5N acetic acid. 3 μl of these solutions were applied in bands on a cellulose plate and developed in solvent system 112E[2]. The amount of insulin formed was determined by comparison with serial dilutions of human insulin. The results are reported in Table 3.

b) In 0.1N ammonia: 50 μg samples were dissolved in 10μl of 0.1N ammonia and left for 2 and 20 h. The samples were evaporated and then treated further as in a.

c) In 0.1N ammonia containing 0.25 mol of 2-mercaptoethanol per mol insulin or isomer: 50 μg samples were dissolved in 5 μl of a solution containing 2.4 μl of mercaptoethanol in 100 ml of 0.1N ammonia. The solution was prepared immediately before use, as the SH-content decreases quickly. The samples were evaporated after 2 or 60 min and then treated further as in a.

Biological assays

a) Glucose oxidation and lipolysis in fat cells

Fat cells were isolated from rat epididymal fat pads by the collagenase technique of Rodbell[6] with minor modifications. The washed cells (approximately 10 μmol triglyceride per flask) were suspended in Krebs-Ringer hydrogen carbonate buffer (2.5mM $Ca^{2\oplus}$) containing bovine serum albumin and glucose (2% and 0.01%, respectively, for glucose oxidation experiments and 4% and 0.1%, respectively, for lipolysis experiments). Oxidation of [1-^{14}C]glucose was measured according to Rodbell[6]. Several dose-response curves covering the full range from zero to maximal stimulation with 8 different concentrations each were run for human insulin and the isomers. The concentrations required to elicit 50% of the maximal effect (EC_{50}) were determined graphically from these curves.

Table 3. Stability of I, II and III under different conditions[a].

Conditions	I	II	III
0.5 N acetic acid, 3 d	**unchanged**	**unchanged**	**unchanged**
95% trifluoroacetic acid 6 h	**unchanged**	**unchanged**	**unchanged**
aqueous pyridine acetate pH 7.0, 3 d	**unchanged**	**unchanged**	**unchanged**
0.1 N ammonia 2 h	**unchanged**	~ **80% start. mat.** ~ **4% I** some polymers[b]	~ **80% start. mat.** ~ **4% I** some polymers[b]
20 h	**unchanged**	~ **40% start. mat.** ~ **10% I** traces of by-prod., remainder polymers[b]	~ **70% start. mat.**[c] ~ **4% I** traces of by-prod., remainder polymers[b]
0.1 N ammonia with 2-mercaptoethanol 2 min	~ **90% start. mat.** traces of by-prod., some polymers[b]	~ **15% start. mat.** ~ **20% I** some by-prod., remainder polymers[b]	not carried out
60 min	~ **80% start. mat.** ~ 5% by-prod., remainder polymers[b]	~ **5% start. mat.** ~ **30% I** some by-prod., remainder polymers[b]	~ **20% start. mat.**[c] ~ **20% I** some by-prod., remainder polymers[b]

[a] For experimental details see Methods.
[b] Remaining at the starting point in thin-layer chromatography.
[c] The reaction stops because III forms fibrils. Treatment with trifluoroacetic acid regenerates the starting material.

Lipolytic activity was assayed as described by Müller and Maier[7]. Inhibition of lipolysis by insulins was determined in cells stimulated submaximally with 0.03 µg of Synacthen/m*l* (corticotrophin-(1-24)-tetracosapeptide). Three concentrations each of insulin and the isomers in the range bracketing 50% inhibition were used and EC_{50} values were determined graphically.

b) Hypoglycaemia test in the rat

Human insulin and isomer were injected s.c. in two doses each into groups (N = 6) of rats fasted for 24 h. Blood glucose was assayed enzymatically (glucose oxidase/peroxidase method) on blood obtained retroorbitally in light anaesthesia (oxygen/carbon dioxide 1 : 1) before and 1, 2, 3, and 5 h after injection. The area under the blood glucose curve was used to calculate the potency of the isomer relative to human insulin by a four point parallel assay. The results of two separate experiments were pooled.

c) Mouse convulsion assay

The procedure used was modified from[8,9]. Human insulin and the isomer were injected s.c. in three different doses each into mice (N = 15) fasted for 10 h and kept at 31 - 33 °C during the experiment. Animals showing convulsions were immediately removed from the cages. Calculation of the potency relative to human insulin was done – based on the numbers of mice convulsing at the different doses – by a quantal response assay (probit analysis)[8,9].

d) Binding to anti-insulin serum

Binding of the isomers to an anti-bovine insulin serum from guinea pigs was determined by competition with [^{125}I]insulin (pork) under the experimental conditions used for the radioimmunoassay of insulin[10]. Antiserum at a final dilution of 1 : 1.25 × 10^6 was incubated with 50 pg of tracer [^{125}I]insulin and various concentrations of isomer or human insulin (0.5 - 10 ng/m*l*) for four days at 4 °C. Free and bound insulin were separated by the dextran-coated charcoal method of Herbert et al.[11]. The concentration of insulin or isomer displacing 50% of the labelled insulin from binding to the anti-serum was calculated after logit (B/B_0)/log transformation[12] and used as a measure of the binding affinity.

Results

The structures of II and III are unequivocally determined by the method of synthesis used, since no disulphide exchange occurred during the conversion of the unsymmetrical cystine derivatives **3** and **5** (Fig. 3 and 4) into the disulphide isomers of human insulin or during their purification by countercurrent distribution. II and III gave the expected values in the amino acid analysis, and were eluted during gel filtration on Sephadex G-75 in 1 M acetic acid with the same elution volume as human insulin.

The isomers II and III exhibit remarkable physicochemical differences from the natural hormone I and to a smaller extent also from each other (see Tab. 2). They are considerably more lipophilic than I during thin-layer chromatography, and isoelectric focussing reveals distinctly lower isoelectric points. Attempts to crystallize the isomers in the form of zinc complexes have failed so far.

The isomers II and III are stable under neutral and acidic conditions, as can be seen in Table 3. However, in 0.1 N ammonia (pH $\approx$ 10.5), especially in the presence of small amounts of 2-mercaptoethanol, they are unstable in contrast to insulin. Under these rather drastic and unphysiological conditions mainly unidentified polymeric material is formed. In addition, insulin (I) is also obtained in variable amounts, as indicated in Table 3.

The biological activity of the two insulin isomers was studied in several test systems in vitro and in vivo in comparison with synthetic human insulin. The latter – prepared by total synthesis as described previously[2] – was used as reference compound, since it has been shown elsewhere[13] to be biologically fully equivalent to natural human and porcine insulin.

In isolated rat fat cells in vitro, both isomers caused stimulation of glucose oxidation and inhibition of lipolysis. They were less potent than human insulin. As shown in Table 4, the isomer II had a relative potency of 0.14 and 0.28 in glucose oxidation and lipolysis, respectively, while III had a potency of 0.17 and 0.26 (human insulin = 1). The dose-response curves of II and III were superimposable on that of insulin. Furthermore, high concentrations of the isomers (about 20 times EC_{50}) produced the same maximum response on glucose oxidation as human insulin, thus indicating that II and III are full agonists.

In the hypoglycaemia test in the rat, the activities observed were of the same order of magnitude as those found in vitro. As indicated in Table 4, the isomer III had only about half the potency of II. The time course of the hypoglycaemic effect of III is shown in Fig. 7 to match exactly that of an equipotent dose of human insulin.

Table 4. Biological activity of the isomers II and III in comparison with human insulin (I).

Test	I		II		III	
	EC_{50}[a]	Rel. potency[b]	EC_{50}[a]	Rel. potency[b]	EC_{50}[a]	Rel. potency[b]
Fat cell, glucose oxidation (+)	0.45 ± 0.01 (n = 4)	1	3.29 ± 0.54 (n = 5)	0.14	2.64 ± 0.27 (n = 5)	0.17
Fat cell, lipolysis (–)	0.46 ± 0.08 (n = 5)	1	1.64 ± 0.28 (n = 5)	0.28	1.75 ± 0.38 (n = 4)	0.26
Rat, hypoglycaemia test		1		0.37 (0.27 - 0.50)		0.21 (0.17 - 0.27)
Mouse, convulsion assay		1		0.29 (0.25 - 0.35)		n.d.
Anti-insulin serum binding (–)	0.94 ± 0.04 (n = 15)	1	4.77 ± 0.23 (n = 5)	0.20	7.27 ± 0.19 (n = 5)	0.13

[a] Concentration for half-maximal stimulation (+) or 50% inhibition (–); ng/ml. Mean + S.E.M., number of experiments in parentheses.
[b] Potency relative to human insulin = 1. Mean, 95% confidence limits in parentheses.
n.d. not determined.

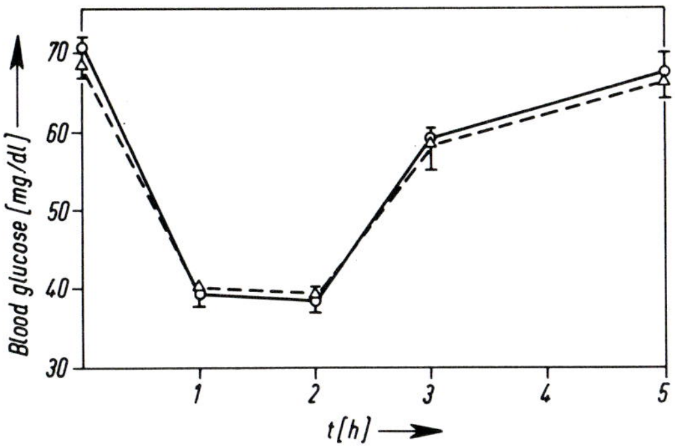

Fig. 7. Time course of the hypoglycaemic effect of human insulin (I) and human [A6-A7,A11-B7-cystine]insulin (III) in the rat.
○——○: I (20 µg/kg); △——△: III (100 µg/kg). Each symbol with the vertical bar represents the mean ± standard error of the mean of 12 animals.

The isomer II was also studied in the mouse convulsion assay and had a relative potency of 0.29, i.e. similar to that in the hypoglycaemia test in the rat.

No major difference was observed between II and III with respect to anti-insulin serum binding in vitro. The potencies relative to human insulin were in the same range as those of the other biological effects (Tab. 4).

Discussion

The synthetic principle of stepwise and selective formation of the three disulphide bridges, which was first used for the synthesis of insulin itself[2], is also applicable for disulphide isomers of the hormone. The formation of the last disulphide bridges of the isomers (A6-B7 in II, and A11-B7 in III) was performed with similar ease and yield as in normal insulin. This had in fact been anticipated, since the disulphide pairing is based on the chemical reaction between two specific derivatives of cysteine, and does not depend on conformational conditions of the precursor molecules **3** and **5**. For the synthesis of III, both unnatural disulphide bridges were formed by oxidation with iodine from two cysteine residues, one being protected by trityl and the other one by acetamidomethyl. The details of this newly developed method will be published separately.

The two isomers II and III can easily be distinguished from insulin I in several analytical procedures. In thin-layer chromatography and countercurrent distribution they display distinctly more lipophilic character. This points to a higher exposition of apolar amino-acid side chains on the outside of the molecules as compared to I. The shift of the isoelectric points of II and III towards lower values is also an indication of conformational changes.

Most noteworthy is the fact that II and III, under conditions rendering possible a partial disulphide exchange, are significantly less stable than I and are partially isomerized to the latter. This instability could be due to an enhanced accessibility of the disulphide bonds in II and III. The cónversion of the isomers to insulin may also be promoted by a higher thermodynamic stability of insulin, if its arrangement of the disulphide bonds allows the molecule to assume a conformation with a lower free energy than that of the isomers. In view of the fact that the three disulphide bonds of I are biosynthetically formed at the stage of the precursor protein preproinsulin[14], this finding was not a priori to be expected. It is, however, in accordance with the chain-recombination experiments of insulin, where higher than statistical yields of the correct cystine pairing have been reported[1].

The biological activity of the insulin isomers II and III is surprisingly high, considering that the modified arrangement of two disulphide bridges induces certain changes in the three-dimensional structure of the molecule. Relative potencies of 0.13 to 0.37 were observed, depending on the test system studied (insulin = 1). This contrasts with some other new insulin analogues, in which minor changes (e.g. replacement of L-cysteine in position A7 by its D-stereoisomer) caused almost complete loss of biological activity (rel. potency around 0.002)*. We had therefore to consider the possibility that the activities observed with the isomers might have been caused by the presence of insulin. A contamination by insulin or a non-enzymatic conversion of the isomers to insulin in the solutions used for biological testing can be clearly ruled out on the basis of the analytical

* To be published.

and stability data presented above. A non-enzymatic formation of insulin was observed only under highly unphysiological conditions as specified above (Table 3). Enzyme-catalyzed isomerisation, however, which might occur in the presence of tissue in the course of the biological assays, cannot be completely excluded, although it appears unlikely. Enzymes catalyzing disulphide exchange have been detected in various animal tissues[15] and isolated from beef liver microsomes[16]. When insulin was used as substrate for the latter enzyme extensive disulphide interchange occurred and insulin immunoreactivity was lost[16], suggesting that in our experiments insulin is unlikely to be formed from one of the isomers by the action of this enzyme. Three additional arguments tend to rule out enzyme-catalyzed isomerisation: 1) Biological potencies of II or III are nearly identical in several in vitro and in vivo tests. If these activities were due to insulin, the rate and extent of the conversion would have to be the same in all tests. 2) When insulin or the isomers are preincubated with the fat cells for 2 h at 37 °C – prior to the assay of glucose oxidation as described above – they exhibit the same potencies as in the usual test without preincubation. Therefore, no enzymatic isomerization of these peptides by fat cells can be detected under the conditions of the assay. 3) The time course of the hypoglycaemic effect of III in the rat does not differ from that of insulin. Isomerisation of III to insulin following injection would most probably result in a delayed onset and a prolonged duration of action compared with insulin itself.

The finding that the biological potencies of both isomers in the various test systems studied, i.e. glucose oxidation and antilipolysis in fat cells, hypoglycaemic effect in the rat, mouse convulsion assay and anti-insulin serum binding, are all remarkably similar, is somewhat unexpected. A comparison of published data on a number of insulin analogues reveals that a dissociation of potencies in different biological tests is rather the rule than the exception; e.g. in the series of insulin analogues studied by Brandenburg et al.[17], potencies in various tests differed by factors of as much as 10 - 15. Particularly great dissociations are usually observed between metabolic effects on the one hand and antibody binding on the other hand[17]. In fact, distinctly different regions of the insulin molecule seem to be involved in receptor binding and in antibody binding[18,19]. For both types of effects, the tertiary structure appears to be critical[18,20,21]. Although both isomers contain the amino acid residues considered by Pullen et al.[18] to constitute the receptor-binding region of insulin, i.e. A1 Gly, A5 Gln, A19 Tyr, A21 Asn, B12 Val, B16 Tyr, B22 Arg, B23 Gly, B24 Phe, B25 Phe and B26 Tyr, it appears unlikely that all these residues have the same spatial arrangement relative to each other as in insulin. So far, a direct investigation of the three-dimensional structure of the isomers by X-ray diffraction analysis is not feasible, as both compounds have not yet been obtained in crystalline form. In the absence of precise information on the topochemical structure, one can only speculate on the influence of deviations from the proposed receptor binding region of insulin on biological activity. The comparatively narrow range of potencies in different tests which we observed with the insulin isomers II and III, together with the fact that only quantitative – but never qualitative – differences from insulin were observed, would suggest that essential regions of the tertiary structure of these isomers may be closely similar to corresponding regions of the insulin molecule. A more detailed knowledge of the three-dimensional structure of these isomers is desirable, as it could contribute to a deeper insight into the structure-activity relationship of insulin.

We thank Mrs. *N. Dietrich*, Mrs. *J. Seeberger*, Mr. *J. Frei* and Mr. *E. Manz* for skilful technical assistance, Mrs. *G. Hostetter*, Ms. *J. Motz* and Mr. *R. Lindecker* for careful performance of the biological assays. The analytical work was kindly performed under the guidance of Mr. *E. von Arx* by Mr. *M. Faupel* and Mr. *W. Morgenthaler*.

Literature

1 Lübke, K. & Klostermeyer, H. (1970) *Adv. Enzymol.* **33**, 445 - 525.

2 Sieber, P., Kamber, B., Hartmann, A., Jöhl, A., Riniker, B. & Rittel, W. (1977) *Helv. Chim. Acta* **60**, 27 - 37.

3 Blundell, T.L., Dodson, G.G., Hodgkin, D.C. & Mercola, D.A. (1972) *Adv. Protein Chem.* **26**, 279 - 402.

4 König, W. & Geiger, R. (1970) *Chem. Ber.* **103**, 788 - 798.

5 Sieber, P., Kamber, B., Eisler, K., Hartmann, A., Riniker, B. & Rittel, W. (1976) *Helv. Chim. Acta* **59**, 1489 - 1497.

6 Rodbell, M. (1964) *J. Biol. Chem.* **239**, 375 - 380.
7 Müller, K. & Maier, R. (1976) *Horm. Metab. Res.* **8**, 271 - 273.
8 British Pharmacopoeia 1963, appendix XV C, pp. 1129 - 1130. London: The Pharmaceutical Press, 1963.
9 Burn, J.H., Finney, D.J. & Goodwin, L.G. (1952) in Biological Standardization, p. 194, Oxford University Press, London.
10 Yalow, R.S. & Berson, S.A. (1960) *J. Clin. Invest.* **39**, 1157 - 1175.
11 Herbert, V., Lau, K.S., Gottlieb, C.W. & Bleicher, S.J. (1965) *J. Clin. Endocrinol. Metab.* **25**, 1375 - 1384.
12 Rodbard, D., Bridson, W. & Rayford, P.L. (1969) *J. Lab. Clin. Med.* **74**, 770 - 781.
13 Märki, F. & Albrecht, W. (1977) *Diabetologia* **13**, 293 - 295.
14 Chan, S.J., Keim, P. & Steiner, D.F. (1976) *Proc. Natl. Acad. Sci. USA* **73**, 1964 - 1968.
15 Fuchs, S., De Lorenzo, F. & Anfinsen, C.B. (1967) *J. Biol. Chem.* **242**, 398 - 402.
16 Givol, D., De Lorenzo, F., Goldberger, R.F. & Anfinsen, C.B. (1965) *Proc. Natl. Acad. Sci. U.S.A.* **53**, 676 - 684.
17 Brandenburg, D., Gattner, H.-G., Weinert, M., Herbertz, L., Zahn, H. & Wollmer, A. (1971) in Diabetes, Proc. Seventh Congr. Int. Diab. Fed., Int. Congr. Series **231**, 363 - 376. Excerpta Medica, Amsterdam.
18 Pullen, R.A., Lindsay, D.G., Wood, S.P., Tickle, I.J., Blundell, T.L., Wollmer, A., Krail, G., Brandenburg, D., Zahn, H., Gliemann, J. & Gammeltoft, S. (1976) *Nature (London)* **259**, 369 - 373.
19 Wilson, S. (1969) in Diabetes, Proc. Sixth Congr. Int. Diab. Fed., Int. Congr. Series **172**, 403 - 405, Excerpta Medica Foundation, Amsterdam.
20 Berson, S.A. & Yalow, R.S. (1961) *Nature (London)* **191**, 1392 - 1393.
21 Berson, S.A. & Yalow, R.S. (1963) *Science* **139**, 844 - 845.

P. Sieber, K. Eisler, B. Kamber, B. Riniker, W. Rittel, F. Märki und M. de Gasparo, Forschungslaboratorien der Division Pharma der Ciba-Geigy AG.,
Klybeck, CH-4002 Basel, Schweiz.

Hoppe-Seyler's Z. Physiol. Chem.
Bd. 359, S. 125 - 127, Januar 1978

SHORT COMMUNICATION

Stringent Factor and Ribosomes are Present in Equimolar Amounts in Cells from a Stringent Strain of *Escherichia coli*

Hartwig SCHMALE, Susanne FEHR and Dietmar RICHTER

Institut für Physiologische Chemie, Abteilung Zellbiochemie, Universität Hamburg

(Received 28 November 1977)

Äquimolares Verhältnis von Pyrophosphotransferase und Ribosomen in Escherichia coli

Zusammenfassung: Zellen des *Escherichia-coli*-Stammes CGSC 2834/a enthalten 10 000 bis 12 000 Moleküle Pyrophosphotransferase; damit liegt dieses Enzym in etwa äquimolaren Verhältnissen zu den Ribosomen vor.

Key words: Stringent factor, ribosomes, stringent factor/ribosomes ratio, (p)ppGpp synthesis.

It is generally accepted that guanosine 5′-diphosphate, 3′-diphosphate (ppGpp) and guanosine 5′-triphosphate, 3′-diphosphate (pppGpp) inhibit synthesis of stable RNA when stringent strains of *Escherichia coli* are deprived of an essential amino acid[1]. In vitro, (p)ppGpp is synthesized on ribosomes when the latter are not actively engaged in the translational cycle. The reaction is catalyzed by the stringent factor, an enzyme that transfers pyrophosphate from ATP to GTP or GDP[1]. Biosynthesis of (p)ppGpp is initiated only when uncharged tRNA is bound to the ribosomes in a codon-specific manner[2], the binding being stringent-factor dependent[3]. It has been calculated that only one molecule of stringent factor is bound per every 500 ribosomes, or that 10 to 20 molecules of stringent factor are present per cell[4], which may reflect that only a few ribosomes carry stringent factor. One might expect a device, known to sense any shortage of charged tRNA immediately, to be present in a larger quantity. It would certainly be more effective if all ribosomes carried stringent factor. Alternatively, limited amounts of stringent factor could exist freely available to any ribosome engaged in guanosine polyphosphate synthesis.

To test one or the other possibility, it was necessary first of all to find appropriate conditions to assay stringent factor in crude cell extracts. So far the reported calculations on stringent factor concentrations per cell were based on numbers obtained from crude ribosomes; isolation steps preceding the ribosome preparation step were not taken into account, mainly because (p)ppGpp synthesizing activity was not detectable in crude cell extracts (ref.[4] and Figure, open squares). However, when complemented with 50S, 30S ribosomal subunits, poly(U) and $tRNA^{Phe}$, (p)ppGpp is synthesized at low protein concentrations of crude cell extract, indicating that the ribosome-mRNA-tRNA complex is a limiting step in this reaction. Less than 95 fmol of ribosomes was calculated per μg protein of cell extract, a number far too low to trigger (p)ppGpp formation in vitro. The inhibition observed at higher protein concentrations is not due to a lack of ATP and/or GTP; although more than 50% of the two nucleotides are hydrolyzed by enzymes present in the crude cell extract, the remaining amount should be still sufficient for (p)ppGpp production (unpublished). It appears more likely that at higher protein concentrations of crude cell extract, tRNA is

Abbreviation: 1 A_{578} unit is the amount of material in 1 m*l* giving an absorbance of 1 in a 1-cm light path at 578 nm.

charged and thus effectively inhibits synthesis of (p)ppGpp by competing with uncharged tRNA for the ribosomal binding site. As shown in the figure, protein concentrations inhibitory to (p)ppGpp synthesis are active in charging tRNA and in poly(Phe) synthesis.

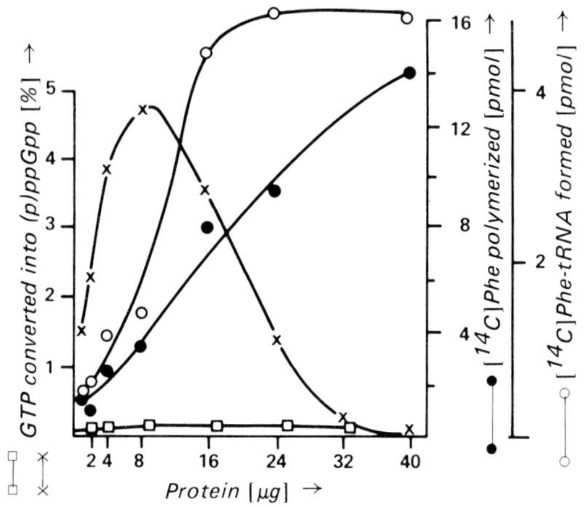

Figure. (p)ppGpp production, aminoacylation of tRNA and protein synthesis in crude cell extracts from *E. coli* (CGSC 2834/a).

Cells were grown at 37 °C to a density of 0.5 A_{578} units/m*l* in 1 *l* medium, poured onto crushed ice buffer, centrifuged and resuspended in 12 m*l* of buffer with 10 µg deoxyribonuclease/m*l*[1]. Cells were disintegrated by two periods of sonication 6 s each with 1 min for cooling. The homogenate was centrifuged for 15 min at 3 200 rpm in a Sorvall SS-34 rotor to remove intact cells. The supernatant fraction was used as crude cell extract and immediately assayed. The (p)ppGpp synthesizing assay in the ribosomal system was the same as described[5]. The assay mixture for aminoacylation and poly(Phe) synthesis was the same as for the (p)ppGpp synthesizing system, except that 550 pmol of [^{14}C]-phenylalanine was present. Incubation was carried out for 1 h at 37 °C; the reaction was stopped with 1 m*l* of 10% trichloroacetic acid. Charged tRNA was isolated by Millipore filtration. Phenylalanine incorporation into protein precipitable with hot trichloroacetic was determined by heating the mixture for 15 min at 92 °C. The precipitate was isolated by filtration.

Formation of (p)ppGpp in the presence of crude cell extract □——□; in the presence of crude cell extract plus 50S, 30S, poly(U), tRNA^{Phe}[5] ×——×; charging of tRNA with [^{14}C]phenylalanine in the presence of crude cell extract plus 50S, 30S, poly(U); tRNA^{Phe}[5] ○——○; poly(Phe) synthesis with crude cell extract plus 50S, 30S, poly(U) and tRNA^{Phe}[5] ●——●.

Table. Molar ratio of stringent factor and ribosomes at various preparation steps.

Crude cell extract was prepared and analyzed as described in the legend to the figure. Stringent factor was released from the ribosomes by dissociation of 70S ribosomes into subunits[5]; stringent factor was purified according to a modified method[5] and was more than 85% pure.

Preparation step	nmol stringent factor per nmol of 70S ribosomes
Crude cell extract	0.92
Released stringent factor	0.024
Purified stringent factor	0.001

Based on the assumption that under our experimental conditions 1 mg of purified stringent factor synthesizes 4000 nmol of (p)ppGpp per minute[4-6], and that crude stringent factor has the same specific activity as the purified preparation, it is calculated that up to 10 to 11 nmol of stringent factor and 12 to 18 nmol of ribosomes exist per gram of cells. This corresponds to 10000 to 12000 molecules of factor per cell (Table). Upon further purification, in particular when stringent factor was released from the ribosomes and further purified, a large decrease in the stringent factor-ribosome ratio was observed. This shift is most likely due to loss of material, a result obtained also by other groups[4-6].

The data presented here indicate that in analogy to elongation factor T and Ts[7], stringent factor is present in an almost equimolar ratio to ribosomes. The equimolar ratio of stringent factor and ribosomes guarantee an effective response to any shortage of charged tRNA immediately.

During preparation of this manuscript, Pedersen and Kjeldgaard reported a stringent factor-ribosome ratio of 1 to 200. Using antibodies against stringent factor that did not interfere with the catalytic site of the enzyme, this group estimated the amount of stringent factor per cell by a combined method of immune precipitation and functional assay rather than by radioimmuno assay[8]. This group also finds a 25 to 100 times higher specific activity of the stringent factor, a result that is in striking conflict with earlier reports[4-6]. Differences in strains could be one explanation; radioimmuno assay with antibodies against stringent factor preparations from various strains should help to resolve these discrepancies.

Literature

1 Haseltine, W. A., Block, R., Gilbert, W. & Weber, K. (1972) *Nature (London)* **238**, 381 - 384.

2 Haseltine, W. A. & Block, R. (1973) *Proc. Natl. Acad. Sci. U.S.A.* **70**, 1564 - 1568.

3 Richter, D. (1976) *Proc. Natl. Acad. Sci. U.S.A.* **73**, 707 - 711.
4 Block, R. & Haseltine, W.A. (1975) *J. Biol. Chem.* **250**, 1212 - 1217.
5 Richter, D., Nowak, P. & Kleinert, U. (1975) *Biochemistry* **14**, 4414 - 4420.
6 Cochran, J.W. & Byrne, R.W. (1974) *J. Biol. Chem.* **249**, 353 - 360.
7 Gordon, J. (1970) *Biochemistry* **9**, 912 - 917.
8 Pedersen, F.S. & Kjeldgaard, N.O. (1977) *Eur. J. Biochem.* **76**, 91 - 97.

Prof. Dr. D. Richter, Dr. H. Schmale and S. Fehr, Institut für Physiologische Chemie, der Universität, Abteilung Zellbiochemie,
Martinistr. 52, D-2000 Hamburg 20.

Hoppe-Seyler's Z. Physiol. Chem.
Bd. 359, S. 129 - 132, Januar 1978

SHORT COMMUNICATION

The Amino Acid Sequences of the Two Main Components of Adult Hemoglobin from Orangutan *(Pongo pygmaeus)*

Tetsuo MAITA, Aiko ARAYA, Morris GOODMAN and Genji MATSUDA

Department of Biochemistry, Nagasaki University School of Medicine, Nagasaki, Japan and
Department of Anatomy, Wayne State University, School of Medicine, Detroit, Michigan, U.S.A.

(Received 13 September 1977)

Aminosäuresequenzen der beiden Hauptkomponenten vom Hämoglobin des adulten Orang-Utan (Pongo pygmaeus)

Zusammenfassung: Die beiden Hauptkomponenten des Hämoglobins aus Orang-Utan-Blut wurden, ohne sie vorher zu trennen, vom Häm befreit; die Globine wurden dann durch Säulenchromatographie an CM-Cellulose getrennt. Dabei erhielt man zwei verschiedene α-Ketten (α-I, α-II) und eine β-Kette. Die tryptischen Peptide der einzelnen *S*-aminoethylierten Ketten wurden isoliert und sequenziert. Die tryptischen Peptide wurden in Homologie zur Primärstruktur des adulten Humanhämoglobins angeordnet, wodurch sich die Primärstrukturen der einzelnen Ketten ableiten ließen.

Im Vergleich zum Humanhämoglobin liegen bei der α-I-Kette drei, bei der α-II-Kette und bei der β-Kette je zwei Aminosäureaustausche vor; außerdem unterscheidet sich die α-I-Kette von der α-II-Kette nur durch den Austausch von Asparaginsäure gegen Glycin in Position 57.

Key words: Orangutan hemoglobin, primary structure, sequence analysis.

We have become interested in the molecular evolution of hemoglobins, in particular, in the rate of evolution and its mechanism. Studies on the primary structures of many proteins, which should shed light on the problem, are in progress. We have taken an interest in primate hemoglobins and have so far reported the primary structures of eight kinds of the hemoglobins[1-8].

Orangutan *(Pongo pygmaeus)* has two main kinds of hemoglobin components[9], although almost all primates have only one main hemoglobin component. In this paper, we present the primary structures of the two main hemoglobin components. The hemoglobin solution obtained from orangutan blood showed two main bands on cellogel electrophoresis with Tris/glycine buffer at pH 8.9. The two main hemoglobin components were dehemed by the method of Anson and Mirsky[10] without separation and the resulting globin subunits were isolated separately using CM-cellulose column chromatography. Three main peaks appeared as shown in Fig. 1, and the mean yields of subunits contained in each peak were 36%, 18% and 16%, respectively, of the amounts of proteins added. We assumed from the fact mentioned above and from the mobilities on cellogel electrophoresis in 8M urea that the first peak contains β chain and the second and third peaks, two α chains, α-I, α-II, corresponding to two main components, respectively.

α-I and α-II chains, after each of them had been *S*-aminoethylated, were digested with trypsin. The tryptic peptides were separated and purified by gel filtration using Sephadex G-50, Chromo Beads P column chromatography and paper chromatography, and their amino acid compositions were analyzed[11]. Of these tryptic pep-

Enzymes:
Chymotrypsin (EC 3.4.21.1); pepsin A (EC 3.4.23.1); subtilisin (EC 3.4.21.14); thermolysin (EC 3.4.24.4); trypsin (EC 3.4.21.4).

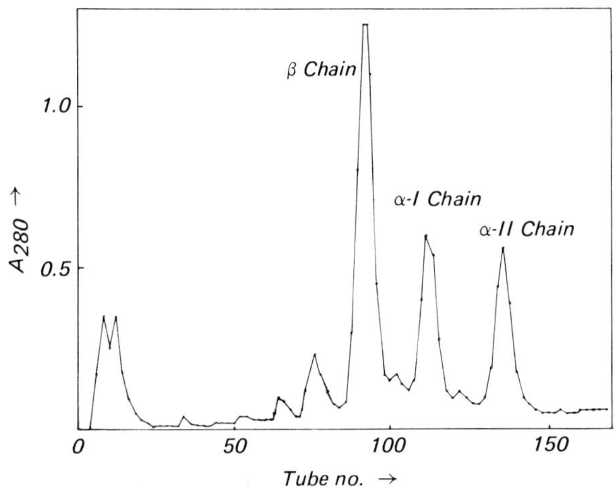

Fig. 1. Separation of the α and β chains of orangutan hemoglobin.

Approx. 400 mg of globin containing two main components of orangutan hemoglobin were dissolved in the first buffer. After incubation at 37 °C for 30 min, they were placed on a CM-52 column (2.5 x 15 cm). Gradient elution was performed with 800 m*l* of the first buffer and 800 m*l* of the second buffer. Fraction size, 10 m*l*/tube. First buffer, 0.015M (in $Na^{\oplus}$) phosphate buffer containing 8M urea, 0.05M 2-mercaptoethanol, pH 6.8; second buffer, 0.07M (in $Na^{\oplus}$) phosphate buffer containing 8M urea, 0.05M 2-mercaptoethanol, pH 6.8.

tides, α-I-T-7 from α-I chain was sequenced as Asp-His-Gly-Lys and α-II-T-7 from α-II, Gly-His-Gly-Lys in three steps of subtractive Edman degradation[12], respectively. All the corresponding tryptic peptides from the α-I and α-II chains, except α-T-7, had the same amino acid compositions and the same mobilities in the various chromatography systems and on paper electrophoresis. The corresponding tryptic peptides from the two α chains were combined. Then the sequences of the smaller tryptic peptides were analyzed by the Edman-dansyl method[13]. The positions of amides were determined from paper electrophoresis at pH 6.4. The longer tryptic peptides were, further, digested with chymotrypsin, pepsin, subtilisin and thermolysin and the sequences of the resulting smaller peptide fragments were analyzed as described above. There were no ambiguous peptides in the sequence analyses. We deduce from the above description that the primary structures of the α-I and α-II chain differ only by one substitution, Asp ↔ Gly at the *N*-terminal of αT-7.

After tryptic digestion of the *S*-aminoethylated β chain, the tryptic peptides were separated and their amino acids were sequenced in the same manner as the α chains. Since no ambiguous peptide was recognized in this case, the β chains in the two components might be the same.

The tryptic peptides described above were ordered by homology with the primary structures of α and β chains of human adult hemoglobin[14], and the primary structures of the α and β chains of orangutan hemoglobin were deduced as shown in Fig. 2. Comparing these with the primary structures of the α and β chains of human adult hemoglobin, three amino acid substitutions have occured in α-I chain: Thr(orangutan) ↔ Ala at position 12, Asp ↔ Glu at position 23, Asp ↔ Gly at position 57; and two substitutions in the β chain: Lys ↔ Thr at position 87 and Gln ↔ Pro at position 125. The α-II chain has Gly at position 57, as the α chain of human hemoglobin does. Two kinds of α chains were recognized in orangutan hemoglobin in this experiment. This suggests that there may be two genes corresponding to the α chain of orangutan hemoglobin.

We wish to thank Miss *M. Ishibashi* for her help in preparing this paper.

Literature

1 Matsuda, G., Maita, T., Takei, H., Ota, H., Yamaguchi, M., Miyauchi, T. & Migita, M. (1968) *J. Biochem.* **64**, 279 - 282.
2 Matsuda, G., Maita, T., Ota, H., Araya, A., Nakashima, Y., Ishii, U. & Nakajima, M. (1973) *Int. J. Protein Res.* **5**, 405 - 418.
3 Matsuda, G., Maita, T., Watanabe, B., Ota, H., Araya, A., Goodman, M. & Prychodko, W. (1973) *Int. J. Protein Res.* **5**, 419 - 421.
4 Matsuda, G., Maita, T., Nakashima, Y., Barnabas, P., Ranjeker, P.K. & Gandhi, N.S. (1973) *Int. J. Protein Res.* **5**, 423 - 425.
5 Matsuda, G., Maita, T., Watanabe, B., Araya, A., Morokuma, K., Goodman, M. & Prychodko, W. (1973) *Hoppe-Seyler's Z. Physiol. Chem.* **354**, 1153 - 1155.
6 Matsuda, G., Maita, T., Watanabe, B., Araya, A., Morokuma, K., Goodman, M., Barnabas, J. & Prychodko, W. (1973) *Hoppe-Seyler's Z. Physiol. Chem.* **354**, 1517 - 1520.
7 Matsuda, G., Maita, T., Suzuyama, Y., Setoguchi, M., Ota, Y., Araya, A., Goodman, M., Barnabas, J. & Prychodko, W. (1973) *Hoppe-Seyler's Z. Physiol. Chem.* **354**, 1517 - 1570.
8 Maita, T., Tanaka, E., Goodman, M. & Matsuda, G. (1977) *J. Biochem.* **82**, 603 - 605.

α	Val	Leu	Ser	Pro	Ala	Asp	Lys	Thr	Asn	Val	Lys	Thr	Ala	Trp	Gly	Lys	Val	Gly	Ala	His	Ala
β	Val	His	Leu	Thr	Pro	Glu	Glu	Lys	Ser	Ala	Val	Thr	Ala	Leu	Trp	Gly	Lys	Val	Asn	Val	Asp
NO.	1	2	3	4	5	6	7	8	9	10	11	12	13	14	15	16	17	18	19	20	21

α	Gly	Asp	Tyr	Gly	Ala	Glu	Ala	Leu	Glu	Arg	Met	Phe	Leu	Ser	Phe	Pro	Thr	Thr	Lys	Thr	Tyr
β	Glu	Val	Gly	Gly	Glu	Ala	Leu	Gly	Arg	Leu	Leu	Val	Val	Tyr	Pro	Trp	Thr	Gln	Arg	Phe	Phe
NO.	22	23	24	25	26	27	28	29	30	31	32	33	34	35	36	37	38	39	40	41	42

α	Phe	Pro	His	Phe	Asp	Leu	Ser	His	Gly	Ser	Ala	Gln	Val	Lys	Asp	His	Gly	Lys	Lys	Val	Ala
β	Glu	Ser	Phe	Gly	Asp	Leu	Ser	Thr	Pro	Asp	Ala	Val	Met	Gly	Asn	Pro	Lys	Val	Lys	Ala	His
NO.	43	44	45	46	47	48	49	50	51	52	53	54	55	56	57	58	59	60	61	62	63

α	Asp	Ala	Leu	Thr	Asn	Ala	Val	Ala	His	Val	Asp	Asp	Met	Pro	Asn	Ala	Leu	Ser	Ala	Leu	Ser
β	Gly	Lys	Lys	Val	Leu	Gly	Ala	Phe	Ser	Asp	Gly	Leu	Ala	His	Leu	Asp	Asn	Leu	Lys	Gly	Thr
NO.	64	65	66	67	68	69	70	71	72	73	74	75	76	77	78	79	80	81	82	83	84

α	Asp	Leu	His	Ala	His	Lys	Leu	Arg	Val	Asp	Pro	Val	Asn	Phe	Lys	Leu	Leu	Ser	His	Cys	Leu
β	Phe	Ala	Lys	Leu	Ser	Glu	Leu	His	Cys	Asp	Lys	Leu	His	Val	Asp	Pro	Glu	Asn	Phe	Arg	Leu
NO.	85	86	87	88	89	90	91	92	93	94	95	96	97	98	99	100	101	102	103	104	105

α	Leu	Val	Thr	Leu	Ala	Ala	His	Leu	Pro	Ala	Glu	Phe	Thr	Pro	Ala	Val	His	Ala	Ser	Leu	Asp
β	Leu	Gly	Asn	Val	Leu	Val	Cys	Val	Leu	Ala	His	His	Phe	Gly	Lys	Glu	Phe	Thr	Pro	Gln	Val
NO.	106	107	108	109	110	111	112	113	114	115	116	117	118	119	120	121	122	123	124	125	126

α	Lys	Phe	Leu	Ala	Ser	Val	Ser	Thr	Val	Leu	Thr	Ser	Lys	Tyr	Arg					
β	Gln	Ala	Ala	Tyr	Gln	Lys	Val	Val	Ala	Gly	Val	Ala	Asn	Ala	Leu	Ala	His	Lys	Tyr	His
NO.	127	128	129	130	131	132	133	134	135	136	137	138	139	140	141	142	143	144	145	146

Fig. 2. The amino acid sequences of the α-I and β chains of adult hemoglobin of the orangutan *(Pongo pygmaeus)*. The α-II chain differs from α-I chain in position 57 (Gly instead of Asp).

9 Buettner-Janusch, J. & Buettner-Janusch, V. (1964) in Evolutionary and Genetic Biology of Primates (Buettner-Janusch, J., ed.) vol. 2, pp. 75 - 88, Academic Press, New York.
10 Anson, M.L. & Mirsky, A.E. (1930) *J. Gen. Physiol.* **13**, 469 - 476.
11 Matsuda, G., Maita, T., Watanabe, B., Ota, H., Araya, A., Goodman, M. & Prychodko, W. (1972) *Int. Peptide Protein Res.* **4**, 405 - 413.
12 Konigsberg, W. & Hill, R.J. (1962) *J. Biol. Chem.* **237**, 2547 - 2561.
13 Gray, W.R. (1967) *Methods in Enzymol.* **11**, 469 - 475.
14 Braunitzer, G., Gehring-Müller, R., Hilschmann, N., Hilse, K., Hobom, G., Rudloff, V. & Wittman-Liebold, B. (1961) *Hoppe-Seyler's Z. Physiol. Chem.* **325**, 283 - 286.

Dr. T. Maita, Dr. A. Araya and Prof. Dr. Genji Matsuda*, Department of Biochemistry, Nagasaki University School of Medicine,
Nagasaki 852, Japan.

Prof. Dr. M. Goodman, Department of Anatomy, Wayne State University, School of Medicine,
Detroit, Mich. 48201, U.S.A.

* To whom all correspondences should be addressed.

Hoppe-Seyler's Z. Physiol. Chem.
Bd. 359, S. 133 - 136, Januar 1978

SHORT COMMUNICATION

A Photometric Estimation of Phospho-L-arginine, Arginine and Octopine Using Homogeneous Octopine Dehydrogenase Isoenzyme 2 from the Squid, *Loligo vulgaris Lam.*

Manfred GRIESHABER, Eva KRONIG and Rudi KOORMANN

Zoologisches Institut der Universität Münster, Lehrstuhl für Tierphysiologie

(Received 14 November 1977)

Photometrische Bestimmung von Phospho-L-arginin, L-Arginin und Octopin mittels homogener Octopin-Dehydrogenase aus dem Kalmar, Loligo vulgaris Lam.

Zusammenfassung: Aus der Mantelmuskulatur des Kalmars *Loligo vulgaris* wurden durch Sephadex-Gel-chromatographie und präparative isoelektrische Fokussierung zwei Isoenzyme der Octopin-Dehydrogenase in homogener Form aufgetrennt. Die isoelektrischen Punkte betrugen pH 8.3 und pH 8.7. Das Isoenzym 2 der Octopin-Dehydrogenase eignet sich sehr gut für die Bestimmung von Phospho-L-arginin, L-Arginin und Octopin im optischen Test.

Key words: Octopine, L-arginine, phospho-L-arginine, octopine dehydrogenase.

Octopine, (N^2-(*d*-1-carboxyethyl)-L-arginine), an unusual amino acid, is synthesized by the reductive condensation of L-arginine and pyruvate in several animal tissues as well as in some crown gall tumors of plants[1,2]. The cosubstrate of the catalyzing enzyme octopine dehydrogenase is NADH in the case of animal tissue[3], and NADPH in plant tumor cells[4].

The elucidation of the biological role of octopine as a temporary endproduct of anaerobic energy metabolism in the mantle muscle of the squid, *Loligo vulgaris*[5], the adductor muscle of the scallop, *Pecten jacobaeus*[6], and the use of octopine as a marker of a possible transfer of genetic material from tumorigenic bacteria into plant cells[7] has sparked some further interest in this compound.

A prerequisite of studies on octopine metabolism is a rapid and specific estimation of this compound in tissue extracts. And since there is a correlation between the formation of octopine and the breakdown of phospho-L-arginine to ATP and L-arginine[5], methods for quantitative determination of these compounds are also desirable. In previous reports, methods estimating phospho-L-arginine[5], arginine[8], and octopine[9] using NAD$^{\oplus}$-dependent octopine dehydrogenase have been described. Octopine dehydrogenase, which must be absolutely free of other dehydrogenases, was purified from adductor muscle of the scallop, *Pecten maximus*, according to the method of Thoai et al.[3]. This procedure, however, required several steps, which were somewhat tedious and time-consuming.

In this paper we report a new purification method for octopine dehydrogenase from the mantle muscle of the squid, *Loligo vulgaris*. This cephalopod is especially well suited as source of octopine dehydrogenase, since it contains high concentrations of the enzyme[5].

Enzymes:
Arginine kinase, ATP:L-arginine N^{ω}-phosphotransferase (EC 2.7.3.3);
Lactate dehydrogenase, L-lactate:NAD$^{\oplus}$ oxidoreductase (EC 1.1.1.27);
Octopine dehydrogenase, N^2-(1-carboxyethyl)-L-arginine:NAD$^{\oplus}$ oxidoreductase (L-arginine-forming) (EC 1.5.1.11).

Material and Methods

Octopine, phospho-L-arginine and arginine kinase were purchased from Sigma, St. Louis. $NAD^{\oplus}$ and NADH (both purity I) were obtained from Boehringer Mannheim GmbH. Ampholine pH 7 - 9 and pH 9 - 11, Ultrodex gel for preparative electrofocusing and polyacrylamide-gel plates for analytical electrofocusing were purchased from LKB, Stockholm. Coomassie Brilliant Blue and 3,3'-(3,3'-dimethoxy-4,4'-biphenylylene)-bis-[2-(4-nitrophenyl)-5-phenyl-2*H*-tetrazolium chloride] were bought from Serva, Heidelberg. All other chemicals were obtained from local distributors in the purest quality possible.

Frozen mantle muscle from *Loligo vulgaris* was shipped in dry ice from the Stazione Zoologica, Naples, Italy.

Enzyme activities were assayed by following the change in A_{339} in a Beckman recording photometer (model Acta V) at 25 °C. One unit (U) was defined as the production or removal of 1 µmol NADH/min.

Octopine dehydrogenase was determined by the method of Thoai et al.[3]. The assay medium contained 50mM triethanolamine/HCl pH 7.0, 0.15mM NADH, 3.5mM sodium pyruvate. To this mixture either crude extracts or preparations of octopine dehydrogenase were added. The assay of octopine dehydrogenase activity was initiated by addition of L-arginine to give a final concentration of 5mM after a constant rate of change of A_{339} was observed.

Arginine kinase was assayed by using the octopine dehydrogenase-mediated formation of octopine as indicator reaction for the formation of L-arginine. The assay mixture contained 50mM glycylglycine, pH 7.5, 5mM $MgCl_2$, 5mM ADP, 0.5mM phospho-L-arginine, 3.5mM sodium pyruvate, 0.15mM NADH and arginine kinase. After a constant change of A_{339} was observed, 4 U of octopine dehydrogenase were added to 3 m*l* of assay mixture.

In some crude extracts with high activities of lactate dehydrogenase and low activities of arginine kinase, the more elaborate method of Leech et al.[10] had to be used for the estimation of arginine kinase.

Protein determination. Protein from 1 m*l* preparations was precipitated by adding trichloroacetic acid to give a final concentration of 5%. The resulting pellet was dissolved in 1.0 m*l* 1N NaOH. The protein content was determined according to Lowry et al.[11].

Preparative gel filtration. The octopine dehydrogenase preparation from the crude extract was applied on top of a Sephadex G-100 column (125 cm × 7.5 cm), equilibrated with 50mM potassium phosphate buffer pH 7.5, containing 0.1mM ethylendiamintetraacetate, 0.1mM dithiothreitol and 10% glycerol. The column was developed at a constant rate of 70 m*l*/h and fractions of 20 m*l* were collected.

Isoelectric focusing. Preparative isoelectric focusing was performed, using the LKB electrofocusing kit for granulated gels, equipped with a power supply which guarantees a constant power during the whole run. The procedure essentially followed LKB application note No. 198[12]

4 g of Ultrodex was suspended in 100 m*l* of bidistilled water containing 2 m*l* each of Ampholine pH 7 - 9 and pH 9 - 11. After the gel plate was poured and dried to the evaporation limit, 3 m*l* of the concentrated octopine dehydrogenase preparation containing 1% of Ampholine pH 7 - 9 and pH 9 - 11 was applied to the center of the gel plate. Electrofocusing was started at 0.3 kV and 13 mA. The power was kept constant at 4 W. After 24 h, the separation was completed at 0.98 kV and 4 mA. The flat gel bed was immediately fractionated into 30 samples with a stainless steel grid. Each fraction was scratched from the glass plate, transferred into a 5 m*l* syringe, resuspended in a minimum amount of potassium phosphate buffer and eluted with the same buffer. During the run and the fractionation the temperature was kept at 2 °C.

Analytical isoelectric focusing was carried out on polyacrylamide plates containing ampholine pH 4 - 10. The enzyme preparations were applied with filter strips. The separation was started at 0.3 kV and 40 mA and stopped at 1.5 kV and 13.4 mA after 90 min. The power was kept constant at 20 W.

The gels were stained with Coomassie Brilliant Blue, following exactly the procedure advised by the distributor of the polyacrylamide plates. Staining for enzymatic activity was performed according to Gaede and Grieshaber[13] after the gels had been rinsed for 3 min in incubation buffer.

Results and Discussion

Purification of octopine dehydrogenase

Mantle muscle of *Loligo vulgaris* which had been stripped of the chromatophore-containing skin, was frozen in liquid nitrogen. After one year of storage at −70 °C only a 10% decrease in the initial enzyme activity was found.

Step 1. 25 g of muscle tissue was cut into thin slices and homogenized in a four-fold volume of cold potassium phosphate buffer (see above) using a Sorvall Omni Mix. Portions of 20 m*l* of this slurry were further homogenized in an Ultra Turrax for 1 min. The resulting homogenate was centrifuged at 50000 × *g* for 60 min.

Step 2. A portion of the supernatant from step 1 containing not more than 1 200 U of octopine dehydrogenase was fractionated on Sephadex G-100. The remaining

Table. Purification of octopine dehydrogenase isoenzyme 2 from the squid, *Loligo vulgaris* Lam.

Fraction	Total act. [U]	[U/ml]	Spec. act. [U/mg prot.]	Purif.
Crude extract as applied to Sephadex G-100	1200	89	10	1
Conc. prep. from Sephadex G-100	780	195	28	2.8
Isoenzyme 2 after isoelectric focusing	126	42	230	23

crude extract could be stored in ice for approximately three weeks.

Octopine dehydrogenase was eluted from the column in a single peak between 2.4 and 3.0 *l*. All fractions containing more than 1 U/m*l* were combined and concentrated to 4 m*l* in an Amicon cell using membrane PM10. The concentrated preparation was centrifuged at 50 000 x *g* for 60 min. At this stage the enzyme is stable for several weeks if bacterial growth is inhibited by the addition of 0.02% NaN_3.

Step 3. 3 m*l* of the concentrated enzyme preparation of step 2 were used for preparative isoelectric focusing. As can be seen from the figure, octopine dehydrogenase separated into two isoenzymes with isoelectric points of pH 8.2 and pH 8.7, designated isoenzyme 1 and isoenzyme 2. The latter usually contained 70% of the total activity recovered.

Analytical isoelectric focusing of isoenzyme 2 on polyacrylamide plates and subsequent staining for protein and enzymatic activity revealed only one band, indicating that it was homogeneous.

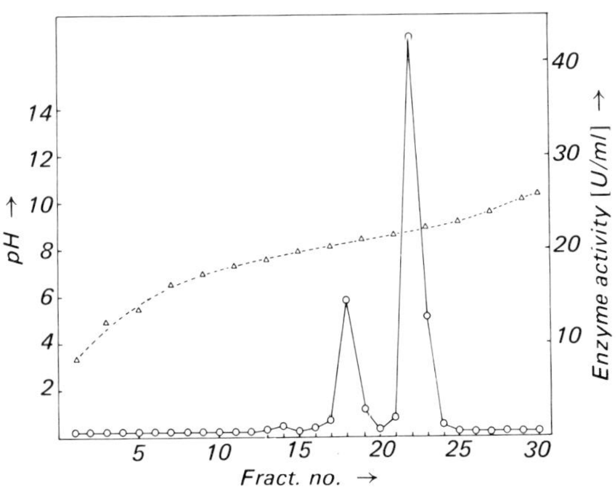

Figure. Fractionation pattern from flat gel bed isoelectric focusing of octopine dehydrogenase from the squid.
(△——△) pH, (○——○) enzymatic activity.

The homogeneous isoenzyme 2 was stabilized by the addition of an equal volume of bidistilled glycerol (89%) and stored at – 10 °C. This preparation, which is stable for at least three months, was used for the substrate determinations.

This preparation of octopine dehydrogenase from *Loligo vulgaris*, which yields a homogeneous isoenzyme after only three purification steps, requires much less time than the previously used method. Every single step is easy to perform. Since gel chromatography and isoelectric focusing give highly reproducible results concerning the fractionation pattern, only a limited number of fractions must be assayed to detect the enzyme once the procedure is standardized. So a continuous supply of octopine dehydrogenase for the estimation of phospho-L-arginine, L-arginine and octopine is possible.

Estimation of substrates

Estimation of phospho-L-arginine and L-arginine. The estimation of these two compounds can be performed in a continuous assay within one reaction mixture. The assay mixture contains 50mM glycylglycine, pH 7.5, 5mM $MgCl_2$, 5mM ADP, 2.5mM sodium pyruvate, 0.4mM NADH and varying amounts of authentic phospho-L-arginine or extracts. The reaction is started by the addition of octopine dehydrogenase isoenzyme 2. The resulting decrease in A_{339} corresponds to the amount of arginine present within the extract.

After all arginine has been condensed to octopine, no further change in absorbance will be observed. The addition of arginine kinase, which of course must be completely free of lactate dehydrogenase and other dehydrogenases, causes the production of ATP and arginine from phospho-L-arginine. L-Arginine, however, is immediately converted to octopine, accompanied by a renewed decrease in A_{339} due to the oxidation of NADH.

The enzymatic breakdown of phospho-L-arginine as well as the enzymatic estimation of L-arginine are very specific methods compared to acid hydrolysis of phospho-L-arginine and colorimetric determination of L-arginine[14].

The described method gives linear results between 0.01 and 0.1mM concentrations of both compounds. The recovery of authentic phospho-L-arginine and L-arginine added to tissues prior to their extraction with 0.6M perchloric acid was 95%. If lower concentrations of phospho-L-arginine must be determined, the more sensitive method of Bigl and Biesold[15] is recommended. In this assay ATP is estimated with the luciferin-luciferase system.

Estimation of octopine. Octopine dehydrogenase isoenzyme 2 can be used for the fluorimetric estimation of octopine as described previously[9], in particular if concentrations lower than 1μM are present. The assay mixture for the photometric estimation contained 0.1M triethanolamine/HCl, pH 9.0, 0.1M hydrazine, 0.3mM $NAD^{\oplus}$ and varying amounts of extract or authentic octopine. When no further change in absorbance is observed, the reaction is started by the addition of isoenzyme 2. The resulting increase in A_{339} is equivalent to the amount octopine present. Recovery, specificity and linearity up to 0.1mM concentrations of octopine are similar to those in the fluorimetric determination.

These determination procedures have now successfully been used for the estimation of phospho-L-arginine, L-arginine and octopine in perchloric acid extracts from animal and plant tissues.

Literature

1 van Thoai, N. & Robin, Y. (1959) *Biochim. Biophys. Acta* **35**, 446 - 453.
2 Ménagé, A. & Morel, M.G. (1964) *C.R. Acad. Sci.* **259**, 4795 - 4796.
3 van Thoai, N., Huc, C., Pho, D.B. & Olomucki, A. (1969) *Biochim. Biophys. Acta* **191**, 46 - 57.
4 Goldmann, A. (1977) *Plant Sci. Lett.* **10**, 49 - 58.
5 Grieshaber, M. & Gaede, G. (1976) *J. Comp. Physiol.* **108**, 225 - 232.
6 Grieshaber, M. & Gaede, G. (1977) *Comp. Biochem. Physiol.*, in press.
7 Kemp, D. (1976) *Biochem. Biophys. Res. Commun.* **69**, 816 - 822.
8 Gaede, G. & Grieshaber, M. (1975) *Anal. Biochem.* **66**, 393 - 399.
9 Grieshaber, M. (1976) *Anal. Biochem.* **74**, 600 - 603.
10 Leech, A.R., Beis, I. & Newsholme, E.A. (1977) *Anal. Biochem.*, in press.
11 Lowry, O.H., Rosebrough, N.J., Farr, A.L. & Randall, R.J. (1951) *J. Biol. Chem.* **193**, 265 - 275.
12 LKB Instrument, Stockholm (1975) Application Note 198.
13 Gaede, G. & Grieshaber, M. (1975) *J. Comp. Physiol.* **102**, 149 - 158.
14 Rosenberg, H., Ennor, A.H. & Morrison, J.F. (1956) *Biochem. J.* **63**, 153 - 159.
15 Bigl, H. & Biesold, D. (1971) *Acta Biol. Med. Ger.* **27**, 245 - 250.

M. Grieshaber, Zoologisches Institut der Westfälischen Wilhelms-Universität, Lehrstuhl für Tierphysiologie, Hindenburgplatz 55, D-4400 Münster.

Berichtigung zu Bd. 358, Heft 12

Direkte immunologische Identifizierung von Proteinen nach elektrophoretischer Fraktionierung in Polyacrylamid-Mikrogradientengelen und Einfluß von Detergentien auf die Immunpräzipitation

Von V. Neuhoff und S. Mesecke

S. 1628: Abb. 4 ist ohne Randbeschriftung um 180° zu drehen.

Redaktionelle Hinweise

(Wichtige Änderungen ab Januar 1976 sind am Rand markiert)

Die Zeitschrift bringt nur Experimantalarbeiten aus den Gebieten der Biochemie und Physiologischen Chemie. Die Manuskripte werden in deutscher oder in englischer Sprache veröffentlicht. Es wird vorausgesetzt, daß alle Koautoren mit der eingereichten Fassung einverstanden sind.

Aktuelle Kurzmitteilungen, deren Umfang 4 Seiten in Kleindruck (ca. 8 Manuskriptseiten) nicht übersteigen darf, werden bevorzugt bearbeitet und in das nächsterreichbare Heft aufgenommen. Die Dringlichkeit zur Prioritätssicherung ist vom Autor in einem Begleitschreiben kurz zu begründen.

Manuskripte sind in zweifacher Ausfertigung an eine der folgenden Anschriften zu senden:

Prof. Dr. A. Butenandt oder Prof. Dr. F. Lynen, Max-Planck-Institut für Biochemie, D-8033 Martinsried b. München;

Prof. Dr. Dr. G. Weitzel, Physiologisch-Chemisches Institut der Universität, Hoppe-Seyler-Str. 1, D-7400 Tübingen;

Dr. A. Dillmann, Hoppe-Seyler-Redaktion, Am Klopferspitz, D-8033 Martinsried b. München, Tel. (089) 85 85-823.

Bei der Abfassung der Manuskripte bitten wir dringend, folgende Regeln zu beachten, die im wesentlichen mit den „Instructions to Authors" von *Biochimica et Biophysica Acta* und *European Journal of Biochemistry* übereinstimmen:

Es soll in Maschinenschrift (2zeilig, auch vorgesehener Kleindruck!) und mit 4 cm breitem Rand geschrieben werden. Vorgesehene Absätze sind deutlich zu markieren.

Seite 1 des Manuskripts soll enthalten (in normaler Schrift, nicht in Großbuchstaben):

a) Titel der Arbeit
b) Autorennamen (jeweils ein Vorname ausgeschrieben)
c) Name des Instituts oder Fachbereichs, in dem die Arbeit ausgeführt wurde (möglichst in der Sprache des Herkunftlandes), vorausgesetzt, daß die Leitung des Instituts oder Fachbereichs der Veröffentlichung zugestimmt hat.
d) Kurztitel, maximal 70 Satzzeichen
e) Adresse für den Versand der Korrekturen

Auf Seite 2 folgen

a) Eventuelle Widmung
b) Vollständige Postadresse der Autoren in der Sprache des Herkunftslandes
c) Ad-hoc-Abkürzungen (nur in seltenen Fällen zugelassen)
d) Enzyme in alphabetischer Reihenfolge der empfohlenen Trivialnamen mit systematischen Namen und EC-Nummern

Auf Seite 3 folgt die *Zusammenfassung* sowie eine Übersetzung des Titels und der (eventuell erweiterten) Zusammenfassung ins Englische (bzw. Deutsche). Diese Übersetzung oder deren sprachliche Überprüfung kann auch von der Redaktion übernommen werden.

Eventuelle *Danksagungen* sind an den Schluß des Manuskriptes zu stellen.

Literaturzitate im Text sind fortlaufend zu numerieren (nicht alphabetisch) und am Schluß des Manuskripts zu bringen. Die in Chemical/Biological Abstracts üblichen Zeitschriftenabkürzungen sind zu verwenden. Zur Anordnung s. IUB Commission of Editors of Biochemical Journals (1973) *Eur. J. Biochem.* **37**, 201 - 202; (1974) *diese Z.* **355**, 1 - 3.

Beispiel: Spackman, D.H., Stein, W.H. & Moore. S. (1958) *Anal. Chem.* **30**, 1190 - 1206.

Zeichnungen sollen einmal in reproduzierbarer Form mit schwarzer Tusche gezeichnet eingereicht werden; Zahlenskalen, gemessene Größen, Einheiten und sonstige Beschriftungen sind jedoch nur leicht mit weichem Bleistift einzutragen.

Die Abbildungen sollen so groß sein, daß sie auf 1/2 oder 1/3 verkleinert werden können. Deutlich beschriftete Kopien für die Referenten sind beizufügen.

Alle *Tabellen* und *Abbildungen* sind mit Legenden zu versehen, so daß die Ergebnisse auch ohne Zuhilfenahme des Textes ersichtlich werden. Es wird gebeten, die Legenden nicht zu unterstreichen und die Abbildungslegenden nicht auf die Abbildungen selbst, sondern auf ein eigenes Blatt zu schreiben.

Symbole für die am häufigsten vorkommenden Größen und ihre Einheiten sind in Tab. 1 zusammengestellt. Wir bitten streng zu unterscheiden, ob bei Ansätzen molare Konzentrationen (mol/l = M; z.B. 0.25M) oder zugesetzte Mengen (z. B. in mol, mmol, μmol oder nmol) gemeint sind.

Hinsichtlich der *Nomenklatur* richte man sich nach der IUPAC-Nomenklatur sowie nach den Empfehlungen der "IUPAC-IUB Commission on Biochemical Nomenclature": Kohlenhydrate, *Eur. J. Biochem.* **21**, 455 - 477 (1971); **25**, 4 (1972); α-Aminosäuren, *Eur. J. Biochem.* **53**, 1 - 14 (1975);

Tab. 1. Symbole für Größen und ihre Einheiten

Größe		Einheit[a]
l	Länge	m, mm, μm, nm, etc.
V	Volumen	m^3, l (dm^3), ml, etc.
λ	Wellenlänge	nm (nicht mμ oder Å)
t	Zeit	h, min, s, ms, μs, etc.
ν	Frequenz	Hz
m	Masse	kg, g, mg, μg, etc.
ρ	Dichte	g/cm^3
s	Sedimentationskoeffizient	$S = 10^{-13}s$
n	Substanzmenge	mol, mmol, etc.
M	molare Masse (m/n)	g/mol
–	Masse eines Moleküls	Dalton[b]
M_r	Molekulargewicht	(dimensionslos)
c_B, [B]	molare Konzentration einer Substanz B	M, mM, μM, etc. (besser als 10^{-3}M, etc.)
T	Temperatur (absolut)	K (nicht °K)
t, θ	Temperatur	°C
q, Q	Wärmemenge	J (besser als cal)[c]
I	Ionenstärke	M (mol/l), mM, etc.
K	Gleichgewichtskonstante	
K_m	Michaeliskonstante	
K_i	Inhibitorkonstante	
k	Geschwindigkeitskonstante	
v	Reaktionsgeschwindigkeit	mol/s (oder mmol, μmol oder nmol/s, etc.), besser als μmol/min
V (nicht V_{max})	Geschwindigkeit der enzymkatalysierten Reaktion bei unendlich großer Substratkonzentration	
I	Lichtintensität	cd
T	Durchlässigkeit (I/I_0)	(dimensionslos)
A	Absorption[d] (– lg T)	(dimensionslos)
ϵ	molarer Extinktionskoeffizient	$M^{-1} \times cm^{-1}$

a Präfixe für Einheiten: T, Tera (10^{12}); G, Giga (10^9); M, Mega (10^6), k, Kilo (10^3); m, Milli (10^{-3}); μ, Mikro (10^{-6}); n, Nano (10^{-9}); p, Piko (10^{-12}); f, Femto (10^{-15}); a, Atto (10^{-18}).

b Das Dalton ist 1/12 der Masse eines Atoms des Nuklids ^{12}C.

c Das Joule ist die SI-Einheit der Energie. Die Kalorie (1 cal = 4.184 J; 1 J = 0.239 cal) ist für eine Übergangszeit noch zugelassen; in jedem Fall sollten aber die Angaben auch in Joule ausgedrückt werden.

d Die Ausdrücke optische Dichte, Extinktion oder Absorbancy sollen nicht mehr verwendet werden.

Cyclite, *diese Z.* **350**, 523 - 535 (1969), *Eur. J. Biochem.* **57**, 1 - 7 (1975); Lipide, *diese Z.* **350**, 279 - 285 (1969); Steroide, *diese Z.* **351**, 663 - 689 (1970); Chinone mit isoprenoiden Seitenketten, Folsäure u. verschiedene andere Verbindungen, *diese Z.* **348**, 266 - 276 (1967); Corrinoide, *Eur. J. Biochem.* **45**, 7 - 12 (1974); Vitamine B_6, *Eur. J. Biochem.* **40**, 325 - 327 (1973); Carotinoide, *Eur. J. Biochem.* **25**, 397 - 408 (1972), **57**, 317 - 318 (1975); synthetische Analoga natürlicher Peptide, *diese Z.* **348**, 262 - 265 (1967); Enzyme Nomenclature, Recommendations (1972), Elsevier Publ. Comp., Amsterdam; multiple Enzymformen, *diese Z.* **353**, 852 - 854 (1972); Eisen-Schwefel-Proteine, *Eur. J. Biochem.* **35**, 1 - 2 (1973); Human-Immunglobuline, *Eur. J. Biochem.* **45**, 5 - 6 (1974); Peptidhormone, *Eur. J. Biochem.* **55**,

Tab. 2 Empfohlene Abkürzungen.

Atome, isotope	=	z. B. [16-^{14}C]Östron
Atome, numeriert	=	z. B. C-5
Curie	=	Ci
Impulse pro Minute	=	Ipm (engl. cpm)
Prozent der Theorie	=	% d. Th.
prozentig	=	proz. (nicht %-ig)
Relative Wanderungsgeschwindigkeit	=	R_F
Schmelzpunkt	=	Schmp. (engl. m.p.)
Siedepunkt	=	Sdp. (engl. b.p.)
Umdrehungen pro Minute	=	U./min (engl. rpm)
Zerfälle pro Minute	=	Zpm (engl. dpm)
Zersetzung	=	Zers. (engl. decomp.)

485 - 486 (1975). Für die Rechtschreibung ist „Jansen/Mackensen, Rechtschreibung der technischen und chemischen Fremdwörter" maßgebend.

Bezüglich der *Abkürzungen und Symbole für chemische Namen,* die in der biologischen Chemie von besonderem Interesse sind, wird auf die Empfehlungen der „IUPAC-IUB Commission on Biochemical Nomenclature" verwiesen (siehe z. B.: Allgemeines, Aminosäuren, Zucker, Phosphate, Coenzyme, *diese Z.* **348**, 245 - 265 (1967); Aminosäurederivate und Peptide, *Eur. J. Biochem.* **27**, 201 - 207 (1972); synthet. Polypeptide, *diese Z.* **349**, 1013 - 1016 (1968); Nucleinsäuren, Polynucleotide und deren Bestandteile, *diese Z.* **351**, 1055 - 1063 (1970). Andere als die dort empfohlenen Abkürzungen werden nur in seltenen Ausnahmefällen zugelassen. Sie müssen am Beginn jeder Arbeit in einer Fußnote erklärt werden.

Sonderdrucke: Von jeder Arbeit werden dem Verfasser 75 Sonderdrucke kostenlos geliefert. Die Bestellung weiterer Sonderdrucke gegen Berechnung ist möglich, muß jedoch spätestens bei der Rücksendung der Korrekturen erfolgen.

Bezugspreis: Der Jahresbezugspreis (1 Jahrg. zu 12 Heften) beträgt DM 590,–, das Einzelheft kostet DM 56,–. Bestellungen nimmt jede Buchhandlung entgegen.

Verlag: WALTER DE GRUYTER & Co.
Genthiner Str. 13, D-1 Berlin 30. Tel. (030) 261 1341
Telegrammadresse: Wissenschaft Berlin
Postscheck: Berlin-West 10307-108

WALTER DE GRUYTER Inc.
3 Westchester Plaza
Elmsford, N.Y. 10523, U.S.A.
Tel. (9 14) 592-5890

Für die nächsten Hefte sind folgende Beiträge vorgesehen:

Alteration in D-glucose uptake by plasma membranes of isolated liver cells, *W. Bachmann, R. Jersild and D. Challoner*

Synthesis of protected decapeptide (B_{21-30}) of human insulin, *V. K. Naithani and J. Föhles*

Effect of $Na^{\oplus}$ on bile acid uptake by isolated rat hepatocytes. Evidences for heterogeneous system, *M. S. Anwer and D. Hegner*

Liver cell heterogeneity: The distribution of fructose-bisphosphatase in fed and fasted rats and in man, *U. Schmidt, H. Schmid, W. G. Guder, B. Funk, A. Mall and J. Weis*

Preparation and physical-chemical characterization of poly-*N*-vinylpyrrolidone-insulin, *B. U. v. Specht, H. J. Kolb, R. Renner, and K. D. Hepp*

The carbohydrate linkage: Site of cow colostrum trypsin inhibitor, *R. Klauser, D. Čechová and H. Tschesche*

On the significance of sialic acid in high affinity 5-hydroxytryptamine uptake by synaptosomes, *G. A. Dette and W. Wesemann*

Stilbene synthases and stilbenecarboxylate synthases, I. Enzymatic synthesis of 3,5,4'-trihydroxystilbene from *p*-coumaroyl coenzyme A and malonyl coenzyme A, *N. Rupprich and H. Kindl*

Studies on bovine insulin with methyl vinyl ketone, *H. E. M. Magojo, D. A. Sutton, B. R. D. Easter and S. E. Drewes*

The stability of A1,B29-Adipoylinsulin to chemical and enzymatic degradation, *S. M. L. Robinson and K. Lübke*

A new organ spectrophotometer for sensitive dual-wavelength absorbance measurement and spectral scanning of intact perfused organs, *H. Schwab and H. Sies*

The elimination in dogs of trypsin-α-macroglobulin complexes inactivated by the Kazal or the Kunitz inhibitor, *A. Eddeland and K. Ohlsson*

The amino acid compositions of the tryptic, chymotryptic and peptic peptides from the L-2 light chain of rabbit skeletal muscle myosin, *G. Matsuda, Y. Suzuyama, M. Setoguchi, T. Maita, B. Watanabe, T. Maekawa and T. Umegane*

Block of cholesterol synthesis in cultured cells by 25-azido-norcholesterol, *W. Stoffel and R. Klotzbücher*

New protein reagents. Synthesis and properties of halogenated maleimides and isomaleimides, *J.-P. Lautié and M. Olomucki*

Neuauflage

Eckhart Buddecke

Grundriß der Biochemie

Für Studierende der Medizin, Zahnmedizin und Naturwissenschaften
5. Auflage

Groß-Oktav. XXXII, 526 Seiten. Mit mehr als 400 Formeln, Tabellen und Diagrammen. 1977. Plastik flexibel DM 39,–
ISBN 3 11 007225 4

Die progressive Zunahme des biochemischen Fachwissens erfordert eine überschaubare und zusammenfassende Darstellung der Biochemie als Hilfsmittel für den Unterricht. Der Wissensstoff ist gegliedert in die Kapitel „Stoffe und Stoffwechsel, Stoffwechselregulation und Funktionelle Biochemie der Organe und Gewebe" mit dem Ziel, durch knappe Darstellung, gesicherte Fakten und gezielte Stoffauswahl dem Leser einerseits eine rasche Information zu bieten, andererseits jedoch auf die vielfältigen Beziehungen und Anwendungsmöglichkeiten der Biochemie zur klinischen Chemie und Molekularpathologie hinzuweisen. Bei der Bearbeitung der 5. Auflage wurden alle Kapitel kritisch redigiert und einzelne Abschnitte entsprechend dem Wissenschaftsfortschritt neugefaßt oder eingefügt. Ebenso wurde die 2. Auflage des Gegenstandskatalogs für die Ärztliche Vorprüfung im Fach Physiologische Chemie eingehend berücksichtigt.

Preisänderung vorbehalten

SN 0018-4888 A 20679 E

Hoppe-Seyler's Zeitschrift für Physiologische Chemie

Mitteilungen der Gesellschaft für Biologische Chemie

WALTER DE GRUYTER · BERLIN · NEW YORK

Hoppe-Seyler's Z. Physiol. Chem. Bd. 359 Heft 2 S. 137 - 250 Februar 1978
HSZPAZ 359 (2) 137 - 250 (1978)

Mitteilungen der Gesellschaft für Biologische Chemie

34. Rundschreiben (Februar 1978)

Einladung zur ordentlichen Mitgliederversammlung

am Donnerstag, dem 6. April 1978, um 17.30 Uhr (im Anschluß an die Nachmittagssitzung) in der Stadthalle in Mosbach (Baden).

Tagesordnung

1. Bericht des Vorstandes
2. Bericht der Kassenprüfer
3. Entlastung des Vorstandes
4. Festsetzung des Beitrages für 1979
5. Wahl der Kassenprüfer für 1978
6. Bekanntgabe der Wahlvorschläge für die Beiratsnachwahl
7. Themen der nächsten Mosbacher Kolloquien
8. Genehmigung einer Studiengruppe „Glykokonjugate"
9. Verschiedenes

Achtung zu Punkt 6:

Mit drei Unterschriften versehene Wahlvorschläge sind beim Schriftführer Prof. Dr. H. Gibian, Müllerstraße 170–172, D-1000 Berlin 65, schriftlich vor oder auf der Versammlung einzureichen. Anzugeben ist, für welches Fach die Nominierung gelten soll: Physiologische Chemie im medizinischen Bereich (Staudinger scheidet turnusmäßig aus), Biochemie im naturwissenschaftlichen Bereich (Fasold scheidet turnusmäßig aus), Spezialfächer (Breuer scheidet turnusmäßig aus). Wiederwahl ist zulässig; Staudinger und Breuer bitten von möglicher Wiederwahl abzusehen.

Freiburg und Berlin, den 10. Januar 1978 — K. Decker, Präsident — H. Gibian, Schriftführer

Wichtige Bitte an unsere Mitglieder

Die Mitgliederzahl der Gesellschaft für Biologische Chemie ist im Laufe der letzten Jahre stark gewachsen, und damit ist der alltägliche Geschäftsgang so umfangreich geworden, daß er die Kapazität der vorhandenen Kräfte zu übersteigen beginnt. Wir wollen deshalb unsere elektronische Buchführung erweitern und nicht nur Mitgliedsdaten speichern, sondern vor allem auch die Beitragskonten elektronisch verwalten.

In der Mitgliederversammlung vom 21. April 1977 ist hierüber schon gesprochen worden. Damals wurde beschlossen, daß wir versuchen sollten, ein an Biochemie interessiertes Unternehmen zu finden, das uns bei dieser Aufgabe behilflich ist. Dieses soll das Recht erhalten, die Adressen unserer Mitglieder auch für eigene Werbezwecke zu nutzen.

Im Laufe des letzten Jahres haben dann Gespräche mit der LKB Instrument GmbH in München-Gräfelfing stattgefunden, die in sehr dankenswerter Weise ihre moderne EDV-Anlage für unsere Zwecke zur Verfügung stellen will. Die Programmierung, die Dateneingabe, die Beitragsverwaltung und die sonstige Bearbeitung werden von LKB vorgenommen. Uns werden lediglich die Selbstkosten dafür in Rechnung gestellt. Wir sind überzeugt, daß damit ein großer Teil der täglichenArbeit für uns wegfallen wird und daß damit zu gegebener Zeit auch ein Wechsel im Amt des Schriftführers oder des Schatzmeisters ganz wesentlich erleichtert wird.

Das erfordert freilich eine Mithilfe unserer Mitglieder. Sie finden beiliegend ein Formular, auf dem Sie nach den Daten gefragt werden, die dann in die EDV eingegeben werden sollen und für die selbstverständlich das Datenschutzgesetz gilt. *Bitte haben Sie die Freundlichkeit, diesen Fragebogen auszufüllen* und die zutreffenden Felder mit einem Kreuz zu versehen! *Schicken Sie ihn dann bitte bis spätestens 1. April 1978 an den Schatzmeister, Prof. Dr. E. Auhagen, Friedrich-Ebert-Str. 217, 5600 Wuppertal 1.* Bitte tragen Sie dazu bei, daß die Rücksendung vollständig erfolgt, damit Nachfragen nicht nötig werden. Wann die neue Buchhaltung ihre Arbeit beginnt, werden wir Ihnen noch rechtzeitig mitteilen.

Bitte erleichtern Sie deren Arbeit, indem Sie sie auf dem Fragebogen zum Einzug Ihrer Mitgliedsbeiträge ermächtigen, auch dann noch einmal, wenn dies bereits erfolgt ist.

Noch einmal die Bitte: Senden Sie Ihren Fragebogen recht bald zurück!

Gesellschaft für Biologische Chemie

K. Decker H. Gibian E. Auhagen

Beitragszahlung

Die Mitgliederversammlung hat am 21.4.1977 beschlossen, den Mitgliedsbeitrag für 1978 unverändert bei DM 30,– (für Studenten bei DM 5,–) zu belassen. Er war am 1. Januar 1978 fällig.

Die Beiträge werden von der Kassenstelle im Laufe des ersten Halbjahrs eingezogen werden, falls sie dazu ermächtigt worden ist oder noch ermächtigt wird. Die übrigen Mitglieder werden hiermit gebeten, ihren Beitrag auf eines der Konten der Gesellschaft für Biologische Chemie (beim Postscheckamt Frankfurt 569 09-608 oder bei der Dresdner Bank AG, Wuppertal 1, Bankleitzahl 330 800 30, Konto 5 049 714) zu überweisen, sofern das noch nicht geschehen ist. Eine gesonderte Aufforderung wird nicht ergehen.

Die Einziehung ausländischer Schecks, auch wenn sie auf DM-Beträge ausgestellt sind, ist mit relativ hohen Bankgebühren belastet. Bitte zahlen Sie deshalb in solchen Fällen DM 5,– zusätzlich, also DM 35,– für 1978.

Noch höhere Gebühren werden uns berechnet, wenn wir vergeblich von Ihrer Einzugsermächtigung Gebrauch machen. Sie können bis zu DM 10,– betragen. Bitte benachrichtigen Sie uns davon, wenn Ihr Konto, das Sie uns angegeben haben, aus irgend einem Grunde gelöscht wird.

Tagungsankündigungen

(neu aufgenommene Ankündigungen sind mit • gekennzeichnet)

Informationen über die zahlreichen *EMBO-Kurse und -Symposien* über molekularbiologische, zellbiologische und virologische Themen gibt das Sekretariat der EMBO, Postfach 10 22 40, D-6900 Heidelberg 1. (Vorläufiges Programm 1978 s. am Schluß dieser Tagungsankündigungen.) Auch mehrere *„NATO Advanced Study Institutes"* betreffen biochemische und molekularbiologische Themen. Information: NATO Scientific Affairs Division, B-1110 Brüssel.

Frühjahrstagung der Gesellschaft für Biologische Chemie

12.-15. März 1978, Frankfurt/Main

Vorläufiges Programm:

Lipoproteine (6 eingeladene Vorträge und 6 - 8 Kolloquiumsvorträge).

Regulatorische Enzyme (4 eingeladene Vorträge und 6 Kolloquiumsvorträge).

Außerdem werden im Rahmen der Tagung die Studiengruppe „Neurobiochemie" der GBCh und die Forschergruppe der DFG „Chemische Modifizierung von Membranen" im Verein mit der Studiengruppe „Membranen" der GBCh je ein eigenes Symposienprogramm durchführen.

Die *3. Adolf-Butenandt-Vorlesung* wird von Prof. U. Westphal, Louisville/Ky., gehalten werden.

Über die Ausbildung naturwissenschaftlicher Studenten in Biochemie soll ein Rundgespräch der Repräsentanten Biochemischer Institute stattfinden. (Siehe hierzu auch die Umfrage „Biochemische Ausbildung naturwissenschaftlicher Studenten in der Bundesrepublik Deutschland" am Ende dieses Rundschreibens.)

Für die Anmeldung wurde ein eigenes Rundschreiben versandt.

Information: Prof. Dr. H. Fasold, Institut für Biochemie der Johann-Wolfgang-Goethe-Universität, Sandhofstraße, Gebäude 75A, D-6000 Frankfurt/Main 71.

International Symposium on
Genetic Engineering: Scientific Developments and Practical Applications
29. - 31. March 1978, Milan

Information: Fondazione Giovanni Lorenzini, Via Monte Napoleone, 23, I-20121 Milano.

FEBS Advanced Course 51 on
New DNA-Sequencing Methods
3. - 7. April 1978, Bristol, England

Information: Dr. J. Hindley, Dept. of Biochemistry, University of Bristol Medical School, University Walk, Bristol BS8 1TD, U.K.

---

29. Mosbacher Kolloquium

6. - 8. April 1978, Mosbach in Baden

Leitung: G. Schäfer, Hannover und E. M. Klingenberg, München.

Anträge auf Reisekostenzuschüsse für junge Biochemiker (siehe 31. und 32. Rundschreiben) sind möglichst rasch, spätestens bis 6. März 1978 beim Schriftführer, Prof. Gibian, einzureichen.

Alle Mitglieder erhalten das endgültige Programm mit diesem Rundschreiben.

Information: Prof. Dr. G. Schäfer, Institut für Klinische Biochemie und Physiologische Chemie der Medizinischen Hochschule, Kurt-Wiechert-Allee 9, D-3000 Hannover 61.

8th International Symposium on
Clinical Enzymology
14. - 16. April 1978, Fondazione Cini, Isola die San Giorgio, Venice, Italy

Topics: The Symposium will consist of Invited Lectures, Communication and Posters on the following subjects: 1) Pathophysiological aspects. 2) Methodology. 3) Clinical aspects.

Information: Prof. Angelo Burlina, Unità di Ricerche Enzimologiche, Lab. di Chimica Clinica, Centro Ospedaliero Borgo Trento, I-37100 Verona.

Europäische Tagung für Biochemische und Instrumentelle Analytik
Biochemische Analytik 78
18. - 21. April 1978, München

Diese Tagung wird in Verbindung mit der *Internationalen Ausstellung* „Analytica 78" stattfinden.

Information: Generalsekretär Dr. Rosmarie Vogel, Nußbaumstr. 20, Postfach 20 03 24, D-8000 München 2.

• 11th Miles International Symposium on
The Mechanisms of Pain and Analgesic Compounds
7.-9. June 1978, Baltimore, Maryland, U.S.A.

Information: Edward G. Bassett, Ph. D., Symposium Coordinator Miles Laboratories, Inc. Elkhart, Ind. 46514 U.S.A. (219) 264-8460

5th Bile Acids Meeting
Biological Effects of Bile Acids (Falk Symposium No. 26)
12.-13. Juni 1978, Freiburg i.Br.

Information: Dr. G. Paumgartner, Dept. of Clinical Pharmacology, University of Bern, Murtenstr. 35, CH-3010 Bern.

• 3rd International Symposium on
Biological Nitrogen Fixation
12.-16. June 1978, Madison, Wisconsin, U.S.A.

Information: W. H. Orme-Johnson, Dept. of Biochemistry, University of Wisconsin-Madison, Madison, Wis. 53706, U.S.A. or W. E. Newton, C. F. Kettering Research Laboratory, Yellow Springs, Ohio 45387, U.S.A.

6th European Workshop on
Drug Metabolism
18.-23. June 1978, Leiden

Information: Prof. D. D. Breimer, Dept. of Pharmacology, Subfaculty of Pharmacy, Wassenaarseweg 72, Leiden, The Netherlands.

• 8th Linderstrøm-Lang Conference on
The Cellular Utilization of Long Chain Length Fatty Acids
27.-30. June 1978, Tønsberg, Norway

Information: Prof. Dr. J. Bremer, Institute for Medical Biochemistry, University of Oslo, Sognsvannsveien 9, Oslo 3, Norway.

12th FEBS Meeting
2.-8. July 1978, Dresden

Symposia

1 DNA-protein interactions
2 Gene expression
3 Protein structure and assembly
4 Structure and function of enzymes
5 Bioenergetics
6 Processing and turnover of proteins and organelles in the cell
7 Cyclic nucleotides in cell regulation
8 Regulation of secondary plant product and hormone metabolism

Colloquia

1 Technology of genetic engineering
2 Molecular immunology
3 Molecular diseases
4 Macromolecular changes and neuronal function
5 Technical enzymology
6 Investigation of biopolymers with scattering methods
7 Cytochrome P-450
8 Biochemical education
9 Xenobiochemistry

Special Colloquium
History of Biochemistry (public evening session).

The Meeting is combined with an *International Exhibition.*

Information: 12th FEBS Meeting Dresden, P. O. B. 313, DDR-806 Dresden.

European Tissue Culture Society, Annual Meeting
3. - 5. July 1978, Glasgow, Scotland

Symposia, workshops and discussion groups.

Information: Dr. I. Freshney, Beatson Institute for Cancer Research, Garscube Estate, Bearsden Road, Glasgow G61, Scotland.

Symposium on
Transport by Proteins
9. - 15. July 1978, Konstanz

The preliminary program includes sections on General Principles; Selected Transport Proteins; Membranes in Transport by Proteins; Metal Transport by Proteins; Molecular Aspects of Transport by Proteins in Physiology and Pharmacology.

Information: Prof. Dr. H. Sund, Fachbereich Biologie, Universität Konstanz, Postfach 77 33, D-7750 Konstanz.

Post-Congress FEBS-Symposium on
Antimetabolites in Biochemistry, Biology and Medicine
10. - 12. July 1978, Prague

Topics:

1) Antimetabolites as Tools in Enzymology
2) Consequences of Analogue Incorporation into Nucleic Acids
3) Aspects of Selective Antiviral Action
4) Rational Approach to the Use of Antimetabolites in Combination Cancer Chemotherapy and in Immunosuppression
5) New Types of Compounds, their Synthesis and Mechanism of Action

Information: Prof. Jan Škoda, Institute of Organic Chemistry and Biochemistry, Czechoslovak Academy of Sciences, Flemingovo 2, CS-16610 Prague 6.
Prof. Peter Langen, Zentralinstitut für Molekulare Biologie der Akademie der Wissenschaften der DDR, Lindenberger Weg 70, DDR-1115 Berlin.

3rd International Symposium on Cytopharmacology
Neurotoxins and their Use as Tools in Neurobiology
10. - 14. July 1978, Venice, Italy.

Information: B. Ceccarelli, F. Clementi - C. N. R. Center of Cytopharmacology, Institute of Pharmacology of Milano, Via Vanvitelli 32, I-20129 Milano.

International Symposium on
Enzyme-Labelled Non-radioimmunoassay of Hormones
10. - 12. July 1978, Ulm

Information: Dr. S. B. Pal, Universität Ulm, Abteilung Innere Medizin I, Postfach 3880, D-7900 Ulm.

• International Conference on
Molecular and Developmental Biology of Insects
13. - 16. July 1978, Heraklion, Crete, Greece

Topics: Genome Organization, Hormones in Development, Intracellular Mechanism of Hormone Action, Model Systems of Development and Differentiation, Molecular Basis of Insect Control.

Information: Dr. E. Fragoulis, Dept. of Biology, University of Athens, Panepistimiopolis Kouponia (621) Athens, Greece.

Deadline: February 15th, 1978.

7th International Congress of Pharmacology
16.- 21. July 1978, Paris

Information: Prof. P. Lechat, Institut de Pharmacologie, Faculté de Médecine, 21 rue de l'Ecole de Médecine, F-75006 Paris.

Meeting on
Conjugation Reactions in Drug Biotransformation
23.- 26. July 1978, Turku, Finnland

Information: Antero Aitio, Dept. of Physiology, Kiinamyllynkatu 10, SF-20520 Turku 52.

4th International Congress of Pesticide Chemistry (IUPAC)
24.- 28. July 1978, Zürich

Topics:
1) Chemical Structure and Biological Activity
2) Natural Products with Biological Activity
3) Biochemistry of Pests and Mode of Action of Pesticides
4) Degradation

Information: Congress Secretariat, 4th International Congress of Pesticide Chemistry, P. O. Box 182, CH-4013 Basle.

• *6th International Forum on Subcellular Methodology*
25.- 28. July 1978, Guildford, U.K.

Topics: Tissue and Protozoan Cell Populations, Markers, and Plant-Cell Organelles

Information: Dr. E. Reid, Wolfson Bioanalytical Centre, University of Surrey, Guildford GU2 5XH, U.K.

3rd International Congress of Plant Pathology
16.- 31. August 1978, München

Information: Dr. E. Fuchs, Biologische Bundesanstalt, Messeweg 11/12, D-3300 Braunschweig.

XIV. Internationaler Kongreß für Genetik
21.- 30. August 1978, Moskau

Grundidee: Genetics and Human Welfare

Information: Nationales Organisationskomitee des XIV. Internationalen Genetik-Kongresses, Fersmana Street 11, n. 4, BOTuC um. N. I. Wawilowa, Moscow 117312, U. S. S. R.

European Society for Neurochemistry, 2nd Meeting
28. August - 1. September 1978, Göttingen

Information: Prof. Dr. Volker Neuhoff, Max-Planck-Institut für Experimentelle Medizin, Hermann-Rein-Str. 3, D-3400 Göttingen.

4th International Congress for Virology
30. August - 6. September 1978, The Hague

Information: 4th International Congress for Virology, Netherlands Congress Centre, P.O. Box 9000, The Hague, The Netherlands.

12th International Congress for Microbiology
3. - 8. September 1978, München
Information: Prof. Dr. K.-H. Schleifer, Arcisstraße 21, D-8000 München 2.

IUPAB, 6th International Biophysics Congress
3. - 9. September 1978, Kyoto, Japan
Information: Prof. Fumio Oosawa, Secretary General, 6th International Biophysics Congress, Dept. of Biophysical Engineering, Osaka University, Toyonaka, Osaka 560, Japan.

International Congress of Bacteriology and Mycology
4. - 12. September 1978, München
Information: Prof. G. Pulverer, Fürst-Pückler-Str. 56, D-5000 Köln 41.

Workshop Course: Modern Techniques in Centrifugation
11. - 16. September 1978, University of Essex
Information: Dr. D. Rickwood, Dept. of Biology, University of Essex, Wivenhoe Park, Colchester, CO4 3SQ, U.K.

Gemeinsame Tagung der Biochemical Society und der Gesellschaft für Biologische Chemie
10. - 13. September 1978, Konstanz

Themen:
1) Biochemie der Nukleinsäuren
2) Membranen, Lipide und Transportprozesse
3) Biochemische Oxidoreduktion, Enzymologie des Redox-Transports

Neben diesen Symposien sind Poster-Präsentationen vorgesehen. Abstracts hierfür müssen bis zum 1. Juli 1978 der Tagungsleitung vorliegen. Anweisungen zur Abfassung der Kurzreferate wurden im Rundschreiben abgedruckt. Das Programm wird allen Mitgliedern der GBCh noch zugestellt.
Anträge auf Reisekostenzuschuß für junge Biochemiker (siehe 31. und 32. Rundschreiben) sind bis zum 10. Juli 1978 beim Schriftführer, Prof. Gibian, einzureichen.

Information: Prof. Dr. Horst Sund, Universität Konstanz, Postfach 7733, D-7750 Konstanz.

- Anläßlich dieser Tagung plant die *Studiengruppe Physikalische Chemie biologischer Makromoleküle* ein erstes Treffen unter dem Thema

  ‚Thermodynamik, Kinetik und Struktur von Biopolymeren'.
  In Form von Kurzvorträgen und Posters sollen im Anschluß an die Tagung am 14.9.78 neue Ergebnisse diskutiert werden. Abstracts vorgesehener Beiträge sind bis zum 15. Juli 1978 zu senden an Prof. Dr. Robert Huber, Max-Planck-Institut für Biochemie, Am Klopferspitz, D-8033 Martinsried bei München.

110. Versammlung der Gesellschaft Deutscher Naturforscher und Ärzte
17. - 21. September 1978, Innsbruck
Generalthema: Das naturwissenschaftliche Weltbild heute – Grundlagen und Bausteine
Information: Prof. Dr. E. Auhagen, Gesellschaft Deutscher Naturforscher und Ärzte, Friedrich-Ebert-Str. 217, D-5600 Wuppertal 1.

- *6th Meeting, International Research Group for Carcino-Embryonic Proteins (I.R.G.C.P.)*
  17.-21. September 1978, Marburg/Lahn
  Themen:
  Carcino-embryonale Proteine:
  1. Klinische Anwendung
  2. Biologie, Immunchemie und Bedeutung in der Embryo- und Carcinogenese
  3. Methoden, Standardisierung

  Information: Prof. Dr. F.-G. Lehmann, Med. Univ.-Klinik, E. Mannkopffstraße 1, D-3550 Marburg/Lahn.

FEBS Advanced Course No. 53
Hormonal Regulation of Specific Gene Expression
19.-22. September 1978, London
Organized by P.N. Campbell and R.K. Craig.

Information: Prof. P.N. Campbell, Courtauld Institute, The Middlesex Hospital Medical School, London, W1P 7PN, England.

- *XX. Symposion der Gesellschaft für Histochemie*
  20.-23. September 1978, Gargellen/Montafon, Österreich
  Hauptthemen:
  1. Cytochemische Merkmale als Ordnungsprinzip in Histologie und Pathohistologie
  2. Pathohistochemie von Magen-Darm-Biopsien

  Information: Prof. Dr. Ch. Pilgrim, Abt. Klinische Morphologie der Universität Ulm, Postfach 4066, D-7900 Ulm/Donau.

- *Tagung der Gesellschaft für Genetik*
  21.-23. September 1978, Innsbruck
  Hauptthema: Molekularbiologische Grundlagen der Humangenetik
  1. Techniken zur Charakterisierung des Genoms (Genomstruktur, Kartierung)
  2. Läsion des Genoms
  3. Biochemische Manifestation von Genläsionen

  Information: Institut für Medizinische und Klinische Chemie der Universität Innsbruck, Fritz-Pegel-Straße 3/VI, A-6020 Innsbruck.
  Voranmeldung bis 30. März 1978.

- *3rd European Symposium on Hormones and Cell Regulation*
  4.-7. October 1978, Bischenberg, near Strasbourg, France
  Topics:
  – Organization and Expression of the Genome – Growth Factors – Kinases and Multisite Phosphorylation –
  – Insulin Secretion – Hormones and Adenylate Cyclase Activities – Steroid Effects on Gene Expression.
  Information: H.J. van der Molen, Dept. of Biochemistry II, Medical Faculty, Erasmus University, P.O. Box 1738, Rotterdam, The Netherlands.

FEBS Special Meeting on
Enzymes
17.-21. April 1979, Dubrovnik-Cavtat, Jugoslavia

Mit diesem Rundschreiben erhalten unsere Mitglieder einen ausführlicheren Prospekt (1st Announcement). Der dort angegebene Anmeldetermin 1. Februar 1978 läßt sich nicht einhalten. Bitte senden Sie trotzdem Ihre Anmeldung möglichst bald ab!

Information: Prof. Dr. P. Mildner, Laboratory of Biochemistry, Technological Faculty, Pierottijeva 6/VI, YU-41000 Zagreb.

- *XII. Acta Endocrinologica Congress*
  26. - 30. Juni 1979, München

  Information: Prof. Dr. Peter C. Scriba, Secretary General, Medizinische Klinik Innenstadt der Universität München, Ziemssenstraße 1, D-8000 München 2.

XIth International Congress of Biochemistry
8. - 14. July 1979, Toronto, Ontario, Canada

Topics:

1) Structure, organization and replication of genetic elements
2) Genetic expression and its regulation
3) Structure and function of proteins
4) Mechanism and regulation of enzyme action
5) Membrane structure and function
6) Bioenergetics and transport
7) Biochemistry of growth and differentiation
8) Biochemical aspects of immunology
9) Neurochemistry
10) Biochemistry of sensory, contractile and motile systems
11) Hormones: biochemistry and molecular mechanisms of action
12) Molecular mechanisms of disease
13) Environmental biochemistry

The Congress is combined with an *International Exhibition.*

Information: R. W. Dolan, Executive Secretary, XIth International Congress of Biochemistry, c/o National Research Council of Canada, Ottawa, Ontario, Canada K1A OR6.

13th FEBS Meeting
24. - 29. August 1980, Jerusalem, Israel

Information: Alisa Tietz-Devir (President of the Israel Biochemical Society), Dept. of Biochemistry, Tel-Aviv University, Ramat Aviv, Tel Aviv.

Nathan Sharon (Chairman of the Organizing Committee), Dept. of Biophysics, The Weizmann Institute of Science, P. O. Box 26, Rehovot.

The Secretariat, 13th FEBS Meeting, P. O. Box 16271, Tel Aviv, Israel.

GDCh-Fortbildungskurse 1978

Die Gesellschaft Deutscher Chemiker veranstaltet auch in diesem Jahr Fortbildungskurse, von denen die folgenden den Fachbereich Biologische Chemie betreffen:

Kurs Nr. 33/78: Fortschritte auf dem Gebiet der physikalisch-chemischen Charakterisierung von Biopolymeren und Phospholipiden
Veranstaltungsort: Freiburg
Leitung: Prof. Dr. Th. Ackermann, Prof. Dr. W. Burchard gemeinsam mit Dr. A. Blume, Dr. H. Klump
Termin*): 25. 9. (v) bis 29. 9. 1978 (n)

Kurs Nr. 41/78: Beziehung zwischen chemischer Struktur und biologischer Aktivität von Wirkstoffmolekülen — Quantitative Struktur-Wirkung-Analysen
Veranstaltungsort: Borstel
Leitung: Prof. Dr. J. K. Seydel
Termin: 2. 10. (v) bis 6. 10. 1978 (m)

Kurs Nr. 42/78: Immunchemie — Immunologische Methoden
Veranstaltungsort: Marburg
Leitung: Prof. K. U. Hartmann, Prof. H. G. Schwick
Termin: 9. 10. (v) bis 13. 10. 1978 (n)

---

*) v = vormittags, m = mittags, n = nachmittags, a = abends.

Kurs Nr. 43/78: Polyacrylamid und Misch-Gele als Träger bei der Analyse von Bio-Molekülen
Veranstaltungsort: Braunschweig
Leitung: Prof. Dr. Stegemann
Termin: 23. 10. (v) bis 27. 10. 1978 (n)

Kurs Nr. 44/78: Einführung in die klinisch-chemische Analytik
Veranstaltungsort: Hannover
Leitung: Prof. Dr. med. A. Delbrück
Termin: 16. 10. (v) bis 20. 10. 1978 (n)

Kurs Nr. 46/78: Hydrophobe und Biospezifische Affinitäts-Chromatographie von Proteinen (Methodenkurs)
Veranstaltungsort: Bochum
Leitung: Prof. Dr. L. M. G. Heilmeyer, Priv.-Doz. Dr. H. Jennissen
Termin: 2. 10. (v) bis 6. 10. 1978 (n)

Kurs Nr. 47/78: Konsequenzen ökologischer und ökologisch-toxikologischer Denkweisen für die Chemie (Übersichtskurs)
Veranstaltungsort: Neuherberg bei München
Leitung: Prof. Dr. F. Korte unter Mitwirkung von Dr. W. Klein, Dr. H. Parlar und Gästen aus den zuständigen Aufsichtsbehörden
Termin: 9. 10. (v) bis 13. 10. 1978 (n)

Informationsblätter und Anmeldekarten sind erhältlich von der Gesellschaft Deutscher Chemiker, Abt. Fortbildungskurse, Postfach 90 04 40, D-6000 Frankfurt 90; Tel. (0611) 7917-356.

EMBO-Kurse und Workshops (Vorläufiges Programm 1978)

Kurse

Thema	Organisator und Informationsadresse	Datum und Ort
Nonsense Mutations and tRNA Suppressors	Dr. J. E. Celis, Kemisk Institut, Aarhus Univ., Langelandsgade 140, DK-8000 Aarhus C	11.-24. Juli Aarhus
Intracellular Ion-sensitive Microelectrodes	Dr. R. C. Thomas, Dept. of Physiology, Univ. of Bristol, The Medical School, University Walk, Bristol BS8 1TD, England	29. Juni - 13. Juli Bristol
Molecular Approaches to Cell Differentiation	Prof. D. R. Newth, Dept. of Zoology, Univ. of Glasgow, Glasgow G12 800, Scotland	2.-15. April Glasgow
Training Course on Hormone Receptors/ Calcium/Cyclic Nucleotides	Dr. J. P. Durham, Dept. of Oncology, Univ. of Glasgow, 1 Horselethill Road, Glasgow G12 9LY, Scotland	20. August - 2. September Glasgow
Reconstruction and Analysis of DNA Molecules	Prof. W. Arber, Abt. Mikrobiologie, Biozentrum d. Univ., Klingelbergstraße 70, CH-4056 Basel	7.-26. August Basel
Structure and Metabolism of Glycoconjugates	Prof. J. Montreuil, Laboratoire de Chimie Biologique, Université des Sciences et Techniques de Lille I, B.P. 36, F-59650 Villeneuve d'Ascq	18.-30. September Villeneuve d'Ascq
New Aspects of Interferon Research	Prof. D. C. Burke, Dept. of Biological Sci., Univ. of Warwick, Coventry CV 4 7AL, England	2.-15. April Warwick
In situ Hybridisation of Nucleic Acids	Dr. J. Bishop, Institute of Animal Genetics, Univ. of Edinburgh, West Mains Road, Edinburgh EH9 3 JN, Scotland	2 Wochen im September Edinburgh
Mammalian Cell Separation	Dr. T. M. Jovin, Abt. Molekulare Biologie, Max-Planck-Institut für Biophysikalische Chemie, Postfach 968, D-3400 Göttingen	4.-13. Mai Göttingen
Mitochondrial Biogenesis in Animal Development	Prof. C. Saccone, Institute of Biological Chemistry, Univ. of Bari, Via Amendola, 165/A, I-70126 Bari	ca. 10.-25. Juni, Kutor, Jugoslawien
Lecture Course: NATO-EMBO Summer School on Molecular and Cell Biology: Control of Gene Expression in Uninfected and virally Infected Cells	Prof. B. F. C. Clark, Kemisk Institut, Aarhus Univ., Langelandsgade 140, DK-8000 Aarhus C	12 Tage Aug./Sept. Spetsai

Workshops

Thema	Organisator und Informationsadresse	Datum und Ort
Molecular Biology of Adenoviruses	Prof. L. Philipson, Dept. of Microbiology, Biomedical Center, Box 581, S-751 23 Uppsala	11.-16. Juni Örenäs
Plasmids and Other Extrachromosomal Genetic Elements	Dr. K. Timmis, Max-Planck-Institut für Molekulare Genetik, Ihnestraße 63 - 73, D-1000 Berlin 33	1.-5. April Berlin
Halophilism	Prof. S. R. Caplan, Dept. of Membrane Research, The Weizman Institute of Science, Rehovot, Israel	14.-19. Mai Arad
Ribosomes	Dr. D. Vázquez, Instituto de Bioquimica de Macromoléculas, Consejo Superior de Investigaciones Cientificas, Velazquez, 144, Madrid 6, Spanien	4 Tage im Juni Salamanca
Intercellular Molecular Communication in the CNS	Prof. M. Cuénod, Institut für Hirnforschung der Univ., August-Forel-Straße 1, CH-8029 Zürich	11.-15. September Schatzalp/Davos
The Fourth European Cell Cycle Workshop	Prof. R. Braun, Institut für Allgemeine Mikrobiologie, Altenbergrain 21, CH-3013 Bern	17.-19. April Rüttihubelbad
Polypeptide Neurotoxins Structure and Function	Prof. M. Lazdunski, Centre de Biochimie, Faculté S P C N I, Université de Nice, Parc Valrose, F-06034 Nice	4 Tage im September oder Oktober Nizza
Incompatibility and Control of Replication of Plasmids	Prof. K. Nordström, Dept. of Molecular Biology, University of Odense, Campusvej 55, DK-5230 Odense M	7.-10. April Odense
Structure and Function of Phospholipases	Prof. L. van Deenen, Biochemisch Laboratorium, Rijksuniversiteit, Padualaan 8, NL-2506 Utrecht	3 1/2 Tage im September oder Oktober Cassis
Vitellogenin	Prof. M. Gruber, Biochemisch Laboratorium, Rijksuniversiteit, Zernikelaan, Groningen, Niederlande	3 Tage im Mai oder September, Groningen
Accuracy	Dr. J. Ninio, I R B M, Tour 43, 2, Place Jussieu, F-75221 Paris Cédex 05	4 Tage im September südl. Paris
Cell Locomotion	Dr. M. Abercrombie, Strangeways Res. Lab., Wort's Causeway, Cambridge CB 1 4RN, England	21.-24. August Cambridge
Regulation of Development in Prokaryotes	Dr. E. Brody, Institut de Biologie Physico-Chimique, 13, Rue P. et M. Curie, F-75005 Paris	4 Tage im September Paris
Viroids and Other Small Replicative Nucleic Acids	Dr. H. L. Sänger, Arbeitsgruppe Pflanzenvirologie der Univ., Schubertstraße 1, D-6300 Gießen/Lahn	24.-26. August Martinsried bei München

Stipendien

EMBO: Short and Long Term Fellowships in Molecular Biology

The European Molecular Biology Organisation intends to award to scientists working within the European area and in Israel both short term (from 1 week to 12 weeks) and long term fellowships for collaborative research or advanced practical training in molecular biology.

The short term fellowships are to support visits to other laboratories for the purpose of carrying out collaborative experiments, especially if these arise at short notice. EMBO short term fellowships are not, however, awarded for exchanges between laboratories within any one country.

The long term fellowships will usually be for a period of up to one year but applications for a renewal for a second year will be considered. To be eligible a candidate must hold a doctors degree. Preference will be given to European and Israeli applicants wishing to work within Europe or Israel. EMBO long term fellowships are not, however, awarded for exchanges between laboratories within any one country. Applications for fellowships to be held outside Europe and Israel are considered but they have a lower priority, as do applications from non-European scientists wishing to work in Europe or Israel.

Application forms and further details may be obtained from Dr. J. Tooze, Executive Secretary, European Molecular Biology Organisation, 69 Heidelberg 1, Postfach 1022.40, Germany.

Applications for *short term* fellowships may be made at any time and will be decided upon soon after receipt of the application. Applications for *long term* fellowships should be made before *February 20* for a decision by *April 30*, and before *August 31* for a decision by *October 31*. Because the selection procedure usually involves an interview these deadlines for applications should be respected.

Zur Umfrage „Biochemische Ausbildung naturwissenschaftlicher Studenten in der Bundesrepublik Deutschland“

Die Umfrage wurde durch Versendung eines Fragebogens in der 1. Hälfte des Jahres 1977 durchgeführt. Nach dem Rücklauf der Fragebögen wurde zunächst eine Tabelle erstellt, die dann an die naturwissenschaftlichen biochemischen Institute mit der Bitte um Korrektur ausgesandt wurde. Insgesamt waren wir bemüht, durch korrekte Interpretation der Beantwortung des Fragebogens und der Rückantworten ein für jede Universität zutreffendes Beantwortungsschema aufzuzeichnen. In manchen Fällen mag dies zu einer gewissen Vereinfachung geführt haben; insgesamt scheint die Tabelle aber einen guten Überblick über die gebräuchliche Ausbildung von Chemikern, Biologen und Studenten anderer Fachrichtungen im Fach Biochemie zu geben.

Das endgültige Ergebnis zeigt die folgende Tabelle. Die einzelnen Spalten seien kurz erläutert: Eine *eigene biochemische Betriebseinheit* bedeutet ein selbständiges Institut, besetzt mit mindestens einem H4- bis H6-Professor, dessen Lehraufgaben ganz vorwiegend darin bestehen, naturwissenschaftliche Studenten im Fach Biochemie auszubilden. Besonders schwierig war es, anhand der beantworteten Fragebögen eine *Definition der Abschlüsse* zu geben. Unter „Diplom-Chemiker-Hauptfach“ sind nur solche Universitäten mit einem Plus-Zeichen aufgeführt, an denen das Fach Biochemie in der Diplom-Chemiker-Hauptprüfung als Prüfungsfach – meist neben den Kernfächern Organische, Anorganische und Physikalische Chemie – gewählt werden kann. Bei dem „Diplom-Biologen-Hauptfach“ gilt analog, daß die Studenten, die im allgemeinen nach dem Vordiplom eine Spezialisierung auf ein Hauptfach mit 2 Nebenfächern durchführen, in der Diplom-Biologen-Hauptprüfung das Fach Biochemie als Hauptfach wählen können. Ungenauer sind die Antworten bei Diplom-Chemiker-, bzw. Diplom-Biologen-*Nebenfach*. Gemeint ist hier, daß im Falle der Chemiker eine ausführlichere Ausbildung durch Lehrveranstaltungen, insbesondere durch ein größeres Praktikum angeboten wird, ohne daß das Fach in der Diplom-Chemiker-Hauptprüfung erscheint. Bei den Diplom-Biologen ist damit gemeint, daß das Fach in der Diplom-Biologen-Hauptprüfung erscheint, und zwar als Nebenfach. Gerade bei dieser letzteren Sparte haben allerdings viele Kollegen auch dann eine positive Antwort gegeben, wenn es nur möglich ist, an der betreffenden Universität eine biochemische Diplomarbeit als Biologe anzufertigen, auch ohne daß für Biologen eine ausführlichere Ausbildung in Form eines Praktikums und mehrerer Vorlesungen im Fach Biochemie angeboten werden kann. Unter der Sparte „Sonstige“ sind meist kürzere Ausbildungen für Lehramtskandidaten, Ingenieure, Ernährungswissenschaftler und landwirtschaftliche Studenten zusammengefaßt. Die Sparte *in chemischen Fachbereichen etabliert* bedeutet, daß eine selbständige biochemische Abteilung, besetzt mit mindestens einem H3-Professor, im chemischen Fachbereich existiert, deren hauptsächliche Aufgabe es ist, Biochemie in Lehrveranstaltungen zu vermitteln. Das gleiche gilt analog für biologische Fachbereiche. In der Rubrik *Studiengang beabsichtigt* sind nur solche Universitäten mit einem Plus-Zeichen aufgeführt, die entweder ein Vollstudium der Biochemie anstreben (Beispiel FU Berlin) oder bis jetzt entweder in den biologischen oder den chemischen Fachbereichen noch keinerlei ausführlichere biochemische Ausbildung vermitteln, dies aber in Kürze tun wollen. In der Rubrik *Wahlausbildung in Chemie* sind solche Universitäten erfaßt, die eine ausführlichere Ausbildung in Biochemie für Chemie-Studenten, in Form mindestens eines Praktikums, anbieten, ohne daß das Fach in der Diplom-Chemiker-Hauptprüfung als Prüfungsfach erscheint. Das gleiche gilt analog für andere Fächer, z.B. für Physik-Studenten. Unter *Wochenstunden* soll angegeben werden, wieviel Gesamtwochenstunden maximal der biochemischen Ausbildung gewidmet werden können, das gleiche gilt analog für die *Semester-Dauer*.

Wenn eine Null in einer Sparte erscheint, so bedeutet dies, daß genaue Auskünfte zu dieser Antwort fehlen und auch nach Rückfrage nicht genau zu ermitteln waren.

Im allgemeinen fällt auf, daß an einer ganzen Reihe von deutschen Hochschulen die Biochemie im naturwissenschaftlichen Bereich entweder in den Biologischen Instituten *oder* in den Chemischen Instituten angesiedelt und auch in den Studiengängen bzw. Prüfungsordnungen verankert ist. Seltener ist der Fall, daß die Biochemie in beiden Fächern vertreten ist. Weiterhin fällt auf, daß die Ausbildung wenig einheitlich angeboten wird. Sie reicht von einer kurzen Einführungsvorlesung (1- bis 2stündig) bis zu zweisemestrigen Praktika mit insgesamt 60 Wochenstunden, kombiniert mit 3 Vorlesungen. Hier erscheint es doch wünschenswert, auf Bundesebene eine gewisse Vereinheitlichung zu erzielen. So sollte ein kürzerer biochemischer Anteil schon in dem Vordiplom-Studium der Biologen wie der Chemiker an jeder Universität vorhanden sein. Nach dem Vordiplom sollte es den Studenten beider Fachrichtungen möglich sein, einen deutlichen Akzent ihrer Ausbildung im Fach Biochemie zu setzen, sei es durch Wahl als Hauptfach, sei es durch die Teilnahme an mindestens einem größeren Praktikum, kombiniert mit einer Vorlesung.

Bei einem ersten Rundgespräch im Rahmen der Frühjahrstagung der Gesellschaft für Biologische Chemie 1977 wurde denn auch als hauptsächliche Forderung das Bestreben erhoben, hier durch Weitergabe einer Empfehlung der

Gesellschaft für Biologische Chemie an die Kultusministerien einerseits und die beteiligten wissenschaftlichen Gesellschaften andererseits zunächst zu einer besseren Vereinheitlichung der biochemischen Lehre in den deutschen naturwissenschaftlichen Fachbereichen zu kommen. Eine solche Empfehlung sollte nach Möglichkeit im Rahmen der Frühjahrstagung der Gesellschaft 1978 in Frankfurt/Main vom 12.–15. März durch Vertreter der naturwissenschaftlichen biochemischen Institute erarbeitet und der Gesellschaft für Biologische Chemie vorgelegt werden.

Prof. Dr. H. Fasold

Hochschule	Eigene Biochemische Betriebseinheit	Biochemie-Vollstudium	Abschlüsse: Dipl.Chem. Hauptfach	Dipl.Chem. Nebenfach	Dipl.Biol. Hauptfach	Dipl.Biol. Nebenfach	Sonstige	Etabliert in: chem. Fachber./Fak.	biol. Fachb./Fak.	Studiengang beabsichtigt	Wahlausbildg. in: Chemie	andere Fächer	Wochenstunden	Semester-Dauer	Betreuung in: Physiol. Chem. Inst. als Diplomanden	Doktoranden	in and. Instituten
Aachen	–	–	–	–	–	+	–	–	–	–	–	+	20- 60	4	+	+	–
Fu Berlin	–	+	–	–	–	–	–	–	–	+	+	+	0	0	+	+	+
TU Berlin	+	–	+	+	–	–	–	+	–	–	–	–	20- 60	2	–	–	+
PH Bielefeld	–	–	–	–	–	–	+	+	–	–	–	–	20	2	–	–	–
Uni Bielefeld	–	–	–	–	–	–	–	–	–	–	–	+	4	1	–	–	–
Bochum	+	–	+	+	+	+	–	+	+	–	–	–	20- 60	4	+	+	+
Uni Bonn	–	–	–	–	–	–	–	–	+	–	–	+	0	0	–	–	–
TU Braunschweig	–	–	+	+	+	+	–	+	+	–	–	–	60-120	2	–	–	–
Bremen	–	–	–	+	–	–	–	+	+	–	+	–	20- 60	3	–	–	–
Darmstadt	+	–	–	+	–	+	+	+	–	–	–	–	20- 60	1	–	–	+
Düsseldorf	–	–	–	–	–	–	–	–	–	–	–	+	20- 60	4	–	+	+
Erlangen	+	–	–	–	+	+	–	+	+	–	+	–	0	0	+	+	+
GH Essen	–	–	–	–	–	–	+	+	–	+	+	–	20	2	+	+	+
Frankfurt	+	–	–	–	+	+	–	–	+	+	+	–	20- 60	3	+	+	+
Freiburg	+	–	–	+	+	+	–	+	+	–	–	–	20- 60	0	+	+	+
Gießen	–	–	+	+	–	+	+	–	–	–	–	–	–	–	+	+	+
Göttingen	–	–	–	–	–	–	+	–	–	+	–	+	–	–	+	+	–
Hamburg	+	–	+	+	–	–	–	+	–	–	+	–	20- 60	2	+	+	+
Hannover	+	+	+	+	+	+	–	+	+	–	+	+	120	0	+	+	+
Heidelberg	+	–	–	+	+	+	–	–	+	–	+	–	20- 60	0	+	+	+
Hohenheim	+	–	+	+	+	+	+	+	+	–	–	+	6- 20	2	+	+	–
Kaiserslautern	–	–	–	–	–	–	–	–	+	+	–	–	0	0	–	–	–
Karlsruhe	–	–	–	–	–	–	–	+	–	–	+	–	20- 60	1	–	–	+
GH Kassel	–	–	–	–	–	–	–	–	–	–	+	+	–	–	–	–	–
Kiel	–	–	–	–	–	+	+	–	–	–	–	+	25	0	+	+	+
Köln	+	–	+	+	+	+	–	+	+	–	–	–	60-120	2	+	+	+
Konstanz	+	–	–	–	+	+	–	–	+	–	+	–	60-120	0	–	–	+
Mainz	+	–	–	+	+	+	–	+	+	–	–	–	20- 60	2	+	+	+
Marburg	+	–	–	+	–	+	+	+	–	+	+	–	20- 60	2	+	+	+
TU München	+	–	–	–	–	+	–	+	–	–	+	–	20- 60	0	+	+	–
Uni München	+	–	+	–	–	+	–	+	–	–	–	–	20	3	+	+	–
Münster	+	–	+	–	–	+	–	+	–	–	+	–	20- 60	1	+	+	–
Oldenburg	–	–	–	+	+	+	–	–	+	–	+	+	20	4	–	–	–
Osnabrück	–	–	–	–	–	–	–	+	–	–	–	+	0	0	–	–	–
Paderborn	–	–	–	+	–	–	–	+	–	–	+	–	10	1	–	–	+
Regensburg	+	–	–	+	+	+	–	–	+	–	+	+	40- 80	6	+	+	+
Saarland	+	–	+	+	–	+	–	+	–	+	–	–	20- 60	4	+	+	+
Stuttgart	+	–	–	+	–	+	+	+	–	–	–	–	12- 20	1 1/2	–	–	+
Tübingen	+	+	–	–	–	+	+	+	+	+	+	+	120	12	+	+	+
Ulm	–	–	–	–	–	+	–	+	+	+	+	–	20	0	+	+	+
Würzburg	+	–	–	–	+	+	–	+	–	–	+	–	20- 60	2	–	+	+

Hoppe-Seyler's Zeitschrift für Physiologische Chemie
Begründet 1877 · Fortgeführt von A. Kossel, F. Knoop und K. Thomas

Bd. 359 Februar 1978 Heft 2

Inhaltsverzeichnis

Fortsetzung nächste Seite

Contents

Continued

Redaktionelle Hinweise siehe gegenüber Seite 250

Indexed in Current Contents

Autoren-Verzeichnis

Diesem Heft liegt ein Prospekt des Verlages Walter de Gruyter bei, den wir zu beachten bitten.

Hoppe-Seyler's Z. Physiol. Chem.
Bd. 359, S. 137 - 146, Februar 1978

Hämoglobine, XXI[1]

Die Sequenzanalyse des Hämoglobins des Schweines

Gerhard BRAUNITZER, Barbara SCHRANK, Anton STANGL und Ute SCHEITHAUER

Max-Planck-Institut für Biochemie, Abteilung Proteinchemie, Martinsried bei München

(Der Schriftleitung zugegangen am 22. Dezember 1977)

Herrn Prof. Dr. Wolfgang Graßmann zum 80. Geburtstag

Zusammenfassung: Das Hämoglobin eines bayerischen Landschweines *(Suidae)* wurde isoliert, die Peptidketten durch Gegenstromverteilung getrennt. Die Ketten wurden mit Trypsin gespalten, die isolierten Peptide mit den hydrophilen Senfölen I und IV oder einem Propinprogramm im Sequenator sequenziert. Die Sequenz der α- und β-Ketten des Schweines wird wiedergegeben. Es werden einige methodische Aspekte diskutiert und der Vergleich der Austausche zum Humanhämoglobin und dem Röntgenmodell wird gegeben. Auf die Regulation der Genexpression und deren Bedeutung für das foetale Leben des Schweines wird eingegangen. Insbesondere wird der Einfluß des 2,3-Bisphosphoglycerats diskutiert.

Hemoglobins, XXI: Sequence Analysis of Porcine Hemoglobin

Summary: The hemoglobin of a bavarian domestic pig *(Suidae)* was isolated. The chains were separated by countercurrent distribution, then cleaved with trypsin. The isolated peptides were sequenced with hydrophilic phenylisothiocyanate I and IV, or with a dimethylaminopropyne program in the sequenator. The sequences of the chains are given. Some methodical aspects of automatic sequencing are discussed and the sequences of human and porcine hemoglobin are compared. The role of adult pig haemoglobin as foetal hemoglobin is discussed.

Key words: Pig hemoglobin, sequence analysis, mammalian.

Wir berichten über die Sequenz der α- und β-Ketten des Hämoglobins des Schweines, nachdem wir schon früher über die tryptischen Peptide berichtet[2,3] und in Kurzmitteilungen die Sequenzen der α- und β-Ketten vorgelegt[4,5] haben. In dieser Arbeit legen wir die Belege vor, die zur Ableitung der Konstitution geführt haben. Diskutiert werden einige methodische Aspekte, d.h., das hydrophobe 3-Dimethyl-[6] oder 3-Diäthylaminopropin-Programm. Weiterhin eine verbesserte Methode zur Identifizierung der Aminosäure-Phenylthiohydantoin-Derivate auf Dünnschichtplatten. Auf

Abkürzungen:
Quadrol: *N,N,N',N'*-Tetrakis(2-hydroxypropyl)äthylendiamin; I = 4-(Isothiocyanato)benzol-sulfonsäure, Natriumsalz. IV = 7-(Isothiocyanato)naphthalin-1,3,5-trisulfonsäure, Trinatriumsalz. HbA = adultes Hämoglobin; HbF = foetales Hämoglobin. TPCK = *N*-Tosyl-L-phenylalanylchlormethan.

einige Aspekte der Primärstruktur dieser Arbeiten wird hingewiesen und die Funktion des adulten Schweinehämoglobins bei der foetalen Atmung wie die Rolle des 2,3-Bisphosphoglycerats diskutiert.

Experimenteller Teil

1. Gewinnung des Schweinehämoglobins und seiner tryptischen Peptide

Blut eines bayerischen Landschweines *(Suidae)* wurde vom Schlachthof erhalten. Damit wurden sämtliche Ergebnisse erarbeitet. Die Gewinnung der Erythrozyten, die Lyse und die Gewinnung des Rohhämoglobins erfolgte nach Standardprozeduren. Die Hämoglobinlösung wurde bei – 30 °C aufbewahrt. Wie schon früher beschrieben, ist in der Elektrophorese nur eine Hämoglobin-Bande sichtbar. Die Gewinnung des Globins erfolgte nach Rossi-Fanelli[7], die Trennung der α- und β-Ketten durch Gegenstromverteilung nach Craig[8] (Glasgerät der Fa. O. Post Inc., New York, N.Y., U.S.A.). Wir arbeiteten nur mit oxidierten Präparaten: die drei Tryptophane der α- und β-Kette haben wir durch Sequenzanalyse der intakten nichtoxidierten Ketten identifiziert. Die Oxidation der Kette erfolgte mit Perameisensäure, die tryptische Spaltung erfolgte mit TPCK-Trypsin der Firma Worthington.

2. Fraktionierung der Peptide

Das tryptische Hydrolysat wurde über eine Sephadex-G-25-Säule (2.5 x 110 cm) vorfraktioniert. Die einzelnen Gipfel wurden lyophilisiert und über Dowex 1X2[9,10] oder Dowex 50X4 rechromatographiert. Auf diese Weise erhielten wir reine Peptide, die anschließend zur Sequenzanalyse gelangten.

3. Sequenzanalyse

Es wurde nach der Filmtechnik[11] gearbeitet. Wir verwendeten zwei Programme: A) das Quadrolprogramm und B) das Dimethyl-[6] oder Diäthylaminopropin-Programm.

Das Quadrolprogramm verwendeten wir an den intakten Ketten. Es wurden meist 45 Schritte analysiert, um die Sequenz der tryptischen, tryptophanhaltigen Peptide zu erhalten. Wir verwendeten dieses Programm auch zur Sequenzierung der Lysylpeptide.

A) Quadrolprogramm

Das Quadrolprogramm[11] wurde in starker Annäherung an die Originalarbeit und stets mit zwei Cyclisierungen angewendet. Die Chemikalien wurden selbst gereinigt, lediglich das 1M Quadrol wurde von der Firma Pierce (Rockfort, Ill., U.S.A.) bezogen.

a) Durchführung der Quadrolmethode

0.5 - 1.0 μmol eines Peptides und 2 - 3 mg I[12] oder IV[13] (8 - 10facher molarer Überschuß) wurden in 200 μl Wasser gelöst in den Sequenator eingebracht und bei niederer Geschwindigkeit getrocknet. Anschließend wurden in den *stehenden* Becher 200 μl 3-Dimethylaminopropin [$(CH_3)_2N\text{-}CH_2\text{-}C{\equiv}CH$: H_2O, 1:1] eingebracht, die Zelle wurde verschlossen, 5 m lang mit N_2 gespült und bei der niedrigen Geschwindigkeit mit den Reagenzien I oder IV 10 min lang bei 50 °C inkubiert. Das überschüssige Propin wurde durch Ausblasen mit Stickstoff entfernt, der Film kurz im Vakuum getrocknet und der Abbau mit Programmpunkt 1 begonnen (Kupplung). Diese Methode ziehen wir heute vor, da die Markierung an der ϵ-Gruppe des C-terminalen Lysins auf diese Weise vollständiger zu sein scheint.

b) Reinigung der Peptide bei automatischer Sequenzanalyse

Sämtliche Peptide wurden unabhängig von der Art der Isolierung über eine kleine Säule (0.9 x 30 cm) Biogel P2 oder Sephadex G-10 rechromatographiert (Fraktionsvol. 1 m*l*). Die Rechromatographie erfolgte entweder in 1N Essigsäure oder 0.1proz. NH_4HCO_3. Das Peptid wurde im ersten Fall in 150 - 200 μl 50proz. Essig- oder Ameisensäure, im zweiten Fall nach Zugabe von 100 mg Harnstoff in 150 - 200 μl H_2O gelöst. Die Fraktionen wurden bei 230 nm (2-mm-Küvette) vermessen. Die Peptide erscheinen zwischen der 5. und der 11. Fraktion.

B) Propingprogramm

a) Herkunft bzw. Reinigung von Dimethyl- und Diäthylaminopropin

Beide Präparate wurden von der Firma Fluka (Buchs, Schweiz) erhalten. 3-Dimethylaminopropin wurde, wie früher beschrieben, gereinigt. 3-Diäthylaminopropin verwendeten wir in jüngster Zeit, da dieses (Siedepunkt 118 °C) konstantere Filme liefert. Die Reinigung mußte allerdings modifiziert werden und wurde wie folgt durchgeführt. 100 m*l* Diäthylaminopropin wurden mit 1 g Maleinsäureanhydrid versetzt, über Nacht stehen gelassen und filtriert. Die Destillation erfolgte in einer Vakuumapparatur unter Stickstoff und Wasserkühlung beim Vakuum der Wasserstrahlpumpe. Um Verluste zu vermeiden, mußten wir das Diäthylaminopropin in einer Aceton-Trockeneis-Falle auffangen. Meist wurde 1/4 der Menge als Vorlauf verworfen und 2/3 vom Rest als Ausgangsmaterial für den Puffer verwendet. Die Reinheit des Präparats wurde spektroskopisch, insbesondere durch NMR getestet. Gute Präparate zeigen auch nach 30 Abbauschritten oder mehr, einen farblosen Film. Schlechte Präparate sind dadurch erkennbar, daß sie gelbe Filme liefern.

b) Einstellung des Puffers

11.2 g Diäthylaminopropin + 35 g n-Propanol + Wasser ad. 100 m*l* mit 1N Essigsäure auf pH 9.2 eingestellt.

c) Abbau von Lysylpeptiden mit dem Propinprogramm

0.5 - 1 μmol Peptid wurden in Wasser suspendiert, kurz am Schüttler kräftig gerührt und mit 1.5 - 2.0 mg I versetzt, geschüttelt und die Suspension bei niederer Gegeschwindigkeit in den Becher gebracht und getrocknet. Anschließend wurden in den stehenden Becher 250 μl Dimethylaminopropin : H_2O (1:1) eingebracht, die Kammer geschlossen und einige Minuten mit N_2 die Luft verdrängt. Die Kopplung von I wurde durch Einschalten der niederen Geschwindigkeit (slow speed) in Gang gesetzt. Nach 15 min wurde kurz mit N_2 gespült und anschließend mit reduziertem Vakuum getrocknet und der Abbau mit Programmpunkt I gestartet. So wird erreicht, daß der erste Schritt, nämlich die Reaktion mit Senföl komplett ist, falls die Kopplung mit I nicht quantitativ verläuft.

d) Konversion und Identifizierung

Früher haben wir mit 1N HCl die Konversion durchgeführt. Später zeigte sich, daß die Salze der Trifluoressigsäure und der Heptafluorbuttersäure mit den Phenylthiohydantoin-Derivaten von Histidin und Arginin und in geringer Menge die Cysteinsäure im Essigester löslich sind. Dies veranlaßte uns, eine einfache Methode zu entwikkeln, um in der Dünnschicht sämtliche Aminosäuren auf einer Platte zu identifizieren. Wir verwendeten meist 20proz. Trifluoressigsäure und behielten die Temperatur der HCl-Konversion (80 °C) bei. Es zeigte sich empirisch aber, daß die Zeit etwas verlängert werden muß. Die Reaktionsbedingungen waren 80 °C, 14 min. Die abgekühlten Fraktionen wurden anschließend im Gerät von Begg durch Gegenstrom extrahiert, nur die Essigesterphase abgedampft und die Thiohydantoine in Essigester/Methanol 1:1 (50 μl) gelöst. In einem aliquoten Teil (5 μl) wurde stets die Ausbeute bestimmt (Messung der Absorption bei 259 nm). Dadurch war es möglich, beim Tüpfeln der Probe auf die Dünnschichtplatte, optimal zu quantitieren. Zur Kontrolle wurde die Menge an Phenylthiohydantoin auch stets noch auf der getüpfelten Platte im UV getestet.

Wie schon früher beschrieben, entwickelten wir die Phenylthiohydantoine in drei Systemen hintereinander an derselben Platte

I. Ester (tert. Butylpropionat),
II. Äthylenchlorid/Essigsäure (60:15) und
III. Äthylenchlorid/Methanol (80:20).

Im ersten System werden die Phenylthiohydantoin-Derivate der Aminosäuren, die hydrophoben Charakter besitzen, sehr gut getrennt. Insbesondere gut ist die Trennung Leu/Ile. Im zweiten System werden die restlichen Aminosäuren bis Gln/Asn getrennt. Dieses System muß eingestellt werden und hängt von der Aktivität der Kieselgelplatten ab. Als Richtmaß kann gelten, daß soviel Essigsäure hinzugegeben wird, daß nach dem Lauf im zweiten System die Flecken Leu/Ile ca. 3 cm von der Front entfernt sind: Weiterhin liegt Threonin zwischen Glutamin- und Asparaginsäure. Das System III dient zur Trennung der Derivate von Histidin und Arginin, die nach den ersten beiden Systemen immer noch am Start verbleiben. Die Trennung der beiden Phenylthiohydantoine, die im allgemeinen als schwierig beschrieben wird, gelingt aber sehr gut und beide wandern als scharfe Flecken und sind einwandfrei zu identifizieren. In Abb. 1 ist das Schema unserer Identifizierung wiedergegeben, indem wir die Testgemische *1*, *2* und *3* nach Edman wiedergeben. Aus der Lage der einzelnen Phenylthiohydantoine ist die eindeutige Identifizierung ohne weiteres ersichtlich. Die Ninhydrinreaktion auf Thiohydantoine als zusätzliche Identifizierung eines Fleckes verwenden wir nur selten.

Die Proben wurden automatisch und quantitativ mit dem TLC-Applikator der Firma Woelm-Pharma GmbH., Eschwege, getüpfelt: es werden besonders scharfe Flekken erhalten.

e) Abbau im Sequenator

Als Sequenator diente der Sequencer der Firma Beckman Instruments, Palo Alto, U.S.A. Wir haben das Gerät auf Anraten und Empfehlung von Dr. *Edman* modifiziert. Diese Modifikation wurde aber ausschließlich für das Propinprogramm verwendet. Ohne das Gerät umzubauen, haben wir zusätzlich einen Synchrongenerator eingebaut, der es erlaubt, mit dem Gerät bei 2700 U./min (Highspeed) und 2000 U./min (Lowspeed) zu arbeiten. Dies bewirkt einerseits, daß der Film dünner ist als bei der alten Version des Sequenators und insbesondere, daß das Waschen der Filme viel effizienter und auch mit kleinen Volumina erfolgen kann. Nach mehrjähriger Erfahrung können wir feststellen, daß der Sequencer ohne Umbauten in dieser Geschwindigkeit arbeiten kann.

S_1 und S_2 und S_3 wurden also stets bei 2700 U in das Gerät gebracht, wodurch das Waschen verkürzt werden kann. Hierzu soll aber später an anderer Stelle berichtet werden.

Ergebnisse

Abbau des intakten Proteins

Die Sequenzanalyse im Sequenator wurde mit einem oxidierten Produkt oder mit einem Carboxymethylderivat durchgeführt und beide Ketten wurden über 45 Schritte analysiert. An beiden Ketten konnten sämtliche Aminosäuren durch Dünnschichtchromatographie allein identifiziert werden.

a) b) c)

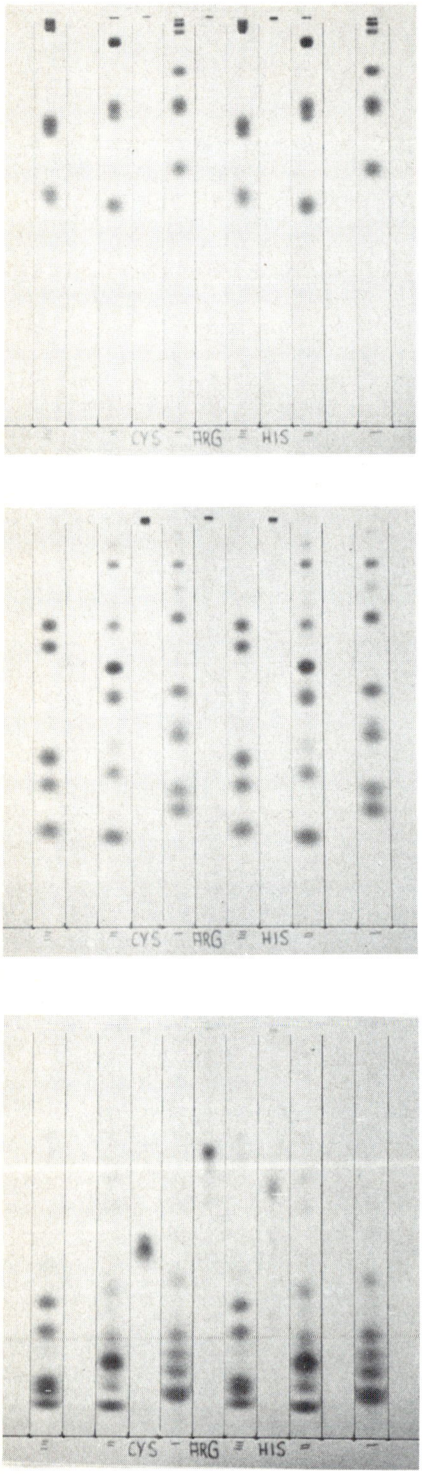

◁ Abb. 1. Auftrennung der Phenylthiohydantoin-Derivate von Aminosäuren mittels Dünnschichtchromatographie. Nach Trifluoressigsäure-Konversion und Extraktion in Essigester werden die eingedampften Fraktionen in Essigester/Methanol (1 : 1, 50 μl) gelöst. Mit 5 μl hiervon wird durch UV-Absorptionsmessung die Ausbeute bestimmt. Entsprechende Aliquoten werden durch den Applikator (s. exptl. Teil) in Paketen zu jeweils 10 getüpfelt. Neben jeder Fraktion läuft eines der Testgemische *1*, *2* oder *3* (in der Abb. mit –, = bzw. ≡ bezeichnet). *Testgemisch 1:* Asn, Ser, Methioninsulfon, Gly, Ala, Pro und Val; *Testgemisch 2:* Gln, Thr, Lys, Tyr, Met und Leu; *Testgemisch 3:* Asp, Glu, Trp, Phe und Ile.

Zur Entwicklung wurden die drei unter Methoden beschriebenen Systeme verwendet. System I trennt die hydrophoben Aminosäuren, System II trennt sämtliche Phenylthiohydantoin-Aminosäuren bis Glutamin und Asparagin und im III. System werden die Phenylthiohydantoin-Derivate von Histidin, Arginin und Cysteinsäure, die nach Trifluoressigsäure-Konversion ebenfalls im Essigester vorgefunden werden, aufgetrennt.

a) Nach Entwicklung in System I.
b) Dieselbe Dünnschichtplatte nach zusätzlicher Entwicklung in System II.
c) Dieselbe Dünnschichtplatte, entwickelt in System I, II und III nacheinander.

Sequenzanalyse der tryptischen Peptide

Diese erfolgte ohne Probleme, da die Anzahl der Aminosäuren der tryptischen Peptide im Vergleich zur Kapazität des Sequenators klein ist.

α-Ketten

Die tryptischen Peptide der α-Ketten sind in Tab. 1 zusammengestellt.

Tp1 (Pos. 1 - 7): Methode: A; Reagenz: IV; Sequenz: Val-Leu-Ser-Ala-Ala-Asp-Lys.

Tp2 (Pos. 8 - 11): Methode: A; Reagenz: IV; Sequenz: Ala-Asn-Val-Lys.

Tp3 (Pos. 12 - 16): nicht isoliert, an der intakten Kette sequenziert. Sequenz: Ala-Ala-Trp*-Gly-Lys.

Tp4 (Pos. 17 - 31): Methode: B; Sequenz: Val-Gly-Gly-Gln-Ala-Gly-Ala-His-Gly-Ala-Glu-Ala-Leu-Glu-Arg.

---

* Oxidierte Form von Tryptophan.

Tab. 1. Tryptische Peptide der α-Kette des Hämoglobins des Schweines.
Hydrolysen: 6N HCl, 110 °C, 20 h. Lediglich die Serin- und Threonin-Werte sind korrigiert.

Lösl. Pept.	Tp1	Tp2	Tp3	Tp4	Tp5	Tp6	Tp7+8	Tp9a	Tp9b	Tp10	Tp11	Tp12	Tp13	Tp14	Summe d. Peptide	Analysen d. Kette
Lys	1.00	1.10	1.01		1.05	0.92	1.01	0.99	0.96		1.00	1.06	1.03		11	11.08
His				1.08		2.03	0.99		2.71			3.80			11	11.40
Arg				1.01						1.02				0.89	3	3.21
Asp	1.05	1.04				2.06		0.98	2.98		1.94	3.81	0.91		15	14.92
Thr					1.92	0.99		1.11				1.01	1.89		7	7.01
Ser	0.96					1.94			1.99			2.88	1.95		10	10.50
Glu				2.88		1.09	0.98								5	5.30
Pro					0.95	0.90			1.15		0.97	1.95			6	5.54
Gly			1.10	3.77	1.21	1.04	0.99		1.95						10	10.10
Ala	2.04	0.94	1.92	3.96			0.99	1.93	4.01			3.20	0.95		20	19.50
Cys												1.06			1	0.90
Val	0.98	0.91		1.20		0.92		0.99	1.04		1.79	2.10	2.00		12	11.30
Met					0.97										1	1.10
Ile															0	0
Leu	0.94			1.08	0.92	0.97		0.95	5.17	0.97		6.03	2.17		19	18.70
Tyr						0.99								1.11	2	2.10
Phe					1.97	1.99					1.20	1.09	1.09		7	7.23
Trp			0.74												1	0.71
AS-Reste	7	4	5	15	9	16	5	7	22	2	7	28	12	2	141	

Tp5 (Pos. 32 - 40): Methode: A; Reagenz: IV; Sequenz:
Met-Phe-Leu-Gly-Phe-Pro-Thr-Thr-Lys.

Tp6 (Pos. 41 - 56):
Methode: A; Reagenz: IV; Sequenz:
Thr-Tyr-Phe-Pro-His-Phe-Asn-Leu-Ser-His-Gly-Ser-Asp-Gln-Val-Lys.

Tp7+8 (Pos. 57 - 61):
Methode: A+B; Reagenz: IV; Sequenz:
Ala-His-Gly-Gln-Lys.

Tp9a (Pos. 62 - 68): Methode: A; Reagenz: IV; Sequenz:
Val-Ala-Asp-Ala-Leu-Thr-Lys.

Tp9b (Pos. 69 - 90): Methode: A; Reagenz: IV; Sequenz:
Ala-Val-Gly-His-Leu-Asp-Asp-Leu-Pro-Gly-Ala-Leu-Ser-Ala-Leu-Ser-Asp-Leu-His-Ala-His-Lys.

Tp10 (Pos. 91 - 92):
Methode: B; Sequenz:
Leu-Arg.

Tp11 (Pos. 93 - 99): Methode: A; Reagenz: IV; Sequenz:
Val-Asp-Pro-Val-Asn-Phe-Lys.

Tp12 (Pos. 100 - 127):
Methode A; Reagenz: IV; Sequenz:
Leu-Leu-Ser-His-Cys-Leu-Leu-Val-Thr-Leu-Ala-Ala-His-His-Pro-Asp-Asp-Phe-Asn-Pro-Ser-Val-His-Ala-Ser-Leu-Asp-Lys.

Tp13 (Pos. 128 - 139): Methode: A; Reagenz: IV; Sequenz:
Phe-Leu-Ala-Asn-Val-Ser-Thr-Val-Leu-Thr-Ser-Lys.

Tp14 (Pos. 140 - 141): Methode: B; Sequenz:
Tyr-Arg.

β-Ketten

Die tryptischen Peptide der β-Ketten sind in Tab. 2 zusammengestellt.

Tp1 (Pos. 1 - 8): Methode: A; Reagenz: IV; Sequenz:
Val-His-Leu-Ser-Ala-Glu-Glu-Lys.

Tp1+2 (Pos. 1 - 17): Methode: A; Reagenz: IV; Sequenz:
Val-His-Leu-Ser-Ala-Glu-Glu-Lys-Glu-Ala-Val-Leu-Gly-Leu-Trp*-Gly-Lys.

---

* Oxidierte Form von Tryptophan.

Tab. 2. Tryptische Peptide der β-Ketten des Hämoglobins des Schweines. Weitere Angaben s. Legende zu Tab. 1.

Lösl. Pept.	Tp1	Tp1+2	Tp3	Tp4	Tp5	Tp6	Tp7	Tp8	Tp9a	Tp9b	Tp10	Tp11	Tp12a	Tp12b	Tp13	Tp14	Tp15	Summe d. Peptide	Analysen d. Kette
Lys	1.07	1.90			1.11	0.99	0.95	0.99	0.95	0.93	1.07				1.05	1.12		11	11.1
His	0.82	0.99					1.02			0.89		1.86			0.92	0.91	1.06	8	8.1
Arg			0.91	1.06								0.96	0.98	0.99				5	5.2
Asp			1.96		4.27				0.99	2.03		2.95	1.36		2.84	1.05		17	16.9
Thr				1.04							0.94							2	2.1
Ser	0.99	1.05			1.99				1.90			1.08						6	6.1
Glu	2.05	3.16	2.05	1.09	1.00				1.09			2.83			1.85			13	13.1
Pro				0.94	0.98							1.14			0.92			4	3.6
Gly		2.10	3.04		1.87		0.96		1.12		1.00		1.01		0.99	1.33		13	13.3
Ala	1.05	1.92	1.04		1.95		1.05				0.98		0.86		2.04	3.78		14	14.2
Cys												0.84						1	0.9
Val	0.84	1.79	2.99	1.71	0.92	1.01			1.05			1.14	4.05		1.12	1.96[a]		19	19.1
Met					0.68													1	0.9
Ile													0.72					1	1.0
Leu	1.17	2.97	1.09	2.16	1.06				1.97	2.09		3.13	2.89		0.99	1.12		19	18.8
Tyr				0.99													0.93	2	2.0
Phe					3.08				0.92		0.98	1.08			2.18			8	8.3
Trp		0.88		0.91														2	1.8
AS-Reste	8	17	13	10	19	2	4	1	10	6	5	17	12	1	15	12	2	146	

[a] bei 200 h Hydrolyse 3 Val.

Tp3 (Pos. 18 - 30): Methode: B; Sequenz:
Val-Asn-Val-Asp-Glu-Val-Gly-Gly-Glu-Ala-Leu-Gly-Arg.

Tp4 (Pos. 31 - 40): Methode: B; Sequenz:
Leu-Leu-Val-Val-Tyr-Pro-Trp*-Thr-Gln-Arg.

Tp5 (Pos. 41 - 59): Methode: A; Reagenz: IV; Sequenz:
Phe-Phe-Glu-Ser-Phe-Gly-Asp-Leu-Ser-Asn-Ala-Asp-Ala-Val-Met-Gly-Asn-Pro-Lys.

Tp6 (Pos. 60 - 61): Methode: B; Sequenz:
Val-Lys.

Tp7 (Pos. 62 - 65): Methode: A + B; Reagenz: IV; Sequenz:
Ala-His-Gly-Lys.

Tp8 (Pos. 66): Lys.

Tp9a (Pos. 67 - 76): Methode: A; Reagenz: IV; Sequenz:
Val-Leu-Gln-Ser-Phe-Ser-Asp-Gly-Leu-Lys.

Tp9b (Pos. 77 - 82): Methode: A; Reagenz: IV; Sequenz:
His-Leu-Asp-Asn-Leu-Lys.

Tp10 (Pos. 83 - 87): Methode: A; Reagenz: IV; Sequenz:
Gly-Thr-Phe-Ala-Lys.

Tp11 (Pos. 88 - 104): Methode: B; Sequenz:
Leu-Ser-Glu-Leu-His-Cys-Asp-Gln-Leu-His-Val-Asp-Pro-Glu-Asn-Phe-Arg.

Tp12a (Pos. 105 - 116): Methode: B; Sequenz:
Leu-Leu-Gly-Asn-Val-Ile-Val-Val-Val-Leu-Ala-Arg.

Tp12b (Pos. 117): Arg.

Tp13 (Pos. 118 - 132): Methode: A; Reagenz: IV; Sequenz:
Leu-Gly-His-Asp-Phe-Asn-Pro-Asp-Val-Gln-Ala-Ala-Phe-Gln-Lys.

Tp14 (Pos. 133 - 144): Methode: A; Reagenz: IV; Sequenz:
Val-Val-Ala-Gly-Val-Ala-Asn-Ala-Leu-Ala-His-Lys.

Tp15 (Pos. 145 - 146): Methode: B; Sequenz:
Tyr-His.

* Oxidierte Form von Tryptophan.

Diskussion

Die Sequenz

Die Sequenz des Hämoglobins des Schweines ist in Abb. 2 wiedergegeben: sie wurde automatisch erarbeitet und noch mit „der Strategie der tryptischen Spaltung" der Ketten durchgeführt: die Anzahl der Methionine (1 oder 2) und Arginine (3) ist gering, so daß auf dem direkten Weg große Spaltprodukte erhalten werden, die mehr Aminosäurereste enthalten als mit den Sequenzmethoden abgebaut werden können. Die Größe der tryptischen Peptide dagegen ist klein gegenüber der Potenz des Sequenators, so daß die Sequenzierung auf dieser Ebene kein Problem darstellt. Wir beschreiben einige Verbesserungen des Sequenators u.a. des Propinprogramms: insbesondere möchten wir auf die Durchführung der Waschung der Filme bei hoher Geschwindigkeit (Umdrehungsgeschw. des Bechers 2700 U./min) hinweisen, da der Film der Waschmittel S_1, S_2, S_3 viel dünner wird und das nötige Waschvolumen der Flüssigkeit umgekehrt proportional der Filmdicke, also der Umdrehungszahl ist. Die α-Ketten unterscheiden sich in 22, die β-Ketten ebenfalls in 22 Aminosäuren vom Hämoglobin des adulten Menschen. Die Durchsicht der Sequenz zeigt, daß von diesen Austauschen weder die $\alpha_1\beta_1$-, $\alpha_1\beta_2$- noch die Hämkontakte[15] betroffen sind, auch nicht, wie weiter unten näher diskutiert, die vier Bisphosphoglycerat-Kontakte, d.h., daß Sauerstofftransport in beiden Fällen gleich ist, was auch die physiologischen Daten ergeben. Die identischen Austausche z.B. bei Pferd, Lama, Rind, bekräftigen die nahe Verwandtschaft dieser Spezies.

Zur Regulation der Expression des foetalen Gens des Schweines

Die Entdeckung[16,17] des foetalen Hämoglobins beim Menschen durch Körber 1866[18] wurde erst seit den 30er Jahren weiter bearbeitet und durch den Nachweis[19], Isolierung[20], Sequenzanalyse[21], Charakterisierung seiner physiologischen Eigenschaften[22] sowie durch die vor einigen Monaten erfolgte Röntgenanalyse[23] zu Ende geführt. Während die Verhältnisse beim Menschen aufgeklärt sind, haben analoge Versuche bei einigen Tieren, z.B. Pferd und Schwein, insofern negative Ergebnisse erbracht, als der Nachweis eines foetalen Proteins nie glückte. Beim Schwein haben wir besonders sorgfältige Untersuchungen vorliegen, insbesondere von Kleihauer et al.[24,25]. Im praenatalen Leben des Schweines werden zwar embryonale Hämoglobine nachgewiesen, der postembryonale Foetus hat jedoch nur noch eine Komponente, die in ihren physiologischen und elektrophoretischen Eigenschaften dem adulten Hämoglobin entspricht. Die endgültige Klärung dieses Befundes ist erst in jüngster Zeit erfolgt. Zwar zeigte sich, daß das foetale Blut des Schweines eine höhere Sauerstoffaffinität besitzt, jedoch war es bis vor kurzem nicht möglich, diesen Befund mit anderen experimentellen Daten in Übereinstimmung zu bringen. Die Physiologie des foetalen Schweinehämoglobins wurde durch die Entdeckung des 2,3-Bisphosphoglycerats[26] in den Erythrozyten und insbesondere in den letzten Jahren durch die Beschreibung von dessen Funktion als allosterischem heterotropem Effektor[27] und weiterhin durch die Bestimmung der Kontaktstellen[28] im Hämoglobin selbst aufgeklärt. Die Untersuchungen ergaben, daß das *Sauerstofftransportsystem nicht aus 2 sondern aus 3 Komponenten besteht*, nämlich Sauerstoff (Kohlendioxid), Hämoglobin und Phosphat.

Bisphosphoglycerat-freies Hämoglobin (stripped Hb) besitzt eine höhere O_2-Affinität (intrinsische Affinität) als Blut. 1 mol Tetramer bindet stöchiometrisch 1 mol Bisphosphoglycerat, wobei die Affinitätskonstante zur Desoxy-Form um annähernd 2 Größenordnungen größer ist. Die Kontaktstellen sind nur die der β-Ketten. Bisphosphoglycerat erniedrigt daher die Affinität des Moleküls zu O_2, was eine regulierte und leichtere Abgabe des Sauerstoffs im Gewebe ermöglicht.

Foetales Hämoglobin = adultes Hämoglobin minus Bisphosphoglycerat

Die 2,3-Bisphosphoglycerat-Kontakte wurden am Hämoglobin des Menschen durch Röntgenanalyse erarbeitet[28] und sind: β1α-NH_2, β2-Histidin, β82-Lysin und β142-Histidin. Die Kontakte des Bisphosphoglycerats sind also, wie ein Vergleich mit den Sequenzen beim Schwein zeigt, bei beiden Species identisch. Die Physiologie wurde in den letzten Jahren gleichzeitig und unabhängig von Duhm et al.[29] und Bartels et al.[30] geklärt und deren Ergebnisse stehen in ausgezeichneter Übereinstimmung mit der hier vorgelegten Sequenz. Die Messung des Sauerstoffpartialdrucks in situ in der Placenta des Schweines hat die höhere Sauerstoffaffinität bestätigt, auch das Blut des Neu-

```
H                Pro          Thr     10                          Ala-His  21     Glu-Tyr
S α  Val-   -Leu-Ser-Ala-Ala-Asp-Lys-Ala-Asn-Val-Lys-Ala-Ala-Trp-Gly-Lys-Val-Gly-Gly-Gln-Ala-Gly-Ala-His-Gly-Ala-Glu-Ala-Leu-
S β  Val-His-Leu-Ser-Ala-Glu-Glu-Lys-Glu-Ala-Val-Leu-Gly-Leu-Trp-Gly-Lys-Val-Asn-      -Val-Asp-Glu-Val-Gly-Gly-Glu-Ala-Leu-
H              Thr-Pro          Ser  10    Thr-Ala                         20

   30              Ser              40                        Asp        50                  Ala
  α -Glu-Arg-Met-Phe-Leu-Gly-Phe-Pro-Thr-Thr-Lys-Thr-Tyr-Phe-Pro-His-Phe-   -Asn-Leu-Ser-His-              -Gly-Ser-Asp-
  β -Gly-Arg-Leu-Leu-Val-Val-Tyr-Pro-Trp-Thr-Gln-Arg-Phe-Phe-Glu-Ser-Phe-Gly-Asp-Leu-Ser-Asn-Ala-Asp-Ala-Val-Met-Gly-Asn-Pro-
       30                          40                           49 Thr

            Gly        Lys  61                    Asn       70 Ala     Val        Met     Asp      80
  α -Gln-Val-Lys-Ala-His-Gly-Gln-Lys-Val-Ala-Asp-Ala-Leu-Thr-Lys-Ala-Val-Gly-His-Leu-Asp-Asp-Leu-Pro-Gly-Ala-Leu-Ser-Ala-Leu-
  β -Lys-Val-Lys-Ala-His-Gly-Lys-Lys-Val-Leu-Gln-Ser-Phe-Ser-Asp-Gly-Leu-Lys-His-Leu-Asp-Asn-Leu-Lys-Gly-Thr-Phe-Ala-Lys-Leu-
       60                        Gly-Ala- 71              Ala          80                  Thr

                   90                            100                           110       Leu
  α -Ser-Asp-Leu-His-Ala-His-Lys-Leu-Arg-Val-Asp-Pro-Val-Asn-Phe-Lys-Leu-Leu-Ser-His-Cys-Leu-Leu-Val-Thr-Leu-Ala-Ala-His-His-
  β -Ser-Glu-Leu-His-Cys-Asp-Gln-Leu-His-Val-Asp-Pro-Glu-Asn-Phe-Arg-Leu-Leu-Gly-Asn-Val-Ile-Val-Val-Val-Leu-Ala-Arg-Arg-Leu-
       90              Lys              100                          Leu 111 Cys          His-His-Phe

     Ala-Glu     Thr     Ala 121                         130 Ser                   140
  α -Pro-Asp-Asp-Phe-Asn-Pro-Ser-Val-His-Ala-Ser-Leu-Asp-Lys-Phe-Leu-Ala-Asn-Val-Ser-Thr-Val-Leu-Thr-Ser-Lys-Tyr-Arg
  β -Gly-His-Asp-Phe-Asn-Pro-Asp-Val-Gln-Ala-Ala-Phe-Gln-Lys-Val-Val-Ala-Gly-Val-Ala-Asn-Ala-Leu-Ala-His-Lys-Tyr-His
     Lys-Glu 122 Thr     Pro              Tyr 131                  140
```

Abb. 2. Die Sequenz der α- und β-Ketten des Hämoglobins des adulten Schweines in homologer[14] Weise gegenübergestellt. Die 22 bzw. 21 Austausche der α- bzw. β-Ketten gegenüber Humanhämoglobin wurde dadurch gekennzeichnet, daß die im Human-Hämoglobin vorkommenden, unterschiedlichen Reste der Sequenz übergestellt wurden. Die für die foetale Atmung erforderliche höhere Affinität wird mit demselben Molekül, (das alle vier Bisphosphoglycerat-Kontakte enthält) jedoch mit geringeren Reduktionen der intrinsischen Affinität des Hämoglobins erzielt, was durch eine um ca. 60% niederere Bisphosphoglycerat-Konzentration gegenüber dem Werte des adulten Schweines erreicht wird (H = Human, S = Schwein) (vgl. Diskussion).

geborenen besaß noch eine erhöhte Sauerstoffaffinität ($p_{50}O_2$ = 3.06 kPa), die jedoch nach wenigen Tagen den Normalwert des adulten Schweines ($p_{50}O_2$ = 4.40 kPa) annahm. Die Lösung dieser anscheinend widersprechenden Daten ergab erst die Messung des Bisphosphoglycerat-Spiegels. Es konnte gezeigt werden, daß die Verminderung der Sauerstoffaffinität nach der Geburt *parallel* zur Erhöhung des Bisphosphoglycerat-Spiegels verläuft, der innerhalb von wenigen Tagen den Pegel des adulten Tieres erreicht (vor der Geburt 2.40mM, einige Tage danach ca. 8mM). Dieses Ergebnis ist überraschend, erklärt aber endgültig die Placentaatmung: das adulte Hämoglobin des Schweines hat wie das des Menschen 4 Kontaktstellen für Bisphosphoglycerat und dessen Konzentration reicht aus, um alle Kontaktstellen zu besetzen. Die Folge ist eine maximale Reduktion der O_2-Affinität. Die Bisphosphoglycerat-Konzentration im Foetus (1/4 der im Erwachsenen) ist aber zu gering, um die intrinsische O_2-Affinität voll zu reduzieren, d.h. alle vier Kontaktstellen zu besetzen, was in einer *höheren* O_2-Affinität des Moleküls, wie sie in der Placenta notwendig ist, um eine schnelle Diffusion des Sauerstoffs aus dem mütterlichen Kreislauf zu ermöglichen, resultiert. Die Regulation der Sauerstoffaffinität wird daher durch die Regulation der intrinsischen Affinität des Hämoglobins bewirkt und hier durch eine verminderte Bisphosphoglycerat-Biosynthese eingestellt. Die Expression des foetalen Gens wird also kompensiert durch die Regulation der Biosynthese des heterotropen, allosterischen Effektors, d.h. durch die Menge an Bisphosphoglycerat. Es entsteht eine ähnliche Situation wie bei der Höhenatmung des adulten Lamas, wo jedoch bei *äquimolarer Bisphosphoglycerat-Menge* eine der vier Kontaktstellen durch eine *Mutation* in $\beta 2$ ($\beta 2$ His → Asn) keine Wasserstoffbrücke zum Phosphat mehr bilden kann[5]. Da das Bisphosphoglycerat durch die Glykolyse der Erythrozyten (Rapoport-Luebering-Zyklus) synthetisiert wird[31], entsteht die Frage, auf welcher Ebene der Sauerstoff mit der Regulation des Bisphosphoglycerats gekoppelt ist[32].

Die Regulation der Sauerstoffaffinität des Hämoglobins unterliegt schon bei Vertebraten einer großen Mannigfaltigkeit. Nicht nur, daß das Phosphat als allosterischer heterotroper Effektor bei verschiedenen Arten verschieden vorliegt, als (Bisphosphoglycerat bei Säugern, als Inosinpentaphosphat z.B. bei Vögeln und als Adenosintriphosphat bei Fischen), sondern daß die Regulation über den Spiegel des allosterischen Effektors den optimalen Partialdruck des Sauerstoffes einstellt und so das Fehlen einer Genexpression kompensiert.

Diese Arbeiten waren im wesentlichen im Jahre 1975 abgeschlossen.

Unser Dank gilt Herrn Dr. *Lutz Heck*, Direktor am Tierpark Hellabrunn. Besonderen Dank schulden wir Frl. *Cornelia Mayr* für ausgezeichnete Mitarbeit. Herrn Dr. *Jörg Pfletschinger* danken wir für die Weiterentwicklung der Chromatographie der Phenylthiohydantoin-Derivate des Histidins und Arginins.

Literatur

1 XX. Mitteilung: Kleinschmidt, T. & Braunitzer G. (1977) *Hoppe-Seyler's Z. Physiol. Chem.* **358**, 765 - 769.
2 Braunitzer, G., Hilse, K., Rudloff, V. & Hilschmann, N. (1964) *Adv. Protein Chem.* **19**, 1 - 71.
3 Braunitzer, G. & Köhler, H. (1966) *Hoppe-Seyler's Z. Physiol. Chem.* **343**, 290 - 293.
4 Braunitzer, G., Schrank, B. & Stangl, A. (1977) *Hoppe-Seyler's Z. Physiol. Chem.* **358**, 409 - 412.
5 Braunitzer, G., Schrank, B., Stangl, A. & Bauer, C. (1977) *Hoppe-Seyler's Z. Physiol. Chem.* **358**, 921 - 925.
6 Braunitzer, G. & Schrank, B. (1970) *Hoppe-Seyler's Z. Physiol. Chem.* **351**, 417.
7 Rossi-Fanelli, A. & Antonini, E. (1958) *Biochim. Biophys. Acta* **30**, 608 - 615.
8 Hill, R. J. & Craig, L. C. (1959) *J. Am. Chem. Soc.* **81**, 2272 - 2273.
9 Hilse, K. & Braunitzer, G. (1959) *Z. Naturforschung* **14b**, 604 - 606.
10 Rudloff, V. & Braunitzer, G. (1961) *Hoppe-Seyler's Z. Physiol. Chem.* **323**, 129 - 144.
11 Edman, P. & Begg, G. (1967) *Eur. J. Biochem.* **1**, 80 - 91.
12 Braunitzer, G., Schrank, B. & Ruhfus, A. (1970) *Hoppe-Seyler's Z. Physiol. Chem.* **351**, 1589 - 1590.
13 Braunitzer, G., Schrank, B., Ruhfus, A., Petersen, S. & Petersen, U. (1972) *Hoppe-Seyler's Z. Physiol. Chem.* **352**, 1730 - 1732.
14 Braunitzer, G., Gehring-Müller, R., Hilschmann, N., Hilse, K., Hobom, G., Rudloff, V. & Wittmann-Liebold, B. (1961) *Hoppe-Seyler's Z. Physiol. Chem.* **325**, 283 - 286.
15 Perutz, M. F., Muirhead, H., Cox, J. M. & Goaman, L.C.G. (1968) *Nature (London)* **219**, 131 - 139.
16 Künzer, W. (1951) Über den Blutfarbstoffwechsel gesunder Säulinge und Kinder. S. Karger, Basel, Bibliotheka Paediatrika, Fasc. 51.

17 Betke, K. (1954) Der menschliche rote Blutfarbstoff bei Foetus und reifem Organismus, Springer-Verlag, Heidelberg.
18 Körber, E. (1866) Über Differenzen des Blutfarbstoffes. Diss. Dorpat, Zbl. Med. Wiss. **1867**, 117.
19 Jonxis, I. H. P. (1937) *Monatsschr. Kindergeneeskd.* **6**, 356 - 373. Vgl. *Zbl. Kinderheilkd.* **34**, 108 (1938).
20 Rhinesmith, H. S., Schroeder, W. A. & Martin, N. (1958) *J. Am. Chem. Soc.* **80**, 3358 - 3361.
21 Allen, D. W. & Schroeder, W. A. (1958) *J. Am. Chem. Soc.* **80**, 1628 - 1634.
22 Haselhorst, G. & Stromberger, K. (1931) *Z. Geburtshilfe* **100**, 48 - 100.
23 Grier, J.H. & Perutz, M. F. (1977) *J. Mol. Biol.* **112**, 97 - 112.
24 Kleihauer, E. & Stöffler, G. (1968) *Mol. Gen. Genet.* **101**, 59 - 69.
25 Tautz, C. H. & Kleihauer, E. (1972) *Res. Exp. Med.* **159**, 44 - 49.
26 Rapoport, S. & Guest, G.M. (1941) *J. Biol. Chem.* **138**, 269 - 282.
27 Benesch, R. E. & Benesch, R. (1974) *Adv. Protein Chem.* **1974**, 211 - 237.
28 Arnone, A. (1972) *Nature (London)* **237**, 146 - 149.
29 Kim, H. D. & Duhm, J. (1974) *Am. J. Physiol.* **1974** 1001 - 1006.
30 Baumann, R., Teischel, F., Zoch, R. & Bartels, H. (1973) *Respir. Physiol.* **19**, 153 - 161.
31 Rapoport, S. & Luebering, J. (1950) *J. Biol. Chem.* **183**, 507 - 516.
32 Duhm, J. & Gerlach, E. (1974) in The Human Red Cell in Vitro (Greenwalt, T.J. & Jamieson, G. A., eds.) pp. 111 - 152. Grune & Stratton, Inc. London, New York.

Prof. Dr. Gerhard Braunitzer, Max-Planck-Institut für Biochemie, D-8033 Martinsried.

Hoppe-Seyler's Z. Physiol. Chem.
Bd. 359, S. 147 - 153, Februar 1978

Zur Synthese von [11-Leucin]Human-Minigastrin I

I. Mitteilung: Darstellung der Gesamtsequenz

Luis MORODER, Friedrich DREES, Ernst JAEGER und Erich WÜNSCH

Max-Planck-Institut für Biochemie, Abteilung für Peptidchemie, Martinsried bei München

(Der Schriftleitung zugegangen am 23. Dezember 1977)

Herrn Prof. Dr. W. Grassmann, emerit. Direktor des ehem. Max-Planck-Instituts für Eiweiß- und Lederforschung zu seinem 80. Geburtstag in steter Verehrung gewidmet

Zusammenfassung: Die Synthese des 11-Leucin-Analogons von Human-Minigastrin I, eines linearen Tridecapeptids mit voller Gastrin-Wirkung, wird beschrieben. Hierzu wurde ein geeignet geschütztes Hexapeptid-Derivat der Sequenz 1 - 6 aufgebaut und dieses mit den früher beschriebenen Peptid-Derivaten der Sequenz 7 - 9 und 10 - 13 zum allseits geschützten analytisch reinen Tridecapeptidamid Boc-Leu-[Glu(OBu^t)]$_5$-Ala-Tyr(Bu^t)-Gly-Trp-Leu-Asp(OBu^t)-Phe-NH$_2$ verknüpft.

Synthesis of Human [11-Leucine]Minigastrin I. – I. Preparation of the Fully Protected Tridecapeptide Amide

Summary: The synthesis of the tridecapeptide amide corresponding to an 11-leucine analog of human minigastrin I is described. The fully protected peptide derivative was obtained by assembly of three suitably protected fragments corresponding to the sequences 1 - 6, 7 - 9 and 10 - 13 by the carbodiimide/*N*-hydroxysuccinimide method.

Key words: Synthetic peptide hormones, human [Leu^{11}]minigastrin I.

Rehfeld und Stadil[3] gelang 1973 der Nachweis einer bislang unbekannten immunreaktiven Gastrin-Komponente niederen Molekulargewichts im menschlichen Blutserum. Dieses neue Mitglied der Gastrinfamilie wurde schließlich 1974 von Gregory und Tracy[4] aus gastrinproduzierenden Tumoren des Zollinger-Ellison-Typs in reiner Form isoliert und als lineares Tridecapeptidamid beschrieben*, in der Primärstruktur dem C-terminalen Sequenzbereich des „Little-Gastrin" entsprechend. Wiederum tritt dieser Peptidwirkstoff,

* Eine gesicherte Identifizierung der Primärstruktur steht noch aus, vgl. dazu Grossman[5]

Abkürzungen: Boc = tert.-Butyloxycarbonyl; Bu^t = tert.-Butyl; OBu^t = tert.-Butylester; DCCD = Dicyclohexylcarbodiimid; ONSu = Succinimidester; OMe = Methylester; Z = Benzyloxycarbonyl.
Aufgrund der Vereinbarungen biochemischer Zeitschriften werden hier die von der IUPAC-IUB Commission on Biochemical Nomenclature empfohlenen Abkürzungen für Aminosäuren und Schutzgruppen verwendet. Die Autoren bevorzugen aber wegen ihrer universellen Anwendbarkeit die im Houben-Weyl[1] verwendeten und eingehend erläuterten Abkürzungen. Die Abkürzungen HG-17 I für „Human Little Gastrin I" und HG-13 I für „Human Minigastrin I" wurden von Walsh et al.[2] vorgeschlagen.

Human-Minigastrin (HG-13) genannt, in den für diese Hormonklasse charakteristischen zwei Formen auf, d.h. entweder mit einem Tyrosin-Rest (Form I) oder einem Tyrosin-*O*-Sulfat-Rest (Form II) in der Position 8 der Sequenz. Gastrische Wirkung, Plasma-Halbwertzeit und -Verteilungsvolumen von HG-13 I wurden von Debas et al.[6] ermittelt.

Die biologischen Eigenschaften und der unzweifelhaft schon der kürzeren Sequenz wegen „aufwendungs-geringere" Syntheseweg offerieren HG-13 I als brauchbaren Ersatz des „Normal-Hormons", d.i. HG-17 I, für den biologisch-medizinischen Einsatz.

Die Sequenzverkürzung am Aminoende erspart unter anderem dem Peptidchemiker den Einbau des zweiten Tryptophan-Restes in die Peptidkette, wodurch ein bedeutender Störfaktor in der Synthese gemindert wird: in unseren Versuchen zur Totalsynthese von [15-Leucin]Human-Gastrin I[7] hatten wir lange Zeit mit einem in größerer Menge anfallenden Nebenprodukt zu kämpfen, dessen Entstehung im Zuge der „letzten" Schutzgruppen-Abspaltung alsbald als tryptophanbedingt bewiesen werden konnte (s. hierzu auch die folgende Arbeit[8]). Würde es im Falle von HG-13 I im Hinblick auf dessen biologische Aktivität möglich sein, in der Sequenzposition 11 wiederum Methionin durch Leucin erfolgreich zu ersetzten (vgl. die positiven Erfahrungen mit [Leu^{15}]HG-17 I), wäre vom Standpunkt der synthetischen Zugänglichkeit und der Substanz-Stabilität, wie z.B. Aktivitätsverlust bei Oxidation des Methionin- zum Methionin-*S*-oxid-Rest[9], eine Ideallage geschaffen.

Der Aufbau des Tridecapeptidamids erfolgte durch Erstellung von drei in geeigneter Form mas-

Z-Glu(OBu^t)-ONSu **[5]** + H-Glu(OBu^t)-OMe **[6]**

↓

Z-Glu(OBu^t)-Glu(OBu^t)-OMe **[5 - 6a]**

↓ $OH^\ominus$

Z-Glu(OBu^t)-Glu(OBu^t)-OH **[5 - 6b]**

↓ H_2/Pd

Z-Glu(OBu^t)-ONSu **[4]** + H-Glu(OBu^t)-Glu(OBu^t)-OH **[5 - 6c]**

↓

Z-Glu(OBu^t)-Glu(OBu^t)-Glu(OBu^t)-OH **[4 - 6a]**

↓ H_2/Pd

Z-Glu(OBu^t)-ONSu **[3]** + H-Glu(OBu^t)-Glu(OBu^t)-Glu(OBu^t)-OH **[4 - 6b]**

↓

Z-Glu(OBu^t)-Glu(OBu^t)-Glu(OBu^t)-Glu(OBu^t)-OH **[3 - 6a]**

↓ H_2/Pd

Z-Glu(OBu^t)-ONSu **[2]** + H-Glu(OBu^t)-Glu(OBu^t)-Glu(OBu^t)-Glu(OBu^t)-OH **[3 - 6b]**

↓

Z-Glu(OBu^t)-Glu(OBu^t)-Glu(OBu^t)-Glu(OBu^t)-Glu(OBu^t)-OH **[2 - 6a]**

↓ H_2/Pd

Boc-Leu-ONSu **[1]** + H-Glu(OBu^t)-Glu(OBu^t)-Glu(OBu^t)-Glu(OBu^t)-Glu(OBu^t)-OH **[2 - 6b]**

↓

Boc-Leu-Glu(OBu^t)-Glu(OBu^t)-Glu(OBu^t)-Glu(OBu^t)-Glu(OBu^t)-OH **[1 - 6a]**

Schema 1. Teilsequenz 1 - 6 (Fragment III).

kierten Fragmenten, d.s. Fragment I: H-Trp-Leu--Asp(OBu^t)-Phe-NH_2 · HCl [**10 - 13**], Fragment II: Z-Ala-Tyr(Bu^t)-Gly-OH [**7 - 9**] und Fragment III: Boc-Leu-[Glu(OBu^t)]$_5$-OH [**1 - 6**] und deren sukzessive Verknüpfung nach der Wünsch-Weygandschen Methode[10,11].

Für die Fragmente I und II konnte auf die von uns bereits früher erarbeiteten Synthesewege zurückgegriffen werden[12]. Die Synthese von Fragment III (s. Schema 1) erfolgte ausgehend von H-[Glu(OBu^t)]$_2$-OH [**5 - 6c**] – zugänglich nach abgeänderten Vorschriften von Marlborough und Rydon[13] in ca. 80proz. Ausbeute über drei Stufen – durch üblichen stufenweisen Anbau der Aminosäurereste 1 - 4 in „Sequenz-Folge" mittels Z-Glu(OBu^t)-ONSu[14] und Boc-Leu-ONSu[15]. Das in 56proz. Ausbeute (bezogen auf [**5 - 6c**]) anfallende Hexapeptidderivat war chromatographisch und analytisch rein. Zur Erstellung der Gesamtsequenz (s. Schema 2) wurden Z-Ala--Tyr(Bu^t)-Gly-OH ([**7 - 9**], Fragment II) und H-Trp-Leu-Asp(OBu^t)-Phe-NH_2 ([**10 - 13**], Fragment I) – freigesetzt in bekannter Manier aus dem Hydrochlorid – nach dem Carbodiimid/-Hydroxysuccinimid-Verfahren zum seitenkettenmaskierten Benzyloxycarbonyl-heptapeptidamid [**7 - 13a**] verknüpft; dessen hydrogenolytische Entacylierung zu H-Ala-Tyr(Bu^t)-Gly-Trp-Leu--Asp(OBu^t)-Phe-NH_2 [**7 - 13b**] gelang über Isolierung als Hydrochlorid durch Titrieren mit methanolischer Salzsäure bei pH 5.0 während der Abspaltung. Der Anbau von Fragment III, d.i. Boc-Leu-[Glu(OBu^t)]$_5$-OH [**1 - 6**] an das erhaltene Heptapeptidamid-Derivat [**7 - 13b**] – freigesetzt aus dem Hydrochlorid – verlief nach dem geschilderten Verknüpfungsverfahren in Dimethylacetamid/Phosphorsäure-tris[dimethylamid]-Lösung ebenfalls mit Erfolg: Das allseits geschützte Tridecapeptidamid konnte in hoher analytischer und chromatographischer Reinheit und in 67proz. Ausbeute (bezogen auf alle Stufen) erhalten werden.

Wie üblich führte die Abspaltung aller Schutzgruppen mittels wasserfreier Trifluoressigsäure in Anwesenheit von Anisol als Kationenfänger zum Rohprodukt des [11-Leucin]Human-Minigastrins I.

Unserem technischen Assistenten Herrn *J. Musiol* danken wir für die ausgezeichnete Mitarbeit.

Z-Ala-Tyr(Bu^t)-Gly-OH [**7 - 9**] + H-Trp-Leu-Asp(OBu^t)Phe-NH_2 [**10 - 13**]

↓ DCCD/HONSu

Z-Ala-Tyr(Bu^t)-Gly-Trp-Leu-Asp(OBu^t)-Phe-NH_2 [**7 - 13a**]

↓ H_2/Pd

Boc-Leu-Glu(OBu^t)-Glu(OBu^t)-Glu(OBu^t)-Glu(OBu^t)-Glu(OBu^t)-OH [**1 - 6**] + H-Ala-Tyr(Bu^t)-Gly-Trp-Leu-Asp(OBu^t)Phe-NH_2 [**7 - 13b**]

↓ DCCD/HONSu

Boc-Leu-Glu(OBu^t)-Glu(OBu^t)-Glu(OBu^t)-Glu(OBu^t)-Glu(OBu^t)-Ala-Tyr(Bu^t)-Gly-Trp-Leu-Asp(OBu^t)-Phe-NH_2 [**1 - 13a**]

↓ F_3AcOH/Anisol

H-Leu-Glu-Glu-Glu-Glu-Glu-Ala-Tyr-Gly-Trp-Leu-Asp-Phe-NH_2 [**1 - 13b**] (Rohprodukt)

Schema 2. Gesamtsequenz 1 - 13.

Beschreibung der Versuche

Aminosäureanalysen nach saurer Hydrolyse mit 6M HCl (20 h) unter Thioglykolsäure-Zusatz für Tryptophan-haltige Peptide[16] wurden am „Amino Acid Analyzer" mit einem „Automatic Digital Integrator" der Fa. Beckman Instruments (Modell 120B und 125) ausgewertet. Die Schmelzpunkte (nicht korrigiert) wurden im Apparat nach Dr. Tottoli, die spezifischen Drehwerte im lichtelektrischen Polarimeter (Modell 241MC) der Fa. Perkin Elmer ermittelt. Die Dünnschichtchromatographie erfolgte auf DC-Fertigplatten Kieselgel 60 der Fa. E. Merck, Darmstadt.

1) Benzyloxycarbonyl-L-glutamyl(γ-tert.-butylester)-L-glutaminsäure-(γ-tert.-butylester)-methylester **[5-6a]**

71.0 g (280 mmol) *H-Glu(OBu^t)-OMe·HCl* **[6]**[17] und 39.2 m*l* (280 mmol) Triäthylamin in 1 *l* Dioxan werden bei 10 °C mit 121.5 g (280 mmol) *Z-Glu(OBu^t)--ONSu* **[5]** versetzt. Die Reaktionsmischung wird nach 12stdg. Rühren im Vak. eingedampft, der verbleibende Rückstand zwischen Essigester und Wasser verteilt, die abgetrennte Essigesterphase nacheinander mit 10% Kaliumhydrogencarbonat-, 5% Kaliumhydrogensulfat-Lösung und Wasser gewaschen. Nach dem Trocknen über Natriumsulfat wird das Lösungsmittel im Vak. eingedampft und der Rückstand aus Essigester/Diisopropyläther umgefällt. Schmp. 48 - 50 °C; $[\alpha]_D^{20}$: – 18.1° bzw. $[\alpha]_{546}^{20}$: – 21.5° (*c* = 1, in Methanol) [Lit.[13]: Schmp. 46 - 47 °C; $[\alpha]_D^{23}$: – 19.4° (*c* = 1, in Methanol)]; chromatographisch rein in n-Butanol/Essigsäure/Wasser 3:1:1, n-Butanol/Essigsäure/Wasser/Pyridin 60:6:24:20 und n-Heptan/tert.-Butanol/Essigsäure 45:45:10; Ausb. 133.0 g (89% d.Th.).

$C_{27}H_{40}N_2O_9$ (536.6)

Ber.	C 60.43	H 7.51	N 5.22
Gef.	C 60.36	H 7.51	N 5.06

2) Benzyloxycarbonyl-L-glutamyl(γ-tert.-butylester)-L-glutaminsäure(γ-tert.-butylester) **[5-6b]**

125 g (233 mmol) *Z-[Glu(OBu^t)]$_2$-OMe* **[5-6a]** in 750 m*l* Dioxan/Wasser 9:1 werden unter Rühren mit 233 m*l* 1M NaOH titrimetrisch bei 20 °C verseift. Nach Ansäuern der Lösung mit der äquiv. Menge 1M HCl dampft man das Dioxan im Vak. weitgehend ab, nimmt den erhaltenen Rückstand in Essigester auf, extrahiert die organische Phase erschöpfend mit verd. Kaliumhydrogencarbonat-Lösung und führt schließlich nach dem Ansäuern der vereinigten Wasser-Extrakte erneut in Essigester über. Die organische Phase wird mit Wasser säurefrei gewaschen, über Natriumsulfat getrocknet und dann im Vak. eingedampft; Öl; chromatographisch rein in n-Butanol/Essigsäure/Wasser 3:1:1, n-Butanol/Essigsäure/Wasser/Pyridin 60:6:24:20 und Cyclohexan/Chloroform/Essigsäure 45:45:10; Ausb. 115.0 g (94% d.Th.).

Eine analytische Probe wurde aus Diäthyläther als Dicyclohexylammonium-Salz gefällt und aus Methanol/Diäthyläther umkristallisiert. Schmp. 145 - 147 °C; $[\alpha]_D^{20}$: – 2.2° bzw. $[\alpha]_{546}^{20}$: – 2.6° (*c* = 2, in Methanol) [Lit.[13]: Schmp. 143 - 144 °C; $[\alpha]_D^{21}$: – 3.8° (*c* = 1, in Methanol)]; Ausb. 88% d.Th.

$C_{38}H_{61}N_3O_9$ (703.9)

Ber.	C 64.84	H 8.74	N 5.97
Gef.	C 64.73	H 8.77	N 6.20

3) L-Glutamyl(γ-tert.-butylester)-L-glutaminsäure-(γ-tert.-butylester) **[5-6c]**

115 g (220 mmol) *Z-[Glu(OBu^t)]$_2$-OH* **[5-6b]** in 2 *l* 95proz. Methanol werden in Gegenwart von Palladiumschwarz hydriert. Nach beendeter Reaktion wird vom Katalysator abfiltriert und das Filtrat im Vak. zur Trockene eingedampft. Der Rückstand wird aus Methanol/Diäthyläther umkristallisiert. Schmp. 194 - 195 °C; $[\alpha]_D^{20}$: + 25.2° bzw. $[\alpha]_{546}^{20}$: + 29.2° (*c* = 1, in Methanol) [Lit.[13]: Schmp. 180 - 181 °C; $[\alpha]_D^{19}$: + 25.3° (*c* = 0.7, in Methanol)]; chromatographisch rein in n-Butanol/Essigsäure/Wasser 3:1:1 und n-Butanol/Essigsäure/Wasser/Pyridin 60:6:24:20; Ausb. 81.0 g (95% d.Th.).

$C_{18}H_{32}N_2O_7$ (388.5)

Ber.	C 55.65	H 8.30	N 7.21
Gef.	C 55.04	H 8.26	N 7.26

4) Benzyloxycarbonyl-L-glutamyl(γ-tert.-butylester)-L-glutamyl(γ-tert.-butylester)-L-glutaminsäure(γ-tert.-butylester) **[4-6a]**

Zu 77.7 g (200 mmol) *H-[Glu(OBu^t)]$_2$-OH* **[5-6c]** in 1 *l* Dimethylformamid werden bei 0 °C 28 m*l* (200 mmol) Triäthylamin, 86.9 g (200 mmol) *Z-Glu(OBu^t)-ONSu* **[4]** und anschließend 16.2 m*l* (200 mmol) Pyridin gegeben. Die Reaktionsmischung wird 2 h bei 0 °C und anschließend 12 h bei Raumtemperatur gerührt und im Vak. eingedampft. Der feste Rückstand wird zwischen Essigester und verd. Kaliumhydrogensulfat-Lösung verteilt, die abgetrennte Essigesterphase mit Wasser sulfatfrei gewaschen und über Natriumsulfat getrocknet. Die Lösung wird im Vak. zur Trockene gebracht und der Rückstand aus Essigester/Diisopropyläther kristallisiert. Schmp. 107 - 109 °C; $[\alpha]_D^{20}$: – 8.5° bzw. $[\alpha]_{546}^{20}$: – 9.7° (*c* = 1, in Dimethylformamid); chromatographisch rein in Cyclohexan/Chloroform/Essigsäure 45:45:10 und Chloroform/Methanol 9:1; Ausb. 129.0 g (91% d.Th.).

$C_{35}H_{53}O_{12}N_3$ (707.8)

Ber.	C 59.39	H 7.55	N 5.94
Gef.	C 59.25	H 7.72	N 5.82

5) L-Glutamyl(γ-tert.-butylester)-L-glutamyl(γ-tert.-butylester)-L-glutaminsäure(γ-tert.-butylester) **[4-6b]**

120 g (170 mmol) *Z-[Glu(OBu^t)]$_3$-OH* **[4-6a]** in 2 *l* 80proz. Essigsäure werden in Gegenwart von Palladium-

schwarz wie üblich katalytisch hydriert. Nach beendeter Reaktion wird das Filtrat im Vak. eingeengt und das Produkt mit Wasser gefällt, abfiltriert, mit Wasser gewaschen und zum Schluß im Vak. bei 40 °C über KOH getrocknet. Schmp. 149 - 151 °C; $[\alpha]_D^{20}$: + 9.3° bzw. $[\alpha]_{546}^{20}$: + 11.6° (*c* = 1, in 80proz. Essigsäure); chromatographisch rein in Methanol/Chloroform 1: 9 und n-Heptan/tert.-Butanol/Essigsäure 3: 2:1; Ausb. 95.0 g (98% d.Th.).

$C_{27}H_{47}N_3O_{10}$ (573.7)

Ber.	C 56.53	H 8.26	N 7.33
Gef.	C 55.40	H 8.35	N 7.21

6) Benzyloxycarbonyl-L-glutamyl(γ-tert.-butylester)-L-glutamyl(γ-tert.-butylester)-L-glutamyl(γ-tert.-butylester)-L-glutaminsäure(γ-tert.-butylester) **[3-6a]**

90.0 g (157 mmol) *H-[Glu(OBu^t)]$_3$-OH* **[4-6b]** in 1.5 *l* Dimethylformamid werden nacheinander unter Rühren mit 22.0 m*l* (157 mmol) Triäthylamin, 68.2 g (157 mmol) *Z-Glu(OBu^t)-ONSu* **[3]** und nach 2stdg. Rühren bei 0 °C mit 12.7 m*l* Pyridin versetzt. Nach weiteren 12 h Reaktionsdauer bei Raumtemperatur wird das Lösungsmittel weitgehend abgedampft und der Rückstand zwischen Essigester und verd. Kaliumhydrogensulfat-Lösung verteilt. Die abgetrennte organische Phase wird anschließend mit verd. Kaliumhydrogencarbonat-, Kaliumhydrogensulfat-Lösung und Wasser gut gewaschen und letztlich über Natriumsulfat getrocknet. Die Lösung wird im Vak. eingedampft und der Rückstand aus Essigester/Diisopropyläther umgefällt. Schmp. 128 - 130 °C; $[\alpha]_D^{20}$: – 10.1° bzw. $[\alpha]_{546}^{20}$: – 11.9° (*c* = 1, in Dimethylformamid); chromatographisch rein in n-Heptan/tert.-Butanol/Essigsäure 3: 2:1 und Chloroform/Methanol 9:1; Ausb. 123.0 g (88% d.Th.).

$C_{44}H_{68}O_{15}N_4$ (893.1)

Ber.	C 59.17	H 7.68	N 6.27
Gef.	C 59.07	H 7.65	N 6.12

7) L-Glutamyl(γ-tert.-butylester)-L-glutamyl(γ-tert.-butylester)-L-glutamyl(γ-butylester)-L-glutaminsäure-(γ-tert.-butylester) **[3 - 6b]**

110 g (123 mmol) *Z-[Glu(OBu^t)]$_4$-OH* **[3 - 6a]** in 5 *l* 95proz. Methanol werden in Gegenwart von Palladiumschwarz als Katalysator hydriert. Nach beendeter Reaktion wird das Filtrat im Vak. eingedampft und der Rückstand aus Methanol/Essigester gefällt. Schmp. 175 - 177 °C; $[\alpha]_D^{20}$: – 8.5° bzw. $[\alpha]_{546}^{20}$: – 9.9° (*c* = 1, in Dimethylformamid); chromatographisch rein in n-Heptan/tert.-Butanol/Essigsäure 3: 2:1 und Chloroform/Methanol 9:1; Ausb. 92.0 g (98% d.Th.).

$C_{36}H_{62}O_{13}N_4$ (758.9)

Ber.	C 56.98	H 8.23	N 7.38
Gef.	C 56.67	H 8.32	N 7.26

8) Benzyloxycarbonyl-L-glutamyl(γ-tert.-butylester)-L-glutamyl(γ-tert.-butylester)-L-glutamyl(γ-tert.-butylester)-L-glutamyl(γ-tert.-butylester)-L-glutaminsäure(γ-tert.-butylester) **[2-6a]**

85.0 g (112 mmol) *H-[Glu(OBu^t)]$_4$-OH* **[3-6b]** und 15.7 m*l* (112 mmol) Triäthylamin in 3 *l* Dimethylformamid werden unter Rühren bei 0 °C mit 48.6 g (112 mmol) *Z-Glu(OBu^t)-ONSu* **[2]** und nach 2 h mit 9.1 m*l* Pyridin versetzt. Das Reaktionsgemisch wird 24 h bei 0 °C und 24 h bei Raumtemperatur gerührt. Nach Abdestillieren des Lösungsmittels im Vak. wird der erhaltene ölige Rückstand wie für **[3-6a]** beschrieben, aufgearbeitet. Schmp. 169 - 171 °C; $[\alpha]_D^{20}$: – 10.6° bzw. $[\alpha]_{546}^{20}$: – 12.6° (*c* = 1, in Dimethylformamid); chromatographisch rein in n-Heptan/tert.-Butanol/Essigsäure 3:2:1, Cyclohexan/Chloroform/Essigsäure 45:45:10 und Chloroform/Methanol 9:1; Ausb. 109.0 g (90% d.Th.).

$C_{53}H_{83}O_{18}N_5$ (1078.5)

Ber.	C 59.04	H 7.76	N 6.49
Gef.	C 59.10	H 7.76	N 6.55

9) L-Glutamyl(γ-tert.-butylester)-L-glutamyl(γ-tert.-butylester)-L-glutamyl(γ-tert.-butylester)-L-glutamyl-(γ-tert.-butylester)-L-glutaminsäure(γ-tert.-butylester) **[2-6b]**

35.0 g (32.5 mmol) *Z-[Glu(OBu^t)]$_5$-OH* **[2-6a]** in 4 *l* 95proz. Methanol werden wie üblich in Gegenwart von Palladiumschwarz als Katalysator hydrogenolytisch entacyliert. Nach beendeter Reaktion wird das bei der Hydrierung zum Teil ausgefallende Produkt durch Erwärmen gelöst und die Lösung vom Katalysator abfiltriert. Das Filtrat wird im Vak. zur Trockene gebracht und das Produkt aus Methanol/Essigester gefällt. Schmp.194 - 195 °C; $[\alpha]_D^{20}$: – 7.2° bzw. $[\alpha]_{546}^{20}$: – 8.2° (*c* = 1, in 80proz. Essigsäure); chromatographisch rein in n-Heptan/tert.-Butanol/Essigsäure 3: 2:1 und Chloroform/Methanol 9:1; Ausb. 28.7 g (93.5% d.Th.).

$C_{45}H_{77}O_{16}N_5$ (944.1)

Ber.	C 57.25	H 8.22	N 7.42
Gef.	C 57.19	H 8.27	N 7.43

10) tert.-Butyloxycarbonyl-L-leucyl-L-glutamyl(γ-tert.-butylester)-L-glutamyl(γ-tert.-butylester)-L-glutamyl-(γ-tert.-butylester)-L-glutamyl(γ-tert.-butylester)-L-glutaminsäure(γ-tert.-butylester) **[1-6a]**

2.83 g (3.0 mmol) *H-[Glu(OBu^t)]$_5$-OH* **[2-6b]** in 100 m*l* Dimethylformamid werden mit 0.42 m*l* (3.0 mmol) Triäthylamin, 1.08 g (3.3 mmol) *Boc-Leu-ONSu* **[1]** und anschließend mit 0.24 m*l* (3.0 mmol) Pyridin versetzt. Die Reaktionsmischung wird 12 h bei Raumtemperatur gerührt und letztlich wie für **[3-6a]** beschrieben aufgearbeitet. Schmp. 225 - 227 °C (Zers.); $[\alpha]_D^{20}$: – 23.4° bzw. $[\alpha]_{546}^{20}$: – 27.9° (*c* = 1, in Methanol); chromatographisch rein in n-Heptan/tert.-Butanol/Essigsäure 3: 2:1 und Chloroform/Methanol 9:1; Ausb. 3.0 g (86% d.Th.).

$C_{56}H_{96}N_6O_{19}$ (1157.4)

Ber.	C 58.12	H 8.36	N 7.26
Gef.	C 58.20	H 8.53	N 7.22

11) Benzyloxycarbonyl-L-alanyl-O-tert.-butyl-L-tyrosyl-glycyl-L-tryptophyl-L-leucyl-L-aspartyl(β-tert.-butylester)-L-phenylalaninamid **[7 - 13a]**

37.0 g (55 mmol) *H-Trp-Leu-Asp(OBu^t)-Phe-NH₂·HCl* [**10 - 13**-HCl]* und 27.5 g (55 mmol) *Z-Ala-Tyr(Bu^t)-Gly-OH* [**7 - 9**][12] in 1 *l* Dimethylformamid werden nach dem Abkühlen auf – 10 °C nacheinander mit 6.05 m*l* (55 mmol) *N*-Methylmorpholin, 6.9 g (60 mmol) *N*-Hydroxysuccinimid und 12.4 g (60 mmol) Dicyclohexylcarbodiimid versetzt. Die Reaktionsmischung wird noch 12 h bei 0 °C und anschließend weitere 36 h bei Raumtemperatur gerührt. Nach dem Abkühlen wird vom Harnstoff abfiltriert und das Filtrat im Vak. eingedampft. Der Rückstand wird in 90proz. Methanol aufgeschlämmt und die Suspension bei 60 °C 15 min gerührt. Das Produkt wird abgesaugt, mit warmen Methanol und letztlich mit Diäthyläther gut gewaschen. Schmp. 223 °C (Zers.); $[\alpha]_D^{20}$: – 29.8° bzw. $[\alpha]_{546}^{20}$: – 35.6° (*c* = 1, in Dimethylformamid); chromatographisch rein in n-Butanol/Essigsäure/Wasser 3:1:1; Ausb. 51.0 g (83% d.Th.).

$C_{60}H_{77}N_9O_{12}$ (1116.34)

Ber.	C 64.55	H 6.95	N 11.29
Gef.	C 64.70	H 7.00	N 11.19

12) L-Alanyl-O-tert.-butyl-L-tyrosyl-glycyl-L-tryptophyl-L-leucyl-L-aspartyl(β-tert.-butylester)-L-phenylalaninamid-monohydrat **[7 - 13b]**

23.0 g (20.6 mmol) *Z-Ala-Tyr(Bu^t)-Gly-Trp-Leu-Asp(OBu^t)-Phe-NH₂* [**7 - 13a**] in 1 *l* Dimethylformamid werden wie üblich in Gegenwart von Palladiumschwarz als Katalysator unter gleichzeitigem Titrieren mit 0.27M methanol. HCl bei pH 5.0 hydriert. Nach beendeter Reaktion wird das Filtrat im Vak. eingedampft und der feste Rückstand mit Diäthyläther verrieben. Das Produkt wird abfiltriert, im Vak. über KOH getrocknet und anschließend in 250 m*l* Dimethylformamid gelöst, mit 2.9 m*l* (20.6 mmol) Triäthylamin und 1 *l* Wasser versetzt. Der dabei gebildete Niederschlag wird abgesaugt, gut mit Wasser und Diäthyläther gewaschen und schließlich im Vak. getrocknet. Schmp. 203 - 204 °C; $[\alpha]_D^{20}$: – 21.0° bzw. $[\alpha]_{546}^{20}$: – 25.1° (*c* = 1.1, in Dimethylformamid); chromatographisch rein in n-Butanol/Essigsäure/Wasser 3:1:1; Ausb. 19.8 g (98% d. Th.).

$C_{52}H_{71}N_9O_{10} \cdot 1\,H_2O$ (1000.2)

Ber.	C 62.40	H 7.35	N 12.60
Gef.	C 62.56	H 7.22	N 12.60

13) tert.-Butyloxycarbonyl-L-leucyl-L-glutamyl(γ-tert.-butylester)-L-glutamyl(γ-tert.-butylester)-L-glutamyl(γ-tert.-butylester)-L-glutamyl(γ-tert.-butylester)-L-glutamyl(γ-tert.-butylester)-L-alanyl-O-tert.-butyl-L-tyrosyl-glycyl-L-tryptophyl-L-leucyl-L-aspartyl(β-tert.-butylester)-L-phenylalaninamid **[1 - 13a]**

0.6 g (0.61 mmol) *H-Ala-Tyr(Bu^t)-Gly-Trp-Leu-Asp(OBu^t)-Phe-NH₂* [**7 - 13b**] und 1.06 g (0.92 mmol) *Boc-Leu-[Glu(OBu^t)]₅-OH* [**1 - 6**] in 10 m*l* Dimethylacetamid und 3 m*l* Phosphorsäure-tris[dimethylamid] werden nach Abkühlen auf – 10 °C nacheinander mit 0.116 g (1.01 mmol) *N*-Hydroxysuccinimid und 0.190 g (0.92 mmol) Dicyclohexylcarbodiimid versetzt. Die Reaktionsmischung wird 48 h bei 4 °C und anschließend 48 h bei Raumtemperatur gerührt und letztlich in 90 m*l* Methanol gegossen. Der gebildete Niederschlag wird abfiltriert und in 200 m*l* Methanol aufgeschlämmt. Die Suspension wird 5 min bei 60 °C und 1 h bei Raumtemperatur gerührt. Das abfiltrierte Produkt wird mit Methanol und Diäthyläther gewaschen und im Vak. getrocknet. Schmp. 255 °C; $[\alpha]_D^{20}$: – 13.5° bzw. $[\alpha]_{546}^{20}$: – 16.2° (*c* = 1, in Phosphorsäure-tris[dimethylamid]); chromatographisch rein in Chloroform/Trifluoräthanol/Methanol 18:2:1 und Chloroform/Trifluoräthanol 3:1; Ausb. 1.07 g (83% d.Th.).

Aminosäureanalyse:

	Asp	Glu	Gly	Ala	Leu	Tyr	Phe	Trp
Ber.	1	5	1	1	2	1	1	1
Gef.	0.94	5.24	0.99	0.97	1.88	0.93	1.02	–

$C_{108}H_{165}N_{15}O_{28}$ (2121.6)

Ber.	C 61.14	H 7.84	N 9.90
Gef.	C 60.91	H 7.75	N 9.86

14) L-Leucyl-L-glutamyl-L-glutamyl-L-glutamyl-L-glutamyl-L-glutamyl-L-alanyl-L-tyrosyl-glycyl-L-tryptophyl-L-leucyl-L-aspartyl-L-phenylalaninamid **[1 - 13b]**

0.99 g (0.466 mmol) *Boc-Leu-[Glu(OBu^t)]₅-Ala-Tyr(Bu^t)-Gly-Trp-Leu-Asp(OBu^t)-Phe-NH₂* [**1 - 13a**] werden unter Argonatmosphäre mit 50 m*l* Trifluoressigsäure und 5 m*l* Anisol übergossen und 90 min bei Raumtemperatur stehengelassen; hierbei tritt Lösung ein. An-

* Das Tetrapeptid-Derivat konnte durch Umfällen aus Methanol/Äther lösungsmittelfrei in 89proz. Ausbeute gewonnen werden. Schmp. 175 - 177 °C (Zers.); $[\alpha]_D^{20}$: – 30.4° bzw. $[\alpha]_{546}^{20}$: – 36.4° (*c* = 1, in Dimethylformamid) [Lit.[12]: H-Trp-Leu-Asp(OBu^t)-Phe-NH₂·HCl·Dimethylformamid; Schmp. 180 °C (Zers.); $[\alpha]_D^{20}$: – 29.6° bzw. $[\alpha]_{546}^{20}$: – 35.1° (*c* = 1.0, in Dimethylformamid].

$C_{34}H_{46}N_6O_6 \cdot HCl$ (671.2)

Ber.	C 60.84	H 7.06	N 12.52
Gef.	C 60.75	H 6.97	N 12.54

schließend wird die Reaktionsmischung mit viel Diäthyläther übergossen und der dabei gebildete Niederschlag abfiltriert, gut mit Diäthyläther gewaschen und im Vak. über KOH getrocknet. Das farblose amorphe Pulver wird in 0.5M NH_4OH aufgenommen und lyophylisiert. Ausb. 0.763 g (Rohprodukt).

Literatur

1 Houben-Weyl, Methoden der Organischen Chemie, 4. Aufl., Bd. 15/1, S. 20 - 27, Verlag G. Thieme, Stuttgart 1974
2 Walsh, J. H., Debas, H. T. & Grossman, M. I. (1974) *J. Clin. Invest.* **54**, 477 - 485.
3 Rehfeld, J. F. & Stadil, F. (1973) *Gut* **14**, 369 - 373.
4 Gregory, R. A. & Tracy, H. J. (1974) *Gut* **15**, 683 - 685.
5 Grossman, M. I. (1976) *Gastroenterology*, **71** 166 - 168, Fußnote S. 167.
6 Debas, H. T., Walsh, J. H. & Grossman, M. I. (1974) *Gut* **15**, 686 - 689.
7 Wünsch, E., Jaeger, E., Deffner, M. & Scharf, R. (1972) *Hoppe-Seyler's Z. Physiol. Chem.* **353**, 1716 - 1720.
8 Jaeger, E., Thamm, P., Schmidt, I., Knof, S., Moroder, L. & Wünsch, E. (1978) *Hoppe-Seyler's Z. Physiol. Chem.* **359**, 155 - 164, nachstehend.
9 Morley, J. S., Tracy, H. J. & Gregory, R. A. (1965) *Nature (London)* **207**, 1356 - 1359.
10 Wünsch, E. & Drees, F. (1966) *Chem. Ber.* **99**, 110 - 120.
11 Weygand, F., Hoffmann, D. & Wünsch, E. (1966) *Z. Naturforsch.* **21b**, 426 - 428.
12 Wünsch, E. & Deimer, K.-H. (1972) *Hoppe-Seyler's Z. Physiol. Chem.* **353**, 1246 - 1254.
13 Marlborough, D. I. & Rydon, H. N. (1972) *J. Chem. Soc. Perkin I*, 1 - 5.
14 Nefkens, G. H. L. & Nivard, R. J. F. (1964) *Trav. Chim.* **83**, 199 - 207.
15 Anderson, G. W., Zimmerman, J. E. & Callahan, F. M. (1964) *J. Am. Chem. Soc.* **86**, 1839 - 1842.
16 Matsubara, H. & Saraki, R. M. (1969) *Biochem. Biophys. Res. Commun.* **35**, 175 - 181.
17 Schröder, E. & Klieger, E. (1964) *Liebigs Ann. Chem.* **673**, 196 - 207.

Prof. Dr. L. Moroder, Dr. F. Drees, Dr. E. Jaeger und Prof. Dr. E. Wünsch, Max-Planck-Institut für Biochemie, Abteilung für Peptidchemie,
D-8033 Martinsried bei München.

Hoppe-Seyler's Z. Physiol. Chem.
Bd. 359, S. 155 - 164, Februar 1978

Zur Synthese von [11-Leucin]Human-Minigastrin I

II. Mitteilung: Reindarstellung des synthetischen Tridekapeptidamids sowie Isolierung und Strukturaufklärung eines Synthese-Nebenproduktes[1],*

Ernst JAEGER, Paul THAMM, Irmgard SCHMIDT, Siegward KNOF, Luis MORODER und Erich WÜNSCH

Max-Planck-Institut für Biochemie, Abteilung für Peptidchemie, Martinsried bei München

(Der Schriftleitung zugegangen am 23. Dezember 1977)

Herrn Prof. Dr. W. Grassmann, emerit. Direktor des ehem. Max-Planck-Instituts für Eiweiß- und Lederforschung zu seinem 80. Geburtstag in steter Verehrung gewidmet

Zusammenfassung: Die Reindarstellung des 11-Leucin-Analogons von Human-Minigastrin I, einem linearen Tridekapeptidamid mit voller „Gastrin-Aktivität", aus dem durch Trifluoressigsäure-Demaskierung des allseits geschützten Tridekapeptidamid-Derivats erhaltenen Rohprodukt gelingt durch Ionenaustauschchromatographie an DEAE-Sephadex A-25 und anschließende Verteilungschromatographie an Sephadex G-25. Die gleichzeitige Isolierung eines Synthese-Nebenproduktes und dessen Identifizierung als [10-(N^{in}-tert.-Butyl-tryptophan),11-Leucin]-Human-Minigastrin I mittels spektroskopischer (UV, Fluoreszenz, ^{1}H-NMR und MS), enzymatischer und chromatographischer Methoden wird beschrieben. Die reinen Tridekapeptidamide [Leu^{11}]Human-Minigastrin I und [Trp(1'-Bu^t)^{10}, Leu^{11}]Human-Minigastrin I werden in einer Gesamtausbeute von 37.4% bzw. 4.6% (Mol-% bezogen auf allseits geschütztes Tridekapeptidamid-Derivat) isoliert.

Synthesis of Human [11-Leucine]Minigastrin I – II. Purification of the Synthetic Tridecapeptide Amide and Isolation of a Side-Product

Summary: The preparation of the pure 11-leucine analogue of human minigastrin I, a linear tridecapeptideamide with full "gastrin activity", from the crude synthetic material obtained after deblocking of the overall protected tridecapeptide amide derivative by means of trifluoroacetic acid, is achieved by ion-exchange chromatography on DEAE-Sephadex A-25 and subsequent partition chromatography on Sephadex G-25. The isolation of a synthetic side-product and its identification as human [10-(N^{in}-tert-butyl-tryptophan),11-leucine]minigastrin I by means of spectroscopic (UV, fluorescence, ^{1}H NMR, MS), enzymatic and chromatographic methods is described. The pure tridecapeptide amides human [Leu^{11}]minigastrin I and human [Trp(1'-Bu^t)^{10},Leu^{11}]minigastrin I are isolated in overall yields of 37.4% and 4.6%, respectively (mol-% with regard to the overall protected tridecapeptide amide derivative).

Abkürzungen:
Aufgrund der Vereinbarungen biochemischer Zeitschriften werden hier die von der IUPAC-IUB Comission on Biochemical Nomenclature empfohlenen Abkürzungen für Aminosäuren und Schutzgruppen verwendet. Die Autoren bevorzugen aber wegen ihrer universellen Anwendbarkeit die im Houben-Weyl[3] verwendeten und eingehend erläuterten Abkürzungen. Die Abkürzungen HG-17 I für Human Little Gastrin I und HG-13 I für Human Minigastrin I wurden von Walsh, J. H. et al.[4] vorgeschlagen.

* Auszugsweise als Vortrag gehalten am ersten „USSR/FRG-Symposium on Peptide and Protein Chemistry", Dushanbe/Tadjikistan, UdSSR, März 1976[2].

Key words: Synthetic peptide hormones, human [11-leucine]minigastrin I, side-reaction, tert-butyl-tryptophan.

In der vorausgehenden Mitteilung[1] hatten wir die Herstellung von rohem [11-Leucin]Human-Minigastrin I (Produkt A) durch Trifluoressigsäure-Demaskierung von Boc-Leu-[Glu(OBu^t)]$_5$--Ala-Tyr(Bu^t)-Gly-Trp-Leu-Asp(OBu^t)-Phe-NH$_2$ beschrieben. Dessen Reinigung versuchten wir zunächst durch Ionenaustauschchromatographie an DEAE-Sephadex A-25 – in Anlehnung an die u.a. mittels Säulenchromatographie an AE-Cellulose erzielte Isolierung des reinen natürlichen Human-Minigastrins I[5]. Es gelang zwar das Rohmaterial in eine Hauptfraktion B$_1$ und eine Nebenfraktion B$_2$ aufzutrennen, eine dünnschichtchromatographische Überprüfung machte jedoch sichtbar (s. Abb. 1), daß diese Operation nur zur Abtrennung eines Teils der vorliegenden Nebenprodukte befähigt war. Veränderungen der Ionenaustauschchromatographie-Bedingungen (z.B. Gradientenelution an QAE-Sephadex A-25 oder AE-Cellulose) brachten keine Verbesserung der Auftrennung. Interessant war das unterschiedliche Farbverhalten der dünnschichtchromatographisch aufgetrennten beiden Komponenten der Hauptfraktion B$_1$ nach Anfärbung mit dem „Tryptophan-Reagens nach Ehrlich"[7]: Komponente 2 (s. Abb. 1) violett, Komponente 3 blau bis blaugrau, gleichbedeutend mit einem ersten Hinweis auf das Vorliegen eines Oligopeptids mit „verändertem" Tryptophan-Rest in der Sequenz.

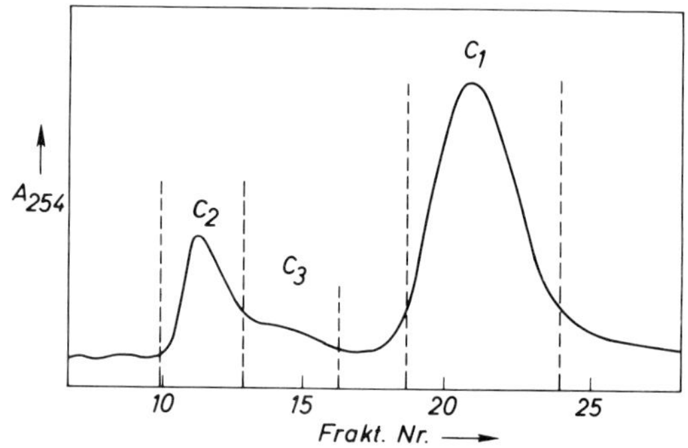

Abb. 2. Verteilungschromatographie von vorgereinigtem [Leu^{11}]Human-Minigastrin I an Sephadex G-25 im Zweiphasensystem 2-Butanol/Äthanol/0.1M Ammoniumacetat (2 : 0.1 : 2), pH 6.0 (i.d. Unterphase). Elutionskurve.

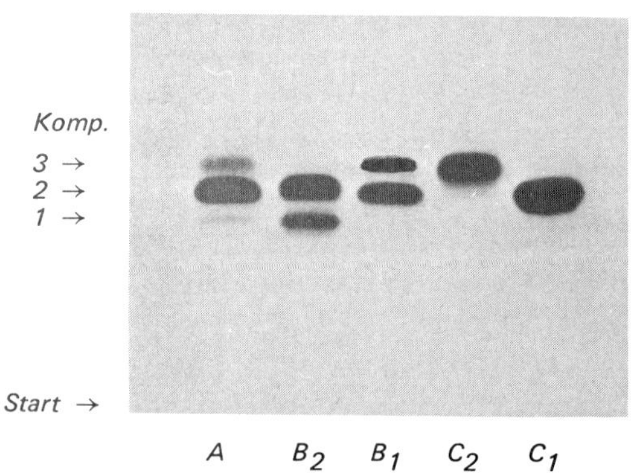

Abb. 1. Dünnschichtchromatographie verschiedener Fraktionen des synthetischen [Leu^{11}]Human-Minigastrins I auf Kieselgel-60-Fertigplatte (Merck) in n-Butanol/Essigsäure/Pyridin/Wasser 60 : 6 : 20 : 24. Anfärbung mit Chlor/Tolidin-Reagens[6].

Aufgetragen wurde (von links nach rechts): Rohes Syntheseprodukt A, Ionenaustauschchromatographie-Frakt. B$_2$, Frakt. B$_1$, reines [Trp(1'-Bu^t)^{10},Leu^{11}]Human-Minigastrin I (= Frakt. C$_2$) und reines [Leu^{11}]Human-Minigastrin I nach Verteilungschromatographie (= Frakt. C$_1$).

Gestützt auf unsere Erfahrungen bei der Reindarstellung des [15-Leucin]Human-Gastrins I[8] durch Verteilungschromatographie an Sephadex G-25 gelang eine weitere Auftrennung der Fraktion B$_1$ in die Fraktionen C$_1$ und C$_2$ sowie eine kleine, im Elutionsdiagramm als „Schulter" erscheinende Zusatzfraktion C$_3$ (s. Abb. 2).

Die dünnschichtchromatographisch (s. Abb. 1) und -elektrophoretisch gut auftrennbaren, einheitlich erscheinenden Substanzen C$_1$ und C$_2$ – letztere nach Rechromatographie an Sephadex G-25 zur Abtrennung geringer Mengen der „überlappenden" Fraktion C$_3$ – zeigten nahezu identische, für [Leu^{11}]Human-Minigastrin I zu erwartende Aminosäurezusammensetzungen einschließlich der Tryptophan-Werte nach saurer Hydrolyse unter Zusatz von Thioglykolsäure (s. Tab. 1, a). Enzymatische Hydrolyse mit Aminopeptidase M aber lieferte Hydrolysate, die nur im Falle der Frakt. C$_1$ die geforderten Analysendaten hatten; in der Aminosäurezusammensetzung der Frakt. C$_2$ fehlte Tryptophan vollständig (s. Tab. 1, b).

Tab. 1. Aminosäureanalysen der Komponenten C_1 und C_2 des [Leu^{11}]Human-Minigastrins I.
In Klammern sind die berechneten Werte für [Leu^{11}]Human-Minigastrin I angegeben.

	a) Nach saurer Hydrolyse mit 6M HCl[a]		b) Nach Hydrolyse mit Aminopeptidase M[b]	
	C_1	C_2	C_1	C_2
Asp (1)	1.02	1.03	0.94	0.97
Glu (5)	5.18	5.15	5.06	5.21
Gly (1)	0.97	1.00	0.99	1.00
Ala (1)	0.98	1.06	0.96	1.01
Leu (1)	0.99	1.00	0.98	0.97
Tyr (1)	1.00	0.99	0.98	0.99
Phe (1)	1.00	1.00	0.96	0.96
Trp (1)	1.01	1.02	0.94	–

[a] Hydrolysebedingungen: Zusatz von 2.5% Thioglykolsäure[9], 24 h, 110 °C;
[b] Inkubationsbedingungen: Tris-Puffer, pH 7.8, 37 °C, 10000 U Enzym/mg Substanz, 6 h.

Die Substanz C_1 erwies sich somit als das gewünschte Tridekapeptidamid

H-Leu-$(Glu)_5$-Ala-Tyr-Gly-Trp-Leu-Asp-Phe-NH_2,

dessen Homogenität zusätzlich durch mikropräparative trägerfreie Hochspannungselektrophorese gesichert wurde: keinerlei Nebenprodukte waren sichtbar (Abb. 3); alle Analysendaten des isolierten Peptids waren identisch mit denen der Frakt. C_1. Die biologische Aktivität dieses [Leu^{11}]Human-Minigastrins I wurde von Konturek et al.[10]

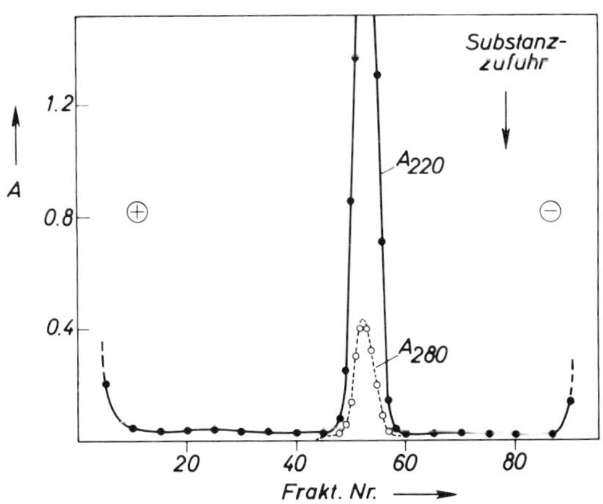

Abb. 3. Trägerfreie kontinuierliche Elektrophorese von [Leu^{11}]Human-Minigastrin I (Frakt. C_1) in 14mM Ammoniumacetatlösung (pH 8.25); 2100 V, 100 mA.

im Vergleich mit der von [Leu^{15}]Human-Gastrin I zu etwa 94% ermittelt.

Substanz C_2 konnte letztlich durch spektroskopische und chromatographische Untersuchungen, sowohl an der Gesamtsequenz [**1 - 13**] als auch an „enzymatischen Spaltstücken“ vorgenommen, als Synthese-Nebenprodukt eindeutig charakterisiert werden: entscheidend war der Austausch des Tryptophans in Position 10 durch einen N^{in}-tert.-Butyl-tryptophan-Rest. Im einzelnen gelang die Aufklärung der Primärstruktur von Frakt. C_2 mittels folgender Methoden (stets vorgenommen im Vergleich mit der Reinsubstanz C_1):

1) Das Ultraviolett-Spektrum a) in Wasser (Abb. 4a) offenbarte eine Absorptionsverschiebung nach größeren Wellenlängen hin (Maximum bei 282 nm gegenüber 279 nm), b) in 0.1M NaOH (Abb. 4b) zeigte nicht mehr das Verhältnis von etwa 1 : 1 für Tryptophan und Tyrosin*, sondern in Anwendung der Beaven-Holiday-Methode[12,13] ein Verhältnis von etwa 0.2 : 1 (eine Auswertung nach den Verfahren von Bencze und Schmid[14] war wegen des Vorliegens nur eines Absorptionsmaximums nicht möglich).

* Die jeweils gemessenen Werte unterscheiden sich aber stets von den berechneten; eine gegenseitige Beeinflussung der aromatischen Systeme der beiden Aminosäuren muß aufgrund des Abstands voneinander (Sequenzpositionen) angenommen werden.

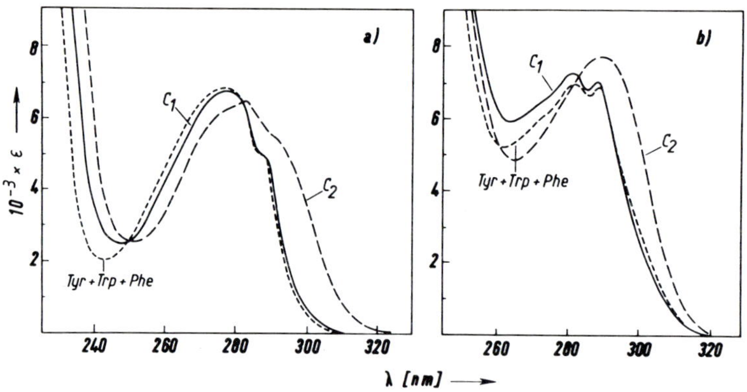

Abb. 4. Ultraviolett-Absorptionsspektren zweier Fraktionen des synthetischen [Leu^{11}]Human-Minigastrins I im Bereich zwischen 230 und 330 nm im Vergleich mit dem Spektrum einer Mischung von L-Tyrosin, L-Tryptophan und L-Phenylalanin (Molverh. 1:1:1).

Konzentrationen: a) 27.5μM in Wasser; b) 108μM in 0.1M NaOH.
Die gemessenen Extinktionswerte für die Tyr + Trp + Phe-Mischung stimmen mit den nach Literaturangaben[11] berechneten Werten weitestgehend überein.

Die Fluoreszenz-Spektren in Wasser zeigten für Tryptophan bei einer etwa doppelt so hohen Intensität sowohl in der Anregung (Abb. 5a) als auch in Emission (Abb. 5b) ebenfalls eine deutliche Rotverschiebung.

2) Im Protonenresonanz-Spektrum von Frakt. C_2 zeigte sich ein signifikantes, scharfes Singulett im „aliphatischen Bereich" bei $\delta = 1.70$ ppm. Mit dem zusätzlich ermittelten Integrationswert für neun Protonen war das Vorliegen einer tert.-Butyl-Gruppierung gesichert; die gleichzeitig relativ hohe chemische Verschiebung sprach außerdem für die Fixierung des tert.-Butyl-Restes am Indolstickstoff (Abb. 6).

3) Enzymatische Hydrolyse mit Chymotrypsin untermauerte die spektroskopisch gefundene Veränderung am Tryptophan-Rest: bei der bekannten Chymotrypsin-Spezifität können Spaltungen der Peptid- bzw. Amid-Bindungen zwischen Pos. 8 und 9 sowie 10 und 11 und an Pos. 13 sowie als

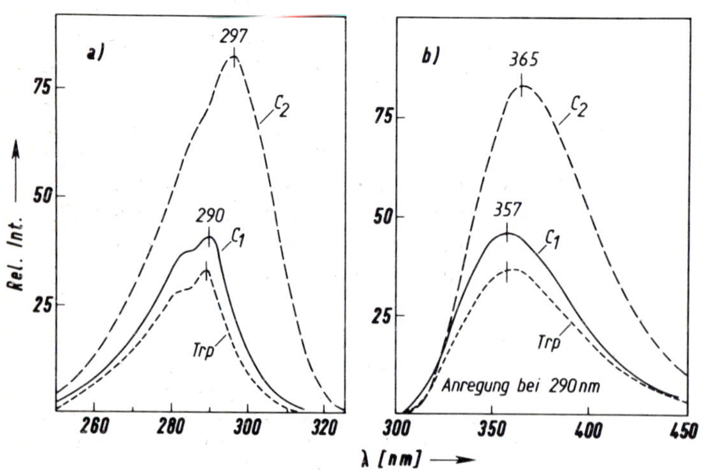

Abb. 5. Fluoreszenz-Anregungs- und Emissionsspektren zweier Fraktionen des synthetischen [Leu^{11}]Human-Minigastrins I im Vergleich mit den Spektren von L-Tryptophan.

a) Anregungsspektren; 27.5μM wäßr. Lösungen. [Messung bei 355 nm (C_1 und Trp) bzw. 370 nm (C_2)].
b) Emissionsspektren (Anregung bei 290 nm); 27.5μM wäßr. Lösungen.

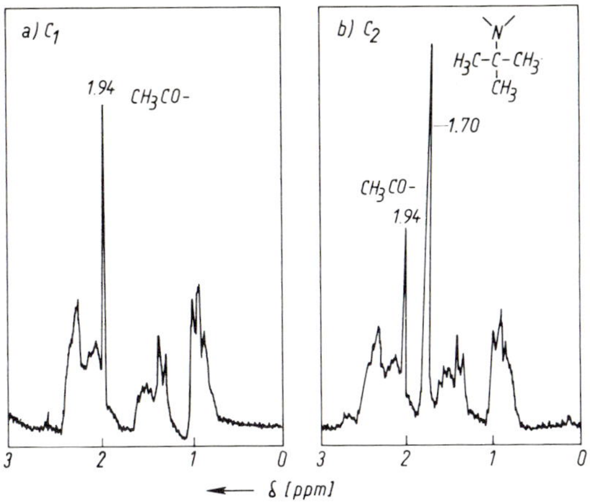

Abb. 6. ^{1}H-NMR-Spektren zweier Fraktionen des synthetischen [Leu][11]Human-Minigastrins I im Bereich zwischen 0 und 3 ppm bei 90 MHz.

Lösungsmittel: 1% ND_3 in D_2O; Acetat als interner Standard: δ = 1.94 ppm.

„Nebenspaltung“ zwischen Pos. 11 und 12 erwartet werden. Diese chymotryptische Aufspaltung der Peptidkette verläuft nur im Falle von Frakt. C_1 so; der Sequenzbereich 10 - 12 widersetzt sich aber im Falle von Frakt. C_2 dem enzymatischen Angriff. So werden hier vorzugsweise nur die Sequenzteilstücke [**1** - **8**] und [**9** - **13**] gefunden (s. Abb. 7 und Tab. 2).

4) Die massenspektrometrische Charakterisierung der chymotryptischen Spaltstücke von Frakt. C_1 und C_2, insbesondere der isolierten Teilsequenzen [**9** - **13**] – durch Acetylierung und Permethylierung* in die geeigneten flüchtigen Derivate von 4 ([**9** - **13**] aus C_1) und von 4′ ([**9** - **13**] aus C_2) überführt –, erhärtete die aus 1. bis 3. bereits resultierenden Folgerungen. Während für das Derivat von 4 neben den sequenzcharakteristischen Ionen auch die intensive Spitze des *N*-methylierten 3-Indolylmethyl-Ions mit m/e = 144 zu finden war (s. Abb. 8a), wies das Spektrum des 4′-Derivats eine jeweils um 42 Masseneinheiten nach höheren Massen verschobene Serie von Sequenzionen auf

* Bei der Permethylierung der *N*-Acetyl-peptide verwendeten wir in Abänderung der üblichen Hakamori-Methode[15] mit Vorteil Kalium-tert.-butylat[16] in Dimethylsulfoxid.

und anstelle des Signals m/e = 144 zwei Peaks mit m/e = 186 bzw. m/e = 130, identifizierend ein *N*-tert.-butyliertes 3-Indolylmethyl-Ion bzw. das nach Abspaltung der tert.-Butyl-Gruppierung daraus entstehende N^{in}-unsubstituierte „Folge-Ion“ (s. Abb. 8b).

Eine ohne Derivatisierungsmaßnahmen mögliche massenspektrometrische Strukturüberprüfung der nach enzymatischer Totalhydrolyse mittels Aminopeptidase M isolierten „indolhaltigen Aminosäuren“ (isoliert mittels Ionenaustausch-

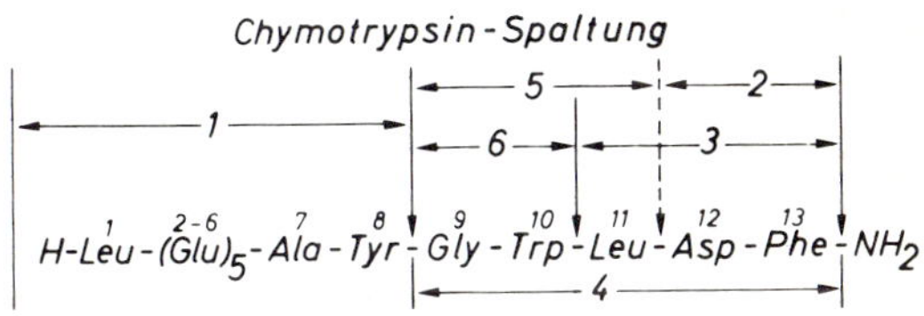

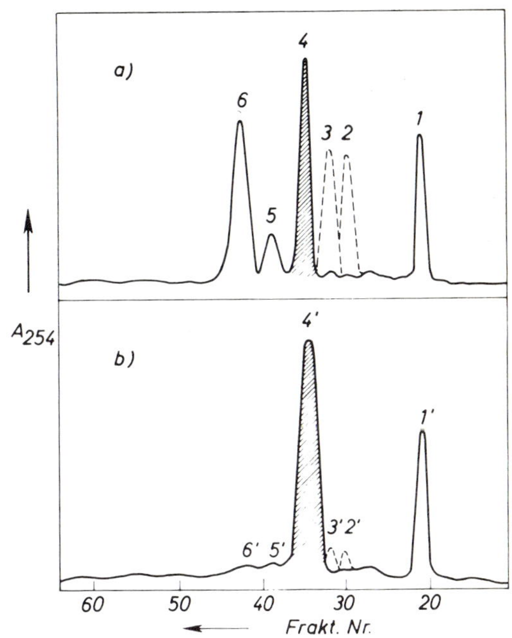

Abb. 7. Gelchromatographische Trennung der chymotryptischen Spaltstücke zweier Fraktionen des synthetischen [Leu[11]]Human-Minigastrins I an Sephadex G-25 in Wasser. Elutionskurven.

a) Fragmente der Frakt. C_1 (Peak Nr. 1 bis 6).

b) Fragmente der Frakt. C_2 (Peak Nr. 1′ bis 6′)

Extinktionsmessungen bei 254 nm (——) und bei 206 nm (– – –; nur im Bereich der Peaks 2 und 3 angegeben).

Tab. 2. Aminosäureanalysen der chymotryptischen Spaltstücke zweier Fraktionen des [Leu^{11}]Human-Minigastrins I nach saurer Hydrolyse mit 6M HCl (24 h; 110 °C) unter Zusatz von Thioglykolsäure[9] (2.5 %). (Vgl. auch Abb. 7).

	Spaltstücke aus Frakt. C_1						aus Frakt. C_2	
	1	2	3	4	5	6	1′	4′
Asp	–	0.96	1.06	1.09	–	–	–	1.06
Glu	4.96	–	–	–	–	–	4.92	–
Gly	–	–	–	1.02	0.85	1.04	–	1.03
Ala	1.00	–	–	–	–	–	1.00	–
Leu	1.01	–	1.00	1.00	1.00	–	0.98	1.00
Tyr	0.98	–	–	–	–	–	0.97	–
Phe	–	1.00	0.94	1.06	–	–	–	1.02
Trp	–	–	–	0.93	0.93	0.93	–	0.95

chromatographie an SP-Sephadex C-25 und als Trp^{10}-C_1 bzw. Trp^{10}-C_2 bezeichnet) bestätigte zusätzlich das Vorliegen einer tert.-Butylierung des Indolringes: Trp^{10}-C_2 offenbarte sich mit einem um 56 Masseneinheiten höheren Molekülionen-Peak und „Fragmentierungsspitze" bei m/e = 186, sprechend für ein (tert.-Butyl-3-indolyl)-methyl-Ion. Demgegenüber zeigte sich Trp^{10}-C_1 als identisch mit authentischem Tryptophan*. Somit ist offenbar das Synthese-Nebenprodukt C_2 als [10-(N^{in}-tert.-Butyl-tryptophan),11-Leucin]Human-Minigastrin I identifiziert*.

Wir danken der *Deutschen Forschungsgemeinschaft* für die uns (E. W., E. J. und L. M.) gewährten, großzügigen Sachbeihilfen.

Beschreibung der Versuche

Die *Aminosäureanalysen* wurden am „Amino Acid Analyzer" mit einem „Automatic Digital Integrator" der Firma Beckman Instruments, (Modell 120 B und 125) ausgewertet. Die *Absorptionsmessungen* an den bei chromatographischen Trennungen erhaltenen Lösungen erfolgten mit Durchflußphotometern der Firma LKB-Produkter (Stockholm), Modelle Uvicord I bzw. Uvicord III. Für die Fraktionierung der Elutionslösungen wurden Fraktionssammler der Firma Serva-Technik, Heidelberg, Modell Linear II, verwendet. Die *chromatographischen und elektrophoretischen Reinheitsteste* erfolgten nach den üblichen Verfahren der Dünnschichtchromatographie und -elektrophorese teils auf Kieselgel-60-Fertigplatten der Fa. Merck, Darmstadt, teils auf frisch bereiteten Platten unter Verwendung von Kieselgel G (nach E. Stahl). Für die analytische Dünnschichtelektrophorese diente ein Gerät der Firma Desaga, für die mikropräparative Hochspannungselektrophorese ein Gerät der Firma Bender & Hobein, München, Modell Elphor VaP_2. Chymotrypsin (EC 3.4.21.1) wurde von der Firma Serva, Feinchemikalien, Heidelberg, Aminopeptidase M (EC 3.4.11.2) von der Firma Carl Roth, Darmstadt, (Batch Nr. 0930139) bezogen.

Zur Aufnahme der *UV-Absorptionsspektren* diente ein registrierendes Doppelstrahl-Spektralphotometer der Firma Varian, Modell Cary 118 C. Die Messungen wurden bei allen Wellenlängen unter Spülung des Gerätes mit Stickstoff durchgeführt; der Probenraum war auf 25 °C thermostatisiert. Es wurden rechteckige Halbmikro-Küvetten aus Quarz (Schichtlänge 1.000 cm) der Fa. Hellma, Mühlheim/Ruhr, verwendet.

Die *Konzentrationen der Peptid-Lösungen* wurden aus den Einwaagen an Peptid (Peptidgehaltsbestimmung der wasserhaltigen Substanzen erfolgte mittels quantitativer Aminosäureanalyse) und an Lösungsmittel sowie aus den berechneten (d_{Peptid} = 1.4; ber. nach Lit.[17]) bzw. bekannten ($d^{25}_{H_2O}$ = 0.99705; $d^{25}_{0.1M\ NaOH}$ = 1.0015) Dichte-Werten bestimmt. Vor den Messungen wurden die wäßrigen Lösungen durch Millipore-Filter GSTF (0.22 μ) filtriert. Die Korrektur der nach Ultrafiltration verbleibenden Streuung und unspezifischen Absorption erfolgte graphisch durch Extrapolation der oberhalb 320 nm (keine Absorption der Peptide) gemessenen Absorptionswerte.

Die *UV-Fluoreszenzspektren* wurden mittels eines Fluoreszenzansatzes ZFM 4 am Spektralphotometer PMQ II der Firma Carl Zeiss, Oberkochen/Württ. aufgenommen. Zur Vermeidung störender, fluoreszierender Substanzen (aus Ionenaustauscherwasser) wurde ein über eine 80 cm lange Glasfüllkörper-Kolonne destilliertes Wasser als Lösungsmittel benutzt.

* Zur weiteren Absicherung siehe eine nachfolgende Arbeit: Jaeger, E., Thamm, P., Knof, S. & Wünsch, E., in Vorbereitung.

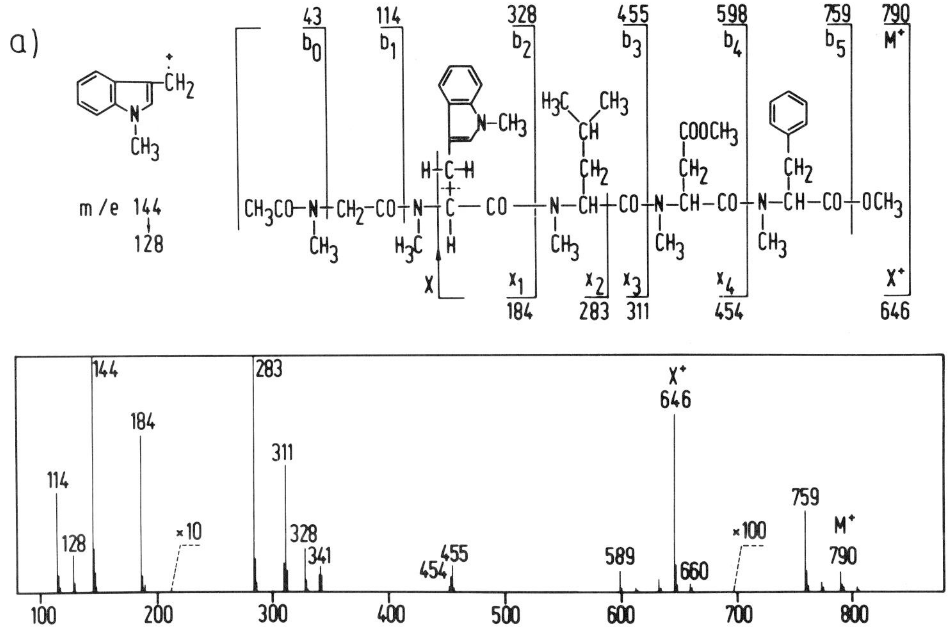

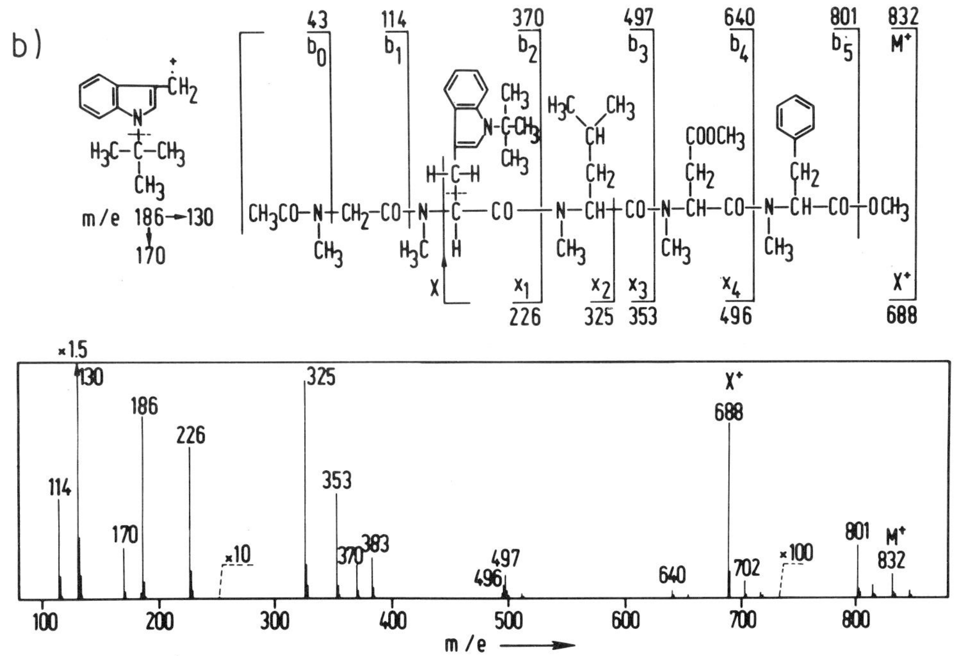

Abb. 8. Massenspektren der nach Acetylierung und Permethylierung zweier chymotryptischer Spaltstücke (Nr. 4 und Nr. 4′ aus den [Leu[11]]Human-Minigastrin-I-Fraktionen C_1 und C_2) erhaltenen Pentapeptid-Derivate.

Elektronenstoß-Ionisation bei 70 eV.

a) Ac-(Me)Gly-(Me)Trp(1′-Me)-(Me)Leu-(Me)Asp(OMe)-(Me)Phe-OMe;

b) Ac-(Me)-Gly-(Me)Trp(1′-Bu^t)-(Me)Leu-(Me)Asp(OMe)-(Me)Phe-OMe.

Zur Messung der *kernmagnetischen Resonanz* diente ein 90-MHz-Spektrometer, Modell WH-90, der Firma Bruker-Physik, Karlsruhe-Forchheim.

Die *Massenspektren* wurden mit einem Gerät MS-902 der Firma AEI, Manchester, bei einer Elektronenenergie von 70 eV und einer Beschleunigungsspannung von 8 KV unter Anwendung der Elektronenstoß-Ionisationstechnik aufgenommen. Die Probeneingabe erfolgte mittels eines Direkteinlaßsystems bei Ionenquellentemperaturen zwischen 200 und 230 °C.

I) REINIGUNG VON ROH-[11-LEUCIN]HUMAN-MINIGASTRIN I

1) Durch Ionenaustauschchromatographie an DEAE-Sephadex A-25

In eine Chromatographiesäule von 28 cm Länge und 3 cm Durchmesser wird ein wie üblich mit Startpuffer (0.5M Ammoniumacetat-Lösung, pH 8.5) äquilibrierter Ionenaustauscher DEAE-Sephadex A-25 (Acetat-Form) gefüllt. Man bringt eine Lösung von 720 mg des rohen [11-Leucin]Human-Minigastrins I (Peptidgehalt: 83.5%; ermittelt durch quantitative Aminosäureanalyse) in 12 m*l* 0.5M Ammoniak-Lösung auf die Säule und eluiert mit dem Startpuffer bei einer Durchflußgeschwindigkeit von 166 m*l*/h. Das Eluat fängt man in Portionen von je 125 m*l* in den Zellen eines Fraktionssammlers auf, wobei die Peptidverteilung durch kontinuierliche Registrierung der Extinktion der Lösung bei 278 nm (LKB-Uvicord III) ermittelt wird. Die so erhaltene Elutionskurve zeigt eine Auftrennung in zwei Hauptkomponenten, B_1 und B_2. Nach dreimaligem Lyophilisieren (zweimal aus 0.5proz. Ammoniak) erhält man aus entsprechenden Fraktionen (Frakt.-Nr. 48 - 82 für B_1 bzw. 20 -32 für B_2) 577 mg (Ausb. 80.1% d. Th.) vorgereinigtes [Leu^{11}]Human-Minigastrin I (B_1) und 35 mg (4.9% d. Th.) des Nebenproduktes B_2 (Gewichtsangaben bezogen auf lyophilisierte, wasserhaltige Fraktionen). Peptidgehalt der Frakt. B_1: 77.8% (durch quantitative Aminosäureanalyse ermittelt). Chromatographisch inhomogen (s. Abb. 1) in n-Butanol/Essigsäure/Wasser/Pyridin (60 : 6 : 24 : 20); Anfärbung mit Chlor/Tolidin-Reagens[6] oder mit Ehrlich-Reagens[7].

2) Durch Verteilungschromatographie an Sephadex G-25

a) Lösungsmittelsystem: Zu einer Mischung von 2 *l* speziell gereinigtem (vgl. Lit.[8]) 2-Butanol, 100 m*l* absol. Äthanol und 2 *l* 0.1M Ammoniumacetat-Lösung (hergestellt aus 0.1M Essigsäure durch Zugabe von Ammoniak, bis pH 5.4 erreicht ist) wird solange Essigsäure gegeben, bis die Unterphase nach Schütteln einen scheinbaren pH-Wert von 6.0 aufweist.

b) Durchführung der Trennung: In eine Säule von 143 cm Länge und 3 cm Durchmesser werden 200 g in der Unterphase des obigen Zweiphasensystems mehrere Stunden gequollenes Sephadex G-25 (fine) gefüllt; nach Äquilibrieren mit 160 m*l* Oberphase bringt man eine Lösung von 200 mg des vorgereinigten [Leu^{11}]Human-Minigastrins I (Frakt. B_1) in einer Mischung von 10 m*l* Oberphase und 0.5 m*l* 3proz. Ammoniak auf die Säule (geringe Mengen unlöslicher Anteile trennt man vorher durch Zentrifugieren ab) und eluiert dann mit Oberphase bei einer Durchflußgeschwindigkeit von 50 m*l*/h. Das Eluat fängt man in Portionen von je 50 m*l* in den Zellen eines Fraktionssammlers auf, wobei die Peptidverteilung durch kontinuierliche Registrierung der Extinktion der Lösung bei 254 nm (LKB-Uvicord I) ermittelt wird. Die Elutionskurve (Abb. 2) zeigt eine Auftrennung in eine Hauptkomponente C_1 und 2 Nebenkomponenten C_2 bzw. C_3. Nach Vereinigen der Fraktionen 18 - 24 (für C_1) dampft man die Lösungsmittel im Vak. ab, löst danach den verbleibenden Rückstand in 20 m*l* 5proz. Ammoniak und isoliert das Peptidmaterial durch Lyophilisieren. Nach 2maligem Wiederholen des Gefriertrocknens aus 5proz. Ammoniak (unlösliche Anteile werden vorher jeweils mittels Filtration durch eine G-4-Fritte abgetrennt) erhält man 110 mg (55% d. Th.) reines, wasserhaltiges [Leu^{11}]Human-Minigastrin I (C_1). In gleicher Weise isoliert man aus den Fraktionen 10 - 12 die Komponente C_2 und erhält daraus nach Rechromatographie unter gleichen Bedingungen 15 mg (7.5% d. Th.) reines [10-(N^{in}-tert.-Butyl-tryptophan),11-Leucin]Human-Minigastrin I (C_2) (Gewichtsangaben bezogen auf lyophilisierte, wasserhaltige Fraktionen). Aminosäureanalysen siehe Tab. 1 [dabei wird ein Peptidgehalt von 84.4% für C_1 (M_r = 1628.75) und von 78.4% für C_2 (M_r = 1684.85) ermittelt]. Chromatographisch rein (Abb. 1) bei den unter I, 1) angegebenen Bedingungen. Elektrophoretisch rein bei Dünnschichtelektrophorese auf Kieselgel-G-Platten in 2.4M Ameisensäure (pH 2.0) und in Pyridin/Essigsäure Puffer (pH 6.0); Anfärbung mit Ninhydrin bzw. Chlor/Tolidin-Reagens.

UV(H_2O); C_1 : λ_{max} = 279 nm (ϵ = 6760); C_2 : λ_{max} = 282 nm (ϵ = 6460); Testmischung [Tyr + Trp + Phe, molar. Verh. 1 : 1 : 1]: λ_{max} = 277 nm (ϵ = 6920) [berechnet[11]: λ_{max} = 278 nm (ϵ = 6820)].

UV(0.1 N NaOH); C_1 : λ_{max} = 281.5 nm (ϵ = 7320); C_2 : λ_{max} = 290 nm (ϵ = 7750); Testmischung [Tyr + Trp + Phe, molar. Verh. 1:1:1]: λ_{max} = 281.5 nm (ϵ = 7080) [berechnet[11]: λ_{max} = 281.5 nm (ϵ = 6980)].

^{1}H-NMR [1% ND_3 in D_2O, Me_4Si = 0]; C_1 : kein Signal zwischen 1.65 ppm und 1.80 ppm; C_2 : Singulett bei δ = 1.70 ppm [9 H, $> NC(CH_3)_3$].

3) Durch trägerfreie kontinuierliche Elektrophorese

Durch die Trennkammer einer VaP_2-Elektrophoreseapparatur wird 14mM Ammoniumacetat-Lösung (pH 8.25) als Elektrolytlösung mit einer Durchlaufgeschw. von 200 m*l*/h gepumpt. Bei 2100 V Klemmenspannung (entsprechend einer Feldstärke von 42 V/cm) und einem bei den gegebenen Bedingungen resultierenden Strom von 100 mA pumpt man mittels der Dosiereinrichtung eine Lösung von 7 mg der Frakt. C_1 in 2.6 m*l* derselben Ammoniumacetat-Lösung innerhalb von 2 h in die über Auffanggläschen Nr. 78 liegende Zuführöffnung ein. Nach der Substanzeingabe wird der Durchfluß der Elektrolytlösung noch ca. 3 h fortgesetzt, bis das Volumen in jedem der insgesamt 95 Gläschen etwa 10.5 m*l* beträgt. Man ermittelt die Peptidverteilung durch Extinktionsmessung bei 220 bzw. 280 nm (PMQ-II, s. Abb. 3). Durch Lyophilisieren von Frakt. 43 - 58 werden 6.8 mg eines Peptids erhalten (Frakt. D_1), dessen sämtliche Analysendaten mit denen von C_1 übereinstimmen (vgl. unter 2. und Tab. 1).

II) ENZYMATISCHE HYDROLYSE DER [11-LEUCIN]HUMAN-MINIGASTRIN-I-FRAKTIONEN C_1 UND C_2

1) Chymotrypsin-Spaltung

20 mg der Frakt. C_1 (bzw. C_2) werden in 6 m*l* Ammoniumacetat-Lösung, pH 7.9, gelöst und nach Zugabe von 4 mg Chymotrypsin (Enzym:Substrat wie 1:5) 24 h bei 25 °C inkubiert. Nach Ansäuern mit 1M Essigsäure wird der Ansatz eingefroren und dann lyophilisiert.

2) Gelchromatographische Trennung der Spaltstücke an Sephadex G-25

In eine Säule von 186 cm Länge und 1.5 cm Durchmesser füllt man wie üblich in 0.1M NH_4HCO_3 mehrere Stunden gequollenes Sephadex G-25 (fine); darauf bringt man eine Lösung des unter 1. erhaltenen Lyophilisates in 3 m*l* 0.1M NH_4HCO_3 und eluiert wie üblich mit demselben Lösungsmittel bei einer Durchflußgeschwindigkeit von 19.5 m*l*/h. Das Eluat fängt man in Portionen von jeweils 6.5 m*l* in den Zellen eines Fraktionssammlers auf und ermittelt die Peptidverteilung durch kontinuierliche Messung der Extinktion der Lösung bei 254 bzw. 206 nm (s. Abb. 7). Durch Lyophilisieren entsprechender Fraktionen werden die Spaltstücke Nr. 1 bis 6 (aus C_1) bzw. Nr. 1′ und 4′ (aus C_2) isoliert. Aminosäurezusammensetzung s. Tab. 2.

III) DERIVATISIERUNG DER TEILSEQUENZEN [9 - 13] VON FRAKT. C_1 UND C_2 ZUR MASSENSPEKTROMETRISCHEN STRUKTURERMITTLUNG

Die Überführung der freien Pentapeptide (Spaltstücke Nr. 4 aus C_1 bzw. 4′ aus C_2) in die für die Massenspektrometrie geeigneten, verdampfbaren N^{α}-Acetyl-*N,O*-permethyl-Derivate wird wie folgt durchgeführt:

a) Acetylierung: Die Lösung von ca. 5 µmol Peptid in einer Mischung von 1 m*l* Wasser und 1 m*l* Acetanhydrid wird mittels eines kleinen Magnetrührers 1 h bei Raumtemperatur gerührt; nach erneuter Zugabe derselben Mengen an Wasser und Acetanhydrid wird das Rühren noch 1 h fortgesetzt. Nach Abdestillieren der Lösungsmittel im Vak. versetzt man den verbleibenden Rückstand noch dreimal mit je 2 m*l* Wasser und bringt jeweils erneut durch Abdestillieren des Lösungsmittels im Vak. zur Trockne. Das verbleibende *N*-Acetyl-peptid wird im Vakuum-Exsiccator über Phosphorpentoxid und Kaliumhydroxid getrocknet.

b) Permethylierung: Zu einer Lösung des unter a) erhaltenen *N*-Acetyl-peptids in 0.5 m*l* trockenem Dimethylsulfoxid (Reinheitsgrad "Uvasol" der Fa. Merck, Darmstadt; zusätzlich getrocknet durch Abdestillieren im Vak. von Calciumhydrid) werden bei Raumtemperatur nacheinander unter Rühren zuerst eine frisch bereitete Lösung von 100 mg Kalium-tert.-butylat in 2 m*l* trockenem Dimethylsulfoxid (hergestellt durch Rühren der Mischung bei 70 °C für 1 h und anschließendes Abkühlen auf Raumtemperatur) und anschließend 0.5 m*l* Methyljodid gegeben. Nach 1 h wird die Mischung mit 3 m*l* 1M Essigsäure versetzt und danach mit etwa 20 m*l* Chloroform erschöpfend extrahiert. Nach 3maligem Waschen der Chloroformlösung mit Wasser wird das Lösungsmittel im Vak. abdestilliert. Der verbleibende Rückstand wird nach Trocknen im Vakuum-Exsiccator über Phosphorpentoxid und Kaliumhydroxid in die Quarzsonde des Direkteinlaßsystems des Massenspektrometers gebracht.

Literatur

1 I. Mitteil.: Moroder, L., Drees, F., Jaeger, E. & Wünsch, E. (1978) *Hoppe-Seyler's Z. Physiol. Chem.* **359**, 147 - 153, voranstehend.

2 Jaeger, E., Knof, S., Schmidt, I., Thamm, P. & Wünsch, E. (1976) Abstracts, 1st USSR/FRG-Symposium on Peptide and Protein Chemistry, Dushanbe/Tadjikistan, März 1976, printed in Moscow, S. 4 - 6.

3 Houben-Weyl, Methoden der Organischen Chemie, 4. Aufl., Bd. 15/1, S. 20 - 27, Verlag G. Thieme, Stuttgart 1974.

4 Walsh, J.H., Debas, H.T. and Grossman, M.I. (1974) *J. Clin. Invest.* **54**, 477 - 485.

5 Gregory, R.A. & Tracy, H.J. (1974) *Gut* **15**, 683 - 685.

6 Reindel, F. & Hoppe, W. (1954) *Chem. Ber.* **87**, 1103 - 1107.

7 Siehe z.B.: Smith, I. (1953) *Nature (London)* **171**, 43 - 44.

8 a) Wünsch, E., Jaeger, E., Deffner, M. & Scharf, R. (1972) *Hoppe-Seyler's Z. Physiol. Chem.* **353**, 1716 - 1720.
b) Newton, G.G.F. & Abraham, E.R. (1950) *Biochem. J.* **47**, 257 - 267.

9 Matsubara, H. & Sasaki, R.M. (1969) *Biochem. Biophys. Res. Commun.* **35**, 175 - 181.

10 Konturek, J., Tasler, J., Cieszkowski, M., Dobrzanska, M. & Wünsch, E. (1976) *Am. J. Physiol.* **233** (3) E170 - E174.
11 Sober, H.A. (1970) Handbook of Biochemistry B-75.
12 Beaven, G.H. & Holiday, E.R. (1952) *Adv. Protein Chem.* **7**, 319 - 386.
13 Goodwin, T.W. & Morton, R.A. (1946) *Biochem. J.* **40**, 628 - 632.
14 Bencze, W.L. & Schmid, K. (1957) *Anal. Chem.* **29**, 1193 - 1196.
15 Vilkas, E., Lederer, E. & Massot, J.C. (1968) *Tetrahedron Lett.* 3089 - 3092.
16 Eagles, J., Laird, W.M., Self, R. & Synge, R.L.M. (1974) *Biomed. Mass Spectrom.* **1**, 43 - 48.
17 Cohn, E.J. & Edsall, J.T. (1943) Proteins, Amino Acids and Peptides as Ions and Dipolar Ions, Reinhold Publ. Corp., New York.

Dr. E. Jaeger, Dr. P. Thamm, Dr. I. Schmidt, Dr. S. Knof, Prof. Dr. L. Moroder und Prof. Dr. E. Wünsch, Max-Planck-Institut für Biochemie, Abteilung für Peptidchemie,
D-8033 Martinsried bei München.

Hoppe-Seyler's Z. Physiol. Chem.
Bd. 359, S. 165 - 172, Februar 1978

Stilbene Synthases and Stilbenecarboxylate Synthases, I

Enzymatic Synthesis of 3,5,4′-Trihydroxystilbene from *p*-Coumaroyl Coenzyme A and Malonyl Coenzyme A

Norbert RUPPRICH and Helmut KINDL

Biochemie (Fachbereich Chemie), Universität Marburg

(Received 24 October 1977)

Summary:

1) Cell-free extracts prepared via acetone powder from rhizome of *Rheum rhaponticum* were found to be capable of converting *p*-coumaroyl-CoA and [2-^{14}C]malonyl-CoA into a 3,5,4′-trihydroxystilbene, resveratrol. The product was characterized by repeated chromatography and by recrystallization to constant specific radioactivity.

2) The mode of condensation of the phenylpropanoid moiety with C_2 units was determined by chemical degradation of the enzymatically formed product.

3) The enzyme system which catalyses the condensation of a *p*-coumaroyl moiety with 3 malonyl units, the intramolecular aldol reaction and the subsequent decarboxylation was classified as stilbene synthase. It could be activated by dithiothreitol.

4) When the extraction of the rhizome was performed with aqueous grinding medium, a high percentage of this stilbene synthase activity was found to be membrane-bound.

Stilben-Synthasen und Stilbencarboxylat-Synthasen, I.
Enzymatische Synthese von 3,5,4′-Trihydroxystilben aus p-Cumaroyl-CoA und Malonyl-CoA

Zusammenfassung:

1) Aus dem Rhizom von *Rheum rhaponticum* wurde über Acetontrockenpulver ein zellfreies System gewonnen, das die Kondensation von *p*-Cumaroyl-CoA und [2-^{14}C]Malonyl-CoA zu dem 3,5,4′-Trihydroxystilben, Resveratrol, katalysiert. Das Reaktionsprodukt wurde durch wiederholte Chromatographie und Kristallisation zur konstanten spezifischen Radioaktivität identifiziert.

Abbreviation: HEPES = 4-(2-Hydroxyethyl)-1-piperazineethanesulfonic acid.
Enzymes:
Acetyl-CoA carboxylase, acetyl-CoA:carbon-dioxide ligase (ADP-forming) (EC 6.4.1.2);
Fatty acid synthase;
Flavanone synthase;
Fructose-bisphosphatase, D-fructose-1,6-bisphosphate 1-phosphohydrolase (EC 3.1.3.11);
6-Methylsalicylate synthase;
Ribonucleoside-diphosphate reductase, 2′-deoxyribonucleoside-diphosphate:oxidized-thioredoxin 2′-oxidoreductase (EC 1.17.4.1);
Stilbenecarboxylate synthase;
Stilbene synthase.

2) Die Charakterisierung der Art der Kondensation erfolgte durch chemischen Abbau des Stilbens zu Benzoesäure-Derivaten. Es wurde gezeigt, daß nur Ring A des Stilbens aus Malonyl-CoA-Einheiten aufgebaut wurde.

3) Das Enzymsystem, welches sowohl die Kondensation von Malonyl-Einheiten an das *p*-Cumaroyl-CoA als auch eine intramolekulare Aldolreaktion und eine Decarboxylierung katalysiert, wurde als Stilben-Synthase bezeichnet. Sie benötigt für die maximale Aktivität Dithiothreit.

4) Wurde das Rhizom in einem wäßrigen Aufschlußmedium homogenisiert, zeigte sich, daß – nach einer Fraktionierung des Extraktes durch Differentialzentrifugation – der Großteil der Stilben-Synthase-Aktivität mit einer Membranfraktion assoziiert war, die sich bei 15000 × *g* sedimentieren ließ.

Key words: Stilbene synthase, resveratrol, *Rheum rhaponticum*, *p*-coumaroyl-CoA, membrane enzyme.

More than twenty different stilbenes have been found in nature. Frequently, they are constituents of heartwood or roots of plants[1]. They have a fungistatic function[2] or are repellent towards insects[3]. Recently, several examples of stilbenes that act as phytoalexines were reported. They are produced by plants as a consequence of an attack by parasites[4-6].

Stilbenes are formed in vivo from cinnamic acids and acetic acid units (cf. ref.[1,7]).

The biosynthesis of resveratrol (3,5,4′-trihydroxystilbene) was investigated in leaves of *Eucalyptus sideroxylon*[8] by means of incorporation of phenylalanine and acetic acid. That indicated that the biosynthetic sequence proceeds by combination of the shikimate and polyacetate pathways.

When one compares both structure and biosynthesis in vivo of a stilbene carboxylic acid (scheme 1) with 6-methylsalicylic acid, which is produced by *Penicillium patulum*, a high degree of homology becomes perceptible. In analogy to the 6-methylsalicylic acid synthase reaction[9], where acetic acid is substituted for *p*-coumaric acid, we propose a hypothetical sequence for the enzymic synthesis of stilbenes (scheme 1).

The sequent steps in the formation of stilbenes are postulated to take place at the level of thioesters, with *p*-coumaroyl-CoA and three units of malonyl-CoA being the substrates of a stilbene synthase. A polyketo derivative may function as an enzyme-bound intermediate which could give rise, formally, to stilbenes (scheme 1) by aldol reaction (between C atoms ∗) or to flavanone by intramolecular acylation (between the two carbon atoms ▲).

While the respective enzyme in the flavonoid biosynthesis, the flavanone synthase, has already been detected[10] and characterized[11] from cell suspension cultures and plant tissues, no successful demonstration of stilbene synthase activities in soluble, cell-free systems has so far been reported. Much effort was devoted to test the hypothetical sequence as outlined in scheme 1 in the case of stilbenes. Using organelle fractions from *Hydrangea macrophylla* leaves[12], one of us was able to demonstrate a conversion of acetate into a stilbene carboxylic acid.

In this communication we report, for the first time, the formation of a stilbene from *p*-coumaroyl-CoA and malonyl-CoA by a soluble enzyme preparation.

Materials and Methods

Plant materials

Seeds of *Rheum rhaponticum* (Holsteiner Blut) were supplied by Küpper, Eschwege.

After imbibation for 24 h, plants were grown on vermiculite at 20 °C under continuous white light (10000 lux). Rhizomes of 8 week-old plants were used for the preparation of acetone powders.

Roots of *Veratrum album* were obtained from Caesar & Loretz, Hilden.

▷

Scheme 1. Hypothetical reaction sequence for the formation of stilbenes or stilbenecarboxylic acids, respectively.

Reactions 1 + 2: stilbene synthase
Reactions 3 + 4: stilbenecarboxylate synthase
Reactions 1 + 5: flavanone synthase.

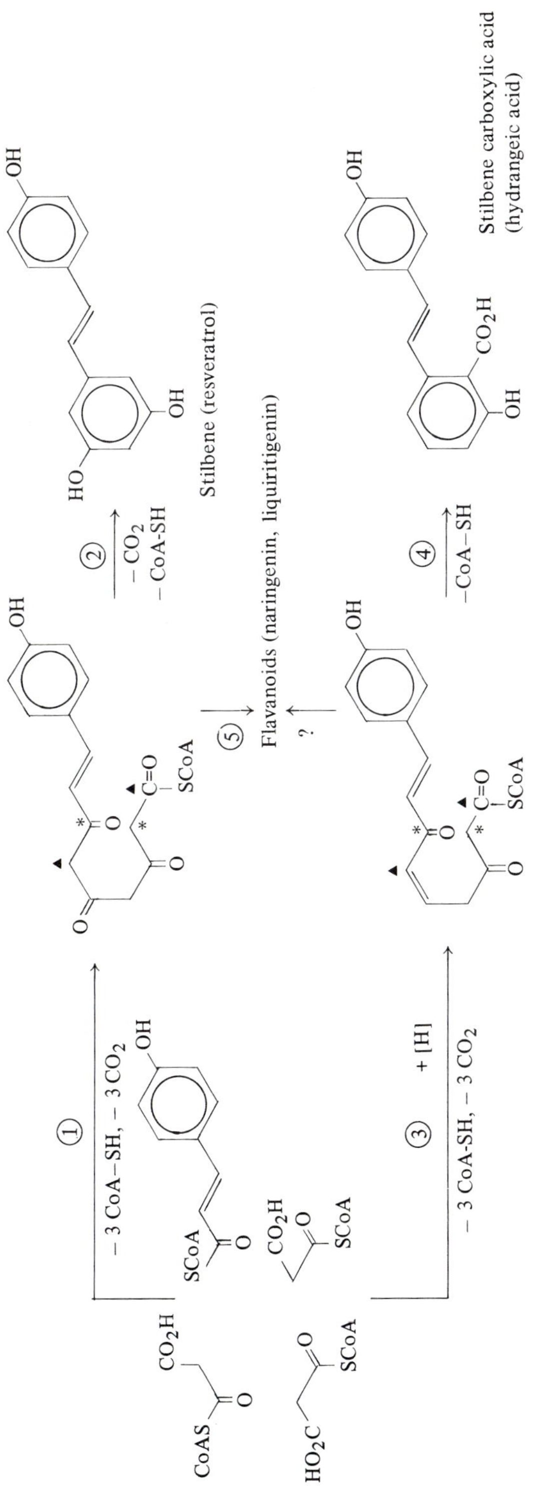

Compounds used

[2-^{14}C]Malonyl-CoA (32.8 mCi/mmol) and [1-^{14}C]acetyl-CoA (11 mCi/mmol) were purchased from NEN Chemicals, Dreieichenhain. *p*-Coumaroyl-CoA was prepared from the respective *N*-hydroxysuccinimide ester according to Stöckigt and Zenk[13]. Resveratrol (3,5,4′-trihydroxystilbene) was prepared from roots of *Veratrum album*. 2 kg of roots were extracted with ethanol[14] and the extract concentrated in vacuo. A preparative column chromatography on silica gel (Woelm, Aktivitätsstufe II) was performed using ethyl acetate/chloroform (4 : 1, v/v) as eluent. Fractions at R_F 0.8 – 0.4 were tested for resveratrol by thin-layer chromatography[15] and pooled in such a way that no contamination with oxyresveratrol (3,5,2′,4′-tetrahydroxystilbene) resulted from fractions with more slowly migrating compounds. Resveratrol was recrystallized from water (mp.: 261 °C[14]). UV-spectrum[15]: λ_{max} = 323 nm (in ethanol); λ_{max} = 338 nm (in ethanol + sodium ethylate). Mass spectrum: *m/e* = 228.1 (100%), 181.1 (20%). Generally, resveratrol was characterized by thin-layer chromatography using 3 different solvent systems, L1, L2, L3.

Enzyme preparation

30 g acetone powder (prepared from 100 g rhizome) was extracted with 100 m*l* medium for 15 min. Medium: 150mM HEPES, pH 7.6, 5mM $MgCl_2$, 1mM dithiothreitol. The extract was filtered through Miracloth and centrifuged at 10000 × *g* for 10 min. The supernatant was desalted by chromatography on Sephadex G-50 which was previously equilibrated with 150mM HEPES, pH 7.6 (v_t = 15 m*l*, v_0 = 6 m*l*).

Membrane preparation

10 g of 8 week-old rhizome was cut in small pieces and extracted with 10 m*l* medium, using a scalpel. Medium: 10% (w/w) sucrose, 150mM HEPES, pH 7.6, 1mM dithiothreitol, 5mM $MgCl_2$. The extract was filtered through Miracloth and subjected to differential centrifugation. The pellets thus obtained were suspended in the grinding medium without sucrose and assayed for stilbene synthase.

Enzyme assays

The standard assay for stilbene synthase contained in a total volume of 2.0 m*l* the following compounds: 15μM [2-^{14}C]malonyl-CoA (1.0 μCi) 100μM *p*-coumaroyl-CoA, 150mM HEPES, pH 7.6, 1mM dithiothreitol, 5mM $MgCl_2$. The reaction was started at 30 °C with the enzyme preparation and was terminated after 30 min by adding 1 m*l* 0.1N HCl. 100 μg inactive resveratrol was added and the aqueous solution was extracted 3 times with ethyl acetate. The organic phase was concentrated in vacuo and the product separated first by thin-layer chromatography on silica gel using L1 as solvent system. The zones corresponding to resveratrol were scraped off, and counted in a scintillation counter. The formation of 1 nmol resveratrol corresponds to the conversion of 3 nmol malonyl-CoA.

Selective degradation of resveratrol

22 mg of resveratrol was dissolved in 0.3 m*l* methanol and methylated with dimethylsulfate (25 mg) and 0.1 m*l* 4N NaOH at room temperature. After 30 min of intensive stirring the reaction mixture was somewhat concentrated in vacuo and extracted with diethyl ether. The ether solution was shaken with 0.5N NaOH and then with water. After concentration of the ether phase, the resulting trimethyl derivative was dissolved in acetone (2 m*l*) and treated with 90 mg powdered $KMnO_4$. The mixture was allowed to react for 1 h and was then treated with $NaHSO_3$. That yields a clear colourless solution which was extracted with ether. The ether solution was separated and reextracted with 1M sodium carbonate. The sodium carbonate solution was acidified and extracted with ether, which was then concentrated in vacuo and the products further separated on paper chromatography in solvent system L5. The respective zones were eluted with methanol and the *p*-methoxybenzoic acid or 3,5-dimethoxybenzoic acid, respectively, recrystallized from water/methanol. *p*-Methoxybenzoic acid, m.p.: 184 °C. 3,5-Dimethoxybenzoic acid, m.p.: 181 °C.

Chromatography

The following solvent systems were used for thin-layer chromatography on silica gel (0.2 mm thick).

L1: chloroform/ethyl acetate/formic acid (5 : 4 : 1, v/v/v)[15];
R_F (resveratrol) = 0.71.

L2: benzene/dioxane/acetic acid (90 : 25 : 4, v/v/v);
R_F (resveratrol) = 0.51.

L3: toluene/ethylformiate/formic acid (5 : 4 : 1, v/v/v);
R_F (resveratrol) = 0.59.

For paper chromatography the following systems were used:

L4: benzene/acetic acid/water (4 : 2 : 1, v/v/v)[16];
R_F (resveratrol) = 0.05.

L5: n-butanol/NH_4OH (conc.)/ethanol/benzene (5 : 3 : 2 : 1, v/v/v/v)[16];
R_F (3,5-dimethoxybenzoic acid) = 0.70;
R_F (*p*-methoxybenzoic acid) = 0.62.

Determination of radioactivity

Radioactive samples were counted in a scintillation spectrometer (Packard Tricarb 3380/544). Absolute radioactivity was calculated by a program based on channel ratio methods. Chromatograms were scanned on a chromatogram scanner (Berthold LB-2723, 4 π-geometry, 15% efficiency for ^{14}C).

Determination of protein

Samples of enzyme preparations or membrane fractions were precipitated with 15% trichloroacetic acid, the precipitate separated by centrifugation and then boiled with acetone for 10 min. After precipitation by centrifugation the protein was determined according to Lowry et al.[17]. This procedure was necessary to remove the large amounts of phenols which we encountered frequently.

Results

1. Identification of the enzymatically formed product

200 nmol *p*-coumaroyl-CoA and 30 nmol malonyl-CoA (1.0 μCi) were incubated with the enzyme preparation (0.92 mg protein) in a total volume of 2.0 m*l*. The resveratrol obtained in this reaction was extracted and chromatographed in 2 different solvent systems (L1 and L4). The profile of radioactivity on the second chromatography showed no other zone of radioactivity except resveratrol.

In addition, a sample of the purified resveratrol was eluted from the chromatogram, diluted with 100 mg inactive material, and recrystallized 3 times from water. The results obtained after the crystallizations are outlined in Table 1.

It can be seen that the resveratrol was already pure after the second chromatography, because no significant variation in specific activity, either in the mother liquor or in the crystals, could be detected.

For a further characterization of the enzymatically obtained resveratrol, a selective degradation of the stilbene should prove whether the enzymatic

Table 1. Recrystallization of rechromatographed resveratrol.

Operation	Fraction	Spec. act. [dpm/mg]
First cryst.	Mother liquor	67.3
	Crystals	76.9
Second cryst.	Crystals	81.8
Third cryst.	Crystals	82.3

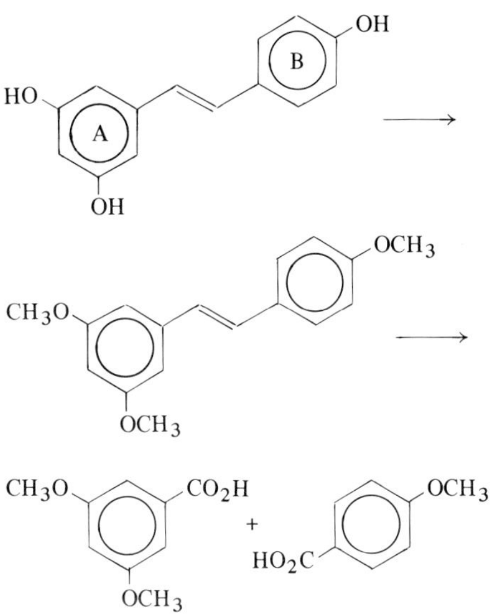

Scheme 2. Chemical degradation of resveratrol.

condensation of *p*-coumaroyl-CoA and malonyl-CoA took place in the way proposed. Therefore, the hydroxystilbene was methylated and ring A and B of the stilbene were separated from each other by oxidation with $KMnO_4$ (Scheme 2). The cleavage of the double bond leads to *p*-methoxybenzoic acid (equivalent to ring B of the stilbene) and 3,5-dimethoxybenzoic acid (equivalent to ring A).

A specific incorporation of [2-^{14}C]malonyl units into ring A was established as it was found that more than 99% of the label within the carbon skeleton of resveratrol was recovered in the 3,5-dimethoxybenzoic acid (Table 2).

Table 2. Balance of radioactivity obtained by chemical degradation of enzymatically formed resveratrol (cf. scheme 2).

Compound	Spec. act. [nCi/mmol]	C atoms
Resveratrol	156	Ring A and B, C_α, C_β
3,5-Dimethoxy-benzoic acid	164	Ring A and C_α
p-Methoxybenzoic acid	< 1	Ring B and C_β

Table 3. Factors affecting the enzymatic resveratrol formation.

An enzyme preparation which was obtained from acetone powder and centrifuged at 10 000 x *g* was either used directly or purified by gel filtration. In the latter case the elution buffer only contained HEPES. Then, dithiothreitol (final concentration 1mM) or $Mg^{2\oplus}$ (final concentration 5mM), respectively, were added to the incubation mixture.

Exp. No.	Enzyme prep.	Substrates	Additions		Resveratrol formed [nmol/(mg prot. x 30 min)]
			Dithiothreitol	$Mg^{2\oplus}$	
1	Prior to gel filtr.	*p*-Coumaroyl-CoA, [2-^{14}C]malonyl-CoA	+	+	0.96
2	Prior to gel filtr.	[2-^{14}C]Malonyl-CoA	+	+	0.05
3	Prior to gel filtr.	*p*-Coumaroyl-CoA, [1-^{14}C]acetyl-CoA	+	+	< 0.01
4	After gel filtr.	*p*-Coumaroyl-CoA, [2-^{14}C]malonyl-CoA	–	–	0.15
5	After gel filtr.	*p*-Coumaroyl-CoA, [2-^{14}C]malonyl-CoA	–	+	0.40
6	After gel filtr.	*p*-Coumaroyl-CoA, [2-^{14}C]malonyl-CoA	+	–	1.30
7	After gel filtr. boiled	*p*-Coumaroyl-CoA, [2-^{14}C]malonyl-CoA	+	+	< 0.01

Table 4. Different cell fractions capable of catalyzing resveratrol formation.
An aqueous extract prepared from rhizome was separated from cell debris by centrifugation at 200 × *g* for 3 min and then fractionated by 3 successive centrifugations.

Operation	Fraction	Protein [mg]	Total activity [pmol resveratrol/30 min]	Spec. act. [pmol resveratrol/(mg prot. × 30 min)]	Activity [%]
Centrifugation at 1200 × *g* for 10 min	Pellet	1.5	29	19.3	10.7
Centrifugation at 15 000 × *g* for 15 min	Pellet	1.6	211	131.8	77.6
Centrifugation at 100 000 × *g* for 1 h	Pellet	0.6	11	18.3	4.0
	Supernat.	1.2	21	17.5	7.7

2. Characterization of the enzymic reaction

A series of experiments (Table 3) showed that (a) the incorporation of radioactivity from malonyl-CoA into resveratrol was abolished when *p*-coumaroyl-CoA was omitted, and that (b) acetyl-CoA could not substitute for malonyl-CoA. Furthermore, this stilbene synthase activity was very much dependent on dithiothreitol.

Mercaptoethanol could not substitute for dithiothreitol. Incubation of the enzyme for 15 min with 1mM iodoacetamide prior to the enzyme test resulted in 90% inhibition of the enzyme activity; the actual stilbene synthase reaction was, here too, followed in the presence of 1mM dithiothreitol.

It was found that, under standard conditions, the reaction was strictly linear with time for at least 20 min.

3. Membrane-associated form(s) of stilbene synthase

One of the reasons why we used acetone powders as starting material for enzyme preparations was that upon extraction in aqueous grinding medium, most of the stilbene synthase activity turned out to be membrane-bound. A differential centrifugation (Table 4) of the homogenate prepared from young rhizome indicates that a high percentage of the extractable activity can be attributed to membranes which are sedimentable at 15 000 × *g*. Data as given in Table 5 demonstrate that the membrane-associated form of stilbene synthase, too, is sensitive to activation by dithiothreitol.

Discussion

Chain-lengthening processes which involve malonyl-CoA as a substrate have been described for the synthesis of fatty acids, 6-methylsalicyclic acid, and flavanones. Also, a microsomal chain-elongation system of fatty acids has been investigated[18]. In the course of stilbene synthesis, too, a chain elongation by repeated condensation of a starter acyl derivative with malonyl-CoA is now evident. As in the formation of flavanones, a *p*-coumaroyl ester is the compound with which

Table 5. Effects on membrane-associated stilbene synthase.
0.91 mg of a membrane preparation suspended in HEPES was incubated with *p*-coumaroyl-CoA, [2-^{14}C]malonyl-CoA, dithiothreitol and $Mg^{2\oplus}$, respectively. As membrane preparation either a crude 1500 × *g* pellet (cf. Table 4) or a purified sediment washed twice in HEPES was used.

Exp. No.	Membrane preparation	Additions: Dithiothreitol	Additions: $Mg^{2\oplus}$	Resveratrol formed [nmol/(30 min × mg prot.)]
1	Untreated	+	+	0.34
2	Washed twice	–	–	0.23
3	Washed twice	+	–	0.75
4	Washed twice	+	+	0.68

the malonyl-CoA is acylated. But otherwise, the stilbene synthesis resembles more the conversion of acetyl-CoA and 3 malonyl-CoA into 6-methylsalicylic acid[9], if one compares the mode of formation of a new intramolecular C-C bond within the polyketo derivative prior to aromatization (cf. Scheme 1).

In presenting evidence for an enzyme which catalyses the stilbene synthesis from CoA-ester of *p*-coumaric acid, we must recall that Hillis and Ishikura[19] reported an enzyme or factor which mediated the conversion of cinnamoyl triacetic acid to 3,5-dihydroxystilbene (pinosylvin). We do not have any clue that corresponding intermediates, e.g. *p*-coumaroyl triacetic acid, are formed by our preparations. Furthermore, such an extremely reactive compound as a 3,5,7-triketoacid is very unlikely as substrate, if not tightly bound to the enzyme. Additional work and comparison of substrate specificity, dependence on active thiol groups, molecular weight, etc., should clarify these discrepancies.

The lack of acetyl-CoA incorporation into the stilbene by our enzyme preparation enables us clearly to define malonyl-CoA as the substrate of chain lengthening. An acetyl-CoA carboxylase activity, leading to malonyl-CoA, did not interfere. Such an enzyme would, if present together with ATP, render it more difficult to distinguish which of the two CoA-esters, malonyl-CoA or acetyl-CoA, is the real substrate.

The drastic influence of dithiothreitol (Table 3) on the stilbene synthase may be an intriguing feature. Such a stimulation was also observed with the membrane-associated form of the enzyme (compare Exp. 2 and 3 in Table 5). Addition of $Mg^{2\oplus}$ did not further stimulate. The phenomenon of stimulation by dithiothreitol is not so much a preservation as an activation of the enzyme, if we compare the stilbene synthase activity before and after washing the membranes with dithiothreitol-free buffers (Table 5, Exp. 1 and 2).

The simplest explanation would be a very sensitive thiol group in the active site of the enzyme. This is substantiated by the significant inhibitory effect of iodoacetamide.

But the selective reduction by dithiothreitol, with mercaptoethanol not being an effective substitute, may point to a thiol group not located in the active center.

It is not unlikely that we have to ascribe a control function to this phenomenon. As it is known that thioredoxins are present even in roots[20], and may exert control by reducing certain enzymes, e.g. fructose-bisphosphatase, a similar mechanism with stilbene synthase cannot be excluded. In several cases, e.g. with ribonucleoside-diphosphate reductase, thioredoxins can be reduced in vitro by dithiothreitol[21]. But so far, we do not have a clue how such a signal can be passed on from one cell (leaf) to another (rhizome).

The specific activity of our stilbene synthase preparation (0.6 μkat/kg) is comparable with the value for an unpurified flavanone synthase from illuminated parsley cell suspension cultures (0.3 μkat/kg)[11].

A more detailed investigation on the properties of stilbene synthase and the question of the enzymes intracellular location is in progress.

This investigation was supported by the *Deutsche Forschungsgemeinschaft*. Financial support from the *Fonds der Chemischen Industrie* is also acknowledged.

Literature

1 Billek, G. (1964) *Fortschr. Chem. Org. Naturst.* **22**, 115 - 152.

2 Shain, L. (1967) *Phytopathology* **57**, 1034 - 1045.

3 Hillis, W. E. & Inoue, T. (1968) *Phytochemistry* **7**, 13 - 22.

4 Keen, N. T. & Ingham, J. L. (1976) *Phytochemistry* **15**, 1794 - 1795.

5 Langcake, P. & Pryce, R. J. (1976) *Physiol. Plant Pathol.* **9**, 77 - 86.

6 Ward, E. W. B., Unwin, C. H. & Stoessl, A. (1975) *Can. J. Bot.* **63**, 964 - 971.

7 Kindl, H. (1971) *Naturwissenschaften* **58**, 554 - 563.

8 Hillis, W. E. & Hasegawa, M. (1962) *Chem. Ind. (London)* 1330 - 1331.

9 Dimroth, P., Walter, H. & Lynen, F. (1970) *Eur. J. Biochem.* **13**, 98 - 110.

10 Kreuzaler, F. & Hahlbrock, K. (1972) *FEBS Lett.* **28**, 69 - 72.

11 Kreuzaler, F. & Hahlbrock, K. (1975) *Eur. J. Biochem.* **56**, 205 - 213.

12 Kindl, H. (1971) *Hoppe-Seyler's Z. Physiol. Chem.* **352**, 767 - 768.

13 Stöckigt, J. & Zenk, M. H. (1975) *Z. Naturforsch.* **30c**, 352 - 358.

14 Takaoka, M. (1940) *J. Faculty Sci., Hokkaido Imp. Univ., Ser. III,* **3**, 1 - 16.

15 Hillis, W. E. & Ishikura, N. (1968) *J. Chromatogr.* **32**, 323 - 336.

16 Billek, G. & Kindl, H. (1962) *Monatsh. Chem.* **93**, 85 - 98.
17 Lowry, O. H., Rosebrough, N. J., Farr, A. L. & Randall, R. J. (1951) *J. Biol. Chem.* **193**, 265 - 275.
18 Podack, E. R., Saathoff, G. & Seubert, W. (1974) *Eur. J. Biochem.* **50**, 237 - 243.
19 Hillis, W. E. & Ishikura, N. (1969) *Phytochemistry* **8**, 1079 - 1088.
20 Buchanan, B. B. & Wolosiuk, R. A. (1976) *Nature (London)* **264**, 669 - 670.
21 Wagner, W. & Follmann, H. (1977) *Biochem. Biophys. Res. Commun.* **77**, 1044 - 1051.

N. Rupprich and H. Kindl, Biochemie (FB Chemie), Universität Marburg, Lahnberge, D-3550 Marburg.

Hoppe-Seyler's Z. Physiol. Chem.
Bd. 359, S. 173 - 180, Februar 1978

The Carbohydrate Linkage Site of Cow Colostrum Trypsin Inhibitor*

Rainer KLAUSER**, Dana ČECHOVÁ*** and Harald TSCHESCHE

** Organisch-Chemisches Institut der Technischen Universität München, Lehrstuhl für Organische Chemie und Biochemie and
*** Institute of Molecular Genetics, Czechoslovak Academy of Sciences, Prague.

(Received 12 September 1977)

Herrn Professor Dr. Georg Wittig zum 80. Geburtstag gewidmet

Summary: Cow colostrum trypsin inhibitor (chromatographic form A IV[7]) was subjected to basic conditions that favour β-elimination of carbohydrates from *O*-glycosidic linkages. The unchanged carbohydrate composition and the unchanged values for serine and threonine indicate the presence of an alkali-stable *N*-glycosidic bond to asparagine. From a tryptic digest of *S*-aminoethylated inhibitor the glyco-decapeptide (residues 24 through 33) was isolated. The carbohydrate composition was identical with that of the *S*-aminoethylated inhibitor. Further degradation of this peptide by carboxypeptidase C (and aminopeptidase) produced the glycopeptide Asn(CHO)-Ser-(Thr) with the same carbohydrate composition. Thus, a single carbohydrate moiety is bound by a *N*-glycosidic linkage to asparagine-27 of the colostrum inhibitor. It is located opposite to the reactive site at the base of a pear-shaped molecule.

Die Kohlenhydrat-Anknüpfungsstelle des Trypsin-Inhibitors aus Rindcolostrum

Zusammenfassung: Colostrum-Trypsin-Inhibitor vom Rind (Chromatographische Form A IV[7]) wurde basischen Bedingungen unterworfen, die eine β-Eliminierung *O*-glykosidisch gebundener Kohlenhydrate begünstigen. Die unveränderte Kohlenhydrat-Zusammensetzung und die unveränderten Werte für Serin und Threonin deuten auf eine alkalistabile *N*-glykosidische Bindung an Asparagin hin. Nach tryptischer Verdauung von *S*-aminoäthyliertem Inhibitor wurde das Glykodecapeptid (Reste 24 bis 33) isoliert. Die Kohlenhydratzusammensetzung war identisch mit der von *S*-aminoäthyliertem Inhibitor. Weiterer Abbau dieses Peptids mit Carboxypeptidase C (und Aminopeptidase) lieferte das Glykopeptid Asn(CHO)-Ser-(Thr) mit der gleichen Kohlenhydrat-Zusammensetzung. Damit ist ein Kohlenhydratrest über eine *N*-glykosidische Bindung an Asparagin-27 des Colostrum-Inhibitors gebunden. Er ist gegenüber dem reaktiven Zentrum an der Basis des birnenförmigen Moleküls angeordnet.

Enzymes: Carboxypeptidase C, peptidyl-L-amino-acid(-L-proline)hydrolase (EC 3.4.12.1);
Trypsin (EC 3.4.21.4).

Abbreviations: Man, mannose; Gal, galactose; Glc, Glucose; Fuc, fucose; GalN, galactosamine; GlcN, glucosamine; NeuAc, *N*-acetylneuraminic acid; TPCK, 1-chloro-(N-*p*-tosyl-L-phenylalanyl)methane; CTI, colostrum trypsin inhibitor.

* This paper is part of the Ph. D. thesis of R. Klauser, Technical University, Munich, 1977.

Key words: Colostrum inhibitor, glycoprotein, carbohydrate attachment, asparagine.

The trypsin inhibitor of bovine colostrum is secreted by the lactogenic system of the cow and has been subject to several studies[1-4]. The amino acid sequence of the inhibitor has been determined[5,6]. A sequence homology of almost 50% has been found with the intracellular trypsin kallikrein inhibitor (Kunitz). However, the colostrum inhibitor is a glycoprotein. The carbohydrate moiety which amounts to 54% of the total molecular weight has been found to be microheterogeneous[7]. In the present paper we want to report on the site of attachment of the carbohydrates to the polypeptide backbone.

Materials and Methods

Alkaline incubation

Two slightly different methods were applied. (a) According to Anderson et al.[8] 1 mg colostrum inhibitor (CTI) was dissolved in 0.5 m*l* 0.5N NaOH and kept at 4 °C for 24 h. (b) The alternative method was to dissolve 1 mg CTI in 0.5 m*l* of a solution of 0.3N NaOH and 1% $NaBH_4$. This solution was kept at 4 °C for 7 days as described by Schiffman et al.[9]. After the incubation 0.5 m*l* acetic acid was added to the solution which then was lyophilized. The inhibitor was desalted by gel filtration on a column 50 x 0.9 cm of BioGel P-2 (200 - 400 mesh, BioRad Laboratories, Munich) in 0.1N acetic acid and then lyophilized.

Amino acid and carbohydrate analysis

Amino acid analyses were performed on a Technicon Autoanalyzer by the method of Spackman et al.[10] as described in[11]. Monosaccharides were determined as silylated methyl glycosides formed by methanolysis with 1.5N HCl in methanol in sealed evacuated ampoules according to Clamp et al.[12]. The reaction was allowed to proceed for 11 h at 95 °C. Bis(trimethylsilyl)acetamide was used for silylation as described by Tschesche et al.[7].

The mixture of silylated methylglycosides was separated with a gas chromatograph (Hewlett Packard 7620 A, Böblingen) fitted with a glass column 200 x 0.3 cm containing 5% methylsilicone gum SE-30 on chromosorb W-AW-DMCS, 80 - 100 mesh (Hewlett Packard). The carrier gas was helium at a flow rate of 60 m*l*/min. Flame ionization detection was used. Injection port temperature was 320 °C and detector temperature was 340 °C. The temperature program for each run was: starting at 130 °C for 2 min then rate 2 °C/min for 30 min (190 °C) and 4 °C/min for 15 min (250 °C) finally 20 min at 250 °C. Identification of the monosaccharides was based on relative retention times with mannitol as internal standard and by comparison with a standard mixture of monosaccharides. The following sugars were used a standards: α-D-galactose (Schuchardt AG, München), D-glucose, D-mannose, D-galactosamine hydrochloride and L-fucose (Merck AG, Darmstadt), D-glucosamine hydrochloride (Fluka AG, Buchs, Switzerland), and *N*-acetylneuraminic acid (Serva AG, Heidelberg).

The amount of each monosaccharide was calculated on the basis of molar response factors as described[12]. The peak areas of the various isomers were added up and compared to the peak area of the internal standard.

Aminoethylation of colostrum inhibitor

Colostrum inhibitor was reduced and aminoethylated in liquid ammonia according to the method of Kuromizu and Meienhofer[13]. 26.5 mg CTI (2.15 μmol) was brought into a two-necked round bottom flask fitted with a condenser for the ammonia. The condenser and the flask were cooled with a mixture of methanol and solid CO_2. The temperature was – 60 °C. The stream of ammonia was maintained until 25 m*l* had condensed into the flask. Then dry nitrogen was flushed through the flask via the second neck and the condenser was removed. Denaturation of the inhibitor was allowed to proceed for 1 h. Then, under a barrier of nitrogen, mercaptoethanol (22 μ*l*, 320 μmol) was added in a 15-fold excess. The reaction was allowed to proceed for half an hour and ethylenimine (Fluka AG, Buchs) (6.4 mmol, 0.33 m*l*) was added. The temperature of the mixture was then allowed to rise and the ammonia evaporated spontaneously The residue was lyophilized. The modified inhibitor was extracted three times with absolute methanol in order to remove by-products, subjected to centrifugation, redissolved in 0.1N acetic acid and desalted on a column 50 x 0.9 cm of BioGel P-2 in 0.1N acetic acid.

Homogeneity of the material was checked by ion exchange chromatography on a column 50 x 0.9 cm of DEAE-Sephadex A-25 equilibrated with 0.02M borate buffer, pH 10.0. The *S*-aminoethylated inhibitor was eluted by a gradient of increasing pH and ionic strength. The inhibitor appeared as a single peak at pH 7.5 and 0.1M borate concentration. After desalting the inhibitor was subjected to amino acid analysis (see Table 2).

Tryptic digestion of S-aminoethylated inhibitor

15 mg *S*-aminoethylated inhibitor was dissolved in 3 m*l* triethanolamine buffer, 0.2M, pH 7.8 and 0.01M $CaCl_2$.

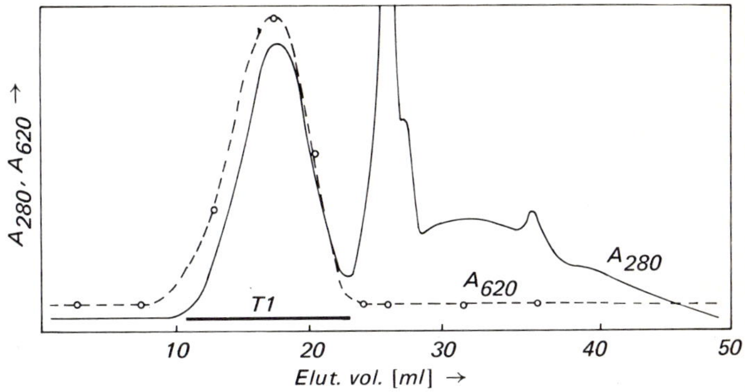

Fig. 1. Chromatography of the tryptic digest of aminoethylated colostrum inhibitor on Biogel P6.

The column (70 x 0.9 cm) was equilibrated with 0.1N acetic acid. Elution was carried out with the same solution. Flow rate was 8 ml/h. Fract. vol.: 2 ml. Full line: absorbance at 280 nm; dashed line: absorbance at 620 nm using the anthrone test (see Methods). The bar indicates pooled fractions.

TPCK-treated bovine trypsin (E. Merck, Darmstadt) (0.624 mg, 26 nmol) was added to the mixture which was digested for 24 h at 25 °C according to Greene and Giordano[14]. The solution was then acidified with 1 ml glacial acetic acid and lyophilized.

Separation of tryptic peptides

A column 70 x 0.9 cm of BioGel P-6 (100 - 200 mesh, BioRad Laboratories, Munich) was equilibrated with 0.1N acetic acid. The peptide mixture was dissolved and eluted in the equilibration buffer. The elution profile is shown in Fig. 1.

The first peak, designated,T-1, was collected as indicated by the bar and lyophilized. This peak was further chromatographed on a column 50 x 0.9 cm of SE-Sephadex C-25 (50 - 150 mesh, Pharmacia, Frankfurt/M.) equilibrated with 0.1M ammonium acetate, pH 6.8. The elution profile is shown in Fig. 2. The major peak containing the carbohydrates was rechromatographed under the same conditions and was then desalted by repeated lyophilization. This fraction designated as peak T-1b was used for all further investigations.

Detection of glycopeptides

Glycopeptides were detected by the anthrone method of Roe[15].

Aliquots of 0.3 ml of each chromatographic fraction were diluted to 1 ml, mixed with 5 ml of the reagent, heated for 20 min on a boiling water bath and cooled to room temperature. The absorption at 620 nm was read against a blank of water in a Shimadzu double beam spectrophotometer. Reagent solution: 50 g anthrone dissolved in a mixture of 280 ml water, 720 ml concentrated sulfuric acid and 10 g thiourea. After cooling to room temperature this solution was ready for use. A standard curve was taken with galactose in a concentration range from 1 - 150 µg/ml.

Carboxypeptidase digestion of decapeptide T-1b

0.85 mg of peptide T-1b was incubated with 2.5 U carboxypeptidase C (Röhm GmbH, Darmstadt) in 1 ml 0.05M sodium citrate buffer, pH 5.3, containing 0.001M sodium azide, according to[16]. Digestion was allowed to proceed for 48 h at 30 °C and was stopped by acidification with glacial acetic acid (0.5 ml) and lyophilization.

The digest was then chromatographed on a 50 x 0.9 cm column of Sephadex G-25 equilibrated with 0.1N acetic acid. The elution profile is shown in Fig. 3. The peptide designated as C-1 was collected as indicated by the bar and lyophilized. Rechromatography was performed on

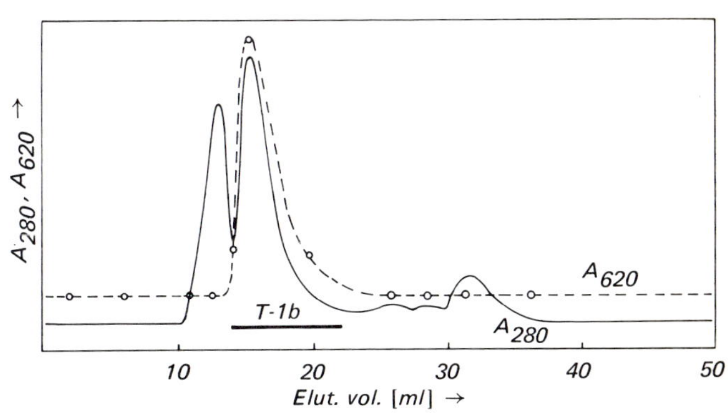

Fig. 2. Chromatography of tryptic peptide mixture T-1 on SE Sephadex C-25.

The column (50 x 0.9 cm) was equilibrated and eluted with 0.1M ammonium acetate, pH 6.8. Flow rate: 5 ml/h. Fract. vol.: 2.5 ml. Full line: absorbance at 280 nm, dashed line: absorbance at 620 nm, using the anthrone test. The bar indicates pooled fraction.

a 30 × 0.9 cm column of Merkogel 2000 (E. Merck, Darmstadt) equilibrated with 0.1N acetic acid. The elution profile is shown in Fig. 4. The peptide designated C-1c was collected as indicated by the bar, lyophilized and subjected to amino acid and carbohydrate analysis.

Results and Discussion

The isolation of bovine colostrum inhibitor and its separation into several components differing in their carbohydrate composition has been described previously[7,17]. The material designated as chromatographic form A IV was used for the present investigations.

Carbohydrate attachment to the polypeptide backbone is possible either by *O*-glycosidic linkage to one of the monohydroxyamino acids, i.e. serine or threonine, or by *N*-glycosidic linkage to asparagine[18]. Both types of linkages are established in many mammalian secretory glycoproteins. The *O*-glycosidic linkage was e.g. found in the mucoproteins of the submaxillary gland of sheep[19] and cattle[20], in the human blood group substances of A-, B- and H-type[9,21], in some glycopeptides of human sputum[22] and in the porcine seminal plasma acrosin inhibitor[23]. It is a characteristic feature of the *O*-glycosidic linkage to be labile in alkaline medium, since, upon β-elimination, the carbohydrates are lost e.g. by formation of dehydroalanine from the serine-carbohydrate bond.

In contrast to the *O*-glycosidic linkage the *N*-glycosidic bond to asparagine is stable in dilute alkali. This type of linkage is formed from glucosamine and presents the more common type of carbohydrate attachment to proteins. It has e.g. been found in ovalbumin[24,25], human γ-globulin[26], thyroglobulin[27], fetuin[28], porcine ribonuclease[29], and orosomucoid[30].

Several different procedures have been in use for alkaline β-elimination of carbohydrates from glycoproteins. We applied both procedures described by Anderson et al.[8] and by Schiffman et al.[9], respectively, in order to determine whether the carbohydrate linkage in colostrum inhibitor is of the *O*- or *N*-glycosidic type. Amino acid and monosaccharide contents of the inhibitor were determined before and after the alkaline treatment. The results are presented in Table 1.

Obviously no decrease in either serine or threonine takes place with either procedure of alkaline treatment. The cysteine content is drastically lowered upon incubation in dilute alkali due to the β-elimination of H_2S. The resulting dehydroalanine is partially hydrogenated to yield alanine in the presence of sodium borohydride as is obvious from the amino acid composition, Table 1.

The carbohydrate composition remains almost unchanged during alkaline incubation, Table 1. Only loss of two residues of glucosamine and one residue of mannose is obvious after the basic treatment that favours β-elimination. A still greater loss of carbohydrates is observed after *S*-aminoethylation in liquid ammonia, Table 2. However, β-elimination of *O*-glycosidically bound carbohydrates give rise to formation of dehydroalanine from serine (or 2,3-dehydro-α-aminobutyric acid from threonine) and a corresponding decrease of the amount of hydroxyamino acids in the amino acid composition. On the basis of the unchanged amounts of either serine or threonine we are in favor of the interpretation that loss of these three monosaccharides occurs from the carbohydrate moiety and *not* from the polypeptide backbone. From these data we conclude that in bovine colostrum trypsin inhibitor the carbohydrate moiety(ies) is (are) linked to the polypeptide backbone by an alkali-stable-*N*-glycosidic bond to asparagine.

In the amino acid sequence of the bovine colostrum trypsin inhibitor, as established by Čechová et al.[5,6], only two potential carbohydrate attachment sites to asparagine are in general agreement with the hypothetical recognition sequence -Asn-X-Thr or -Asn-X-Ser, respectively,[25,29,30-33] (for review see[34]. If we assume that identification of the residue Asp or Asn in the positions 27 or 65 is still questionable then these two sites are in the partial sequences:

$$\text{-Tyr-}\overset{27}{\text{Asx}}\text{-Thr-Ser-} \quad \text{and} \quad \text{-}\overset{65}{\text{Asp}}\text{-Lys-Ser-.}$$

The question whether both asparagine residues or just one, and which one bear carbohydrates could only be answered by isolation and investigation of the proper partial peptides.

The minimum amount of inhibitor available required a highly specific degradation procedure in order to obtain a sufficient yield of the respective

Table 1. Amino acid and carbohydrate composition of the colostrum inhibitor CTI A IV before and after alkaline treatment.

Values are given as mol amino acid/mol protein.

	CTI A IV	Integer	After treatment with	
			0.5N NaOH	0.3N NaOH + 3% $NaBH_4$
Asp	8.90	8	8.27	7.56
Thr	5.30	6	5.72	5.18
Ser	3.46	3	3.75	3.63
Glu	10.86	10	11.06	10.86
Pro	7.52	7	7.06	7.32
Gly	4.06	4	3.91	4.02
Ala	4.67	4	4.62	5.48[a]
Val	0.00	0	0.00	0.00
Cys	4.10	6	0.99[b]	0.90[b]
Met	0.71	1	0.67	0.69
Ile	0.76	1	0.86	0.84
Leu	5.89	5	5.24	6.50
Tyr	2.62	3	2.59	2.85
Phe	3.58	4	3.60	3.83
Lys	2.29	2	2.96	2.97
His	0.00	0	0.00	0.00
Arg	2.57	3	2.87	2.98
Fuc	0.8	1	0.8	0.7
Man	5.6	6	5.6	4.8
Gal	1.6	2	1.7	1.4
Glc	0.8	1	0.5	0.4
GalN	4.4	4	4.0	3.4
GlcN	7.8	8	5.9	5.4
NeuAc	0.9	1	0.6	0.5

[a] High value due to alanine formed via dehydroalanine from cysteine.
[b] Cystine eliminates H_2S in alkaline medium.

glycopeptide(s) for further studies. Trypsin as the most specific protease was chosen to digest *S*-aminoethylated colostrum inhibitor. *S*-Aminoethylation of the three cysteine bridges was carried out after reduction by β-mercaptoethanol in liquid ammonia according to Kuromizu and Meienhofer[13]. Thus, all derivatized cystine peptide bonds were rendered sensitive to tryptic hydrolysis[35,36]. Though the rate of tryptic cleavage is slower than that of lysyl bonds[37], six additional sites of tryptic cleavage could be introduced into the inhibitor. The *S*-aminoethylated inhibitor obtained was eluted as a single peak after DEAE-Sephadex chromatography and

Table 2. Carbohydrate composition of native colostrum inhibitor A IV, *S*-aminoethylated inhibitor, glycopeptide T-1b and glycopeptide C-1c.

Values in mol monosaccharide per mol peptide.

	CTI A IV	Integer	*S*-Aminoethyl-CTI A IV	Integer	Peptide T-1b	Integer	Peptide C-1c	Integer
Fuc	0.8	1	0.5	1	0.2	1	0.8	1
Man	5.6	6	1.5	2	1.9	2	1.8	2
Gal	1.5	2	0.5	1	0.7	1	0.9	1
Glc	0.8	1	0.4	1	0.2	1	0.3	1
GalN	4.4	5	1.5	2	2.3	2	2.2	2
GlcN	7.4	8	1.6	2	2.1	2	1.7	2
NeuAc	0.8	1	0.3	1	0.7	1	0.3	1

Table 3. Amino acid analyses of *S*-aminoethylated inhibitor, glycopeptide T-1b, and glycopeptide C-1c. Values in mol amino acid per mol peptide.

	S-Aminoethyl-CTI	Integer	T-1b		C-1c	
Asp	7.88	8	1.73	2	1.00	1
Thr	5.47	6	0.94	1	0.31	–
Ser	3.39	3	2.23	2	0.84	1
Glu	10.93	10	0.17	–	–	–
Pro	7.77	7	–	–	–	–
Gly	4.10	4	0.15	–	–	–
Ala	4.79	4	1.12	1	–	–
Val	0.00	0	–	–	–	–
Met	0.98	1	–	–	–	–
Ile	0.92	1	–	–	–	–
Leu	4.17	5	–	–	–	–
Tyr	2.51	3	1.94	2	–	–
Phe	3.90	4	1.03	1	–	–
Lys	1.97	2	–	–	–	–
His	0.00	0	–	–	–	–
Arg	3.05	3	–	–	–	–
$[CH_2]_2NH_2$[a] – Cys	4.85	6	2.81	1	–	–

[a] *S*-Aminoethyl-Cys ($[CH_2]_2NH_2$ – Cys) shows up in the amino acid chromatogram just before lysine.

showed apparent homogeneity. The amino acid composition is given in Table 3.

Tryptic digestion of *S*-aminoethylated bovine colostrum trypsin inhibitor was carried out for 24 h[14]. The tryptic peptides were separated by gel filtration on BioGel P-6. Only one carbohydrate-containing peak, T-1, could be detected (Fig. 1). The fractions of this peak were combined and subjected to further chromatography on SE-Sephadex C-25. Again, only one peak, T-1b, gave a positive anthrone reaction for carbohydrates[15] (Fig. 2). The amino acid analysis of the glycopeptide T-1b is given in Table 3. The following amino acid sequence is in accordance with the amino acid composition of this peptide and with the primary structure of the bovine colostrum trypsin inhibitor[5,6]:

```
  25                 30
-Tyr-Phe-Tyr-Asn-Ser-Thr-Ser-Asn-Ala-Cys-.
             |
            CHO
```

In *S*-aminoethylated bovine colostrum trypsin inhibitor there is just this one and no other carbohydrate-protein attachment site. This results from the amino acid composition of the inhibitor prior to and after alkaline treatment and was proven by determination of the monosaccharide compositions of *S*-aminoethylated inhibitor, and of the glycopeptides T-1b and C-1c (Table 2). All contain almost the same amounts of fucose, mannose, galactose, glucose, galactosamine, glucosamine, and neuraminic acid. The glycopeptide T-1b, thus, contains the complete carbohydrate moiety of the *S*-aminoethylated inhibitor.

Peptide T-1b was subjected to carboxypeptidase C treatment in order to obtain an even shorter fragment containing the carbohydrate moiety. This enzyme was chosen because of its combined specificity of both carboxypeptidase A and B. Contamination of the enzyme with aminopeptidase activity was reported by the producer (Röhm). After 48 h incubation at 30 °C the digest was chromatographed on Sephadex G-25 in order to remove the enzyme and to separate the glycopeptide. The elution profile is shown in Fig. 3. Only one carbohydrate containing peak C-1 was detectable. Rechromatography of this material was performed on Merckogel 2000 as shown in Fig. 4.

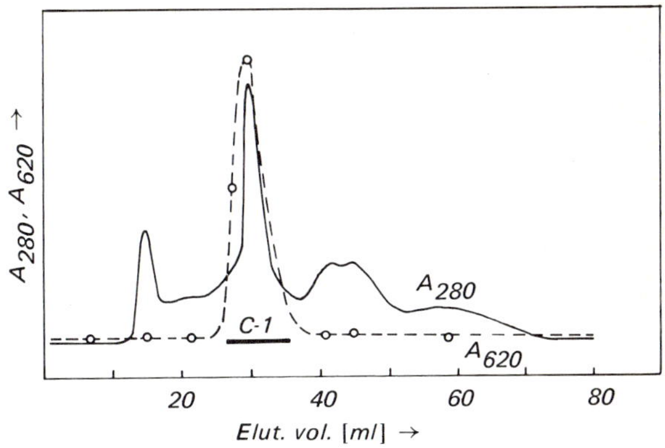

Fig. 3. Chromatography of carboxypeptidase C digest of peptide T-1b on Sephadex G-25.

The column (50 × 0.9 cm) was equilibrated and eluted with 0.1N acetic acid. Flow rate: 10 m*l*/h. Fract. vol.: 1.5 m*l*. Full line: absorbance at 280 nm, dashed line: absorbance at 620 nm, using the anthrone test. The bar indicates pooled fractions.

Two minor peaks could be separated from the single carbohydrate containing peptide C-1c. The amino acid composition of this peptide is given in Table 3. It is in accordance with the amino acid sequence

$$\overset{27}{\text{-Asn}}\text{-Ser-(Thr).}$$
$$\text{|}$$
$$\text{CHO}$$

Besides cleavage of the *C*-terminal residues from peptide T-1b the three *N*-terminal residues Tyr--Phe-Tyr have been removed by the contaminating aminopeptidase.

The carbohydrate composition of the glycodipeptide C-1c is almost identical to that of *S*-aminoethylated colostrum inhibitor and tryptic decapeptide T-1b (Table 2).

Though only one of the chromatographic forms, i.e. component A IV, of bovine colostrum trypsin inhibitor was investigated, it is highly likely that the result is representative for all other forms. A microheterogeneity of the several components with respect to the carbohydrate moiety has been demonstrated previously[7]. We conclude that the colostrum inhibitor has one big moiety of carbohydrates attached to asparagine 27. This position is fairly distant from the reactive site lysine 17 and is compatible with the three-dimensional structure where this region extends to the surface of the molecule. It is noteworthy that the side chain of the neighbouring tyrosine residue 26 rotates more slowly than in the structurally homologous inhibitors of bovine or snail tissue[38]. A detailed NMR-spectral investigation of this phenomenon which for the first time documents intramolecular

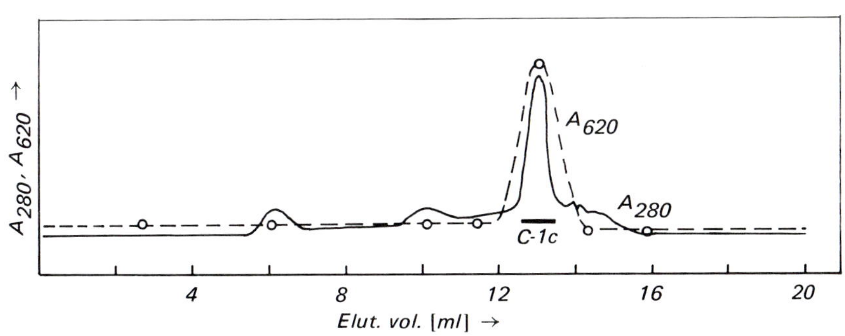

Fig. 4. Chromatography of carbohydrate containing fraction C-1 an Merkogel 2000.

The column (30 × 0.9 cm) was equilibrated and eluted with 0.1N acetic acid. Flow rate: 5 m*l*/h. Fract. vol.: 1.5 m*l*. Full line: absorbance at 280 nm, dashed line: absorbance at 620 nm using the anthrone test. The bar indicates pooled fractions.

interaction on a glycoprotein will be the subject of a separate paper*.

We are grateful to Miss *C. Frank* for her expert performing of the amino acid analyses. We are indebted to the *Stiftung Volkswagenwerk* for providing us with a Hewlett-Packard gas chromatograph. This work was supported by the *Deutsche Forschungsgemeinschaft.*

Literature

1 Laskowski, M., Jr. & Laskowski, M., Sr. (1951) *J. Biol. Chem.* **190**, 563 - 573.
2 Laskowski, M., Jr., Mars, P.H. & Laskowski, M., Sr. (1952) *J. Biol. Chem.* **198**, 745 - 752.
3 Pospišilová, D., Dlouha, V. & Šorm, F. (1968) in Federation Abstracts, 5th FEBS Meeting, Prague, Abstr. No. 971, Academic Press, London and New York.
4 Čechová-Pospišilová, D., Svestková, V. & Šorm, F. (1969) in Federation Abstracts, 6th FEBS Meeting, Madrid, Abstr. No. 225, Academic Press, London and New York.
5 Čechová, D., Jonáková, V. & Šorm, F. (1971) *Collect. Czech. Chem. Commun.* **36**, 3342 - 3357.
6 Čechová, D., Jonáková, V. & Šorm, F. (1971) in Proc. Int. Res. Conf. on Proteinase Inhibitors (Fritz, H. & Tschesche, H., eds.) pp. 105 - 107, W. de Gruyter, Berlin.
7 Tschesche, H., Klauser, R., Čechová, D. & Jonáková, V. (1974) *Hoppe-Seyler's Z. Physiol. Chem.* **356**, 1759 - 1764.
8 Anderson, B., Seno, N., Sampson, P., Riley, J.G., Hofman, P. & Meyer, K. (1964) *J. Biol. Chem.* **239**, PC 2716 - 2717.
9 Schiffman, G., Kabat, E.A. & Thompson, W. (1964) *Biochemistry* **3**, 113 - 120.
10 Spackman, D.H., Stein, W.H. & Moore, S. (1958) *Anal. Chem.* **30**, 1190 - 1206.
11 Tschesche, H., Frank, C. & Ebert, H. (1973) *J. Chromatogr.* **85**, 35 - 43.
12 Clamp, J.R., Bhatti, T. & Chambers, R.E. (1972) in Glycoproteins (Gottschalk, A., ed.) 2nd edn. Part A, pp. 300 - 321, Elsevier Publ. Co., Amsterdam, London, New York.
13 Kuromizu, K. & Meienhofer, J. (1972) *J. Biol. Chem.* **247**, 5646 - 5647.
14 Greene, L.J. & Giordano, J.S., Jr. (1969) *J. Biol. Chem.* **244**, 285 - 298.

* Wagner, G., Wüthrich, K. & Tschesche, H. (1978) Eur. J. Biochem., submitted.

15 Roe, J.H. (1955) *J. Biol. Chem.* **212**, 335 - 343.
16 Tschesche, H. & Kupfer, S. (1972) *Eur. J. Biochem.* **26**, 33 - 36.
17 Čechová, D., Jonáková-Svestková, V. & Šorm, F. (1970) *Collect. Czech. Chem. Commun.* **35**, 3085 - 3091.
18 Spiro, R.G. (1960) *Methods Enzymol.* **VIII**, 3 - 26.
19 Bhavanadan, V.P., Buddecke, E., Carubelli, R. & Gottschalk, A. (1964) *Biochem. Biophys. Res. Commun.* **16**, 353 - 357.
20 Gottschalk, A. & Graham, E.R.B. (1959) *Biochim. Biophys. Acta* **34**, 380 - 391.
21 Lloyd, K.O., Kabat, E.A., Layug, E.J. & Gruezo, F. (1966) *Biochemistry* **5**, 1489 - 1501.
22 Roberts, G.P. (1974) *Eur. J. Biochem.* **50**, 265 - 280.
23 Tschesche, H., Kupfer, S., Klauser, R., Fink, E. & Fritz, H. (1976) in Protides of the Biol. Fluids, 23rd Colloquium (Peeters, H., ed.) pp. 255 - 266, Pergamon Press Oxford and New York.
24 Nuenke, R.H. & Cunningham, L.W. (1961) *J. Biol. Chem.* **236**, 2452 - 2460.
25 Johansen, P.G., Marshall, R.D. & Neuberger, A. (1961) *Biochem. J.* **78**, 518 - 527.
26 Rosevear, J.W. & Smith, E.L. (1961) *J. Biol. Chem.* **236**, 425 - 435.
27 Arima, T. & Spiro, R.G. (1972) *J. Biol. Chem.* **247**, 1836 - 1848.
28 Spiro, R.G. (1962) *J. Biol. Chem.* **237**, 382 - 388.
29 Jackson, R.L. & Hirs, C.H.W. (1970) *J. Biol. Chem.* **245**, 624 - 636.
30 Eylar, E.H. (1962) *Biochem. Biophys. Res. Commun.* **8**, 195 - 199.
31 Satake, M., Okuyama, T., Ishihara, K. & Schmid, K. (1965) *Biochem. J.* **95**, 749 - 757.
32 Iwanaga, S., Blombäck, B., Gröndahl, N.J., Hessel, H. & Wallén, P. (1968) *Biochim. Biophys. Acta* **160**, 280 - 283.
33 Clamp, J.R. & Putnam, F. (1964) *J. Biol. Chem.* **239**, 3233 - 3240.
34 Neuberger, A., Gottschalk, A., Marshall, R.D. & Spiro, R.G. (1972) in Glycoproteins, 2nd edn. (Gottschalk, A., ed.) Part A, pp. 450 - 490, Elsevier Publ. Co., Amsterdam, London, New York.
35 Raftery, M.A. & Cole, R.D. (1963) *Biochem. Biophys. Res. Commun.* **10**, 467 - 472.
36 Lindley, H. (1956) *Nature (London)* **178**, 674 - 678.
37 Wang, S.S. & Carpenter, F.H. (1968) *J. Biol. Chem.* **243**, 3702 - 3710.
38 Wüthrich, K., Wagner, G. & Tschesche, H. (1976) in Protides of the Biological Fluids, 23rd Coll. (Peeters, H., ed.) pp. 201 - 204, Pergamon Press, Oxford and New York.

Dr. R. Klauser, Dept. of int. Medicine, University of Texas, Health Science Center at Dulles, Dulles, Tex. 75235, U.S.A.

Dr. Dana Čechová, Institute of Molecular Genetics, Czechoslovak Academy of Science, Femingovo náměsti 2, CSSR-16610 Praha 6.

Prof. Dr. H. Tschesche, Lehrstuhl für Biochemie, Fakultät für Chemie, Universität Bielefeld, Universitätsstraße, D-4800 Bielefeld 1 (new address to which requests for reprints should be addressed).

Hoppe-Seyler's Z. Physiol. Chem.
Bd. 359, S. 181 - 192, Februar 1978

Effect of $Na^{\oplus}$ on Bile Acid Uptake by Isolated Rat Hepatocytes

Evidence for a Heterogeneous System

M. Sawkat ANWER and Dietmar HEGNER

Institut für Pharmakologie, Toxikologie und Pharmazie
Fachbereich Tiermedizin der Universität München

(Received 8 August 1977)

Summary: The effect of $Na^{\oplus}$ on cholate and taurocholate uptake was studied in isolated rat hepatocytes. Uptake of both bile acids can be divided into three components: a) a non-saturable component independent of $Na^{\oplus}$, b) a $Na^{\oplus}$-independent saturable and c) a $Na^{\oplus}$-dependent saturable component. The two saturable components were inhibited by KCN, oligomycin and various other bile acids, which suggests carrier-mediated, energy-requiring processes. The relationship between $Na^{\oplus}$ concentration and the $Na^{\oplus}$-dependent uptake rate was sigmoidal. The maximal uptake rate, rather than the K_m, was affected by $Na^{\oplus}$. These results suggest that more than one $Na^{\oplus}$ (probably two) are cosubstrates in the transport of one bile acid molecule and $Na^{\oplus}$ exerts a stimulatory effect on the translocation of the bile acid-carrier complex across the membranes. The $Na^{\oplus}$-dependent uptake was inhibited by ouabain, suggesting that this uptake process is linked to the activity of membrane bound ($Na^{\oplus}$-$K^{\oplus}$)ATPase. Inhibition and competition studies suggest that cholate and taurocholate share a common $Na^{\oplus}$-independent transport system while there are at least two $Na^{\oplus}$-dependent transport systems for taurocholate, one of which is shared by cholate.

Der Einfluß von $Na^{\oplus}$ auf die Gallensäureaufnahme in isolierte Leberparenchymzellen der Ratte. Hinweise auf ein heterogenes Transportsystem

Zusammenfassung: Der Einfluß von $Na^{\oplus}$ auf die Cholsäure- und Taurocholsäureaufnahme in isolierte Hepatozyten der Ratte wurde untersucht. Die Aufnahme der beiden Gallensäuren kann in drei Komponenten unterteilt werden: a) einen nicht sättigbaren, $Na^{\oplus}$-unabhängigen Anteil, b) einen $Na^{\oplus}$-unabhängigen, sättigbaren Anteil und c) einen $Na^{\oplus}$-abhängigen, sättigbaren Anteil. Die beiden sättigbaren Komponenten waren durch KCN, Oligomycin und verschiedene andere Gallensäuren hemmbar. Diese Anteile werden als carrier- und energieabhängige Prozesse beschrieben. Die Beziehung zwischen der $Na^{\oplus}$-Konzentration und der $Na^{\oplus}$-abhängiger Aufnahmegeschwindigkeit zeigt einen sigmoiden Verlauf. Die maximale Aufnahmegeschwindigkeit war durch $Na^{\oplus}$ beeinflußbar, nicht jedoch der K_m Wert. Die Ergebnisse weisen darauf hin, daß mehr als ein $Na^{\oplus}$-Ion (wahrscheinlich zwei) für den Transport eines Gallensäuremoleküles erforderlich sind. $Na^{\oplus}$ hat einen stimulierenden Effekt auf den Transport des Gallensäure-Carrier-Komplexes durch die Membran. Die $Na^{\oplus}$-abhängige Aufnahme war durch Ouabain hemmbar. Das weist darauf hin, daß dieser Aufnahmeprozeß von der Aktivität der membrangebundenen ($Na^{\oplus}$-$K^{\oplus}$)ATPase abhängig ist. Untersuchungen über Hemmung und Kompetition weisen darauf hin, daß Cholsäure und Taurocholsäure ein gemeinsames $Na^{\oplus}$-unabhängiges Transportsystem besitzen, während zwei $Na^{\oplus}$-abhängige Transportsysteme für Taurocholsäure postuliert werden. Eines davon transportiert auch Cholsäure.

Key words: Bile acid uptake, rat hepatocytes, $Na^{\oplus}$ dependency, heterogeneous system.

The total bile acid pool of the body is confined almost exclusively in the enterohepatic cicula-tion because of very efficient uptake of bile acids by the liver. Moreover other organic anions and drugs are also transported by the liver. Hepatic diseases and drugs could affect hepatic uptake of bile acids and thereby could affect the enterohepatic circulation of bile acids. Thus an understanding of the mechanism of hepatic bile acid uptake process and the influence of drugs and organic anions on this process are of clinical as well as theoretical interest.

Hepatic uptake of bile acids has been shown to be carrier-mediated in isolated rat hepatocytes[1,2] and in isolated perfused rat liver[3,4]. The uptake of bile acids has been reported to be $Na^{\oplus}$-dependent[2-4]. These studies suggested that the uptake is $Na^{\oplus}$-specific, since replacement of $Na^{\oplus}$ by $K^{\oplus}$, $Li^{\oplus}$, choline, Tris and sucrose failed to restore the effect of $Na^{\oplus}$. No quantitative relationship of bile-acid uptake with $Na^{\oplus}$ was, however, investigated in these studies. Replacement of $Na^{\oplus}$ by $K^{\oplus}$ or sucrose decreased taurocholate uptake up to 75% in isolated rat hepatocytes[2]. Cholate uptake seemed to be less $Na^{\oplus}$-dependent than taurocholate uptake in isolated perfused rat liver[4]. These results could indicate that hepatic uptake of bile acids may involve both $Na^{\oplus}$-dependent and $Na^{\oplus}$-independent processes as shown for amino acids[5]. From our previous studies on the inhibitory effect of different bile acids[6] and bile-acid analogues[7] on cholate and taurocholate uptake, we have postulated that more than one carrier is involved in the hepatic uptake of bile acids. $Na^{\oplus}$ dependency was not investigated in those studies. In the present study we have investigated the effect of $Na^{\oplus}$ on cholate and taurocholate uptake by isolated rat hepatocytes, in order to further characterize the $Na^{\oplus}$ dependency of hepatic bile-acid uptake and to test the multiple carrier hypothesis. We have used isolated rat hepatocytes since this system allowed independent determination of uptake[1,2]. The two bile acids, cholate and taurocholate, represent the major unconjugated and conjugated bile acids in rat.

Materials and Methods

Chemicals

All inorganic chemicals were obtained from Merck (Darmstadt, Germany). Tauro[*carbonyl*-^{14}C]cholic acid (54 mCi/mmol), [*carboxyl*-^{14}C]cholic acid (59.5 mCi/mmol) and [^{3}H]insulin were purchased from Amersham Buchler, Braunschweig, Germany. All unlabeled bile acids (Calbiochem, San Diego, California) were the best grade available. Ouabain and oligomycin were products of Sigma Chemical Co., Munich, Germany. Fusidate and taurofusidate were generous gifts of Drs. *W. von Dahne* and *W. O. Godtfredsen* of Leo Pharmaceutical Products, Denmark.

Preparation of hepatocytes

Hepatocytes were isolated from male Wistar rats (250 - 300 g) fed on standard laboratory diet as previously described[1]. The cells showing the following viability criteria were used for uptake studies: less than 10% stainable by trypan blue, intracellular $K^{\oplus}$ concentration above 85mM and less than 15% loss of lactate dehydrogenase. The cells were washed twice with the desired buffer before resuspending in the same buffer. Cells suspended in $Na^{\oplus}$-free buffer showed no significant alteration of viability criteria. The composition of the basic buffer (pH 7.4) was as follows (concentrations in mmol/*l*): NaCl (135); KCl (2); KH_2PO_4 (3); $CaCl_2$ (0.3); $MgSO_4$ (1) and Tris (10). Unless otherwise mentioned, NaCl was replaced by choline chloride to obtain desired concentrations of $Na^{\oplus}$.

Uptake studies

Bile acid uptake by isolated rat hepatocytes was studied at 37 °C in duplicate using a centrifugal filtration technique[6]. The initial uptake rate was calculated from the time dependent uptake of bile acids[6] and expressed as nmol × min^{-1} × mg protein^{-1}. The protein content of the cells was determined using a modified biuret method[8]. To obtain the $Na^{\oplus}$-dependent uptake rate, bile acid uptake rates were determined in $Na^{\oplus}$-free buffer (NaCl replaced by choline chloride) and in buffers containing desired amounts of $Na^{\oplus}$. The difference in the uptake rate represented $Na^{\oplus}$-dependent uptake rate. The $Na^{\oplus}$-independent uptake rate was obtained from the uptake rate in $Na^{\oplus}$-free buffer.

Results

Effect of $Na^{\oplus}$ replacement by other cations

Uptake rates of both cholate and taurocholate were decreased significantly when $Na^{\oplus}$ was replaced by $K^{\oplus}$, $Li^{\oplus}$ or choline (Table 1). Considering similar results from other studies where $Na^{\oplus}$ was replaced by Tris[3] or sucrose[2], this result would suggest that a part of the bile acid uptake is specifically dependent on $Na^{\oplus}$. The result shows that the $Na^{\oplus}$-independent uptake rate of cholate (50 - 60%) is about three-fold higher than that of taurocholate (15 - 25%). Since cations used to replace $Na^{\oplus}$ did not affect the $Na^{\oplus}$-independent uptake rate significantly (Table 1) and since $K^{\oplus}$ and $Li^{\oplus}$ may have other cellular effects[9], $Na^{\oplus}$ was replaced by choline in all other experiments.

Bile acid uptake as a function of the external concentration

The effect of $Na^{\oplus}$ on concentration-dependent uptake of cholate and taurocholate was next studied (Fig. 1). Above 200μM the initial uptake rates for cholate and taurocholate increased linearly both in the presence and in absence of $Na^{\oplus}$ The slope of this "non-saturable" component showed no $Na^{\oplus}$ dependency in three separate experiments. Thus it was considered valid to calculate the $Na^{\oplus}$-dependent uptake rate by subtracting the $Na^{\oplus}$-independent uptake rate from the uptake rate in the presence of $Na^{\oplus}$. For both bile acids hyperbolic curves were obtained for the $Na^{\oplus}$-dependent uptake (Fig. 1). Subtraction of the "non-saturable" component from the $Na^{\oplus}$-independent uptake rate also resulted in hyperbolic curves for both cholate and taurocholate (data not shown). Thus both cholate and taurocholate are transported into hepatocytes by a) a non-saturable uptake process, b) a $Na^{\oplus}$-independent saturable uptake process and c) a $Na^{\oplus}$-dependent saturable uptake process.

These two saturable processes could be fitted by a single straight line using the Eadie-Hofstee plot from which the kinetic parameters, apparent Michaelis constant and apparent maximum uptake rate, were calculated for three separate experiments (Table 2). Both of these constants for the $Na^{\oplus}$-dependent uptake of cholate differed significantly from those of taurocholate.

Effect of external $Na^{\oplus}$ concentrations on bile acid uptake

The relationship between external $Na^{\oplus}$ concentration ($[Na^{\oplus}]_0$) and bile acid uptake rate was sigmoidal for all three bile acid concentrations studied (Fig. 2). The result would indicate that more than one $Na^{\oplus}$ may interact with the bile acid transport system[10-12]. Straight lines were obtained when the data were plotted as 1/($Na^{\oplus}$-dependent uptake rate) versus $1/([Na^{\oplus}]_0)^2$ (Fig. 3).

Table 1. Effect of $Na^{\oplus}$ replacement on bile acid uptake.

Hepatocytes were washed twice before resuspending in the buffer containing equimolar amounts of choline chloride, KCl or LiCl instead of NaCl (135 mmol/*l*). After 15 min of preincubation at 37 °C, the uptake of cholate and taurocholate was studied by incubating hepatocytes (0.8 – 1.0 mg cell protein) with radiolabeled bile acids for 15, 30 and 45 s. The uptake rate was calculated from the time dependent uptake as described before[1]. The values represent mean ± SEM for four determinations.

NaCl replaced by	Uptake rates [nmol × min^{-1} × mg protein^{-1}]			
	Cholate		Taurocholate	
	10μM	50μM	10μM	50μM
None	0.25 ± 0.021	0.71 ± 0.061	0.63 ± 0.041	1.48 ± 0.082
Choline chloride	0.13 ± 0.013	0.36 ± 0.041	0.10 ± 0.016	0.37 ± 0.052
LiCl	0.14 ± 0.012	0.34 ± 0.025	0.12 ± 0.011	0.32 ± 0.018
KCl	0.14 ± 0.013	0.38 ± 0.035	0.09 ± 0.007	0.33 ± 0.021

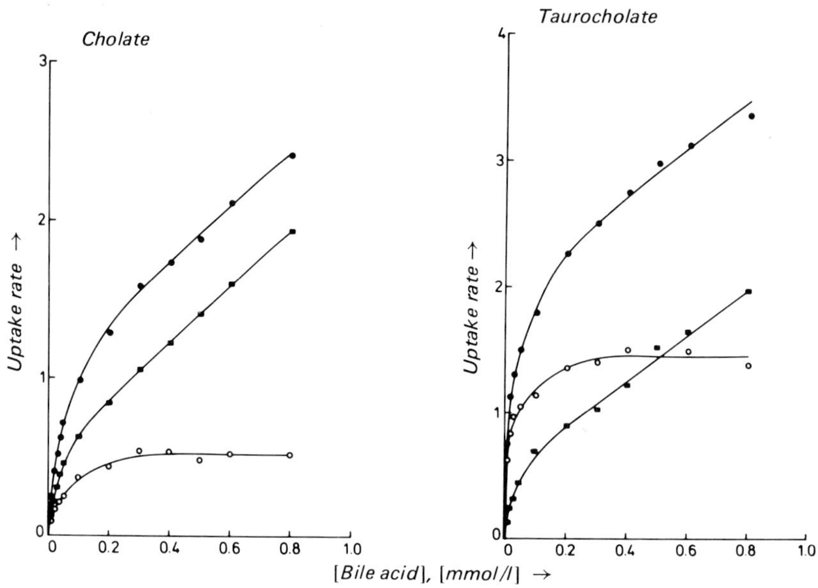

Fig. 1. Effect of $Na^{\oplus}$ on concentration dependent uptake of bile acids.

Uptake rates of cholate and taurocholate were determined in presence of 135mM $Na^{\oplus}$ (●——●) and in the absence of $Na^{\oplus}$ (■——■). The $Na^{\oplus}$-dependent uptake rates (○——○) were obtained by subtracting the uptake rates in the absence of $Na^{\oplus}$ from the corresponding uptake rates in the presence of $Na^{\oplus}$. The uptake rates are expressed in nmol × min^{-1} × mg $protein^{-1}$.

Table 2. Kinetic parameters of bile acid uptake.

The kinetic parameters, apparent Michaelis constant K_m(app) and apparent maximum uptake rate V(app), were calculated from Eadie-Hofstee plots using the data in Fig. 1. The $Na^{\oplus}$-independent uptake parameters were obtained by using the curve obtained after subtracting the non-saturable component. The $Na^{\oplus}$-dependent uptake curve was obtained by subtracting the $Na^{\oplus}$-independent uptake rate from the uptake rate in presence of 135mM $Na^{\oplus}$. The values represent mean ± SEM of three separate experiments.

Bile acid	Uptake components	Kinetic parameters	
		K_m(app) [μM]	V(app) [nmol × min^{-1} × mg $prot.^{-1}$]
Cholate	$Na^{\oplus}$-independent	39 ± 1.5	0.60 ± 0.074
	$Na^{\oplus}$-dependent	58 ± 5.4	0.57 ± 0.039
Taurocholate	$Na^{\oplus}$-independent	57 ± 3.8	0.74 ± 0.138
	$Na^{\oplus}$-dependent	15 ± 0.9	1.34 ± 0.044

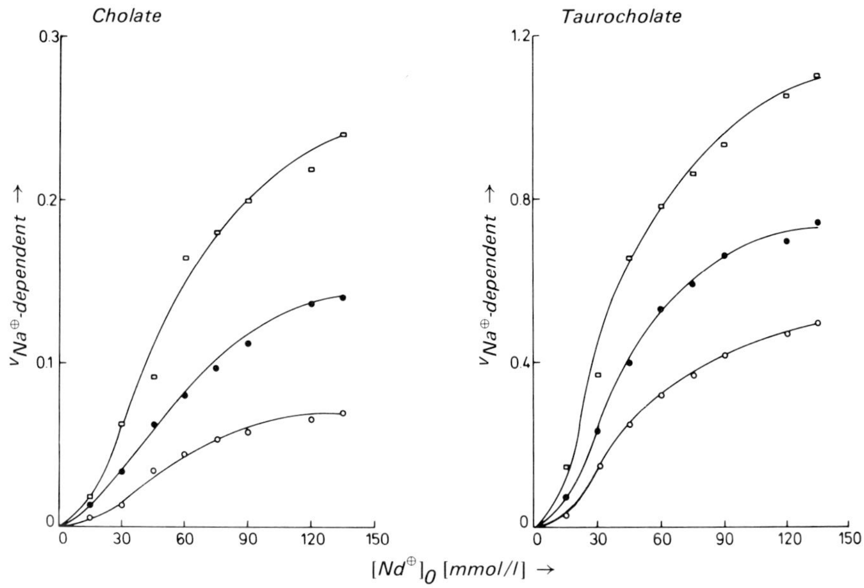

Fig. 2. Effect of various $Na^{\oplus}$ concentrations on bile acid uptake.

Cholate and taurocholate uptake rates were determined in presence of various external concentrations of $Na^{\oplus}$ ($[Na^{\oplus}]_0$). The $Na^{\oplus}$-dependent uptake rate $v_{Na^{\oplus}\text{-dependent}}$) at any value of $[Na^{\oplus}]_0$ was calculated by subtracting the uptake rate in the absence of $Na^{\oplus}$ from the uptake rate at a particular $[Na^{\oplus}]_0$. Bile acid concentrations are 10 (○——○), 20 (●——●) and 50 (□——□) μM. The uptake rates are expressed in nmol × min^{-1} × mg $protein^{-1}$.

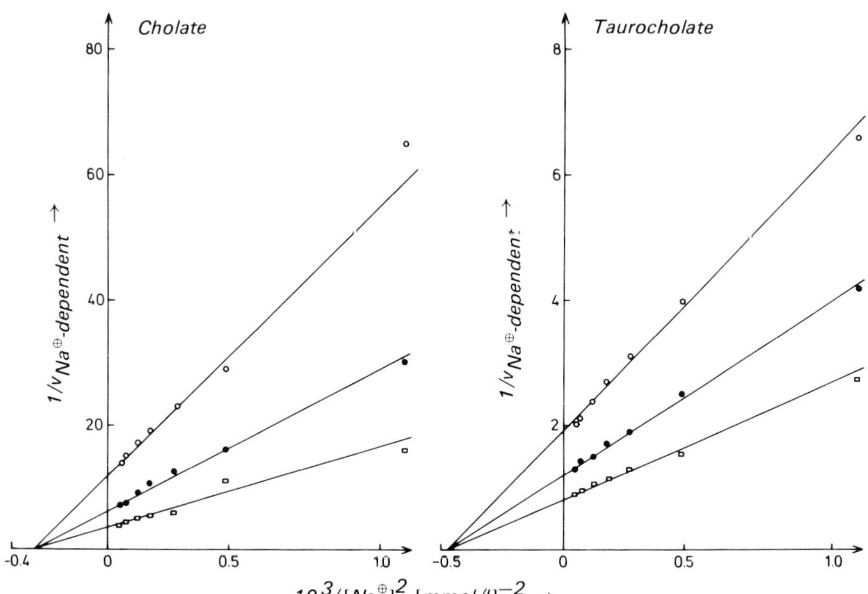

Fig. 3. Double reciprocal plot of $Na^{\oplus}$-dependent uptake rate and square of the external $Na^{\oplus}$ concentration.

The data and other conditions are the same as in Fig. 2. The affinity of $Na^{\oplus}$ for the bile acid uptake system ($K_{Na^{\oplus}}$) was calculated from the intercept on the abscissa.

A Hill plot of these data ($\log[v/(V-v)]$ versus $\log[Na^{\oplus}]_0$; cf. see below) had a least square slope (Hill coefficient) of 1.99 ± 0.03 for cholate and 2.07 ± 0.36 for taurocholate (plots not shown). This would indicate that two $Na^{\oplus}$ are probably involved in the transport of one bile acid molecule. If this is the case, the bile acid uptake rate (v) in the presence of different external $Na^{\oplus}$ concentrations can be described by the following equation[13].

$$v = \frac{V}{1 + \left(\frac{K_{Na^{\oplus}}}{[Na^{\oplus}]_0}\right)^2}$$

where V is the maximal $Na^{\oplus}$-stimulated bile acid uptake rate and $K_{Na^{\oplus}}$ is the apparent Michaelis constant for the interaction of $Na^{\oplus}$ with the transport system. Since straight lines intersected on the abscissa (Fig. 3), $K_{Na^{\oplus}}$ was not affected by bile acids. The $K_{Na^{\oplus}}$ value for cholate uptake (54 ± 4.6mM, n = 3) was slightly higher than that of taurocholate uptake (40 ± 4.2mM, n = 3). The effect of various values of $[Na^{\oplus}]_0$ on the uptake rate of bile acids as a function of external bile acid concentrations was studied to find out the effect of $[Na^{\oplus}]_0$ on bile acid uptake parameters. A Lineweaver-Burk plot of the data (Fig. 4) showed that $Na^{\oplus}$ did not affect the K_m(app) value for the uptake of taurocholate. A similar result was obtained with cholate (data not shown). This would suggest that neither $Na^{\oplus}$ nor bile acid affected the affinity of the other for the transport system. Thus the mechanism of binding of $Na^{\oplus}$ and bile acid to the carrier would be of random type. This finding agrees well with our previous finding in isolated rat liver plasma membranes[14], i.e. that cholate binding to a high affinity binding site, postulated to be a bile acid carrier, was not affected by $Na^{\oplus}$. On the other hand, the V(app) values of both bile acids were increased by $Na^{\oplus}$. This would suggest that $Na^{\oplus}$ exerts a stimulatory effect on the translocation of bile acid-carrier complex across the membranes.

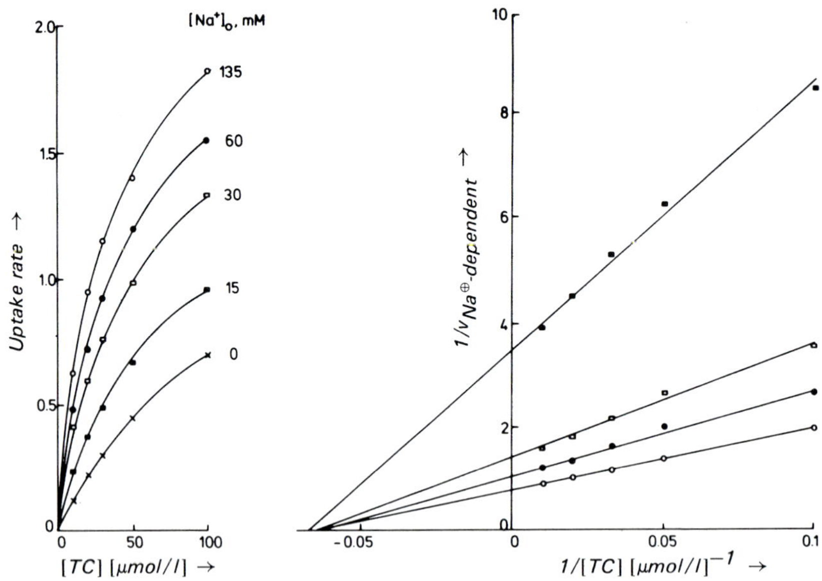

Fig. 4. Effect of $Na^{\oplus}$ on taurocholate uptake.
Left: uptake rate of taurocholate as a function of its concentrations ([TC]) at various concentrations of external $Na^{\oplus}$ ($[Na^{\oplus}]_0$). Right: double reciporcal plot of $Na^{\oplus}$-dependent uptake rate ($v_{Na^{\oplus}\text{-dependent}}$) against taurocholate concentrations at 15 (■ – – – ■), 30 (□——□), 60 (● – – – ●) and 135 (○——○) mM $[Na^{\oplus}]_0$.

Table 3. Effect of metabolic inhibitors on bile acid uptake.

Hepatocytes were incubated with different inhibitors for 15 min at 37 °C before uptake studies were done. $Na^{\oplus}$-independent uptake rate was measured in $Na^{\oplus}$-free buffer ($Na^{\oplus}$ was replaced by choline). $Na^{\oplus}$-dependent uptake rate was obtained from the uptake rate in presence of 135mM $Na^{\oplus}$ minus the $Na^{\oplus}$-independent uptake rate. The values are expressed as percent of control uptake rate and each value represents mean ± SEM for four determinations.

Inhibitors (conc.)		Uptake rates [% of control]			
		Cholate (50μM)		Taurocholate (50μM)	
		$Na^{\oplus}$-independent	$Na^{\oplus}$-dependent	$Na^{\oplus}$-independent	$Na^{\oplus}$-dependent
Ouabain	(0.1mM)	101 ± 5	63 ± 4	101 ± 6	65 ± 5
Ouabain	(1.0mM)	96 ± 4	59 ± 5	100 ± 5	56 ± 4
KCN	(2mM)	60 ± 4	55 ± 4	65 ± 6	57 ± 4
Oligomycin	(20 μg/ml)	57 ± 5	30 ± 2	44 ± 3	29 ± 3

Effect of metabolic inhibitors on bile acid uptake

The $Na^{\oplus}$-dependent uptake but not the $Na^{\oplus}$-independent uptake of both cholate and taurocholate was inhibited by ouabain (Table 3). This would suggest that the $Na^{\oplus}$-dependent uptake of both bile acids is linked to the activity of membrane bound ($Na^{\oplus}$-$K^{\oplus}$)ATPase. KCN and oligomycin inhibited both $Na^{\oplus}$-dependent and $Na^{\oplus}$-independent uptake of both bile acids (Table 3). Similar results were obtained when the bile acid concentration was 10μM. This result would indicate that both $Na^{\oplus}$-independent and $Na^{\oplus}$-dependent uptake of cholate and taurocholate are dependent on metabolic energy supplies.

Inhibition studies with bile acid and bile acid analogues

Different bile acids and bile acid analogues (fusidate and taurofusidate) inhibited both $Na^{\oplus}$-independent and $Na^{\oplus}$-dependent uptake of cholate and taurocholate (Table 4). Considering that both of these uptake processes are saturable (Fig. 1) this result would suggest that there are specific transport systems for $Na^{\oplus}$-independent and $Na^{\oplus}$-dependent uptake of bile acids. It is possible that both cholate and taurocholate are transported by a common $Na^{\oplus}$-independent and by a common $Na^{\oplus}$-dependent uptake system. If this is the case, the inhibitory activities of different bile acids for

Table 4. Effect of bile acids and their analogues on cholate and taurocholate uptake.

After 15 min of preincubation at 37 °C uptake of cholate or taurocholate (20μM) by hepatocytes was studied in the presence of 100μM of various bile acids or their analogues (fusidate and taurofusidate). The $Na^{\oplus}$-independent and the $Na^{\oplus}$-dependent uptake rates were determined as described in Table 3. The values represent the average of two determinations from four different cell preparations.

Bile acids or analogues	Uptake rates [% inhibition]			
	Cholate		Taurocholate	
	$Na^{\oplus}$-independent	$Na^{\oplus}$-dependent	$Na^{\oplus}$-independent	$Na^{\oplus}$-dependent
Cholate	51	28	57	40
Taurocholate	36	64	36	76
Glycocholate	38	84	50	48
Chenodeoxycholate	45	100	34	77
Taurochenodeoxycholate	58	60	60	98
Glycochenodeoxycholate	64	55	67	86
Deoxycholate	54	67	50	62
Taurodeoxycholate	33	100	30	92
Fusidate	42	34	38	42
Taurofusidate	76	100	60	60

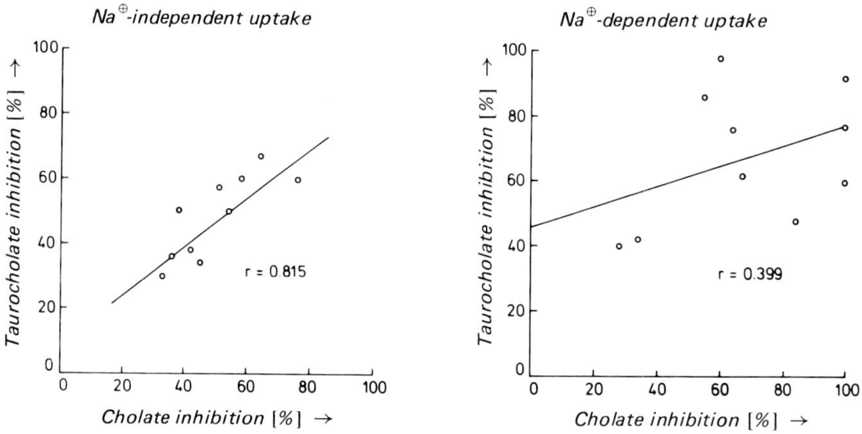

Fig. 5. Correlation between the inhibitory activities of different bile acids and their analogues on the $Na^{\oplus}$-independent and the $Na^{\oplus}$-dependent uptake of cholate and taurocholate.

The data are from the same experiment as in Table 4.

the inhibition of cholate and taurocholate uptake would be correlated for both uptake processes[15]. The data from Table 4 were then plotted as shown in Fig. 5 to test this possibility. There is a good correlation for the $Na^{\oplus}$-independent uptake (Fig. 5, left side) but the correlation is poor for $Na^{\oplus}$-dependent uptake (Fig. 5, right side). This may indicate that both cholate and taurocholate are transported by a common $Na^{\oplus}$-independent uptake system. On the other hand the $Na^{\oplus}$-dependent uptake system seems to be heterogeneous (more than one transport system).

Competition studies with cholate and taurocholate

To further test the heterogeneity of the $Na^{\oplus}$-dependent uptake system and the homogeneity of the $Na^{\oplus}$-independent uptake system we have investigated in detail the inhibitory effect of cholate and taurocholate on taurocholate and cholate uptake. The effects of various concentrations of cholate and taurocholate on the $Na^{\oplus}$-dependent and the $Na^{\oplus}$-independent uptake of cholate and taurocholate are shown in Fig. 6 and 7. The data were analyzed (plot of $1/v_0 - v_t$) versus 1/[inhibitor] as shown on the right hand side of Fig. 6 and 7) according to the method of Inui and Christensen[16]. From this plot the reciprocal of the rate of substrate uptake subject to inhibition at infinite concentration of the inhibitor ($1/V_i$) can be calculated from the intercept with the y axis. Since no inhibitor for a particular bile acid uptake system is known, we used cholate and taurocholate as their own inhibitors. Cholate uptake at a particular concentration would be completely inhibited by the infinite concentration of cholate. Now if another bile acid could inhibit cholate uptake to the same extent, it would indicate that this particular bile acid shares the cholate uptake system(s). However, this would not indicate that this particular bile acid is not also transported by other carrier(s). If the latter is the case, cholate would be unable to inhibit the uptake of this particular bile acid completely. Thus to evaluate the uptake systems for cholate and taurocholate we have used competition studies with both bile acids. Calculation of V_i (Fig. 6) revealed that both cholate and taurocholate inhibited the $Na^{\oplus}$-independent uptake of cholate to the same extent. The residual uptake was comparable to non-saturable uptake. A similar result was obtained for the inhibition of $Na^{\oplus}$-independent uptake of taurocholate by cholate and taurocholate (Fig. 7). However, the effect on $Na^{\oplus}$-dependent uptake is different. Both cholate and taurocholate inhibited $Na^{\oplus}$-dependent uptake of cholate completely (Fig. 6). On the other hand, $Na^{\oplus}$-dependent uptake of taurocholate was completely inhibited by taurocholate but only 68% by cholate (Fig. 7). The same sets of experiments

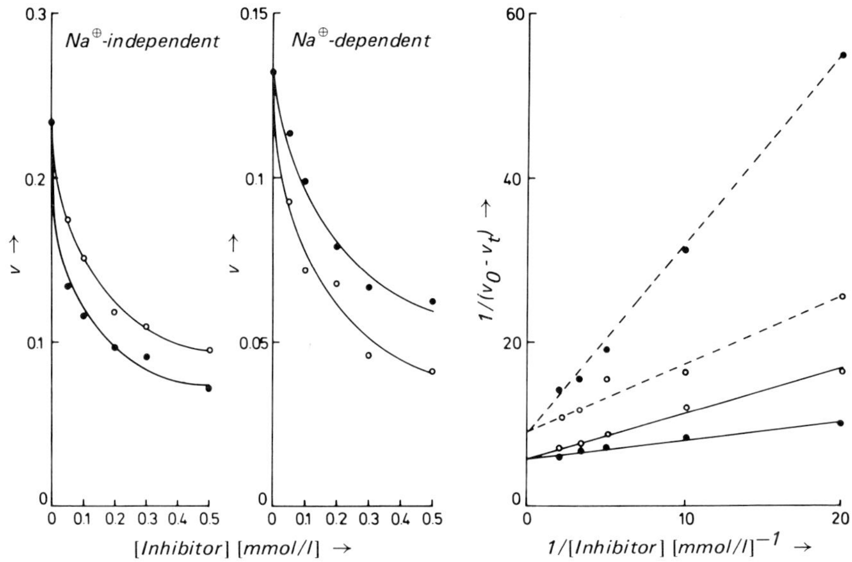

Fig. 6. Effect of cholate and taurocholate on $Na^{\oplus}$-independent (left) and $Na^{\oplus}$-dependent uptake rate (middle) of cholate.

The rate of cholate (20μM) uptake was determined in the absence (v_0) and in the presence of different concentrations of cholate and taurocholate (v_t). v = uptake rate in nmol × min^{-1} × mg $protein^{-1}$; [Inhibitor] = concentrations of cholate (•) and taurocholate (○). In the $1/(v_0 - v_t)$ versus 1/[inhibitor] plot (right) the solid line represents the $Na^{\oplus}$-independent uptake and the broken line represents the $Na^{\oplus}$-dependent uptake.

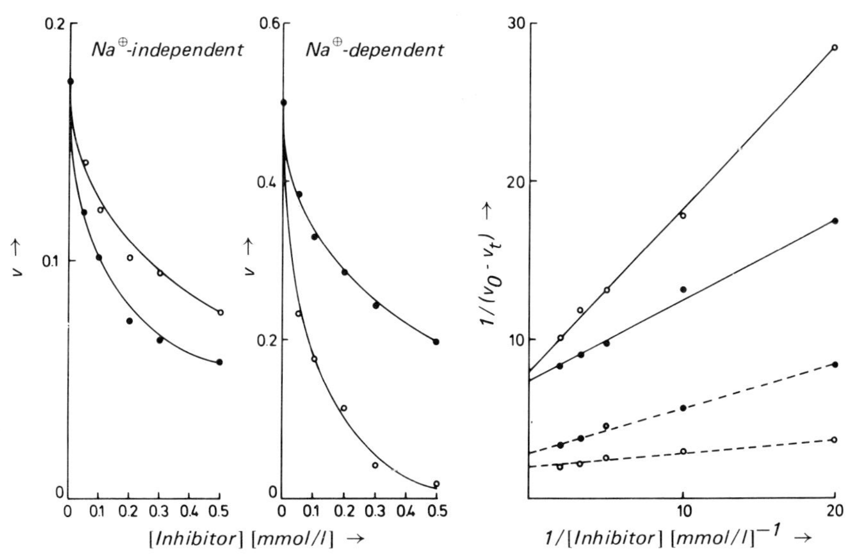

Fig. 7. Effect of cholate and taurocholate on the $Na^{\oplus}$-independent and the $Na^{\oplus}$-dependent uptake rate of taurocholate (20μM). Other conditions are as described in Fig. 6.

Table 5. $Na^{\oplus}$-dependent and $Na^{\oplus}$-independent uptake of bile acids subject to cholate and taurocholate inhibition.

The results were obtained from similar experiments described in Fig. 6 and 7. The uptake rate subject to bile acid inhibition was calculated from the reciprocal of the Y axis intercept of the $1/(v_0 - v_t)$ versus 1/[inhibitor] plot and is expressed as a percentage. The values thus obtained represent the percent inhibition of the uptake rate by an infinite concentration of the inhibitors (cholate or taurocholate). The $Na^{\oplus}$-independent uptake rate is corrected for non-saturable uptake. The values are the average of two determinations at each concentration.

Bile-acid conc.	Percent of cholate uptake subject to taurocholate inhibition		Percent of taurocholate uptake subject to cholate inhibition	
[μM]	$Na^{\oplus}$-independent	$Na^{\oplus}$-dependent	$Na^{\oplus}$-independent	$Na^{\oplus}$-dependent
10	103	96	97	75
20	100	99	108	68
50	97	98	90	58

were then done with two other concentrations (10 and 50μM) of bile acids. Again cholate failed to completely inhibit the $Na^{\oplus}$-dependent uptake of taurocholate (Table 5). A Dixon plot of the data (10 and 50μM bile acids) showed that both $Na^{\oplus}$-dependent and $Na^{\oplus}$-independent uptake of cholate (data not shown) and taurocholate (Fig. 8) were competitively inhibited by taurocholate and cholate, respectively. If these two transport systems are not heterogeneous, the K_i values (competitive inhibition constant) of cholate and taurocholate would be comparable to their respective K_m values. The K_i values of both cholate (40μM) and taurocholate (65μM) for the inhibition of

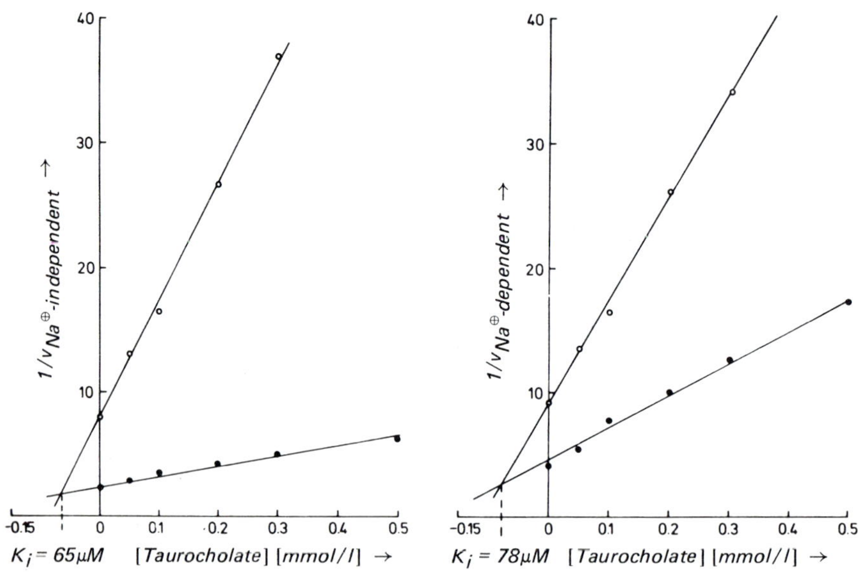

Fig. 8. Dixon plot for the inhibition of $Na^{\oplus}$-independent (left) and $Na^{\oplus}$-dependent (right) uptake of cholate by taurocholate.

$v_{Na^{\oplus}\text{-independent}}$ and $v_{Na^{\oplus}\text{-dependent}}$ represent $Na^{\oplus}$-independent and $Na^{\oplus}$-dependent uptake rates, respectively, and are expressed in nmol × min^{-1} × mg $protein^{-1}$. Cholate concentrations are 10 (○) and 50 (●) μM.

$Na^{\oplus}$-independent uptake were comparable to the respective K_m values for this system (Table 2). However, the K_i values of both cholate (170μM) and taurocholate (78μM) for the inhibition of $Na^{\oplus}$-dependent uptake were considerably higher than their respective K_m values for this system. These results would suggest that both cholate and taurocholate are transported by a common $Na^{\oplus}$-independent uptake system while the $Na^{\oplus}$-dependent uptake system is heterogeneous.

Discussion

Bile acid uptake processes

The total uptake rate of cholate and taurocholate can be divided into three components: a) non-saturable uptake, b) $Na^{\oplus}$-independent saturable uptake and c) $Na^{\oplus}$-dependent saturable uptake. The non-saturable uptake is not $Na^{\oplus}$ dependent (Fig. 1). We have previously shown that this component of cholate uptake is practically insensitive to temperature and to a depletion of cellular energy supplies[1]. Thus this process may represent a passive diffusion phenomenon. However, bile acid uptake by isolated perfused rat liver did not reveal this component[3,4]. Hence the question arises as to whether such a process would be operative in vivo or whether this would result from the cell isolation procedure. In any case, the contribution of this component is so small (about 2% of the total uptake rate at 50μM) that under physiological conditions this process would not play a significant role.

The $Na^{\oplus}$-independent saturable uptake process is sensitive to depletion of cellular energy supplies but insensitive to ouabain (Table 3). This component is inhibited by other bile acids and bile acid analogues (Table 4) suggesting a carrier-mediated system. This process has also been reported in isolated perfused rat liver for both cholate and taurocholate where it has been shown to be concentrative[4]. These results would suggest that an active transport system independent of $Na^{\oplus}$ is responsible for a part of bile acid uptake into hepatocytes. A similar transport system has been postulated for cycloleucine uptake by isolated rat hepatocytes[5]. At low concentrations about 50 - 60% of cholate and 15 - 25% of taurocholate uptake takes place by this system.

The $Na^{\oplus}$-dependent uptake process is saturable (Fig. 1), energy dependent (Table 3) and inhibited by other bile acids and bile acid analogues (Table 4). This uptake process has been reported to be concentrative for both cholate and taurocholate in isolated perfused rat liver[5]. Thus a part of cholate and taurocholate uptake can be considered to be active and $Na^{\oplus}$-dependent. The $Na^{\oplus}$-dependent uptake represents about 40 - 50% of cholate and 75 - 85% of taurocholate uptake. The $Na^{\oplus}$-dependent uptake is linked to the activity of membrane bound ($Na^{\oplus}$-$K^{\oplus}$)ATPase (Table 3). Analysis of $Na^{\oplus}$ dependency of bile acid uptake (see Results) suggested that two $Na^{\oplus}$ are involved in the transport of one bile acid molecule and $Na^{\oplus}$ stimulates the translocation of bile acid-carrier complex across the membrane. Considering the "gradient hypothesis" and its applicability in various transport systems[17,18] it can be postulated that a quarternary complex is formed among two $Na^{\oplus}$, bile acid and the carrier. This complex is then translocated across the membrane taking advantage of the $Na^{\oplus}$-electrochemical gradient. A similar mechanism has been proposed for glycine uptake into pigeon erythrocytes[12]. Considering negative potential and a higher bile-acid concentration inside the cell, the anionic bile acids have to be transported against an electrochemical potential gradient. Thus coupling of this transport with two $Na^{\oplus}$ would help to circumvent part of this gradient barrier. This may also explain why taurocholate, a stronger anion than cholate, would be mostly transported by this system. If two $Na^{\oplus}$ are cotransported with each molecule of bile acid (containing one negative charge), the overall transport would be electrogenic.

Heterogeneity of transport systems

The results discussed so far would indicate that cholate and taurocholate are transported by two active transport systems; of which one is $Na^{\oplus}$-independent and the other is $Na^{\oplus}$-dependent. Since kinetic data do not generally permit detection of heterogeneity of transport systems[16], inhibition and competition studies were performed to find out if both cholate and taurocholate are transported by a common $Na^{\oplus}$-independent and a common $Na^{\oplus}$-dependent transport system. The results from these studies (Fig. 5 - 8 and Table 5) suggested that both cholate and taurocholate are transported by a common Na$^{\oplus}$-independent transport system. Since Na$^{\oplus}$-dependent uptake of cholate and taurocholate is competitively inhibited by taurocholate and cholate, respectively, both cholate and taurocholate should share a part of the Na$^{\oplus}$-dependent uptake system. The total Na$^{\oplus}$-dependent uptake of cholate should be shared by taurocholate since cholate uptake by this system is subject to complete inhibition by an infinite concentration of taurocholate (Table 5). Since Na$^{\oplus}$-dependent uptake of taurocholate is only partly inhibited by an infinite concentration of cholate (Table 5), total Na$^{\oplus}$-dependent uptake of taurocholate is not shared by cholate. This would indicate that there are at least two Na$^{\oplus}$-dependent transport systems for taurocholate, one of which is shared by cholate.

It is possible that the Na$^{\oplus}$-independent uptake system and the Na$^{\oplus}$-dependent uptake system shared by cholate and taurocholate represent the same carrier which has separate binding sites for bile acid and Na$^{\oplus}$. The carrier is able to transport bile acid when no Na$^{\oplus}$ is present. In the presence of Na$^{\oplus}$, the carrier also binds Na$^{\oplus}$, and the complex is translocated faster. If this is the case, the K_m(app) value for the uptake of bile acid would be the same, both in the presence and in the absence of Na$^{\oplus}$, since K_m(app) was not affected by Na$^{\oplus}$ (Fig. 4). This seems to be the case for cholate uptake but not for taurocholate uptake (Table 2). However, since Na$^{\oplus}$-dependent uptake of taurocholate represents at least two transport systems, the K_m(app) value would not be representative of either system. Hence it cannot be said definitely whether these two transport processes represent the same carrier or different carriers. Other bile acids may share all of these or only a part of these transport systems for their transport into hepatocytes. It is also possible that there are other transport systems for other bile acids.

The present study was supported by a grant from *Deutsche Forschungsgemeinschaft*. The authors wish to thank Drs. *R. Kinne* and *H. Murer* of Max-Planck-Institut für Biophysik, Frankfurt, for helpful discussion and Mrs. *U. Rüberg* for excellent technical assistance.

Literature

1 Anwer, M.S., Kroker, R. & Hegner, D. (1976) *Hoppe-Seyler's Z. Physiol. Chem.* **357**, 1477 - 1486.
2 Schwarz, L.R., Burr, R., Schwenk, M., Pfaff, E. & Greim, H. (1975) *Eur. J. Biochem.* **55**, 617 - 623.
3 Reichen, J. & Paumgartner, G. (1976) *Am. J. Physiol.* **231**, 734 - 742.
4 Dietmaier, A., Gassel, R., Graf, J. & Peterlik, M. (1976) *Biochim. Biophys. Acta* **443**, 81 - 91.
5 LeCam, A. & Freychet, P. (1977) *J. Biol. Chem.* **252**, 148 - 156.
6 Anwer, M.S., Kroker, R. & Hegner, D. (1976) *Biochem. Biophys. Res. Commun.* **73**, 63 - 71.
7 Anwer, M.S. & Hegner, D. (1977) *Naunyn-Schmiedeberg's Arch. Pharmacol.* (abstr.) **297**, R8.
8 Szarkowska, L. & Klingenberg, M. (1963) *Biochem. Z.* **338**, 674 - 697.
9 Schultz, S.G., Frizzell, R.A. & Nellans, H.N. (1974) *Annu. Rev. Physiol.* **36**, 51 - 91.
10 Hoffmann, N., Thees, M. & Kinne, R. (1976) *Pflügers Arch.* **362**, 147 - 156.
11 Blaustein, M.P. & Christie Ector, A. (1976) *Biochim. Biophys. Acta* **419**, 295 - 308.
12 Vidaver, G.A. (1964) *Biochemistry* **3**, 662 - 667.
13 Blaustein, M.P. (1974) *Rev. Physiol. Biochem. Pharmacol.* **70**, 33 - 82.
14 Anwer, M.S., Kroker, R., Hegner, D. & Petter, A. (1977) *Hoppe-Seyler's Z. Physiol. Chem.* **358**, 543 - 553.
15 Christensen, H.N.((1969) *Adv. Enzymol.* **32**, 1 - 20.
16 Inui, Y. & Christensen, H.N. (1966) *J. Gen. Physiol.* **50**, 203 - 224.
17 Crane, R.K. (1977) *Rev. Physiol. Biochem. Pharmacol.* **78**, 99 - 159.
18 Schultz, S.G. & Curran, P.E. (1970) *Physiol. Rev.* **50**, 637 - 718.

M.S. Anwer and Prof. D. Hegner, Institut für Pharmakologie, Toxikologie und Pharmazie, Fachbereich Tiermedizin der Universität München,
Veterinärstraße 13, D-8000 München 22.

Hoppe-Seyler's Z. Physiol. Chem.
Bd. 359, S. 193 - 198, Februar 1978

Liver Cell Heterogeneity

The Distribution of Fructose-Bisphosphatase in Fed and Fasted Rats and in Man

Udo SCHMIDT[a,b], Heide SCHMID[a] and Walter G. GUDER[c] with the technical assistance of Barbara FUNK[b], Adelheid MALL[a] and Johanna WEIS[b]

[a] Pathologisches Institut der Universität, Tübingen
[b] Enzymlaboratorium der Medizinischen Poliklinik der Universität, Basel
[c] Klinisch-Chemisches Institut des Städtischen Krankenhauses München-Schwabing, München

(Received 8 August/13 December 1977)

Summary: Periportal and perivenous hepatocytes were isolated by microdissection from lyophilized liver slices (16 μm) from fed and fasted rats and from a human patient. $NADP^{\oplus}$/NADPH cycling was used to determine fructose-1,6-bisphosphatase activity in the isolated hepatocytes (10 ng dry weight). The periportal hepatocytes contain 3 times as much fructose-1,6-bisphosphatase activity as the perivenous hepatocytes. A 24 h fast led to a two-fold increase in the activity in the periportal hepatocytes and a four-fold increase in the perivenous hepatocytes. Fructose-1,6-bisphosphatase parallels closely with the key enzyme phospho*enol*pyruvate carboxykinase, and therefore can be considered a suitable marker for gluconeogenic capacity.

Heterogenität des Lebergewebes: Die Verteilung der Fructose-bisphosphatase bei gefütterten und hungernden Ratten sowie beim Menschen

Zusammenfassung: Mit Hilfe der Mikrodissektionstechnik wurden aus lyophilisierten Leberschnitten (16 μm) von gefütterten und hungernden Ratten sowie vom Menschen Hepatocyten aus dem periportalen und perivenösen Feld isoliert. Die Fructose-bisphosphatase wurde mit dem $NADP^{\oplus}$/NADPH cycling in den isolierten Hepatocyten (10 ng Trockengewicht) bestimmt. Die Hepatocyten im periportalen Feld weisen eine dreimal höhere Fructose-bisphosphatase Aktivität im Vergleich zu den Hepatocyten im perivenösen Feld auf. Eine 24 h Hungerperiode steigert die Fructose-bisphosphatase Aktivität um das zweifache im periportalen und um das vierfache im perivenösen Feld. Die Fructose-bisphosphatase verhält sich wie das Schlüsselenzym, die Phospho*enol*pyruvat-Carboxykinase, und ist daher als Maß zur Abschätzung der gluconeogenetischen Aktivität geeignet.

Enzymes:
Fructose-bisphosphatase, D-fructose-1,6-bisphosphate 1-phosphohydrolase (EC 3.1.3.11);
Glucose-6-phosphatase, D-glucose-6-phosphate phosphohydrolase (EC 3.1.3.9);
Glucose-6-phosphate dehydrogenase, D-glucose-6-phosphate:$NADP^{\oplus}$ 1-oxidoreductase (EC 1.1.1.49);
Glucosephosphate isomerase, D-glucose-6-phosphate ketolisomerase (EC 5.3.1.9);
Glutamate dehydrogenase ($NAD(P)^{\oplus}$), L-glutamate:$NAD(P)^{\oplus}$ oxidoreductase (deaminating) (EC 1.4.1.3);
Phospho*enol*pyruvate carboxykinase, GTP:oxaloacetate carboxy-lyase (transphosphorylating) (EC 4.1.1.32);
Phosphogluconate dehydrogenase (decarboxylating), 6-phospho-D-gluconate:$NADP^{\oplus}$ 2-oxidoreductase (decarboxylating) (EC 1.1.1.44);
Pyruvate kinase, ATP:pyruvate 2-*O*-phosphotransferase (EC 2.7.1.40).

Key words: Quantitative histochemistry, fructose-1,6-bisphosphatase, liver, rat, human.

In a recent microdissection study we were able to show that the activity of phospho*enol*pyruvate carboxykinase – thought to be rate-limiting in the gluconeogenic pathway – was three times higher in the periportal than in the perivenous hepatocytes from the fed rat[1]. The glycolytic enzyme pyruvate kinase (EC 2.7.1.40) revealed an inverse distribution pattern within the liver parenchyme: the activity in perivenous hepatocytes was twice that of the periportal cells[1].

Katz et al. found that glucose-6-phosphatase activity of periportal liver cells was more than twice as high as in the perivenous region, when microdissected from fed rats[2].

These quantitative findings confirmed the assumption of a "metabolic zonation" of liver parenchyme with regard to gluconeogenesis and glycolysis, as postulated on the basis of histochemical studies[3].

The present study deals with the quantification of a further enzymatic step in the gluconeogenic pathway, i.e. the fructose-bisphosphatase within the periportal and perivenous liver tissue obtained by microdissection from rat and human liver slices.

Methods

Male rats (Ivanovas strain) of 200 g body weight, fed or fasted for 24 h, with free access to water, were used. The preparation of liver slices for the microdissection procedure was done as reported previously in detail[1]. Periportal and perivenous specimens of 90×60 μm were separated from a tissue section of 16 μm thickness. The scheme of a liver acinus according to Rappaport[4] served as a guide for identification of the periportal and perivenous specimens. The dry weight of the specimens was measured with a quartz-fiber balance of 6×10^{-9} g/mm sensitivity.

Fructose-bisphosphatase activity was measured in single specimens of 5 to 15 ng using a modification of the assay conditions described by Latzko and Gibbs[5] combined with an enzymatic cycling procedure[6] (see Scheme).

To achieve the necessary sensitivity, the first three enzymatic steps were performed in small volumes of reagents using the oil-well technique[6].

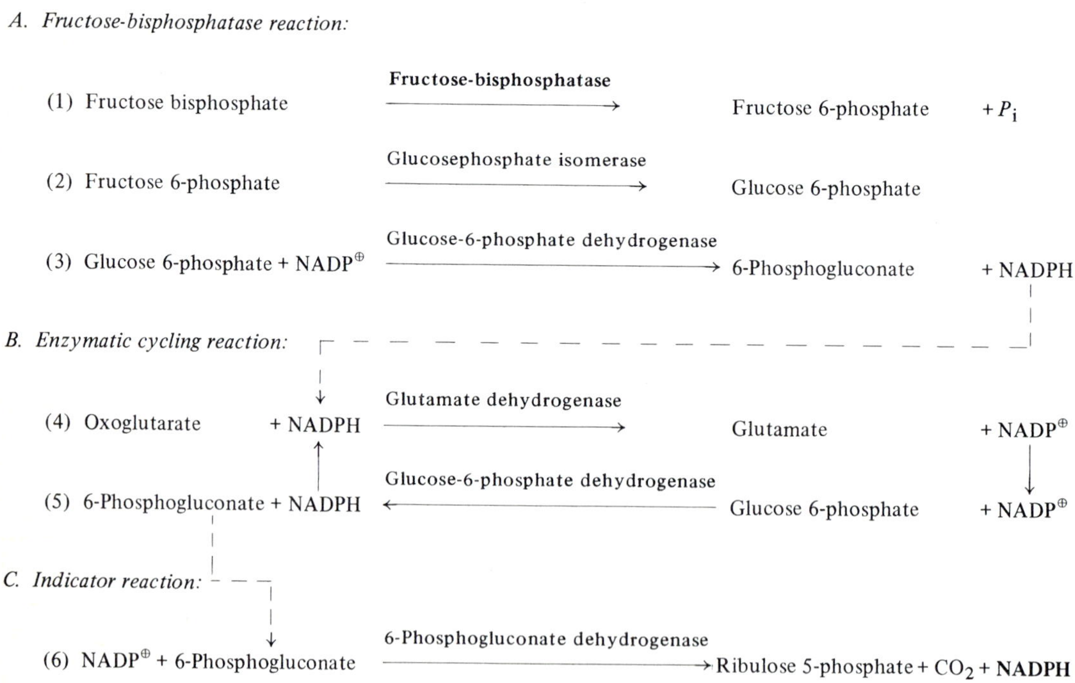

A specimen was brought into 400 n*l* of 0.1M Tris/HCl buffer, pH 8.0, containing 0.05% bovine serum albumin, which was pipetted into the well of a teflon block and covered immediately with two droplets of oil to prevent evaporation[6]. The loaded teflon rack was placed on an aluminium block in an ice bath for 10 min; then 400 n*l* of reagent containing 0.1M Tris/HCl, pH 8.0, 10mM $MgCl_2$, 1mM $NADP^{\oplus}$, 48mM mercaptoethanol, 1.2mM fructose bisphosphate, 40 μg glucosephosphate isomerase/m*l*, 20 μg glucose-6-phosphate dehydrogenase/m*l* and 0.05% bovine serum albumin was added. The teflon block was incubated in a water bath at 37 °C for 32 min, including a 2 min rewarming time.

After cooling in an ice bath, 400 n*l* of the incubated mixture was pipetted into 200 μ*l* 0.04N NaOH. After an incubation at 60 °C for 10 min, an aliquot of 6 μ*l* was added to 25 μ*l* of an enzymatic cycling reagent, containing 0.1M Tris/HCl, pH 8.0, 5mM oxoglutarate, 1mM glucose 6-phosphate, 0.3mM ADP, 25mM NH_4 acetate, 35 μg glucose-6-phosphate dehydrogenase/m*l*, 150 μg glutamate dehydrogenase/m*l* and 0.02% bovine serum albumin. After incubation for 1 h at 37 °C, the reaction was stopped by boiling for 2 min, then 200 μ*l* of indicator reagent containing 0.02M Tris/HCl, pH 8.0, 90μM $NADP^{\oplus}$, 3.3mM $MgCl_2$, 0.1mM EDTA, 0.05% bovine serum albumin, and 100 μg 6-phosphogluconate dehydrogenase/m*l* was added. NADPH fluorescence was recorded after 30 min at room temperature in a spectrofluorometer. NADPH standards of 10, 5 and 2.5 × 10^{-11} mol were carried throughout the procedure. The activity is expressed in U/g dry weight.

In some experiments, 0.5mM 5′-AMP was added as an inhibitor of fructose-bisphosphatase to evaluate possible unspecific phosphatase activities in the samples tested.

Results

Assay conditions

Optimal conditions for fructose-bisphosphatase were deduced from studies with homogenates of lyophilized liver slices taken after microdissection. Under the conditions described, fructose 6-phosphate was formed at a constant rate for at least 40 min by a homogenate of lyophilized liver corresponding to 50 ng dry weight. Fructose 6-phosphate production was also linearly dependent on the amount of incubated liver homogenate in the range of 6 to 100 ng dry weight, dissolved in 10 μ*l* of reagent.

Fructose-bisphosphatase activity in liver homogenates of fed rats was in the range of 32 to 68 U/g dry weight, which lies within the range reported by others[7].

Liver tissue contains 6-phosphogluconate dehydrogenase, which could increase the NADPH formed/fructose bisphosphate used during incubation. From the kinetics of the reactions, we calculated that only 10% of the NADPH formed in an incubation period of 40 min results from 6-phosphogluconate dehydrogenase activity[5]. This small amount was neglected in the calculation of fructose-bisphosphatase activity.

Our assay for fructose-bisphosphatase was sensitive enough to measure the activity in samples containing only 5 hepatocytes or even less[8].

Table 1. Distribution of fructose-1,6-bisphosphatase activity in rat liver.

Fructose-1,6-bisphosphatase was measured in microdissected samples from the periportal and perivenous region of the liver lobule, as described in the Methods section. Corrected activities were obtained by subtraction of residual activities measured after the addition of 0.5mM 5′-AMP. All values are given as means ± standard deviation with the number of samples in parentheses.

	Periportal (zone 1) [U/g dry wt.]	Perivenous (zone 3) [U/g dry wt.]	Zone 1 / Zone 3
Fed rat	35.0 ± 5.0 (12)	10.0 ± 3.3 (12)	3.5
+ AMP	5.0	3.3	
Corrected	30.0	6.7	4.5
Starved rat	80.0 ± 8.3 (13)	35.0 ± 4.0 (13)	2.3
+ AMP	6.7	3.3	
Corrected	73.3	31.7	2.3
"Starved" divided by "fed"	2.3	3.5	
Corrected	2.4	4.7	

Thus we were able to isolate relatively homogeneous cell populations from the periportal or perivenous zone. From histochemical studies with glucose-6-phosphatase[3], it is evident that there is a gradual transition from one metabolic zone to another, and that the diameter of a single zone changes throughout the acinus. Therefore we decided to assay very small samples to avoid cells from the borderline of the specific zone.

Distribution of fructose-bisphosphatase activity and effects of fasting

Table 1 demonstrates that fructose-bisphosphatase activity is more than three times higher in the periportal hepatocytes than in the perivenous cells of the fed rat. The ratio of periportal to perivenous activity of 3.5 is increased to 4.5 when the corrected values were taken into consideration. After 24 h fasting, fructose-bisphosphatase rose more than two-fold in the periportal and more than four-fold in the perivenous zone when the corrected values were compared (Table 1). An activity increase of similar magnitude was also found in the homogenates of liver slices from fasted rats. This increase is further evidence for the specificity of the assay.

With AMP in the assay, a significant residual activity was measured in each type of hepatocyte (Tabel 1). However, no difference was found between the two cell types with respect to the magnitude of the residual activity. In addition, no change in the residual activity was detected in either cell type after fasting, which is evidence for the unspecificity of the residual activity.

Table 2. Distribution of fructose-1,6-bisphosphatase in human liver.

A liver sample obtained during laparatomy from a woman who had fasted for 18 h was treated as described for rat liver in the methods section. Enzyme activities are given as means ± standard deviation with the number of samples in parentheses. Activities are expressed as U/g dry weight and were obtained in the absence of 5′-AMP.

	Periportal (zone 1)	Perivenous (zone 3)	$\frac{\text{Zone 1}}{\text{Zone 3}}$
Human liver	70.0 ± 11.7 (5)	28.3 ± 3.3 (5)	2.5

In a human liver biopsy, obtained from a woman during laparatomy after an 18-h fast, fructose-bisphosphatase exhibited a distribution pattern within the acinus which corresponds closely to the pattern found in the liver of fasted rats (Table 2).

Discussion

The present finding of a 3.5-fold higher fructose-bisphosphatase activity in the periportal than in the perivenous zone fits with the periportal/perivenous activity ratios of phospho*enol*pyruvate carboxykinase[1] and of glucose-6-phosphatase[2]. The correspondence of all three gluconeogenic enzymes supports the concept that the liver acinus (or lobule) is heterogeneous with respect to carbohydrate metabolism, i.e. with a more gluconeogenic periportal area (zone 1 according to Rappaport) and a more glycolytic perivenous area (zone 3)[1-3]. Katz et al. have introduced the term "metabolic zonation" of the liver parenchyme[2].

Fasting leads to a two-fold increase in fructose-bisphosphatase in the periportal and a four-fold increase in the perivenous area (Table 1); an increment corresponding to that of phospho*enol*-pyruvate carboxykinase[1] and glucose-6-phosphatase[2] in the same areas. The increase in fructose-bisphosphatase observed after a 24 h fast seems of special interest in view of the negative results reported in the literature. Pontremoli et al. even described a decrease in fructose-bisphosphatase in rabbit liver during the first 30 h of starvation, followed by an increase of at least twofold up to 100 h[9]. In addition, Messner et al. were not able to detect an increase in fructose-bisphosphatase in the rat liver during a 16-h fast[10].

The former authors explained the biphasic behaviour of the activity of this enzyme during a long fast by assuming a proteolytic modification of the enzyme in the fasted animal[9]. Species differences might also be responsible for the different results. In addition, Scholz[11] found that in the rat kidney, the increase in fructose-bisphosphatase during fasting was only detectable in the microdissected proximal tubule, but not in the homogenate of the slice in which microdissection was performed. This might be further evidence for the assumption of a modification of the catalytic properties of the enzyme, depending on the kind of preparation[7].

The marked increase in fructose-bisphosphatase activity in the perivenous area during starvation underlines the concept of Sasse et al.[3] that the "metabolic zonation" of carbohydrate metabolism in the liver acinus is a dynamic one. This concept was supported by results which showed an increase in phospho*enol*pyruvate carboxykinase and glucose-6-phosphatase within the periportal area and a more pronounced increase in the perivenous area of liver parenchyme during 24 h starvation[1,2]. The dynamic behaviour of fructose-bisphosphatase distribution in the liver parenchyme in different feeding states is in contrast to the enzyme behaviour within the rat nephron. Here fructose-bisphosphatase was found to be mainly located in the proximal convolution. The segments of the distal tubule revealed only a low but significant activity[11]. Starvation led to a two-fold increase in fructose-bisphosphatase in the proximal convolution, whereas the activity of the distal part of the nephron remained unchanged. This supports the conclusion that metabolic heterogeneity of carbohydrate metabolism in the nephron is stable, despite a change in the nutritional state. The difference between liver and kidney with regard to metabolic heterogeneity might be caused by the difference in the function of liver and kidney gluconeogenesis. Gluconeogenesis in liver tissue seems to be important mainly for regulating blood glucose concentration; in kidney, however, this pathway might be responsible for the reabsorption of organic acids, i.e. lactate, glutamine and proline[12].

On the basis of our present data and previous reports[1-3], gluconeogenic capacity seems to decrease along the liver sinusoidal blood flow, whereas glycolytic pyruvate kinase was found to exhibit an inverse distribution pattern[1]. Although the functional significance of the "metabolic zonation" in liver parenchyme is still obscure, this heterogeneity seems to be an integral part of the mechanism by which this organ is able to synthesize and degrade glucose simultaneously.

Addendum: During the printing of this article a paper by Katz et al.[13] appeared in which data on fructose-bisphosphatase activity in microdissected periportal and perivenous liver samples were published. These findings, like those reported here, reveal an activity ratio of 2:1 between the periportal and perivenous area in the fed rat. However, they failed to establish a significant increase in fructose-bisphosphatase activity in the two liver areas from the fasted rat.

We assume that the discrepancy between Katz's results and ours is due to the difference in sensitivity of the two methods. In their measurements, at least 250 ng of dry weight was used (this sample size represents the total periportal or perivenous zone), whereas in our study samples of only 5 ng were sufficient for measurement; this sample size corresponds to only a limited part of the liver zone.

It might be that the activity change during starvation becomes evident only when defined parts of the various zones are compared.

This project was supported by the *Deutsche Forschungsgemeinschaft*, Bonn-Bad Godesberg; *Schweizerischer Nationalfonds, Barell-Stiftung Hoffmann-La Roche* and *Cilag-Chemie.*

Literature

1 Guder, W.G. & Schmidt, U. (1976) *Hoppe-Seyler's Z. Physiol. Chem.* **357**, 1793 - 1800.
2 Katz, N., Teutsch, H.F., Sasse, D. & Jungermann, K. (1977) *FEBS Lett.* **76**, 226 - 230.
3 Sasse, D., Katz, N. & Jungermann, K. (1975) *FEBS Lett.* **57**, 83 - 88.
4 Rappaport, A.M. (1976) *Beitr. Pathol.* **157**, 215 - 243.
5 Latzko, E. & Gibbs, M. (1974) in Methoden der enzymatischen Analyse (Bergmeyer, H.U., ed.) pp. 914 - 917, Verlag Chemie, Weinheim.
6 Lowry, O.H. & Passonneau, J.V. (1972) A flexible system of enzymatic analysis, Academic Press, New York, London.
7 Horecker, B.L., Melloni, E. & Pontremoli, S. (1975) *Adv. Enzymol.* **42**, 193 - 226.
8 Zahlten, R.N. & Stratman, F.W. (1974) *Arch. Biochem. Biophys.* **163**, 600 - 608.
9 Pontremoli, S., Melloni, E., Salamino, F., de Flora, A. & Horecker, B.L. (1974) *Proc. Natl. Acad. Sci.* **71**, 1776 - 1779.
10 Messner, B., Berndt, J. & Still, J. (1976) *Nature (London)* **263**, 599 - 600.
11 Scholz, M. (1978) Thesis, Universität Tübingen, (in preparation).
12 Guder, W.G. & Schmidt, U. (1976) 6th Int. Congr. Nephrology, pp. 187 - 195, Karger Verlag, Basel.
13 Katz, N., Teutsch, H.F., Jungermann, K. & Sasse, D. (1977) *FEBS Lett.* **83**, 272 - 276.

Prof. Dr. U. Schmidt and Dr. H. Schmid, Enzymlaboratorium der Medizinischen Poliklinik der Universität Basel, Hebelstr. 1, CH-5046 Basel,
and Pathologisches Institut der Universität Tübingen,
Liebermeisterstr. 8, D-7400 Tübingen.

PD. Dr. W.G. Guder, Klinisch-Chemisches Institut des Städtischen Krankenhauses München-Schwabing, Kölnerplatz 1, D-8000 München 40.

Hoppe-Seyler's Z. Physiol. Chem.
Bd. 359, S. 199 - 209, Februar 1978

Inhibition of Cholesterol Synthesis in Cultured Cells by 25-Azidonorcholesterol

Wilhelm STOFFEL and Rainer KLOTZBÜCHER

Institut für Physiologische Chemie der Universität Köln

(Received 14 November 1977)

Summary: 25-Azidonorcholesterol, a side chain analogue of cholesterol, inhibits cholesterol biosynthesis in BHK 21 cells in tissue culture; but 3-azido-5-cholestene, the azido analogue of cholesterol has no inhibitory effect. Cell growth was unimpaired by the azido analogue, which showed only a minimal uptake by the cell, compared with that of cholesterol or 25-azidonorcholesterol. 25-Azidonorcholesterol is incorporated into the membranes of BHK 21 cells in place of cholesterol as demonstrated by isotope experiments. The incorporation of radioactive acetate is strongly inhibited by the side chain azido substituted cholesterol derivative; the incorporation of radioactive mevalonate is, however, unimpaired. We conclude that the inhibition occurs at the hydroxymethylglutaryl-CoA-reductase step. Cell growth is not significantly altered.

Hemmung der Cholesterin-Synthese in Zellen in der Gewebekultur durch 25-Azidonorcholesterin

Zusammenfassung: Das Seitenketten-Cholesterin-analoge 25-Azidonorcholesterin führt in Zellen in der Gewebekultur (BHK 21) zur Hemmung der Cholesterin-Synthese, während das 3-Azido-5-cholesten, eine Verbindung mit der Azidogruppe in der Ringstellung, keine Hemmung verursacht. In Isotopenversuchen wurde gezeigt, daß 25-Azidonorcholesterin an Stelle des Cholesterins in die Membranen der BHK-Zellen inkorporiert wird. Isotopenexperimente mit radioaktivem Acetat und Mevalonat zeigen auf, daß die Hemmung auf der Stufe der 3-Hydroxymethylglutaryl-CoA-Reduktase erfolgt. Das Zellwachstum wird nicht signifikant verändert.

Key words: 25-Azidonorcholesterol, 3α-azido-5-cholestene, inhibition of hydroxymethylglutaryl-CoA reductase, cell growth, membrane synthesis.

Enzymes:
Hydroxymethylglutaryl-CoA reductase (NADPH), mevalonate:NADP$^{\oplus}$ oxidoreductase (CoA-acylating) (EC 1.1.1.34); Sterol 24-reductase (not yet registered).

Abbreviations:
Mesyl = methylsulfonyl; DBED = *N,N*-dibenzylethylenediammonium.

An attempt has been made to incorporate azido-labelled lipid precursors into the membranes of eucaryotic cells, which act as host cells for enveloped viruses. We recently demonstrated that phospholipids, labelled in their fatty acyl moieties with the photosensitive azido group, are suitable chemical probes for the determination of the nearest neighbour to the carbon segment of the lipid molecule because the aliphatic nitrene group generated by UV irradiation has a very short lifetime ($\sim 10^{-11}$ s), during which inter- or intra-molecular reactions occur. Intermolecular reactions (insertion into C-H-bonds and addition to double bonds) with adjacent fatty acyl chains and with polypeptide chains have been demonstrated[1]. During experiments aimed at the incorporation of the two photosensitive azido-labelled cholesterol derivatives ([25-^{3}H]25-azidonorcholesterol and [26-^{14}C]3α-azido-5-cholestene) into the membranes of baby hamster kidney (BHK 21) cells, it became apparent that the 25-azido compound inhibits the de novo synthesis of cholesterol, and that 3α-azido-5-cholestene lacks these properties. Isotope experiments suggest that 3-hydroxy-3-methylglutaryl-CoA synthesis is blocked. Concommittantly both compounds are incorporated into the membranes, but cell growth is barely inhibited.

Material and Methods

The syntheses of [26-^{14}C]3α-azido-5-cholestene and [25-^{3}H]25-azidonorcholesterol are described below. [2-^{14}C]Na-acetate (specif. radioactiv. 40 - 60 mCi/mmol) and [5-^{3}H]D,L-mevalonate (DBED salt, 5 mCi/mmol) were purchased from New England Nuclear, [5-^{14}C]-mevalonate from Amersham Buchler, Braunschweig and 26-norcholesten-3β-ol-25-one from Steraloids, Inc. Wilton, N.H.. BHK 21 cells were grown in Dulbecco's medium supplemented with 10% Bacto tryptose phosphate broth (Difco Laboratories, Inc. Detroit, Michigan) and 5% normal or lipoprotein free (d = 1.22 g/ml) fetal calf serum[2]. Monolayer cultures were grown in 12-cm Petri dishes, in a 5% CO_2 atmosphere at 37 °C in a humidified incubator.

Radioactive pulse experiments

The azidocholesterol derivatives, dissolved in 95% ethanol (5 mg/ml) and aqueous solutions of radioactive sodium acetate and mevalonate, were added as indicated in the legends to figures. The medium was decanted at appropriate times, the cell layer washed once with 5 ml medium containing 2% albumin, once with medium containing 5% serum and four times with phosphate buffered saline. Cells were released from the bottom of the dish with a rubber policeman, harvested in 4 ml phosphate buffered saline and sedimented by centrifugation at 250 × g for 10 min. The pellet was sonicated for 1 min and protein determined. The rest of the suspension was lyophilized in Sovirel tubes, and lipids were saponified with 2 ml 0.5N methanolic KOH for 1 h at 70 °C. Sterols were extracted three times with petroleum ether (30 - 60 °C). The extracts were taken to dryness in a stream of nitrogen at 40 °C, and the residue dissolved in 1 ml chloroform/methanol (1 : 1). Sterols were precipitated as digitonides[3]. The digitonides were pelleted, washed twice with 2 ml diethyl ether and dissolved in 0.5 ml soluene 350 and 10 ml toluene scintillator solution.

Gas liquid chromatographic analyses

A Packard gas-chromatograph model 805 was used. This was equipped with U-shaped 2-m glass columns packed with 1% OV 17 and operated at 250 °C. The instrument was connected to a Hewlett-Packard recorder integrator, model 3380 A. The effluent was split in a 1:1 ratio for mass and radioactivity determination of cholesterol, desmosterol, lanosterol and dihydrolanosterol. The latter and β-cholestane were used as standards. The Varian MAT CH5 mass spectrometer was used with the direct inlet and combined with gas-chromatographic separation (Varian model 1840).

Syntheses of radioactive azido-sterols

a) [25-^{3}H]25-Azidonorcholesterol (norcholest-5-en-3β-ol)

3β-Acetoxy-5-cholesten-25-one (I)

586.3 mg (1.5 mmol) *26-nor-5-cholesten-3β-ol-25-one* in 15 ml dry pyridine and 1.6 ml freshly distilled *acetic-anhydride* were heated at 60 - 70 °C for 1 h and left overnight at room temperature. The solution was poured into 500 ml ice water, the precipitate isolated by filtration, washed thoroughly with distilled water and dried over P_2O_5 in a desiccator. Yield: 587 mg (1.4 mmol, 93% of th.). The product proved to be pure in thin-layer chromatography (solvent system: petroleum ether/ether/methanol 70:30:7).

[25-^{3}H]3β-Acetoxy-5-cholesten-25-ol (II): 420 mg (1 mmol) *3β-acetoxy-5-cholesten-25-one (I)* was dissolved in 30 ml tetrahydrofuran/0.2M phosphate buffer pH 8.5 (8:1 v/v). A solution of 25 mCi (0.08 mmol) *NaB^{3}H$_4$* in 1 ml phosphate buffer pH 7 was added dropwise to the ketone. After 1 h additional 35.1 mg $NaBH_4$ completed the reaction within 2 h. The reaction mixture was adjusted to pH 6.0 with acetic acid and the solution poured into 400 ml ice water. The tritiated

product was extracted with ether, the ether solutions washed with water, dried over Na_2SO_4 and taken to dryness under reduced pressure. Yield: 420 mg (0.98 mmol) 98% of th.; specif. radioactiv. 7.7 mCi/mmol. The product proved to be homogeneous in thin-layer chromatography using the solvent system mentioned above.

[25-^{3}H]3β-Acetoxy-25-methylsulfonyloxy-5-cholestene-(III): To 200 mg (0.46 mmol) *[25-^{3}H]3β-acetoxy-5-cholesten-25-ol (II)* dissolved in 8 m*l* dry pyridine, 64 mg (0.55 mmol) *mesylchloride* was added and the mixture stirred at room temperature for 12 h. The solvent was evaporated under vacuum and the residue extracted six times with 50 m*l* diethyl ether. The ether extracts were washed with water, dried over Na_2SO_4 and then taken to dryness. Yield: 180 mg (0.36 mmol), 77% of th. Thin-layer chromatography showed, that the reaction had run essentially to completion.

[25-^{3}H]3β-Acetoxy-25-azido-5-norcholestene (IV): 180 mg (0.36 mmol) of the *mesyl derivative (III)* was dissolved in 20 m*l* dimethylformamide/water 10:1 (v/v) and 117 mg (1.8 mmol) sodium azide was added to this solution, which was subsequently stirred in the dark for 24 h. The mixture was concentrated at 70 °C under vacuum and the residue extracted five times with 50 m*l* chloroform. The combined extracts were washed with water, dried over Na_2SO_4 and taken to dryness. Yield: 165 mg (0.36 mmol). The product proved to be pure in thin-layer chromatography using the above system. IR-spectroscopy revealed the typical azido-absorption band at 2100 cm^{-1} and the ester band at 1730 cm^{-1}.

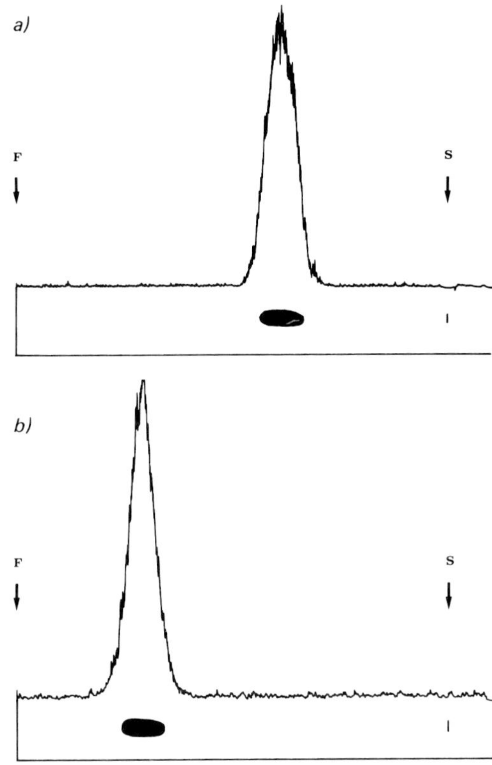

Fig. 1. Radio thin-layer chromatogram of a) [25-^{3}H]25-azidonorcholesterol, b) [26-^{14}C]3α-azido-5-cholestene. Solvent system: petroleum ether/ether/methanol 70:30:15 (V/V); S = start; F = front.

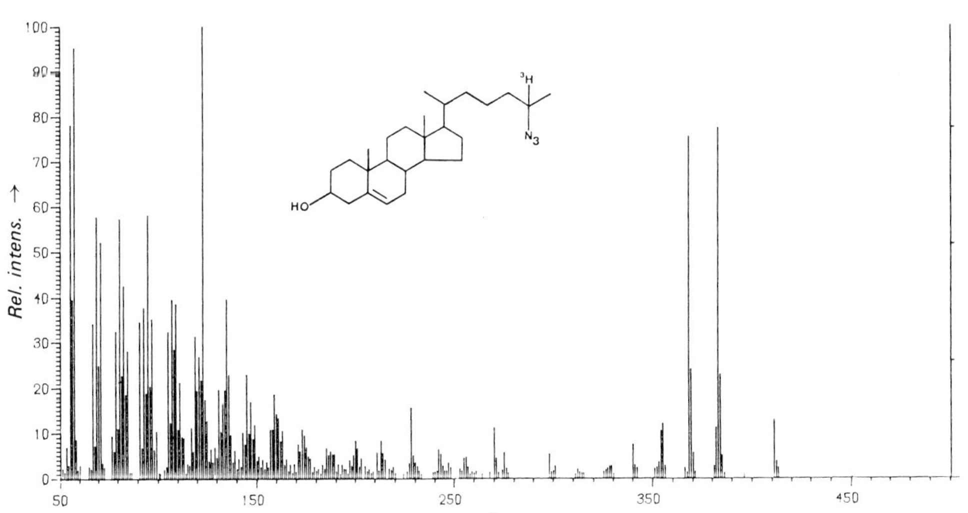

Fig. 2. Mass spectrum of 25-azidonorcholesterol.

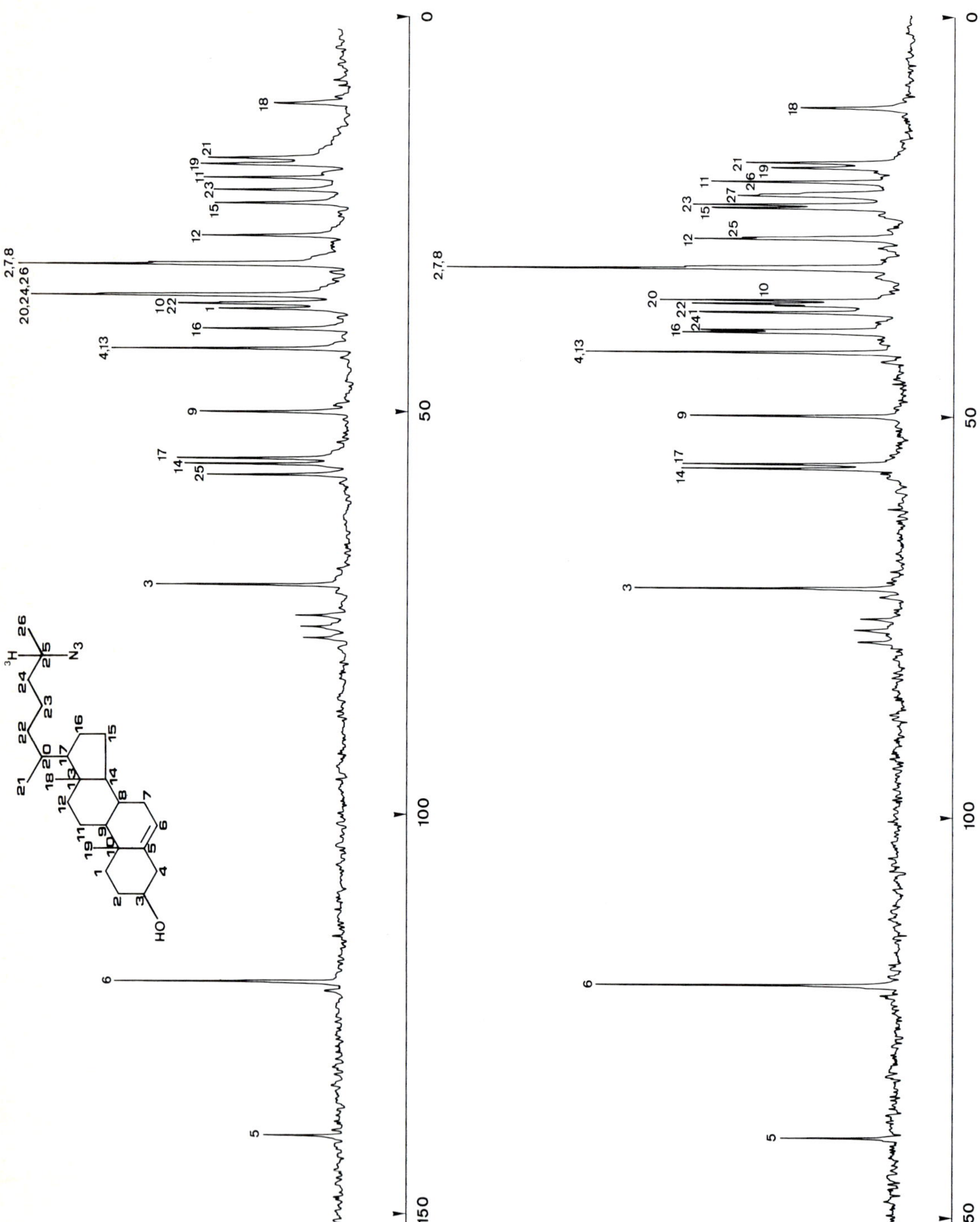

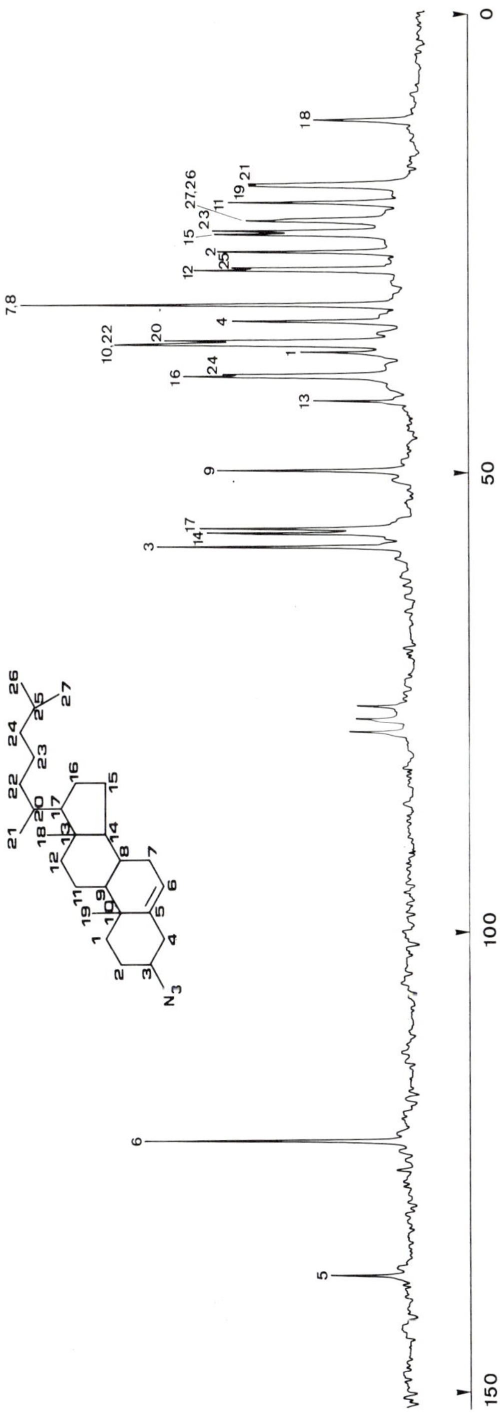

Fig. 3. ^{13}C-NMR spectra of a) 25-azidonorcholesterol, b) cholesterol, c) 3α-azido-5-cholestene.

[25-^{3}H]25-Azidonorcholesterol (V): 165 mg of *IV* was saponified with 60 m*l* 0.5N methanolic KOH at room temperature overnight. The solution was diluted with 40 m*l* of water, neutralized with 6N HCl and extracted three times with 50 m*l* chloroform. The extracts were dried over Na_2SO_4 and concentrated under vacuum. Yield: 144 mg (0.35 mmol), 97% of th., mp. 119.5 °C. The product proved to be pure in thin-layer chromatography (Fig. 1a). IR spectrum: absorption bands at 2100 cm^{-1} -N_3; 3 600 – 3 200 cm^{-1} -OH group. Specif. radioactiv. 7.7 mCi/mmol. The compound deteriorates during gas-chromatography. The mass spectrum (Fig. 2) reveals characteristic peaks at: 413 M$^{\oplus}$; $^m/e$ M$^{\oplus}$-(H_2+N_2) 383; $^m/e$ M$^{\oplus}$-(CH_3+H_2+N_2) 368. The chemical shift of the ^{13}C resonance lines of every carbon atom relative to TMS (tetramethylsilane) are compared with those of cholesterol in Fig. 3 and Table 1.

Table 1. Differences in chemical shifts of ^{13}C resonance lines (ppm relative to TMS) of cholesterol, 3α-azido-5-cholestene and 25-azidonorcholesterol.

C atom	Cholesterol	3α-Azido-5-cholestene	25-Azidonor-cholesterol
2	31.58	25.80	31.45
3	71.37	57.86	71.29
4	41.98	33.31	41.89
24	39.18	39.18	35.20
25	27.83	27.61	57.62
26	22.39	22.39	35.20

b) [26-^{14}C]3α-Azido-5-cholestene

81 mg (0.21 mmol) *[26-^{14}C]cholesterol*[4] (specif. radioactiv. 0.198 mCi/mmol (purified by bromination and debromination) was dissolved in 4 m*l* dry pyridine, 30 *μl mesylchloride* was added and the mixture stirred for 2 h at room temperature. The solvent was evaporated under reduced pressure, the residue extracted with ether, the combined ether extracts washed with cold 2N HCl and water, dried over Na_2SO_4 and taken to dryness. Yield: 101.7 mg (100% of th.).

The *mesyl derivative* was dissolved in 5 m*l* hexamethylphosphotriamide (HMPT)[5], 200 mg *Na-azide* was added and the reaction mixture kept at 80 °C for 3 h under an atmosphere of argon. 10 m*l* water was then added to the cold mixture, which was extracted several times with hexane. The combined hexane extracts were washed with water, dried over Na_2SO_4 and concentrated. 3α-Azido-5-cholestene was purified by preparative thin-layer chromatography, mp. 107 - 108 °C.

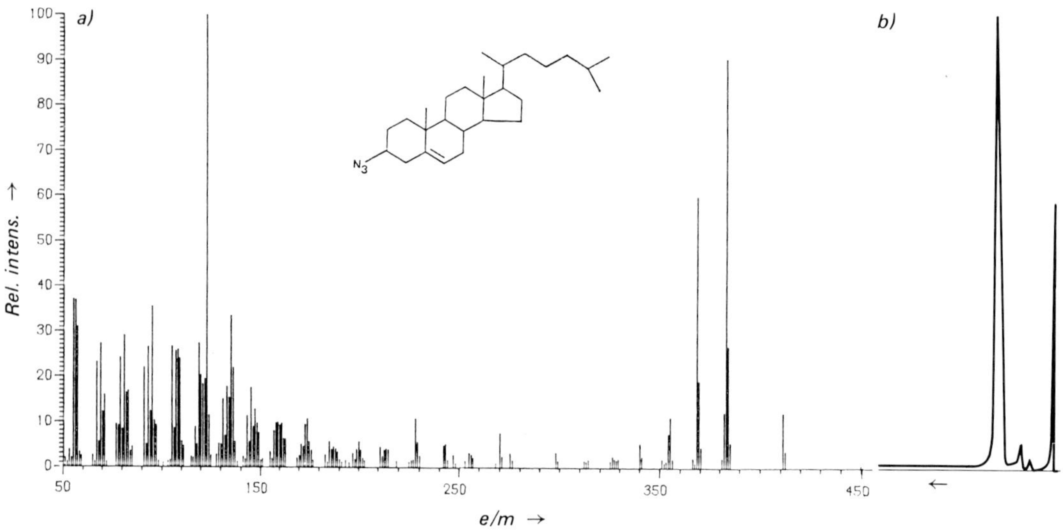

Fig. 4. Combined gas liquid chromatography-mass spectroscopy of 3α-azido-5-cholestene.

36.5 mg [26-^{14}C]3α-azido-5-cholestene (42% of th.) was separated from 3,5-cholestadiene and obtained in pure form. The product was characterized by gas-chromatography-mass spectroscopy (Fig. 4). The IR spectrum revealed the characteristic absorption bands at 2080 - 2120 cm^{-1} -N_3, C=C: 800 cm^{-1}. The chemical shifts of the ^{13}C resonance lines can be taken from Fig. 3 and Table 1. It proved to be radiochemically pure in radio thin-layer chromatography, (Fig. 1b).

Results

Incorporation of [25-^{3}H]25-azidonorcholesterol into BHK 21 cells

BHK 21 cells were grown for 24 h in media supplemented with either lipoprotein-containing serum or with lipoprotein-free serum, and the uptakes of [26-^{14}C]cholesterol and [25-^{3}H]25-azidonorcholesterol were compared during this period. Cells take up both sterols linearly, depending on the concentration in the medium (Fig. 5). In the concentration range up to 10 μg of cholesterol and 25-azidonorcholesterol/m*l* the incorporation of both sterols was distinctly higher, when lipoprotein-free serum was used as a supplement. [25-^{3}H]25-Azidonorcholesterol, at a concentration of 5 μg/m*l* medium, was also added to BHK 21 cell cultures growing in the logarithmic phase in medium supplemented with 5% normal fetal calf serum, and its incorporation was measured at different time intervals. Whereas the total amount of digitonin precipitable sterols (cholesterol + 25-azidonorcholesterol) remained constant after 24 h, the ratio of cholesterol to azidonorcholesterol did not change after confluency was

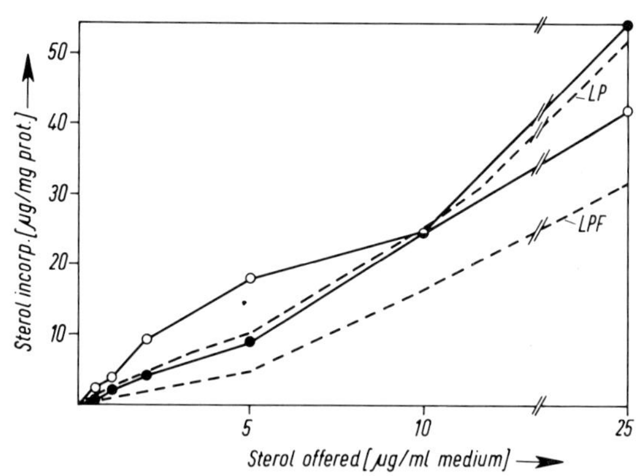

Fig. 5. Comparative study of the incorporation of [25-^{3}H]25-azidonorcholesterol and [26-^{14}C]cholesterol into BHK 21 cells, showing the dependence on their concentration in the medium.

LP = lipoprotein containing medium, LPF = lipoprotein free medium. —— 25-Azidonorcholesterol, - - - cholesterol.

Table 2. Incorporation of [25-^{3}H]25-azidonorcholesterol into BHK 21 cells grown on lipoprotein-free medium in the presence of 5 μg/m*l* 25-azidonorcholesterol.

Time	+ 25-Azidonorcholesterol				Control	
	Protein	Digitonin-precipitable sterols	25-Azidonor-cholesterol	Cholesterol	Protein	Cholesterol
[h]	[mg]	[μg/mg prot.]	[μg/mg prot.]	[μg/mg prot.]	[mg]	[μg/mg prot.]
24	2.92	16.4	9.7	6.7	1.83	6.6
48	3.74	14.3	10.8	3.5	2.61	8.8
72	2.92	17.1	11.4	5.7	1.53	11.6
96	2.31	17.2	10.4	6.8	1.25	13.5

reached. The cholesterol concentration does not increase within 72 to 96 h, whereas in control cells the cholesterol concentration doubles. Apparently 25-azidonorcholesterol can substitute for cholesterol and inhibit cholesterol synthesis at the same time (Table 2). Evidence for the inhibition of cholesterol biosynthesis at an early stage was obtained by the determination of cholesterol and its precursors in cells after exposure to a constant concentration of 25-azidonorcholesterol (5 μg/m*l* medium for different time intervals). Cholesterol, desmosterol and dihydrolanosterol were quantified by gas liquid chromatography. The results are summarized in Table 3. In order to localize the inhibitory action within the biosynthetic pathway of cholesterol biosynthesis sodium-[2-^{14}C]acetate and [5-^{14}C]- and [5-^{3}H]-mevalonate were used as precursors in cholesterol biosynthesis. Their incorporation into digitonin-precipitable sterols was measured. The inhibition of 25-azidonorcholesterol was compared to that of cholesterol (Fig. 6). The regulatory function of cholesterol at the level of 3-hydroxymethylglutaryl-CoA reductase is well established[6].

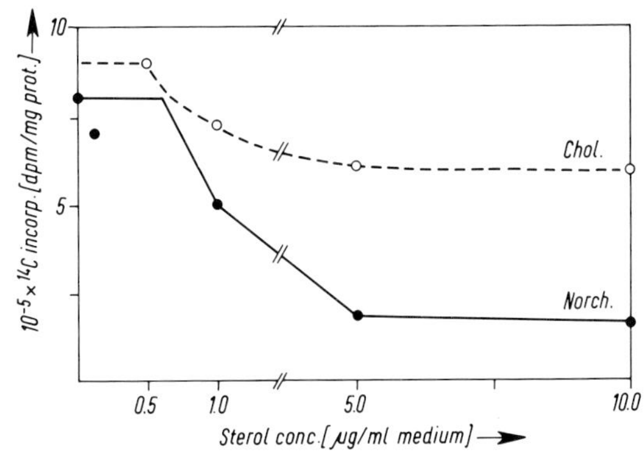

Fig. 6. [1-^{14}C]Acetate incorporation into digitonin-precipitable sterols of BHK 21 cells in the presence of 25-azidonorcholesterol (•——•) and cholesterol (○---○).

Table 3. Inhibition of cholesterol synthesis in BHK 21 cells in monolayer culture.

The medium was supplemented with 5 μg 25-azidonorcholesterol/m*l* medium. Cells were grown in 10-cm Petri dishes with 10 m*l* medium. They were harvested and the sterol fraction isolated as described under Methods and Materials. They were separated by gas liquid chromatography (2 m 1% OV-1 column; temperature 245 °C; 30 m*l*/argon/min. Values in μg/mg protein.

[h]	+ 25-Azidonorcholesterol			Control		
	Cholesterol	Desmosterol	Dihydro-lanosterol	Cholesterol	Desmosterol	Dihydro-lanosterol
24	14.4	1.2	1.3	16.1	2.1	–
48	11.2	1.9	0.3	18.8	1.4	–
72	8.6	1.7	–	21.5	1.3	–

Fig. 6 indicates that 25-azidonorcholesterol is a strong inhibitor of cholesterol biosynthesis, causing 50% inhibition at a concentration of 1 - 2 μg/m*l* medium (2.5 x 10^{-6}M). Whereas cholesterol, at a concentration of 10 μg/m*l* medium (2.5 x 10^{-5}M), reduces the incorporation of [^{14}C]acetate by 40% of the control experiment, 25-azidonorcholesterol, at the same concentration, inhibits cholesterol biosynthesis by more than 85%. The inhibition by 25-azidonorcholesterol did not lead to the accumulation of sterol intermediates such as desmosterol, dihydrolanosterol and lanosterol, as can be seen in the gas-chromatographic analyses of the sterol fraction. Therefore radioactive mevalonate was supplied as precursor of cholesterol synthesis to the medium of BHK 21 cells. Cells were incubated in the presence and absence of the inhibitory 25-azidonorcholesterol (5 μg/m*l*) and the medium supplemented with [5-^{14}C]mevalonate (0.2 μCi/m*l*). Cells were harvested at time intervals indicated in Fig. 7. The incorporation of radioactive

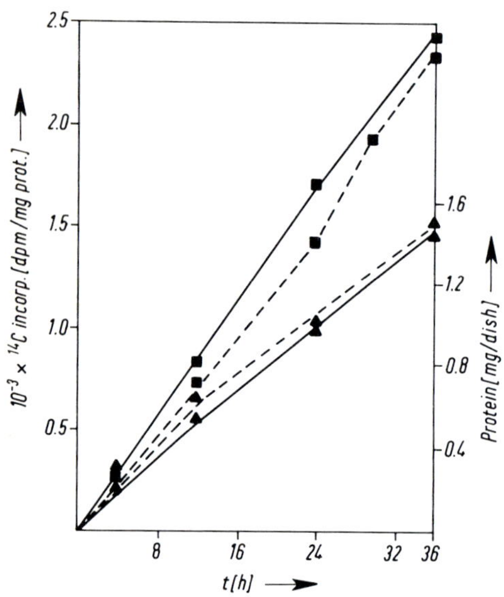

Fig. 7. [5-^{14}C]Mevalonate incorporation into digitonin-precipitable sterols of BHK 21 cells.

Digitonin-precipitable sterols: ■——■ Control, ■– – –■ in the presence of 25-azidonorcholesterol (5 μg/m*l*). Protein: ▲——▲ control, ▲– – –▲ in the presence of 25-azidonorcholesterol (5 μg/m*l*).

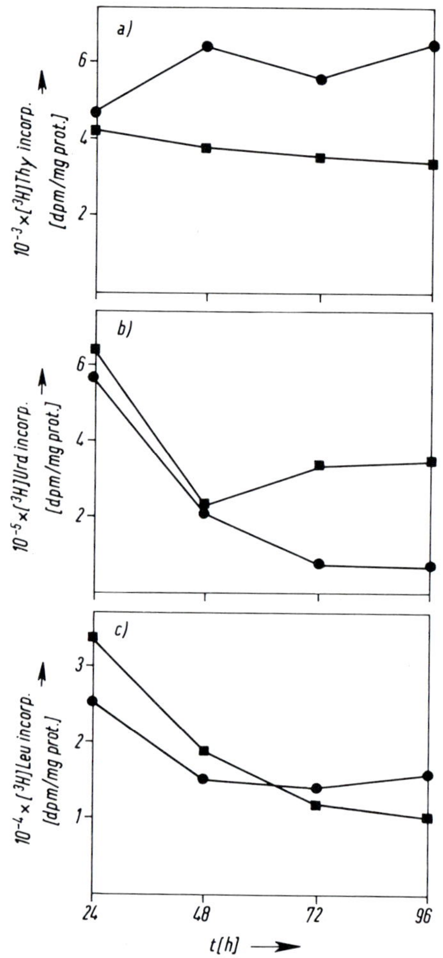

Fig. 8. Incorporation of a) [^{3}H]thymine (3 μCi/m*l*), b) [^{3}H]uridine (3 μCi/m*l*), c) [^{3}H]leucine (2 μCi/m*l*) into BHK 21 cells growing in the absence ●——● and presence ■——■ of 25-azidonorcholesterol (5 μg/m*l*).

Fig. 9. Morphology of BHK 21 cells growing 48 h (a, c) and 72 h (b, d), respectively, in the absence (a, b) and presence (c, d) of 25-azidonorcholesterol. ▷

Lipoprotein-free medium was used. Magnification approx. 1000-fold.

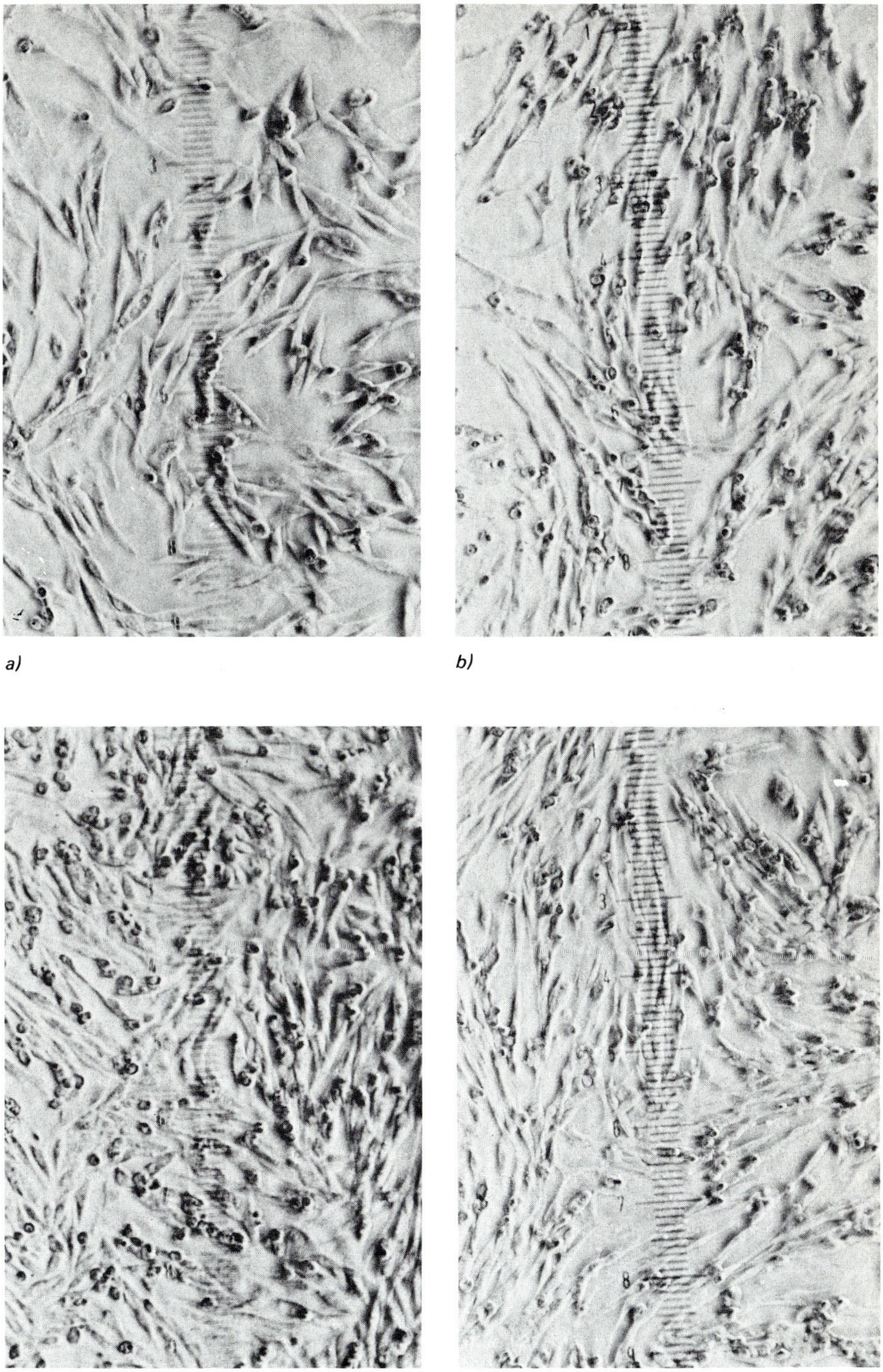

a) b) c) d)

mevalonate into the digitonin precipitable sterol fraction was identical in control cells and those growing on the azido derivative. Therefore the inhibition must occur at the level of hydroxymethylglutaryl-CoA reductase. The inhibition of cholesterol synthesis by 25-azidonorcholesterol does not influence cell growth. Growth curves of control cells (Fig. 7) and those growing in the presence of 25-azidocholesterol were identical, as indicated by the protein determination.

[^{3}H]Thymine, [^{3}H]uridine and [^{3}H]leucine were also added to BHK 21 cells with inhibited and non-inhibited cholesterol synthesis. The precursors of DNA, RNA and protein synthesis were incorporated at very similar rates in both sets of experiments (Fig. 8). The morphological appearance of the cells is normal and unaltered even after 96 h (Fig. 9a - d).

Incorporation of 3α-azido-5-cholestene into BHK 21 cells

If the 3-hydroxy group of cholesterol is substituted by the azido group a derivative is obtained, which, in the concentration range between 0 and 6×10^{-5} M, has no inhibitory action on the incorporation of Na-[^{14}C]acetate into cholesterol. Its uptake into the cell is minimal, as compared to cholesterol and 25-azidonorcholesterol. Cell growth remained unimpaired by this compound.

Discussion

The studies reported here indicate that 25-azidonorcholesterol is a very effective inhibitor of cholesterol biosynthesis in BHK 21 cells. Isotope experiments prove that the biosynthetic sequence is inhibited at the level of hydroxymethylglutaryl-CoA reductase, because mevalonate releases the block of cholesterol synthesis. 25-Azidonorcholesterol not only inhibits cholesterol synthesis but is also incorporated in the membranes of the cell linearly with time. Kandutsch et al.[7,8] have demonstrated that cholesterol-derived sterols, with additional oxygen containing functional groups, depress cholesterol synthesis at the level of hydroxymethylglutaryl-CoA reductase. These additional hydroxyl or carbonyl groups may either be in the sterane ring system (C-6, C-7, C-15) or in the side chain (C-20, C-22, C-25). 7-Oxocholesterol has been studied in detail by Erickson et al.[9] in perfused rat liver and intact rat liver. These authors demonstrated that hydroxymethylglutaryl-CoA reductase is inhibited only for a brief period of time. The synthesis of oxygenase is then enhanced and the inhibitor is quickly transformed into more polar products, which are excreted in the bile. The azido group, as a substituent in the side chain of cholesterol, lends the sterol molecule properties which lead to an effective inhibition of the regulatory enzyme of cholesterol biosynthesis, hydroxymethylglutaryl-CoA reductase. No polar derivatives could be detected during the period of incubation with BHK 21 cells. By following DNA, RNA and protein synthesis with appropriate precursors, we observed that a minor deviation in thymine incorporation into DNA could be correlated with a slightly slower growth of the cells growing on 25-azidonorcholesterol. [^{3}H]Uridine and [^{3}H]-leucine incorporation resulted in specific activities of RNA and protein (dpm/mg protein) which were quite comparable in cells with and without inhibition of cholesterol synthesis. The appearance of both types of cells in light microscopy showed no morphological differences. This was also observed for 3α-azido-5-cholestene. Therefore the two azido derivatives, one of which is labelled in the side chain (25-azidonorcholesterol), the other in the nucleus (3α-azido-5-cholestene), are not toxic for cell growth in the concentration range studied here.

The low incorporation of 3α-azido-5-cholestene into the membranes of BHK 21 cells underlines the importance of the 3-hydroxy group in the 3 position. On the other hand substitution of a CH_3 group of the side chain of cholesterol by an azido group does not reduce its uptake compared with that of cholesterol. Whereas the 3α-azido-5-cholestene does not decrease cholesterol biosynthesis, 25-azidonorcholesterol has an inhibitory effect, similar to that of cholesterol on the de novo synthesis of cholesterol.

On the basis of isotope studies ([^{14}C]acetate and [^{14}C]mevalonate) this inhibition occurs at the level of hydroxymethylglutaryl-CoA reductase. Further in vitro studies are planned for the elucidation of the inhibition mechanism (inhibition of enzyme or enzyme synthesis). Photoaffinity labelling with this azido-labelled sterol may further help to determine the site of inhibition.

We acknowledge gratefully the support of this study by the *Deutsche Forschungsgemeinschaft*.

Literature

1 Stoffel, W., Därr, W. & Salm, K.P. (1977) *Hoppe-Seyler's Z. Physiol. Chem.* **358**, 453 - 462.
2 Cham, B.E. & Knowles, B.R. (1976) *J. Lipid Res.* **17**, 176 - 181.
3 Knauss, H.J., Porter, J.W. & Wasson, G. (1959) *J. Biol. Chem.* **234**, 2835 - 2840.
4 Organic Syntheses with Isotopes (1958) (Murray, A., III, & Williams, D.L., eds.) Part I, pp. 1056 - 1059, Interscience Publishers, New York.
5 Cavé, A., Jarreau, F.-X., Khuong-Huu, Q., Leboeuf, M., Serban, N. & Goutarel, R. (1967) *Bull. Soc. Chim. Fr.* 701 - 706.
6 Siperstein, M.D. (1970) *Curr. Top. Cell Regul.* **2**, 74 - 76.
7 Kandutsch, A.A. & Chen, H.W. (1974) *J. Biol. Chem.* **249**, 6057 - 6061.
8 Kandutsch, A.A. & Chen, H.W. (1973) *J. Biol. Chem.* **248**, 8108 - 8117.
9 Erickson, S.K., Cooper, A.D., Matsui, S.M. & Gould, R.G. (1977) *J. Biol. Chem.* **252**, 5186 - 5193.

Prof. Dr. Dr. W. Stoffel and Dipl.-Chem. R. Klotzbücher, Institut für Physiologische Chemie der Universität Köln, Joseph-Stelzmann-Str. 52, D-5000 Köln 41.

Hoppe-Seyler's Z. Physiol. Chem.
Bd. 359, S. 211 - 218, Februar 1978

New Protein Reagents

Synthesis and Properties of Halogenated Maleimides and Isomaleimides

Jean-Paul LAUTIÉ and Martin OLOMUCKI

Laboratoire de Biochimie Cellulaire, Collège de France, Paris

(Received 9 December 1977)

Summary: The title compounds were investigated as examples of potential hetero-bifunctional protein reagents having an electrophilic moiety in addition to a maleimide or isomaleimide ring. The synthesis of such products by cyclization of the corresponding maleamic acids gives lower yields in the maleimide than in the isomaleimide series, especially when the second function is labile, owing to the more drastic reaction conditions necessary to form the symmetric isomer. On the other hand, a comparison of the isomeric imides showed that despite some differences in reactivity, isomaleimides can be used instead of maleimides for chemical modification of proteins. Therefore, if the synthesis of new maleimide reagents containing a second fragile functional group is difficult, they can be designed in the isoimide form.

Neue Proteinreagenzien: Synthese und Eigenschaften halogenierter Maleinimide und Isomaleinimide

Zusammenfassung: Die im Titel genannten Verbindungen wurden als potentielle hetero-bifunktionelle Proteinreagenzien, die eine elektrophile Gruppe und einen Maleinimid- oder Isomaleinimid-Ring enthalten, untersucht. Die Synthese solcher Produkte durch Ringschluß der entsprechenden Maleinsäuremonoamide gibt niedrigere Ausbeuten in der Maleinimid- als in der Isomaleinimid-Reihe, besonders wenn die zweite Funktion labil ist, weil die Bildung der symmetrischen Isomeren schärfere experimentelle Bedingungen erfordert. Andererseits zeigt ein Vergleich der isomeren Imide, daß, trotz gewisser Unterschiede in der Reaktivität, Isomaleinimide ebenfalls für die chemische Modifizierung der Proteine verwendet werden können. Wenn also die Synthese eines neuen Maleinimidreagenzes Schwierigkeiten bereitet, sollte die Isoimid-Form in Betracht gezogen werden.

Key words: Protein reagents, cross-linking, maleimides, isomaleimides.

N-Substituted maleimides are commonly used as relatively specific thiol reagents in proteins[1]. Several compounds containing two maleimide rings have been used as homo-bifunctional -SH

Enzymes:
Alcohol dehydrogenase (yeast), alcohol:NAD$^{\oplus}$ oxidoreductase (EC 1.1.1.1.); papain (EC 3.4.22.2).

cross-linking reagents in studies of tertiary and quaternary structure of proteins[1,2]. However, compounds including a maleimide moiety and an unspecific alkylating group able to detect the presence of another nucleophile in the vicinity of a thiol in proteins are so far almost unknown[3]. Still, such versatile reagents, which could perform various cross-links, could be useful in biochemistry, e.g. for exploring the microenvironment of -SH groups in active sites of enzymes or other biological receptor sites. Compounds containing two different functional groups often also offer the practical advantage that the cross-linking may be carried out in two steps. It therefore seemed interesting to synthesize and investigate the properties of some maleimide derivatives of this type as potential hetero-bifunctional protein reagents.

All the maleimides studied in the present work contain on the nitrogen atom a halogenated substituent able to react with different functional groups. To facilitate the identification of the modified amino acid residues, two types of substituents were chosen: halogenoacetyl groups giving well known carboxymethylated amino acids after total hydrolysis of the modified protein, and the chromogenic chloro-*o,o'*-dinitrophenyl group which, after substitution of chlorine by nucleophiles such as amines or thiols, exhibits characteristic UV absorptions[4]. Some of these compounds were synthesized in both the imide and isoimide form to compare their properties and determine to what extent isomaleimides can also be applied for chemical modification of proteins. Actually, despite certain advantages of the synthesis of isomaleimides, the possibility of using them for that purpose has up to now seldom been envisaged[5,6].

Experimental

Organic syntheses

N-Acyl-N'-(carboxyacryloyl)hydrazines (Ia, b, c)

A solution of 0.1 mol of hydrazine in 20 m*l* of chloroform was added dropwise in 30 min to a stirred ice-cold solution of 0.11 mol of *p*-nitrophenylacetate or halogenoacetate in 120 m*l* of tetrahydrofuran. After 20 min, 150 m*l* of anhydrous ether was added and a small amount of solid material was filtered. A solution of 0.11 mol of maleic anhydride in 70 m*l* of chloroform was then added dropwise. After 30 min the precipitate was collected and recrystallized in acetone.

The products thus obtained were:

Ia, yield 70%, m.p. 200° - 202 °C (Lit.[7]: m.p. 202 °C)

Ib, yield 70%, m.p. 184 °C; IR (nujol) 3170, 1715, 1688, 1220 cm^{-1}. NMR $(C^2H_3)_2SO$: δ 11.2, 10.45, 6.22, 4.05 ppm.

$C_6H_7N_2O_4Cl$ (206.58)
Calc.: Cl 17.16.
Found: Cl 16.9.

Ic, yield 70%, m.p. 182 °C.

$C_6H_7N_2O_4Br$ (251.04),
Calc.: Br 18.80.
Found: Br 19.1.

N-(Acylamino)maleimides (II a, b, c)

A solution of 0.02 mol of compound Ia, b or c in 60 m*l* of acetic acid was refluxed for 2 h, the solvent evaporated in vacuo and the white residual solid recrystallized in benzene. The products thus obtained are indicated in the Table.

N-(Acylamino)isomaleimides (III a, b, c)

Method A: A solution of 0.07 mol of compound Ia, b or c in 250 m*l* of acetic anhydride was kept 1 - 2 days at room temperature (or heated to 80 °C for 15 min), evaporated and the residue recrystallized in benzene.

Method B: A solution of 0.02 mol of dicyclohexyl carbodiimide in 20 m*l* of methylene chloride was added dropwise to a stirred solution of 0.02 mol of product I in 10 m*l* of methylene chloride. After 3 h the suspension was filtered, the filtrate evaporated and the residue recrystallized in benzene (see Table).

N-(4-Chloro-3,5-dinitrobenzoyloxymethyl)maleimide II d

Method A: 4 g (15 mmol) of 4-chloro-3,5-dinitrobenzoyl chloride[8] in 5 m*l* of acetone and 1.3 m*l* of pyridine were added to 1.9 g (15 mmol) of *N*-hydroxymethylmaleimide[9] in 5 m*l* of acetone and the mixture was refluxed for 1 h. The precipitated pyridine hydrochloride was filtered and the filtrate treated with an equal volume of water. The brown solid was collected and recrystallized in benzene giving yellow cristals, m.p. 172 °C. A second recrystallization in benzene with treatment with Norit yielded 1.3 g (25%) of white crystals, m.p. 184 °C. IR (KBr): 3100, 1730, 1615, 1350, 1320, 1280, 840 cm^{-1}.

$C_{12}H_6N_3O_8Cl$ (355.65)

Calc.:	C 40.52	H 1.70	N 11.81	Cl 9.97
Found:	C 40.5	H 1.7	N 11.5	Cl 10.2

Method B: A solution of 1 g (4 mmol) of 4-chloro-3,5-dinitrobenzoic acid in 10 m*l* of tetrahydrofuran was added dropwise to a solution of 0.51 g (4 mmol) of *N*-hydroxymethylmaleimide and 0.83 g (4 mmol) of dicyclohexylcarbodiimide in 20 m*l* of tetrahydrofuran

Table. Maleimides and isomaleimides.

Product	m.p. [°C]	Yield [%]	Empirical formula	M_r	Analysis[a] C	H	N	O	Hal
Maleimides									
IIa	164-6[b]	60							
IIb	115	40	$C_6H_5N_2O_3Cl$	188.57	38.22 38.1	2.67 2.8	14.86 15.2	25.46 25.3	18.81 18.8
IIc	130-2	35	$C_6H_5N_2O_3Br$	233.03	30.92 30.6	2.16 2.2	12.02 11.9	20.60 20.3	34.29 34.0
Isomaleimides									
IIIa	184[c]	86							
IIIb	148[d,e]	70	$C_6H_5N_2O_3Cl$		38.3	2.9	14.7	25.5	18.8[f]
IIIc	150[d]	50	$C_6H_5N_2O_3Br$		31.1	2.2	12.5	20.1	34.0[f]

[a] Found under calculated values, [b] Lit.[7]: m.p. 164-6 °C, [c] Lit.[7]: m.p. 184 °C; [d] Prepared by method A; [e] Prepared by method B, [f] Found values.

and the mixture was stirred overnight. The precipitated dicyclohexylurea was filtered, the filtrate evaporated, the residue was taken off with acetone and a small part of undissolved solid was filtered. The precipitate formed by addition of petroleum ether was filtered and recrystallized in benzene, yielding 1.06 g (75%) of product having the same m.p. and mixed m.p. as above.

N-(Acylamino)-2-(benzylthio)succinimides

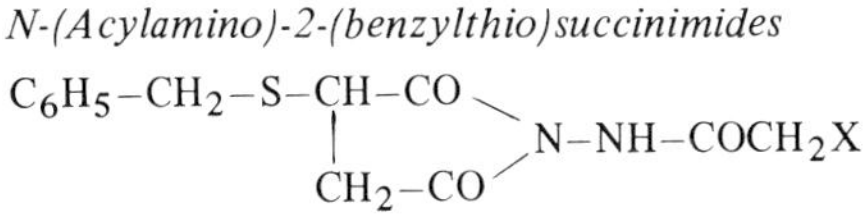

To a stirred solution of 3.7 mmol of *N*-(acylamino)-maleimide or the corresponding isoimide in 15 m*l* of acetone was added 3.7 mmol of benzylmercaptan in 15 m*l* of acetone followed by 3.7 m*l* of 1N alcoholic NaOH. After 15 min the reaction mixture was evaporated, the remaining benzylmercaptan extracted with petroleum ether and the residue crystallized in benzene. In each case both isomers led to the same product (m.p. and mixed m.p.; IR (KBr): 3280, 1700 cm^{-1}; NMR $[(C^2H_3)_2SO]$: δ 7.27, 4.26, 3.93, 3.20 ppm; ^{13}C NMR data indicated in ref.[10]). The products thus obtained were: X = H, yield 75%, m.p. 136 °C. X = Cl, yield 70%, m.p. 110 °C.

$C_{13}H_{13}N_2O_3SCl$ (312.77)
Calc.: C 50.04 H 4.19 N 8.97 S 10.27 Cl 11.36
Found: C 49.8 H 4.1 N 9.1 S 10.2 Cl 11.6

X = Br, yield 70%, m.p. 138 °C.

$C_{13}H_{13}N_2O_3SBr$ (357.23)
Calc.: C 43.71 H 3.67 N 7.84 S 8.97 Br 22.37
Found: C 43.4 H 3.6 N 7.8 S 9.4 Br 22.8

2-(Carboxymethylthio)-N-phenylsuccinimide

This product was obtained either according to Marrian[11], by treating *N*-phenylmaleimide with aqueous thioacetic acid in the presence of alkali or, under the same conditions, starting from phenylisomaleimide, m.p. and mixed m.p. 154 °C. Lit.[11]: m.p. 153.5 - 154.5 °C.

N-[Tetrahydro-3-oxo-2H-1,4-thiazin-2-yl)acetyl]-N'-(halogenoacetyl)hydrazines IVb, c

The reactions were performed starting from chloro- and bromoacetylamino maleimides and from the corresponding isoimides. To 5 mmol of product in 20 m*l* of acetone was added 5 mmol of aminoethanethiol hydrochloride in 5 m*l* of water, the apparent pH adjusted to 8 with 1N NaOH and the solution was stirred for 10 h. The white precipitate was then filtered and crystallized from water. For the four syntheses the yields were 70 - 80%.

IVb, m.p. 212 - 214 °C. IR (KBr): 3220, 3080, 1690, 1615 cm^{-1} (amide). NMR $[(C^2H_3)_2SO]$: δ 10.14 (CO-NH-NH-CO), 7.85 (4-NH), 4.05 (CH_2Cl), 3.77 (CH), 3.38 (5-CH_2), 2.78 (6-CH_2), 2.27 (CH-CH_2-CO) ppm.

$C_8H_{12}N_3O_3SCl$ (265.72)
Calc.: C 36.16 H 4.55 N 15.81 O 18.06 S 12.06 Cl 13.34
Found: C 36.2 H 4.5 N 15.5 O 17.7 S 12.0 Cl 13.1

IVc, m.p. 134 - 136 °C.

$C_8H_{12}N_3O_3SBr$ (310.17)
Calc.: C 30.97 H 3.89 N 13.54 O 15.47 S 10.33 Br 25.76
Found: C 31.6 H 4.0 N 13.4 O 15.7 S 10.2 Br 25.6

2-(Benzylthio)-N-(benzylthioacetylamino)succinimide

To a stirred solution of 5 mmol of 2-(benzylthio)-*N*-(bromoacetylamino)succinimide or the corresponding *N*-chloroacetylamino compound in 20 m*l* of acetone were added 7 mmol of benzylmercaptan then, dropwise, 5 m*l* of 1N NaOH. After 5 h the mixture was concentrated and the reaction product was precipitated by adding water, filtered and crystallized in ethyl acetate, m.p. 140 °C. Yield 70% in both cases. NMR $[(C^2H_3)_2SO]$: δ 7.27, 3.93, 3.77, 3.20 ppm.

$C_{20}H_{20}N_2O_3S_2$ (400.52)

Calc.:	C 60.05	H 5.04	N 7.02	S 16.03
Found:	C 60.0	H 4.8	N 6.9	S 16.0

Reaction of N-(bromoacetylamino)maleimide (IIc) with papain

Papain (Sigma) was activated with cysteine as described in ref.[12]. Enzyme concentrations were determined from the extinction at 278 nm[13].

0.4 m*l* of 0.1M acetate buffer, pH 5.6 and 0.1 m*l* of a 10mM solution of compound IIc in acetone were added to 0.5 m*l* of a 0.2mM solution of papain. After 15 min at room temperature, the excess of reagent was eliminated by dialysis against 0.1M acetate buffer, pH 5.6, the -SH groups titrated by Ellman's method[14] and histidine residues determined using diazobenzene sulphonic acid according to Barnard and Stein[15].

Reaction of N-(4-chloro-3,5-dinitrobenzoyloxymethyl)-maleimide (IId) with yeast alcohol dehydrogenase

Enzyme concentrations were determined using $A_{1\,cm}^{1\%}$ = 12.6 at 280 nm[16].

A mixture of 0.5 m*l* of 72μM solution of alcohol dehydrogenase (Boehringer Mannheim), 0.1 m*l* of a 5.4mM solution of IId in dimethylformamide and 2.4 m*l* of 0.1M acetate buffer, pH 5.1, was incubated 30 min at 20 °C, then dialyzed against the same buffer to eliminate the excess of reagent. Thiol groups were titrated in aliquots by the method of Ellman[14] and the protein solution was dialyzed against 0.1M carbonate buffer, pH 9, for 24 h. The number of $-NH_2$ groups substituted by the chlorodinitrophenyl part of the reagent was determined by the absorbance at 430 nm (ϵ_M = 7 000).

Results and Discussion

Synthesis of halogenated maleimides

To prepare maleimides containing a halogenoacetyl group on the nitrogen atom we took hydrazine as a starting point. Its reaction with a *p*-nitrophenyl halogenoacetate yields the corresponding halogenoacetyl hydrazines, unstable compounds decomposing even at room temperature with evolution of nitrogen. These compounds are therefore generally isolated in the form of hydrochlorides[17]. However, by adding maleic anhydride directly to the reaction mixture containing the freshly formed acylhydrazine, we obtained the expected *N*-acyl-*N'*-carboxyacryloylhydrazines Ib and Ic in 70% yields.

Heating these compounds in acetic acid converted them to the symmetrical imides II. To obtain the corresponding isoimides III, we treated products I at room temperature either with acetic anhydride or with dicyclohexylcarbodiimide (DCCI) in a neutral solvent. For comparison all these syntheses were also performed with the acetyl derivative a (X = H).

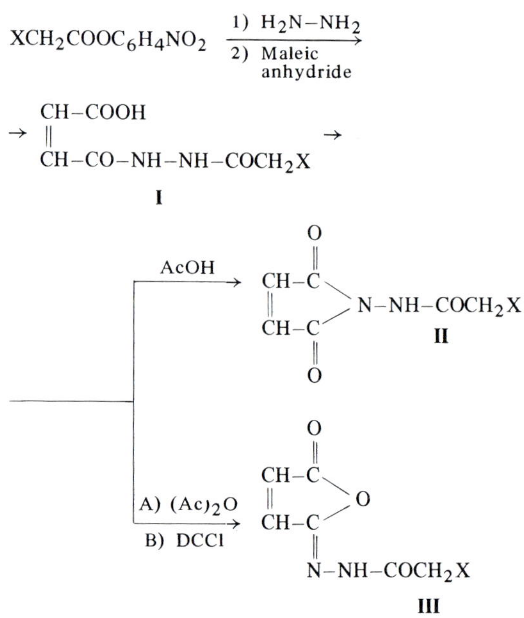

a, X = H; b, X = Cl; c, X = Br

To obtain a maleimide containing a chloro-*o,o'*-dinitrophenyl substituent on the nitrogen atom, we synthesized *N*-(4-chloro-3,5-dinitrobenzoyloxymethyl)maleimide IId

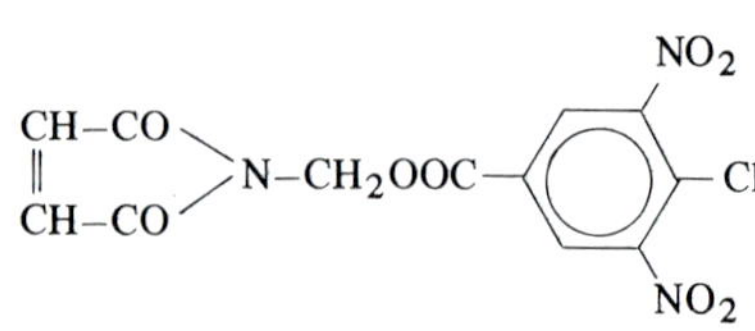

by treating hydroxymethyl maleimide with 4-chloro-3,5-dinitrobenzoyl chloride in presence of pyridine or, better, by 4-chloro-3,5-dinitrobenzoic acid and dicyclohexylcarbodiimide.

Physical properties of halogenated maleimides

Spectral data for maleimides were: UV, 215 nm (ϵ_M 1.4 x 10^4); infra-red, ν_{CO} 1725 and 1695 cm^{-1}; NMR, singlet at δ = 7.05 ppm for the two equivalent ethylenic protons, while for isomaleimides these values were: UV, 310 nm (ϵ_M 2 x 10^4); infra-red, 1790 and 1705 cm^{-1}; NMR, two doublets at 7.77 and 6.77 ppm (J = 6 Hz) for the ethylenic protons. The ^{13}C NMR spectra were published in a preceding paper[10].

In the case of isoimides, the question could also arise of the *syn-anti* isomerism of the C=N bond. This problem was of interest in view of a possible use of these compounds as cross-linking reagents, since the distance between their two functions (the C=C bond and the halogen) depends on the form of the molecule. X-ray crystallographic studies of *N*-(chloroacetylamino)isomaleimide IIIb, chosen as an example, showed that the molecule has an *anti* configuration, the chloroacetic chain being maximally elongated. The distances between chlorine and the olefinic carbon atoms are 6.80 and 7.55 Å. Moreover, ^{13}C NMR spectroscopy has shown that this compound, as well as *N*-acetylamino isomaleimide IIIa are also *anti* isomers in solution[10].

When aromatic amino acids are added to solutions of maleimides, new UV absorption bands are observed, e.g. at 300 nm in mixtures of *N*-(acetylaminomaleimide IIa with tryptophan and at 290 nm in mixtures IIa-tyrosine. These absorptions disappear after addition of a thiol such as *N*-acetylcysteine, which converts the maleimide rings into alkylthiosuccinimides. Thus, it seems that aromatic amino acids form charge transfer complexes with maleimides. To our knowledge the only example so far mentioned in the literature of such a complex formed by a *N*-substituted maleimide was the one given by *N*-(hydroxymethyl)maleimide with resorcinol[9].

Comparison of the reactivity of maleimides and isomaleimides

The Table shows that isomaleimides are obtained in higher yields than the symmetric isomers, probably because of the milder reaction conditions necessary for the synthesis of the former isomers. On the other hand, in both series the yields decrease with increasing lability of the nitrogen substituent (a > b > c). These results illustrate the fact that maleimides, especially those containing a second labile function, are easier to prepare in the isoimide form. It was therefore interesting to check the possibility of using isomaleimides in biochemistry instead of the symmetric isomers.

While the reactivity of maleimides has been sudied by several authors[18,19], few data concerning isomaleimides are available, although the greater lability of this isomer is known. In particular, no systematic comparison of the stability and the reactivity of the two isomers with thiols has so far been made. We were interested in this comparison, especially under the conditions in which maleimides are used in protein chemistry.

We tested the stability of compounds II and III at different pH values by following the variations in UV absorption due to hydrolytic ring opening according to the method of Gregory[19]. As could be expected, the stability of both isomers decreased with increasing pH. On the other hand, however, while the greater instability of isomaleimides appears clearly at acidic pH, in the pH range in which maleimides are normally used (neutral or slightly alkaline), the differences are small (Fig. 1).

The rates of addition of thiols (*N*-acetylcysteine) were studied by following the changes in absorption at 215 nm for maleimides and at 310 nm for isomaleimides. Again, although maleimides react faster in acidic media, the difference in reactivity decreases under neutral conditions (Fig. 2). The nature of the nitrogen substituent has no influence on the reaction rate.

We tried to isolate the adducts of thiols with maleimides and isomaleimides. Since the purification of addition products of *N*-acetylcysteine proved difficult, other thiols such as benzylmercaptan, thioacetic acid or cysteamine were used in reaction with compounds II and IIIa, b, c and also, to enlarge the tested series, with the two isomeric *N*-phenylmaleimides. In each case both isomers led to the same adduct. A ^{13}C NMR study of the reaction products described in a preceding paper[10] showed that all the adducts were symmetric alkylthiosuccinimides.

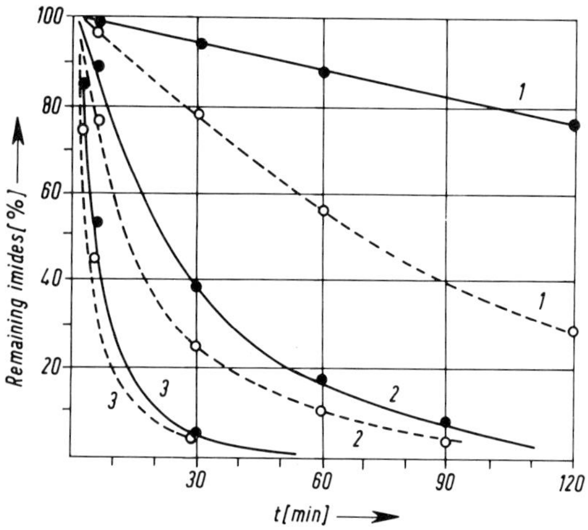

Fig. 1. Hydrolysis of *N*-(acylamino)maleimides and -isomaleimides.

The changes in absorbance at 215 and 250 nm were followed in 60μM solutions of *N*-(acylamino)maleimides IIa, b, c (full lines) and at 310 nm in solutions of the same concentration of the corresponding isoimides IIIa, b, c (dotted lines) in the following buffers: (1) 0.05M sodium acetate, pH 5.1; (2) 0.05M sodium phosphate, pH 7.2; (3) 0.05M tris(hydroxymethyl)aminomethane, pH 8.2.

In the case of cysteamine, a side reaction between the amino and the carbonyl group followed the addition of thiol and led to the thiazines IV in a 70-80% yield. Such reactions have already been observed[18,20,21].

When two equivalents of a mercaptan were reacted with compounds IIb and c, the quantity of thiol decreased in the reaction medium at two distinctly different rates corresponding to a first rapid addition on the maleimide double bond, then a much slower substitution of halogen

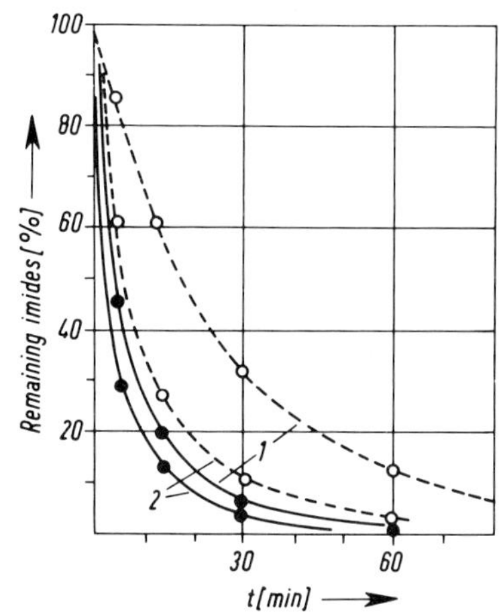

Fig. 2. Reaction of *N*-(acylamino)maleimides and -isomaleimides with *N*-acetylcysteine.

The changes in absorbance at 215 nm were followed in equimolar 50μM solutions of *N*-acetylcysteine and a *N*-(acylamino)maleimide IIa, b or c (full lines) and at 310 nm in solutions of the same concentrations of the corresponding isoimides III a, b or c (dotted lines) in the following buffers: (1) 0.05M sodium acetate, pH 5.1, (2) 0.05M sodium phosphate, pH 6.8.

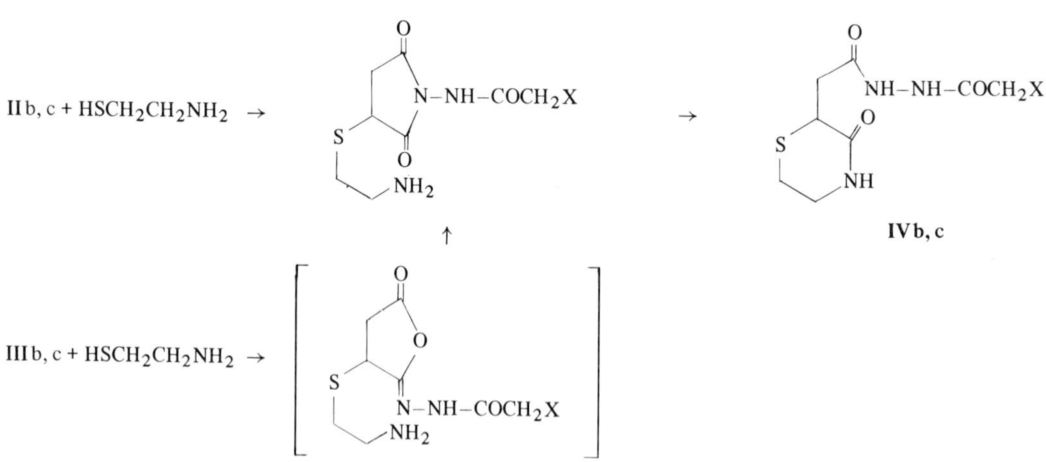

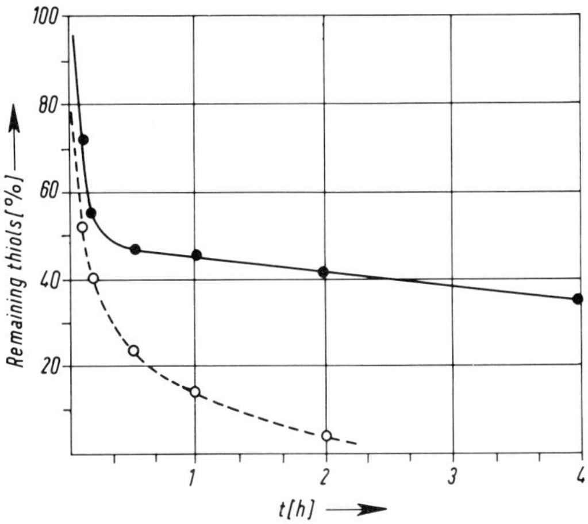

Fig. 3. Reactions of *N*-acetylcysteine with *N*-halogenoacetylaminomaleimides.

The thiol concentration was periodically determined by Ellman's method[14] in solutions 0.021M in *N*-acetylcysteine and 0.02M in *N*-chloroacetylaminomaleimide IIb (full line) or the corresponding bromo derivative IIc (dotted line).

(Fig. 3). It should be noted, however, that in these experiments the latter reaction was intermolecular, but in the case of proteins a higher rate could be expected, since the cross-linking is an intramolecular reaction. Nevertheless, if desired, a two-step cross linking can be favoured by lowering the pH, which hinders the substitution of halogen, while addition of thiol on the maleimide part is still possible.

Reaction of halogenated maleimides with proteins

In order to illustrate by a preliminary experiment the behavior of halogenomaleimides towards proteins, we tested the reaction of *N*-(bromoacetylamino)maleimide IIc with papain. This enzyme, containing in its active site a cysteine and a histidine residue, was treated for 15 min at pH 5.6 with compound IIc. Titration of thiols then showed that one -SH group had reacted per mol of protein. After dialysis against a pH 8 buffer, titration of histidine residues showed that one histidine residue was substituted. Other experiments showed that both cysteine and histidine substitution could also occur at pH 5.6 in 15 min. Thus, in this case the two steps of cross-linking could not be distinguished. This, however, is in agreement with previous results of Husain and Lowe[22] who showed that the histidyl residue vicinal to Cys 25 in papain is particularly reactive towards the -$COCH_2Br$ group at pH 5.6, since the two residues could be very easily crosslinked by 1,3-dibromoacetone.

The longer reagent *N*-(4-chloro-3,5-dinitrobenzoyloxymethyl)maleimide IId was reacted at pH 5.1 with the tetrameric enzyme yeast alcohol dehydrogenase. Four essential SH groups of the enzyme were rapidly substituted with a 90% loss of activity. When, after dialysis, the pH was adjusted to 9, a new absorption band appeared at 430 nm and soon reached values corresponding to the substitution of two -NH_2 groups by the chlorodinitrophenyl part of the reagent. A more thorough investigation of yeast alcohol dehydrogenase made in this laboratory with a similar hetero-bifunctional reagent containing an organomercuric group instead of the maleimide fragment also showed that anchoring of four reagent molecules on -SH groups of the enzyme is followed by cross-linking with only two -NH_2 groups. The proposed explanation was that the four subunits of yeast alcohol dehydrogenase could be arranged as a dimer of dimers[23].

Difficulties can sometimes be encountered in the synthesis of new hetero-bifunctional protein reagents containing a maleimide ring and a second very reactive functional group, since many of the known methods of preparation of *N*-substituted maleimides use rather drastic reaction conditions. One possibility of obviating this difficulty could consist in esterifying *N*-(hydroxymethyl)maleimide with an acid containing the desired function, which can be done under mild conditions. However, the ester linkage in the resulting compound may sometimes prove too labile[3]. On the other hand, equally mild experimental conditions can also be used in the most classical method of synthesis of maleimides: cyclization of the corresponding maleamic acid, but in that case isoimides are obtained instead of the symmetric isomer. As shown in the present paper, differences in stability and reactivity with thiols between the two isomers are small at neutral pH. The rate of addition of thiols to isomaleimides is high enough compared to the rate of hydrolysis or reaction with amines of these compounds. Finally, in both cases reaction with thiols yields the same addition products. It seems, therefore, that if the synthesis of a new maleimide derivative proves difficult, the preparation of the corresponding isoimide can be attempted without major objections concerning its use.

The authors are thankful to *Prof. R. Gay* and *Drs. J.P. Bideau* and *P. Marsau* (Laboratoire de Cristallographie, Université Bordeaux I) for determination of the structure of a isomaleimide by X-ray diffraction, Mrs. *L. Lacombe* (Laboratory of Prof. A. Horeau in Collège de France) for NMR spectra, Mr. *J. Y. Le Gall* for technical assistance and the *Centre National de la Recherche Scientifique* for financial support.

Literature

1 Friedman, M. (1973) The Chemistry and Biochemistry of the Sulfhydryl Group in Amino Acids, Peptides and Proteins, Pergamon Press, New York.

2 Wold, F. (1972) *Methods Enzymol.* **25B**, 629 - 632.

3 Arndt, D.J. & Konigsberg, W. (1971) *J. Biol. Chem.* **246**, 2594 - 2601 and 2602 - 2608.

4 Diopoh, J. & Olomucki, M. (1972) *Biochim. Biophys. Acta* **263**, 220 - 226.

5 Tsou, K.C., Barrnett, R.J. & Seligman, A.M. (1955) *J. Amer. Chem. Soc.* **77**, 4613 - 4616.

6 Barrat, M.D., Davies, A.P. & Evans, M.T.A. (1971) *Eur. J. Biochem.* **24**, 280 - 283.

7 Godin, J. & LeBerre, A. (1968) *Bull. Soc. Chim. Fr.*, 4210 - 4221.

8 Lindemann, H. & Wessel, W. (1925) *Ber. Dtsch. Chem. Ges.* **58B**, 1221 - 1230.

9 Tawney, P.O., Snyder, R.H., Conger, R.P., Leibrand, K.A., Stiteler, C.H. & Williams, A.R. (1961) *J. Org. Chem.* **26**, 15 - 21.

10 Combrisson, S., Lautié, J.-P. & Olomucki, M. (1975) *Bull. Soc. Chim. Fr.* 2769 - 2770.

11 Marrian, D.H. (1949) *J. Chem. Soc.* 1515 - 1516.

12 Sluyterman, L.A. (1967) *Biochim. Biophys. Acta* **139**, 430 - 438.

13 Glazer, A.N. & Smith, E.L. (1961) *J. Biol. Chem.* **236**, 2948 - 2951.

14 Ellman, G.L. (1959) *Arch. Biochem. Biophys.* **82**, 70 - 77.

15 Barnard, E.A. & Stein, W.D. (1959) *J. Mol. Biol.* **1**, 339 - 349.

16 Dickinson, F.M. (1972) *Biochem. J.* **126**, 133 - 138.

17 Buyle, R. (1964) *Helv. Chim. Acta* **47**, 2449 - 2451.

18 Smyth, D.G., Nagamatsu, A. & Fruton, J.S. (1960) *J. Amer. Chem. Soc.* **82**, 4600 - 4604.

19 Gregory, J.D. (1955) *J. Amer. Chem. Soc.* **77**, 3922 - 3923.

20 Clark-Walker, G.D. & Robinson, H.C. (1961) *J. Chem. Soc.*, 2810 - 2812.

21 Wu, C.W., Yarbrough, L.R. & Wu, F.Y.H. (1976) *Biochemistry* **15**, 2863 - 2868.

22 Husain, S.S. & Lowe, G. (1968) *Biochem. J.* **108**, 855 - 859 and 861 - 866.

23 Diopoh, J. & Olomucki, M. (1977) *Eur. J. Biochem.* **75**, 441 - 444.

J.-P. Lautié and M. Olomucki, Laboratoire de Biochimie Cellulaire, Collège de France, 11, Place Marcelin Berthelot, F-75231 Paris Cedex 05.

Hoppe-Seyler's Z. Physiol. Chem.
Bd. 359, S. 219 - 224, Februar 1978

Studies on Bovine Insulin with Methyl Vinyl Ketone

Isolation of Crystalline A1-3-Oxobutyl Derivative of Bovine Insulin

Herbert E. M. MAGOJO, Donald A. SUTTON, Barratt R. D. EASTER and Siegfried E. DREWES

Department of Chemistry, University of Natal

(Received 2 November 1977)

Summary: Reaction of methyl vinyl ketone with bovine insulin affords, amongst others, a crystalline A1-(3-oxobutyl)insulin. This retains full biological activity in the mouse convulsion assay. The presence of zinc ions changes the course of the above reaction, and illustrates the use of zinc as a "blocking function".

Untersuchungen an Rinderinsulin mit Hilfe von Methylvinylketon: Isolierung des kristallisierten A1-3-Oxobutyl-Derivats des Rinderinsulins

Zusammenfassung: Insulin wurde mit Methylvinylketon umgesetzt. Unter anderem wurde ein kristallines A1-3-Oxobutyl-Derivat erhalten, das beim Mauskrampftest volle biologische Aktivität zeigte. In Gegenwart von Zn-Ionen entstehen zum Teil andere Reaktionsprodukte, wodurch die Verwendbarkeit von Zink als „blockierende Funktion" demonstriert wird.

Key words: Bovine insulin, methyl vinyl ketone, A1-3-oxobutyl derivative, zinc addition.

In earlier communications[1-3] we have described the preparation of crystalline derivatives of insulin which preserve the hydrophilic nature of the amino terminus by maintaining the positive charge at this site at physiological pH. Our rationale for this approach was that modification of hydrophilic groups with suitable hydrophilic reagents should result in minimal changes in the conformation of the hormone. Recently, we have come to believe that both the size of the modifying group and the charge it carries play an important role in receptor binding.

With the A1-(2-nitro-4-trimethylammoniophenyl) derivative of insulin[2], the cause of the decreased biological activity could not be assigned with certainty. It could have been due to the shielding of important interactions with the receptor site or to some distortion in conformation. In order to eliminate the effects of distortion resulting from size, the reagent methyl vinyl ketone ($CH_2 = CH\text{-}CO\text{-}CH_3$), was selected to modify insulin. This reagent is hydrophilic in nature, and once attached to an amino group is unlikely to bring about conformational distortion. Furthermore, the group $R\text{-}NH\text{-}CH_2\text{-}CH_2\text{-}CO\text{-}CH_3$ will be positively charged at physiological pH. Since the carbon-nitrogen bond is a strong one, it would also resist enzymatic degradation during studies *in vivo*.

Reaction of methyl vinyl ketone with insulin under mild reaction conditions yielded a mixture of products which was separated by ion-exchange chromatography on DEAE-Sephadex A-25 at pH 7.3.

Materials and Methods

Bovine insulin (Boots Company S.A. (Pty) Ltd, Batch No. NT80) was purified by gel filtration as oultined previously[2]. Methyl vinyl ketone was a product of the British Drug House. Urea was deionized by percolating a 7M solution through a mixed-bed Amberlite MB3 column and using it without delay. All separations on columns were done at 4 °C. Salts and urea were removed by dialysis against distilled water using Union Carbide dialysis tubing. Samples for amino acid analysis were obtained by hydrolysis in constant-boiling 6M HCl in an evacuated, sealed tube for 6h at 120 °C[4].

Cellulose acetate electrophoresis

Runs were carried out on strips (12.5 × 2.5 cm) as described earlier[2]. A 65mM phosphate buffer at pH 6.5 was employed in this case[5].

Sulphitolysis and electrophoretic separation

The conditions used have been described earlier[2].

Crystallization

The "hot box" method[2] was used to grow rhombohedral crystals in a citrate buffer containing zinc ions.

Biological assay

Tests were done in the laboratories of the Boots Company Nottingham, using the standard mouse convulsion assay. Comparison was made with the 4th International Standard for insulin.

Preparation of model compound, N-(3-oxobutyl)-phenylalanine

Phenylalanine (2 g, 0.01 mol) was allowed to react with methyl vinyl ketone (2 m*l*, 0.02 mol) for 1 h at 40 °C in 0.2M $NaHCO_3$ solution (20 m*l*) at pH 8.4. The reaction mixture was neutralized with dilute HCl, the resultant precipitate filtered off and recrystallized from water, m.p. 189 - 190 °C (lit.[6] m.p. 190 - 191 °C).

$C_{13}H_{17}NO_3$ (235.29)	Calcd.	C 66.4	H 7.3
	Found	C 66.2	H 7.3

Mass and NMR spectra confirmed the structure of the compound.

Reaction of methyl vinyl ketone with insulin

Purified insulin (200 mg, 33.4 μmol) was reacted with methyl vinyl ketone (0.03 m*l*, 330 μmol) for 2 h at 40 °C in 0.2M $NaHCO_3$ (40 m*l*) at pH 8.4. Dialysis was against 1% acetic acid first, until precipitation occurred, followed by further dialysis against distilled water. The freeze-dried sample was desalted on Sephadex G-50 using 10% acetic acid, then freeze-dried again (190 mg).

A similar run was carried out at pH 7.5 using 0.2M phosphate buffer.

The "zinc reaction"

Insulin (100 mg, 16.7 μmol) with $ZnCl_2$ (10 mg, 0.07 μmol), dissolved in 0.2M phosphate buffer (20 m*l*), pH 7.5, was reacted with a 10-fold molar excess of methyl vinyl ketone for 2 h at 40 °C. The products were isolated as described above.

Chromatographic separation of products

The reaction products were separated on a column (2.5 × 40 cm) of DEAE-Sephadex A-25 by a modification of Lindsay and Shall's method[7]. The column was equilibrated with the starting buffer containing 0.01M Tris/0.05M NaCl/7M urea adjusted to pH 7.3 with 1M HCl. The sample (200 mg) in this buffer (1 m*l*), was applied to the column. A flow rate of 35 m*l*/h was maintained and 10 m*l* fractions were collected. After 100 m*l* of eluant had been collected, a linear salt gradient was applied. This was obtained by running 0.01M Tris/0.15M NaCl/7M urea, adjusted to pH 7.3 (1 *l*), into a reservoir of the starting buffer (1 *l*). The protein concentration was determined by continuous recording of the absorption at 280 nm. The appropriate fractions at the centre of each peak (Fig. 1) were combined, dialysed, desalted and freeze-dried. In subsequent separations it was found advantageous to use a salt gradient starting at 0.04M and rising to 0.20M in the limiting buffer.

Results and Discussion

The model compound based on phenylalanine was prepared in order to establish the optimum experimental conditions for the specific 3-oxobutylation of free amino groups. Using these conditions (large excess reagent, pH 8.4) for insulin and methyl vinyl ketone, a reaction product substituted at the amino termini as well as at the B-29 lysine residue was obtained (Table 1). This contrasts with the findings of Bosshard and co-workers[8]. These authors found that acrylonitrile, a reagent closely resembling methyl vinyl ketone, reacted with insulin at pH 8.6 to give a derivative substituted at one histidine residue, the two amino termini, and at the lysine position

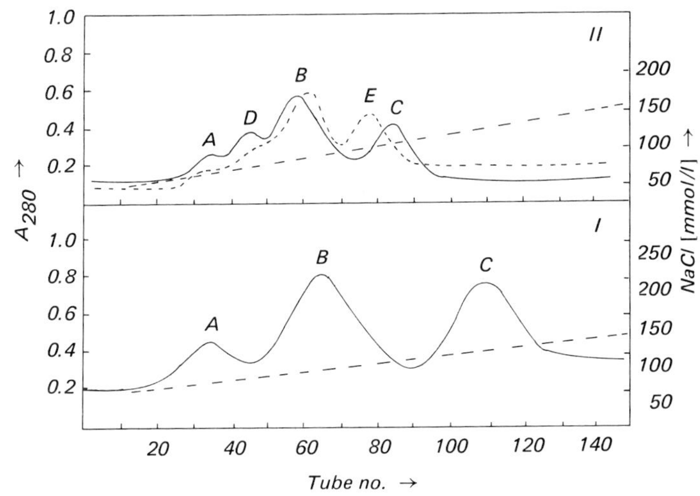

Fig. 1. Separation of insulin reaction mixtures on DEAE-Sephadex A-25.

Details, see text. I: Reaction mixture, pH 8.4; II: —— Reaction mixture, pH 7.5, ------ Reaction mixture, pH 7.5, "zinc reaction".

Table 1. Amino acid analyses of insulin treated with large excess of methyl vinyl ketone at pH 8.4.

Amino acid	Residues in parent insulin		Residues in modified insulin
	Calcd.	Found	
Lys	1	0.98	**0.09**
His	2	2.05	1.98
Arg	1	1.00	1.00
Asp	3	3.03	3.18
Thr	1	1.06	1.12
Ser	3	3.02	2.81
Glu	7	7.56	7.43
Pro	1	1.11	1.21
Gly	4	4.31	**3.18**
Ala	3	3.00	3.00
Cys	6	3.68	4.88
Val*	5	2.52	3.25
Ile*	1	0.23	0.66
Leu	6	5.46	5.88
Tyr	4	3.75	3.92
Phe	3	2.81	**2.17**

* Values low due to incomplete hydrolysis under conditions used.

Separation of insulin treated with a lower excess (10-fold molar) of methyl vinyl ketone at pH 8.4, achieved better control of the reaction. Three peaks, A, B and C (Fig. 1, I) were obtained, where A was unchanged insulin. On cellulose acetate strips A, B and C migrated to the positive pole as discrete bands. B was more negative than A, and C more negative than B. Oxidative sulphitolysis of the peaks followed by electrophoretic separation of the chains (pH 2) showed no difference from insulin. It is suggested that this is due to the protonation of the secondary amine present in the 3-oxobutyl derivatives. Amino acid analyses (Table 2) indicate that peak B is a Gly^{A1}-modified derivative. Under these reaction conditions (pH 8.4) it is contaminated by the Lys^{B29} derivative. Peak C is a Gly^{A1},Phe^{B1}-disubstituted derivative, but not entirely pure.

When the reaction with insulin took place at pH 7.5, an extra peak, D, eluting between A and B was found (Fig. 1, II). In all probability this is a Phe^{B1}-modified derivative. It behaves as the Gly^{A1}-modified derivative (peak B) does on cellulose acetate strip electrophoresis at pH 6.5. The amino acid analyses in Table 3 show that B and C are obtained pure under these conditions.

Reaction of insulin with methyl vinyl ketone in the presence of zinc (the "zinc reaction") led to a different, but not inexplicable, result. On separation of the mixture three peaks resulted (Fig. 1, II, broken line). These were identified (in order of elution) as insulin, A1-(3-oxobutyl)insulin (peak B above) and a A1,A1-bis(3-oxobutyl)insulin derivative (E). On the basis of its amino acid analysis (Table 4), which is the same as for peak B, we conclude that the Gly^{A1} bears *two* 3-oxobutyl

Table 2. Amino acid analyses of peaks B and C obtained from the reaction at pH 8.4.

Amino acid	Residues in parent insulin		Residues in B	Residues in C
	Calcd.	Found		
Lys	1	0.92	**0.53**	**0.69**
His	2	1.78	1.97	1.85
Arg	1	1.00	1.00	1.00
Asp	3	3.00	3.04	2.91
Thr	1	1.06	1.02	0.94
Ser	3	3.02	2.94	2.78
Glu	7	7.10	7.22	6.35
Pro	1	1.01	1.12	1.01
Gly	4	4.26	**3.31**	**3.17**
Ala	3	3.00	3.00	3.00
Cys	6	4.61	4.61	3.67
Val*	5	4.22	3.40	1.86
Ile*	1	0.49	0.42	0.25
Leu	6	5.97	5.83	5.02
Tyr	4	3.60	3.55	3.64
Phe	3	2.87	**2.65**	**1.71**

* Values low due to incomplete hydrolysis under conditions used.

Table 3. Amino acid analyses of peaks B and C obtained from the reaction at pH 7.5.

Amino acid	Residues in parent insulin		Peak B	Peak C
	Calcd.	Found		
Lys	1	1.00	0.90	0.94
His	2	1.81	1.80	1.80
Arg	1	1.00	1.00	1.00
Asp	3	3.04	2.96	3.03
Thr	1	1.04	1.00	1.13
Ser	3	2.82	2.72	3.01
Glu	7	6.95	7.13	6.80
Pro	1	0.99	1.11	1.23
Gly	4	4.13	**3.28**	**3.33**
Ala	3	3.00	3.00	3.00
Cys	6	4.48	4.93	4.54
Val*	5	3.88	4.15	2.92
Ile*	1	0.51	0.61	0.44
Leu	6	5.94	6.22	5.85
Tyr	4	3.63	3.53	3.66
Phe	3	2.74	2.65	**2.08**

* Values low due to incomplete hydrolysis under conditions used.

Table 4. Amino acid analyses of peaks obtained from the "zinc reaction".

Amino acid	Residues in parent insulin		Peak B	Peak E
	Calcd.	Found		
Lys	1	1.00	0.77	0.78
His	2	1.96	2.03	1.99
Arg	1	1.00	1.00	1.00
Asp	3	3.12	3.19	3.09
Thr	1	1.08	0.99	1.11
Ser	3	2.95	2.89	2.89
Glu	7	7.51	7.63	7.48
Pro	1	1.15	1.20	1.21
Gly	4	4.13	**3.39**	**3.31**
Ala	3	3.00	3.00	3.00
Cys	6	5.21	4.80	4.74
Val*	5	3.83	4.25	4.25
Ile*	1	0.42	0.59	0.63
Leu	6	6.26	6.21	6.14
Tyr	4	4.05	3.98	3.86
Phe	3	2.95	2.82	2.73

* Values low due to incomplete hydrolysis under conditions used.

groups. This is a tentative, but plausible explanation, which requires verification. It receives support from the work of Friedman and Finley[9], who isolated N^{ϵ},N^{ϵ}-bis(ethylsulphonylethyl)-lysine from the reaction between ethyl vinyl sulphone and lysine. All the above results are summarized in Table 5.

The results show that the reaction at the Gly^{A1} terminus is favoured at pH 7.5, whereas that of Lys^{B29} is suppressed. This is anticipated, since the ϵ-NH_2 group will be protonated at this pH[10]. The reactivity of the Phe^{B1} residue is far from clear. Judging from the amount of residual insulin at pH 8.4, it appears less reactive than Gly^{A1}. At pH 7.5 it appears equally reactive, however.

The results of the "zinc reaction" deserve further comment. It is known that $Zn^{2\oplus}$ binds to the imidazole ring and hence to the histidines[11]. This has a profound effect on the degree of aggregation of insulin monomers to form hexamers[12,13]. It has also been suggested that zinc may bind to the α-amino groups[14,15]. Our results show that the phenylalanine is unavailable

Table 5. Derivatives from reaction of insulin with methyl vinyl ketone.
% yields are given in brackets.

pH of reaction	Peak A	Modified amino-acid residues			
		Peak D	Peak B	Peak E	Peak C
8.4	Insulin (13)		Gly^{A1} (30)		Gly^{A1}, Phe^{B1} (23)
7.5	Insulin (5)	Phe^{B1} (10)	Gly^{A1} (39)		Gly^{A1}, Phe^{B1} (21)
7.5 (Zn react.)	Insulin (2)		Gly^{A1} (42)	Gly^{A1} (twice) (29)	

for reaction when zinc is present. The non-availability of the phenylalanine may be due to direct chemical blocking. A similar effect was observed by Covelli and Wolff[16] for histidine. These authors demonstrated that the presence of zinc lowered the reactivity of this residue towards iodine. In our case, however, the failure of the phenylalanine to react under the zinc conditions probably results from its inaccessibility during the process of aggregation of insulin from monomer to dimer to hexamer[12]. This appears more feasible than direct chemical blocking by reaction with the zinc.

Crystallization of A1-(3-oxobutyl)insulin (Fraction B) afforded small rhombohedral crystals (Fig. 2), very similar to those of native insulin. The disubstituted derivative (Fraction C) failed to crystallize, however. Lindsay and Shall[17] have previously shown that modification of the Gly^{A1} amino group by small acylating reagents does not prevent crystal formation. By contrast,

Fig. 2. Crystal of A1-(3-oxobutyl)insulin.

modification of the Phe^{B1} terminus does interfere with crystal formation in most instances. Lack of material has so far hampered crystallization studies on A1,A1-bis(3-oxobutyl)insulin.

Tests for biological activity (mouse convulsion) on the A1-3-oxobutyl derivative showed full retention of activity. This could be anticipated because of the small size of the modifying group. The activity is higher than that for the A1-aceto-acetyl (3-oxobutyryl) derivative (89%)[17], although the two groups are of comparable size. The reason for this may be that at physiological pH, the 3-oxobutyl derivative retains a positive charge, whereas the acetoacetyl does not. This, and our earlier contentions on the importance of charged substituents[2], is further supported by recent findings of Friesen[18]: it has been found that A1 amino-substituted insulins with positively charged groups at neutral pH, such as the guanidino group, have relatively high biological activities in studies in vitro, whereas insulins with neutral side chains had relatively lower biological activities.

The authors express their thanks to the *South African Council for Scientific and Industrial Research*, and the *University of Natal*, for financial assistance.

Literature

1 Sutton, D. A., Drewes, S. E. & Welz, U. (1972) *Biochem. J.* **130**, 589 - 595.

2 Sutton, D. A., Easter, B. R. D. & Drewes, S. E. (1976) *Hoppe-Seyler's Z. Physiol. Chem.* **357**, 971 - 976.

3 Drewes, S. E. & Sutton, D. A. (1976) *Hoppe-Seyler's Z. Physiol. Chem.* **357**, 1261 - 1266.

4 Levy, D. (1973) *Biochim. Biophys. Acta* **328**, 107 - 113.

5 Lindsay, D.G. & Shall, S. (1970) *Eur. J. Biochem.* **15**, 547 - 554.
6 Friedman, M. & Romersberger, J.A. (1968) *J. Org. Chem.* **33**, 154 - 157.
7 Lindsay, D.G. & Shall, S. (1969) *Biochem. J.* **115**, 587 - 595.
8 Bosshard, H.R., Jorgensen, K.H. & Humbel, R.E. (1969) *Eur. J. Biochem.* **9**, 353 - 362.
9 Friedman, M. & Finley, J.W. (1975) *Int. J. Pept. Protein Res.* **7**, 481 - 486.
10 Tanford, C. & Epstein, J. (1954) *J. Am. Chem. Soc.* **76**, 2170 - 2176.
11 Edsall, J.T., Felsenfeld, G., Goodman, D.S. & Gurd, F.R.N. (1954) *J. Am. Chem. Soc.* **76**, 3054 - 3061.
12 Blundell, T.L., Dodson, G.G., Hodgkin, D.C. & Mercola, D. (1972) *Adv. Protein Chem.* **26**, 279 - 402.
13 Wood, S.P., Blundell, T.L., Wollmer, A., Lazarus, N.R. & Neville, R.W.J. (1975) *Eur. J. Biochem.* **55**, 531 - 542.
14 Cunningham, L.W., Fischer, R.L. & Vestling, C.S. (1955) *J. Am. Chem. Soc.* **77**, 5703 - 5707.
15 Goldman, J. & Carpenter, F.H. (1974) *Biochemistry* **13**, 4566 - 4574.
16 Covelli, R.W. & Wolff, J. (1967) *J. Biol. Chem.* **242**, 881 - 886.
17 Lindsay, D.G. & Shall, S. (1971) *Biochem. J.* **121**, 737 - 745.
18 Friesen, H. (1976) Thesis, Techn. Univ. Aachen.

Mr. H.E.M. Magojo, Prof. D.A. Sutton, Mr. B.R.D. Easter and Prof. S.E. Drewes, Department of Chemistry, University of Natal,
P.O. Box 375, Pietermaritzburg, South Africa.

Hoppe-Seyler's Z. Physiol. Chem.
Bd. 359, S. 225 - 229, Februar 1978

The Stability of A1,B29-Adipoylinsulin to Chemical and Enzymatic Degradation

Stephen M. L. ROBINSON and Klaus LÜBKE

Aus den Forschungslaboratorien der Schering AG, Berlin/Bergkamen

(Received 28 October 1977)

Summary: The degradation of A1,B29-adipoylinsulin by partial hydrolysis, by disulphide-bond exchange, by chymotrypsin and by a cytosolic enzyme fraction of rat liver was measured and compared with that of A1,B29-diacetylinsulin and insulin. The cross-linked insulin is, in general, more resistant than insulin itself or the diacetylated derivative. However, in the case of enzymatic degradation, A1,B29-diacetylinsulin resembles more the cross-linked derivative than insulin. This provides evidence that the properties of A1,B29-adipoylinsulin could not in all circumstances be attributed to its 'proinsulin-like' structure.

Die Stabilität von A1,B29-Adipoylinsulin gegen chemischen und enzymatischen Abbau

Zusammenfassung: Der Abbau von A1,B29-Adipoyl-insulin durch Partialhydrolyse, durch Disulfid-Austausch, durch Chymotrypsin und durch eine cytoplasmatische Enzymfraktion aus Rattenleber wurde bestimmt und mit dem von A1,B29-Diacetylinsulin und Insulin verglichen. Im allgemeinen ist das vernetzte Insulin stabiler als Insulin oder sein Diacetyl-Derivat. In Hinblick auf einen enzymatischen Abbau ist A1,B29-Diacetylinsulin dem vernetzten Derivat wesentlich ähnlicher als Insulin. Daraus folgt, daß die Eigenschaften von A1,B29-Adipoylinsulin nicht nur durch seine 'Proinsulin-artige' Struktur zu erklären sind.

Key words: Insulin, derivatives, degradation, chemical, enzymatic.

The A1,B29 intramolecularly cross-linked insulins exhibit considerably decreased biological activity in vitro and an apparent retention of activity in vivo[1]. This discrepancy may be attributed, in part, to their comparatively slower rate of degradation than insulin in vivo.

The aim of this study was to investigate the resistance of A1,B29-adipoylinsulin to various forms of chemical and enzymatic degradation. The results were compared with those of insulin and A1,B29-diacetylinsulin. A1,B29-diacetylinsulin, which can be considered chemically as a seco-crosslinked insulin derivative, shows similarities in conformation to A1,B29-adipoylinsulin, measured by circular dichroism[2,3]. In the blood-glucose depression assay in the rat, A1,B29-diacetyl-

Enzyme: Chymotrypsin (EC 3.4.21.1).

insulin exhibits full biological potency[4] whereas A1,B29-adipoylinsulin possesses about 40% - 50% of the activity of insulin. In the in vitro fat-cell assay of Gliemann and Gammeltoft[5], A1,B29-diacetylinsulin has approximately 26% - 32%, and A1,B29-adipoylinsulin, less than 6% activity.

The following conditions of degradation were chosen: partial acid hydrolysis, disulphide bond exchange, digestion with chymotrypsin and with an enzyme fraction from the cytosol of rat liver[6]. On the assumption that degradation would result in altered binding to insulin antibodies, the kinetics of degradation were measured using an insulin radioimmunoassay.

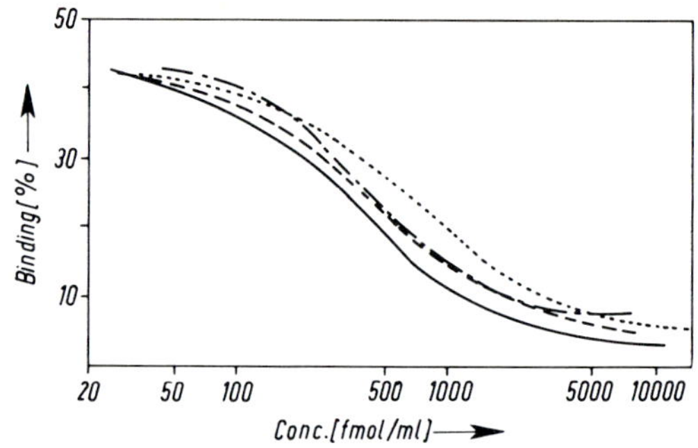

Fig. 1. Representative antibody binding curves of insulin (——), A1,B29-diacetylinsulin (– – –), A1,B29-adipoylinsulin (– · –) and proinsulin (·········).

At the relative insulin binding of 50%, the cross reactivities of the insulin compounds were 75%, 78% and 48%, respectively. Standard deviations have been omitted for clarity. The maximum standard deviation was 3%.

Materials and Methods

Chymotrypsin was purchased from Worthington (Freehold, N.J.). All other chemicals were of analytical grade.

Bovine insulin (Fa. Ch. Brunnengräber, Lübeck) was purified to homogeneity by the method of Schmidt and Arens[7]. With this procedure proinsulin could be isolated from the insulin and purified to homogeneity as well.

A1,B29-adipoylinsulin and A1,B29-diacetylinsulin were prepared by treating insulin, dissolved in aqueous dioxane/sodium hydrogencarbonate mixtures, with the *N*-hydroxysuccinimide esters of adipic and acetic acid, respectively (cf. ref.[4]). The derivatives were purified by countercurrent distribution (n-butanol/acetic acid/methanol/water; 20:1:5:20).

For the radioimmunoassay, the double antibody method of Hales and Randle[8] was used (Insulin Immunoassay Kit, Radiochemical Centre, Amersham).

Reproducible standard binding curves were established for insulin, proinsulin, A1,B29-diacetylinsulin and A1,B29-adipoylinsulin. Proinsulin and the insulin derivatives showed a high cross-reactivity with insulin in this assay (Fig. 1). The concentration of insulin, proinsulin and the insulin derivatives and of their immunoreactive products of degradation were estimated from the logit-log-model of Rodbard et al.[9] for binding curves.

Partial acid hydrolysis of insulin, A1,B29-diacetylinsulin and A1,B29-adipoylinsulin

166 nmol of insulin or of its derivative were dissolved in 1 m*l* of 11.6N hydrochloric acid and incubated at 37 °C under nitrogen. At given times, 50-μl samples were pipetted into 950 μl of a cold solution of 1M Tris buffer, pH 7.5. A portion of this mixture was diluted and taken for immunoassay. Control experiments showed that degradation was negligible in the buffer used for dilution and in the cold hydrochloric acid solution.

Reduction of insulin, A1,B29-diacetylinsulin and A1,B29-adipoylinsulin with 2-mercaptoethanol

166 nmol of insulin or of its derivative were dissolved in 800 μl of 0.1M Tris/HCl buffer, pH 8.3. 1.0 μmol of 2-mercaptoethanol in 200 μl of the same buffer was added. Reduction was allowed to proceed at room temperature under nitrogen. At given times, 100-μl samples of the reduction mixture were reacted with an excess of 2,2'-dithiopyridine (50-fold on a molar basis), dissolved in 900 μl of a mixture of Tris buffer and methanol (7:3, v/v), in order to stop the reaction[10].

The total thiol content of the reduction mixture was constant for periods of up to 18 h as determined spectroscopically[10]. None of the reagents used was found to affect the standard antibody binding curve of insulin after dilution.

Degradation of insulin, proinsulin, A1,B29-diacetylinsulin and A1,B29-adipoylinsulin with chymotrypsin

166 nmol of insulin, of proinsulin or of the insulin derivative was dissolved in 900 μl of 1% ammonium hydrogencarbonate buffer, pH 8.6. The digestion was started by the addition of 0.38 nmol of α-chymotrypsin dissolved in 100 μl of 1mM hydrochloric acid and was allowed to proceed at 30 °C. At given times, the digestion was terminated by adding 10 m*l* of ethanol containing 100 μl of concentrated hydrochloric acid and heating the sealed reaction vessels in a water bath at 65 °C for 25 min. Controls were included to examine the effects of this

heat treatment on insulin, proinsulin and the insulin derivatives. Decreases in their immuno-reactivities were not observed.

Degradation of insulin, A1,B29-diacetylinsulin and A1,B29-adipoylinsulin with the cytosolic enzyme fraction of rat liver

The enzyme fraction was prepared from rat liver cytosol and diluted according to the method of Burghen et al.[6]. The protein content of the diluted enzyme fraction was determined according to the method of Lowry et al.[11] by interpolation from a standard curve of bovine serum albumin.

Insulin or its derivative was dissolved at a concentration of 300 ng/m*l* in 0.13M sodium borate buffer, pH 7.4, containing 0.5% bovine serum albumin. 1.0 m*l* of the insulin solution was diluted with 1.0 m*l* of the borate buffer and allowed to reach 37 °C in a shaking water-bath. The digestion was started by the addition of 0.25 m*l* of the enzyme preparation and was allowed to proceed at 37 °C. At given times, the digestion was stopped by adding 0.25 m*l* of a borate buffer solution, containing 0.01M *N*-ethylmaleimide.

Results and Discussion

Partial hydrolysis

The results of the partial hydrolysis of insulin, A1,B29-diacetylinsulin and A1,B29-adipoylinsulin are shown in Fig. 2. Large differences in the rates of hydrolysis of insulin and its derivatives are not observed.

It has been shown that disulphide bond exchange under similar conditions would be too slow to account for the losses in immunoreactivity[12]. Therefore, the principle degradative pathway involves the hydrolytic fission of a peptide or an amide bond which was influenced neither by substitution nor by cross-linking of the A1 and B29 amino groups.

Disulphide bond exchange

Preliminary experiments have shown that the immunoreactivities of insulin and its derivatives were unchanged after 90 min when 2 mol mercaptoethanol were present per mol insulin. When the concentration of mercaptoethanol was raised to 6 mol, insulin and its derivatives were inactivated (Fig. 3). The decrease in the immunoreactivity of insulin and of A1,B29-diacetylinsulin was greater than that of A1,B29-adipoylinsulin.

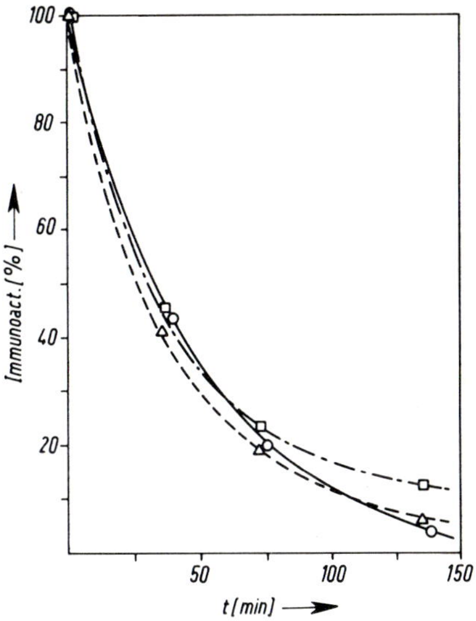

Fig. 2. Partial hydrolysis of insulin (○——○), A1,B29-diacetylinsulin (△– – –△) and A1,B29-adipoylinsulin (□–.–.–□) in 11.6N hydrochloric acid at 37 °C.

Standard deviations have been omitted for clarity. The maximum standard deviation was 7%.

Within three hours equilibrium was reached. The equilibria were not changed in presence of 1mM EDTA or of 1 - 10μM cupric ions. Thus, the influence of heavy metal ions on disulphide bond exchange and on oxidation could be excluded. At equilibrium, the reaction mixtures will be populated with various disulphide bond and thiol-

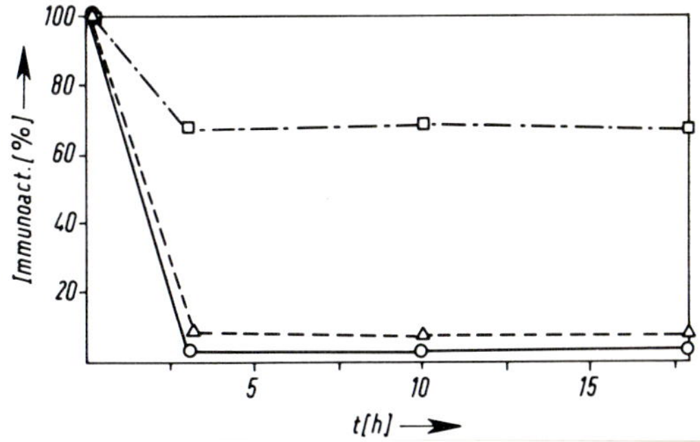

Fig. 3. The effect of disulphide bond exchange on insulin (○——○), A1,B29-diacetylinsulin (△– – –△), and A1,B29-adipoylinsulin (□–.–.–□).

Standard deviations have been omitted for clarity. The maximum standard deviation was 12%.

containing isomers. Insulin immunoreactivity should be due to those molecules in which the native bonds have remained intact and those which have reformed the correct disulphide bond structure. Therefore, in the case of insulin and the diacetylinsulin, the mixture may contain 2 - 5% of the original molecules. In contrast, more than 60% immunoreactivity was found with the cross-linked insulin derivative.

It is noteworthy that studies concerning the reoxidation of reduced insulin or insulin derivatives support these results. In these recombination experiments, approximately the same yields are obtained[13-16].

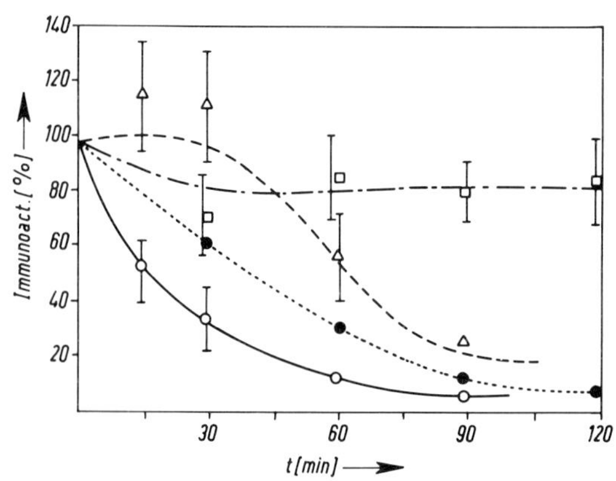

Fig. 4. Degradation of insulin (○——○), proinsulin (●........●), A1,B29-diacetylinsulin (△– – –△), and A1,B29-adipoylinsulin (□–·–·–□) by chymotrypsin at 30 °C.

Only standard deviations greater than 10% are shown.

Chymotryptic degradation

The degradation of insulin, proinsulin, A1,B29-diacetylinsulin and A1,B29-adipoylinsulin by chymotrypsin is shown in Fig. 4. The relative order or resistance to degradation is A1,B29-adipoylinsulin ≫ A1,B29-diacetylinsulin > proinsulin > insulin.

The rate of inactivation of A1,B29-adipoylinsulin by chymotrypsin is considerably decreased in relation to that of insulin and proinsulin. In order to establish unequivocally that degradation was occurring, A1,B29-adipoylinsulin was digested for a further 8 and 16 h, at which times the immunoreactivities were 45 ± 8% and 6 ± 4%, respectively. A1,B29-diacetylinsulin was completely inactivated at these times. Clearly, modification of the A1 and B29 amino groups in insulin increases its resistance to hydrolysis by chymotrypsin. The effect of cross-linking these groups is more pronounced than that of their substitution.

Cytosolic enzyme fraction of rat liver

The cytosolic enzyme fraction of rat liver has been shown to contain a protease highly specific for insulin and insulin derivatives[6]. The results of this experiment are shown in Fig. 5. The relative order of resistance to degradation is A1,B29-adipoylinsulin > A1,B29-diacetylinsulin > insulin.

Modification of the A1 and B29 amino groups in insulin increases the resistance to enzymatic degradation. This is probably more a consequence of modification of the A1 than of the B29 amino group, since studies with derivatives of proinsulin, having fragments of the connecting peptide of various lengths attached to Gly[A1], indicate that

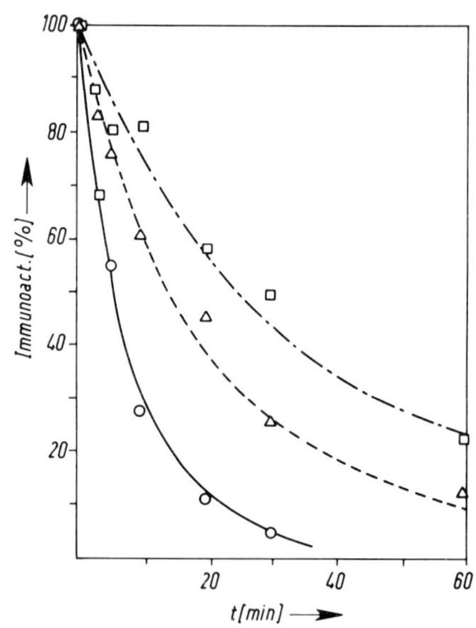

Fig. 5. Degradation of insulin (○——○), A1,B29-diacetylinsulin (△– – –△), and A1,B29-adipoylinsulin (□–·–·–□) by the cytosolic enzyme fraction of rat liver at 37 °C.

Standard deviations have been omitted for clarity. The maximum standard deviation was 10%.

the free A chain *N*-terminus in insulin is necessary for optimal substrate activity[17]. Cross-linking of the A1 and B29 amino groups with a substituent of comparable steric properties to the two acetyl groups renders insulin even more resistant to degradation.

Conclusion

A1,B29-adipoylinsulin is, in general, more resistant than A1,B29-diacetylinsulin and insulin to the various forms of chemical and enzymatic degradation. The effects of the adipoyl residue on the stability of the molecule, however, could not be attributed only to cross-linking, but also to the modification of the A1 and B29 amino groups because A1,B29-diacetylinsulin is also more stable than insulin, in particular, towards chymotrypsin and the enzymes of the cytosolic fraction of rat liver. This provides evidence that the properties of A1,B29-adipoylinsulin could not in all circumstances be attributed to its 'proinsulin-like' structure.

Literature

1 Freychet, P., Brandenburg, D. & Wollmer, A. (1974) *Diabetologia* **10**, 1 - 5.

2 Brandenburg, D., Gattner, H.-G. & Wollmer, A. (1972) *Hoppe-Seyler's Z. Physiol. Chem.* **353**, 599 - 617.

3 Brandenburg, D. (1972) *Hoppe-Seyler's Z. Physiol. Chem.* **353**, 869 - 873.

4 Lindsay, D. G. & Shall, S. (1971) *Biochem. J.* **121**, 737 - 745.

5 Gliemann, J. & Gammeltoft, S. (1974) *Diabetologia* **10**, 105 - 113.

6 Burghen, G. A., Kitabchi, A. E. & Brush, J. S. (1972) *Endocrinology* **91**, 633 - 642.

7 Schmidt, D. D. & Arens, A. (1968) *Hoppe-Seyler's Z. Physiol. Chem.* **349**, 1157 - 1168.

8 Hales, C. N. & Randle, P. J. (1963) *Biochem. J.* **88**, 137 - 146.

9 Rodbard, D., Bridson, W. & Rayford, P. L. (1969) *J. Lab. Clin. Med.* **74**, 770 - 781.

10 Grassetti, D. R. & Murray, J. F. (1967) *Arch. Biochem. Biophys.* **119**, 41 - 49.

11 Lowry, O. H., Rosebrough, N. J., Farr, A. L. & Randall, R. J. (1951) *J. Biol. Chem.* **193**, 265 - 275.

12 Ryle, A. P. & Sanger, F. (1955) *Biochem. J.* **60**, 535 - 540.

13 Robinson, S. M. L., Beetz, I., Loge, O., Lindsay, D. G. & Lübke, K. (1973) *Tetrahedron Lett.* **12**, 985 - 988.

14 Brandenburg, D. & Wollmer, A. (1973) *Hoppe-Seyler's Z. Physiol. Chem.* **354**, 613 - 627.

15 Du, Y. C., Zhang, Y. S., Lu, Z. X. & Tsou, C. L. (1961) *Sci. Sin.* **10**, 84 - 86.

16 Brandenburg, D., Gattner, H.-G., Herbertz, L., Kreil, G. & Wollmer, A. (1971) in Proc. Congr. Int. Diabetes Fed., 7th Buenos Aires, 1970, Excerpts medica Found. Int. Congr. Ser. No. 231 (Margolies, M. ed.) pp. 363 - 376, Amsterdam.

17 Baskin, F. K. & Kitabchi, A. E. (1973) *Eur. J. Biochem.* **37**, 489 - 496.

Dr. K. Lübke and Dr. S.M.L. Robinson, Dept. Allg. Biochemie, Schering AG, Müllerstraße 170 - 178, D-1000 Berlin 65.

Hoppe-Seyler's Z. Physiol. Chem.
Bd. 359, S. 231 - 238, Februar 1978

Preparation and Physical-Chemical Characterization of Poly-*N*-vinylpyrrolidone-Insulin

Berndt U. v. SPECHT, Helmut J. KOLB, Rolf RENNER and K. Dieter HEPP

Institut für Diabetesforschung, München and
Institut für Chirurgische Forschung der Universität München

(Received 13 August 1977)

Summary: In an attempt to modify the structure of insulin, the hormone was coupled by covalent linkage to poly-*N*-vinylpyrrolidone (PVP). PVP-insulin is soluble in aqueous solution and was purified by column chromatography. Special care was taken to check for a possible leakage of insulin from the synthetic polymer. The amount of bound insulin was determined by quantitative amino acid analysis after acid hydrolysis. After protection of the *N*-terminal glycine residue of the A-chain by citraconylation, a biologically active PVP-insulin was obtained.

The molecular weight determined by analytical ultracentrifugation and column chromatography was 50 000 - 60 000. The biological activity of the coupled hormone was between 0.5 and 7% when compared with native insulin in the system in vitro. The immunological activity was about 20 - 50%.

Darstellung und physikalisch-chemische Charakterisierung von Poly-N-vinylpyrrolidon-Insulin

Zusammenfassung: Im Rahmen von Struktur-Wirkungsstudien wurde Insulin kovalent an Poly-*N*-vinylpyrrolidon (PVP) gekoppelt. PVP-Insulin ist wasserlöslich und wurde säulenchromatographisch gereinigt. Besonders eingehend wurde die Stabilität des gekoppelten Insulins untersucht. Die Menge von polymergebundenem Insulin wurde nach Säurehydrolyse durch quantitative Aminosäureanalyse bestimmt. Die Verwendung von Citraconsäure als Schutzgruppe für das *N*-terminale Glycin der A-Kette ergab ein biologisch aktives PVP-Insulin.

Das durch analytische Ultrazentrifugation und Säulenchromatographie gemessene Molekulargewicht betrug 50 000 - 60 000. Im Vergleich mit nativem Insulin lag die in vitro ermittelte biologische Aktivität des gekoppelten Hormons im Bereich zwischen 0.5 und 7%. Die immunologische Aktivität belief sich auf etwa 20 - 50%.

Key words: Insulin modification, poly-*N*-vinylpyrrolidone-coupled insulin, biological insulin activity, immunological insulin activity, isolated fat cell.

During the past years attempts have been made to couple insulin to a matrix in order to gain more information on the mechanism of action in vivo and in vitro. Experiments with Sepharose-coupled

---

Abbreviation: PVP = Poly-*N*-vinylpyrrolidone.

insulin were interpreted as experimental proof that the hormone does not enter the intact fat cell for its action[1,2]. However, these studies have been seriously questioned[3], and it has been reported that the preparation released free hormone[4-7], which could have been entirely responsible for the biological effect[7,8]. In face of the difficulty of linking insulin to an insoluble matrix, attempts were made to attach insulin to a soluble macromolecule.

Poly-*N*-vinylpyrrolidone, a water-soluble polymer, has been used previously as a matrix for proteolytic enzymes[9,10]. The modified enzyme could be extensively purified by column chromatography. Kallikrein covalently attached to polyvinylpyrrolidone not only retained esterase activity in vitro, but was also able to reduce the blood pressure in dogs[10]. β-Hexosaminidase from human placenta was also coupled to polyvinylpyrrolidone. The modified enzyme showed an increased stability towards proteolysis and prolonged clearance from rabbit blood[11].

Therefore, polyvinylpyrrolidone was considered to be a suitable matrix and in this communication, the coupling of insulin to the polymer is reported. Particular care was taken to investigate a possible leakage of "free" insulin from the matrix. The first products were biologically inactive. After protection of the *N*-terminal residues of the hormone with citraconic acid during the coupling reaction, a polyvinylpyrrolidone-coupled insulin was obtained which was immunologically and biologically active. Preliminary results have been reported elsewhere[12].

Material and Methods

Poly-*N*-vinylpyrrolidone K 15, mean molecular weight 10 000, was obtained from Fluka, Switzerland. Citraconic acid, dicyclohexylcarbodiimide and *N*-hydroxysuccinimide were from Merck, Darmstadt. A protein marker kit, for the determination of the molecular weight was purchased from Serva, Heidelberg. Bovine insulin was a gift of Farbwerke Hoechst.

Coupling of insulin to polyvinylpyrrolidone

Polyvinylpyrrolidone K 15 was hydrolyzed by potassium hydroxide and the hydroxysuccinimide ester prepared with dicyclohexylcarbodiimide and *N*-hydroxysuccinimide as described previously[9,10]. The "activated" ester was dried and stored in the desiccator. 25 mg insulin was dissolved in 10% sodium hydrogencarbonate and 200 mg "activated" polyvinylpyrrolidone ester added. The solution was stirred overnight at 4 °C. The formed precipitate was removed by centrifugation and the clear supernatant dialyzed against 20 vol. of dist. water at 4 °C. The dialyzed solution was concentrated by evaporation and further purified by column chromatography.

Modified synthesis of PVP-insulin

In a slightly modified synthesis the *N*-terminal amino acid residues of insulin were protected by citraconic acid as described by Zahn et al.[13]. 150 mg insulin was dissolved in a mixture of 23 m*l* dimethylsulfoxide and 40 m*l* 0.5M phosphate buffer, pH 7.0. 10 μ*l* re-distilled citraconic acid was dissolved in 1 m*l* dry dioxane and the solution was added to the insulin solution in portions of 50 μ*l*. The pH was kept between 7.0 and 7.1 by adding 0.1N NaOH. After 30 min the pH was adjusted to 9.5 with NaOH and a 10-fold excess of dry (w/w) "activated" polyvinylpyrrolidone ester added and stirred overnight. After dialysis against distilled water at 4 °C, the sample was concentrated by evaporation and further purified by column chromatography.

Purification of PVP-insulin on Biogel P-30

In order to remove any native "free" insulin, the reaction mixture was passed over a Biogel P-30 column (5 × 50 cm). The elution solvent was 3M acetic acid. The fractions were recorded at 280 nm. The front part of the first peak was collected, concentrated by evaporation and rechromatographed on another Biogel P-30 column (2 × 100 cm) equilibrated with 3M acetic acid.

Molecular weight of PVP-insulin

The molecular weight of PVP-insulin was estimated by chromatography on a Sephadex G-200 column (2 × 100 cm) equilibrated with 50mM phosphate buffer, pH 7.0. The column had been calibrated with the usual globular protein markers. The elution volume was determined and the molecular weight calculated as described[14]. In addition, the molecular weight was obtained from ultracentrifugation data using a Beckman-Spinco Model E instrument, equipped with schlieren and absorption optics. The apparent sedimentation[15] and diffusion coefficients[16] were determined as outlined and corrected to standard conditions[17]. The mean molecular weight was calculated according to the equation given by Svedberg and Peterson[18]

$$M_r = \frac{S \times R \times T}{D \times (1 - \bar{V}\rho)},$$

where S and D are the values of the sedimentation and diffusion coefficients, R is the gas constant and ρ the density of the solution. The partial specific volume $\bar{V}$

was calculated from the two components: PVP, $\bar{V}$ = 0.788 m*l*/g[10]; and insulin, $\bar{V}$ = 0.720 m*l*/g[19].

Chemical analysis

The carboxyl groups present on the polymer after hydrolysis of the lactams and methylation of the secondary amino groups formed were titrated with 0.05M sodium methylate in dimethylformamide[20]. Care was taken to prevent uptake of CO_2 from the atmosphere. Thymol blue was used as indicator. Because of the slight yellow colour of the polyvinylpyrrolidone solution, it was titrated to a green endpoint. An average of about 8 - 10 carboxyl groups per molecule of polyvinylpyrrolidone (M_r 10 000) was obtained.

The insulin content of PVP-insulin was estimated by quantitative amino acid analysis after hydrolysis in 6N HCl for 24 h at 110 °C according to Spackman et al.[21]. Before hydrolysis the sample was evaporated to remove excess oxygen and sealed under vacuum. For stabilization of tyrosine a few crystals of phenol were added. After hydrolysis the hydrochloric acid was drawn off, the amino acids were dissolved in buffer and the sample put on a Beckman Multichrom amino acid analyzer. The polyvinylpyrrolidone concentration in solution was quantitatively estimated by iodination as described[22].

Bioassay with isolated fat cells

The assays for biological insulin activity was performed with isolated fat cells in triplicate. Isolated fat cells from epididymal fat pads were prepared by collagenase treatment[23]. The stimulation of lipogenesis from [2-^{3}H]glucose was measured essentially as outlined[24]. The biological activity of PVP-insulin was estimated from dose-response curves. The concentration of coupled hormone which yielded 50% of the maximal stimulation of lipogenesis was determined and compared with native insulin standard.

Radioimmunoassay

Radioimmunological insulin activity was determined according to Herbert et al.[25] where the antibody-bound and free insulin are separated by dextran-coated charcoal. The assay was carried out in triplicate in 7mM barbital/sodium acetate buffer pH 7.4, containing 5 mg bovine serum albumin/m*l*. Details of the assay are described elsewhere[26].

Results and Discussion

Preparation and purification of PVP-insulin

PVP-Insulin was purified and separated from native "free" insulin by chromatography on a Biogel P-30 (5 × 50 cm) column, equilibrated with 3M acetic acid. The optical density of the eluate was recorded at 280 nm and a typical elution pattern is depicted in Fig. 1a. As can be seen, two major peaks are obtained. The first peak near the hold-up volume of the column is well separated from the second peak of native insulin. The front part of the first peak was collected, concentrated by evaporation and re-chromatographed on another Biogel P-30 column (2 × 100 cm) in 3M acetic acid. Fig. 1b shows a typical chromatogram of purified PVP-insulin.

Only one rather narrow peak is eluted near the hold-up volume of the column. The arrow marks the position at which native insulin would be eluted. As seen from the data in Table 1, 6 preparations of PVP-insulin have been purified by the above mentioned method. Preparations 1 - 3 were performed without protection of the terminal amino groups. In preparations 4 - 6 the terminal amino groups were protected with citraconic anhydride according to Zahn et al.[13]. In no case was there any indication of leakage of insulin from the polymer backbone.

The validity of the data obtained from studies of PVP-insulin depends on the stability of the compound. Previously, with Sepharose-coupled insulin[1,2], different groups reported[4,5] that "free" insulin was leaking from the insoluble carrier. Therefore, in the case of PVP-insulin, additional controls were carried out.

In one of the preparations of PVP-insulin (prep. No. 6), besides native insulin, ^{125}I-insulin was added as a tracer and the coupling reaction was carried out as described under Methods (modified procedure).

Fig. 1c shows the elution pattern of PVP-insulin containing PVP-^{125}I-insulin. The radioactivity and optical density at 280 nm are plotted versus the tube number. Again, only one peak near the hold-up volume of the column can be seen. The arrow marks the position where ^{125}I-insulin would be eluted. No radioactivity can be detected at this position, indicating that PVP-insulin is stable in 3M acetic acid at room temperature.

Insulin content and amino acid composition of PVP-insulin

The insulin content of purified PVP-insulin was analyzed after acid hydrolysis in 6N HCl at 110 °C.

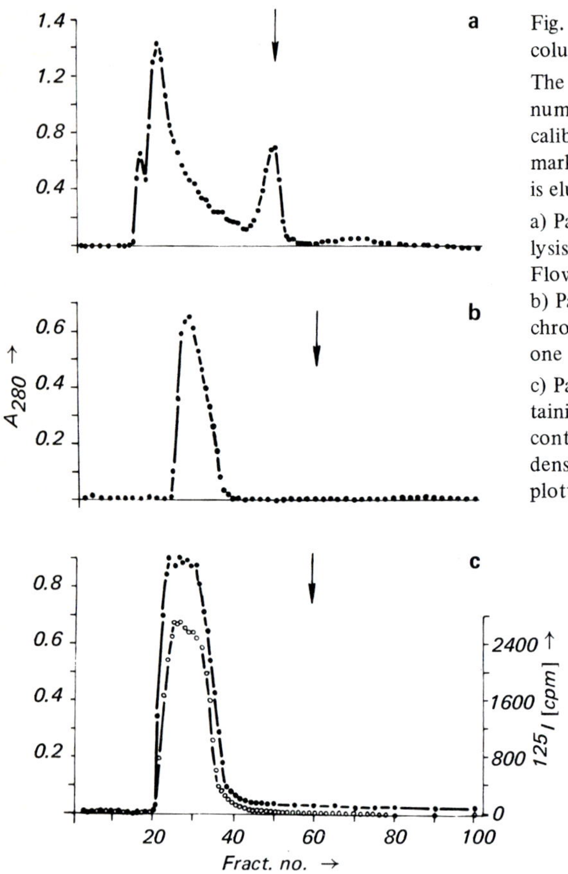

Fig. 1. Chromatography of PVP-insulin on Biogel P-30 columns in 3M acetic acid.

The optical density at 280 nm is plotted versus the tube number (•——•). The columns had previously been calibrated with native bovine insulin and the arrow marks the tube where the maximum of native insulin is eluted.

a) Pattern of crude PVP-insulin (Prep. No. 4) after dialysis. Column size 5 x 50 cm; one tube contained 8 m*l*. Flow rate: 25 m*l*/h.

b) Pattern of purified PVP-insulin (Prep. No. 4) after rechromatography on Biogel P-30. Column size 2 x 100 cm; one tube contained 4 m*l*. Flow rate: 8 m*l*/h.

c) Pattern of purified PVP-insulin (Prep. No. 6) containing ^{125}I-insulin. Column size 2 x 100 cm; one tube contained 4 m*l*. Flow rate: 8 m*l*/h. Besides the optical density (•——•) on the left ordinate, radioactivity is plotted on the right ordinate (○——○).

Table 1. Hydrodynamic properties and molecular weight of PVP-insulin.

Prep. No.	µg Insulin / mg PVP	Part. spec. vol. ($\bar{V}$) [m*l*/g]	Sedim. coeff. *S*	Diff. coeff. *D*	M_r (accord. to Svedberg[18])	M_r (Sephadex chromatogr.)
1	129	0.780	n.d.	n.d.	–	300000
2	240	0.774	4.3	1.6	275000	300000
3	176	0.777	4.0	1.7	260000	n.d.
4	155	0.778	1.9	n.d.	–	n.d.
5	172	0.777	1.7	3.9	48000	60000
6	190	0.776	1.6	3.4	51000	60000

n.d. = not determined.

Table 2. Amino acid composition of PVP-insulin after hydrolysis with 6N HCl at 110 °C (Prep. No. 4).

Amino acid	nmol	Residues	Native bovine insulin
Asp	191.3	2.7	3
Thr	44.3	0.6	1
Ser	178.6	2.5	2
Glu	502.2	7.1	7
Pro	77.5	1.1	1
Gly	250.6	3.6	4
Ala	179.8	2.6	3
Val	280.4	4.0	5
Ile	42.5	0.6	1
Leu	331.3	4.8	6
Tyr	131.7	2.0	4
Phe	186.7	2.7	3
Lys	77.1	1.1	1
His	141.0	2	2
Arg	78.2	1.1	1

The amino acid concentration in the hydrolysate was determined by use of the amino acid analyzer. Table 2 shows the amino acid composition of PVP-insulin (Prep. No. 4). A similar pattern was observed for the other preparations of PVP-insulin.

Although precautions were taken to protect tyrosine during hydrolysis (Methods) the yield for this amino acid was only about 50 - 60% in all the preparations. The reason for the apparent instability of tyrosine in PVP-insulin is not known. Despite this fact, the amino acid pattern can be regarded as additional evidence that insulin has indeed been coupled to the synthetic polymer.

Molecular weight of PVP-insulin

The molecular weight of polymer-bound insulin was determined by chromatography on a Sephadex G-200 column (2 × 100 cm). The column had previously been calibrated with the usual globular protein marker and a straight line was obtained when the elution volume of the individual protein markers was plotted versus the logarithm of the molecular weight. The optical density of the tubes was recorded at 280 nm. The first two preparations of PVP-insulin were eluted as a rather broad peak near the hold-up volume of the column, even in front of the marker enzyme catalase (M_r 240 000), suggesting a molecular weight of about 300 000 (Table 1).

However, it is not very likely that PVP-insulin has a globular conformation; it rather has a random-coil structure.

Therefore, the molecular weight of PVP-insulin was also determined by analytical ultracentrifugation, which is independent of the conformation of the polymer. From the scanner patterns the apparent sedimentation and diffusion coefficients were determined and the mean molecular weight calculated using the equation given by Svedberg[18]. The results of the ultracentrifugation runs are listed in Table 1. Preparation No. 2 had a mean molecular weight of about 275 000. Although the molecular weight is somewhat lower than that obtained by column chromatography, the ultracentrifugation experiments indicate that cross-links have been formed and that different polyvinylpyrrolidone chains are linked via insulin molecules to each other.

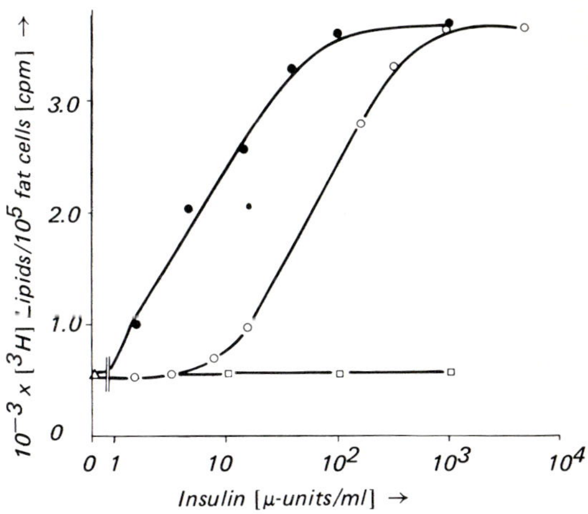

Fig. 2. Stimulation of lipogenesis by insulin and PVP-insulin in isolated fat cells.

The incorporation of [2-^{3}H]glucose into the lipid fraction of isolated fat cells (10^5 cells/m*l*) was determined in Tris/albumin buffer, pH 7.4 at 37 °C. The total volume was 1.25 m*l*, the incubation time 30 min. The assay was carried out in triplicate. Radioactivity (cpm) in the lipid fraction is plotted versus the insulin concentration as determined by amino acid analysis.

●——● Native insulin; □——□ PVP-insulin (Prep. No. 2); ○——○ PVP-insulin (Prep. No. 4).

Biological and immunological activity of insulin linked to polyvinylpyrrolidone

The biological activity of polymer-bound insulin was measured with isolated fat cells. The stimulation of lipogenesis from [2-^{3}H]glucose served as a parameter for insulin activity. Preparation No. 1 - 3 had no insulin activity in this system. Even a 1 000-fold excess of PVP-insulin compared with native insulin promoted no stimulation of lipogenesis in isolated fat cells (Fig. 2). Essentially the same negative result was obtained from radioimmunological studies. PVP-Insulin (Prep. No. 2) could not displace ^{125}I-insulin from the antibody (Fig. 3).

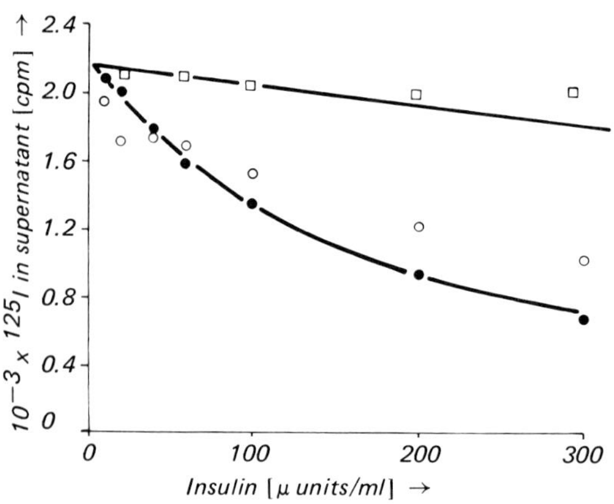

Fig. 3. Comparison of immunoreactivity of native insulin and PVP-insulin with insulin antiserum.

Different concentrations of insulin and PVP-insulin were incubated with insulin antiserum and ^{125}I-insulin (5 micro-units) in 0.5 m*l* barbital/albumin buffer, pH 7.4 at 37 °C for 3 h. Afterwards, 0.5 m*l* Dextran-coated charcoal was added, centrifuged and a portion of the supernatant counted. Radioactivity is plotted versus insulin concentration.

●——● Native insulin; □——□ PVP-insulin (Prep. No. 2); ○ PVP-insulin (Prep. No. 4).

Due to the work of Weber et al.[27] and Zahn et al.[13] it is known that the *N*-terminal part of the insulin molecule, particularly the α-amino group of glycine of the A-chain, is essential for the biological activity of the hormone. Therefore, the α-amino groups of the hormone were protected by citraconic acid[11] and a modified preparation was worked out (Methods). The product was purified and a molecular weight of about 50 000 (ultracentrifugation) to 60 000 (Sephadex chromatography) was obtained (Table 1). In the isolated fat cell assay the PVP-insulin preparations No. 4 - 6 were biologically active.

Fig. 2 compares the biological activity of polymer-bound insulin (Prep. No. 4) with native insulin. The two hormones cause the same maximal stimulation of lipogenesis, but the dose-response curve of PVP-insulin is shifted to the right. 100 Micro-units of polymer-bound insulin exhibits about the same biological effect as 10 micro-units of the native hormone. PVP-Insulin preparations No. 4 and 6 were also immunologically active. Increasing concentrations of polymer-bound insulin displaced ^{125}I-insulin from the antibody. The immunological activity was between 30 and 50% (Table 3). It should be mentioned that controls with polyvinylpyrrolidone, which served as a matrix for insulin, have been performed. A control preparation was carried through the coupling purification procedure without the addition of insulin. No interference with the bioassay or radioimmunoassay was observed, when comparable concentrations of the control and the untreated polyvinylpyrrolidone were added.

An important and as yet unsolved problem regarding the biological action of insulin is the question, whether the hormone acts on the cell surface or if it has to enter the cell for its biological function.

One approach to the problem has been the coupling of insulin to the insoluble matrix Sepharose[1,2]. However, it was shown that the observed biological insulin activity of Sepharose-bound insulin was due to free insulin leaking from the matrix[7,8].

Using two different preparation procedures the problem of insulin leaking from the matrix could be excluded, because it was found that only the preparations in which the terminal amino groups of insulin had been protected by citraconication were biologically active. Moreover, extensive re-chromatography together with ^{125}I-labeled insulin showed that no "free" insulin is released from the polymer under these conditions.

Table 3. Biological and immunological properties of different preparations of PVP-insulin.

Prep. No.	Insulin content [a]		Biological activity		Immunol. act.
	[μg/ml]	[(m-units/ml)] [b]	[m-units/ml]	[%]	[%]
1	168	(4.2)	0 (n = 3)	0	0
2	652	(16.3)	0 (n = 2)	0	0
3	600	(15.0)	0 (n = 2)	0	n.d.
4	264	(6.6)	47.5 ± 9 (SEM; n = 3)	7.2 ± 1.3	30
5	728	(18.2)	9.1 ± 2 (SEM; n = 5)	0.5 ± 0.1	n.d.
6	860	(21.5)	44.1 ± 4.5 (SEM; n = 7)	2.1 ± 0.2	50

[a] According to amino acid analysis after acid hydrolysis.
[b] The numbers in brackets indicate the theoretical insulin activity calculated from the insulin content, assuming 1 μg insulin = 25 m-units.

The biologically active preparations of PVP-insulin had a mean molecular weight of about 50 - 60 000, as can be derived from analytical ultracentrifugation and column chromatography on Sephadex G-200. Preparations with higher molecular weights were biologically and immunologically inactive. This could be due to the formation of cross-links, where different polyvinylpyrrolidone chains are linked via insulin molecules. This idea is supported by the fact that one mole of polyvinylpyrrolidone contains about 8 - 10 carboxyl groups after hydrolysis with base. This means that the expected molecular weight of PVP-insulin, without cross-link formation, would be about 50 000 - 60 000.

In the other preparations the biological activity is rather low as compared to identical concentrations of the native hormone. The reason could be that despite protection by citraconylation, the *N*-terminal glycine residue of the A-chain has been linked to the polymer, which would yield inactive products, as has been shown previously [13,27]. It also seems possible that, since the molecule of PVP-insulin is rather bulky, steric hindrance might impair interaction with the receptor at the plasma membrane, thus leading to the decreased biological activity.

For future studies on the structure-function relationship of insulin, polyvinylpyrrolidone-bound insulin with its high molecular weight and the possibilities of labelling the polymer backbone could serve as a powerful tool.

The skillful technical assistance of Ms. *Jutta Drechsel*, Ms. *Sigrid Haßlinger* and of Ms. *Roswitha Anding* is greatly appreciated. The work was supported by the *Deutsche Forschungsgemeinschaft*.

Literature

1 Cuatrecasas, P. (1969) *Proc. Natl. Acad. Sci. U.S.A.* **63**, 450 - 457.
2 Cuatrecasas, P. (1969) *Biochem. Biophys. Res. Commun.* **35**, 531 - 537.
3 Katzen, H.M. & Vlahakes, G.J. (1973) *Science* **179**, 1142 - 1143.
4 Davidson, M.B., Van Herle, A.J. & Gerschenson, L.E. (1973) *Endocrinology* **92**, 1442 - 1446.
5 Schwartz, J., Nutting, D.F., Goodman, H.M., Kostyo, J.L. & Fellows, R.E. (1973) *Endocrinology* **92**, 439 - 445.
6 Kolb, H.J., Weiss, L. & Wieland, O.H. (1974) *Diabetologia* **10**, 373 - 374.
7 Renner, R., Kolb, H.J. & Hepp, K.D. (1974) *Hoppe-Seyler's Z. Physiol. Chem.* **355**, 1242.
8 Kolb, H.J., Renner, R., Hepp, K.D., Weiss, L. & Wieland, O.H. (1975) *Proc. Natl. Acad. Sci.* U.S.A. **72**, 248 - 252.
9 von Specht, B.U., Seinfeld, S. & Brendel, W. (1973) *Hoppe-Seyler's Z. Physiol. Chem.* **354**, 1659 - 1660.
10 von Specht, B.U., Wahl, M., Kolb, H.J. & Brendel, W. (1975) *Arch. Int. Pharmadyn. & Thér.*, **213**, 242 - 250.
11 Geiger, B., von Specht, B.U. & Arnon, R. (1977) *Eur. J. Biochem.* **73**, 141 - 147.
12 Kolb, H.J., von Specht, B.U., Renner, R. & Hepp, K.D. (1976) *Diabetes* **25**, Suppl. 1, 370.
13 Zahn, H., Brandenburg, D. & Gattner, H.G. (1972) *Diabetes* **21**, 468 - 475.
14 Andrews, P. (1965) *Biochem. J.* **96**, 595 - 606.
15 Schachman, H.K. & Edelstein, S.J. (1966) *Biochemistry* **5**, 2681 - 2705.
16 Schumaker, V.N. & Schachman, H.K. (1957) *Biochim. Biophys. Acta* **23**, 628 - 634.

17 Schachman, H.K. (1957) *Methods Enzymol.* **4**, 32 - 103.
18 Svedberg, T. & Peterson, K.O. (1940) The Ultracentrifuge, Clarendon, Oxford.
19 Jeffrey, P.D. & Coates, J.H. (1966) *Biochemistry* **5**, 3820 - 3824.
20 Patchornik, A. & Rogozinsky, S.E. (1959) *Anal. Chem.* **31**, 985 - 989.
21 Spackman, D.H., Stein, W.H. & Moore, S. (1958) *Anal. Chem.* **30**, 1190 - 1206.
22 Poullain, P. & Piette, M. (1948) *Bull. Soc. Chim. Biol.* **30**, 496 - 501.
23 Rodbell, M. (1964) *J. Biol. Chem.* **239**, 375 - 380.
24 Moody, A.J., Stan, M.A., Stan, M. & Gliemann, J. (1974) *Horm. Metab. Res.* **6**, 12 - 16.
25 Herbert, V., Lau, K.S., Gottlieb, S.W. & Bleicher, S.J. (1965) *J. Clin. Endocrinol. Metab.* **25**, 1375 - 1381.
26 Löffler, G. & Weiss, L. (1975) in Methoden der Hormonbestimmung (Breuer, H., Hamel, D. & Krüskemper, H.L., eds.), pp. 75 - 88, Thieme-Verlag Stuttgart.
27 Weber, U., Hörnle, S., Grieser, G., Herzog, K.H. & Weitzel, G. (1967) *Hoppe-Seyler's Z. Physiol. Chem.* **348**, 1715 - 1717.

B. U. v. Specht, Institut für Chirurgische Forschung der Universität München, Schillerstraße 37, D-8000 München 2.

H. J. Kolb*, R. Renner and K. D. Hepp, Institut für Diabetesforschung, Kölner Platz 1, D-8000 München 40.

* To whom correspondence should be addressed.

Hoppe-Seyler's Z. Physiol. Chem.
Bd. 359, S. 239 - 242, Februar 1978

SHORT COMMUNICATION

Alteration in D-Glucose Uptake by Plasma Membranes of Isolated Liver Cells

Werner BACHMANN, Ralph JERSILD and David CHALLONER

Indiana University School of Medicine, Departments of Medicine and Anatomy, Indianapolis, Indiana, U.S.A.

(Received 24 March 1977/5 September 1977)

Geänderte D-Glucose-Aufnahme durch Plasmamembranen von isolierten Leberzellen

Zusammenfassung:

1) Plasmamembranen wurden nach der Neville'schen Methode in der Modifikation nach Ray aus isolierten Leberzellen angereichert. Diese für Ganzleber ausgearbeitete Methode hat sich auch zur Präparation von Plasmamembranen aus isolierten Leberzellen als geeignet erwiesen (Solyom, A. et al. (1972) *Biochim. Biophys. Acta* **274**, 631 - 637).

2) Plasmamembranen aus isolierten Leberzellen nahmen im Vergleich zu Plasmamembranen aus Ganzleber um 50% weniger D-Glucose auf.

3) Plasmamembranen aus Gesamtleber, die mit Kollagenase in gleicher Konzentration wie sie zur Isolation der Leberzellen verwendet wurde, inkubiert wurden, zeigten eine um 65% reduzierte D-Glucoseaufnahme.

4) Aus den Ergebnissen wird geschlossen, daß für die verminderte D-Glucoseaufnahme der Plasmamembranen aus isolierten Leberzellen die Vorbehandlung der Zellen mit Kollagenase verantwortlich zu machen ist.

Key words: Isolated liver cells, plasma membranes, D-glucose uptake.

Isolated liver cells are widely used in metabolic studies. Solyom et al.[1] have shown that plasma membranes of isolated hepatocytes are dissimilar in some details to those prepared from whole liver, i.e. lower cholesterol content and low specific activities of 5'-nucleotidase and leucyl-β-naphthylamidase. Therefore it was of interest to determine whether the first step in the metabolism of glucose, its interaction with the liver cell membrane, is altered in plasma membranes of isolated cells. In this study D-[^{3}H]glucose uptake experiments with plasma membranes from isolated hepatocytes were performed and compared to those of plasma membranes from whole liver[2]. The uptake of D-glucose by the plasma membranes was defined as binding to the membrane and uptake into the internal space of the membrane vesicles, the latter being the predominant morphological feature of the plasma membrane fraction of whole liver and of isolated liver cells.

This paper should be understood as a final report on studies dealing with glucose transport in plasma membranes of rat liver which were performed several years ago. Although some questions could not be answered by these experiments, we believe that the data are of value to those continuing research in this area.

Enzymes:
Adenosinetriphosphatase, ATP phosphohydrolase (EC 3.6.1.3);
Glucose-6-phosphatase, D-glucose-6-phosphate phosphohydrolase (EC 3.1.3.9);
5'-Nucleotidase, 5'-ribonucleotide phosphohydrolase (EC 3.1.3.5);
Succinate dehydrogenase, succinate:(acceptor) oxidoreductase (EC 1.3.99.1).

Materials and Methods

D-[1-^{3}H]Glucose (spec. act. 57 Ci/mol) was purchased from New England Nuclear, Boston, Mass. PCS solubilizer was obtained from Amersham/Searle Corp., Arlington Heights, Ill. Adenosine 5'-monophosphate disodium salt, glucose 6-phosphate and collagenase (Type I, 125 U/mg) were from Sigma Chemical Comp., St. Louis, Miss.

Precoated thin-layer chromatography cellulose sheets and all other chemicals were obtained from Merck AG, Darmstadt.

Fed male Wistar rats weighing 140 - 160 g were used. The isolated cells were prepared by Drs. *N.R. Ingebretsen* and *S. R. Wagle*, Indiana University, Indianapolis. The isolation medium consisted of $Ca^{2\oplus}$-free Hanks' solution containing 20 mg/100 m*l* collagenase and no hyaluronidase. For detail see Ingebretsen and Wagle[3].

Preparation of plasma membranes from isolated cells

The isolation of plasma membranes was carried out according the method of Neville[4] as modified by Ray[5], which was described previously for plasma membranes of the whole liver[2]. 4 to 5 grams of isolated cells (wet weight) were used for one preparation. The purity of the plasma membrane fraction was checked by electron microscopy and by several enzymes: 5'-nucleotidase[6], $Mg^{2\oplus}$-ATPase[7], glucose-6-phosphatase[8] and succinate dehydrogenase*[9]. The protein concentration was determined by the method of Lowry et al.[10], with bovine serum albumin as the standard.

Glucose uptake experiments

The HA Millipore filter (pore size 45 μm) assay procedure for D-glucose uptake has been described in detail[2]. The specific activity for D-[^{3}H]glucose was adjusted to 40 μCi/μmol. The concentration of the labeled sugar was 1mM, except in the glucose-dependent uptake experiment. The membrane fractions (12.5 to 100 μg of membrane protein/sample) were preincubated at 20 °C in 50mM Tris/HCl, pH 7.4, for 15 min in a volume of 0.2 m*l*.

The reaction was started by adding 0.05 m*l* of the labeled sugar. After 5 min of incubation with D-glucose in concentrations up to 100mM, steady state conditions were reached. To identify the radioactive compounds taken up or later released by the membranes into the incubation medium, thin-layer chromatography was performed according to Clark[11] as described before[2]. No metabolic products of D-[^{3}H]glucose were found in the membranes or in the incubation medium after 20 min of incubation at 20 °C.

Incubation of plasma membranes from whole liver with collagenase containing buffer

One to 2 mg of plasma membrane protein, stored in distilled water, was centrifuged at 30000 × *g* for 20 min. The protein-free supernatant was discarded. The membrane pellet was resuspended in 2.0 m*l* of $Ca^{2\oplus}$-free Hanks buffer, either with or without 0.02% collagenase and incubated at 37 °C for 15 min. The membrane pellet was washed twice with distilled water by resuspending and centrifugation. The last pellet was suspended in distilled water. Protein, enzyme activities, and D-[^{3}H]-glucose uptake were measured.

Results and Discussion

Plasma membranes from isolated cells

The yield of plasma membranes was 1.25 mg to 2.0 mg membrane protein per g liver cells (wet weight), corresponding to 0.7 to 1.5% of the total protein of the homogenate. As shown in Table 1, the specific activity of 5'-nucleotidase was only 6 - 7 fold higher in the plasma membrane fraction than in the cell homogenate. This corresponds with results of Solyom et al.[1], who found a 3 - 4 fold increase in the specific activities of this enzyme. This lower activity of 5'-nucleotidase in the plasma membranes of isolated cells was not due to greater contaminations of other cell organelles, since the specific activities of glucose-6-phosphatase and succinate dehydrogenase were comparable to those from plasma membranes of whole liver[2]. In addition, electron microscopic examination showed no evidence for significant organelle contamination. Therefore it seemed possible that the collagenase might have a direct effect on the plasma membrane enzymes. As shown in Table 2, a loss of 50% of membrane protein, as soluble protein, occured during incubation of plasma membranes from whole liver. If one calculates the total activity of 5'-nucleotidase before and after the incubation, 62% of the activity was lost. Control incubations without collagenase revealed no change in protein content or enzyme activities. Thus 5'-nucleotidase may be removed from the plasma membrane by collagenase and be lost after cell disruption during the isolation procedure. If the enzyme were inactivated it should also be drastically diminished in the homogenate of the isolated cells (Table 1).

Uptake and release of D-glucose by plasma membranes of isolated liver cells

As shown in Fig. 1, the uptake of D-glucose by plasma membranes isolated from isolated hepatocytes is about 40 - 50% lower than the uptake by plasma membranes of whole liver. Similar properties of the glucose uptake of plasma membranes of whole liver and the glucose uptake by plasma membranes of isolated cells were: (a) rapidity, the half maximal uptake taking about 30 s (Fig. 1), (b) the linear increase in uptake with increase in

---

* Measured with 2-(4-iodophenyl)-3-(4-nitrophenyl)-5-phenyltetrazolium iodide as electron acceptor.

Table 1. Enzyme activities in cell homogenate and in plasma membranes from isolated hepatocytes.

Results are given as means ± standard deviations with number of plasma membrane preparation examined (in parentheses). Specific activities are given as μmol substrate/(mg protein × h). Recovery and relative specific activity values refer to the homogenate taken as 100% and unity, respectively.

Enzyme		Cell homogenate	Plasma membrane fraction		
		Spec. act. [units/mg prot.]	Spec. act. [units/mg prot.]	Recovery [%]	Rel. spec. act.
5'-nucleotidase	(7)	3.1 ± 0.32	20.4 ± 2.41	7 - 10	6.7
$Mg^{2\oplus}$-ATPase	(7)	7.1 ± 0.83	22.5 ± 2.34	3 - 4	3.2
Glucose-6-phosphatase	(7)	5.0 ± 1.12	0.77 ± 0.32	23 - 27 × 10^{-4}	0.15
Succinate: INT-dehydrogenase	(5)	2.4 ± 0.14	0.03 ± 0.03	35 - 67 × 10^{-5}	0.013

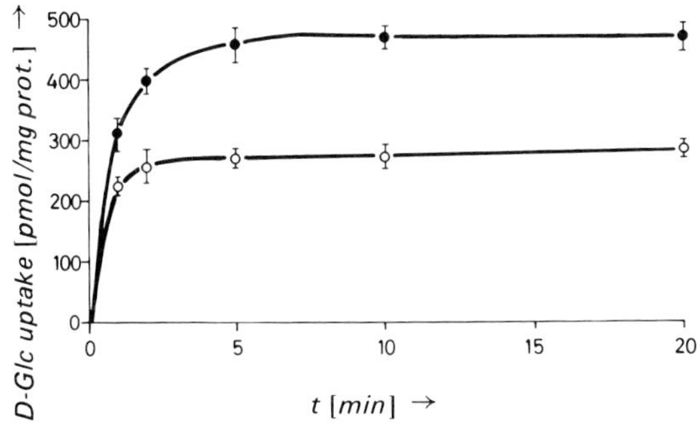

Fig. 1. Kinetics of D-glucose uptake by plasma membranes.

Plasma membranes of whole liver (56 μg membrane protein/sample) and of isolated cells (47 μg membrane protein/sample) were preincubated for 15 min. At zero time D-[^{3}H]glucose (spec. act. 40 μCi/μmol) in a final concentration of 1mM was added and at varying times later the samples were removed and processed as described[2]. Each point is the mean value of 3 membrane preparations. (Means ± standard deviations) • whole liver, ○ isolated cells.

the glucose concentration in the incubation medium (Fig. 2), and (c) the linear relation of D-glucose uptake into the plasma membranes to the plasma membrane concentration (12 μg to 100 μg of membrane protein). In contrast to those common properties, only the release of D-glucose from previously loaded plasma membranes was inhibited in membranes from isolated cells (Fig. 3).

Incubation of plasma membranes from whole liver with a medium containing collagenase in the same concentration used for liver perfusion diminished the D-[^{3}H]-glucose uptake capacity to 35% of the controls (Table 2). That is even more than observed in plasma membranes of isolated cells. In control studies without collagenase in the incubation medium, no change in D-glucose uptake could be detected. The results reported show that treatment of plasma membranes with collagenase, either during liver cell isolation or by incubating of plasma membranes of whole liver with collagenase, lead to an inhibited uptake and release of D-glucose, probably due to a loss of membrane proteins by collagenase itself or by contaminating proteases of the collagenase preparation used. Nevertheless, the results presented raise the question whether conclusions from studies on glucose

Table 2. Enzyme activities and D-glucose uptake by rat liver plasma membranes incubated in the presence of collagenase.

1 mg of plasma membranes isolated from whole liver was incubated for 15 min at 37 °C in $Ca^{2\oplus}$-free Hanks' solution containing 20 mg/100 m*l* collagenase. After washing with distilled water, the enzyme activities were tested and D-[^{3}H]glucose uptake measured as indicated under "Material and Methods". Specific activities of enzymes are given as μmol P_i/(mg protein × h). The D-[^{3}H]glucose uptake is expressed as pmol/(mg protein × 5 min). Means ± standard deviations.

		Protein [mg]	5'-nucleotidase	$Mg^{2\oplus}$ ATPase	D-[^{3}H]glucose uptake
Before incubation	(3)	1.010 ± 0.102	59.5 ± 3.8	30.3 ± 2.3	561 ± 63
After incubation	(3)	0.484 ± 0.053	47.8 ± 4.1	28.0 ± 2.0	195 ± 21

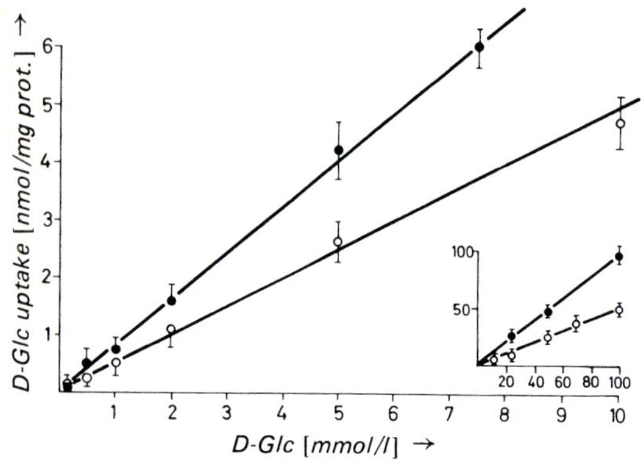

Fig. 2. Effect of D-glucose concentration on D-glucose uptake by plasma membranes.

Plasma membranes of whole liver (50 μg/sample) and of isolated cells (53 μg/sample) were incubated for 15 min. Then D-[^{3}H]glucose of different concentrations was added (spec. act. 40 μCi/μmol and 0.8 μCi/μmol). After 5 min of further incubation, the reaction was stopped as described[2]. Each point is the mean value of 3 determinations. (Means ± standard deviations) • whole liver, ○ isolated cells.

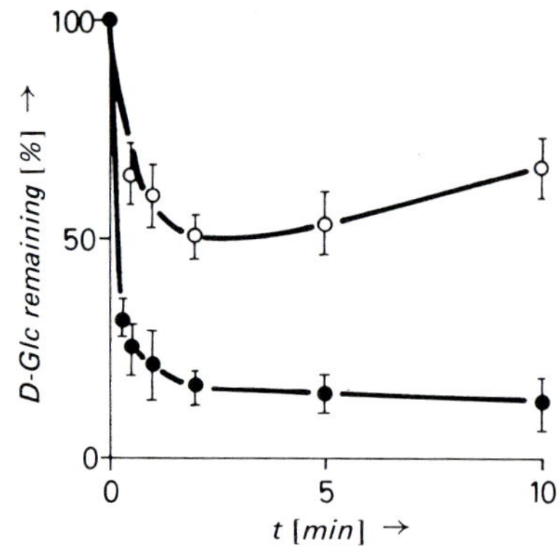

Fig. 3. D-Glucose release from plasma membranes.

Plasma membranes from whole liver (65 μg membrane protein/sample) and from isolated cells (47 μg membrane protein/sample) were preloaded with 1mM D-[^{3}H]-glucose (spec. activity 40 μCi/μmol) by incubating for 5 min at 20 °C. At zero time the incubation mixture was diluted 10-fold with buffer and the rate of release measured as described[2]. The results are expressed as percentage of sugar remaining in the plasma membranes. The amount of labeled sugar taken up before dilution was 100%. Each point is the mean value of 3 membrane preparations. (Means ± standard deviations) • whole liver, ○ isolated cells.

metabolism in isolated liver cells must be qualified by the demonstrated abnormalities in glucose uptake by the isolated membranes.

Literature

1 Solyom, A., Lauter, C. J. & Trams, E. G. (1972) *Biochim. Biophys. Acta* **274**, 631 - 637.
2 Bachmann, W. & Challoner, D. (1976) *Biochim. Biophys. Acta* **443**, 254 - 266.
3 Ingebretsen, W. R., Jr. & Wagle, S. R. (1972) *Biochem. Biophys. Res. Commun.* **47**, 403 - 410.
4 Neville, D. M. (1960) *J. Biophys. Biochem. Cytol.* **8**, 413 - 422.
5 Ray, T. K. (1970) *Biochim. Biophys. Acta* **196**, 1 - 9.
6 Michell, R. H. & Hawthorne, F. N. (1965) *Biochem. Biophys. Res. Commun.* **21**, 333 - 338.
7 Emmelot, P. & Bos, C. J. (1966) *Biochim. Biophys. Acta* **120**, 369 - 382.
8 DeDuve, C., Pressmann, B. C., Gianetto, R., Wattiaux, R. & Appelmans, O. O. (1955) *Biochem. J.* **60**, 604 - 617.
9 Pennington, F. (1961) *Biochem. J.* **80**, 649 - 654.
10 Lowry, O. H., Rosebrough, N. J., Farr, A. L. & Randall, R. J. (1951) *J. Biol. Chem.* **193**, 265 - 275.
11 Clark, C. M., Jr. (1973) *Diabetes* **22**, 41 - 49.

W. Bachmann, Forschergruppe Diabetes und III. Med. Abteilung des Krankenhauses München-Schwabing
Kölner Platz 1, D-8000 München 40

Hoppe-Seyler's Z. Physiol. Chem.
Bd. 359, S. 243 - 245, Februar 1978

SHORT COMMUNICATION

Assessment of Glycerol as Non-Aqueous Isolation Medium for Cell Nuclei

Günther SIEBERT and Rosemarie HANNOVER

Institut für Biologische Chemie der Universität Hohenheim

(Received 28 November 1977/12 January 1978)

Bewertung von Glycerin als nicht-wäßriges Isolierungsmedium für Zellkerne

Zusammenfassung: Die hohe Affinität von Glycerin zur Absorption von Wasserdampf wie auch die Bildung flüssiger binärer Glycerin-Wasser-Gemische bei – 19 °C sind ein ernstes Hindernis für die Verwendung von Glycerin als Isolierungsmedium für Zellorganellen. Die Löslichkeit von Alkalichloriden, Adenosin-5′-monophosphat und mehreren Proteinen in wasserfreiem Glycerin wird bestimmt; auf Grund der erhaltenen Werte sollte Glycerin außer in seltenen Spezialfällen als Isolierungsmedium vermieden werden. Die funktionelle Bedeutung des Begriffes „nicht-wäßrig" wird kurz diskutiert.

Key words: Nuclear isolation, glycerol, non-aqueous media, protein solubility in glycerol.

Since the first description of nuclear isolation by Behrens almost half a century ago[1], the principle of using lyophilized cells in the presence of organic solvents, like light petroleum, carbon tetrachloride and cyclohexane, has been increasingly widely used. With the introduction of freeze-clamping[2] for preservation of the original biochemical state of cells and tissues, numerous investigations have demonstrated the suitability of this procedure in studies of nuclear constituents and their metabolism (for reviews see ref.[3-6]).

However, almost complete loss of lipids and concomitant destruction of lipoprotein complexes are an inevitable feature of the non-aqueous isolation procedure for nuclei and other cell organelles, and have therefore led to renewed efforts to replace water-nonmiscible solvents by non-aqueous liquids which maintain anhydrous conditions and do not extract or alter any nuclear constituent. Silicon oil[7] as well as glycerol[8] have been proposed, the latter being used more widely in recent years.

The present author, mainly by correspondence with colleagues from abroad, has become aware of rather serious sources of error incurred by the uncritical use of glycerol; it is the aim of this paper to demonstrate the limited applicability of glycerol for experiments in which biochemically unaltered nuclei are to be isolated.

Materials and Methods

Glycerol, water-free, d 1.26, most pure according to DAB 7, BP 1973, Ph. Helv. VI, USP XIX, was obtained from E. Merck AG., Darmstadt, and used without further purification. Crystalline bovine hemoglobin was purchased from Worthington Biochemical Corp., Freehold, New Jersey, U.S.A., bovine serum albumin from Behringwerke, Marburg-Lahn, casein (alkali-soluble) from Serva Feinbiochemica, Heidelberg, and adenosine 5′-monophosphate from Boehringer Mannheim GmbH. All other chemicals were from E. Merck AG., Darmstadt.

Analyses for sodium and potassium were performed, at proper dilution with redistilled water, by atomic absorption spectrometry in a Beckman Atomic Absorption Spectrophotometer No. 1233. Hemoglobin, proteins, and 5′-AMP were analyzed spectrophotometrically at 450, 280, and 260 nm respectively, with aqueous standard solutions for calibration, in a Zeiss spectrophotometer PMQ 2. Glycerol, in these measurements, was diluted with redistilled water between 40- and 2000-fold and did not exert any influence on the absorption measurements.

Results

Glycerol is very well known for its avid sorption of water vapor. A simple experiment demonstrates that a centrifuge cup, filled with 50 m*l* of glycerol and kept in the atmosphere of a cold-room at + 2 °C, gains enough weight within 4 - 5 h to indicate a water content of 3 - 5% (v/v).

It should also be pointed out that low temperatures do not prevent glycerol-water interaction. Frozen pieces of redistilled water and glycerol are kept separately and well stoppered for 24 h in the deep-freeze at – 19 °C before some pieces of ice are quickly transferred to the top of the glycerol, which is immediately stoppered again. Within 6 h at – 19 °C, a binary fluid phase forms which contains, according to weight determinations of measured volumes, 52% water in glycerol (v/v). Data on equilibria of glycerol-water mixtures are available in literature[9].

Glycerol, when used as a suspension medium for cell organelles, is expected not to extract any constituents from subcellular particles but in reality dissolves all tested compounds shown in the table. Solubility data of inorganic salts in glycerol are tabulated[10], e.g. 7.55% NaCl or 6.43% KCl (w/v) in 100% glycerol at 25 °C, and indicate about 1/5 of the solubility in water. If hemoglobin and 5′-AMP are kept in anhydrous glycerol at – 19 °C instead, their solubilities are diminished sharply. The conclusion is justified from these measurements that glycerol will dissolve inorganic salts, nucleotides, and proteins from lyophilized cells during homogenization and centrifugation, which may last more than 18 h[13].

Table. Solubility of cell constituents in anhydrous glycerol.

Known amounts of some proteins, adenosine 5′-monophosphate, or inorganic salts are placed into small glass tubes with tight covers; the proteins and 5′-AMP have been stored before in the deep-freeze in the presence of a desiccant. The glass tubes are filled with anhydrous glycerol and immediately restoppered. The amount of glycerol is determined by weighing. After 5 days at room temperature, the contents of the tubes are centrifuged at 105 000 x *g* for 2 h and the supernatants are diluted with redistilled water for analyses (see Methods).

Compound	Solubility [g/100 m*l*]
NaCl	0.9
KCl	0.4
$MgCl_2$	Interphase indicates solubility
$CaCl_2$	No interphase formed
5′-AMP	0.57
Hemoglobin	1.35
Casein	0.007
Serum albumin	0.74

Discussion

All available enzymological evidence indicates that a few percent of water – whether in solvents or in so-called dry or in deep-frozen foods – suffice to allow enzymatic hydrolyses to proceed[11,12]. Low temperatures, as e.g. used in ref.[13], do not prevent the existence of fluid water in glycerol mixtures. In laboratory practice, unusual efforts and sophisticated instrumentation would be required if truly anhydrous conditions were to be maintained during homogenization and centrifugation steps in the isolation of cell organelles in the presence of glycerol.

Glycerol is capable of dissolving (Table) a variety of cell constituents, which should exclude glycerol from studies on inorganic and low molecular weight compounds. Even in the case of proteins and soluble enzymes, extensive control experiments are obviously needed in every individual case if glycerol is used as an isolation medium. This holds true in spite of the fascinating morphology of nuclei isolated in glycerol[8], and despite the protective action of poly-hydroxy compounds like glycerol, mannitol, or sucrose in shielding proteins from denaturation.

Whereas organic solvents may easily be mixed for obtaining density gradients in equilibrium banding experiments[14] and for zonal centrifugation[15], it is more difficult to conceive of density gradients based on glycerol, whose viscosity poses an additional technical problem in keeping centrifugation times short.

The term "non-aqueous", generally used for isolation procedures based on the Behrens principle[1], seems to be formally correct if applied to anhydrous glycerol. But besides the problems involved in the prevention of access of water vapor, the functional meaning of "non-aqueous" also implies the absence of any solubility of water-soluble compounds in a non-aqueous solvent, a requirement obviously not met by glycerol. In consequence, glycerol cannot be recommended as an isolation medium, except in very restricted conditions which need careful counterchecks for applicability. Experiments like those described above may be required before different alternatives for solvents like polyethylene or polypropylene glycols are explored for practical laboratory use.

The authors are grateful for financial support by *Fonds der Chemie* and *Universitätsbund Hohenheim.*

Literature

1 Behrens, M. (1932) *Hoppe-Seyler's Z. Physiol. Chem.* **209**, 59 - 74.
2 Hohorst, H.J., Kreutz, F.H. & Bücher, T. (1959) *Biochem. Z.* **332**, 18 - 46.
3 Siebert, G. & Humphrey, G.B. (1965) *Adv. Enzymol.* **27**, 239 - 288.
4 Siebert, G. (1968) *Compreh. Biochem.* **23**, 1 - 37.
5 Siebert, G. & Langendorf, H. (1970) *Naturwissenschaften* **57**, 119 - 124.
6 Siebert, G. (1972) *Sub-Cell. Biochem.* **1**, 277 - 292.
7 Elbers, R., Heldt, H.W., Schmucker, P., Soboll, S. & Wiese, H. (1974) *Hoppe-Seyler's Z. Physiol. Chem.* **355**, 355 - 378.
8 Kirsch, W.M., Leitner, J.W., Gainey, M., Schulz, D., Lasher, R. & Nakane, P. (1970) *Science* **168**, 1592 - 1595.
9 Lane, L.B. (1925) *Ind. Eng. Chem., Ind. Ed.* **17**, 924.
10 Landolt-Börnstein (1923) Physikalisch-Chemische Tabellen (Roth, W.A. & Scheel, K., eds.) 5th edit., vol. 1, pp. 724 - 739, Springer-Verlag, Berlin.
11 Acker, L. (1961) *Adv. Food Res.* **11**, 263 - 330.
12 Drapon, R. & Guilbot, A. (1962) *An. Technol. Agric.* **11**, 175 - 218; 275 - 317.
13 Foster, D.N. & Gurney, T., Jr. (1976) *J. Biol. Chem.* **251**, 7893 - 7898.
14 Siebert, G. & Hannover, R. (1977) in Methodol. Surveys in Biochem. (Reid, E., ed.) Vol. 6, pp. 189 - 196, Horwood, Chichester.
15 Shakoori, A.R., Romen, W., Oelschläger, W., Schlatterer, B. & Siebert, G. (1972) *Hoppe-Seyler's Z. Physiol. Chem.* **353**, 1735 - 1748.

Hoppe-Seyler's Z. Physiol. Chem.
Bd. 359, S. 247 - 250, Februar 1978

SHORT COMMUNICATION

Cold-Insoluble Globulin (Fibronectin), IV[1-3] Affinity to Soluble Collagen of Various Types

Franz JILEK and Helmut HÖRMANN

Max Planck-Institut für Biochemie, Martinsried bei München

(Received 12 January 1978)

Kälte-unlösliches Globulin (Fibronektin), IV. Affinität zu löslichem Kollagen verschiedener Typen

Zusammenfassung: Die Affinität zwischen Kälte-unlöslichem Globulin (Fibronektin) und nativem sowie hitzedenaturiertem löslichem Kollagen der Typen I, II und III wurde in folgenden Systemen untersucht: 1. Adsorption von ^{125}I-markierten löslichen Kollagenderivaten an immobilisiertes Fibronektin. 2. Komplexbildung zwischen ^{125}I-Kollagen und Fibronektin in Lösung und anschließende Fällung des Komplexes mit Anti-Fibronektin-Antiserum. 3. Hemmung der Komplexbildung von löslichem denaturiertem ^{125}I-Kollagen, Typ III, und Fibronektin durch nichtmarkierte lösliche Kollagenderivate. Auch in diesem Falle wurden die gebildeten Komplexe durch Anti-Fibronektin-Antiserum gefällt. In allen drei Systemen zeigte Fibronektin die stärkste Affinität zu denaturiertem löslichem Kollagen, Typ III. Es folgten mit etwa gleicher Wirksamkeit denaturiertes Kollagen, Typ I und Typ II. Von den nativen löslichen Kollagenpräparaten wies nur Typ III in allen drei Tests eine gewisse Affinität auf, die deutlich geringer war als die der denaturierten Präparate. Im Inhibitionstest konnte auch für natives Kollagen, Typ I, eine Affinität nachgewiesen werden, die noch geringer war als die von nativem Kollagen, Typ III.

Key words: Collagen, fibronectin, cold-insoluble globulin, complex formation, affinity behavior.

Cold-insoluble globulin is a high molecular weight plasma protein representing the soluble derivative of a fibroblast surface protein[4] (LETS-protein[5]). It also exists as fibrils on the cell surface[6] as well as in the intercellular space[7]. Recently, all the various forms were uniformly denoted fibronectin[8,9].

In many tissues fibronectin is associated with collagen, e.g. in basement membranes[10] or in the neighbourhood of fibroblasts[11]. A derivative isolated from serum was found to be identical with a gelatin-binding protein (antigelatin factor)[12] whose existence had been postulated by immunologists for several years. It was able to mediate binding of denatured collagen to macrophages[13], which probably initiates an intracellular degradation. In addition, fibronectin was recognized as the effective component of a cell-attachment factor[14] isolated from serum[15], which mediates the attachment of fibroblasts and other cells to collagen.

The outlined data indicate an affinity between fibronectin and collagen; the region in which the binding sites of Type I and II collagen lie has already been located[16]. However, it was not clear whether the fibronectin affinity extended only to denatured collagen or whether the globulin was also bound by some of the native proteins. Engvall and Ruoslahti[17] were able to demonstrate that fibronectin was considerably more

---

Abbreviation: LETS-protein = Large external transformation sensitive protein.

strongly adsorbed by denatured Type I collagen than by native Type III collagen.

In the present investigation, the binding of native and denatured soluble collagen, Type I, II and III, by fibronectin was measured in various systems using ^{125}I-labelled collagen preparations. Soluble Type I and II collagens had been extracted from fetal calf skin and were fractionated by NaCl precipitation as outlined by Timpl et al.[18]. Soluble Type II collagen was prepared from joint cartilage of fetal calves by methods including pepsin digestion at low temperature[19]. The collagen preparations were labelled with ^{125}I by the Chloramine T method[20] resulting in a specific activity between 20000 and 25000 cpm per 100 ng. If desired, the samples were denatured by incubating a 0.07% collagen solution in 0.05% acetic acid at 50 °C for 30 min. Fibronectin (cold-insoluble globulin) was prepared from human plasma as outlined by Mosesson and Umfleet[21].

1) Batchwise adsorption of soluble ^{125}I-collagen onto immobilized fibronectin (Fibronectin-Sepharose)

Fibronectin-Sepharose was prepared by coupling cold-insoluble globulin (1.5 mg/m*l*) to cyanogen bromide-activated Sepharose 4B (Deutsche Pharmacia GmbH) by the method of Cuatrecasas[22]. It contained 80 - 90% of the maximum load of immobilized globulin. Batchwise adsorption experiments were done in plastic tubes by mixing 0.5 m*l* of a standard suspension of ca. 50 mg Fibronectin-Sepharose in 0.1M Tris/HCl, pH 7.4, containing 0.1M NaCl, 5mM $CaCl_2$, 0.5% bovine serum albumin and 0.04% Tween 20, with 100 ng ^{125}I-collagen in 0.1 m*l* 0.005% acetic acid followed by incubation at 4 °C for 4 or 20 h, respectively, with periodic swirling. Subsequently, the Fibronectin-Sepharose was spun down, washed twice each time with 0.5 m*l* buffer at 4 °C and counted in a Beckman scintillation spectrometer Gamma 300. Controls were run with ^{125}I-labelled collagen and Sepharose 4B as well as without any adsorbents to check unspecific binding to the vessel wall.

The table shows that of the denatured collagen samples, Type III was most effectively adsorbed by Fibronectin-Sepharose. Considerable affinity was also found for Types I and II. In all cases the adorbed quantity was nearly equal after 4 and 20 h incubation. In contrast, the only native sample adsorbed in reasonable amounts was Type III collagen. The binding of this material was not complete after 4 h. After 20 h, the fixed quantity was about half of that of denatured collagen, Type I. In contrast to denatured Type III collagen, the native sample also showed a non-negligible affinity to unloaded Sepharose 4B. When considering this result, it became clear that native Type III collagen was less firmly bound than denatured Type I and II collagen. Native Type I and II collagens were essentially not adsorbed by Fibronectin-Sepharose.

Table. Batchwise adsorption of soluble ^{125}I-collagen (100 ng/0.6 m*l*) onto Fibronectin-Sepharose (ca. 50 mg). Controls in the absence of adsorbant always yielded 0.3 - 0.4% adsorption of ^{125}I-collagen onto the vessel wall.

^{125}I-Collagen	% adsorbed onto			
	Fibronectin-Sepharose		Sepharose 4B	
	4 h	20 h	4 h	20 h
Type I denatured	38.3	39.7	0.9	1.2
Type II denatured	49.6	49.5	1.2	1.7
Type III denatured	71.2	72.4	2.8	5.1
Type I native	4.4	5.5	0.7	0.8
Type II native	6.6	8.8	1.2	1.6
Type III native	28.6	37.6	8.3	11.8

2. Radio-collagen binding assay

In this assay a constant amount of ^{125}I-collagen was complexed in solution by various quantities of fibronectin and the aggregates obtained were precipitated by monospecific anti-fibronectin antiserum. Decreasing portions of 0.05% solutions of fibronectin (constant volume 100 *μl*) in 0.1M Tris/HCl, pH 7.4, 0.1M NaCl were diluted to 300 *μl* by the same buffer which, in addition, contained 0.5% bovine serum albumin and 0.04% Tween 20. 100 ng ^{125}I-collagen in 100 *μl* 0.05% acetic acid was added, followed by 20 h incubation at 4 °C. Subsequently, 200 *μl* rabbit anti-human fibronectin antiserum was added, and after 20 h the precipitate formed was centrifuged 20 min at 3000 × *g* and 4 °C. It was thrice washed with incubation buffer and counted. The values of controls without fibronectin were subtracted.

Fig. 1 shows the amounts of soluble ^{125}I-collagen precipitated in presence of increasing concentrations of fibronectin. The most active material, giving rise to a final precipitation of nearly 80%, was denatured Type III collagen. Half-maximum precipitation of 100 ng collagen was achieved by ca. 2 *μ*g fibronectin. Denatured Type I and II collagen required somewhat greater amounts. The affinity of these two collagen samples to fibronectin was nearly equal. Considerably less native than denatured collagen was precipitated under the conditions of the assay. A recognizable affinity was only observed for Type III collagen, which, however, resulted in a precipitation of less than 20% of the material applied, even with the highest concentration of fibronectin. Native Type I and II collagen did not show any affinity to fibronectin.

3. Inhibition of radio-collagen binding assay

The radio-assay outlined for binding of ^{125}I-collagen by fibronectin in the preceeding section suffered from the

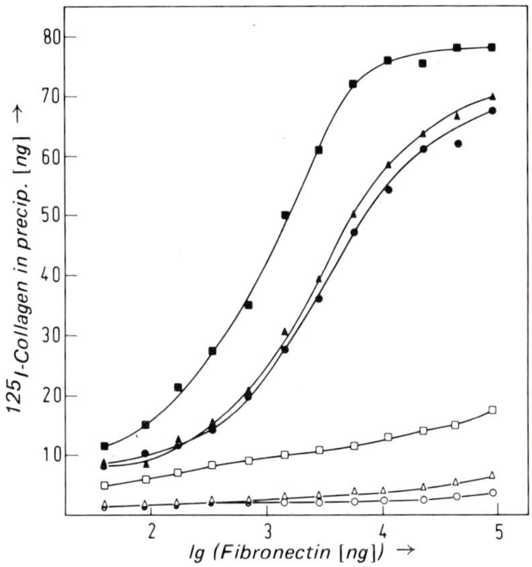

Fig. 1. Complex formation of ^{125}I-collagen (100 ng) with fibronectin followed by precipitation of the complex with anti-fibronectin antiserum.

Reaction volume 0.4 m*l*, pH 7.4.
Collagen: Type I (●, ○), Type II (▲, △), Type III (■, □); Filled symbols: denatured collagen; open symbols: native collagen.

drawback that the ratio of fibronectin and antiserum was different in the various test combinations, resulting in a variable precipitability of fibronectin and its complexes. This inconsistancy was overcome by an inhibition assay using a constant amount (100 ng) of denatured ^{125}I-collagen, Type III, as well as a constant amount of fibronectin and inhibiting the formation of radioactive complexes by the addition of unlabelled soluble collagen of various types. A quantity of fibronectin was chosen which precipitated 72% of the applied denatured Type III ^{125}I-collagen under the conditions of the experiment shown in Fig. 1. First fibronectin was incubated with the unlabelled collagen preparation for 20 h at 4 °C in the system of the previous experiment. Subsequently, 100 ng denatured ^{125}I-collagen, Type III, was added and incubation was continued for another 20 h at 4 °C, followed by precipitation of the complex with anti-fibronectin antiserum.

The data shown in Fig. 2 indicate that complex formation of fibronectin and denatured ^{125}I-collagen, Type Type III, was most strongly inhibited by denatured soluble collagen, Type III. Denatured collagen, Type I and Type II, produced a comparable inhibition at somewhat higher concentrations. The two preparations were nearly equally effective. Native soluble Type III collagen had to be applied in a ca. 100-fold higher concentration than denatured Type III to achieve a similar degree of inhibition. When native Type I collagen was used, a 500-fold higher concentration was necessary, and native Type II collagen inhibited incompletely. When evaluating these data, it must be kept in mind that the experiments with native Type I and II collagen were performed with such high concentrations of inhibitor that side reactions like fibril formation and unspecific precipitation could not be excluded. Therefore, the results obtained in this region must be interpreted with caution.

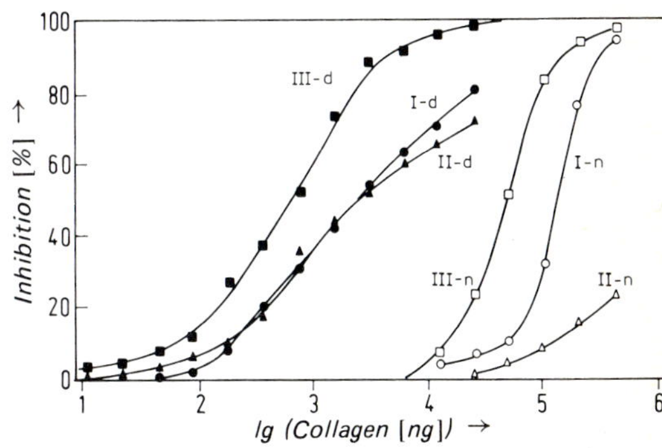

Fig. 2. Inhibition of complex formation of 100 ng denatured soluble collagen, Type III, and 2.8 μg fibronectin by soluble collagens of various types. Conditions and symbols as in Fig. 1.

All experiments outlined above agreed that among the samples investigated, denatured Type III collagen showed the strongest affinity to fibronectin. It was followed by the denatured Types I and II, which were nearly equally effective. The affinity of fibronectin to native soluble collagen was demonstrated by the adsorption assay and the radioactive binding assay only for the Type III preparation. It was considerably lower than those of the denatured samples. The affinity of native Type I and Type III collagen was only detectable by the inhibition assay, using very high concentrations. Native Type II collagen gave unclear results in this assay.

The moderate affinity of native Type III collagen and fibronectin might be of importance for processes taking place in atherosclerosis. Type III collagen is found in blood vessels immediately beneath the epithelium layer and, if this layer is degraded, it comes into contact with blood[23].

The expert technical assistance of Mrs. *Jelinić* and Mrs. *M. Hable* is gratefully acknowledged. We are indebted to Dr. *B. C. Adelmann* for kindly providing us with various soluble collagen preparations. The investigations were supported by funds of the *Deutsche Forschungsgemeinschaft*, Sonderforschungsbereich 51 and project Ho 740/1.

Literature

1 1. Commun.: Stemberger, A. & Hörmann, H. (1976) *Hoppe-Seyler's Z. Physiol. Chem.* **357**, 1003 - 1005.
2 2. Commun.: Jilek, F. & Hörmann, H. (1977) *Hoppe-Seyler's Z. Physiol. Chem.* **358**, 133 - 136.
3 3. Commun.: Jilek, F. & Hörmann, H. (1977) *Hoppe-Seyler's Z. Physiol. Chem.* **358**, 1165 - 1168.
4 Ruoslahti, E. & Vaheri, A. (1975) *J. Exp. Med.* **141**, 497 - 501.
5 Hynes, R. O. & Bye, J. M. (1974) *Cell* **3**, 113 - 120.
6 Wartiovaara, J., Linder, E., Ruoslahti, E. & Vaheri, A. (1974) *J. Exp. Med.* **140**, 1522 - 1533.
7 Muir, L. W., Bornstein, P. & Ross, R. (1976) *Eur. J. Biochem.* **64**, 105 - 114.
8 Vaheri, A. (1976) in Virus transformed cell membranes (Nicolau, C., ed.) Academic Press New York and London, in press, cited by ref.[9].
9 Keski-Oja, J. (1976) *FEBS-Lett.* **71**, 325 - 329.
10 Linder, E., Vaheri, A., Ruoslahti, E. & Wartiovaara, J. (1975) *J. Exp. Med.* **142**, 41 - 49.
11 Bornstein, P. & Ash, J. F. (1977) *Proc. Natl. Acad. Sci. U.S.A.* **74**, 2480 - 2484.
12 Dessau, W., Jilek, F., Adelmann, B. C. & Hörmann, H. (1978) *Biochim. Biophys. Acta*, in press.
13 Hopper, K. E., Adelmann, B. C., Gentner, G. & Gay, S. (1976) *Immunology* **30**, 249 - 259.
14 Pearlstein, E. (1976) *Nature (London)* **262**, 497 - 500.
15 Klebe, R. J. (1974) *Nature (London)* **250**, 248 - 251.
16 Dessau, W., Adelmann, B. C., Timpl, R. & Martin, G. R. (1977) *Biochemical J.* **169**, 55 - 59.
17 Engvall, E. & Ruoslahti, E. (1977) *Int. J. Cancer* **20**, 1 - 5.
18 Timpl, R., Glanville, R. W., Nowack, H., Wiedemann, H., Fietzek, P. P. & Kühn, K. (1975) *Hoppe-Seyler's Z. Physiol. Chem.* **356**, 1783 - 1792.
19 Miller, E. J. (1972) *Biochemistry* **11**, 4903 - 4909.
20 Adelmann, B. C., Gentner, G. J. & Hopper, K. (1973) *J. Immunol. Methods* **3**, 319 - 336.
21 Mosesson, M. W. & Umfleet, R. A. (1970) *J. Biol. Chem.* **245**, 5728 - 5736.
22 Cuatrecasas, P. (1970) *J. Biol. Chem.* **245**, 3059 - 3065.
23 Gay, S., Balleisen, L., Remberger, K., Fietzek, P. P., Adelmann, B. C. & Kühn, K. (1975) *Klin. Wochenschr.* **53**, 899 - 902.

Dipl.-Chem. F. Jilek and Prof. Dr. H. Hörmann, Max Planck-Institut für Biochemie, Am Klopferspitz, D-8033 Martinsried bei München.

Titel Nr. 3 109 002 689

Redaktionelle Hinweise

(Wichtige Änderungen ab Januar 1976 sind am Rand markiert)

Die Zeitschrift bringt nur Experimantalarbeiten aus den Gebieten der Biochemie und Physiologischen Chemie. Die Manuskripte werden in deutscher oder in englischer Sprache veröffentlicht. Es wird vorausgesetzt, daß all Koautoren mit der eingereichten Fassung einverstanden sind

Aktuelle Kurzmitteilungen, deren Umfang 4 Seiten in Kleindruck (ca. 8 Manuskriptseiten) nicht übersteigen darf, werden bevorzugt bearbeitet und in das nächsterreichbare Heft aufgenommen. Die Dringlichkeit zur Prioritätssicherung ist vom Autor in einem Begleitschreiben kurz zu begründen.

Manuskripte sind in zweifacher Ausfertigung an eine der folgenden Anschriften zu senden:

Prof. Dr. A. Butenandt oder Prof. Dr. F. Lynen, Max-Planck-Institut für Biochemie, D-8033 Martinsried b. München;

Prof. Dr. Dr. G. Weitzel, Physiologisch-Chemisches Institut der Universität, Hoppe-Seyler-Str. 1, D-7400 Tübingen;

Dr. A. Dillmann, Hoppe-Seyler-Redaktion, Am Klopferspitz, D-8033 Martinsried b. München, Tel. (089) 85 85-823.

Bei der Abfassung der Manuskripte bitten wir dringend, folgende Regeln zu beachten, die im wesentlichen mit den „Instructions to Authors" von *Biochimica et Biophysica Acta* und *European Journal of Biochemistry* übereinstimmen:

Es soll in Maschinenschrift (2zeilig, auch vorgesehener Kleindruck!) und mit 4 cm breitem Rand geschrieben werden. Vorgesehene Absätze sind deutlich zu markieren.

Seite 1 des Manuskripts soll enthalten (in normaler Schrift, nicht in Großbuchstaben):

a) Titel der Arbeit
b) Autorennamen (jeweils ein Vorname ausgeschrieben)
c) Name des Instituts oder Fachbereichs, in dem die Arbeit ausgeführt wurde (möglichst in der Sprache des Herkunftlandes), vorausgesetzt, daß die Leitung des Instituts oder Fachbereichs der Veröffentlichung zugestimmt hat.
d) Kurztitel, maximal 70 Satzzeichen
e) Adresse für den Versand der Korrekturen

Auf Seite 2 folgen

a) Eventuelle Widmung
b) Vollständige Postadresse der Autoren in der Sprache des Herkunftslandes
c) Ad-hoc-Abkürzungen (nur in seltenen Fällen zugelassen)
d) Enzyme in alphabetischer Reihenfolge der empfohlenen Trivialnamen mit systematischen Namen und EC-Nummern

Auf Seite 3 folgt die *Zusammenfassung* sowie eine Übersetzung des Titels und der (eventuell erweiterten) Zusammenfassung ins Englische (bzw. Deutsche). Diese Übersetzung oder deren sprachliche Überprüfung kann auch von der Redaktion übernommen werden.

Eventuelle *Danksagungen* sind an den Schluß des Manuskriptes zu stellen.

Literaturzitate im Text sind fortlaufend zu numerieren (nicht alphabetisch) und am Schluß des Manuskripts zu bringen. Die in Chemical/Biological Abstracts üblichen Zeitschriftenabkürzungen sind zu verwenden. Zur Anordnung s. IUB Commission of Editors of Biochemical Journals (1973) *Eur. J. Biochem.* **37**, 201 - 202; (1974) *diese Z.* **355**, 1 - 3.

Beispiel: Spackman, D.H., Stein, W.H. & Moore. S. (1958) *Anal. Chem.* **30**, 1190 - 1206.

Zeichnungen sollen einmal in reproduzierbarer Form mit schwarzer Tusche gezeichnet eingereicht werden; Zahlenskalen, gemessene Größen, Einheiten und sonstige Beschriftungen sind jedoch nur leicht mit weichem Bleistift einzutragen.

Die Abbildungen sollen so groß sein, daß sie auf 1/2 oder 1/3 verkleinert werden können. Deutlich beschriftete Kopien für die Referenten sind beizufügen.

Alle *Tabellen* und *Abbildungen* sind mit Legenden zu versehen, so daß die Ergebnisse auch ohne Zuhilfenahme des Textes ersichtlich werden. Es wird gebeten, die Legenden nicht zu unterstreichen und die Abbildungslegenden nicht auf die Abbildungen selbst, sondern auf ein eigenes Blatt zu schreiben.

Symbole für die am häufigsten vorkommenden Größen und ihre Einheiten sind in Tab. 1 zusammengestellt. Wir bitten streng zu unterscheiden, ob bei Ansätzen molare Konzentrationen (mol/*l* = M; z.B. 0.25M) oder zugesetzte Mengen (z. B. in mol, mmol, μmol oder nmol) gemeint sind.

Hinsichtlich der *Nomenklatur* richte man sich nach der IUPAC-Nomenklatur sowie nach den Empfehlungen der "IUPAC-IUB Commission on Biochemical Nomenclature": Kohlenhydrate, *Eur. J. Biochem.* **21**, 455 - 477 (1971); **25**, 4 (1972); α-Aminosäuren, *Eur. J. Biochem.* **53**, 1 - 14 (1975);

Tab. 1. Symbole für Größen und ihre Einheiten

Größe		Einheit[a]
l	Länge	m, mm, μm, nm, etc.
V	Volumen	m^3, l (dm^3), ml, etc.
λ	Wellenlänge	nm (nicht mμ oder Å)
t	Zeit	h, min, s, ms, μs, etc.
ν	Frequenz	Hz
m	Masse	kg, g, mg, μg, etc.
ρ	Dichte	g/cm^3
s	Sedimentationskoeffizient	$S = 10^{-13} s$
n	Substanzmenge	mol, mmol, etc.
M	molare Masse (m/n)	g/mol
–	Masse eines Moleküls	Dalton[b]
M_r	Molekulargewicht	(dimensionslos)
c_B, [B]	molare Konzentration einer Substanz B	M, mM, μM, etc. (besser als 10^{-3} M, etc.)
T	Temperatur (absolut)	K (nicht °K)
t, θ	Temperatur	°C
q, Q	Wärmemenge	J (besser als cal)[c]
I	Ionenstärke	M (mol/l), mM, etc.
K	Gleichgewichtskonstante	
K_m	Michaeliskonstante	
K_i	Inhibitorkonstante	
k	Geschwindigkeitskonstante	
v	Reaktionsgeschwindigkeit	mol/s (oder mmol, μmol oder nmol/s, etc.), besser als μmol/min
V (nicht V_{max})	Geschwindigkeit der enzymkatalysierten Reaktion bei unendlich großer Substratkonzentration	
I	Lichtintensität	cd
T	Durchlässigkeit (I/I_0)	(dimensionslos)
A	Absorption[d] (– lg T)	(dimensionslos)
ϵ	molarer Extinktionskoeffizient	$M^{-1} \times cm^{-1}$

[a] Präfixe für Einheiten: T, Tera (10^{12}); G, Giga (10^9); M, Mega (10^6), k, Kilo (10^3); m, Milli (10^{-3}); μ, Mikro (10^{-6}); n, Nano (10^{-9}); p, Piko (10^{-12}); f, Femto (10^{-15}); a, Atto (10^{-18}).

[b] Das Dalton ist 1/12 der Masse eines Atoms des Nuklids ^{12}C.

[c] Das Joule ist die SI-Einheit der Energie. Die Kalorie (1 cal = 4.184 J; 1 J = 0.239 cal) ist für eine Übergangszeit noch zugelassen; in jedem Fall sollten aber die Angaben auch in Joule ausgedrückt werden.

[d] Die Ausdrücke optische Dichte, Extinktion oder Absorbancy sollen nicht mehr verwendet werden.

Cyclite, *diese Z.* **350**, 523 - 535 (1969), *Eur. J. Biochem.* **57**, 1 - 7 (1975); Lipide, *diese Z.* **350**, 279 - 285 (1969); Steroide, *diese Z.* **351**, 663 - 689 (1970); Chinone mit isoprenoiden Seitenketten, Folsäure u. verschiedene andere Verbindungen, *diese Z.* **348**, 266 - 276 (1967); Corrinoide, *Eur. J. Biochem.* **45**, 7 - 12 (1974); Vitamine B_6, *Eur. J. Biochem.* **40**, 325 - 327 (1973); Carotinoide, *Eur. J. Biochem.* **25**, 397 - 408 (1972), **57**, 317 - 318 (1975); synthetische Analoga natürlicher Peptide, *diese Z.* **348**, 262 - 265 (1967); Enzyme Nomenclature, Recommendations (1972), Elsevier Publ. Comp., Amsterdam; multiple Enzymformen, *diese Z.* **353**, 852 - 854 (1972); Eisen-Schwefel-Proteine, *Eur. J. Biochem.* **35**, 1 - 2 (1973); Human-Immunglobuline, *Eur. J. Biochem.* **45**, 5 - 6 (1974); Peptidhormone, *Eur. J. Biochem.* **55**, 485 - 486 (1975). Für die Rechtschreibung ist „Jansen/Mackensen, Rechtschreibung der technischen und chemischen Fremdwörter" maßgebend.

Tab. 2 Empfohlene Abkürzungen.

Atome, isotope	=	z.B. [16-^{14}C]Östron
Atome, numeriert	=	z.B. C-5
Curie	=	Ci
Impulse pro Minute	=	Ipm (engl. cpm)
Prozent der Theorie	=	% d. Th.
prozentig	=	proz. (nicht %-ig)
Relative Wanderungsgeschwindigkeit	=	R_F
Schmelzpunkt	=	Schmp. (engl. m.p.)
Siedepunkt	=	Sdp. (engl. b.p.)
Umdrehungen pro Minute	=	U./min (engl. rpm)
Zerfälle pro Minute	=	Zpm (engl. dpm)
Zersetzung	=	Zers. (engl. decomp.)

Bezüglich der *Abkürzungen und Symbole für chemische Namen,* die in der biologischen Chemie von besonderem Interesse sind, wird auf die Empfehlungen der „IUPAC-IUB Commission on Biochemical Nomenclature" verwiesen (siehe z.B.: Allgemeines, Aminosäuren, Zucker, Phosphate, Coenzyme, *diese Z.* **348**, 245 - 265 (1967); Aminosäurederivate und Peptide, *Eur. J. Biochem.* **27**, 201 - 207 (1972); synthet. Polypeptide, *diese Z.* **349**, 1013 - 1016 (1968); Nucleinsäuren, Polynucleotide und deren Bestandteile, *diese Z.* **351**, 1055 - 1063 (1970). Andere als die dort empfohlenen Abkürzungen werden nur in seltenen Ausnahmefällen zugelassen. Sie müssen am Beginn jeder Arbeit in einer Fußnote erklärt werden.

Sonderdrucke: Von jeder Arbeit werden dem Verfasser 75 Sonderdrucke kostenlos geliefert. Die Bestellung weiterer Sonderdrucke gegen Berechnung ist möglich, muß jedoch spätestens bei der Rücksendung der Korrekturen erfolgen.

Der Verlag erwirbt mit der Annahme des Manuskriptes das Verlagsrecht und das ausschließliche Recht der Vervielfältigung, Verbreitung und Übersetzung. Der Verlag behält sich vor, den Inhalt dieser Zeitschrift für fremdsprachige Ausgaben zu verwenden.

Bezugspreis: Der Jahresbezugspreis (1 Jahrg. zu 12 Heften) beträgt DM 590,–, das Einzelheft kostet DM 56,–. Bestellungen nimmt jede Buchhandlung entgegen.

Verlag: WALTER DE GRUYTER & Co.
Genthiner Str. 13, D-1 Berlin 30. Tel. (030) 261 1341
Telegrammadresse: Wissenschaft Berlin
Postscheck: Berlin-West 10307-108

WALTER DE GRUYTER Inc.
3 Westchester Plaza
Elmsford, N.Y. 10523, U.S.A.
Tel. (914) 592-5890

Für die nächsten Hefte sind folgende Beiträge vorgesehen:

Synthesis of protected decapeptide (B_{21-30}) of human insulin, *V.K. Naithani and J. Föhles*

Preparative polyacrylamide gel electrophoresis of nuclear RNA, *R. Böcker and W. Kersten*

On the significance of sialic acid in high affinity 5-hydroxytryptamine uptake by synaptosomes, *G.A. Dette and W. Wesemann*

Comparison of tRNA-nucleotidyltransferase from *Escherichia coli* and *Lactobacillus acidophilus*, *M. Leineweber and G.R. Philipps*

On the linkages between chromophore and protein in biliproteins, VII: Amino acid sequence in the chromophore regions of C-phycoerythrin from *Pseudanabaena* W 1173 and *Phormidium persinicum*, *G. Muckle, J. Otto and W. Rüdiger*

Substrate specificity of neuraminidase from *Erysipelothrix rhusiopathiae*, *H. von Nicoali, H.E. Müller and F. Zilliken*

A new organ spectrophotometer for sensitive dual-wavelength absorbance measurement and spectral scanning of intact perfused organs, *H. Schwab and H. Sies*

The elimination in dogs of trypsin-α-macroglobulin complexes inactivated by the Kazal or the Kunitz inhibitor, *A. Eddeland and K. Ohlsson*

The amino acid compositions of the tryptic, chymotryptic and peptic peptides from the L-2 light chain of rabbit skeletal muscle myosin, *G. Matsuda, Y. Suzuyama, M. Setoguchi, T. Maita, B. Watanabe, T. Maekawa and T. Umegane*

A new method for the preparation of band 3, the main integral protein of the human erythrocyte membrane, *D. Schubert and B. Domning*

Immunoreactive trypsin in serum and peritoneal fluid in acute pancreatitis, *A. Borgström and K. Ohlsson*

ıd 359 Heft 3 März 1978

ʒN 0018-4888 A 20679 E

Ioppe-Seyler's Zeitschrift für 'hysiologische Chemie

Frühjahrstagung der Gesellschaft für Biologische Chemie

Herausgegeben von

A. Butenandt, F. Lynen, G. Weitzel

unter Mitwirkung von K. Bernhard, K. Decker, J. Engel, H. Fritz, B. Hamprecht,
H. Hanson, E. Helmreich, H. Herken, B. Hess, N. Hilschmann, H. Hilz, P.W. Jungblut,
ʒarlson, H.L. Kornberg, K. Kühn, F. Leuthardt, D. Oesterhelt, K. Rajewsky, J. Seelig, G. Siebert,
imon, Hj. Staudinger, W. Stoffel, H. Tuppy, H. Wiegandt, H.G. Wittmann, H.G. Zachau, H. Zah

Redaktion A. Dillmann, G. Peters

WALTER DE GRUYTER · BERLIN · NEW YORK

Hoppe-Seyler's Z. Physiol. Chem. Bd. 359 S. 251 - 429 März 1978
HSZPAZ 359 (3) 251 - 429 (1978)

Hoppe-Seyler's Zeitschrift für Physiologische Chemie
Begründet 1877 · Fortgeführt von A. Kossel, F. Knoop und K. Thomas

Bd. 359 März 1978 Heft 3

Inhaltsverzeichnis / Contents

Fortsetzung nächste Seite *Continued*

Redaktionelle Hinweise siehe gegenüber Seite 430

Indexed in Current Contents

Autoren-Verzeichnis

Autoren-Verzeichnis der Kurzreferate für die Frühjahrstagung

Hoppe-Seyler's Z. Physiol. Chem.
Bd. 359, S. 251 - 343, März 1978

Frühjahrstagung der Gesellschaft für Biologische Chemie

12. - 15. März 1978, Frankfurt am Main

Autorenreferate der angemeldeten Vorträge und Poster

Die Referate sind nach der alphabetischen Reihenfolge der *Erstautoren* geordnet.

H. Abberger, H.-P. Buscher, W. Gerok, W. Kramer and G. Kurz

Identity of Bile Acids Conjugating Enzymes

The number and the specificity of enzymes which catalyze the transfer of the acyl group from hydroxycholanoyl-SCoA to appropriate acceptors in bile acid metabolism is still a controversial subject[1-3].

In order to gain insight into the formation and transport of conjugated bile acids, we have purified the hydroxycholanoyl transferase activity (EC 2.3.1.?) from rat liver 300-fold. In general, the stop-time assay was used:

hydroxycholanoyl-SCoA + acceptor = conjugated hydroxycholanic acid + CoASH

CoASH + Nbs_2 = CoAS-Nbs + Nbs-H

[Nbs_2 = 5,5′-dithiobis(2-nitrobenzoic acid)].

Experiments were done using glycine, taurine or L-arginine[4,5] as acceptor. Whereas L-arginine proved not to act as acceptor at all, the following results were obtained with glycine and taurine:

1) Throughout the purification procedure (ammonium sulfate fractionation, DEAE-chromatography, CM-chromatography, gel permeation chromatography) the ratio of the reaction velocities with glycine and taurine did not change.
2) This ratio of reaction velocities also remained constant during heat inactivation; the same heat inactivation curve was obtained with both acceptors.
3) The apparent molecular weight of the transferase acitivity was the same with glycine as with taurine, as determined by gel permeation chromatography (preliminary M_r = 60000).
4) After electrophoresis in polyacrylamide gels at pH 7.4 and 8.3, the transferase activity appeared at the very same position with glycine as with taurine.
5) After isoelectric focusing in polyacrylamide gels as well as in sucrose gradients (pH range 3.5 to 10.0 and 6.0 to 8.0, respectively) the transferase activity behaves identically with both acceptors. The isoelectric point of the enzyme is 7.05 ± 0.05.

These results demonstrate clearly that the conjugation of bile acids, starting out from the hydroxycholanoyl-SCoA intermediates, is catalyzed by the same enzyme.

1 Bremer, J. (1956) *Acta Chem. Scand.* **10**, 56 - 71.
2 Killenberg, P. C. (1976) *Gastroenterology* **71**, A-23/916.
3 Vessey, D. A., Crissey, M. H. & Zakim, D. (1977) *Biochem. J.* **163**, 181 - 183.
4 Myher, J. J., Marai, L., Kuksis, A., Yousef, I. M. & Fisher, M. M. (1975) *Can. J. Biochem.* **53**, 583 - 590.
5 Yousef, I. M. & Fisher, M. M. (1975) *Can. J. Physiol. Pharmacol.* **53**, 880 - 887.

Hubert Abberger, Werner Kramer, Gerhart Kurz, Lehrstuhl Biochemie, Chemisches Laboratorium der Universität, Albertstraße 21, D-7800 Freiburg i. Br.
Hans-Peter Buscher, Wolfgang Gerok, Medizinische Klinik der Universität, Hugstetter Straße 55, D-7800 Freiburg i. Br.

V. K. Antonov

Specificity and Mechanism of Acid Proteinases

One of the best known acid proteinases, porcine pepsin (EC 3.4.23.1), catalyzes peptide hydrolysis in acidic medium, transpeptidation of the acyl- and amino-transfer types, and also carboxyl oxygen exchange in acylamino acids. These properties had served as basis for regarding pepsin catalysis as proceeding by a nucleophilic mechanism with formation of covalent "acyl-enzyme" and "amino-enzyme" intermediates[1]. We have shown, however, that in pepsin catalysis the intermediates are non-covalent complexes of the enzyme with the hydrolysis products, and that pepsin functions by the mechanism of general base catalysis. This is substantiated by the following experimental data: a) Leucyl-leucine, a product of the pepsin-catalyzed transpeptidation of leucyltyrosine amide, exchanges its amide oxygen for ^{18}O, when the reaction is carried out in heavy-oxygen water; b) The non-reacted leucyl-tyrosine amide undergoes oxygen exchange in the peptide bond to the extent of ca. 20%; c) The rate of oxygen exchange in the carboxyl group of the acylamino acid increases in the presence of some tripeptides due to pepsin-catalyzed synthesis-hydrolysis of the corresponding tetrapeptides; d) No oxygen exchange in the carboxyls of the pepsin active site occurs in the course of the substrate hydrolysis in $H_2{}^{18}O$.

Therefore pepsin (and, presumably, other acid proteinases cf. ref.[2]) acts as a general base catalyst in the functioning of which water participates at the stage immediately following formation of the Michaelis complex. A kinetic scheme is proposed for the pepsin-catalyzed transfer reactions which involves a substrate-induced conformational change of the enzyme with a subsequent chemical stage and relaxation of the "activated" enzyme-product complex. It is with this complex that the receptor is assumed to react. Such a mechanism is in accord with the kinetic specificities of pepsin catalysis with respect to peptides of different chain length.

1 Fruton, J.S. (1976) *Adv. Enzymol.* **44**, 1 - 36.
2 James, M.N.G., Hsu, I.-N. & Delbeare, L.T.J. (1977) *Nature (London)* **267**, 808 - 813.

Prof. Dr. V. Antonov, Shemyakin Institute of Chemistry of Natural Products, USSR Academy of Sciences, Vavilova Street 32, USSR.

J. Arendes, R. K. Zahn und W. E. G. Müller

Sialyltransferase assoziiert mit dem Species-spezifischen Aggregationsfaktor in Schwämmen

Die Untersuchungen von Zell-Zell-Wechselwirkungen am biologischen Modell des Kieselschwamms *Geodia cydonium* zeigen, daß die Reaggregation dissoziierter Zellen in 3 Phasen unterteilt werden kann:

1) Bildung primärer Aggregate in Anwesenheit von $Ca^{2\oplus}$;
2) Bildung sekundärer Aggregate und
3) Ausbildung eines funktionellen Systems.

Die beiden letzten Phasen erfordern die Anwesenheit eines Aggregationsfaktors, der bei Schwämmen species-spezifisch ist[1].

Der Aggregationsfaktor aus *Geodia* ist als Untereinheit mit großen, extrazellulären Partikeln assoziiert. Als weiteres funktionelles Molekül tragen diese Partikel eine Sialyltransferase (= Glykoprotein-Sialinsäure-Transferase), die am Reaggregationsprozeß beteiligt ist. Die Experimente wurden in intakten Zellen und isolierten Enzymsystemen durchgeführt.

Während der Aggregation wird die *Geodia*-Zellmembran enzymatisch sialysiert. Desialysierte Zellen haben eine niedrigere Aggregationskapazität als die Kontrollen. Der Sialysierungsprozeß ist temperaturabhängig und kann durch *N*-Äthylmaleinimid inhibiert werden. Die Sialysierung erfolgt hauptsächlich an einer Komponente der Zelloberfläche, dem Aggregationsrezeptor[2].

Die Sialyltransferase kann durch folgende Schritte isoliert und gereinigt werden: Sepharose 4B, CM-Cellulose, Nonidet-Behandlung und Sephadex 100. Die erzielte Reinigung ist 680fach mit einer Ausbeute von 31%. Das Enzym katalysiert den Transfer von Sialinsäure (*N*-Acetylneuraminsäure) auf den Aggregationsrezeptor. Das Molekulargewicht der Sialyltransferase beträgt 52000, die gereinigte Transferase hat ein pH-Optimum von pH 7.75 und benötigt 200mM NaCl zur maximalen Aktivität. $Mg^{2\oplus}$ oder $Ca^{2\oplus}$ sind nicht erforderlich. Kinetische Untersuchungen zeigen keine lag-Phase und eine Abhängigkeit von der Enzymkonzentration. Die Reaktion wird inhibiert durch 10μM *N*-Äthylmaleimid[3].

1 Müller, W. E. G. & Zahn, R. K. (1973) *Exp. Cell. Res.* **80**, 95 - 104.
2 Müller, W. E. G., Müller, I., Zahn, R. K. & Kurelec, B. (1976) *J. Cell. Sci.* **21**, 227 - 241.
3 Müller, W. E. G., Arendes, J., Kurelec, B., Zahn, R. K. & Müller, I. (1977) *J. Biol. Chem.* **252**, 3836 - 3842.

J. Arendes, R. K. Zahn und W. E. G. Müller, Institut für Physiologische Chemie der Universität Mainz, Duesbergweg, D-6500 Mainz.

F. J. Barrantes

Ligand-Induced Fluorescence Changes in the Membrane-Bound Acetylcholine Receptor from Torpedo marmorata.

Conformational transitions in the state of the acetylcholine receptor have been postulated to account for the pharmacological effects exerted by cholinergic ligands. Considerable effort has been devoted in recent years to the search for model systems and appropiate techniques to identify such conformational states and transitions.

In order to tackle the above problem, we have applied fluorescence techniques, with their inherent high sensitivity and time resolution, to the study of the cholinergic receptor in its membrane environment using a) synthetic fluorescent probes whose pharmacological specificity is assessed in vivo using electrophysiological methods (see e.g. ref.[1]), and b) non-fluorescent cholinergic ligands of recognised pharmacological activity and a physical signal arising from intrinsic reporter groups in the receptor itself[2,3].

The most relevant question addressed via the above approach concerns the identification of discrete conformational states in the membrane-bound receptor from *Torpedo marmorata*, the state transitions induced by cholinergic ligands, and their kinetic pathways. An initial step consistent with the binding of the agonist to a "low" affinity form of the receptor has been recently observed with suberyldicholine*. The on- and off- rate constants for this step ($R + L \rightleftharpoons RL$) were determined to be $9.8 \times 10^6 M^{-1}s^{-1}$ and $10 s^{-1}$ respectively. A larger fluorescence change follows in the time scale of seconds, consistent with the isomerisation of the liganded state to a "high" affinity state ($RL \rightleftharpoons DL$). This third state (DL) differs from both the free (R) and liganded (RL) conformers in its natural fluorescence and other properties.

The minimal reaction mechanism accounting for fluorescence changes, α-toxin inhibition and other parameters is consistent with the cyclic scheme postulated from studies in vivo[4], the most conspicuous physical change being attributed to the phenomenon of "desensitisation"[4].

* Barrantes, F. J., submitted for publication.

1 Barrantes, F. J., Sakmann, B., Bonner, R. F., Eibl, H. & Jovin, T. M. (1975) *Proc. Natl. Acad. Sci. U.S.A.* 72, 3097 - 3101.
2 Barrantes, F. J. (1976) *Biochim. Biophys. Res. Commun.* 72, 479 - 488.
3 Bonner, R. F., Barrantes, F. J.,& Jovin, T. M. (1976) *Nature (London)* 263, 429 - 431.
4 Katz, B. & Thesleff, S. (1957) *J. Physiol. (London)* 138, 63 - 80.

F. J. Barrantes, Max-Planck-Institut für Biophysikalische Chemie, Postfach 968, am Faßberg, D-3400 Göttingen-Nikolausberg.

C. A. Barth and I. Hillmar

Gluconeogenesis and Glycogen Levels in Primary Cultures of Adult Rat Hepatocytes

We have further investigated whether hepatocytes in primary culture will preserve organotypic metabolic features. For this purpose glycogen metabolism and gluconeogenesis were studied in this system.

Isolation of hepatocytes and culture conditions were as described by us in an earlier communication[1]. Cultured hepatocytes proved to be glucocorticoid-sensitive, as they reacted with a considerable rise of tyrosine aminotransferase (EC. 2.6.1.5) to the addition of 10^{-5}M dexamethasone. Therefore gluconeogenesis was measured after a 24-h preincubation in 10^{-5}M dexamethasone by incubating hepatocyte monolayers in Krebs-Ringer hydrogencarbonate supplemented with 10^{-5}M adrenalin. Glucose liberation in presence of 10mM alanine plus 10mM lactate plus 1mM pyruvate was 0.65 μmol/(min × *g* wet wt.) and in presence of 10mM fructose amounted to 1.42 μmol/(min × *g* wet wt.). As the simultaneous glycogenolysis proved to be insignificant (less than 6% of glucose liberation) we conclude that gluconeogenesis proceeds in cultured hepatocytes as in the perfused liver or in vivo.

In conclusion, gluconeogenesis is a further metabolic pathway, besides ketogenesis and urea synthesis[2], that is preserved quantitatively in hepatocytes kept in culture.

1 Barth, C. A., Willershausen, B., Walther, B. & Weis, E. E. (1977) *Hoppe-Seyler's Z. Physiol. Chem.* 358, 211.
2 Barth, C. A. & Willershausen, B. (1977) XIth FEBS Meeting, Copenhagen, A 1-5052 3/4.

Dr. Christian Barth, Forschergruppe Ernährung an der Medizinischen Poliklinik, Pettenkoferstraße 8a, D-8000 München 2.

P. Bartholmes und H. Balk

Kinetik der Bindung von Pyridoxal-5'-phosphat an Tryptophan-Synthase aus Escherichia coli

Wie bereits beschrieben[1], bindet Pyridoxal-5'-phosphat, das Coenzym von Tryptophan-Synthase [L-Serin-Hydro-Lyase (Indol addierend) (EC 4.2.1.20)][2,3], kooperativ an die isolierte β_2-Untereinheit mit einem Hill-Koeffizienten $n_H = 1.7$. Die mikroskopischen Dissoziationskonstanten, K_d, wurden unter Anwendung der Adair-Gleichung zu 8.7μM und 0.23μM bestimmt. Im Gegensatz dazu bindet Pyridoxal-5'-phosphat nicht kooperativ mit $K_d = 1.0 \mu M$ an den nativen Komplex $\alpha_2\beta_2$. Diese Beobachtungen legen nahe, daß die β_2-Untereinheit in wenigstens drei verschiedenen Konformationszuständen existiert, von welchen einer durch die α-Untereinheit stabilisert wird.

Dementsprechend werden Verstärkung der Fluroeszenzemission, Fluoreszenz-Energie-Transfer und optisches Differenzspektrum der Bindung des Koenzyms an β_2 sowie an den nativen $\alpha_2\beta_2$-Komplex in mehreren, voneinander unterscheidbaren Prozessen aufgebaut. Deren kinetische Analyse zeigt für β_2, daß nach Auftreten eines Primärkomplexes die Bildung des Azomethins von einer langsamen Konformationsänderung begleitet wird.

Es folgt ein weiterer, konzentrationsunabhängiger Isomerisierungsprozeß, der dem Auftreten der vollen enzymatischen Aktivität parallel verläuft. Mit $\alpha_2\beta_2$ wird neben dem schnellen Bindungsschritt nur ein langsamer Isomerisierungsvorgang beobachtet, der die Azomethinbildung begleitet.

Entsprechende Mechanismen für die Bindung des Koenzyms an Tryptophan-Synthase werden diskutiert.

1 Barholmes, P., Kirschner, K. & Geschwind, H.-P. (1976) *Biochemistry* **15**, 4712 - 4717.
2 Yanofsky, C. (1960) *Bacteriol. Rev.* **24**, 221 - 245.
3 Yanofsky, C. & Crawford, I. P. (1972) *Enzymes* **7**, 1 - 31.

Dr. Peter Bartholmes, Institut für Biophysik und Physikalische Biochemie, Universität Regensburg, Universitätsstr. 31, D-8400 Regensburg.

R. Baur, H. Wohlert and H. Kröger

Methylation of DNA in Developing Embryos of the Sea Urchin Psammechinus miliaris

It has often been suggested that 5-methylcytosine might affect gene expression and differentiation[1-3]. The experimental background giving rise to this hypothesis comes from observations that DNA from various tissues or from various developmental stages of the same organism shows different concentrations of 5-methylcytosine[4-7].

When we worked with various developmental stages of *Xenopus laevis* in connection with the problem of methylation of DNA, we were able to show that there was no significant difference in the concentration of 5-methylcytosine detectable[8]. The assay we used in order to detect 5-methylcytosine involved base separation by medium pressure liquid chromatography and quantitative detection by ultraviolet absorption. The method is quite sensitive and detects as little as 100 pmol of 5-methylcytosine.

In this presentation we describe some experiments on embryos of the sea urchin *Psammechinus miliaris* taking advantage of this method of quantification of 5-methylcytosine. In additional assays we controlled the 5-mCyt/(5-mCyt + Cyt) ratios by labeling the embryos with [6-^{3}H]uridine. As uridine is metabolised to thymidine and cytidine, we could separate [6-^{3}H]cytosine from [6-^{3}H]5-methylcytosine and calculate the ratios mentioned above by counting the disintegrations per minute. This technique avoids the problem of pool size inherent in radioactive labeling with methionine.

Embryos of the sea urchin *Psammechinus miliaris* were cultured in filtered sea water. At the desired developmental stage, i.e. 32 cells, blastula, gastrula and pluteus stage, the embryos were harvested. DNA was extracted and hydrolysed. The nuclear bases of all various developmental stages were separated as mentioned earlier. The ratios of molarities of 5-methylcytosine divided by (5-methylcytosine plus cytosine) and the ratios of disintegrations per minute of [6-^{3}H]5-methylcytosine divided by ([6-^{3}H]-5-methylcytosine plus [6-^{3}H]cytosine) were calculated. Both determinations yielded 4.7% ± 0.3 for all developmental stages and did not show significant differences. Even the ratio for DNA from sperm of *Psammechinus miliaris* did not show significant deviation from the ratio determined for the embryos.

These results are in contradiction to various earlier reports, which showed delayed methylation of DNA during developmental processes. The hypothesis that methylation of DNA may play an important role in differentiation cannot be supported by quantifying the 5-methylcytosine of total DNA at various developmental stages of different organisms.

1 Holliday, P. & Pugh, I.E. (1975) *Science* **187**, 226 - 233.
2 Sager, R. & Kitchin, R. (1975) *Science* **189**, 426 - 433.
3 Riggs, A.D. (1975) *Cytogenet. Cell Genet.* **14**, 9 - 25.
4 Vanyushin, B.F., Mazin, A.L., Vasilyev, V.K. & Belozersky, A.N. (1975) *Biochim. Biophys. Acta* **299**, 397 - 403.
5 Comb, D.G. (1965) *J. Mol. Biol.* **11**, 851 - 855.
6 Scarano, E., Iaccarino, M., Grippo, P. & Parisi, E. (1967) *Proc. Natl. Acad. Sci. U.S.A.* **57**, 1394 - 1398.
7 Adams, R.L.P. (1973) *Nature (London), New Biol.* **244**, 27 - 29.
8 Baur, R. & Kröger, H. (1976) *Hoppe-Seyler's Z. Physiol. Chem.* **357**, 308.

Dr. Roland Baur, Robert-Koch-Institut, Abteilung Biochemie, Nordufer 20, D-1000 Berlin 65.

H. Blum, B. Weber, B. Deus, H. Ueberberg und W. Gerok

Nachweis eines mitochondrialen Isoenzyms der Kreatin-Kinase in menschlichem Herzmuskel

Im Rahmen von Studien über das Vorkommen von Isoenzymen der Kreatin-Kinase (EC 2.7.3.2) in verschiedenen Muskeltypen des Menschen wurde die subzelluläre Lokalisation dieses Enzyms im Herzmuskel untersucht.

Die Aktivität der Kreatin-Kinase-Isoenzyme wurde quantitativ mit dem optischen Test unter Zusatz von spezifischen inhibierenden Antikörpern sowie nach elektrophoretischer Auftrennung auf Celluloseacetat-Folie durch Aktivitätsfärbung in Gegenwart der spezifischen inhibierenden Antikörper bestimmt. Qualitativ wurden die Isoenzyme der Kreatin-Kinase mit spezifischen präzipitierenden Antikörpern durch Doppel-Immundiffusion nach Ouchterlony nachgewiesen.

Mit diesen Methoden ließen sich im Homogenat des menschlichen Herzmuskels die folgenden Isoenzyme der Kreatin-Kinase identifizieren:

CK-MM (M = Muscle)
CK-BB (B = Brain)
CK-MB (Hybrid)
CK-Mitochondrial.

Durch Differential-Zentrifugation des Herzmuskel-Homogenats erhielten wir vier subzelluläre Fraktionen, deren Reinheit durch Messung der Aktivität der entsprechenden „Marker"-Enzyme sowie elektronenoptisch nachgewiesen wurde.

Im Sarkoplasma fanden sich die Isoenzyme CK-MM, CK-BB und CK-MB. In den Myofibrillen wurde nur CK-MM nachgewiesen, ebenso im sarkoplasmatischen Retikulum. In der gereinigten Mitochondrien-Fraktion fand sich ausschließlich das mitochondriale Isoenzym der Kreatin-Kinase. Auch nach Solubilisierung war keines der anderen Isoenzyme der Kreatin-Kinase nachweisbar.

Dieses mitochondriale Isoenzym der Kreatin-Kinase zeigt keine Antigen-Gemeinschaft mit inhibierenden oder präzipitierenden Antikörpern gegen die M- bzw. B-Untereinheit der Kreatin-Kinase. Im Unterschied zu den Isoenzymen CK-MM, CK-BB und CK-MB wandert das mitochondriale Isoenzym bei pH 8.6 kathodenwärts.

Daraus schließen wir, daß das mitochondriale Isoenzym nicht durch Assoziation der bisher bekannten Untereinheiten CK-M bzw. CK-B entsteht, sondern eine besondere molekulare Variante der Kreatin-Kinase darstellt.

Weiterführende Untersuchungen haben ergeben, daß das mitochondriale Isoenzym der Kreatin-Kinase nicht nur im Herzmuskel, sondern auch in menschlichen Skelettmuskeln mit überwiegend oxidativer Stoffwechsel-Leistung (z.B. M. Soleus, M. Gastrocnemius) nachweisbar ist.

Dr. Hubert Blum, Zentrum Innere Medizin der Universität Freiburg, Hugstetter Straße 55, D-7800 Freiburg.

V. Braun, K. Hantke, A. Hartmann, S. Hussein und H. J. Krieger-Brauer

Multiple Funktionen der Rezeptorproteine bakterieller Zelloberflächen

Als Beispiele für die multifunktionellen Eigenschaften von Proteinen der Zellhülle von *Escherichia coli* werden Rezeptoren für Phagen und Colicine beschrieben, die ebenfalls für den Durchtritt von Komplexen des dreiwertigen Eisens mit Ferrichrom, Enterochelin und Citrat durch die äußere Membran essentiell sind[1]. Diese Proteine werden unter Eisenmangel gebildet. Zur Induktion des Eisencitrat-Transportsystems muß Eisencitrat von außen angeboten werden. Intracelluläres Citrat hat keine induzierende Wirkung. Zur Aufrechterhaltung des Eisencitrat-Transportsystems bedarf es viel geringerer Citratkonzentrationen als zur primären Bildung. Fluorcitrat induziert das Eisentransportsystem, ist aber kein Eisen-Carrier und gelangt nicht in wachstumshemmender Konzentration in die Zelle. Im Falle der Ferrichromaufnahme kann der Komplexbildner wiederholt zum Transport benutzt werden[2]. Das strukturanaloge Antibiotikum Albomycin wird über das gleiche Transportsystem aufgenommen. Die ursprünglich paradoxe Beobachtung, daß kaum [^{3}H]Acetat-markiertes Albomycin in der Zelle zu finden ist, konnte durch Markierung mit ^{35}S insoweit geklärt werden, als von dem Molekül ein Teil abgespalten wird, der in der Zelle verbleibt. Somit ist es möglich, daß der Eisenkomplex nur als Träger für das Eindringen des Antibiotikums in die Zelle dient und erst nach Hydrolyse ein Spaltstück antibiotisch aktiv ist.

Die Aufnahmegeschwindigkeit von Nucleosiden, außer Cytidin und Desoxycytidin, wird in Mutanten (*tsx*), denen das gemeinsame Rezeptorprotein für den Phagen T6 und das Colicin K in der äußeren Membran fehlt, um 60 - 70% reduziert. Die Aufnahme der freien Basen und der Nucleosidmonophosphate wird nicht vom *tsx*-Protein beeinflußt. Mutanten in einem weiteren Hauptprotein der äußeren Membran (*ompA*) zeigen einen um 20 - 30% verringerten Nucleosidtransport (ausgenommen Thymidin) und die Zellen sind Colicin-K-tolerant, was auf eine funktionelle Interaktion der Proteine *tsx* und *ompA* schließen läßt.

Es gibt Hinweise, daß der Funktionszustand der äußeren Membranproteine von der Energetisierung der Cytoplasmamembran abhängt. Im Gegensatz zu den Rezeptorsystemen bei Eukaryonten werden sehr verschiedene Substanzen an ein Rezeptorprotein gebunden und können die Zelle in verschiedener Weise beeinflussen.

1 Braun, V. & Hantke, K. (1977) in Microbial Interactions (Receptors and Recognition, Series B, Vol. 3), (Reissig, J. L., Hrsg.) S. 101 - 137.
2 Braun, V., Hancock, R. E. W., Hantke, K. & Hartmann, A. (1976) *J. Supramol. Struct.* **5**, 37 - 58.

Prof. Dr. V. Braun, Institut für Biologie II, Universität Tübingen, Auf der Morgenstelle 28, D-7400 Tübingen 1.

H. Breer and S. J. Morris

Ions and Lectins Binding to Purified Cholinergic Synaptic Vesicles from the Electromotor Synapse of Torpedo marmorata

The release of transmitters stored in synaptic vesicles is induced by the influx of extracellular $Ca^{2\oplus}$, but the molecular action of the cations in the presynaptic terminal is still unknown, although some interaction with vesicle membranes is suggested. Lanthanides such as $Eu^{3\oplus}$ and $Tb^{3\oplus}$ have been used as fluorescent and electronmicrographic probes for $Ca^{2\oplus}$-binding sites on erythrocytes[1] and chromaffin granule membranes[2].

Vesicles were isolated and purified by isoosmotic density gradients followed by chromatography on CPG-10-3000 porous glass beads. This separation method yields vesicles of high purity which are free from soluble proteins and other membrane structures[3]. The titration of vesicle membranes resuspended in 2mM HEPES buffer with $TbCl_3$ resulted in an enhancement of fluorescence at 543 nm accompanied by a depression of tryptophan emmission (330 nm) at an excitation wavelength of 290 nm. A binding constant (K_d) of about 2.56μM for $Tb^{3\oplus}$ was determined. The binding of terbium to vesicle membranes is competitively inhibited by $Ca^{2\oplus}$ ions ($K_i \approx 1.00$mM), demonstrating that $Tb^{3\oplus}$ is binding to calcium-binding sites of the membrane proteins. $Mg^{2\oplus}$ cations compete slightly less effectively ($K_i \approx 1.55$mM).

It has been demonstrated that Ruthenium Red inhibits the binding of $Ca^{2\oplus}$ ions to plasma membranes and calcium binding proteins[4]. At low concentration of Ruthenium Red, a marked inhibition of $Tb^{3\oplus}$ binding to the vesicle membranes was shown; a dye concentration of 1μM inhibits the maximal binding capacity by more than 50%. In contrast to the divalent cations the interaction of Ruthenium Red with $Tb^{3\oplus}$ binding could be characterized as a non-competitive inhibition.

Ruthenium Red is known to react selectively with mucopolysaccharides and related substances in biological tissue[5]; therefore the membrane of water-lysed synaptic vesicles was explored for carbohydrates by measuring the binding capacity for various lectins with the agglutination method of Meyer and Burger[6]. Vesicle membranes were agglutinated by concanavalin A and by *Ricinus communis* agglutinin. No agglutination could be demonstrated with wheat germ agglutinin.

1 Mikkelsen, R. B. & Wallach, D. F. H. (1974) *Biochim. Biophys. Acta* **363**, 211 - 218.
2 Morris, S. J. & Schober (1977) *Eur. J. Biochem.* **75**, 1 - 12.
3 Breer, H., Morris, S. J. & Whittaker, V. P. (1977) *Eur. J. Biochem.* **80**, 313 - 318.
4 Kamino, K., Ogawa, M., Uyesaka, N. & Inouye, A. (1976) *J. Membr. Biol.* **26**, 345 - 356.
5 Luft, J. H. (1971) *Anat. Rec.* **171**, 347 - 368.
6 Meyer, D. J. & Burger, M. M. (1976) *Biochim. Biophys. Acta* **443**, 428 - 436.

H. Breer and S. J. Morris, Abteilung Neurochemie, Max-Planck-Institut für Biophysikalische Chemie, Postfach 968, D-3400 Göttingen.

R. Brossmer and D. Roelcke

Inhibition of Human Anti-Gd cold Agglutinins by Sialyl-Lactose

Human cold agglutinins with the serological specificities anti-I/-i, anti-$Pr_{1\text{-}3}$, and anti-Gd are autoantibodies directed against red blood cells with a temperature optimum of 0 °C. High-titer cold agglutinins occur transiently in the course of virus and *Mycoplasma pneumoniae* infections, or persistently in chronic lymphoproliferative diseases. Two examples of IgM κ-monotypic cold agglutinins, occurring in patients with Non-Hodgkin lymphomas, reacted with erythrocyte autoantigens, which were protease resistant but were inactivated by neuraminidase. The cold agglutinins, termed anti-Gd, were inhibited by the trisaccharide sialyl-lactose [NeuAc(α,2→3) Gal(β,1→4) Glc], while anti-I/-i and anti-$Pr_{1\text{-}3}$ cold agglutinins were not inhibited, although $Pr_{1\text{-}3}$ determinants are inactivated by neuraminidase, like Gd determinants. NeuAc and lactose alone did not inhibit anti-Gd. Therefore at least the disaccharide NeuAc (α,2→3) Gal seems to be essential for Gd antigen determinants. The benzyl α-ketoside of NeuAc had only a slight inhibitory effect. This points also to a significant role of a subterminal sugar molecule for Gd antigen activity. The use of other NeuAc-containing oligosaccharides should help to determine the definite structure of Gd and other antigen determinants (I/i, $Pr_{1\text{-}3}$), which correspond to cold agglutinins.

Reinhard Brossmer, Institut für Biochemie II (med. Fak.), Universität Heidelberg, Im Neuenheimer Feld 328, D-6900 Heidelberg, und Dieter Roelcke, Institut für Immunologie und Serologie, Sektion für Immunhämatotologie, Universität Heidelberg, Im Neuenheimer Feld 305, D-6900 Heidelberg.

D. E. Büchel, P. H. Schreier and R. W. Davies

DNA-Sequence of a Mutant Attachment-site (b508) of Escherichia coli Phage λ

Bacteriophage λ integrates into its host chromosome by recombination between specific sites of the bacterial and viral genome, the so called attachment sites (*att*-sites). We previously determined the nucleotide sequence of the *att*-site of λ[1,2]. Landy and Ross[3] found a sequence of 15 base pairs which is common to both bacterial and viral *att*-sites. To get more detailed information about the recombination process we determined the DNA sequence of the λ deletion mutant b508, which shows a decreased frequency of site-specific recombination[4]. λ b508 carries a deletion of 3.9% of the λ genome, including the right part of *att* and the genes *int* and *xis* necessary for integration and excision. We used a new sequencing technique (Sanger, F., Nicklen, S. & Coulson, A.R. (1977) in press) priming on a λ single strand with restriction fragment HinfI 1 (γ)[1,5]. The DNA sequences of λ b508 and λ wild type are homologous to the left of the common core; within the common core three bases at positions 11 - 13 are different. No homology can be detected further to the right of the common core (below in brackets).

5′ GTTCA[GCTTTTTTATACTAA]GTTGG 3′
3′ CAAGT[CGAAAAAATATGATT]CAACC 5′
λ wild type

5′ GTTCA[GCTTTTTTAT*GTA*AA]*GAAAC* 3′
3′ CAAGT[CGAAAAAATA*CAT*TT]*CTTTG* 5′
λ b508

1 Davies, R.W., Schreier, P.H. & Büchel, D.E. (1977) *Nature (London)* **270**, 757 - 760.
2 Davies, R.W., Schreier, P.H. & Büchel, D.E. (1977) *Hoppe-Seyler's Z. Physiol. Chem.* **358**, 1191 - 1192.
3 Landy, A. & Ross, W. (1977) *Science* **197**, 1147 - 1160.
4 Parkinson, J.S. (1971) *J. Mol. Biol.* **56**, 385 - 401.
5 Schreier, P.H., Wayne Davies, R., Büchel, D.E., Gronenborn, B., Fanning, T.G. & Wilcken, von B. (1977) *Nature (London)* **267**, 555 - 557.

D.E. Büchel and P.H. Schreier, Institut für Genetik der Universität zu Köln, Weyertal 121, D-5000 Köln 41. R.W. Davies, Department of Biology, University of E Essex, Wivenhoe Park, Colchester CO4 3SQ UK.

Z. I. Cabantchik, H. Ginsburg and M. Barzilay

Transport of Nucleosides across Mammalian Cell Membranes: Kinetic and Chemical Properties

Nucleosides penetrate mammalian cell membranes via systems of facilitated diffusion. These systems in general, and that of red blood cells in particular, were studied in order to characterize transport in terms of simple steady-state kinetic schemes and to probe chemically the underlying molecular mechanisms. The erythrocyte system is the first known example of an asymmetric transport system which is fully consistent with a simple carrier model[1,2]. Although the kinetic parameters (K_m, J_m) differed with respect to the direction of flux measurement by a factor of 4 - 5, they did so in a compensatory manner, such that J_m/K_m was the same, regardless of the procedure used to characterize transport. Differential resistance to the inward and outward movement of free carrier (or carrier conformation) accounted for the asymmetry of the system[3].

The nucleoside systems differed in their response to SH group modification, but shared common susceptibilities to S-substituted or R-Hg derivatives of 6-mercaptopurine ribosides, which inhibited transport in a competitive 1:1 fashion, displaying K_i values in the nanomolar range[4]. The high affinity inhibitors were used:

1) on the basis of kinetic profiles, to estimate the number of carriers (10 000 - 50 000 per cell, depending on the system)[4];
2) in the form of radioactive photoreactive analogs, e.g. (*p*-azidophenacyl)-6-[^{35}S]mercaptopurine riboside, to identify functional components;
3) in the form of non-penetrating analogs of *S*-benzyl-6-mercaptopurine riboside coupled to dextran to study sidedness of inhibitory effects in erythrocyte ghosts and its relationship to the transmembrane (kinetic) asymmetry of the system, and
4) to establish that stimulatory and/or inhibitory effects of either free or dextran-coupled organomercurials resulted from conformational changes induced by ligand-carrier interactions[5,6].

1 Cabantchik, Z.I. & Ginsburg, H. (1977) *J. Gen. Physiol.* **69**, 75 - 96.
2 Lieb, W.R. & Stein, W.D. (1974) *Biochim. Biophys. Acta* **373**, 178 - 196.
3 Ginsburg, H. & Cabantchik, Z.I. (1977) *J. Gen. Physiol.* **70**, 679 - 680.
4 Eilam, Y. & Cabantchik, Z.I. (1977) *J. Cell Physiol.* **92**, 185 - 202.
5 Heichal, O., Bibi, O., Katz, J. & Cabantchik, Z.I. (1978) *J. Membr. Biol.* **39**, in press.
6 Bibi, O., Schwartz, J., Eilam, Y., Shohami, E. & Cabantchik, Z.I. (1978) *J. Membr. Biol.* **39**, in press.

Z. Ioav Cabantchik, Hagai Ginsburg and Moshe Barzilay, LKEM/NHLBI/NIH, 10/6N320, 9000 Rockville Pike, Bethesda, Md. 20014 U.S.A, 301-496-3032.

E. Conzelmann and K. Sandhoff

G_{M2}-Gangliosidosis: Evidence for the Deficiency of a Factor Stimulating the Enzyme-Catalyzed Degradation of Ganglioside G_{M2} in the AB-Variant

Human kidney extracts heated to 60 °C and devoid of hexosaminidase activity (EC 3.2.1.30) stimulate more than 20-fold the hexosaminidase A-catalyzed degradation of ganglioside G_{M2} and of glycolipid G_{A2}, the neuronal storage compounds of G_{M2} gangliosidosis. The stimulating factor of this extract, which is labile at temperatures above 60 °C, is also present in kidney extracts from patients of infantile G_{M2} gangliosidosis having a deficiency of hexosaminidase A (Tay-Sachs disease, variant

B) and a deficiency of hexosaminidase A and B (variant O). Evidence is presented that this factor is defective in the AB-variant of infantile G_{M2} gangliosidosis, which is characterized by an accumulation of glycolipids G_{M2} and G_{A2} despite the fact that the degrading enzymes, hexosaminidase A and B, retain normal activity levels. Variant AB is the first example of a fatal lipid storage disease which is not caused by a defect of a degrading enzyme but rather by a defective factor necessary for the interaction of lipid substrates and the water-soluble hydrolase.

E. Conzelmann and K. Sandhoff, Max-Planck-Institut für Psychiatrie, Neurochemische Abteilung, Kraepelinstraße 2, D-8000 München 40.

C. Daub and H. U. Weltzien

Mechanism of Hemolysis by Saturated and Unsaturated Lysolecithin Analogs

Both the number of carbon atoms in the apolar part of the molecule[1] and the number of *cis*-double bonds play an important role in determining physico-chemical, cytolytic, and adsorption properties of lysolecithin analogs (LLA). We have investigated the interactions of 1-linoleoyl- (18 : 2) and 1-stearoyl-propandiol-3-phosphorylcholine (18 : 0) with human erythrocytes, using 0.5% cell suspensions and ^{14}C-labeled analogs, and determining $K^{\oplus}$ and hemoglobin (Hb) in the supernatant as indicators for hemolysis.

At 0 °C, Hb release by 18 : 2 is much slower than by 18 : 0, the rate of hemolysis being only slightly dependent on the concentration of the analog and the extent of lysis not exceeding 70% even after 4 h with 40 nmol/m*l* of 18 : 2. 18 : 0 lyses the cells totally at a concentration of 8 nmol/m*l* within 40 min.

Hb release does not take place with 18 : 0 unless more than 50% of the $K^{\oplus}$ content has been released, whereas $K^{\oplus}$ and Hb efflux begin simultaneously with 18 : 2, but continue at different rates.

Adsorption kinetics of the two substances show the reverse order. Adsorption equilibrium is attained within less than a minute by 18 : 2, whereas 18 : 0 takes 10 to 20 min to reach equilibrium. The ratio LLA_{bound}/LLA_{free} is always higher for 18 : 0 than it is for 18 : 2.

We conclude from these data that hemolysis by 18 : 0 and 18 : 2, respectively, takes place by different mechanisms. Lysis by 18 : 0 probably occurs by action of the colloid-osmotic pressure after equilibration of the small molecules across the membrane. Incorporation of 18 : 2, however, seems to render the cell membrane more permeable but more resistant to osmotic pressure differences, so that the diffusion of the marker molecules across the membrane seems to be the rate-limiting step, whereas in the case of 18 : 0, adsorption or incorporation is rate-limiting.

1 Weltzien, H. U., Arnold, B. & Reuther, R. (1977) *Biochim. Biophys. Acta* **466**, 411 - 421.

C. Daub and H. U. Weltzien, Max-Planck-Institut für Immunbiologie, Stübeweg 51, D-7800 Freiburg.

B. Deus, J. Schmitz, H. Becher und W. Gerok

Regulatorische Eigenschaften von Glutamin-Synthetase aus menschlichem Gehirn

Glutamin-Synthetase (EC 6.3.1.2) wurde aus menschlichem Gehirn durch Chromatographie an DEAE-Cellulose, Filtration durch Sephadex G-100 und Dichtegradienten-Zentrifugation gereinigt. Das Enzym ergab in der Polyacrylamidgel-Disk-Elektrophorese eine einzige scharfe Bande.

Für die Aktivitätsmessung der Glutamin-Synthetase wurde ein radiochemischer Test in geeigneter Weise modifiziert.

Die Sättigungskonzentration der Glutamin-Synthetase für $NH_4^{\oplus}$ folgt einer Michaelis-Menten-Kinetik ohne Hemmung des Enzyms bei Substrat-Überschuß. Abhängig von der ATP-Konzentration werden optimale Konzentrationen für $Mg^{2\oplus}$- bzw. $Mn^{2\oplus}$-Ionen gefunden; bei weiterer Steigerung der Metallionen-Konzentrationen kommt es zu einer ausgeprägten Hemmung der Glutamin-Synthetase. Die $Mg^{2\oplus}$-abhängige Enzym-Aktivität wird durch Zusatz von $Mn^{2\oplus}$-Ionen stark vermindert. Bei Überschreiten einer optimalen ATP-Konzentration wird die Aktivität der Glutamin-Synthetase nahezu vollständig gehemmt. Für eine maximale Enzym-Aktivität muß ein bestimmtes Verhältnis zwischen ATP- und Metallionen-Konzentration im Testansatz vorhanden sein. Die Sättigungs-Konzentrationen des Enzyms für Glutamat sind in Gegenwart optimaler $Mg^{2\oplus}$- bzw. $Mn^{2\oplus}$-Konzentrationen identisch.

Die Aminosäuren Histidin und Tryptophan beeinflussen die Aktivität der Glutamin-Synthetase nicht; dagegen führen Serin, Alanin und Glycin konzentrationsabhängig zu einer starken nicht-kompetitiven Hemmung der Enzym-Aktivität.

2-Oxoglutarat steigert die Aktivität der Glutamin-Synthetase ohne Veränderung der Michaelis-Konstanten für Glutamat, während Citrat das Enzym kompetitiv hemmt. Succinat dagegen hat keinen Einfluß auf die Glutamin-Synthetase-Aktivität.

Durch Zusatz von anorganischem Phosphat wird die Aktivität der Glutamin-Synthetase gehemmt, und zwar nicht-kompetitiv. Carbamoylphosphat hemmt die Enzym-Aktivität erheblich stärker als äquimolare Mengen anorganischen Phosphats; die Hemmung ist kompetitiv.

Durch $NAD^{\oplus}$, CTP und AMP wird die Glutamin-Synthetase aus menschlichem Gehirn nicht beeinflußt.

Da Glutamin-Synthetase zu $NH_4^{\oplus}$-Ionen eine etwa 10mal so hohe Affinität besitzt wie zu Glutamat und die Enzymaktivität durch Überschuß an $NH_4^{\oplus}$-Ionen nicht gehemmt wird, könnte die Glutamin-Synthetase für die Assimilation des für das menschliche Gehirn toxischen Ammoniaks von großer Bedeutung sein.

Prof. Dr. Bruno Deus, Zentrum Innere Medizin der Universität Freiburg, Hugstetter Straße 55, D-7800 Freiburg.

G. Dickneite, H. P. Jennissen and L. M. G. Heilmeyer, Jr.

Phosphorylation of Troponin and Phosphorylase b by Two Different Sites of Phosphorylase Kinase

Phosphorylase kinase (EC 2.7.1.38), a large oligomer with the subunit structure $\alpha_4 \beta_4 \gamma_4$ (M_r 1.25×10^6) is known to phosphorylate the inactive phosphorylase *b* (EC 2.4.1.1), thus forming the active *a* form. This kinase also phosphorylates the *T* subunit of troponin, a regulatory protein of the actomyosin ATPase. The latter phosphorylation occurs ca. 500-fold more slowly than that of phosphorylase *b*, which could be due to a lower catalytic effectiveness for troponin, *T*, or to a contamination of phosphorylase kinase with another protein kinase. The present study shows that both the phosphorylase and troponin kinase activities are enriched together from a crude extract up to the homogeneous enzyme. Therefore, it is likely that one protein carries both activities. However, when the enzyme is eluted from a DEAE-column, the activity ratio between phosphorylase *b* and troponin phosphorylation is not constant in all fractions. Concomitantly a small shift in the subunit composition of phosphorylase kinase is observed. This might suggest that both activities are located on different subunits. Antibodies against phosphorylase kinase inhibit phosphorylase *b* phosphorylation noncompetitively, whereas troponin phosphorylation is competitively inhibited. The phosphorylation of phosphorylase *b* is not inhibited by troponin. If one catalytic center for both enzymes is assumed, a competitive inhibition by the second substrate would be expected.

These results lead to the conclusion that phosphorylase kinase is a multienzyme complex with at least two different catalytic centers.

Dipl.-Biol. Gerhard Dickneite, Ruhr-Universität Bochum, Institut für Physiologische Chemie, Lehrstuhl I, Postfach 102148, D-4630 Bochum 1.

C. M. Dobson, R. G. Ratcliffe and R. J. P. Williams

The Study of Biological Compartments In Vivo by NMR

Biological molecules have been successfully studied by high resolution NMR in the range of very small to very large molecular weights. Their conformational mapping often uses magnetic perturbations. The application of low resolution NMR to the mapping of space, zeugmatography using field gradients, is also under rapid development, but here only the total signal from all protons, mainly water, is observed. We are making a series of exploratory experiments on the use of *high resolution NMR* in the study of biological systems in vivo. This is at present very largely an analytical approach, but there are new experimental ways of attacking the macroscopic mapping of space using both molecular and macroscopic probe techniques.

As an example I shall illustrate the way in which some components of the adrenal gland have been studied[1]. The method is applicable to systems of interest to neurobiologists, since we can study the adrenal vesicles in considerable detail in the intact organ. Manganese(II) can be used as a mapping probe. The ability to follow manganese distribution could also be of interest to neurobiologists.

In the study of other biological objects line position and line shape may be affected by bulk susceptibility, and it could well be possible to distinguish the environment of a molecule on this basis alone. Chemical shifts and line widths will not be easy to interpret, however, since the situation can be confused by pH, and ionic strength differences in different compartments.

The bulk susceptibility in a particular region of space can be altered by the insertion of a macroscopic probe; e.g. glass-beads or a thin metal wire. All resonances from nuclei close to the wire are perturbed. The magnetic field in a particular region of space can also be changed by current flowing through the wire when the perturbation produced is proportional to the current flowing and dependent upon the geometry of the current-carrying insert, e.g. a straight wire or a coil. Some of these possibilities will be described using model systems. The methods can be used with a wide range of nuclei. High fields and large sample space are a great advantage. The prospect of mapping chemicals in biological compartments using NMR merits intensive further study.

1 Daniels, A., Korda, A., Tanswell, D., Williams, A. R. & Williams, R. J. P. (1974) Proc. Royal Society, London B187, 353 - 361.

C. M. Dobson, R. G. Ratcliffe and R. J. P. Williams, Inorganic Chemistry Laboratory, Oxford University, South Parks Road, Oxford, England.

J. Eckel, M.L. Rao, G.S. Rao and H. Breuer

Kinetics of Thyroid Hormone Transport through Cell Membranes: Comparison of the Transport of L-Triiodothyronine by Isolated Rat Liver Cells with Binding to Rat Liver Cytosol

After entering the cell the thyroid hormone binds to intracellular binding proteins which carry the hormone to the nucleus, the site of biological action. The role of such proteins for the action of triiodothyronine (T_3) is not fully understood. The present studies were initiated to assess the part played by these cytosol proteins in the uptake of T_3 by the intact liver cell. Viable rat liver cells suspended in a medium containing [^{125}I]triiodothyronine take up the hormone by a saturable process; the uptake exhibits positive cooperativity[1]. Cellular energy, in the form of ATP, is involved in this process. Treatment of cells with metabolic inhibitors does not lead to significant leakage of intracellular proteins, or intracellular T_3 binding components. The transport of T_3 into the cell proceeded 20-fold faster at 23 °C than at 1.5 °C. The binding of T_3 to cytosol proteins was 1.6-fold faster at 23 °C than at 1.5 °C. During the initial phase of transport, cells take up 4.4 fmol of T_3 per mg cell protein in 15 s, whereas 0.13 fmol of T_3 is bound per mg cytosol protein. The binding of T_3 by cytosol proteins is insensitive to inhibitors of metabolic energy, indicating that cellular ATP is not required.

These results show that T_3 is actively accumulated by liver cells, and that the transport is sensitive to energy blocking agents. The amount of T_3 bound to cytoplasmic proteins is 1/30 compared to transport of T_3 into liver cells, and is not inhibited by energy blocking agents. In conclusion, transport of T_3 into the cell is mediated by protein components of the plasma membrane and does not involve intra-cellular binding proteins.

1 Rao, G. S., Eckel, J., Rao, M. L. & Breuer, H. (1976) *Biochem. Biophys. Res. Commun.* **73**, 98 - 104.

Prof. Dr. Govind S. Rao, Institut für Klinische Biochemie der Universität Bonn, Venusberg, D-5300 Bonn 1.

E. Eigenbrodt, B. Schering, H. Brunn und W. Schoner

Studien zur Phosphorylierung und Inaktivierung der Pyruvat-Kinase M_2

Inkubation von Nierenrindenschnitten hungernder Ratten mit 20mM Lactat und 1mM Dibutyryl-cAMP führt zu einer Stimulierung der Gluconeogenese um 25% [Kontrolle: 10.48 ± 0.4 µmol Glucose/(g Frischgew. × 30 min)] und zu einer Abnahme des Pyruvat-Kinase-Fluxes um 30% [Kontrolle: 10.3 ± 0.8 µmol/(g Frischgew. × 30 min)]. Die Stimulierung der renalen Gluconeogenese durch 2µM Parathormon war ebenfalls mit einer Abnahme des Pyruvat-Kinase-Fluxes verbunden. Da in der Niere 90% der Pyruvat-Kinase-Aktivität (ATP:Pyruvat-Phosphotransferase, EC 2.7.1.40) aus Pyruvat-Kinase Typ M_2 besteht[1], Pyruvat-Kinase M_2 aber nicht durch eine cAMP-abhängige Kinase phosphoryliert werden soll[2], wurde untersucht, wie die Aktivität der Pyruvat-Kinase Typ M_2 reguliert werden kann.

Gereinigte Pyruvat-Kinase Typ M_2 aus Hühnerleber wird durch eine mit cAMP nicht stimulierbare Protein-Kinase phosphoryliert und inaktiviert. Der allosterische Effektor Fructose-1,6-bisphosphat verlangsamt die Phosphorylierung und Inaktivierung, Alanin beschleunigt sie. Vollständige Inaktivierung erfolgt bei einem Einbau von 1 mol Phosphat/mol Untereinheit der tetrameren Pyruvat-Kinase. Die Phosphopeptidbindung ist gegenüber Kochen mit 2N HCl bei 100 °C für 20 h stabil. Phosphorylierte inaktive Pyruvat-Kinase M_2 kann durch eine Phosphatase dephosphoryliert und reaktiviert werden.

Die cAMP-unabhängige Protein-Kinase vom Typ III[3] wird bei der Reinigung der Pyruvat-Kinase an das Enzym gebunden angereichert. Die partiell gereinigte Protein-Kinase zeigt in der Gelchromatographie ein Mol.-Gew. von 170 000. Der K_m-Wert für Pyruvat-Kinase Typ M_2 beträgt 6×10^{-10}M. Das Enzym hat eine breite Nucleotidspezifität und phosphoryliert auch Pyruvat-Kinase Typ L. Das gegenüber Phosphopyruvat-Kinase L andersartige kinetische Verhalten phosphorylierter Pyruvat-Kinase Typ M_2 und das Fehlen von Arginin-Resten im CNBr-Phosphopeptid aus phosphorylierter Pyruvat-Kinase sprechen gegen eine Übereinstimmung der Phosphatakzeptorstellen bei der Phosphorylierung mit cAMP-abhängiger Protein-Kinase[4,5] und der cAMP-unabhängigen Protein-Kinase.

1 Osterman, J., Fritz, P. J. & Wuntsch, T. (1973) *J. Biol. Chem.* **248**, 1011 - 1018.
2 Humble, E., Berglund, L., Titanji, V., Ljungström, O., Edlund, B., Zetterquist, Ö. & Engström, O. (1975) *Biochem. Biophys. Res. Commun.* **66**, 614 - 621.
3 Traugh, J. A., Asby, C. D. & Walsh, D. A. (1974) *Methods Enzymol.* **38**, 290 - 299.
4 Hjelmquist, G., Anderson, J., Edlund, B. & Engström, L. (1974) *Biochem. Biophys. Res. Commun.* **61**, 559 - 563.
5 Edlund, B., Anderson, J., Titanju, V., Dahlquist, K., Ekman, P., Zetterquist, Ö. & Engström, L. (1975) *Biochem. Biophys. Res. Commun.* **67**, 1516 - 1521.

Erich Eigenbrodt, Beate Schering, Hubertus Brunn und Wilhelm Schoner, Institut für Biochemie und Endokrinologie, FB 18, Frankfurter Straße 100, D-6300 Giessen.

B. Emmerich, H. Schuster, R. Baier und J. Rastetter

Analyse der relativen Synthese von Immunglobulin- und Globinpolypeptidketten im zellfreien System aus Plasmazelltumoren und Retikulozyten unter Einfluß von Aurintricarbonsäure

Beim Vergleich der Empfindlichkeit der Gesamtproteinsynthese im zellfreien System aus Mausplasmazelltumoren und Kaninchenretikulozyten fanden wir, daß das Retikulozytensystem gegenüber Aurintricarbonsäure 100fach empfindlicher ist als das Plasmazelltumorsystem[1]. Es konnte gezeigt werden, daß Aurintricarbonsäure in beiden Zellsystemen die Messenger-RNA-Bindung an die 40S-Ribosomenuntereinheit hemmt[2]. Kürzlich wurde berichtet, daß in Kulturen von Plasmazelltumoren bei einer Hemmung der Proteinsynthese-Initiation die Immunglobulinsynthese im Vergleich zur Synthese der übrigen endogenen Proteine relativ resistent ist[3,4]. Als Ursache dafür wurden unterschiedliche Bindungsaffinitäten zur 40S-Ribosomenuntereinheit diskutiert. Dies war für uns Anlaß zu untersuchen, ob der unterschiedlichen Empfindlichkeit gegenüber Aurintricarboxylsäure im Retikulozyten- und Plasmazelltumorsystem ein ähnlicher Mechanismus zugrunde liegt.

Methode: Auf Natriumdodecylsulfat-Polyacrylamidgelen wurden nach Inkubation mit und ohne Hemmstoff die endogenen Proteine des TEPC 15 (IgA, K) Plasmazelltumor- und Retikulozytensystems analysiert und ihre relative Synthesegeschwindigkeit durch Messung der Radioaktivität in den einzelnen Banden bestimmt. Die Hemmstoffkonzentration wurde so gewählt, daß in beiden Zellsystemen ca. 50% der Gesamtproteinsynthese blockiert war.

Resultate: Im TEPC-15-Plasmazelltumorsystem ist unter Einfluß von Aurintricarbonsäure die Einbaugeschwindigkeit in die leichte Immunglobulinpolypeptidkette geringer vermindert als die der übrigen Plasmazelltumorproteine. Im Retikulozytensystem dagegen ist die Einbaugeschwindigkeit in die Globinketten, die 80% der Gesamtproteinsynthese ausmacht, im gleichen Ausmaß wie die der restlichen Proteine vermindert.

Dieser Befund untersützt die Vorstellung, daß Immunglobulin-Messenger-RNAs im Vergleich zu anderen Messenger-RNAs eine besonders hohe Affinität zum Initiationskomplex haben.

1 Emmerich, B., Preis, R., Hoffmann, H., Erben, V. & Rastetter, J. (1976) *Hoppe-Seyler's Z. Physiol. Chem.* **357**, 291 - 292.

2 Emmerich, B., Preis, R., Baier, R. & Rastetter, J. (1977) *Hoppe-Seyler's Z. Physiol. Chem.* **358**, 230.

3 Nuss, D.L. & Koch, G. (1976) *J. Mol. Biol.* **102**, 601 - 612.

4 Sonensheim, G.E. & Brawerman, G. (1976) *Biochemistry* **15**, 5497 - 5501.

Dr. B. Emmerich, Hämatolog. Abteilung der 1. Med. Klinik der Technischen Universität München, Ismaninger Str. 22, D-8000 München 80.

H. A. Ensinger, H. E. Pauly and G. Pfleiderer

The Role of Zn(II) in Calf Intestinal Alkaline Phosphatase as Studied by the Influence of Chelating Agents and the Chemical Modification of Histidine Residues

Incubation of alkaline phosphatase (EC 3.1.3.1) with ethylenediamine tetraacetic acid at pH 8.0 results in a reversibly inactivated apoenzyme. At pH 9.8 this inactivation is irreversible. The Zn(II) and Mg(II) contents of native alkaline phosphatase and the reversibly and irreversibly inactivated apoenzymes were determined by atomic absorption spectroscopy, showing a direct relation between zinc content and residual activity of the enzyme. Azocoupling of the acetamidinated enzyme with diazonium-1-*H*-tetrazol results in a complete loss of enzymatic activity, after modification of histidine residues. Zinc analyses by atomic absorption spectroscopy of the azocoupled enzyme showed that these histidine residues are involved in binding of $Zn^{2\oplus}$ ions.

The kinetics of inactivation by chelating agents or by azocoupling of the histidine residues show a biphasic first order reaction with a change in the inactivation constant at 50% residual activity.

The conformation of the holoenzyme and the reversibly and irreversibly inactivated apoenzymes were investigated by means of optical rotatory dispersion and UV and fluorescence spectroscopy.

Helmut Alfred Ensinger, Institut für Organische Chemie, Biochemie und Isotopenforschung, Universität Stuttgart, Pfaffenwaldring 55, D-7000 Stuttgart 80.

U. Essig, J. Neeff and D. Mecke

Catabolite Degradation in Yeast: A Selective Process of Enzyme Regulation and of Cellular Economy

In the yeast *Saccharomyces cerevisiae*, some glucose-repressible enzymes [cytoplasmic malate dehydrogenase (EC 1.1.1.37), fructosebisphosphatase (EC 3.1.3.11), phospho*enol*pyruvate carboxykinase (EC 4.1.1.49), galactozymase, maltose uptake system] are rapidly inactivated after addition of glucose to derepressed cells[1]. At least for cytoplasmic malate dehydrogenase, there is now strong evidence that this inactivation is due to a rapid degradation of the enzyme protein[2,3].

In this paper we present experiments which show that glucose addition to radioactively labeled, derepressed cells causes a significant degradation of labeled proteins. The data are consistent with the interpretation that the addition of glucose not only initiates the degradation of enzymes that are no longer necessary or are even harmful, but also leads to a degradation of other derepressible proteins. It is therefore concluded that catabolite degradation does not function exclusively in enzyme regulation, but plays a more general role in cellular economy.

1 Holzer, H. (1976) *Trends Biochem. Sci.* **1**, 178 - 181.
2 Hägele, E., Neeff, J. & Mecke, D. (1978) *Eur. J. Biochem.* **83**, 67 - 76.
3 Neeff, J., Hägele, E., Nauhaus, J., Heer, U. & Mecke, D. (1978) *Biochem. Biophys. Res. Commun.*, in press.

Dr. Jürgen Neeff, Physiologisch-Chemisches Institut der Universität Tübingen, Hoppe-Seylerstraße 1, D-7400 Tübingen.

K. Feldmann, M. Hörl and K. Schnackerz

^{31}P NMR Studies on the Structure and Function of Glycogen Phosphorylase

Pyridoxal-*P* is indispensible for the catalytic activity of α-glucanphosphorylases (EC 2.4.1.1). The rabbit skeletal muscle glycogen phosphorylase can be activated in the presence of glycogen and anionic substrates either by addition of AMP or by conversion of phosphorylase *b* to phosphorylase *a* via phosphorylation of a specific seryl residue.

We performed ^{31}P NMR measurements at 72.8 MHz in 20 mm tubes with a Bruker WH 180 spectrometer in order

1) to clarify the role of pyridoxal-*P* and characterize its binding site in the dimeric and monomeric form of the enzyme[1];

2) to follow the well documented heterotropic interactions between the binding sites for AMP and substrates[2] and

3) the conformational transition on interconversion; and

4) to establish the function of the inhibitors ATP and glucose-6-*P*.

Adenosine-5′-*O*-(γ-thio)triphosphate (ATP[S]) was used to prepare an active phosphorylase *a* containing thiophosphoserine residues[3]. As effector, the analogue adenosine-5′-*O*-monothiophosphate (AMP[S]) activates phosphorylase *b* even more efficiently[4]. We substituted arsenate for inorganic phosphate[2] and reconstituted apo-phosphorylase *b* with pyridoxal 5′-deoxymethylene-phosphonate to an active enzyme[5]. Thus it was possible to separate the various phosphate signals.

Distinct forms of enzyme-bound pyridoxal-*P* have been identified[6]. The transition of the inactive native form of phosphorylase *b* to the active enzyme-effector complex in the presence of arsenate generates, via deprotonation, a dianionic phosphate group of the cofactor in a protected binding site. This form is already found in native active phosphorylase *a*. The effector or the phosphorseryl residue, interacting with two different groups of the protein, trigger this transition.

Phosphorylase monomers produced by limited succinylation exhibit the same pyridoxal-*P* resonance as native phosphorylase *b*. This shows that pyridoxal-*P* is shielded from the interaction with the solvent within the single subunit. Therefore the cofactor is not located at the interprotomeric contact surface.

ATP is a competitive inhibitor with respect to AMP and does not affect the pyridoxal-*P* resonance of the native enzyme. Glucose-6-*P* weakens the tight phosphorylase *b*-effector complex.

In contrast to phosphorylase, the pyridoxal-*P* of D-serine dehydratase (EC 4.2.1.14) shows a typical pH dependence of the ^{31}P resonance as compared with a model Schiff base. With the D-serine dehydratase-iso-serine complex, only the signal of the dianionic form of pyridoxal-*P* is observed, probably being fixed through a salt bridge to an arginine.

1 Feldmann, K., Hörl, M., Klein, H. & Helmreich, E. (1977) in 11th FEBS – Meeting Kopenhagen. Pergamon Press, in press.
2 Helmreich, E. & Cori, C. F. (1964) *Proc. Natl. Acad. Sci. U.S.A.* **51**, 131 - 138.
3 Gratecos, D. & Fischer, E. H. (1974) *Biochem. Biophys. Res. Commun.* **58**, 960 - 967.
4 Murray, A. W. & Atkinson, M. R. (1968) *Biochemistry* **7**, 4023 - 4029.
5 Vidgoff, J. M., Pocker, A., Hullar, T. L. & Fischer, E. H. (1974) *Biochem. Biophys. Res. Commun.* **57**, 1166 - 1174.
6 Feldmann, K. & Hull, W. H. (1977) *Proc. Natl. Acad. Sci.* U.S.A. **74**, 856 - 860.

Dr. Knut Feldmann, Physiologisch-Chemisches Institut der Universität Würzburg, Koellikerstraße 2, D-8700 Würzburg.

H. Fiedler und W. Kleinow

Auftrennung der Proteine aus mitochondrialen Ribosomen von Locusta migratoria durch zweidimensionale Gelelektrophorese

Die Ribosomen aus den Mitochondrien der Wanderheuschrecke *Locusta migratoria* haben eine Sedimentationskonstante von nur 60S und ribosomale-RNA-Moleküle von M_r = 520 000 und 280 000. Sie zeigen eine Morphologie, die der anderer Ribosomen entspricht[1,2]. Für die Ribosomen aus Vertebratenmitochondrien – die in den genannten Merkmalen mit den Ribosomen aus *Locusta*-Mitochondrien weitgehend übereinstimmen – wurde durch zweidimensionale Gelelektrophorese eine hohe Zahl an ribosomalen Proteinen festgestellt (Zusammenstellung s.l.c.[3]). Unsere Versuche sollen prüfen, ob dieses Merkmal ebenfalls für die mitochondrialen Ribosomen aus Invertebraten zutrifft.

Ribosomen wurden aus gradientengereinigten Mitochondrien gewonnen, das Lysieren der Mitochondrien geschah in Anwesenheit hoher Magnesiumkonzentrationen, um Proteinverunreinigungen der Ribosomen zu vermindern[4]. Die Ribosomen wurden anschließend durch Zentrifugieren in einem Gradienten, der 0.5M Ammoniumchlorid und 0.001M Magnesiumchlorid enthielt, weiter gereinigt und in ihre Untereinheiten zerlegt. Die Freisetzung der ribosomalen Proteine erfolgte durch Inkubation der Untereinheiten mit RNA-spaltenden Enzymen, da die Abtrennung der RNA mit 66proz. Essigsäure bei diesem Ribosomentyp anscheinend nicht vollständig erreicht werden kann. Eine befriedigende Auftrennung der Proteine durch die zweidimensionale Gelelektrophorese nach l.c.[5,6] gelang nur, wenn die mitochondrialen Ribosomen unmittelbar nach dem Präparieren analysiert wurden.

Die Auszählung der Flecke nach Auftrennung der ribosomalen Proteine in der zweidimensionalen Gelelektrophorese ergab bisher folgende Resultate:
Große Untereinheit – 24 stärkere und 16 schwächere Flecken,
kleine Untereinheit – 20 stärkere und 24 schwächere Flecken,
undissoziierte Ribosomen – 42 stärkere und 37 schwächere Flecken.
Hieraus läßt sich die Zahl 79 als obere Grenze für die mögliche Anzahl verschiedener ribosomaler Proteine bei *Locusta* ableiten.

In Übereinstimmung mit den Befunden an Wirbeltiermitochondrien zeigte ein relativ großer Anteil der aus der mitochondrialen Ribosomenfraktion freigesetzten Proteine einen isoelektrischen Punkt unter 8.6.

1 Kleinow, W., Neupert, W. & Miller, F. (1974) *J. Cell Biol.* **62**, 860 - 875.
2 Kleinow, W. (1977) *Insect Biochem.* **7**, 121 - 124.
3 Faye, G. & Sor, F. (1977) *Mol. Gen. Genet.* **155**, 27 - 34.
4 Kleinow, W. (1974) *Hoppe-Seyler's Z. Physiol. Chem.* **355**, 1027 - 1034.
5 Kaltschmidt, E. & Wittmann, H.G. (1970) *Anal. Biochem.* **36**, 401 - 412.
6 Howard, G.A. & Traut, R.R. (1973) *FEBS-Lett.* **29**, 177 - 180.

Prof. Dr. W. Kleinow, Zoologisches Institut der Universität zu Köln, Weyertal 119, D-5000 Köln 41.

I. Filipovic and E. Buddecke

Metabolism of Homologous Lipoproteins in Cultured Human and Bovine Aortic Smooth Muscle Cells as Influenced by Hypoxia

Cultured human and bovine aortic smooth muscle cells internalize homologous ^{125}I-labelled low density lipoproteins (LDL) under a gas phase of 95% O_2/5% CO_2 by two processes: (1) by a specific, probably receptor-mediated process saturable at 30 - 40 µg lipoprotein-protein/m*l* incubation medium and (2) by a nonspecific endocytosis which is not saturable up to 100 µg lipoprotein-protein/m*l* medium. High density lipoproteins (HDL) are internalized only by nonspecific endocytosis. Both human and bovine smooth muscle cells take up more LDL than HDL, the amount of internalized LDL and HDL per mg cell protein being 3 times higher in bovine than in human cells.

During a 3 h chase experiment, the cells released 25 - 30% of the internalized radioactivity into the chase medium, more than 1/3 of which was precipitable by specific antisera and approximately 10% as trichloroacetic acid-soluble material. The distribution pattern of the labelled lipid subfractions eliminated into the chase medium did not reflect that of the cells, thus suggesting that the elimination of lipids from the cells is the result of a selective export process.

In hypoxia (95% N_2/5% CO_2 as gas phase), the uptake of lipoproteins by the cells proceeds by the same mechanisms as in aerobiosis, but the total LDL or HDL radioactivity found in the cells at the end of the incubation period is 35 - 40% higher than that found in aerobiosis. Pulse-chase experiments revealed that the increase in the cellular radioactivity was caused by a diminished regurgitation of internalized lipoproteins in hypoxia. Under these conditions, the lipid-bound radioactivity of lipoproteins was retained to a higher extent than apoprotein-bound radioactivity. This finding is interpreted as a regulatory mechanism in hypoxia in which the insufficient endogenous synthesis of polyunsaturated long chain fatty acids and phospholipids might be compensated by an increased retention of internalized lipoprotein lipids.

Ivan Filipovic and Eckhart Buddecke, Institut für Physiologische Chemie, Universität Münster, Waldeyerstr. 15, D-4400 Münster.

G. Fischer und H. Jatzkewitz

Funktion des Aktivators der Sulfatase A

Beim enzymatischen Abbau von Sphingolipiden in vitro müssen im allgemeinen Detergentien bzw. spezifische Proteine (z.B. Lit.[1]) zugesetzt werden, die als Aktivatoren oder Effektoren bezeichnet werden.

Der Aktivator der Sulfatase A (Cerebrosid-Sulfatase) (EC 3.1.6.8) ist ein niedermolekulares hitzestabiles Protein (molare Masse ca. 20000), das den enzymatischen Abbau von Sulfogalakto-sphingolipiden und -glycerolipiden aktiviert. Diese Aktivierung zeigt ein pH-Optimum bei pH 4.8 - 5.0 und nimmt mit zunehmender Ionenstärke ab. Wie am Beispiel des enzymatischen Abbaus der Sulfatide (Cerebrosid-3-sulfat) gezeigt werden konnte, beruht diese Aktivierung auf der Bildung von Aktivator-Lipid Komplexen der Stöchiometrie 1 : 1[2], die dem Enzym als Substrat dienen[3]. Dies konnte durch Untersuchungen wahrscheinlich gemacht werden, die zeigten, daß

1) die Abbaurate der Sulfatide mit der Konzentration des Aktivator-Lipid-Komplexes zunimmt
2) die Komplexbildung durch Cerebroside und Phosphatidylserin im selben Ausmaß gehemmt wird wie der enzymatische Abbau der Sulfatide
3) die Komplexbildung zwischen Aktivator und Sulfatiden reversibel ist.

Da sowohl Sulfatase A als auch der Aktivator in den Lysosomen lokalisiert sind[4] könnte dieser Aktivierungsmechanismus auch auf In-vivo-Bedingungen beim Abbau der Sulfatide übertragbar sein. Der Abbau anderer Sphingolipide durch saure, lysosomale Sphingolipid-Hydrolasen verläuft möglicherweise analog.

1 Li, S.-C. & Li, Y.-T. (1976) *J. Biol. Chem.* **251**, 1159 - 1163.
2 Fischer, G. & Jatzkewitz, H. (1977) *Biochim. Biophys. Acta* **481**, 561 - 572.
3 Fischer, G. & Jatzkewitz, H. (1978) *Biochim. Biophys. Acta* **528**, 69 - 76.
4 Mraz, W., Fischer, G. & Jatzkewitz, H. (1976) *Hoppe-Seyler's Z. Physiol. Chem.* **357**, 1181 - 1191.

Dr. G. Fischer, Max-Planck-Institut für Psychiatrie, Abteilung Neurochemie, Kraepelinstr. 2, D-8000 München 40.

E.-Ch. Foerster, B. Wühr und F. Lynen

Eine empfindliche Methode zur Bestimmung von Malonyl-CoA im pmol-Bereich

Die Malonyl-CoA-Konzentration der Rattenleber unter verschiedenen Stoffwechselbedingungen

Zur Beantwortung der Frage, ob die Synthese von Malonyl-CoA den Regulationsschritt bei der Fettsäurebiosynthese darstellt, wurde die Malonyl-CoA-Konzentration der Rattenleber unter verschiedenen Stoffwechselbedingungen untersucht: Haltungsdiät, Hunger, Fettfütterung, Kohlenhydratfütterung, Alloxan-Diabetes.

Mit folgender radiochemischen Methode können Malonyl-CoA-Mengen noch unter 10 pmol quantitativ bestimmt werden:
Der vom Eiweiß befreite Gewebeextrakt wird mit [B-^{3}H]NADPH, Acetyl-CoA und Fettsäure-Synthetase aus Hefe versetzt, um das anwesende Malonyl-CoA quantitativ in radioaktive Fettsäuren umzuwandeln. Die gebildeten Fettsäuren werden extrahiert und ihre Radioaktivität gemessen. Dies bedeutet eine Steigerung der Nachweisempfindlichkeit von Malonyl-CoA um den Faktor 10^3 gegenüber dem photometrischen Test[1].

Die Beobachtungen legen nahe, daß die Acetyl-CoA-Carboxylase (EC 6.4.1.2) in Rattenleberzellen die Fettsäurebiosynthese reguliert. Dieses Ergebnis steht im Einklang mit Bestimmungen des Acetyl-CoA-Carboxylase-Gehaltes der Leberzellen unter verschiedenen metabolischen Bedingungen[2-5].

1 Guynn, R. W. & Veech, R. L., *Methods Enzymol,* **XXXV**, (Lipids) Part. B, 312 - 315.
2 Nakanishi, S. & Numa, S. (1970) *Eur. J. Biochem.* **16**, 161 - 173.
3 Nakanishi, S. & Numa, S. (1971) *Proc. Natl. Acad. Sci. U.S.A.* **68**, 2288 - 2292.
4 Nishikori, K., Iritani, N. & Numa, S. (1973) *Febs-Lett.* **32**, 19 - 21.
5 Numa, S., Nakanishi, S., Hashimoto, T., Iritani, N. & Okazaki, T. (1970) *Vitam. Horm. (Leipzig)* **28**, 213 - 243.

E.-Ch. Foerster, Max-Planck-Institut für Biochemie, Abteilung Prof. Dr. F. Lynen, D-8033 Martinsried/ München.

H. Follmann und W. Ludwig

Spezifität, Regulationsmuster und Struktur bakterieller Ribonucleotid-Reduktasen

Die zwei bekannten Typen der Ribonucleotid-Reduktasen [Nicht-Häm-Eisen-haltige Diphosphat-Reduktasen (EC 1.17.4.1) und Coenzym-B_{12}-abhängige Triphosphat-Reduktasen (EC 1.17.4.2)][1] zeichnen sich durch ungewöhnlich vielseitige allosterische Regulation des Umsatzes ihrer vier natürlichen Substrate aus[2,3]; Effektoren sind Desoxyribonucleotide oder/und ATP. Mit basenmodifizierten Purinnucleotiden ist jedoch zu zeigen, daß die Enzyme aus *Escherichia coli* (Typ I) und *Lactobacillus leichmannii* (Typ II) generell wenig substratspezifisch sind (nur Ribonucleotide mit *syn*-Konformation oder

der anionischen Xanthinbase sind inaktiv), und daß es unter Substraten und Effektoren keine spezifischen Kombinationen gibt. Vielmehr modifizieren die meisten Nucleosidtriphosphate der Desoxyribose, Ribose und Arabinose die Enzymaktivität in je nach Substrat unterschiedlichem Maße, von Stimulierung bis zu nichtkompetitiver Hemmung. Unter unphysiologischen Nucleotiden sind z.B. 8-Brom-dATP, 2,6-Diaminopurinribotid, *ara*ATP und *ara*CTP wirksame Effektoren. Substituenteneffekte sprechen für H-brückenartige Beziehungen zwischen Nucleobasen und Aminosäureresten im System Substratnucleotid/Protein/Effektornucleotid(e). Ein „zweiseitiger induced fit"[4] vermag die Mehrzahl der individuellen kinetischen Daten zu erklären. Das sehr flexible Verhalten des Enzymproteins – das sich von der klassischen allosterischen Regulation unterscheidet – scheint Vorbedingung für vergleichbare Reduktionsgeschwindigkeiten aller Pyrimidin- und Purinnucleotide zu sein. Weitgehende Ähnlichkeiten in Spezifität und Regulation der beiden in Protein- und Cofaktorstruktur sehr verschiedenen Enzymtypen lassen auf gleiche katalytische Prinzipien der Ribonucleotidreduktion schließen, eine ähnliche Grundeinheit der bakteriellen Reduktasen mit drei Nucleotidbindungszentren (Molekulargewicht ca. 80000) wird erkennbar. Als ein solches, einfaches Enzym der Eukaryonten kann die Diphosphat-Reduktase der Grünalge *Scenedesmus obliquus*[5] gelten.

1 Follmann, H. (1974) *Angew. Chemie* **86**, 624 - 634; *Angew. Chemie Int. Ed.* **13**, 569 - 579.
2 Brown, N. C. & Reichard, P. (1969) *J. Mol. Biol.* **46**, 39 - 55.
3 Vitols, E., Brownson, C., Gardiner, W., & Blakley, R. L. (1967) *J. Biol. Chem.* **242**, 3035 - 3041.
4 Ludwig, W. & Follmann, H. (1978) *Eur. J. Biochem.* **82**, 393 - 403.
5 Feller, W. & Follmann, H. (1976) *Biochem. Biophys. Res. Commun.* **70**, 752 - 758.

Prof. Dr. H. Follmann und Dr. W. Ludwig, Fachbereich Chemie der Universität, Arbeitsgruppe Biochemie, D-3550 Marburg.

W. W. Franke

Membran-assoziierte, unlösliche Skelettstrukturen

Nach präparativen Kriterien wird das Material von Membranfraktionen gemeinhin eingeteilt in Komponenten, die (a) mit Lösungen niedriger bzw. hoher Ionenstärke oder (b) mit nicht-denaturierenden Detergentien in Lösung gebracht werden können. Bei vielen Membranfraktionen gibt es aber auch Komponenten, die unter solchen Bedingungen unlöslich sind; dabei werden häufig bestimmte, eng membran-assoziierte, cytoplasmatische Skelettstrukturen angereichert. Einige typische Vertreter solcher resistenter Teilfraktionen werden vorgestellt bzw. diskutiert:

(1) die Strukturelemente der Porenkomplexe der Zellkernhülle und der „annulate lamellae";
(2) der Clathrin-haltige Besatz der „coated vesicles";
(3) der Binnenbesatz der Membran der Milchfetttropfen;
(4) die Präkeratin-haltigen Filamente, die an Desmosomen inserieren;
(5) die Desmosomen-Matrix selbst;
(6) der dichte Besatz der postsynaptischen Membranen.

Die Struktur, das präparative Verhalten und die biochemische Zusammensetzung solcher resistenter Komponenten wird beschrieben; Funktionen solcher Strukturen werden diskutiert.

Prof. Dr. Werner W. Franke, Abteilung für Membranbiologie und Biochemie, Institut für Experimentelle Pathologie, Deutsches Krebsforschungszentrum, Im Neuenheimer Feld 280, D-6900 Heidelberg 1.

G. Garweg, D. von Rehren und U. Hintze

Regionale Verteilung der Pyrrolin-2-carboxylat-Reduktase im Säugergehirn

Bei der Metabolisierung von D-Prolin bzw. D-Lysin über die D-Aminosäure-Oxidase (EC 1.4.3.3) entstehen im Säugerorganismus als Zwischenprodukte 1-Pyrrolin- bzw. 1-Piperidein-2-carbonsäure, die zu L-Prolin bzw. L-Pipecolsäure umgesetzt werden. Die Reaktion wird durch die 1-Pyrrolin-2-carboxylat-Reduktase (EC 1.5.1.1) katalysiert. L-Pipecolsäure wird in der Proteinsynthese nicht verwendet.

Autoradiographische Untersuchungen an der Maus haben gezeigt, daß isomerisiertes L-Prolin in die Nervenzellproteine des Gehirns mit auffallender regionaler Unterschiedlichkeit aufgenommen wird[1]. Für das hirntopographische Einbaumuster ist die ebenfalls regional unterschiedliche Enzymaktivität der D-Aminosäure-Oxidase[2] nicht verantwortlich, da im Zentral-Nervensystem Enzymverteilung und Einbauverteilung nicht übereinstimmen: Maximale Einbauaktivität und Fehlen der D-Aminosäure-Oxidase im Telencephalon steht der nicht vorhandenen D-Prolin-Utilisation trotz maximaler D-Aminosäure-Oxidase-Aktivität im Cerebellum gegenüber[3].

In der vorliegenden Studie wird die hirntopographische Verteilung der Pyrrolin-2-carboxylat-Reduktase-Aktivität, die Biogenese von Pipecolsäure und das regionspezifische Bindungsverhalten der Pipecolsäure im Zentral-Nervensystem untersucht und mit dem regionalen Verwertungsmuster von D-Prolin im Gehirn der Maus verglichen.

Die Ergebnisse weisen darauf hin, daß mit Hilfe der [^{3}H]D-Prolin-Markierung das Enzym Pyrrolin-2-carboxylat-Reduktase autoradiographisch lokalisiert werden kann. Die in der 50000 × *g*-Überstandsfraktion gemessenen Aktivitätsunterschiede dieser Reduktase in verschiedenen Hirnarealen von Maus, Hund und Affe (maximale Umsatzgeschwindigkeiten für 1-Piperidein-2-carbonsäure und NADH in Rinde und subcorticalen Stammganglien des Großhirns, kein Umsatz im Kleinhirn) stützen die Annahme, daß die Pyrrolin-2-carboxylat-Reduktase und die Pipecolsäure im Säugergehirn einem Wirkungsprinzip zuzuordnen sind, über dessen funktionelle Bedeutung noch keine ausreichenden Kenntnisse vorliegen.

1 Garweg, G. & Dahnke, H. G. (1973) *Anat. Anz.* **139**, 297 - 303.
2 Goldstein, D. B. (1966) *J. Neurochem.* **13**, 1011 - 1016.
3 Garweg, G. (1974) Habilitationsschrift, Med. Fak. Univ. Bonn.

Prof. Dr. med. Gerhard Garweg, Anatomisches Institut der Universität, Martinistr. 52, D-2000 Hamburg 20.

P. Geck, B. Pfeiffer and C. Henschel

The Influence of Ascorbate + Phenazine Methosulfate (PMS) on the Amino Acid Transport in Ehrlich Cells

The $Na^{\oplus}$-dependent accumulation of amino acids is best explained by Crane's gradient hypothesis[1], which postulates that the amino acids are brought into the cells via cotransport with $Na^{\oplus}$, and that the energy stored in the $Na^{\oplus}$ gradient is used to drive the amino acid uphill into the cell. Recently, however, Garcia-Sancho et al.[2] have reported that ascorbate + PMS stimulates the accumulative amino acid transport into Ehrlich cells, and concluded that amino acid transport is coupled directly to a redox reaction as an additional energy source. The following experiments were designed to decide whether a redox reaction stimulates the transport directly or whether this stimulation is a consequence of an alteration in the electrochemical potential difference of $Na^{\oplus}$, and so compatible with Crane's hypothesis.

Kinetically, the stimulation of amino acid uptake is described by a reduced Michaelis constant of transport, while the maximum velocity remains unchanged. This result corresponds to the behaviour of the transport at increased membrane potential[3]. Simultaneous measurements of amino acid transport and membrane potential (via the distribution of the lipid-soluble cation tetraphenylphosphonium) show a parallel influence of the redox system on both parameters, i.e. ascorbate + PMS hyperpolarizes the cell membrane. The mechanism of this hyperpolarization is unclear; it is not exclusively a consequence of an activation of the sodium pump, since an elevation of the membrane potential is also observed in ouabain inhibited cells. The inhibition by 2-deoxyglucose is not relieved by ascorbate + PMS while oligomycin, valinomycin, FCCP (mesoxalodinitril [*p*-(trifluoromethoxy)phenyl]hydrazone) and ouabain inhibition is partially relieved. PMS alone reverses the inhibition by antimycin A, while with antimycin A, ascorbate + PMS cause stimulation of transport. The simultaneous determination of the membrane potential shows that under all conditions tested, the changes in transport are paralleled by changes in membrane potential.

Since ascorbate + PMS influences the membrane potential and as a consequence the electrochemical potential difference of $Na^{\oplus}$, it is not necessary to postulate a direct coupling of the transport process to a redox reaction to explain the effects of ascorbate + PMS. By which mechanism ascorbate + PMS affects the membrane potential is not known.

1 Crane, R. A. (1965) *Fed. Proc.* **24**, 1000 - 1006.
2 Garcia-Sancho, J., Sanchez, A., Handlogten, M. E. & Christensen, H. N. (1977) *Proc. Natl. Acad. Sci. U.S.A.* **74**, 1488 - 1491.
3 Heinz, E. & Geck, P. (1978) in The Peter Curran Memorial Symposium, (Hoffman, J. F., ed.) Raven Press, in press.

Dr. Peter Geck, Gustav-Embden-Zentrum der Biologischen Chemie, Abteilung für Physikalische Biochemie der Universität Frankfurt am Main, Theodor-Stern-Kai 7, D-6000 Frankfurt 70.

P. Geck, C. Pietrzyk, E. Heinz and B. Pfeiffer

Electroneutral $K^{\oplus}$-$Cl^{\ominus}$-Cotransport in Ehrlich Ascites Tumor Cells

In Ehrlich cells which have been depleted of $K^{\oplus}$, there is a metabolism-dependent $K^{\oplus}$-uptake, even after the $Na^{\oplus}$-$K^{\oplus}$ pump is completely blocked by ouabain. This ouabain-insensitive $K^{\oplus}$ uptake is accompanied by stoichiometric uptake of $Cl^{\ominus}$ ions. Even though both $K^{\oplus}$ and $Cl^{\ominus}$ flows appear to depend on the presence of cellular ATP, there is no evidence of an extra ATP utilization or oxygen consumption. Since the electrical potential across the plasma membrane is not noticeably influenced by this KCl uptake, it is assumed that these ions enter the cell in an electroneutral form rather than as single ions. Two hypothetical mechanisms would explain the observations:

1) An electroneutral cotransport of KCl or
2) an active $K\text{-}H^{\oplus}$ exchange pump which in the presence of a $Cl^{\ominus}\text{-}OH^{\ominus}$ antiportsystem would cause a net uptake of KCl.

This second hypothesis, however, is not supported by any experimental evidence. The addition of $K^{\oplus}$ to a $K^{\oplus}$-free medium does not induce $H^{\oplus}$ extrusion, nor any detectable change of intracellular pH. Also the $Cl^{\ominus}\text{-}OH^{\ominus}$ antiport, though existent, is of minor importance in Ehrlich cells. Hence the first hypothesis of KCl symport is the only explanation at hand. Since the product $[K^{\oplus}]\cdot[Cl^{\ominus}]$tends to be higher inside the cell than outside, energy appears to be required for this process, even though it does not seem to derive from ATP hydrolysis. The KCl uptake described is completely inhibitable by various diuretics, including furosemide, bumetanide, piretanide and ethacrynic acid. Since these inhibitors also inhibit the isotopic $K^{\oplus}$ exchange fluxes in the steady state, they appear to affect the KCl transport system directly. The concentration of furosemide required for half-maximal inhibition is 10^{-4}M, and for the other diuretics mentioned, 10^{-5}M. At these low concentrations, these drugs have hardly any other effect.

Dr. Peter Geck, Gustav-Embden-Zentrum der Biologischen Chemie, Abteilung für Physikalische Biochemie, Universität-Frankfurt am Main, Theodor-Stern-Kai 7, D-6000 Frankfurt 70.

J. Gerschitz, R. Rudolph and R. Jaenicke

Reconstitution of Horse-Liver Alcohol Dehydrogenase after Dissociation and Deactivation in Guanidinine Hydrochloride

Horse-liver alcohol dehydrogenase (EC 1.1.1.1) is a dimer requiring 2 $Zn^{2\oplus}$ per subunit for full enzymatic function. Denaturation in 6M guanidine · HCl leads to inactive monomers; removal of the denaturant under optimum solvent conditions causes partial reactivation (~ 70%). The concentrations of the enzyme, the coenzyme, $NAD^{\oplus}$, and $Zn^{2\oplus}$ affect the reactivation significantly, since high concentrations promote the formation of wrong aggregates.

Analysis of the sigmoidal kinetics of reconstitution suggests a superposition of first order transconformation and second order association reactions; the latter are corroborated by the concentration dependence of the reactivation reaction. $NAD^{\oplus}$ has no influence on the kinetics of reactivation. $Zn^{2\oplus}$ affects both the rate and yield of reconstitution. In its absence no reactivation is detectable, excess leads to inactive aggregates, and optimum reactivation occurs at 10μM $Zn^{2\oplus}$ after short incubation in the denaturant.

The refolding of the Zn-free enzyme obeys a first order process which may be separated from second order dimerization started by the addition of the essential metal ion. This association reaction is superimposed by a reversible complex formation of inactive monomers with $Zn^{2\oplus}$

R. Jaenicke, Institut für Biophysik und Physikalische Biochemie, Universität Regensburg, Universitätsstraße 31, D-8400 Regensburg.

M. Giesel

Affinitätsmarkierung von Proteinen mit Guanosinnucleotid-Analoga

Zur Untersuchung Guanosinnucleotid-abhängiger Proteine wurden verschiedene chemisch reaktive Analoga von Guanosinnucleotiden synthetisiert. Diese besitzen als Abgangsgruppe einen Dinitrophenylrest, der über eine Thioätherbindung mit der Position 6 des Purinringes verknüpft ist. Unter schwach alkalischen Bedingungen (pH 7.2) bindet das Reagenz unter Abspaltung von 2,4-Dinitrothiophenol kovalent an SH-Seitenketten im Nucleotid-Bindungszentrum des Proteins.

Das Mikrotubuli-formende Protein Tubulin zum Beispiel benötigt zur Polymerisation GTP, das dabei zu GDP hydrolysiert wird. Durch das GTP-Analoge 6-[*S*-(2,4-Dinitrophenyl)mercapto]-2-aminopurinnucleosid-triphosphat (N_2ph-S-GTP) läßt sich die Polymerisation nahezu ebensogut stimulieren wie durch GTP selbst; allerdings nur so lange, wie es noch nicht kovalent gebunden ist.

Gleichzeitig wird aber die Polymerisation von Tubulin verhindert, wenn die SH-Gruppen mit *N*-Äthylmaleinimid blockiert wurden. Stöchiometrische Markierung mit $[^{14}C]$*N*-Äthylmaleinimid zeigt zwei radioaktiv markierte Peptide im Fingerprint.

Die gleichen, offensichtlich im GTP-Bindungszentrum liegenden, SH-Gruppen werden auch durch $[^{32}P]N_2$ph-S-GTP markiert.

Michael Giesel, Institut für Biochemie der Universität Frankfurt, Sandhofstraße, Gebäude 75 A, D-6000 Frankfurt/M.-Niederrad 1.

M. Giesing, K. Wemhöner and F. Zilliken

A Comparative Study on the Action of Acetylcholine and Adrenalin on Glycerolipid Biosynthesis in Cortex Grey Matter Cultures

Biosynthesis of cell glycerolipids was studied with exogenous $[1\text{-}^{14}C]$stearate and $[2\text{-}^{3}H]$glycerol at two stages of development: first in 8-day-old cultures prior to glia maturation, and second, in 16-day-old cultures, after the biogenesis of synapses had been completed. Incorporation of glycerol into phosphoglycerides was stimulated during development, whereas stearate uptake was reduced. Significant changes were observed in the presence of either acetylcholine or adrenaline in concentrations of 10^{-5} and 10^{-4}M. Acetylcholine and adrenalin reduced the incorporation rate and acylation with stearate, except in phosphatidic acid. The inhibitory effect was concentration dependent. The maximum effect was observed at 8 days. Acetylcholine was more effective in 16-day cultures than adrenalin. The effect of acetylcholine and adrenalin on glycerol incorporation into cell lipid classes was similar to that of stearate. However adrenalin led rather to a stimulation of phosphatidylcholine formation – this only at the 8-day stage. Other individual lipids showed a transmitter-specific reduction of radioactivity. The results indicate that the adrenergic system is developed prior to the cholinergic system.

It was expected that the subcellular distribution of radiolabelled lipids would give further information on the specificity of neurotransmitter action. The majority of subcellular lipids isolated from stimulated cultures, aside from microsomal diglyceride, phosphatidylinositol and phosphatidate, exhibited a reduced radioactivity. The ability of subcellular fractions to synthesize lipids from [^{14}C]glycerol-3-phosphate was found to be characteristic. Synaptosomal diglyceride formation was enhanced some 30-fold, microsomal biosynthesis of phosphatidylcholine, -inositol, -ethanolamine and phosphatidate between 2 and 3-fold. It is concluded that diglycerides are directly involved in synapse function, whereas microsomal phosphoglycerides rather seem to serve as material that is resonant to neurotransmitter action. The results also may explain studies on the specificity of pre- and postganglionic glycerolipid biosynthesis[1].

1 Giesing, M. & Zilliken, F. (1977) *Hoppe-Seyler's Z. Physiol. Chem.* **358**, 236 - 237.

Dr. Michael Giesing, Institut für Physiologische Chemie der Universität Bonn, Nussallee 11, D-5300 Bonn.

H. Glossmann

Regulation of Adenylate and Guanylate Cyclase in Bovine Adrenal Cortex

Cyclic AMP and cyclic GMP are believed to control steroid secretion, cell growth and differentiation in the adrenal cortex. Adenylate cyclase (EC 4.6.1.1) is bound to the plasma membrane and its activity is controlled by hormone receptors and cytosol-oriented guanyl nucleotide sites. The cyclase-coupled guanyl nucleotide sites have affinity for GTP > Gpp(NH)p. The intrinsic activity of GTP is low, due to rapid turnover of GTP at the sites. The turnover is considerably slowed down by modification of the plasma membranes with cholera toxin. Another set of guanyl nucleotide sites interacts with the angiotensin receptor in the same plasma membrane. For the latter sites, which are most likely to be involved in resensitization, the affinity for GTP equals that for Gpp(NH)p. The two nucleotides have the same intrinsic activity, indicating that there is no turnover of GTP.

Although angiotensin II increases cyclic AMP in intact bovine adrenal cortex cells, stimulation of adenylate cyclase in broken cell preparations cannot be observed. Divalent cations, especially $Mn^{2\oplus}$ and $Mg^{2\oplus}$, also control adenylate cyclase activity. In the presence of corticotropin (ACTH), the affinity of the cyclase for $Mn^{2\oplus}$ or $Mg^{2\oplus}$ increases 100-fold. $Ca^{2\oplus}$ competes with $Mn^{2\oplus}$ at the divalent cation sites. The importance of the guanyl nucleotide sites in regulating hormonal excitation of adenylate cyclase in intact tissue is exemplified by experiments with rat and bovine adrenal cells and the respective particulate adenylate cyclases using corticotropin analogues.

In contrast to adenylate cyclase, little is known about the regulation of guanylate cyclase (EC 4.6.1.2) in adrenal cortex. The enzyme is found in both particulate (plasma-membrane-bound) and soluble fractions. Although corticotropin, Nps-corticotropin and angiotensin II increase cyclic GMP levels in isolated bovine adrenal cells, neither the particulate nor the soluble enzyme is activated in vitro by hormones. Sodium nitroprusside and oncogenic nitrosoamines increase cyclic GMP levels dramatically in isolated adrenal cells. The cyclic AMP level remains unchanged. Since guanylate cyclase has very low activity with $Mg^{2\oplus}$ as divalent cation, the rise in cyclic GMP levels by the non-physiological stimulators was obscure until it was found that the same agents stimulate soluble guanylate cyclase dramatically and convert the enzyme into a form where $Mg^{2\oplus}$ is the preferred cation. Sodium nitroprusside and nitrosoamines activate adenylate cyclase in bovine adrenal homogenates when $Mn^{2\oplus}$ is present. The activatable form of adenylate cyclase is almost exclusively found in the 100 000 x g supernatant and shares a number of properties with soluble guanylate cyclase. We conclude that one protein is responsible for both enzymic activites.

The non-hormonal stimulants apparently convert guanylate cyclase into a form which not only has affinity for ATP, but converts it into cyclic AMP at 1/6 - 1/10 of the rate which is seen with GTP, when $Mn^{2\oplus}$ is present. Since only cyclic GMP levels are increased in isolated cells, this probably reflects a low intracellular $Mn^{2\oplus}$ concentration.

A possible clue to the regulation of guanylate cyclase in vivo was the finding that hydrophobic ligands, like unsaturated fatty acids, lysolecithin and non-ionic detergents, inhibit the effects of sodium nitroprusside and nitrosoamines on guanylate cyclase, but stimulate the enzyme in the absence of the SH-oxidizing agents.

Adenylate cyclase activity is controlled, among many other factors, by GTP. Guanylate cyclase activity, on the other hand, must operate (unless compartimentization is postulated) under ATP control.

Prof. Dr. Hartmut Glossmann, Pharmakologisches Institut der Universität Giessen, Frankfurter Straße 107, D-6300 Lahn-Giessen 1.

F. Göbber und H. Lachmann

Mehrwellenlängen-Analyse chemischer und biochemischer Titrationssysteme: Ein universelles Computerprogramm zur Auswertung von spektrometrischen Titrationen

Zur Auswertung von ein-, zwei- oder mehrstufigen spektrometrischen Säure-Base-Titrationen wurde in den letzten Jahren eine Anzahl neuer Methoden entwickelt [1-5,*], die jeweils zwei oder drei Absorptions-pH-Meßreihen A_λ (pH) verschiedener Wellenlänge mittels Absorptions- (= A-) Diagrammen oder Absorptionsdifferenzen-Quotienten- (= ADQ-) Diagrammen verknüpfen. Mit diesen Diagrammen kann zunächst die Zahl s der (spektroskopisch beobachtbaren) Titrationsstufen ermittelt werden. Anschließend werden mittels ausgewählter A_λ (pH)-, A- und/oder ADQ-Diagramme die pK-Werte des Titrationssystems ermittelt. Für ein-, zwei- und mehrstufige sequentielle Titrationen sowie für Simultantitrationen stehen jeweils mehrere Methoden zur absoluten und relativen pK-Wertbestimmung zur Verfügung[1-5,*].

Vor kurzem wurde ein universelles iteratives Fitting-Programm entwickelt, das bei beliebigen spektrometrischen Titrationssystemen anwendbar ist und sämtliche A_λ (pH)- Meßreihen *zugleich* zur Auswertung heranzieht. Dadurch wird eine gekoppelte Auswertung der gesamten Information von Mehrwellenlängen-Messungen ermöglicht. Das Programm berechnet neben den pK-Werten die Absorptionen $A_{\lambda 0}$ bzw. Absorptionskoeffizienten ϵ_λ der (reinen) völlig protonierten bzw. deprotonierten Formen AH_s bzw. $A^{s\ominus}$ sowie sämtlicher Ampholyte, auch wenn diese in reiner Form nicht existieren. Es wird daher nicht notwendig vorausgesetzt, daß das saure bzw. alkalische Ende der Titration quantitativ erreicht wird.

Die Güte und Signifikanz der Messung und Auswertung kann nicht nur aus der Summe der Fehlerquadrate, sondern auch durch Aufprägen von Zufallsfehlern bei der Kontrollsimulation geprüft werden (Monte-Carlo-Methode, vgl.[4]). Als Startwerte für die Iteration genügen sehr grobe Schätzungen der pK-Werte (etwa auf 1 pH-Einheit genau).

Das Verfahren ist nicht auf absorptionsspektrometrische Messungen beschränkt, es können auch spektralfluorimetrische, -polarimetrische, NMR-Titrationen etc. ausgewertet werden, vgl.[6,7].

Falls der Titrationsmechanismus gesichert ist, kann auch bei komplizierteren Titrationssystemen eine einzige Titrationskurve zur Bestimmung aller pK-Werte ausreichen, falls sich sämtliche Titrationsstufen des Systems hinreichend in den Meßwerten bemerkbar machen.

Die Leistungsfähigkeit der Methode wird an verschiedenen biochemisch relevanten Beispielen mit bis zu vier (spektroskopisch beobachtbaren) Titrationsstufen demonstriert: z.B. *o*- und *p*-Nitrophenol-Chromophore (s = 1), *cis*- und *trans*-Urocaninsäure (s = 2), Verbindungen der Vitamin-B_6-Gruppe (s = 2 bis 4).

---

* Lachmann, H., Polster, J., Übersichtsarbeit in Vorbereitung.

1 Blume, R., Lachmann, H., Mauser, H. & Schneider, Fr. (1974) *Z. Naturforsch.* **29b**, 500 - 513.

2 Blume, R., Lachmann, H. & Polster, J. (1975) *Z. Naturforsch.* **30b**, 263 - 276.

3 Polster, J. (1975) *Z. Phys. Chem. NF* **97**, 55 - 68, 113 - 126.

4 Göbber, F. & Polster, J. (1976) *Anal. Chem.* **48**, 1546 - 1550.

5 Lachmann, G. & Lachmann, H. (1977) *Hoppe-Seyler's Z. Physiol. Chem.* **358**, 1237.

6 Lachmann, H. (1976) *Hoppe-Seyler's Z. Physiol. Chem.* **357**, 270 - 271.

7 Lachmann, H. (1978) *Z. Anal. Chem.* **290**, in Druck.

Dr. H. Lachmann, Institut für Physikalische und Theoretische Chemie der Universität, Auf der Morgenstelle 8, D-7400 Tübingen.

B. Goertz

Regulation of Protein Synthesis during Postnatal Maturation of Mouse Brain

It has been reported by several investigators that the rate of protein synthesis significantly decreases during neural development[1-3]. This maturation-dependent loss in protein synthesizing activity seemed to be the result of an alteration of either ribosomes, or pH-5 enzymes, or both of them, but its actual nature has not been fully resolved[4].

On a DNA basis, we find in the developing mouse brain (NMRI strain) a twofold higher concentration of polysomes than in the adult brain, but there is no maturation-dependent decrease in tRNA content. The amino-acid-incorporating activity of cell-free systems with polysomes of newborn mouse brain exceeds that of the adult controls about 100%. This is, on the one hand, a polysomal effect: addition of newborn polysomes stimulates in vitro systems with adult tRNA and enzymes to an extent of 40%, whereas adult polysomes decrease the activity of newborn systems to a similar degree. On the other hand, aminoacylation of newborn tRNA exceeds the in vitro loading of the adult tRNA ($tRNA^{Phe}$ 25%, $tRNA^{Lys}$ 30%, $tRNA^{Leu}$ 34%, $tRNA^{Meth}$ 129%). The velocities of aminoacylation of the newborn tRNA, too, are 4 times higher than in the adult loading systems. Unloaded newborn tRNA increases the incorporating activity of tRNA-dependent adult polysomal systems to 173% of the controls with adult tRNA. Precharged newborn tRNA stimulates the activity of these systems to 134%. In contrast to polysomes and tRNA the newborn enzymes do not show higher activities than the adult controls.

These experiments indicate that the different protein synthesizing activity in developing and'adult mouse brain is the result of higher biological activities of the newborn messenger ribosome complexes and, most important, of a stimulated aceeptance and transfer function of the newborn tRNA.

1 Lerner, M. P. & Johnson, T. C. (1970) *J. Biol. Chem.* **245**, 1388 - 1393.
2 Zomzely, C. E., Roberts, S., Peache, S. & Brown, D. M. (1971) *J. Biol. Chem.* **246**, 2097 - 2103.
3 Johnson, D. E. & Sellinger, O. Z. (1971) *J. Neurochem.* **18**, 1445 - 1460.
4 Johnson, T. C. (1976) *J. Neurochem.* **27**, 17 - 23.

Dr. Bernd Goertz, C. & O. Vogt-Institut für Hirnforschung der Universität, Himmelgeister Str. 300/5, D-4000 Düsseldorf.

H. Greten, J. Augustin und O. Schrecker

Regulatorische Enzyme für den Katabolismus menschlicher Plasmalipoproteine

Exogen aufgenommene und endogen synthetisierte Triglyceride werden im menschlichen Plasma als Chylomikronen (Chylos) und very-low density Lipoproteine (VLDL) transportiert. Die intravasale Hydrolyse dieser Makromoleküle wird im wesentlichen durch zwei membrangebundene, am Kapillarendothel lokalisierte Enzyme reguliert: Die extrahepatische Lipoprotein-Lipase (LPL) (EC 3.1.1.34) und die der Leber entstammende hepatische Triglycerid-Lipase (HTGL). Hierbei handelt es sich um zwei Isoenzyme mit einem Molekulargewicht von ungefähr 67000. Die Reindarstellung beider Enzyme ermöglichte Untersuchungen zu ihrer Zusammensetzung und Funktion sowohl bei Gesunden als auch unter pathologischen Bedingungen. Beide Enzyme sind Glykoproteine, die sich bei gleichem Proteinanteil durch ihre Kohlenhydratzusammensetzung unterscheiden. Im Gegensatz zum Leberenzym benötigt die Lipoprotein-Lipase ein Apoprotein, das Apo C II der VLDL-Dichteklasse, als Cofaktor. Das Reaktionsoptimum für beide Enzyme unterscheidet sich außerdem durch pH-Wert und Ionenstärke. Es gelang monospezifische Antikörper gegen beide Enzyme herzustellen, die den selektiven Enzymnachweis aus dem Plasma von Gesunden und von Patienten mit primären und sekundären Hyperlipämien ermöglichen. Sowohl bei bestimmten primären als auch bei einigen sekundären Fettstoffwechselstörungen bestehen selektive Mängel an einem der beiden Enzyme, so daß eine unterschiedliche Wirkungsweise dieser Isoenzyme in vivo wahrscheinlich ist. Erste kinetische Untersuchungen mit isolierten Lipoproteinen und gereinigten Enzymfraktionen in vitro deuten darauf hin, daß *beide* Enzyme triglyceridtransportierende Lipoproteine hydrolysieren. In-vitro-Inkubationen von künstlichen Lipid/Protein-Komplexen aus Triolein und isoliertem Apo C II mit gereinigter Lipoprotein-Lipase geben Aufschluß über den möglichen Wirkungsmechanismus dieses Cofaktors in vivo.

H. Greten, J. Augustin und O. Schrecker, Klinisches Institut für Herzinfarktforschung, Medizinische Universitätsklinik Heidelberg, Bergheimerstr. 58, D-6900 Heidelberg 1.

E. Grundmann and E. Görlitz

Studies with 4,4'-Diisothiocyanato-2,2'-(1,2-ethanediyl)-bis(benzenesulfonic acid) on Phalloidin-Sensitive Hepaptocytes and on Phalloidin-Insensitive Ascites Hepatoma Cells

Phalloidin treatment of isolated rat hepatocytes results in the development of characteristic protrusions on the cell surface. In contrast, phalloidin does not affect AS 30D ascites hepatoma cells.

Both 4,4'-diisothiocyanato-2,2'-(1,2-ethanediyl)-bis-(benzenesulfonic acid) (I) and [^{3}H]I inhibit the development of protrusions irreversibly*. Because the above dihydrostilbens do not penetrate the cell membrane[1], it has been concluded that they act on proteins on the surface of the cell. This conclusion was substantiated by comparison of the specific radioactivity of cell proteins after labeling intact cells with [^{3}H]I with that of a [^{3}H]I-labeled cell homogenate. The specific radioactivity of the labeled cell homogenate was about 80 times that of the specific radioactivity of intact cells labeled with [^{3}H]I.

For understanding of the different action of phalloidin on hepatocytes and on hepatoma cells, we studied the binding of [^{3}H]I to the membrane proteins of both cell types, using sodium dodecylsulfate electrophoresis* and subsequent fluorography[2].

Marked differences in the distributions of labeled protein were observed between the two cell types. In isolated hepatocytes, six proteins were significantly labeled (molecular weights of 25000, 32000, 50000, 56000, 170000, and 230000). In contrast, only two hepatoma proteins exhibit a detectable amount of the tracer, corresponding to molecular weights of 135000 and 230000 Propably one of the labeled proteins on the surface of hepatocytes, not present in the membrane of hepatoma cells, is involved in the recognition and uptake of phalloidin.

The inhibition of the phalloidin response by [^{3}H]I cannot be due to an interaction with actin, because this contractile protein was not labeled after incubation of intact cells.

---

* Petzinger, E., Grundmann, E., Veil, L. B., Frimmer, M. & Fasold, H., in preparation.

1 Lepke, S., Fasold, H., Pring, M. & Passow, H. (1976) *J. Membr. Biol.* **29**, 147 - 177.
2 Bonner, W. M. & Laskey, R. A. (1974) *Eur. J. Biochem.* **46**, 83 - 88.

Dr. Ewald Grundmann, Institut für Pharmakologie und Toxikologie im Fachbereich 18 der Universität Gießen, Frankfurter Straße 107, D-6300 Lahn-Gießen.

G. Gruner, D. Richter, E. Rohm and W. Groß

Extracellular Dextranase – Its Role in the Utilization of Dextrans for Growth of Streptomyces hydrogenans

Dextrans can be used as carbon sources for the growth of *Strept. hydrogenans* cells. Since compounds of high molar mass cannot be taken up as such, they have to be split into smaller products: the α-1,6-linked glucans (dextran 60 and 200) were degraded by an extracellular endo-dextranase (EC 3.2.1.11). Sephadex particles, too, were attacked and deformed, as microscopic studies revealed. Therefore, gel filtration of *Strept.* extracts on Sephadex columns may yield erroneous results.

Growth in synthetic media with dextrans as sole carbon sources is pH-dependent, suggesting that the dextranase is more active in a slightly acid medium. During cell growth there is an increase in anthranol-positive dialysable products. Free glucose cannot be detected (as judged by the glucose oxidase/peroxidase method). The amount of dialysable products (molar mass < 1 200; maximally 5 - 6 glucose units) is higher in dextran 60-media than in those containing dextran 200. The viscosity of the dextran media decreases with time, concomitantly with the decrease in the total amount of carbohydrates in the media and indirectly linked to the increase in oligosaccharides. However, the extent of the oligosaccharide increase is the result of two processes: their liberation from glucans by the dextranase and their consumption by the growing *Strept.* cells; maximal values were reached in cell suspensions between the 2nd and 4th day of growth.

Prof. Dr. W. Groß, Gustav Embden-Zentrum der Biologischen Chemie, Universität Frankfurt am Main, Theodor Stern-Kai 7, D-6000 Frankfurt 70.

W. Gutensohn

5'-Nucleotidase aus menschlicher Plazenta-Plasmamembran als Rezeptor für Concanavalin A und Lens-culinaris-Lectin

5'-Nucleotidase (EC 3.1.3.5) und alkalische Phosphatase (EC 3.1.3.1) wurden aus Mikrosomen oder Plasmamembranfraktionen von menschlicher Plazenta durch Detergensextraktion gewonnen, durch DEAE-Cellulose-Chromatographie voneinander getrennt und durch Affinitätschromatographie an *Lens-culinaris*-Lectin-Sepharose weiter angereichert. Die Wirkung einer Reihe von Pflanzen-Lectinen auf beide Enzymaktivitäten in der intakten Plasmamembran, sowie auf die isolierten Enzyme wurde studiert.

Dabei erweist sich auf Grund der Hemmung der Enzymaktivität 5'-Nucleotidase eindeutig als Rezeptor für Concanavalin A und *Lens-culinaris*-Lectin. Die für 50proz. Hemmung erforderliche Lectin-Konzentration liegt für das isolierte Enzym um 2 Zehnerpotenzen niedriger, als für die 5'-Nucleotidase im Membranverband. Zugabe des spezifischen Zuckers α-Methylmannose verhindert die Hemmung, bzw. hebt sie wieder auf. Die Hemmung des Enzyms in der intakten Membran ist somit höchstwahrscheinlich nicht auf unspezifische Membranperturbation zurückzuführen.

Die Ergebnisse der Affinitätschromatographie lassen darauf schließen, daß auch die alkalische Phosphatase als Rezeptor für *Lens-culinaris*-Lectin fungiert, jedoch ist hier und auch mit Concanavalin A keine Hemmung der Enzymaktivität zu beobachten.

Dr. Wolf Gutensohn, Institut für Anthropologie und Humangenetik der Universität München, Arbeitsgruppe Biochemische Humangenetik, Schillerstr. 42, D-8000 München 2.

C. W. M. Haest, D. Kamp, G. Plasa and B. Deuticke

Erythrocyte Membrane Protein Modification by Reversible Formation of Inter- and Intramolecular Disulfide Bonds and of Mixed Disulfides between Proteins and Glutathione

The major *ex*trinsic protein of the erythrocyte membrane, spectrin, is cross-linked via disulfide bonds upon incubation of erythrocytes or ghosts with SH-oxidizing agents (tetrathionate, diamide). In ghosts, the crosslinking can be suppressed by glutathione (GSH). The major *in*trinsic membrane protein (Band 3, 100000 dalton) is easily dimerized by the agents in ghosts, but reacts in intact cells only after selective blockade of intracellular GSH by iodoacetate. Formation of mixed disulfides between membrane proteins and GSH probably prevents intermolecular crosslinking. Indeed, about 30% of the membrane SH-groups were found, by tracer techniques and chemical analysis, to be crosslinked with GSH in erythrocytes treated with diamide, demonstrating that these groups are readily accessible to GSH. The mixed disulfides were distributed over most of the membrane protein fractions.

In the absence of GSH, the oxidizing agents oxidize up to 80% of the membrane SH-groups, suggesting that pairs of membrane SH-groups are close enough to each other to build up disulfide bonds. After reduction of these disulfide bonds by sulfite, which releases one SH-group while the other one remains blocked as $S\text{-}SO_3^{\ominus}$ group, the released SH-groups can be reoxidized by a second treatment with diamide. This demonstrates that even more than 2 SH-groups are arranged very closely.

The oxidation of the SH-groups is reversible, although to a limited extent, upon incubation of the cells with adenosine, while treatment of cells or ghosts (e.g. dithioerythritol, GSH) brings about a complete reversal. The limited metabolic cleavage is probably due to increasing blockade of glycolysis. The formation of disulfide bonds and its metabolic reversal could have important implications for the proper function of biological membranes (e.g. the protection against radicals and oxidative stress).

C. W. M. Haest, D. Kamp, G. Plasa and B. Deuticke, Abt. Physiologie, RWTH Aachen, Melatener Str. 211, D-5100 Aachen.

H.-H. Hamm and K. Decker

Cell-free Synthesis of 6-Hydroxy-D-nicotine Oxidase Containing Covalently Bound FAD

6-Hydroxy-D-nicotine oxidase (EC 1.5.3.6) with covalently bound FAD is synthesized in *Arthrobacter oxidans* during induction by D-nicotine. A riboflavin-requiring mutant was prepared to separate the translation of the apoprotein from the reaction(s) by which the coenzyme molecule becomes attached to the apoprotein. The maximal level as well as the rate of synthesis of the enzyme is dependent on the concentration of riboflavin in the medium. At any extracellular riboflavin level, however, enzyme synthesis was blocked at once by addition of chloramphenicol or streptomycin, with or without simultaneous riboflavin supplements. This indicates that even at suboptimal riboflavin levels, an apoprotein does not accumulate in vivo, and that the flavin-binding reaction may be a cotranslational event.

For studies in vitro of the synthesis of 6-hydroxy-D-nicotine oxidase, polysomes from cells isolated during induction of this enzyme were combined with the translation system of *E. coli* MRE 600. Since FAD and riboflavin had no influence on the general translation capacity of ribosomes, the products of translation in the presence of flavins (25μM riboflavin + 25μM FAD) were analyzed using sodium dodecylsulfate polyacrylamide gel electrophoresis.

A major peak indicated the preferential synthesis of one protein, whose identity with 6-hydroxy-D-nicotine oxidase was substantiated by the following evidence:

1) During dodecylsulfate gel electrophoresis, the protein migrated with the same velocity as purified [^{14}C]-riboflavin-labelled 6-hydroxy-D-nicotine oxidase, indicating a molecular weight (about 50 000) similar to this enzyme.

2) Treatment of the translation product with anti-6-hydroxy-D-nicotine oxidase antiserum resulted in a specific precipitate which yielded the same labelled peak in dodecylsulfate gel electrophoresis.

3) Omission of flavins in the translation assay led to the preferential synthesis of a protein which migrates faster than 6-hydroxy-D-nicotine oxidase in dodecylsulfate gel electrophoresis.

4) ^{14}C-labelled riboflavin added to an unlabelled translation assay led to the incorporation of radioactivity into this protein.

5) With polysomes from cells which were grown without inducer, the cell-free translation system failed to synthesize any protein of 50000 daltons and no incorporation of ^{14}C radioactivity was observed after incubation in the presence of [^{14}C]riboflavin.

H.-H. Hamm and K. Decker, Biochemisches Institut an der Medizinischen Fakultät der Universität Freiburg, Hermann-Herder-Straße 7, D-7800 Freiburg i. Br.

S. Harada, D. P. Agarwal and H. W. Goedde

Alcohol-Metabolizing Enzymes in Human Tissues and Cultured Skin Fibroblasts: Methods of Isozyme Separation

Alcohol is primarily metabolized in the liver by alcohol dehydrogenase (EC 1.1.1.1) and aldehyde dehydrogenase (NAD(P)$^{\oplus}$) (EC 1.2.1.5). Isozyme variation in both the enzymes is controlled by three structural gene loci[1-3,*]. Most of the reported data are based on studies with tissue autopsy samples.

Cultured skin fibroblasts have proved helpful as diagnostic tools for studying several metabolic defects. In this preliminary report, we present methods which allow the demonstration of alcohol dehydrogenase and aldehyde dehydrogenase isozymes in cultured skin fibroblasts from normal individuals. Extracts from biopsy specimens (about 2.5 mm diameter) were also found suitable for the identification of alcohol dehydrogenase types. The implication of these methods is obvious, as they eliminate the need of liver biopsies for phenotype determination. Investigations on the frequency of alcohol dehydrogenase and aldehyde dehydrogenase variants in cultured skin fibroblasts and liver autopsy samples of alcoholic subjects and normal controls are under study and will be presented.

* Harada, S., Agarwal, D. P. & Goedde, H. W. (1978), unpublished results.

1 Smith, M., Hopkinson, P. A. & Harris, H. (1973) *Ann. Hum. Genet.* **37**, 49 - 67.

2 Harada, S., Agarwal, D. P. & Goedde, H. W. (1977) *Hoppe-Seyler's Z. Physiol. Chem.* **358**, 1173 - 1307.

3 Harada, S., Agarwal, D. P. & Goedde, H. W. (1978) *Humangenetik* **40**, 215 - 220.

Dr. S. Harada, Priv.-Doz. Dr. D. P. Agarwal and Prof. Dr. H. W. Goedde, Institut für Humangenetik der Universität Hamburg, Butenfeld 32, D-2000 Hamburg 54.

S. Hartmann, J. Neeff and D. Mecke

Arenaemycin: an Antibiotic which Inhibits Glycolysis

Under appropriate culture conditions the strain Tü 469 of *Streptomyces arenae* produces the new antibiotic Arenaemycin[1]. This substance has been shown to act on procaryotic as well as on eucaryotic cells only if the organisms grow on carbon compounds which are metabolized via glycolysis. Carbon compounds which enter the metabolism "below" pyruvate antagonize the inhibitory effect of the antibiotic. Theoretical considerations suggested that arenaemycin inhibits pyruvate kinase[1,2]. In the experiments in vivo presented here, it will be shown that:

1) The uptake of the glycolytically metabolized substrates is not inhibited (no transport – antibiotic);
2) In the presence of the antibiotic, the level of fructose-1,6-bisphosphate is remarkably increased;
3) The concentrations of lactate and pyruvate decrease;
4) The inhibitory effect is fully reversible in vivo.

These measurements show that the antibiotic is really an inhibitor of glycolysis. Preliminary results obtained in vitro suggest that the antibiotic is an inhibitor of glyceraldehyde-3-phosphate dehydrogenase (EC 1.2.1.12).

1 Keller-Schierlein, W., Lemke, J., Nyfeler, R. & Zähner, H. (1972) *Arch. Mikrobiol.* **84**, 301 - 316.

2 Lemke, J. (1971) Doctoral Thesis at the University of Tübingen.

Jürgen Neeff, Physiologisch-Chemisches Institut der Universität Tübingen, Hoppe-Seylerstraße 1, D-7400 Tübingen.

J. Hausdörfer and W. Heller

High Doses of Glucocorticoid Protecting the Metabolism of Mechanically Traumatized Central Nervous Tissue

Glucocorticoid applied in experimental head injury demonstrates a stabilizing effect on central nervous tissue. Following a recent trend in clinical neurotraumatology high doses of dexamethasone are given in this study.

210 male guina pigs (700 g average body weight) are divided into three groups. 140 anaesthetized animals are subjected to defined head injuries. In a special trauma apparatus three consecutive though not lethal blows add up to contusions of the cortex and pial bleedings. 12 h prior to trauma half (70) the animals receive an initial dose of 2 mg dexamethasone/100 g body weight intraperitoneally. A repeat dose of 1 mg dexamethasone/ 100 g body weight is injected three times (every 6 h). 12 h after the injury, brain tissue is recovered by freeze stop technique. The study also calls for 70 anaesthetized animals without trauma and therapy to constitute a control group.

Compared to material from damaged but untreated guinea pigs, dexamethasone-treated brain tissue reveals significantly higher concentrations of ATP. The values approach control group results. Metabolites of glycolysis (especially lactate and pyruvate) are found to be significantly reduced, but are not reduced to control group levels. The metabolism of mechanically traumatized central nervous tissue thus shows a distinct improvement with dexamethasone.

Prof. Dr. W. Heller, Chirurgische Universitätsklinik, Calwer Straße 7, D-7400 Tübingen.

M. v. Hedenström und M. Höfer

Die Wirkung von Nystatin auf intakte Zellen und Mitochondrien der Hefe Rhodotorula glutinis

Die obligat aerobe Hefe *Rhodotorula glutinis* (CBS 6681) akkumuliert Monosaccharide im Zellinnern durch einen Membrancarrier[1]. Der die Energie liefernde Schritt ist noch nicht eindeutig bekannt. Experimente zeigten[2], daß ein elektrochemisches Potential über der Membran aufgebaut und ein Protonengradient errichtet werden, durch die das Substrat mittels eines $H^{\oplus}$-Substrat-Symports akkumuliert werden kann.

Das Polyen-Antibiotikum Nystatin beeinflußt den Substrattransport in sterolhaltigen Membranen von Eukaryonten[3]. D-Xylose-Aufnahmeversuche zeigen, daß Nystatinkonzentrationen um 3.5×10^{-10} mol/mg Trockengewicht der Hefe den Transport gerade hemmen. Höhere Nystatinkonzentrationen unterbinden einen Substrattransport völlig. Ebenso wird der Sauerstoffverbrauch reduziert. Im Grenzbereich der Nystatinkonzentration wird die vorübergehende Alkalinisierung des Mediums bei Substratakkumulation um das zweifache verstärkt, ohne aber die Geschwindigkeit des Eintransports von D-Xylose zu mindern.

Die Wirkung von Nystatin könnte auf einer direkten Beeinflussung der Transportkomponenten – das Protonensystem wird bereits bei niedrigerer Konzentration gestört – der Membran beruhen oder auf der Hemmung des Energietransfers von den Mitochondrien zur Plasmamembran und den Energie verbrauchenden Transportsystemen. Aus Protoplasten[4] der Hefe isolierte Mitochondrien zeigen jedoch, daß Nystatin weder auf die Bildung von ATP noch auf den Sauerstoffverbrauch Einfluß hat. Seine Wirkung ist auf die Plasmamembran beschränkt. Da es dort mit Sterolen reagiert[5], scheinen Sterole eine wesentliche Rolle bei der Funktion von aktiven Transportsystemen in der Zellmembran von *Rhodotorula glutinis* zu spielen.

1 Höfer, M. (1970) *J. Membr. Biol.* **3**, 73 - 82.
2 Misra, P. C. & Höfer, M. (1975) *FEBS Lett.* **52**, 1 95 - 99.
3 Komor, B., Komor, E. & Tanner, W. (1974) *J. Membr. Biol.* **17**, 231 - 238.
4 v. Hedenström, M. & Höfer, M. (1974) *Arch. Microbiol.* **98**, 51 - 57.
5 Demel, R. A., Van Deenen, L. L. M. & Kinsky, S. (1965) *J. Biol. Chem.* **240**, 2749 - 2753.

Dr. Michael v. Hedenström, Biologisches Institut der Universität Stuttgart, Abteilung Botanik, Ulmer Str. 227, D-7000 Stuttgart 60.

C. Heilmann, T. Lindl and W. Müller

Biochemical and Ultrastructural Characterization of Microsomal Fractions from Normal and Hypertrophied Rat Hearts

Hypertension leads to hypertrophy of the myocardium, which is accompanied by substantial alterations of heart contractility[1]. In order to elucidate the possible role of the sarcoplasmic reticulum in this process, isolated microsomes of the left ventricles of spontaneously hypertensive rats (Okamoto-Aoki strain, SHR)[2] and normotensive Wistar rats (NR) were characterized biochemically and electron-microscopically. The weight of the left ventricle relative to body weight was 1.9 (NR) and 3.3 (SHR). $Mg^{2\oplus}$-dependent (basal) and $Mg^{2\oplus}$, $Ca^{2\oplus}$-dependent (extra) ATPase (EC 3.6.1.3) activities were reduced by 56% and 45%, respectively, as compared to NR. The initial rate and capacity of $Ca^{2\oplus}$ uptake in the presence of oxalate were found to be depressed by about 20% and 30%, respectively. $Ca^{2\oplus}$ binding (without oxalate added) was not significantly altered. $K_{0.5(Ca^{2\oplus})}$ for ATPase and $Ca^{2\oplus}$ uptake activities did not differ significantly. Phosphorylation experiments revealed a delayed dephosphorylation of the $Ca^{2\oplus}$ transport ATPase with 30% higher (in the presence of $Mg^{2\oplus}$ and $Ca^{2\oplus}$) than in NR. In the presence of $Ca^{2\oplus}$ alone, $t_{1/2}$ was higher by a factor of 2.7 for SHR. Steady state levels of phosphoprotein formation in the presence of $Mg^{2\oplus}$ and $Ca^{2\oplus}$ were also reduced by about 30%. Sodium dodecylsulfate gel electrophoreses revealed a substantial reduction of the concentration of a polypeptide (M_r 115 000), presumably representing the $Ca^{2\oplus}$ pump protein. The latter was confirmed by radioautographic analysis of phosphorylated compounds.

Additionally, negative staining and freeze fracturing analyses were performed. The following conclusion was drawn: severe hypertrophy of the myocardium is paralleled by a reduction in the $Ca^{2\oplus}$ pumping sites in the sarcoplasmic reticulum and a delayed dephosphorylation of the phosphorylated intermediate of the $Ca^{2\oplus}$ pump, which both lead to a disturbed intracellular $Ca^{2\oplus}$ metabolism. It is postulated that these alterations are at least partially responsible for the altered contractility of hypertrophied myocardium.

1 Katz, A. M. (1970) *Physiol. Rev.* **50**, 63 - 158.
2 Aoki, K., Keda, N., Yamashita, K. & Hotta, K. (1974) *Jpn. Heart J.* **15**, 475 - 484.

Dr. Claus Heilmann, Fachbereich Biologie, Universität Konstanz, Postfach 7733, D-7750 Konstanz.

B. Hess, R. Korenstein and D. Kuschmitz

On the Coupling of the Photochemical Cycle and Proton Transfer in Bacteriorhodopsin

Bacteriorhodopsin molecules are arranged in a two-dimensional array, forming an almost perfect crystal lattice of space group P3[1]. Thus, three protein molecules are in direct contact, forming a trimeric cluster and imposing a strong protein-protein interaction within the lipid texture of the bacterial membrane. In each bacteriorhodpsin molecule the transition dipole moment of the retinal chromophore, as analyzed at 568 nm, forms an angle of $\leqslant 27°$ with the plane of the membrane as shown by analysis of linear dichroism[2,3]. The chromophore is immobilized within the purple membrane matrix as shown by photoselection experiments in the time domain of minutes[3]. Since the microviscosity of the lipid domains in the purple membrane was found to be 5 poise in previous experiments[4] it can be concluded that the viscosity that bacteriorhodopsin experiences is mainly due to protein-protein interactions and not due to lipid-protein interactions.

Steady illumination of bacteriorhodopsin in open membrane sheets generates a photochemical state, which is characterized by a steady level of a component absorbing at 412 nm (BR-412) with a half time in the millisecond range and coupled to a steady release and re-uptake of protons[1]. With a quantum yield of approximately 0.8 a ratio of $H^{\oplus}$/BR-412 of 0.5 - 3.0 is found depending on the ionic strength as well as on the light intensity. At least two intrinsic proton pools are indicated. Presently it is not known, to what extent the pools are within the pathway of vectorial proton transfer through the bacterial membrane[5].

Light pulse excitation of the system reveals a series of photocycle intermediates (in the pico-, nano-, micro- and milliseconds range) which can be correlated to a fast proton release (in the microsecond range) and relatively slow proton re-uptake (in the millisecond range)[1].

The interaction between the chromophore through the various events of the photocycle and the protein is indicated in tryptophan fluorescence changes[1] and the chemical reactivity of retinal under illumination[1]. Recently, investigation of the coupling mechanism between the photo- and proton cycle demonstrate conformational changes of the protein more directly[6]. In accordance with the sensitivity of the $H^{\oplus}$/BR-412 ratio to the ionic strength, pH titrations of the purple membrane in the dark demonstrate a heterogeneous population of acidic groups, the p*K* values of which are a function of ionic strength in agreement with the Gouy-Chapman theory. A light triggered conformational change of the protein can be detected by titrations of the membrane in light compared to dark: at high ionic strength we discovered an increase in the buffer capacity at pH > 6.8 and a decrease at pH < 6.8 coupled to the appearance of an apparent p*K* around 7.8 as a function of light. It is interesting to note that the latter p*K*, not detectable in dark bacteriorhodopsin, can also be detected in an apomembrane preparation. These results allow us to present a p*K* cascade through the purple membrane molecule as triggered by light relevant to a recent structural model of bacteriorhodopsin[7].

The lattice organization of the purple membrane suggests a strong cooperativity between the three bacteriorhodopsin molecules in each set. Spectroscopic evidence of such an interaction was shown by circular dichroic spectrum of the membrane[1,8]. Recently double excitation experiments have demonstrated that the photocycle kinetics of a bacteriorhodopsin molecule is dependent on the conformational state of its nearest neighbours. The decay rates of the photochemical cycle as analyzed in terms of a sum of exponentials increase along with the increase of the photostationary concentration of the BR-412 intermediate of the cycle. These results fit a cooperative model based on subunit-subunit interaction within the trimer.

1 Henderson, R. (1977) *Annu. Rev. Biophys. Bioeng.* 6, 87 - 109.
2 Heyn, M. P., Cherry, R. J. & Müller, U. (1977) *J. Mol. Biol.* **117**, 607 - 620.
3 Korenstein, R. & Hess, B., in preparation.
4 Korenstein, R., Sherman, W. V. & Caplan, S. R. (1976) *Biophys. Struct. Mechanism* **2**, 267 - 276.
5 Kuschmitz, D. & Hess, B. (1977). Abstract of the 11th. FEBS Meeting Copenhagen, A4-13/708.
6 Kuschmitz, D. & Hess, B. in preparation.
7 Ovchinnikov, Yu. A., Abdulaev, N. G., Feigina, M. Yu., Kiselev, A. V. & Lobanov, N. A. (1977) *FEBS Lett.* **84**, 1 - 4.
8 Reed, T. & Hess, B. (1976) *Fed. Proc., Fed. Am. Soc. Exp. Biol.* **35**, 1360.

Professor Dr. Benno Hess, Max-Planck-Institut für Ernährungsphysiologie, Rheinlanddamm 201, D-4600 Dortmund.

C. C. Heuck

Nephelometrische Untersuchungen immunologischer Reaktionen an Lipoproteinen des menschlichen Serums

Immunologische Reaktionen an Serumlipoproteinen gestatten qualitative und quantitative Aussagen über den Gehalt bestimmter Apolipoproteine[1,2]. Nephelometrische Untersuchungen ermöglichen zusätzliche Informationen über die Oberfläche dieser Partikel. Anhand einer Studie der Reaktion zwischen Anti-Apolipoprotein B und Apolipoprotein B in isolierten Lipoproteinen wurde der Einfluß von physiologischerweise vorhandenen Faktoren auf die Immunkomplexbildung abgeklärt.

An isolierten Fraktionen von Very-Low-Density-Lipoproteinen (VLDL) und Low-Density-Lipoproteinen (LDL) wurde die Trübungsentwicklung nach Zusatz von Anti-Apolipoprotein B in Gegenwart von Albumin, Fettsäuren und Kationen untersucht. Ähnliche Untersuchungen wurden an mit verschiedenen lipolytischen Enzymen vorbehandelten Lipoproteinen durchgeführt.

Die Entwicklung der Trübung bei der Immunkomplexbildung zwischen Anti-Apolipoprotein B und Apolipoprotein B in VLDL verläuft langsamer als die gleiche Reaktion zwischen Anti-Apo-B und Apo-B in LDL. Als Ursache hierfür können verschiedene Faktoren in Betracht gezogen werden. Nach einer enzymatischen Hydrolyse von Triglyceriden und Cholesterinestern beobachtet man eine verminderte Trübungsentwicklung. Die Vorbehandlung mit Phospholipase A_2 bewirkt eine Hemmung der Immunreaktion bei gleichzeitiger Trübungsverminderung. Ähnliches wird auch nach Zusatz von Laurinsäure zu nativen VLDL beobachtet. Albumin klart die Eigentrübung von Lipoproteinen und die Trübung bei der Immunkomplexbildung auf. Spezifische Effekte werden durch Calcium hervorgerufen.

Die Versuchsergebnisse lassen sich in folgender Weise deuten:

1) Die durch eine Immunreaktion hervorgerufene Trübung wird neben dem Gehalt der antigenen Proteinkomponente auch durch den Gehalt bestimmter Lipide beeinflußt.

2) Der verzögerte Verlauf der Trübungsentwicklung bei der Reaktion zwischen Anti-Apo-B und Apo-B in VLDL weist auf eine stärkere Abschirmung der antigenen Determinanten an der Oberfläche der Partikel hin. Teilweise kann die Abschirmung durch die Abspaltung von Neuraminsäure und der polaren Gruppen von Phospholipiden aufgehoben werden.

1 Schonfeld, G., Lees, R. S., George, P. K. & Pfleger, B. (1974) *J. Clin. Invest.* **53**, 1458 - 1467.
2 Lees, R. S. (1970) *Science* **169**, 493 - 495.

C.C. Heuck, Klinisches Institut für Herzinfarktforschung, Bergheimer Straße 58, D-6900 Heidelberg.

H. Hilz, P. Adamietz, R. Bredehorst und K. Wielckens

Kovalente Modifikation von Proteinen durch ADP-Ribosylierung

Die Übertragung des ADP-Ribose(ADPR)-Teiles von NAD auf Proteine unter gleichzeitiger Freisetzung von Nicotinamid stellt eine kovalente Modifikation dar, vergleichbar der Phosphorylierung oder Acetylierung von Proteinen. Mehrere Toxine und Phagen-Enzyme katalysieren solche Transferreaktionen, die in den meisten Fällen mit einer Inaktivierung der modifizierten Akzeptorproteine verbunden sind (vgl. l.c.[1]). Im Zellkern von Eukaryonten ist ein Enzymsystem lokalisiert, welches die Bildung von proteingebundenen oligomeren und polymeren $(ADPR)_n$-Ketten katalysiert. Zahlreiche Analysen über die Veränderung der poly(ADPR)-Synthase-Aktivität unter verschiedenen metabolischen Zuständen lieferten keine zuverlässigen Hinweise auf eine Beteiligung an den bekannten Funktionen des Chromatins (für eine Übersicht vgl. l.c.[2]). Die Entwicklung eines empfindlichen Nachweisverfahrens für ADPR ermöglichte neuerdings die Quantifizierung der proteingebundenen ADPR-Spiegel in intakten Geweben[3]. Seine Anwendung erbrachte bisher folgende Befunde: In allen untersuchten Geweben beträgt die Menge an ADPR-Resten weniger als 1% des DNA-Gehaltes; die Masse der ADPR-modifizierten Proteine ist nicht mit den polymeren $(ADPR)_n$-Ketten, sondern mit mono(ADPR)-Resten besetzt; $(ADPR)_n$-Ketten finden sich nur in untergeordnetem Maße; ADPR-Reste sind über mindestens zwei verschiedene Bindungstypen an Proteine geknüpft, die sich durch ihr Verhalten gegenüber neutralem Hydroxylamin unterscheiden und so quantifizieren lassen; die Masse der ADPR-Reste befindet sich nicht an den Histonen, sondern an Nicht-Histon-Proteinen.

Im Verlauf des Zellzyklus bei *Physarum polycephalum* (Zusammenarbeit mit W. Sachsenmaier) ändert sich der Spiegel der proteingebundenen ADPR-Reste nur geringfügig. Dies indiziert entweder, daß ADP-Ribosylierungsreaktionen keine wesentliche Regulationsfunktion beim Zellzyklus ausüben oder daß nur eines von den ADPR-tragenden Proteinen während des Zellzyklus modifiziert wird, wegen seiner geringen Konzentration aber nicht stark in Erscheinung tritt.

Die Wichtigkeit von Analysen an $(ADPR)_n$-Proteinkonjugaten aus intakten Geweben anstelle von isolierten Kernen wird dokumentiert durch Befunde an ADP-ribosyliertem Histon H1, das aus [^{3}H]Adenosin-markierten HeLa-Zellen gewonnen wurde*. Bezüglich der Art der Bindungstypen und dem Anteil an mono(ADPR)- und poly(ADPR)-Resten zeigte es fundamentale Unterschiede zu Histon H1-$(ADPR)_n$-Konjugaten, die durch Inkubation von isolierten Kernen mit [^{3}H]NAD gebildet wurden.

* Adamietz, P., Bredehorst, R. & Hilz, H., in preparation.

1 Hayaishi, O. (1976) *Trends Biochem. Sci.* **1**, 9 - 10.
2 Hilz, H. & Stone, P. R. (1976) *Rev. Physiol. Biochem. Pharmacol.* **76**, 1 - 59.
3 Bredehorst, R., Ferro, A. & Hilz, H. (1978) *Eur. J. Biochem.* **82**, 115 - 121.

Helmuth Hilz, Peter Adamietz, Reinhard Bredehorst und Klaus Wielckens, Institut für Physiologische Chemie der Universität Hamburg, Martinistr. 52, D-2000 Hamburg 20.

J. Hoppe

ATP Binding to cAMP-Dependent Protein Kinase Type I

Studies on the substrate specificity of the catalytic subunit of rabbit muscle adenosine 3':5'-monophosphate-dependent protein kinase (EC 2.7.1.37, ATP:protein phosphotransferase) have been made using 24 different ATP analogues modified at various positions of the nucleotide. These experiments show that position N^6 is the dominant factor in the affinity between the enzyme and the nucleotide, since replacement of the NH_2-group at this position by -OH dramatically increases the K_i. Modification of the N^6 by methylation or dimethylation significantly decreases the interaction affinity, indicating a strict steric requirement at this locus of the ATP binding site. N^6-benzyl-ATP was found to be inactive. Increasing K_i-values are found with increasing van der Waals radii of the respective substituents at position C-8. The acceptance of substituents at C-8 suggests a *"syn"* conformation of the bound ATP. The fact that Tubercidin triphosphate, which does not form an *"anti"* complex with Mg ions, binds with the same affinity as ATP does supports the assumption of a *"syn"* complex. Chemical modifications of the sugar moiety cause little or no increase in K_i. Comparison of the data with binding constants of the same series of ATP derivatives to the holoenzyme R_2C_2 previously reported[1] reveals almost identical structural requirements of both sites for the adenine moiety. Remarkable differences are found with the sugar part and the phosphates of ATP. Thus it is suggested that both sites take the same adenine binding domain, but ribose and phosphates are bound in different states, depending on whether the enzyme is associated (R_2C_2) or dissociated by cAMP ($R_2(cAMP)_2$ + 2 C (active)).

1 Hoppe, J., Marutzky, R., Freist, W. & Wagner, K. G. (1977) *Eur. J. Biochem.* **80**, 369 - 372.

Dr. Jürgen Hoppe, Abteilung Molekularbiologie, Gesellschaft für Biotechnologische Forschung, Mascheroder Weg 1, D-3300 Braunschweig-Stöckheim.

J. Huber and W. Rüdiger

Subcellular Localisation and Properties of 3-Hydroxy-3-methylglutaryl-Coenzyme A Reductase in Plant Cell Suspension Cultures of Anise

Phytosterols, carotenoids and the phytyl side chain of chlorophylls are major products of the isoprenoid pathway of plants. They fulfill various functions, especially those related to plant growth and development. That variety of functions necessitates a system coordinating and controlling isoprenoid biosynthesis. The existence of two separate pathways was postulated (1): 1) A pathway leading to phytosterols at a cytoplasmic site. 2) A pathway leading to isoprenoid pigments at a plastidic site. Therefore, plastids are expected to contain the complete set of enzymes required for pigment formation from Acetyl-CoA.

3-Hydroxy-3-methylglutaryl-coenzyme A (HMG-CoA) reductase [mevalonate:NADP$^{\oplus}$ oxidoreductase (CoA-acylating) (EC 1.1.1.34)] is the regulatory enzyme of sterol biosynthesis of mammals[2] and yeast[3]. Its occurrence in higher plants has been demonstrated[4]. The enzyme was shown to be associated with subfractions of chloroplasts and mitochondria, as well as with a microsomal fraction. There is no conclusive proof that HMG-CoA reductase is present in different compartments, however.

In this communication we now wish to report:

1) Subcellular localisation of HMG-CoA reductase.
2) Characterization of particle fractions showing reductase activity using specific marker enzymes.
3) Regulatory properties of HMG-CoA reductase from plastidic and microsomal fractions.

As model system cell suspension cultures of anise *Pimpinella anisum*, L. (cell strain An 1175) were used[5]. This cell strain may be induced to produce somatic embryos which will turn green on illumination.

1 Goodwin, T. W. & Mercer, E. I. (1963) *Symp. Biochem. Soc.* **24**, 37.
2 Bucher, N. L. R., Overath, P. & Lynen, F. (1960) *Biochem. Biophys. Acta* **40**, 491.
3 Berndt, J., Boll, M., Löwel, M. & Gaumert, R. (1973) *Biochem. Biophys. Res. Commun.* **51**, 843 - 848.
4 Brooker, J. D. & Russell, D. W. (1975) *Arch. Biochem. Biophys.* **167**, 730 - 737.
5 Huber, J., Constabel, F. & Gamborg, O. L. (1978) *Plant Sci. Lett.*, submitted for publication.

Dr. J. Huber, Botanisches Institut der Universität München, Phytochemie III, Menzinger Straße 67, D-8000 München 19.

F. Hucho

The Acetylcholine Receptor as a Protein Complex of the Postsynaptic Membrane

Membrane fragments, probably of postsynaptic origin, can be obtained from the electric organ of *Torpedo californica* which contain four main types of polypeptide chains, α (M_r 42000), β (M_r 48000), γ (M_r 62 000), and δ (M_r 68000). We propose that these chains form a protein complex, with an unknown stoichiometry, which regulates the ion permeability of the postsynaptic membrane. The α-chains have been shown to contain the binding site for an SH-group-directed affinity label[1] and also for α-neurotoxins from snake venom[2]. They are assumed to represent the acetylcholine receptor, defined as the transmitter-binding moiety. The function of the other chains is unknown, but they are closely attached to the α-chains: They are co-purified by affinity chromatography from detergent extracts of the membrane; their ratio remains constant in membrane fractions with different specific activities, and they can be cross-linked to a molecule of M_r 270000. Using a cleavable cross-linker, we have shown that the M_r 270000 particle and a M_r 500000 molecule, probably a dimer, contain the four chains α-δ. The purified complex can be reconstituted into artificial lipid vesicles. There it regulates the ion permeability of the vesicle membrane in response to cholinergic effector molecules. (For a recent review see ref.[3]).

The site of action of the local anesthetic quinacrine has been investigated. The alkylating derivative quinacrine mustard reacts in a covalent manner, preferentially with the α-chains, but the other membrane components, including the lipid, incorporated the label also. Alkylation blocks irreversibly the carbachol-induced sodium efflux from receptor-membrane vesicles. Acetylcholine still binds to the modified membranes.

1 Karlin, A. (1974) *Life Sci.* **14**, 1385 - 1415.
2 Hucho, F., Layer, P., Kiefer, H.R. & Bandini, G. (1976) *Proc. Natl. Acad. Sci. U.S.A.* **73**, 2624 - 2629.
3 Hucho, F. & Schiebler, W. (1978) *Mol. Cell. Biochem.*, in press.

F. Hucho, Fachbereich Biologie der Universität Konstanz, Postfach 77 33, D-7750 Konstanz.

F. Hucho, S. Stengelin, W. Rathmayer and L. Béress

Neurotoxins ATX I and ATX II: Radioactive Labeling and Binding to Axonal Membranes

The neurotoxic action of the polypeptides ATX I and II from the sea anemone *Anemonia sulcata* is based on a prolongation of action potentials by blocking of sodium channel inactivation[1,2]. Because of their selectivity and high affinity these toxins may be useful tools for the biochemical characterization of components of the voltage-dependent sodium channels. We intend to use them in a binding assay for these components in vitro.

We have labeled ATX I and ATX II and have shown that the radioactive derivatives retained their toxicity. ATX I was iodinated by the lactoperoxidase method with ^{125}I. Even after incorporation of more than one ^{125}I per polypeptide chain, the toxic activity remained unchanged. The iodinated toxin is bound by the intact walking nerve of the crayfish *Astacus leptodactylus.* Specificity of the binding was demonstrated by its saturability and by competition experiments with excess unlabeled toxin. Reciprocal binding plots extrapolated to 43.5 pmol ^{125}I-ATX I bound per gram nerve. TTX did not compete with ATX binding. Binding to nerve homogenates or purified membrane fragments have not been obtained so far.

ATX II, which does not contain a tyrosine residue as a site for iodination, was labeled by reductive alkylation via a Schiff's base intermediate with pyridoxal phosphate and sodium [^{3}H]borohydride. Pyridoxal phosphate introduces an additional charge into the polypeptide, allowing separation of monosubstituted from more highly modified and from unreacted toxin by ion-exchange chromatography. The pyridoxamine toxins retained their toxicity but changes in effectiveness and reversibility were observed.

Further radioactive derivatives were obtained by catalytic exchange of iodine for tritium (ATX I) and by reductive methylation of aminogroups (ATX II).

1 Bergman, C., Dubois, J.M., Rojas, E. & Rathmayer, W. (1976) *Biochim. Biophys. Acta* **455**, 173 - 184.
2 Romey, G., Abita, J.P., Schweitz, H., Wunderer, G. & Lazdunski, M. (1976) *Proc. Natl. Acad. Sci. U.S.A.* **73**, 4055 - 4059.

Ferdinand Hucho, Fachbereich Biologie der Universität Konstanz, Postfach 77 33, D-7750 Konstanz.

L. Jaenicke und J.-C. Jésior

Die Kinetik der oktameren Glutamin-Synthetase: Deutung der Sigmoid-Kinetik und Halbzentren-Reaktivität

Glutamin-Synthetase (EC 6.3.1.2) läßt sich mittels Cibacronblau-Sepharose ausgezeichnet reinigen[1]. Das läßt auf eine Affinität des Farbstoffs eher zu einer allgemeinen Nucleotid-Faltung als zu einer speziellen NAD$^\oplus$-Bindestelle[2] schließen.

Mit dem hoch gereinigten Enzym aus Schweinehirn wurde die Rolle von $Mg^{2\oplus}$ und ATP für die Bildung von Glutamin untersucht. Die Reaktionsgeschwindigkeit hängt bei einem gegebenen pH-Wert nicht so sehr von der absoluten Konzentration der beiden Reaktionspartner ab als von ihrem jeweiligen stöchiometrischen Verhältnis. Er läßt sich zeigen und durch Computer-Simulation nachvollziehen, daß die Reaktion von der Konzentration des Enzym-Mg/ATP-Komplexes als einer der verschiedenen im Gleichgewicht stehenden Komplex-Species abhängt. Das Enzym hat zwei benachbarte, aber voneinander verschiedene Bindungsstellen, eine für das

Metallion, die andere für das Nucleotid. Zwischen dem Enzym-Mg- und dem Enzym-ATP-Komplex stellt sich rasch ein Gleichgewicht ein.

Aus der jeweiligen Beteiligung der beiden Bindungsstellen erklärt sich die Sigmoidizität und Glockenform der $v/Mg^{2\oplus}$-Kurven, die scheinbare „halfsite-reactivity" des Enzyms gegenüber Glutamin[3,4] aus den experimentellen Bedingungen. Diese Deutungen[5] lassen sich vermutlich auch auf andere Ligasen und Synthetasen anwenden.

1 Jaenicke, L. & Berson, W. (1977) *Hoppe-Seyler's Z. Physiol. Chem.* **358**, 883 - 889.
2 Thompson, S.T., Gass, K.H. & Stellwagen, E. (1975) *Proc. Natl. Acad. Sci. U.S.A.* **72**, 669 - 672.
3 Meister, A. (1974) in The Enzymes (Boyer, P.D., ed.) vol. 10, pp. 699 - 754, Academic Press, New York.
4 Seydoux, F., Malhotra, O.P. & Bernhard, S.A. (1974) *CRC Critical Rev. Biochem.* **2**, 227 - 257.
5 London, W.P. & Steck, T.L. (1969) *Biochemistry* **8**, 1767 - 1779.

Prof. Dr. L. Jaenicke, Institut für Biochemie der Universität zu Köln, An der Bottmühle 2, D-5000 Köln 1.

A. Jakob and S. Diem

Role of ATP and $Mg^{2\oplus}$ in the Regulation of the Activity of Hepatic Phosphorylase Phosphatase

Adenine nucleotides and divalent cations have been known for some time to influence the activity of hepatic phosphorylase phosphatase (EC 3.1.3.17) but from earlier studies it was concluded that ATP and $Mg^{2\oplus}$ are not of regulatory importance in vivo. The following experiments indicate that this conclusion was premature.

Three preparations of phosphorylase phosphatase were used. The enzyme was preincubated with adenine nucleotides and divalent cations and subsequently assayed by measuring the inactivation of purified phosphorylase *a*. With equimolar concentrations of ATP plus $Mg^{2\oplus}$, we found that the inhibition was not as predicted from data obtained earlier with each agent alone. Therefore the enzyme activity was measured at different levels of uncomplexed ATP that were kept virtually constant throughout the experiment, while $ATPMg^{2\ominus}$ was increased from 1µM to 5mM. Total amounts of ATP and $Mg^{2\oplus}$ necessary to obtain these concentrations were calculated from pH, dissociation and stability constants. As $ATPMg^{2\ominus}$ was increased, the inhibitory effect of uncomplexed ATP was suppressed. In addition, an increasing degree of complexation of ATP with $Mg^{2\oplus}$ altered the sigmoidicity of the curves obtained by plotting total ATP against enzyme activity. At concentrations of total ATP around 1mM, inhibition was complete in the absence of $Mg^{2\oplus}$ ($ATPMg^{2\ominus}$ 0%). Inhibition was progessively less pronounced as the percentage of Mg-complexed ATP was increased to 90% of total ATP. Total ATP and $Mg^{2\oplus}$ concentrations in the intact cell are 1 - 5mM, and it is therefore possible that a decrease in Mg-complexed ATP produces an inhibition of phosphorylase phosphatase in vivo.

PD Dr. A. Jakob, Metabolic Unit, Dept. of Medicine, Universitätsspital, CH-8091 Zürich.

F. Jakob und R. Zwilling

Flußkrebs-Carboxypeptidase: Affinitäts-Chromatographie, Charakterisierung und N-terminale Sequenz

Die Carboxypeptidase (EC 3.4.12.3) aus dem Flußkrebs *Astacus fluviatilis* ist im Hinblick auf die frühe Entwicklung der Carboxypeptidasen von besonderem Interesse.

Zur Isolierung des Enzyms wurde eine Affinitäts-Chromatographie an einem Inhibitor aus der Kartoffel[1] entwickelt, wobei aus 100 m*l* Rohextrakt bis zu 25 mg reine *Astacus*-Carboxypeptidase gewonnen werden können. Die isolierte Fraktion zeigt Carboxypeptidase-B-Spezifität (Substrat: Hippuryl-L-arginin) und ist in der Acetatfolien-, Disk- und Dodecylsulfat-Elektrophorese absolut homogen.

Das Enzym wurde durch Doppeldiffusions-Experimente und Laurell-Elektrophoresen immunologisch untersucht. Durch Chelatbildner und Neutronenaktivierungs-Analysen konnte gezeigt werden, daß dieses Wirbellosen-Enzym ebenso wie die pankreatischen Säuger-Carboxypeptidasen ein Zinkatom in seinem Katalyse-Mechanismus enthält[2]. Von der *C*-terminalen Sequenz der A-Kette des oxidierten Insulins -Cys-Asn wird kein Rest, von der B-Ketten-Sequenz -Thr-Pro-Lys-Ala werden die Reste Alanin und Lysin freigesetzt.

Krebs-Carboxypeptidase ist gegen die synthetischen Hemmstoffe β-Hydroxyzimtsäure und ε-Aminocapronsäure 10mal sensitiver als Schweine-Carboxypeptidase B. Das pH-Optimum des Flußkrebs-Enzyms liegt bei 6.5 und ist gegenüber den Säuger-Carboxypeptidasen nach dem Sauren hin verschoben. Durch isoelektrische Fokussierung wurde der isoelektrische Punkt bei pH 4.0 gefunden.

Die Aminosäuren-Summenanalyse, die durch einen hohen Gehalt an sauren Resten auffällt, und vor allem die *N*-terminale Sequenz der Flußkrebs-Carboxypeptidase lassen deutlich erkennen, daß dieses Invertebraten-Enzym und die pankreatischen Säuger-Carboxypeptidasen aus einem gemeinsamen Vorläufer-Gen hervorgegangen sind.

1 Ryan, C.A., Hass, G.M. & Kuhn, R.W. (1974) *J. Biol. Chem.* **249**, 5495 - 5499.
2 Bauer, Hch., Diplomarbeit (1977), Heidelberg.

Prof. Dr. R. Zwilling, Zoologisches Institut, Fachrichtung Physiologie, Universität Heidelberg, Im Neuenheimer Feld 230, D-6900 Heidelberg.

K. H. Jakobs and G. Schultz

Adrenalin-Induced Inhibition of Cholera Toxin-Stimulated Platelet Adenylate Cyclase

In cell-free preparations from human platelets, adrenalin inhibits adenylate cyclase (EC 4.6.1.1) through α-adrenergic receptors[1]. Whereas prostaglandin-induced stimulation is not or is only slightly increased by GTP, inhibition by adrenalin requires GTP[2]. In contrast, the effect of adrenalin is reversed or abolished in the presence of GMP-P(NH)P, GMP-P(CH_2)P or GTPγS, stable GTP analogues, which cause time-dependent and irreversible activation of the enzyme. Like these GTP analogues, cholera toxin causes adenylate cyclase stimulation in various systems, presumably by inhibition of a GTPase[3]. The effect of cholera toxin on adenylate cyclase was studied in human platelet particles. Cholera toxin (0.3 to 30 μg/m*l*) caused enzyme stimulation up to about 5-fold; after a time-lag of 2 to 3 min, constant activities were reached. Addition of GTP (1 to 30μM), which had very little effect on the untreated enzyme, caused further stimulation (up to 2-fold) of the cholera toxin-treated enzyme. Adrenalin (0.3 to 30μM, with 10μM pindolol and 30μM GTP) reduced adenylate cyclase activity by maximally 50 to 60% in the absence and presence of cholera toxin (up to 30 μg/m*l*). The extent of inhibition of the cholera toxin-treated enzyme was unchanged over at least 20 min. These findings, which are in contrast to those obtained with the stable GTP analogues, suggest that different stimulated states are reached with GTP analogues and with cholera toxin.

1 Jakobs, K.H., Saur, W. & Schultz, G. (1976) *J. Cycl. Nucl. Res.* **2**, 381 - 392.
2 Jakobs, K.H., Saur, W. & Schultz, G. (1978) *FEBS Lett.* **85**, 167 - 170.
3 Cassel, D. & Selinger, Z. (1977) *Proc. Natl. Acad. Sci. U.S.A.* **74**, 3307 - 3311.

Dr. K. H. Jakobs, Pharmakologisches Institut der Universität Heidelberg, Im Neuenheimer Feld 366, D-6900 Heidelberg.

J. Janecki, A. Fijalkowska und H. Tomaszewska

Geschlechtsgebundene, östrogenabhängige Subfraktionen der kleinmolekularen Serumlipoproteide (SMDL) des Menschen

Mit Hilfe der 3-schichtigen Polyacrylamidgel-Disk-Elektrophorese[1] lassen sich die kleinmolekularen Serumlipoproteide[2] des Menschen reproduzierbar in 12 Subfraktionen differenzieren[3-5]. Wird die Methodik streng eingehalten, so sind die Ergebnisse auch quantitativ, auf Grund der Extinktionsmessungen bei der Densitometrie, auswertbar.

Es bestehen sichtbare Unterschiede zwischen den Subfraktionen-Mustern bei verschiedenen Personen. Die auffälligsten Änderungen treten in Verbindung mit manchen pathologischen Zuständen auf. Man kann aber auch mit sehr hoher Signifikanz ($p < 0.001$) das Muster, das bei jungen Frauen vorkommt von dem der jungen Männern unterscheiden[6]: die Mittelwerte der von uns mit den Nrn. 6 und 7 bezeichneten Fraktionen liegen bei den Frauen viel höher, als bei den Männern (85.9 gegenüber 45.4 für Frakt. 7). Wir haben entsprechende Messungen auch im Entwicklungsalter (bei Kindern) durchgeführt und fanden fast ständig, geringfügig höhere Werte bei Mädchen.

Bei Menstruationsstörungen[7] findet man meistens niedrige Frakt.-7-Werte ($\bar{x} = 68.0$). Behandelt man diese Störungen mit Östrogen-Präparaten[8], kann man meistens einen Anstieg dieser Fraktion beobachten ($\bar{x}_\Delta = 25$) besonders, wenn die Ausgangswerte niedrig lagen ($\bar{x} = 55$, wenn $\bar{x}_A = 33$). Wir haben immunologisch im Serum derselben Frauen Apo-A-Lipoproteide quantitativ bestimmt und die Werte (vor der Behandlung $\bar{x} = 94.5\%$ der Norm, nach der Behandlung $\bar{x} = 122.5\%$) mit denen der Frakt. 7 verglichen. Es fand sich eine positive Korrelation ($r = 0.72$), aber auffällig war die Parallelität des Anstieges beider Parameter infolge der Behandlung. Auf Grund mancher Erwägungen könnte man vermuten, daß im Schnitt die Hälfte des Apo-A-Lipoproteins mit der Frakt. 7 zusammen auf die Östrogen-Therapie reagiert.

1 Wollenweber, J. & Kahlke, W. (1970) *Clin. Chim. Acta.* **29**, 411 - 420.
2 Dangerfield, W.G. & Pratt, J.J. (1970) *Clin. Chim. Acta.* **30**, 273 - 278.
3 Fijalkowska, A. (1977) Promotionsarbeit, Medizinische Akademie Warszawa, 1 - 84.
4 Feliste, R., Dousset, N. & Douste-Blazy, L. (1973) *Clin. Chim. Acta.* **47**, 329 - 333.
5 Utermann, G. (1972) *Clin. Chim. Acta.* **36**, 521 - 529.
6 Reider, H.P. & Peyer, Ch. (1970) *Clin. Chim. Acta.* **30**, 305 - 310.
7 Barclay, M., Skipski, V.P., Trebus-Kekish, O., Müller, C.H., Shoh, E. & Elkins, W.L. (1965) *Biochem. J.* **96**, 205 - 209.

8 Gustafson, A. & Svanborg, A. (1972) *Clin. Endocrinol. Met.* **35**, 203 - 207.

J. Janecki, Zakład Diagnostyki Laboratoryjnej Instytutu Pediatrii Akademii Medycznej, 00-576 Warszawa, Marszałkowska 24.

B. Jastorff, T.M. Konijn, J. Mato, J. Hoppe and K.G. Wagner

Comparison of the Molecular Interactions between cAMP and Its Receptor Protein in Dictyostelium discoideum and Protein Kinase Type I

Binding of effector molecules like cAMP to their specific protein receptors depends on chemical interaction of distinct atoms or atom groups of the small molecule with single amino acid side chains or with the backbone of the protein. Several X-ray studies of proteins with effector or substrate molecules have shown that these chemical interactions are either hydrogen bonding, ionic or hydrophobic interactions.

We have developed a method for determining these essential interactions using specifically modified analogs of the effector molecule as molecular probes in different protein systems. Using cAMP and two different biological receptors for it as a model system, we elucidated

i) the exact atom or atom group of cAMP essentially involved in binding.
ii) the type of chemical interaction with the receptor amino acids which led us to
iii) proposals for the kind of side chains involved in binding.

A comparison of the binding sites of the receptor proteins of the amoeba *Dictyostelium discoideum** and of different protein kinanases type I** showed that there are five distinct chemical interactions between cAMP and the receptors, but only three of them are common to both namely one ionic interaction with the negative charge, one hydrogen donating bond with the 3′-oxygen, and one hydrophobic interaction with the heterocyclic base. Protein kinase further binds to the 2′-hydroxyl group and to the 5′-oxygen, while the *Dictyostelium receptor* binds to the 6-amino group and N-7.

* Konijn, T.M., Mato, J., Morr, M. & Jastorff, B. (1978) *Proc. Natl. Acad. Sci. U.S.A.*, submitted for publication.
** Hoppe, J., Morr, M. & Jastorff, B., in preparation.

Prof. B. Jastorff, Fachbereich Biologie-Chemie, Universität Bremen, Leobenerstr., D-2800 Bremen 33.

H.P. Jennissen

Off-Rates of Phosphorylase b Bound to Hydrophobic Agaroses

The multivalent binding of phosphorylase *b* (EC 2.4.1.1) to immobilized butyl residues on Sepharose 4B (unless otherwise stated: 30 μmol/m*l* packed gel, 5 °C) in the presence of 1.1M ammonium sulfate has been shown to be an endothermic, non-independent reaction[1]. At an initial concentration (c_0) of 0.3 mg/m*l*, the adsorption (for method see ref.[1]) of reduced, ^{3}H-labelled phosphorylase *b* ([^{3}H]phosphorylase b_r) is a second-order reaction with an apparent rate constant of k'_{+1} = 0.74 m*l*/(mg × min).

The dissociation rate studied over 60 min after 10 - 300-fold dilution of the [^{3}H]phosphorylase b_r-agarose complex (for method see ref.[2]) does not follow a single first-order reaction. Between 0.4 and 2 min, however, one apparent first-order rate constant (k'_{-1}) is observed. At an initial concentration of adsorbed labelled enzyme (a_0) of 4.5 mg/m*l* packed gel $k'_{-1} = 4.2 \times 10^{-3}$ min^{-1}. This rate constant is independent of the dilution (20 - 300-fold) and the stirring rate (400 - 1200 rpm) but is a function of the fractional gel saturation (e.g. at a_0 = 23 mg/ml packed gel, $k'_{-1} = 13.1 \times 10^{-3}$ min^{-1}) and a function of the degree of substitution (e.g. at 12 μmol/m*l* packed gel, $k'_{-1} = 14.4 \times 10^{-3}$ min^{-1}, a_0 = 3.7 mg/m*l* packed gel).

On dilution of the [^{3}H]phosphorylase b_r-agarose complex (a_0 = 4.5 mg/m*l* packed gel, see above) in the presence of free, unlabelled phosphorylase b_r (1 mg/m*l*) k'_{-1} is increased ca. 5-fold to 20.5×10^{-3} min^{-1}. This enhancement of k'_{-1} increases with the initial concentration of unlabelled enzyme in the bulk but decreases to ca. 1.5-fold on gels with a low degree of substitution (e.g. 12 μmol/m*l* packed gel) or at high fractional gel saturation (e.g. a_0 = 23 mg/m*l* packed gel).

Tentatively the described increase in k'_{-1} in the presence of unlabelled enzyme could be explained by a decrease in the affinity (and valence) of a *binding unit* (i.e. unit encompassing the number of binding sites necessary for the multivalent adsorption of a protein ligand), to which a labelled enzyme molecule is adsorbed, by the sequential binding of an unlabelled molecule to an adjacent, overlapping binding unit on the agarose.

The association-dissociation reaction of [^{3}H]phosphorylase b_r on butyl-Sepharose appears to be a hysteretic process in finite times, probably of kinetic origin. This is illustrated by a desorption isotherm (desorption 60 min, a_0 = 3.3 mg/m*l* packed gel) which is described according to Freundlich by the two constants α = 4.1 mg/m*l* packed gel and $1/n$ = 0.047, both of which are ca. 10-fold lower than those of the corresponding adsorption isotherms (see ref.[1]).

The described experiments serve the understanding of multivalent interactions between proteins and structurally immobilized components e.g. also of cellular membranes. Furthermore, hysteresis may be important as a mode of information transfer in multivalent effector-receptor interactions (see ref.[2]).

1 Jennissen, H.P. (1976) *Biochemistry* **15**, 5683-5692.
2 Jennissen, H.P. (1977) *Hoppe-Seyler's Z. Physiol. Chem.* **358**, 255.

Dr. Herbert P. Jennissen, Institut für Physiologische Chemie der Ruhr-Universität Bochum, Postfach 10 21 48, D-4630 Bochum 1.

T.F. Joos und F.A. Anderer

Charakterisierung der Proteinphosphorylierung in isolierten Plasmamembran- und Reticulumfraktionen SV 40-transformierter Mäusefibroblasten

Die endogene Phosphorylierung von hochgereinigten Fraktionen der Plasmamembran und des endoplasmatischen Reticulum wurde mit $[\gamma\text{-}^{32}P]$ATP und $[\gamma\text{-}^{32}P]$-GTP als radioaktiven Vorläufern untersucht.

Es zeigte sich, daß Plasmamembranen ATP und GTP, Reticulum aber nur ATP zur Proteinphosphorylierung verwerten. Bei Phosphorylierung mit ATP erfolgt maximaler ^{32}P-Gesamteinbau in beiden Membranfraktionen bei pH 7.8 und 10mM $MgCl_2$ nach 1 min, bei Phosphorylierung mit GTP in der Plasmamembran bei pH 7.8 und 10mM $MgCl_2$ nach 3 min bei 37 °C.

In der Natriumdodecylsulfat-Polyacrylamidgel-Elektrophorese zeigt die Plasmamembran rund 15 Phosphoproteine, während das Reticulum im wesentlichen nur eine einzige phosphorylierte Komponente mit M_r 35 000 enthält.

Der ATP-abhängige ^{32}P-Einbau in die Plasmamembran führt zu einer wesentlich höheren spezifischen Aktivität als der GTP-abhängige mit einem signifikanten Unterschied: Die GTP-spezifische Hauptkomponente hat ein Molekulargewicht von 100 000, die ATP-spezifische Hauptkomponente dagegen eines von 110 000. Bei Erhöhung der Konzentration der Plasmamembranproteine von 0.2 auf 2 mg/m*l* wird die ATP-spezifische Hauptkomponente bei M_r 110 000 nicht mehr phosphoryliert, dagegen mit GTP im wesentlichen nur noch die Hauptkomponente bei M_r 100 000. Die relative Markierung der verschiedenen Plasmamembran-Phosphoproteine ist außerdem vom pH-Wert abhängig, nicht aber von der $MgCl_2$-Konzentration und der Inkubationszeit. Durch ATP und durch GTP werden die Phosphoproteine bei M_r 110000.bzw. 100000 erst ab pH 7.8 als Hauptkomponenten markiert. Inkubationszeit, pH-Wert, $MgCl_2$-Konzentration sowie Proteinkonzentration haben keinen qualitativen Einfluß auf das Phosphorylierungsmuster des Reticulums.

Beide Membranfraktionen zeigen eine unterschiedliche Sensitivität gegenüber graduellem Einfrieren: Während die endogene Phosphorylierung des Reticulums intakt bleibt, lassen sich danach Plasmamembranen nicht mehr mit ATP phosphorylieren.

Diese Befunde lassen auf mindestens 2 Kinase-Typen in der Plasmamembran schließen, von denen einer ATP-spezifisch, der andere GTP-spezifisch ist, und auf mindestens eine Kinase im Reticulum, die vom ATP-spezifischen Kinase-Typ in der Plasmamembran verschieden ist.

Dr. Thomas F. Joos, Friedrich-Miescher-Laboratorium der Max-Planck-Gesellschaft, Abteilung Prof. Dr. F. Alfred Anderer, Spemannstraße 37 - 39, D-7400 Tübingen.

R. Kaleja and M. Schütrumpf

Active-Site-Directed Inhibition of Myokinase by an AMP-Derivative

Myokinase from hog muscle is inhibited by N^6-(2,3-diiodo-1-propyl)adenosine monophosphate in the following manner:

I. a) At 34 °C myokinase is inhibited in a time-dependent step.
   b) Myokinase is protected by AMP.
   c) ATP slightly accelerates the inhibition reaction.

II. a) At 26 °C inhibition is slow.
   b) Acceleration of inhibition by ATP is fast.

It is concluded that myokinase possesses two distinct binding sites – one for AMP and one for ATP – which exert a mutual influence on each other.

R. Kaleja and M. Schütrumpf, Institut für Biochemie der Universität Frankfurt, Sandhofstraße, Gebäude 75 A, D-6000 Frankfurt.

O. Kandler und H. König

Chemie der peptidoglycanfreien Zellwände der Methanbakterien

Die schon früher auf Grund biochemischer Besonderheiten vermutete phylogenetische Sonderstellung der methanbildenden Bakterien ist neuerdings durch den Vergleich der Oligonucleotidmuster der ribosomalen 16S-RNA[1] mit den 16S-RNA-Mustern anderer Prokaryonten und die Auffindung neuartiger Zellwandpolymerer besonders deutlich geworden. Wie wir zeigen konnten, besitzen von den 4 Gattungen der Methanbakterien nur die beiden gram-positiven Gattungen *Methanobacterium* und *Methanosarcina* einen typischen Sacculus. Die gram-negative Gattung *Methanococcus* weist dagegen nur eine fragile, aus leicht desintegrierenden Untereinheiten bestehende äußere Proteinhülle auf, und die ebenfalls gram-negative Gattung *Methanospirillum* hat zwar eine sich über viele Zellen erstreckende Scheide aus fibrillärem und gegen Na-Dodecylsulfat und Protease resistentem Protein entwickelt, die Einzelzellen besitzen aber keinen Sacculus.

Die Analyse der Sacculi von *Methanosarcina* ergab kein Peptidoglykan, sondern ein Heteropolysaccharid, das aus Neutralzuckern, Uronsäuren und Aminozuckern besteht[2]. Die Sacculi der Gattung *Methanobacterium* bestehen dagegen überwiegend aus L-Lys, L-Ala, L-Glu und GlcAc in einem Molverhältnis von rund 1 : 1.2 : 2 : 1. Außerdem enthalten die Sacculi der einzelnen Arten unterschiedliche Mengen von Neutralzuckern, die aber durch Formamidextraktion ohne Formverlust entfernt werden können. Sie gehören daher nicht zum formgebenden Zellwandpolymer. Die Analyse der durch sauere Partialhydrolyse der Sacculi erhaltenen überlappenden Peptide ergibt folgende vorläufige Primärstruktur (I) der Peptidkomponente, wobei die Art der Bindung des *N*-Acetylglucosamins noch ungeklärt bleibt.

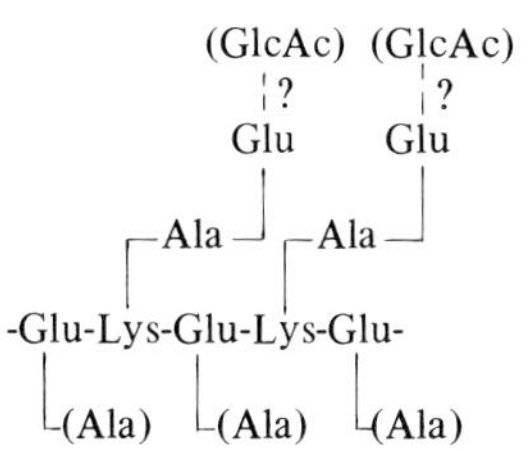

(I)

Der in Klammern gesetzte Alaninrest ist in der Zellwand zu etwa 80% abgespalten. Da etwa 10 bis 15% der Lysinreste *C*-terminal vorliegen, ist zu entnehmen, daß die aus wiederkehrenden Peptideinheiten aufgebauten Polypeptidstücke nur 6 bis 10 Einheiten umfassen. Die Vernetzung dieser Stücke zum hochpolymeren Sacculus könnte durch die Glucosaminylreste erfolgen, deren Bindungsart noch unklar ist. Dieses Zellwandpolymer der Methanbakterien erinnert nur wegen der Art der 3 Aminosäuren und des Aufbaus aus wiederkehrenden Peptideinheiten an das Peptidoglykan[3] typischer Bakterien, unterscheidet sich aber davon prinzipiell durch das Fehlen von Muraminsäure und der D-Aminosäuren. Die auch für diese Zellwände notwendige Resistenz gegenüber normalen Proteasen ist vermutlich durch die Häufung ungewöhnlicher Peptidbindungen gewährleistet.

1 Fox, G. E., Magrum, L. J., Balch, W. E., Wolfe, R. S. & Woese, C. R. (1977) *Proc. Natl. Acad. Sci. U.S.A.* **74**, 4537 - 4541.
2 Kandler, O. & Hippe, H. (1977) *Arch. Microbiol.* **113**, 57 - 60.
3 Schleifer, K. H. & Kandler, O. (1972) *Bact. Rev.* **36**, 407 - 477.

Prof. Dr. Otto Kandler und Helmut König, Botanisches Institut der Universität München, Menzinger Straße 67, D-8000 München 19.

W. Keil, H.-D. Klenk und R. T. Schwarz

Charakterisierung des Bindungstyps zwischen dem Kohlenhydrat- und Proteinanteil von Influenzavirus-Glykoproteinen

Die Hydrolyse von Influenzavirus-Glykoproteinen mit Pronase ergibt zwei verschiedene Typen von Glykopeptiden. Typ I, der als komplexer Typ bezeichnet wird, enthält Glucosamin, Mannose, Galaktose und Fucose. Der Kohlenhydratanteil von Typ II besteht dagegen nur aus Mannose und Glucosamin[1].

Es war möglich, Typ I weiter aufzutrennen in Typ Ia und Typ Ib. Die individuellen Glykopeptide wurden hinsichtlich der Verknüpfungsart zwischen dem Kohlenhydrat- und Proteinanteil untersucht. Durch milde saure Hydrolyse[2] und nachfolgende Derivatisierung der Spaltprodukte war in allen Fällen Asn-GlcNAc nachweisbar, was das Vorliegen einer *N*-glykosidischen Bindung bei allen Oligosaccharidresten beweist.

Versuche mit β-Elimination ergaben keine Hinweise für das Vorliegen einer *O*-glykosidischen Bindung. Es kann demnach als gesichert gelten, daß das Kohlenhydratkomplement von Influenzavirus-Glykoproteinen über *N,N'*-Diacetylchitobiose an Asparagin gebunden ist.

1 Schwarz, R. T., Schmidt, M. F. G., Anwer, U. & Klenk, H.-D. (1977) *J. Virol.* **23**, 217 - 226.
2 Monsigny, M., Adam-Chosson, A. & Montreuil, J. (1968) *Bull. Soc. Chim. Biol.* **50**, 857 - 886.

Walter Keil, Ralph T. Schwarz und Hans-Dieter Klenk, Institut für Virologie, Frankfurter Straße 107, D-6300 Lahn-Gießen.

M. A. Kempfle, R. Müller und H. A. Winkler

Glutamat-Dehydrogenase-Tetrajodfluorescein-Komplexe: Verdrängung des spezifisch gebundenen Farbstoffs durch Adeninnucleotide

Es ist schon länger bekannt, daß 2,4,5,7-Tetrajodfluorescein mit bestimmten Enzymen spezifische Komplexe bilden kann. Dabei wurde gezeigt[1], daß dieses Molekül an die Bindungsstelle für den Adeninring von Adeninnucleotiden gebunden wird, und somit mit diesen Nucleotiden um diese Bindungsstelle konkurriert.

Aufgrund dieser Eigenschaft wurde eine einfache Methode entwickelt, mit der sich aus der Verdrängung des Farbstoffs durch Nucleotide die Dissoziationskonstanten von Enzym-Coenzym-Komplexen bestimmen lassen[2]. Dabei haben sich auch Einblicke in die Struktur der Bindungsstelle ergeben, da nicht alle Nucleotide den Farbstoff gleich gut verdrängen.

Bei der Bindung von Tetrajodfluorescein an Glutamat-Dehydrogenase (NAD(P)$^{\oplus}$) (EC 1.4.1.3) aus Rinderleber treten zwei Bindungsstellen auf mit Dissoziationskonstanten von 6.5μM und 30μM; sie unterscheiden sich durch ihre optischen Eigenschaften.

Die affinere Bindungsstelle ist identisch mit der Adenosinbindungsstelle, da nur von dieser der Farbstoff durch Adeninnucleotide verdrängt werden kann. Setzt man verschiedene Nucleotide ein, so ergeben sich folgende Dissoziationskonstanten $K_{D,EL}$ für die Enzym-Liganden-Komplexe (bei 0.1M Na-Phosphat-Puffer, pH 7.4, 25.0 °C):

Adenin (keine Verdrängung), Adenosin (270μM), 5'-AMP (17μM), 3'(2')-AMP (460μM), 3',5'-cAMP (530μM), 5'-ADP (3μM), Adenosin-5'-*O*-(2-thiodiphosphat) (3μM), 5'-ATP (40μM), 5'-GDP und 5-GTP (jeweils keine Verdrängung);

NADH (12μM – enzymatisch aktiv), α-NADH (12μM – aber *nicht* aktiv), NAD$^{\oplus}$ (60μM – aktiv), α-NAD$^{\oplus}$ (60μM – aber *nicht* aktiv), ε-NAD$^{\oplus}$ (keine Verdrängung – aber aktiv), NADP$^{\oplus}$ und NADPH (keine Verdrängung – aber aktiv).

Diese Ergebnisse zeigen, daß eine Anzahl von Wechselwirkungen zur Ausbildung einer „Bindung" erforderlich ist:

1) eine freie exozyklische NH_2-Gruppe am Adeninsystem,
2) unbesetzte 2' und 3'-OH-Gruppen der Ribose und
3) 5'-Phosphatgruppen[3].

Dies wird deutlich für die Nicotinsäureamid-dinucleotide, die nur unter den genannten Voraussetzungen an der Adenosinbindungsstelle gebunden werden können. Daß auch ohne eine solche Bindung das Enzym katalytische Aktivität aufweist, ist ein weiterer Hinweis auf das Vorhandensein mehrerer Bindungsstellen der Glutamat-Dehydrogenase für NAD$^{\oplus}$ bzw. NADH.

Ferner zeigen diese Ergebnisse, daß die Farbstoffverdrängungsmethode auch zur Untersuchung von Struktur-Funktions-Zusammenhängen herangezogen werden kann.

1 Wassarman, P.L. & Lentz, P.J. (1971) *J. Mol. Biol.* **60**, 509 - 522.
2 Kempfle, M.A., Müller, R.F., Sonnenberg, A. & Winkler, H.A. (1977) *Z. Anal. Chem.* **286**, 232 - 238.
3 Subramanian, S., Stickel, D.C. & Fisher, H.F. (1975) *J. Biol. Chem.* **250**, 5885 - 5889.

Prof. Dr. M. Kempfle, Physiologisch-Chemisches Institut der Universität Bonn, Nußallee 11, D-5300 Bonn.

N. Q. Khanh, R. Linde, R. Lipecky und H. G. Gassen

Inaktivierung des ribosomalen Proteins S1 durch reduktive Methylierung

Das ribosomale Protein S1, das auch als Untereinheit der Qβ-Replicase fungiert, spielt für die Übersetzung einer natürlichen mRNA eine wichtige Rolle. Dieses Protein soll das 3'-Ende der 16S-RNA in einer Konformation halten, die für eine Doppelstrangbildung mit der Intercistronsequenz einer mRNA optimiert ist[1]. Entfernt man S1 von 70S-Ribosomen (30S(– 1) + 50 S), so wird auch die poly(U)-abhängige Phe-tRNA-Bindung und die poly(Phe)-Synthese auf 20% des ursprünglichen Wertes reduziert. Da das Protein S1 mit Homopolynucleotiden Komplexe bildet, deren Assoziationskonstanten zwar von der Kettenlänge aber nicht von der Natur der Base abhängig sind[2], könnten die ε-Ammoniumgruppen der Lysine an einer elektrostatischen Wechselwirkung zu den Phosphatgruppen der Nucleinsäuren beteiligt sein. Reduktive Methylierung mit Formaldehyd-$NaBH_4$ modifiziert 6 - 7 Lysine pro Molekül S1. Dieses modifizierte Protein S1* assoziiert noch mit 30S Ribosomen, bildet aber keinen Komplex mehr mit poly(U). Rekonstituierte Ribosomen (30S(– 1)·S1* + 50S) sind inaktiv in der poly(U)-codierten Phe-tRNA-Bindung und der poly(Phe)-Synthese. Die Beteiligung der Lysinseitenketten an der biochemischen Funktion des Proteins S1 wird auch durch die pH-Abhängigkeit der Komplexbildung zwischen poly(U) und S1 bestätigt.

1 Dahlberg, A.E. & Dahlberg, J.E. (1975) *Proc. Natl. Acad. Sci. U.S.A.* **72**, 2940 - 2944.
2 Lipecky, R., Kohlschein, J. & Gassen, H.G. (1977) *Nucl. Acids Res.* **4**, 3627 - 3642.

Prof. Dr. H. G. Gassen, Institut für Organische Chemie und Biochemie, Technische Hochschule Darmstadt, Petersenstr. 22, D-6100 Darmstadt.

P. Kleine-Katthöfer and R.E. Zimmermann

The Aggregation Mechanism of Fibrin Monomers

When fibrin monomers are formed by the thrombin-induced removal of A- and B-fibrinopeptides, the aggregation of the intermediates occurs in such a way that intact fibrinogen molecules form dimers with fibrin monomers[1]. This hypothesis is strongly supported by the fact that the clotting time of fibrinogen seems to be correlated with the protein concentration[2]. When the reaggregation of solubilized fibrin monomers is studied the opposite effect is observed: fibrinogen increases the rate of this reaction, whereas fragments D and E have an inhibitory effect. One must conclude that different intermediates are formed when either fibrinogen or the fibrin monomer is present in higher concentration than its reaction partner.

1 Brass, E.P., Forman, W.B., Edwards, R.V. & Lindan, O. (1976) *Thromb. Haemost.* **36**, 37 - 48.
2 Bang, N.U. & Chang, L. (1974) *Semin. Thromb. Haemost.* **1**, 91 - 123.

Prof. Dr. R.E. Zimmermann, Physiologisches Institut II, Robert-Koch-Straße 28, D-4400 Münster.

L. Klungsøyr, D.A. Harris and G.K. Radda

Nucleotide Specificity and Dinitrophenol Effects in Mitochondrial ATPase

The classical uncoupler of oxidative phosphorylation, 2,4-dinitrophenol, stimulates the ATPase (EC 3.6.1.3) of intact mitochondria and submitochondrial particles, possibly by a mechanism involving proton transport.

Phenolic uncouplers also influence the activity of the isolated mitochondrial coupling factor F_1. Low concentrations of dinitrophenol stimulate the ATPase of F_1, while inhibition occurs at higher concentrations. On the other hand, F_1 ITPase is inhibited at all levels of dinitrophenol[1].

We have studied the nucleotide specificity of dinitrophenol stimulation or inhibition of F_1 from ox heart, and found that the nucleoside triphosphates can be divided into two classes: When F_1 acts on ITP, GTP, ATP-N^1-oxide and etheno-ATP, inhibition is seen with dinitrophenol at all levels. When F_1 acts on ATP, 2′-deoxy-ATP and iso-GTP, initial stimulation followed by inhibition is observed with increasing dinitrophenol concentrations.

This is the same specificity as that observed for several other functions on the isolated F_1 or in submitochondrial particles, such as binding to F_1 "tight nucleotide sites", reversal of electron flow in submitochondrial particles, ATP-P_i exchange in particles, and nucleoside triphosphate synthesis from the corresponding diphosphates in particles[2].

Preincubation of F_1 with high levels of ATP, iso-GTP and 2′-deoxy-ATP converted the enzyme to an activated form which has an increased rate of hydrolysis of both ATP and ITP. Dinitrophenol stimulated ITP hydrolysis by the activated F_1, in contrast to the untreated enzyme.

1 Stockdale, M. & Selwyn, M.J. (1971) *Eur. J. Biochem.* **21**, 416 - 423.
2 Harris, D.A., Gomez Fernandez, J.C., Klungsøyr, L. & Radda, G.K. (1977) in Structure and Function of Energy Transducing Membranes (van Dam, K. & van Gelder, B.F., eds.) pp. 319 - 327 Elsevier, Amsterdam.

Leiv Klungsøyr, Department of Biochemistry, University of Bergen, Norway.
David A. Harris and George K. Radda, Department of Biochemistry, University of Oxford, U.K.

I. Knack and K.-H. Röhm

A General-Purpose Computer Method for the Evaluation of Enzyme Kinetic Experiments

Although the advantages of the electronic processing of kinetic data are obvious (improved accuracy and objectiveness of calculations, simplified documentation and representation of results, saving of time), most enzyme kinetic studies are still interpreted by means of conventional graphical methods, which are rather painstaking and inevitably affected with a considerable subjective bias. Therefore we have developed a computer program which, starting from progress curves (i.e. the time dependences of absorption or another quantity related to the substrate or product concentration), automatically calculates initial velocities and from these, the best values of the kinetic parameters[1].

If the experimental values are available in a digital form on punched paper tape or another data carrier, the necessary calculations may be performed within a few seconds by any Fortran-programmable computer. No special knowledge of computer languages is required from the user of the program, which may be easily accomodated to a variety of experimental situations. Examples illustrating the application and efficiency of the method will be presented.

1 Herrmann, V., Knack, I. & Röhm, K.-H. (1978) *Hoppe-Seyler's Z. Physiol. Chem.* **359**, 47 - 61.

Ingo Knack and Klaus-H. Röhm, Institut für Physiologische Chemie der Universität, Lahnberge, D-3550 Marburg/L.

J. Knappe

Pyruvate Formate-Lyase, Mechanism and Regulation by Interconversion

Escherichia coli and various other bacteria may obtain their energy from fermentation processes which involve the production of acetyl-CoA from pyruvate by the non-oxidative reaction

pyruvate + CoA ⇌ acetyl-CoA + formate.

The pertinent enzyme, pyruvate formate-lyase (formate acetyltransferase, EC 2.3.1.54), operates through a double displacement mechanism according to:

a) ESH + pyruvate ⇌ ES-acetyl + formate,
b) ES-acetyl + CoA ⇌ ESH + acetyl-CoA[1];

the acetyl group being transferred to and from a cysteinyl residue.

Pyruvate formate-lyase occurs in an active (E_a) and in an inactive (E_b) form which are reversibly convertible. E_b into E_a conversion is catalyzed by an "activase", depends on pyruvate as an allosteric effector and is coupled to the reaction of adenosylmethionine and dihydroflavodoxin to yield 5'-deoxyadenosine and methionine[2]. Back conversion of E_a into E_b is inhibited by pyruvate and possibly involves a distinct "deactivase". Circumstantial evidence suggests an intramolecular modification, i.e. rearrangement, of the protein structure as the interconversion mechanism of pyruvate formate-lyase.

1 Knappe, J., Blaschkowski, H.P., Gröbner, P. & Schmitt, T. (1974) *Eur. J. Biochem.* **50**, 253 - 263.
2 Knappe, J. & Schmitt, T. (1976) *Biochem. Biophys. Res. Commun.* **71**, 1110 - 1117.

Prof. Dr. J. Knappe, Institut für Biologische Chemie der Universität, Im Neuenheimer Feld 501, D-6900 Heidelberg.

K. Knobloch

Role of Ionophorous Antibiotics on Energy Transformation in Purple Photosynthetic Bacteria

It has been shown that energy transformation occurs when an electrochemical proton gradient ($\Delta\tilde{\mu}_{H^+}$) is formed across a closed membrane vesicle upon the input of energy. The electrochemical gradient formed is composed of both a proton gradient (Δ pH) and an electric membrane potential ($\Delta\psi$). Energy transformation will be inhibited when both the ionophorous antibiotics nigericin (plus $K^{\oplus}$), which collapses a proton gradient, and valinomycin, which will nullify the membrane potential, are added[1]. This has been demonstrated, e.g., for closed chromatophore vesicles from *Rhodopseudomonas capsulata*[2], *Rps. spheroides*[3], and *Rhodospirillum rubrum*[4,5].

The purple bacteria mentioned above develop a vesicular membrane system[6]. However, there are photosynthetic bacteria, such as *Rps. palustris*, which develop an intracytoplasmic membrane system by invagination of the plasma membrane in such a way that uniform closed vesicles are not pinched off. As a result, the *Rps. palustris* cell maintains one single spherical membrane with the intracytoplasmic membrane system remaining connected to the plasma membrane[6].

Chromatophores isolated from *Rps. palustris* are of different size and shape[7]. They do not form closed vesicles but represent open global sheets[8]. For this reason, the contents of both the cell compartments are collected within the chromatophore-free supernatant fraction after ultracentrifugation. The supernatant, which still contains membrane particles[9], phospholipids[10], and complete electron transport chains[11], as well as the ATPase protein[7], does catalyze energy transformations[7-12], in contrast to the empty chromatophore fragments. The energy-transforming steps in the supernatant fraction from *Rps. palustris* will likewise occur in the presence of nigericin plus valinomycin, at concentrations which would nullify an energy-rich gradient across a closed membrane vesicle from other purple photosynthetic bacteria.

One has to conclude that in spite of the experimentally proved energy coupling in the chemiosmotic sense[13], there must be energy transformation in the system presented here which is catalyzed apart from a vesicular level, where energy may be transmitted by an interlocking protein network.

1 von Stedingk, L.-V. (1969) in Progress in Photosynthesis Research (Metzner, H., ed.) vol. III, pp. 1410 - 1419, Tübingen.
2 Casadio, R., Baccarini-Melandri, A., Zannoni, D. & Melandri, B.A. (1974) *FEBS Lett.* **49**, 203 - 207.
3 Crofts, A.R. & Jackson, J.B. (1970) in Electron Transport and Energy Conservation (Tager, J.M. et al., eds.) pp. 383 - 408, Adriatrica Editrice, Bari.
4 Schuldiner, S., Padan, E., Rottenberg, H., Gromet-Elhanan, Z. & Avron, M. (1974) *FEBS Lett.* **49**, 174 - 177.
5 Leiser, M. & Gromet-Elhanan, Z. (1977) *Arch. Biochem. Biophys.* **178**, 79 - 88.
6 Oelze, J. & Drews, G. (1972) *Biochim. Biophys. Acta* **265**, 209 - 239.
7 Knobloch, K. (1974) *Ber. Dtsch. Bot. Ges.* **87**, 119 - 128.
8 Knobloch, K. & Acker, G. (1977) *Hoppe-Seyler's Z. Physiol. Chem.* **358**, 1232 - 1233.

9 Knobloch, K. & Mayer, F. (1973) Ninth Internatl. Congr., Biochem., Stockholm, 4 h 6.
10 Gemeinhardt, F. & Knobloch, K. (1977) *Hoppe-Seyler's Z. Physiol. Chem.* **358**, 1205.
11 Knobloch, K. (1975) *Z. Naturforsch.* **30c**, 342 - 348.
12 Appelt, N. & Knobloch, K. (1977) *Hoppe-Seyler's Z. Physiol. Chem.* **358**, 1174 - 1175.
13 Mitchell, P. (1968) Chemiosmotic Coupling and Energy Transduction, Bodmin, England, Glynn Research.

Priv. Doz. Dr. K. Knobloch, Institut für Botanik und Pharmazeutische Biologie der Universität Erlangen-Nürnberg, Schloßgarten 4, D-8520 Erlangen.

H. E. Knoell

Ubiquinone Synthesis in vitro: Studies with Sonicates of Escherichia coli charged with 2-octaprenylphenol in the Cytoplasmic Membrane

The membrane-bound synthesis of ubiquinone comprises especially the introduction of hydroxy and methyl groups into 2-octaprenylphenol[1]. A more detailed study of this process requires a labeled precursor of ubiquinone which is correctly positioned in the membrane phase of the cells. A method was developed to accumulate [U-^{14}C]2-octaprenylphenol in situ in the cytoplasmic membrane. The conversion of 2-octaprenylphenol into ubiquinone is absolutely dependent on molecular oxygen[2].

Cells charged with 2-octaprenylphenol under argon should be very suitable for investigations in vitro of the subsequent hydroxylation and methylation reactions. An experimental advantage is that the conversion rate of 2-octaprenylphenol into ubiquinone is strongly dependent on the temperature (transition temperature, $T_t \sim$ 25 °C; an upper end, $T_h \sim 32$ °C; and a lower end, $T_l \sim$ 16 °C), and is absolutely zero at 0 °C. This makes it possible to break cooled suspensions of cells charged with 2-octaprenylphenol by sonication and to centrifuge the sonicate at 40 000 × *g* without loss of the precursor by conversion into ubiquinone.

If the supernatant (50 mg protein; 10 nmol [U-^{14}C]2-octaprenylphenol, 20 000 dpm) was shaken at 30 °C for 1 h under air with *S*-adenosyl-L-[*methyl*-^{3}H]methionine (100 nmol, 2 μCi), and heptane-soluble components were subsequently fractionated by high-pressure liquid chromatography, the double label was detected stoichiometrically in ubiquinone (2 - 4% conversion of the 2-octaprenylphenol pool). In a second experiment performed under argon, no labeled ubiquinone was detectable. This in vitro synthesis of ubiquinone by a crude cell fraction is partly inhibited by well known inhibitors of hydroxylases (metyrapone, 5mM; SKF-525A, 3mM).

Also, the capacity to synthesize ubiquinone is lost after filtration of the cell-fraction through Sephadex G-25 and is restored by NADPH.

These experiments suggest the involvement of an enzyme system working as an external monooxygenase.

Abbreviations:
Metyrapone,1,2-Di-(3-pyridyl)-2-methyl-1-propanone; SKF-525A, 2-Dimethylaminoethyl-2,2-diphenyl valerate · HCl.

1 Young, J.G., Stroobant, P., MacDonald, C.G. & Gibson, F. (1973) *J. Bacteriol.* **114**, 42 - 52.
2 Knoell, H.E. & Knappe, J. (1977) 11th FEBS Meeting, Copenhagen, August 1977, Abstract book, C-9, 811.

Dr. H.E. Knoell, Institut für Biologische Chemie der Universität Heidelberg, Im Neuenheimer Feld 501, D-6900 Hiedelberg.

J. Köhrle, D. Auf dem Brinke and R. D. Hesch

Microsomal Localization and Solubilization of L-Thyroxine-5'-deiodinase of Rat Liver

5'-Deiodination of L-thyroxine (T_4), forming one bioactive iodothyronine L-3,3',5-triiodothyronine (T_3), is an enzymatic process[1,2]. We investigated 5'-deiodination of T_4 in vitro using subcellular fractions of rat liver homogenate, prepared by differential and density gradient centrifugation[3,4], and characterized by determination of marker enzymes. T_4-5'-deiodinase was solubilized by incubation of microsomes with various salts and detergents. Solubilized protein was separated from microsomal fragments by centrifugation at 100 000 × *g* for 90 min.

T_4-5'-deiodinase activity was linear with incubation time up to 20 min and with protein concentration (50 - 500 μg/m*l*). The pH optima of T_4-5'-deiodinase were at pH 6.8 in all subcellular fractions and T_4-5'-deiodinase activity correlates only with microsomal marker enzymes, NADPH-cytochrome-*c* reductase (EC 1.6.2.4) and glucose-6-phosphatase (EC 3.1.3.9). All three enzymes were enriched 4 - 6-fold in the microsomal fractions, compared to homogenate.

Enzyme kinetic studies in microsomes gave sigmoidal v vs. [S] plots at pH < 7.0 [$[S]_{0.5V}$=2-5μM T_4 and V = 10 pmol T_3/(mg × min)], but Michaelis-Menten plots above this pH.

We succeeded in solubilizing T_4-5'-deiodinase from microsomal fractions by treating them with 1% cholate or

4M KCl for 1 h. Solubilization did not cause any loss in specific activity of T_4-5'-deiodinase, and 70% of the activity was recovered.

These results indicate that T_4-5'-deiodinase is solely localized in microsomes of rat liver. We succeeded in saturating the enzyme with respect to its substrate, L-T_4. T_4-5'-deiodinase can be solubilized by cholate or KCl. Non-Michaelis-Menten kinetics of T_4-5'-deiodinase at pH < 7.0 might be caused by other regulatory active iodothyronines derived from T_4.

1 Hesch, R.D., Brunner, G. & Söling, H.D. (1975) *Clin. Chim. Acta* **59**, 209 - 213.
2 Visser, Th.J., Van der Does-Tobé, I., Docter, R. & Hennemann, G. (1976) *Biochem. J.* **157**, 479 - 482.
3 de Duve, C. (1964) *J. Theor. Biol.* **6**, 33 - 59.
4 Trouet, A. (1974) *Methods Enzymol.* **31**, 323 - 329.

J. Köhrle, Abteilung für Klinische Endokrinologie, Medizinische Hochschule Hannover, Karl-Wiechert-Allee 9, D-3000 Hannover 61.

W. Koester-Eiserfunke, A.M. Gressner, D. Bruns and H. Greiling

A Method for the Isolation of Proteoglycans from Rat Liver

The biosynthesis of proteoglycans in normal liver and the molecular mechanisms of their accelerated production in injured hepatic tissue are poorly understood[1]. As is suggested by studies on cartilage[2], their overall synthesis might be regulated at the level of translation, depending on the amount of core protein synthesized [3,4]. Detailed studies on core protein formation in vivo and in cell-free systems necessitate profound knowledge of its structure which, however, is not yet available. This prompted us to develop an efficient and quite simple procedure for the isolation of proteoglycans, omitting extraction by guanidinium chloride.

Liver homogenate from rats previously injected with [^{35}S]sulphate was used to prepare postmitochondrial supernatant (20 min, 17 000 × *g*, 2 °C), which was subsequently centrifuged for 6 h at 213 000 × *g*, 2 °C. The sediment containing more than 90% of ^{35}S-labeled total liver proteoglycans was resuspended in buffer A (0.02M Tris, pH 7.6, 0.5M KCl, 3mM $MgCl_2$) with 0.5% (w/v) deoxycholate and centrifuged in a linear 10 - 30% (w/v) sucrose gradient in the same buffer to remove ribosomes. The upper third of the gradient, containing nearly 95% of [^{35}S]proteoglycans, was concentrated by diafiltration, applied to a column (2 × 120 cm) of Sepharose CL-4B and eluted with buffer A to eliminate contaminating RNA and protein. The proteoglycans recovered 110 m*l* behind V_0 were further purified 250-fold by isopycnic CsCl gradient centrifugation (60 h, 155 000 × *g*, 20 °C). The ^{35}S material banding at d = 1.69 g/m*l* had the same composition of labeled specific proteoglycans as in the liver homogenate (60% heparan sulphate, 12.5% dermatan sulphate, 2.5% chondroitin sulphate), indicating their nonselective recovery. Proteoglycans were characterized by amino acid analysis, analytical ultracentrifugation and enzymatic treatment.

1 Gressner, A.M., Pazen, H. & Greiling, H. (1977) *Hoppe-Seyler's Z. Physiol. Chem.* **358**, 825 - 833.
2 Telser, A., Robinson, H.C. & Dorfman, A. (1965) *Proc. Natl. Acad. Sci. U.S.A.* **54**, 912 - 919.
3 Gibson, K.D., Segen, B.J. & Audhya (1977) *Biochem. J.* **162**, 217 - 233.
4 Elders, M.J., McNatt, M.L., Kilgore, B.S. & Hughes, E.R. (1977) *Biochem. Biophys. Res. Commun.* **77**, 557 - 565.

Priv.-Doz. Dr. A.M. Gressner, Klinisch-Chemisches Zentrallaboratorium der Medizinischen Fakultät der RWTH, Goethestr. 27 - 29, D-5100 Aachen.

S. Komor und L.M.G. Heilmeyer, Jr.

Vergleichende Untersuchungen an den 42 000-Molekulargewichtskomponenten Isoaktin und Aktin aus Kaninchenskelettmuskel

Eine Proteinkomponente mit einem Molekulargewicht von 42 000, Isoaktin, kann im Gegensatz zu Aktin bei niedriger Ionenstärke (1mM $NaHCO_3$) aus Kaninchenskelettmuskel extrahiert und durch 11% Ammoniumsulfatfraktionierung in reiner Form gewonnen werden. Die Ausbeute beträgt ~ 0.7 g/100 g Muskel, die mit der des Aktins (~ 1 g/100 g Muskel) vergleichbar ist. Der Vergleich der molekularen Eigenschaften des Isoaktins und Aktins ergibt, daß Isoaktin und Aktin von verschiedenen Strukturgenen codiert werden.

Im Gegensatz zu Aktin enthält Isoaktin kein ADP gebunden und geht mit Myosin keine Wechselwirkung ein. Die Geschwindigkeit des intrazellulären Auf- und Abbaus des Isoaktins ist 1.5fach höher als die des Aktins. Isoaktin ist wie Aktin polydispers, bildet jedoch bei Erhöhung der Ionenstärke keine filamentösen Polymere. Die Bildung filamentöser Aktin-Polymere wird von Isoaktin zu ca. 75% gehemmt. Isoaktin zeigt mit Aktin Ähnlichkeiten hinsichtlich der Aminosäure-Zusammensetzung, einschließlich 1 mol 3-Methylhistidin/42 000 g Protein. Die CD-Spektren von Isoaktin, G- und F-Aktin sind identisch, nicht dagegen das des Apoaktins. Beide, Isoaktin und Apoaktin, binden maximal 6 mol $Ca^{2\oplus}$/ 42 000 g Protein. Im Scatchard-Diagramm zeigt sich sowohl in Abwesenheit als auch in Gegenwart von 0.1mM $Mg^{2\oplus}$ ein nach unten gekrümmter Kurvenverlauf. Die Bindung von $Ca^{2\oplus}$ an Isoaktin und Apoaktin ist negativ kooperativ. Die Hill-Koeffizienten ergeben Werte von 0.70 bis 0.84. $Ca^{2\oplus}$ wird von $Mg^{2\oplus}$ kompetetiv verdrängt. In Aktin können mit cAMP-abhängiger Protein-Kinase 0.003 bis 0.2 mol Phosphat/42000 g Protein und in Isoaktin maximal 0.02 mol Phosphat/42000 g Protein inkorporiert werden. Der Phosphorylierungsgrad ist nicht mit dem endogenen Phosphatgehalt korrelierbar.

Sibylle Komor, Institut für Physiologische Chemie der Ruhr-Universität, Lehrstuhl I, Postfach 2148, D-4630 Bochum.

P. Kraus

Partikelgebundene Glutathion-S-Transferasen

Die von zahlreichen Autoren durchgeführte Erforschung von Glutathion-S-Transferasen (EC 2.5.1.18) hat sich fast ausschließlich auf die lösliche, im Cytosol enthaltenen Enzyme dieser Art konzentriert[1-6]. Während unserer Untersuchung des Stoffwechsels von Hexachlorcyclohexan haben wir aber festgestellt[7,8], daß diese lipophile Substanz in Gegenwart von Glutathion und Mikrosomen bzw. Mitochondrien als Enzymquelle in wasserlösliche Produkte umgewandelt wird. Daraus wurde der Schluß gezogen, daß die Glutathion-S-Transferasen auch in den korpuskulären Zellfraktionen enthalten sind und die nachstehend beschriebenen Versuche haben diese Annahme bestätigt.

Die zu diesen Experimenten benutzten Zellpartikeln wurden so lange gewaschen, bis die in ihnen enthaltene Glutathion-S-Transferase ein konstantes Aktivitätsniveau erreicht hat. Auf diese Weise wurde sichergestellt, daß das benutzte Material frei von cytoplasmatischer Kontamination ist.

Die Bestimmung der Enzymaktivität erfolgte entweder aufgrund der gaschromatographisch gemessenen Abnahme des elektrophilen Substrates oder aufgrund der photometrisch ermittelten Abnahme von Glutathion[9].

Die Mitochondrien setzten 1,2-Dichlor-4-nitrobenzol, 1,2-Epoxy-3-phenoxypropan, Methyljodid und β-1,3,4,-5,6-Pentachlorcyclohex-1-en um; die Mikrosomen, zusätzlich zu diesen Substraten, auch noch 1-Menaphthylsulphat und Bromsulphthalein. Bei der Mitochondrien- oder Mikrosomen-katalysierten Umsetzung von β-1,3,4,-5,6-Pentachlorcyclohex-1-en entsteht 2,4-, 2,5-, 2,6- und 3,4-isomeres Dichlorphenylglutathion. Die prozentuale Zusammensetzung der Produktenmischung richtet sich nach der Art des benutzten Enzymmaterials.

Alle Versuche, die mikrosomale Glutathion-S-Transferase zu solubilisieren, waren erfolgslos. Dagegen konnte das mitochondriale Enzym durch die Fällung mit Amoniumsulfat und anschließende Extraktion mit Puffer (pH 8.5) löslich gemacht werden. Die solubilisierte mitochondriale Glutathion-S-Transferase setzt die gleichen Substrate um wie die nativen Mitochondrien auch, und hat ein Molekulargewicht von ca. 45 000.

1 Habig, W.H., Pabst, M.J. & Jakoby, W.B. (1974) *J. Biol. Chem.* **249**, 7130 - 7139.

2 Habig, W.H., Pabst, M.J. & Jakoby, W.B. (1976) *Arch. Biochem. Biophys.* **175**, 710 - 716.

3 Gilham, B. (1971) *Biochem. J.* **121**, 667 - 672.

4 Fjellsted, T.A., Allen, R.A., Duncan, B.K. & Jakoby, W.B. (1973) *J. Biol. Chem.* **248**, 3702 - 3707.

5 Askerlöf, P., Guthenberg, C., Jakobson, I. & Mannervik, B. (1975) *Biochem. J.* **147**, 513 - 522.

6 Kamisaka, K., Habig, W.H., Ketley, J.N., Arias, I.M. & Jakoby, W.B. (1975) *Eur. J. Biochem.* **60**, 153-161.

7 Kraus, P., Noack, G. & Portig, J. (1973) *Naunyn-Schmiedeberg's Arch. Pharmacol.* **279**, 199 - 202.

8 Kraus, P. (1975) *Naunyn-Schmiedeberg's Arch. Pharmacol.* **291**, 79 - 87.

9 Ellman, G.L. (1959) *Arch. Biochem. Biophys.* **82**, 70 - 77.

P. Kraus, Institut für Toxikologie und Pharmakologie der Universität Marburg, Pilgrimstein 2, D-3550 Marburg/L.

G.-B. Kresze und L. Steber

Wirkungsmechanismus der Pyruvat-Dehydrogenase-Inaktivase

Die Gesamtaktivität des aus tierischem Gewebe gewonnenen Multienzymkomplexes Pyruvat-Dehydrogenase (EC 1.2.4.1) wird irreversibel durch Pyruvat-Dehydrogenase-Inaktivase, ein lysosomales Enzym der Rattenleber, zerstört[1]; alle Teilaktivitäten des Komplexes (Pyruvat-Dehydrogenase-Komponente, Dihydrolipoyl-Transacetylase, Dihydrolipoyl-Dehydrogenase) bleiben jedoch unverändert meßbar.

Mit verschiedenen Methoden (Gelchromatographie, Gradientenzentrifugation, Elektronenmikroskopie) konnten wir zeigen, daß der Verlust der Pyruvat-Dehydrogenase-Gesamtaktivität auf der Dissoziation des Multienzymkomplexes in die einzelnen Teilenzyme beruht, die – wiewohl jedes für sich aktiv – nicht mehr in der Lage sind, den koordinierten Ablauf der Gesamtreaktion zu vermitteln.

Dieser Zerfall des Pyruvat-Dehydrogenase-Komplexes wird hervorgerufen durch proteolytische Spaltung mindestens einer Peptidbindung in der Dihydrolipoyl-Transacetylase, die das „core" des Enzymkomplexes darstellt. Trotz dieses partiellen Abbaus kann die Transacetylase als Aggregat mit hohem Molekulargewicht und voller enzymatischer Aktivität isoliert werden. Die peripheren Teilenzyme Dihydrolipoyl-Dehydrogenase und Pyruvat-Dehydrogenase werden hingegen durch die Inaktivase nicht angegriffen.

1 Lynen, A., Sedlaczek, E. & Wieland, O.H. (1978) *Biochem. J.* **169**, im Druck.

Dr. Georg-Burkhard Kresze, Institut für Biochemie der Universität München, Karlstr. 23, D-8000 München 2.

B. de Kruijff and E.J.J. van Zoelen

Protein-Mediated Transbilayer Movements of Phospholipids

Transbilayer movements of lipids in biomembranes are in general faster than in lipid vesicles[1]. Half times of transbilayer movements of lipids in vesicles under equilibrium conditions are on the order of days or more. Recently, however, using NMR techniques several conditions have been found in which the "flip-flop" rate is significantly increased. Asymmetrical incorporation of dioleoyllecithin in dimyristoyllecithin bilayers resulted in a relatively rapid transbilayer movement of dioleoyllecithin[2]. Similarly, phosphatidic acid formed in the outer monolayer of lecithin vesicles by phospholipase D (EC 3.1.4.4) is translocated to the inner monolayer with a half time of 40 min or less[3]. Both observations indicate that a non-equilibrium transbilayer distribution of the lipid molecules can lead to transbilayer movements of lipids without loss of barrier properties of the bilayer. Because biological membranes are highly dynamic structures in which lipid molecules are continuously renewed by metabolic processes or exchange, transient non-equilibrium states must exist at the sites of synthesis, degradation or exchange. At these sites transmembrane movements of lipids could possibly occur. Another major difference between lipid vesicles and biological membranes is the absence of membrane-proteins in the former. Therefore the influence of the incorporation of glycophorin, a membrane spanning protein from the human erythrocyte membrane, on the transbilayer movements of lysolecithin in vesicles was investigated[4]. It was found that glycophorin increased the rate of transbilayer movement of lysolecithin by two orders of magnitude.

1 Rothman, J.E. & Lenard, J. (1977) *Science* **195**, 743 - 753.

2 de Kruijff, B. & Wirtz, K.W.A. (1977) *Biochim. Biophys. Acta* **468**, 318 - 326.

3 de Kruijff, B. & Baken, P. (1978) *Biochim. Biophys. Acta*, in press.

4 Van Zoelen, E.J.J., de Kruijff, B. & van Deenen, L.L.M. (1978) *Biochim. Biophys. Acta*, in press.

B. de Kruijff and E.J.J. van Zoelen, Department of Molecular Biology and Department of Biochemistry, University of Utrecht, Transitorium III, Padualaan 8, De Uithof, Utrecht, The Netherlands.

J. Krumsiek, H. Reiber und V. Neuhoff

Die Aktivität lipidabhängiger Enzyme als Funktion der Testbedingungen und deren biologische Relevanz am Beispiel der Galaktosyltransferase

Der Aktivitätstest membrangebundener, wasserunlöslicher Enzyme bereitet häufig Schwierigkeiten, besonders dann, wenn auf Detergenzien verzichtet werden muß.

Die Begriffe „Aktivierung" und „Hemmung" wie sie für wasserlösliche Enzyme verständlich sind, sind für wasserunlösliche Enzyme oftmals mehrdeutig. Dies ist vor allem dann der Fall, wenn die Zu- oder Abnahme des Umsatzes durch Lipide bewirkt wird und die Substrate selbst wasserunlöslich sind.

Bei der UDP-Galaktose:Ceramid Galaktosyltransferase (EC 2.4.1.45) wird ein maximaler Umsatz des Substrats in Anwesenheit von ungesättigtem Phosphatidyläthanolamin (Eikephalin) gefunden. Dabei wird gefunden, daß Enzym, Ceramid als Substrat und Phosphatidyläthanolamin in derselben Phase (Lipidphase) sein müssen, um einen maximalen Umsatz des Ceramids zu erreichen. Diese maximale Aktivität ist sowohl durch gemeinsame Adsorption der drei Komponenten an die Glaswand des Reaktionsgefäßes[1] als auch durch homogene Suspension der drei Komponenten im Reaktionspuffer mit Ultraschall erreichbar.

Durch die Variation der Testbedingungen wird deutlich, daß der Einfluß des Phosphatidyläthanolamins auf den Umsatz durch die Bildung einer einheitlichen Phase für Enzym und Substrat erklärbar ist.

Der Einfluß des Lipids ist in den alternativen Testsystemen quantitativ unterschiedlich. Dies ist für die Beurteilung der biologischen Relevanz der im Test gefundenen Lipidabhängigkeiten[2] von Bedeutung. Die gefundene starke Hemmung des Substratumsatzes durch Cholesterin[2] ist eine Konsequenz der Testbedingungen[1]. Erst der Vergleich der genannten alternativen Testbedingungen gibt Aufschluß über die mögliche Relevanz dieses Befundes im biologischen System.

Die Gegenwart von Lipid-Doppelschichten (Bilayer-Vesikel oder multilamelläre Strukturen) im Testsystem wasserunlöslicher Enzyme kann zur Kompartimentierung des Systems führen und dadurch bei voller Aktivität des Enzyms einen Umsatz des Substrats verhindern.

1 Brenkert, A. & Radin, N. S. (1972) *Brain Res.* **36**, 183 - 193.
2 Krumsiek, J., Reiber, H. & Neuhoff, V. (1977) *Hoppe-Seyler's Z. Physiol. Chemie* **358**, 1236.

J. Krumsiek, H. Reiber und V. Neuhoff, Forschungsstelle Neurochemie, Max-Planck-Institut für Experimentelle Medizin, D-3400 Göttingen.

J. Kubicki, G. Buse und H. D. Ohlenbusch

Assoziationszustand von Cytochrom-c-Oxidase in Monoschichten: Vergleich mit analogen Befunden im kristallinen Zustand bei Vesikeln

πc-Isothermen von Cytochrom-*c*-Oxidase (EC 1.9.3.1) wurden nach dem Verfahren der Oberflächenmanometrie[1] bei verschiedenen Sublösungsbedingungen gemessen. Aus der Konzentrationsabhängigkeit des reduzierten Oberflächendrucks wurde das mittlere Molekulargewicht im Bereich kleiner Oberflächendrucke ($\pi < 0.1$ dyn/cm) nach der Volmerschen Gleichung bestimmt.

Zwischen pH 3 und pH 8 resultiert ein Molekulargewicht von 290 000; auf der Grundlage des Lipid- und Detergenzgehaltes des Präparates[2] entspricht dies einem Molekulargewicht des Proteinanteils von 229 000. Dieser Wert stimmt mit Ergebnissen überein, die für Cytochrom-*c*-Oxidase einerseits im gelösten und andererseits im vesikulären Zustand gefunden wurden[3,4]; er entspricht nach heutiger Kenntnis dem Dimeren der aktiven Enzymeinheit.

Oberhalb von pH 8, bei hohen Natriumchloridkonzentrationen und bei Zusatz von Natriumdithionit dissoziiert das Protein zunehmend in seine Untereinheiten. Dieses Verhalten ist korrelierbar mit entsprechenden elektronenmikroskopischen Beobachtungen an den zweidimensionalen Gitterstrukturen der Cytochrom-*c*-Oxidase in Vesikeln[5], die sich bei gleichen Bedingungen aufzulösen beginnen. Auf Grund dieser Ergebnisse wird vermutet, daß die Kräfte, die den Enzymkomplex stabilisieren, auch für die Ausbildung dieser zweidimensionalen Gitterstruktur verantwortlich sind.

In Gegenwart von zweiwertigen Kationen in der Sublösung beginnt das Enzym über seinen dimeren Zustand hinaus zu assoziieren. Für dieses Verhalten sind entsprechende Befunde für den Zustand in Lösung nicht bekannt. Hierbei könnte es sich um eine Ausbildung von Salzbrücken handeln, die durch die Orientierung der Teilchen in der Grenzphase begünstigt und durch zweiwertige Kationen vermittelt würden. Diese Assoziation scheint außerdem durch den Phospholipidgehalt des Präparates gefördert zu werden.

1 Kubicki, J. & Ohlenbusch, H. D. (1977) *Hoppe-Seyler's Z. Physiol. Chem.* **358**, 267 - 268.
2 Steffens, G. & Buse, G. (1976) *Hoppe-Seyler's Z. Physiol. Chem.* **357**, 1125 - 1137.
3 Seki, S. & Oda, T. (1970) *Arch. Biochem. Biophys.* **138**, 122 - 134.
4 Rubin, M. S. & Tzagoloff, A. (1973) *J. Biol. Chem.* **248**, 4269.
5 Vanderkooi, G. (1974) *Biochim. Biophys. Acta* **344**, 307 - 345.

Dr. Jörn Kubicki, Abteilung Physiologische Chemie der RWTH-Aachen, Melatener Str. 211, D-5100 Aachen.

D. Kübler, W. Pyerin und V. Kinzel

Protein-Kinase und Substrate an der Oberfläche von Gewebekulturzellen

Die Bedeutung der Phosphorylierung von Proteinen bei der Regulation sehr verschiedener Vorgänge wird durch die nahezu universelle Rolle zyklischer Nucleotide unterstrichen, die ihre Wirksamkeit vor allem über diese Form der postsynthetischen Proteinmodifikation, nämlich durch Aktivierung von Protein-Kinasen – in fast allen Zellkompartimenten nachgewiesen – ausübt. Neuland ist diesbezüglich die Zelloberfläche.

Nachdem die Existenz phosphorylierbarer Oberflächenproteine bei HeLa-Zellen wahrscheinlich gemacht werden konnte[1], berichten wir nun über Protein-Kinase-Aktivität auf der Plasmamembran von Gewebekulturzellen.

Die Inkubation von HeLa-Monolayer-Zellen unter physiologischen Bedingungen mit ATP (γ-^{32}P-markiert), von dem bekannt ist, daß es intakte Membranen nicht penetriert, führt zur Inkorporation radioaktiven Phosphats in säureunlösliches Material. Die Reaktion ist abhängig von Inkubationszeit, Temperatur und Zahl eingesetzter Zellen; sie wird weder durch cAMP noch einen hitzestabilen Protein-Kinase-Inhibitor beeinflußt. Elektrophoretische Auftrennung von Zellproteinen in Gegenwart von Natriumdodecylsulfat ergibt reproduzierbare, für jede Zelllinie charakteristische Muster ^{32}P-markierter Proteine. Nach partieller Säurehydrolyse wird hauptsächlich [^{32}P]Phosphoserin, in geringerer Menge [^{32}P]Phosphothreonin gefunden. Behandlung phosphorylierter Zellen mit Trypsin führt zum Verlust ihrer Markierung. Wie für alle Verfahren zur selektiven Modifizierung oder Markierung von Zelloberflächenbestandteilen ist eine intakte Membranschranke Voraussetzung. In Autoradiographien von Zellkulturen werden abgestorbene Zellen besonders stark phosphoryliert gefunden. Bei Verwendung von Kulturen mit extra hohen Trypanblau-Indices wird die Phosphorylierungsreaktion durch cAMP und den Inhibitor beeinflußbar – ein Hinweis für die Beteiligung zellinterner Protein-Kinasen und Substrate. Die Verhältnisse werden besonders klar, wenn nach Phosphorylierung intakte und abgestorbene Zellen getrennt (z.B. nach mikrofluorometrischer Zellsortierung) untersucht werden. Die auf die Zellzahl bezogene spezifische Radioaktivität ist bei abgestorbenen Zellen über 10fach höher als bei intakten; die elektrophoretischen Muster phosphorylierter Proteine unterscheiden sich signifikant.

Die zellphysiologische Bedeutung der Oberflächen-Proteinkinase ist noch unbekannt. Sie wird in Phänomenen zu suchen sein, die durch exogenes ATP bei intakten Zellen ausgelöst werden.

1 Kinzel, V. & Mueller, G.C. (1973) *Biochim. Biophys. Acta* **322**, 337 - 351.

Dieter Kübler, Walter Pyerin und Volker Kinzel, Institut für Experimentelle Pathologie, Deutsches Krebsforschungszentrum, Im Neuenheimer Feld 280, D-6900 Heidelberg.

H. Kühn

Light-dependent, Reversible Binding of Rhodopsin Kinase and (An)other Protein(s) to the Photoreceptor Membrane

Rhodopsin, the major protein in photoreceptor membranes, is phosphorylated by ATP (GTP) and an endogenous protein kinase subsequent to light absorption. The enzyme can be readily extracted from dark-adapted, but not from illuminated rod outer segments; kinase activity in dark extracts was up to 17 times higher than in corresponding light extracts. Binding of the enzyme to the bleached membranes is reversible in the dark and decays with a half time of about 25 min at 20 °C. This was used to purify the enzyme partially by specific binding to freshly bleached membranes. When thoroughly washed rod outer segment disc membranes were bleached in the presence of added crude extract, most of the protein of the extract could be washed out by centrifugation, whereas a few proteins were transiently bound to the membranes and could later be desorbed in the dark. One of these selectively bound proteins is the rhodopsin kinase. Its molecular weight, as determined by molecular sieve chromatography of the active enzyme and by SDS gel electrophoresis, is about 67000. The other most prominent transiently bound protein, the function of which is not yet known, has a molecular weight of about 45000 - 50000. The decay of binding capacity in the dark after bleaching corresponds in its time course roughly to the decay of phosphorylation capacity of rhodopsin after bleaching[1]. Neither $Mg^{2\oplus}$ nor ATP is needed for the light-dependent binding; EDTA, which abolishes the phosphorylation reaction, has no influence on the binding of the enzyme to its substrate membrane.

1 McDowell, J. H. & Kühn, H. (1977) *Biochemistry* **16**, 4054 - 4060.

Dr. H. Kühn, Institut für Neurobiologie, Kernforschungsanlage Jülich GmbH, Postfach 1913, D-5170 Jülich 1.

P. Laggner

Die molekulare Architektur der Plasma-Lipoproteine

Die Plasma-Lipoproteine umspannen ein breites Spektrum makromolekularer Teilchen mit variabler chemischer Zusammensetzung (Triglyceride, Cholesterin-Fettsäureester, Cholesterin, Phospholipide und spezifische Apolipoproteine) im Größenbereich von ca. 10 nm (High-Density-Lipoproteine, HDL, d 1.063 - 1.21 g/ml) bis etwa 1 μm (Chylomikronen, $d < 0.95$ g/ml). Zur Aufklärung der molekularen Bauprinzipien wurden bisher, vor allem wegen der relativ leichten Isolierbarkeit monodisperser Fraktionen, die Dichteklassen der Low-Density-Lipoproteine LDL (d 1.006 - 1.063 g/ml) und der High-Density-Lipoproteine mit Hilfe der Röntgenkleinwinkelstreuung untersucht. Diese Methode ermöglicht die Bestimmung von Größe, Gestalt und innerer Struktur von Makromolekülen in verdünnter Lösung (Übersichtsartikel zur Methodik s. Ref.[1,2]).

Alle bisher untersuchten HDL-Fraktionen (LpA von HDL_3 und HDL_2, und LpC von HDL_2) sind annähernd kugelförmig mit charakteristischen Ähnlichkeiten in ihrer radialen Elektronendichteverteilung: ein Kern niedriger Elektronendichte, entsprechend den Kohlenwasserstoffanteilen der Lipide, von Radien zwischen 3.7 nm und 5.5 nm ist umgeben von einer äußeren Schale hoher Elektronendichte, entsprechend den polaren Phospholipid-Kopfgruppen und dem Proteinanteil, mit einer konstanten Dicke von ca. 1.5 nm. Daraus läßt sich als gemeinsames Strukturmodell für die HDL eine radial-micellare Anordnung der Cholesterinester im Kern der Teilchen ableiten, mit welchem die amphiphilen Bestandteile (Phospholipide, Cholesterin) von der Schale her ineinandergreifen[3]. Modellrechnungen sowie Rekonstitutionsversuche haben gezeigt, daß der überwiegende Teil des Proteins in der äußeren, 1.5 nm dicken Schale lokalisiert ist.

Auch die LDL-Struktur ist annähernd kugelsymmetrisch, sie zeigt jedoch eine völlig andere radiale Elektronendichteverteilung[4]. Hier ist ein Kern von ca. 9.5 nm Radius mit durchschnittlich niedriger Elektronendichte von einer etwa 2 nm dicken Schale hoher Elektronendichte umgeben. Die Struktur des Kerns zeigt eine reversible, thermische Umwandlung zwischen 15 und 30 °C, wobei sich bei niedrigen Temperaturen eine Trennung hoher und niedriger Elektronendichte zeigt, die beim Erwärmen über 30 °C verschwindet. Die Analogien zum flüssig-kristallinen Verhalten reiner Cholesterinester und die Tatsache, daß das Volumen des temperaturempfindlichen Kerns mit dem theoretischen Volumen der Cholesterinester pro Teilchen gut übereinstimmt, läßt darauf schließen, daß im LDL die Cholesterinester in Kern angeordnet sind, und die polaren Lipide zusammen mit dem Protein die Schale in Form einer monomolekularen Oberflächenschicht bilden. Dieses Modell wurde durch Untersuchungen an teilweise trypsinierten LDL-Präparaten erhärtet.

Ein völlig anderes Bauprinzip wurde für das pathologische Lipoprotein-X (Lp-X) gefunden[5], dessen chemische Zusammensetzung sich vor allem durch das Fehlen von Cholesterinestern von den normalen Lipoproteinen abhebt. Hier stehen die Röntgenkleinwinkelergebnisse mit einer einfachen Lipid-Doppelschichtstruktur (Vesikel) in guter Übereinstimmung.

1 Kratky, O. (1963) *Prog. Biophys.* **13**, 107 - 173.
2 Kratky, O. & Pilz, I. (1972) *Quart. Rev. Biophys.* **5**, 481 - 537.
3 Laggner, P. (1978) in Protides of the Biological Fluids, Proc. 25th Colloquium Brugge (Peeters, H., Hrsg.) S. 59 - 66, Pergamon Press, Oxford.
4 Laggner, P., Degovics, G., Müller, K.W., Glatter, O., Kratky, O., Kostner, G. & Holasek, A. (1977) *Hoppe-Seyler's Z. Physiol. Chem.* **358**, 771 - 778.
5 Laggner, P., Glatter, O., Müller, K., Kratky, O., Kostner, G. & Holasek, A. (1977) *Eur. J. Biochem.* **77**, 165 - 171.

Dr. Peter Laggner, Institut für Röntgenfeinstrukturforschung der Österreichischen Akademie der Wissenschaften und des Forschungszentrums Graz, Steyrergasse 17, A-8010 Graz.

R. J. Laib, H. Ottenwälder and H. M. Bolt

Formation of Etheno Derivatives of Nucleic Acid Bases in vivo by Metabolites of Vinyl Chloride

The detection of mutagenic and carcinogenic effects of vinyl chloride greatly stimulated research on the molecular mechanism of vinyl chloride disease.

Most probably, vinyl chloride is initially metabolized to its epoxide derivative (chloroethylene oxide), which may alkylate cellular components. Furthermore, chloroethylene oxide undergoes rearrangement to chloroacetaldehyde.

Chloroacetaldehyde and chloroethylene oxide may interact with adenosine and lead to formation of 1,N^6-ethenoadenosine, as described by Barbin et al.[1].

Rat liver microsomes were incubated with NADPH, [1,2-^{14}C]vinyl chloride and polyadenylic acid or polycytidylic acid. The latter were re-isolated from the incubation mixtures and hydrolyzed. The radioactivity originating from [^{14}C]vinyl chloride, which was irreversibly bound to the polyadenylic or polycytidylic acid, was confined to 1,N^6-ethenoadenosine or 3,N^4-ethenocytidine.

When rats were exposed to [1,2-^{14}C]vinyl chloride, part of the radioactivity was incorporated into the liver RNA. Analysis of hydrolysates of liver RNA showed that all natural nucleosides of RNA were labelled. Besides, small amounts of radioactivity could be detected which were confined to 1,N^6-ethenoadenosine and 3,N^4-ethenocytidine[2]. DNA from livers of rats exposed to [1,2-^{14}C]-vinyl chloride also contained significant amounts of radioactivity.

When DNA was incubated with rat liver microsomes, NADPH and [^{14}C]vinyl chloride, radioactivity was also incorporated into the DNA. Re-isolation of the DNA, hydrolysis and separation of the nucleosides showed that radioactive 1,N^6-etheno-2'-deoxyadenosine and 3,N^4-etheno-2'-deoxycytidine were formed.

The experiments support the theory that vinyl chloride metabolites react with adenine or cytosine moieties of nucleic acids to form etheno analogues.

1 Barbin, A., Brésil, A., Croisy, P., Jacquignon, P., Malaville, C., Montesano, R. & Bartsch, H. (1975) *Biochem. Biophys. Res. Commun.* **67**, 596 - 602.
2 Laib, R. J. & Bolt, H. M. (1977) *Toxicology* **8**, 185 - 195.

R. J. Laib, H. Ottenwälder and H. M. Bolt, Institut für Toxikologie der Universität Tübingen, Wilhelmstr. 56, D-7400 Tübingen.

O. Leiß, H. Egge und U. Murawski

Lecithin-Cholesterin-Acyltransferase-Aktivität in Abhängigkeit von verschiedenen Lipoproteinparametern

Die Lecithin-Cholesterin-Acyltransferase (EC 2.3.1.43), die im Plasma die Übertragung einer Fettsäure von Lecithin auf Cholesterin katalysiert, hat eine bedeutende Funktion im Cholesterinstoffwechsel[1]. Mit den bisherigen Methoden zur Aktivitätsbestimmung („common substrate methods", „self substrate methods", gaschromatographischen und kolorimetrischen Methoden) kann jedoch schlecht zwischen Einflüssen von Enzym und Lipoproteinsubstrat differenziert werden. Es wurde deshalb ein Verfahren entwickelt, das es erlaubt, den Einfluß der einzelnen Lipoproteinfraktionen sowie deren Bestandteile auf die Acyltransferase-Aktivität zu untersuchen. Dieses Verfahren beruht auf einer schrittweisen Fällung der very low density-(VLDL), low density-(LDL) und high density-Lipoproteine (HDL) aus 50 μl Serum jeweils vor und nach 5stdg. Inkubation bei 37 °C, sowie anschließender quantitativer Lipidfraktionierung in Cholesterinester (CHE), Triglyceride (TG), freies Cholesterin (CH) und Phospholipide (PL). Die erhaltenen mg/*l*-Werte für die einzelnen Lipidklassen in den VLDL, LDL, HDL,und VHDL (= Überstand nach HDL-Fällung) werden zur Summe (= Gesamtserum) aufaddiert. Die Veresterungsrate wird nach Korrektur für die Wiederfindung durch Differenzbildung aus den Summenwerten des freien Cholesterins vor und nach Inkubation ermittelt. Bei 19 männlichen Probanden wurden im Durchschnitt 20.6 ± 4.5 mg Cholesterin/(l × h) (Bereich 14.6 - 29.1) verestert. Signifikante Korrelationen der Acyltransferase-Aktivität betrugen gegen mg Gesamtlipid VLDL $r = -0.530$, ($p < 0.05$), gegen % Gesamtlipid in den VLDL $r = -0.517$ ($p < 0.05$), gegen % CHE in den VLDL $r = -0.572$ ($p < 0.05$), gegen den Quotienten TG/CH + CHE in den VLDL $r = 0.576$ ($p < 0.01$), gegen den Quotienten HDL/VLDL $r = 0.554$ ($p < 0.05$), gegen den Quotienten VHDL/VLDL $r = 0.544$ ($p < 0.05$), während keine bzw. nur schwache Korrelationen gegen mg Gesamtlipid HDL $r = 0.063$, gegen mg CH in den HDL $r = 0.068$ und gegen den Quotienten PL/CH in den HDL $r = 0.153$ bestanden.

1 Glomset, J. A. (1968) *J. Lipid Res.* 9, 155 - 167.

Prof. Dr. H. Egge, Physiologisch-Chemisches Institut, Nußallee 11, D-5300 Bonn.

O. Leiß, U. Murawski und H. Egge

Mikromethode zur quantitativen Lipoproteinanalyse

Durch Kombination einer bewährten Mikromethode für Lipidanalysen[1] mit bekannten Präzipitationsverfahren zur Lipoproteinisolierung[2] und ihre Adaptation an den Mikromaßstab wurde eine Methode entwickelt, die es erlaubt, aus 50 μl Serum (oder Plasma) einen kompletten Lipidstatus der einzelnen Lipoproteinklassen zu erstellen. Hierbei werden schrittweise in der angegebenen Reihenfolge die very low density-(VLDL), low density-(LDL) und high density-Lipoproteine (HDL) mittels Polyanionen und zweiwertigen Kationen (Heparin/$Mg^{2\oplus}$; Dextransulfat 500/$Mg^{2\oplus}$; Dextransulfat 500/$Mn^{2\oplus}$) gefällt und durch Zentrifugation abgetrennt. Aus den gefällten Lipoproteinen sowie aus dem Überstand nach HDL-Fällung (very high density-Lipoproteine = VHDL) werden die Lipide extrahiert und der jeweilige Lipidgehalt gravimetrisch ermittelt. Die quantitative Bestimmung des veresterten (CHE) und des freien Cholesterins (CH), der Triglyceride (TG) und der Phospholipide (PL) erfolgt densitometrisch nach dünnschichtchromatographischer Trennung und Veraschung. Die Werte wurden verglichen mit den entsprechenden Werten aus den mittels Ultrazentrifugation erhaltenen Lipoproteinfraktionen. Hierbei ergaben die Fraktionen der Dichte $d < 1.063$ m*l*, $d > 1.063$ m*l* und $d = 1.063$ - 1.21 m*l* gute Übereinstimmung in der relativen Lipidverteilung, während die Fraktionen der VLDL ($d < 1.006$ m*l*) und der VHDL ($d > 1.21$ m*l*) weniger gut mit den entsprechenden Fraktionen nach Fällung übereinstimmten ($n = 12$). Verglichen mit den Werten aus dem Gesamtserum betragen die Wiederfindungsraten (in %) der einzelnen Lipidklassen aus allen Lipoproteinfraktionen für CHE = 102.63 ± 9.13, für CH = 100.41 ± 10.83, für TG = 101.56 ± 9.13, für PL = 91.91 ± 9.14 und für das Gesamtlipid = 99.67 ± 4.43. Die Variationskoeffizienten für die einzelnen Lipidklassen liegen in den VLDL bei 10%, in den LDL und HDL zwischen 5 und 8% und in den VHDL bei 10% ($n = 10$).

1 Egge, H., Murawski, U., Müller, J. & Zilliken, F. (1970) *Z. Klin. Chem. Klin. Biochem.* 8, 488 - 491.
2 Burstein, M. & Scholnick, H. (1973) *Adv. Lipid Res.* **11**, 67 - 108.

U. Murawski, Institut für Physiologische Chemie, Nußallee 11, D-5300 Bonn.

C. Lemmen und Ch. Barth

Regulation von lipogenetischen Enzymen in menschlichen diploiden Fibroblasten

In tierischen Geweben weisen die Aktivitäten lipogenetischer Enzyme Schwankungen auf, die durch wechselnde Ernährungsbedingungen sowie Hormone ausgelöst werden[1].

In dieser Mitteilung berichten wir über den Einfluß, den Hormone und Substrate auf die Aktivitäten der Acetyl-CoA-Carboxylase (EC 6.4.1.2) und Glucose-6-phosphat-Dehydrogenase (EC 1.1.1.49) von menschlichen Fibroblasten haben.

Die spezifische Aktivität der Acetyl-CoA-Carboxylase wurde mittels Aufschluß durch Frieren und Tauen nach Jacobs und Majerus[2], die Aktivität der Glucose-6-phosphat-Dehydrogenase im optischen Test nach Warburg bestimmt. Die Zellen befanden sich zwischen der 16. - 20. Passage und im konfluenten Wachstumszustand. Zunächst wurde die Aktivität der Acetyl-CoA-Carboxylase in Gegenwart von Eagle's Minimalmedium + 10% fötalem Kälberserum bestimmt (Basisaktivität); sie betrug 0.23 ± 0.02 mU/mg Protein ($\bar{x}$ ± S.E.M.; n = 8) (37 °C). Wurde das Serum durch fettsäurefreies Albumin plus 10 Millieinheiten Insulin/m*l* ersetzt, so fand sich ein Anstieg der Acetyl-CoA-Carboxylase-Aktivität nach 72 h auf 150% der Kontrolle. Kürzere Inkubationszeiten hatten keine Aktivitätsänderungen zur Folge. Ein paralles Verhalten wurde für die Dehydrogenase beobachtet. 0.5mM Dibutyryl-cAMP hatte einen 30proz. Abfall der Acetyl-CoA-Carboxylase-Aktivität zur Folge. Die Aktivität der Glucose-6-phosphat-Dehydrogenase fiel dabei um 43% ab.

Kontrollmessungen ergaben, daß die Aktivität des konstitutiven Enzyms Lactat-Dehydrogenase unter den obigen Bedingungen keine Änderungen aufwies.

1 Lane, M. D., Moss, Y. & Polakis, S. E. (1974) Curr. Top. Cell. Reg. (Horecker, B. L. & Stadtman, E. R., eds.) vol. 8, pp. 139 - 195, Acad. Press, New York.

2 Jacobs, R. A., Sly, W. S. & Majerus, Ph. W. (1973) *J. Biol. Chem.* **248**, 1268 - 1276.

Dr. Christa Lemmen, Forschergruppe Ernährung an der Medizinischen Poliklinik, Pettenkoferstraße 8a, D-8000 München 2.

W. Lenk

Hill Plots and Hill Profiles of the Kinetic Data from Microsomal ω- and (ω – 1)-Hydroxylations

The kinetics of ω- and (ω – 1)-hydroxylation of the propionic residue, integrated into *N*-(4-chlorophenyl)-propionamide (I), has been studied with the microsomal fraction of liver homogenate from untreated, phenobarbital-, and 3-methylcholanthrene-treated rabbits as the enzyme source.

Enzyme saturation curves, obtained by plotting the initial velocities v versus substrate concentrations S, had a rectangular hyperbolic form for ω-hydroxylation, and an intermediary plateau region[1] for (ω – 1)-hydroxylation, indicative of apparent negative cooperativity[2]. Hill plots of the kinetic data gave straight lines with the slope of unity for ω-hydroxylation and right-hand deviations (kinks) for (ω – 1)-hydroxylation. Hill plots, modified by Watari and Isogai[3], permitted the determination of the Hill coefficient n_H, which was always 1 for ω – and between 0.48 and 0.69 for (ω – 1)-hydroxylation. Since n_H is substrate-dependent, profiles of the Hill slope were recommended by Endrenyi et al.[4] as a characterisation of non-hyperbolic enzymes. Such profiles, constructed according to Kurganov[5], revealed sudden and slow changes of n_H as a function of the relative (ω – 1)-hydroxylation velocity, depending on the degree of fractional saturation of the mixed function oxygenase (EC 1.14.14.1). Thus, the specific form of cytochrome *P*-450 which is inducible by phenobarbital and catalyses the ω-hydroxylation of I, displays normal kinetic behaviour, whereas the 3-methylcholanthrene-inducible form, which catalyses the (ω – 1)-hydroxylation of I, displays apparent negative homotropic cooperativity.

1 Teipel, J. & Koshland, D.E., Jr. (1969) *Biochemistry* 8, 4656 - 4663.

2 Levitzki, A. & Koshland, D.E., Jr. (1969) *Proc. Natl. Acad. Sci. U.S.A.* **64**, 1121 - 1128.

3 Watari, H. & Isogai, Y. (1976) *Biochem. Biophys. Res. Commun.* **69**, 15 - 18.

4 Endrenyi, L., Fajszi, C. & Kwong, F.H.F. (1975) *Eur. J. Biochem.* **51**, 317 - 328.

5 Kurganov, B.I. (1973) *Acta Biol. Med. Germ.* **31**, 181 - 201.

Priv.-Doz. Dr. Werner Lenk, Pharmakologisches Institut der Universität München, Nußbaumstr. 26, D-8000 München 2.

R. Lichtner and H. U. Wolf

Characterization of the Phosphorylated High-affinity $Ca^{2⊕}$-ATPase from Human Erythrocyte Membranes

High-affinity $Ca^{2⊕}$-ATPase (Adenosinetriphosphatase, EC 3.6.1.3) has been solubilized and partially purified as described before[1], and phosphorylated by 2μM [γ-^{32}P]-ATP (1.25 Ci/mmol) in the presence of 50mM $K^⊕$ or $Na^⊕$ and varying concentrations of $Mg^{2⊕}$ and $Ca^{2⊕}$. The rate of phosphorylation was dependent on [$Ca^{2⊕}$], with a maximum at 10mM, and independent of [$Mg^{2⊕}$]. There was no significant difference between the effects of $Na^⊕$ and $K^⊕$.

Phosphorylation in the presence of 1mM $Ca^{2⊕}$ and 0.1mM $Mg^{2⊕}$ and subsequent addition of 10mM EDTA resulted in a release of 30% of the bound $^{32}P_i$ after 30 s, in agreement with the data of Knauf et al.[2]. Phosphorylation in the presence of 10mM $Ca^{2⊕}$ and 0.01mM $Mg^{2⊕}$ and subsequent addition of 10 - 25mM EDTA resulted in a release of 80% of the bound $^{32}P_i$ after 30 s. Addition of 2.5mM $Mg^{2⊕}$ accelerated the dephosphorylation.

Hydroxylamine (0.2M, 10 min, pH 5.5) split the phosphorylated enzyme nonenzymatically. The pH-dependent hydrolysis at 40 °C and 2 min showed characteristics of an acylphosphate bond.

Phosphorylation was not increased in the presence of cAMP, indicating the absence of cAMP-stimulated kinases.

The amount of phosphorylated protein in the 145000 band obtained after sodium dodecylsulfate-gel electrophoresis at pH 2.4 and 2 °C was dependent on the addition of MgATP or treatment of the phosphorylated enzyme with hydroxylamine as described above.

1 Wolf, H. U., Dieckvoß, G. & Lichtner, R. (1977) *Acta Biol. Med. Ger.* **36**, 847 - 858.
2 Knauf, Ph. A., Proverbio, F. & Hoffmann, J. F. (1974) *J. Gen. Physiol.* **63**, 324 - 336.

Prof. Dr. H. U. Wolf, Abteilung Pharmakologie und Toxikologie der Universität Ulm, Oberer Eselsberg, D-7900 Ulm/Donau.

T. Lindl

Calcium Ions and their Relations to Extracellularly Recorded Action Potentials and to Drug-Induced Cyclic AMP Accumulations in Rat Superior Cervical Ganglion in vitro

Calcium ions play a prominent role in the excitability of nervous cells[1]. It was hypothesized that increases in intracellular calcium content lead to an increase of cyclic AMP (cAMP)[2]. But drug-induced accumulation of cAMP seemed to be virtually unaffected by prior incubation of brain slices in calcium-free incubation medium[3].

The superior cervical ganglia (SCG) have proved to be an important model for studying nerve cell function[4]. It used the rat SCG in vitro for investigations of effects of calcium ions on the transmission through the ganglion and on accumulation of cAMP in the ganglion. Omission of calcium in the bathing fluid abolished action potentials recorded after preganglionic electrical stimulation. Threshold concentrations for the appearence of potentials were lower than 1.0mM. Higher concentrations of calcium (20 - 40mM) resulted in a complete reduction of all potentials recorded (fast and slow). Raising the calcium concentration in the superfusion fluid to 60mM led to an increase of the slow excitatory potential. This effect was calcium-specific and it could be neither antagonized nor mimicked by the alkali-earth ions $Ba^{2\oplus}$ and $Sr^{2\oplus}$.

On the other hand, amine-induced cAMP accumulations in the ganglion occurred whether $Ca^{2\oplus}$-ions are present in the incubation medium or not. Even in the presence of a high external calcium concentration (60mM) drug-induced cAMP accumulation was virtually unaffected. This was shown by incubation of ganglia with low concentrations of isoproterenol, adrenalin, dopamine, histamine and prostaglandin E_1. The β-adrenergic-induced rise of cAMP and its interaction with calcium ions were further investigated. A high external calcium concentration (60mM) did not affect the amount of isoproterenol-induced accumulation of cAMP in the ganglion but the ability of the β-blocker propranolol to antagonize the drug-induced rise of cAMP was affected to some extent.

It was concluded that calcium ions are necessary for the transmission process through the ganglion. This event seemed to be independent of drug-induced changes in the ganglionic cAMP content.

1 Reuter, H. (1963) *Prog. Biophys. Mol. Biol.* **26**, 1 - 43.
2 Roth, R.H., Morgenroth, V.H. & Salzman, P.M. (1975) *Naunyn-Schmiedeberg's Arch. Pharmacol.* **289**, 327 - 343.
3 Daly, J.W. (1977) Cyclic Nucleotides in the Nervous System. Plenum Press, New York.
4 Nathanson, J.A. (1977) *Physiol. Rev.* **57**, 157 - 256.

Toni Lindl, Fachbereich Biologie, Universität Konstanz, Postfach 77 33, D-7750 Konstanz.

G. Löffler, M. Lucas, R. Kromas and G. Schmid

Ca-Metabolism of Isolated Pancreatic Exocrine Cells

Enzyme secretion of isolated pancreatic exocrine cells of the guinea pig can be stimulated by carbamoylcholine or by pancreozymine and is independent of the extracellular $Ca^{2\oplus}$ concentration. After preloading the cells with $^{45}CaCl_2$, a marked ^{45}Ca efflux occurs together with the stimulated enzyme secretion, which can be explained by Ca liberation from intracellular depots, such as mitochondria or microsomes. Both organelles are able to accumulate $^{45}Ca^{2\oplus}$ in an ATP-dependent reaction. Mitochondrial ^{45}Ca accumulation can also be driven by respiration in the presence of ADP, and substrate and is diminished by mitochondrial inhibitors. Microsomal $^{45}Ca^{2\oplus}$ accumulation is stimulated several-fold by addition of oxalate and is strongly inhibited by thiol-reagents. Pancreatic microsomes are not only able to accumulate $^{45}Ca^{2\oplus}$, but contain a Mg-dependent Ca-ATPase, which has been characterized with respect to cation and nucleotide specificity, activation by phospholipids, thiol reagents and inhibition by NaF.

G. Löffler, M. Lucas, R. Kromas and G. Schmid, Fachbereich Biologie und Vorklinische Medizin der Universität Regensburg, Universitätsstraße 1, D-8400 Regensburg.

J. Lohmann, N. Schulze-Waltrup, T. Kerényi, H. Schriewer and W. H. Hauss

Investigations on the Biosynthesis of Phosphatidylcholine in Human Atherosclerotic Aorta

In order to get more information about the de novo synthesis of phosphatidylcholine in human aorta, incorporation of [*methyl*-^{14}C]choline – into phosphorylcholine, CDP-choline and phosphatidylcholine was studied in homogenates and slices of human aorta. In normal aortic tissue, the highest radioactivity was measured in the phosphatidylcholine fraction, and the lowest in the CDP-choline fraction. This proportion confirms the supposition that CTP:choline phosphate cytidylyltransferase (EC 2.7.7.15) is the rate-limiting step in the synthesis of phosphatidylcholine, with choline as precursor, in vitro.

Therefore the activity of the cytidylyltransferase was studied. The apparent K_m for phosphorylcholine was estimated to be 0.32 mmol/*l*. The pH optimum was about pH 7.0. 50% of the enzyme activity was lost when aortic homogenates were stored at 4 °C for 3 days. The enzyme activity in atherosclerotic tissue of human aorta was only enhanced if intimal proliferation was accompanied with cellular infiltration. In regions of the same aorta with less cellular infiltration we found lower activities of cytidylyltransferase, independent of the degree of atherosclerosis. These findings indicate that the endogenous synthesis of phosphatidylcholine may be correlated more with the activity of the atherosclerotic process than with the degree of intimal thickening.

Dr. Jörg Lohmann, Institut für Arteriokleroseforschung an der Universität Münster, Westring 3, D-4400 Münster.

N. B. Madsen, P. J. Kasvinsky, J. Sygusch and R. J. Fletterick

Crystallographic Insights into Regulatory Mechanisms in Glycogen Metabolism

Availability of the crystal structure of phosphorylase (EC 2.4.1.1) to a resolution of 2.5 Å enables us to locate ligand binding sites in three-dimensional relationship to each other and to relate occupation of these sites with kinetic studies of the enzyme in solution. These studies plus the finding of new sites have suggested new hypotheses for the molecular mechanisms of control associated with this enzyme. Thus a glycogen storage site on each monomer was found 30 Å from the active site but on the same concave face of the dimer which contains the entrances to the active sites. Kinetics indicate tight binding of glucose chains at the storage site but loose binding at the active site. On the opposite, convex, face of the dimer, near the subunit interface, are found the serine-phosphate and the AMP (activator) sites. Thus the enzyme can bind tightly to the glycogen particle while leaving the site for the interconvertible control (serine^{14}-phosphate) available for the binding of the specific phosphatase and kinase.

The high resolution structure of phosphorylase was obtained from crystals grown in the presence of the allosteric inhibitor, glucose, and therefore depicts the enzyme in its inhibited (T) conformation. Addition of a virtual substrate, 5-thioglucose-1-*P*, or the substrates causes the crystals to shatter and then reanneal in the same space group. X-Ray crystallography shows a 2 Å increase in unit cell dimensions, indicating the monomers in each dimer have moved apart slightly. A long α-helix at the sub-unit interface between the monomers has shifted inward by 5 Å along its entire length, pulling with it the serine-phosphate attached to its outer surface. The shift of this helix provides a means of communicating between the effector (AMP binding) sites of the dimer, as well as exerting the glucose control over the action of phosphorylase phosphatase on the serine-phosphate and hence an effect on glycogen metabolism in general. Structural changes leading to the active site (by the pyridoxal-phosphate) and the glycogen storage site may also be seen.

The secondary nucleotide site has been found 10 Å from the active site where glucose-1-*P* binds, and 40 Å from the nucleotide activator site. The secondary nucleotide site shows a preference for nucleosides and their derivatives such as caffeine or theophylline. These compounds show competitive inhibition with respect to glucose-1-*P*, while synergism with glucose inhibition is exhibited. Crystallographic studies of the binding of these ligands indicate that a nucleoside bound in its site prevents glucose-1-*P* from binding in the active site because the shift of a protein loop, necessary to accommodate the phosphate of the substrate, is physically prevented by the nucleoside inhibitor. The relationship of the glucose and caffeine inhibition to carbohydrate metabolism in the liver will be discussed.

N. B. Madsen, P. J. Kasvinsky, J. Sygusch and R. J. Fletterick, Dept. of Biochemistry, & M. R. C. Group on Protein Structure and Function University of Alberta, Edmonton, Alberta, T6G 2H7, Canada.

E.-M. Mandelkow and E. Mandelkow

Structures Resembling Flagellar Doublets and Centriolar Triplets from Brain Tubulin in vitro

Although tubulins from different sources are chemically similar, they assemble into distinct structures such as single tubes in the cytoplasm, doublet tubes in flagellar axonemes, triplets in centrioles, etc. The question arises, whether the bonds generating these aggregates are inherent in the tubulin molecule or in other factors, e.g. some microtubule associated proteins (MAP).

Singlet microtubules can be formed from tubulin alone, so the bonds forming microtubule walls are a property of the tubulin molecules themselves. Their subunit arrangement is very similar in flagellar and cytoplasmic microtubules and polymorphic forms[1-4]. However, doublet or triplet tubules require junctions between wall segments whose nature is not well understood,

We have produced a variety of tubular forms from brain tubulin in vitro under conditions similar to those generating hoops[4]*. They range from incomplete walls (C-tubules) to two C's joined into an S, microtubules with one or two C's attached, doublet tubules resembling those of flagellae, and even centriolar triplet-like structures. The relative curvatures of walls are best observed in micrographs of thin sections of pelleted protein, while the substructures of the walls are deduced from negatively stained samples, using optical diffraction and computer image processing.

The surface lattice of tubulin monomers is similar to that in microtubules, sheets, or hoops (40 Å axial repeat, 50 Å protofilament separation, protofilament stagger 10 Å). The pairing of monomers in the α-β-dimers is weak so the 80 Å layer line indicating the dimer arrangement is rarely visible. Available evidence suggests that it is like that in flagellar B subfibers[1] which is compatible with the observations on cytoplasmic microtubules and hoops[2,4]. The junction between wall segments itself is often blurred, presumably due to distortions and flattening during drying on the grid, but it appears that the protofilament stagger is also around 10 Å, although the angle between walls may vary from 0° to 90°.

Since the wall junctions can be formed in the presence or absence of MAPs they reflect an intrinsic bonding property of tubulin, distinct from that leading to walls or tubules. It is possible that even the more complex microtubular structures in the cell are based simply on tubulin-tubulin interactions, and extra factors present in situ may be active in selecting which of the various possibilities of assembly is realised.

* Mandelkow, E.-M. & Mandelkow, E. (1978), manuscript submitted for publication.

1 Amos, L.A. & Klug, A. (1974) *J. Cell. Sci.* **14**, 523 - 549.

2 Mandelkow, E., Thomas, J. & Cohen, C. (1977) *Proc. Natl. Acad. Sci. U.S.A.* **74**, 3370 - 3374.

3 Erickson, H.P. (1974) *J. Cell Biol.* **60**, 153 - 167.

4 Mandelkow, E.-M., Mandelkow, E., Unwin, N. & Cohen, C. (1977) *Nature (London)* **265**, 655 - 657.

Eva-Maria Mandelkow and Eckhard Mandelkow, Max-Planck-Institut für medizinische Forschung, Abteilung Biophysik, Jahnstr. 29, D-6900 Heidelberg 1.

K. Mann

Mikrotubuli-assoziierte Proteine

Mikrotubuli können aus Schweinehirnen nativ, d.h. ohne Depolymerisation gewonnen werden.* Bei der Reinigung dieser Mikrotubuli zeigt sich, daß an das Tubulin-Gerüst noch eine Reihe anderer Proteine mit höherem Molekulargewicht als Tubulin assoziiert ist. Diese Proteine können nach ihrer Wanderungsgeschwindigkeit in Natriumdodecylsulfat-Harnstoff-Polyacrylamidgelen bei pH 8.5 in sieben Gruppen eingeteilt werden:

I entspricht dem von mehreren Autoren beschriebenen τ-Faktor, der die Polymerisation von Tubulin beschleunigen soll.[1,2] II ist ein Protein mit einem Molekulargewicht von etwa 85 000. III ist eine Gruppe von drei Proteinen mit einem Molekulargewicht zwischen etwa 100 000 und 130 000. IV beinhaltet mindestens drei dicht aufeinanderfolgende Banden zwischen etwa 140 000 und 160 000. V besteht aus zwei dicht beieinanderliegenden Banden sehr ungleicher Stärke um etwa 200 000. Noch höhere Molekulargewichte haben die Proteine der Gruppen VI, einer Doppelbande, und VII, die die schon mehrfach in der Literatur beschriebenen hochmolekularen Mikrotubuli-assoziierten Proteine (Molekulargewichte bis 300 000) enthält.[3,4]

* Fasold, H., Giesel, M., Haase, W. & Mann, K. (1978) FEBS Lett., in Vorbereitung.

1 Weingarten, M.B., Lockwood, A.H., Hwo, S.Y. & Kirschner, M.W. (1975) *Proc. Natl. Acad. Sci. U.S.A.* **72**, 1858 - 1862.

2 Fellous, A., Francon, J., Lennon, A.-M. & Nunez, J. (1977) *Eur. J. Biochem.* **78**, 167 - 174.

3 Sloboda, R.D., Dentler, W.L. & Rosenbaum, J.L. (1976) *Biochemistry* **15**, 4497 - 4505.

4 Gaskin, F., Kramer, S.B., Cantor, C.R., Adelstein, R. & Shelanski, M.L. (1974) *FEBS Lett.* **40**, No. 3, 281 - 285.

Karlheinz Mann, Institut für Biochemie der Universität Frankfurt am Main, Sandhofstraße, Gebäude 75A, D-6000 Frankfurt/71.

R. Merz, M. Löffler and Fr. Schneider

Phosphatase and Nucleosidtriphosphatase Activity of Intact Ehrlich Ascites Tumor Cells

The substrate specificity and the effects of nucleotides and SH-blocking agents on the *p*-nitrophenylphosphatase activity of intact Ehrlich ascites tumor cells were studied. DL-β-Glycerophosphate, *O*-phosphoethanolamine, phosphocholine, glucose 6-phosphate, *o*-carboxyphenylphosphate, phospho*enol*pyruvate and AMP were not attacked by intact cells. ATP > GTP > UTP > PP_i >

p-nitrophenyl-*P* were cleaved with decreasing velocity. A stimulation of the cleavage of *p*-nitrophenyl-*P* by the following nucleotides was observed with decreasing effectivity: ATP > ADP > GTP > UTP; AMP was ineffective. The phosphatase activity was not affected by malate, tartrate or oxidized glutathione. The SH blocking agents diamide, thimerosal and 2,2'-dicarboxy-4,4'-bis(iodoacetamido)azobenzene were more effective inhibitors of the *p*-nitrophenylphosphatase than of the ATPase activity, whereas the hydrolysis of ATP is more affected by the ATP analog adenylylimidodiphosphate. The present data are best compatible with a double headed enzyme[1]: Both acitve sites interact with ATP, only one is active against *p*-nitrophenyl-*P* and sensitive against SH-blocking agents.

1 Robinson, J. D. (1976) *Biochim. Biophys. Acta* **429**, 1006 - 1019.

Reiner Merz, Physiologisch-Chemisches Institut II, Universität Marburg, Lahnberge, D-3550 Marburg.

W. Meyer und H. Follmann

Adenosincarbonsäuren als Modellnucleotide zum Studium von 5'-AMP-Aminohydrolase und 5'-Nucleotidase

Zur Untersuchung Nucleotid-umsetzender und/oder durch Nucleotide regulierter Enzyme sind Substrate oder Effektoren mit strukturell modifiziertem 5'-Ende erforderlich; solche Verbindungen waren bisher nur in Einzelfällen bekannt[1,2]. Wir beschreiben als neuartige Modellnucleotide die aus Adenosin synthetisch zugänglichen Adenosincarbonsäuren des Typs 9-Adeninyl-$C_4H_6O_3$-$[CH_2]_nCOOH$ (n = 0: Adenosin-5'-uronsäure, I; n = 1: 5'-Desoxyadenosin-5'-carbonsäure[3], II; n = 2: 5'-Desoxyadenosin-5'-essigsäure[4], III). Sie sind hydrolysebeständig, jedoch bei physiologischen pH-Werten dem 5'-AMP in Ladung und Molekülkonformation[5-6] gleichende Analoge, die durch Variation der Kettenlängen Aussagen über die Bedeutung der relativen Position von negativer Ladung und Adeninbase in Enzymreaktionen erlauben. So setzt die spezifische 5'-AMP-Aminohydrolase (EC 3.5.4.6) aus Kaninchenmuskel auch die Carbonsäuren II und III um (relative Reaktionsgeschwindigkeiten: AMP = 100; III = 21; II = 1), die Uronsäure I ist inaktiv. 5'-Nucleotidase (EC 3.1.3.5) aus Schlangengift wird durch die Adenosincarbonsäuren kompetitiv gehemmt (II > III ≫ I), doch ist wiederum die 5'-Uronsäure am wenigsten wirksam. Dem Abstand der Ladung von anderen Molekülteilen scheint daher in Nucleotiden eine kritische Bedeutung zuzukommen, und die starke intramolekulare Fixierung der 5'-Uronsäuren über eine H-Brücke zu Purin- oder Pyrimidinbasen[5-6] trägt zur enzymatischen Inaktivität dieser Substanzklasse bei.

1 Halford, M.H. & Jones, A.S. (1968) *J. Chem. Soc.* 2667 - 2670.
2 Hampton, A., Howgate, P., Harper, P.J., Perini, F., Kappler, F. & Preston, R.K. (1973) *Biochemistry* **12**, 3328 - 3332.
3 Meyer, W., Böhnke, E. & Follmann, H. (1977) *Angew. Chem.* **88**, 512 - 513; *Angew. Chem. Int. Ed.* **15**, 499 - 500.
4 Walker, T.E., Follmann, H. & Hogenkamp, H.P.C. (1973) *Carbohydr. Res.* **27**, 225 - 234.
5 Follmann, H., Kuntz, I. & Zacharias, W. (1975) *Eur. J. Biochem.* **58**, 31 - 41.
6 Follmann, H., Pfeil, R. & Witzel, H. (1977) *Eur. J. Biochem.* **77**, 451 - 461.

Dipl.-Chem. W. Meyer und Prof. Dr. H. Follmann, Fachbereich Chemie der Universität, Arbeitsgruppe Biochemie, D-3550 Marburg.

G. Mohn and M. Weber

The Earliest Inducing Effects of Cyclohexane in Vivo and in Vitro in Female Mice

In female mice, one of the earliest inducing effects of cyclohexane was measured in vivo by determining the hexobarbital sleeping time. It was decreased by 50% between the 12th and 24th hour of cyclohexane inhalation, whereas the additional decrease after the 3rd day of inhalation was only 15%. The early steep decrease in the sleeping time was completely prevented by actinomycin D.

As early as 6 h after beginning of the inhalation, the type II-binding spectrum of aminopyrine changed to a type I-spectrum in the liver microsomes in vitro. $\Delta A_{max}/\mu M$ cyt. P-450 increased during the following 18 h, and remained constant thereafter. K_S was well reproducible and independent of the concentration range of substrate and solvent.

Obviously, the *N*-demethylation of aminopyrine and the hydroxylation of hexobarbital were preferably induced by cyclohexane during the first day, whereas the main increase in the ethylumbelliferone dealkylation was not seen before 1 - 3 days of induction. This may be the cause of the fast decrease of the relative dealkylase activity during the first day of inhalation of cyclohexane or DL-camphor[1,2].

In the liver microsomes of female mice, the change in the aminopyrine binding spectrum from type II to Type I was the most sensitive test for demonstrating the earliest and weakest inducing effects of cyclohexane, camphor and phenobarbital.

1 Mohn, G. & Philipp, E.-M. (1974) *Hoppe-Seylers Z. Physiol. Chem.* **355**, 564 - 575.

2) Mohn, G., Niederlaender, P. & Kessler, C. (1977) *Biochem. Pharmacol.* **26**, 1885 - 1892.

G. Mohn, Fachrichtung für Physiologische Chemie, Universität des Saarlandes, D-6650 Homburg/Saar.

H. Mohr, H. Jüppner and R.D. Hesch

Binding and Degradation of Parathormone in the Kidney Cortex

Parathyrin (parathormone), which influences the metabolism of calcium in higher vertebrates, binds to specific receptors at plasma membranes, e.g. of the kidney cortex[1]. This membrane also contains a proteolytic activity capable of degrading the hormone[2]. Our observations show that the contraluminal part of the plasma membrane (the basal lateral membrane) of bovine kidney cortex contains receptor sites as well as "parathyrinase" activity. The protease is not part of the hormone receptor, which can be extracted from the matrix with chaotropic salts or urea, and is also digested by trypsin (EC 3.4.21.4)[3]. This indicates that the hormone binding protein is located on the membrane surface. The parathyrinase seems to be anchored more inside the lipid bilayer. It is not extractable and remains intact after tryptic digestion of membranes. Furthermore, it is heat stable when membrane bound. This property is lost after solubilization with Triton X-100. *p*-Nitrophenylformiate and *p*-nitrophenylhydrazine inhibit the PTHase but not the binding of the hormone to its receptor. Phenylmethane sulfonylfluoride, an inhibitor of serine proteases[4], has no influence on the parathyrinase. Binding as well as cleavage of hormone are unaffected by SH-reagents.

There seems to be no direct relationship between attachment of parathyrin to its receptor and degradation of the hormone. Oxidized parathyrin has lost its ability to combine with the binding protein[5], but is degraded as efficiently as intact hormone.

1 Sutcliffe, H.S., Martin, T.J., Eisman, J.A. & Pilczyk, R. (1973) *Biochem. J.* **134**, 913 - 921.
2 Chu, L.L., Forte, L.R., Anast, C.S. & Cohn, D.C. (1975) *Endocrinology* **97**, 1014 - 1023.
3 Chase, L.R. (1974) *Endocrinology* **96**, 70 - 76.
4 Shaw, E. (1970) in The Enzymes 1, (Boyer, P.D. ed.) 3rd edn., p. 99, Academic Press, New York and London.
5 DiBella, F.P., Doûsa, T.P., Miller, S.S. & Arnaud, C.D. (1974) *Proc. Natl. Acad. Sci. U.S.A.* **71**, 723 - 726.

H. Mohr, Abteilung für Klinische Endokrinologie, Medizinische Hochschule Hannover, Karl-Wiechert-Allee 9, D-3000 Hannover 61.

R. Müller, M. Giesing and F. Zilliken

Taurine in Developing Brain Cultures: Biosynthesis, Cellular and Subcellular Distribution

The function of taurine in brain is not yet completely understood. It was, therefore, of considerable interest to study the metabolism of taurine in explant cultures. Taurine content was determined in grey matter of cortex (CC), glial cells from the neuropil (NP), diencephalon (DE) and mesencephalon (ME) prior to cultivation and during development in vitro. Significant changes among the tissues were observed immediately after dissection from rat brain: Glia contained 1010 nmol taurine per mg cell protein, followed by ME (710), DE (480) and CC (340). Development of the cells in vitro was paralleled by a decrease in intracellular taurine as known from studies in vivo. The rate of decline was found to be tissue specific: after 8 days in vitro, glia contained only 2% of the original amount, ME 6%, DE 8% and CC still 53%. Since CC seemed to play an exceptional role (lowest taurine content, highest intracellular maintenance), the tissue was chosen to investigate subcellular distribution and biosynthesis. More than 80% of the intracellular taurine was discovered in the cytosol fraction, whereas nerve endings contained only some 4%. Microsomes showed the highest amount of the particulate fractions. Specific activities, as measured from incorporated [^{14}C]taurine, differed only slightly among subcellular fractions. Biosynthesis of taurine was investigated by using [^{14}C]cysteine. The pattern of radiolabelled metabolites showed that the formation of taurine during development was substantially enhanced. The specific activities of taurine between the 4th and the 15th day in vitro was increased some 6-fold. The results shed light on two facets of taurine action in brain. First, it seems unlikely that taurine is a neurotransmitter, and second, since glia cells show opposite results in comparison with cortex neurons, taurine may function in the interaction process between component cells during a certain stage of brain development. However, the mechanism remains completely unknown so far.

Dr. Robert Müller, Institut für Physiologische Chemie der Universität Bonn, Nußallee 11, D-5300 Bonn.

E. Neumann

Die Kinetik der Acetylcholin-Bindung an Rezeptor und Esterase in Beziehung zur neuralen Informationsübertragung

Relaxationskinetische Studien an isolierten, hochgereinigten Acetylcholinesterasen (EC 3.1.1.7) und Acetylcholinrezeptoren (in 0.1% nichtionischem Detergens) haben zu folgenden Ergebnissen geführt[1-3]:

a) Die Geschwindigkeitskonstante k für die Assoziation des Acetylcholins an die Esterase ist $\geqslant 2 \times 10^8\ l \times mol^{-1} \times s^{-1}$ (0.1M NaCl, pH 7 und 293 K)[1]. Die k-Werte für nicht-hydrolysierbare Liganden wie *N*-Methylacridinium und 1-Methyl-7-hydroxichinolinium sind für gleiche Bedingungen $2 \times 10^9\ l \times mol^{-1} \times s^{-1}$. Aus der Ionenstärkeabhängigkeit sowohl der relaxationskinetisch bestimmten k-Werte als auch der Katalyseparameter (k_{cat} und K_{app}) folgt, daß das aktive Zentrum durch 6 (± 1) negative Ladungen charakterisiert ist und daß innerhalb weiter Grenzen der Substratkonzentration, der sehr schnelle bimolekulare Schritt der Ligandenassoziation zu dem noch geschwindigkeitsbestimmend für die katalytische Zwischenstufe der Enzymacylierung ist[3]. Die Zahlenwerte für k und für die effektive Ladungszahl lassen vermuten, daß nicht nur die aktiven Zentren sondern größere Oberflächenbereiche des Enzyms Substrat einfangen und durch Oberflächendiffusion an die Bindungstaschen weiterleiten können.

b) Isolierte Acetylcholinrezeptoren binden zwar Acetylcholin ebenfalls schnell, die Assoziationsgeschwindigkeit $k = 3 \times 10^7\ l \times mol^{-1} \times s^{-1}$ (0.1M NaCl, pH 8.5, 293 K) ist aber etwa 10- (vielleicht auch 100-)mal kleiner als bei der Esterase. Wegen des Fehlens geeigneter optischer Indikation wurde die Bindung des Acetylcholins an den Rezeptor über die angekoppelte Ca-Freisetzung mittels Farbindikator (Murexid) gemessen. Die weitere Analyse zeigt, daß isolierte Rezeptoren zwischen mindestens drei Konformationen equilibrieren können und daß mindestens eine der angezeigten Strukturänderungen in charakteristischer Weise von der Konzentration des Acetylcholins abhängig ist. Mit fluoreszierenden kationischen Liganden ergeben sich k-Werte von $10^8\ l \times mol^{-1} \times s^{-1}$ und eine effektive Ladungszahl von – 3 (± 1) für den Rezeptor. Die Ähnlichkeit zwischen dem k-Wert für das Acetylcholin von $3 \times 10^7\ l \times mol^{-1} \times s^{-1}$ und dem Wert $k_{on} = 10^7\ l \times mol^{-1} \times s^{-1}$, der aus elektrophysiologischen Stromrelaxationen hergeleitet wurde[4], laßt den Schluß zu, daß die Geschwindigkeit der Ca-Freisetzung nicht nur die Geschwindigkeit der Acetylcholinbindung widerspiegelt, sondern auch den geschwindigkeitsbestimmenden Schritt für den Leitfähigkeitsanstieg in der Membran darstellt.

Wegen des großen Unterschieds in der Geschwindigkeit der Acetylcholinassoziation und des daraus resultierenden Nachteils für den Rezeptor (im Vergleich zur schnelleren Esterase) erhebt sich nun die Frage, wie der funktionell sequentielle Ablauf, erst die Rezeptorreaktion dann der enzymatische Abbau des Acetylcholins, in der Struktur der Zellmembranen und der Synapse gewährleistet ist. Alternativ zu dem Konkurrenzmodell (gleichzeitiger Zugang des freigesetzten Acetylcholins zu beiden Proteinen), wie es in der Neurotransmissions-Hypothese implizit ist, wird eine sequentielle Reaktionsfolge diskutiert, die eine Trennung der Reaktionsräume für die Rezeptorreaktion innerhalb der Membran und der Acetylcholinhydrolyse nahe an der Membran beinhaltet.

1 Rosenberry, T. L. & Neumann, E. (1977) *Biochemistry* **16**, 3870 - 3878.
2 Neumann, E. & Chang, H. W. (1976) *Proc. Natl. Acad. Sci. U.S.A.* **73**, 3994 - 3998.
3 Nolte, H. J., Rosenberry, T. L. & Neumann, E. (1978) *Biochemistry*, in press.
4 Sheridan, R. Z. & Lester, H. A. (1977) *J. Gen. Physiol.* **70**, 187 - 219.

Prof. Dr. E. Neumann, Max-Planck-Institut für Biochemie, D-8033 Martinsried bei München.

H. v. Nicolai and F. Zilliken

Substrate Specificity of Neuraminidase from Arthrobacter sp.

Different soil bacteria belonging to the genus *Arthrobacter* are known to produce neuraminidase (EC 3.2.1.18) when cultivated in glycoprotein- or colominic acid-containing media[1,2].

The specificity of an *Arthrobacter* neuraminidase (Nakarai Kagaku Ltd. Japan, kindly supplied by Boehringer Mannheim GmbH) has been investigated, using several NeuAc-containing oligosaccharides, glycolipids, and glycoproteins as substrates.

The enzyme exhibits its highest activity towards the $\alpha2 \rightarrow 6$ linkage of II^6NeuAc-Lac. Considering this cleavage rate as 100, it is 60 for the $\alpha2 \rightarrow 3$ linkage of II^3NeuAc-Lac, and 15 for the $\alpha2 \rightarrow 8$ linkage of II^3NeuAc$_2$-Lac. Among more complex oligosaccharides, the cleavage rate is 80 for III^6NeuAc-LcOse$_4$, 47 for IV^6NeuAc-nLcOse$_4$, and 95 for IV^3NeuAc, III^6NeuAc-LcOse$_4$ respectively. Colominic acid and bovine brain gangliosides are cleaved in the range of 5 to 10 only. With glycoproteins as substrates, the enzyme reaches values of 65 for orosomucoid (α_1-acid glycoprotein) and fetuin, 16 for human urinary glycoprotein (Tamm and Horsfall mucin), and 5 for bovine submaxiallary mucin.

In general, the enzyme shows a specificity pattern similar to that found for neuraminidases from other non-pathogenic bacteria[3], with exception of its remarkably high affinity to $\alpha2 \rightarrow 6$ linked NeuAc. Under appropriate conditions, *Arthrobacter* neuraminidases may therefore be useful as a tool in structural analysis as well as for specific modifications of cells and membrane surfaces.

1 Uchida, Y., Tsukada, Y. & Sugimori, T. (1977) *J. Biochem. (Tokyo)* **82**, 1425 - 1433.

2 Flashner, M., Wang, P., Hurley, J.B. & Tanenbaum, S.W. (1977) *J. Bacteriol.* **129**, 1457 - 1465.
3 Nicolai, v., H. (1976) Habilitationsschrift, Bonn.

H. v. Nicolai, Institut für Physiologische Chemie der Universität, Nußallee 11, D-5300 Bonn.

Dr. Hans Nohl, Institut für Pharmakologie, Fakultät für Veterinärmedizin, Universität München, Veterinärstraße 13, D-8000 München 22.

H. Nohl

Evidence for the Presence of Catalase in Rat-Heart Mitochondria

We have recently shown that the mitochondria of living rat-heart cells accumulate products of lipid peroxidation resulting from a radical chain reaction involoving $O_2^{\ominus}$ radicals and H_2O_2[1]. The presence of superoxide-dismutase (EC 1.15.1.1) and glutathione peroxidase (EC 1.11.1.9) in heart mitochondria may indicate a physiological control of lipid peroxidation by keeping deleterious metabolites at a low level. The apparent lack of catalase (EC 1.11.1.6) activity in preparations from heart mitochondria[2-4] has led to conflicting reports of the way in which heart cells regulate and dispose of mitochondrial H_2O_2. Therefore the existence of catalase in crude fractions of heart mitochondria was intensively investigated, using various techniques. Evidence for the existence of catalase in heart mitochondria comes from (a) the sensitivity of H_2O_2 metabolism towards haem-binding inhibitors, (b) the steady state of catalase-H_2O_2 intermediate in H_2O_2-producing mitochondria and (c) the disappearance of these phenomena after pretreatment of the animals with 3-amino-1,2,4-triazole. From the latency of catalase activity, it was concluded that the mitochondrial membrane imposes an imperfect permeability barrier to H_2O_2. Further evidence for this concept comes from the observation that intact motochondria destroy exogenous H_2O_2 at a very much slower rate than do disrupted particles. The cytochemical detection of catalase by the 3,3′-diaminobenzidine technique supports not only the biochemical findings but also the interpretation that catalase is indeed present in the mitochondria, and not as a contaminant from peroxisomes.

1 Nohl, H. & Hegner, D. (1978) *Eur. J. Biochem.* **82**, 563 - 567.
2 Loschen, G., Azzi, A. & Flohé, L. (1973) in Glutathione, Proc. 16th Conf. Germ. Soc. Biol. Chem., Tübingen (Flohé, L., Benöhr, H.Ch., Sies, H., Waller, H.D. & Wendel, A., eds.) pp. 228 - 236, Georg Thieme, Stuttgart.
3 Loschen, G. (1975) Thesis, Universität Tübingen.
4 Boveris, A. & Chance, B. (1973) *Biochem. J.* **134**, 707 - 716.

D. Nonnenmacher, B. Schätzler and R. Brossmer

Chemical modification studies of Concanavalin, Lens culinaris and Peanut lectins

The following modifications of mitogenic lectins such as concanavalin A (Con A), *Lens culinaris* hemagglutinin (LcH) and peanut agglutinin (PNA) have been investigated.

1) Derivatization of the tyrosyl residues:
– nitration with tetranitromethane
– binding by azolinkage of the 4′-diazobenzyl-α-ketoside of *N*-acetyl-D-neuraminic acid
2) Derivatization of the free amino groups:
– reaction with 1-isothiocyanato-4-benzenesulfonic acid
– succinylation (= 3-carboxypropionylation)

We obtained modification of about 10 - 20% of the tyrosyl residues and 60 - 90% of the free amino groups.

The influence of the incorporation of nitro and anionic groups on the biological properties, such as agglutination of human erythrocytes and stimulation of human peripheral lymphocytes, and physicochemical properties is compared with the properties of the nitrated-Con A and succinyl-Con A. The hemogglutinating activity of all the derivatives of Con A and LcH is reduced, whereas it remains unchanged for PNA.

The dose-response curve of LcH in the mitogenic stimulation of human lymphocytes is similar to that of Con A. However, the modified LcH and Con A show a response which is diminished at high concentrations compared to succinyl-Con A. An exception is the isothiocyanate derivative of LcH.

For PNA the maximal stimulation of neuraminidase-treated human lymphocytes is found at a concentration of 150 µg/m*l*; the response to the succinyl and isothiocyanate derivatives is strongly reduced.

We observed that these chemical modifications of Con A, LcH and PNA alter the biochemical properties, but not the structure, of LcH and PNA, in contrast to tetrameric Con A, which is converted by succinylation to a dimeric molecule. Moreover, it seems that an anionic label is not the only factor for the particular mitogen properties of the succinyl-Con A at high doses.

Danielle Nonnenmacher, Bernd Schätzler und Reinhard Brossmer, Institut für Biochemie II (Med. Fak.), Universität Heidelberg, Im Neuheimer Feld 328, D-6900 Heidelberg.

A. Paessens and H. Rüdiger

Purification of the Tetrahydropteroylglutamate Methyltransferase by Hydrophobic and Affinity Chromatography

The vitamin B_{12}-dependent methionine synthesizing enzyme tetrahydropteroylglutamate methyltransferase (5-methyltetrahydropteroyl-L-glutamate: L-homocysteine *S*-methyltransferase, EC 2.1.1.13) from *Escherichia coli*[1]:

$$\begin{matrix} \text{5-CH}_3\text{-H}_4\text{folate} \\ + \text{Hcy} \end{matrix} \xrightarrow[\text{Met(Ado)}]{\text{B}_{12}\text{-enzyme, red.flavin}} \begin{matrix} \text{H}_4\text{folate} \\ + \text{Met} \end{matrix}$$

may be purified by gels containing aminoalkyl residues bound to Sepharose[2,3]. Aminoethyl, aminopropyl and aminobutyl gels do not adsorb the enzyme. The enzyme is retained on gels containing aminopentyl up to amino-octyl groups, and may be desorbed by increasing salt concentrations (0.3 - 0.9M NaCl). Aminodecyl and aminododecyl gels adsorb the enzyme irreversibly. Neither salt solutions (up to 5M NaCl) nor chaotropic reagents (ethylene glycol, dioxane) are able to remove enzyme activity or protein from these gels.

Alkyl gels without the terminal amino group retain the enzyme in generally the same way. In these gels, however, shorter alkyl chains (6 C atoms) are able to adsorb the enzyme irreversibly. n-Alkyl residues bind the enzyme more strongly than branched ones.

At a given chain length, the precise salt concentration necessary for elution of the enzyme is dependent on the amount of alkyl groups bound per g gel and may thus be varied by employing different amounts of cyanogen bromide in gel activation[4].

Affinity chromatography was studied with gels containing cobalamine or 5-methyltetrahydrofolate. Cobalamine gels do not retain the enzyme, presumably because the immobilized cobalamine does not compete with the tightly bound B_{12} cofactor of the enzyme. Methyltetrahydrofolate gels adsorb the enzyme. Due to the presence of other folate-binding proteins, however, no significant purification could be achieved.

Hydrophobic chromatography, in connection with ion-exchange and gel filtration, were successfully applied to purify the enzyme to homogeneity.

1 Rüdiger, H. & Jaenicke, L. (1973) *Mol. Cell. Biochem.* **1**, 157 - 168.
2 Axén, R., Porath, J. & Ernback, S. (1967) *Nature (London)* **214**, 1302 - 1304.
3 Shaltiel, S. (1970) *Methods Enzymol.* **34B**, 126 - 140.
4 Jennissen, H.P. (1976) *Hoppe-Seyler's Z. Physiol. Chem.* **357**, 1201 - 1203.

Prof. Dr. H. Rüdiger, Institut für Pharmazie und Lebensmittelchemie der Universität Würzburg, Am Hubland, D-8700 Würzburg.

D. Palm, H. Schächtele and E. Schiltz

Escherichia coli Phosphorylase as a Model of Glycogen Phosphorylases. Comparison of the Pyridoxalphosphate-Binding Peptides

Maltodextrin phosphorylase (EC 2.4.1.1) from *E. coli* shows an absolute dependence on pyridoxal-5'-phosphate, as the glycogen phosphorylases of higher organisms do. The lack of control by phosphorylation or allosteric effectors makes the bacterial enzyme an ideal model to study the functions of pyridoxalphosphate[1].

The sequences of the pyridoxalphosphate-binding peptides of some eukaryotic phosphorylases are known from protease digests of "reduced" phosphorylase. The deeply buried Schiff's-base-linked pyridoxalphosphate-lysyl residue of maltodextrin phosphorylase could only be reduced in the presence of detergents. Due to incomplete tryptic digestion we obtained two pyridoxylphosphate-containing peptides, which on subsequent cleavage by thermolysin yield one unique octapeptide. The amino acid composition and partial sequence determination demonstrates a high degree of homology with the sequence preceeding the pyridoxalphosphate-binding lysyl residue in rabbit muscle glycogen phosphorylase[2]. Similarly, homologous sequences are also seen at the amino terminus.

1 Palm, D., Thanner, F. & Schwenk, L. (1976) in Metabolic Interconversion of Enzymes 1975 (Shaltiel, S. ed.) pp. 37 - 43, Springer Verlag, Heidelberg.
2 Titani, K., Koide, A., Hermann, J., Ericsson, L.H, Kumor, S., Wade, R.D., Walsh, K.A., Neurath, H. & Fischer, E.H. (1977) *Proc. Natl. Acad. Sci. U.S.A.* 74, 4762 - 4766.

Prof. Dr. D. Palm, Physiologisch-Chemisches Institut, Universität Würzburg, Koellikerstr. 2., D-8700 Würzburg.

H. Pelzer

Isolierung eines Teilpeptids der Muskel-Phosphorylase b aus der Bindungsstelle des Adenosin-5'-monophosphats

Von 5'-AMP befreite Phosphorylase *b* (EC 2.4.1.1) wird durch 6-(3-Carboxy-4-nitrophenylthio)purinribosid-5'-phosphat (I) aktiviert[1]; die Aktivierung gelingt ebenso mit 6-(2,4-Dinitrophenylthio)purinribosid-5'-phosphat. Zur Synthese von I wird zunächst 5,5'-Dithiobis(2-nitrobenzoesäure) in wasserfreiem Methanol mit Natriumborhydrid reduziert. Die entstehende 5-Mercapto-2-nitrobenzoesäure wird mit isopropylidiertem 6-Chlorpurinribosid zum Nucleosid-thioäther umgesetzt, der mit radioaktivem $^{32}POCl_3$ phosphoryliert wird[1,2]. Nach Inkubation des gereinigten Enzyms mit ^{32}P-markiertem I kann der radioaktiv markierte Phosphorylase-Anteil isoliert werden: das durch enzymatische Hydrolyse der Phosphorylase *b* entstehende Peptidgemisch wird durch eine Kombination von präparativen Papierelektrophoresen, präparativer Papierchromatographie und anschließende Fingerprints aufgetrennt. Aus der erhaltenen „Peptidkarte" kann ein radioaktiv markiertes Peptid eluiert werden, welches nach saurer Hydrolyse im Analysator untersucht wird.

1 Hulla, F.W. & Fasold, H. (1972) *Biochemistry* **11**, 1056 - 1061.
2 Yoshikawa, M., Kato, T. & Takenishi, T. (1967) *Tetrahedron Lett.* **50**, 5065 - 5068.

Hermann Pelzer, Institut für Biochemie der Universität Frankfurt a. Main, Sandhofstr., D-6000 Frankfurt/71.

E. Perzborn und K. Gersonde

Reinigung und Charakterisierung der Sorbit-Dehydrogenase aus Hammel-Leber

Die Sorbit-Dehydrogenase (L-Iditol-Dehydrogenase) (EC 1.1.1.14) wurde in einer elektrophoretisch einheitlichen Form in einer Ausbeute von 18% und mit einer 1000fachen Anreicherung durch Anwendung der folgenden Trennverfahren aus Hammel-Leber isoliert: Fraktionierte Fällung mit Ammoniumsulfat oder Polymin p, Ionenaustauscher-Chromatographie auf CM-Sephadex und Affinitätschromatographie auf 5'-AMP-Sepharose 4B.

Die Sorbit-Dehydrogenase ist ein tetrameres Protein mit einem scheinbaren Molekulargewicht von 40000 pro Untereinheit. Sie enthält 8 $Zn^{2\oplus}$-Ionen/Molekül, die durch Dialyse gegen EDTA unter Verlust der enzymatischen Aktivität entfernt werden können. Durch Zugabe von $Mn^{2\oplus}$-Ionen wird die Enzym-Aktivität wieder hergestellt. NADH-Bindung verhindert den durch Metall-Komplexbildner bewirkten Aktivitätsverlust[1].

Zur Aufrechterhaltung der enzymatischen Aktivität sind 11 freie SH-Gruppen, die in der Sorbit-Dehydrogenase durch Reaktion mit 5,5'-Dithiobis(2-nitrobenzoesäure) nachgewiesen wurden, erforderlich. Diese Zahl der freien SH-Gruppen entspricht der Gesamtzahl der Cysteinreste, die mittels quantitativer Aminosäureanalyse nachgewiesen wurden. Ein Überschuß an Zink-Ionen hemmt die Enzym-Aktivität. NADH-Bindung verhindert auch die Schwermetall-Hemmung.

Das Temperaturoptimum der enzymatischen Reduktion der Fructose liegt bei 54 °C. Die Reaktion verläuft optimal bei pH 6.3. Das pH-Optimum der enzymatischen Oxidation des Sorbits liegt bei pH 10.1. Die Michaelis-Konstanten der enzymatischen Reduktion der Fructose bzw. Oxidation des Sorbits betragen: $K_{m,Fructose} = 0.46$ mol/*l*; $K_{m,NADH} = 3.95 \times 10^{-5}$ mol/*l*; $K_{m,Sorbit} = 2.51 \times 10^{-3}$ mol/*l*; $K_{m,NAD^{\oplus}} = 3.43 \times 10^{-5}$ mol/*l*. Diese Ergebnisse stehen in guter Übereinstimmung mit den in der Literatur beschriebenen Werten[2-3]. Hohe Fructose-, Sorbit- und $NAD^{\oplus}$-Konzentrationen hemmen die Enzym-Aktivität.

Die kinetischen Ergebnisse werden im Sinne eines sequenziellen Reaktionsmechanismus diskutiert.

1 Schult, R., Cullmann, W., Ohlenbusch, H.D., Kubicki, J. (1973) *Hoppe Seyler's Z. Physiol. Chem.* **354**, 212.
2 Holzer, H., Goedde, H.W. (1960) *Biochem. Biophys. Acta* **40**, 297 - 308.
3 Smith, M.G. (1962) *Biochem. J.* **83**, 135 - 144.

E. Perzborn und Prof. Dr. K. Gersonde, Abteilung Physiologische Chemie, Sektion Physikalische Chemie der Proteine, Rheinisch-Westfälische Technische Hochschule Aachen, Melatener Str. 213, D-5100 Aachen.

H.W. Peter and T. Günther

Influence of Ionic Strength on Polyamine- and Phospholipid Content of Several Marine Bacteria

In the marine bacteria MB 3, MB 28 and MB 45 we found the polyamines spermidine and putrescine in amounts of ca. 30 nmol per mg of cell protein. When the bacteria were transferred to a medium of low ionic strength (dilution 1 : 10) we observed a rapid increase in the extinction during the first 10 min, indicating a swelling of the cells caused by an influx of water. At the same time, higher amounts of putrescine and spermidine were synthesized. A possible explanation for the function of the polyamines could be that the increased polyamine concentration causes a stabilisation of the membrane and reduces passive permeability. The increase in polyamines was followed by a decrease in putrescine, and especially in spermidine. After 30 - 60 min, this decrease came to an end, and a new increase of polyamines was observed which, however, was slower than the one of the initial phase.

Further investigations were concerned with possible changes in the amounts and composition of membrane-phospholipids during growth in a medium diluted 1 : 10. In MB 3 and MB 28 we identified phosphatidylethanolamine as the main phospholipid. We found minor amounts of phosphatidylglycerol and cardiolipin and traces of lysophosphatidylethanolamine. In the strain MB 45, cardiolipin was lacking. We found that there were indeed changes in the amounts of phospholipids when the bacteria were transferred to the diluted medium. These changes were in accordance with the time-dependent increase and decrease in polyamines as described above. An explanation for this phenomenon could be that polyamines possibly are able to stimulate the synthesis of phospholipids as described by Jamdar[1] for microsomes from rat adipose tissue.

1 Jamdar, S.C. (1977) *Arch. Biochem. Biophys.* **182**, 723 - 731.

Priv.-Doz. Dr. H.W. Peter, Institut für Tierphysiologie der Freien Universität Berlin, Grunewaldstr. 34, D-1000 Berlin 41.

J. Peters und R. Peters

Bakteriorhodopsin: Inkorporation von „Brauner Membran" in Lipidvesikeln und Untersuchung des lichtabhängigen Protonentransports

„Braune Membran" wurde aus *Halobacterium halobium* R_1M_1 isoliert, dessen Retinalsynthese durch Nicotin[1] weitgehend gehemmt war (s. nachfolgendes Referat). Um zu untersuchen, ob das in der „Braunen Membran" enthaltene Bakteriorhodopsin ebenso wie das der Purpurmembran[2] in der Lage ist, Lichtenergie für den Transport von Protonen zu verwenden, wurden die folgenden Experimente durchgeführt:

1) „Braune Membran" (ohne Retinalzugabe) wurde mit Vesikeln aus gereinigtem Sojabohnenlecithin gemischt; durch Beschallung bei pH 6.0 wurden Proteoliposomen hergestellt und die Größe des lichtabhängigen Protonentransports als Funktion der Beschallungszeit gemessen; analoge Messungen wurden
2) mit „Brauner Membran" durchgeführt, die mit Retinal gesättigt war, und
3) mit Purpurmembran; in einer weiteren Meßserie wurde
4) „Braune Membran" mit Lipidvesikeln gemischt und beschallt, der Protonentransport wurde dann vor und nach Sättigung mit Retinal in derselben Probe gemessen.

Es ergab sich, daß „Braune Membran" (ohne Retinalzugabe) in Proteoliposomen eine kleine Transportkapazität für Protonen besitzt (das Maximum der pH-Veränderung bei Belichtung entsprach der Aufnahme von 10 - 20 neq $H^{\oplus}$/mg Opsin). Nach Sättigung mit Retinal hat „Braune Membran" etwa die gleiche Transportkapazität wie Purpurmembran (80 - 100 neq $H^{\oplus}$/mg Bakteriorhodopsin). Werden „Braune Membran" und Liposomen erst beschallt und dann mit Retinal gesättigt, beträgt die maximale Transportkapazität etwa 40 - 50 neq $H^{\oplus}$/mg Opsin. Wir schließen aus diesen Versuchen, daß Bakteriorhodopsin auch in der Form, wie es in der „Braunen Membran" vorliegt, Protonen transportieren kann. Die Anordnung des Bakteriorhodopsins in einem Kristallgitter, wie es in der Purpurmembran vorliegt, scheint keine Voraussetzung für diese Funktion zu sein.

1 Sumper, M., Reitmeier, H. & Oesterhelt, D. (1976) *Angew. Chem.* **88**, 203 - 210.
2 Racker, E. & Stoeckenius, W. (1974) *J. Biol. Chem.* **249**, 662 - 663.

J. Peters und R. Peters, Gustav-Embden-Zentrum der Biologischen Chemie, Abteilung für Mikrobiologische Chemie, Klinikum der Universität Frankfurt am Main, Theodor-Stern-Kai 7, D-6000 Frankfurt 70.

R. Peters und J. Peters

Bakteriorhodopsin: Hell-Dunkel-Adaptation der „Braunen Membran"

Sumper et al.[1] haben kürzlich eine Fraktion der Zellmembran von Halobakterien beschrieben, die sie „Braune Membran" nannten. Die „Braune Membran" wird als biosynthetische Vorstufe der Purpurmembran betrachtet und enthält (neben anderen Bestandteilen) Bakteriorhodopsin. Wir haben die Hell-Dunkel-Adaptation des Bakteriorhodopsins in „Brauner Membran" untersucht und sie mit der des Bakteriorhodopsins in Purpurmembran verglichen. Hierzu wurde „Braune Membran" aus *Halobacterium halobium* R_1M_1 isoliert, dessen Retinalsynthese durch Nicotinzugabe (1 mM) gehemmt war. Es zeigte sich, daß die Hemmung der Retinalsynthese inkomplett war und daß ca. 5% des gesamten Bakterio-Opsins der „Braunen Membran" mit Retinal kombiniert waren und als Bakteriorhodopsin vorlagen. Die Hell-Dunkel-Adaptation verläuft qualitativ ähnlich in Purpurmembran und „Brauner Membran". Quantitativ bestehen jedoch große Unterschiede. Bei der Purpurmembran verläuft der Übergang von der hell- zur dunkeladaptierten Form mit einer Halbwertszeit von $t_{1/2}$ = 41 min (28 °C), die Kinetik entspricht der einer Reaktion 1. Ordnung. Bei „Brauner Membran" ist $t_{1/2}$ = 125 s vor Retinalzugabe und $t_{1/2}$ = 7 min nach kompletter Sättigung mit Retinal. Vor Retinalzugabe zeigt die Reaktion einen Verlauf 1. Ordnung, nach Retinalsättigung liegt eine höhere Ordnung vor. Wir schließen daraus, daß in „Brauner Membran" nicht eine einzige, sondern mehrere Bakteriorhodopsin-Species vorliegen, die sich in ihrer Affinität zu Retinal und in photochemischen Eigenschaften unterscheiden. Bei der Hell-Dunkel-Adaptation des Bakteriorhodopsins ist eine *trans-cis*-Isomerisierung des Retinal-Chromophors beteiligt[2]. Die physiologische Bedeutung dieses Prozesses ist noch unbekannt. Wir erwarten deshalb, daß weitere vergleichende Untersuchungen der Photochemie des Bakteriorhodopsins in „Brauner Membran" und Purpurmembran wichtige Aufschlüsse über die Photophysiologie von Halobakterien ergeben werden.

1 Sumper, M., Reitmeier, H. & Oesterhelt, D. (1976) *Angew. Chem.* **88**, 203 - 210.
2 Oesterhelt, D., Meentzen, M. & Schumann, L. (1973) *Eur. J. Biochem.* **40**, 453 - 463.

R. Peters und J. Peters, Gustav-Embden-Zentrum der Biologischen Chemie, Abteilung für Mikrobiologische Chemie, Klinikum der Universität Frankfurt am Main, Theodor-Stern-Kai 7, D-6000 Frankfurt 70.

G. Pfleiderer, M. Baier, M. Boll, H. Ensinger, E. Mössner and H. E. Pauly

New Investigations of the Structure-Function Relationship of Mammalian Alkaline Phosphatase

Many contradictory results have been published about mammalian alkaline phosphatases (EC 3.1.3.1), concerning specific activity, molecular weight, number of subunits, zinc content, etc.

We have developed modified isolation procedures to obtain very highly active enzyme preparations, especially from kidney. One step is the intensive incubation with free neuraminidase in order to reduce the number of multiple forms. With pure preparations from calf liver, kidney and intestine we have re-investigated various biochemical, physical and immunological properties in order to clear some discrepancies in the literature, like molecular weight by means of gel filtration and ultracentrifugation, number of subunits in sodium dodecylsulfate electrophoresis, amino acid analysis, zinc content by atomic absorption spectroscopy and the question whether real isozymes exist. Some contradictory results in the literature are caused by ignoring the influence of buffer ions on the catalytic activity. In some cases, a time dependent increase in specific activity is observed. Another aspect of our study concerned the function of the carbohydrate moiety, which is characteristic for this and other membrane-bound enzymes, like γ-glutamylcyclotransferase (EC 2.3.2.4) or alanine amino peptidase (EC 3.4.11.2), and which protects the enzymes against proteolytic attack. The importance of zinc ions for structural and catalytic effectiveness was also studied. A detailed report will be given by Ensinger in a short communication. Finally we have summarized the current concept of the structure and function of mammalian alkaline phosphatases.

Professor Dr. G. Pfleiderer, Institut für Organische Chemie, Biochemie und Isotopenforschung, Universität Stuttgart, Pfaffenwaldring 55, D-7000 Stuttgart 80.

C. Pietrzyk and G. Werner

Stimulation of the $Na^{\oplus}$,$K^{\oplus}$-ATPase concomitant with the uptake of amino acids in Ehrlich cells

Ehrlich ascites tumor cells accumulate amino acids by co-transport with $Na^{\oplus}$[1]; the electrochemical potential gradient of $Na^{\oplus}$ energizes this uptake[2]. Metabolizing cells show little change in $Na^{\oplus}$ content as they take up amino acid and with it by way of co-transport $Na^{\oplus}$, while in cells with inhibited $Na^{\oplus}$,$K^{\oplus}$ pump, the uptake of $Na^{\oplus}$ co-transported with amino acid is reflected by the increased $Na^{\oplus}$ content[3]. A stimulation of the $Na^{\oplus}$,$K^{\oplus}$ pump, known to be activated by internal $Na^{\oplus}$[4], might be responsible for the constant low $Na^{\oplus}$ content of uninhibited cells.

The ouabain-inhibitable $^{86}Rb^{\oplus}$ influx, as a measure of the activity of the $Na^{\oplus}$,$K^{\oplus}$ pump, was stimulated by addition of amino acid. Addition (10mM) of glycine, L-proline, L-2-aminobutyric acid, L-serine, L-histidine increased the $^{86}Rb^{\oplus}$ uptake by about 50%, 2-aminoisobutyric acid, L-alanine, L-valine, L-leucine, L-methionine by about 30%, while D-alanine, D-methionine, D-serine, L-arginine, L-lysine, L-glutamate, L-tryptophan and 3-aminobutyric acid showed no effect on the $Rb^{\oplus}$ uptake. The stimulation was concentration-dependent. For glycine the half-maximally effective concentration was found to be about 3mM, which is close to the K_M reported for glycine uptake[5]

This stimulation of the $Na^{\oplus}$,$K^{\oplus}$ pump should increase ATP consumption. No differences in metabolic activity have been reported between cells transporting amino acid and controls not transporting[6]. Upon inhibition of oxidative phosphorylation and addition of glucose, cells taking up amino acids showed an increase in the small fraction of lactate production, which was inhibitable by ouabain and was attributed, therefore, to the $Na^{\oplus}$,$K^{\oplus}$ pump, while the total lactate production remained unchanged.

The mechanism, by which the $Na^{\oplus}$,$K^{\oplus}$-ATPase is stimulated, is presently not known. Three possibilities can be considered:

1) a direct stimulation of the enzyme by the external amino acid,
2) an increase in the internal $Na^{\oplus}$ available to the pump,
3) the depolarization of the membrane upon $Na^{\oplus}$ dependent amino acid transport.

1 Christensen, H.N. (1970) in Membranes and Ion Transport, (Bittar, E.E., ed.) vol. 1, pp. 365 - 394, Wiley-Interscience, New York.
2 Heinz, E., Geck, P. & Pietrzyk, C. (1975) *Ann. N. Y. Acad. Sci.* **264**, 428 - 441.
3 Eddy, A.A. (1968) *Biochem. J.* **108**, 195 - 206.
4 Post, R.L., Merrit, C.R., Kinsolving, C.R. & Albright, C.D. (1960) *J. Biol. Chem.* **235**, 1796 - 1802.
5 Eddy, A.A., Mulcahy, M.F. & Thompson, P.J. (1967) *Biochem. J.* **103**, 863 - 876.
6 Geck, P., Heinz, E. & Pfeiffer, B. (1974) *Biochim. Biophys. Acta* **339**, 419 - 425.

C. Pietrzyk, Gustav-Embden-Zentrum der Biologischen Chemie, Abteilung für Physikalische Biochemie, Klinikum der Universität Frankfurt am Main, Theodor-Stern-Kai 7, D-6000 Frankfurt 70.

I. von Pusch, N. Paul und W. Wesemann

Veränderungen im Adeninnucleotidstoffwechsel menschlicher Thrombozyten durch Induktion der Release-Reaktion

Um die Release-Reaktion zu verfolgen, bei der aus den Speichergranula der Thrombozyten hauptsächlich Serotonin, ATP und ADP unter Energieverbrauch freigesetzt werden[1], wurden menschliche Blutplättchen mit radioaktiv markiertem Serotonin und Adenin inkubiert. Das in die Thrombozyten inkorporierte Serotonin fand sich nach Dichtegradientenzentrifugation der lysierten Zellen überwiegend in den osmiophilen Speichergranula, Adenin vornehmlich im zytoplasmatischen Raum. Diese Verteilung wurde durch autoradiographische Untersuchungen bestätigt. Durch elektrophoretische Auftrennung wurde gezeigt, daß [^{14}C]Adenin in ATP (33%), ADP (15%), IMP (5%), AMP (1%) inkorporiert bzw. zu Hypoxanthin/Inosin (11%) metabolisiert wird.

Im Gegensatz zu Holmsen und Mitarb., die durch Thrombin, Kollagen[2], ADP und Adrenalin[3] eine Release-Reaktion mit nachfolgender Aggregation auslösten, wurden in eigenen Versuchen Reaktionsbedingungen und Substanzen so gewählt, daß der Energiestoffwechsel nur während der Release-Reaktion untersucht werden konnte. Hierbei nahm die Radioaktivität in den zuvor mit [^{3}H]Adenin inkubierten Blutplättchen ab und zwar stärker nach Induktion der vollständigen Release-Reaktion (Freisetzung von ATP, ADP und Serotonin durch Reserpin, Imipramin, 1-Amino-3,5-dimethyladamantan) als nach Inkubation mit Substanzen, die nur Serotonin ausschütten (Tyramin, Tryptamin, 1-Aminoadamantan). Im Nucleotid-/Nucleosidmuster waren im Vergleich zur Kontrolle radioaktiv markiertes ATP und ADP vermindert, die Monophosphate, sowie Hypoxanthin/Inosin und Adenin vermehrt. Nach Ablauf der Release-Reaktion wird verstärkt Adenin aufgenommen und vor allem in ATP und ADP inkorporiert.

Die Ergebnisse legen nahe, daß auch bei der Release-Reaktion ohne nachfolgende Aggregation Energie in Form von ATP verbraucht wird, das anschließend durch vermehrte Aufnahme der Vorstufe Adenin und erhöhte Synthese regeneriert wird.

1 Holmsen, H. (1975) in Biochemistry and Pharmacology of Platelets (Elliott, K. & Knight, J., eds.), Ciba Found. Sympos. 35, new series, pp. 175 - 196, Elsevier/Excerpta Medica/North Holland, Amsterdam, Oxford, New York.
2 Holmsen, H., Day, H.J. & Storm, E. (1969) *Biochim. Biophys. Acta* **186**, 254 - 266.
3 Holmsen, H., Day, H.J. & Setkowsky, C.A. (1972) *Biochem. J.* **129**, 67 - 82.

W. Wesemann, Institut für Physiologische Chemie II, Universität Marburg, Lahnberge, D-3550 Marburg/Lahn.

H. Rahmann, R. Hilbig und H. Rösner

Hirngangliosid-Muster bei Vertebraten und deren Temperatur-abhängige Variabilität

Die Adaptation von Vertebraten an Änderungen der Außentemperatur soll aufgrund elektrophysiologischer Befunde mit funktionellen Veränderungen im Nervensystem, speziell im Bereich der Synapsen korreliert sein[1,2]. Der molekulare Mechanismus der Temperatur-Adaptation ist bislang noch unbekannt, doch wird vermutet, daß er möglicherweise auf biochemischen Änderungen von spezifischen Membranstrukturen beruht. Es stellt sich die Frage, ob neuronale Ganglioside, denen in zunehmendem Maße funktionelle Bedeutung am neuronalen Geschehen zugesprochen wird[3], wegen ihrer 10fach höheren Konzentration im Synapsenbereich am Prozeß der Temperatur-Adaptation beteiligt sein könnten. Eine Zusammenstellung bisher beschriebener Hirngangliosid-Muster verschiedenster Vertebraten-Species ermöglicht keine Katalogisierung etwa im Sinne einer Chemotaxonomie dieser Sialo-Glykolipide. Erst unter Heranziehung der Temperatur als wichtigstem Ökofaktor lassen sich Korrelationen zwischen der Ausgestaltung neuronaler Gangliosid-Muster und der ökophysiologisch-systematischen Stellung der Vertebraten aufzeigen:

1) Bei *homoiothermen adulten Vertebraten* (mit Thermoregulation: Vögel, Säuger) besteht das Hirngangliosid-Muster aus wenig polaren Fraktionen mit eindeutiger Dominanz des Disialogangliosides $II^3(NeuAc)_2$-$GgOse_4Cer$ ($\hat{=}$ G_{D1a}).

2) *Heterotherme Phasen von Homoiothermen* (Thermoregulation noch nicht voll ausgeprägt):
a) während der *perinatalen Entwicklung:* Beispiel Maus bzw. Küken;
b) während des *Winterschlafs:* (Beispiel Goldhamster) sind charakterisiert durch ein temporär polareres Muster der Hirnganglioside.
3) Bei *poikilothermen Vertebraten* (Beispiel: versch. Teleosteer, Amphibien) ist die Adaptation an erniedrigte Außentemperaturen im Falle
a) der *Akklimatisation* (Adaptation an saisonale Temperatur-Änderungen) sowie
b) auch der *Akklimation* (experimentell ausgelöste Temperatur-Änderungen) gekennzeichnet durch eine reversible „Poly-Sialisierung" der Hirnganglioside.

Erste Untersuchungen der Frage, inwieweit die aufgezeigten Hirngangliosid-Musterausprägungen einen molekularen Mechanismus zur Temperatur-Adaptation darstellen, werden im Referat von Rösner et al.[4] angesprochen.

1 Hazel, J., Prosser, C. L. (1974) *Physiol. Rev.* **54**, 620 - 677.
2 Lagerspetz, K. Y. H. (1974) *Biol. Rev.* **49**, 477 - 514.
3 Rahmann, H., Rösner, H. & Breer, H. (1976) *J. Theor. Biol.* **57**, 231 - 237.
4 Rösner, H., Probst, W. & Rahmann, H. (1978) *Hoppe-Seyler's Z. Physiol. Chem.* **359**, 311.

Prof. Dr. H. Rahmann, Dr. R. Hilbig und Priv. Doz. Dr. H. Rösner, Institut für Zoologie der Universität Hohenheim, Postfach, D-7000 Stuttgart 70.

M. Reddington and E. Mehl

The Phosphorylation of Synaptic Membrane Proteins by Endogenous cAMP-Stimulated Protein Kinase Activity in Different Regions of Rat Brain

It is now well established that synaptosomal plasma membrane fractions from mammalian brain exhibit cAMP-stimulated protein kinase activity which phosphorylates a number of endogenous synaptic membrane proteins in vitro[1-3]. Since cAMP synthesis in nervous tissue can be stimulated by a number of putative neurotransmitter substances[4] it is reasonable to suppose that as least some of the effects of neurotransmitters might be mediated through the phosphorylation of synaptic membrane proteins. However, virtually nothing is known about the nature of the synaptic phosphoproteins. One of a number of possibilities would be that the neurotransmitter receptors are themselves phosphorylated[5,6]. We have therefore examined by polyacrylamide gel electrophoresis the distribution of substrates for cAMP-stimulated protein phosphorylation in synaptosomal membranes from various brain regions containing different populations of neurotransmitter receptors, with the following results:

1) Synaptosomal membrane fractions from whole brain contain at least ten proteins (molecular weights 47 000 and above) whose phosphorylation is stimulated by cAMP, although the phosphorylation pattern obtained was dependent upon the incubation conditions used. None of the synaptosomal membrane phosphoproteins were present in myelin or mitochondria, demonstrating that they are not derived from contamination of the synaptosomal membranes by these fractions.

2) No differences in phosphoprotein pattern were observed in synaptosomal membranes from cerebral cortex, corpus striatum (enriched in dopamine receptors) and cerebellum (lacking high affinity LSD-serotonin receptors). It is therefore unlikely that any of the main phosphoproteins detected under our conditions represent proteins associated with the presence of these specific neurotransmitter receptors.

1 Weller, M. & Rodnight, R. (1970) *Nature (London)* **225**, 187 - 188.
2 Ueda, T., Maeno, H. Greengard, P. (1973) *J. Biol. Chem.* **248**, 8295 - 8305.
3 Routtenberg, A. & Ehrlich, Y. (1975) *Brain Res.* **92**, 415 - 430.
4 Bloom, F. E. (1975) *Rev. Physiol. Biochem. Pharmacol.* **74**, 1 - 103.
5 Gordon, A. S., Davis, C. G., Milfay, D. & Diamond, I. (1977) *Nature (London)* **267**, 539 - 540.
6 Teichberg, V. I. & Changeux, J.-P. (1977) *FEBS Lett.* **74**, 71 - 76.

Dr. M. Reddington, Max-Planck-Institut für Psychiatrie, Kraepelinstr. 2, D-8000 München 40.

K. Reichlmeier, K. Citherlet and M. Ermini

Some Aspects of the Protein Kinase System in the Ageing Human Brain Cortex

Earlier studies have shown an age-dependent decrease in cAMP-activated protein kinase (EC 2.7.1.37) activity in human brain cortex as measured at an optimal cAMP concentration in vitro of 5 μM[1,2]. It was therefore of interest to perform more detailed investigations. Significantly less cAMP-dependent protein kinase activity was found in the brain cortex of old persons (> 70 years, $n = 6$) than in younger subjects (< 50 years, $n = 6$) over a broad range of cAMP concentrations (0.05 - 50.0 μM). This effect was significant for most of the cAMP concentrations used.

The effect of age on the cAMP-activated protein kinase was observed with several substrates such as different histones, protamine, and endogenous substrate contained in the homogenate. Calf thymus histone was subject to the highest degree of phosphorylation, whereas casein and albumin were not phosphorylated.

In different subcellular fractions, the effect of age was found to be associated with the particulate enzyme. In the crude nuclear fraction, there was a clear, age-related decrease in cAMP-activated protein kinase, which disappeared after separation into neuronal and glial nuclei. Preliminary experiments show an age-dependent increase in cAMP-activated protein kinase activity in the soluble fraction after sonication and centrifugation of the original homogenate, and a diminished activity in the remaining pellet. From this observation, we assume a shift of the particulate enzyme to a soluble form with increasing age, but this idea has to be proved in further experiments.

1 Reichlmeier, K., Schlecht, H.P. & Iwangoff, P. (1977) *Experientia* **33**, 798.
2 Reichlmeier, K., Ermini, M. & Schlecht, H.P. (1978) *Aktuel. Gerontol.*, in press.

Dr. K. Reichlmeier, Basic Med. Res., Sandoz Ltd., CH-4002 Basel.
Dr. M. Ermini, Institut für Pharmakologie und Biochemie, Veterinär-Med. Fakultät der Universität Zürich, CH-8057 Zürich.

K. Ring, P. van Dijck, J. Op den Kamp und J. de Gier

Korrelationen zwischen Wachstumstemperatur-induzierten Veränderungen der Membranpermeabilität, des Lipidmusters und thermotropen Lipid-Phasenübergängen bei Streptomyces hydrogenans

Durch Herabsetzen der Inkubationstemperatur von 30 °C auf 18 °C wird die Zellmembranpermeabilität bei *Streptomyces hydrogenans* und anderen Streptomyceten drastisch erhöht, so daß niedermolekulare, hydrophile Substanzen die Membran in beiden Richtungen rasch penetrieren können[1]. Der für diese Permeabilitätserhöhungen kritische Temperaturbereich hängt von der Züchtungstemperatur der Zellen ab. Herabsetzung der Züchtungstemperatur verschiebt den kritischen Temperaturbereich zu niedrigeren Temperaturen hin[2]. Die Erhöhung der Membranpermeabilität ist reversibel; durch Wiedererwärmen der Zellen auf T > 20 °C erhält die Membran ihre ursprüngliche Barriere-Funktion wieder zurück. Exogene Faktoren, wie die Protonen-, Calcium- oder Magnesiumkonzentration, sowie die Ionenstärke verändern die Temperatursensitivität der Membran signifikant.

Untersuchungen der Membranlipide von *Streptomyces hydrogenans* ergaben einen hohen Gehalt an verzweigtkettigen Fettsäuren der iso- und anteiso-Serie; monoungesättigte Fettsäuren sind nur in geringer Menge vorhanden. Unter den Phospholipiden dominiert mit 70 - 85% Phosphatidyläthanolamin; daneben sind vor allem Cardiolipin (10 - 20%) und Phosphatidylinosit (2 - 5%) vorhanden. Änderung der Züchtungstemperatur verändert sowohl das Phospholipidmuster als auch das Spektrum der Fettsäuren. Differentialcalorimetrische Untersuchungen an isolierten Lipidfraktionen zeigten, daß scharfe thermotrope Phasenübergänge im gleichen Temperaturbereich erfolgen, in dem auch die Änderungen der Membranpermeabilität zu beobachten sind. Durch Erniedrigung der Züchtungstemperatur der Zellen wird auch die Phasenübergangstemperatur herabgesetzt; gleichzeitig ist eine Verminderung der Enthalpieänderung festzustellen. Wie bisher nur bei tierischen Phospholipiden beschrieben[3], verhindert Cholesterin in genügend hoher Konzentration (ca. 25 mol-%) thermotrope Phasenübergänge auch bei diesen Lipiden vollständig.

1 Ring, K. (1965) *Biochem. Biophys. Res. Commun.* **19**, 576 - 581.
2 Ring, K. (1965) *Biochim. Biophys. Acta* **94**, 598 - 600.
3 Demel, R.A., Jansen, J., van Dijck, P., van Deenen, L.L.M. (1977) *Biochim. Biophys. Acta* **465**, 1 - 10.

Klaus Ring, Biochemisch Laboratorium, Rijksuniversiteit Utrecht/Niederlande und Gustav-Embden-Zentrum der Biologischen Chemie, Abteilung für Mikrobiologische Chemie, Klinikum der Universität Frankfurt am Main, Theodor-Stern-Kai 7, D-6000 Frankfurt 70.

K. Ring, B. Foit-Minuth, Ch. Meyer und H. Fasold

Hemmung der Nucleosidaufnahme in Ehrlich-Maus-Ascites-Tumorzellen durch 6-(2,4-Dinitrophenylthio)-purinribosid

6-(2,4-Dinitrophenylthio)purinribosid (I) ist ein spezifischer und effizienter Hemmstoff des Nucleosidtransports in Ehrlich-Maus-Ascites-Tumorzellen. Trotz seines Puringrundgerüstes hemmt es den Transport der Pyrimidinnucleoside stärker als den der Purinnucleoside. Die Wirkung ist nach kurzfristiger Inkubation der Zellen bei pH 7.4 kompetitiv, wobei der Cytidin-Influx erheblich stärker gehemmt wird als der des Uridins. Die Inhibitorkonstanten für den Uridintransport hängen vom pH-Wert und der Inkubationstemperatur ab; sie liegen im Bereich von 40 - 120μM. Beim Cytidintransport beträgt K_i für das wenig affine System ca. 100μM, für das hochaffine System ca. 7μM.

Eine Reihe weiterer kinetischer Befunde läßt darauf schließen, daß I nicht nur das für die Cytidintranslokation verantwortliche Zentrum am Transportsystem besetzt, sondern ebenfalls – wenn auch mit geringer Affinität – ein regulatorisches, dessen Besetzung die Aktivität des Systems vermindert. Dieses regulatorische Zentrum hat offenbar eine reaktive SH-Gruppe, die mit *p*-Chlormercuriphenylsulfonat umgesetzt werden kann. Gleichzeitige Inkubation der Zellen mit I und *p*-Chlormercuriphenylsulfonat führt zu einer Vervielfachung der Transporthemmung. Dies mag darauf hindeuten, daß durch Anlagerung von I an das Transportzentrum das Transportprotein eine Konformation erhält, die die rasche und vollständige Umsetzung des regulatorischen Zentrums mit *p*-Chlormercuriphenylsulfonat begünstigt.

Durch Änderung der Inkubationsbedingungen (Erhöhung des pH-Wertes auf 8 und Verlängerung der Inkubationsdauer) wird die bei pH 7.4 reversible Hemmung durch I praktisch vollständig irreversibel. Wir nehmen an, daß unter diesen Bedingungen I kovalent an sein Rezeptorprotein gebunden wird.

Klaus Ring, Gustav-Embden-Zentrum der Biologischen Chemie, Abteilung für Mikrobioloigsche Chemie, Klinikum der Universität Frankfurt am Main, Theordor-Stern-Kai 7, D-6000 Frankfurt 70;
Hugo Fasold, Biochemisches Institut der Universität Frankfurt.

H. Rink

Water-Soluble Proteins of Lens Epithelial Cells during Ageing in vitro

The lens, normally characterized by its transparency, contains some lens-specific proteins, called α-, β- and γ-crystallins. γ-Crystallin is said to be synthesized if lens epithelial cells become differentiated to lens fiber cells. This event happens in vivo behind the equatorial zone, when the nucleus disappears and some kind of mRNA becomes stabilized[1]. The elongated fiber cells form the mass of the lens body.

Bovine lens epithelial cells were cultured in vitro (the procedure will be described) in order to check the definite lifespan of diploid mammalian cells according to Hayflick[2] and to investigate the mechanisms involved[3]. The system is described by the possible generation numbers per day, as well as by the cumulative cell numbers, the occurence of the cell types I, II and III and the alteration of this pattern during ageing. The crystallins were analyzed and characterized by means of isoelectric focussing[4]. The occurence of γ-crystallin in older cultures is discussed as a biochemical marker, indicating that the in-vitro ageing of bovine lens epithelial cells is due to a genetic program working in vitro as well as in vivo.

1 Papaconstantinou, J. (1967) *Science* **56**, 338 - 346.
2 Hayflick, L. (1975) *Fed. Proc.* **34**, 9 - 13.
3 Bayreuther, K. (1975) *Verh. Dtsch. Ges. Pathol.* **59**, 101 - 108.
4 Bours, J. (1975) in Progress in Isoelectricfocussing and Isotachophoresis (Rhigetti, P.G., ed.) pp. 235 - 256, North Holland Publ., Amsterdam.

Priv.-Doz. Dr. Hermann Rink, Institut für Biophysik, Venusberg, Sigmund-Freud-Str. 15, D-5300 Bonn 1

H. Rink and H.-J. Partke

Alterations of the Protein Synthesis in X-Irradiated Yeast Cells (Saccharomyces uvarum)

X-irradiation is known to inhibit the colony-forming ability. Obviously the most sensitive target is the DNA molecule. But DNA lesions are randomly spread in the various DNA molecules of the irradiated population. Thus each damaged cell may give rise to more or less severe changes in the newly synthesized proteins. There is no reason for the assumption that the one or the other altered protein – occuring in one single cell – will be accumulated in the whole population. Thus there may be many different causes, all of them based on altered proteins, for the observable endpoint.

Under defined growth conditions (limitation of the C-source), X-irradiated yeast cells (100 krad) grow to giant cells[1] in which the DNA content remains constant (no polyploidy), whereas the RNA and protein contents as well as the average fresh weight of a cell increases 2-fold. In this special case it was of interest to check the rate of protein synthesis by pulse label experiments with [^{14}C]leucine and to analyze the spectra of the synthesized proteins by means of isoelectricfocussing[2].

The results show that, depending on the culture time, the rate of protein synthesis is greater than the control values. The characterization of the water-soluble proteins show some small quantitative differences, but in addition, one protein band is lacking at distinct times in culture in the X-irradiated samples. If no cell division takes place (lag-phase, stationary phase), the same protein band is also lacking in the control samples, This finding indicates that a special protein which is engaged in the cell division mechanism is lost in yeast cells growing in suspension culture after X-irradiation.

1 Rink, H., Partke, H.-J. (1975) *Radiat. Environ. Biophys.* **12**, 119 - 125.
2 Bours, J. (1975) in Progress in Isoelectric Focusing and Isotachophoresis (Rhigetti, P.G. ed.) pp. pp. 235 - 256, North Holland Publ., Amsterdam.

Priv.-Doz. Dr. Hermann Rink, Institut für Biophysik, Venusberg, Sigmund-Freud-Str. 15, D-5300 Bonn 1.

K. Rittinghaus, U. Schwuléra and K. Dose

Mechanism of Substrate Conversion at the Active Site of an F_0-F_1-ATPase from Micrococcus luteus ATCC 4698

The present investigations have been carried out with an F_0-F_1-ATPase complex (EC 3.6.1.3) from *Micrococcus luteus* solubilized with Triton X-100 as described elsewhere[1,2].

In order to obtain information about the function of $Mg^{2\oplus}$ ions in ATP hydrolysis, the concentration of free $Mg^{2\oplus}$ was varied at several fixed ATP concentrations and vice versa. At low concentrations of $Mg^{2\oplus}$ or ATP, the ATPase activity is increased, whereas higher concentrations inhibit the activity of the complex. An excess of free $Mg^{2\oplus}$ or free ATP decreases the reaction velocity. MgATP binds more readily to the enzyme than ATP. In the presence of $Mg^{2\oplus}$-free ATP, no F_0-F_1-ATPase activity is observed. The K_m value for MgATP is about 0.16mM.
A Hill-coefficient for MgATP of 1.0 indicates that one molecule MgATP is bound per enzyme molecule.

The kinetic results show that MgATP is the primary substrate in the reaction and that free $Mg^{2\oplus}$ or ATP are competitive inhibitors of the F_0-F_1-complex.

Only the enzyme-substrate-$Mg^{2\oplus}$ complex undergoes conversion into the end products.

1 Rittinghaus, K., Schmitt, M., Schwuléra, U. & Dose, K. (1977) *Hoppe-Seyler's Z. Physiol. Chem.* **358**, 1265.
2 Schmitt, M., Rittinghaus, K., Scheurich, P., Schwuléra, U. & Dose, K. (1978) *Biochem. Biophys. Acta*, in press.

Prof. Dr. K. Dose, Institut für Biochemie der Universität Mainz, Joh.-Joachim-Becher-Weg 30, D-6500 Mainz.

H. Rösner, W. Probst und H. Rahmann

Interaktionen zwischen neuronalen Gangliosiden und Kationen und deren Temperaturabhängigkeit

Veränderungen des Hirngangliosidmusters im Verlauf der Akklimatisation sowie Akklimation poikilothermer Vertebraten an unterschiedliche Temperaturen[1,2] führten zu der Frage, ob diese Veränderungen einen Mechanismus einer molekularen Adaptation, vor allem im Hinblick auf $Ca^{2\oplus}$-abhängige Prozesse an neuronalen Membranen, einschließlich der synaptischen Transmission, darstellen könnten. Es wurden daher in vitro die Interaktionen von Gangliosiden mit $Ca^{2\oplus}$ sowie weiteren Kationen, und deren Temperaturabhängigkeit *indirekt* anhand der gangliosidinduzierten Toluidinblau-Metachromasie[3] (a) und *direkt* durch ionenselektive Bindungsanalysen (b) untersucht.

a) Kationen vermindern die durch Gangliosidmicellen induzierte Metachromasie von Toluidinblau in der Sequenz: $Ca^{2\oplus} > Mg^{2\oplus} > K^{\oplus} = Na^{\oplus}$, wobei $K^{\oplus}$ und $Na^{\oplus}$ im Gegensatz zu $Ca^{2\oplus}$ und $Mg^{2\oplus}$ erst in „unphysiologischen" Konzentrationen (> 5mM) wirken. Die bzgl. $Ca^{2\oplus}$ untersuchte Temperaturabhängigkeit dieses ionalen Effektes läßt auf eine erheblich verminderte $Ca^{2\oplus}$-Wirkung bei abnehmender Temperatur (50 °C bis 4 °C) schließen.
b) Im Gegensatz zur freien *N*-Acetylneuraminsäure (NeuAc) sowie De-Sphingo-Gangliosiden (DeS-G) ändert sich die $Ca^{2\oplus}$-Affinität heteropolarer Gangliosidmicellen (0.1mM Gangl.-NeuAc) in Abhängigkeit von der $Ca^{2\oplus}$-Konzentration. Die Temperaturabhängigkeit der Interaktion von $Ca^{2\oplus}$ mit heteropolaren Gemischen, Gangliosideinzelfraktionen sowie DeS-G wird anhand isomolarer Testansätze von Gangl.-NeuAc-$Ca^{2\oplus}$ aufgezeigt.

Die in vitro gefundene ausgeprägte Temperatursensibilität der Interaktion von Gangliosiden mit $Ca^{2\oplus}$ sowie diesbezügliche Unterschiede einzelner Gangliosidfraktionen unterstützen die Annahme, daß den in vivo beobachteten Veränderungen im Hirngangliosidmuster[1] im Verlauf der Akklimatisation und Akklimation funktionelle Bedeutung als *ein* molekularer Mechanismus der neuronalen Temperaturadaptation zukommt.

1 Rahmann, H., Hilbig, R. & Rösner, H. (1978) *Hoppe-Seyler's Z. Physiol. Chem.* **359**, 307 - 308.
2 Rahmann, H. (1978) *J. Jap. Exp. Med.*, in press.
3 Sylvén, B. (1954) *Q. J. Microsc. Sci.* **95**, 327 - 358.

Priv.-Doz. Dr. H. Rösner, W. Probst und Prof. Dr. H. Rahmann, Institut für Zoologie der Universität Hohenheim, Postfach, D-7000 Stuttgart 70.

W. Rommel

Synthese photoaktiver Enkephalin-Analoga

Die Wechselbeziehungen der endogenen Morphine wie Endorphin und Enkephalin mit Opiatrezeptoren beruhen in vivo notwendigerweise auf leicht spaltbaren Bindungen. Dies erschwert jedoch die Lokalisation und Identifikation der Rezeptoren erheblich, so daß Möglichkeiten gesucht wurden, diese Reagenzien kovalent zu binden. Hier bietet sich der Einbau von photoaktiven Substanzen in Enkephalin-Analoga besonders an, da sie im Dunkeln chemisch inert sind und dadurch unter Standardbedingungen getestet werden können.

Von uns wurden verschiedene Enkephalin-Analoga mit wechselnder Aminosäure-Sequenz mit folgendem Ziel synthetisiert:

1) Bei erhaltener biologischer Aktivität Aminosäuren in Enkephalin einzuführen, die sich als Photoaffinitätsmarkierung einsetzen lassen.
2) Erreichen einer hohen spezifischen Aktivität (20 Ci/mmol) bei radioaktiver Markierung, da die optimale Konzentration des Reagenzes am Rezeptor bei 10^{-8} mol liegt.
3) Kopplung der photoaktiven Analoga über einen hydrophilen Spacer an Fluoreszenzfarbstoffe.

Bei der Synthese der Analoga wurden die Positionen Arg^{60}-Tyr^{61}-Gly^{62}, Phe^{64}-Leu^{65}-Thr^{66} variiert und die Position Tyr^{61}, Phe^{64}, Leu^{65}, Thr^{66} durch *p*-Azidophenylalanin (AzPhe) bzw. Gly-AzPhe ersetzt und auf biologische Aktivität getestet. Die hohe spezifische Radioaktivität wurde durch selektive Acetylierung der ϵ-Aminogruppe des Lysins erhalten. Die Fluoreszenzmarkierung wurde mit Hilfe von $(Pro)_6$-Asp als Spacer an 2-(*p*-Benzoesäurehydrazid)-5-phenyloxazol erreicht.

Wolfgang Rommel, Institut für Biochemie der Universität Frankfurt, Sandhofstraße, Gebäude 75 A, D-6000 Frankfurt/M.-Niederrad 1.

G. Rosenfelder und P. F. Mühlradt

Radioaktive Glykosphingolipide als biochemische Marker von Lymphozyten-Populationen

[1-^{14}C]Galaktose wird in vitro in die Plasmamembran Concanavalin-A-stimulierter Mausthymozyten eingebaut[1] und markiert dort neben Glykoproteinen auch Glykosphingolipide[2]. Diese Markierungsmethode ermöglicht prinzipiell die selektive Untersuchung der Glykosphingolipide bestimmter Immunzellpopulationen, soweit diese sich entweder selektiv stimulieren oder rein gewinnen lassen. Zur selektiven Stimulierung von T-Zell-Populationen dienten Phytohämagglutinin und Concanavalin A, zur B-Zell-Stimulierung *Salmonella minnesota* R595-Lipopolysaccharid oder *E.-coli*-Lipoprotein. Zum Vergleich wurden Milz-T-Zellen mit Nylonsäulen gewonnen und stimuliert oder aus Milzen athymischer nu/nu-Mäuse B-Zellen markiert. Ein Vergleich der Neutralglykolipid- und Gangliosid-Markierungsmuster ergab sowohl Unterschiede zwischen T-Lymphozyten aus verschiedenen Organen (Thymus und Milz) wie auch zwischen B- und T-Zellen der Milz.

1 Van Eijk, R.V.W. & Mühlradt, P.F. (1977) Eur. J. Biochem. 78, 41 - 54.
2 Rosenfelder et al. (1978) *Eur. J. Biochem.*, im Druck.

Günter Rosenfelder und Peter F. Mühlradt, Gesellschaft für Biotechnologische Forschung mbH, Mascheroder Weg 1, D-3300 Braunschweig-Stöckheim.

H. Rotering und V. Braun

Biosynthese des kovalent gebundenen Diglycerids im Murein-Lipoprotein von E. coli

An einer *E.-coli*-K-12-*lpm*-Mutante mit freier Thiolgruppe[1], deren Lipoprotein anscheinend kein kovalent gebundenes Diglycerid enthielt, wurde die Biosynthese des Lipidteils von Murein-Lipoprotein untersucht. Durch Einbau von [2-^{3}H]Glycerin in Lipoprotein und quantitative Fettsäureanalyse konnten wir zeigen, daß in dieser Mutante Glycerin kovalent an die Thiolgruppe des Amino-terminalen Cysteins des Lipoproteins gebunden ist, die estergebundenen Fettsäuren (FS) des Diglycerids jedoch fehlen. Wir bestimmten nur eine Fettsäure pro Polypeptidkette. Die Fettsäurezusammensetzung und die fehlende Esterbande im Infrarotspektrum zeigen, daß es sich hierbei um die amidgebundene Fettsäure des Lipoproteins handeln muß. Die freie Thiolgruppe im Lipoprotein der Mutante ist auf einen Aminosäureaustausch von Arginin gegen Cystein an der zweitletzten Position des Carboxylendes des Proteins zurückzuführen[2]. Das Lipoprotein aus einer merodiploiden Mutante *lpm*/F'*lpm*$^+$, die ein F'-Plasmid mit intaktem Lipoprotein-Gen enthält, besteht aus Wildtyp- und Mutanten-Polypeptidketten, jedoch mit 3 Fettsäuren pro Polypeptidkette. Es werden also das Wildtyp-Gen auf dem Plasmid und das mutierte chromosomale Gen exprimiert, wobei die intakte Fettsäuren-Transferase die mutierte ersetzt. Dies zeigt, daß die fehlenden Diglycerid-Fettsäuren nicht auf einen pleiotrophen Effekt des Gens, das den Aminosäureaustausch verursacht, zurückzuführen sind, sondern daß eine Doppelmutation in zwei nahe benachbarten Genen, dem Strukturgen *lpp* des Peptidteils und dem Gen *lpm* für die Funktion der Fettsäure-Transferase(n) vorliegt. Auf Grund dieser Beobachtungen nehmen wir folgenden Biosyntheseweg für das kovalent gebundene Diglycerid im Lipoprotein an:

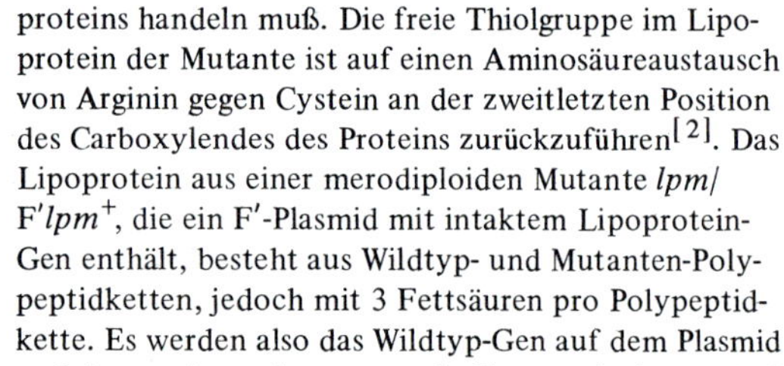

1 Suzuki, H., Nishimura, Y., Iketani, H., Campisi, J., Hirashima, A., Inonye, M. & Hirota, Y. (1976) *J. Bacteriol.* 127, 1494 - 1501.

2 Rotering, H. & Braun, V. (1977) *FEBS Lett.* 83, 41 - 44.

Dipl.-Biol. H. Rotering und Prof. Dr. V. Braun, Lehrstuhl für Mikrobiologie II, Universität, Auf der Morgenstelle 28, D-7400 Tübingen.

R. Rudolph, I. Heider and R. Jaenicke

Reassociation and Reactivation of Malate Dehydrogenase

The analysis of the reassociation and reactivation of mitochondrial malate dehydrogenase (EC 1.1.1.37) from pig heart is of interest in connection with previous studies on the reconstitution of tetrameric enzymes, e.g. lactate dehydrogenase[1,2], or D-glyceraldehyde-3-phosphate dehydrogenase[3]. All three enzymes are NAD-dependent dehydrogenases. However, unlike the latter two, malate dehydrogenase is a dimer which is characterized by a close structural relationship to lactate dehydrogenase. Applying the nomenclature introduced by Rossmann[4] with an orthogonal right-handed coordinate system, the "Q-axis" turns out to be accurately maintained in both lactate dehydrogenase and malate dehydrogenase. On the basis of this evidence, obtained from crystallographic data, it is of considerable interest to investigate whether the kinetics of reactivation and reassociation for the dimeric enzyme are comparable to the reconstitution reactions which were found to be rate-limiting in the reactivation of tetrameric dehydrogenases[1-3].

The kinetic analysis of the reassociation processes are based on two necessary requirements. Firstly, the native enzyme must not show concentration-dependent dissociation under the conditions of reconstitution. As shown by gel chromatography, the dimeric state (M_r = 66 100) is clearly established in the concentration range from 100 to 0.2 μg/m*l*. Secondly, the assumption has to be made that the product of reconstitution and the enzyme in its initial native state are identical. Evidence from physico-chemical and enzymological experiments which corroborates earlier findings for other systems[5] proves this assumption to be true.

The kinetics of reactivation of malate dehydrogenase is closely similar to the reactivation of lactate dehydrogenase from pig heart. Both enzymes are characterized by sigmoidal reactivation relaxations, the rates of which depend on enzyme concentration. For both enzymes, the reactivation is determined by consecutive first and second order reactions which may be described by an irreversible uni-bimolecular reaction scheme (cf. ref.[2]).

The coenzyme, $NAD^{\oplus}$, does not influence either the rate or the yield of the reactivation reaction. It may be concluded that the folded monomers which are produced as intermediates in the overall reconstitution cannot be fully active.

1 Rudolph, R. & Jaenicke, R. (1976) *Eur. J. Biochem.* 63, 409 - 417.
2 Rudolph, R., Heider, I., Westhof, E. & Jaenicke, R. (1977) *Biochemistry* 16, 3384 - 3390.
3 Rudolph, R., Heider, I. & Jaenicke, R. (1977) *Eur. J. Biochem.* 81, 563 - 570.
4 Rossmann, M.G., Liljas, A., Brändén, C.-I. & Banaszak, L.J. (1975) in The Enzymes (Boyer, P.D. ed.), 3rd edn., vol. XI, pp. 61 - 102, Academic Press, New York,
5 Jaenicke, R., Engelhard, M., Kraus, E. & Rudolph, R. (1975) *Trans. Biochem. Soc.* 3, 1051 - 1054.

R. Rudolph, I. Heider and R. Jaenicke, Institut für Biophysik und Physikalische Biochemie der Universität Regensburg, Universitätsstraße 31, D-8400 Regensburg.

H. Rüterjans

Die Untersuchung der Struktur und Funktion von Proteinen mit Hilfe der KMR-Spektroskopie

Es wird ein Überblick gegeben über neuere Fortschritte in der Anwendung von KMR-Methoden zur Untersuchung von Struktur und Funktionsbeziehungen in Proteinen.

^{1}H-, ^{13}C- und ^{15}N-Kerne können als Sonden genutzt werden für das Studium des Ionisierungsverhaltens von Aminosäureseitengruppen. Aus dem nicht normalen Ionisierungsverhalten kann auf Wechselwirkungen der entsprechenden Aminosäureseitengruppen mit der Umgebung geschlossen werden. Diese Seitengruppenwechselwirkungen sind in vielen Fällen Voraussetzung einer enzymatischen Aktivität.

Wie an mehreren Beispielen gezeigt wird, werden solche Wechselwirkungen durch Ligandenbindung und/oder allosterische Umwandlungen beeinflußt.

Besonders die Möglichkeiten der Anwendung der ^{15}N-KMR-Spektroskopie sollen erläutert werden.

H. Rüterjans, Institut für Physikalische Chemie der Universität Münster, Schloßplatz 4, D-4400 Münster.

G. Sander

Interaction of 30S Ribosomal Subunits with 50S Subunits Exposes Ribosomal RNA for Interaction with Elongation Factor G

Bacterial polypeptide chain elongation factor G (EF-G) exhibits a GTPase activity (EC 3.6.1.-) dependent either upon the 50S subunit alone in the absence of monovalent cations, or upon the simultaneous presence of 50S and 30S subunits when monovalent cations are present. A steady fall of activity with increasing salt concentration in the former case contrasts strongly with the bell-shaped curve found for 50S plus 30S subunits[1]. Only in the presence of the 30S subunit is this GTPase activity modulated by changes in the pH; as the pH increases, the bell-shaped curve appoaches the steadily descending curve observed with the 50S subunit alone. An interpretation in terms of the polyelectrolyte theory[2] suggests that in the presence of both ribosomal subunits, there are immobile negative charges near the binding site for the factor and/or the catalytic center; the relatively weak response to changes in pH, when compared with similar changes in the case of a direct interaction between RNA and RNAase[1], implies that these negative charges do not themselves partake in catalytic activity. This then leads to the following model: the 50S subunit interacts weakly with EF-G · GTP[3], with no apparent participation of ribosomal RNA; upon association with the 30S subunit, a critical region of the ribosomal RNA is exposed, leading to a much stronger interaction with EF-G · GTP, which is resistant to high salt concentration. GTP is hydrolyzed in either case as the result of specific activation of the catalytic center, probably situated on the elongation factor[3].

1 Voigt, J., Sander, G., Nagel, K. & Parmeggiani, A. (1974) *Biochem. Biophys. Res. Commun.* **57**, 1279 - 1286.
2 Douzou, P. & Maurel, P. (1977) *Proc. Natl. Acad. Sci. U.S.A.* **74**, 1013 - 1015.
3 Marsh, R.C., Chinali, G. & Parmeggiani, A. (1975) *J. Biol. Chem.* **250**, 8344 - 8352.

Dr. Gernot Sander, Laboratoire de Biochimie-Ecole Polytechnique, Laboratoire Associé n° 240 du CNRS, F-91128 Palaiseau Cédex.

N. R. Saunders

Axonal Transport in Peripheral Cholinergic Neurons

Axonal transport is the mechanism by which parts of a neuron distant from the cell body are supplied by certain materials which cannot be obtained directly from the extracellular environment and which cannot be synthesized by the distant parts of the neuron.

Axonal transport in cholinergic neurons will be considered mainly from the point of view of its possible contribution to synaptic function. This has two aspects (I) the contribution of structural materials, such as protein[1], phospholipid[2] and glycoprotein[3], which are important for the maintenance of the structure of nerve axons and terminals; and (II) the more specific contribution of enzymes and transmitter substance to synaptic function[4,5].

Transport of the enzymes acetylcholinesterase (EC 3.1.1.7)[6], choline acetyltransferase (EC 2.3.1.6) [7,8] and of the transmitter acetylcholine[5,9] has been studied in experiments involving axotomy of peripheral nerves and estimation of the subsequent redistribution of enzyme or transmitter on either side of the site of nerve injury. There is evidence that these enzymes and the transmitter acetylcholine may be localized in several different subcellular compartments[5,10]; uncertainties about the proportions of material in the different compartments has given rise to variable estimates of the rates of axonal transport of these materials. More direct experiments on acetylcholine transport have been carried out by intracellular injection of [^{3}H]choline in *Aplysia*[11] and in cat spinal motor neurons[12].

It has been shown that nerve terminals are dependent upon a continued supply of materials such as protein and phospholipid from the cell bodies. It has also been shown that the presence of cholineacetyltransferase and acetylcholinesterase is dependent upon supply of these enzymes from the cell body. Both axotomy and [^{3}H]-choline injection experiments suggest that acetylcholine may also be transported from cell body to periphery. However, the amount transported is small compared with the turnover of acetylcholine at the nerve terminals[13]. Possibly the significance of axonal transport of acetylcholine may be more in relation to an association with the supply of subcellular organelles involved in transmitter release at the nerve terminals.

An important aspect of axonal transport which has been little studied is the mechanism by which the periphery is able to influence the output of material transported from the cell body. This will be considered in the context of changes in the transport of labelled phospholipids in cholinergic neurons following nerve injury; experiments will be discussed which suggest that the cell body is able to respond to a distant injury within one hour, by considerably increasing the amount of phospholipid (mainly lecithin) which is transported to the periphery[14].

1 Ochs, S. (1972) *J. Physiol.* **227**, 627 - 645.
2 Miani, N. (1964) *Progr. Brain. Res.* **13**, 115 - 126.
3 Droz, E., Koenig, H. & Di Gamberdino, L. (1973) *Brain Res.* **60**, 93 - 127.
4 Lubinska, L. (1975) *Int. Rev. Neurobiol.* **17**, 241 - 269.
5 Saunders, N.R. (1975) in Cholinergic Mechanisms (Waser, P.G., ed.) pp. 177 - 185, Raven Press, New York.
6 Lubinska, L. & Niemierko, S. (1971) *Brain Res.* **27**, 329 - 342.
7 Saunders, N.R., Dziegielewska, K., Häggendal, J. & Dahlström, A. (1975) *J. Neurobiol.* **4**, 95 - 103.

8 Tuček, S. (1975) *Brain Res.* **86**, 259 - 270.
9 Dahlström, A., Evans, C.A.N., Häggendal, C.J., Heiwall, P.O. & Saunders, N.R. (1974) *J. Neural. Transm.* **35**, 1 - 11.
10 Whittaker, V.P. (1969) *Progr. Brain Res.* **31**, 211 - 222.
11 Koike, H., Eisenstadt, M. & Schwartz, J.H. (1972) *Brain Res.* **37**, 152 - 159.
12 Schubert, P., Lux, H.D. & Kreutzberg, G.W. (1971) *Acta Neuropathol. Suppl.* **V**, 179 - 186.
13 Häggendal, C.J., Dahlström, A.B. & Saunders, N.R. (1973) *Brain Res.* **58**, 494 - 499.
14 Dziegielewska, K.M., Evans, C.A.N. & Saunders, N.R. (1977) *J. Physiol.* **263**, 200 - 201 p.

Norman R. Saunders, Reader in Physiology, Department of Physiology, University College London, Gower Street, London WC1E 6BT.

B. Schade, H.-D. Lüdemann und R. Jaenicke

Dissoziation und Strukturänderungen von Tabakmosaikvirus-Protein unter hohem hydrostatischem Druck

Tabakmosaikvirus-Protein assoziiert bei pH 6.5, $I = 0.1$ M in endothermer Reaktion vom Zustand des „A-Proteins" zu helikalen Stäbchen[1]. Im Konzentrationsbereich von $c \sim 2$ mg/m*l* beträgt die charakteristische Übergangstemperatur $t_m = 20 \pm 1$ °C. Wie aus dilatometrischen und gravimetrischen Experimenten[2] hervorgeht, ergibt sich ein positives Reaktionsvolumen und ein der Assoziation parallel-verlaufender „Water release"-Effekt von 0.033 ± 0.0003 g/g Protein; das partielle spezifische Volumen erfährt eine Zunahme von $\Delta v_2 = 0.0049 \pm 0.0003$ cm^3/g. Die erwähnten Daten lassen einen signifikanten Einfluß hohen hydrostatischen Druckes auf die endotherme Assoziation in dem Sinne erwarten, daß hohe Drucke der Assoziation entgegenwirken[3]. Dementsprechend sollten die Isobaren der Assoziationsreaktion mit zunehmendem Druck zu höheren Übergangstemperaturen t_m hin verschoben sein. Experimente im Druckbereich von $1 < p < 1500$ bar bestätigen die Voraussagen mit $\Delta t_m = 14$ °C für den Übergang bei 700 bar. Die Assoziation erfolgt unter den gegebenen Bedingungen vollständig reversibel. Die Produkte der Assoziation bei unterschiedlichen Drucken und Temperaturen legen verschiedene Konformationszustände des assoziierenden Proteins nahe. Tatsächlich läßt ein Reaktionszyklus gemäß

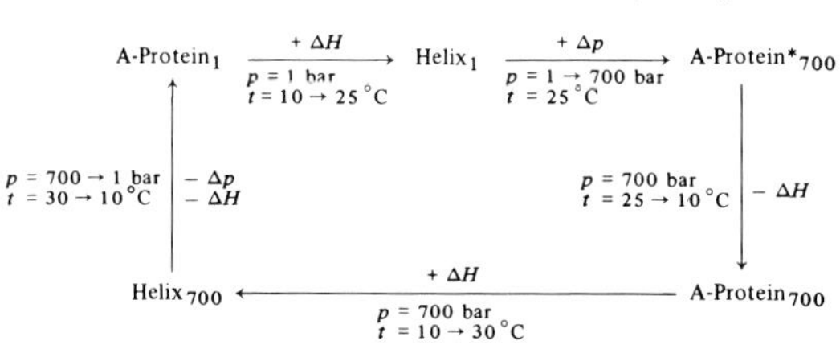

anhand von Messungen der Trübung bzw. des Circulardichroismus Strukturvarianten erkennen.

1 Smith, C.E. & Lauffer, M.A. (1967) *Biochemistry* **6**, 2457 - 2465.
2 Jaenicke, R. & Lauffer, M.A. (1969) *Biochemistry* **8**, 3083 - 3092.
3 Lauffer, M.A. (1975) Entropy-Driven Processes in Biology, Springer, Berlin, Heidelberg, New York.

R. Jaenicke, Institut für Biophysik und Physikalische Biochemie, Universität Regensburg, Universitätsstraße 31, D-8400 Regensburg.

H.-J. Schäfer, P. Scheurich, G. Rathgeber and K. Dose

8-Azido-1,N^6-etheno-ATP – a possible photoaffinity label for F$_1$ ATPases

8-Azido-ATP has been used as a photoaffinity label for the identification of nucleotide binding sites of F_1 ATPases[1,2]. The detection of the label after binding poses some problems if nonradioactive agents are used. Therefore we have synthesized 8-azido-1,*N*^{6}-etheno-ATP – a fluorescent photoaffinity label[3]. This ATP-analog has been characterized by elementary analysis, thin-layer chromatography, and IR, NMR and UV spectroscopy[3]. In aqueous solution the photoreaction of the analog is pH-dependent. The photolysis products obtained at pH 2.0 differ from those obtained at pH 7.0 and 10.0, as indicated by UV spectroscopy. The occurrence of isosbestic sections in all experiments shows that the progress of the photoreaction is largely uniform. The fluorescence emission and excitation spectra are also pH-dependent.

A few preliminary results on labeling F_1 ATPase (EC 3.6.1.3) from *Micrococcus luteus* are reported here. In the presence of divalent metal ions, the enzyme hydrolyzes to their corresponding diphosphates the nucleoside triphosphates ATP, 1,*N*^{6}-etheno-ATP, 8-bromo-ATP, 8-azido-ATP and 8-azido-1,*N*^{6}-etheno-ATP. The K_m values for ATP (and 1,*N*^{6}-etheno-ATP) and its 8-substituted analogs differ by a factor of 10. These differences probably depend on differences between the conformation of the nucleotides. Whereas ATP (and 1,*N*^{6}-etheno-ATP) is more stable in the *anti* conformation, 8-bromo-ATP[4] and very likely also 8-azido-ATP and 8-azido-1,*N*^{6}-etheno-ATP preferentially attain the *syn* conformation, due to the presence of bulky substituents in position 8. Irradiation of F_1 ATPase in the presence of 8-azido-1,*N*^{6}-etheno-ATP and $Mg^{2\oplus}$ ions causes a decrease in enzymic activity. This inhibition is of the noncompetitive type. The rate of inactivation is decreased if ATP or ADP is added prior to the irradiation. AMP affords no protection, inasmuch as it is not specifically bound by F_1 ATPase. The present results suggest the suitability of 8-azido-1,N^6-etheno-ATP as a fluorescent photoaffinity label for ATP-dependent enzymes.

1 Wagenvoord, R.J., van der Kraan, I. & Kemp, A. (1977) *Biochim. Biophys. Acta* **460**, 17 - 24.
2 Scheurich, P., Schäfer, H.-J. & Dose, K. (1977) *Hoppe-Seyler's Z. Physiol. Chem.* **358**, 298.
3 Schäfer, H.-J., Scheurich, P., Rathgeber, G. & Dose, K. (1977) *Hoppe-Seyler's Z. Physiol. Chem.* **358**, 1270 - 1271.
4 Ikehara, M., Uesugi, S. & Yoshida, K. (1972) *Biochemistry* **11**, 830 - 836.

Dr. Hans-Jochen Schäfer, Institut für Biochemie der Universität Mainz, J.J.-Becher-Weg 30, D-6500 Mainz.

R. Schmidt

Sequential Release of ATP and Acetylcholine from Cholinergic Synaptic Vesicles Modulated by $Ca^{2\oplus}$ and Synaptic Membranes in vitro

Cholinergic synaptic vesicles isolated from the electric organ of *Torpedo marmorata*[1] contain 0.2 molecules of ATP per molecule of acetylcholine (ACh)[2]. When the electric organ is stimulated in situ, both constituents are released concomitantly from the vesicles[3]. It is generally accepted that $Ca^{2\oplus}$ ions are the trigger for transmitter release. Therefore we investigated whether this metal cation can induce changes in vesicle composition during in vitro incubation, either on its own or on addition of isolated presynaptic membranes.

Synaptic vesicles were purified by a combination of several differential centrifugation steps and separation on a discontinuous isoosmolar sucrose/glycine gradient[4], then incubated in 0.8M glycine for 1 min at 4 °C. The intrinsic Ca concentration in the incubation medium is 50μM. The addition of further $Ca^{2\oplus}$ ions to the incubation medium induced a marked release and hydrolysis of ATP (– 50% at 700μM Ca), presumably by intrinsic ATP-ase activity of the vesicles[5] or of contaminating membranes.

When presynaptic membranes are isolated from a nerve terminal sac fraction (T-sacs[6]), and added to the incubation medium, the release of ATP is drastically enhanced. However, membranes are ineffective after boiling. The presence of high concentrations of metal ions is not essential for membrane induced release of vesicular ATP in vitro.

Under these conditions only slight changes in the ACh content of the vesicles are observed (– 10%).

The significance of these results for the understanding of the physiological release mechanisms will be discussed.

1 Whittaker, V.P., Essman, W.B.,& Dowe, G.H.C. (1972) *Biochem. J.* **128**, 833 - 846.
2 Dowdall, M.J., Boyne, A.F. & Whittaker, V.P. (1974) *Biochem. J.* **140**, 1 - 12.
3 Zimmermann, H. & Whittaker, V.P. (1974) *J. Neurochem.* **22**, 435 - 450.
4 Schmidt, R. (1975) Diplomarbeit, Johannes Gutenberg-Universität, Mainz, pp. 28 - 35.
5 Breer, H., Morris, S.J. & Whittaker, V.P. (1977) *Eur. J. Biochem.* **80**, 313 - 318.
6 Dowdall, M.J. & Zimmermann, H. (1977) *Neuroscience* **2**, 405 - 421.

R. Schmidt, Abteilung für Neurochemie, Max-Planck-Institut für Biophysikalische Chemie, Postfach 968, D-3400 Göttingen-Nikolausberg.

R. Schmidt

Release of $Ca^{2\oplus}$ from postsynaptic mitochondria in the electric organ of Torpedo during nerve activity

The cytoplasmic concentration of $Ca^{2\oplus}$ is essential for the coupling of stimulation and contraction or secretion in a variety of excitable tissues. Based on studies of isotope fluxes, possible extracellular as well as intracellular Ca sources have been taken into consideration. However, little is known on the Ca content of subcellular organelles. The electric organ of *Torpedo* has been frequently used as a model system for the neuromuscular junction, since its electrocytes develop from myoblasts and are innervated by cholinergic electromotor neurons.

Mitochondria are isolated from this tissue by a combination of several homogenization and centrifugation steps and purified on discontinuous sucrose, glycine density gradients. Cytochrome *c* oxidase (EC 1.9.3.1) is used as the marker enzyme for the mitochondria. Ca concentrations are analysed by atomic absorption spectrophotometry.

The Ca concentration of the mitochondrial fraction is much higher than that of any other subcellular fraction (900 nmol/mg protein). When the tissue is stimulated with 5 000 pulses via the electromotor nerve[1] before separation, the Ca concentration of the mitochondrial fraction is reduced by 80%. However, the Ca content of the mitochondria is unaffected by the stimulation, when tissue blocks which have been perfused[2] with d-tubucurarine are used, until the response of the electrocytes to single test pulses is almost abolished. Acetylcholine is still released from the presynaptic terminal under these conditions.

It is concluded that postsynaptic mitochondria release their Ca during nerve activity, and this result will be discussed in relation to similar findings in other excitable tissues.

1 Zimmermann, H. & Whittaker, V.P. (1974) *J. Neurochem.* **22**, 435 - 450.

2 Zimmermann, H. & Denston, C.R. (1977) *Neuroscience* **2**, 695 - 714.

R. Schmidt, Abteilung für Neurochemie, Max-Planck-Institut für Biophysikalische Chemie, Postfach 968, D-3400 Göttingen-Nikolausberg.

M. Schmitt, S. Risi and K. Dose

F_1 ATPase from Micrococcus sp. ATCC 398: Artificial Immobilization to Sepharose CL-6B

Highly purified F_1 ATPase from *Micrococcus sp.*[1] was immobilized on Sepharose CL-6B by the cyanogen bromide reaction[2]. F_1 ATPase loses activity when coupled to the gel matrix in the cold, but it retains its active conformation when coupled in the presence of bovine serum albumin. The Sepharose-F_1 ATPase conjugate can be stored moist in the cold at least 6 months without appreciable loss of activity and it displays increased resistance to thermal inactivation.

The artificially immobilized F_1 ATPase was filled into a column and used in a system of automated analysis. Kinetic measurements make evident that the transport of substrate to the active site of the enzyme becomes a rate-limiting step of the reaction and that diffusional hindrance influences all kinetic parameters determined for the immobilized F_1 ATPase. The $K_{m\,(app)}$ for both F_1 ATPase forms shows no significant change from the native for the immobilized enzyme. The pH optimum for the reaction with respect to $Ca^{2\oplus}$-ATP as substrate is 8.4 for both F_1 ATPase forms, whereas, in the presence of $Mg^{2\oplus}$-ATP, the pH optimum for the Sepharose-bound enzyme shows a shift from 7.8 for the soluble F_1 ATPase to 6.6 for the matrix-bound enzyme. The Arrhenius plots for both F_1 ATPase forms present discontinuities near 17.5 and 30 °C, at pH 8.0, indicating that both F_1 ATPase forms can exist in at least three different active conformations as a result of the transition between a low- and a high-temperature protein structure.

1 Risi, S., Höckel, M., Hulla, F.W. & Dose, K. (1977) *Eur. J. Biochem.* **81**, 103 - 109.

2 Cuatrecasas, P. (1972) in Advances in Enzymology 36 (Meister, A., ed.) pp. 29 - 89, Interscience Publishers, New York.

Prof. Dr. Klaus Dose, Institut für Biochemie der Universität Mainz, Joh.-Joachim-Becher-Weg 30, D-6500 Mainz.

D. Schneider

Das System Phosphodiesterase I – alkalische Phosphatase in Herpes-simplex-Virus (Typ 1)-transformierten Hamsterzellen. Biochemische und biologische Eigenschaften

Die Zelloberflächenmembran der Säugetierzelle ist an physiologischen Prozessen wie Zellwachstum, Zellteilung, Differenzierung, Zellbewegung und Zell-Zell-Kommunikation unmittelbar beteiligt. Änderungen in der Zusammensetzung oder im Aufbau der Membran wirken sich auf den „normalen" Ablauf dieser Prozesse aus und sind besonders deutlich bei der Umwandlung der normalen Säugetierzelle in eine Tumorzelle nachweisbar[1].

Besonders dramatische transformationsabhängige Veränderungen treten bei den für die normale Zellmembran typischen abbauenden Enzymen auf[1]. So konnte gezeigt werden, daß in Hamsterzellen, die durch Viren transformiert wurden, die Hydrolyse von Nucleosiddiphosphatzuckern zu den freien Zuckern, die über einen Zweistufenmechanismus erfolgt

$$\text{Nucleosiddi-phosphatzucker} \xrightarrow[\text{(EC 3.1.4.1)}]{\text{Phosphodiesterase I}} \text{Zuckerphosphat}$$

$$\xrightarrow[\text{(EC 3.1.3.1)}]{\text{alkal. Phosphatase}} \text{Zucker,}$$

auf der Stufe der alkalischen Phosphatase blockiert ist[2].

Die Inaktivierung der alkalischen Phosphatase wurde auch nach Transformation durch das Rous-Sarkom-Virus[3] und in durch Herpes-simplex-Virus(Typ 1)-transformierten Hamsterzellen (HSV_1-Zellen)[4] gefunden. Möglicherweise handelt es sich bei dieser Inaktivierung des Enzyms um einen Vorgang, der für die Transformation von Säugetierzellen durch Viren typisch ist.

Die Eigenschaften der alkalischen Phosphatase und deren Induzierbarkeit durch Prednisolon (11β,17α,21-Trihydroxy-1,4-pregnadien-3,20-dion) sind in Zellen humanen Ursprungs umfassend untersucht worden[5], ohne daß die biologische Funktion der alkalischen Phosphatase und ihrer Isoenzyme bisher eindeutig aufgeklärt werden konnte. Möglicherweise spielt das Enzym eine wesentliche Rolle bei der Regulation des Zellwachstums[6].

Da die alkalische Phosphatase auch in HSV_1-Zellen durch Prednisolon induzierbar ist[4], sind diese Zellen ein geeignetes Modellsystem, um Wechselwirkungen verschiedener Parameter wie virusbedingte Transformation, Aktivität von Plasmamembranenzymen, Enzyminduktion durch Hormone in transformierten Zellen und Zellwachstum zu untersuchen.

Durch Affinitätschromatographie von Mikrosomenfraktionen bzw. gereinigten Plasmamembranfraktionen aus HSV_1-Zellen an Linsenagglutinin (LcH)-Sepharose und anschließende präparative Flachgel-Polyacrylamid-Gelelektrophorese der an LcH-Sepharose gebundenen Membranfraktion ließ sich ein Glykoprotein (M_r = 120 000 bis 130 000) isolieren, das als alkalische Pyrophosphatase identifiziert wurde und dessen biochemische Eigenschaften mitgeteilt werden. Werden nach der gleichen Methode Plasmamembranfraktionen bzw. mikrosomale Fraktionen aus primären Hamsterembryofibroblasten oder aus Hamsterembryonen untersucht, so enthält die an LcH-Sepharose bindende Fraktion neben der alkalischen Pyrophosphatase ein Enzym, das in der Gelelektrophorese als Banden mit M_r = 150 000 und 170 000 auftritt und als alkalische Phosphatase identifiziert wurde. Die von der LcH-Sepharose nicht gebundene Fraktion enthält einen hohen Anteil an alkalischer Phosphatase, der auch nach Rechromatographie nicht an LcH-Sepharose gebunden wird und möglicherweise ein Isoenzym mit modifzierter oder fehlender Kohlenhydratseitenkette enthält. In der an LcH-Sepharose bindenden Fraktion aus HSV_1-Zellmembranen lassen sich in der Gelelektrophorese keine Banden der alkalischen Phosphatase nachweisen. Werden die HSV_1-Zellen in einem Medium gezüchtet, das 1% Kälberserum anstelle der üblicherweise verwendeten 10% enthält, läßt sich durch Zugabe von Prednisolon (5 μg/m*l*) zum Medium eine alkalische Phosphatase induzieren.

Vorläufige Ergebnisse der Versuche, das induzierte Enzym anzureichern und zu charakterisieren, werden mitgeteilt und diskutiert.

1 Nicolson, G.L. (1976) *Biochim. Biophys. Acta* **458**, 1 - 72.
2 Sela, B. & Sachs, L. (1973) *FEBS Lett.* **30**, 100 - 104.
3 Bader, M., Bigelow, M. & Kondratik, J. (1975) Abstr. FASEB Ann. Meeting, p. 249.
4 Schneider, D., unveröffentlichte Ergebnisse.
5 Fishman, W.H. (1974) *Am. J. Med.* **56**, 617 - 650.
6 Sela, B. & Sachs, L. (1974) *J. Cell. Physiol.* **83**, 27 - 34.

Dr. Dietmar Schneider, Abteilung für Exptl. Virologie, Institut für medizinische Mikrobiologie der Universität Mainz, Hochhaus am Augustusplatz, D-6500 Mainz.

E. Schneider, K. Rittinghaus, V. Thiele, U. Schwuléra and K. Dose

Adenosine Triphosphatase Complex from Rhodospirillum rubrum: Alternative Method for Solubilization and Further Properties

Recently[1,2] we solubilized and purified an oligomycin-sensitive ATPase complex (EC 3.6.1.3) from *Rhodospirillum rubrum* chromatophores by the method of Tzagoloff and Meagher[3].

In order to obtain the pure enzyme in higher yield, the F_0 - F_1 ATPase was solubilized in a different way, essentially following a method described by Sone at al.[4]. According to this method, the chromatophores are first washed with 1% cholate for removal of loosely bound enzymes and then the ATPase complex is solubilized with 2% Triton X-100. In agreement with the preparation described previously[1,2], the enzyme is sensitive to oligomycin and dicyclohexylcarbodiimide.

The K_m value is about 0.06mM for Mg-ATP and about 0.4mM for Ca-ATP. Furthermore, free $Mg^{2\oplus}$ or $Ca^{2\oplus}$ ions and free ATP are competitive inhibitors.

1 mg of the purified ATPase complex contains about 0.14 mg phospholipids and at least 0.1 mg Triton X-100.

Analytical ultracentrifugation revealed a single boundary. The s_{20} value obtained from these experiments was only 1.45, although the molecular weight of the F_0 - F_1 complex is about 480 000[1]. The phospholipid content and bound Triton X-100 might account for the low s_{20} value.

1 Schneider, E., Müller, H.W., Schwuléra, U. & Dose, K. (1977) *Hoppe-Seyler's Z. Physiol. Chem.* **358**, 1276 - 1277.
2 Schneider, E., Schwuléra, U., Müller, H.W. & Dose, K. (1978) *FEBS-Lett.*, in press.
3 Tzagoloff, A. & Meagher, P. (1971) *J. Biol. Chem.* **246**, 7328 - 7336.
4 Sone, N., Yoshida, M., Hirata, H. & Kagawa, Y. (1975) *J. Biol. Chem.* **250**, 7917 - 7823.

Udo Schwuléra, Institut für Biochemie der Universität Mainz, Joh.-Joachim-Becher-Weg 30, D-6500 Mainz.

H. Schneider-Bernlöhr, H. Dietrich, W. Maret, I. Andersson and M. Zeppezauer

Analytical and Molecular Characterization of Sepharose-Bound Liver Alcohol Dehydrogenase

Active site titration of liver alcohol dehydrogenase (EC 1.1.1.1), using the ternary complex formation with $NAD^{\oplus}$/pyrazole and NADH/ isobutyramide, respectively, has been used to characterize enzyme preparations bound to agarose by the cyanogen bromide technique of Axén and Ernback[1]. Sensitivity is enhanced by means of the dual-wavelength technique[2]. By monitoring the active site contents of the preparations, the coupling procedure could be optimized. Conditions have been worked out where more than 90% of the active sites are preserved during immobilization. The apparent specific activity of the immobilized enzyme is decreased substantially, however, which indicates the effects of the microenvironment and diffusion on the observed reaction rate. Preparations containing more than 90% active sites lend themselves to a more penetrating molecular characterization. For example, the intrinsic zinc ions of the enzyme could be replaced by cobalt with relative ease; the spectra of these preparations are reported.

1 Axén, R. & Ernback, S. (1971) *Eur. J. Biochem.* 18, 351 - 360.

2 Einarsson et al. (1976) *Anal. Lett.* 9, 815 - 823.

Prof. M. Zeppezauer, Fachbereich 15, Fachrichtung Biochemie, Universität des Saarlandes, D-6600 Saarbrücken.

K. F. Schnell, W. Flossmann and R. Vangala

Transport of Anionic Spin-Labels across the Red Cell Membrane and Modification of the Red Cell Permeability by Covalently Bound Spin-Labels

The transport of the anionic spin-label 2,2,5,5-tetramethyl-3-pyrrolin-1-oxyl-3-carboxylate (I) across the red cell membrane was studied in red cell ghosts. I is transported across the red cell membrane. The ESR spectra exhibited an isotropic mobility of the label. The transport of I is temperature dependent (E_A = 20 kcal/mol) and is inhibited by 4-acetamido-4′-isothiocyanostilbene-2,2′-disulfonic acid which specifically inhibits the anion transport system of the red blood cell. 4-Maleiimido-2,2,6,6-tetramethylpiperidinooxyl (II) and 4-isothiocyanato-2,2,6,6-tetramethylpiperidinooxyl (III) react covalently with the red cell membrane. The ESR spectra can be interpreted as a superposition of a mobile and an immobile spectrum. The relative concentration of the immobilized label is increased by repeatedly washing the red cell ghosts in 165mM KCl solutions. The effect of these agents on the anion transport, on the sugar transport and on the passive cation permeability was studied by pretreating the red cell ghosts for 30 min with 5mM II and III at 37 °C and pH 7.2. The sulfate self-exchange (37 °C, pH 7.2), the chloride self-exchange (0 °C, pH 7.2) and the $K^{\oplus}$ loss (37 °C, pH 7.2) were not affected by these labels. The transport of L-arabinose (37 °C, pH 7.2) was slightly inhibited. The results of our studies suggest that II and III are mainly bound to membrane groups which are not involved in the regulation of the anion transport and the sugar transport across the red cell membrane.

K. F. Schnell, W. Flossmann and R. Vangala, Institut für Physiologie und Institut für Biophysik und Physikalische Biochemie, Universität Regensburg, Postfach 397, D-8400 Regensburg.

E. Schott und E. Schlimme

β-Nicotinamid-α-adenin-dinucleotid: Synthese und Eigenschaften eines Coenzymanalogons

Über die Konformation des hydridübertragenden Coenzyms $NAD^{\oplus}$ in Lösung sowie in Holoenzymkomplexen wurden in den letzten Jahren detaillierte Kenntnisse erarbeitet[1,2]. So liegt das Coenzym im aktiven Zentrum von Dehydrogenasen, z.B. der Lactat-Dehydrogenase (EC 1.1.1.27), in gestreckter Konformation vor, wobei die beiden Heterocyclen Nicotinamid, N, und Adenin, A, senkrecht zueinander stehen und etwa 1.2 nm voneinander entfernt sind. Dabei sind N und A in hydrophoben Taschen des Enzyms fixiert[1]. Fehlt der Adeninring wie im Nicotinamidmononucleotidphosphoribosid, $NMNPR^{\oplus}$, fällt die Coenzymaktivität (Maximalgeschwindigkeit *V*) auf etwa 3% ab[3].

Uns interessierte deshalb die Frage, wie sich eine Epimerisierung des Adenosins am C1′ vom β- zum α-Anomeren auf die Coenzymaktivität auswirkt.

α-Adenosin-5′-*O*-monophosphat, α-AMP, wurde durch chemische Phosphorylierung von α-Adenosin mikrosynthetisch präpariert und chromatographisch sowie enzymatisch charakterisiert. Einzelheiten werden an anderer Stelle ausführlich beschrieben[4].

β-Nicotinamid-α-adenin-dinucleotid, $N\alpha AD^{\oplus}$, wurde in wässerigem Pyridin durch Kondensation von α-AMP und β-NMN mit *N*, *N*′-Dicyclohexylcarbodiimid dargestellt. Die Aufreinigung erfolgte über eine DE-52-Cellulose-Säule mit einem Triethylammonium-hydrogencarbonat-Gradienten (I), von 0 - 0.2M. $N\alpha AD^{\oplus}$ wurde bei 0.1M I eluiert und nach enzymatischer Spaltung chromatographisch sowie UV- und ^{1}H-NMR-spektroskopisch (δ = 6.25, H1′ α-A; δ = 5.90, H1′ β-A; ppm) identifiziert.

$N\alpha AD^{\oplus}$ wird von den Lactat-Dehydrogenasen aus Schweinemuskel, 4M aus Schweinemuskel und 4H aus Schweineherz (Boehringer Mannheim GmbH) als Coenzym akzeptiert, wobei die K_m-Werte sich etwa um den Faktor 5 - 10, verglichen mit $NAD^{\oplus}$, erhöhen, *V* allerdings nur bei M-haltigen Isoenzymen vermindert ist. 4M: K_m = 1.74mM ($NAD^{\oplus}$: K_m = 0.31mM), *V* = 36%; 4H: K_m = 0.9mM ($NAD^{\oplus}$: K_m = 0.1mM), *V* = 95%.

Da die Heterocyclen N und A im aktiven Holoenzymkomplex gebunden vorliegen müssen und α-$NAD^{\oplus}$ inaktiv ist, muß eine Rotation um den Torsionswinkel (C4′ - C5′ O5′ - P1) des A-Teils vom $N\alpha AD^{\oplus}$ während des Bindungsvorganges von *trans* nach *cis* gefordert werden. Damit liegen alle Molekülteile des $N\alpha AD^{\oplus}$ in einer $NAD^{\oplus}$-äquivalenten Konformation im Holoenzym vor mit Ausnahme der Ribose des α-Adenosins, so daß deren 2′- und 3′-Hydroxylgruppen nicht wie im unmodifizierten $NAD^{\oplus}$ zur Bindung beitragen können. Die veränderten Michaelis-Menten-Werte lassen sich auf dieser Basis erklären.

1 Holbrook, J. J., Liljas, A., Steindel, S. J. & Rossmann, M.G. (1975) in The Enzymes XI, Oxidation-Reduction (Boyer, P. D., Hrsg.) S. 191 - 292, Academic Press, New York.
2 Saenger, W., Reddy, B. S., Mühlegger, K. & Weimann, G. (1977) *Nature (London)* **267**, 225 - 229.
3 a Pfleiderer, G., Sann, E. & Ortanderl, F. (1963) *Biochim. Biophys. Acta* **73**, 39 - 49.
b Woenckhaus, C. & Scherr, D. (1973) *Hoppe-Seyler's Z. Physiol. Chem.* **354**, 53 - 59.
4 Boos, K. S. & Schlimme, E. (1978) *Eur. J. Biochem.*, in Vorbereitung.

E. Schott und E. Schlimme, Laboratorium für Biologische Chemie im Fachbereich Naturwissenschaften II der Universität – Gesamthochschule, Warburger Str. 100, Bauteil J, D-4790 Paderborn.

H.-U. Schröder, V. Neuhoff und N. N. Osborne

Effekt von Pharmaka und Tryptamin-Analogen auf den Influx von radioaktivem 5-Hydroxytryptamin, Dopamin und Tryptamin in Schneckennervengewebe (Helix pomatia)

Frühere Arbeiten mit ganzen Schneckenganglien haben gezeigt, daß es besondere High-Affinity-Aufnahme-Mechanismen sowohl für 5-Hydroxytryptamin als auch für Dopamin gibt, aber nicht für Tryptamin[1,2]. Die hier vorgelegten Experimente dienten der Aufklärung der Wirkung verschiedener Pharmaka und Tryptaminanaloga auf den Influx von 0.1μM radioaktiv markiertem 5-Hydroxytryptamin, Tryptamin oder Dopamin nach den oben beschriebenen Methoden[1]. Von den geprüften Pharmaka [Veratridin, *N*-Äthylmaleimid, 2,4-Dinitrophenol, KCN, Chlorpromazin, Chloramphenicol, Amantadin, 2-Desoxyglucose, Methylenblau, Probenecid und Lilly 110140 (3-*p*-Trifluormethylphenoxy-*N*-methyl-3-phenoxypropylamin-hydrochlorid)] waren Lilly 110140 (ID_{50}-Wert 5×10^{-6}M) und Chlorpromazin (ID_{50}-Wert 1×10^{-5}M) die stärksten Hemmstoffe der 5-Hydroxytryptamin-Aufnahme. Alle übrigen Substanzen hemmten die 5-Hydroxytryptamin-Aufnahme mit ID_{50}-Werten im Bereich von 10^{-3} bis 10^{-5}M, mit Ausnahme von 2,4-Dinitrophenol, das selbst bei 10^{-3}M Konzentration kaum einen Einfluß hatte. Lilly 110140 und Chlorpromazin beeinflußten den Dopamin- und Tryptamin-Influx bei der Konzentration, die die Aufnahme von 5-Hydroxytryptamin drastisch auf 50% senkte, nur um etwa 5 - 10%. Alle übrigen Substanzen hemmten die Dopamin-Aufnahme tendenziell ähnlich stark wie die von 5-Hydroxytryptamin (mit Ausnahme des Veratridins, das einen schwächeren Einfluß, und des KCN, das einen stärkeren Einfluß auf die Dopamin-Aufnahme hat). Demgegenüber übten die gleichen Konzentrationen auf die Tryptamin-Akkumulation einen Einfluß von maximal 15% aus. Die Untersuchung verschiedener Tryptamin-Analoga (6-Hydroxytryptamin, 5,6-Dihydroxytryptamin, *N*, *N'*-Dimethyltryptamin, 5,7-Dihydroxytryptamin, Tryptamin, 5-Methoxytryptamin und 5-Hydroxyindol) auf den Influx von 0.1μM Dopamin, 5-Hydroxytryptamin und Tryptamin zeigte, daß die Konzentration des Strukturverwandten, die die 5-Hydroxytryptamin-Aufnahme auf die Hälfte reduzierte (ID_{50}-Werte lagen im Bereich zwischen 2.4×10^{-6}M für 6-Hydroxytryptamin und 8×10^{-4}M für 5-Hydroxyindol), keinen Einfluß auf den Dopamin- oder Tryptamin-Influx hatte – mit Ausnahme des 5-Methoxytryptamins. 5-Methoxytryptamin hemmte die Tryptamin-Akkumulation um 40%, hatte jedoch keinen signifikanten Einfluß auf die Dopamin-Aufnahme.

Die vorliegenden Daten bestätigen früherer Ergebnisse[1,2] und werden im Sinne eines hoch spezifischen energieabhängigen Aufnahmemechanismus für den Influx von 0.1μM 5-Hydroxytryptamin in Schneckennervengewebe interpretiert. Zudem läßt sich sagen, daß die Dopamin-Aufnahme ebenfalls energieabhängig ist, wohingegen der Tryptamin-Influx weitestgehend unabhängig zu sein scheint sowohl von Pharmaka, die die verschiedenen energieproduzierenden Systeme beeinflussen, als auch von eng strukturverwandten Substanzen.

1 Osborne, N. N., Hiripi, L. & Neuhoff, V. (1975) *Biochem. Pharmacol.* **24**, 2141 - 2148.
2 Schröder, H.-U., Neuhoff, V. & Osborne, N. N. (1977) in Proc. 6th Europ. Malac. Congress.

H.-U. Schröder, V. Neuhoff und N. N. Osborne; Max-Planck-Institut für Experimentelle Medizin, Forschungsstelle Neurochemie, D-3400 Göttingen.

D. Schubert und B. Domning

Isolierung des „Band 3"-Proteins aus Erythrozytenmembranen durch präparative Elektrophorese in einer organischen Säure

„Band 3", das häufigste integrale Protein der menschlichen Erythrozytenmembran (M_r 100 000, das „Anionentransport-Protein"), wurde durch eine neuartige Präparationsmethode in detergenzfreier Form isoliert:

1) Die peripheren Membranproteine wurden durch Behandlung der Membran mit 10proz. Essigsäure aus ihr herausgelöst.
2) Der zurückbleibende Lipoprotein-Komplex wurde in 92proz. Essigsäure solubilisiert und dann durch Zonenelektrophorese in einem Gradienten aus Essig-

säure (mittlere Konzentration: 80%), Wasser und Saccharose in seine Komponenten getrennt (hierzu wurde die Elektrofokussiersäule LKB 8100-2 benutzt).

Die Ausbeute an „Band 3"-Protein betrug etwa 25 mg pro Lauf (60 - 70% des eingesetzten Proteins 3), die Reinheit etwa 95%. – Nach Abtrennung des organischen Lösungsmittels durch Dialyse gegen Wasser ist das isolierte Protein in wäßrigen Puffern löslich. Es erscheint in dieser Form als geeignetes Ausgangsmaterial für physikalisch-chemische Untersuchungen, die detergenzfreies Protein erfordern.

Prof. D. Schubert, Max-Planck-Institut für Biophysik, Abteilung Zellphysiologie, Heinrich-Hoffmann-Str. 7, D-6000 Frankfurt/M. 71.

E. Schütz und K. Radsak

Nucleäre und mitochondriale DNA-Polymerase γ: Untersuchungen zur Unterscheidung

Nucleäre und mitochondriale γ-Polymerase[1-3] aus Mauszellkulturen wurden chromatographisch partiell gereinigt mit dem Ziel, diese beiden zellulären DNA-Polymerasen auf biochemische Unterscheidungsmerkmale zu untersuchen. Zwei Drittel der zellulären Gesamt-γ-Polymerase-Aktivität wurden in der mitochondrialen Fraktion gefunden. Beide Enzyme wiesen annähernd identische Sedimentationskoeffizienten auf und zeigten auch bei der Re- und Co-Chromatographie auf Phosphocellulose ein vergleichbares Verhalten. Auch die Empfindlichkeit gegenüber Inhibitoren wie Äthidiumbromid, Phosphonoessigsäure und *ara*-CTP ließ keine wesentlichen Unterschiede erkennen. Die K_m-Werte für das nucleäre als auch das mitochondriale Enzym lagen zwischen 1 und 2μM. Bei Einzympräparationen aus Chloramphenicol-behandelten Zellkulturen wurde ein signifikant erniedrigter K_m-Wert ausschließlich für die mitochondriale γ-Polymerase bestimmt. Ergänzende Untersuchungen werden an Präparationen mitochondrialer γ-Polymerase aus synchronisierten Zellkulturen durchgeführt.

1 Bolden, A., Pedrali Noy, G. & Weissbach, A. (1977) *J. Biol. Chem.* **252**, 3351 - 3356.
2 Bertazzoni, U., Scovassi, A. & Brun, G. (1977) *Eur. J. Biochem.* im Druck.
3 Radsak, K. & Seidel, G. (1976) *Biochem. Biophys. Res. Commun.* **71**, 1259 - 1268.

Dr. K. Radsak, Hygiene-Universität der Universität Marburg, Pilgrimstein 2, D-3350 Marburg.

H.-P. Schultheiß, G. Bispink, V. Neuhoff und H.-D. Bolte

Die Isoenzyme der Lactat-Dehydrogenase im Herzmuskel: Unterschiedliche Verteilungsmuster innerhalb verschiedener Anteile und Wandschichten des Myocards beim Meerschweinchen

Eine der wesentlichen Funktionen der Lactat-Dehydrogenase (LDH) (EC 1.1.1.27) ist die Regulation des $NAD^{\oplus}$/NADH-Quotienten[1]. Bisher wurde angenommen, daß die H-Anteile des Enzyms im Myocard bei weitem überwiegen[2,3]. Da genaue Untersuchungen über die Verteilung der Isoenzyme in verschiedenen Myocardanteilen nicht vorliegen, bestimmten wir das Isoenzymverteilungsmuster mit Hilfe der Isoelektrischen Fokussierung in Mikrogelen (10 *μl*)[4].

Jeweils fünf Proben bei fünf verschiedenen Meerschweinchen wurden aus beiden Vorhöfen sowie aus den subendocardialen, mittleren und subepicardialen Schichten beider Ventrikel entnommen. Dabei kann gezeigt werden, daß keineswegs die H-Anteile in allen Myocardregionen überwiegen. Sowohl im rechten als auch im linken Vorhof überwiegen die Isoenzyme LDH_3, LDH_4 und LDH_5 – insgesamt also die M-Anteile (M-Anteil rechter Vorhof 65.5%, M-Anteil linker Vorhof 59%) (p = 0.01).

In den Ventrikeln findet man gegenüber den Vorhöfen eine deutliche Zunahme der Isoenzyme LDH_1 und LDH_2 – also der H-Anteile (H-Anteil rechter Ventrikel 49.5%, H-Anteil linker Ventrikel 55%) (p = 0.001). Zusätzlich zeigt sich im Querschnitt beider Ventrikelwände ein unterschiedliches Isoenzymmuster. Die H-Anteile nehmen von den subepicardialen zu den subendocardialen Schichten deutlich zu – Beispiel linker Ventrikel – (subepicardial 54%, subendocardial 61%) (p = 0.001), die M-Anteile gleichzeitig ab (subepicardial 46 %, subendocardial 39%) (p = 0.001). Die Konzentration der LDH_1 bei sich entsprechenden Lokalisationen und Wandschichten ist jeweils innerhalb des linken Herzens höher als innerhalb des rechten Herzens. Das Verhalten der LDH_5 ist dem entgegengesetzt. Die Unterschiede sind jeweils statistisch signifikant (p = 0.01).

Die Ergebnisse zeigen deutlich die energetischen und metabolischen Unterschiede sowohl in den verschiedenen Anteilen des Herzens als auch in den unterschiedlichen Wandschichten des Ventrikelmyokards.

1 Nadal-Ginard B & Markert, C. L. (1975) Isoenzymes II, Physiological Functions, N. Y. Academic Press, 45 - 67.
2 Kowalewski S. L. (1972) Die Isoenzyme der Lactat-Dehydrogenase, S. 57 - 67, Georg Thieme Verlag, Stuttgart.
3 Hammond, G. L., Nadel-Ginard, B., Talner, N. S. & Markert, C. L. (1976) *Circulation* **53**, 637 - 643.
4 Quentin, C. D. & Neuhoff, V. (1972) *Int. J. Neurosci.* **4**, 17 - 24.

Dr. H.-P. Schultheiß und Prof. Dr. H.-D. Bolte, Medizinische Klinik I der Universität München, Klinikum Großhadern, Marchioninistr. 15, D-8000 München 70.
Dr. G. Bispink und Prof. Dr. Neuhoff, Max-Planck-Institut für Experimentelle Medizin, Hermann-Rein-Str. 3, D-3400 Göttingen.

R. T. Schwarz und R. Datema

Beeinflussung der lipidabhängigen Glykosylierung durch 2-Desoxy-D-glucose

Die Glykosylierung von Glykoproteinen wird durch 2-Desoxy-D-glucose (dGlc) in vivo spezifisch gehemmt[1,2]. Der Mechanismus der Hemmung wurde in vivo und in vitro näher charakterisiert. Nach Markierung Influenzavirus-infizierter Hühnerembryofibroblasten mit [^{3}H]Glucosamin und [2-^{3}H]Mannose in vivo in Gegenwart von 2-Desoxy-D-glucose wurde der Einbau in verschieden extrahierbaren Fraktionen bestimmt. Es zeigte sich, daß die Radioaktivität in der Fraktion, die die Oligosaccharid-Lipide enthält, am stärksten gesenkt ist.

Inkubationen von UDP-[^{14}C]Glucosamin oder GDP-Mannose mit partikulären Präparationen aus Hühnerembryofibroblasten in vitro in Gegenwart von GDP-2-Desoxy-D-glucose, dem für die Hemmung in vivo verantwortlichen Metaboliten[3], zeigten, daß aberrante Formen von Dolichylpyrophosphoryl-oligosacchariden wie z.B. Dol-p-p(GlcNAc)$_2$dGlc auftreten. Es kann deshalb angenommen werden, daß 2-Desoxy-D-glucose durch Beeinflussung der lipidabhängigen Glykosylierung seine hemmende Wirkung entfaltet.

1 Scholtissek, C (1975) *Curr. Top. Microbiol. Immunol.* **70**, 101 - 119.
2 Schwarz, R. T. & Klenk, H.-D. (1974) *J. Virol.* **14**, 1023 - 1034.
3 Schmidt, M. F. G., Schwarz, R. T. & Scholtissek, C. (1976) *Eur. J. Biochem.* **70**, 55 - 62.

Ralph T. Schwarz, Institut für Virologie, Fachbereich Veterinärmedizin, Frankfurter Str. 107, D-6300 Lahn-Gießen.

I. von Schwarzenfeld

Falsche Transmitter im cholinergen System

Da Neurotransmitter synthetisierende Enzyme nicht streng substratspezifisch sind, ist es möglich, durch Gabe nichtphysiologischer Vorstufen die Bildung falscher Transmitter hervorzurufen. Im adrenergen System hat die Einbeziehung der falschen Transmitter in biochemische und pharmakologische Untersuchungen die Kenntnisse über die Aminaufnahme, -speicherung und -freisetzung wesentlich bereichert[1]. Obwohl Burgen[2] bereits 1956 für Triäthylcholin die mögliche Bildung eines falschen Transmitters (Acetyltriäthylcholin) vermutete, eröffnete erst die gezielte Synthese differenter, nicht homologer, Cholinanaloga durch Barker und Mittag[3] neue Wege, um offene Probleme mittels falscher Transmitter auch im cholinergen System zu untersuchen. Die differente Spezifität bestimmter metabolischer Schritte bei einem generell dem physiologischen Transmitter analogen Metabolismus dürfte es ermöglichen, die bestehenden Kontroversen bezüglich des Ursprungs von freigesetztem Acetylcholin, die insbesondere auf der metabolischen Heterogenität der synaptischen Vesikel basieren, zu lösen. Unter diesem Aspekt durchgeführte Versuche mit Pyrrolidincholin am elektrischen Organ von *Torpedo marmorata*[4] sowie am Meerschweinchengehirn[5] unterstützen die Hypothese der vesikulären Acetylcholin-Freisetzung.

Am elektrischen Organ von *Torpedo marmorata* konnte darüber hinaus die Aufnahme der Vorstufe Homocholin in synaptische Vesikel und dessen Freisetzung nach elektrischer Reizung demonstriert werden*. Kombinierte Untersuchungen mit falschen Transmittern und synaptisch aktiven Pharmaka sollten es ermöglichen, den subzellulären Angriffspunkt von im cholinergen System wirkenden Substanzen zu spezifizieren.

* Sudlow, G., persönl. Mitteilung

1 Muscholl, E. (1972) in Handb. d. Exp. Pharmak., Bd. 33 (Blaschko, H. & Muscholl, E., Hrsg.) S. 618 - 660, Springer-Verlag, Berlin.
2 Burgen, A. S. V., Burke, G. & Desbarats-Schonbaum, M.-L. (1956) *Brit. J. Pharmacol.* **11**, 308 - 312.
3 Barker, L. A. & Mittag, T. W. (1975) *J. Pharmacol. Exp. Ther.* **192**, 86 - 94.
4 Zimmermann, H. & Dowdall, M. J. (1977) *Neuroscience* **2**, 731 - 739.
5 v. Schwarzenfeld, I. (1978) *Naunyn-Schmiedeberg's Arch. Pharmacol. Suppl. I*, R 216.

I. von Schwarzenfeld, Pharmakologisches Institut der Universität Mainz, Obere Zahlbacher Str. 67, D-6500 Mainz.

M. Schwenk and V. Lopez del Pino

Transport and Metabolism of Estrone Sulfate in Isolated Rat Liver Cells

Estrone sulfate is considered to be a major circulating estrogen in the human. It undergoes enterohepatic circulation and is subject to hepatic uptake, metabolism in the liver, and biliary secretion[1]. These biochemical steps may be of importance with regard to the pharmacokinetics and bioavailability of estrogens.

Methods: Transport was studied by the use of isolated rat liver cells, as previously described[2]. Radioactive metabolites were analyzed in both intra- and extracellular compartments by thin-layer chromatography in order to identify chemical forms and amounts.

Results: Estrone sulfate uptake by hepatocytes is mediated by an energy-dependent anion carrier. This conclusion is based on the following observations of initial rates of uptake: uptake is saturable, with a K_m of approximately 1μM. It is inhibited by such -SH reagents as mersalyl. Taurocholate and estrone sulfate compete for uptake. The activation energy is 67 kJ/mol. There is a pH optimum in the physiological range. Inhibition of mitochondrial respiration lowers uptake immediately by 93%. Estrone sulfate is accumulated inside the cells, where it is partly cleaved to the free steroid. This is conjugated with glucuronic acid (about 30% at an external estrone sulfate concentration of 0.2μM). In this form it is finally secreted from the cells.

Conclusion: Hepatic uptake of estrone sulfate is mediated by an anion carrier. Cleavage to the free steroid within the hepatocytes may be of importance with regard to the bioavailability of estrogens. The results are consistent with data obtained with the isolated perfused liver.

1 Höller, M., Grochtmann, W., Napp, M. & Breuer, H. (1977) *Biochem. J.* **166**, 363 - 371.

2 Schwenk, M., Burr, R., Schwarz, L. & Pfaff, E. (1976) *Eur. J. Biochem.* **64**, 189 - 197.

Michael Schwenk, Institut für Toxikologie, Universität Tübingen, Wilhelmstr. 56, D-7400 Tübingen.

J. Seeger, H. Murer and R. Kinne

Substrate-Induced Sodium Movements Mediated by Sodium-Cotransport Systems in Renal and Intestinal Brush Border Membrane Vesicles

During the past several years, sodium-dependent substrate uptake has been studied extensively in isolated renal and intestinal brush border microvilli by a rapid filtration technique[1]. Due to the large contribution of "substrate-independent" sodium movement to the net sodium movement across the membrane it was, however, difficult to demonstrate unequivocally a substrate-dependent movement of the cosubstrate sodium.

To reduce the net sodium movement across the membranes the experiments described below were performed with brush border vesicles preequilibrated with 5mM ^{22}Na. The change in intravesicular sodium was followed after addition of different substrates to the extravesicular space. In order to avoid any membrane potentials, the experiments were carried out in the presence of valinomycin, and of KSCN in equal concentrations at both sides of the vesicle membrane.

Addition of 5mM D-glucose to the extravesicular space led to a transient 10 - 15% increase in intravesicular sodium concentration in renal as well as intestinal brush border membrane vesicles. The glucose-induced sodium movement was not observed in the presence of 0.1mM phlorrhizin or after addition of L-glucose instead of D-glucose. Addition of 5mM L-phenylalanine produced similar changes in the intravesicular sodium content, and again the non-transported stereoisomer was ineffective. The same results were obtained with L-glutamate. The effect of substrates on the sodium movement was prevented by the presence of monactin, which increases the sodium conductivity of the membranes.

These results indicate that under appropriate conditions, sodium movements mediated by sodium-solute cotransport systems can be detected. Since, by using different labels, it should be possible to measure sodium and substrate movement simultaneously, such studies should allow direct determination of the coupling coefficients in sodium cotransport systems.

1 Murer, H. & Kinne, R. (1977) in Biochemistry of Membrane Transport, FEBS-Symposium No. 42 (Semenza, G. & Carafoli, E., eds.) pp. 292 - 304, Springer-Verlag, Berlin, Heidelberg, New York.

J. Seeger, H. Murer und R. Kinne, Max-Planck-Institut für Biophysik, Kennedyallee 70, D-6000 Frankfurt 70.

D. Seidel

Der Stoffwechsel und die klinische Bedeutung der Plasmalipoproteine

In Deutschland leiden 20 - 30% der Bevölkerung an der Erhöhung einer oder mehrerer Plasmalipidfraktionen; damit ist die Hyperlipämie häufiger als der Diabetes mellitus und die häufigste Stoffwechselerkrankung überhaupt. Die große klinische Bedeutung des Fettstoffwechsels ergibt sich durch die vielfältigen Hinweise und Beweise, daß den Lipiden eine besondere Rolle in der Ätiologie degenerativer vasculärer Erkrankungen, speziell der Atherosklerose, zukommt. Eine erste Beziehung zwischen Cholesterin und Atherosklerose stellte die Entdeckung von Windaus her, daß die atheromatösen Veränderungen in den Arterien hauptsächlich Cholesterin und dessen Ester enthalten. Heute, 65 Jahre später, wissen wir, daß dieses Cholesterin zum allergrößten Teil von den im Plasmapool zirkulierenden löslichen Lipoproteinen stammt. Die Ergebnisse genetischer sowie biochemischer Forschung der letzten Jahre auf diesem Gebiet zeigen in guter Übereinstimmung, daß die pathobiochemischen Zusammenhänge von Störungen des Fettstoffwechsels nur dann erkannt und therapeutisch erfolgreich angegangen werden können, wenn sich die Analytik auf wohl definierte Lipoproteineinheiten und deren Stoffwechselregulation im Plasma und den verschiedenen Geweben konzentriert. Unterschiede in physikochemischen und chemischen Eigenschaften der verschiedenen Plasmalipoproteine, die festgelegt sind durch die Natur und Konzentration ihrer einzelnen Protein- und Lipidkomponenten, bedingen eine unterschiedliche metabolische Beziehung zwischen einer Lipidfraktion und den am Lipidstoffwechsel beteiligten Organen. Eingeleitet durch Untersuchungen an Fibroblastenkulturen, die später auf andere Zellsysteme ausgedehnt wurden, ist nunmehr gesichert, daß es zwischen den Lipid-transportierenden Lipoproteinen des Plasmas und bestimmten Organen zu einer durch den Proteinanteil gesteuerten Wechselwirkung kommt. Hierdurch erhalten Unterschiede in der Apoproteinkomposition einzelner Lipoproteinfraktionen die größte biologische Bedeutung. Obgleich im Plasma zwischen den einzelnen Lipoproteinfraktionen ein steter Austausch sowohl ihrer Protein- wie ihrer Lipidkomponenten stattfindet, und alle Lipoproteinfraktionen Regelmechanismen unterliegen, die eng miteinander verknüpft sind, ist es wichtig zu erkennen, daß sich die einzelnen Lipoproteinfraktionen aufgrund ihrer unterschiedlichen Komposition in ihrer klinischen Wertigkeit unterscheiden, ja zum Teil sogar gegensätzlich verhalten können. Dies gilt sowohl für die primären, genetisch determinierten Formen als auch für die sekundären Dyslipoproteinämien und setzt zur Beurteilung eine zuverlässige und aussagekräftige Analytik des Lipoproteinsystems voraus.

Dietrich Seidel, Klinisch-Chemisches Laboratorium, Medizinische Universitätsklinik, Bergheimerstraße 58, D-6900 Heidelberg.

H. Seifert, W. Northemann and P. C. Heinrich

Low Molecular Weight RNAs Are Components of Nuclear Ribonucleoprotein Particles

Ribonucleoprotein (RNP) particles were isolated from EDTA-washed rat liver nuclei after gentle sonication and centrifugation on a continuous (15 - 35%) sucrose gradient. In the presence of RNAase inhibitor, purified from rat liver cytosol, RNP particles of 50 - 150S (polyparticles) were isolated.

These particles contain, in addition to the rapidly labeled heterogeneous nuclear RNA (hnRNA), several small RNA (snRNA) species[1,2]. In order to exclude the possibility that the snRNAs are degradation products contaminating the polyparticles, the following experiments were carried out.

(1) Polyparticles were treated with 0.14, 0.25, 0.5, 1.0 and 2.0M NaCl and centrifuged through a 1.0M sucrose layer containing the respective NaCl concentration. The RNA of the sedimented NaCl-treated particles was phenol-extracted and analyzed by formamide polyacrylamide micro gel electrophoresis (linear gradient of 1% to 20%). Particle snRNAs were resolved into 8 sharp bands on the gradient gels. All 8 species were present in particles treated with 2M NaCl, although 63% of particle proteins were lost under these conditions. There was a decrease in intensity of only one snRNA band with increasing NaCl concentrations. A tight association of the snRNA species with the ribonucleoprotein complex must therefore exist.

(2) Polyparticles were chromatographed on an oligo(dT) column. The RNP particle fraction, which was bound to oligo(dT) and eluted from the column with 50% formamide, still contained snRNAs.

(3) In contrast to hnRNA, which is readily labeled by [^{3}H]orotate as precursor, the snRNA did not incorporate radioactivity, even after repeated [^{3}H]orotate injections for several days.

1 Northemann, W., Scheurlen, M., Gross, V., Heinrich, P.C. (1977) *Biochem. Biophys. Res. Commun.* **76**, 1130 - 1137.

2 Deimel, B., Louis, C., Sekeris, C.E. (1977) *FEBS-Lett.* **73**, 80 - 84.

Hans Seifert, Wolfgang Northemann and Peter C. Heinrich, Biochemisches Institut der Universität Freiburg, Hermann-Herder-Str. 7, D-7800 Freiburg i. Br.

S. Shaltiel and G. Halperin

Probing the Surface of Cells with Homologous Series of Hydrocarbon-Coated Agaroses

Cells can be efficiently adsorbed onto and desorbed from homologous series of hydrocarbon-coated agaroses under isotonic conditions[1]. Such column materials (prepared by reaction of agarose with CNBr and subsequent coupling with an α-aminoalkane, n carbon atoms long) are suitable for the reversible binding of erythrocytes, since they cause essentially no physical entrapment or lysis, and since the erythrocytes eluted from such columns and those yet to be applied are not only morphologically indistinguishable, but exhibit an identical osmotic fragility profile, suggesting that they are not significantly damaged by the adsorption-desorption process.

At neutral pH (where the adsorption is carried out), the columns are positively charged, while the erythrocyte surface is negatively charged. Nevertheless, hydrophobic interactions play a key role in this adsorption[2] since (a) columns with equal or even decreasing net charge exhibit an increased capacity for erythrocytes with increasing hydrocarbon chain length, (b) even after acetylation of the columns (which was reported to cancel their positive charge) there is a chain-length dependent gradation in their capacity for erythrocytes, (c) enzymatic removal of sialic acid residues from the surface of guinea pig erythrocytes, with concomitant decrease in the negative charge of the cell surface, does not significantly affect the adsorption profile of these cells.

Erythrocytes from different sources exhibit different affinities for members of such column series (DBA/1 mice > C57BL/6J mice > random bred rabbits > Heston guinea pigs > humans), illustrating their potential use in probing the surface of cells.

1 Halperin, S. & Shaltiel, S. (1976) *Biochem. Biophys. Res. Commun.* **72**, 1497 - 1503.
2 Halperin, G. & Shaltiel, S. (1978) in Affinity Chromatography (Hoffmann Ostenhof et al., eds.) Pergamon Press, Oxford, in press.

Shmuel Shaltiel and Gideon Halperin, The Weizmann Institute of Science, P.O.B. 26, Rehovot, Israel.

A.E. Shamoo, T.L. Scott and J.J. Abramson

Molecular Mechanism of $Ca^{2\oplus}$ Transport in Sarcoplasmic Reticulum $Ca^{2\oplus}$ + $Mg^{2\oplus}$ ATPase

The energy for $Ca^{2\oplus}$ transport is derived from the hydrolysis of ATP by $Ca^{2\oplus}$ + $Mg^{2\oplus}$ ATPase (EC 3.6.1.3, ATP dephosphohydrolase). The intact enzyme (M_r = 100 000) has been shown to contain a $Ca^{2\oplus}$-dependent and selective ionophore in artificial lipid bilayers[1]. Two fragments of 55 000 and 45 000 daltons have been purified from the first tryptic digest (TD1) of the enzyme. Subsequent digestion (TD2) cleaves the 55 000 into 30 000 and 20 000 dalton fragments. The 55 000 and 30 000 fragments have been shown to contain the site of phosphorylation and of *N*-ethylmaleimide binding, indicative of the hydrolytic site of the enzyme. The 55 000 and 20 000 dalton fragments have been shown to contain the $Ca^{2\oplus}$-dependent and selective ionophoric activity inhibited by ruthenium red and by mercuric chloride, but not by methylmercury chloride, an inhibitor of the hydrolytic site of the enzyme[2]. Further, we have data to indicate that the 45 000 fragment serves as large non-selective channel. The reduction of the disulfide bridges in this fragment causes the disruption of its ionophoric activity and consequently the loss of the ionophoric activity of the intact enzyme. Thus, the 45 000 fragment appears to be in series with the 55 000 fragment in the overall ATPase system.

The decrease in $Ca^{2\oplus}$ transport in sarcoplasmic reticulum vesicles parallels the cleavage of 55 000 (TD2) and the appearance of the 30 000 and 20 000 fragments. This cleavage (TD2) also results in the conversion of the high affinity sites for $Ca^{2\oplus}$ to low affinity sites with the total number of binding sites remaining constant. These and other evidence suggest that $Ca^{2\oplus}$ transport is driven by the hydrolysis of ATP at the 30 000 fragment, which changes the $Ca^{2\oplus}$ ion transport site (20 000) so that it transports $Ca^{2\oplus}$ through the large channel (45 000)[3].

1 Shamoo, A.E. & MacLennan, D.H. (1974) *Proc. Natl. Acad. Sci. U.S.A.* **71**, 3522 - 3526.
2 Shamoo, A.E., Ryan, T.E., Stewart, P.S. & MacLennan, D.H. (1976) *J. Biol. Chem.* **251**, 4147 - 4154.
3 Shamoo, A.E. & Goldstein, D.E. (1977) *Biochim. Biophys. Acta* **472**, 13 - 53.

Dr. Adil E. Shamoo, Department of Radiation Biology and Biophysics, University of Rochester Medical Center, Rochester, N.Y. 14642, U.S.A.

G.G. Shipley

Lipid Structure and Lipid-Protein Interactions

Although detailed information on both the structure of cell membranes and plasma lipoproteins and the lipid-protein interactions responsible for their stability remains fragmentary, precise information on the structure of biological lipids (phospholipids, sphingolipids, cholesterol, cholesterol esters and triglycerides) in the crystalline state is now available. The molecular conformation and intermolecular arrangement of these lipids in the crystalline state will be used as a starting point for understanding lipid organization in less ordered lipid-water, lipid-lipid, and lipid-protein systems. The structural information derived from these model systems using x-ray diffraction/scattering, calorimetry and nuclear magnetic resonance techniques will be compared with structural data derived from similar studies on both membranes and plasmam lipoproteins.

G. Graham Shipley, Biophysics Division, Department of Medicine, Boston University School of Medicine, 80 East Concord Street, Boston, Mass. 02118, U.S.A.

E. A. Siess, D. G. Brocks und O. H. Wieland

Zur Differenzierung der Glucagonwirkung auf die Gluconeogenese und Ketogenese in isolierten Leberzellen der Ratte

In isolierten Leberzellen hungernder Ratten zeigt die Glucagonwirkung auf die Gluconeogenese aus Lactat die gleiche Dosisabhängigkeit wie die Steigerung der Ketogenese, die Erhöhung des mitochondrialen Redoxzustandes und die Erniedrigung des zellulären Glutamat- und 2-Oxoglutaratgehalts. Senkt man durch Verabreichung einer Biotinmangeldiät die Aktivität der Pyruvat-Carboxylase (EC 6.4.1.1) auf ein Zehntel des Normalwertes, so ist Glucagon nicht mehr in der Lage, die Glucosebildung aus Lactat (wohl aber aus Glutamat) zu stimulieren. Dennoch steigert das Hormon im Biotinmangel unvermindert die Ketonkörperbildung und senkt die Spiegel der C_5-Dicarbonsäuren. Diese Ergebnisse zeigen nicht nur, daß Gluconeogenese und Ketogenese gleich sensitiv auf Glucagon reagieren, sondern auch, daß beide Prozesse unabhängig voneinander durch Glucagon stimuliert werden.

E. A. Siess, D. G. Brocks und O. H. Wieland, Forschergruppe Diabetes und Klinisch-Chemisches Institut, Krankenhaus München-Schwabing, Kölner Platz 1, D-8000 München 40.

S. Soboll, K. Werdan, E. Erdmann und H. W. Heldt

Messung der mitochondrialen und cytosolischen ATP/ADP-Quotienten in kultivierten Rattenherz-Fibroblasten

Herzzellen ein bis drei Tage alter Ratten wurden nach Harary et al.[1] isoliert. Nach Abtrennung der Zellen vom muskulären Typ[2] wurden die fibroblastoiden Zellen auf Plastik gezüchtet, nach der zweiten Subkultivierung mit flüssigem Stickstoff gefriergestoppt und lyophilisiert. Mit einer Modifikation der von Elbers et al.[3] beschriebenen Methode der nichtwäßrigen Fraktionierung erfolgte die Auftrennung des Zellmaterials in eine cytosolische und mitochondriale Fraktion. In beiden Kompartimenten wurden die Gehalte von ATP, ADP, Kreatin und Kreatinphosphat gemessen. Kreatinphosphat und Kreatin konnten nur im Cytosol gefunden werden, wobei der Quotient Kreatinphosphat/Kreatin etwa 4 betrug.

Die Mitochondrien machten 20% des Zellproteins aus und enthielten 8% des zellulären ATP- und 14% des zellulären ADP-Gehaltes. Daher lag der mitochondriale ATP/ADP-Quotient (im Mittel 1.9; $n = 5$) niedriger als der cytosolische Wert (im Mittel 5.2; $n = 5$). Ähnliche Verhältnisse findet man in der perfundierten Rattenleber*, jedoch sind die Unterschiede der subzellulären ATP/ADP-Quotienten in Herzfibroblasten wesentlich geringer als in Lebergewebe.

Die Ergebnisse zeigen, daß die für perfundierte Leber entwickelte Fraktionsmethode in nichtwäßrigen Medien auch an Zellkulturen anwendbar ist. Auf diese Weise besteht die Möglichkeit, den Energiestoffwechsel des Herzgewebes durch Bestimmung der subzellulären ATP/ADP-Quotienten in den beiden Zellarten – Herzmuskelzellen und fibroblastoide Zellen – getrennt zu untersuchen.

* Soboll, S., Scholz, R. & Heldt, H. W. (1978) *Eur. J Biochem.*, zur Veröffentlichung eingereicht.

1 Harary, I., Hoover, F. & Farley, B. (1974) *Methods Enzymol.* **32**, 740 - 745.
2 Polinger, I. S. (1970) *Exp. Cell. Res.* **63**, 78 - 82.
3 Elbers, R., Heldt, H. W., Schmucker, P., Soboll, S. & Wiese, H. (1974) *Hoppe-Seyler's Z. Physiol. Chem.* **335**, 378 - 393.

S. Soboll und H. W. Heldt, Institut für Physiologische Chemie, Physikalische Biochemie und Zellbiologie der Universität München, Goethestraße 33, D-8000 München 2.

K. Werdan und E. Erdmann, Medizinische Klinik I der Universität München, Klinikum Großhadern, Marchioninistraße 15, D-8000 München 70.

H. Stadler

Reinvestigation of Vesiculin and the Presence of Glycosamionglycans in Cholinergic Synaptic Vesicles

Vesiculin, released after water dialysis of hypo-osmotically shocked vesicle preparations from the electric organ of *Torpedo marmorata* and isolated by gel filtration, was interpreted earlier as a protein that constitutes part of the core of synaptic vesicles[1]. A chemical re-examination revealed that most of the amino acids detected are not present in peptide linkage. The material complexing them, with a molecular weight of about 10 000, was identified as a glycosaminoglycan by its characteristic molar ratio of hexosamine to uronic acid of 1 : 1, and showed a strong binding property for free amino acids, especially acidic ones. Since glucosamine is the predominant amino sugar in vesiculin, and sulphate was detected, it might be a carbohydrate of the heparin type.

If highly purified vesicle preparations are analysed for total glycosaminoglycan content[2], preliminary values indicate an amount of not less than 30 to 40% per mg vesicle protein.

Vesiculin is thus regarded as a breakdown of a probably high molecular weight proteoglycan; a localisation within the vesicular core has not been demonstrated yet. It is concluded that cholinergic synaptic vesicles do not contain a low molecular weight core protein, but a highly negatively charged glycosaminoglycan in an amount that could play a functional role in the storage of acetylcholine within vesicles.

1 Whittaker, V. P., Dowdall, M. J., Dowe, G. H. C., Facino, R. M. & Scotto, J. (1974) *Brain Res.* **75**, 115 - 131.
2 Margolis, R. U. & Margolis, R. K. in Research Methods in Neurochemistry (1972) (Marks, N. & Rodnight, R., eds.) pp. 249 - 253, Plenum Press, New York.

Dr. Herbert Stadler, Max-Planck-Institut für Biophysikalische Chemie, Abteilung Neurochemie, Postfach 968, D-3400 Göttingen-Nikolausberg.

G. Steffens und G. Buse

Teilsequenzen der Polypeptidkette II von Cytochrom-c-Oxidase (EC 1.9.3.1) aus Rinderherzen

Cytochrom-*c*-Oxidase wird nach bekannten Verfahren aus Rinderherzmuskeln isoliert. Die Auftrennung in einzelne Polypeptidkomponenten läßt sich durch Gelfiltration über Bio-Gel-P-60 in 2proz. Natriumdodecysulfat oder Craig-Verteilung im System (2-Butanol/4proz. Essigsäure) erreichen. Die Charakterisierung[1] der Polypeptidketten erfolgt durch Natriumdodecylsulfat-Gelelektrophorese, Aminosäureanalyse und *N*- und *C*-terminale Endgruppenbestimmung.

Die Aminosäuresequenz der Polypeptidkette II beginnt *N*-terminal mit *N*-Formylmethionin. Dies bestätigt die Befunde[2] eines mitochondrialen Ursprungs dieser Kette. Als *C*-Terminus wird Leucin gefunden.

Die Sequenzanalyse wird hauptsächlich über große Fragmente durchgeführt, die durch überlappende Spaltung mit BrCN oder Trypsin dargestellt werden. Wir erhalten nach Trennung über Bio-Gel P-6 vierzehn BrCN-Peptide und nach Trennung über Bio-Gel P-10 elf lösliche tryptische Peptide. Damit gelingt eine eindeutige Zuordnung von dreißig Sequenzen im *N*-terminalen und von über hundert im *C*-terminalen Bereich.

Für die automatische Sequenzierung der Spaltpeptide mit dem Beckman-Sequencer wurde ein neues Peptid-Programm entwickelt, das einen vollständigen Abbau – auch hydrophober Peptide – ohne vorherige chemische Modifizierung oder Ankupplung an eine Festphase erlaubt.

Nach Aufarbeitung von drei weiteren schwerlöslichen tryptischen Peptiden sollte es gelingen, die vollständige Primärstruktur der Polypeptidkette II aus Cytochrom-*c*-Oxidase zu erhalten.

1 Steffens, G. & Buse, G. (1976) *Hoppe-Seyler's Z. Physiol. Chem.* **357**, 1125 - 1137.
2 Yatscoff, R. W. & Freeman, K. B. (1977) *FEBS Lett.* **81**, 7 - 9.

Dr. Gerd Steffens and Prof. Dr. G. Buse, Abteilung Physiologische Chemie, RWTH Aachen, Melatener Str. 211, D-5100 Aachen.

W. Stender

Quaternary Structure of Escherichia coli RNA Polymerase I. Measurement of Distances by Fluorescence Energy Transfer

The elucidation of the quaternary structure of *E. coli* RNA polymerase (RNA nucleotidyltransferase, EC 2.7.7.6) together with the knowledge of subunit function will be important prerequisites for complete understanding of the transcriptional process in bacteria. Recently the arrangement of subunits has been presented in 3 models by protease digestion[1] and by crosslinking experiments[2,3] but quantitative spatial relationships between subunits or other fixed points on RNA polymerase have only been reported in one case[4] (σ-subunit – rifamycin binding site). We employed the latter approach together with enzyme reconstitution techniques to determine the distances between the fluorescent rifamycin analogue 3-[2-(7-nitro-2,1,3-benzoxadiazol-4-ylamino)ethylthio]rifamycin SV (NBD-Rif) and the subunits labelled with 5-[2-(iodoacetamido)ethylamino]-1-naphthalenesulfonate (1,5-J-AEDANS)[5]. Experiments using differently labelled subunits and mixed reconstitution techniques are in progress.

1 Lill, U. I. & Hartmann, G. R. (1975) *Eur. J. Biochem.* **54**, 45 - 53.
2 Coggins J. R., Lumsden, J. & Malcolm, A. D. B. (1977) *Biochemistry* **16**, 1111 - 1116.
3 Hillel, Z. & Wu. C.-W. (1977) *Biochemistry* **16**, 3334 - 3342.
4 Wu, C.-W., Yarbrough, C. R., Wu, F. Y.-H. & Hillel, Z. (1976) *Biochemistry* **15**, 2097 - 2104.
5 Hudson, E. N. & Weber, G. (1973) *Biochemistry* **12**, 4154 - 4161.

Dr. Walter Stender, Physiologisch-Chemisches Institut, Abteilung Molekulare Enzymologie, Universität Würzburg, Koellikerstr. 2, D-8700 Würzburg.

R. Stermann and K. Decker

Changing the Intracellular P_i Content Influences the Rate of Glycogenolysis in Isolated Rat Hepatocytes

Glucose output and rates of glycogenolysis in isolated rat hepatocytes were found to be decreased in the presence of 2mM D-galactosamine. The amino sugar is known to trap inorganic phosphate by intracellular accumulation of D-galactosamine 1-phosphate. This suggested that the intracellular concentration of inorganic phosphate could influence the rate of glycogen breakdown.

Addition of D-galactosamine to isolated hepatocytes lowered the intracellular P_i level within 20 min of incubation from 3.6 to 2.3 μmol/g wet cells, concomitant with an intracellular decrease of the ATP content from 2.34 to 1.86 μmol/g wet cells. At the same time, the rate of glycogenolysis and of glucose output was reduced to about 50% of controls. Addition of inorganic phosphate to the incubation medium raised the intracellular phosphate concentration, reaching a plateau at an extracellular concentration of about 20mM. At this level of inorganic phosphate, the intracellular P_i content rose by about 56% to 6.4 μmol/g wet cells, whereas the intracellular ATP concentration increased to 2.96 μmol/g wet cells. Under these conditions, a stimulation of glycogenolysis and glucose release of more than 100% was measured. Addition of sulfate ions had no effects on the intracellular phosphate concentration and on glycogenolysis and glucose output. 2mM D-galactosamine and 20mM potassium phosphate supplied simultaneously counteracted their individual effects.

Under all conditions tested, a close correlation was found between the rate of glycogenolysis and the intracellular phosphate concentration (r = 0.83). It is concluded that in isolated hepatocytes, the rate of glycogen breakdown is limited by the intracellular P_i level. These findings are substantiated by studies on the changes of P_i in subcellular fractions of normal, GalN-treated and P_i supplemented hepatocytes.

R. Stermann and K. Decker, Biochemisches Institut an der Medizinischen Fakultät der Universität Freiburg, Hermann-Herder-Str. 7, D-7800 Freiburg i. Br.

F. Stöckmann, L. Phivan and H.-D. Söling

The Possible Role of a Protein-Carboxymethylase in Stimulus-Secretion Coupling in the Exocrine Pancreas

Diliberto et al.[1-3] have recently shown a regional and subcellular distribution of protein-carboxymethylase activity in brain and other tissues, including the adrenal medulla and pancreas. They discuss the possibility that a carboxymethylation of the vesicle membranes might be necessary for membrane fusion during secretion, because carboxymethylation led to a release of methylated proteins from isolated vesicles[4].

When studying the activity of a protein-carboxymethylase in isolated rat pancreas lobules and in vivo, we found a 2 - 3-fold increase in enzyme activity following the administration of pancreocymin or carbachol, while secretin is ineffective. The increase in protein-carboxymethylase activity occured within 2 min after the addition of the secretagogue and lasted for at least 40 min. The mechanism of this rapid enzyme activity change is under investigation. The rapid stimulation could not be abolished by cycloheximide. Isolated plasma membranes and granules from rat pancreas are specific substrates for the protein-carboxymethylase, whereas no carboxymethylation of proteins in the 100 000 x g supernatant occurs. It is proposed that carboxymethylation is involved in exocytosis of various types of secreting cells by facilitating the membrane fusion process.

1 Diliberto, E. J. & Axelrod, J. (1974) *Proc. Natl. Acad. Sci. U.S.A.* **71**, 1701 - 1704.
2 Diliberto, E. J. & Axelrod, J. (1976) *J. Neurochem.* **26**, 1159 - 1165.
3 Diliberto, E. J., Viveros, O. H. & Axelrod, J. (1976) *Proc. Natl. Acad. Sci. U.S.A.* **73**, 4050 - 4054.
4 Diliberto, E. J., Viveros, O. H. & Axelrod, J. (1977) 11th FEBS Meeting, Copenhagen, Abstracts, p. 326.

Prof. Dr. H.-D. Söling, Abteilung Klinische Biochemie, Medizinische Universitätsklinik, Humboldtallee 1, D-3400 Göttingen.

W. Stoffel, W. Därr, P. Metz, K. P. Salm und B. Tunggal

Untersuchungen über die molekulare Organisation der Lipide und Apoproteine in menschlichen Serum-High-Density-Lipoproteinen

Untersuchungen über die Anordnung der am Aufbau der High-Density-Lipoprotein-Partikel des menschlichen Serums beteiligten Lipide und Lipoproteine mit Hilfe der ^{13}C-NMR-Spektroskopie führten zu dem Vorschlag eines Modells für diese Partikel[1].

In der Oberfläche sind danach die Apoproteine A-I und A-II mit den polaren Seiten ihrer amphiphilen α-helikalen und β-Strukturen zusammen mit den polaren Köpfen der Phospholipide und der 3-Hydroxylgruppe des Cholesterins angeordnet.

Diesen Folgerungen liegen neben stöchiometrischen Überlegungen, Berechnungen der Sekundärstruktur nach Chou und Fasman[2] und Corey-Prelog-Koltun-Modellen, vorwiegend Rekombinationsexperimente zugrunde. Hierzu wurde eine Methode ausgearbeitet, die zu reproduzierbaren Stöchiometrien und Ausbeuten führt. Eine detaillierte Analyse der Wechselwirkungen, die die Apoproteine mit den Lipiden eingehen, wurde an Komplexen der gereinigten Apoproteine A-I und A-II mit in den Fettsäureketten und den polaren Gruppen ^{13}C angereicherten Lysolecithin-, Phosphatidylcholin- und Sphingomyelinspecies durchgeführt. Aus Änderungen der Rotationskorrelationszeiten der mit ^{13}C angereicherten Segmente der Lipidmoleküle, die über Spin-Gitter-Relaxationszeiten bestimmt wurden, kann geschlossen werden, daß im wesentlichen hydrophobe Wechselwirkungen die Stabilität der Lipoproteinkomplexe garantieren. Die Bindung der Bis(acyl)lipide führt zu einer erheblichen Induktion von α-Helix-Struktur.

Komplexe, die aus Apo A-I und zwei hydrophoben Alkanketten-haltigen Lipiden bestehen, bilden scheibenförmige Partikeln (50 x 150 Å). Die am Aufbau beteiligten vier Apo-A-I-Moleküle werden durch Dimethylsuberimidat zu Dimeren quervernetzt. Die Form dieser Lipoproteinkomplexe entspricht der der nativen, neu assoziierten HDL-Partikeln beim Verlassen der Leberzelle[2,3]*.

Es wird eine neue Methode beschrieben, die helfen soll, die Bindungsart und die molekulare Organisation zwischen Lipiden und Apoproteinen zu studieren. Dazu wurden Lipide mit photosensitiven Gruppen rekombiniert. Durch Photolyse wurde aus der Azidogruppe eine hochreaktive Nitrengruppe gebildet, die mit den nächsten Nachbarn quervernetzt. Abbau der kovalent gebundenen Lipoproteine führt zu Lipopeptiden, deren Analyse einen weiteren Beitrag zur Topochemie der HDL-Partikel liefern soll[4,5]**.

Ferner wird über eine Lipidaustausch-Methode berichtet***. Die nativen Lipide werden gegen solche mit Markierung mit ^{13}C oder photosensitiven Gruppen ausgetauscht. Hierbei bleibt die Apoproteinkonformation erhalten. Die Bedeutung der Methode für unsere physikalischen und chemischen Analysemethoden wird durch erste Ergebnisse unterstrichen.

* Stoffel, W., Därr, W. & Salm, K. P. (1978), in preparation.

** Stoffel, W., Salm, K. P. & Körkemeier, U. (1978) in preparation.

*** Stoffel, W., Salm, K. P. & Melzner, I. (1978), in preparation.

1 Stoffel, W., Zierenberg, O., Tunggal, B. & Schreiber, E. (1974) *Proc. Natl. Acad. Sci. U.S.A.* **71**, 3696 - 3700.

2 Chou, P. Y. & Fasman, G. D. (1974) *Biochemistry* **13**, 222 - 245.

3 Stoffel, W., Därr, W. & Salm, K. P. (1977) *Hoppe-Seyler's Z. Physiol. Chem.* **358**, 1 - 11.

4 Felker, T. E., Fainaru, M., Hamilton, R. L. & Havel, R. J. (1977) *J. Lipid Res.* **18**, 465 - 473.

5 Stoffel, W., Därr, W. & Salm K. P. (1977) *Hoppe Seyler's Z. Physiol. Chem.* **358**, 453 - 462.

W. Stoffel, Physiol.-Chem. Institut der Universität zu Köln, Joseph-Stelzmannstr. 52, D-5000 Köln 41.

Ch. Stournaras and G. Kurz

An Exceptional 6-Phospho-D-gluconate Dehydrogenase from Pseudomonas fluorescens

In the catabolism of D-glucose 6-phosphate in *Pseudomonas fluorescens* two pathways are operative, the Entner-Doudoroff pathway and the hexose monophosphate pathway, both starting with the dehydrogenation of D-glucose 6-phosphate. They are assumed to be regulated by two distinct D-glucose-6-phosphate dehydrogenases [D-glucose-6-phosphate:NAD(P)$^{\oplus}$ 1-oxidoreductase (EC 1.1.1.49)][1]. The dehydrogenase appointed to the Entner-Doudoroff pathway, an inducible enzyme, has a molecular weight of 220 000, and is composed of four identical subunits, each with a molecular weight of 55 000. It shows homotropic effects with D-glucose 6-phosphate and is strongly inhibited by ATP. The other enzyme, for which no regulatory properties have been demonstrated so far, is also active with both coenzymes and has been assigned to the hexose monophosphate pathway. It has a molecular weight of 260 000 and is also composed of four polypeptide chains, each with a molecular weight of 65 000.

In order to detect further possibilities for regulation of the two pathways and to study the evolutionary relationship between the dehydrogenases belonging to them, we have isolated the 6-phospho-D-gluconate dehydrogenase [(6-phospho-D-gluconate:NAD(P)$^{\oplus}$ oxidoreductase (decarboxylating) (EC 1.1.1.44)] from *Ps. fluorescens*. This enzyme is active with NAD$^{\oplus}$ and with NADP$^{\oplus}$, showing normal kinetics and prefering NAD$^{\oplus}$; it is inhibited by ATP and weakly by ADP.

Based on molecular structure, the *Ps. fluorescens* enzyme is the first exception in the series of 6-phospho-D-gluconate dehydrogenases so far investigated. The native enzyme has a molecular weight of 128 000, as determined by gel permeation chromatography and sedimentation equilibrium studies, and it is composed of four polypeptide chains, each with a molecular weight of 32 000, as determined by sodium dodecylsulfate gel electrophoresis. No association-dissociation processes could be observed with the native enzyme. In contrast to these results, all the 6-phospho-D-gluconate dehydrogenases isolated so far from a variety of other sources seem to be dimeric molecules, containing subunits with molecular weights about 50 000.

The 6-phospho-D-gluconate dehydrogenase from *Ps. fluorescens* differs from the two enzymes from *Ps.*

multivorans[2,3], which are both composed of two subunits; the subunits of one *Ps. multivorans* enzyme have a molecular weight of 60 000 each, and those of the other, 40 000. Contrary to our expectations, the 6-phospho-D-gluconate dehydrogenase from *Ps. fluorescens* is in no respect similar to the D-glucose-6-phosphate dehydrogenases from the same organism.

1 Lessmann, D., Schimz, K.-L. & Kurz, G. (1975) *Eur. J. Biochem.* **59**, 545 - 559.
2 Lessie, T. G. & Vander Wyk, J. C. (1972) *J. Bacteriol.* **110**, 1107 - 1117.
3 Lee, Y. N & Lessie, T. G. (1974) *J. Bacteriol.* **120**, 1043 - 1057.

Christos Stournaras and Gerhart Kurz, Lehrstuhl Biochemie, Chemisches Laboratorium der Universität, Albertstraße 21, D-7800 Freiburg i. Br.

B. Stumpf und H. Kraus

Zur Regulation des Glucose- und Ketonkörperstoffwechsels im Gehirn ketotischer Tiere

Wenn im Gehirn die Ketonkörperoxidation an Bedeutung gewinnt, nimmt gleichzeitig die cerebrale Glucoseoxidation ab[1-3]. Möglicherweise wird der Glucoseabbau auf mehreren Stufen gedrosselt. Regulationen werden diskutiert beim Einstrom von Glucose in die Zelle, auf dem Niveau der Hexokinase- und der Phosphofructokinase-Reaktion sowie schließlich auf der Ebene der Pyruvatoxidation. Um die Teilschritte zu charakterisieren, an denen der Stoffwechsel der Glucose und Ketonkörper im Gehirn erwachsener ketotischer Ratten reguliert wird, wurden bei hungernden streptozotocindiabetischen sowie phlorrhizindiabetischen Tieren charakteristische Hirnmetabolite und der Interkonvertierungszustand der Pyruvat-Dehydrogenase (Lipoat) (EC 1.2.4.1) im gefriergestoppten Gewebe[4] gemessen.

Typische Veränderungen im Sinne eines Crossovers auf der Ebene der Phosphofructokinase (EC 2.7.1.11) mit einem Anstieg von Glucose-6-phosphat und Fructose-6-phosphat sowie einem Abfall von Fructose-1,6-bisphosphat fanden sich nur im Streptozotocindiabetes. Von den möglichen Effektoren der Phosphofructokinase zeigte Citrat ansteigende Tendenz. Messungen von Phospho*enol*pyruvat, Pyruvat und Lactat wiesen auf eine Drosselung der Pyruvatoxidation bei allen drei Ketoseformen hin. Gleichzeitig stieg unter diesen Bedingungen die Lactatproduktion durch das Gehirn drastisch an[1-3]. Obwohl diese Verhältnisse eindeutig für eine Hemmung der Pyruvatoxidation sprechen, änderte sich im Gegensatz zu anderen Geweben aber der Interkonvertierungszustand der Pyruvat-Dehydrogenase im Gehirn der ketotischen Tiere nicht. Andere Mechanismen, die möglicherweise die Pyruvatoxidation auf mitochondrialer Ebene kontrollieren, werden zur Zeit untersucht.

1 Owen, O. E., Morgan, A. P., Kemp, H. G., Sullivan, J. M., Herrera, M. G. & Cahill, G. F. Jr. (1967) *J. Clin. Invest.* **46**, 1589 - 1595.
2 Ruderman, N. B., Ross, P. S., Berger, M. & Goodman, M. N. (1974) *Biochem. J.* **138**, 1 - 10.
3 Kraus, H., Schlenker, St. & Schwedesky, D. (1974) *Hoppe-Seyler's Z. Pysiol. Chem.* **355**, 164 - 170.
4 Veech, R. L., Harris, R. L., Veloso, D. & Veech, E. H. (1973) *J. Neurochem.* **20**, 183 - 188.
5 Wieland, O., Siess, E., v. Funcke, H. G., Patzelt, G., Schumann, A., Löffler, G. & Weiss, L. (1972) in Metabolic Interconversion of Enzymes (Wieland, O., Helmreich, E. & Holzer, H., Hrsg.) S. 293 - 309, Springer-Verlag, Heidelberg.

Dr. Bruno Stumpf, Universitäts-Kinderklinik, Humboldtallee 38, D-3400 Göttingen.

W.-H. Thomas and D. Chmelevsky

Determination of the Configuration of Two Proteins from Central and Peripheral Nerves

Myelin protein A_1 prepared from the central nervous system is known to be antigenic for the allergic encephalomyelitis, while myelin protein P_2 from the peripheral nervous system is antigenic for the experimental allergic neuritis. It is, accordingly, of interest to investigate the structure of these two myelin proteins. An attempt has therefore been made to determine the relative contribution of the three possible protein conformations to the proteins A_1 and P_2. This has been done by a circular-dichroism (CD) analysis of the samples. CD data for the α, β and random coil conformation of poly-L-lysine have been used as calibration spectra.

It can be assumed that the observed CD spectrum is a linear superposition of three components corresponding to the three basic configurations:

$$f(\lambda) = a_1 f_1(\lambda) + a_2 f_2(\lambda) + a_3 f_3(\lambda) \quad (1)$$

a_1, a_2, and a_3 are the fractions of the protein corresponding to the α, β, and random coil configuration, $f_j(\lambda)$ (j = 1, 2, 3) are the molar ellipticities of the three basic conformations at the wave length λ, and $f(\lambda)$ is the observed molar ellipticity of the sample. Eq(1) is valid only insofar as non-protein chromophores do not influence the results.

The three coefficients a_1, a_2, and a_3 are linked by the condition:

$$a_1 + a_2 + a_3 = 1 \quad (2)$$

Systematic as well as statistical deviations may affect both the calibration spectra and the observed values. This implies that Eq.(1) is not strictly valid. However one can formulate it in the modified form:

$$f(\lambda) = a_1 f_1(\lambda) + a_2 f_2(\lambda) + a_3 f_3(\lambda) + \epsilon_\lambda \qquad (3)$$

$$\int_{190\,\mathrm{nm}}^{240\,\mathrm{nm}} \epsilon_\lambda^2 \, d\lambda = \min \qquad (4)$$

As an additional aspect, one has to consider a possible wave-length-dependent weight factor for the deviations. With a suitable choice of this weight factor one obtains the following solution of the least squares problem for protein A_1:

where the term ϵ_λ accounts for the fact that even a suitable choice of the three parameters a_1, a_2, and a_3 will not lead to complete agreement between the experimental and the theoretical curve. In this formulation the analysis is reduced to the least squares problem, i.e. one has to find those coefficients which minimize the integral of the square of the deviations:

α-helix: $a_1 = 0.24$
β-structure: $a_2 = 0.02$
random coil: $a_3 = 0.74$

For the P_2 protein the following values were obtained:

α-helix: $a_1 = 0$
β-structure: $a_2 = 0.70$
random coil: $a_3 = 0.30$

In the case of the P_2 protein, one finds that the least squares estimate of the three parameters actually corresponds to a negative value of a_1. The best estimate has therefore been derived under the additional constraint that no negative parameters occur.

In addition to the estimates of the parameters, one may consider the ranges of those values which are still consistent with the data. It is difficult to separate, in the present case, systematic errors from statistical uncertainties. A rigorous error analysis can therefore not be obtained. Nevertheless one can make meaningful statements on the possible range of variation of the three parameters. To this purpose one may consider those combinations of the three parameters which lead to mean square deviations that exceed the least squares value by less than a specified factor. This condition defines ellipsoids in the space of the three parameters a_1, a_2, and a_3.

Dr. W.-H. Thomas and Dr. D. Chmelevsky, Institut für Med. Strahlenkunde der Universität Würzburg, Versbacher Landstr. 5, D-8700 Würzburg.

A. Thrun

Quervernetzung von Actin mit Tartrylaziden und die Verwertung von Actin-Dimeren

Das kontraktile Protein Actin kann im filamentären Zustand bei pH 7.5 - 8 und 6 °C in Phosphat-Puffer quervernetzt werden, wobei die beteiligten Aminosäuren ausschließlich Lysine sein sollten[1,2].

Je nach Reagenz ergibt die Quervernetzung geringere oder größere Ausbeuten an Dimeren: Die Vernetzung unterhalb der Denaturierungsgrenze des Proteins mit Tartrylbis-(glycylazid) (I), 13 Å lang, ergibt ca. 0.5 - 1% Dimere, bezogen auf die eingesetzte Proteinmenge; die Vernetzung unterhalb der Denaturierungsgrenze des Proteins mit Tartrylbis(hexanoylazid) (II), 23 Å lang, ergibt ca. 3.0% Dimere, bezogen auf die eingesetzte Proteinmenge.

Bei Quervernetzung mit beiden Reagenzien können die Actin-Dimeren nicht mehr polymerisieren[1].

Die Actin-Dimeren eignen sich für Myosin-ATPase-Tests (EC 3.6.1.3), die der physiologischen Actin-aktivierten Myosin-ATPase-Aktivität möglicherweise ähneln. Darauf weisen mehrere Versuche mit I-Actin-Dimeren[1] sowie mehrere Versuche mit II-Actin-Dimeren hin.

Azid-Actin-Dimere könnten auch eine besondere Rolle bei der Nucleation spielen, wenn sich monomeres Actin zu F-Actin polymerisiert. In diesem Zusammenhang sind die Modellbetrachtungen von Engel und Fasold[3] zur Actinpolymerisation interessant. Jedenfalls scheinen sich Actin-Dimere, die selbst nicht mehr in der Lage sind zu polymerisieren, mit polymerisationsfähigem Monomeren-Actin zu F-Actin einzupolymerisieren.

1 Fasold, H., Bäumert, H & Bender, N. (1975) *Biochem. Soc. Trans.* **3** 935.
2 Peters, K. & Richards, F. M. (1977) *Ann. Rev. Biochem.* **46**, 523 - 51.
3 Engel, J., Fasold, H., Hulla, F. W., Waechter, F. & Wegner, A. (1978) *Mol. Cell. Biol.*, im Druck.

Alexander Thrun, Institut für Biochemie, Sandhofstr. 75 A, D-6000 Frankfurt/Main.

E. Töpfer-Petersen and A. Henschen

Primary Structure of Bovine Fibrin γ-Chain

Mammalian fibrinogen is an assembly of three pairs of peptide chains, the different chains being denoted Aα, Bβ and γ. The primary structure analysis of human fibrinogen is nearing its completion. A structural study on fibrin(ogen) from other species seemed of great interest for several reasons. First, the fibrinopeptides A and B have been shown to have the highest evolutionary mutation rates of all proteins studied[1] and the question arises if the fibrin parts, α, β and γ, also show high rates. Furthermore, information about fibrinogen sequences in other species would shed light on structure-function relationships in this protein.

For the present investigation, the amino acid sequence of bovine fibrin was selected. In the γ-chain only short sequences from the *N*-terminus[2,3], around the carbohydrate linkage site[4] and from the *C*-terminus[5] are known.

The bovine γ-chain was isolated by CM-cellulose chromatography after mercaptolysis and alkylation. The chain was then cleaved in a way to produce large fragments suitable for automatic sequencing, e.g. with cyanogen bromide at the methionyl bonds and with trypsin after citraconylation at the arginyl bonds. The fragments obtained were isolated by gel-filtration and CM-cellulose chromatography and then analysed for *N*-terminal sequence. It turned out that the positions of several methionine and arginine residues were different in bovine and human γ-chain. Thus, the fragmentation patterns were different. However, a large part of the bovine γ-chain sequence could be determined. The chain is homologous with the corresponding human chain to an extent of about 85%.

1 McLaughlin, P. J. & Dayhoff, M. O. (1972) in Atlas of Protein Sequence and Structure (Dayhoff, M. O., ed.) pp. 47 - 52, National Biomedical Research Foundation, Washington.
2 Pirkle, H. & Henschen, A. (1968) *Biochemistry,* 7, 1362 - 1366.
3 Timpl, P., Fietzek, P. P., Wachter, E. & van Delden, V. (1977) *Biochim. Biophys. Acta* **490**, 420 - 429.
4 Töpfer-Petersen, E. & Henschen, A. (1977) *Thromb. Res.* **11**, 881 - 892.
5 Sharp, J. J., Cassman, K. G. & Doolittle, R. F. (1972) *FEBS Lett.* **25**, 334 - 336.

Dr. E. Töpfer-Petersen and A. Henschen, Max-Planck-Institut für Biochemie, Am Klopferspitz, D-8033 Martinsried bei München.

J. Veser, P. Geywitz und H. Thomas

Anreicherung einer Catechol-Methyltransferase aus Candida tropicalis

Die Catechol-Methyltransferase[1,2] (EC 2.1.1.6) überträgt eine Methylgruppe vom *S*-Adenosylmethionin auf eine der beiden vicinalen Hydroxylgruppen von Adrenalin[1], Noradrenalin[2], Dopamin[2] und anderen Dihydroxyphenyl-Verbindungen[2,3]. Als Substrate dienen so z.B. auch 2-Hydroxyöstrogene[4] und ein Cumarin-Derivat wie 6,7-Dihydroxycumarin (Äsculetin)[5,6].

Das Enzym kommt in vielen Organen aller bisher untersuchten Tierarten und auch in einigen Pflanzen vor[3]. Kürzlich konnte auch in den Sproßpilzen *Candida albicans, Candida tropicalis* CBS 94 und *Candida stellatoidea* CBS 1905 enzymatische Aktivität nachgewiesen werden, durch die eine Methylgruppe des *S*-Adenosylmethionins auf die meta- bzw. para-ständige OH-Gruppe von Adrenalin und 3,4-Dihydroxybenzoesäure übertragen wird[7]. In diesen Sproßpilzen kann auch eine Methylierung von Äsculetin zu 7-Hydroxy-6-methoxycumarin (Scopoletin) und 6-Hydroxy-7-methoxycumarin (Isoscopoletin) erfolgen.

Es wird nunmehr über die Anreicherung des Enzyms, das diese Methylierungen katalysiert, aus *Candida tropicalis* berichtet. Kulturen von *C. tropicalis* wurden auf Sabouraud-Glucose 2% Bouillon gezüchtet. Nach 2 - 3 Tagen wurde das durch Zentrifugation gewonnene Zellsediment im Zellhomogenisator mit Glasperlen lysiert. Der 100 000 x *g*-Überstand des so gewonnenen Homogenats enthielt 92% der Gesamt-Enzymaktivität. Die weitere Anreicherung erfolgte über die folgenden Schritte: Säurefällung bei pH 5.0, Säulenchromatographie an DEAE-Cellulose, Säulenchromatographie an DEAE-Sepharose CL-6B und Ultrogel-AcA44-Filtration.

Bezogen auf das Homogenat ergab sich eine etwa 900fache Anreicherung des Enzyms. Das Molekulargewicht ließ sich mittels Gelfiltration an Sephadex G-100 zu 35 000 ± 5 000 bestimmen. Durch Isoelektrofokussierung an Sephadex G-75 wurde ein isoelektrischer Punkt von 5.0 ± 0.1 ermittelt. Es fand sich über den gesamten pH-Gradienten (pH 3.0 - 10) nur ein Aktivitätspeak. Das pH-Optimum liegt bei pH 7.6, das Temperaturoptimum bei 37.5 °C. Während Cytosol und Säuresediment nach wiederholtem Einfrieren und Auftauen über 2 Monate keinen Aktivitätsverlust zeigten, fiel die Aktivität der AcA44-Fraktion nach einmaligem Einfrieren und Auftauen auf 20 - 40% ab.

1 Axelrod, J. (1957) *Science* **126**, 400.
2 Axelrod, J. & Tomchick, R. (1958) *J. Biol. Chem.* **233**, 702 - 705.
3 Guldberg, H. C. & Marsden, C. A. (1975) *Pharmacol. Rev.* **27**, 135 - 206.
4 Knuppen, R., Lubrich, W., Haupt, O., Ammerlahn, U. & Breuer, H. (1969) *Hoppe-Seyler's Z. Physiol. Chem.* **350**, 1067 - 1075.
5 Müller-Enoch, D., Seidel, E. & Thomas, H. (1976) *Z. Naturforsch.* **31c**, 280 - 284.
6 Thomas, H., Veser, J. & Müller-Enoch, D. (1976) *Hoppe-Seyler's Z. Physiol. Chem.* **357**, 1347 - 1351.
7 Müller-Enoch, D., Thomas, H., Streng, W., Wildfeuer, W. & Haferkamp, O. (1976) *Z. Naturforsch.* **31c**, 509 - 513.

Dr. J. Veser, P. Geywitz und Prof. Dr. H. Thomas, Abteilung Biochemie I der Universität Ulm, Oberer Eselberg N 26, D-7900 Ulm.

Prof. Dr. Hildegard Debuch, Institut für Physiologische Chemie, Lehrstuhl II, Postfach 41 05 80, D-5000 Köln 41.

M. Vierbuchen und H. Debuch

Plasmalogenbiosynthese im Rattengehirn: Subzelluläre Lokalisation der Enzymaktivität und ihr Verhalten während der Gehirnentwicklung

Es wird angenommen, daß die Plasmalogene im Säugetierorganismus durch Desaturierung aus den Alkylglycerolipiden entstehen. Für die Eigenschaften dieser Desaturierungsreaktion im Rattengehirn findet man allerdings z.T. wiedersprechende Angaben[1,2]. Bei unseren Untersuchungen über die Plasmalogenbiosynthese verwendeten wir die Gehirne 14 Tage alter Ratten, weil um die Zeit der Myelinisierung eine rasche Vermehrung der ethanolaminhaltigen Plasmalogene zu beobachten ist.

Sowohl das Hirnhomogenat, als auch die daraus gewonnen (rohen) Zellkerne, Myeline, Synaptosomen, Mitochondrien, Mikrosomen und das Cytosol[3], wurden im Hinblick auf die Fähigkeit zur Plasmalogenbiosynthese untersucht. Als Vorläufer diente 1-(9,10-$^3H_2$)Hexadecyl-*sn*-glycero-3-phosphoethanolamin, welches in vivo in hohem Maße in die Plasmalogene eingebaut wurde[4]. Die höchste Enzymaktivität [hier angegeben in pmol/(mg Prot. × h)] fanden wir im Homogenat (264) bzw. in der „Kern"-Fraktion (286). (Die Kerne waren noch mit Neuronen und Gliazellen verunreinigt, wodurch vielleicht die Enzymaktivität erklärt werden kann.) Im Myelin und im Cytosol lag die Enzymaktivität an der Grenze der Nachweisbarkeit, während sie in den Mikrosomen 68, in den Synaptosomen 75 und in der mitochondrialen Fraktion 110 pmol Plasmalogen/(mg Prot. × h) betrug. Die Enzymaktivität der mikrosomalen Fraktion ließ sich durch Zusatz von Cytosol nicht steigern.

In einer 2. Versuchsreihe verwendeten wir Hirnhomonat – wegen der hohen Einbaurate des Etherlipids – von Versuchstieren verschiedenen Alters (in Tagen) 7, 11, 14, 21, 365 und fanden folgende Enzymaktivitäten: 268, 308, 274, 122, 11.

Wie auch von den anderen Autoren beschrieben[2] war in unserem System stets O_2 und NADH oder NADPH notwendig.

1 Horrocks, L. A. & Radominska-Pyrek, A. (1972) *FEBS Lett.* **20**, 190 - 192.
2 Wykle, R. L. & Schremmer Lockmiller, J. M. (1975) *Biochim. Biophys. Acta* **380**, 291 - 298.
3 Mangan, J. L. & Whittaker, V. P. (1966) *Biochem. J.* **98**, 128 - 137.
4 Gunawan, J. & Debuch, H. (1977) *Hoppe-Seyler's Z. Physiol. Chem.* **358**, 537 - 548.

J. Voss, E. Reichwein, B. Stoll and W. Groß

Utilization of L- and D-Glutamate in Streptomyces hydrogenans

Streptomyces hydrogenans can be grown in the presence of L-glutamate as a sole source of carbon and nitrogen. However, a mutant of *Strept. hydrogenans* isolated in our laboratory (DGA-cells) has been shown to utilize D-glutamate almost with the same efficiency as the L-enantiomer[1]. The DGA-cells can utilize L-glutamate, since this enantiomer is easily transported into the cells by two carriers, as shown by kinetic analysis, and by exchange experiments. However, there is no evidence for a transport of D-glutamate across the cell membrane. This different behaviour of the two enantiomorphic forms of glutamate was the subject of our further studies. Marked differences could be observed in competition experiments: whereas L-glutamate is a potent inhibitor of D-glutamate utilization in DGA-cells, the reciprocal effect is missing. D-glutamate, even at high concentrations, does not influence L-glutamate uptake. Further differences were established with respect to the sensitivity to monovalent cations: L-glutamate transport is stimulated by protons and by potassium ions; but with equivalent ion concentrations, the D-glutamate utilization is insensitive. Divalent cations like $Ca^{2\oplus}$ and $Mg^{2\oplus}$ also affect differently; at concentrations of about 10 mmol/*l* the L-glutamate uptake is stimulated 2- to 3-fold, whereas the utilization of D-glutamate is only slightly enhanced (about 15%). From these and additional results, it is concluded that the initial step in D-glutamate utilization is the formation of a racemate from which the L-from is taken up.

1 Groß, W., Voss, J. & Reichwein, E. (1976) FEBS Symposium Zürich, P 125.

Dr. J. Voss, c/o Zentrum der Biologischen Chemie der Universität Frankfurt, Theodor-Stern-Kai 7, D-6000 Frankfurt.

K. G. Wagner, J. Hoppe, E. Rieke and B. Jastorff

Mechanism of Activation of Protein Kinase I

cAMP-dependent protein kinases (EC 2.7.1.37) which have a tetrameric quarternary structure differ in their reaction with ATP; while type II undergoes autophosphorylation, type I binds ATP with high affinity. In a survey the properties of these two kinases and the different influence of ATP on their activation is discussed[1,2].

Data on the ligand binding sites of type I enzyme, i.e. the cAMP site, the ATP site at the holoenzyme (high affinity ATP site) and the ATP site at the free catalytic subunit, are presented. They indicate the different nature of the adenine locus for the cAMP and the ATP sites[3], whereas this locus is identical for the two ATP sites. However, the ribose and triphosphate moieties of ATP are bound in different states, depending on whether the enzyme is associated (R_2C_2) or dissociated (C).

The geometry of the cAMP site is based on results obtained with spin-labeled cAMP derivatives[4]. A conformational change in this site is observed upon addition of ATP. Data on the chemical features of the cAMP site are presented and an approach is outlined which allows the characterization of different cAMP binding proteins using a minimal number of selected cAMP analogues.

The activation scheme:

$$R_2C_2 \text{ (inactive)} + 2\text{ cAMP} \rightleftharpoons R_2(\text{cAMP})_2 + 2\text{ C (active)}$$

represents an intricate interrelation between ligand interactions and changes in the protein quarternary structure. The influence of the effectors cAMP and ATP is discussed on the basis of results obtained with agarose-immobilized cAMP derivatives[5] and with free ligand binding. Models are presented which allow computer calculations to simulate the ligand binding curves obtained under various conditions. From these studies the magnitude of the equilibrium constants are estimated and the cooperative behaviour of the kinase is explained.

1 Beavo, J. A., Bechtel, P. J. & Krebs, E. G. (1975) *Adv. Cyclic Nucleotide Res.* **5**, 241 - 251.
2 Rosen, O. M., Erlichman, J. & Rubin, C. S. (1975) *Adv. Cyclic Nucleotide Res.* **5**, 253 - 263.
3 Hoppe, J., Marutzky, R., Freist, W. & Wagner, K. G. (1977) *Eur. J. Biochem.* **80**, 369 - 372.
4 Hoppe, J., Rieke, E. & Wagner, K. G. (1978) *Eur. J. Biochem.*, in press.
5 Rieke, E., Hoppe, J. & Wagner, K. G. (1978) *Eur. J. Biochem.*, in press.

K. G. Wagner, Abteilung Molekularbiologie, Gesellschaft für Biotechnologische Forschung mbH., Mascheroder Weg 1, D-3300 Braunschweig-Stöckheim.

W. Wagner und H. Follmann

Thioredoxine aus Scenedesmus und die multiplen Funktionen von Thioredoxin in Enzymsystemen

Die ein -SH ⇄ -S-S-Redoxsystem enthaltenden Thioredoxine wurden in Bakterien, Tieren und Pflanzen vorwiegend als reduzierendes Substrat der enzymatischen Ribonucleotidreduktion charakterisiert, besitzen jedoch auch Aktivität bei der Licht-Regulation von Chloroplasten-Enzymen[1] und in der assimilatorischen Sulfatreduktion [2,3]. Es ist zur Zeit ungewiß, wie weit sie in der Zelle jeweils spezifische Funktionen ausüben oder universell verwendet werden. Quantitative Kreuzexperimente (in Zusammenarbeit mit A. Schmidt, Botanisches Institut, München) mit reduziertem Thioredoxin aus der Grünalge *Scenedesmus obliquus*[4], der blaugrünen Alge *Synechococcus* 6301, *E. coli*, Hefe und Spinat belegen weitgehende Austauschbarkeit dieser Polypeptide in drei völlig verschiedenen Enzymsystemen und sprechen gegen eine hohe Spezialisierung. So stimulieren gleiche Mengen der Thioredoxine die Ribonucleosid-diphosphat-Reduktase aus *E. coli* (EC 1.17.4.1) und 3′-Phosphoadenosin-5′-phosphosulfat-Sulfotransferase (EC 2.8.2.-) aus *Synechococcus*[3] in sehr ähnlichem Maße, Fructose-1,6-bisphosphatase (EC 3.1.3.11) aus Spinat wird durch das homologe Thioredoxin besser aktiviert als durch die Proteine aus Mikroorganismen. Bei der chromatographischen Reinigung von *Scenedesmus*-Thioredoxin ist neben der Hauptfraktion (M_r 12 600) eine kleine Menge höhermolekularen Thioredoxins (M_r 23 500) zu isolieren, das die Fructose-bisphosphatase stärker stimuliert als die Ribonucleosid-disphosphat-Reduktase; hierdurch scheint erstmals der Nachweis eines cytoplasmatischen und eines chloroplastidären Thioredoxins nebeneinander gelungen. Thioredoxin-abhängige Enzyme sind daher allgemein unspezifisch für ihr gemeinsames Substrat (Cofaktor), individuelle Aktivitäten werden durch strukturbedingte Varationen der K_m-Werte und/oder Kompartimentierung bestimmt. Dagegen setzt Thioredoxin-Reduktase (NADPH) (EC 1.6.4.5) aus *E. coli* in Gegenwart von NADPH heterologe oxidierte Thioredoxine in keinem Fall um, in *Scenedesmus* haben wir aber auch keine analoge NADPH-Thioredoxin-Reduktase gefunden. In Mikroorganismen und Pflanzen gibt es offenbar zur Regeneration des multifunktionellen reduzierten Thioredoxins streng spezialisierte Enzyme.

1 Wolosiuk, R. A. & Buchanan, B. B. (1977) *Nature (London)* **266**, 565 - 567.
2 Tsang, M. L. S. & Schiff, J. A. (1976) *J. Bacteriol.* **125**, 923 - 933.
3 Schmidt, A. & Christen, U. (1978) *Planta*, im Druck.
4 Wagner, W. & Follmann, H. (1977) *Biochem. Biophys. Res. Commun.* **77**, 1044 - 1051.

Dipl.-Chem. Wolfgang Wagner und Prof. Dr. Hartmut Follmann, Fachbereich Chemie der Universität, Arbeitsgruppe Biochemie, D-3550 Marburg.

G. Warren and B. Dobberstein

Protein Transfer across Microsomal Membranes: Functional Reassembly from Separated Membrane Components

The synthesis of secretory proteins is initiated on ribosomes in the cell cytoplasm, and the first part of the polypeptide chain to appear is a short sequence termed the signal sequence[1]. This signal is thought to direct the ribosomal complex to the endoplasmic reticulum membrane, where the synthesis of the rest of the secretory protein is tightly coupled to its transfer across the membrane into the cisternal space. In all but one case, it is clear that the polypeptide chain is processed during transfer to remove the signal sequence[2,3].

Little is known about the membrane proteins that are involved in conveying the growing polypeptide chain across the membrane. Some must be involved in recognising and binding the signal sequence and ribosomal complex to the membrane; others might be needed to form a channel across the membrane. Proteins binding the ribosome to the membrane have been identified, but no function in protein transfer has been demonstrated[4]. What has been lacking is a dissection of the microsomal membrane into inactive components that can be reassembled subsequently into a functional entity. This would allow the purification of components involved in protein transfer. Here we report the first such reassembly. Microsomal vesicles washed with high salt are incapable of transfering and processing secretory proteins, but readdition of the salt extract restores full activity. Membrane proteins are responsible for the restoration of activity and they may be involved in the binding of the signal peptide and ribosomal complex to the membrane; the salt-washed vesicles still retain proteins that are essential for protein transfer and processing.

1 Blobel, G. & Dobberstein, B. (1975) *J. Cell Biol.* **67**, 835 - 851.
2 Blobel, G. & Dobberstein, B. J. (1975) *J. Cell. Biol.* **67**, 852 - 867.
3 Campbell, P. N. & Blobel, G. (1976) *FEBS Lett.* **72**, 215 - 226.
4 Palmiter, R. D., Thibodeau, S. N., Gagnon, J. & Walsh, K. A. 11th FEBS Meeting, Copenhagen, 1977, Pergamon Press.
5 Kreibich, G., Pereyra, B. & Sabatini, D. D. (1976) Abstract of the Tenth International Congress of Biochemistry, Hamburg 297.

Graham Warren and B. Dobberstein, European Molecular Biology Laboratory, Postfach 10 22 09, D-6900 Heidelberg.

N. Weiner und W. Wesemann

Vergleichende Untersuchungen zur Serotoninbindung an mitochondriale und mikrosomale Membranfraktionen der Nervenendigungen

Neurophysiologische Versuche lassen auf die Präsenz prä- und postsynaptischer Serotoninrezeptoren schließen[1]. Während die Serotoninbindung an die mikrosomale Membranfraktion (P_3) als postsynaptisch charakterisiert wird[2], sind nach Lysis der mitochondrialen Fraktion (P_2), die die Synaptosomen enthält, vorwiegend präsynaptische Rezeptoren zu erwarten. – Zur Differenzierung der beiden Bindungsstellen werden mikrosomale und mitochondriale Membranfraktionen aus Rattenhirnhomogenaten elektronenmikroskopisch, gelelektrophoretisch und hinsichtlich der Serotoninbindung vergleichend untersucht. Beide Membranfraktionen unterscheiden sich im Proteinmuster. Im Bereich hoher Affinität wird in der mitochondrialen Membranfraktion für die Serotoninbindung ein K_D-Wert (5×10^{-9}M) gefunden, während die Serotoninbindungskurve der mikrosomalen Fraktion einen biphasischen Verlauf aufweist (K_D: 4×10^{-9}M; K_{D2}: 2×10^{-8}M). Die Bindung an Membranen wird durch Filtration über Glasfaserfilter, die Bindung an Detergenzextrakte der Membranen (Mulgofen) durch Gleichgewichtsdialyse und Molekularfiltration bestimmt. Die spezifische Bindung wird aus der Differenz der Gesamt-[^{3}H]Serotoninbindung und der Bindung in Gegenwart eines 1000fachen Überschusses an unmarkiertem Liganden ermittelt. Die spezifische Bindung der mitochondrialen Membranfraktion [450 fmol/mg Protein] und des entsprechenden Mulgofenextraktes [15 pmol/mg Protein] ist ca. doppelt so hoch wie die Serotoninbindungskapazität der mikrosomalen Membranen (230 fmol/mg Protein) bzw. des Extraktes dieser Fraktion (7.5 pmol/mg Protein). In beiden Membranfraktionen wird Serotonin aus seiner Bindung durch Tryptamin, 5-Methoxytryptamin, Bufotenin, Lysergsäurediäthylamid und dessen Bromderivat, allerdings mit unterschiedlichen Werten für die ED_{50}, verdrängt. – Die Ergebnisse können durch die Präsenz differenter Akzeptormoleküle in mitochondrialen und mikrosomalen Membranfraktionen interpretiert werden, obwohl eine eindeutige Zuordnung der Bindungsstellen zu prä- und postsynaptischen Rezeptoren noch nicht möglich ist.

1 Haigler, H. J. & Aghajanian, G. K. (1974) *J. Pharmacol. Expt. Ther.* 188, 688 - 699.
2 Snyder, S. H. & Bennett, Jr., J. P. (1975) in Pre- and Postsynaptic Receptors (Usdin, E. & Bunney, W. E. Jr., eds.) pp. 191 205, Marcel Dekker, Inc., New York.

W. Wesemann, Institut für Physiologische Chemie II, Universität Marburg, D-3550 Marburg/Lahn.

D. G. Weiss, G. W. Gross and G. W. Kreutzberg

Subcellular Distribution of Rapidly Transported Material in the Olfactory Nerve of the Pike (Esox lucius)

[^{3}H]Leucine applied to the olfactory epithelium of the pike results in rapid precursor uptake, incorporation into proteins and export of labelled material by axoplasmic transport. The various regions of the resulting characterisitc isotope distribution along the nerve have been subjected to subcellular fractionation, yielding the following fractions: homogenate, total soluble material (100 000 $\times g$, 1 h – supernatant) and total particulate material (100 000 $\times g$, 1 h – pellet). The soluble material was separated into trichloroacetic acid-soluble and precipitable material. The particulate material was subdivided into a nuclear-like pellet (1000 $\times g$, 10 min), a mitochondrial-like pellet (15 000 $\times g$, 20 min) and the microsomal pellet (100 000 $\times g$, 1 h).

In the most rapidly moving profile region the labelled material was found to be predominantly associated with the microsomal particulate fraction, and to a lesser extent with the soluble macromolecules. Only a small amount was represented by free amino acids in the acid-soluble fraction. Fractions containing particles larger than microsomes do not contain much rapidly transported radioactive material. The more slowly moving labelled material consists mainly of soluble macromolecules, whereas the most slowly moving material which can be studied by this procedure,is mainly acid-soluble.

This can be interpreted to mean that particulate material is transported more rapidly than soluble macromolecules and these, in turn, more rapidly than low molecular weight molecules. Therefore these results give, in addition to our earlier report[1], further evidence for a size-dependent molecule separation during axonal transport. It will be discussed whether these results together with the results of the ultrastructural characterization of this nerve[2] may support the chromatographic hypothesis of axoplasmic transport[3].

1 Gross, G. W. & Weiss, D. G. (1977) *Neurosci. Lett.* **5**, 15 - 20.

2 Kreutzberg, G. W. & Gross, G. W. (1977) *Cell Tiss. Res.* **181**, 443 - 457.

3 Gross, G. W. (1975) in Advances in Neurology, (Kreutzberg, G. W., ed.) vol. 12, pp. 283 - 296, Raven Press, New York.

Dr. D. G. Weiss, Zoologisches Institut der Universität München, Luisenstr. 14, D-8000 München 2.

P. Wendt, K. Zänker, W. Tölle and G. Blümel

On the Nature of the Complex Between Protein and Phospholipid in Pulmonary Surfactant

Studies of the interactions between proteins and phospholipids in the alveolar lining layer are fundamental to our understanding of the physiological functioning of pulmonary surfactant. To maintain the alveoli in an expanded state during expiration, a surface-active material is secreted into the fluids that line the alveoli to reduce their surface tension. Ample evidence has been provided that surfactant material is lipo-protein material[1]; it is well etablished that dipalmitoyl lecithin is an integral part of the surfactant molecule. The following study was designed to determine whether protein-phospholipid complex formation is due to hydrophobic or electrostatic interactions.

Incubation of [^{14}C]phosphatidyl choline with a protein preparation derived from the alveolar fluid revealed the existence of proteins which bind tightly to phosphatidylcholine. Alveolar lining proteins were pre-incubated with a water-soluble carbodiimide in order to mask carboxyl groups. Plots of phosphatidylcholine binding to various concentrations of carboxyl-group-modified surfactant protein suggest an involvement of carboxyl group(s) in protein-phospholipid interactions. There is a significant decrease in phosphatidylcholine binding capacity as compared to native surfactant protein. At saturation, the binding of label to modified protein is at least 50% lower.

1 Goerke, J. (1974) *Biochem. Biophys. Acta* **344**, 241 - 261.

Dr. P. Wendt, Institut für Experimentelle Chirurgie der Technischen Universität München, Ismaningerstr. 22, D-8000 München 80.

M. Wenzel

Stabilisierung biologischer Systeme durch D_2O

Der Vortrag befaßt sich mit den Wirkungen, die in biologischen Systemen ausgelöst werden, bei Ersatz des normalen Lösungsmittels H_2O durch D_2O.

Beim Austausch von H gegen D in Wasserstoffbrücken von Makromolekülen werden diese Makromoleküle „stabiler". Dies liegt an der oft um den Faktor 2 - 5 verlangsamten Reaktionsgeschwindigkeit einer X-D-Bindung im Vergleich zu einer X-H-Bindung.

Man kann diesen Effekt zur Konservierung von Makromolekülen z. B. von Enzymen oder von Membran-Proteinen in Zellen ausnutzen. Es gelingt z. B. durch D_2O-Zugabe, Enzyme vor ihrer thermischen Denaturierung zu schützen, wie anhand der katalytischen Aktivität nachweisbar. Durch D_2O läßt sich auch eine Stabilisierung der Zellen erreichen. So zeigten Human-Erythrozyten, die in D_2O-haltigem Medium suspendiert sind, bei Hyperthermie oder osmotischer Schädigung eine geringere Hämolyse als Erythrozyten, die stets in H_2O-Medium gehalten wurden. Auch andere Gewebe sind durch D_2O stabilisierbar.

Auf praktische Schlußfolgerungen wird eingegangen.

Dr. Martin Wenzel, Biochem. Abteilung des Pharmazeutischen Instituts, Freie Universität Berlin, Königin-Luise-Straße 2 - 4, D-1000 Berlin 33.

1 Whittaker, V. P., Essman, W. B. & Dowe, G. H. C. (1972) *Biochem. J.* **128**, 833 - 846.
2 Nagy, A., Baker, R. R., Morris, S. J. & Whittaker, V. P. (1975) *Brain Res.* **109**, 285 - 309.
3 Uvnäs B. (1976) *Acta Physiol. Scand.*, Suppl. **440**.
4 Winkler, H. & Smith, A. D. (1975) *Hdb. Physiol.* **6**, 321 - 339.

Prof. Dr. V. P. Wittaker, MPI f. Biophysikalische Chemie, Abteilung Neurochemie, Postfach 968, D-3400 Göttingen.

V. P. Whittaker

The Structure of the Cholinergic Synaptic Vesicle: Recent Results

The electric organ of *Torpedo marmorata*, with its profuse cholinergic innervation, provides an excellent source material for the bulk isolation of very pure, homogeneous cholinergic synaptic vesicles using high-resolution density gradient separation in a zonal rotor followed by gel filtration[1,2]. Recent work in this laboratory* shows that the acetlycholine: phospholipid ratio is the most useful criterion of vesicle purity. The vesicle membranes bind actin and other proteins which can be detached by EGTA [ethyleneglycolbis(2-aminoethyl ether)tetraacetic acid] treatment without loss of acetylcholine, but with a decrease in vesicle stability. The vesicle is also rich in glycosaminoglycan. On prolonged osmotic shock, mucopolysaccharides are split off and are recovered in the vesiculin fraction.** The strongly negative change of this polyanion may be involved in the binding of acetylcholine in much the same way as heparin is supposed to bind histamine in the mast cell granule[3].

Density measurements made by iso-pycnic centrifugation of vesicles into gradients of varying osmotic pressure made with non-penetrating and penetrating solutes permit the derivation of a multi-compartment model of the vesicle*** consisting of a flexible, but inelastic semipermeable membrane, hydrated core and osmotically active water space containing the diffusible small molecular weight solutes acetylcholine and ATP. Such a model is reasonably consistent with the chemical analyses and leads to the conclusion that the *Torpedo* vesicle does not contain a condensed phase of biogenic amine analogous to that postulated for adrenalin in the chromaffin granule[7].

* Tashiro, T. & Stadler, H., to be published.
** Stadler, H. & Whittaker, V. P., to be published.
*** Breer, H., Morris, S. J. & Whittaker, V. P., submitted for publication.

A. J. Wieczorek und B. Walther

Opaleszente High-Density-(HDL) Plasmalipoproteinfraktionen bei malignen Lymphomen

Koppel et al.[1] berichteten über einen erhöhten Gehalt an glykolipidgebundener Neuraminsäure in Seren von Patienten mit malignen Tumoren. Wir fanden einen um den Faktor 2.5 bis 3.5 erhöhtem Plasmagehalt an Hämatosid (*N*-Acetylneuraminyl-galaktosyl-glucosylceramid) bei vier untersuchten Nicht-Hodgkin-Lymphomen durch photometrische Neuraminsäurebestimmung[2] nach dünnschichtchromatographischer Trennung. Keine Erhöhung wurde bei zwei Hodgkin-Lymphomen beobachtet. Indessen wurden bei der Suche nach den Trägern des vermehrten Hämatosids im Plasma nach KBr-Dichtegradientenultrazentrifugation[3] bei allen sechs untersuchten malignen Lymphomen opaleszente HDL-Fraktionen beobachtet, die z.T. unmittelbar über den regulären HDL-Fraktionen flotierten.

Die Messung des Gehaltes an hämatosidgebundener Neuraminsäure bei allen Lipoproteinfraktionen zeigte, daß der vermehrte Plasmagehalt an Hämatosid den opaleszenten HDL-Fraktionen zuzuordnen war. Bei den Hodgkin-Lymphomen kam es aufgrund niedriger Konzentrationen der regulären Lipoproteine zu keinem Bruttoanstieg.

Opaleszente HDL-Fraktionen wurden weder bei Gesunden noch bei Patienten mit Fettstoffwechselstörungen beobachtet.

Die Lipide dieser Fraktionen wurden nach dünnschichtchromatographischer Trennung photometrisch bestimmt [2,4,5] und zeigten ein den regulären HDL ähnliches Verteilungsmuster. Die durch Dodecylsulfat-Polyacrylamidelektrophorese[6] aufgetrennten Apoproteine zeigten dagegen abweichende Muster. Elektronenmikroskopische Studien wurden durchgeführt.

1 Kloppel, Th. M., Keenan, Th. W., Freeman, M. J & Morrè, D. J. (1977) *Proc. Natl. Acad. Sci. U.S.A.* 74, 3011 - 3013.
2 Aminoff, D. (1961) *Biochem. J.* 81, 384 - 392.

3 Redgrave, T. G., Roberts, D. C. K. & West, C. E. (1975) *Anal. Biochem.* **65**, 42 - 49.
4 Zlatkis, A., Zak, B. & Boyle, A. J. (1957) *J. Lab. Clin. Med.* **41**, 486 - 492.
5 Bartlett, G. R. (1959) *J. Biol. Chem.* **234**, 466 - 471.
6 Weber, K. & Osborn, M. (1969) *J. Biol. Chem.* **244**, 4406 - 4411.

Dr. Andreas, J. Wieczorek, Forschergruppe Ernährung an der Medizinischen Poliklinik der Universität, Pettenkoferstr. 8a, D-8000 München 2.

O. H. Wieland

Regulation des Pyruvat-Dehydrogenase-Komplexes durch Phosphorylierung und Dephosphorylierung

Der mitochondriale Pyruvat-Dehydrogenase-Komplex (EC 1.2.4.1) wird sowohl durch Produkthemmung (Acetyl-CoA, NADH), als auch durch kovalente Modifizierung (Phosphorylierung-Dephosphorylierung) reguliert. Phosphorylierung eines Serins der α-Kette der Untereinheit Pyruvat-Dehydrogenase durch ATP-abhängige PDH-Kinase führt zur inaktiven Form des Komplexes (PDH_b) und umgekehrt wird durch Abspaltung des Phosphatrestes durch eine $Mg^{2\oplus}$- und $Ca^{2\oplus}$-abhängige Phosphatase die aktive Form (PDH_a) wiederhergestellt. Die Kinase ist fest, die Phosphatase locker im Enzymkomplex gebunden, für beide Enzyme besteht strenge Substratspezifität. In vivo befinden sich PDH_a und PDH_b in einem Fließgleichgewicht, dessen Lage von den relativen Aktivitäten von Kinase und Phosphatase abhängt. Aus Untersuchungen isolierter PDH-Komplexe kennt man eine Vielzahl von Substanzen, die die Eigenschaften der beiden regulatorischen Untereinheiten beeinflussen. An der Kinase wirkt ADP (kompetitiv zu ATP) hemmend, so daß mit steigendem Verhältnis von ATP/ADP die Phosphorylierungsgeschwindigkeit zunimmt und umgekehrt. Von ähnlichem Einfluß scheinen die mitochondrialen Quotienten von $NADH/NAD^{\oplus}$ und von Acetyl-CoA/CoA zu sein. Nicht-kompetitive Hemmstoffe der Kinase sind Pyruvat und Dichloracetat. Die Phosphatase benötigt neben Spuren von $Ca^{2\oplus}$ hohe Mengen von $Mg^{2\oplus}$ (10mM) und wird durch $F^{\ominus}$ gehemmt. Eine Beteiligung des Cyclo-AMP-Systems kann mit ziemlicher Sicherheit ausgeschlossen werden. Untersuchungen intakter Tiere, perfundierter Organe, isolierter Zellen und Mitochondrien lassen den Schluß zu, daß im allgemeinen erhöhte Fettsäureverbrennung mit einer Aufphosphorylierung des Pyruvat-Dehydrogenase-Systems einhergeht, und vice versa. Folglich fällt im Hunger oder Diabetes der Anteil an PDH_a auf weniger als ein fünftel der Gesamtaktivität ab. Am Fettgewebe besitzt Insulin die bemerkenswerte Eigenschaft, den Enzymkomplex durch Umwandlung von PDH_b in PDH_a zu aktivieren. Am Skelettmuskel ist das gleiche nach elektrischer Reizung zu beobachten. Die Frage, welche Faktoren letztlich in vivo an der Regulation des Pyruvat-Dehydrogenase-Komplexes beteiligt sind, ist schwer zu beantworten, solange es nicht gelingt, die aktuellen Stoffkonzentrationen innerhalb der Mitochondrien zu bestimmen. Hier zeichnen sich durch neue Methoden der Fraktionierung von Leber- und Fettgewebszellen gewisse Fortschritte ab.

Literatur

Reed, L. J., Pettit, F. H., Roche, I. E., Pelley, J. W. & Butterworth, P. J. (1976) in Metabolic Interconversion of Enzymes (Shaltiel, S., Hrsg.) S. 121 - 124, Springer, Berlin, Heidelberg, New York.
Wieland, O. H., Löffler, G., Siess, E. A. & Bard, S. (1976) ebenda, S. 125 - 135.
Randle, P. J. & Denton, R. M. (1976) ebenda, S. 136-141.

Professor Dr. O. H. Wieland Klinisch-Chemisches Institut des Städt. Krankenhauses München-Schwabing und Forschergruppe Diabetes, Kölner Platz 1, D-8000 München 40.

A. H. Wilhelm

Variation des Lactat-Dehydrogenase- und Kreatin-Kinase-Isoenzymverteilungsmusters in menschlichen Herz- und Skelettmuskeln

Untersuchung der Kreatin-Kinase-(EC 2.7.3.2) Isoenzymverteilung in menschlichen Skelettmuskeln zeigte eine starke Variation des Isoenzym-MB-Anteils von 2% (Musculus rectus abdominis) bis 29% (Musculus gluteus maximus). In 20 Herzmuskelproben (linker Ventrikel) wurde eine Variation der relativen Aktivität der Kreatin-Kinase MB von 1 bis 38% gefunden. In drei Fällen eines frischen Herzinfarktes wurden deutliche Unterschiede zwischen infarzierten und nicht infarzierten Bezirken beobachtet, die Werte aus den nicht infarzierten Bezirken in Klammern: 28 (22); 26 (11); 27 (16).

Das Maximum des Lactat-Dehydrogenase-(EC 1.1.1.27) Isoenzymverteilungsmusters schwankt bei den 20 Herzmuskelproben zwischen Isoenzym 1 und 3. In vier Fällen frischen Herzinfarktes wurde eine kathodische Verschiebung des Lactat-Dehydrogenase-Isoenzymmusters in den infarzierten gegenüber den nicht infarzierten Bezirken gefunden, wie aus den folgenden prozentualen Anteilen der A-Untereinheiten[1] entnommen werden kann, die Werte der nicht infarzierten Kontrollen in Klammern: 53 (20); 44 (20); 34 (20); 39 (28).

Schließlich wurden die Isoenzyme aus 13 Proben eines einzigen Herzens bestimmt, das starke ischämische Veränderungen aufwies. Die Kreatin-Kinase-MB-Werte schwankten innerhalb dieses Herzes zwischen 4 und 36%, die der Lactat-Dehydrogenase-A-Untereinheiten zwischen 24 und 44%. Der Korrelationskoeffizient zwischen Kreatin-Kinase-MB (%) und Lactat-Dehydrogenase-A-Untereinheiten (%) betrug 0.88. Autolyse als Ursache für die beobachtete Isoenzymvariation wurde durch Lagerungsversuche ausgeschlossen.

Diese Befunde bestätigen ähnliche Ergebnisse bezüglich der Lactat-Dehydrogenase-Isoenzyme[1] und unterstützen die Interpretation von Isoenzymverschiebungen als Kompensationsmechanismen gegenüber Änderungen von Bedarf und Angebot an metabolischen Substraten in Abhängigkeit zirkulatorischer und funktioneller Parameter.

1 Hammond, G. L., et al. (1976) *Circulation* **53**, 637 - 643.

Andreas H. Wilhelm, Medizinische Universitätsklinik, Westring 3, D-4400 Münster.

E.-L. Winnacker, Th. Reiter und B. Weingärtner

Ursprungspunkte der Adenovirus-DNA-Replikation

Dieser Bericht beschreibt eine Methode zur Isolierung der Startpunkte der Adenovirus-DNA-Replikation. Adenovirus-Typ-2-(AD2-) DNA, ein lineares, doppelsträngiges Molekül vom Molekulargewicht 22×10^6, repliziert in Zellkernen infizierter menschlicher HeLa-Zellen. Auf Grund kinetischer, elektronenoptischer und genetischer Untersuchungen wurde von uns festgestellt[1], daß sowohl Start- als auch Endpunkte der DNA-Replikation sich an den molekularen Enden des DNA-Moleküls befinden. Durch Isolierung replizierender Formen auf BND-Cellulose und nachfolgende, sequentielle Verdauung der DNA mit Restriktionsendonucleasen unterschiedlicher Spezifizität konnten die Startpunkte angereichert und charakterisiert werden. Aufbauend auf den besonderen funktionellen und strukturellen Eigenschaften der beiden molekularen Enden, die einerseits ein am 5'-Ende der DNA kovalent gebundenes Protein enthalten sowie andererseits eine „umgekehrte terminale Repetition" darstellen, wird ein Modell für den Ablauf der Adenovirus-DNA-Replikation vorgeschlagen und das Problem der Erhaltung der genetischen Information an den molekularen Enden während der Replikation diskutiert.

1 Weingärtner, B., Winnacker, E. L., Tolun, A., Pettersson, U., (1976) *Cell* **9**, 259.

Ernst-L. Winnacker, Th. Reiter und B. Weingärtner, Institut für Biochemie der Universität München, Karlstraße 23, D-8000 München 2.

J. H. Wissler

Fluidity (Microviscosity) of Surface Lipid Hydrocarbon Layers of Membranes of Chemokinetically and Chemotactically Stimulated Neutrophil Polymorphonuclear Leucocytes

Different forms of motion responses and biological activities (adhesion, chemokinesis, chemotaxis, chemoattractive trapping, phagocytosis, extrusion of lysosomal constituents, etc.) are shown by the various leucocyte types upon exposure to selected blood proteins and serum-derived inflammatory mediators in the form of either diffusion equilibrium or nonequilibria of solutes (gradients and boundaries between different solute concentrations; for survey, see[1,2]). They represent equivalents in vitro for the description of the phenomenon of leucocyte accumulation at inflammatory sites in vivo, consisting of recognition and removal of host's own damaged and intruded foreign components, and initiation of tissue repair.

So far there have been no investigations on the physical alterations of the membrane of leucocytes upon their stimulation by purified chemokinetic and chemotactic solutes. The problem was unsolved as to whether some of these cellular activities can be attributed to differences in the properties of cellular membranes and how these are interrelated. The determination and comparison of the fluidity (microviscosity) of the surface lipid hydrocarbon layers of membranes of chemokinetically and chemotactically stimulated leucocytes is one partial approach to the problem, and is a way to gain a more detailed understanding of the cellular recognition process[1,3]. Microviscosity η (fluiditity, respectively) may be defined as the harmonic mean of the effective viscosities, given in poise units, along the symmetry axes of a system with a reporter probe embedded (for survey and methods, see ref.[4-7]). The fluid state of the surface membrane lipid hydrocarbon layers of normal, chemokinetically and chemotactically stimulated neutrophils was studied by total intensity and polarization of fluorescence on a quantum basis through pulse excitation of 1,6-diphenyl-1,3,5-hexatriene, perylene and *N*-phenyl-1-naphthylamine as fluidity and structural probes.

Normal peripheral neutrophils of guinea pigs have an average η of 1.5 poise when kept in Gey's balanced salt solution at 37 °C, pH 7.10. In the biologically active dose range of 10^{-9} to 10^{-6}M of pig serum-derived chemotactic solutes which induce cellular direction finding only but no locomotion, if isolated and crystallized by standard procedures[1,8-10], the neutrophils have an unchanged η.

However, surface lipid hydrocarbon regions are strongly shielded from the aqueous surroundings, as shown by the kinetics of the absolute signal pulse series. Highly purified, non-defatted serum albumin, as chemokinetic stimulus[2] for the cells which induces increased locomotion only but no direction finding, above an average threshold of about 10^{-5}M enhances η to a range of 2 - 5 poise and competes for the probes with the cell surface. Intimate contact and exchange of probe between the cell surface lipid hydrocarbon regions and the serum albumin can be demonstrated. Exposure of cells to both, chemotactic and chemokinetic solutes, to elicit chemotaxis in form of directional locomotion, induces both effects together.

Chemotactic solutes act directly and independently of serum albumin (chemokinetic solutes) on the cell surface. They apparently bind to the cell surface, forming larger hydrophilic surface areas, without altering the fluidity of the lipid hydrocarbon layer of the membrane. The driving force for this interaction represents the processing at the cellular surface of the information (recognition process in chemotaxis of leucocytes) that resides in the metastable chemotactic solute-water phase[1,3]. In terms of the statistical mechanical model of water structure, this information can be visualized by a 5-fold coordinated water structure at the chemotactic solute[3]. The binding of chemotactic solutes to the cell surface itself follows the transition to four-fold coordinated, fully hydrogen bonded water clusters and separation of the chemotactic solute from the water phase. On the basis of the data presented here this binding can be best described by the model of multivalent adsorption[11], which may lend additional bits of information (and specificity) to the recognition of basic information[3]. On the other hand, the present investigation and others[2,8,9,12] can clearly invalidate the notion drawn from biological studies[13] that serum-derived chemotactic solutes[1,8-10] exert there biological activity on the cell only when complexed by serum albumin, but not directly and alone.

1 Wissler, J.H. (1975) in Proc. Immunosymposium Wien 1973: Gram-negative Bacterial Infections and Mode of Endotoxin Actions, Pathophysiological, Immunological and Clinical Aspects (Urbaschek, B., Urbaschek, R. & Neter, E., eds.) pp. 91 - 105, Springer-Verlag, Wien.

2 Keller, H.U., Wissler, J.H., Hess, M.W. & Cottier, H. (1977) in Movement, Metabolism and Bactericidal Mechanismsm of Phagocytes (Rossi, F., Patriarca, P.L. & Romeo, D., eds.) pp. 15 - 20, Piccin Medical Books, Padova.

3 Wissler, J.H. (1976) *Hoppe-Seyler's Z. Physiol. Chem.* **357**, 286 - 287.

4 Weber, G. (1953) *Adv. Protein Chem.* **8**, 415 - 459.

5 Overath, P., Thilo, L. & Träuble, H. (1976) *Trends Biochem. Sciences* **1**, 186 - 189.

6 Shinitzky, M. & Barenholz, Y. (1974) *J. Biol. Chem.* **249**, 2652 - 2657.

7 Shinitzky, M. & Inhbar, M. (1974) *J. Mol. Biol.* **85**, 603 - 615.

8 Wissler, J.H. (1971) *Experientia* **27**, 1447 - 1448.

9 Wissler, J.H. (1972) *Eur. J. Immunol.* **2**, 73 - 83.

10 Wissler, J.H. (1972) *Eur. J. Immunol.* **2**, 84 - 89.

11 Jennissen, H.P. (1976) *Biochemistry* **15**, 5683 - 5692.

12 Keller, H.U., Wissler, J.H., Hess, M.W. & Cottier, H. (1978) *Eur. J. Immunol.*, in press.

13 Wilkinson, P.C. (1976) *Exp. Cell Res.* **103**, 415 - 418.

Dr. Josef H. Wissler, Arbeitsgruppe Biochemie, Max-Planck-Institut für Experimentelle Kardiologie, Sprudelhof 11, D-6350 Bad Nauheim.

U. Zabel und K. Ring

Kinetik der Aufnahme von Pyrimidinnucleosiden in Ehrlich-Maus-Ascites-Tumorzellen.

Cytidin und Uridin werden mit hoher Geschwindigkeit in Ehrlich-Maus-Ascites-Tumorzellen aufgenommen. Die Aufnahme wird durch Inhibitoren des Energiesstoffwechsels gehemmt; sie ist unabhängig von extrazellulären Natriumionen. Obwohl der überwiegende Anteil der aufgenommenen Nucleoside während der Membranpassage oder unmittelbar im Anschluß daran zu den Mono-, Di- und Triphosphaten aufphosphoryliert wird, werden insbesondere bei physiologisch älteren Zellen auch freie Nucleoside im Zellpool akkumuliert. Der Transport ist praktisch unidirektional; Efflux durch freie, erleichterte oder Austauschdiffusion ist nicht nachweisbar gewesen.

Sowohl der Uridin- als auch der Cytidininflux erfolgt durch multiple Systeme mit überlappenden Substratspezifitäten. Bei 37 °C und pH 7.4 wird Uridin durch eine im Lineweaver-Burk-Diagramm homogen erscheinende Transportkomponente aufgenommen. Bei Temperaturen unterhalb von 28 °C, sowie bei erniedrigtem pH-Wert wird jedoch eine zweite Transportkomponente mit höherer Substrataffinität nachweisbar. Die scheinbare Michaelis-Konstante für das erste System liegt bei 100 - 200μM, die für das zweite System bei 10 - 17μM. 0.5mM *p*-Chlormercuriphenylsulfonat hemmt den Transport zu mehr als 50%; dextrangebundenes *p*-Chlormercuribenzoat ist hingegen wirkungslos. Vorinkubation der Zellen mit unmarkiertem Uridin hemmt den [^{14}C]-Uridininflux signifikant.

Für Cytidin lassen sich bei pH 7.4 und 37 °C zwei Transportfunktionen stark unterschiedlicher Affinität nachweisen, deren scheinbare Michaelis-Konstanten im Bereich von 5 - 25μM, bzw. 200μM, liegen. Veränderung des pH-Wertes oder der Temperatur ergeben keine substantiellen Änderungen. Wie der Uridintransport wird auch der Cytidintransport durch *p*-Chlormercurisulfonat oder durch Vorbeladung der Zellen gehemmt. Verschiedene kinetische Beobachtungen deuten darauf hin, daß die Cytidinaufnahme durch eine regulatorische Bindungsstelle am Transportsystem gesteuert wird, die über eine mit *p*-Chlormercurisulfonat reaktive SH-Gruppe verfügt.

Ulrich Zabel und Klaus Ring, Gustav-Embden-Zentrum der Biologischen Chemie, Abteilung der Mikrobiologischen Chemie der Universität Frankfurt am Main, Theodor Stern Kai 7, D-6000 Frankfurt 70.

K. Zechel

Actin from the Smooth Muscle of Chicken Gizzard Is Different from Cytoplasmic Actin and from Skeletal Muscle Actin

Cytoplasmic actins and actin from skeletal muscle can be distinguished by isoelectric focusing in polyacrylamide gels in the presence of 9M urea and the nonionic detergent Nonidet P40[1], and by sodium dodecylsulfate/urea gel electrophoresis[2]. The actin isolated from the smooth muscle of chicken gizzard is resolved by isofocusing under denaturing conditions into two isoelectric species: one major band that focuses at slightly higher pH than the most basic cytoplasmic actin (γ-actin), and one minor band that focuses at the same apparent pI as the acidic cytoplasmic actin (β-actin). In dodecylsulfate/urea polyacrylamide gels, the chicken gizzard γ-like component has the same mobility as actin from skeletal muscle (α-actin). The γ-like component of chicken gizzard is, according to sequence data of Vandekerckhove and Weber[3] more closely related to skeletal muscle α-actin than to cytoplasmic actin. However, the aminoterminal part of chicken gizzard actin is more similar to cytoplasmic γ-actin than to muscle actin. The aminoterminal sequence therefore appears to determine the isoelectric position of the chicken gizzard γ-like component while the mobility of the polypeptide in dodecylsulfate/urea polyacrylamide gels is more dependent on the rest of the sequence. However, since the γ-like component focuses differently from cytoplasmic γ-actin, it must contain an undetected difference that is responsible for its different isoelectric point. This contention is supported by differences noted during a comparative analysis of partial tryptic digests of chicken gizzard actin and rabbit skeletal muscle actin by dodecylsulfate gel electrophoresis.

1 Whalen, R.G., Butler-Browne, G.S. & Gros, F. (1976) *Proc. Natl. Acad. Sci. U.S.A.* **73**, 2018 - 2022.
2 Storti, R.V. & Rich, A. (1976) *Proc. Natl. Acad. Sci. U.S.A.* **73**, 2346 - 2350.
3 Vandekerckhove, J. & Weber, K. (1978) *Proc. Natl. Acad. Sci. U.S.A.*, in press.

Dr. Kasper Zechel, Abteilung 09-Biochemie, Max-Planck-Institut für Biophysikalische Chemie, Am Faßberg, D-3400 Göttingen-Nikolausberg.

R. Zecher and H. U. Wolf

Characterization and Purification of Phosphatase Activity from Human Erythrocytes

In order to confirm our hypothesis that high-affinity $Ca^{2\oplus}$-ATPase (EC 3.6.1.3) and $Ca^{2\oplus}$-dependent *p*-nitrophenylphosphatase (EC 3.1.3.1) of human erythrocyte membranes are not identical totally, or in part, we tried to isolate and characterize the phosphatase activity of human erythrocytes.

Human blood (group 0, Rh^+) was washed three times with isotonic KCl solution. The packed erythrocytes were diluted 1:1 with 10mM $CaCl_2$ and 40mM morpholinopropanesulfonic acid (I), pH 7.5, and hemolyzed by freezing at − 80 °C. After thawing, the hemolyzed erythrocytes were diluted 1:1 with 20mM I, pH 7.5, and the membranes were sedimented at 37 400 × *g* for 40 min. Then they were washed several times in 20mM I and 10mM $K^\oplus$ and frozen at − 80 °C. Both membranes and the first supernatant were used for characterization of phosphatase activity.

Phosphatase activity was measured under the following conditions: 40mM Tris/HCl or Tris/maleate (variable pH-values), 5mM $Mg^{2\oplus}$, 40μM $Ca^{2\oplus}$, 100mM $K^\oplus$ and 5mM *p*-nitrophenylphosphate, $T = 37$ °C.

Phosphatase activities from membranes and the first supernatant were similar with respect to the dependence of the activity on the pH value (main activity peaks at pH 6.0 and 8.6) and on the concentration of monovalent cations such as $Na^\oplus$ or $K^\oplus$ up to 100mM concentrations.

Phosphatase activity from the first supernatant was purified on DEAE-Sepharose CL-6B using a linear gradient of 20 - 250mM KCl (+ 2mM $Mg^{2\oplus}$, 10mM I, pH 6.8, $T \leq 4$ °C). After precipitation of the most active fractions with 80% $(NH_4)_2SO_4$, further purification was carried out on Sephadex G-100 at 150mM $K^\oplus$, 2mM $Mg^{2\oplus}$, 0.2% mercaptoethanol, and 20mM I, pH 6.8. The phosphatase activity could be purified 2100-fold up to 11.6 U/mg protein. As shown by sodium dodecylsulfate gel electrophoresis, the M_r of the active material is ≤ 30 000, indicating that the phosphatase activity is not identical in part or totally with the high-affinity $Ca^{2\oplus}$-ATPase of human erythrocyte membranes (M_r = 145 000, 115 000, and 105 000[1]).

1 Wolf, H.U. (1977) *Acta Biol. Med. Ger.* **36**, 847 - 858.

Prof. Dr. H.U. Wolf, Abteilung Pharmakologie und Toxikologie der Universität Ulm, Oberer Eselsberg, D-7900 Ulm.

D. Ziegler, G. Keilich and R. Brossmer

Isolation and Characterization of Two Neuraminidases from Vibrio cholerae

Two neuraminidases (EC 3.2.1.18) could be isolated from the culture filtrate of *Vibrio cholerae**. Both enzymes were adsorbed to *N*-(4-nitrophenyl)oxamic acid coupled to Sepharose 4B containing 1,6-diaminohexane as the spacer group (I)[1]. Separation of the enzymes was achieved by elution with 50mM sodium acetate buffer, pH 5.0, containing 1.0M sodium chloride. The two neuraminidases show different optimum pH values, which were determined in 50mM sodium acetate buffer (2mM $Ca^{2\oplus}$ ions) to be pH 4.8 and pH 5.4, respectively. Furthermore, their substrate specificities were different. The kinetic data, which were obtained with a series of synthetic as well as natural substrates, were compared and discussed with respect to the structure of the aglycon. The two enzymes differ significantly in their behaviour against the synthetic substrate benzyl α-ketoside of neuraminic acid. For the neuraminidase with the K_m value 3.6mM, I proved to be a more specific adsorbent than for the neuraminidase with the K_m value 1.4mM. Although the nature of the adsorption effect is not clearly understood, these findings indicate a correlation between the substrate specificity of the neuraminidases and their binding characteristics to the affinity column.

* Lyophilized ultrafiltrate from *Vibrio cholerae*, Behringwerke AG, Marburg.

1 Brossmer, R., Ziegler, D. & Keilich, G. (1977) *Hoppe-Seyler's Z. Physiol. Chem.* **358**, 397 - 400.

Dietmar Ziegler, Gunda Keilich und Reinhard Brossmer, Institut für Biochemie II (Med. Fak.), Universität Heidelberg, Im Neuenheimer Feld 328, D-6900 Heidelberg.

G. Zimmer, L. Lacko and B. Wittke

Diethylpyrocarbonate Inhibits Glucose Transport in Human Red Cells in a Temperature-Dependent Manner

Diethylpyrocarbonate has been found to react with histidine-imidazole residues in different enzymes at pH values below 7.0[1-4].

Different biophysical approaches have led to the suggestion that the reagent interferes with a polar lipid-protein interaction site at the red cell membrane[5].

We have, therefore, centered our interest on the question of further interaction of diethylpyrocarbonate with glucose transport in human red cells.

It was found that diethylpyrocarbonate at 1 - 3mM concentrations inhibited glucose transport in a noncompetitive manner (K_i = 5.4mM at 20 °C). The dissociation constant was found to decrease in the temperature range of 10 - 20 °C, while it increased from 20 - 30 °C. The enthalpy change (ΔH) for the association of the inhibitor with the transport system was 5.9 kcal/mol below 20 °C and − 3.9 kcal/mol above 20 °C. There was a large increase in entropy for the association of the inhibitor of 30.5 cal/(Mol x degree) below 20 °C, while above that temperature, the change in entropy was about 1/10 of that value. These data indicate a major structural rearrangement of the red cell membrane about 20 °C, which intimately affects the glucose transport system.

1 Mühlrad, A., Hegyi, G. & Toth, G. (1967) *Acta Biochim. Biophys. Acad. Sci. Hung.* **2**, 19 - 29.
2 Huc. C., Olomucki, A., Le-thi-Lan, D. B. P. & van Thoai, N. (1971) *Eur. J. Biochem.* **21**, 161 - 169.
3 Holbrook, J. J. & Ingram, V. A. (1973) *Biochem. J.* **131**, 729 - 738.
4 Burstein, Y., Walsh, K. A. & Neurath, H. (1974) *Biochemistry* **13**, 205 - 210.
5 Bastian, P. & Zimmer, G. (1976) Biochemistry of Membrane Transport, FEBS Symposium, Zürich, Abstract No. 334.

Dr. Guido Zimmer, Gustav-Embden-Zentrum der Biologischen Chemie, Universität Frankfurt am Main, Theodor-Stern-Kai 7, D-6000 Frankfurt 70.

H. Zimmermann and M. J. Dowdall

Uptake of [^{3}H]Adenosine into Nerve Endings and Incorporation of Adenine Nucleotides into Synaptic Vesicles in the Torpedo Electric Organ.

For uptake studies of [2-^{3}H]adenosine, synaptosomes were isolated from the *Torpedo* electric organ[1]. [^{3}H]-Adenosine is taken up into isolated nerve endings by a saturable uptake mechanism with a K_m value of ≙ 1 μM. This uptake can be inhibited by 2-deoxyadenosine. The incorporated adenosine is rapidly metabolized, with ADP and ATP constituting half of the metabolic products.

On application of [^{3}H]adenosine to perfused blocks of electric tissue[2], radiolabel is taken up into synaptic vesicles. It can be shown by ion-exchange chromatography that about 80% of the ^{3}H-labelled products in the isolated synaptic vesicle fraction are in the form of ATP. The remaining radioactivity was in the form of ADP and traces of AMP. On stimulation of tissue blocks at 0.1 Hz, two populations of synaptic vesicles can be isolated, one of which (fraction VP_2) can be shown to contain the bulk of newly synthesized transmitter[3].

In the present investigations, newly synthesized ATP was distributed as newly synthesized ACh was, with fraction VP_2 containing vesicles of high specific ATP radioactivity.

The data suggest that adenosine which can be taken up into the nerve terminal by a high affinity uptake mechanism is an effective precursor for vesicular ATP. On stimulation, synaptic vesicles become heterogeneous. Density centrifugation methods allow the separation of a population of synaptic vesicles which have a high turnover rate for both ACh and ATP.

1 Morel, N., Israël, M., Manaranche, R. & Mastour-Frachon, P. (1977) *J. Cell. Biol.* **75**, 43 - 55.
2 Zimmermann, H. & Denston, C.R. (1977) *Neuroscience* **2**, 695 - 714.
3 Zimmermann, H. & Denston, C.R. (1977) *Neuroscience* **2**, 715 - 730.

H. Zimmermann and M.J. Dowdall, Max-Planck-Institut für Biophysikalische Chemie, Abteilung Neurochemie, Am Faßberg, D-3400 Göttingen.

R.E. Zimmermann and M. Vohle

Conformation Changes of Fibrinogen Induced by Thrombin

Since fibrinogen is adsorbed to immobilized fibrin momers by its α-chains, and, in addition, the aggregation of fibrinogen with formaldehyde or carbodiimide leads only to α-chain polymerisation, we must assume that the surface of the native fibrinogen molecule is mainly composed of α-chains[1]: During the physiological process of fibrin formation, another mechanism comes into play: the γ-chain dimerisation is the initial event, followed by α-chain polymerisation[2].

When native fibrinogen is treated with increasing amounts of mercaptoethanol, the β- and the γ-chains are released together. When they are separated by isoelectric focusing, a very high degree of structure modification is demonstrated. Chain separation on CM 52 cellulose supports this result and leads to the conclusion that after the proteolytic attack of thrombin, the α-chain and γ-chain interactions stabilize the aggregate in such a manner that particularly the γ-chains are attached to the clot and are not very easily separable by mercaptoethanol treatment. Furthermore, as an additional effect, a certain proportion of the γ-chains shows structure modifications caused by the thrombin action. When fibrinogen is aggregated with leucine amino peptidase, chain fragments of various kinds are formed which continue to respond to thrombin with an increased polymerisation rate.

1 Blombäck, B. & Blombäck, M. (1972) *Ann. N. Y. Acad. Sci.* **202**, 77 - 95.
2 Seelich, Th., Furlan, M. & Beck, E.A. (1976) *Thromb. Haemost.* **35**, 620 - 627.

Prof. Dr. R.E. Zimmermann, Physiologisches Institut II, Robert-Koch-Straße 28, D-4400 Münster.

R. Zwilling

Ist das Flußkrebs-Trypsin älter als die Thrombin/Elastase/-Chymotrypsin/Trypsin-Divergenz?

Flußkrebs-Trypsin ist ein Invertebraten-Enzym, welches viele Merkmale einer frühen Entwicklungsstufe der Serin-Proteasen bewahrt hat.

Zur effektiveren Reinigung wurde eine Affinitäts-Chromatographie an Sojabohnen-Trypsin-Inhibitor entwickelt, wobei Flußkrebs-Trypsin ein gegensätzliches Verhalten im Vergleich zu Rinder-Trypsin zeigt. Es denaturiert irreversibel unter pH 4.0 und läßt sich im Sauren nicht mehr von der Inhibitorsäule eluieren. Eine quantitative Zerlegung des Komplexes wird jedoch durch eine NH_3-Lösung von pH 11.0 erreicht, wofür gleichzeitig eine äußerst geringe Ionenstärke Voraussetzung ist. Auf dieser Basis können nach Entfernung des flüchtigen Elutionsmittels durch Gefriertrocknung aus 50 m*l* Cardia-Flüssigkeit 150 mg des reinen Enzyms gewonnen werden*.

Ein Vergleich der *N*-terminalen Sequenz des Flußkrebs-Trypsins mit anderen Serin-Proteasen zeigt, daß zumindest in diesem Teil des Moleküls keineswegs die größte Übereinstimmung mit den Säuger-Trypsinen, sondern eher mit dem Bakterien-Trypsin aus *Streptomyces griseus* besteht[1]. Das Flußkrebs-Trypsin besitzt ebenso wie *Str.-griseus*-Trypsin nur 3 Disulfidbrücken, welche sich bei allen Serin-Proteasen wiederfinden. Zu diesen 3 ursprüngli chen kommen im weiteren Verlauf der divergenten Entwicklung bei Thrombin, Elastase, Chymotrypsin und Trypsin der Wirbeltiere weitere, neuerworbene Disulfidbrücken hinzu.

Es wird diskutiert, ob das Flußkrebs-Trypsin den Typ einer Serin-Protease repräsentiert, welche noch vor der Divergenz von Thrombin/Elastase/Chymotrypsin/Trypsin eine eigenständige Entwicklung genommen hat.

* Zwilling, R., (1978), in Vorbereitung.

1 Zwilling, R., Neurath, H., Ericsson, L.H. & Enfield, D.L. (1975) *FEBS Lett.* **60**, 247 - 249.

Prof. Dr. R. Zwilling, Zoologisches Institut, Fachrichtung Physiologie, Universität Heidelberg, Im Neuenheimer Feld 230, D-6900 Heidelberg.

Hoppe-Seyler's Z. Physiol. Chem.
Bd. 359, S. 345 - 355, März 1978

On the Linkages between Chromophore and Protein in Biliproteins, VII[1]

Amino Acid Sequence in the Chromophore Regions of C-Phycoerythrin from *Pseudanabaena* W 1173 and *Phormidium persicinum*

Günther MUCKLE*, Joachim OTTO** and Wolfhart RÜDIGER*

* Botanisches Institut and ** Institut für Physiologische Chemie, Physikalische Biochemie und Zellbiologie, Universität München

(Received 21 October 1977)

Dedicated to Professor Dr. Dr. h.c. mult. A. Butenandt on the occasion of his 75th birthday

Summary: Bilipeptides from all chromophore regions were prepared by trypsin digestion of C-phycoerythrin from *Pseudanabaena* W 1173 and *Phormidium persicinum*. Analytical separation and quantitative determination of bilipeptides was achieved by isoelectric focusing, preparative isolation by gel chromatography and ion-exchange chromatography. Amino acid analysis revealed cysteine as the only amino acid common to all chromopeptides. Amino acid sequences were determined by Edman degradation and the dansyl-Edman technique. Sequences are different in all 5 and 6 chromophore regions, respectively. Possible homologies are discussed. A thioether linkage between ring A of the chromophore and cysteine was found in the bilipeptides (as before in biliproteins). A second linkage (serine ester) was found in only one peptide (CM 4.I from *Pseudanabaena* W 1173). This peptide absorbs as cation at a longer wavelength (559 nm) than the other bilipeptides (542 - 550 nm).

Über die Bindungen zwischen Chromophor und Protein in Biliproteiden, VII: Aminosäuren-Sequenz der Chromophorbereiche von C-Phycoerythrin aus Pseudanabaena W 1173 und Phormidium persicinum

Zusammenfassung: Mittels Trypsins wurden Bilipeptide aus allen Chromophorbereichen von C-Phycoerythrin aus *Pseudanabaena* W 1173 und *Phormidium persicinum* gewonnen. Die analytische Trennung und quantitative Bestimmung der Bilipeptide gelang durch isoelektrische Fokussierung; für die präparative Isolierung wurde die Gelchromatographie und Ionenaustausch-Chromatographie verwendet. Die Aminosäureanalysen zeigten, daß Cystein als einzige Aminosäure allen Chromopeptiden gemeinsam ist. Die Aminosäuresequenzen wurden durch Edman-Abbau und die Dansyl-Edman-Technik ermittelt. Die Sequenzen sind für alle 5 bzw. 6 Chromophorbereiche verschieden. Mögliche Homologien werden diskutiert. Eine Thioätherbindung zwischen Ring A des Chromophors und Cystein wurde auch in den Bilipeptiden (wie zuvor in Biliproteiden) gefunden. Eine zweite Bindung (Serinester) wurde nur in einem Peptid (CM 4.I aus *Pseudanabaena* W 1173) gefunden, das als Kation längerwellig (559 nm) absorbiert als die anderen Bilipeptide (542 - 550 nm).

Enzyme: Trypsin (EC 3.4.21.4).
Abbreviations: Dansyl, 5-dimethylamino-naphthalene 1-sulphonyl; > PhNCS, phenylthiohydantoin.

Key words: C-phycoerythrin, bilipeptide, cyanobacteria, chromic acid degradation.

The biliproteins C-phycoerythrin and C-phycocyanin are the major accessory pigments of photosynthesis in cyanobacteria (blue-green algae)[2,3]. They can account for up to 24% of the dry weight[4] or about half of the total soluble protein[5] of the algal cells. C-Phycoerythrin and C-phycocyanin are both composed of α- and β-subunits[6,7]. The amino-terminal sequences of C-phycoerythrin and C-phycocyanin subunits exhibit a remarkable degree of conservation amongst all cyanobacteria tested so far[8,9]. Sufficient homology can still be detected between the amino-terminal sequences of α- and β-subunits to postulate a common ancestral gene for both subunits[9,10]. However, the homology within C-phycoerythrin and C-phycocyanin α-subunits from various algae – or within all β-subunits – is strikingly greater than that between α- and β-subunits, even if these are the components of a single biliprotein. The homology of amino-terminal sequences between C-phycoerythrin and C-phycocyanin contrasts to the finding that these two biliproteins are immunologically unrelated[11,12].

In the course of our investigation of chromophore-protein linkages in biliproteins, the question of the peptide sequences around the chromophore-bearing amino acid arose. C-Phycoerythrins from various cyanobacteria carry 5 or 6 chromophores (2 on the α-subunit, 3 or 4 on the β-subunit, respectively)[13]. Is the linking amino acid in all of these cases cysteine, as had been demonstrated earlier[14] for one chromophore region in B-phycoerythrin from a red alga, and are the sequences in all of these chromophore regions different? We describe here the preparation and Edman degradation of C-phycoerythrin bilipeptides from *Pseudanabaena* W 1173 (5 chromophores) and *Phormidium persicinum* (6 chromophores). A similar approach has been described for 2 bilipeptides of C-phycocyanin from *Mastigocladus laminosus*[15].

Material and Methods

C-Phycoerythrin was isolated from *Pseudanabaena* W 1173 (obtained from R. Wagenmann, München) and *Phormidium persicinum* (obtained from Prof. Chapman, Los Angeles). Frozen algal cells (10 g fresh weight) were disrupted by grinding with glass powder (7 g, diam. 0.25 mm) in a Vibrogen Zellmühle (Bühler, Tübingen) for 0.5 min *(Ph. persicinum)* or 2 - 5 min (*Ps.* W 1173). All subsequent procedures were performed at 0 - 4 °C. Pigments were extracted with 0.1M sodium phosphate buffer (3 x 5 m*l*, pH 7.0); after centrifugation, C-phycoerythrin was precipitated with ammonium sulfate (50% saturation), dissolved in 1 - 2 m*l* 0.1M Tris/HCl buffer (pH 7.0) and applied to gel chromatography (Sephadex G-100, elution with 0.01M Tris/HCl buffer, pH 7.0, containing 0.1M KCl, 1mM EDTA and 1mM NaN_3). The main C-phycoerythrin fraction was then subjected to ion-exchange chromatography (Servacel DEAE 23SS, equilibrated with 0.1M Tris/HCl buffer, pH 8.0, elution with the same buffer and 0 - 0.3M KCl gradient). Fractions with $A_{560}/A_{280} > 6.3$ were pooled, and C-phycoerythrin obtained by ammonium sulfate precipitation (40 - 60% saturation), dialysis (12 - 14 h against 1mM Tris/HCl buffer, pH 7.0 and then 3 - 4 h against bidistilled water), and lyophilisation was pure according to disc gel electrophoresis (7.5% polyacrylamide gel, Tris/glycine buffer, pH 8.8, electrophoresis for 8 h at 4 °C and 150 - 200 V) and sodium dodecylsulfate gel electrophoresis[6]. C-Phycoerythrin α- and β-subunits were separated by ion-exchange chromatography with a linear urea gradient as described by Muckle and Rüdiger[13].

Trypsin digestion for analyses was performed with 4 mg C-phycoerythrin (or 1 - 2 mg of subunits) in 5 m*l* (3 m*l*) 0.1M NH_4HCO_3, pH 7.9. After heat denaturation (2 min at 80 °C), the solution was incubated at 37 °C under nitrogen with 40 µg (10 - 20 µg) trypsin (treated with L-1-tosylamido-2-phenylethylchloromethyl ketone, Fa. Merck/Darmstadt, fresh solution 1 mg/m*l* H_2O). The same amount trypsin was added every 30 min. After 2 h the pH was adjusted to 3 - 4 with acetic acid and the mixture was lyophilized. Preparative digestion was performed with larger amounts (e.g. 150 mg) of C-phycoerythrin with the same substrate/enzyme ratio 100:1 (w/w).

Purification of chromopeptides on a preparative scale was achieved by the following steps:
1) gel chromatography on Sephadex G-25, equilibrated with 5% acetic acid, elution with the same solvent 2 m*l*/h. On a column 2.5 x 80 cm, 4 red zones with the following elution volumes were found in a digest from 150 mg C-phycoerythrin: G.1, 100 - 150 m*l*; G.2, 160 - 180 m*l*; G.3, 190 - 225 m*l*; G.4, 230 - 300 m*l*. A rough calibration of the column was achieved with dextran-blue (void volume) and the isolated peptide CM 4.I (see below). 2) The separated fractions were then further fractionated on carboxymethyl-cellulose

(Serva, Germany), column 2.5 x 25 cm equilibrated with 0.01M sodium acetate buffer, either pH 4.5 (G.4 and part of G.3) or pH 4.2 (G.2 and G.3), elution with a gradient 0.01 - 0.05M ("acidic" peptides) or 0.01 - 0.1M ("basic" peptides) sodium acetate of the same pH.
3a) "Basic" chromopeptide fractions ($p_I > 5$) were further purified on a silica gel column (1.6 x 15 cm) by adsorption from the above mentioned acetate buffer, washing with 250 m*l* water, elution with 2% pyridine/water. 4a) The last step for "basic" chromopeptides was thin-layer chromatography (silica gel G, Merck, Germany) with n-butanol/acetic acid/water (3:3:1). Coloured zones were eluted with methanol.
3b) "Acidic" chromopeptides ($p_I < 5$) were purified on a Bio-Rex 70 column (1.6 x 25 cm), equilibrated with 10% acetic acid in water and eluted with a gradient 20 - 60% acetic acid. The main chromopeptide fractions were concentrated under reduced pressure. This was repeated 1 - 2 times after addition of 5 - 10 m*l* water until most of the acetic acid was removed

All the chromopeptides, the "basic" as well as the "acidic" were then adsorbed on silica gel, washed with water and eluted with 2% pyridine/water. Purification of chromopeptides was checked by isoelectric focusing with the method of Wrighley[16], by amino acid analysis and by determination of the *N*-terminus after dansylation, hydrolysis and two-dimensional thin-layer chromatography[17]. Fraction B 2.I from *Ph. persicinum* C-phycoerythrin which gave multiple bands in isoelectric focusing, was dissolved in 5% formic acid and digested with pepsin (enzyme/substrate = 1:10) for 9 h at 37 °C. Thin-layer chromatography of the digest as above (see 4a) yielded one chromopeptide band ("B 2.I pepsin") ($R_F = 0.78$) and colourless peptides ($R_F = 0.54$ and 0.46).

Amino acid analysis was carried out after hydrolysis of peptides with 6N HCl for 24 h under nitrogen, either with an analyzer working at the nanomole range[18] (*Pseudanabaena* W 1173 peptides) or by dansylation and analysis of dansylamino acids by thin-layer chromatography[17] (*Ph. persicinum* peptides).

Chromic acid degradation and CrO_3/NH_3 degradation were performed as previously described[1,19-21]. Electronic spectra were recorded with a DMR 22 spectrophotometer (Zeiss, Germany), gels were scanned with a densitometer TLD 100 with recorder UR 400 and integrator (Vitatron, Germany), slit width 0.25 mm, filter 559 nm.

Results

Trypsin digestion, number and distribution of chromopeptides

Chromopeptides from all chromophore regions can only be obtained if trypsin digestion is complete and if no bleaching of chromophores occurs. This can be checked by spectrophotometry (Fig. 1). Native C-phycoerythrin has a high absorption at 560 nm, which is reduced by about 50% by unfolding of the peptide chain or proteolytic digestion[13]. Digestion for 1 - 2 h reduces the absorption to this level, which stays constant afterwards if anaerobic conditions are used. A 2-h incubation was chosen for the preparation of chromopeptides because prolonged incubation with trypsin leads to unspecific cleavage of peptides.

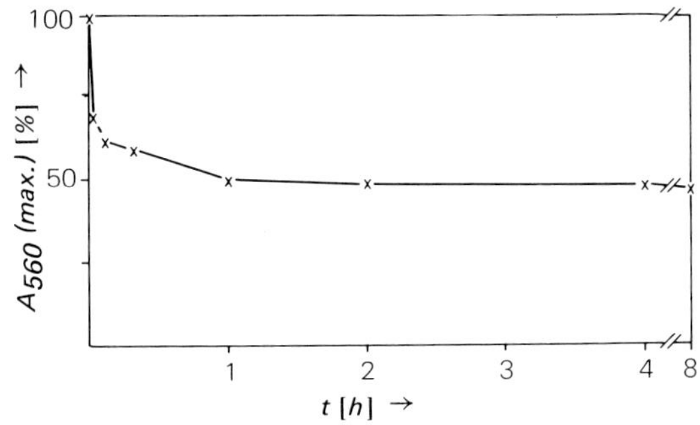

Fig. 1. Decrease in absorption of C-phycoerythrin from *Pseudanabaena* W 1173 during trypsin digestion.
Pure C-phycoerythrin in 0.1M Tris/HCl buffer (pH 8.0) was incubated with trypsin (C-phycoerythrin: enzyme ratio 100:1) at 37 °C under nitrogen. After various incubation times, samples were acidified with HCl to pH 2 (formation of the chromophore cation) and the extinction at the maximum (560 nm) was determined. Time zero = extinction of native C-phycoerythrin without acidification.

A convenient and rapid method for separation and determination of chromopeptides is isoelectric focusing in gel tubes[16]. Gradients of pH 2 - 11, 4 - 6 and 5 - 7 were used. No staining was necessary because of the inherent colour of the chromopeptides. Scanning was possible after incubation of gels with trichloroacetic acid. This treatment yielded a uniform colour (cation of the chromophore) independent of the p_I of single zones. Two groups of chromopeptides were distinguishable by isoelectric focusing (Fig. 2): "acidic chromopeptides" with $p_I < 5$ and "basic chromopeptides" with $p_I > 5$. In the region $p_I = 4.3 - 4.4$, varying numbers of coloured zones appeared in different digestion experiments.

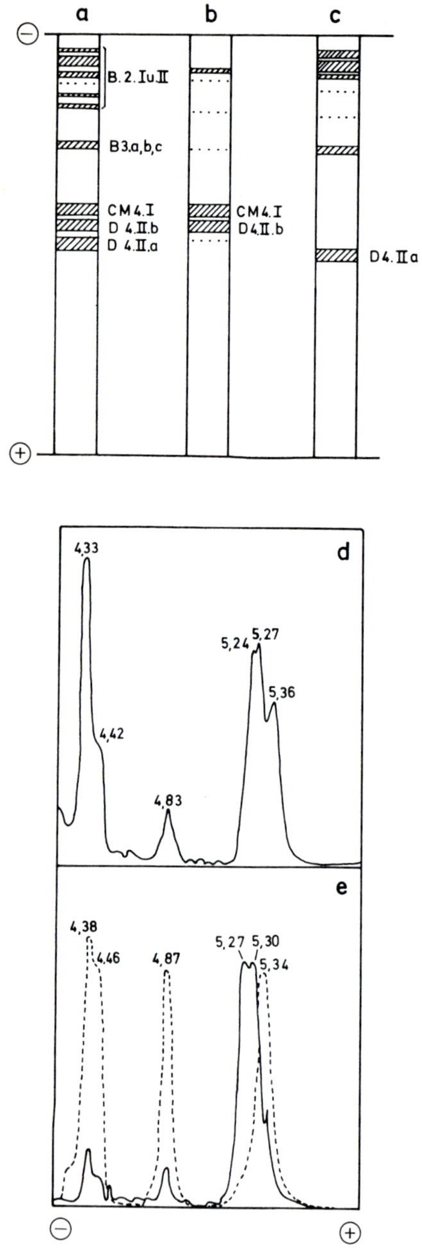

Fig. 2. Isoelectric focusing of tryptic bilipeptides from *Pseudanabaena* W 1173 with a pH gradient 4.0 - 7.5.

a) Peptides from C-phycoerythrin; b) peptides from α-subunit; c) peptides from β-subunit; d) densitogram of gel a; e) densitograms of gels b (——) and c (– – –). The single isolated bilipeptides were run in control tubes for identification of the bands as indicated. Numbers in d) and e) are isoelectric points.

These zones were so close to each other, that only one peak appeared upon scanning. Whereas the intensities of the single zones of this region varied, the amount of all chromophore-containing peptides in this region was constant (Table 1). The "basic region", p_I = 5.2 - 5.4, exhibited 3 bands of equal intensity. Scanning of single bands was sometimes difficult in this region because of rapid diffusion. The ratio "basic": "acidic" chromopeptides is 3.02 : 2.32 for *Ps.* W 1173 and 3.0 : 3.2 for *Ph. persicinum.* This corresponds to the previously determined content of 5 and 6 chromophores, respectively, for C-phycoerythrin from these algae[13]. The higher value of acidic chromopeptides (calculated 3 : 2 and 3 : 3) is due to products of incomplete digestion (G-1 peptides, see below) which migrate at p_I = 4.5 - 4.7.

Tryptic digestion of the subunits revealed that the α-subunit yields only 2 "basic" and no "acidic" chromopeptides. The β-subunit yields one "basic" and two (*Ps.* W 1173) or three *(Ph. persicinum)* "acidic" chromopeptides. This again corresponds to the previously[13] determined chromophore ratio of 2 : 3 and 2 : 4 for α : β subunits.

Amino acid composition and sequence of chromopeptides

Whereas preparation of minute amounts of chromopeptides by elution from isoelectric focusing gels allowed determination of *N*-terminal amino acids, a different procedure was applied for chromopeptide isolation on a preparative scale (Fig. 3). This included gel chromatography on Sephadex G-25, which yielded 4 chromopeptide fractions with a typical chromophore ratio G-1 : G-2 : G-3 : G-4 = 31.3 : 14.6 : 22.5 : 31.6 for *Ps.* W 1173 and 17.1 : 29.1 : 26.7 : 27.1 for *Ph. persicinum.* G-1 peptides were products of incomplete proteolysis: further trypsin digestion of the G-1 fraction yielded chromopeptides identical with those from G-2, G-3 and G-4 as identified by isoelectric focusing. Further separation was achieved with carboxymethyl cellulose: the G-2 and G-3 fractions, which contained "acidic" chromopeptides, were chromatographed at pH 4.2. Traces of "basic" peptides which were present in some experiments in fraction G-3

Table 1. Densitometry of trypsin digests of C-phycoerythrin and subunits after isoelectric focusing. Values related to the "basic" chromopeptide.

Chromopeptides	"Acidic"		"Basic"		
	B 2	B 3	CM 4.I	D 4.II.b	D 4.II.a
Pseudanabaena W 1173					
C-Phycoerythrin	1.98	0.32	1.03	1.03	1.00
α-Subunit	0.20	0.08	1.00	1.00	–
β-Subunit	1.02	0.59	–	–	1.00
Phormidium persicinum					
C-Phycoerythrin	2.6	0.6	1.0	1.0	1.0

remained at the top of the column. They were then eluted at pH 4.5. G-4, which contained "basic" chromopeptides, was further fractionated at pH 4.5. Whereas "acidic" chromopeptides were further separated on Bio-Rex, "basic" chromopeptides were fractionated by thin-layer

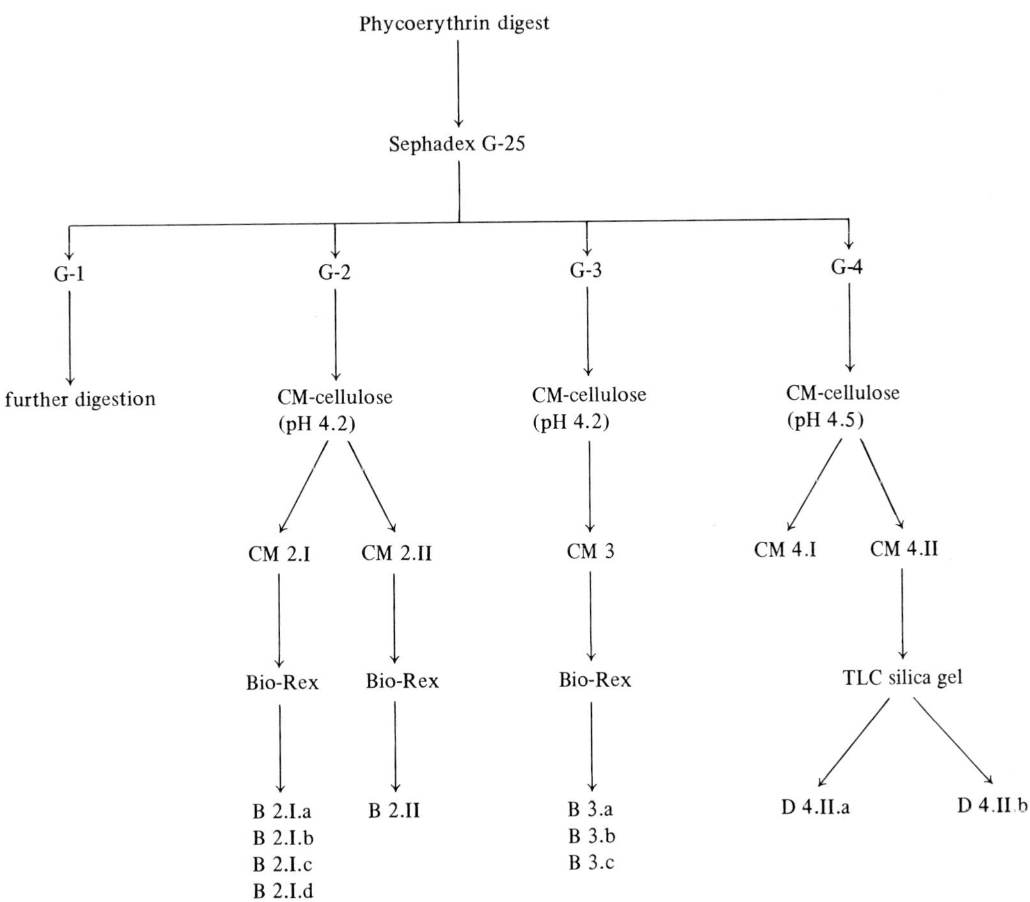

Fig. 3. Scheme of purification of tryptic bilipeptides on a preparative scale. TLC = thin-layer chromatography.

chromatography. The chromopeptide fractions thus obtained were pure according to thin-layer chromatography and isoelectric focusing.

Amino acid analyses of chromopeptide fractions (Table 2) support the view that separation on Sephadex G-25 was achieved according to the molecular weight of the peptides: chromopeptides derived from G-2 (peak on a calibrated column at 2700 - 3200 daltons) contain 14 - 20 amino acids; those derived from G-3 (peak at 1900 - 2500 daltons) and G-4 (peak at 1100 - 1800 daltons) contain only 3 - 6 amino acids. If the amino acid analysis revealed the presence of contaminating (colourless) peptides, a subsequent purification step (thin-layer chromatography) was applied before sequencing. The only amino acid present in all chromopeptides was cysteine (analyzed either as cystine or cysteic acid). This fact led to the assumption that Cys is the amino acid linked to the chromophore in all chromophore regions. This agrees with the analysis on chromopeptides from R-phycoerythrin[22], C-phycocyanin[23] and cryptomonad phycoerythrin[24], in which cysteine was always present.

Sequence determinations of purified chromopeptides were carried out using the dansyl-Edman procedure described by Hartley[17], in some cases using in addition the classical Edman degradation[25]. An important point in sequencing of chromopeptides was the identification of Cys. This amino acid did not appear as free Cys > PhNCS or dansyl-Cys, but remained covalently linked to the chromophore. During extraction of the > PhNCS derivatives, Cys was extracted into the organic phase in form of a PhNCS < Cys-chromophore compound[14] from which it was recovered by hydrolysis and identified either as Cys or as dansyl-Cys by thin-layer chromatography. Generally, extraction of the bulk of chromophore was observed at the same step of Edman degradation at which no amino acid was directly detectable. This correlation could be observed only for some degradation steps because of chromophore bleaching. At first it changed its colour from red to greenish-yellow during 4 - 5 Edman steps. After 6 - 7 steps, the peptides were colourless. Localization of the chromophore-bearing amino acid was therefore performed with smaller peptides, if necessary.

Table 2. Amino acid composition of chromopeptides from *Pseudanabaena* W 1173 phycoerythrin. Trypsin digestion, isolation of peptides as in Fig. 3. All values in molar ratio, next integer in brackets.

	B 2.I.a	B 2.I.b	B 2.I.c	B 2.I.d	B 2.II	B 3.a		B 3.b	B 3.c	CM 4.I	CM 4.II.a	CM 4.II.b
Cys		0.70 (1)	0.77 (1)	0.61 (1)		0.76 (1)		0.76 (1)	0.56 (1)	0.81 (1)		0.56 (1)
O_3H-Cys	0.30 (1)				0.76 (1)		0.98 (1)				1.00 (1)	
Asx	1.61 (2)	1.88 (2)	1.91 (2)	1.91 (2)	1.68 (2)	0.15 (0)	0.12 (0)	0.15 (0)	0.29 (0)			
Thr	1.21 (1)	0.57 (1)	0.20 (0)	0.36 (0)	0.32 (0)	0.05 (0)			0.20 (0)		0.61 (0)	
Ser	1.76 (2)	1.92 (2)	1.48 (1)	1.19 (1)	1.31 (1)	0.90 (1)	0.95 (1)	0.94 (1)	1.06 (1)	1.00 (1)	0.59 (0)	
Glx	2.06 (2)	2.11 (2)	1.88 (2)	1.66 (2)	1.36 (1)	0.33 (0)	0.18 (0)	0.65 (1)	0.56 (0 - 1)			
Pro	0.96 (1)	0.80 (1)	0.77 (1)							1.06 (1)	+ (0)	
Gly	2.41 (2 - 3)	3.48 (3)	2.05 (2)	1.95 (2)	1.72 (2)	0.30 (0)	0.36 (0)	0.32 (0)	0.82 (1)	1.06 (1)	1.00 (1)	
Ala	1.63 (2)	3.12 (3)	2.99 (3)	2.45 (2)	2.38 (2)	1.34 (1)	1.07 (1)	2.11 (2)	2.03 (2)	0.15 (0)	0.59 (0)	
Val	1.16 (1)	2.00 (2)	1.05 (1)	0.83 (1)	0.93 (1)	0.12 (0)		0.15 (0)	0.20 (0)			
Ile	1.13 (1)	1.18 (1)	1.17 (1)	1.01 (1)	1.00 (1)	0.21 (0)		0.59 (1)	0.47 (0 - 1)			
Leu	1.31 (1)	1.45 (1)	0.94 (1)	1.01 (1)	0.83 (1)*	0.21 (0)	0.32 (0)	0.15 (0)	0.30 (0)	0.12 (0)	1.00 (1)	0.98 (1)
Tyr	0.18 (0)	0.19 (0)	0.28 (0)	0.50 (1)	+ (0)	0.07 (0)		0.13 (0)	0.10 (0)			
Phe	0.45 (0 - 1)					0.10 (0)		0.09 (0)	0.15 (0)			
Lys	0.25 (0)				0.64 (0 - 1)							
His	0.13 (0)											
Arg	1.31 (1)	0.80 (1)	0.20 (0)		1.02 (1)					1.06 (1)	1.00 (1)	1.45 (1)

* Low value presumably because of short hydrolysis; sequence indicates 2 Leu.

Table 3. Results of dansyl-Edman degradation of chromopeptides.
Peptides marked with an asterisk(*) were sequenced with the classical Edman technique in addition. Bold-face type indicates the positions of chromophore linkage. Residues in brackets are derived from amino acid analysis.

Peptide	Position 1	2	3	4	5	6	7	8	9	10	
Pseudanabaena W 1173											
B 2.I.a	Phe	Pro	Gly	Val	Gly	Asx	Leu	(Gly, Cys, Ala_2, Ser_2, Ile, Glx_2, Asx, Thr, Arg)			
B 2.I.b	Pro	Gly	Val	Gly	(Asx, Leu, Gly, Cys, Ala_3, Ser_2, Ile, Glx_2, Val, Asx, Thr, Arg)						
B 2.I.c	(Pro)	Gly	Val	Gly	(Pro, Asx, Leu, Cys, Ala_3, Ser, Ile, Glx_2, Asx)						
B 2.II	Ile	Leu	Asx	**(Cys)**	Leu	Tyr Gly	Asx	Val	(Gly, Glx, Ser, Ala_2, Arg)		
B 3.a*	**Cys**	Ala	Ser								
B 3.b	**Cys**	Ala	Ser	Ile	(Ala, Glx)						
B 3.c	**Cys**	Ala	Ser	(Ile, Ala, Glx, Gly)							
CM 4.I*	Gly	**Cys**	**Ser**	Pro	Arg						
D 4.II.a*	Gly	Leu	**Cys**	Arg							
D 4.II.b*	**Cys**	Leu	Arg								
Phormidium persicinum											
"B 2.I pepsin"	Phe Leu	Leu	Gly Val	(Gly, Leu, Cys, Ala, Ser, Ile, Pro, Thr)							
B 2.II	Ile	Leu	Glx	**(Cys)**	Leu	Gly	Asx	Val Leu	(Pro, Gly, Ile, Ser, Ala, Arg)		
B 3.a	**(Cys)**	Leu	Gly	Asx	(Val, Pro, Gly_{1-2}, Ile)						
B 3.b	**Cys**	Thr Leu	Gly	Pro	Leu	Asx	Glx	Gly	Gly	Ile	(Val, Ser, Lys)
B 3.c	Gly	Ala	Gly	O_3H Cys	Leu Thr	Gly	Pro	Leu Gly	Asx Pro Leu	(Glx, Ile, Val)	
CM 4.I	Gly	**Cys**	Ala	Pro	Arg						
D 4.II.a	Gly	Ala	**(Cys)**	Leu	(Arg)						
D 4.II.b	**Cys**	Leu	Arg								

The following deductions may be based on the results of sequencing (Table 3): "acidic" chromopeptides are either large peptides containing Asx and Glx with a *C*-terminal Arg (B 2-peptides) or Lys (B 3.b from *Ph. persicinum*) or peptides without a *C*-terminal basic amino acid (B 2.I.c and all B 3-peptides except B 3.b from *Ph. persicinum*). The latter are products of unspecific peptide cleavages. "Basic" chromopeptides are small peptides with a *C*-terminal Arg which has a positive charge that is not compensated by any acidic amino acid but only by the carboxylic group of the chromophore (CM 4 I, D 4 II.a and D 4 II.b). Some chromopeptides are derived from the same peptide region: this is true for B 2.I.a, b and c, for B 3.a, b and c from *Pseudanabaena* W 1173, and for B 3.b/c and B 2.II/B 3.a from *Ph. persicinum*. The varying chain length of these peptides allowed the deduction of the probable position for many of the amino acids beyond the sequenced part of the peptide (given in brackets in Tab. 3 and 4). Amino acid analysis suggested furthermore that B 3.a, b and c from *Pseudanabaena* are partial sequences of B 2.I.a. This was supported by prolonged treatment of B 2.I.a with trypsin, which yielded a product chromatographically identical with B 3.b.

In summary, 5 peptide sequences of the chromophore regions were derived for *Pseudanabaena* W 1173 C-phycoerythrin (Tab. 4). For the 6 chromophore regions of *Ph. persicinum* C-phycoerythrin, 5 peptide sequences could be derived for certain. The sixth region was expected in fraction B 2.I, which gave multiple coloured bands on isoelectric focusing, each in small yield. Subsequent pepsin digestion yielded one main chromo-

Table 4. Sequences of chromopeptides isolated from the trypsin digest of C-phycoerythrin, ⇁ identified by dansyl-Edman degradation, ⇁ (crossed) Cys identified by hydrolysis of PhNCS < Cys-chromophore, - - ⇁ position of amino acid linked to the chromophore.

Subunit	*Pseudanabaena* W 1173		*Phormidium persicinum*	
α	CM 4.I	Gly-Cys-Ser-Pro-Arg	CM 4.I	Gly-Cys-Ala-Pro-Arg
α	D 4.II.b	Cys-Leu-Arg	D 4.II.b	Cys-Leu-Arg
β	D 4.II.a	Gly-Leu-Cys-Arg	D 4.II.a	Gly-Ala-Cys-Leu-Arg
β	B 2.I.a	Phe-Pro-Gly-Val-Gly-Asx-Leu-(X*)- Cys-Ala-Ser-Ile-(Ala, Glx)-(Y)* ←B 3.a→ ←B.3.b→	"B 2.I. pepsin"	(Phe/Leu)-Leu-(Gly/Val)-(Gly, Leu, Cys, Ala, Ser, Ile, Pro, Thr)
β	B 2.II	Ile-Leu-Asx-Cys-Leu-(Tyr/Gly)-Asx-Val- (Ser, Glx, $Gly_{1\text{-}2}$, Ala_2, Arg)	B 2.II	Ile-Leu-Glx-Cys-Leu-Gly-Asx-(Val/Leu)-(Pro, ←B 3.a Gly, Ile)-(Ser, Ala, Arg) →
β	–		B 3.c/B 3.b	Gly-Ala-Gly-Cys-(Leu/Thr)-Gly-Pro-Leu-Asx- ←B 3.c ←B 3.b Glx-Gly-Gly-Ile-Val-Ser-Lys → →

* X and Y together constitute Gly, Glx, Asx, Ser, Thr and Arg. The position of these amino acids and their distribution between X and Y is unknown.

peptide fraction which was purified by thin-layer chromatography. The material was sufficient only for two Edman steps, but the amino acid composition indicated that this peptide is derived from a chromophore region that is the homologue of B 2.I.a from *Pseudanabaena* W 1173. It does not fit into any of the other five chromophore regions of *Ph. persicinum.* Previously[13] 5 chromophores had been determined for *Pseudanabaena* W 1173 C-phycoerythrin and 6 chromophores for *Ph. persicinum* C-phycoerythrin. The number of chromophore regions derived here agree with this independent determination of the chromophore content in these two algae. The distribution of the chromopeptides over the subunits was determined by isoelectric points of single chromopeptides (see Fig. 2 and Table 1).

Although it is difficult to consider sequence homologies with short peptide sequences (e.g. D-4-IIa and b), some conclusions can be drawn from the sequences of the chromophore regions as known at present.

1) Chromophore regions are extremely conservative. This can easily be seen by comparison of the data from *Pseudanabaena* W 1173 with those from *Phormidium persicinum.*
2) The sixth chromophore region of *Ph. persicinum* (B 3.b), which has no chromophore-bearing counterpart in *Ps.* W 1173 yields the only Lys-containing bilipeptide.
3) All chromophore regions of one C-phycoerythrin are distinct from each other, so that each can easily be recognized.
4) The sequence homology between α and β chains, mainly derived from *N*-terminal sequences of phycocyanins[9,10], cannot be recognized here with confidence because of short sequences. But D 4.II.a may be homologous to D 4.II.b in both algae.
5) Whereas *N*-terminal sequences of phycocyanins and phycoerythrins are very similar[8,9], no correlation of their chromopeptide sequences can be detected. The only reported sequences in the chromophore region of a phycocyanin are Leu-Ile-Cys-Gly-Ala-Ala for the α-chain and Lys-Ser-Lys-Cys-Arg-Asp-Ile-His-Gly-Tyr for the β-chain[15] (remote amino acid residues omitted here). These sequences do not fit into any of the C-phycoerythrin sequences determined here. This difference corresponds to the immunological behaviour: C-phycoerythrin and C-phycocyanin do not cross-react in spite of the similarity of their *N*-terminal sequences[8,9,11,12].

The differences in the chromophore regions may be an adaptation to different chromophores (phycoerythrobilin in C-phycoerythrin, phycocyanobilin in C-phycocyanin) which certainly require different peptide conformations for optimal spatial arrangement of the tetrapyrrole moiety.

Chromophore-amino acid-linkages

The question of covalent linkages between chromophore and amino acids of the peptide chain was investigated by oxidative degradation, by determination of spectral data and by Edman degradation. "Nonhydrolytic" and "hydrolytic" chromic acid oxidation[19] and "CrO_3-NH_3-degradation"[20,21,26] were applied to peptides CM 4.I and CM 4.II from *Pseudanabaena* W 1173, leading to essentially the same results as with biliproteins[1,27]: chromic acid degradation of peptide CM 4.I at 20 °C yielded 4-methyl-2,5-dioxo-3-pyrroline-3-propionic acid (hematinimide = **1**) and 2-methyl-3-vinyl-maleimide (**2**) (see Fig. 4). Subsequent heating of the acid solution liberated more **1** and 2-ethylidene-3-methyl succinimide (**3**). The yield of **3** was higher when the solution – after CrO_3 degradation at 20 °C – was treated with dilute ammonia at 20 °C. This is consistent with the assumption of a thioether at the side chain of ring A (see Fig. 4) because model thioether compounds of the same basic structure are at first oxidized to a sulfone and then eliminate to **3** under the applied conditions[20,21,26]. Although the results concerning ring A and the imides **2** and **3** were the same with peptide CM 4.II, imide **1** was obtained in full yield with CrO_3 at 20 °C – no increased yield was observed by subsequent treatment with CrO_3 at 100 °C or with ammonia in contrast to CM 4.I.

This difference in the degradation results parallels the spectral properties of the two peptides (as the cations): CM 4.I absorbs as cation at 559 nm, CM 4.II at 548 nm (Tab. 5). All other peptides absorb at shorter wavelength, like CM 4.II. No differences in the absorption of CM 4.I and CM 4.II were found in the spectra of their zinc complexes. We interpret the difference in the cation absorption with a conformational change

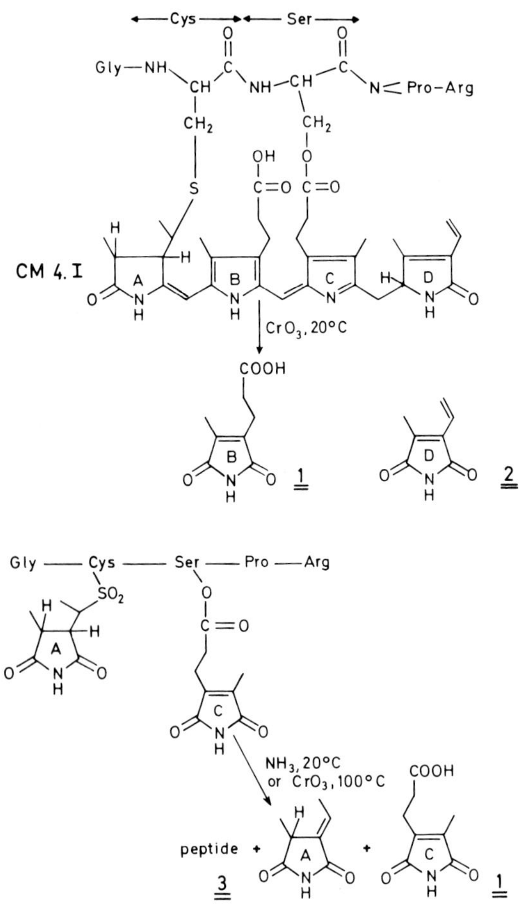

Fig. 4. Chromic acid degradation and combined chromic acid-ammonia degradation of bilipeptide CM 4.I.

Table 5. Absorption maxima of chromopeptides [nm], shoulders in brackets.

Peptide	Cation (2% HCl/H_2O) λ_{max} [nm]	Zinc complex ($ZnCl_2$/ 1% NH_3/H_2O) λ_{max} [nm]
Pseudanabaena W 1173		
Mixture of chromopeptides	550	582 (538, 510)
CM 4.I	559	585 (542, 515)
CM 4.II	548	585 (509)
B 3.a	546	584
B 2.I.a	548	
B 2.II	542	
Phormidium persicinum		
CM 4.I	550	
CM 4.II	550	

induced in CM 4.I by a second covalent linkage between chromophore and peptide. This second linkage follows from the oxidative degradation experiments and even better from the results of Edman degradation.

For a comparison of the Edman degradation, peptides B 3.a and CM 4.I were chosen. Whereas nearly all chromophore was extracted into the organic phase together with cysteine during the first Edman step of B 3.a, the result was different with CM 4.I (see Fig. 5). With CM 4.I only some of the chromophore was extracted during the first step together with glycine. This was due to the unspecific chromophore cleavage that always accompanies the Edman reaction. Free chromophore (ethylidene group at ring A) can be distinguished from bound chromophore (substituted ethyl group at ring A) by the absorption maximum of the zinc complex which is at 600 - 607 nm for the free and at 584 - 585 nm for the bound chromophore[27]. The zinc complex of the first Edman product of B.3.a absorbed at 605 nm,

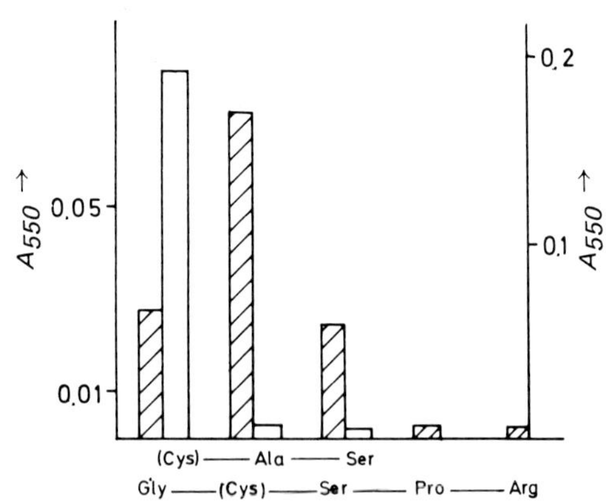

Fig. 5. Chromophore extraction into the organic phase during Edman degradation of bilipeptides B 3.a and CM 4.I. Extinction at 550 nm determined in ethyl acetate (3 m*l*), B 3.a = right scale and empty columns, CM 4.I = left scale and hatched columns.

that of CM 4.I (containing glycine > PhNCS) at 585 nm. More than half of the chromophore was extracted together with cysteine in the second Edman step of peptide CM 4.I, but a substantial portion was extracted together with serine in the third step. This is compatible with one linkage between cysteine and the chromophore and a second linkage between serine and the chromophore. The results of chromophore degradation (cleavage by acid at 100 °C or ammonia at 20 °C yielding free imide **1**) suggest an ester linkage between a propionic acid side chain of the chromophore and the serine hydroxyl group. The significance of this ester linkage is not clear. It could be a native linkage or an artefact formed during trypsin digestion. If it is a native linkage, it cannot be essential for the function of all phycoerythrins. The corresponding CM 4.I peptide from *Phormidium persicinum* has an alanine residue instead of serine and all data speak against a second linkage in this chromopeptide.

We thank *Prof. Dr. F. Wagner* and *Dr. Koppenhagen*, Stöckheim, for mass cultivation of *Phormidium persicinum*. The work was supported by the *Deutsche Forschungsgemeinschaft* and the *Fonds der Chemischen Industrie*.

Literature

1 VI. Commun.: Klein, G., Grombein, S. & Rüdiger, W. (1977) *Hoppe-Seyler's Z. Physiol. Chem.* **358**, 1077 - 1079.
2 Rüdiger, W. (1975) *Ber. Dtsch. Bot. Ges.* **88**, 125 - 139.
3 O'Carra, P. & O'hEocha, C. (1976) in Chemistry and Biochemistry of Plant Pigments (Goodwin, T.W., ed.) 2. edn., vol. 1, pp. 328 - 376, Academic Press, London and New York.
4 Myers, J. & Kratz, W. A. (1955) *J. Gen. Physiol.* **39**, 11 - 22.
5 Bennett, A. & Bogorad, L. (1973) *J. Cell. Biol.* **58**, 419 - 435.
6 Glazer, A. N. & Cohen-Bazire, G. (1971) *Proc. Natl. Acad. Sci. U.S.A.* **68**, 1398 - 1401.
7 O'Carra, P. & Killilea, S. D. (1971) *Biochem. Biophys. Res. Commun.* **45**, 1192 - 1197.
8 Harris, J. U. & Berns, D. S. (1975) *J. Mol. Evol.* **5**, 153 - 163.
9 Glazer, A. N., Apell, G. S., Hixson, C. S., Bryant, D. A., Rimon, S. & Brown, D. M. (1976) *Proc. Natl. Acad. Sci. U.S.A.* **73**, 428 - 431.
10 Frank, G., Zuber, H. & Lergier, W. (1975) *Experientia* **31**, 23 - 26.
11 Bennett, A. & Bogorad, L. (1971) *Biochemistry* **10**, 3625 - 3634.
12 Glazer, A. N., Cohen-Bazire, G. & Stanier, R. Y. (1971) *Proc. Natl. Acad. Sci. U.S.A.* **68**, 3005 - 3008.
13 Muckle, G. & Rüdiger, W. (1977) *Z. Naturforsch.*, **32c**, 957 - 962.
14 Köst-Reyes, E., Köst, H.-P. & Rüdiger, W. (1975) *Liebigs Ann. Chem.* **1975**, 1594 - 1600.
15 Byfield, P. G. H. & Zuber, H. (1972) *FEBS Lett.* **28**, 36 - 40.
16 Wrighley, C. (1968) *Sci. Tools* **15**, 17 - 23.
17 Hartley, B. S. (1970) *Biochem. J.* **119**, 805 - 822.
18 Krejci, K. & Machleidt, W. (1969) *Hoppe-Seyler's Z. Physiol. Chem.* **350**, 981 - 993.
19 Rüdiger, W. (1969) *Hoppe-Seyler's Z. Physiol. Chem.* **350**, 1291 - 1300.
20 Klein, G. (1977) Dissertation Universität München.
21 Klein, G. & Rüdiger, W. (1977) *Liebigs Ann. Chem.*, in press.
22 Fujiwara, T. (1957) *J. Biochem. (Tokyo)* **44**, 723 - 733.
23 Crespi, H. L. & Smith, U. H. (1970) *Phytochemistry* **9**, 205 - 212.
24 Brooks, C. & Chapman, D. J. (1972) *Phytochemistry* **11**, 2663 - 2670.
25 Edman, P. & Henschen, A. (1975) in Protein Sequence Determination, Molecular Biology, Biochemistry and Biophysics (Needleman, S. B., ed.) vol. 8, pp. 232 - 279, Springer-Verlag Berlin, Heidelberg, New York.
26 Schoch, S., Klein, G., Linsenmeier, U. & Rüdiger, W. (1976) *Liebigs Ann. Chem.* **1976**, 549 - 558.
27 Köst, H.-P., Rüdiger, W. & Chapman, D. J. (1975) *Liebigs Ann. Chem.* **1975**, 1582 - 1593.

Günther Muckle and Prof. Dr. Wolfhart Rüdiger, Botanisches Institut der Universität München, Menzinger Straße 67, D-8000 München 19.

Dr. Joachim Otto, Institut für Physiologische Chemie, Physikalische Biochemie und Zellbiologie, Goethestr. 33, D-8000 München 2.

Hoppe-Seyler's Z. Physiol. Chem.
Bd. 359, S. 357 - 368, März 1978

The Amino Acid Compositions of the Tryptic, Chymotryptic and Peptic Peptides from the L-2 Light Chain of Rabbit Skeletal Muscle Myosin

Genji MATSUDA, Yoji SUZUYAMA, Masayuki SETOGUCHI, Tetsuo MAITA, Bunji WATANABE, Tomoyuki MAEKAWA and Toshiyo UMEGANE

Department of Biochemistry, Nagasaki University School of Medicine, Nagasaki, Japan

(Received 11 November 1977)

Dedicated to Prof. Dr. Adolf Butenandt on the occasion of his 75th birthday

Summary: The light chain fraction was separated from rabbit skeletal muscle myosin and four kinds of light chains, L-1, L-2, L-3 and L-4 in the fraction were further isolated by column chromatography using DEAE-cellulose DE-52. After aminoethylation, the L-2 light chain was digested with trypsin. It was also digested with chymotrypsin and pepsin, respectively, after carboxymethylation. Each of the tryptic, chymotryptic and peptic peptides thus obtained was separated and purified and their amino acid compositions were analyzed.

Die Aminosäure-Zusammensetzung der tryptischen, chymotryptischen und peptischen Peptide der leichten Kette L-2 des Myosins aus Kaninchenskelettmuskel

Zusammenfassung: Aus dem Myosin des Kaninchenskelettmuskels wurden die leichten Ketten L-1, L-2, L-3 und L-4 durch Säulenchromatographie an DEAE-Cellulose DE-52 isoliert. Die L-2-Kette wurde entweder nach Aminoethylierung mit Trypsin oder nach Carboxymethylierung mit Chymotrypsin oder Pepsin gespalten. Alle so erhaltenen tryptischen, chymotryptischen und peptischen Peptide wurden gereinigt und ihre Aminosäure-Zusammensetzung wurde bestimmt.

Key words: Skeletal muscle, myosin, L-2 light chain, amino acid composition.

The structure of myosin, a main component of muscle proteins, is of interest in connection with the mechanism of muscle contraction. Myosin separated from rabbit skeletal muscle contains two heavy chains with molecular weights about 200 000 and four light chains with molecular weights about 20 000[1-5]. In this light chain fraction, there are four recognizable components, which migrate toward the anode in electrophoresis on thin-layer starch gels (pH 8.6) containing 8M urea. They are designated L-1, L-2, L-3 and L-4 in the order of their increasing anionic mobil-

---

Enzymes: Chymotrypsin (EC 3.4.21.1); Pepsin A (EC 3.4.23.1); Trypsin (EC 3.4.21.4).

Abbreviations: DTNB = 5,5′-dithiobis(2-nitrobenzoic acid), $\overset{\displaystyle CH_2CO_2H}{\overset{|}{Cys}}$ = carboxymethylcysteine; $\overset{\displaystyle EtNH_2}{\overset{|}{Cys}}$ = aminoethylcysteine.

ities. L-1 and L-4 light chains are also called alkali light chain 1 (A1) and alkali light chain 2 (A2), respectively and L-2 and L-3 light chains, DTNB light chains. L-3 is a phosphorylated L-2 light chain[6]. The role of the L-2 light chain of the skeletal muscle myosin in muscle contration is not yet clear. We determined the primary structure of the L-2 light chain of rabbit skeletal muscle myosin. The amino acid sequence of this chain has already been published in a preliminary communication[7].

Material and Methods

Isolation and purification of the light chains of rabbit skeletal muscle myosin

The forty or so rabbits used in this experiment were all the native Japanese white race. Myosin was extracted from skeletal muscles by the method of Weeds and Hartley[8] and purified. The isolation and purification of the light chains in this myosin were carried out according to the method of Perrie and Perry[9], with modifications. That is, solid guanidine-HCl and 2-mercaptoethanol were added to myosin paste, and after mixing them thoroughly, cold water and ethanol were added and the heavy chains were precipitated. The supernatant containing the light chains was evaporated under reduced pressure and demineralized by gel filtration through Sephadex G-25. The four components in this light chain fraction were isolated by column chromatography on DEAE cellulose DE-52 (Whatman). Elution was carried out with a gradient between 4M urea/0.035M KCl/25mM Tris (pH 7.6) and 4M urea/0.30M KCl/25mM Tris (pH 7.0).

Thin-layer starch gel electrophoresis

To 9 g of the hydrolyzed potato starch was added 60 m*l* of 0.1 M Tris/glycine buffer, pH 8.3, containing 8M urea and 0.1% 2-mercaptoethanol. The starch gel was boiled for some minutes in a water bath and installed in a frame (20 × 20 × 0.1 cm), covered with a plate of glass, and stored overnight in the cold before use. Electrophoresis of the light chains of rabbit skeletal muscle myosin was carried out at 250 V for 15 h. The buffer used for the electrode compartments was 0.1 M Tris/glycine buffer, pH 8.3. After electrophoresis, the light chains were stained with amido black.

Aminoethylation and carboxymethylation of L-2 light chain

L-2 light chain was aminoethylated by Cole's method[10]. That is, 500 mg of the L-2 light chain was dissolved in 60 m*l* of 0.1 M Tris/HCl buffer (pH 8.6) containing 8M urea and 0.2% EDTA and was reduced by adding 0.5 m*l* of mercaptoethanol under N_2 gas at room temperature for 4 h. After that, the L-2 was aminoethylated by adding 0.1 m*l* of ethylenimine for 1 h. Finally, the products were purified by gel filtration using Sephadex G-25. L-2 light chain was carboxymethylated by the method of Crestfield et al.[11]. That is, 400 mg of the L-2 light chain was dissolved into 60 m*l* of 0.5M Tris/HCl buffer (pH 8.6) containing 8M urea and 0.3% EDTA and was reduced by adding 0.5 m*l* of 2-mercaptoethanol under N_2 gas at room temperature for 4 h. After that, the L-2 was carboxymethylated by adding 1.5 g of monoiodoacetic acid for 1 h. Finally, the products were purified by gel filtration using Sephadex G-25.

Tryptic, chymotryptic and peptic digestion of L-2 light chain

The aminoethylated L-2 light chain was dissolved in water, adjusted to pH 8.5 by adding ammonia water and digested with trypsin at 37 °C, pH 8.0 - 8.5 for 3 h under N_2 gas. The carboxymethylated L-2 light chain was digested with chymotrypsin under the same conditions. The carboxymethylated L-2 light chain was also dissolved in water, adjusted to pH 2.0 by adding 1N HCl, and digested with pepsin at 37 °C for 1 h under N_2 gas. In all digestions, the ratio of the L-2 light chain to enzyme was 50:1.

Isolation and purification of tryptic, chymotryptic and peptic peptides

Column chromatography of tryptic, chymotryptic and peptic peptides with strongly acidic cation exchange resin was performed using a column (0.9 × 20 cm) of Chromo Beads P (Technicon Co.) connected to an automatic peptide analyzer[12]. The peptides were eluted by a gradient made by mixing three kinds of pyridine acetate buffers in a varigrad with six chambers, as shown in Table 1. Fraction volume was 5 m*l*, temperature 55 °C, flow rate to the column 120 m*l*/h, and flow rate to the direct ninhydrin reaction 3 m*l*/h. In chromatography on a DE-52 column (0.9 × 50 cm), the material was eluted with a gradient between 0.05M NH_4HCO_3 and 0.4M NH_4HCO_3. Fraction volume was 3 m*l* and the absorbance at 230 nm was measured. In gel filtration on Sephadex G-25 (3 × 50 cm) and Sephadex G-50 (5 × 130 cm), the gels were eluted by 0.05M NH_4HCO_3. With Sephadex G-25, the fraction volume was 3 m*l* and with Sephadex G-50, it was 12 m*l*. The peptides were detected by measurement of the absorbance at 230 nm.

Paper chromatographs were developed with the upper phase of 1-butanol, acetic acid and water (4:1:5). The peptides were identified by ninhydrin and other specific color reactions, and were eluted from the filter paper with 0.5% acetic acid.

Table 1. Composition of buffers used for a varigrad with six chambers for the column chromatography on Chromo Beads P.

Buffer A, 0.05 M pyridine/acetate, pH 3.1; buffer B, 0.2 M pyridine/acetate, pH 3.1; buffer C, 2.0 M pyridine/acetate, pH 5.0.

Chamber no.	Composition		
	A [ml]	B [ml]	C [ml]
1	150	0	0
2	0	150	0
3	0	150	0
4	0	150	0
5	0	100	50
6	0	0	150

Results and Discussion

Myosin was extracted from skeletal muscles of about forty adult rabbits of the Japanese white native race and was purified. The light chain fraction in this myosin was separated from the heavy chain fraction. In the light chain fraction, as Fig. 1 shows, there were four recognizable components, which migrate toward the anode in thin-layer starch gel electrophoresis. They are designated L-1, L-2, L-3 and L-4 light chains in the order of their increasing anionic mobilities. As Fig. 2 shows, four components in the light chain fraction were separated by column chromatography on DE-52. That is, first unadsorbed protein and then L-1, L-2, L-3 and L-4 were successively eluted. Each of the eluted fraction was collected separately and lyophilized. Thin-layer starch gel electrophoresis showed that there were four components, as shown in Fig. 1.

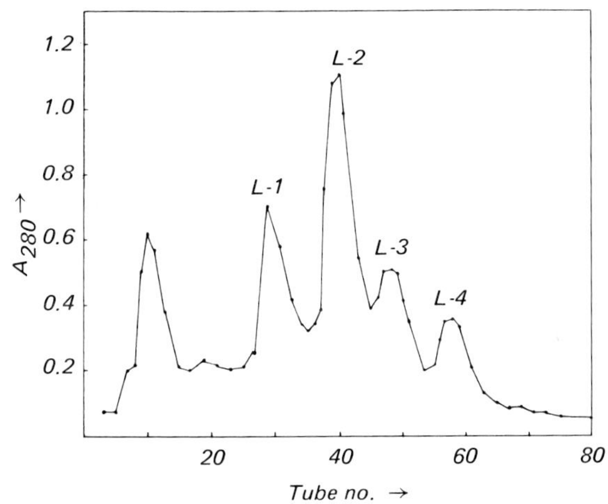

Fig. 2. Column chromatogram of the four light chain components of rabbit skeletal muscle myosin.

Light chain, 500 mg; column, DE-52, 3.6 × 30 cm; flow rate, 100 ml/h; fraction volume, 12 ml; detection, absorbance at 280 nm.

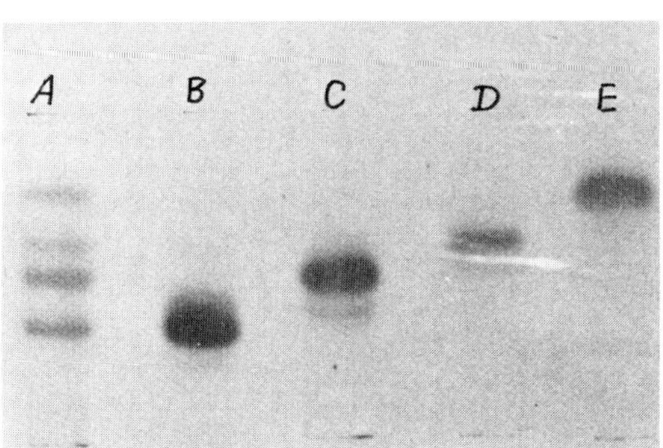

Fig. 1. Thin-layer starch gel electrophoresis of the light chain components of rabbit skeletal muscle myosin.

A: mixture of four light chain components; B: L-1 light chain; C: L-2 light chain; D: L-3 light chain; E: L-4 light chain.

The L-2 light chain was, after aminoethylation, digested with trypsin under N_2 gas at 37 °C, pH 8.0 - 8.5 for 3 h. The peptides obtained were dissolved in 0.05 M pyridine/acetate buffer (pH 3.1). However, as some of the peptides were insoluble, they were separated into soluble and insoluble fractions by centrifugation. The soluble tryptic peptides were separated by column chromatography using Chromo Beads P. As a result, fourteen peaks were obtained, as shown in Fig. 3. The tubes containing each peak were collected separately, and the effluent was lyophilized and further tryptic peptides in the peaks were identified by paper chromatography. As a result, twenty-four spots were recognized, as shown in Fig. 4. Then peptides in peaks which contain more than one peptide were further purified by paper chromatography and gel filtration using G-25.

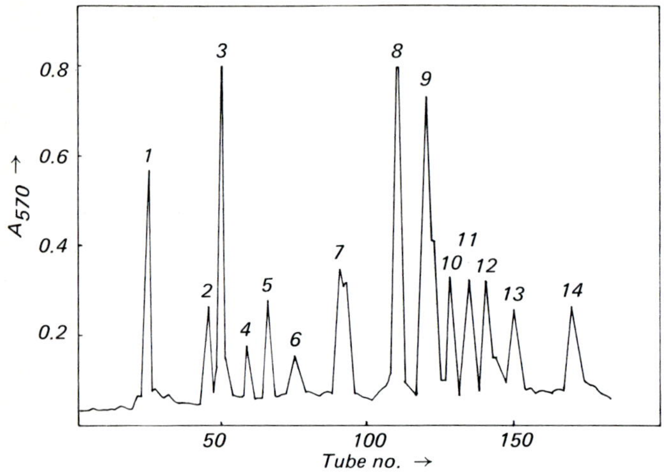

Fig. 3. Column chromatogram of the soluble tryptic peptides from the aminoethylated L-2 light chain of rabbit skeletal muscle myosin.

Insoluble tryptic peptides were first separated by gel filtration on Sephadex G-50. Two peaks were found, as shown in Fig. 5. The effluent in the tubes of each peak were collected separately and lyophilized. Further tryptic peptides in the two peaks were purified by column chromatography on DE-52. The amino acid compositions of twenty-six tryptic peptides thus obtained were analyzed. Table 2 shows the result.

After carboxymethylation, the L-2 light chain was digested with α-chymotrypsin under N_2 gas at 37 °C, pH 8.0 - 8.5 for 3 h. The peptides thus obtained were separated into three groups, Chy-I, Chy-II and Chy-III, by gel filtration on Sephadex G-50. Five main peptides in the Chy-I fraction were further purified by column chromatography on Chromo Beads P and by paper chromatography. The Chy-II fraction was dissolved in 0.05M pyridine/acetate buffer (pH 3.1). However, as part of it was insoluble, it was separated into soluble and insoluble fractions at pH 3.1. The former was designated Chy-II-S, the latter, Chy-II-I. Four main peptides in Chy-II-S were further purified by column chromatography using Chromo Beads P and by paper chromatography. A main peptide in Chy-II-I was purified by gel

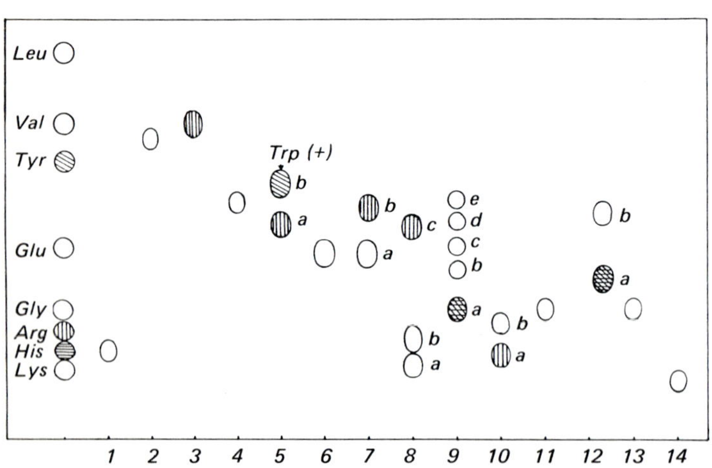

Fig. 4. Paper chromatogram of the soluble tryptic peptides from the aminoethylated L-2 light chain of rabbit skeletal muscle myosin. The numbers 1 - 14 refer to the peaks in Fig. 3 and the spots to the left in the figure are the authentic amino acids.

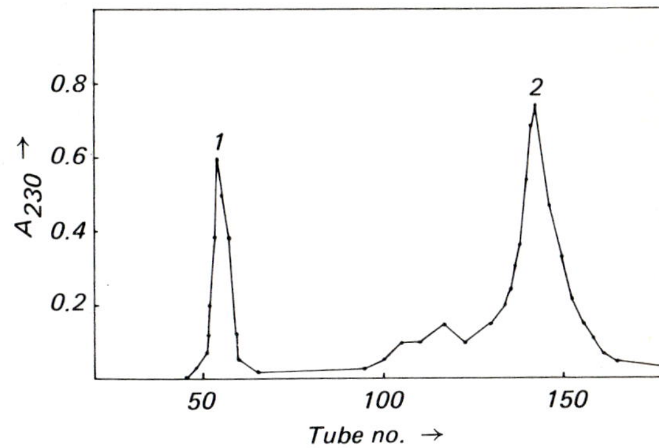

Fig. 5. Gel filtration of the insoluble tryptic peptides from the aminoethylated L-2 light chain of rabbit skeletal muscle myosin.

Column, Sephadex G-50, 4.8 × 130 cm; developer, 0.05M NH_4HCO_3; flow rate, 100 ml/h; fraction volume, 12 ml; detection, absorbance at 230 nm.

filtration using Sephadex G-50. Eleven main peptides in the Chy-III fraction were further purified by column chromatography on Chromo Beads P and by paper chromatography. The amino acid compositions of twenty-one main chymotryptic peptides thus obtained were analyzed. Table 3 shows the result.

The L-2 light chain, after carboxymethylation, was digested with pepsin under N_2 gas at 37 °C, pH 2.0 for 1 h. The peptic peptides obtained were fractionated into two groups, Pep-I and Pep-II by gel filtration on Sephadex G-50. Fourteen main peptides in the Pep-I fraction were further purified by column chromatography on Chromo Beads P and by paper chromatography. Five main peptides in Pep-II were also purified in the same manner as Pep-I. The amino acid compositions of the nineteen main peptic peptides thus obtained were analyzed. Table 4 shows the result.

In the experiments mentioned above, we separated and purified the tryptic, chymotryptic and peptic peptides from the L-2 light chain of rabbit skeletal muscle myosin and further analyzed their amino acid compositions.

Table 2. Amino acid compositions of the tryptic peptides from the aminoethylated L-2 light chain of rabbit skeletal muscle myosin.
The figures under (x) represent residues per mole of peptide in the 24 h hydrolysate, the figures under (y) the assumed integers. Only 8c was hydrolysed for 40 h.

Peak no.	Try-S																											
	1		2		3		4		5a		5b		6		7a		7b		8a		8b		8c		9a		9b	
	x	y	x	y	x	y	x	y	x	y	x	y	x	y	x	y	x	y	x	y	x	y	x	y	x	y	x	y
Lys			1.12	1			1.00	1			1.08	1	1.00	1	1.02	1			1.00	1	1.08	1	1.09	1	1.09	1		
His																									1.08	1		
Arg					0.91	1			1.19	1							1.10	1					1.07	1				
EtNH$_2$-Cys					0.73	1																					0.83	1
Asx	0.93	1	2.08	2	0.98	1	2.07	2	2.27	2	4.21	4	1.13	1			1.07	1					3.21	3	2.11	2	1.13	1
Thr			0.97	1	2.08	2			0.90	1							1.01	1							1.07	1		
Ser															0.78	1												
Glx	1.07	2	1.21	1	4.18	4	2.13	2	2.37	2			1.19	1	3.27	3							1.08	1	2.17	2		
Pro			1.09	1							2.21	2	0.98	1							0.92	1						
Gly			2.16	2							1.92	2	0.91	1			1.03	1					1.02	1	1.10	1		
Ala			2.08	2			1.08	1	0.98	1	2.08	2					2.09	2							1.03	1		
Val			0.63	1					0.53	1	1.96	2	0.87	1											0.71	1		
Met							1.64	2			0.64	1					0.59	1										
Ile			0.61	1					0.64	1					0.83	1							1.62	2	0.72	1	1.04	1
Leu					3.00	3	1.09	1					0.92	1									1.01	1				
Tyr											0.88	1													0.83	1		
Phe			1.07	1	1.12	1			1.02	1	1.12	1			1.10	1	1.11	1										
Trp											+	1																
Total		3		13		13		9		10		17		7		7		8		1		2		10		12		3
Final designation (ref.[7])	C-terminus		T-15		T-19ab		T-12		T-7		T-21		T-16		T-20		T-10		T-2 T-18		T-1		T-8, 9		T-22b		T-22a	

Peak no.	Try-S																				Try-I			
	9c		9d		9e		10a		10b		11		12a		12b		13		14		1		2	
	x	y	x	y	x	y	x	y	x	y	x	x	x	y	x	y	x	y	x	y	x	y	x	y
Lys	1.10	1	1.16	1	1.07	1					1.07	1	1.18	1	0.97	1	2.22	2	2.10	2	1.12	1	1.10	1
His													1.09	1										
Arg							1.12	1	1.00	1														
EtNH$_2$-Cys					0.73	1																		
Asx			0.99	1			0.88	1					1.19	1							1.22	1	2.08	2
Thr	0.92	1			1.88	2							1.07	1			0.89	1			1.93	2	1.07	1
Ser																					1.02	1	2.72	3
Glx					4.18	4															2.38	2	5.31	5
Pro																					0.92	1		
Gly	1.18	1											1.19	1			1.18	1			1.82	2	2.08	2
Ala											0.93	1	1.20	1					0.93	1	1.11	1	2.90	3
Val			0.92	1									0.53	1							0.83	1	1.02	1
Met																					0.82	1	0.63	1
Ile	0.81	1											0.72	1			0.71	1			0.94	1	1.01	1
Leu			0.93	1	2.99	3									1.03	1					1.04	1		
Tyr													0.83	1										
Phe					1.08	1															2.83	3	3.10	3
Trp																								
Total		4		4		12		2		1		2		9		2		5		3		18		23
Final designation (ref.[7])	T-17		T-11		T-18, 19a		T-19b		T-4 T-5		T-3		pos. 157 ~ 165		T-14		T-17, 18		T-2, 3		T-13		T-6	

Table 3. Amino acid compositions of the chymotryptic peptides from the carboxymethylated L-2 light chain of rabbit skeletal muscle myosin. The figures under (x) represent residues per mole of peptide in the 24 h hydrolysate, the figures under (y) the assumed integers.

Peak no.	Chy-I										Chy-II-S								Chy-II-I	
	1		2		3a		3b		4		1		2		3		4		1	
	x	y	x	y	x	y	x	y	x	y	x	y	x	y	x	y	x	y	x	y
Lys	1.00	1	1.88	2	0.97	1	0.83	1	2.62	3			0.92	1			2.98	3		
His					0.92	1														
Arg							1.62	2							0.89	1				
CM-Cys															0.61	1				
Asx	2.09	2	4.18	4	2.29	2	5.60	6	1.01	1	3.28	3	1.28	1	1.19	1			1.12	1
Thr	0.97	1			1.08	1	1.83	2	0.91	1					1.80	2			1.01	1
Ser			0.82	1									0.82	1					1.01	1
Glx	1.11	1	3.12	3	2.10	2	2.27	2	1.28	1			3.27	3	1.30	1			4.28	4
Pro	0.82	1	0.91	1					0.81	1	2.02	2					0.90	1		
Gly	2.18	2	1.18	1	1.12	1	1.28	1	2.08	2	2.00	2								
Ala	2.18	2	2.11	2	0.97	1					1.88	2					1.12	1		
Val			1.18	1	0.71	1	1.07	1	1.19	1	2.11	2								
Met	0.73	1	1.62	2									0.53	1					0.42	1
Ile			0.99	1	0.82	1	3.19	3	0.97	1			1.18	1					1.07	1
Leu	0.82	1	1.03	1			1.20	1	1.13	1										
Tyr											0.72	1								
Phe	1.10	1	0.98	1			1.11	1			0.99	1			1.21	1			2.09	2
Trp													+	1						
Total		13		20		11		20		12		13		9		7		5		11
Final designation (ref.[7])	pos. 91 ~ 103		pos. 60 ~ 79		pos. 158 ~ 168		pos. 34 ~ 53		pos. 104 ~ 115		pos. 140 ~ 152		pos. 131 ~ 139		pos. 124 ~ 130		pos 1 ~ 5		pos 19 ~ 29	

Peak no.	Chy-III																					
	1		2		3		4		5a		5b		6		7		8		9		10	
	x	y	x	y	x	y	x	y	x	y	x	y	x	y	x	y	x	y	x	y	x	y
Lys											1.10	1			1.02	1	0.92	1			1.13	1
His																						
Arg									0.72	1			1.00	1					0.89	1		
CM-Cys																					0.72	1
Asx	1.02	1							1.10	1											1.18	1
Thr							0.89	1			1.12	1										
Ser	2.10	2							1.63	2												
Glx	0.93	1			1.89	2			1.20	1	1.17	1			0.97	1	1.17	1				
Pro																						
Gly	2.09	2							2.11	2	1.11	1							1.18	1		
Ala	2.80	3	2.19	2					3.09	3					1.09	1						
Val	1.17	1					1.10	1	1.12	1												
Met			0.81	1							0.43	1										
Ile																					1.28	1
Leu					3.11	3					2.17	2							0.93	1		
Tyr																					0.69	1
Phe	0.89	1					1.01	1	1.03	1	1.00	1			0.92	1	0.91	1				
Trp																						
Total		11		3		5		3		12		8		1		4		3		3		5
Final designation (ref.[7])	pos. 8 ~ 18		pos. 54 ~ 56		pos. 119 ~ 123		pos. 80 ~ 82		pos. 7 ~ 18		pos. 83 ~ 90		pos. 6		pos. 30 ~ 33		pos. 116 ~ 118		pos. 57 ~ 59		pos. 153 ~ 157	

Table 4. Amino acid compositions of the peptic peptides from the carboxymethylated L-2 light chain of rabbit skeletal muscle myosin. The figures under (x) represent residues per mole of peptide in the 24 h hydrolysate, the figures under (y) the assumed integers.

Peak no	Pep-I																			
	1		2		3		4		5		6		7		8		9a		9b	
	x	y	x	y	x	y	x	y	x	y	x	y	x	y	x	y	x	y	x	y
Lys									0.72	1			0.97	1	2.11	2	0.99	1	0.83	1
His																	0.83	1		
Arg											0.89	1							1.10	1
CM-Cys											0.63	1								
Asx	0.73	1	3.27	3					1.21	1	1.10	1	3.22	3	2.17	2	1.98	2	5.16	5
Thr	0.92	1			0.97	1					2.09	2			0.63	1	1.00	1	0.99	1
Ser									0.91	1	1.01	1								
Glx	4.38	4					2.27	2	1.30	1	3.37	3	2.09	2	2.28	2	2.27	2	1.83	2
Pro			1.99	2					1.10	1					1.10	1				
Gly			2.08	2	1.02	1			1.21	1					2.20	2	1.01	1	1.08	1
Ala			1.72	2	1.12	1			1.28	1			0.80	1	0.72	1	0.93	1		
Val			1.91	2	0.98	1							1.01	1			0.82	1	0.81	1
Met									0.43	1					0.54	1				
Ile	0.97	1			0.81	1			1.12	1							1.17	1	3.28	3
Leu							0.73	1			0.91	1	0.91	1	2.32	2			1.01	1
Tyr																				
Phe			1.03	1					0.82	1	1.00	1			1.13	1			0.92	1
Trp																				
Total		7		12		5		3		10		11		9		15		11		17
Final designation (ref.[7])	pos. 22 ~ 28		pos. 140 ~ 151		pos. 98 ~ 102		pos. 120 ~ 122		pos. 70 ~ 79		pos. 123 ~ 133		pos. 60 ~ 68		pos. 83 ~ 97		pos. 158 ~ 168		pos. 33 ~ 49	

Peak no	Pep-I								Pep-II									
	10		11		12		13		1		2a		2b		3		4	
	x	y	x	y	x	y	x	y	x	y	x	y	x	y	x	y	x	y
Lys					3.73	4	2.64	3	1.10	1	0.91	1	0.72	1				
His																		
Arg	1.21	1	0.82	1			1.90	2										
CM-Cys											0.71	1						
Asx			1.29	1	1.01	1	1.37	1	1.19	1	1.29	1						
Thr			0.99	1	0.82	1									1.18	1		
Ser							1.88	2									1.28	1
Glx					1.29	2	1.29	1	1.37	1			1.19	1				
Pro					1.18	1	0.81	1										
Gly	1.29	1			2.38	2	2.18	2										
Ala	1.71	2					4.09	4					0.89	1				
Val					0.91	1	0.92	1							0.99	1		
Met	0.62	1							0.53	1							0.54	1
Ile					1.00	1			0.81	1	1.21	1						
Leu	1.17	1			1.99	2												
Tyr											1.88	2						
Phe			0.90	1	1.51	2	0.93	1					1.20	1	0.83	1	1.18	1
Trp									+	1								
Total		6		4		17		18		6		6		4		3		3
Final designation (ref.[7])	pos. 54 ~ 59		pos. 50 ~ 53		pos. 103 ~ 119		pos. 1 ~ 18		pos. 134 ~ 139		pos. 152 ~ 157		pos. 29 ~ 32		pos. 80 ~ 82		pos. 19 ~ 21	

We wish to thank Miss *Michiko Ishibashi* for her help in preparing this paper.

Literature

1 Gershman, L.C., Stracher, A. & Dreizen, P. (1969) *J. Biol. Chem.* **244**, 2726 - 2736.
2 Gazith, J., Himmelfarb, S.H. & Harrington, W.F. (1970) *J. Biol. Chem.* **245**, 15 - 22.
3 Weeds, A.G. & Lowey, S. (1971) *J. Mol. Biol.* **61**, 701 - 725.
4 Yagi, K., Okamoto, Y. & Yazawa, Y. (1975) *J. Biochem. (Tokyo)* 77, 333 - 342.
5 Hayashi, Y. (1972) *J. Biochem. (Tokyo)* **72**, 83 - 100.
6 Perrie, W.T., Smillie, L.B. & Perry, S.V. (1972) *Cold Spring Harbor Symp. Quant. Biol.* **37**, 17 - 18.
7 Matsuda, G., Maita, T., Suzuyama, Y., Setoguchi, M. & Umegane, T. (1977) *J. Biochem. (Tokyo)* **81**, 809 - 811.
8 Weeds, A.G. & Hartley, B.S. (1968) *Biochem. J.* **107**, 531 - 548.
9 Perrie, W.T. & Perry, S.V. (1970) *Biochem. J.* **119**, 31 - 38.
10 Cole, R.D. (1967) *Methods Enzymol.* **11**, 315 - 317.
11 Crestfield, A.M., Moore, S. & Stein, W.H. (1963) *J. Biol. Chem.* **238**, 622 - 627.
12 Matsuda, G., Maita, T., Watanabe, B., Ota, H., Araya, A., Goodman, M. & Prychodko, W. (1972) *Int. J. Pept. Protein Res.* **4**, 405 - 413.

Prof. Dr. G. Matsuda*, Dr. Y. Suzuyama, Dr. M. Setoguch, Dr. T. Maita, Dr. B. Watanabe, Dr. T. Maekawa and Mrs. T. Umegane: Department of Biochemistry, Nagasaki University School of Medicine, Nagasaki 852, Japan.

* To whom all correspondence should be addressed.

Hoppe-Seyler's Z. Physiol. Chem.
Bd. 359, S. 369 - 378, März 1978

Bradykinin and Human Forearm Metabolism: Inhibition of Endogenous Prostaglandin Synthesis

Günther DIETZE, Matthias WICKLMAYR, Lothar MAYER, Ingo BÖTTGER and Hansjoachim v. FUNCKE

III. Medizinische Klinik, Abteilung Stoffwechsel und Endokrinologie, and Forschergruppe Diabetes, Städtisches Krankenhaus München-Schwabing

(Received 16 January 1978)

Summary: The metabolism of the human forearm was studied in healthy volunteers by the registration of arterial and deep-venous concentrations of oxygen and glucose and of resting forearm blood flow during the intrabrachial arterial infusion of 0.9% NaCl (n = 4), papaverine (n = 5), bradykinin (n = 9) and bradykinin after the oral pretreatment with indomethacin, an inhibitor of prostaglandin synthesis (n = 10). The blood flow was not affected during the infusion of NaCl (0.2 ml/min) but was increased, depending on the concentrations, 2- and 3-fold during infusion of papaverine (2 or 3 μg/(kg × min)), and 2-fold during infusion of bradykinin (0.2 ng/(kg × min)).

The latter effect on the blood flow was significantly smaller when endogenous synthesis of prostaglandins was impaired by indomethacin pretreatment. Arterial oxygen and glucose concentrations were maintained during the individual infusions. This was also true for the arterial–deep-venous concentration differences of oxygen and glucose during NaCl infusions. Corresponding to the acceleration of the blood flow caused by papaverine and bradykinin, arterial–deep-venous concentration oxygen differences declined, indicating that oxygen consumption of the muscle was not changed. A similar behaviour was observed for muscular extraction and consequently uptake of glucose when the acceleration of the blood flow was caused by papaverine. In contrast, doubling of the blood flow caused by the infusion of bradykinin was accompanied with enlargement of the arterial–deep-venous concentration difference and thus of muscular glucose uptake. There was no change in the arterial and forearm deep-venous levels of insulin under these conditions. Also, this effect of bradykinin was nearly abolished after oral pretreatment with indomethacin. From these data one may suggest that bradykinin, besides its well-known action on blood flow, also affects the glucose metabolism of the human forearm apparently via endogenously synthetized prostaglandins.

Bradykinin und Unterarmstoffwechsel: Hemmung der endogenen Prostaglandin-Synthese

Zusammenfassung: An gesunden Probanden wurde der Stoffwechsel des Unterarms studiert. Dazu wurden die arteriellen und tiefvenösen Konzentrationen von Sauerstoff und Glucose sowie der Unterarmdurchfluß gemessen. Dies geschah während einer Infusion von 0.9% NaCl (n = 4), von Papaverin (n = 5), von Bradykinin (n = 9) sowie von Bradykinin nach oraler Vorbehandlung mit Indomethacin, einem Inhibitor der Prostaglandin-Biosynthese (n = 10). Infusion von NaCl änderte den Unterarmdurchfluß nicht, während Papaverin (2 und 3 $\mu g \times kg^{-1} \times min^{-1}$) ihn konzentrationsabhängig auf das 2- und 3-fache ansteigen ließ und Bradykinin ($0.2\,ng \times kg^{-1} \times min^{-1}$) verdoppelte ihn.

Dieser Bradykinin-Effekt war signifikant kleiner nach Indomethacin-Vorbehandlung. Während der Infusionen änderten sich die arteriellen Spiegel von Sauerstoff und Glucose nicht. Dies traf auch für die arterio-tiefvenöse Differenz der Sauerstoff- und Glucosekonzentrationen während der NaCl-Infusion zu. Unter Papaverin- und Bradykinin-Infusionen fiel die Konzentrationsdifferenz von Sauerstoff entsprechend der Steigerung der Durchblutung. Das gleiche galt unter der Infusion von Papaverin für die arterio-tiefvenöse Konzentrationsdifferenz sowie die Aufnahme von Glucose. Dagegen war die Verdopplung des Unterarmdurchflusses durch Bradykinin von einer Erweiterung der Konzentrationsdifferenz und damit der Aufnahme von Glucose in den Unterarm begleitet. Dabei verhielten sich die arteriellen und tiefvenösen Konzentrationen von Insulin unauffällig. Auch dieser Effekt von Bradykinin wurde durch Indomethacin-Vorbehandlung signifikant reduziert.

Nach den vorliegenden Daten scheint Bradykinin neben seiner bekannten Wirkung auf den Durchfluß auch einen Effekt auf den Stoffwechsel des Unterarms zu besitzen, der durch endogen synthetisierte Prostaglandine übertragen wird.

Key words: Arterial and deep-venous concentrations of oxygen and glucose, forearm blood flow, papaverine, bradykinin, indomethacin.

Werle and coworkers[1,2] were the first to recognize that kallikrein acts on blood pressure and blood flow[3,4] via the liberation of kinins from kininogen. Later, Rocha e Silva and coworkers isolated bradykinin and showed that it produced almost identical effects on blood flow as kallikrein[5]. 45 years ago, Frey and his coworkers presented evidence that kallikrein, besides its hemodynamic effects, can also cause a lowering of the blood glucose[6]. However, this observation was not followed up.

In principle, in the fasted state, a decrease in the blood glucose level can result either from reduction of hepatic glucose output or from acceleration of peripheral, mainly muscular, glucose uptake. Data concerning an involvement of kallikrein or kinins in either mechanism were not previously available. We became interested in this possibility during our studies on the metabolism of the human forearm and could indeed present evidence that endogenously liberated kinins are involved in the physiological adaptation of muscular glucose metabolism during work load[7] and hypoxia[8]. The following investigations were performed in order to study the effect of bradykinin on the glucose metabolism of the human forearm during rest. Furthermore, we have investigated the possible role of prostaglandins, which were suggested to be involved in the hemodynamic effects of the kinins[9,10]. For that purpose studies were carried out after the pretreatment with indomethacin, which prevents endogenous prostaglandin synthesis in man[11].

Material and Methods

Subjects

34 healthy volunteers were recruited from medical students of the Munich University School of Medicine. All were informed about the aim and the risks of the study and gave their consent. The protocol for this study was reviewed and approved by the Investigation and Ethical Committee of the Sonderforschungsbereich 51 of the Deutsche Forschungsgemeinschaft according to the Code of Ethics of the World Medical Association[12]. Physical examination as well as laboratory tests excluded internal diseases. All the subjects fasted overnight and received no special premedication. 4 subjects received intrabrachial-arterial infusion of physiological saline as described in more detail in the next section; another 5, infusions of papaverine; 9, bradykinin; and 10, bradykinin after the oral pretreatment with indomethacin (100 mg three times daily for two days and another 100 mg 1 h prior to the test). Their clinical data are listed below (mean ± SEM).

Saline infusion: age, 22.0 ± 1.6 years; height, 177.0 ± 4.3 cm; weight, 72.8 ± 3.1 kg (103.8 ± 9.9% of ideal weight according to the Metropolitan Life Insurance Tables, 1959); 5.26 ± 0.21mM postprandial glucose (2 h after 100 g oral glucose).

Papaverine infusion: age, 22.8 ± 0.7 years; height, 175.3 ± 2.1 cm; weight, 68.8 ± 2.0 kg (102.0 ± 1.0% of ideal weight); 5.33 ± 0.15mM postprandial glucose.

Indomethacin + bradykinin: age, 23.7 ± 1.0 years; height, 179.8 ± 2.2 cm; weight, 73.9 ± 2.2 kg (103.8 ± 4.8% of ideal weight); 5.08 ± 0.13mM postprandial glucose. The clinical data of the 10 volunteers in the study on the precision of blood flow determinations were as follows: 25.3 ± 2.1 years, 176.3 ± 1.8 cm, 72.1 ± 2.6 kg (104.2 ± 4.1% of ideal weight).

Catheterization, blood sampling and registration of blood flow

All studies were performed between 8:30 a.m. and 11:30 a.m. in a recumbent position. Special care was taken that the arm rested comfortably and was positioned above heart level to secure free venous drainage. Under sterile conditions and local anaesthesia, one of the cubital veins apparently draining deep tissue of the forearm was cannulated and a teflon catheter (Code 1.512, Vygon, France) inserted retrograde as deep as possible. This procedure was thought to be successful when the tip of the catheter could no longer be palpated and when the artieral–deep-venous oxygen difference exceeded 6 vol.%. In all except four, the success of the procedure was controlled by X-rays. The catheter was flushed continuously with physiological saline containing small amounts of heparin, the rate of which [0.1 U/(kg × min)] guaranteed there would be no significant effect on lipoprotein lipase activity[13]. Then, a Cournand-needle (Pe 160, Kifa, Sweden) was inserted into the right femoral artery and a thin Luer-Lok cannula (A 150, Henke-Sass Wolf, Tübingen) into the brachial artery and flushed continuously with physiological saline (0.2 m*l*/min).

Arterial and deep-venous blood samples were collected simultaneously at 5-min intervals throughout a 15-min basal period, followed by a 25-min test period for chemical analysis.

This was performed in 5 subjects during the intrabrachial-arterial infusion of physiological saline (0.2 m*l*/min) for the whole test period. 5 subjects received identical infusions of saline containing papaverine in a concentration that resulted in a dose of 2 μg/(kg × min) during the first 10 min and of 3 μg/(kg × min) thereafter. 9 subjects, and 10 subjects pretreated with indomethacin, were infused with bradykinin instead of papaverine at a rate of 0.2 ng/(kg × min). Forearm blood flow was usually estimated by two independent methods. One of them was a venous occlusion plethysmography using a modification[14] of the Whitney[15] "mercury in rubber strain gauge" plethysmography. Forearm volume changes were registered by changes in the electrical resistance of a double mercury straingauge due to lengthening. The gauges were applied on the forearm 6 - 8 cm distal to the elbow at adequate tension without causing congestion and were calibrated before each measurement. The venous occlusion cuff just proximal to the elbow was inflated automatically to 60 mmHg for 20 s. The precision of the method based on serial determinations is given under Results. Blood flow measurements by this technique were usually performed immediately after the collection of blood samples.

The other method for blood flow determination was accomplished by the brachial arterial injection of 0.3 Ci ^{133}Xe dissolved in 0.5 m*l* saline. The deep-venous disappearance of ^{133}Xe was registered by blood sampling at 30-s intervals for 5 min and then at minute intervals for further 5 min. The rate constant for the washout of ^{133}Xe from muscle was calculated as previously described[16]. Two measurements were done, one between 5 and 15 min during the basal period, and one between 10 and 20 min during the drug infusion period (see Table 4). Data on human forearm blood flow obtained with the two different methods are given under Results.

Analyses

For glucose determinations, 0.1 m*l* of blood was mixed immediately after the withdrawal with 1 m*l* of ice-cold 0.3M perchloric acid, and centrifuged at 4 °C. Glucose was determined in specimens stored overnight at – 20 °C. Blood samples for gas analysis were taken in heparinized syringes and analyzed promptly. Glucose was determined enzymatically and oxygen oximetrically as described elsewhere[17]. Serum insulin was assayed according to a modification[18] of the Yalow-Berson immunoassay[19]. From each blood sample, determinations of the individual substrates were performed at least in duplicate. Basal values represent the mean of 4 determinations obtained at 5-min intervals during the basal period averaged for each subject. For statistical calculation the student's *t*-test for paired and unpaired samples was used when applicable[20]. Mean values are given with the standard error of the mean (SEM). Substrate uptake was calculated from arterial–deep-venous concentration differences and the corresponding blood flow rate estimated by the venous occlusion plethysmography[21].

Materials

Bradykinin was used as a sterile solution of synthetic bradykinin (Sandoz AG,-Basel) diluted in physiological saline to give a concentration of 70 nmol/*l*. ^{133}Xe was purchased as a sterile solution in 0.9% sodium chloride (10 mCi/m*l*) with a specific activity of 44.8 Ci/mmol (Radiochemical Center, Amersham). Heparin was used as a sterile solution of sodium heparinate containing 2500 USP/m*l* (Hoffmann La Roche, Grenzach). Papaverine was used as a sterile solution of papaverine hy-

Table 1. Arterial concentrations of oxygen and glucose.

	n	Basal[a]	Infusion[b]	
			15 min	25 min
Oxygen[c]				
Saline	5	20.4 ± 0.2	20.5 ± 0.3	20.3 ± 0.2
Papaverine	5	19.9 ± 0.4	19.6 ± 0.3	19.8 ± 0.3
Bradykinin	8	19.9 ± 0.4	19.7 ± 0.4	19.7 ± 0.4
Indomethacin + bradykinin	8	19.4 ± 0.4	19.3 ± 0.4	19.4 ± 0.4
Glucose[d]				
Saline	5	4.70 ± 0.13	4.78 ± 0.14	4.76 ± 0.15
Papaverine	5	4.76 ± 0.21	4.72 ± 0.20	4.74 ± 0.24
Bradykinin	9	4.75 ± 0.24	4.74 ± 0.23	4.72 ± 0.25
Indomethacin + bradykinin	10	5.31 ± 0.18[e]	5.41 ± 0.18[e]	5.44 ± 0.18[e]

[a] From 4 determinations at 5 min intervals averaged for each subject; [b] dose under materials and methods; [c] mean ± SEM in m*l*/100 m*l*; [d] mean ± SEM in μmol/100 m*l*; [e] significant at $P < 0.05$ as compared to the residual values.

drochloride (Department of Pharmacy, Schwabing City Hospital, Munich) diluted in physiological saline to a concentration of 5 mmol/*l*. Indomethacin* was used as Amuno-capsules (Sharp & Dohme, Munich), containing 50 mg of indomethacin.

Results

1) Arterial concentrations of oxygen and glucose

Arterial concentrations of oxygen and glucose (Table 1) were in good agreement with previously published data[7,8,17,22]. Oxygen concentrations did not differ between the individual groups. In agreement with earlier observations[23], the mean arterial glucose concentration was significantly elevated after indomethacin pretreatment ($p < 0.05$; unpaired t-test). During infusion of the different substances, arterial levels of oxygen and glucose remained unchanged. The arterial and deep-venous concentrations of insulin, determined in 3 subjects, showed only small fluctuations during the infusion of bradykinin:

Basal: arterial, 4.33 ± 0.22 μU/m*l*; deep-venous, 4.36 ± 0.23 μU/m*l*.

15 min bradykinin: arterial, 3.83 ± 0.05 μU/m*l*; deep-venous 3.9 ± 0.09 μU/m*l*.

20 min bradykinin: arterial, 3.83 ± 0.29 μU/m*l*; deep-venous, 3.67 ± 0.22 μU/m*l*.

25 min bradykinin: arterial,3 .37 ± 0.47 μU/m*l*; deep-venous, 3.47 ± 0.47 μU/m*l*.

2) Arterial–deep-venous concentration differences of oxygen and glucose

The arterial–deep-venous concentration differences of oxygen and glucose observed in the four experimental groups during the basal period were in good accord with otherwise published data [7,8,22] and did not differ significantly from one group to the other (Table 2). These values remained constant during the infusion of 0.9% NaCl, indicating that the experimental procedure by itself did not influence the results. During the infusion of papaverine, both values for oxygen and glucose were significantly reduced within 5 min, corresponding to the increase in blood flow (Table 3), and a further reduction was caused by further acceleration of blood flow when concentration of papaverine was increased once more. During the infusion of bradykinin, the arterial–deep-venous concentration difference of oxygen was also reduced, corresponding to the doubling of blood flow, but that of glucose was enlarged significantly, in contrast to the values obtained during papaverine infusion (Table 2, Fig. 1).

---

* Indomethacin = [1-(p-chlorobenzoyl)-5-methoxy-2-methyl-1-indolyl]acetic acid.

Table 2. Arterial–deep-venous concentration differences of oxygen and glucose.

	n	Basal[a]	Infusion[b]				
			5 min	10 min	15 min	20 min	25 min
Oxygen[c]							
Saline	4	7.75 ± 0.59	7.71 ± 0.33	7.71 ± 0.33	7.44 ± 0.34	7.73 ± 0.32	7.98 ± 0.40
Papaverine	3	7.83 ± 0.91	–	4.57 ± 1.50[f]	4.67 ± 1.52[e,f]	4.13 ± 1.30[e,f]	4.17 ± 0.90[e,f]
Bradykinin	6	9.00 ± 0.66	7.48 ± 0.66	6.10 ± 0.60[f]	6.32 ± 0.73	4.83 ± 0.51[e,f]	5.13 ± 0.54[e,f]
Indomethacin + bradykinin	8	8.21 ± 0.44	7.26 ± 0.54	6.98 ± 0.96	6.95 ± 0.89	6.71 ± 0.48[e,h]	6.83 ± 0.49[e,h]
Glucose[d]							
Saline	4	28.2 ± 4.0	27.2 ± 3.4	28.4 ± 3.1	28.2 ± 3.0	27.8 ± 3.5	28.0 ± 3.9
Papaverine	5	34.1 ± 5.2	25.3 ± 3.3[e]	25.1 ± 3.3[e]	18.0 ± 3.1[e,f]	18.2 ± 3.0[e,f]	16.3 ± 4.2[e,f]
Bradykinin	9	25.4 ± 4.3	34.3 ± 4.1[e]	35.0 ± 5.2	37.7 ± 4.1[e,g]	36.4 ± 4.2[e,g]	41.1 ± 3.3[e,f,g]
Indomethacin + bradykinin	10	34.0 ± 2.8	30.5 ± 2.7	30.5 ± 3.9	26.5 ± 4.4[h]	26.0 ± 4.1[e,h]	24.0 ± 4.8[e,h]

[a] Mean values ± SEM averaged for each subject from 4 determinations at 5 min intervals; [b] dosage is given under materials and methods; [c] mean values ± SEM are given in m*l* oxygen/100 m*l*; [d] mean values ± SEM are given in μmol/100 m*l*; [e] significant at $P < 0.05$ as compared to basal; [f] to saline; [g] to papaverine; [h] to bradykinin.

After indomethacin pretreatment, as could be expected from the smaller blood flow response (Table 3), only a small diminution of the oxygen balance was observed. In contrast to the findings obtained before indomethacin treatment there was a reduction of the arterial–deep-venous difference of glucose (Table 2).

3) *Forearm blood flow*

Our data on blood flow derived by two different techniques (Table 3) are in good agreement with those reported in the literature[14-16]. The precision of the plethysmographic method was assessed from duplicate determinations in 10 individuals; the ratio of the first (2.86 ± 0.16 m*l*/(100 g × min)) to the second value (3.03 ± 0.19 m*l*/(100 g × min)) was 0.952 ± 0.035. The ^{133}Xe technique yielded a ratio of the first value (10.7 ± 0.85 m*l*/(100 g × min)) to the second value (10.3 ± 1.05 m*l*/(100 g × min)) of 1.032 ± 0.032. In agreement with other findings, the ^{133}Xe procedure yielded much higher values than the plethysmographic procedure[16].

During saline infusion, blood flow rates estimated by the two methods did not change (Table 3). Infusion of papaverine resulted in a two-fold and, at higher concentrations, in an almost three-fold increase in blood flow. Similarly, bradykinin stimulated forearm blood flow by about 100%, an effect persisting during the whole test period. After indomethacin pretreatment, the effect of bradykinin was significantly smaller (Table 3).

4) *Uptake of oxygen and glucose*

Uptake of substrates was estimated only in those subjects whose blood flow and arterial and deep-venous oxygen and glucose concentrations were measured simultaneously. The rates of oxygen

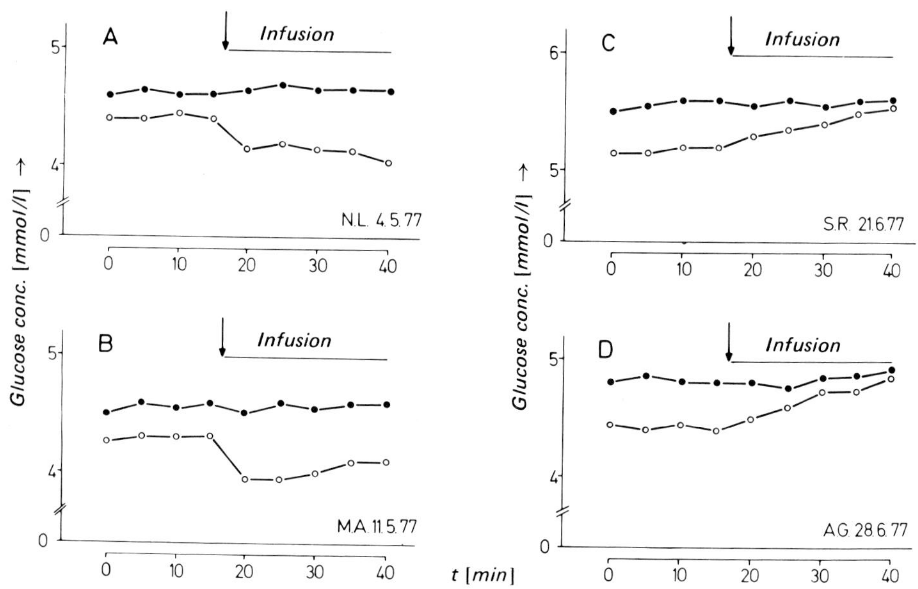

Fig. 1. Arterial (●) and deep-venous (○) concentrations of glucose in four individuals during the infusion of bradykinin (A, B) and papaverine (C, D).

While the arterial–deep-venous concentration difference of glucose was immediately enlarged during the infusion of bradykinin, it was continuously reduced during that of papaverine.

uptake were in good agreement with earlier findings[7,8,22] and did not differ significantly either during baseline measurements or drug infusion. The mean ± SEM of oxygen uptake in ml/(100 g tissue × min) was as follows:

Saline infusion (n = 4): basal 0.24 ± 0.13; after 15 min, 0.24 ± 0.04; after 25 min, 0.25 ± 0.05. Papaverine (n = 3): basal 0.27 ± 0.05; after 15 min, 0.33 ± 0.05; after 25 min, 0.36 ± 0.05. Bradykinin (n = 4): basal 0.29 ± 0.06; after 15 min, 0.34 ± 0.06; after 25 min, 0.30 ± 0.07. Bradykinin after indomethacin pretreatment (n = 5): basal 0.27 ± 0.04; after 15 min, 0.28 ± ± 0.04; after 25 min, 0.31 ± 0.05. The effect of the individual infusions on glucose uptake is illustrated in Fig. 2.

Discussion

For a long time, insulin was generally considered to represent the only physiologically important stimulator of glucose uptake in muscle. This ef-

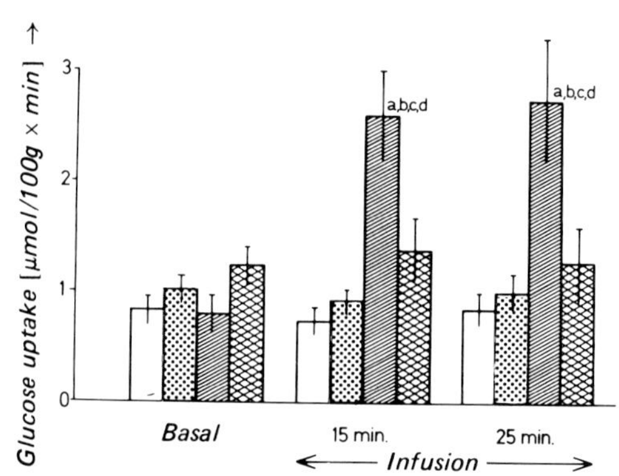

Fig. 2. Glucose uptake of the human forearm.

The columns represent the mean ± SEM in μmol per 100 g tissue and min obtained during the intrabrachial-arterial infusion of physiological saline (unshaded bars; n = 4), of papaverine (shaded bars; n = 3) of bradykinin (hatched bars; n = 4) and of bradykinin after the oral pretreatment with indomethacin (chequered bars; n = 7).

[a] Significant difference to basal; [b] to NaCl; [c] to papaverine; [d] to bradykinin after indomethacin pretreatment.

Table 3. Forearm blood flow[a]

	Subjects	Venous occlusion plethysmography			Xenon technique	
		Basal[b]	Infusion[c]		Basal	Infusion[c] (15 min)
			15 min	25 min		
Physiol. saline						
	D.R.	2.9	2.8	2.9	8.3	7.1
	B.M.	2.5	2.9	2.9	12.3	11.0
	P.K.	2.8	2.8	2.8	10.3	9.9
	L.G.	2.5	2.4	2.3	9.7	9.7
	$\bar{x} \pm$ SEM	2.88 ± 0.18	2.73 ± 0.11	2.73 ± 0.14	10.15 ± 0.88	9.43 ± 0.82
Papaverine						
	S.R.	3.5	9.2	12.4	9.6	26.2
	A.G.	3.7	7.3	11.0	13.5	24.3
	V.O.	2.9	5.5	7.5	11.6	16.2
	$\bar{x} \pm$ SEM	3.36 ± 0.30	7.33 ± 1.1[d,e]	10.30 ± 1.5[d,e]	11.56 ± 0.88	22.23 ± 3.12[d,e]
Bradykinin						
	M.A.	3.3	7.5	7.3	13.7	34.8
	N.L.	2.8	8.0	7.5	14.5	29.0
	Z.P.	3.9	6.2	6.0	4.7	7.2
	K.D.	2.4	3.5	3.2	9.7	12.2
	$\bar{x}$ + SEM	3.10 ± 0.32	6.31 ± 1.0[d,e]	6.00 ± 0.9[d,e]	10.65 ± 2.22	20.81 ± 6.50[d]
Indomethacin + bradykinin						
	W.R.	2.9	3.9	3.5		
	S.F.	3.8	4.5	4.5		
	U.R.	3.5	4.5	5.5		
	G.H.	2.5	3.8	3.9		
	F.J.	4.8	5.5	5.5		
	$\bar{x} \pm$ SEM	3.50 ± 0.30	4.44 ± 0.24[d,e,f]	4.58 ± 0.32[d,e,f]		

[a] m*l*/(100 g of tissue × min); [b] mean of 4 determinations at 5 min intervals; [c] for dosage see material and methods section; [d] significant to basal; [e] to saline; [f] to bradykinin.

fect of insulin has been well established in experiments in vitro as well as in vivo including the model of the human forearm[24]. Further studies revealed that muscular work, anaerobiosis and also some chemical agents can mimick the effect of insulin on glucose transport[25], but the mechanisms of these interrelationships have remained obscure. First evidence that the kallikrein-kinin system, besides its known hemodynamic actions [3-5,26,27], may be involved, came from studies in this laboratory on the metabolism of the human forearm. It was shown that infusion of a kallikrein-trypsin inhibitor suppressed the increase in the blood flow and of glucose uptake which normally occurs during muscular work[7] or in hypoxia[8] of the forearm. Moreover, this effect was overcome by intrabrachial-arterial application of physiological doses of bradykinin. These observations, together with other reports, such as an activation of the prekallikrein-kallikrein conversion[28], a lowering of plasma kininase activity by a fall in pH[29], and an increase and decrease of the venous concentrations of kinins and kininogen, respectively[30], led us to suggest that the kinins, might be related to the postulated "muscular activity factor"[31], and may play a physiological role in the regulation of blood and substrate supply for the working muscle.

The present work provides experimental evidence that bradykinin also exerts its effects on glucose metabolism in the resting muscle. At a brachial arterial perfusion rate of about 20 m*l*/min[32], the expected bradykinin concentration in our experiments would amount to 0.5 - 1.0 ng/m*l* blood, thus falling well within the physiological range as measured in man[30]. These small concentrations led to an immediate acceleration of blood flow by about 100% (Table 3) and of muscular glucose uptake by about 350% (Fig. 2). The latter effect was not only caused by the acceleration of blood flow, but also by an enlargement of the arterial–deep-venous glucose difference (Table 2, Fig. 1).

Control experiments with papaverine disproved that the effect of bradykinin on glucose uptake was merely due to the accompanying acceleration of blood flow: although it was able to accelerate forearm blood flow to a similar extent, the arterial–deep-venous concentration difference of glucose was simultaneously reduced, indicating unchanged glucose uptake (Fig. 1, 2). These effects of bradykinin on the metabolism of the human forearm have also been confirmed by similar studies in maturity onset diabetics[33]. That the glucose taken up was utilized in the muscle cell was indicated by the finding that no efflux of glucose was detectable after the termination of the experiments. An increase of glycogen synthetase I activity and of muscular glycogen content in insulin-deprived juvenile diabetics after local muscular work[34] would favour the view that at least part of the glucose taken up is channeled into glycogen under the influence of kinins. Measurements of the arterial and deep-venous insulin concentrations, which remained fairly constant, excluded the possibility that bradykinin stimulated insulin secretion or released bound insulin within the capillary bed[35].

According to studies on fibroblasts, bradykinin, after binding to specific receptors of the plasma membrane, stimulates prostaglandin synthesis by increasing the deacylation of phospholipids[36]. That bradykinin might act on blood flow via prostaglandin formation was already suggested from studies on kidney and heart muscle, where the action of bradykinin on perfusion was diminished after suppression of endogenous prostaglandin synthesis[9,10]. Moreover, prostaglandin E_1 applied into the brachial artery in man accelerated forearm blood flow[37] as bradykinin did, and conversely, inhibition of endogenous prostaglandin synthesis was shown to delay reactive and functional hyperaemia[38]. In the present studies, indomethacin was used as an inhibitor of prostaglandin synthesis[11] in a dose sufficient to decrease the urinary excretion of the main prostaglandin metabolite by 77 - 98%[39]. Our findings that the action of bradykinin on both blood flow and arterial–deep-venous glucose concentration difference was significantly smaller after the pretreatment with indomethacin lends further support on the view that the actions of bradykinin on muscle may be mediated by prostaglandins.

As demonstrated by others[38], there was no effect of indomethacin on the basal rates of forearm blood flow (Table 3). However, indomethacin pretreatment caused significant elevation of basal arterial glucose concentrations, an effect ascribed to be due to a two- to three-fold rise of serum human growth hormone[23]. Since a much higher elevation of human growth hormone concentration is required for a reduction of glucose uptake of the human forearm[40], it is not very likely that human growth hormone plays a significant role in the effect of indomethacin treatment in our experiments. Finally, it should be noted that the prostaglandins themselves were shown to produce an accelerating effect on muscle glucose uptake[41,42] and oxidation[43], and a reduction of the rate of muscle fatty acid extraction[41,42], furthermore suggesting that the glucose entering the cell under the influence of prostaglandins was indeed utilized.

The demonstration that kinins are involved in the control of reactive and functional hyperemia and of glucose uptake in muscle, and that prostaglandins may participate therein has important implications for the normal function of skeletal muscle. Our working hypothesis evolving from the present and earlier results is presented in Fig. 3. It is proposed that in addition to the hormonal control which coordinates metabolism of all the tissues within the whole organism, a defined control system must exist in each individual tissue which, in cases of emergency, as in muscle during work or hypoxia, secures the rapid and selective supply of oxygen and glucose.

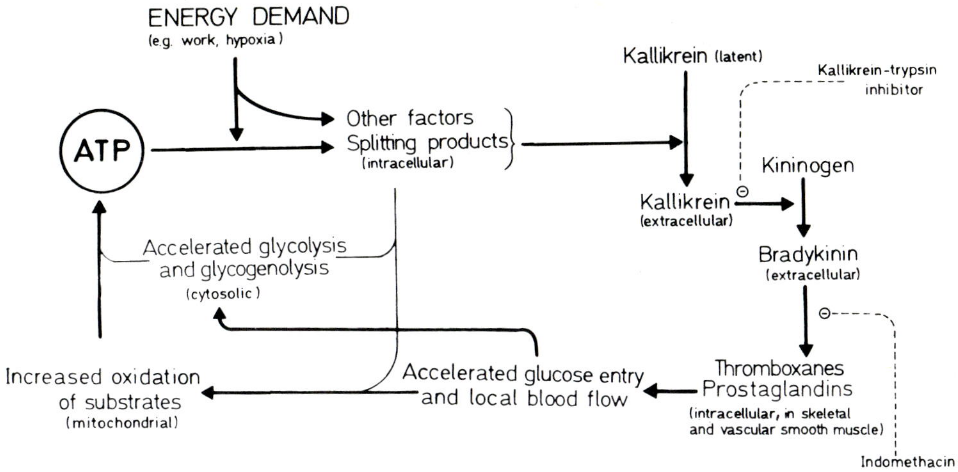

Fig. 3. Hyopthetical scheme on the role of the kallikrein-kinin system during adaptation of metabolism to local muscular work or hypoxia.

Under these circumstances, the demand for energy may exceed the endogenous substrate supply by glycogenolysis and glycolysis[44], thus making local control essential. It is proposed that products of ATP degradation, i.e. adenosine, inorganic phosphate, etc. and/or other factors, such as calcium ions or pH changes, will induce kallikrein formation, which liberates bradykinin from kininogen within the extracellular space. This in turn will stimulate thromboxane[45] and prostaglandin synthesis inside the membrane in skeletal and vascular smooth muscle. The following acceleration of glucose entry at the cell surface and the pericellular vasodilation make glucose and oxygen available for glycolysis and oxidation. Thus, energy demand may be covered and local muscular work continued. Since kinins are produced locally and are rapidly inactivated[46], accelerated glucose entry and blood flow would be restricted to the required site.

The authors are indebted to *Prof. Dr. E. K. Frey*, who has stimulated this work, to *Prof. Werle* and *Prof. Kraut* for valuable discussions, to *Prof. Fritz, Prof. Mehnert* and *Prof. Wieland* for their advice and critical review of this manuscript, and to *Dr. P. H. Wright* for a gift of the antibody GPAIS III. Furthermore, they wish to acknowledge the technical assistance of *E. A. Bauer, A. Bammert, E. Maerker, H. Kirschner* and *G. Trotleff*. Thanks are also due to *Sandoz AG* for a gift of synthetic bradykinin.

Literature

1 Werle, E., Götze, W. & Keppler, A. (1937) *Biochem. Z.* **289**, 217 - 224.

2 Werle, E. & Berek, U. (1948) *Angew. Chem.* **60**, 53 - 57.

3 Frey E. K. (1926) *Arch. Klin. Chir.* **142**, 663 - 669.

4 Frey, E. K. & Kraut, H. (1926) *Hoppe-Seyler's Z. Physiol. Chem.* **157**, 32 - 61.

5 Rocha e Silva, M., Bevaldo, W. T. & Rosenfeld, G. (1949) *Am. J. Physiol.* **156**, 261 - 269.

6 Frey, E. K. Kraut, H. & Werle E. (1932) *Klin. Wochenschr.* **11**, 1 - 9.

7 Dietze, G. & Wicklmayr, M. (1977) *FEBS Lett.* **74**, 205 - 208.

8 Dietze, G., Wicklmayr, M. & Mayer, L. (1977) *Hoppe-Seyler's Z. Physiol. Chem.* **358**, 633 - 638.

9 McGiff, J. C., Terragno, N, A., Malik, K. U. & Lonigro, A. J. (1972) *Circ. Res.* **31**, 36 - 43.

10 Needleman, P., Key, S. L., Denny, S. E., Isahron, P. C. & Marshall, G. R. (1975) *Proc. Natl. Acad. Sci. U.S.A.* **72**, 2060 - 2063.

11 Vane, J. R. (1971) *Nature (London) New Biol.* **231**, 232 - 235.

12 World Medical Association (1964) *Br. Med. J.* ii. 177 - 180.

13 Boberg, J. (1972) *Acta Med. Scand.* **191**, 97 - 102.

14 Gutmann, J., Kachel, V. & Bründl, G. (1969) *Elektromedizin*, Sonderh. 87.

15 Whitney, R. J. (1953) *J. Physiol. (London)* **121**, 1 - 9.

16 Tønnesen, V. H. & Sejrsen, P. (1967) *Circ. Res.* **20**, 552 - 564.

17 Dietze, G., Wicklmayr, M., Hepp, K. D., Bogner, W., Mehnert, H., Czempiel, H. & Henftling, H. G. (1976) *Diabetologia* **12**, 555 - 561.

18 Herbert, V., Kam-Seng Lau, Gottlieb, C. W. & Bleicher, S. (1965) *J. Endocrinol.* **25**, 1375 - 1379.
19 Yalow, R. S. & Berson, S. A. (1960) *J. Clin. Invest.* **39**, 1157 - 1175.
20 Snedecor, G. W. & Cochran, W. G. (1967) Statistical Methods. 6th edn., Ames, Iowa, Iowa State University Press.
21 Fick, A. (1870) *Physiol. Med. Ges., Würzburg* **2**, 16 - 28.
22 Jorfeldt, L. & Wahren, J. (1970) *Scand. J. Clin. Lab. Invest.* **22**, 73 - 81.
23 Syvälahti, E. K. G. (1974) *Int. J. Clin. Pharmacol.* **10**, 111 - 116.
24 Andres, R., Baltzan, M. A., Cader, G. & Zierler, K. L. (1962) *J. Clin. Invest.* **41**, 108 - 115.
25 Randle, P. J. & Smith, H. G. (1958) *Biochem. J.* **70**, 501 - 508.
26 Fox, R. H., Goldsmith, R., Kidd, D. J. & Lewis, G. P. (1961) *J. Physiol. (London)* **157**, 589 - 602.
27 Gersmeyer, E. F. & Spitzbarth, H. (1961) *Klin. Wochenschr.* **39**, 1227 - 1233.
28. Frey, E. K. (1930) *Arch. Klin. Chir.* **162**, 334 - 340.
29 Edery, H. & Lewis, G. P. (1962) *Br. J. Pharmacol.* **19**, 299 - 305.
30 Armstrong, D. & Stewart, I. W. (1960) *Nature (London)* **188**, 1193 - 1195.
31 Goldstein, M. S. (1961) *Am. J. Physiol.* **200**, 67 - 70.
32 Huldická, O. (1973) Muscle Blood Flow, Its Relation to Muscle Metabolism and Function, Swets & Zeitlinger B. U., Amsterdam.
33 Wicklmayr, M. & Dietze, G. (1978) in Kininogenases (Haberland, G. L., Rohen, J. W. & Suzuki, R. O., eds.) F. K. Schattauer-Verlag, Stuttgart.
34 Roch-Norlund, A. E., Bergström, J. & Hultman, E. (1972) *Scand. J. Clin. Lab. Invest.* **30**, 77 - 84.
35 Rasio, E., Whichelow, M. J., Butterfield, W. J. H. & Hicks, B. M. (1972) *Diabetologia* 8, 244.
36 Su-Chen, L. H. & Levine, L. (1976) *J. Biol. Chem.* **251**, 5814 - 5816.
37 Bevegard, S. & Orö, L. (1969) *Scand. J. Clin. Lab. Invest.* **23**, 347 - 353.
38 Kilbom, A. & Wennmalm, A. (1976) *J. Physiol. (London)* **257**, 109 - 121.
39 Hamberg, M. (1972) *Biochem. Biophys. Res. Commun.* **49**, 720 - 726.
40 Rabinowitz, D., Klassen, G. A. & Zierler, K. L. (1965) *J. Clin. Invest.* **44**, 51 - 61.
41 Maxwell, G. M. (1967) *Br. J. Pharmacol.* **31**, 162 - 168.
42 Glaviano, V. & Masters, T. (1968) *Circulation* **38**, Suppl. VI, 83 - 97.
43 Willebrands, A. F. & Tasseron, S. J. A. (1968) *Am. J. Physiol.* **215**, 1089 - 1095.
44 Helmreich, E. & Cori, C. F. (1965) *Adv. Enzyme Regulat.* **3**, 91 - 105.
45 Hamberg, M., Svensson, I. & Samuelson, B. (1975) *Proc. Natl. Acad. Sci. U.S.A.* 72, 2994 - 2998.
46 McCarthy, D. A., Potter, D. E. & Nicolaides, E. D. (1965) *J. Pharmacol. Exp. Ther.* **148**, 117 - 122.

Priv. Doz. Dr. G. Dietze, III. Medizinische Klinik, Krankenhaus München-Schwabing, Kölner Platz 1, D-8000 München 40.

Hoppe-Seyler's Z. Physiol. Chem.
Bd. 359, S. 379 - 384, März 1978

The Elimination in Dogs of Trypsin-α-Macroglobulin Complexes Inactivated by the Kazal or the Kunitz Inhibitor

Allan EDDELAND and Kjell OHLSSON

Departments of Surgery and Clinical Chemistry, University of Lund, Malmö General Hospital, Malmö, Sweden.

(Received 7 November 1977)

Summary: The elimination of trypsin-α-macroglobulin complexes and similar complexes with the trypsin inactivated by low-molecular weight inhibitor was studied in anesthetized dogs. The complex was inactivated either by the Kazal (pancreatic secretory trypsin inhibitor, PSTI) or the Kunitz inhibitor (Trasylol BE). The inhibitors were labelled with ^{125}I and in the case of the trypsin-α-macroglobulin complex the trypsin was labelled with ^{125}I. All of the inactivated complexes exhibited a half-life of about 5 min in the dog. The elimination in plasma was exponential until 80 - 85% of the initial dose was cleared in 30 min and nearly negligible thereafter as seen by radioactivity measurements. Simultaneously increasing amounts of dialyzable radioactive substances with a lower molecular weight than the inhibitors were recovered in the urine. No significant differences in the elimination of trypsin-α-macroglobulin complexes were detected in plasma or in the urine before and after inactivation with the Kazal inhibitor (PSTI) or the Kunitz inhibitor (Trasylol BE).

Zur Eliminierung von Trypsin-α-Makroglobulin-Komplexen, die durch den Kazal- bzw. den Kunitz-Inhibitor gehemmt sind

Zusammenfassung: Die Eliminierung von „freien" und durch niedermolekulare Inhibitoren inaktivierten Trypsin-α-Makroglobulin-Komplexen wurde in narkotisierten Hunden untersucht. Die Hemmung erfolgte entweder durch den sekretorischen Panreas-Inhibitor (Kazal-Typ, PSTI) oder durch den Kunitz-Typ-Inhibitor (Trasylol). Die Inhibitoren waren dafür zuvor mit ^{125}I markiert worden, im Falle des freien Komplexes war ^{125}I-markiertes Trypsin verwendet worden. Die inaktivierten Komplexe wurden mit einer Halbwertzeit von 5 min aus dem Blut eliminiert. Die Eliminierung verlief exponentiell, bis 80 - 85% der injizierten Radioaktivität nach 30 min aus dem Plasma ausgeschieden waren, danach verlief die weitere Eliminierung praktisch nur noch sehr langsam. Gleichzeitig wurden im Urin zunehmend radioaktive Substanzen nachweisbar, deren Molekulargewichte niedriger waren als das der Inhibitoren. Signifikante Unterschiede im Eliminierungsverhalten der freien und durch den Kazal- bzw. Kunitz-Inhibitor inaktivierten Trypsin-α-Makroglobulin-Komplexe waren nicht erkennbar.

Abbreviations:
PSTI = pancreatic secretory trypsin inhibitor; BzArgNHNp = benzoyl-DL-arginine-*p*-nitroanilide-HCl.

Key words: α-Globulins; macroglobulins; trypsin; trypsin inhibitors.

The pancreatic secretory trypsin inhibitor (PSTI), also called the Kazal inhibitor, is a specific trypsin inhibitor secreted from the acini of the pancreas into the pancreatic juice. It can normally be found in minute amounts in the circulation and in varying amounts in cases of pancreatic disease.* On intravenous administration it is eliminated rapidly from the circulation, the kidneys being considered as primary organ of catabolism for low molecular weight proteins like these[1-5]. Another trypsin inhibitor, the Kunitz inhibitor, commercially available as Trasylol BE, resembles the Kazal inhibitor in molecular weight (M_r 6500). It is a strong trypsin inhibitor which has recently been shown to be able to prevent the shock regularly seen in dogs on saturation of the α-macroglobulin in the circulation by trypsin[6].

α_2-Macroglobulin** is thought to be extremely important in the endogenous defence against proteolytic enzymes released in vivo[7,8]. α_2-Macroglobulin binds all types of endoproteases, and proteases initially bound by α_1-antitrypsin seem to be transferred to α_2-macroglobulins[7]. The protease-α-macroglobulin complexes are then rapidly cleared by the reticuloendothelial cells in the liver, spleen and bone marrow[9].

Electron microscopic studies of α_2-macroglobulin suggest that the binding of trypsin results in a profound structural change of the inhibitor molecule[10]. The trypsin-α_2-macroglobulin complex exhibits hydrolytic activity against low molecular weight substrates but no proteolysis of larger molecules[11]. Ganrot[12] showed that the hydrolytic activity of the complex could be inhibited by the Kunitz inhibitor and we have shown similar inhibition by the Kazal inhibitor[13]. Such PSTI-trypsin-α-macroglobulin complexes are furthermore formed on intravenous injection of PSTI-trypsin complexes[5].

The exact mechanism of elimination of the trypsin-α-macroglobulin complexes is unknown. Their elimination may depend either on the recognition of an altered tertiary structure by the cells of the reticuloendothelial system[9,10] or on the hydrolytic activity of a protease-α-macroglobulin complex[11]. To study the latter possibility the elimination of trypsin-α-macroglobulin complexes inactivated by either Trasylol or PSTI was compared to the elimination of active trypsin-α-macroglobulin complexes. The metabolism of PSTI and Trasylol when injected in complex with trypsin-α-macroglobulin was also studied.

Materials and Methods

Materials

Sephadex G-25 Fine and *Sephadex G-200* were obtained from Pharmacia Fine Chemicals AB, Uppsala, Sweden.
Agarose, batch 520 A, was purchased from Miles-Seravac, Maidenhead, England.
Freund's complete adjuvant was a product of the Difco Laboratories, Detroit, Michigan, U.S.A.
Kodak Tri-X-Pan Professional Film, 4164, thick, was purchased from Eastman Kodak Company, Rochester, N.Y. 14650, U.S.A.
Benzoyl-DL-arginine-p-nitroanilide · HCl (BzArgNHNp) was a product of Fluka AG, Switzerland and Mebumal vet. 60 mg/m*l* of ACO Läkemedel AB, Solna, Sweden.
Trasylol BE (aprotinin) was graciously provided by Bayer AG, Leverkusen, West Germany.
Anionic dog trypsin[14] and the *dog pancreatic secretory trypsin inhibitor* (PSTI)[13] were purified as previously reported.
Rabbit antisera. The same antiserum against dog α-macroglobulin was used as earlier described[15] and antiserum against Trasylol was previously produced in our laboratory.
Experimental animals. Eleven fully grown mongrel dogs of either sex weighing 8 - 39 kg were used for this study.

Operative procedure

The animals were anesthetized with Mebumal vet. 60 mg/m*l*, initially 30 mg/kg body weight followed by repeated maintenance doses. Cuffed tube endotracheal intubation was maintained during the experiment. A 16 gauge polyethylene catheter was inserted into a femoral vein for blood sampling, another into a femoral artery for monitoring blood pressure and a third into a brachial vein for injection. Blood samples were drawn at regular intervals after the injection of complexes. An indwelling catheter was placed in the urinary bladder and urine was collected at half hour intervals.

* Eddeland, A. & Ohlsson, K., to be published.

** In dog plasma unlike in human plasma there are two α-macroglobulins [Ohlsson, K. (1971) *Biochim. Biophys. Acta* **236**, 84 - 91.

Analytical methods

Electrophoretic method. The electroimmuno assay technique of Laurell[16] was used to identify α-macroglobulin containing fractions as well as the trypsin-α-macroglobulin complexes. The antiserum against the protein to be studied was added to a 0.8% (w/v) agarose gel that contained 0.075 mol barbital buffer, pH 8.6/*l* with 0.002 mol calcium lactate/*l* and the agarose gel plate was made 1 mm thick.

Autoradiography was used in combination with the electroimmuno assay. Sheets of Tri-X-Pan Professional Film were cut to the size of the electroimmuno assay plates and exposed to them in total darkness for 48 h and then developed to locate the radioactive samples.

Gel filtration was done in Sephadex G-25 Fine (columns 0.9 × 55 cm or 1.6 × 15 cm) and in Sephadex G-200 (column 2.5 × 35 cm).

Protein concentration in fractions from gel filtrations was followed by measurement of optical density at 280 nm.

Ultrafiltration of protein fractions from columns was done in an Amicon cell with either a Diaflo membrane UM-2 (PSTI and Trasylol) or UM-10 (trypsin-α-macroglobulin).

Radioactive labelling. A lactoperoxidase method[17] was used for the labelling of the trypsin, PSTI and Trasylol with ^{125}I giving an average specific activity of 0.8 mCi/mg protein. Gel filtration on a Sephadex G-25 column (1.6 × 16 cm) equilibrated with a 0.04 mol Tris/HCl buffer, pH 7.6,/*l*, containing 0.002 mol $CaCl_2$ and 0.2 mol NaCl/*l* was used to separate the unbound ^{125}I from the labelled protein.

Trypsin activity was measured with BzArgNHNp as a substrate according to Erlanger et al.[18].

Preparations of complexes in vitro

Complexes of ^{125}I-trypsin-α-macroglobulin were prepared by Sephadex G-200 gel filtration (0.04 mol Tris-HCl buffer, pH 7.6,/*l*, containing 0.002 mol $CaCl_2$ and 0.2 mol NaCl/*l*, column 2.5 × 35 cm) of 5 m*l* dog serum to separate the α-macroglobulins and refiltration on the same column of the pooled α-macroglobulin fractions to which was added 2.0 mg radioactively labelled anionic dog trypsin. The complex-containing fractions were pooled and utilized for studies in two dogs.

The radioactivity, which was measured in a well-type scintillation detector was eluted in two peaks, one with free trypsin and one confined to the fractions containing α-macroglobulin, by electroimmuno assay with antiserum against α-macroglobulin. When the electroimmuno assay precipitates were autoradiographed the radioactivity was observed in the same fractions as the trypsin-α-macroglobulin. Furthermore, all the radioactivity could be precipitated from solution by antibodies against α-macroglobulin. These tests all showed that the labelled enzyme injected was in the form of a complex with α-macroglobulin.

Complexes of ^{125}I-PSTI-trypsin-α-macroglobulin were obtained by first preparing trypsin-α-macroglobulin complexes as described above using unlabelled trypsin. The pool of these complexes was allowed to react with 0.8 mg of PSTI labelled with ^{125}I and the mixture was gel filtered on Sephadex G-200 (same column and buffer as above). The radioactivity measured in a scintillation detector or by autoradiography of electroimmuno assay precipitates appeared in the same fractions as those containing trypsin-α-macroglobulin complexes when measured by electroimmuno assay. The radioactivity was precipitable by antiserum against α-macroglobulin.

Complexes of ^{125}I-Trasylol-trypsin-α-macroglobulin were prepared and localized in the same fashion described above for ^{125}I-PSTI-trypsin-α-macroglobulin, substituting Trasylol for PSTI.

Intravenous administration of α-macroglobulin complexes prepared in vitro

The ^{125}I-trypsin-α-macroglobulin complexes corresponding to 2 m*l* dog serum were injected intravenously in each dog in less than 10 s (2 dogs) and the time at the beginning of the injection was designated zero. Blood samples were drawn at measured intervals until 3.5 - 4 h thereafter and urine was sampled every half hour. Identical administration and sampling methods were used for ^{125}I-PSTI-trypsin-α-macroglobulin (5 dogs) and ^{125}I-Trasylol-trypsin-α-macroglobulin (4 dogs).

Results

General observations of the experimental animals

During continuous monitoring of pulse, blood pressure or respiratory rate no change was detectable after administration of the complexes.

Elimination in vivo of ^{125}I-trypsin-α-macroglobulin

The elimination of the labelled material is shown in Fig. 1. The curve appears exponential until 80% of the administered dose is eliminated and the half-life in the circulation is calculated to be 6 min (see table). The decrease of the plasma radioactivity became markedly slower after the first 30 min. The initial urine specimen contained a radioactive substance which increased in concentration in later samples and was dialyzable. The urinary radioactivity could not be precipitated by antibodies against trypsin.

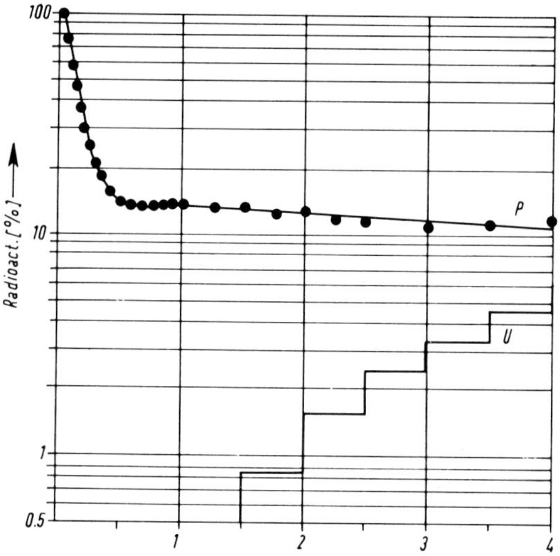

Fig. 1. Elimination of plasma radioactivity in dog after i.v. administration of dog ^{125}I-trypsin-α-macroglobulin complexes corresponding to 2.5 m*l* of dog serum.

Ordinate: the radioactivity in plasma (P) is expressed as percentage of the value at 2 min after the injection. The cumulative radioactivity in the urine (U) is expressed as percentage of the total amount injected. Abscissa: time in hours.

Table. Half-life of radioactive substance in the circulation of dogs after intravenous injection of trypsin-α-macroglobulin complexes and such complexes inactivated by the pancreatic secretory trypsin inhibititor or Trasylol.

Complex injected	Dog		$t_{1/2}$ [min]
	No.	Weight [kg]	
^{125}I-Trypsin-α-macroglobulin	259/77	39	6
	263/77	32	6.2
^{125}I-PSTI-trypsin-α-macroglobulin	119/74	17	5
	173/74	8	5.5
	189/74	20	4.5
	192/74	17	4.3
	122/74	17	4
^{125}I-Trasylol-trypsin-α-macroglobulin	128/74	25	4.2
	119/75	18	3.0
	120/75	16	4.0
	56/76	26	4.0

Elimination in vivo of ^{125}I-PSTI-trypsin-α_2-macroglobulin

A typical curve for the elimination of the radioactive substance of these complexes is seen in Fig. 2. The curve is exponential until 80% of the initial radioactive dose is eliminated, then becomes level just as in the case of trypsin-α-macroglobulin. The half-life is about 5 min (see table). Urinary radioactivity increased and was dialyzable. A plasma sample drawn 2 h after the administration of the complex was subjected to gel filtration on Sephadex G-50 (column 0.9 × 55 cm). The radioactivity was eluted with a lower K_{av} value than native PSTI.

Elimination in vivo of ^{125}I-Trasylol-trypsin-α-macroglobulin

The elimination curve for radioactivity of this complex exhibited a striking resemblance to the corresponding elimination of the other complexes. The elimination is exponential until about 83%

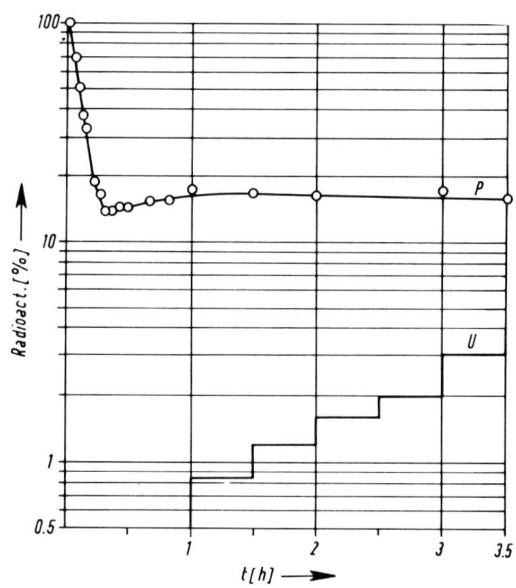

Fig. 2. Elimination of plasma radioactivity in dog after i.v. administration of dog ^{125}I-PSTI-trypsin-α-macroglobulin complexes corresponding to 2.5 m*l* of dog serum.

Ordinate: the radioactivity in plasma (P) is expressed as percentage of the value at 2 min after the injection. The cumulative radioactivity in the urine (U) is expressed as percentage of the total amount injected. Abscissa: time in hours.

of the initially administered radioactive dose is eliminated, and thereafter the level is only slightly reduced. The half-life is about 4 min (see table). Dialyzable radioactivity was recovered in the urine. Antibodies against Trasylol could not precipitate the radioactivity in plasma drawn 2 h after the injection of the complex nor could the radioactivity in the urine be precipitated by these antibodies. The plasma radioactivity was eluted with a lower K_{av} than that of native Trasylol when plasma was gel filtered on Sephadex G-50 (column 0.9 × 55 cm).

Discussion

This study confirms earlier results[19] that intravenous trypsin-α-macroglobulin complexes are cleared very rapidly from the circulating blood and in an earlier investigation[9] it was shown that these complexes are eliminated by the reticuloendothelial system.

In vitro studies show that Trasylol can inhibit most of the hydrolytic activity of the trypsin-α_2-macroglobulin complex[12] and PSTI demonstrates the same effect[13] representing a general tendency of low molecular weight substrates to gain access to the protease active sites. The present work indicates a rapid elimination plus short half-life of 4 to 5 min for trypsin-α-macroglobulin complexes with inhibited hydrolytic activity on low molecular substrates after reaction with Trasylol or PSTI (Fig. 2). This suggests that the hydrolytic activity of the trypsin-α_2-macroglobulin complexes against low molecular weight substrates is relatively unimportant for the elimination of the complexes.

When free PSTI is injected into the circulation some can be revealed in the urine as an active inhibitor[1,5] and when Trasylol is injected in the same way, about 10% can be found as a biologically active inhibitor in the urine[1,20]. Here no free inhibitor could be detected, but radioactivity bound to material of lower molecular weight than PSTI and Trasylol appeared in plasma and urine after the injection of ^{125}I-PSTI or ^{125}I-Trasylol complexes with trypsin-α-macroglobulin complexes. The cumulative appearance of radioactivity in urine is similar to that found during elimination of ^{125}I-trypsin-α-macroglobulin (Fig. 1, 2). This suggests that the two labelled inhibitors may be degraded by the reticuloendothelial system along with the entire trypsin-α-macroglobulin[9]. In contrast, the free low molecular weight inhibitors seem to be metabolized in the kidneys[1-4]. This is also the fate of the major part of PSTI injected as trypsin complexes as most of the trypsin is transferred to the plasma protease inhibitors and PSTI thus released[5]. Only a small percentage of the PSTI is rebound by the trypsin-α-macroglobulin complexes formed and thus probably degraded in the reticuloendothelial system.

Neither Trasylol nor PSTI in complex with the protease of trypsin-α-macroglobulin complexes has any observable effect on the elimination of the complex by the reticuloendothelial system. Further studies in vivo of these two low molecular weight protease inhibitors must take into consideration their elimination by the reticuloendothelial system when bound to protease-α-macroglobulin complexes.

This investigation was supported by the Swedish Medical Research Council (project no. B77-17X-03910-05B), the Medical Faculty, University of Lund and Bayer AG.

Literature

1 Fritz, H., Oppitz, K.-H., Meckl, D., Kemkes, B., Haendle, H., Schult, H. & Werle, E. (1969) *Hoppe-Seyler's Z. Physiol. Chem.* **350**, 1541 - 1550.

2 Just, M., Török, P. & Habermann, E. (1973) in Kininogenases-Kallikrein (Haberland, G.L. & Rohen, J.W., eds.) pp. 123 - 127, F.K. Schattauer Verlag, Stuttgart, New York.

3 Wunderer, G., Kummer, K., Fritz, H., Beress, L. & Machleidt, W. (1974) in Proteinase Inhibitors – 2nd Int. Conf. (Fritz, H., Tschesche, H., Greene, L.J. & Truscheit, E., eds.) pp. 277 - 281, Springer-Verlag, Berlin, Heidelberg, New York.

4 Strober, W. & Waldman, T.A. (1974) *Nephron* **13**, 35 - 66.

5 Eddeland, A. & Ohlsson, K. (1978) *Scand. J. Clin. Lab. Invest.*, in press.

6 Balldin, G. & Ohlsson, K. (1975) in X. Scand. Conf. Gastroent., Aarhus, Denmark.

7 Ohlsson, K., Ganrot, P.-O. & Laurell, C.-B. (1971) *Acta Chir. Scand.* **137**, 113 - 121.

8 Dyce, B. & Haverback, B.J. (1960) *Am. J. Gastroenterol.* **34**, 481 - 486.

9 Ohlsson, K. (1971) *Acta Physiol. Scand.* **81**, 296 - 272.

10 Bloth, B., Chesebro, B. & Svehag, S.-E. (1968) *J. Exp. Med.* **127**, 749 - 763.

11 Haverback, B.J., Dyce, B. & Bundy, H.F. (1962) *J. Clin. Invest.* **41**, 972 - 980.

12 Ganrot, P.-O. (1967) *Arkiv Kemi* **26**, 583 - 591.
13 Eddeland, A. & Ohlsson, K. (1976) *Scand. J. Clin. Lab. Invest.* **36**, 815 - 820.
14 Ohlsson, K. & Tegner, H. (1973) *Biochim. Biophys. Acta* **317**, 328 - 337.
15 Ohlsson, K. & Eddeland, A. (1975) *Gastroenterology* **69**, 668 - 675.
16 Laurell, C.-B. (1972) *Scand. J. Clin. Lab. Invest.* Suppl. (29) **124**, 21 - 37.
17 Thorell, J.I. & Johansson, B.G. (1971) *Biochim. Biophys. Acta* **251**, 363 - 369.
18 Erlanger, B.F., Kokowsky, N. & Cohen, W. (1961) *Arch. Biochem. Biophys.* **95**, 271 - 278.
19 Ohlsson, K. (1971) *Scand. J. Clin. Lab. Invest.* **28**, 219 - 223.
20 Vogel, R., Trautschold, I. & Werle, E. (1968) in Natural proteinase inhibitors, Academic Press, New York, London.

Dr. Allan Eddeland, Kirurgiska kliniken, Malmö allmänna sjukhus,
S-214 01 Malmö, Sverige.

Doc. Kjell Ohlsson, Kliniskt kemiska centrallaboratoriet, Malmö allmänna sjukhus,
S-214 01 Malmö, Sverige.

Hoppe-Seyler's Z. Physiol. Chem.
Bd. 359, S. 385 - 392, März 1978

A New Organ Spectrophotometer for Sensitive Dual-Wavelength Absorbance Measurement and Spectral Scanning of Intact Perfused Organs

Helmut SCHWAB and Helmut SIES

Institut für Physiologische Chemie, Physikalische Biochemie und Zellbiologie der Universität München

(Received 7 November 1977)

Summary:
1) A new spectrophotometer was designed and constructed for application in high sensitivity dual-wavelength absorbance and fluorescence measurement and spectral scanning (non corrected) from intact perfused organs, combining advantages of other instruments described in the literature.

2) The time-sharing principle was applied to gain high stability. All optical and electronic components except the light paths of the wavelength modulation system are common to both wavelengths. For this purpose the logarithmic stage was introduced before demodulation. Stability is about 0.001 A/h and noise about 0.001 A.

3) The performance of the instrument with a biological object, the isolated perfused rat liver, is demonstrated in the dual-wavelength mode and in the spectral scanning mode. Limitations in stability and in time resolution in such an application result from the properties of the biological object, not from the instrument.

Ein neues Organspektrophotometer zur Messung von Extinktionsdifferenzen im Zweiwellenlängenverfahren und von Spektren bei intakten perfundierten Organen

Zusammenfassung:
1) Ein neues Spektrophotometer, welches Vorteile von in der Literatur beschriebenen Geräten vereinigt, wurde zur Anwendung in der empfindlichen Zweiwellenlängen-Extinktions- und -Fluoreszenzmessung sowie der Differenz-Spektrenregistrierung (nicht korrigiert) von intakten perfundierten Organen entwickelt.

2) Zur Erzielung hoher Stabilität wurde das Time-Sharing-Prinzip eingesetzt. Alle optischen und elektronischen Teile außer den Lichtwegen des Wellenlängenmodulationssystems sind beiden Wellenlängen gemeinsam. Zu diesem Zweck wurde die Logarithmierstufe vor der Demodulation eingefuhrt. Die Stabilität ist etwa 0.001 A/h, das Rauschen 0.001 A.

3) Die Eigenschaften des Instruments bei Messungen mit einem biologischen Objekt, der isoliert durchströmten Rattenleber, werden im Zweiwellenlängenverfahren und bei Spektrenmessung demonstriert. Die Begrenzung bei Stabilität und Zeitauflösung in solch einer Anwendung liegt begründet in den Eigenschaften des Objekts, nicht in denen des Instruments.

Key words: Organ spectrophotometry; organ spectrofluorometry; perfused organs; perfused liver.

The increasing knowledge on properties of biological components in purified systems and in isolated subcellular organelles has been paralleled by the development, in recent years, of methods providing for measurements from intact organs and organisms. Sensitive detection of reactants and intermediates in the intact organ brings our understanding to a new level both with respect to their properties in situ and to their interrelations with other components. In this regard, optical methods have proven most useful, and indeed following the pioneering developments of suitable instrumentation for highly sensitive and rapid response measurement of light absorbance by Chance[1], there has been a continual development of specialized optical equipment, e.g. in the groups of Chance[2-4], Lübbers[5] and Bücher[6], Brauser[7], Hess[8] and Hollaway[9].

While emphasis was placed on the rapidity of measurement in some of these photometers, allowing for experiments in the range of ms and below, high time resolution is of little importance in the analysis of steady states in intact organs. For example, the transit time of extracellular markers in the isolated perfused liver is of the order of a few s. Based on the special requirements for an organ spectrophotometer and spectrofluorometer for selective steady state analysis, we have designed and constructed a new instrument capable of sensitive dual wavelength absorbance measurements as well as spectral scanning, combining advantages of the above mentioned designs.

A short account of this material has been presented[10].

A) Requirements for an organ spectrophotometer

As organs consist of a heterogeneous array of intracellular and extracellular compartments containing numerous different components (multicomponent system), the problem of assigning measured signals to transitions of specific components is solved (a) by biochemical information on the selectivity of the transition, and (b) by use of the dual wavelength technique and (c) absorbance and fluorescence difference spectra[10a], as has been exemplified in the organ photometry of catalase Compound I in the hemoglobin free perfused rat liver[11].

The requirements regarding dual wavelength measurement from the organ include stability of the apparatus for extended time periods of the order of an hour, sensitivity at an acceptable signal/noise ratio to detect small absorbance difference changes so that the necessity to measure at isosbestic points in the multicomponent system is avoided, and suitable mechanical and optical design to operate the photometer in conjunction with a perfusion system.

Further requirements in spectral scanning are reproducible spectral resolution and an on-line facility of spectral processing and presentation of results, which permit the experimenter to direct the course of the experiment on the basis of online responses. This requirement will become of particular importance in more refined experimentation, e.g. in steady-state-relaxation experiments.

B) Design

The principal arrangement (Fig. 1) of the photometer allows for the application of absorbance – in contrast to reflectance – spectrophotometry through the highly turbid object, based on the initial proposal of Brauser[7]. Short optical pathways of incident light were accomplished by flexible light guides and by use of end-on photomultipliers beneath the turbid object.

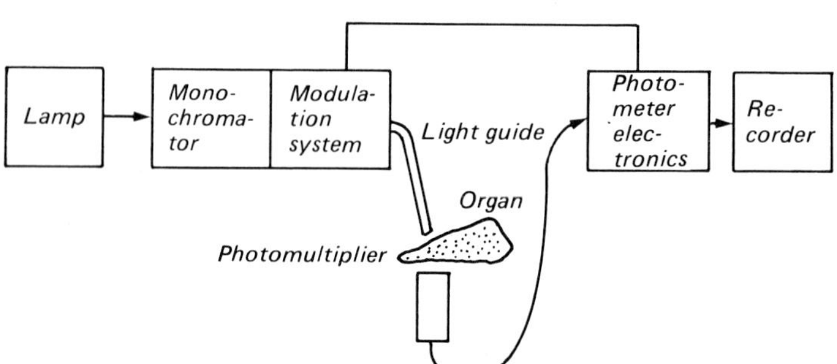

Fig. 1. Principal arrangement of the organ photometer.

1) Optical unit

A light source illuminates the entrance slit of a monochromator generating light of selectable wavelengths, λ_1 and λ_2, which is alternately transmitted to the organ by the time-sharing modulation system.

The range of wavelengths from 340 - 700 nm is covered by a 100 W tungsten iodine light source (power supply: model HR20-10, Trygon Electronics), that between 220 and 400 nm by a 150 W xenon arc lamp (Osram XBO 150 W/1). The xenon arc lamp power supply is a constant current supply of our own construction, the lamp current variations being $< 10^{-3}$ for 200 - 240 V line voltage. A quartz lens (f = 50 mm; 30 mm diameter; Infrasil, Schott) focuses light onto the entrance slit of the grating monochromator (Farrand No. 117, 182; aperture $f/3.5$, f = 160 mm, 220 - 700 nm; 570 lines/mm). The air-cooled lamp house and other optical parts are mounted on a sturdy cast iron machine table to achieve long-term stability.

2) Modulation system (Fig. 2)

A double exit slit at the monochromator selects two wavelengths, λ_1 and λ_2. The corresponding light energies enter two quartz rods of pyramidal form which direct them to two flexible quartz light guides. A chopper disc is interpositioned between rods and flexible guides to provide for light modulation at 120 Hz. A common light path is formed by a solid quartz rod of 5 mm diameter which directs light onto the surface of the organ.

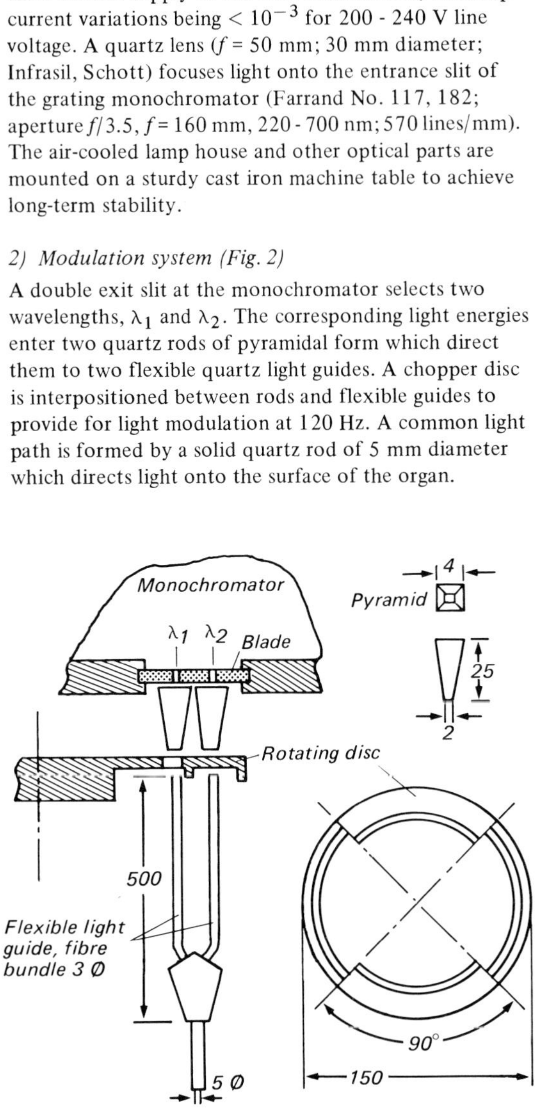

Fig. 2. Modulation System.
Lengths are given in mm.

The wavelength difference $\lambda_1 - \lambda_2$ as well as the spectral bandwidths are determined by a metal blade carrying appropriate slits. By this arrangement, a change of wavelength difference is simple and reproducible by exchange of the blade; blades are available from 10 nm up to 50 nm at spectral bandwidths of 1, 3, and 5 nm. Maximum wavelength difference is limited by the width of the entrance areas of the pyramidal rods, corresponding to 75 nm. Light intensities may be balanced by varying the length of the exit slits. Light modulation by the slits in the chopper disc is 100% complete. This is an advantage over systems using reflecting surfaces; modulation noise is negligible.

3) Photometer unit

The electronics of the photometer allow for three modes of operation as is indicated in the block diagram of Fig. 3.

a) Dual-wavelength mode

The dual-wavelength mode is the routine mode of operation of the photometer. While resembling the time-sharing apparatus of Chance et al.[3] by use of the rotating chopper disc, the difference is that a change of wavelength is rapid and almost unlimited, since interference filters are not required. However, the stability advantage of such a system is retained, since only one channel is required in the time-sharing system. Instabilities in the electronics and the optical system are canceled in a first-order approximation.

In the dual wavelength mode, the photomultiplier (EMI 9592B for visible spectral region; EMI 9601B for near UV and visible, in measurements at wavelengths less than 390 nm in conjunction with a Schott UG2 guard filter) is illuminated sequentially by the light penetrating the organ at wavelengths λ_1 and λ_2. The photomultiplier current enters a log stage (Analog Devices AD 752P log module with operational amplifier) which produces a rectangular waveform alternating between $\log I_1$ and $\log I_2$. The peak-to-peak amplitude of this waveform then is $\log I_1 - \log I_2 = \log (I_1/I_2)$, i.e. the absorbance difference ΔA to be measured. The introduction of the logarithmic stage before demodulation leads to a gain in stability since there is only one electronic channel for both wavelengths. Because this is a non-linear operation, a slight measurement error is introduced which depends on noise amplitude; however, this error may be dismissed from consideration. On the average, noise levels of A did not exceed 0.001 at the system output after bandwidth limitation to about 1 Hz; the bandwidth of the signal entering the logarithmic amplifier is about 1 kHz, so the noise at the log stage was 30-fold. Under such conditions, when measuring A = 0.1, the error was calculated to be about 0.1%, which is negligible.

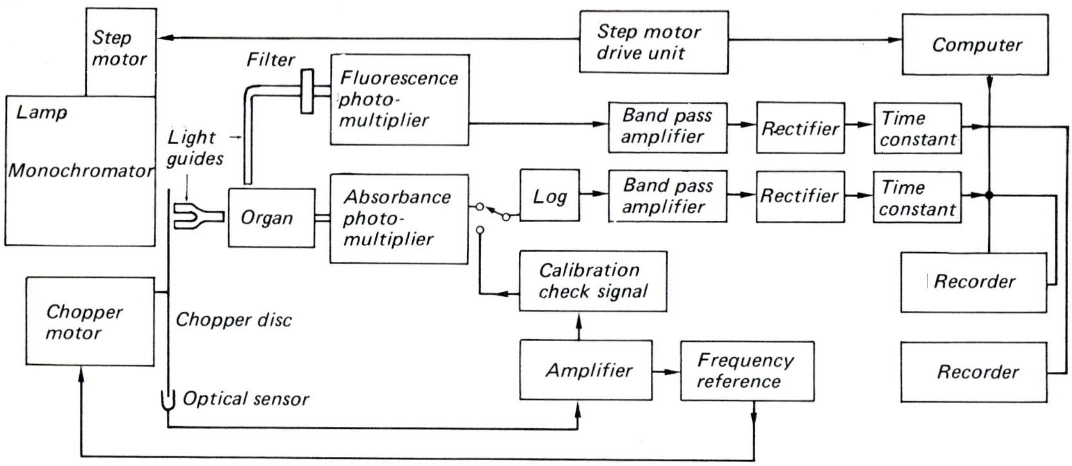

Fig. 3. Block diagram of organ spectrophotometer and fluorometer[10].

The log stage output signal is amplified by a band pass amplifier of 10 Hz bandwidth and is rectified. Chopping frequency is 120 Hz to exclude interference of line frequency (50 Hz) and its harmonics. It is stabilized at the filter mean frequency by feedback control.

Proper operation of the electronics can be checked during an experiment by application of a check signal. This signal has a constant amplitude at the actual chopping frequency as derived from the disc by an optical sensor. When applied to the electronics input, it results in a response of $A = 0.020$.

b) Spectral scanning mode

In the spectral scanning mode a single slit is employed at the monochromator exit, so that illumination of the object now alternates between light energy at the selected wavelength and zero energy. The log amplifier is placed after the time constant block in this mode. Spectral sweeps are performed by means of a step motor so that the photometer output delivers a signal of absorbance as a function of wavelength, containing the spectral characteristics of the lamp, the monochromator and other components of the optical system. In the presence of the organ in the light path, the total spectrum is recorded in the memory of an instrument computer (Nicolet, Model 1072) and is stored at 12 bit resolution. Following an experimental change, another spectrum representing the new condition is recorded, and the difference spectrum between this new spectrum and the initial (reference) spectrum is obtained by digital subtraction. The difference spectrum may be viewed on the screen of an oscilloscope or may be plotted by an xy-recorder or recorded on computer tape.

Since no reference beam is used in this mode, stability of the instrument is lower than in the dual wavelength mode; instabilities, however, were found to be less than 1% of the measured intensity. Light attenuation in the lobe of a perfused rat liver, for example, is on the order of 10^2 to 10^5. Nevertheless, the apparatus provides a signal of sufficiently low noise even in the ultraviolet so that single scans proved adequate. Thus, in routine use with the perfused organ, the sweep speed is 1 - 2 nm/s, and spectra are usually scanned twice to ensure steady state conditions.

When a blade with a double slit is used while scanning the wavelength, an approximation of the first derivative of a spectrum is obtained; only the absolute value of the derivative will be indicated, however.

c) Use as an organ spectrofluorometer

As the relative fluorescence yield of compounds provides information e.g. on their state of binding, the facility of recording the fluorescence intensity emitted from the surface of the organ at the site of the incident light beam is a useful addition. As is indicated in Fig. 3, a flexible light guide is used to feed the fluorescence intensity to a second photometer unit after passing appropriate cut-off filters. This allows for the simultaneous recording of fluorescence and absorbance, either in the dual wavelength mode or the spectral scanning mode.

Since its introduction in a single wavelength arrangement by Chance and Jöbsis[12], the surface fluorescence of pyridine nucleotides has been monitored from a variety of organs. Studies on the enhancement of pyridine nucleotide fluorescence in liver have been performed [13,14] and striking differences became evident in metabolic transitions involving the mitochondrial and cytosolic compartments. In other words, transitions comprising similar changes of reduced pyridine nucleotides in terms of nmol/g tissue may be represented by largely different changes in fluorescence intensity.

C) Performance

1) Characteristic properties of the apparatus

Stability measurements without an organ in the light path were performed using stained glass or wire mesh to introduce absorbance. Fig. 4 shows typical photometer outputs, the deviation of A being within $\pm 2 \times 10^{-4}$ for several hours with the tungsten iodine lamp (Fig. 4a). With the xenon arc lamp there are spikes of statistical occurrence (Fig. 4b), probably caused by arc fluctuations on the cathode surface. Thus, stability with the xenon arc lamp is about an A value of 10^{-3}. Although inferior in this regard to the tungsten iodine lamp, the xenon lamp is preferred in the UV range and at high turbidity because of its higher light intensity. When the xenon lamp is used with low values of absorbance, the noise contributed by the lamp exceeds that contributed by the photomultiplier. At an A value of 6

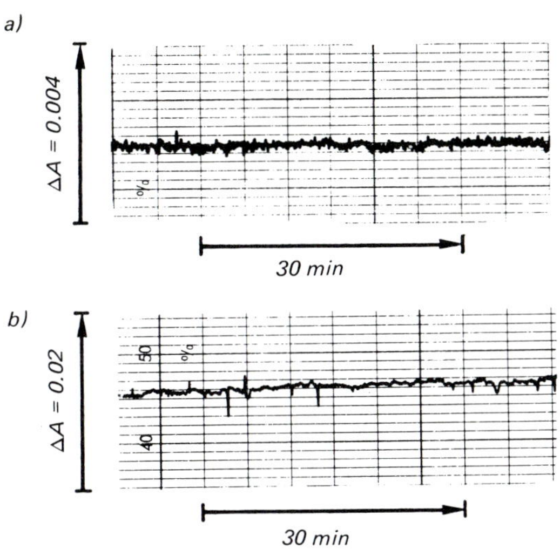

Fig. 4. Stability measurement of organ spectrophotometer at 490 - 510 nm, with tungsten iodine lamp (a) and xenon arc lamp (b).

Bandwidth, 5 nm; time constant, 1 s; wire mesh absorbance, 3.

(see below), however, both contributions to instability approach each other.

The second major parameter is photomultiplier noise which at the photocathode surface is determined by the shot noise relationship[15],

$$I_n = \sqrt{2\,eIB}$$

where

I_n = noise current,
e = charge of electron (1.6×10^{-19} As),
I = cathode current,
B = electrical bandwidth of the system.

The noise amplitude as a fraction of mean cathode current is

$$I_n/I = \sqrt{2\,eB/I} = \sqrt{2\,eB\;10^A/I_{max}}$$

or

$$\log\,(I_n/I) = 1/2\,\log\,(2\,eB/I_{max}) + 1/2\,A$$

where I_{max} is the maximum photomultiplier cathode current obtained with a fictitious object of zero absorbance and A is the absorbance of a given object. Thus, on a double logarithmic scale there is a linear relationship between noise amplitude and A, valid for any photometer with I_{max} and B being the parameters. I_{max} depends on the optical properties of the instrument. Measurements of the noise level at different values of object absorbance were performed, and in the table the data are presented as the value of object absorbance yielding a noise level of 0.001 A. Commercial spectrophotometers allow less absorbance than the organ photometer, thus limiting the possibility of adapting a commercial instrument to

Table. Absorbance of objects that will result in a (photomultiplier) noise value of 0.001 A.

Measurements were carried out using absorbance by wire mesh or tinted glass; spectral bandwidth, $\Delta\lambda$ = 3 nm, electrical bandwidth, B = 1 Hz, and measurement wavelength, λ = 500 nm.

Photometer	Object absorbance A at 0.001 A noise
This organ photometer, tungsten iodine lamp	5
Commercial photometers Shimadzu UV 300	4
Cary 118	2

the application of organ photometry at high absorbance.

The above consideration applies to the dual wavelength mode with an assumed bandwidth of 1 Hz. The noise level will be higher by a factor of 3 in the spectral scanning mode where B = 10 Hz.

2) Performance of the organ photometer with isolated hemoglobin free perfused rat liver

a) Dual wavelength recordings

The instrument is capable of detecting small experimental changes of absorbance difference in the presence of a high basal absorbance. This has been documented in applications extending from the near-infrared to the near-ultraviolet regions of the spectrum[10,14,16,17] with selective recordings of catalase-H_2O_2 (Compound I), the cytochromes of the respiratory chain, and the reduced nicotinamide nucleotides, respectively. Basal absorbance exhibited by a lobe of perfused liver of approx. 4 mm thickness rises from an A value of about 1 - 2 in the red to 4 - 5 in the region of the dihydroband of nicotinamide nucleotides. Actual noise is approx. 0.001 A in the latter case, and actual drift is determined not by the instrument (0.001 A/h) but rather by slight changes in the positioning of the organ on the perfusion table; however, usually the actual drift observed is approaching the instrument drift.

An example of a dual wavelength recording from a lobe of perfused liver is presented in Fig. 5. The degree of reduction of the terminal oxidase of the endoplasmic reticulum monooxygenase system, cytochrome P-450, is recorded against time at the wavelength pair 450 - 465 nm in the presence of indicator amounts (3.5 vol %) of carbon monoxide in the gas mixture equilibrating with the perfusate. An increased steady state reduction of cytochrome P-450 ensues upon addition of a drug substrate, aminopyrine, to the perfusate. Following the addition of an inhibitor, 6-chloro-1,2,3-benzothiadiazole, an insecticide synergist, the aminopyrine effect is no longer observable. The recording demonstrates the time-course of the inhibitor effect.

The performance is similar to other types of organ photometer mentioned above, the advantage of this apparatus being the versatility of wavelengths due to the monochromator, the output being directly proportional to absorbance difference, and the stability of the system (no phasing or amplitude problems as known in other systems).

b) Spectral scanning

In spectral scanning in the presence of the organ, a storage and processing facility is employed, and usually difference spectra representing the absorbance difference observed between two metabolic states are of interest, although first derivative spectra and other special types of display may be obtained as well. A difference spectrum obtained from the perfused liver in the presence of sodium cyanide (reference: before and after the addition) in the α, β-band region of the cytochromes is shown in Fig. 6; the components represented are, from the red towards the blue, catalase-HCN, cytochrome oxidase, cytochromes b, c_1 and c. A

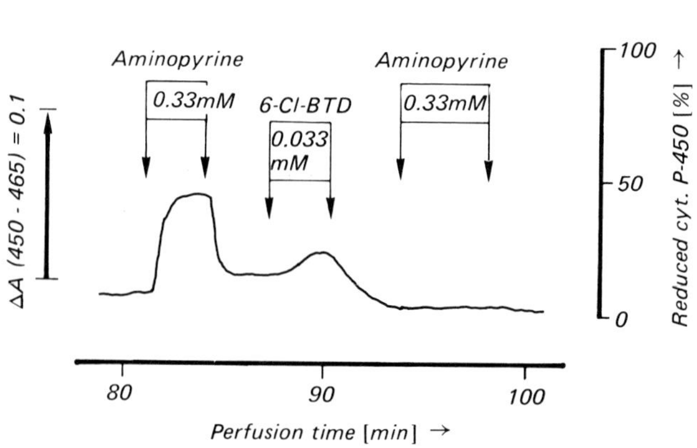

Fig. 5. Absorbance difference ΔA (450 - 465), recorded from the lobe of a perfused liver from a phenobarbital-pretreated rat.

Changes are observed upon addition of a drug substrate, aminopyrine, and an inhibitor, 6-chloro-1,2,3-benzothiadiazole. The signal represents the degree of reduction of cytochrome P-450. Carbon monoxide was present in gas mixture at 3.5 vol. %. Full reduction (100%) was obtained by N_2 anoxia and infusion of 0.2mM sodium dithionite (experiment in collaboration with Dr. *L. Gil*).

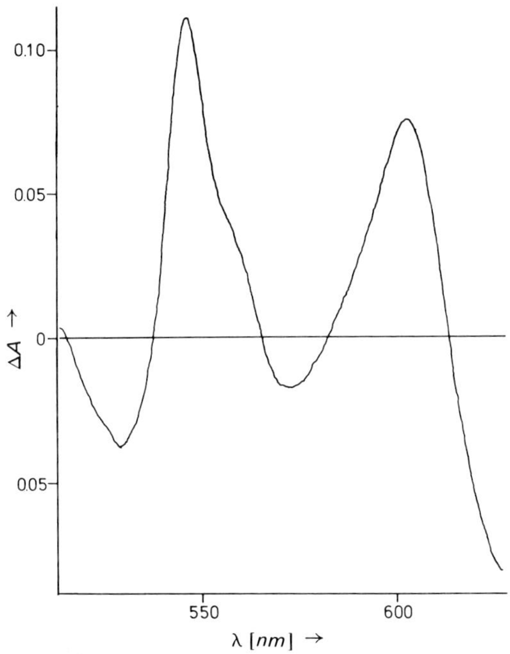

Fig. 6. Liver absorbance difference spectrum 2 min after addition of 0.8mM sodium cyanide to perfusate[10]. Spectra before and after the cyanide addition (not recorded) were located at the line of zero absorbance difference. Signal corresponds to approx. 20×10^6 cells.

sequence of difference spectra representing progressive steady states of hypoxia is shown in Ref.[16].

It is clear that spectral identification of a component in a metabolic transition of the organ is an indispensable requisite before application of the dual wavelength method. Thus, a time-consuming systematic scanning with the dual wavelength mode as was applied in identification of the catalase-H_2O_2 complex[18] is now unnecessary, as has recently been shown for identification of the cytochrome P-450 halothane complex in the intact perfused liver[19].

Since in this instrument a second compensating channel is not included, baseline drifts resulting from slight changes in the positioning of the liver on the perfusion table as well as from energy output of the system are reflected by approximately parallel lines in the difference spectrum because drift is usually small during the time required for one scan.

c) Calibration and linearity

A check of the calibration of the instrument is performed as indicated above. A special problem in measurement with an organ, however, is the calibration of a measured change in terms of optical pathlength, or ultimately of nmol/g tissue. A solution to this problem is given by the use of absorbent (and in determination of fluorescence yields, also fluorescent) compounds known to be non-penetrant, so that a measure of the extracellular space is obtained; the latter may be determined independently by use of labeled compounds such as inulin. We have used dansylated serum albumin for optical calibration of the absorbance and fluorescence signals, and a detailed description of the procedure has been given[14]. For a theoretical analysis of pathlength determination, see Ref.[20].

An intrinsic check of linearity is given by infusion of compounds known to bind to intracellular sites. An example is the binding curve of cyanide to catalase heme, resulting in a linear double reciprocal plot and in a dissociation constant similar to that of isolated catalase[21].

D) Conclusions and outlook

The development of a compact and high intensity spectrophotometer suitable for the special needs of organ spectrophotometry has opened a new level of analytical approach, i.e. the investigation of cellular constituents during the actual occurrence of metabolic processes like biosyntheses etc. The photometer properties are such that the limitation of the method resides in the stability and reproducibility of the organ and, except for small organs with rapid transit times and during non-stationary conditions, the two modes of performance are applicable. Thus, future development of the wavelength modulation system used here, i.e. a step-motor, into a more rapid wavelength scanning system, will permit spectral resolution in transition states, too. The basic goal, the provision of a high stability and low noise readout of optical signals from intact perfused liver or other organs, has been satisfactorily achieved as shown in the examples of Figs. 5, 6 and the cited references.

The expert electronic and mechanical work in constructing the instrument was provided by *H. Strobel*, *H. Erk* and *K. Metzner*. Thanks are extended to *Th. Bücher* and *B. Brauser* for their continued interest and encouragement. Supported by *Deutsche Forschungsgemeinschaft*, Sonderforschungsbereich 51 „Medizinische Molekularbiologie und Biochemie“, Grants F/1 and D/8.

Literature

1 Chance, B. (1951) *Rev. Sci. Instrum.* **22**, 634 - 638.
2 Chance, B., Graham, N., Sorge, J. & Legallais, V. (1972) *Rev. Sci. Instrum.* **34**, 62 - 71.
3 Theorell, H., Chance, B., Yonetani, T. & Oshino, N. (1972) *Arch. Biochem. Biophys.* **151**, 434 - 444.
4 Chance, B., Legallais, V., Sorge, J. & Graham, N. (1975) *Anal. Biochem.* **66**, 498 - 514.
5 Niesel, W., Lübbers, D. W., Schneewolf, D., Richter, J. & Botticher, W. (1964) *Rev. Sci. Instrum.* **35**, 578 - 581.
6 Schnitger, H., Scholz, R., Bücher, Th. & Lübbers, D.W. (1965) *Biochem. Z.* **341**, 334 - 339.
7 Brauser, B. (1968) *Z. Anal. Chem.* **237**, 8 - 17.
8 Hess, B., Kleinhans, H. & Schlüter, H. (1970) *Hoppe-Seyler's Z. Physiol. Chem.* **351**, 515 - 531.
9 Hollaway, M. R. & White, H. A. (1975) *Biochem. J.* **149**, 221 - 231.
10 Sies, H. & Schwab, H. (1977) *Adv. Exp. Med. Biol.* **94**, 119 - 126.
10a Brauser, B., Bücher, Th., Sies, H. & Versmold, H. (1972) in Molecular Basis of Biological Activity (Gaede, K., Horecker, B. L. & Whelan, W. J., eds.) pp. 197 - 219, Academic Press, New York.
11 Sies, H. (1974) *Angew. Chem.* **86**, 789 - 801; *Int. Ed. Engl.* **13**, 706 - 718.
12 Chance, B. & Jöbsis, F. F. (1959) *Nature (London)* **184**, 195 - 196.
13 Bücher, T., Brauser, B., Conze, A., Klein, F., Langguth, O. & Sies, H. (1972) *Eur. J. Biochem.* **27**, 301 - 317.
14 Sies, H., Häussinger, D. & Grosskopf, M. (1974) *Hoppe-Seyler's Z. Physiol.* **355**, 305 - 318.
15 RCA Technical Manual PT-60 (1963) Phototubes and Photocells, RCA, Lancaster, PA.
16 Sies, H. (1977) *Hoppe-Seyler's Z. Physiol. Chem.* **358**, 1021 - 1032.
17 Sies, H. (1978) *Methods Enzymol.* **51**, 48 - 59.
18 Sies, H. & Chance, B. (1970) *FEBS Lett.* **11**, 172 - 175.
19 Nastainczyk, W., Ullrich, V. & Sies, H. (1978) *Biochem. Pharmacol*, in press.
20 Wodick, R. & Lübbers, D. W. (1973) *Pflügers Arch.* **342**, 29 - 40.
21 Sies, H., Bücher, Th., Oshino, N. & Chance, B. (1973) *Arch. Biochem. Biophys.* **154**, 106 - 116.

Dr. H. Schwab and Priv.-Doz. Dr. H. Sies, Inst. f. Physiologische Chemie, Physikalische Biochemie u. Zellbiologie, Goethestr. 33, D-8000 München 2.

Hoppe-Seyler's Z. Physiol. Chem.
Bd. 359, S. 393 - 398, März 1978

Substrate Specificity of Neuraminidase from *Erysipelothrix rhusiopathiae*

Hubertus VON NICOLAI, Hans E. MÜLLER und Fritz ZILLIKEN

Institut für Physiologische Chemie der Universität Bonn und Staatliches Medizinaluntersuchungsamt Braunschweig

(Received 26 October 1977)

Summary: *Erysipelothrix rhusiophatiae*, a non-sporulating gram-positive bacillus, causes swine erysipelas and sometimes local inflammations in man. Depending on the culture conditions, it produces several neuraminidase fractions of different molecular size. This paper deals with the isolation of the lower molecular main fraction from the culture medium and its purification by gel chromatography on Sephadex G-100. The enzyme possesses a molecular weight of about 48 000 and a pH optimum of 6.2. The K_m values are 9×10^{-4}M for the $\alpha 2 \rightarrow 3$ linkage in II^3NeuAc-Lac and 5×10^{-3}M for the $\alpha 2 \rightarrow 6$ linkage in II^6NeuAc-Lac.

In order to obtain model substrates of known structure with various types of *O*-glycosidic linkages, a simple and time saving method for the isolation of NeuAc containing oligosaccharides from human and cow's milk is described.
The *Erysipelothrix* enzyme exhibits its highest activity on II^3NeuAc-Lac. Taking the rate of cleavage of this substrate as 100, the corresponding rate for the isomeric II^6NeuAc-Lac is 20 and the value for the $\alpha 2 \rightarrow 8$ linkage of $II^3(NeuAc)_2$-Lac is also 20. Gangliosides and colominic acid are very poor substrates, their rates of cleavage being below 5. Depending on their individual structures, the more complex oligosaccharides are cleaved in the range of 10 to 50. Similarly, high molecular glycoproteins reach values of 25 to 50. Only the synthetic substrate NeuAc$\alpha 2 \rightarrow$(3′-methoxyphenyl)-glycoside is hydrolyzed to the same extent as II^3NeuAc-Lac. In general, the enzyme shows a specificity pattern similar to that found in other bacteria of low pathogenicity towards man.

Substratspezifität einer Neuraminidase aus Erysipelothrix rhusiopathiae

Zusammenfassung: *Erysipelothrix rhusiopathiae* ist ein nicht-sporenbildendes grampositives mikroaerophiles Stäbchenbakterium, das bei Schweinen Rotlauf erzeugt und beim Menschen gelegentlich lokal begrenzte und spontan abheilende Infektionen (Erysipeloide) verursacht. Je nach Kulturbedingungen und Inkubationsdauer produziert *Erysipelothrix* in vitro mehrere Neuraminidasen

---

Enzyme: Neuraminidase, acylneuraminosyl hydrolase (EC 3.2.1.18).

Abbreviations: NeuAc = N-Acetylneuraminic acid; II^3NeuAc-Lac = NeuAc$\alpha 2 \rightarrow 3$Gal$\beta 1 \rightarrow 4$Glc; II^6NeuAc-Lac = NeuAc$\alpha 2 \rightarrow 6$Gal$\beta 1 \rightarrow 4$Glc; $II^3(NeuAc)_2$-Lac = NeuAc$\alpha 2 \rightarrow 8$NeuAc$\alpha 2 \rightarrow 3$Gal$\beta 1 \rightarrow 4$Glc; IV^3NeuAc-$LcOse_4$(Pentaose a) = NeuAc$\alpha 2 \rightarrow 3$Gal$\beta 1 \rightarrow 3$GlcNAc$\beta 1 \rightarrow 3$Gal$\beta 1 \rightarrow 4$Glc; III^6NeuAc-$LcOse_4$(Pentaose b) = Gal$\beta 1 \rightarrow 3$GlcNAc-(6$\leftarrow$2αNeuAc)$\beta 1 \rightarrow 3$Gal$\beta 1 \rightarrow 4$Glc; IV^6NeuAc-n$LcOse_4$(Pentaose c) = NeuAc$\alpha 2 \rightarrow 6$Gal$\beta 1 \rightarrow 4$GlcNAc$1 \rightarrow 3$Gal$\beta 1 \rightarrow 4$Glc; IV^3NeuAc, III^6NeuAc-$LcOse_4$ = NeuAc$\alpha 2 \rightarrow 3$Gal$\beta 1 \rightarrow 3$GlcNAc(6$\leftarrow$2αNeuAc)$\beta 1 \rightarrow 3$Gal$\beta 1 \rightarrow 4$Glc.

unterschiedlicher Molekülgröße und Aktivität, von denen die Hauptkomponente der niedrigermolekularen Fraktion näher untersucht wurde. Die Reinigung des Enzyms erfolgte nach Isolierung aus dem Kulturmedium durch Gelfiltration an Sephadex G-100. Das Präparat besitzt ein Molekulargewicht von etwa 48000 und ein pH-Optimum bei 6.2. Die Michaeliskonstanten (K_m) betragen für II^{3}NeuAc-Lac 9×10^{-4}M und für II^{6}NeuAc-Lac 5×10^{-3}M.

Von allen untersuchten Substraten wird II^{3}NeuAc-Lac am leichtesten gespalten. Setzt man die Spaltungsgeschwindingkeit dieses Substrates gleich 100, so ergeben sich für II^{6}NeuAc-Lac und für II^3(NeuAc)$_2$-Lac Werte um 20, für III^{6}NeuAc-LcOse$_4$ etwa 8, für IV^{6}NeuAc-nLcOse$_4$ ungefähr 28 und für IV^{3}NeuAc, III^{6}NeuAc-LcOse$_4$ annähernd 50. Hingegen werden Colominsäure und Ganglioside mit < 5 wenig angegriffen. Hochmolekulare Glykoproteine und Mucine erreichen ebenfalls Werte zwischen 25 und 50. Lediglich das synthetische Substrat NeuAcα2→(3'-methoxyphenyl)-glykosid wird fast ebenso schnell gespalten wie II^{3}NeuAc-Lac. Das Enzym zeigt somit ein Spezifitätsprofil, wie es auch bei anderen, wenig menschenpathogenen Keimen gefunden wurde.

Key words: Neuraminidase, purification, substrate specificity, neuraminosyl oligosaccharides, *Erysipelothrix rhusiopathiae.*

Neuraminidases cleave α-ketosidic linkages between *N*-acetylneuraminic acid and the carbohydrate residues of oligosaccharides, glycolipids, and glycoproteins of body fluids and cell tissues. These enzymes have been found in viruses[1], bacteria[2], protozoa[3], and vertebrates[4].

Depending on their origin and the analytical conditions, neuraminidases show differences in specificity towards α2→3, α2→6, and α2→8 glycosidic linkages of NeuAc to the adjacent carbohydrate molecule. Even the structures of the oligosaccharide residues and type and size of the aglykons (i.e. lipid or protein) play a role in the enzyme-substrate interactions[5,6].

Neuraminidases have been discussed as pathogenic factors in microbial infections[7]. They are also known to unmask antigenic sites on tumor glycoproteins and are already used in cancer immunotherapy[8]. In order to elucidate the molecular basis of neuraminidase-antigen interactions, it seems to be reasonable to investigate the different neuraminidases for their substrate specificities towards model substrates of known structure.

Milk of many mammalian species, especially human milk, contains a great number of more or less complex NeuAc containing oligosaccharides[9-12]. Human milk is therefore an excellent source for substrates of neuraminidases. A simple and rapid isolation procedure for these compounds is described in this communication.

Erysipelothrix rhusiopathiae (insidiosa) is a gram-positive nonsporulating microaerophilic bacillus. In man it sporadically causes local inflammations known as erysipeloid. Neuraminidase of *E. rhusiopathiae* was first discovered by Müller and coworkers[13,14]. Depending on the culture medium and the cultivation time, they observed two neuraminidase fractions with molecular weights of 200000 and about 50000.

The present paper deals with the low molecular form. However, further isoenzymes or subunits of both fractions seem to exist.

Materials and Methods

Isolation of Erysipelothrix rhusioparthiae neuraminidase

The cultivation of the bacteria has been described elsewhere[15]. The supernatant of 120-h cultures of *E. rhusiopathiae* (Strain 119) was dialyzed 3 times for 12 h against 0.1M sodium phosphate buffer pH 6.5. The undialyzable residue was concentrated in a collodium filter tube to 1/10 of its initial volume. The solution was fractionated by gel filtration on a Sephadex G-75 column (50 × 2.6 cm). Neuraminidase activity was identified by incubation of 100 μl 3.0×10^{-3}M II^{3}NeuAc-Lac in water with 100 μl of each column fraction for 14 h. The amount of NeuAc released was estimated by the thiobarbiturate method according to Aminoff[16]. A high molecular weight fraction of low enzyme activity, which was probably particle-bound, was discarded. The main peak was concentrated again and rechromatographed in barbital acetate buffer (pH 6.0) according to Michaelis[17] on a Sephadex G-100 column (90 x 2.6 cm). The estimation of the molecular weight of the enzyme was performed by loading the column with calibration standards including cytochrome *c*, chymotrypsinogen A, egg white albumin, serum albumin, aldolase and ferritin.

Isolation of NeuAc containing oligosaccharides

About 1 *l* of milk (i.e. human or bovine colostrum) was centrifuged at 3000 x*g* for 30 min at 4 °C to remove the fat. The aqueous layer was treated with an equal volume of precooled acetone. The denatured protein was allowed to settle overnight and was then removed by centrifugation at 3000 x*g* for 1 h. The clear yellowish supernatant was concentrated on a rotary evaporator to 1/10 of its original volume. The acetone-free residue was centrifuged again, applied to a Sephadex G-15 column (30 x 10 cm), and eluted with H_2O. NeuAc-containing fractions were identified by the direct Ehrlich reaction, pooled and dialyzed 3-times against 5 *l* H_2O for 12 h at 4 °C. Both the dialyzable and the non-dialyzable solutions were collected, concentrated, and lyophilized.

In the case of human milk, the dialyzable part contains oligosaccharides ranging from hexaoses. More complex NeuAc- (and partly L-fucose-) containing glycosides and glycopeptides remain in the undialyzable residue. Bovine colostrum contains dialyzable oligosaccharides only, i.e. II^3NeuAc-Lac, II^6NeuAc-Lac, and II^3(NeuAc)$_2$-Lac, respectively.

About 10 g of the lyophilized white powder was dissolved in 50 m*l* of 0.05M Tris buffer, pH 7.2, and applied to a column of DEAE-Sephadex (30 x 8 cm). The column was eluted with a linear gradient of 3.0 *l* of 0.0 - 0.5M sodium chloride in Tris buffer solution. The NeuAc-containing fractions were identified as described, desalted on Sephadex G-10, and lyophilized again.

The dialyzable part of human milk shows three Ehrlich positive fractions, Fract. I contains IV^3NeuAc-LcOse$_4$, III^6NeuAc-LcOse$_4$, and IV^6NeuAc-nLcOse$_4$. Fract. II represents mainly II^6NeuAc-Lac with II^3NeuAc-Lac as minor constituent. Fract. III is part of the semi-dialyzable IV^3NeuAc, III^6NeuAc-LcOse$_4$ which is generally found in the undialyzable fraction.

The undialyzable part consists of at least seven different NeuAc-containing fractions[18]. The first three seem to be more or less complex glycopeptides. Fract. IV, V, and VI represent mixtures of more than three different oligosaccharides not yet identified but possibly identical with some described by Grimmonprez[11] or Kobata et al.[10,12]. Fract. VII is identical with IV^3NeuAc, III^6NeuAc-LcOse$_4$, first isolated also by Grimmonprez and coworkers[19]. The separation of the NeuAc-containing oligosaccharides was performed by preparative paper chromatography on Whatman No. 3 in a solvent mixture of pyridine/ethyl acetate/acetic acid/water 5 : 5 : 1 : 3 (v/v) for 36 or 48 h, respectively. The substances were identified by staining with alkaline triphenyltetrazolium chloride and by comparative chromatography with authentic samples, kindly provided by Dr. *A. Gauhe*, Heidelberg.

Compared with lactose (R_F = 1.00), the R_F values are 0.61 for II^3NeuAc-Lac, 0.38 for II^6NeuAc-Lac, 0.32 for II^3(NeuAc)$_2$-Lac, 0.24 for IV^3NeuAc-LcOse$_4$, 0.21 for III^6NeuAc-LcOse$_4$, 0.17 for IV^6NeuAc-nLcOse$_4$, and 0.11 for IV^3NeuAc,III^6NeuAc-LcOse$_4$, respectively.

After proper localization, the paper chromatograms were cut into strips containing the separated oligosaccharides, which were then eluted with water. The effluents were concentrated, further purified on small Sephadex G-15 columns, and finally lyophilized and stored in a desiccator.

Substrates of other origin

Colominic acid was a generous gift of Dr. *Y. Uchida*, Uji, Japan, and purified fetuin was kindly provided by Dr. R. Drzeniek, Hamburg. Human α_1-acid glycoprotein was a gift of the Behringwerke, Marburg. Human urinary glycoprotein was isolated as described by Tamm and Horsfall[20]. Carboxymethylated II^3NeuAc-Lac was prepared as described elsewhere[21].

NeuAc2→(3'-methoxyphenyl)-glycoside, NeuAcβ2-methyl glycoside, and bovine submaxillary glycoprotein (mucin) were purchased from Boehringer Mannheim GmbH. Bovine brain ganglioside mixture was from Sigma Chemical Co.

Neuraminidase assay

Neuraminidase activity was determined by measuring the amount of NeuAc released according to a slightly modified procedure of Aminoff[16]. Substrates were dissolved in barbital/acetate buffer, pH 6.0, so as to give a total of ≙50 μg glycosidically bound NeuAc in each test. Additional buffer and 10 μ*l* of enzyme preparation were added to a final volume of 200 μ*l*. Incubation times were from 15 min up to 24 h at 37 °C. With NeuAcα2→(3'-methoxyphenyl)-glycoside as substrate, the enzyme activity was tested according to Palese et al.[22].

Results and Discussion

In contrast to other bacterial neuraminidases, the enzyme of *Erysipelothrix rhusiopathiae* shows a non-homogenous elution pattern of at least three active peaks on Sephadex G-100, indicating the existence of isoenzymes or partial dissociation

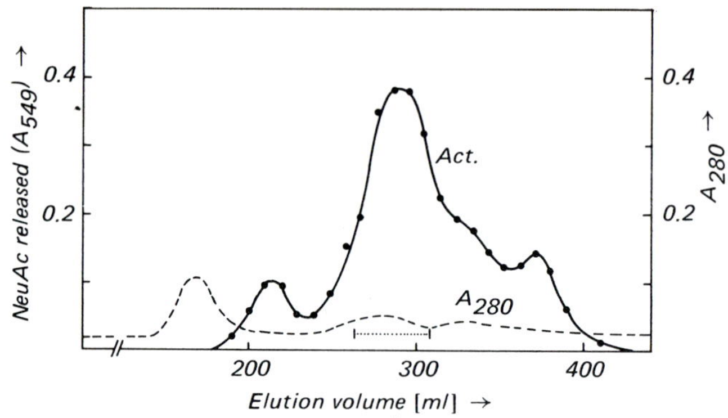

Fig. 1. Elution pattern of neuraminidase from *Erysipelothrix rhusiopathiae* on Sephadex G-100.

Column 50 × 2.5 cm, elution buffer 0.03M barbital/acetate pH 6.0. Determination of NeuAc by the thiobarbiturate method according to Aminoff[16]. Solid line = enzyme activity, i.e. NeuAc released, measured at A_{549}; broken line = protein absorption at 280 nm; dotted line = enzyme fraction used for substrate specificity investigations.

into subunits (Fig. 1). After subtraction of the higher and lower molecular neuraminidase activities, the main fraction represents a 4-fold enrichment of the enzyme by gel chromatography. As determined on a calibrated column, the molecular weight of the main peak is in the region of 48 000. The enzyme shows a pH optimum of 6.0 - 6.5 for different substrates in 0.03M barbital/acetate buffer. As plotted by the method of Lineweaver and Burk[23], the K_m values are 9×10^{-4}M for II³NeuAc-Lac and 5×10^{-3}M for II⁶NeuAc-Lac. The substrates with two NeuAc residues of different linkage type, i.e. II³(NeuAc)₂-Lac and IV³NeuAc, III⁶NeuAc-LcOse₄, show fictitious K_m values of 1.5×10^{-3}M.

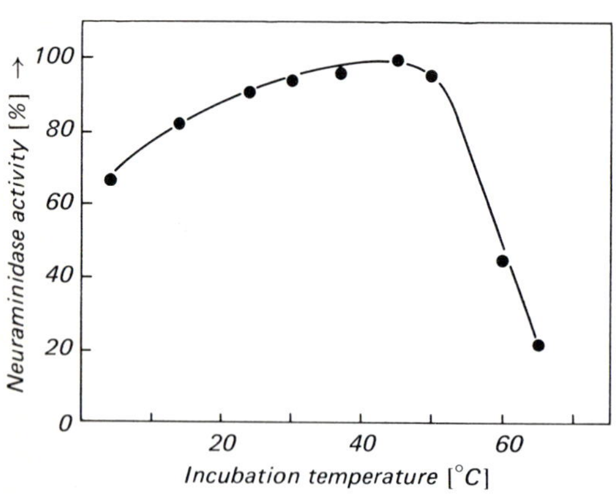

Fig. 2. Temperature dependence of optimum neuraminidase activity.

Experimental conditions: see Materials and Methods. Substrate: human serum; incubation time: 20 min.

Using II^3NeuAc-Lac as substrate, the enzyme preparation has a spec. act. of 65 μkat/g protein; it is therefore in the range of the commercially available neuraminidases from the highly pathogenic bacteria *Clostridium perfringens* and *Vibrio cholerae.*

The enzyme shows a remarkable temperature stability with an optimum efficiency at 45 °C (Fig. 2), a fact that could be of significance for the pathomechanism of erysipeloid inflammations.

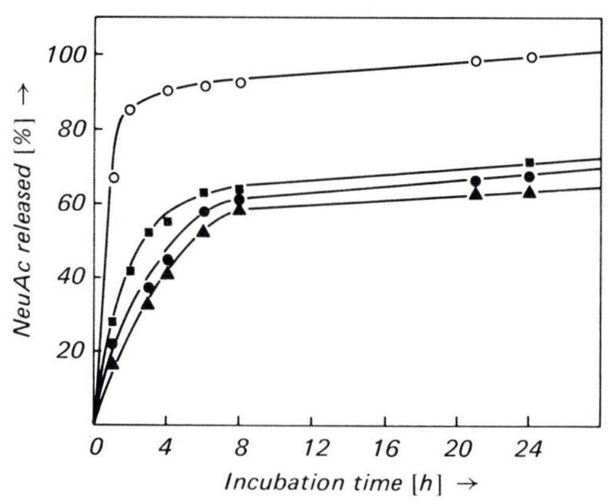

Fig. 3. Release of NeuAc from different substrates, percentage of the total amount of glycosidically bound NeuAc.

Incubation mixture: 190 μl substrate solution (≙50 μg NeuAc) in barbital/acetate buffer pH 6.0, incubated with 10 μl enzyme preparation at 37 °C. Determination of NeuAc as described.

○——○ II^3NeuAc-Lac; ●——● II^6NeuAc-Lac; ▲——▲ $II^3(NeuAc)_2$-Lac; ■——■ IV^3NeuAc, III^6NeuAc-$LcOse_4$.

As shown in Fig. 3, II^{3}NeuAc-Lac is readily cleaved up to 100% whereas II^{6}NeuAc-Lac, II^3(NeuAc)$_2$-Lac, and IV^{3}NeuAc, III^{6}NeuAc-LcOse$_4$ show cleavage rates not higher than 60% within 24 h.

This finding is in close agreement with earlier results on the substrate specificities of Newcastle disease virus and fowl plaque virus neuraminidases, which also cleave $\alpha 2 \rightarrow 3$ linkages preferentially[24].

In order to get a "profile" of the substrate specificity of *Erysipelothrix* neuraminidase on different types of substrates, the relative cleavage rates of NeuAc were plotted together in Fig. 4: II^{3}NeuAc-Lac was set to 100. Compared to this reference substrate, the $\alpha 2 \rightarrow 6$ and $\alpha 2 \rightarrow 8$ linkages of the other NeuAc-Lac type substrates are cleaved to an extent of only 20. A considerable difference can also be seen between IV^{6}NeuAc-LcOse$_4$, III^{6}NeuAc-nLcOse$_4$, and IV^{3}NeuAc, III^{6}NeuAc-LcOse$_4$. III^{6}NeuAc-nLcOse$_4$ is cleaved nearly three times as fast as IV^{6}NeuAc-LcOse$_4$. The close proximity of the inner $\alpha 2 \rightarrow 6$ linked NeuAc residue to the outer one in $\alpha 2 \rightarrow 3$position suggests an inhibitory effect for the cleavage rate of the outer $\alpha 2 \rightarrow 3$ linkage, slowing down the velocity of hydrolysis compared with II^{3}NeuAc-Lac. Colo-

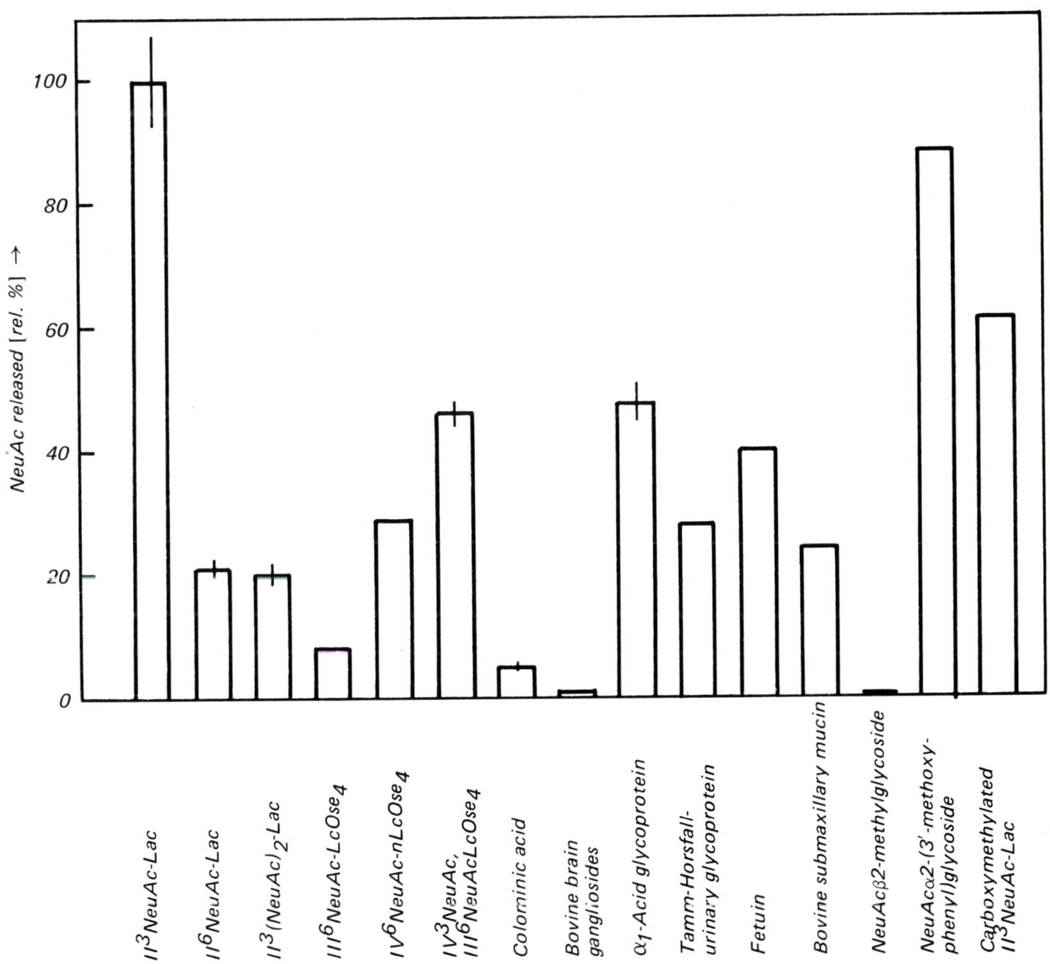

Fig. 4. Substrate specificity pattern of *Erysipelothrix rhusiopathiae* neuraminidase on different substrates, expressed as percentage of the cleaveage rate of II^{3}NeuAc-Lac.

At the abscissa the substrates are indicated. Analytical conditions: see Fig. 3; incubation time: 4 h.

minic acid and gangliosides are fairly resistant. Among high molecular weight substrates, the globular α_1-acid glycoprotein (orosomucoid) and fetuin are cleaved half as fast as II³NeuAc-Lac, mucous substrates i.e. human urinary glycoprotein of Tamm and Horsfall and bovine submaxillary mucin give values in the range of 20. Considering the different modes of determination, the affinity of the enzyme to the synthetic NeuAc2→(3'-methoxyphenyl)glycoside is the same as for II³NeuAc-Lac. The turnover rate of the semi-synthetic carboxymethylated II³NeuAc-Lac is about 60, indicating that the negatively charged free carboxylate group plays a role in the formation of the enzyme-substrate complex.

As shown with NeuAcβ2-methylglycoside, *Erysipelothrix* neuraminidase is, like all the other neuraminidases so far known, strictly specific for α-glycosidic linkages.

For more common substrates, standard deviations have been calculated, which show a highly significant difference ($2p < 0.0001$) between the α2→3- and the α2→6-linkage of the isomeric NeuAc-lactoses. Substrate specificity patterns like these are typical for neuraminidases from low or non-pathogenic bacteria, i.e. *Streptococcus viridans**, *Bifidobacterium bifidum var. pennsylvanicus*[25], which also cleave α2→3 linkages of low molecular oligosaccharide type substrates preferentially. In contrast, neuraminidases from bacteria of higher pathogenicity, i.e. *Vibrio cholerae, Clostridium perfringens,* and *Clostridium septicum*[28], show higher affinities mainly to high molecular weight substrates of body fluids, i.e. mammalian serum glycoproteins like orosomucoid or fetuin, respectively.

This work was supported by the *Deutsche Forschungsgemeinschaft* (Ni 147/3 and Sonderforschungsbereich 54).

* Nicolai, v. H., Müller, H. E. & Zilliken, F., unpublished results.

Literature

1 Drzeniek, R. (1972) *Curr. Top. Microbiol. Immunol.* **59**, 35 - 75.
2 Müller H. E. (1974) *Behring Inst. Mitt.* **55**, 34 - 56.
3 Müller, H. E., Nicolai, v. H. & Zilliken, F. (1974) *Pathol. Microbiol.* **41**, 323 - 333.
4 Gielen, W. (1974) *Behring Inst. Mitt.* **55**, 85 - 88.
5 Gottschalk, A. & Drzeniek, R. (1972) in Glycoproteins, their Composition, Structure and Function (Gottschalk. A., ed.) pp. 381 - 402, Elsevier Publishing Company, Amsterdam-London-New York.
6 Drzeniek, R. (1974) *Behring Inst. Mitt.* **55**, 1 - 10.
7 Müller, H. E. (1976) *Zbl. Bakt. Hyg., I. Abt. Orig. A* **235**, 106 - 110.
8 Sedlacek, H. H., Seiler, F. R. & Schwick, H. G. (1977) *Klin. Wochenschr.* **55**, 199 - 214.
9 Nicolai, v. H. (1971) Dissertation, Bonn.
10 Kobata, A. (1972) *Methods Enzymol.* **XXVIII**, 262 - 271.
11 Grimmonprez, L. & Montreuil, J. (1975) *Biochimie* **57**, 695 - 701.
12 Yamashita, K., Tachibana, Y. & Kobata, A. (1977) *Arch. Biochem. Biophys.* **182**, 546 - 555.
13 Müller, H. E. (1971) *Pathol. Microbiol.* **37**, 241 - 248.
14 Müller, H. E. & Böhm, K. H. (1973) *Zbl. Bakt. Hyg., I. Abt. Orig. A* **223**, 220 - 227.
15 Krasemann, C. & Müller, H. E. (1975) *Zbl. Bakt. Hyg., I. Abt. Orig. A* **231**, 206 - 213.
16 Aminoff, D. (1961) *Biochem. J.* **81**, 384 - 391.
17 Michaelis, L. (1931) *Biochem. Z.* **234**, 139 - 141.
18 György, P., Jeanloz, R. W., Nicolai, v. H. & Zilliken, F. (1974) *Eur. J. Biochem.* **43**, 29 - 33.
19 Grimmonprez, L., Bouquelet, S., Bayard, B., Spik, G., Monsigny, M. & Montreuil, J. (1970) *Eur. J. Biochem.* **13**, 484 - 492.
20 Tamm, I. & Horsfall, F. L. (1952) *J. Exp. Med.* **95**, 71 - 97.
21 Nicolai, v. H. (1976) Habilitationsschrift, Bonn.
22 Palese, P., Bucher, D. & Kilbourne, E. D. (1973) *Appl. Microbiol.* **25**, 195 - 201.
23 Lineweaver, M. & Burk, D. J. (1934) *J. Am. Chem. Soc.* **56**, 658 - 666.
24 Nicolai, v. H. Drzeniek, R. & Zilliken, F. (1971) *Z. Naturforsch.* **26b**, 1049 - 1051.
25 Nicolai, v. H. & Zilliken, F. (1974) *Behring Inst. Mitt.* **55**, 78 - 84.
26 Nicolai, v. H. & Zilliken, F. (1976) Tenth Intl. Congress of Biochemistry, Abstr. 16 - 7 - 280, Hamburg.
27 Nicolai, v. H., Werner, H. & Zilliken, F. (1977) *Hoppe-Seyler's Z. Physiol. Chem.* **358**, 283 - 284.
28 Nicolai, v. H., Müller, H. E. & Zilliken, F. (1977) *Hoppe-Seyler's Z. Physiol. Chem.* **358**, 1252.

Priv.-Doz. Dr. H. v. Nicolai und Prof. Dr. F. Zilliken, Institut für Physiologische Chemie, Nußallee 11, D-5300 Bonn,

Prof. Dr. Dr. H. E. Müller, Staatliches Medizinaluntersuchungsamt, Hallerstraße 1, D-3300 Braunschweig.

Hoppe-Seyler's Z. Physiol. Chem.
Bd. 359, S. 399 - 406, März 1978

On the Significance of Sialic Acid in High Affinity 5-Hydroxytryptamine uptake by Synaptosomes

Gerhard A. DETTE and Wolfgang WESEMANN

Institut für Physiologische Chemie der Universität Marburg

(Received 20 September/8 December 1977)

Summary: Synaptosomes isolated from rat brain cortex incorporated [^{14}C]5-hydroxytryptamine at 37 °C with high affinity. An apparent transport constant of K_t = 50nM was found. The high affinity uptake was decreased by treatment of synaptosomes with neuraminidase from *Vibrio cholerae* or *Clostridium perfringens* prior to incubation with [^{14}C]5-hydroxytryptamine. The inhibition was related to the amount of sialic acid released, with a K_i value of 3.5μM. A non-competitive type of inhibition was observed after treatment with neuraminidase. The inhibition caused by ouabain could not be enhanced by simultaneous treatment with neuraminidase. Neuraminidase did not lower the activity of ($Na^{\oplus}$ + $K^{\oplus}$)-ATPase or $Mg^{2\oplus}$-ATPase. These results suggest that sialic acid is involved in the 5-hydroxytryptamine uptake mechanism without functional linkage to the energy pump of the membrane, which maintains the sodium gradient necessary for 5-hydroxytryptamine transport.

Zur Abhängigkeit der Serotoninaufnahme hoher Affinität vom Sialinsäuregehalt der Synaptosomen

Zusammenfassung: Synaptosomen, die aus Großhirnrinde der Ratte isoliert wurden, zeigten bei 37 °C eine [^{14}C]5-Hydroxytryptamin-Aufnahme mit hoher Affinität (K_t = 50nM). Inkubation der Synaptosomen mit Neuraminidase aus *Vibrio cholerae* oder *Clostridium perfringens* verminderte bei nachfolgender Inkubation mit [^{14}C]5-Hydroxytryptamin die Aufnahme hoher Affinität. Die Aufnahmehemmung war eine Funktion der abgespaltenen Neuraminsäure. Es wurde der Typ der nicht-kompetitiven Hemmung mit einem K_i-Wert von 3.5μM gefunden. Die durch Ouabain hervorgerufene Hemmung konnte durch gleichzeitige Inkubation mit Neuraminidase nicht verstärkt werden. Weder die Aktivität der ($Na^{\oplus}$ + $K^{\oplus}$)-ATPase noch die der $Mg^{2\oplus}$-ATPase wurde durch Neuraminidase erniedrigt. Diese Ergebnisse weisen darauf hin, daß Neuraminsäure am Transportmechanismus des Serotonins beteiligt ist, ohne jedoch die Ionenpumpe der Membranen zu beeinflussen, die den für die Serotoninaufnahme erforderlichen Natriumionengradienten aufrechterhält.

Key words: 5-Hydroxytryptamine, synaptosomes, neuraminidase, rat brain cortex, uptake kinetics.

Enzymes: Adenosinetriphosphatase (ATPase), ATP phosphohydrolase (EC 3.6.1.3.);
Neuraminidase, acylneuraminyl hydrolase (EC 3.2.1.18).

Abbreviation: NeuAc, *N*-acetylneuraminic acid.

The uptake of biogenic amines with putative neurotransmitter function into the presynaptic terminal has been suggested to represent the termination mechanism of synaptic transmission. The specific, saturable, high affinity uptake of 5-hydroxytryptamine by synaptosomes is restricted to low concentrations in the surrounding medium and is masked at high doses by passive processes[1,2]. Thus the uptake process in mammalian brain of amines like serotonin (5-hydroxytryptamine) was investigated using brain slices, homogenates or pinched-off nerve endings (synaptosomes). Although parameters like energy dependence, kinetics, and structural requirements of 5-hydroxytryptamine analogues have been well studied[3-5], little is known about the chemical nature of the 5-hydroxytryptamine uptake mechanism. In studies on this topic, particular attention should be paid to the composition of nerve ending membranes. The membranes of synaptosomes are rich in glycoproteins and glycosphingolipids containing neuraminic acid[6-8]. Because of the low pK value (2.6) of sialic acid, it is conceivable that this negatively charged molecule may be of importance for membrane conformation. Functional roles of sialic acid may include transport processes and binding of the 5-hydroxytryptamine molecule which is positively charged at physiological pH. However, the exact significance of sialic acid in membranes is not known. In the present paper we describe the effects of sialic acid release on the high affinity uptake system of 5-hydroxytryptamine in synaptosomes.

Materials and Methods

5-Hydroxy-3-(2-amino-[1-^{14}C]ethyl)-indole ([^{14}C]5-hydroxytryptamine) creatinine sulfate,(specific activity 58 mCi/mmol), was obtained from Amersham Buchler, Braunschweig, Germany. Unlabeled 5-hydroxytryptamine creatinine sulfate was purchased from Merck, Darmstadt, Germany. Purified neuraminidase from *Vibrio cholerae* (activity 500 U/m*l* according to Mohr and Schramm[9] and *Clostridium perfringens* (activity 0.6 U/m*l*, 1 unit (U) liberates 1/μmol of sialic acid (NeuAc) from mucine per min at 37 °C) was kindly provided by Behringwerke AG, Marburg, Germany, or purchased from Boehringer Mannheim GmbH, Germany. The neuraminidase was free of protease activity as measured with benzoyl-DL-arginin-*p*-nitroanilide as substrate[10]. Tris-ATP was from Sigma Chemicals, St. Louis, U.S.A., and NCS tissue solubilizer from Nuclear Chicago, Heusenstamm, Germany. All other chemicals were of analytical grade.

Animals and dissection

Male Wistar II rats (160 - 180 g) with free access to diet and water were killed by decapitation. The brains were rapidly removed and placed into ice-cold 0.32 M sucrose. After excision of cerebellum and medulla oblongata, the hemispheres were separated by longitudinal section. White matter and basal ganglia were quickly removed from the greyish cerebral cortex by the use of a curved forceps. All preparation steps and the following isolation were performed at + 4 °C.

Preparation of synaptosomes

Purified cortex synaptosomes were isolated from cortex homogenates by differential and discontinuous sucrose density gradient centrifugation as described by Bradford[11]. The synaptosomal fraction was collected from the 0.8 - 1.2 M sucrose layer, dropwise, carefully diluted with an equal volume of ice-cold double distilled water, and centrifuged for 10 min at 34 800 x *g*. The resulting pellet was resuspended in 0.32 M sucrose (freshly prepared) or Krebs phosphate buffer (124 mM NaCl, 5 mM KCl, 20 mM Na_2HPO_4, 1.2 mM KH_2PO_4, 1.3 mM $MgSO_4$, 0.75 mM $CaCl_2$) containing 10 mM glucose as energy source. This synaptosomal suspension (0.6 to 1.0 mg protein/m*l*) was kept in an ice-water bath until further use. The purity of the preparations was tested by electron microscopy.

Uptake studies

The synaptosomal suspensions were preincubated for 15 min at 37 °C in a shaking incubator (shaking rate: 80 strokes per min, amplitude of the stroke: 1.5 cm) in the absence and presence of neuraminidase from *Vibrio cholerae* or *Clostridium perfringens.* Neuraminidase was used at varying concentrations, ranging from 0.012 to 0.12 U/mg of synaptosomal protein (see Results!). In some experiments ouabain (1 mM) was present. During the period of preincubation the suspensions were gassed with carbogen (95% CO_2, 5% O_2) until the pH was 7.2 Then, synaptosomal samples (500 μl) were added to temperature-regulated microtubes containing different amounts of [^{14}C]5-hydroxytryptamine (10 μl) to give final concentrations ranging from 5 to 100 nM. The samples were incubated at 37 °C for 3 min, since previous studies[12] had shown that after 3 min of uptake, the rate was still linear (Fig. 2). – Linearity of the uptake with time could not be assessed for samples treated with neuraminidase, since the enzyme could not be removed completely from the synaptosomal fraction. After two washing procedures by means of ultracentrifugation, NeuAc was still split off as was shown by positive thiobarbituric acid assay[13]. – The uptake was stopped by adding formaldehyde at a final concentration of 1.3% (w/v) and rapid chilling to + 4 °C as described by Costa and Murphy[14]. Chilling alone was shown to be inadequate for rapid inhibition of uptake[15]. Correction for non-specific adsorption ana diffusion was done by incubating samples in parallel at + 4 °C. These blank values were substracted from the total uptake values at 37 °C in order to obtain the specific uptake.

After cooling on ice, the microtubes were centrifuged for 4 min at 12 000 x *g*, + 2 °C. The supernatants were discarded, the pellets were dissolved in 500 *μl* of NCS by incubation for 2 h at + 45 °C, and the solution was transferred to counting vials containing 15 m*l* of Bray's solution[16]. Radioactivity was determined by liquid scintillation spectrometry.

Release studies

For release studies the synaptosomal suspension (20 m*l*, 0.60 to 0.70 mg protein/m*l*) was preincubated at + 37 °C for 5 min and gassed with carbogen. Then [^{14}C]5-hydroxytryptamine (10 *μl*) at a final concentration of 0.1*μ*M was added in the presence of the monoamine oxidase inhibitor iproniazide (25*μ*M) and the incubation continued for a further 5 min. The uptake reaction was stopped by rapid chilling and centrifugation (2 °C, 17 800 x *g*, 5 min). The supernatant was discarded. The pellet was washed twice with the original volume of ice-cold Krebs phosphate buffer and finally resuspended in the same volume of Krebs buffer, containing neuraminidase (0.11 U/mg protein). In some experiments cocaine (0.1mM) was added simultaneously with neuraminidase. Portions of 500 *μl* were incubated at + 4 °C and + 37 °C for 0, 5, 10, 15 min and prepared for counting as described above. Controls without enzyme were treated in the same way.

Determination of sialic acid

The amount of enzymatically released sialic acid was calculated from the total NeuAc content still present in the synaptosomal aliquots (2.5 m*l*) after incubation in presence or absence of neuraminidase. The samples were quickly chilled to + 4 °C in an ice bath and the protein was precipitated by ice-cold trichloroacetic acid at a final concentration of 5% (w/v). After centrifugation at + 4 °C for 10 min (about 6 500 x *g*) the clear supernatant was discarded and the pellet resuspended and hydrolysed in 0.1N H_2SO_4 (4.5 m*l*) for 60 min at 80 °C. Thereafter the samples were centrifuged again and the resulting supernatants were analysed for free NeuAc by the method of Aminoff[13]. Standards of crystalline NeuAc, prepared as described by Wesemann and Zilliken[17] were used as reference. The values given were not corrected for degradation of free NeuAc during hydrolysis by 0.1N H_2SO_4.

ATPase determinations

The activity of ATPase was measured by the amount of inorganic phosphate liberated during 5 min incubations of ATP with synaptosomes. 100-*μl* samples of synaptosomes, resuspended in 0.32M sucrose (0.80 mg protein/m*l*) were added to the following Tris/HCl (50mM, pH 7.2 at 37 °C) buffered medium (final concentrations in 600 *μl*) for assaying total ATPase: 10mM KCl, 115mM NaCl and 5 mM $MgCl_2$. $Mg^{2⊕}$-ATPase was assayed in presence of 1 mM ouabain. ($Na^⊕$ + $K^⊕$)-activated ATPase was calculated as difference between total and $Mg^{2⊕}$-ATPase activity. After a preincubation period of 5 min at 37 °C, the reaction was started with 4mM Tris-ATP. The reaction was stopped by the addition of 100 *μl* trichloroacetic acid (10% final concn., w/v). Inorganic phosphate was measured according to Bonting et al.[18].

Other assays

Binding of [^{14}C]5-hydroxytryptamine to neuraminidase was tested by charcoal assay and equilibrium dialysis as described elsewhere[19].

Thin-layer chromatography of 5-hydroxytryptamine and 5-hydroxyindoleacetic acid were performed according to Kuhar et al.[20] as described elsewhere[21].

Protein was determined by the method of Lowry et al.[22] using bovine serum albumin as a standard.

Electron microscopy

The samples were fixed at + 4 °C for 1 h in glutaraldehyde (3%), postfixed in osmium tetroxide (1%) and embedded in Epon. The sections were examined in a Siemens Elmiskop 101. The authors are grateful to *Dr. N. Paul*, Institut für Cytobiologie, Marburg, for carrying out these examinations.

Results

High affinity uptake of 5-hydroxytryptamine

Synaptosomes isolated from rat brain cortex incorporate [^{14}C]5-hydroxytryptamine by means of a saturable uptake mechanism (Fig. 1). In the concentration range of 10 to 100nM, only a high affinity uptake of 5-hydroxytryptamine is observed. The apparent uptake constant was calculated from a double reciprocal plot [accumulated 5-hydroxytryptamine]$^{-1}$ versus [free 5-hydroxytryptamine]$^{-1}$ according to Lineweaver and Burk[23] and found to be K_t = 50nM. The linear dose-related and non-saturable part found after incubation at + 4 °C represents non-specific binding.

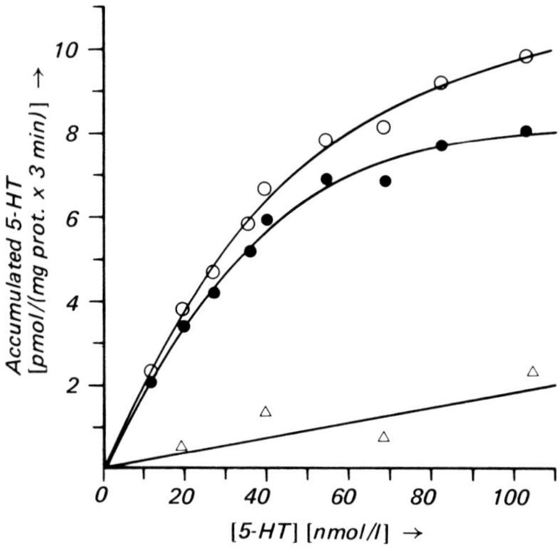

Fig. 1. 5-Hydroxytryptamine (5-HT) uptake as a function of concentration.

Synaptosomal suspensions (0.50 mg protein/m*l*) were incubated at 37 °C for 3 min with final [^{14}C]5-hydroxytryptamine concentrations ranging from 10 to 100nM. Non-specific uptake (△——△) measured at 4 °C was subtracted from total (○——○) to give specific uptake (●——●). The values represent means of triplicate determinations (SD ± within 5% range).

Inhibition of 5-hydroxytryptamine uptake after incubation with neuraminidase

The type of inhibition and the kinetic parameters were evaluated by Lineweaver-Burk and Dixon plots[23,24]. At a given concentration of neuraminidase (0.11 U/mg protein) and varied [^{14}C]5-hydroxytryptamine concentrations, specific 5-hydroxytryptamine uptake was diminished, resembling a non-competitive type of inhibition. Fig. 3 illustrates that the maximal uptake rate V was decreased by neuraminidase treatment to about 50 per cent without any change in uptake affinity as shown by the same K_t value. Binding of 5-hydroxytryptamine to neuraminidase has been excluded by equilibrium dialysis and charcoal assay.

The effect of neuraminidase on 5-hydroxytryptamine uptake was related to the amount of sialic acid released. In Fig. 4 the data are plotted according to Dixon[24] with inhibitor concentrations replaced by molar concentrations of liberated sialic acid. The apparent K_i value for sialic acid release was calculated as 3.5 µM. This graph again confirms the similarity of the decrease in 5-hydroxytryptamine uptake due to neuraminidase treatment to non-competitive inhibition in enzymology.

ATPase activity

Accumulation of [^{14}C]5-hydroxytryptamine was reduced by ouabain (1mM) to 20.6% of control (Fig. 5). This inhibition could not be enhanced by simultaneous treatment with neuraminidase from *Vibrio cholerae* (210 U/mg synaptosomal protein) or *Clostridium perfringens* (0.045 U/mg). Even at higher concentrations of the enzyme (0.12 U/mg protein), which reduced 5-hydroxytryptamine uptake by about 50%, no greater effect could be achieved than with ouabain alone.

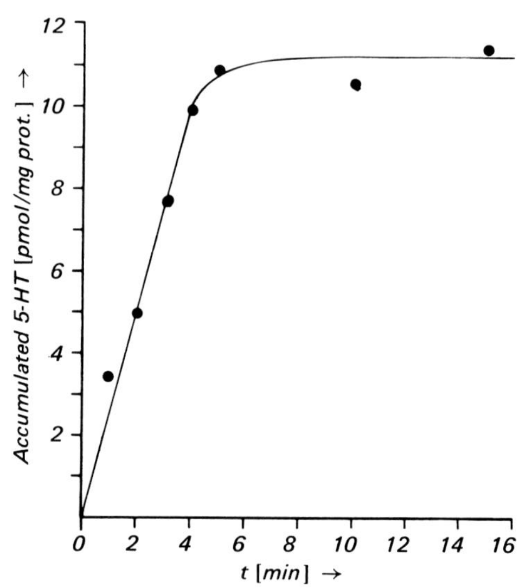

Fig. 2. Specific 5-hydroxytryptamine (5-HT) uptake as a function of time.

Synaptosomal aliquots (0.4 mg/m*l*) were incubated at 37 °C and 4 °C with [^{14}C]5-hydroxytryptamine (final concentration 100nM) for the time periods indicated. Nonspecific uptake measured at 4 °C was subtracted from total (incubation at 37 °C) to obtain the specific uptake given in the figure. The points are means of triplicate (SD ± within the 5% range) determinations.

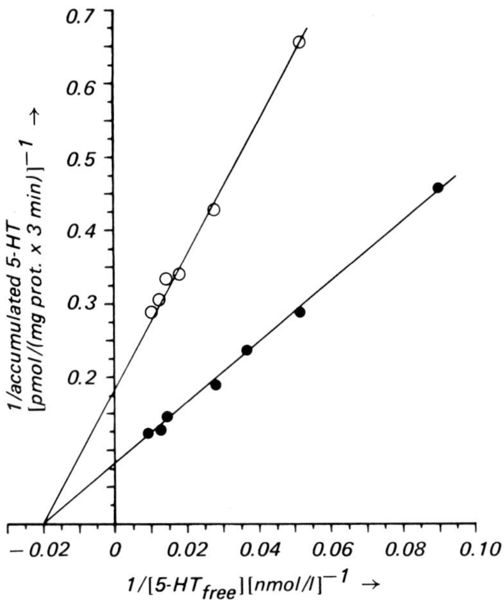

Fig. 3. Inhibition of 5-hydroxytryptamine (5-HT) uptake by neuraminidase treatment.

Lineweaver-Burk plot. Synaptosomes were incubated at 37 °C for 15 min in presence (○——○) or absence (●——●) of *Clostridium perfringens* neuraminidase (0.11 U/mg synaptosomal protein) prior to incubation, 3 min, 37 °C, with various [^{14}C]5-hydroxytryptamine concentrations (10 to 100nM). The values represent means of triplicate determinations (SD ± within the 5% range).

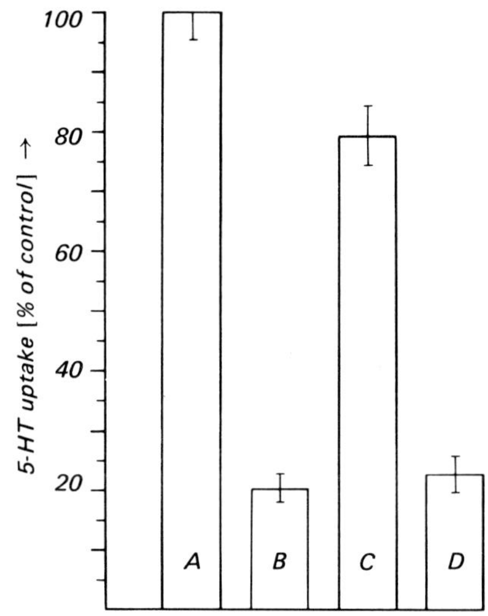

Fig. 5. Inhibition of 5-hydroxytryptamine (5-HT) uptake after incubation with ouabain and neuraminidase.

Samples of synaptosomes were preincubated at 37 °C for 15 min (control, A); with 1mM ouabain (B); neuraminidase (*Vibrio cholerae,* 210 U/mg synaptosomal protein) (C); or simultaneously with the same concentration of ouabain and neuraminidase (D). This pretreatment was followed by a 3 min incubation period at 37 °C with [^{14}C]5-hydroxytryptamine, 0.1μM. The values given are the residual [^{14}C]5-hydroxytryptamine uptake in per cent of control (A) [9.8 pmol/(mg protein x 3 min)] and are the means of 5 - 6 experiments ± SD.

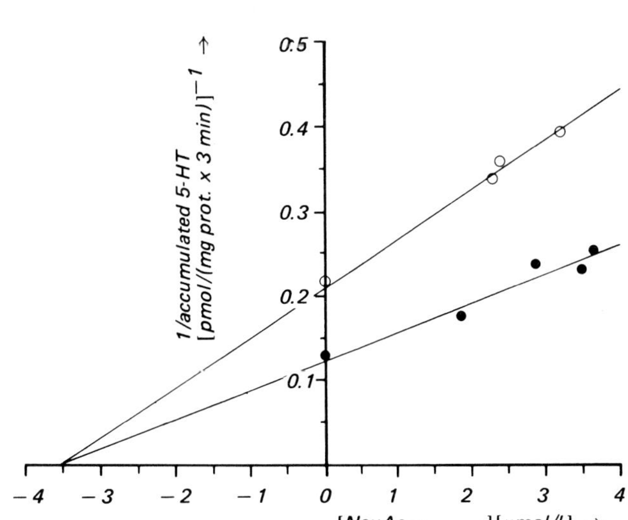

Fig. 4. Inhibition of 5-hydroxytryptamine (5-HT) uptake as a function of released sialic acid.

Dixon plot. Synaptosomes were treated with different concentrations of neuraminidase from *Clostridium perfringens* (0.012 to 0.12 U/mg synaptosomal protein) at 37 °C for 15 min followed by incubation with [^{14}C]5-hydroxytryptamine 30nM (●——●) and 100nM (○——○) for 3 min. The values represent means of triplicate determinations (SD ± within the 5% range).

In a separate set of experiments the ($Na^{\oplus}$ + $K^{\oplus}$)-ATPase and the $Mg^{2\oplus}$-ATPase (in the presence of $Na^{\oplus}$, $K^{\oplus}$, $Mg^{2\oplus}$, and 1mM ouabain) were determined. Neither the activity of the ($Na^{\oplus}$ + $K^{\oplus}$)-ATPase, [6.6 μmol P_i/(mg protein × h)] nor the activity of the $Mg^{2\oplus}$-ATPase [10.1 μmol P_i/(mg protein × h)] was decreased by 15 min preincubation with neuraminidase applied in the same concentrations as given above. Contamination of neuraminidase with free phosphate in the ATPase assay has been excluded by blanks with neuraminidase alone, treated in the same way as the samples.

Effect of neuraminidase on 5-hydroxytryptamine release

Fig. 6 gives the time dependence of spontaneous 5-hydroxytryptamine release. After an incubation time of 15 min about 55% of previously incorporated [^{14}C]5-hydroxytryptamine was set free in the controls. As could be shown by thin-layer chromatography, [^{14}C]5-hydroxytryptamine was not metabolized in the presence of the monoamine oxidase inhibitor iproniazid. As the incubation proceeded, release was slightly increased by neuraminidase treatment and exceeded the liberation from untreated synaptosomes by about 20% at 15 min. This additional release was not affected by cocaine (100μM) present at the same time.

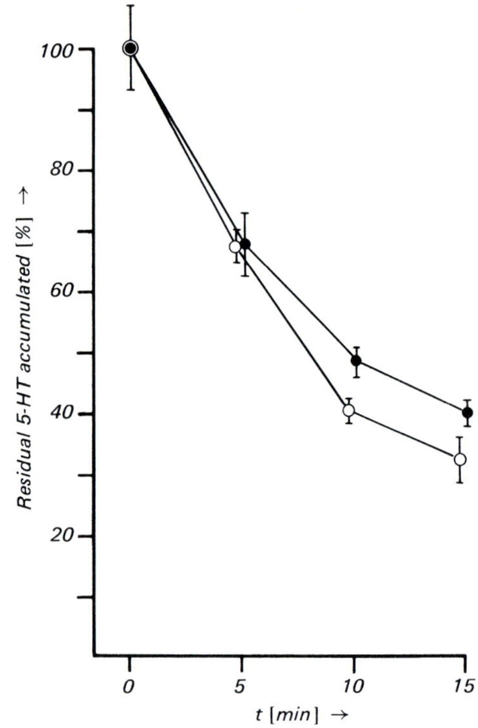

Fig. 6. Release of accumulated [^{14}C]5-hydroxytryptamine (5-HT) as a function of time.
Synaptosomes were incubated with 0.1μM [^{14}C]5-hydroxytryptamine for 10 min at 37 °C and then pelleted at + 2 °C (17 000 × *g*, 5 min). The pellet was washed and resuspended in chilled Krebs-phosphate buffer with (○——○) or without (●——●) *Clostridium perfringens* neuraminidase (0.11 U/mg synaptosomal protein). The samples were incubated at 37 °C for the times indicated (abscissa) and [^{14}C]5-hydroxytryptamine accumulated was expressed as % of 5-hydroxytryptamine at 0 min of incubation [26.8 pmol/(mg protein × 3 min)] (ordinate). The values given represent the means of 5 - 6 experiments ± SD.

Morphological examination

Electron microscopical examination revealed no changes in synaptosomal morphology after treatment with neuraminidase.

Discussion

The kinetic parameters of 5-hydroxytryptamine uptake by cortex synaptosomes used in the present experiments agree well with those reported for high-affinity uptake by other authors[25-29]. The occurrence of two uptake mechanisms with two different K_t values described by Shaskan and Snyder[5] was not observed, as the 5-hydroxytryptamine in the medium was restricted to concentrations below 0.1μM. Only higher concentrations were reported to saturate sites not specific for 5-hydroxytryptamine, e.g. catecholaminergic uptake sites.

Although many analogues of 5-hydroxytryptamine were tested in order to elucidate the structural requirements of the 5-hydroxytryptamine molecule for ligand-receptor interaction, little is known about the significance of the molecular structure of the membrane for 5-hydroxytryptamine uptake and transport. The synaptosomal membrane has been shown to be rich in gangliosides and glycoproteins with the terminal non-reducing position of the oligosaccharides occupied by sialic acid[6,7,30].

Treatment of synaptosomes with phytohemagglutinins like concanavalin A, known as D-manno- or D-glucopyranose determinants, did not change 5-hydroxytryptamine uptake[31]. The same was found after incubation with moderate doses of trypsin[31]. Only a slight decrease in 5-hydroxytryptamine uptake due to elevated concentrations of trypsin was reported[32].

The present paper describes an uptake inhibition due to incubation with neuraminidase from *Vibrio cholerae* and *Clostridium perfringens*. The enzymes from both sources attack the same linkages, but the enzyme derived from *Clostridium perfringens* does not require calcium ions[33]. Under the experimental conditions used, the uptake inhibition was dose-related to enzyme concentration (Fig. 3).

In order to elucidate the mode of action of neuraminidase treatment on 5-hydroxytryptamine uptake, the effect of this enzyme on synaptosomal $(Na^{\oplus} + K^{\oplus})$-ATPase was studied. Since 5-hydroxytryptamine uptake depends on energy supplied by a sodium gradient[26], the $(Na^{\oplus} + K^{\oplus})$-ATPase appears to represent the key enzyme for the uptake mechanism mediated by a hypothetical carrier of limited capacity. The finding that the ATPase inhibitor ouabain reduced 5-hydroxytryptamine uptake by almost 80% agrees well with this concept.

Therefore we first determined whether the neuraminidase-induced decrease in 5-hydroxytryptamine uptake was related to ATPase inhibition. A destruction of the membrane was unlikely, considering the selective decrease in uptake rate without effect on affinity. Moreover, in the case of membrane degradation, one should expect no dose relation, but a rapid levelling off. Inhibition of $(Na^{\oplus} + K^{\oplus})$-ATPase as well as selective changes in the permeation process itself should cause inhibition of the non-competitive type, as was actually found. The comparative experiment with ouabain in the presence and absence of neuraminidase demonstrated that uptake inhibition due to ouabain could not be enhanced by neuraminidase, even when inhibition due to neuraminidase alone reached 50%. As compared with the effect when each inhibitor was tested separately, the simultaneous application of neuraminidase and ouabain did not enhance the inhibitory action.

Moreover, direct measurements of ATPase activities have shown no inhibition due to neuraminidase, either of $(Na^{\oplus} + K^{\oplus})$-ATPase or of $Mg^{2\oplus}$-stimulated ATPase. Although the decrease in 5-hydroxytryptamine uptake obtained after neuraminidase treatment is more pronounced, it parallels the action of trypsin, which inhibits 5-hydroxytryptamine uptake slightly without effect on the $(Na^{\oplus} + K^{\oplus})$-ATPase activity[32].

To avoid the misleading interpretation that decreased uptake is caused by enhancement of release, the possible interrelationship of sialic acid removal and 5-hydroxytryptamine release was studied. The spontaneous release of 5-hydroxytryptamine is relatively high[27,34], as confirmed in the present study.

After 15 min incubation with neuraminidase, the spontaneous release of incorporated 5-hydroxytryptamine was augmented as compared with the controls, but the difference was too small to explain the values found for uptake inhibition. Moreover, considering the short incubation time of 3 min used in the uptake studies, it seems very unlikely that increased efflux should mimick uptake inhibition. On the other hand, the effect of reuptake on the spontaneous release was tested with cocaine, which strongly inhibits the retrieval mechanism[4]. Under these conditions, a certain difference in release rates was still detectable, indicating that neuraminidase treatment has a true release-promoting effect.

The results provide further evidence that 5-hydroxytryptamine transport in nerve endings has some features in common with the uptake mechanism in blood platelets. Like the nerve-ending-system, the thrombocyte uptake system exhibits temperature and sodium dependence and K_t values in the range of $0.1 \mu M$[14,35-37].

Conversely, after incorporation of NeuAc into the outer membrane, platelets took up 5-hydroxytryptamine with increased velocity but without increment of the total amount[38].

At present, the results allow two interpretations: Firstly, sialic acid is an integral part of the 5-hydroxytryptamine transport system, not located at the 5-hydroxytryptamine binding site

(K_t is not affected). Secondly, the effects are due to membrane alteration, e.g. structural changes of the membrane with consecutive interference with transport.

The authors thank Miss *Christiane v. Winterfeld* and Mr. *Ralf Stahl* for skilful and patient technical assistance.

This work was supported by a grant from the *Deutsche Forschungsgemeinschaft.*

Literature

1 Robinson, J. D., Anderson, J. H. & Green, J. P. (1965) *J. Pharmacol. Exp. Ther.* **147**, 236 - 243.
2 Schanberg, S. M. (1963) *J. Pharmacol. Exp. Ther.* **139**, 191 - 200.
3 Blackburn, K. J., French, P. C. & Merrills, R. J. (1967) *Life Sci.* **6**, 1653 - 1663.
4 Ross, S. B. & Renyi, A. L. (1967) *Life Sci.* **6**, 1407 - 1415.
5 Shaskan, E. G. & Snyder, S. H. (1970) *J. Pharmacol. Exp. Ther.* **175**, 404 - 418.
6 Eichberg, J., Whittaker, V. P. & Dawson, R. M. C. (1964) *Biochem. J.* **92**, 91 - 100.
7 Brunngraber, E. G., De Kirmenjian, H. & Brown, B. D. (1967) *Biochem. J.* **103**, 73 - 78.
8 De Kirmenjian, H. & Brunngraber, E. G. (1969) *Biochim. Biophys. Acta* **177**, 1 - 10.
9 Mohr, E. & Schramm, G. (1960) *Z. Naturforsch.* **15b**, 568 - 575.
10 Erlanger, B. F., Kokowsky, N. & Cohen, W. (1961) *Arch. Biochim. Biophys.* **95**, 271 - 278.
11 Bradford, H.F. (1972) in Methods of Neurochemistry, (Fried, R., ed.), vol 3, pp. 155 - 202, Marcel Dekker, New York.
12 Wildenhahn, G. (1976) Dissertation Marburg/L. Promoter: Wesemann, W.
13 Aminoff, D. (1961) *Biochem. J.* **81**, 384 - 392.
14 Costa, J. L. & Murphy, D. L. (1975) *Nature (London)* **255**, 407 - 408.
15 Tuomisto, J. (1974) *J. Pharmacol.* **26**, 92 - 100.
16 Bray, G. A. (1960) *Anal. Biochem.* **1**, 279 - 285.
17 Wesemann, W. & Zilliken, F. (1966) *Liebigs Ann. Chem.* **695**, 209 - 216.
18 Bonting, S. L., Simon, K. A. & Hawkins, N. M. (1961) *Arch. Biochem. Biophys.* **95**, 416 - 423.
19 Dette, G. A. & Wesemann, W. (1975) *J. Neural Transm.* **37**, 281 - 295.
20 Kuhar, M. J., Roth, R. H. & Aghajanian, G. K. (1972) *J. Pharmacol. Exp. Ther.* **181**, 36 - 45.
21 Lathia, D. & Wesemann, W. (1975) *J. Neural Transm.* **37**, 111 - 126.
22 Lowry, O. H., Rosebrough, N. J., Farr, A. L. & Randall, R. J. (1951) *J. Biol. Chem.* **193**, 265 - 275.
23 Lineweaver, H. & Burk, D. (1934) *J. Am. Chem. Soc.* **56**, 658 - 666.
24 Dixon, M. (1953) *Biochem. J.* **55**, 170.
25 Tuomisto, L. & Tuomisto, J. (1973) *Naunyn-Schmiedeberg's Arch. Pharmacol.* **279**, 371 - 380.
26 Bogdanski, D. F., Tissari, A. H. & Brodie, B. B. (1970) *Biochim. Biophys. Acta* **219**, 189 - 199.
27 Kellar, K. J., Elliott, G. R., Holman, R. B., Vernikos-Danellis, J. & Barchas, J. D. (1976) *J. Pharmacol. Exp. Ther.* **198**, 619 - 625.
28 Ciofalo, F. R. (1974) *J. Pharmacol. Exp. Ther.* **189**, 83 - 89.
29 Wesemann, W. & Wildenhahn, G. (1977) *Hoppe-Seyler's Z. Physiol. Chem.* **358**, 323 - 324.
30 Lapetina, E. G. & Soto, E. F. (1967) *Biochim. Biophys. Acta* **135**, 33 - 43.
31 Wang, Y. J., Gurd, J. W. & Mahler, H. R. (1975) *Life Sci.* **17**, 725 - 734.
32 Hitzemann, B.A., Hitzemann, R. J. & Loh, H. H. (1975) *J. Neurochem.* **24**, 323 - 330.
33 Müller, H. E. (1974) *Behring Inst. Mitt.* **55**, 34 - 56.
34 Azzaro, A. J. & Rutledge, C. O. (1973) *Biochem. Pharmacol.* **22**, 2801 - 2813.
35 Sneddon, J. M. (1969) *Brit. J. Pharmacol.* **37**, 680 - 688.
36 Glynn, M. F. X. (1973) *Am. J. Clin. Pathol.* **60**, 636 - 643.
37 Gielen, W. & Viehöfer, B. (1974) *Behring Inst. Mitt.* **55**, 194 - 197.
38 Szabados, L., Mester, L., Michal, F. & Born, G. V. R. (1975) *Biochem. J.* **148**, 335 - 336.

Dr. G. A. Dette and Prof. Dr. W. Wesemann, Abteilung Neurochemie, Institut für Physiologische Chemie der Philipps-Universität Marburg,
Lahnberge, D-3550 Marburg/Lahn.

Hoppe-Seyler's Z. Physiol. Chem.
Bd. 359, S. 407 - 411, März 1978

Synthesis and Application of a New Bifunctional Amidination Reagent

Joachim MÜLLER* and Gerhard PFLEIDERER

Institut für Organische Chemie, Biochemie und Isotopenforschung der Universität Stuttgart.

(Received 23 November 1977)

Summary: Methyl 4-hydroxy-3-nitrobenzimidate hydrochloride (I) is presented as a newly developed amidination reagent. It is synthesized in two steps from commercially available 4-hydroxybenzonitrile. Its incorporation into proteins can be easily determined spectroscopically. The activities of the three enzymes lactate dehydrogenase (pig heart), alkaline phosphatase (calf intestine) and tryptophan synthase *(E. coli)* were not greatly reduced after modification with the reagent.

The nitro groups can be reduced by mild treatment with $Na_2S_2O_4$ without alteration of the enzymatic activity. The newly formed aromatic amino groups can further be made to react at lower pH than the native amino groups.

Synthese und Anwendung eines neuen bifunktionellen Imidoesters

Zusammenfassung: 4-Hydroxy-3-nitrobenzimidsäure-methylester-hydrochlorid ist ein neu entwickeltes Reagenz für Aminogruppen in Proteinen. Die Darstellung erfolgt in zwei Stufen aus käuflichem 4-Hydroxybenzonitril.

Die Bestimmung des Einbaus in Proteine kann spektroskopisch erfolgen. Bei Lactat-Dehydrogenase, alkalischer Phosphatase und Tryptophan-Synthase war nach Modifizierung mit dem Reagenz keine wesentliche Aktivitätsabnahme festzustellen.

Die Nitrogruppen des eingebauten Reagenzes können mit Natriumdithionit zu aromatischen Aminogruppen reduziert werden. Diese können dann bei niedrigerem pH als α- und ϵ-Aminogruppen des Proteins weiter umgesetzt werden.

Key words: Chemical modification of ϵ-amino groups in proteins; bifunctional imidic ester; chromophoric imidic ester.

Enzymes:
Alkaline phosphatase (calf intestine), orthophosphoric-monoester phosphohydrolase (alkaline optimum) (EC 3.1.3.1);
Lactate dehydrogenase (pig heart), L-lactate: $NAD^{\oplus}$ oxidoreductase (EC 1.1.1.27);
Tryptophan synthase (*E. coli*), L-serine hydro-lyase (adding indoleglycerol-phosphate) (EC 4.2.1.20).

* This work is part of the Ph. D. thesis of Joachim Müller

The specific reaction of imidic esters with proteins to form amidines

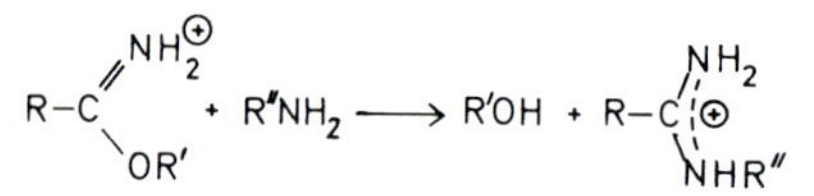

is widely used for modification of essential and non-essential amino groups in enzymes[1-3]. The great advantages of these reagents are the mild reaction conditions and the conservation of the positive charge of the modified group, which in many cases is necessary for stability and activity of the protein[4]. The reagent presented here was initially designed for stepwise immobilisation of enzymes.

In the first modification step, the number of groups capable of binding to the matrix could be varied and followed quantitatively by simple optical measurement. The effect of this modification could be studied.

After reduction of the nitro group with $Na_2S_2O_4$, the macromolecule can be immobilised with matrices binding amino groups at low pH, where only the newly formed aromatic amino groups are able to react.

The great advantage of this stepwise treatment is the separate control of the effects of modification and immobilisation. Effects of different matrices, reactive groups and spacer lengths could be studied in this way, but these results will be presented in a following paper.

Here the synthesis, properties and applications of the newly developed reagent, excluding immobilisation, are presented.

Experimental

Materials

p-Hydroxybenzonitrile was a product of Fluka AG, Buchs SG. Trinitrobenzene sulfonic acid was obtained from Serva, Heidelberg. Lactate dehydrogenase (pig heart) and alkaline phosphatase (calf intestine) were obtained as crystal suspensions from Boehringer Mannheim GmbH. Tryptophan synthase was prepared from *E. coli* W 3110 according to Heilmann[5]. All other reagents were mostly analytically pure.

1) Synthesis of 4-hydroxy-3-nitrobenzonitrile

p-Hydroxybenzonitrile was dissolved in as little ether as possible and cooled in an ice bath. A 3-fold excess of 100% *HNO_3* was added dropwise. After 2 h at 0 °C the mixture was diluted 10-fold with water and extracted 3 times with ether. The combined ether layers were washed with water, dried over Na_2SO_4 and evaporated. The residue was taken up in $CHCl_3$ and purified on a Silica gel/$CHCl_3$ column. The desired material eluted first. Yield after recrystallisation was 25%.

Spectroscopic data:
NMR ($CDCl_3$), chemical shifts in δ: 3 aromatic protons at 7.35; 7.85; 8.55 ppm, 1 O-H proton at 11 ppm.
IR (KBr): -CN at 2240 cm^{-1}, -NO_2 at 1530 and 1330 cm^{-1}.

Elementary analysis: $C_7H_4N_2O_3$ (164.1)

	C	H	N	O
Calcd.	51.22	2.44	17.07	29.27
Found	51.29	2.31	17.02	29.38

2) Synthesis of methyl 4-hydroxy-3-nitrobenzimidate hydrochloride (I)

1 g of *4-hydroxy-3-nitrobenzonitrile* was dissolved by warming in 10 ml of dry $CHCl_3$ and added to a concentrated solution of *HCl* in dry *methanol* in the ice bath. After 12 h at 0 °C light yellow crystals had formed. The solvent was sucked off, and the crystals were dried in vacuo and stored over KOH in the cold. Yield was 80%.

Spectroscopic data:
NMR (D_2O) chemical shifts in δ: 3 aromatic protons at 7.0, 8.0, 8.7 ppm, 3 methyl protons in a singlet at 4.3 ppm. This signal shifts to 3.65 ppm on addition of NaOD.

Elementary analysis: $C_8H_7N_2O_4 \cdot HCl$ (231.6)

	C	H	N	O	Cl
Calcd.	41.49	3.47	12.09	27.63	15.30
Found	41.47	3.76	11.88	27.66	15.23

3) Other methods

Protein was determined by the Biuret method[6]. Amino groups were determined with trinitrobenzene sulfonic acid according to ref.[4].

Amino acid analysis was done on a Biotronik LC 6000 analyser.

Activities of lactate dehydrogenase[7], alkaline phosphatase[8], and tryptophan synthase[9] were determined by previously published methods.

Results

1) Synthesis and properties of methyl 4-hydroxy-3-nitrobenzimidate hydrochloride (I)

Starting with 4-hydroxybenzonitrile the following pathway was used:

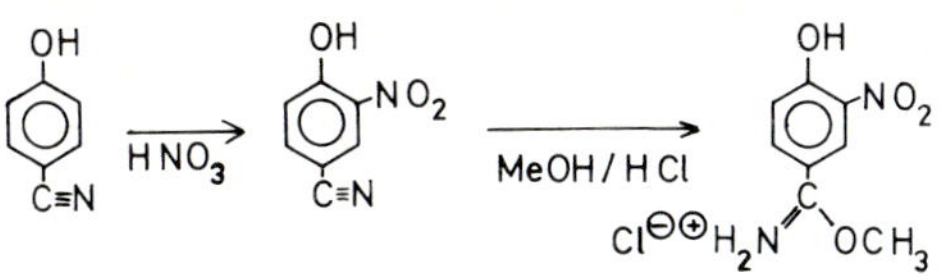

For details of synthesis see "Methods".

In agreement with two dissociable groups the spectra indicate the existence of three different structures which can be represented as follows:

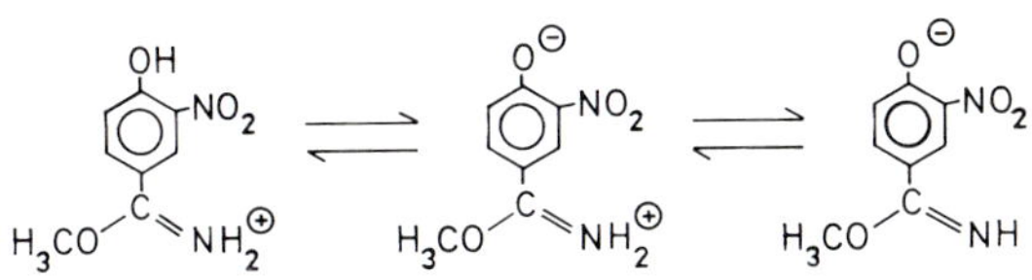

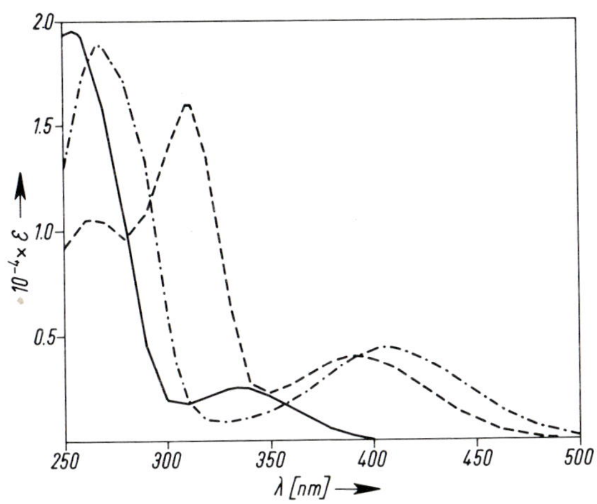

Fig. 1. Spectra of I at various pH.
——: 0.1M HCl; pH 1; ---- : water; pH 5.2;
-·-·-·: 0.5M borate; pH 9.5.
A 1mM stock solution of I in water was diluted 20-fold into the buffer. The spectrum of the stock solution (pH 5) was not altered for at least 1 h, indicating unusually great stability of I.

All these structures are ionic ones; this fact explains the very low solubility of I in organic solvents used for NMR-spectra.

The spectra (Fig. 2) indicate a slow reaction, half of the reagent reacting within about 20 min. After completion of the reaction, extinction coefficients for the 4-hydroxy-3-nitro-benzamidinium cation were calculated from the spectrum and tentatively used for quantitation of incorporation in the modification experiments. The coefficients of ϵ_{400} = 4500 cm^2/mmol and ϵ_{280} = 14500 cm^2/mmol were found to be in good agreement with trinitrobenzene sulfonic acid determination of amino groups and with biuret measurement in the modification experiments.

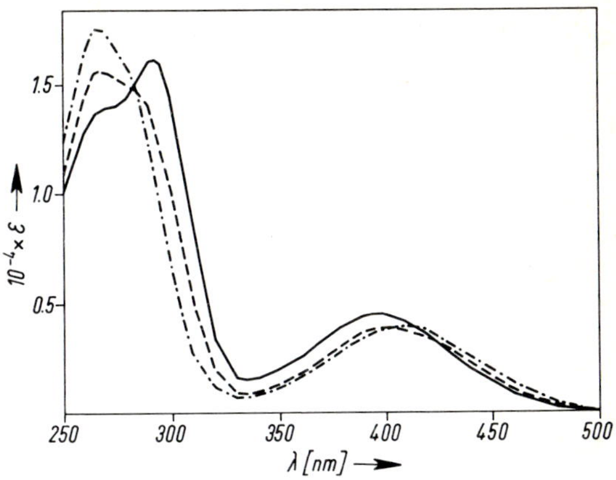

Fig. 2. Spectral changes accompanying reaction of 50μM I with 1M NH_4HCO_3, pH 9.0 at 25 °C.
Each spectral scan was done in 5 min, beginning with the longer wavelengths. -·-·-·: 0 - 5 min after start of reaction; ---- : 13 - 18 min after start of reaction; ——: 12 h after start of reaction.

Amino acid analysis done by *Dr. Jany* was in agreement, too. After 24 h of hydrolysis the expected loss of lysines was observed. With a special buffer the modified lysine could be eluted after arginine.

2) *Modification of enzymes with I*

The usual modification procedure consisted of the incubation of the enzyme (0.5 - 5 mg/m*l*) in 0.5M borate pH 9.5 with 10 mg/m*l* I for 3 h at 25 °C. After addition of I, the pH was readjusted with sodium carbonate. Reaction was termined by passage over Sephadex G-25.

Table: Modification of three different enzymes.

Enzyme	No. of groups incorp. / M_r	Spec. act. [%]
Lactate dehydrogenase	26/144000	80
Alkaline phosphatase	20/144000	100
Tryptophan synthase	36/146000	95

The two essential lysine residues of tryptophan synthase are not attacked by I, whereas methylacetimidate (0.1M, pH 8.7) destroys over 50% of the activity.

For lower incorporations, the same procedure can be used at 0 °C or at pH 8.5. Higher incorporation could be achieved by treatment over night: lactate dehydrogenase with 56 groups/tetramer had 80% of the specific activity of native enzyme (Fig. 3).

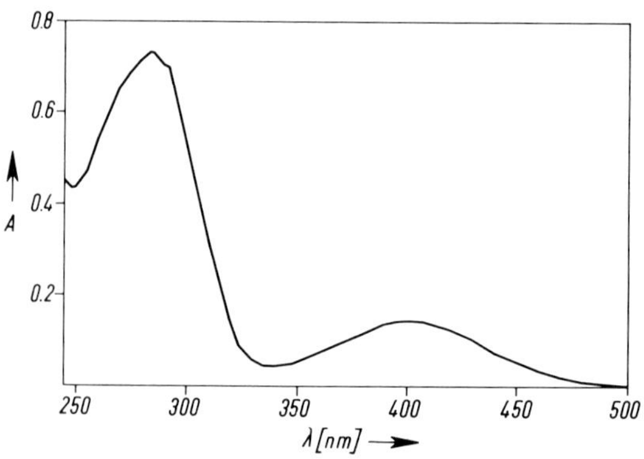

Fig. 3. Spectrum of modified H_4 lactate dehydrogenase. From the spectrum, concentration of incorporated groups is determined as 32.2μM. Protein concentration was determined as follows: from A_{280}, the extinction of incorporated groups at 280 nm is subtracted:

$$A^*_{280} = A^*_{400} \times \frac{\epsilon_{280}}{\epsilon_{400}}.$$

The remainder is the extinction of the protein. In this case, protein content was 0.18 mg/m*l* = 1.25μM. An incorporation of 26 groups is determined in this way. The biuret test gave 0.185 mg/m*l* and reaction with trinitrobenzene sulfonic acid, 28 reacted groups.

3) *Reduction of the NO_2 group*

The aromatic nitro group can be reduced by adding about 1 mg $Na_2S_2O_4$/m*l* to the protein solution. Decolorization is achieved within about 1 min.

No changes in the enzymatic activities were observed. After desalting, the newly formed aromatic amino groups could be made to react with trinitrobenzene sulfonic acid even at pH 6, where native amino groups are known to react only very slowly.

4) *Further applications*

As mentioned above, the aromatic amino groups can react at low pH. Fluorescent labeling is possible by reaction with dansyl chloride at pH 6. Immobilisation of the modified enzymes was performed on several matrices. Results will be presented in a separate paper. Binding of divalent cations by the *o*-aminophenolic structure is demonstrated by a great increase in extinction at ~ 300 nm on addition of $Cu^{2\oplus}$ or $Zn^{2\oplus}$. This could be useful in X-ray structure determinination.

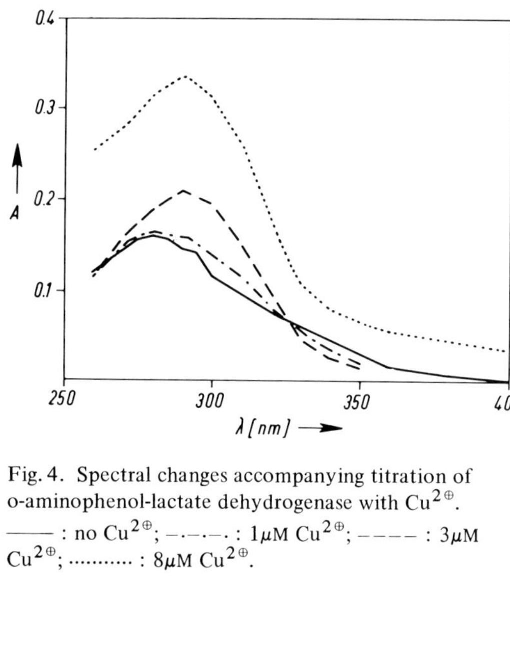

Fig. 4. Spectral changes accompanying titration of o-aminophenol-lactate dehydrogenase with $Cu^{2\oplus}$. —— : no $Cu^{2\oplus}$; –·–·–· : 1μM $Cu^{2\oplus}$; – – – – : 3μM $Cu^{2\oplus}$; ·········· : 8μM $Cu^{2\oplus}$.

Discussion

Even if initially not designed for other applications than immobilisation of enzymes, the reagent offers some advantages.

In simple modification experiments it is a stable, soluble and chromophoric reagent. Its incorporation does not cause greater losses of activity than that of simple imidates like methyl acetimidate, but determination of incorporation is very simple. The reagent is much more soluble and hydrophilic than other chromophoric imidic esters[10,11].

The hydrophobic nature of the 2,4-dinitroaniline chromophore used in ref.[10,11] causes great difficulties in modification and handling of the treated protein[10]. In our opinion, hydrophilisation cannot be achieved by introduction of a carboxylic function as proposed by ref.[11]. In Pinner synthesis (MeOH/HCl) this carboxylic function would be esterified, and water produced by this reaction would destroy the imidic ester.

In fluorescent labeling the reagent offers the same advantages as in immobilisation: the effects of modification and introduction of a highly hydrophobic fluorophore can be controlled separately. Naturally, this is not possible with, for instance, a dansyl imidic ester[12,13].

Finally, the metal chelating properties of the *o*-aminophenolic structure produced by reduction of the nitro group should be mentioned. They are known from 3-aminotyrosine produced by nitration with tetranitromethane and reduction with $Na_2S_2O_4$[14].

In contrast to the known picolinamidine group[15], our reagent binds cations at some distance from the originally modified group.

All these applications of one new reagent make it a valuable alternative to the use of many different reagents.

Literature

1 Plapp, B. V. (1970) *J. Biol. Chem.* **245**, 1727 - 1735.

2 Wood, F. T. Wu, M. M. & Gerhart, J. C. (1975) *Anal. Biochem.* **69**, 339 - 349.

3 Means, G. E. & Feeney, R. E. (1971) Chemical Modification of Proteins, pp. 89 - 93, Holden Day Inc., San Francisco.

4 Tuengler, P. & Pfleiderer, G. (1977) *Biochim. Biophys. Acta* **484**, 1- 8.

5 Heilmann, H.-D. (1976) Habilitationsschrift Universität Bochum.

6 Beisenherz, G., Boltze, H. J., Bücher, Th., Czok, R., Garbade, K. H., Meyer-Arendt, E. & Pfleiderer, G., (1953) *Z. Naturforsch.* **8b**, 555.

7 Pfleiderer, G., Holbrook, J. J., Zaki, L. & Jeckel, D. (1968) *FEBS Lett.* **1**, 129 - 132.

8 Klinisches Labor (1974) p. 207, E. Merck, Darmstadt.

9 Smith, O. H. & Yanofsky, C. (1962) *Methods Enzymol.* **5**, 794 - 806.

10 Kapmeyer, W. & Pfleiderer, G. (1977) *Biochim. Biophys. Acta* **481**, 328 - 339.

11 Schramm, H. J. (1967) *Hoppe-Seyler's Z. Physiol. Chem.* **348**, 289 - 292.

12 Bösterling, B., Engel, J. Steinemann, A., & Schramm, H. J. (1976) *Hoppe-Seyler's Z. Physiol. Chem.* **357**, 1283 - 1296.

13 Schramm, H. J. (1975) *Hoppe-Seyler's Z. Physiol. Chem.* **356**, 1375 - 1379.

14 Sokolovsky, M., Riordan, J. F. & Vallee, B. L. (1967) *Biochem. Biophys. Res. Commun.* **27**, 20.

15 Benisek, W. F. & Richards, F. M. (1968) *J. Biol. Chem.* **243**, 4267 - 4271.

Prof. Dr. G. Pfleiderer and J. Müller, Institut für Organische Chemie, Biochemie und Isotopenforschung der Universität Stuttgart, Pfaffenwaldring 55, D-7000 Stuttgart 80.

Hoppe-Seyler's Z. Physiol. Chem.
Bd. 359, S. 413 - 420, März 1978

Preparative Polyacrylamide Gel Electrophoresis of Nuclear RNA

Ronald Böcker and Walter Kersten

Institut für Physiologische Chemie, Universität Erlangen–Nürnberg

(Received 16 August 1977)

Summary: A method is described for the separation of nuclear RNA by preparative polyacrylamide gel electrophoresis. Separation is performed on a slab gel. The elution chamber system is rectangular. Total nuclear RNA can be fractionated into the prominent RNA species 4 - 7S, 18S and 28S RNA. The less prominent RNA, especially the high molecular weight precursor molecules of ribosomal RNA 45S (M_r = 4.1 x 10^6) and 32S (M_r = 2.1 x 10^6) can be separated from the total nuclear RNA without contamination by other RNA. The time needed for a preparative run is 260 min.

Präparative Elektrophorese von Kern-RNA in Polyacrylamid-Gelen

Zusammenfassung: Es wird eine Methode beschrieben, mit der Kern-RNA durch präparative Gelelektrophorese aufgetrennt werden kann. Die Trennung wird in einem Plattengel durchgeführt. Die Elutionskammer ist rechteckig. Die Gesamt-Kern-RNA kann in die Hauptkomponenten 4 - 7S, 18S und 28S RNA aufgetrennt werden. Die weniger prominenten RNA-Arten, insbesondere die hochmolekularen Vorläufer der ribosomalen RNA 32S (M_r = 2.1 x 10^6) und 45S (M_r = 4.1 x 10^6), können von der Gesamt-Kern-RNA abgetrennt werden, ohne daß sie mit anderen RNA-Arten kontaminiert sind. Die Zeit für eine präparative Trennung beträgt 260 min.

Key words: Preparative gel electrophoresis, separation of nuclear RNA.

Polyacrylamide gel electrophoresis on a gel column in a circular elution chamber system is used for preparative separation of ribonucleic acids[1-4]. With this method only eucaryontic cytoplasmic RNA or bacterial RNA can be separated. The high molecular weight rRNA precursor molecules from the nucleus have as yet only been identified by analytical electrophoresis[5-9] or by gradient centrifugation[10].

This paper describes a preparative polyacrylamide gel electrophoresis system which separates the prominent and less prominent RNA molecules of the nucleus in the range of 4S to 45S (M_r = 3 x 10^4 - 4.1 x 10^6) in one step.

The time needed for a complete separation of nuclear or cytoplasmic RNA by the method described here is shorter than with comparable methods of the literature[1,2,4]. The quality of separation for the short time is better than that in a previously described fast preparative agarose electrophoresis method[3].

The precursor molecules of ribosomal RNA 45S- and 32S RNA can be isolated without contamination by other RNA species. The separation is performed on a gel slab, the elution chamber system is rectangular.

Materials and Methods

Chemicals

Acrylamide, *N,N'*-methylenebisacrylamide (bisacrylamide), *N,N,N',N'*-tetramethylethylenediamine, agarose and sucrose were from Serva, Heidelberg. All other chemicals were of reagent grade.

Cells and isolation of subcellular components

RNA was isolated from Ehrlich ascites tumor cells. The cells were grown in the abdominal cavity of white mice for six days. The cells were harvested by sucking the fluid from the cavity. The cells were washed three times with buffer (0.14M NaCl, 0.027M KCl, 0.065M $Na_2HPO_4 \times 2H_2O$, 0.015M KH_2PO_4, pH 7.4). Nuclei were prepared as described[11]. Supernatants from nuclei preparations were centrifuged at 40000 × *g* for 20 min. The resulting pellets were discarded and the clear supernatants were used to isolate cytoplasmic RNA.

Isolation of nuclear RNA

Nuclear RNA was obtained from the purified nuclei by a sodium dodecylsulfate/hot phenol method. Nuclei from 1.5 g ascites cells (wet weight) were suspended in ice cold 8 m*l* buffer (0.01M CH_3COONa, 0.01M Na_2EDTA, pH 5.1), 0.05 m*l* water saturated phenol was added to inhibit enzymatic activities. The suspension was made 0.5% with respect to sodium dodecylsulfate warmed up to and kept at room temperature for 5 min. Then 8 m*l* of water saturated phenol was added and the mixture heated up to 60 °C and shaken for 3 min. NaCl was added to give a final concentration of 0.2M, the mixture was cooled to 4 °C in an ice bath and centrifuged at 10000 × *g* for 10 min at 4 °C.

The upper cloudy phase was treated with 2 m*l* of chloroform and centrifuged at 10000 × *g* for 10 min. From the clear supernatant the RNA was precipitated with 2 vol. of cold ethanol at – 20 °C over night.

Isolation of cytoplasmic RNA

The cytoplasmic fraction was stirred with an equal volume of water saturated phenol for 30 min at 0 - 4 °C. The concentration of NaCl was increased to 0.2M and 5 m*l* chloroform was added per 20 m*l* water layer. Phenol and chloroform were removed by centrifugation and RNA precipitated as described above.

The precipitated RNA was spun down, dried and dissolved in buffer (0.01M CH_3COONa; 0.1M NaCl; 0.005M Na_2EDTA; pH 5.1)[12] to a maximal concentration of 100 A_{260} units/m*l*.

Gradient centrifugation

RNA, about 0.5 mg, was layered onto a 21 m*l* linear sucrose gradient (10 - 30%) in the pH 5.1 buffer. Centrifugation was carried out in the MSE 65 superspeed centrifuge, 3 × 23 m*l* swinging bucket rotor (26000 rpm, 16 h, 2 °C, 80000 × *g*). The gradient was fractionated with the LKB fractionator sucking the sucrose with a needle from the bottom of a centrifuge tube through a flow cell. The elution profile was automatically recorded with a Zeiss PMQ II photometer connected to a chart recorder. From one tube 32 fractions were collected.

Electrophoresis

For analytical and preparative gel electrophoresis the commercially available apparatus from Isco/Colora (Ultraphor™ Colora Meßtechnik, D-7073 Lorch, Württ.) was used. The power supply was from Desaga, D-6900 Heidelberg, with maximal 1200 V and 200 mA. The temperature of the electrophoresis system was regulated using a cooling circulator (Colora Meßtechnik). The same cell was used for preparative and analytical gel electrophoresis.

Before preparing a gel the glass chamber was cleaned intensively with water, acetone and chloroform. Acrylamide and bisacrylamide were recrystallized according to Loening[13]. The electrophoresis buffer used in all experiments was Tris/EDTA/borate, pH 8.3 (0.089M Tris; 0.089M boric acid, 2.5mM Na_2EDTA)[14].

Analytical gel electrophoresis

Analytical polyacrylamide gel electrophoresis was performed in a 2% acrylamide/bisacrylamide gel (19:1), 0.5% agarose[14]. Acrylamide and bisacrylamide were dissolved in 70 m*l* of water, 20 m*l* of 10 × electrophoresis buffer and 0.1 m*l* tetramethylethylenediamine were added and the solution prewarmed to 40 °C. 1 g agarose was boiled in 100 m*l* of water with stirring and cooled to 40 °C. The solutions were combined and degassed under reduced pressure for 30 s. 10 m*l* of 2% ammonium persulfate in water was added. The mixture was gently shaken and poured into the precooled (14 °C) electrophoresis cell. The slot former (8 slots, each 1.0 × 0.4 cm) was set onto the gel. After 1 h the gel was ready for electrophoresis. The electrophoresis system was filled with buffer and cooled to 4 °C. A maximum of 1.4 A_{260} units of RNA in 0.1 m*l* buffer pH 5.1 was applied to each slot. To increase the density, each sample contained 8% glycerol and was cooled in an ice bath.

Each sample contained 0.01 m*l* of a bromophenol red solution (0.01%) to indicate the buffer front during the electrophoretic run. The RNA enters the gel at 40 mA and 100 V during 30 min. Then the current is held at 100 mA for 120 min. Under these conditions the buffer front migrates about 12 cm. The gel is removed from the chamber and stained with ice-cold 0.2% methylene blue for 2 h[15]. Excess dye is removed in 2 x 10 *l* ice-cold demineralized water with stirring for 24 h.

Preparative gel electrophoresis

The gel for preparative electrophoresis was 2% in respect to acrylamide/bisacrylamide (19:1) and 0.5% in respect to agarose. The gels were prepared by mixing 22.2 m*l* aqueous acrylamide solution (1.14 g acrylamide, 0.06 g bisacrylamide) with 6 m*l* 10 x electrophoresis buffer and 0.05 m*l* tetramethylethylenediamine. This solution was prewarmed to 40 °C. Agarose, 0.31 g, was boiled in 30 m*l* water for 1 min and cooled to 40 °C. The solutions were combined, degassed and 1.8 m*l* of 2% aqueous ammonium persulfate was added and the mixture gently shaken.

A 4-cm gel slab, which was supported by a nylon mesh at the bottom of the precooled (14 °C) electrophoresis cell, was polymerized. To fix this gel to the chamber at the top of the gel a wall with a single slot (1.5 x 13.5 x 0.4 cm) was formed of 11% acrylamide (1 g acrylamide; 0.1 g bisacrylamide; 0.02 m*l* tetramethylethylenediamine; 1 m*l* 10 x electrophoresis buffer; 9 m*l* water; 0.3 m*l* of 2% aqueous ammonium persulfate). After 1 h the slot

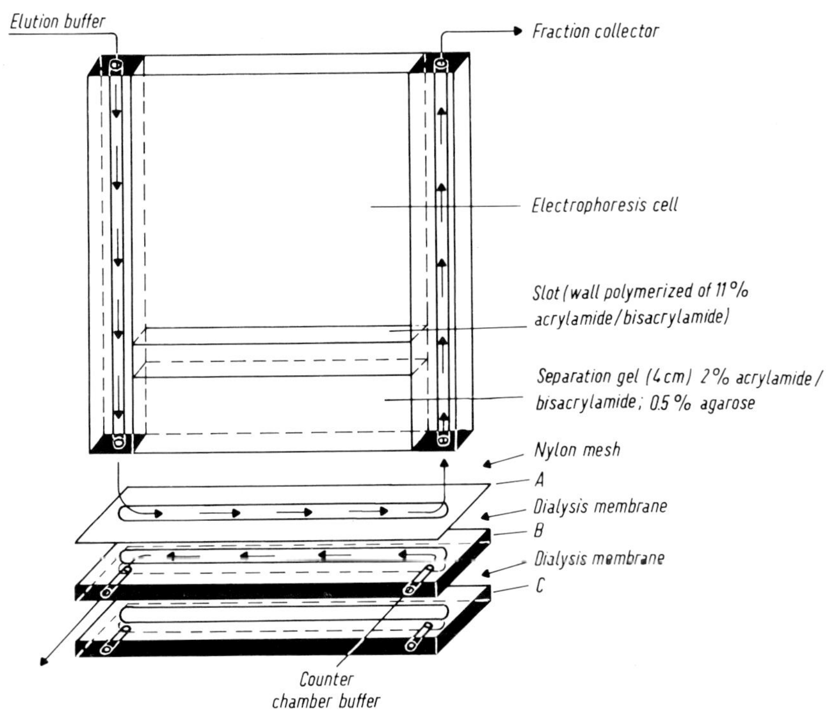

Fig. 1. System for preparative gel electrophoresis.

The separation gel (4 cm, 2% acrylamide/bisacrylamide, 0.5% agarose) is first polymerized into the precooled electrophoresis cell. The slot former is set onto the gel. A wall of 11% acrylamide/bisacrylamide is polymerized around the slot former.

The elution chamber system consists of three main parts A, B and C. The Visking dialysis membrane between A and B separates the elution chamber from the counter chamber. The Visking membrane between B and C separates the counter chamber from the electrophoresis buffer. The elution chamber system is bolted together by Teflon screws and pressed under the electrophoresis cell. Elution buffer and counter chamber buffer flow in counter direction to minimize formation of an electric field gradient.

After filling the system with electrophoresis buffer air bubbles below the counter chamber can be removed through the holes in part C by sucking.

For further details compare Materials and Methods.

former was removed and the rectangular elution chamber system pressed under the gel. The nylon mesh which supported the gel at the bottom was not removed. (See also Fig. 1.)

The elution buffer contained 0.7 x electrophoresis buffer. It was pumped at a flow rate of 42 m*l*/h. The elution chamber (0.1 cm in height) was separated from the counter chamber by a dialysis Visking membrane. The counter chamber buffer contained 3 x electrophoresis buffer and was pumped at a rate of 20 m*l*/h. For the elution buffer a lower ion concentration was chosen compared to the gel buffer to increase the strength of the electric field. Thus RNA molecules were accelerated out of the gel into the elution chamber. The high ion concentration of the buffer in the counter chamber reduces the field strength at the Visking membrane of the elution chamber.

The electrophoresis system was filled with buffer and cooled to 4 °C. Total nuclear RNA up to 1.8 mg in electrophoresis buffer with 8% glycerol was cooled and layered on top of the gel into the slot. Dye was not added. The loading of the gel was visible due to the different refractive indices of the sample and the electrophoresis buffer. The RNA sample entered the gel during 15 min at 40 mA and 120 V. Up to the end of the electrophoresis the current was held at 90 mA. Fractions were collected at 5 min intervals. The absorbance of the effluent was monitored in a LKB Uvicord III at 254 nm and automatically recorded.

When all UV-absorbing material had been eluted from the gel, the power was switched off and the elution continued. In this way RNA sticking at the elution chamber membrane was removed and eluted. To isolate the RNA from the different fractions of a preparative separation NaCl to give a final concentration of 0.2M and two volumes of cold absolute ethanol (– 20 °C) were added to the eluted fractions to precipitate the RNA. The RNA was spun down, dried and dissolved in buffer pH 5.1.

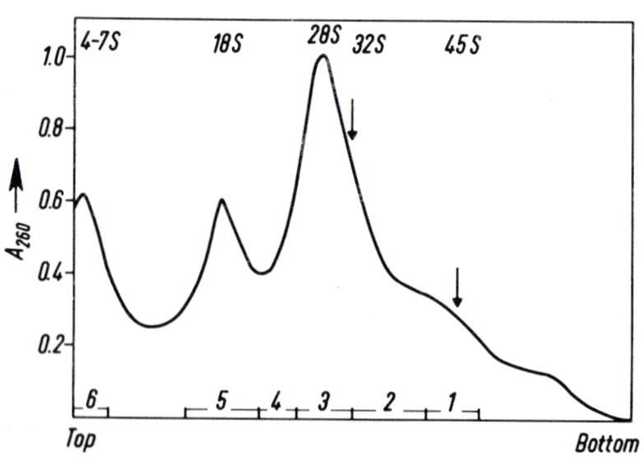

Fig. 2. Sucrose density-gradient centrifugation of nuclear RNA from Ehrlich ascites cells.

0.5 mg RNA in 0.3 m*l* buffer was placed atop a 21 m*l* linear 10–30% sucrose gradient in buffer pH 5.1 and centrifuged at 26 000 rpm for 16 h at 2 °C. For further details see Materials and Methods. The fractions – indicated 1 to 6 – were used for analytical gel electrophoresis (see Plate 1).

Results

The separation of nuclear RNA from Ehrlich ascites cells by gradient centrifugation is shown in Fig. 2. The 4 - 7S, 18S and 28S RNA are eluted

Fig. 3. ▷

Plate I: Analytical gel electrophoresis of nuclear RNA on a composite 2% polyacrylamide-, 0.5% agarose gel. Electrophoretic conditions were described under Materials and Methods. The RNA samples for this electrophoresis were derived from the gradient centrifugation of nuclear RNA shown in Fig. 2. From left to right the samples applied are: Slot 1: unfractionated cytoplasmic RNA as marker for 18S and 28S RNA; Slots 2 - 7 are Fract. 1 - 6 of the gradient centrifugation of Fig. 2. Slot 2: Fract. 1 = 45S RNA; Slot 3: Fract. 2 = 32S RNA; Slot 4: Fract. 3 = 28S RNA; Slot 5: Fract. 4 = 18 - 28S RNA; Slot 6: Fract. 5 = 18S RNA; Slot 7: Fract. 6 = 4 - 7S RNA; Slot 8: unfractionated nuclear RNA. Migration from start of the prominent RNA species: 4S 11.4 cm; 18S 5.6 cm; 28S 3.4 cm.

Plate II: Analytical gel electrophoresis of nuclear RNA of Ehrlich ascites tumor cells.

Fractions were pooled from the preparative gel electrophoresis shown in Fig. 5. From left to right, the samples applied are:

IIa Slot 1: unfractionated cytoplasmic RNA of ascites cells as marker; Slot 2: Fract. 1 = UV absorbing material eluted prior to 4S RNA; Slot 3: Fract. 2 = 4 - 7S RNA; Slot 4: Fract. 3; Slot 5: Fract. 4; Slot 6: Fract. 5 = 18S RNA; Slot 7: Fract. 10 = RNA material adsorbed at the Visking membrane and eluted after the power was switched off; Slot 8: unfractionated nuclear RNA.

IIb Slot 1: unfractionated cytoplasmic RNA; Slot 2: Fraction 5 = 18S RNA; Slot 3: Fract. 6 = intermediate fraction between 18S and 28S RNA; Slot 4: Fract. 7 = 28S RNA; Slot 5: Fract. 8 = 32S RNA; Slot 6: Fract. 9 = 45S RNA; Slot 7: Fract. 10 = RNA material adsorbed at the Visking membrane; Slot 8: unfractionated nuclear RNA. Migration of the different RNA species in plate IIb: 4S 12.6 cm; 18S 5.4 cm; 28S 3 cm; 32S 2.3 cm; 45S 1.3 cm.

Plate I

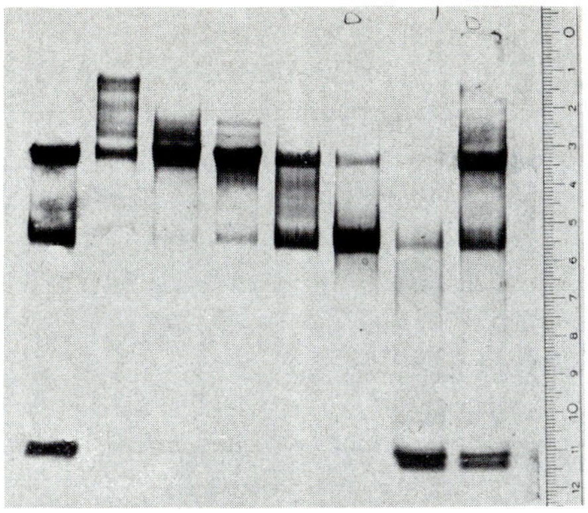

1 2 3 4 5 6 7 8 ← *Slot no*

Plate IIa

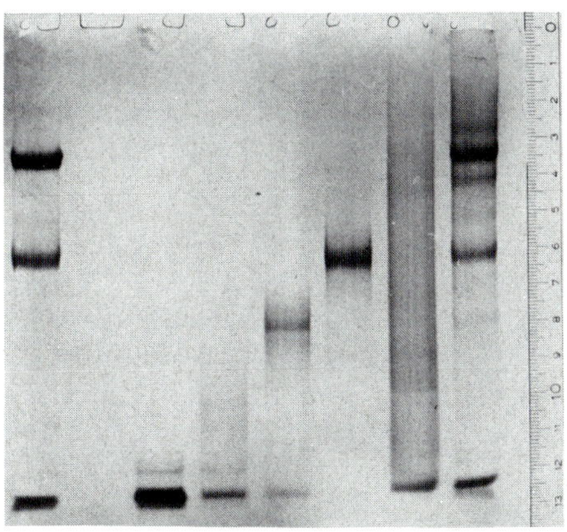

1 2 3 4 5 6 7 8 ← *Slot*

Plate IIb

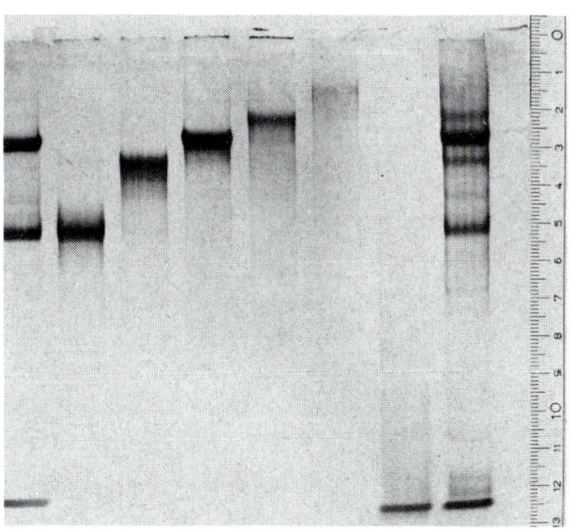

1 2 3 4 5 6 7 8

in the main peaks. RNA molecules sedimenting at 32S and 45S are indicated by the arrows. This elution profile is similar to those described in the literature[10,14]. After separation by gradient centrifugation the different fractions of nuclear RNA were tested by analytical gel electrophoresis (Plate I in Fig. 3). In Fract. 1 of the sucrose gradient, in addition to 45S RNA, other precursor molecules and also 28S RNA are eluted (Plate I Slot 2). In Fract. 3 mainly 28S RNA would be expected, but 18S and 32S can also be identified by electrophoresis. These results show that by gradient centrifugation the different RNA species, isolated from the nuclei of ascites cells, cannot be separated completely. Nuclear 28S and 18S RNA have the same electrophoretic mobilities as the corresponding cytoplasmic RNA species (compare Slot 1 and 8 in Plate I). To achieve a better separation of these RNA molecules we developed and improved conditions for a preparative separation of RNA on a polyacrylamide gel. The final conditions are described under material and methods. The resolution power of the method is shown in Fig. 4 where cytoplasmic RNA of ascites cells was applied to the gel and separated within 3 h. The separation is equivalent to those obtainable with any other preparative separation procedure[1-4].

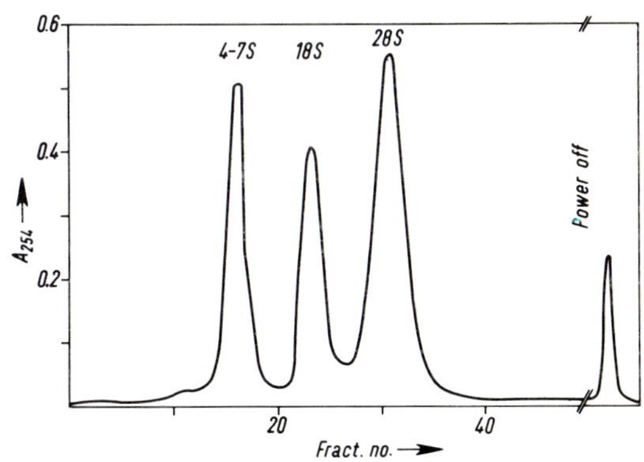

Fig. 4. Preparative polyacrylamide gel electrophoresis of cytoplasmic RNA from Ehrlich ascites tumor cells.

1 mg RNA in 1.1 m*l* gradient buffer, pH 5.1, 8% glycerol, was applied onto a 2% acrylamide-, 0.5% agarose gel. Electrophoresis was performed as described under Materials and Methods. Fractions of 4.2 m*l* (5 min) were collected. UV absorbance was recorded automatically.

Nuclear RNA can be also resolved into three main peaks by preparative gel electrophoresis, representing the prominent RNA 4 - 7S, 18S and 28S, see Fig. 5. Beside these RNA species less prominent RNA compounds were resolved as distinguishable but sometimes overlapping peaks.

To analyze the quality of separation and to control the molecular weights of the different RNA types by their electrophoretic mobilities fractions were pooled as indicated in Fig. 5. RNA was precipitated by ethanol and dissolved in buffer pH 5.1 and submitted to analytical gel electrophoresis, see Plate II a and b in Fig. 3.

UV absorbing material eluted prior to 4 - 7S was not precipitable by ethanol (Fract. 1 does not give stainable material, compare Slot 2 on Plate II a). With the electrophoretic conditions used RNA molecules with low molecular weights, smaller than about 12S, cannot be separated into distinct, separable fractions. This can easily be achieved in gels with higher acrylamide concentrations, however RNAs with high molecular weights are retarded in those gels and the peaks of the eluted RNAs are broadened[4,15]. 18S RNA can be eluted in two fractions as a single entity, see Slot 6 in Plate II a and Slot 2 in Plate II b. In repeated experiments we found a small peak between 18S and 28S RNA. Upon analytical electrophoresis this RNA migrates as a single species. From the mobility we have calculated a molecular weight of 1.4×10^6.

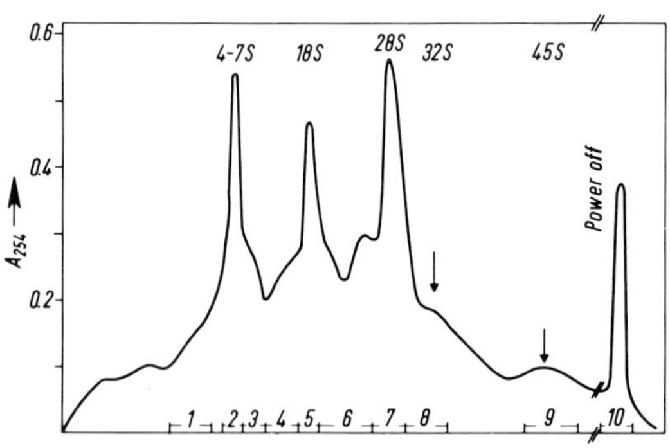

Fig. 5. Preparative polyacrylamide gel electrophoresis of nuclear RNA from Ehrlich ascites tumor cells.

1.75 mg RNA in 1.1 m*l* gradient buffer, 8% glycerol was applied onto the gel. All other conditions were as described under Materials and Methods and in the legend of figure 3. The fractions were pooled as indicated and examined by analytical gel electrophoresis.

Also 28S, 32S and 45S RNA can be separated from each other without cross contamination, see Slot 4, 5 and 6 on plate II b. Thus RNA molecules with sedimentation constants greater than about 12S can clearly be separated by preparative electrophoresis on a 2% acrylamide gel.

The pattern of the analytical separation of the collected fractions does not change when the RNA is separated on gel containing 6M urea, which demonstrates that there are no aggregates.

During electrophoresis some RNA material is adsorbed at the Visking membrane. When the power supply is switched off and elution continued this RNA is eluted (Fract. 10 in Fig. 5). The material is inhomogeneous and migrates predominantly as 4S RNA (Slot 7 in Plate II a and II b).

Usually the amount of the "sticking material" was about 10% of the total eluted RNA when nuclear RNA was separated and about 5 - 7% when cytoplasmic RNA was separated. At a current higher than 100 mA or after a contact of the gel with the Visking membrane, the amount of the "sticking material" increases drastically. Including this sticking material 100% of the applied material was recovered from the gel.

A logarithmic relationship exists between the molecular weight of RNA and its electrophoretic mobility[6,10]. When the logarithms of the molecular weights are plotted versus the mobilities of the RNAs the plot is linear (see Fig. 6a). Using this plot the molecular weight of the RNA eluted in Fract. 6 of the preparative electrophoresis was determined to have an M_r value of 1.4×10^6. The plot of the logarithm of the molecular weights versus the elution position of the RNAs after preparative electrophoresis is not linear as also described by others using a circular elution chamber system for electrophoresis[4]. The mobility of RNA is directly proportional to the distance migrated for a constant time. Here the mobility is measured for a fixed distance. A linear relation exists between the logarithm of molecular weight

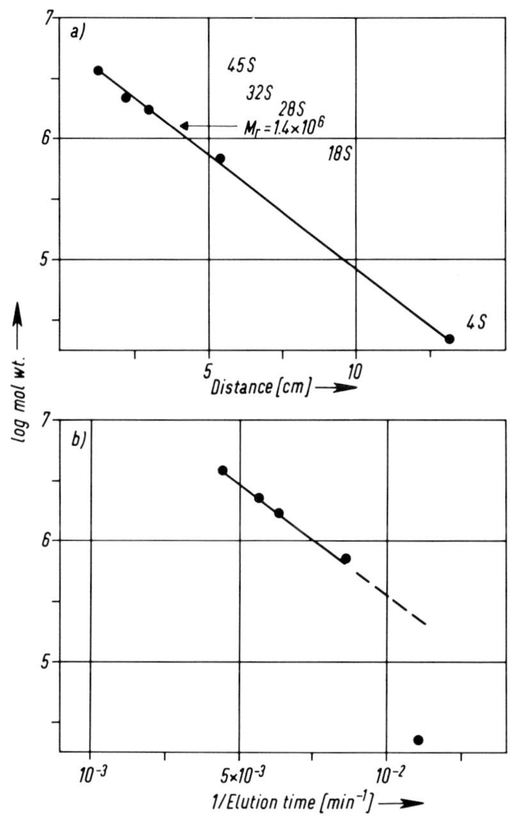

Fig. 6. Electrophoretic mobility of nuclear RNA upon analytical and preparative gel electrophoresis.

a) The logarithms of the molecular weights of nuclear RNA are plotted versus the electrophoretic migration in cm. 4 S, 18 S and 28 S RNA of cytoplasm were used as reference markers.

b) The logarithms of the molecular weights of the eluted RNA species after preparative separation are plotted versus the reciprocal of the elution time.

and the reciprocal of the elution time, but in this system only for molecules greater than 4S (Fig. 6b).

Discussion

Eucaryontic ribosomal RNA 28S ($M_r = 1.75 \times 10^6$), 18S ($M_r = 0.75 \times 10^6$) and 7S ($M_r = 4 \times 10^4$) are synthesized as one precursor molecule of 45S ($M_r = 4.1 \times 10^6$) in the nucleolus. This precursor molecule is modified and processed in several steps to mature ribosomal RNAs and transported into the cytoplasm[6-9].

To investigate labeling or base composition of the intermediates the different RNA species must be separated without contamination by other RNAs. The method described here enables us to isolate the prominent RNA molecules of the nucleus as well as the less prominent precursor molecules and their intermediates during processing. By gradient centrifugation nuclear RNA can be separated into three main peaks. Only some of the less prominent molecules can be separated, but they can not be resolved completely[10]. A better resolution of nuclear RNA can be achieved by analytical polyacrylamide gel electrophoresis. But determination of UV absorbance is sometimes difficult because of the high background absorbance of thick gels. The amount of RNA that can be separated by analytical electrophoresis is small. The preparative electrophoresis described here, resolves total nuclear or cytoplasmic RNA into definite RNA species. The method is a one step separation. The UV absorbance can easily be determined in the effluent. This method combines the advantages of gradient centrifugation and gel electrophoresis. The time needed for a preparative run is shorter than described in the literature [1,2,4]. The identified different types of nuclear RNA are not aggregates of small species. This was proved by analytical electrophoresis of the separated RNA in gels with 6M urea.

Cytoplasmic RNA is nearly completely separated, the UV absorbance reaches the baseline between the different peaks. Nuclear RNA is apparently far more heterogeneous; there is always some material between the distinct peaks of the well characterized prominent RNA peaks. In some part these are the intermediates of the maturation process of ribosomal RNA. To some extent these may be RNA species which are not yet fully characterized.

The composition of nuclear RNA prepared by the method described is very similar to that described in the literature[10,14,16] for other eucaryontic organisms. Purity of preparations of nuclei can easily be detected with the microscope after staining the nuclear preparation with haematoxillin/eosine[16]. No attempts were made to isolate nucleoli and to prepare nucleolar RNA.

On agarose columns[3] a large amount of RNA can be separated but the separation is not better than the separation by gradient centrifugation.

The slab form of the gel and the rectangular construction of the elution chamber guarantees a fast and complete elution of the RNA. A rectangular shaped elution chamber is better and faster eluted than a circular chamber[17].

The small volume of the elution chamber avoids excessive dilution of the RNA. The slab gel technique has another advantage:
the gel can be cooled uniformly over a wide area. By supporting the gel at the bottom with a nylon mesh the gel cannot slide out of the electrophoresis cell and come into contact with the elution chamber membrane. This mesh does not change the electrophoretic conditions but enables one to handle the gel without difficulties over longer periods of electrophoresis. Some RNA sticks at the membrane's surface. Because of the heterogeneity of the material it might be degraded RNA. When the current is increased above 90 mA the amount of intact 18S and 28S RNA decreases and the amount of degraded RNA eluted at the end of electrophoresis increases drastically. This can also be shown for purified 18S or 28S RNA. When 18S and 28S RNA isolated by preparative electrophoresis is re-electrophoresed at too high a current the intact molecules were degraded. A contact of the Visking membrane with the gel has to be avoided, because this also causes degradation of intact RNA.

The degree of resolution of total RNA can be varied by the concentration of acrylamide in the gel. The high molecular weight RNA species up to a molecular weight of 4.1×10^6 can best be separated on a 2% acrylamide gel.

We thank Dr. *S. Frank* from this institute for supplying the ascites cells. For careful technical assistance we thank Mrs. *B. Böcker*.
This work was supported by the *Deutsche Forschungsgemeinschaft* grant Ke 61/13.

Literature

1 Lanclos, K. & Bresnick, E. (1972) *Anal. Biochem.* **45**, 645 - 652.
2 Shimada, K., Sekikawa, K., Fujinaga, K. & Ito, Y. (1973) *Anal. Biochem.* **51**, 456 - 465.
3 Weil, P. A. & Hampel, A. (1973) *Biochemistry* **12**, 4361 - 4367.
4 Hagen, F. S. & Young, E. T. (1974) *Biochemistry* **13**, 3394 - 3400.
5 Dingman, C. W. & Peacock, A. C. (1968) *Biochemistry* **7**, 659 - 668.
6 Weinberg, R. A. & Penman, S. (1970) *J. Mol. Biol.* **47**, 169 - 178.
7 Fujisawa, T. Abe, S., Kawada, T., Satake, M. & Ogata, K. (1973) *Biochim. Biophys. Acta* **324**, 226 - 240.
8 Maden, B. E. H. (1971) *Progr. Biophys.* **22**, 129 - 177.
9 Darnell, J. E., jr. (1968) *Bacteriol. Rev.* **32**, 262 - 290.
10 Grinchishin, V., Savage, H., Nazar, R. & Busch, H. (1972) *Fed. Proc., Fed. Am. Soc. Exp. Biol.* **31**, 639.
11 Grummt, F. & Grummt, I. (1976) *Eur. J. Biochem.* **64**, 307 - 312.
12 Darnell, J. E. & Scherrer, K. (1962) *Biochem. Biophys. Res. Commun.* **7**, 486 - 490.
13 Loening, U. E. (1967) *Biochem. J.* **102**, 251 - 257.
14 Peacock, C. A. & Dingman, C. W. (1968) *Biochemistry* **7**, 668 - 674.
15 Peacock, C. A. & Dingman, C. W. (1967) *Biochemistry* **6**, 1818 - 1827.
16 Böcker, R., Jones, I. K. & Kersten, W. (1977) *J. Nutr.* **107**, 1737 - 1746.
17 Nees, S. (1974) in Electrophoresis and Isoelectric Focusing in Polyacrylamide Gel (Allen, R. L. & Maurer, H. R., eds.) pp. 189 - 198, Walter de Gruyter & Co., Berlin, New York.

Dr. R. Böcker, Institut für Pharmakologie und Toxikologie, Universität Erlangen–Nürnberg, Universitätsstraße 22, D-8520 Erlangen.

Prof. W. Kersten*, Institut für Physiologische Chemie der Universität, Wasserturmstr. 5, D-8520 Erlangen.

* To whom correspondence should be addressed.

Hoppe-Seyler's Z. Physiol. Chem.
Bd. 359, S. 421 - 425, März 1978

Differences in Intracellular Corrective Activity in Cultured Mannosidosis Cells of Acidic α-Mannosidases from Various Sources

Günther MERSMANN, Franz-Josef WILDE and Eckhart BUDDECKE

Physiologisch-Chemisches Institut, Universität Münster

(Received 21 November 1977/13 February 1978)

Summary: The ability of acidic α-mannosidases from pig kidney, bovine liver, human urine and placenta, and jack bean to correct the deranged metabolism of cultured skin fibroblasts from manosidosis patients is studied. One cause for the different corrective abilities are, as shown for other systems, the different rates of endocytosis of the enzyme forms into the cells. Besides this, it is demonstrated for the first time, that there are differences in the intracellular corrective activity of the internalized enzymes, which can be explained by different specificities of the enzymes against the storage material.

Unterschiede in der intrazellulären Korrekturfähigkeit saurer α-Mannosidasen verschiedener Herkunft in kultivierten Mannosidosezellen

Zusammenfassung: Die Fähigkeit der α-Mannosidasen mit saurem pH-Optimum aus Schweinenieren, Rinderleber, menschlichem Urin, menschlicher Plazenta und Schwertbohnen, den gestörten Stoffwechsel von kultivierten Hautfibroblasten von Mannosidosepatienten zu normalisieren, wird untersucht. Es wird gezeigt, daß die gefundenen Unterschiede einmal, wie für andere Systeme bekannt, auf der verschiedenen Aufnahmegeschwindigkeit der Enzymformen in die Zellen beruht. Daneben wird erstmals ein Unterschied in der intrazellulären Korrekturfähigkeit der aufgenommenen Enzyme demonstriert, der auf die unterschiedliche Spezifität der Enzyme für das Speichermaterial zurückgeführt wird.

Key words: α-mannosidase, mannosidosis, corrective activity, enzyme replacement.

The deranged metabolism in cultured cells of patients suffering from enzyme deficiency diseases can, for several lysosomal storage diseases, be corrected by addition of the deficient enzyme to the culture medium[1-8]. There are great differences in the corrective activity of different enzyme forms, and these differences have previously been ascribed to variations in the rate of uptake of the various enzyme forms[2-4]. We here present evidence that differences in the corrective

Enzyme:
α-Mannosidase, α-D-mannoside mannohydrolase (EC 3.2.1.24)

activity of α-mannosidases from pig kidney, bovine liver, human urine and placenta and jack bean in mannosidosis cells are not only due to their different rates of uptake, but also that the internalized enzymes exhibit different intracellular activities towards the storage material.

Experimental

Cell experiments

The methods of cell culturing, harvesting, and homogenizing, and the measurement of radioactivity and α-mannosidase activity were exactly as described[8].

Cells were preincubated with 25 μCi (160 nmol) [2-^{3}H]-mannose per 5 m*l* medium for 48 h, the medium was removed, the cells washed five times with Hanks' balanced salt solution and twice with medium, fed with medium containing the appropriate amount of α-mannosidase activity and incubated for 12 h.

Enzyme preparation

α-Mannosidase was purified from pig kidney as described[9]. A preparation of specific activity 6.7 U/mg protein was used for these experiments.

α-Mannosidase from human urine was purified by ammonium sulfate precipitation and gel filtration as reported for α-*N*-acetylglucosaminidase[10]. α-Mannosidase eluted from Sephadex G-200 slightly later than the latter enzyme. 6.6 U of the enzyme pooled after this step was dialysed against 25mM Tris/HCl, pH 8.0, and put on a column (3 x 15 cm) of ConA-Sepharose (Pharmacia, Freiburg, West Germany) equilibrated with the same buffer. Elution was done first with the starting buffer (20 fractions of 15 m*l*) and then α-mannosidase was eluted with 0.25M methyl α-D-mannoside in the starting buffer. 5.6 U of specific activity 80 U/g protein were pooled from the eluant. α-Mannosidase from bovine liver and human placenta was purified from freshly prepared tissue homogenates (1 : 3 w/v) in 25mM Tris/HCl, pH 8.0 by affinity chromatography on ConA-Sepharose, as described above for the enzyme from human urine, and subsequent gel filtration on Sephadex G-200 in 10mM sodium phosphate buffer, pH 6.0. By this procedure α-mannosidase was purified from bovine liver 46-fold (final specific activity 91 U/g) and from human placenty 90-fold (final specific activity 57 U/g).

Jack bean α-mannosidase was purchased from Boehringer Mannheim GmbH (Mannheim, West Germany).

Corrective units

The deranged metabolism in mannosidosis cells can be judged from the ratio of radioactivity in the trichloroacetic acid supernatants and precipitates of the homogenates of cells preincubated with [2-^{3}H]mannose (for details see ref.[8]). The ratios for uncorrected mannosidosis cells (M_0) and normal cells (N_0) were remeasured for each series of experiments. The ratio for cells incubated in presence of α-mannosidase activity (M_{corr}) was determined and correction was calculated as:

$$\text{Correction} = (M_0\text{-}M_{corr})/(M_0\text{-}N_0)$$

An extracellular (intracellular) corrective unit is that amount of extracellular (intracellular) α-mannosidase activity necessary to achieve half-maximal correction[7]. It was determined as the slope of the lines obtained in plots of 1/correction versus 1/α-mannosidase (extra cellular or intracellular, respectively).

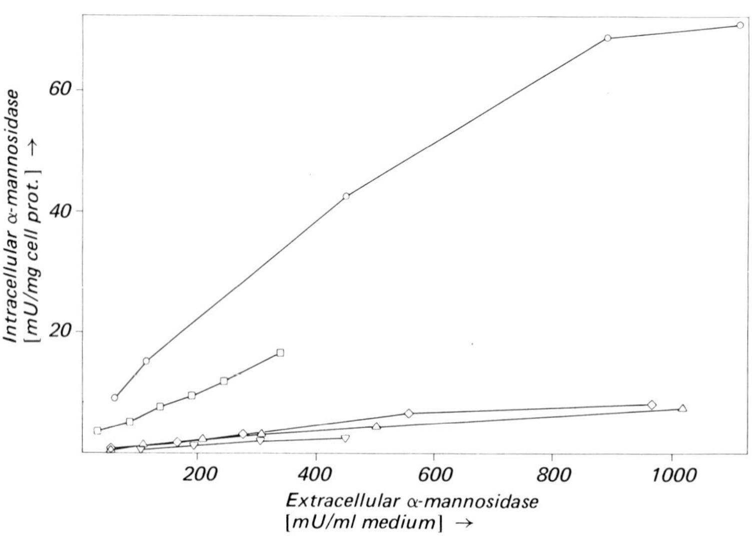

Fig. 1: Uptake of α-mannosidases from different sources into mannosidosis fibroblasts.

The intracellular activity after 12 h of incubation in culture medium containing different amounts of the various α-mannosidase forms is given. For details of incubation and assay conditions see experimental section. α-Mannosidase sources: ○——○, pig kidney; △——△, bovine liver; □——□, human placenta; ▽——▽, human urine; ◇——◇, jack bean.

Results

α-Mannosidases purified from different sources were added to the culture medium of mannosidosis cells preincubated with [2-^{3}H]mannose for 48 h and the amount of endocytosed enzyme was measured after 12 h of incubation. Fig. 1 shows that the enzymes are internalized at different rates.

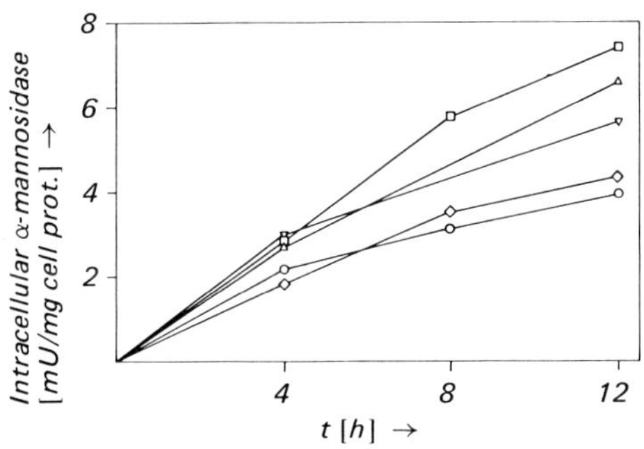

Fig. 2. Time course of intracellular α-mannosidase activity in mannosidosis cells incubated in medium containing α-mannosidase from different sources.

The extracellular activities of the enzymes were: α-mannosidase from pig-kidney, 40 U/*l* (○); bovine liver, 860 U/*l* (△); human placenta, 160 U/*l* (□); human urine, 900 U/*l* (▽); jack bean, 350 U/*l* (◇). The residual intracellular α-mannosidase activity of 0.4 mU/mg was subtracted from the measured values. For details of incubation and assay conditions see experimental section.

The endocytosis of the different enzymes is nearly linear with time over 8 - 12 h (Fig. 2). Less than 1 % of the total extracellular α-mannosidase activity was internalized within 12 h, and a decrease in the extracellular enzymatic activity was not detectable after this period. The α-mannosidases tested were not inhibited or inactivated by compounds present in the cell homogenates of control or mannosidosis cells. The intracellular half-lives of the internalized α-mannosidases from pig kidney and bovine liver were determined in chase experiments as will be described elsewhere. G. Mersmann, R. Becker and E. Buddecke, unpublished results). They are of the same magnitude (18 - 20 and about 23 h, respectively).

The deranged degradation of mannose-containing compounds in mannosidosis cells can be judged from the ratio of radioactivity in the trichloroacetic-acid-soluble and -precipitable fractions of the homogenates from cells preincubated with labelled mannose as has been shown in ref.[8]. For the cell line used for these experiments this value was about 2.0 - 2.8 for mannosidosis cells, compared to approximately 0.5 - 0.7 for normal cells.

Fig. 3 shows that the increased ratio in mannosidosis cells can be reduced to the normal value by addition of sufficient α-mannosidase activity to the culture medium. It is advantageous to quantitate the correction by dividing the difference between the ratios for non-corrected and corrected

Fig. 3. Correction of metabolism in mannosidosis fibroblasts by bovine liver α-mannosidase.

The ratio of radioactivity in trichloroacetic acid supernatant and precipitate obtained after 12 h of incubation of prelabelled cells in presence of different amounts of α-mannosidase in the culture medium. Inset: double reciprocal plot. The slope of the line is equal to the corrective unit for this enzyme. For preincubation, incubation and assay conditions and calculation of correction and corrective unit, see experimental section.

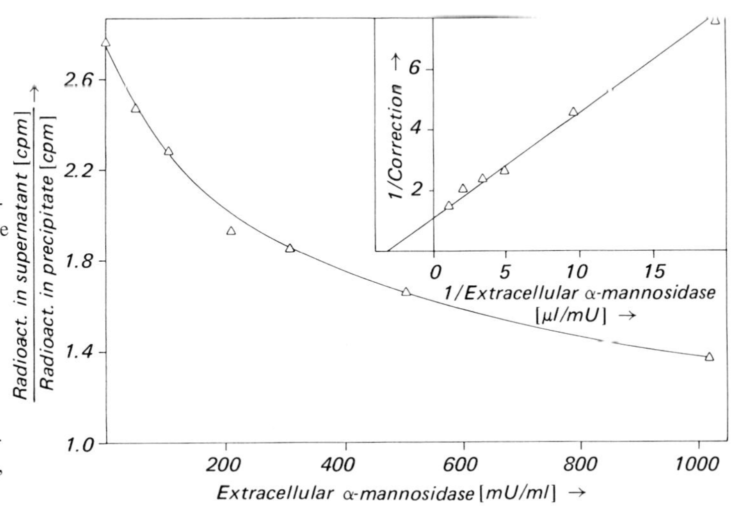

mannosidosis cells by the difference between the ratios for non-corrected mannosidosis and for normal cells (= maximal reduction). Correction is then unity for reduction of the ratio to the normal value. One corrective unit of the added enzyme is that amount of enzyme necessary for half-maximal reduction. There is a linear relation between 1/correction and 1/enzyme activity added to the medium[1,7] (insert in Fig. 3). Different α-mannosidase preparations yield lines with different slopes. The slopes are equal to the corrective units for the respective enzyme form and are given in the table (left column).

Enzyme source	Extracellular α-mannosidase [mU/m*l*]		Intracellular α-mannosidase [mU/mg cell protein]	
Pig kidney	20	(4)	3.3	(0.6)
Bovine liver	334	(11)	4.2	(0.1)
Human placenta	44	(9)	6.7	(0.3)
Human urine	605	(26)	4.8	(0.2)
Jack bean	1573	(184)	11.4	(0.3)

Table. Amounts of extracellular and intracellular α-mannosidases from different sources necessary for half-maximal correction.

For the calculation of the corrective units, see experimental section. The numbers in brackets are the standard deviations of the calculated values for the corrective units.

The right column of this table contains the slopes obtained when the reciprocal value of the enzyme activity endocytosed into the cells is taken as the abscissa.

Discussion

The ability of different preparations of α-mannosidase to correct the deranged metabolism of cultured mannosidosis cells, related to the extracellularly applied enzymatic activity, varies over a wide range (table, left column). 80 times more α-mannosidase activity from jack bean than from pig kidney must be added to the medium to obtain the same degree of correction.

Similar variations in the corrective activity of lysosomal enzymes obtained from different sources or by different methods of purification have been reported for several other lysosomal enzymes and the differences were shown to correlate with the different endocytosis rates of the enzymes[2-4]. Fig. 1 also demonstrates great differences in the uptake rates of the α-mannosidase preparations tested by us. As the uptake rates of pig kidney α-mannosidases of different specific activity are identical (not shown here), these variations are not supposed to be due to the different stages of purity but may be caused by a different content of "uptake form" of α-mannosidase in the different preparations (as shown for pig kidney α-mannosidase, G. Mersmann, R. Becker and E. Buddecke, unpublished results) or possibly by species- or tissue-specific differences in affinity to the endocytosis receptors of the cell surface. The data presented in the table (right column) demonstrate, however, that differences in the rate of uptake are not the exclusive cause for differences in corrective activity, but that, in addition, different amounts of intracellular enzyme activity from the different preparations are needed to obtain the same corrective effect. The corrective activity of the intracellular α-mannosidase activity varied for the different sources tested by a factor of up to 3.5. These differences may be explained by different susceptibility of the intracellular storage material to the degradation by the internalized enzyme forms.

Differences in specificity of multiple α-mannosidase forms against natural substrates have been described[11]. We have demonstrated here for the first time that such differences for α-mannosidases from various sources contribute to the overall corrective ability of these enzymes for α-mannosidase deficient cells. This adds another criterion, besides availability, immunological properties and rate of uptake, to be taken into account in the choice of the best enzyme for enzyme replacement therapy.

The authors thank *Dr. K. von Figura* for helpful discussions and for his advice and support in cell culturing. The excellent technical assistance of Mrs. *G. Scheidgen* is gratefully acknowledged. This work was in part supported by a grant from the *Deutsche Forschungsgemeinschaft* (SFB 104).

Literature

1 Neufeld, E.F. & Cantz, M. (1973) in Lysosomes and Storage Diseases (Hers, H.G. & van Hoof, F., eds.) pp. 262 - 275, Academic Press, New York and London.

2 Brot, F.E., Glaser, J.H., Roozen, K.J., Sly, W.S. & Stahl, P.D. (1974) *Biochem. Biophys. Res. Commun.* **57**, 1 - 8.

3 Shapiro, L.J., Hall, C.W., Leder, I.G. & Neufeld, E.F. (1976) *Arch. Biochem. Biophys.* **172**, 156 - 161.

4 Mersmann, G., von Figura, K. & Buddecke, E. (1974) *Biochim. Biophys. Acta* **364**, 88 - 96.

5 Hickman, S. & Neufeld, E.F. (1972) *Biochem. Biophys. Res. Commun.* **49**, 992 - 999.

6 Kresse, H. & Neufeld, E.F. (1972) *J. Biol. Chem.* **247**, 2164 - 2170.

7 Barton, R.W. & Neufeld, E.F. (1971) *J. Biol. Chem.* **246**, 7773 - 7779.

8 Mersmann, G., von Figura, K. & Buddecke, E. (1976) *Hoppe-Seyler's Z. Physiol. Chem.* **357**, 641 - 648.

9 Mersmann, G. & Buddecke, E. (1977) *Int. J. Biochem.* **8**, 763 - 768.

10 von Figura, K. (1977) *Eur. J. Biochem.* **80**, 525 - 533.

11 Hultberg, B., Lundblad, A., Masson, P.K. & Öckerman, P.A. (1975) *Biochim. Biophys. Acta* **410**, 156 - 163.

Dr. G. Mersmann, stud.med.dent. F.-J. Wilde and Prof. Dr. E. Buddecke, Physiologisch-Chemisches Institut der Universität Münster,
Waldeyerstr. 15, D-4400 Münster.

Hoppe-Seyler's Z. Physiol. Chem.
Bd. 359, S. 427 - 429, März 1978

SHORT COMMUNICATION

The Activity of Glutathione Reductase in A_2-Cell and B-Cell-Rich Pancreatic Islets

Ambjörn ÅGREN, Christian BERNE and Sven E. BROLIN

Department of Histology, University of Uppsala, Sweden

(Received 14 November 1977/23 January 1978)

Aktivität der Glutathion-Reduktase in A_2- und B-Zellen-reichen Langerhans-Inseln

Zusammenfassung: In den Langerhans-Inseln kann Glutathion-Reduktase NADH zur Reduktion von oxidiertem Glutathion verwenden, allerdings nur in geringem Maße, wie In-vitro-Versuche mit NADPH als Vergleich gezeigt haben. Eine hohe Aktivität der Glutathion-Reduktase wurde bisher als Charakteristikum der B-Zellen angesehen und der Umsatz von Glutathion wurde mit der Sekretion von Insulin in Zusammenhang gebracht. Wie bei Versuchen an Mäusen besitzen auch die Langerhans-Inseln aus Meerschweinchen unmittelbar nach Isolierung durch Kollagenase und nach 7tägiger Kultur auf 6.1 M Glucose eine hohe Aktivität des Enzyms, höher als die des frischen Lebergewebes.
Zerstört man die B-Zellen mit Streptozotocin, so erhält man Präparate, in denen A_2-Zellen angereichert sind; auch diese zeigen eine bemerkenswert hohe Aktivität der Glutathion-Reduktase, sowohl unmittelbar nach der Präparation als auch nach 7tägiger Kultur. Sie liegt in derselben Größenordnung wie die der B-Zellen. Die Versuche weisen darauf hin, daß in beiden Zelltypen trotz der verschiedenen Hormonsekretion ähnliche Glutathion-Redox-Systeme vorliegen.

Key words: glutathione reductase, pyridine nucleotides, pancreatic islets, tissue culture

The islets of Langerhans contain considerable amounts of reduced glutathione[1] and display also, as a characteristic feature, a high activity of glutathione reductase exceeding that of exocrine pancreas, liver, muscle tissue and kidney cortex[2]. It has been assumed that the activity of pancreatic islets mainly reflect the B-cells, and the glucagon-producing A_2-cells have not yet been studied. Among possible relationships between insulin secretion and the conversion between oxidized and reduced forms of glutathione[3], maintenance of membrane thiol groups in the reduced state may turn out to be essential [4-6]. Experimental evidence suggests that SH-groups may participate in the recognition of insulin secretagogues[7]. The disulphide form of glutathione is reduced in the reductase reaction by either NADH or NADPH, but the K_m for the latter nucleotide is generally much lower[8,9]. It has also been shown that several species of laboratory rodents contain a glutathione-dependent enzyme for insulin degradation[10].

Materials and Methods

Experiments with NADPH and NADH as a cofactor for glutathione reductase

For comparison between NADPH and NADH as regards hydrogen transfer to the disulphide form of glutathione,

Enzymes:
Collagenase (Clostridiopeptidase A)(EC 3.4.24.3);
Glutathione reductase, NAD(P)H:oxidized-glutathione oxidoreductase (EC 1.6.4.2).

islet samples were isolated by microdissection from freeze-dried sections cut 20 μm thick. These were obtained from two obese hyperglycaemic mice (gene symbol *ob/ob*[11]. The incubation and measuring conditions were as described previously[2] (see Table 1), except for the nucleotide concentrations, which in addition to 0.2 mmol/*l* also were kept at 2.0 mmol/*l*.

Comparisons between A_2-cell and B-cell rich islets of guinea pigs

For comparisons between islets rich in A_2- and B-cells, 30 male guinea pigs about two months of age were used (body weight 243 ± 11 g). Fluorimetric assay and the incubation preceding the measurements were carried out as described previously[2] (see Table 1) with 0.2mM NADPH. Liver specimens from guinea pigs showed an activity of 23 ± 7 μmol/(min × g dry weight), which corresponds well to 20 μmol/(min × g dry weight) found in livers of obese hyperglycaemic mice[2]. In order to obtain A_2-cell rich islets, B-cells were destroyed by a single intraperitoneal injection of streptozotocin in doses of 75 to 80 mg, depending on the body weight. The effect of streptozotocin was checked by microscopical examination of the pancreatic samples. Islets containing about 70% A_2-cells are produced by this procedure[12]. Islets from non-injected animals contain an average of 70% B-cells and are subsequently called B-cell rich islets. Both kinds of islets were isolated by a modified collagenase method[13]. The isolated islets were either immediately freeze-dried or cultured for 7 days and then subjected to freeze-drying before assay of glutathione reductase activity. The culture was carried out at 37 °C in an atmosphere consisting of 5% CO_2 in humidified air and in plastic Petri dishes containing tissue culture medium (Parker's TCM 199 solution) supplemented with 6.1 mmol/*l* glucose, 10% (w/v) calf serum, penicillin (100 U/m*l*) and streptomycin (0.1 mg/m*l*)[14,15]. On the third and fifth day of culture, the entire medium was replaced by fresh, and on the seventh day of culture, the islets were harvested mechanically with a rubber "policeman"[16]. Before the enzyme assay, the weights of the freeze-dried islet samples were estimated on a quartz fiber balance[17].

Results and Discussion

The results of the experiments with NADPH and NADH as cofactors for glutathione reductase are summarized in Table 1 and show that the enzyme activity is several times higher with NADPH as a cofactor than with NADH. When considering the intracellular concentrations of pyridine nucleotides previously reported for the pancreatic islets from these obese hyperglycaemic mice[19,20], our conclusion would be that NADH in vivo may be utilized in the reduction of glutathione, but only to a minor extent. As in other tissues, NADPH thus stands out as the predominating cofactor for the glutathione reductase reaction. It deserves mention that glutathione reductase, being involved in the cytoplasmic turnover of $NADP^{\oplus}$ and NADPH[21], may influence the carbohydrate and lipid metabolism in both A_2- and B-cells.

In accordance with the findings in mice the B-cells of guinea pigs display high activities of glutathione reductase, 2 - 3 times greater than those found in the liver. It is worthy of note that the A_2-cells possess high activities, comparable to those of the B-cells, as evidenced both in the freshly prepared and in the cultured islets, see Table 2.

Table 1. Glutathione reductase activity at different concentrations of the reduced pyridine nucleotides.

6 - 7 islet samples obtained from two mice were used to calculate the means and S.E.M. Islet samples were incubated for 20 min at 38 °C in 2.7 μ*l* at 0.1 mol Tris/*l*, pH 7.6, containing in addition to NADPH and NADH at the two concentrations given in the table, 2 mmol oxidized glutathione/*l*, 0.6 mmol EDTA/*l*, 0.2 mg bovine plasma albumin/m*l*. The formation of $NADP^{\oplus}$ was measured fluorimetrically after treatment with alkali and after destruction of the remaining reduced forms according to the principles of Lowry et al.[18].

Nucleotide	Concentration [mmol/*l*]	Enzyme activity [μmol/ (min × g dry wt.)]
NADPH	0.2	60 ± 3
NADPH	2.0	52 ± 7
NADH	0.2	2 ± 0.3
NADH	2.0	10 ± 2

Table 2. Enzyme activities of glutathione reductase in B-cell and A_2-cell-rich guinea pig pancreatic islets.

	A_2-cell-rich	B-cell-rich
Isolation with collagenase	58 ± 12 (n = 10)[a]	30 ± 4 (n = 20)[b]
Isolation with collagenase followed by culture	61 ± 15 (n = 7)	61 ± 10 (n = 12)

[a] Values are given as μmol/(min × g dry weight). Numbers of animals are given within parentheses.
[b] $p < 0.01$ for the difference in activity between freshly isolated and cultured B-cell-rich islets.

The long-term culture used in the present study increased the glutathione reductase activity of the B-cells, which may be of interest for possible connexions between redox reactions involving SH-groups in the B-cells and the insulin release process. The B-cells, however, are known to maintain their secretory response to glucose under these experimental conditions[16].

Financial support from the *Helge Ax:son Johnson Foundation,* the *Swedish Diabetes Association* and the *Swedish Medical Research Council* is gratefully acknowledged.

Literature

1 Havu, N. (1969) *Acta Endocrinol. (Kbh.) Suppl.* **139**, 61 - 92.
2 Berne, C. (1975) *J. Histochem. Cytochem.* **23**, 660 - 665.
3 Ammon, H.P.T. & Steinke, J. (1972) *Diabetes* **21**, 143.
4 Hellman, B. (1974) 8. Karlsburger Symposium über Diabetesfragen (Bibergeil, H., Fiedler, H. & Poser, U., eds). pp. 327 - 345, Zentralinstitut für Diabetes, Gerhard Katsch, Karlsburg, DDR.
5 Hellman, B., Idahl, L.-Å., Lernmark, Å., Sehlin, J. & Täljedal, I.-B. (1974) in Proc. 8th Congr. Int. Diab. Fed. (Malaisse, W.J. & Pirart, J., eds.). Excerpta Med. Congr. Series No 132, pp. 65 - 78.
6 Ammon, H.P.T., Akhtar, M.S. & Niklas, H. (1976) *Diabetologia* **12**, 376.
7 Bloom, G.D., Hellman, B., Idahl, L.-Å., Sehlin, J. & Täljedal, I.-B. (1972) *Biochem. J.* **129**, 241 - 254.
8 Scott, E.M., Duncan, I.W. & Ekstrand, V. (1963) *J. Biol. Chem.* **238**, 3928 - 3933.
9 Icén, A. (1967) *Scand. J. Clin. Lab. Invest.* **20**, *Suppl.* 96, 1 - 67.
10 Kohnert, K.-D., Hahn, H.J., Zühlke, H., Schmidt, S. & Fiedler, H. (1974) *Biochem. Biophys Acta* **338**, 68 - 77.
11 Lowry, O.H. & Passonneau, J.V. (1972). A Flexible System of Enzymatic Analysis. Academic Press, New York.
12 Petersson, B., Hellerström, C. & Gunnarsson, R. (1970) *Horm. Metab. Res.* **2**, 313 - 317.
13 Asplund, K. & Hellerström, C. (1972) *Horm. Metab. Res.* **4**, 153 - 163.
14 Andersson, A. & Hellerström, C. (1972) *Diabetes* **21**, *Suppl.* 2, 546 - 554.
15 Andersson, A., Petersson, B., Westman, J., Edwards, J., Lundqvist, G. & Hellerström, C. (1973) *J. Cell. Biol.* **57**, 241 - 247.
16 Andersson, A. (1976) *Acta Endocrinol. (Kbh). Suppl.* **205**, 283 - 293.
17 Lowry, O.H. (1953) *J. Histochem. Cytochem.* **1**, 420 - 428.
18 Lowry, O.H., Roberts, N.R. & Kapphahn, J.I. (1957) *J. Biol. Chem.* **224**, 1047 - 1063.
19 Brolin, S.E., Berne, C. & Isacsson, U. (1972) *Anal. Biochem.* **50**, 50 - 55.
20 Berne, C., Brolin, S.E. & Ågren, A. (1973) *Horm. Metab. Res.* **5**, 141 - 142.
21 Veech, R.L., Eggleston, L.V. & Krebs, H.A. (1969) *Biochem. J.* **115**, 609 - 619.

Dr. A. Ågren, Doc. Ch. Berne and Prof. S.E. Brolin, Histologiska Institutionen, Uppsala Universitet, Box 571, S-751 23 Uppsala, Sverige

Titel Nr. 3 109 002 690

 – Printed in Germany – Satz: R. & J. Blank, München; Druck: Grafik + Druck, München; Bindearbeiten: F. Steinberger & Sohn, München.

Redaktionelle Hinweise

(Wichtige Änderungen ab Januar 1976 sind am Rand markiert)

Die Zeitschrift bringt nur Experimantalarbeiten aus den Gebieten der Biochemie und Physiologischen Chemie. Die Manuskripte werden in deutscher oder in englischer Sprache veröffentlicht. Es wird vorausgesetzt, daß alle Koautoren mit der eingereichten Fassung einverstanden sind.

Aktuelle Kurzmitteilungen, deren Umfang 4 Seiten in Kleindruck (ca. 8 Manuskriptseiten) nicht übersteigen darf, werden bevorzugt bearbeitet und in das nächsterreichbare Heft aufgenommen. Die Dringlichkeit zur Prioritätssicherung ist vom Autor in einem Begleitschreiben kurz zu begründen.

Manuskripte sind in zweifacher Ausfertigung an eine der folgenden Anschriften zu senden:

Prof. Dr. A. Butenandt oder Prof. Dr. F. Lynen, Max-Planck-Institut für Biochemie, D-8033 Martinsried b. München;

Prof. Dr. Dr. G. Weitzel, Physiologisch-Chemisches Institut der Universität, Hoppe-Seyler-Str. 1, D-7400 Tübingen;

Dr. A. Dillmann, Hoppe-Seyler-Redaktion, Am Klopferspitz, D-8033 Martinsried b. München, Tel. (089) 85 85-823.

Bei der Abfassung der Manuskripte bitten wir dringend, folgende Regeln zu beachten, die im wesentlichen mit den „Instructions to Authors" von *Biochimica et Biophysica Acta* und *European Journal of Biochemistry* übereinstimmen:

Es soll in Maschinenschrift (2zeilig, auch vorgesehener Kleindruck!) und mit 4 cm breitem Rand geschrieben werden. Vorgesehene Absätze sind deutlich zu markieren.

Seite 1 des Manuskripts soll enthalten (in normaler Schrift, nicht in Großbuchstaben):

a) Titel der Arbeit
b) Autorennamen (jeweils ein Vorname ausgeschrieben)
c) Name des Instituts oder Fachbereichs, in dem die Arbeit ausgeführt wurde (möglichst in der Sprache des Herkunftlandes), vorausgesetzt, daß die Leitung des Instituts oder Fachbereichs der Veröffentlichung zugestimmt hat.
d) Kurztitel, maximal 70 Satzzeichen
e) Adresse für den Versand der Korrekturen

Auf Seite 2 folgen

a) Eventuelle Widmung
b) Vollständige Postadresse der Autoren in der Sprache des Herkunftslandes
c) Ad-hoc-Abkürzungen (nur in seltenen Fällen zugelassen)
d) Enzyme in alphabetischer Reihenfolge der empfohlenen Trivialnamen mit systematischen Namen und EC-Nummern

Auf Seite 3 folgt die *Zusammenfassung* sowie eine Übersetzung des Titels und der (eventuell erweiterten) Zusammenfassung ins Englische (bzw. Deutsche). Diese Übersetzung oder deren sprachliche Überprüfung kann auch von der Redaktion übernommen werden.

Eventuelle *Danksagungen* sind an den Schluß des Manuskriptes zu stellen.

Literaturzitate im Text sind fortlaufend zu numerieren (nicht alphabetisch) und am Schluß des Manuskripts zu bringen. Die in Chemical/Biological Abstracts üblichen Zeitschriftenabkürzungen sind zu verwenden. Zur Anordnung s. IUB Commission of Editors of Biochemical Journals (1973) *Eur. J. Biochem.* **37**, 201 - 202; (1974) *diese Z.* **355**, 1 - 3.

Beispiel: Spackman, D.H., Stein, W.H. & Moore. S. (1958) *Anal. Chem.* **30**, 1190 - 1206.

Zeichnungen sollen einmal in reproduzierbarer Form mit schwarzer Tusche gezeichnet eingereicht werden; Zahlenskalen, gemessene Größen, Einheiten und sonstige Beschriftungen sind jedoch nur leicht mit weichem Bleistift einzutragen.

Die Abbildungen sollen so groß sein, daß sie auf 1/2 oder 1/3 verkleinert werden können. Deutlich beschriftete Kopien für die Referenten sind beizufügen.

Alle *Tabellen* und *Abbildungen* sind mit Legenden zu versehen, so daß die Ergebnisse auch ohne Zuhilfenahme des Textes ersichtlich werden. Es wird gebeten, die Legenden nicht zu unterstreichen und die Abbildungslegenden nicht auf die Abbildungen selbst, sondern auf ein eigenes Blatt zu schreiben.

Symbole für die am häufigsten vorkommenden Größen und ihre Einheiten sind in Tab. 1 zusammengestellt. Wir bitten streng zu unterscheiden, ob bei Ansätzen molare Konzentrationen (mol/*l* = M ; z.B. 0.25M) oder zugesetzte Mengen (z. B. in mol, mmol, µmol oder nmol) gemeint sind.

Hinsichtlich der *Nomenklatur* richte man sich nach der IUPAC-Nomenklatur sowie nach den Empfehlungen der "IUPAC-IUB Commission on Biochemical Nomenclature": Kohlenhydrate, *Eur. J. Biochem.* **21**, 455 - 477 (1971); **25**, 4 (1972); α-Aminosäuren, *Eur. J. Biochem.* **53**, 1 - 14 (1975);

Tab. 1. Symbole für Größen und ihre Einheiten

Größe		Einheit[a]
l	Länge	m, mm, µm, nm, etc.
V	Volumen	m^3, l (dm^3), ml, etc.
λ	Wellenlänge	nm (nicht mµ oder Å)
t	Zeit	h, min, s, ms, µs, etc.
ν	Frequenz	Hz
m	Masse	kg, g, mg, µg, etc.
ρ	Dichte	g/cm^3
s	Sedimentationskoeffizient	$S = 10^{-13} s$
n	Substanzmenge	mol, mmol, etc.
M	molare Masse (m/n)	g/mol
–	Masse eines Moleküls	Dalton[b]
M_r	Molekulargewicht	(dimensionslos)
c_B, [B]	molare Konzentration einer Substanz B	M, mM, µM, etc. (besser als 10^{-3} M, etc.)
T	Temperatur (absolut)	K (nicht °K)
t, θ	Temperatur	°C
q, Q	Wärmemenge	J (besser als cal)[c]
I	Ionenstärke	M (mol/l), mM, etc.
K	Gleichgewichtskonstante	
K_m	Michaeliskonstante	
K_i	Inhibitorkonstante	
k	Geschwindigkeitskonstante	
v	Reaktionsgeschwindigkeit	mol/s (oder mmol, µmol oder nmol/s, etc.), besser als µmol/min
V (nicht V_{max})	Geschwindigkeit der enzymkatalysierten Reaktion bei unendlich großer Substratkonzentration	
I	Lichtintensität	cd
T	Durchlässigkeit (I/I_0)	(dimensionslos)
A	Absorption[d] (– lg T)	(dimensionslos)
ϵ	molarer Extinktionskoeffizient	$M^{-1} \times cm^{-1}$

[a] Präfixe für Einheiten: T, Tera (10^{12}); G, Giga (10^9); M, Mega (10^6), k, Kilo (10^3); m, Milli (10^{-3}); µ, Mikro (10^{-6}); n, Nano (10^{-9}); p, Piko (10^{-12}); f, Femto (10^{-15}); a, Atto (10^{-18}).

[b] Das Dalton ist 1/12 der Masse eines Atoms des Nuklids ^{12}C.

[c] Das Joule ist die SI-Einheit der Energie. Die Kalorie (1 cal = 4.184 J; 1 J = 0.239 cal) ist für eine Übergangszeit noch zugelassen; in jedem Fall sollten aber die Angaben auch in Joule ausgedrückt werden.

[d] Die Ausdrücke optische Dichte, Extinktion oder Absorbancy sollen nicht mehr verwendet werden.

Cyclite, *diese Z.* **350**, 523 - 535 (1969), *Eur. J. Biochem.* **57**, 1 - 7 (1975); Lipide, *diese Z.* **350**, 279 - 285 (1969); Steroide, *diese Z.* **351**, 663 - 689 (1970); Chinone mit isoprenoiden Seitenketten, Folsäure u. verschiedene andere Verbindungen, *diese Z.* **348**, 266 - 276 (1967); Corrinoide, *Eur. J. Biochem.* **45**, 7 - 12 (1974); Vitamine B_6, *Eur. J. Biochem.* **40**, 325 - 327 (1973); Carotinoide, *Eur. J. Biochem.* **25**, 397 - 408 (1972), **57**, 317 - 318 (1975); synthetische Analoga natürlicher Peptide, *diese Z.* **348**, 262 - 265 (1967); Enzyme Nomenclature, Recommendations (1972), Elsevier Publ. Comp., Amsterdam; multiple Enzymformen, *diese Z.* **353**, 852 - 854 (1972); Eisen-Schwefel-Proteine, *Eur. J. Biochem.* **35**, 1 - 2 (1973); Human-Immunglobuline, *Eur. J. Biochem.* **45**, 5 - 6 (1974); Peptidhormone, *Eur. J. Biochem.* **55**, 485 - 486 (1975). Für die Rechtschreibung ist „Jansen/Mackensen, Rechtschreibung der technischen und chemischen Fremdwörter" maßgebend.

Tab. 2 Empfohlene Abkürzungen.

Atome, isotope	=	z.B. [16-^{14}C]Östron
Atome, numeriert	=	z.B. C-5
Curie	=	Ci
Impulse pro Minute	=	Ipm (engl. cpm)
Prozent der Theorie	=	% d. Th.
prozentig	=	proz. (nicht %-ig)
Relative Wanderungsgeschwindigkeit	=	R_F
Schmelzpunkt	=	Schmp. (engl. m.p.)
Siedepunkt	=	Sdp. (engl. b.p.)
Umdrehungen pro Minute	=	U./min (engl. rpm)
Zerfälle pro Minute	=	Zpm (engl. dpm)
Zersetzung	=	Zers. (engl. decomp.)

Bezüglich der *Abkürzungen und Symbole für chemische Namen,* die in der biologischen Chemie von besonderem Interesse sind, wird auf die Empfehlungen der „IUPAC-IUB Commission on Biochemical Nomenclature" verwiesen (siehe z.B.: Allgemeines, Aminosäuren, Zucker, Phosphate, Coenzyme, *diese Z.* **348**, 245 - 265 (1967); Aminosäurederivate und Peptide, *Eur. J. Biochem.* **27**, 201 - 207 (1972); synthet. Polypeptide, *diese Z.* **349**, 1013 - 1016 (1968); Nucleinsäuren, Polynucleotide und deren Bestandteile, *diese Z.* **351**, 1055 - 1063 (1970). Andere als die dort empfohlenen Abkürzungen werden nur in seltenen Ausnahmefällen zugelassen. Sie müssen am Beginn jeder Arbeit in einer Fußnote erklärt werden.

Sonderdrucke: Von jeder Arbeit werden dem Verfasser 75 Sonderdrucke kostenlos geliefert. Die Bestellung weiterer Sonderdrucke gegen Berechnung ist möglich, muß jedoch spätestens bei der Rücksendung der Korrekturen erfolgen.

Der Verlag erwirbt mit der Annahme des Manuskriptes das Verlagsrecht und das ausschließliche Recht der Vervielfältigung, Verbreitung und Übersetzung. Der Verlag behält sich vor, den Inhalt dieser Zeitschrift für fremdsprachige Ausgaben zu verwenden.

Bezugspreis: Der Jahresbezugspreis (1 Jahrg. zu 12 Heften) beträgt DM 590,–, das Einzelheft kostet DM 56,–. Bestellungen nimmt jede Buchhandlung entgegen.

Verlag: WALTER DE GRUYTER & Co.
Genthiner Str. 13, D-1 Berlin 30. Tel. (030) 261 1341
Telegrammadresse: Wissenschaft Berlin
Postscheck: Berlin-West 10307-108

WALTER DE GRUYTER Inc.
3 Westchester Plaza
Elmsford, N.Y. 10523, U.S.A.
Tel. (914) 592-5890

Für die nächsten Hefte sind folgende Beiträge vorgesehen:

Synthesis of protected decapeptide (B_{21-30}) of human insulin, *V. K. Naithani and J. Föhles*

Comparison of tRNA-nucleotidyltransferase from *Escherichia coli* and *Lactobacillus acidophilus, M. Leineweber and G. R. Philipps*

A new method for the preparation of band 3, the main integral protein of the human erythrocyte membrane, *D. Schubert and B. Domning*

Purification and characterization of aromatic-amino-acid-glyoxylate aminotransferase from monkey and rat liver, *I. Harada, T. Noguchi and R. Kido*

Immunoreactive trypsin in serum and peritoneal fluid in acute pancreatitis, *A. Borgström and K. Ohlsson*

Studies on the involvement of lysosomes in estrogen action, I: Isolation and enzymatic properties of pig endometrial lysosomes, *W. Sierralta, A. J. Truitt and P. W. Jungblut*

Localization of kallikrein in porcine pancreas and submandibular gland as revealed by the indirect immunofluorescence technique, *T. Dietl, J. Kruck and H. Fritz*

Multiple protein kinases from *Trypanosoma gambiense, R. D. Walter*

Adenosine 3',5'-cyclic monophosphate binding proteins from *Trypanosoma gambiense, R. D. Walter*

The binding of lysolecithin to human serum high density apoprotein A-I. A ^{13}C-NMR-study, *W. Stoffel, P. Metz and B. Tunggal*

Aggregation of sponge cells. XVIII. Glycosyltransferase linked with the aggregation factor, *W. E. G. Müller, R. K. Zahn, H. J. Rohde, B. Kurelec and I. Müller*

Investigations on the functional role of rat liver carboxylesterase isoenzymes, *R. Arndt, H.-E. Schlaak, D. Uschtrin, D. Südi, K. Michelssen and W. Junge*

Immunohistological studies on the distribution of cytochrome b_5 reductase in rat liver microsomes, *H.-U. Schulze, B. Fleischer and L. Pop*

Physiological ligands of putative LSD-serotonin receptors. Heterogeneity of LSD-displacing factors in human body fluids and nervous tissue, *E. Mehl and L. Guiard*

Release of granulocyte eleastase in lethal canine endotoxin shock, *A. O. Aasen and K. Ohlsson*

Uptake of thymidine into isolated rat hepatocytes. Evidences for two transport systems, *R. F. Ungemach and D. Hegner*

Vitamin-B_6-Biosynthese in *Bacillus subtilis, W. Pflug and F. Lingens*

Enhanced phosphorylation of ribosomal protein S6 and other cytoplasmic proteins after mengovirus infection of Ehrlich ascites tumor cells, *I. Rosnitschek, U. Traub and P. Traub*

E. Buddecke

Pathobiochemie

Ein Lehrbuch für Studierende und Ärzte

17,0 cm × 24,0 cm. Etwa XXXV, 446 Seiten. Mit 247 Abbildungen und zahlreichen Tabellen. 1978. Plastik flexibel DM 34,–
ISBN 3 11 007526 1

Das Lehrbuch **Pathobiochemie** beschreibt die biochemischen Grundlagen genetischer und erworbener Störungen des Stoffwechsels, Abweichungen in der chemischen Struktur der Bausteine des menschlichen Körpers, fehlerhafte biochemische Prozesse, soweit sie sich als Symptome mit Krankheitswert manifestieren.

Die **Pathobiochemie** umfaßt die Hauptabschnitte: Stoffwechsel, Stoffwechselregulation, Zellen, Gewebe und Organe und Dynamische Systeme.

Die **Pathobiochemie** liefert der Medizin neue theoretische Grundlagen, zeigt, in welchen Bereichen ein erfolgreicher Vorstoß bis zu einer „molekularen Krankheitslehre" möglich ist, und vermittelt als fachgebietsübergreifende Wissenschaft auch Beziehungen zu zahlreichen Nachbargebieten wie z. B. zur Pathologie, Immunologie, Pharmakologie, Klinischer Chemie und Innerer Medizin.

Bei der Auswahl des Stoffes wurde die revidierte und neugegliederte **2. Auflage (1978)** des **Gegenstandskataloges „Pathophysiologie** und **Pathobiochemie"** eingehend berücksichtigt.

Ein entsprechendes **Korrelationsregister** zum **GK** ist dem Buch beigefügt.

Preisänderung vorbehalten

d 359 Heft 4 April 1978

N 0018-4888 A 20679 E

oppe-Seyler's Zeitschrift für hysiologische Chemie

3. Adolf-Butenandt-Vorlesung
29. Mosbacher Kolloquium und 24. Konferenz der Gesellschaft für Biologische Chemie

Herausgegeben von

A. Butenandt, F. Lynen, G. Weitzel

unter Mitwirkung von K. Bernhard, K. Decker, J. Engel, H. Fritz, B. Hamprecht,

H. Hanson, E. Helmreich, H. Herken, B. Hess, N. Hilschmann, H. Hilz, P. W. Jungblut,

arlson, H. L. Kornberg, K. Kühn, F. Leuthardt, D. Oesterhelt, K. Rajewsky, J. Seelig, G. Siebert,

mon, Hj. Staudinger, W. Stoffel, H. Tuppy, H. Wiegandt, H. G. Wittmann, H. G. Zachau, H. Zah

Redaktion A. Dillmann, G. Peters

WALTER DE GRUYTER · BERLIN · NEW YORK

Hoppe-Seyler's Z. Physiol. Chem. Bd. 359 S. 431 - 546 April 1978
HSZPAZ 359 (4) 431 - 546 (1978)

Hoppe-Seyler's Zeitschrift für Physiologische Chemie
Begründet 1877 · Fortgeführt von A. Kossel, F. Knoop und K. Thomas

Bd. 359 April 1978 Heft 4

Inhaltsverzeichnis

Fortsetzung nächste Seite

Contents

Continued

Redaktionelle Hinweise siehe gegenüber Seite 546

Indexed in Current Contents

Autoren-Verzeichnis

Hoppe-Seyler's Z. Physiol. Chem.
Bd. 359, S. 431 - 447, April 1978

Bindung von Progesteron und anderen Steroidhormonen an Proteine des Blutserums

DRITTE ADOLF-BUTENANDT-VORLESUNG
gehalten vor der Gesellschaft für Biologische Chemie in Frankfurt/Main am 14. März 1978

Ulrich WESTPHAL

Biochemistry Department, University of Louisville School of Medicine

Herrn Professor Butenandt zum 75. Geburtstag gewidmet. Dieser Bericht soll zugleich ein Ausdruck der Verehrung und Dankbarkeit sein für alles, was der Jubilar dem Verfasser für seine wissenschaftliche Laufbahn – und für sein Leben – mitgegeben hat.

Die ehrenvolle Einladung der Deutschen Gesellschaft für Biologische Chemie, die dritte Adolf Butenandt-Vorlesung zu halten, war für mich eine große Freude, wie ich mit besonderer Dankbarkeit zum Ausdruck bringen möchte. Sie gibt mir einmal eine schöne Gelegenheit, Herrn Professor Butenandt meine Wünsche zu seinem 75. Geburtstag selbst zu überbringen. Zum andern erlaubt sie mir noch einmal, wie so oft vor langen Zeiten, meinem alten Chef einen Arbeitsbericht vorzulegen über das, was ich mit meinen Mitarbeitern in den letzten Jahren getan habe – ich hoffe daß er das Resultat mit Milde beurteilen wird.

Fast auf den Tag vor 50 Jahren begann ich das Studium der Chemie im Institut von Professor Windaus in Göttingen. Da war ein Saalassistent im Organischen Praktikum, gerade erst 25 Jahre alt; sein Name war Adolf Butenandt. Ich höre noch wie heute die Stimme des Volksmundes, das heißt, der weisen älteren Semester des Laboratoriums, wie sie zu mir sagten: „Das ist der kommende Mann"! Niemand von uns ahnte, wie wahr sich dies Urteil erfüllen sollte.

Die Welt kennt heute das Forschungswerk Butenandts. Generationen von Studenten haben Professor Butenandt als akademischen Lehrer erlebt und sind durch seine formvollendeten Vorlesungen für die Biochemie als treibendes Grundelement des Lebens begeistert worden. Diejenigen aber, die den Vorzug hatten, Adolf Butenandt als ihren Doktorvater zu haben – eine Verwandtschaftsbezeichnung, die in der Sprache meiner neuen Heimat unbekannt ist – wissen, wieviel er uns, zusätzlich zu dem wissenschaftlichen, für unser Leben mitgegeben hat. Seine Persönlichkeit mit ihrem das Gesamtleben des Menschen umfassenden Interessenkreis war es, die die Atmosphäre für den „Arbeitskreis Butenandt" bestimmte. Für jeden, der empfänglich war, war diese Erfahrung ein Leitstern für sein ganzes Leben. Ich bin mir dessen bei unzähligen Gelegenheiten, im wissenschaftlichen wie im menschlichen Bereich, dankbar bewußt geworden.

Bald nach Isolierung der Keimdrüsenhormone durch Butenandt und seine Mitarbeiter[1-3] wurden mancherlei Überlegungen angestellt, wie diese Wirkstoffe wohl ihre spezifische Aktivität entfalten könnten. Unter uns Kollegen im Butenandtschen Arbeitskreis war es besonders Hans Friedrich-Freksa, der unsere Gedanken mit

---

Abkürzungen:
AAG, α_1-saures Glykoprotein oder Orosomucoid;
CBG, Corticosteroid-bindendes Globulin oder Transcortin;
HSA, Humanserumalbumin;
PBG, Progesteron-bindendes Globulin;
SBP, Sexualsteroid-bindendes Protein = TeBG, Testosteron-Östradiol-bindendes Globulin;
HNBB, 2-Hydroxy-5-nitrobenzylbromid.

seinen Vorstellungen über mögliche Mechanismen der Hormonwirkung am Erfolgsorgan stimulierte und bereicherte. Wer Paul Ehrlichs Lehre Glauben schenkte: „Corpora non agunt nisi fixata", mußte überzeugt sein, daß der erste Schritt der physiologischen Funktion eines Hormons die Assoziation an die spezifischen Zellen des Zielorgans sei. Die stoffliche Grundlage der lebenden Substanz stellten im Wissen der Zeit die Proteine dar; die fundamentale Bedeutung der Nucleinsäuren für das Lebensgeschehen war noch nicht im einzelnen erkannt.

Durch die Verlagerung des Kaiser Wilhelm-Instituts für Biochemie von Berlin-Dahlem nach Tübingen während des Zweiten Weltkrieges kamen Professor Butenandt und seine Arbeitsgruppe in engen Kontakt mit Professor Bennhold, dem Direktor der Medizinischen Universitätsklinik. Bennhold hatte durch seine Arbeiten über „Die Vehikelfunktion der Serumeiweißkörper"[4] ein bedeutsames neues Gebiet erschlossen, das in seinem Heimatlande unverdientermaßen nur geringe Beachtung fand. Er zeigte in einer Reihe klassischer Arbeiten, daß die verschiedenartigsten niedrigmolekularen Stoffe – saure und basische Farbstoffe, Vitamine und Hormone, natürliche Pigmente und synthetische Arzneimittel, Sterine und Gallensäuren – an Albumin und andere Serumproteine gebunden werden, und daß diese Bindung durch Dissoziation des Komplexes in einfacher Weise gelöst werden kann[5]. Die noch weiter zurückliegende Auffindung einer Serumeiweißbindung herzwirksamer Steroide durch Oppenheimer[6], sowie die Beobachtung der Bindung östrogener Steroide an Plasmaproteine durch Brunelli[7] blieben ebenfalls unbeachtet.

Als ich nach dem Kriege im chemischen Laboratorium der Bennholdschen Klinik den Problemen der Proteinbindung niedrigmolekularer Stoffe, wie Bilirubin und Azorubin, näher kam, kehrten meine Gedanken zu den früheren Überlegungen der Bindung eines Hormons an spezifische Eiweißstrukturen des Erfolgsorgans zurück. Offenbar synthetisierte ein Organismus, in dem Hormone wirksam sind, Proteine mit der spezifischen Eigenschaft, Hormone an sich zu binden; dies mußte der erste Schritt einer biologischen Wirkung sein. Solche Proteine werden heute Rezeptoren genannt. Sie waren damals unbekannt, und ihre chemische Natur ist bis heute nicht aufgeklärt.

Die Fähigkeit des tierischen Organismus, Proteine mit spezifischen Bindungsaffinitäten zu synthetisieren war jedoch in der Immunchemie, in der Bildung von Antikörpern gegen körperfremde Proteine als Antigene, bekannt. Darüber hinaus hatten die Arbeiten Landsteiners und anderer Autoren gezeigt, daß das Immunsystem auch Antikörper gegen niedrigmolekulare Stoffe herstellen kann, wenn diese als Haptene an immunogene Proteine gekoppelt werden[8]. Auf dieser Basis versuchte ich vor 30 Jahren, gemeinsam mit meinem Doktoranden Erich Meinzinger, spezifische Antikörper gegen Östradiol zu gewinnen. In gleicher Weise sollten Antikörper gegen Diäthylstilböstrol bereitet werden, um das bindende Protein für das natürliche Hormon mit dem Antikörper gegen das synthetische Östrogen zu vergleichen. Das schien von Interesse, da wir die spezifischen Antikörper als Modelle für Bindungsproteine der Zielorgane betrachteten, und da Diäthylstilböstrol weitgehend die gleiche biologische Wirkung entfaltet wie Östradiol. Dieses Arbeitsprogramm ist in der Dissertation von E. Meinzinger[9] dargelegt. Ein 4- (oder 2-) Aminoöstradiol wurde synthetisiert (Ref.[10], S. 446 - 448) und sollte über die Azobindung an Eiweiß gekoppelt werden. Die Arbeit wurde unterbrochen, als ich, vor fast 30 Jahren, meine Tätigkeit in die Vereinigten Staaten verlegte. Erst jetzt haben wir die Untersuchung spezifischer, gegen Progesteron gerichteter Antikörper wieder aufgegriffen.

Anläßlich der ersten Butenandt-Vorlesung hat vor diesem Gremium Professor Elwood Jensen von der Universität Chicago über den Stand der Erforschung der Steroid-Rezeptorproteine gesprochen, an deren Entdeckung und Bearbeitung er mit seinen Mitarbeitern bahnbrechend gewirkt hat [11,12]. Diese Proteine sind es, die die Steroidhormone im Zielorgan aufnehmen und in den Zellkern einschleusen, so daß die biologische Wirkung am Genort realisiert werden kann. Wir wollen heute einen Schritt zurückgehen und das Steroidhormon von seinem Eintritt aus seiner Bildungsstätte in das zirkulierende Blut, und auf dem langen Weg in der Blutbahn verfolgen, bis es schließlich von den Zellen des Erfolgsorgans aufgenommen wird. Hier ist es, wo die Steroid-bindenden Serumproteine ihre Funktion erfüllen.

Diese Funktion beginnt unmittelbar wenn das neu synthetisierte Hormon seinen Bildungsort verläßt und durch die Wand der Blutkapillaren in die Zirkulation gelangt. Vor einigen Jahren stellte ich die Hypothese auf, daß die Steroid-bindenden Serumproteine diese Passage erleichtern[13]. Dieser Vorschlag gründete sich auf Versuche (Ref.[10], S. 188), in denen Corticosteron mittels Dialyse durch Cellulosemembranen aus Rattenserum entfernt werden sollte. Tab. 1 zeigt daß die Dialysegeschwindigkeit durch die Gegenwart von Serumalbumin in der Außenflüssigkeit erhöht wird; Tierkohle hat einen ähnlichen, obwohl geringeren, Effekt. Die wirksamste Entfernung von Corticosteron wurde erreicht, wenn die Dialyse gegen das hochaffine Corticosteroid-bindende Globulin (CBG oder Transcortin) erfolgte, obwohl das Verhältnis Außen: Innenvolumen in diesem Falle 10mal kleiner war[14]. Wir nehmen an, daß die Serumproteine den Durchtritt der Steroidhormone durch das Endothel der Kapillarwände erleichtern, indem sie die (vornehmlich hydrophoben) Steroide binden und somit ihre chemische Aktivität an der Durchtrittsstelle verringern.

Tab. 1. Entfernung von Corticosteron aus verdünntem Rattenserum durch Dialyse (40 °C, 24 h).

Expt.	Außenflüssigkeit[a]	Entfernt [%]
1	Phosphatpuffer, pH 7.6	25
2	Puffer-Tierkohle, 2.5 proz.	38
3	HSA-Lösung, 1 proz.	46
4	CBG im Serum nebennierenloser Ratten, 1 : 5 verdünnt	51

[a] Außenvolumen; Innenvolumen war 13 : 1 in Exp. 1 - 3, und 1.3 : 1 in Exp. 4. Nach Ref.[14].

Diese Annahme ist neuerdings experimentell bewiesen worden[15]. Ewing und Mitarbeiter in Baltimore fanden in Perfusionsversuchen am Kaninchenhoden, daß die Sekretion von Testosteron auf das vierfache erhöht wird, wenn der Perfusionsflüssigkeit 3% Rinderserumalbumin zugesetzt wird. Zusatz von 3% Dextran, das den osmotischen Druck des Mediums aufrechterhält, aber keine nennenswerte Bindungsaffinität für das Steroidhormon hat, läßt die Testosteronsekretion unverändert. Die Autoren deuten diesen Befund im Sinne der oben angeführten Hypothese.

Die Bindung der Steroidhormone an die Proteine des Serums hat eine Unterdrückung der biologischen Aktivität zur Folge. Der Wirkstoff wird in einer inaktiven Form im Blut transportiert und stellt somit eine indifferente Hormonreserve dar. Da die Eiweißbindung jedoch nichtkovalenter Natur ist, kann der Komplex jederzeit am Erfolgsorgan dissoziieren und somit das freie, biologisch aktive Hormon verfügbar gemacht werden. Die Proteinbindung schützt das Steroidhormon weiterhin vor chemischem und enzymatischem Angriff, vor Aufnahme durch die Leberzellen, metabolischer Umwandlung und Ausscheidung. Das Resultat ist eine Konservierung des lebenswichtigen Wirkstoffes.

Die wichtigsten Steroid-bindenden Proteine des menschlichen Blutserums sind in Tab. 2 zusammengestellt. Albumin, das in der weitaus höchsten Konzentration vorkommt, bindet Progesteron mit der geringsten Affinität; das Sexualsteroid-bindende Protein (SBP), das in geringster Menge vorhanden ist, hat die höchste Bindungsstärke für die androgenen und östrogenen Hormone. Da die Verteilung der Steroidhormone im Bindungsgleichgewicht von Proteinmenge und Affinität abhängt, spielt das Albumin wegen seiner hohen Konzentration als Transportprotein eine erhebliche Rolle – eine Tatsache, die oftmals übersehen wird.

Die allgemeinen chemischen Eigenschaften der Steroid-bindenden Serumproteine sollen hier nicht im einzelnen behandelt werden. Humanserumalbumin (HSA) weist die geringste Spezifität in der Bindung seiner Liganden auf[21], und die Affinität für Steroidhormone folgt der Polaritätsregel, nach der das Steroid mit stärkstem hydrophobem Charakter am festesten gebunden wird (Ref.[10], S. 101 - 132). Verglichen mit HSA ähnelt das α_1-saure Glykoprotein (AAG) in seiner Bindungsfähigkeit für Steroide und in anderen Eigenschaften (Ref.[10], S. 375 - 433) mehr den hochaffinen Glykoproteinen CBG und SBP. Nach chemischen Modifikationsversuchen sind Tryptophan, Lysin und Tyrosin an der Progesteronbindung durch AAG beteiligt[22]. Die molekularen Eigenschaften des CBG sind in verschiedenen Laboratorien erarbeitet worden (Ref.[10], S. 315 - 355)[23-25]; die Gegenwart von Histidin [18,26,27], Methionin[27], Tryptophan[18] und Arginin[18] nahe der Bindungsstelle ist wahrscheinlich gemacht worden. Diese neueren Ergeb-

Tab. 2. Steroidbindende Proteine im Humanserum.

Protein	M_r	Kohlenhydrat [%]	Konz. [µM]	Gebund. Steroid[a]	Assoziationskonstante [μM^{-1}] 4 °C	37 °C	Lit.
HSA	69000	–	550	P	0.36	0.18	16
AAG	41000	42	18	P	1.5	0.6	17
CBG	52000	26	0.7	P	700	26	18
				C	790	40	18
SBP	94000	32	0.04	T	1100	350	19
				DHT	2400	990	19
				E	600	220	19
PBG[b]	88000	71	13	P	2200	350	20

[a] P, Progesteron; C, Cortisol; T, Testosteron; DHT, Dihydrotestosteron = 17β-Hydroxy-5α-androstan-3-on; E, Östradiol.
[b] Vom Serum des trächtigen Meerschweinchens.

nisse wurden an CBG-Präparaten gewonnen, die mittels der Affinitätschromatographie gereinigt worden waren. Dieselbe Methodik hat auch die Herstellung des reinen SBP erleichtert[19,28].

Butenandt hat vor Jahren auf das Ökonomieprinzip hingewiesen, das die Natur bei der Synthese der Steroidhormone anwendet, um eine Grundstruktur durch geringfügige chemische Veränderungen in zahlreiche Wirkstoffe mit völlig verschiedenartigen biologischen Spezifitäten zu verwandeln. In ähnlicher Weise erfüllen auch die Steroid-bindenden Glykoproteine des Serums mehr als eine Funktion. Das Corticosteroid-bindende Globulin assoziiert das Ovarialhormon Progesteron mit derselben hohen Affinität wie die Nebennierenrindenhormone Cortisol oder Corticosteron; SBP spielt eine Doppelrolle indem es sowohl die androgenen Steroide, Testosteron und Dihydrotestosteron, als auch Östradiol bindet (Ref.[10], S. 356 - 374). Das Sexualsteroid-bindende Protein wird daher auch Testosteron-Östradiol-bindendes Globulin, TeBG, genannt. Diese Eiweißstoffe haben eine einzige hochaffine Bindungsstelle für die gegebenen Steroidhormone in ihrem Molekül.

Dieses sind die Proteine mit denen die Steroidhormone unmittelbar nach Eintritt in die Blutbahn feste Komplexe bilden. Wie schnell erfolgt diese Assoziation? Wie hoch ist die Dissoziationsgeschwindigkeit der Komplexe, wenn die freien Hormone am Zielorgan benötigt werden? Was passiert mit der Proteinstruktur, wenn das Steroid gebunden wird? Was wissen wir über die chemische Natur der Bindungsstelle? Können wir uns ein ungefähres Bild von dem Bindungsort machen, an dem die verschiedenen Steroidhormone mit hoher Affinität gebunden werden? Das sind die Fragen, denen wir nun nachgehen wollen.

1) Kinetik der Assoziation und Dissoziation

Unsere Untersuchungen über den Mechanismus der Steroidbindung sind bisher meist mit dem Progesteron-bindenden Globulin (PBG) des trächtigen Meerschweinchens durchgeführt worden. Wir entdeckten dies Protein vor etwa 10 Jahren[29]; es wurde in mehreren Laboratorien rein dargestellt, wobei neben anderen Methoden auch die Affinitätschromatographie angewendet wurde[30]. PBG eignet sich in idealer Weise für die Bearbeitung grundlegender Probleme der Interaktionen zwischen Steroiden und Proteinen; es findet sich im Serum des trächtigen Meerschweinchens in der relativ hohen Konzentration von etwa 1 mg m*l*, und ist stabil bei erhöhter Temperatur und extremen pH-Werten. Zu den ungewöhnlichen Eigenschaften des PBG gehören der hohe Kohlenhydratgehalt von etwa 70%, und der niedrige isoelektrische Punkt von 2.8. PBG ist ein polydisperses Glykoprotein mit einem Polypeptidkern des Molekulargewichtes 27000. Durch Affinitätschromatographie ist eine Molekülform mit einem Molekulargewicht von 88 000 in reiner Form erhalten worden; die Polydispersität zeigt sich jedoch im Vorkommen einer lückenlosen Reihe von Molekulargewichten zwischen 70000 und 120000, von denen einzelne Vertreter isoliert worden sind. Das reine PBG besitzt eine einzige hochaffine Bindungsstelle für Progesteron und andere Steroide; die Affinitätskonstante, K_a, des Progesteronkomplexes beträgt ungefähr $2 \times 10^9 \text{M}^{-1}$ bei 4 °C und pH 7.4.

Die Bestimmung der Assoziations- und Dissoziationsgeschwindigkeiten von PBG-Steroid-Komplexen wurde sehr erleichtert durch die Entdeckung[31], daß die Eigenfluoreszenz des PBG durch die Bindung von Progesteron und anderen Δ^4-3-Ketosteroiden gelöscht wird (Abb. 1). Bei der Assoziation von Progesteron mit reinem PBG tritt ein Verlust der Fluoreszenzintensität von über 80% ein. Diese Fluoreszenzlöschung erfolgt hauptsächlich durch strahlungslose Energieübertragung vom angeregten Tryptophan des Proteins zu dem nah-ultravioletten Absorptionsband des Δ^4-3-Ketochromophors. Diese Resonanzübertragung ist möglich, weil das Absorptionsspektrum des 3-Oxo-4-en-steroids dem Fluoreszenzemissionsspektrum des PBG überlagert ist[31]. Steroide ohne Chromophor mit solch überlagerter Absorption bewirken keine Fluoreszenzlöschung, auch wenn sie mit gleich hoher Affinität an das PBG gebunden sind.

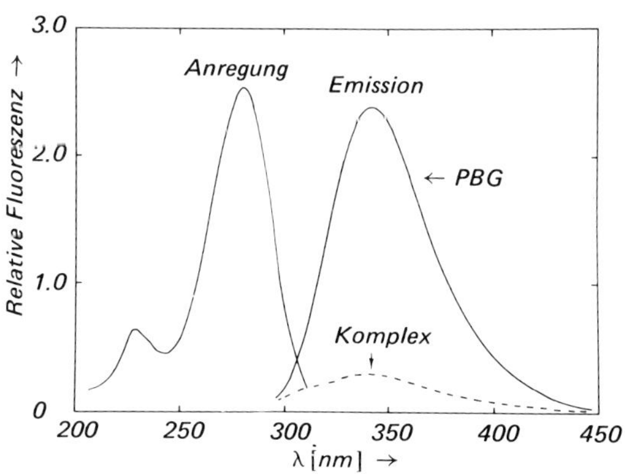

Abb. 1. Fluoreszenzanregungs- und Emissionsspektren von PBG und dem PBG-Progesteron-Komplex.

Die Emission des Anregungsspektrums wurde bei 340 nm gemessen; die Anregung der Emissionsspektren erfolgte bei 280 nm.

Da maximale Löschung der Fluoreszenz eintritt, wenn ein mol Steroid einen stöchiometrischen Komplex mit einem mol PBG bildet, kann man die Fluoreszenzänderung zur Grundlage einer titrimetrischen Konzentrationsbestimmung der Bindungsstellen in einer PBG-Lösung machen[31]. Kleine Anteile der Steroidlösung bekannter Konzentration werden der Proteinlösung zugesetzt und die relative Fluoreszenz wird gemessen (Abb. 2). Der Äquivalenzpunkt wird graphisch erhalten und definiert die Konzentration der Bindungsstellen in PBG. Aus dem Kurvenanteil nahe dem Äquivalenzpunkt läßt sich die Assoziationskonstante, K_a, berechnen.

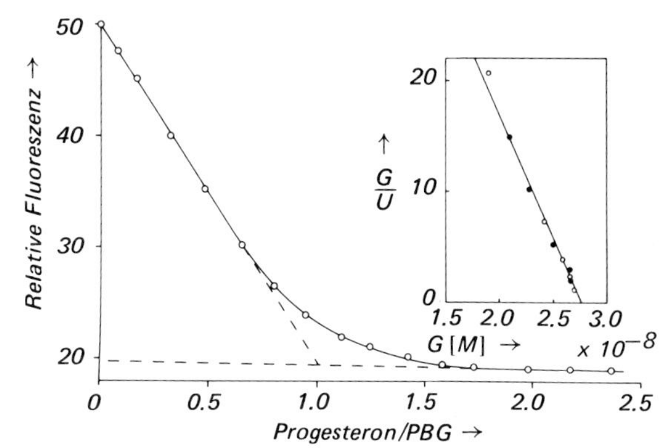

Abb. 2. Fluoreszenzlöschungstitration von PBG mit Progesteron.

Die Nebenfigur zeigt die Auswertung der Daten der Hauptfigur nach der graphischen Methode von Scatchard. G = gebunden, U = ungebunden. Nach Ref.[31].

Wird die Fluoreszenzlöschung in einem Stop-Flow-Spektrophotometer gemessen, so kann die Assoziationsgeschwindigkeit zwischen PBG und Steroid quantitativ bestimmt werden[20]. Die gleiche Methodik läßt sich auch zur Messung der Dissoziationsgeschwindigkeit der Komplexe verwenden. In diesem Falle nutzen wir die Tatsache aus, daß ein „transparentes" Steroid, das keinen zur Energieaufnahme geeigneten Chromophor besitzt, aber fest vom PBG gebunden wird, das Δ^4-3-Ketosteroid vom PBG verdrängt. Diese Verdrängung bewirkt eine Umkehr der Fluoreszenzlöschung, und die Fluoreszenz des PBG nimmt wieder ihren ursprünglichen Wert an (Abb. 3). Die

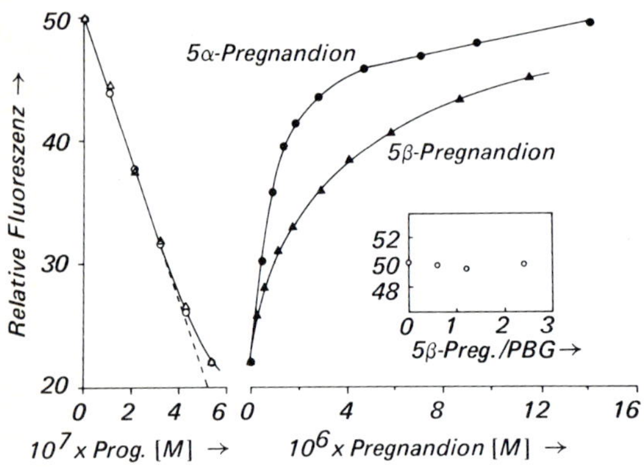

Abb. 3. Fluoreszenzlöschungstitration von PBG mit Progesteron und anschließende Zugabe von überschüssigem 5α-Pregnan-3,20-dion oder 5β-Pregnan-3,20-dion, worauf Progesteron verdrängt und die ursprüngliche Fluoreszenz wiederhergestellt wird.

Die Nebenfigur zeigt, daß die PBG-Fluoreszenz durch Zugabe von mehr als einem mol 5β-Pregnandion pro mol PBG nicht weiter verändert wird. Nach Ref.[31].

Geschwindigkeit der Fluoreszenzzunahme ist durch die Dissoziationsgeschwindigkeit des PBG-Δ^4-3-Ketosteroid-Komplexes bestimmt[20].

Die Geschwindigkeitskonstanten der Assoziation und Dissoziation, k_1 und k_{-1}, für den Progesteron-PBG-Komplex im Temperaturbereich von 4 - 45 °C sind in Tab. 3 zusammengestellt. Der k_1-Wert steigt in diesem Bereich auf das 10fache an, während k_{-1} zwischen denselben Temperaturen 100mal größer wird. Die Affinitätskonstante, die durch k_1/k_{-1} gegeben ist, fällt daher von 4 bis 45 °C um den Faktor 10. Die Temperaturabhängigkeit der Affinität wird also hauptsächlich durch Änderung der Dissoziationsgeschwindigkeit bestimmt. Tab. 3 zeigt weiterhin, daß die kinetischen Affinitätskonstanten, k_1/k_{-1}, mit den durch Gleichgewichtsdialyse erhaltenen K_a-Werten gut übereinstimmen.

Das Primat der Dissoziationsgeschwindigkeit über die Assoziationsgeschwindigkeit in der resultierenden Affinitätskonstanten findet man auch bei den PBG-Komplexen anderer Steroide (Tab. 4). Die Assoziationsgeschwindigkeit der Steroide mit K_a-Werten $> 10^8 \text{M}^{-1}$ (20°) ist innerhalb der Fehlergrenzen die gleiche. Die Dissoziationsgeschwindigkeiten dieser Komplexe sind jedoch sehr verschieden, so daß k_1/k_{-1} im Verhältnis bis zu 1 : 20 variiert. Wiederum ist es die Dissoziationsgeschwindigkeit, die die Affinität der Bindung zwischen PBG und den Steroiden bestimmt[20].

Der kontrollierende Einfluß der Dissoziationsgeschwindigkeit ist jedoch nicht von allgemeiner Gültigkeit. Bestimmung der Kinetik von Steroidkomplexen mit einem anderen Serumprotein, dem menschlichen CBG, ergibt ein andersartiges Bild[18]. Die Dissoziationsgeschwindigkeit des CBG-Komplexes mit Progesteron ist 5mal größer als die des CBG-Cortisol-Komplexes, obwohl die Assoziationskonstanten nicht sehr verschieden sind (Tab. 5). Dies ist nur möglich, wenn sich auch die Assoziationsgeschwindigkeiten unterscheiden; die Berechnung ergibt einen 4fach größeren k_1-Wert für den Progesteronkomplex, verglichen mit dem Cortisolkomplex. Direkte Bestimmung der Assoziationsgeschwindigkeiten mittels Stop-Flow-Fluoreszenz bestätigt die berechneten Konstanten (Tab. 5).

Tab. 3. Progesteron-PBG-Komplex. Assoziationskonstanten von kinetischen Daten und Gleichgewichtsdialyse.
Nach Ref.[20].

Temperatur [°C]	$10^{-7} \times k_1$ $[\text{M}^{-1}\,\text{s}^{-1}]$	k_{-1} $[\text{s}^{-1}]$	$10^{-9} \times k_1/k_{-1}$ $[\text{M}^{-1}]$	$10^{-9} \times K_a$ (Gleichgewicht) $[\text{M}^{-1}]$
4	2.2	0.0095	2.4	2.2
14	4.3	0.031	1.4	1.4
23	7.4	0.083	0.89	1.0
37	16.2	0.39	0.41	0.35
45	24.6	1.05	0.23	0.22

Tab. 4. Kinetik der PBG-Komplexe mit verschiedenen Steroiden bei 20 °C. Nach Ref.[20].

Steroid	$10^{-7} \times k_1$ [M^{-1} s^{-1}]	k_{-1} [s^{-1}]	$10^{-8} \times k_1/k_{-1}$ [M^{-1}]	$10^{-8} \times K_a$ [M^{-1}]
Progesteron	8.8	0.060	14.5	20
Desoxycorticosteron	8.3	0.12	6.7	10
Testosteron	7.8	0.43	1.8	2.9
Testosteronacetat	9.3	0.15	6.1	9.2
Medrogeston*	8.5	0.024	35.3	45.5
Corticosteron	~ 5	1.4	0.4	0.2
Cortisol	~ 5	90	–	0.02

* 6,17α-Dimethyl-4,6-pregnadien-3,20-dion.

Tab. 5. Kinetik einiger Steroid-CBG-Komplexe bei 20 °C.

	$10^{-8} \times K_a$ (Gleichgewichts-dialyse) [M^{-1}]	k_{-1} (Fluorometrie) [s^{-1}]	$10^{-7} \times k_1$ (ber. als $K_a \times k_{-1}$) [M^{-1} s^{-1}]	$10^{-7} \times k_1$ (gemessen durch Stop-Flow) [M^{-1} s^{-1}]
Cortisol	3.9	0.031	1.2	1.4
Corticosteron	2.8	0.047	1.3	–
Progesteron	2.8	0.16	4.5	5.8

Ein weiterer Unterschied zwischen den PBG- und CBG-Komplexen zeigt sich im Vergleich der Dissoziationsgeschwindigkeiten von Steroiden mit kontinuierlich zunehmender Polarität (Hydroxylgruppen). Bei den Komplexen mit CBG[18] hat Progesteron den höchsten, Cortisol den niedrigsten Wert für k_{-1} (Tab. 6). Das umgekehrte Verhältnis trifft für die PBG-Komplexe zu[20]: Der k_{-1}-Wert des Cortisolkomplexes ist 1500mal größer als der des Progesteronkomplexes. Wir nehmen an, daß dieser grundlegende Unterschied dadurch bedingt ist, daß die Steroidbindung am

Tab. 6. Dissoziationsgeschwindigkeitskonstanten von Steroidkomplexen mit CBG und PBG bei 20 °C, pH 7.4.

Steroid	CBG k_{-1} [s^{-1}]	PBG k_{-1} [s^{-1}]
Cortisol	0.031	90
Corticosteron	0.047	1.4
Desoxycorticosteron	0.10	0.12
Progesteron	0.16	0.06

CBG mehr hydrophilen Charakter hat, was typisch ist für „polare" Steroide wie Cortisol, während die charakteristische Affinität des PBG auf mehr hydrophobe Steroide wie Progesteron gerichtet ist.

Wenn wir die physiologische Bedeutung der Dissoziations- und Assoziationsgeschwindigkeiten der Steroid-Proteinkomplexe verstehen wollen, so ist es zweckmäßig, diese Werte für verschiedene Typen von Bindungsproteinen zu vergleichen. In Tab. 7 sind die k_1- und k_{-1}-Werte für die Steroidkomplexe mit 5 Transportproteinen des Serums und mit verschiedenen Rezeptoren der Zielorgane zusammengestellt. Es ist offenbar, daß die k_{-1}-Werte für alle Serumproteine um 1, 2 oder 3 Größenordnungen höher sind als die k_{-1}-Werte für die Rezeptorproteine. Eine hohe Dissoziationsgeschwindigkeit macht das freie Hormon den Zellen der Erfolgsorgane unmittelbar verfügbar, so daß es von den Rezeptoren des Cytoplasmas aufgenommen werden kann. Demgegenüber hat die Bindung des Steroidhormons an die intrazellulären Rezeptorproteine eine andere Funktion: das Hormon bleibt an den Rezeptor gebunden, wäh-

Tab. 7. Assoziations- und Dissoziationsgeschwindigkeitskonstanten für steroidbindende Proteine bei 0 - 4 °C.

Protein	Steroid[a]	$10^{-4} \times k_1$ $[M^{-1} s^{-1}]$	$10^5 \times k_{-1}$ $[s^{-1}]$	$10^{-8} \times k_1/k_{-1}$ $[M^{-1}]$	Lit.
PBG, Meerschweinchen	P	2200	950	24	20
CBG, Mensch	C	–	230	–	32
CBG, Mensch	C	220	280	7.9	18
CBG, Mensch	P	840	1200	7.0	18
CBG, Ratte	C	–	50	–	33
SBP, Mensch	T	–	90	–	34
SBP, Mensch	DHT	–	15	–	34
SBP, Kaninchen	DHT	–	220	–	35
Huhn Ovidukt Rezeptor A	P	28	1.9	150	36
Huhn Ovidukt Rezeptor B	P	63	2.4	260	36
Huhn Ovidukt Rezeptor, gereinigt	P	–	1.6	–	37
Rattenleber Rezeptor	D	1.4	2.0	7.3	33
Rattenuterus Rezeptor	E	–	3.9[b]	–	38

[a] P, Progesteron; C, Cortisol; T, Testosteron; DHT, Dihydrotestosteron; D, Dexamethason (9α-Fluor-11β,17α,21-trihydroxy-16α-methyl-1,4-pregnadien-3,20-dion); E, Östradiol.
[b] 25 °C.

rend Konformationsänderungen und Reaktionen zur Translokalisierung in den Zellkern statthaben. Die Geschwindigkeitskonstanten für die verschiedenen Steroidkomplexe in Tab. 7 stehen daher in Übereinstimmung mit den geforderten Funktionen.

2) *Konformationsänderungen*

Vor vielen Jahren beobachtete ich, daß die molare Absorptivität des Progesterons durch Bindung an HSA erniedrigt wurde (Abb. 4); dieser hypochrome Effekt ging mit einer geringfügigen Blauverschiebung des Absorptionsmaximums, λ_{max}, einher[39]. Die Spektraländerungen waren durch Transferierung des Steroids in eine mehr hydrophobe Umgebung, und durch Perturbation aromatischer Aminosäuren bei der Komplexbildung bewirkt. Als wir entsprechende Spektralmessungen an dem Progesteronkomplex mit PBG durchführten, fanden wir[40], daß der Absorptionskoeffizient des Progesterons durch die Bindung an das Meerschweinchenprotein erhöht wurde (Abb. 5). Eine ähnliche Hyperchromie sahen wir auch bei der Assoziation von PBG mit anderen Δ^4-3-Ketosteroiden. Wie bei der Bindung an HSA trat auch hier eine Blauverschiebung von λ_{max} um einige nm auf.

Zur Erklärung der Hyperchromie bei der Steroidbindung an PBG, und der Hypochromie bei der

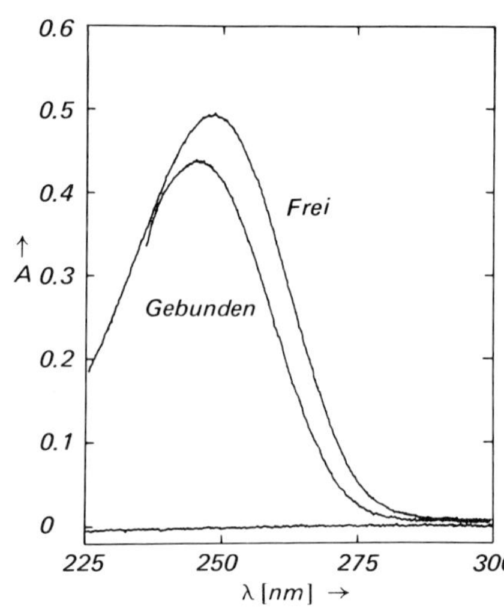

Abb. 4. Absorptionsspektrum des Progesterons, frei in wäßriger Lösung und gebunden an HSA.

Albuminbindung, haben wir die Exciton-Theorie der Hypochromie[41] herangezogen. Obwohl diese Theorie ursprünglich entwickelt wurde, um die Absorptionserniedrigung von Polynucleotiden zu erklären, kann sie auch zur qualitativen Interpretation der Spektralbeobachtungen an proteingebundenen Steroiden verwendet werden. Die Hy-

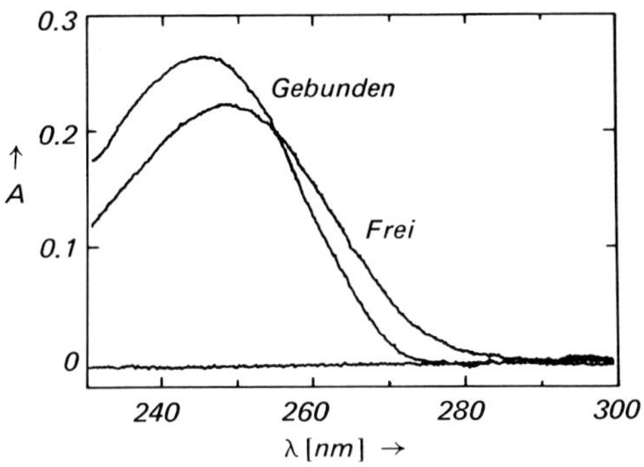

Abb. 5. Absorptionsspektrum des Progesterons, frei in wäßriger Lösung und gebunden an PBG.
Nach Ref.[40].

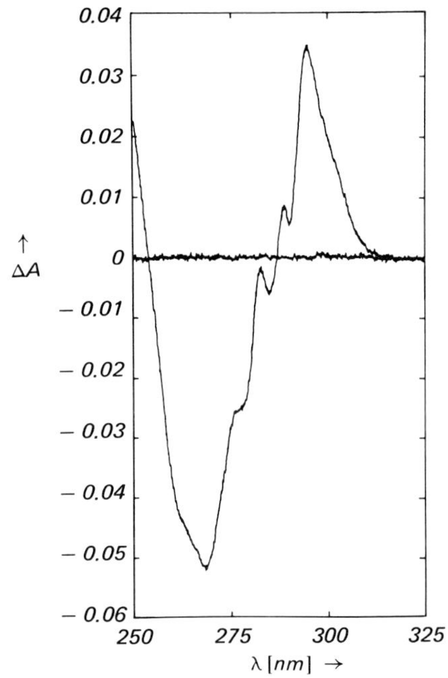

Abb. 6. Differenzspektrum des PBG-Progesteron-Komplexes.
Nach Ref.[40].

pochromie im Falle der Bindung an HSA könnte dann davon herrühren, daß der ungeordnete Chromophor (Steroid) durch die Bindung in solcher Weise orientiert wird, daß sein elektrisches Dipolmoment parallel zum Dipolmoment des Chromophors im Protein gelagert wird. Die Hyperchromie bei der Assoziation mit PBG würde verständlich wenn man annimmt, daß die Übergangsmomente der Chromphore von Steroid und Protein durch die Komplexbildung in einer Kopf-Schwanz-Anordnung ausgerichtet werden[40].

Zur genaueren Analyse der bei der Steroidbindung auftretenden Spektraländerungen wurden Differenzspektren der PBG-Komplexe mit einer Reihe von Steroiden aufgenommen[40]. Abb. 6 zeigt das Differenzspektrum des PBG-Progesteron-Komplexes. Das tiefe Minimum bei 268 nm ist dem Steroid mit seinem 3-Oxo-4-en-Chromophor zuzuschreiben; bei den PBG-Komplexen mit gesättigten Steroiden gleichstarker Bindungsaffinität ist es nicht vorhanden. Das negative Signal bei 268 nm rührt in erster Linie von einer Verschärfung des Absorptionsbandes des PBG-gebundenen Steroids her; geringere Beiträge werden von der Blauverschiebung des Absorptionsmaximums und von der Proteinperturbation geliefert. Das prominente Maximum bei 294 nm stammt von der Perturbation eines Tryptophanrestes; die kleineren Gipfel bei 288 und 283 nm zeigen Perturbation von Tyrosin und Tryptophan an[42]. Konformationsänderungen im PBG-Molekül als Folge der Steroidbindung wurden auch durch Circulardichroismus und Dispersion der optischen Drehung nachgewiesen[42].

Mit der Bindung der Steroide an PBG gehen tiefgreifende Veränderungen in den Absorptionsspektren der Steroide und des Proteins einher. Diese Perturbationen sind das Resultat starker und spezifischer Interaktionen zwischen Steroidmolekül und Aminosäureresten in der Bindungsstelle. Welche Aminosäuren finden sich hier?

3) *Chemische Natur der Bindungsstelle in PBG*

Die Teilnahme des *Tryptophanrestes* an der Assoziation der Steroide mit PBG folgt aus den besprochenen spektralen Perturbationen und aus der Fluoreszenzlöschung, deren Lage im Spektrum (Abb. 1) charakteristisch für Tryptophan ist. Die Bedeutung dieses Aminosäurerestes läßt sich auch durch chemische Modifikationsversuche aufzeigen[43]. Läßt man PGB mit 2-Hydroxy-5-nitrobenzylbromid (NHBB) reagieren, einem Reagens, das in spezifischer Weise Tryptophan substitutiert*, so wird die Bindungsstelle für

* PBG ist frei von Sulfhydrylgruppen.

Tab. 8. Inaktivierung von PBG durch HNBB und Schutz der Bindungsstelle durch Progesteron.
Nach Ref.[30].

Reagentien [mol pro mol PBG]	Trp modif. [mol pro mol PBG]	Bindungsanalyse durch Gleichgewichtsdialyse, 4 °C	
		% Bindungsstellen	K_a (M^{-1})
0	–	100	1.1×10^9
20 HNBB	1.2	10	1.1×10^9
10 Progest., 20 HNBB	0.7	80	0.8×10^9

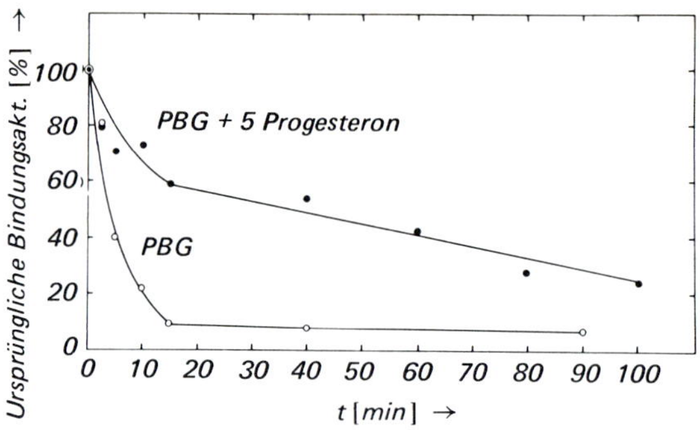

Abb. 7. Inaktivierungsgeschwindigkeit von ungeschütztem PBG und von PBG-Progesteron-Komplex durch Trinitrobenzolsulfonsäure.

Die Bindungsaktivität wurde durch multiple Gleichgewichtsdialyse bestimmt. Nach Ref.[43].

Progesteron inaktiviert (Tab. 8); unter bestimmten Bedingungen mit einem 20fachen molaren Überschuß an HNBB waren etwa 90% der Bindungsstellen nicht länger nachweisbar. Bei gleichen Reaktionsbedingungen in Gegenwart von 10 mol Progesteron pro mol PBG wurden nur 20% der Bindungsstellen inaktiviert. Wir nehmen an, daß die Assoziation des Progesterons an der Bindungsstelle den Tryptophanrest vor der Reaktion mit HNBB schützt. In einer anderen Versuchsreihe wurde Parallelität gefunden zwischen der Anzahl eingesetzter mol HNBB pro mol PBG, der Anzahl der eingeführten Hydroxynitrobenzylreste und der Abnahme der aktiven Bindungsstellen, bestimmt durch Fluoreszenzlöschungstitration oder Gleichgewichtsdialyse[30].

Modifikation der *Lysinreste* in PBG durch Trinitrobenzolsulfonsäure führte ebenfalls zum Verlust der Affinität für Progesteron[30]. Auch hier hinderte die Gegenwart von Progesteron die chemische Reaktion, so daß eine kleinere Zahl von Lysinresten durch Trinitrophenylreste substituiert wurde. Zugleich mit der Behinderung der Substitution wurde die Inaktivierung der Steroidaffinität verlangsamt (Abb. 7).

Einführung von Nitrogruppen in die *Tyrosinreste* von PBG mittels Tetranitromethan reduzierte die Zahl der Bindungsstellen für Progesteron auf 3 - 10% des ursprünglichen Wertes. Bildung des Progesteronkomplexes schützte das PBG vor dem Angriff durch das Tetranitromethan unter sonst gleichen Bedingungen, sodaß die Nitrierung der Tyrosinreste gehemmt wurde und etwa zwei Drittel der Bindungsstellen in aktiver Form erhalten blieben[43].

Die Ergebnisse dieser chemischen Umwandlungsversuche am PBG zeigen, daß Tryptophan, Lysin, und Tyrosin an der Bindung des Progesterons beteiligt sind. Covalente Substitution inaktiviert die Bindungsstelle. Komplexbildung mit Progesteron schützt die Seitenketten dieser Aminosäuren vor chemischer Modifikation, und die Bindungsstelle vor Inaktivierung. Die veränderte chemische Reaktionsbereitschaft der Aminosäurereste ist entweder die Folge einer direkten Interaktion mit Progesteron am Bindungsort, oder resultiert von Konformationsänderungen, die mit der Komplexbildung einhergehen.

Auf einem anderen Wege zur Aufklärung der chemischen Natur der Steroidbindungsstelle haben wir PBG der Affinitätsmarkierung mit radioaktivem 11α-Bromacetoxyprogesteron unterworfen, um die Aminosäuren in der Nähe des gebundenen Steroids identifizieren zu können[44]. Diese Versuche sind noch nicht abgeschlossen.

4) *Einfluß der Steroidstruktur auf die Bindungsaffinität für PBG*

Eine indirekte Methode, über die Natur der Bindungsstelle in PBG Aufschluß zu erhalten, ist die Untersuchung des Einflusses struktureller und sterischer Änderungen im Steroidmolekül auf die Bindungsstärke. Wir haben daher die Assoziationskonstanten, K_a, der PBG-Komplexe von über hundert Steroiden der Pregnan- und Androstan-

Tab. 9. Bindungsstärke der PBG-Komplexe: Einführung von Hydroxylgruppen in das Progesteronmolekül. Nach Ref.[45].

Steroid	=O	–OH	$10^{-8} \times K_a$ [M^{-1}]	ΔG° [a] [kcal/mol]	$\delta\Delta G^{\circ}$ [b] [kcal/mol]
Progesteron	2	0	18.5	– 12.55	–
Desoxycorticosteron	2	1	8.6	– 12.11	+ 0.4
Corticosteron	2	2	1.4	– 11.04	+ 1.5
Cortisol	2	3	0.02	– 8.54	+ 4.0

[a] Obwohl die 2. Dezimale nicht signifikant ist, wird sie zur zuverlässigeren Berechnung von $\delta\Delta G^{\circ}$ angegeben.
[b] Beitrag der eingeführten Gruppe(n) zur freien Bindungsenergie.

reihe bestimmt[45]; hierfür wurde die Fluoreszenzlöschungstitration, oder, wenn das Steroid keinen geeigneten Chromophor besaß, eine Gleichgewichtsdialyse mit Bindungskonkurrenz benutzt. Durch Vergleich jeweils zweier Steroide, die sich nur in einem Substituenten oder einem anderen strukturellen Merkmal unterscheiden, kann der positive oder negative Effekt einer bestimmten molekularen Änderung auf die Bindungsaffinität abgeschätzt werden. Auf diese Weise läßt sich der Einfluß von „Polarität" (Besitz polarer Gruppen), Hydrophobie, Raumerfüllung, und anderer sterischer und konformationeller Eigenschaften untersuchen. Für einen besseren Vergleich haben wir auch die freie Energie (Gibbs) der Bindung, ΔG°, der Komplexe berechnet, sodaß die jeweiligen Beiträge der strukturellen Änderungen auf die thermodynamische Stabilität der PBG-Steroidkomplexe quantitativ erfaßbar worden[45]. Da die relativ hohen Bindungsaffinitäten vieler Steroide für PBG eine nahe Assoziation von Ligand und Protein voraussetzen[31], kann man auf diese Weise ein Komplementärbild der wahrscheinlichsten Form der Bindungsstelle ableiten, und Gebiete hydrophober und hydrophiler Bindung definieren. Weiterhin können wir Auskunft erhalten über den in der Bindungsstelle verfügbaren Raum, über die Lokalisierung von Wasserstoffdonatoren oder -akzeptoren für die Wasserstoffbrücken, und über andere Charakteristika, die auf die Bindungsaffinität Einfluß haben.

Die Einführung von Hydroxylgruppen in das Progesteronmolekül führt zu einer Herabsetzung der Bindungsstärke, und zwar proportional zu der Zahl der OH-Gruppen (Tab. 9). Die drei Hydroxylgruppen in Cortisol erniedrigen K_a 1000fach, und tragen 4 kcal/mol zur freien Bindungsenergie bei. Dies kann nicht einfach als additiver Effekt erklärt werden, denn die Beispiele in Tab. 10 zeigen, daß der Einfluß einer einzelnen Hydroxylgruppe sehr von ihrer Stellung im Steroidmolekül und dessen übrigen strukturellen Eigenschaften abhängt. Der Befund, daß Einführung von Hydroxylgruppen ausnahmslos die K_a-Werte der Steroid-PBG-Komplexe herabsetzt, zeigt, daß die Interaktion der Polaritätsregel[10] folgt, d.h., daß die Bindung vornehmlich hydrophob ist. Eintritt von Oxogruppen hat in den meisten Fällen eine Reduktion der Assoziationskonstante zur Folge, obwohl auch hier die Stärke dieses Effektes durch die Stellung im Steroidgerüst bedingt ist (Tab. 11).

Tab. 10. Bindungsstärke der PBG-Steroid-Komplexe: Einführung einzelner Hydroxylgruppen. Nach Ref.[45].

Steroid	$10^{-8} \times K_a$ [M^{-1}]	$\delta\Delta G^{\circ}$ [a] [kcal/mol]
Progesteron	18.5	–
21-Hydroxy-	8.6	+ 0.4
11β-Hydroxy-	1.8	+ 1.4
2α-Hydroxy-	0.02	+ 4.6
Androstendion	2.3	–
11β-Hydroxy-	1.1	+ 0.4
17,21-Dihydroxy-progesteron	3.3	–
11β-Hydroxy-	0.02	+ 3.0

[a] Beitrag der eingeführten Gruppe zur freien Bindungsenergie.

Tab. 11. Bindungsstärke der PBG-Steroid-Komplexe: Einführung von Oxogruppen. Nach Ref.[45].

Steroid	$10^{-8} \times K_a$ [M^{-1}]	$\delta\Delta G^{\circ}$ [a] [kcal/mol]
Progesteron	18.5	–
21-Oxo-	3.8	+ 0.9
11-Oxo-	0.07	+ 3.3
Androstendion	2.3	–
11-Oxo-	0.03	+ 3.1

[a] Beitrag der eingeführten Gruppe zur freien Bindungsenergie.

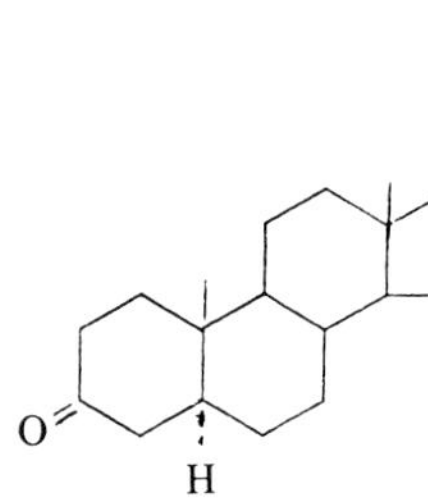

III

$K_a = 21.0 \times 10^8 M^{-1}$
$\Delta G^{\circ} = -11.8$ kcal/mol

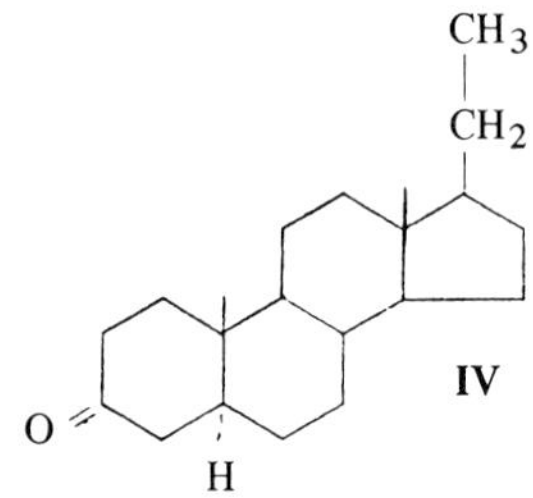

IV

$K_a = 16.9 \times 10^8 M^{-1}$
$\Delta G^{\circ} = -11.7$ kcal/mol

Wir beobachteten zwei Ausnahmen von der Polaritätsregel bei den Ketosteroiden[45]. Die Oxogruppe an C-3 ist wesentlich für hohe Affinität zu PBG. Die Steroide der Formel I und II unterscheiden sich nur durch Gegenwart oder Abwesenheit der 3-Oxogruppe; ihre Entfernung (Formel II) führt zu einer 150fachen Erniedrigung der Assoziationskonstanten.

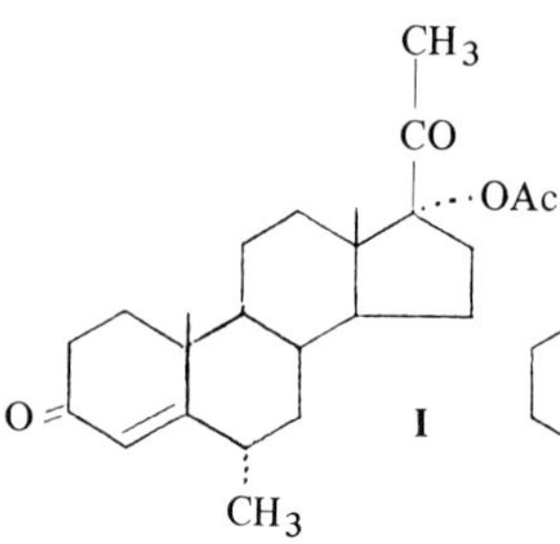

I

$K_a = 6.6 \times 10^8 M^{-1}$
$\Delta G^{\circ} = -11.1$ kcal/mol

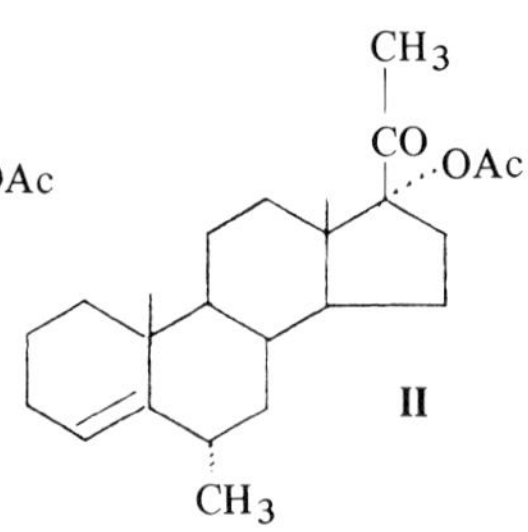

II

$K_a = 0.04 \times 10^8 M^{-1}$
$\Delta G^{\circ} = -8.4$ kcal/mol

Im Gegensatz zu der Bedeutung der 3-Oxogruppe ist die Oxogruppe an C-20 für die Bindungsaffinität unwesentlich. Die Assoziationskonstante von 5α-Pregnan-3,20-dion (III) ist innerhalb der Fehlergrenze die gleiche wie die des 5α-Pregnan-3-ons (IV).

Ein sterischer Effekt auf die Bindungsaffinität zeigt sich beim Vergleich der PBG-Komplexe 6α- und 6β-substituierter Steroide (Abb. 8). In allen untersuchten Beispielen hat der äquatoriale 6α-Substituent den stärkeren Einfluß: die 6α-OH-Gruppe erniedrigt K_a mehr als die 6β-OH-Gruppe; die 6α-Methylderivate des Androstendions, Testosterons, und 17-Methyltestosterons werden etwa 3 - 4mal stärker gebunden als die entsprechenden 6β-Derivate[45].

Ein interessantes Beispiel für die Bedeutung der Raumerfüllung stellt der PBG-Komplex des Testosteronacetats dar. Testosteron (V) unterscheidet sich strukturmäßig von Progesteron (VII) nur durch den Besitz einer Hydroxylgruppe an C-17β statt einer Acetylgruppe; die Assoziationskonstan-

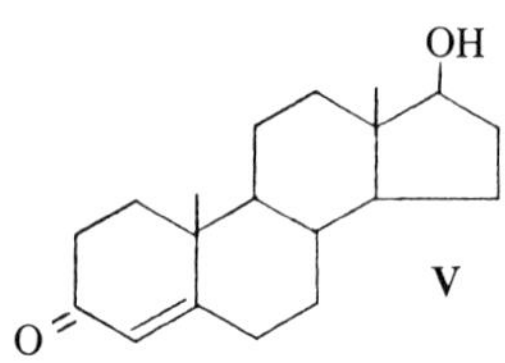

V

Testosteron
$K_a = 2.9 \times 10^8 M^{-1}$
$\Delta G^{\circ} = -11.5$ kcal/mol

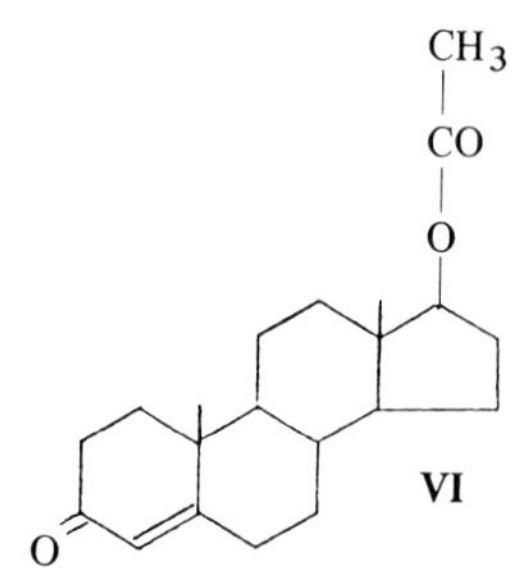

VI

Testosteronacetat
$K_a = 9.2 \times 10^8 M^{-1}$
$\Delta G^{\circ} = -12.2$ kcal/mol

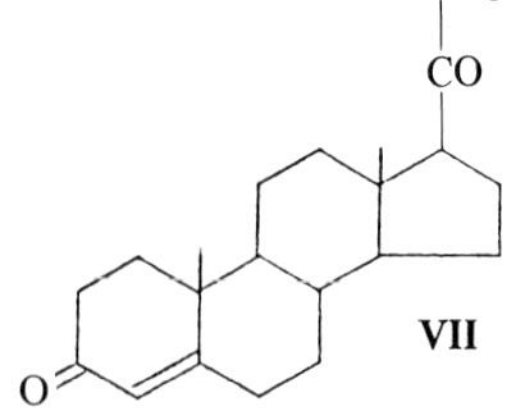

VII

Progesteron
$K_a = 18.5 \times 10^8 M^{-1}$
$\Delta G^{\circ} = -12.6$ kcal/mol

ten liegen um einen Faktor von mehr als 6 auseinander. Wird die Hydroxylgruppe des Testosterons acetyliert (VI), so steigt K_a um mehr als das dreifache und erreicht die Hälfte des Progesteronwertes. Diese Annäherung der Bindungsaffinität ist ein Ausdruck für die ähnliche Raumerfüllung der Steroide VI und VII, was besonders gut an Stuartmodellen sichtbar ist[45]. Der Unterschied zwischen den beiden Molekülen ist lediglich die Einfügung eines Estersauerstoffes zwischen C-17 und die Acetylgruppe.

Die optimale Einpassung des Steroids in die Bindungsstelle des PBG hängt nicht nur von den Substituenten an den verschiedenen Kohlenstoffatomen ab, sondern auch von der Gesamtkonformation des Steroidmoleküls. Es ist seit langem bekannt, daß zwischen 5α- und 5β-Steranen ein Konformationsunterschied besteht, der in einer starken Abbiegung des A-Ringes um die C(5)-C(10)-Bindung in dem β-Isomeren zum Ausdruck kommt[46]. Die Assoziationskonstante des PBG-Komplexes mit 5α-Pregnan-3,20-dion ist 6mal größer als die des Komplexes mit 5β-Pregnan-3,20-dion[45]. Abb. 9 enthält neben diesen Daten die Konformationsbilder wie sie sich aus der kristallographischen Röntgenstrukturanalyse von 5α- und 5β-Steroiden, am Beispiel zweier Androstanderivate, ergeben[47]. Die 5α-Struktur ist fast eben, während die 5β-Konformation die charakteristische Abbiegung des A-ringes erkennen läßt. Als Maß für den Grad der Abknickung ist der Winkel (– 65.1°) angegeben, den die Ebene des A-Ringes mit der horizontalen Ebene der Ringe B, C, und D bildet[47]. Diese Abbiegung kann auch durch die Entfernung der 3-Oxogruppe von der horizontalen B-C-D-Ringebene (3 Å) ausgedrückt werden (Abb. 9).

Einführung einer C(1,2)-Doppelbindung in ein 3-Oxo-4-en-steroid hat ebenfalls eine Abbiegung des A-Ringes zur Folge; ein Biegungswinkel von 49.1°, und eine Distanz von 3.2 Å zwischen der 3-Oxogruppe und der B-C-D-Ringebene ist für das 17β-Hydroxy-1,4-androstadien-3-on gefunden worden[47]. Beim Vergleich der PBG-Komplexe von 11β-Hydroxy-4-androsten-3,17-dion (K_a = 1.1 × 10^8M^{-1}) und 11β-Hydroxy-1,4-androstadien-3,17-dion (K_a = 0.02 × 10^8M^{-1}), die sich nur durch Ab- oder Anwesenheit der C(1,2)-Doppelbindung unterscheiden, hat die 1,4-Dienstruktur mit dem abgebogenen A-Ring eine 50mal kleinere Assoziationskonstante[45]. Die Differenz der ΔG°-Werte beträgt 2.9 kcal/mol.

Der Eintritt einer 6-en-Doppelbindung in das Molekül der Δ^4-3-Ketosteroide, ebenso wie die Entfernung der C(19)-Methylgruppe, führt zu einer mehr geebneten, flacheren Konformation[47]. Das Resultat ist eine Zunahme der Assoziationskonstanten[45], wie in Tab. 12 für einige Beispiele ersichtlich ist.

Sterischer Einfluß
$10^{-8} \times K_a\,[M^{-1}]$

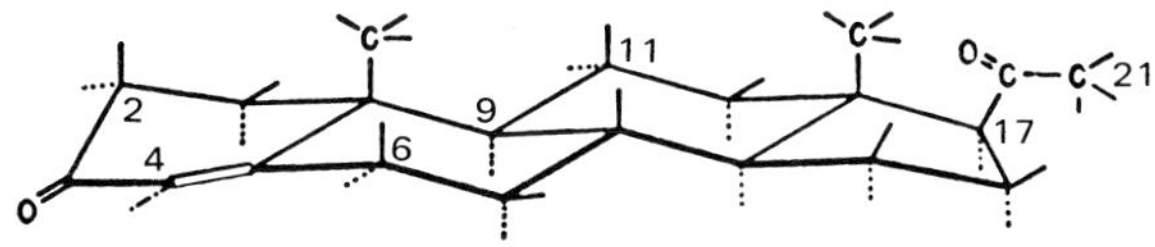

-Progesteron

6α-OH-	1.6
6β-OH-	3.2

	-Androstendion	-Testosteron	-17-Methyltestosteron
6α-CH_3-	9.9	8.5	13.0
6β-CH_3-	3.2	3.6	4.1

Abb. 8. Sterischer Effekt auf die Assoziationskonstanten von PBG-Steroid Komplexen.
Der äquatoriale 6α-Substituent bindet mit vornehmlich hydrophober Region. Nach Ref.[45]

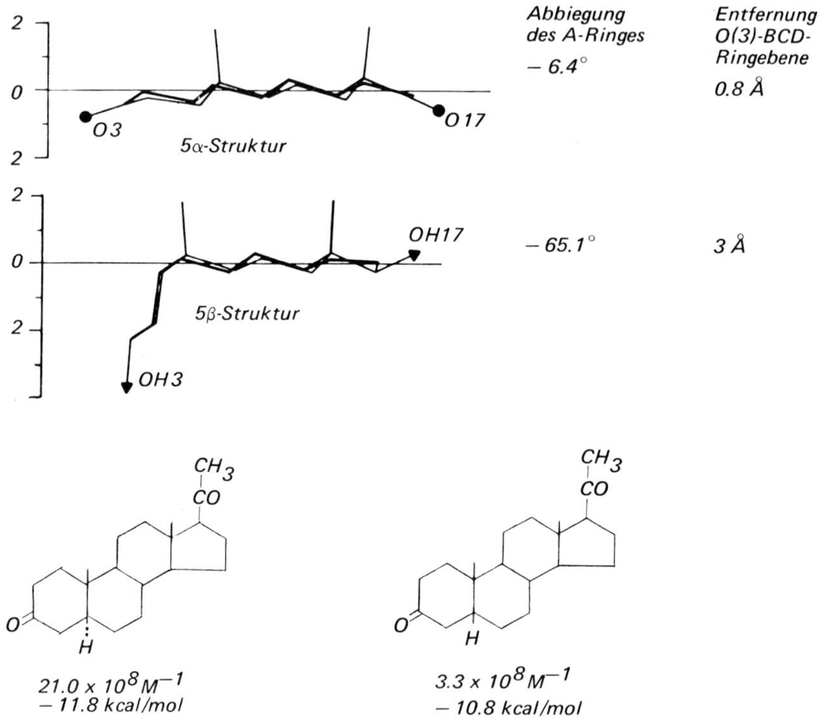

Abb. 9. Einfluß der Steroidkonformation auf die Assoziationskonstante. Nach Ref.[45,47].

All diese Ergebnisse sind mit PBG, dem Serumprotein des trächtigen Meerschweinchens, erhalten worden. Es soll jedoch betont werden, daß die Befunde der Beziehung zwischen Steroidstruktur und Bindungsstärke nicht auf PBG beschränkt sind. Wir haben die relativen Bindungsaffinitäten von 17 Steroiden für 7 Progesteron-bindende Proteine des Wirbeltierreiches zusammengestellt[45] und gefunden, daß die typischen, an PBG beobachteten Änderungen von K_a durch Einführung

Tab. 12. Einfluß der Verflachung der Steroidkonformation. Nach Ref.[45].

Steroid	$10^{-8} \times K_a$ [M^{-1}]	$\Delta G^{\circ\,a}$ [kcal/mol]	$\delta\Delta G^{\circ\,b}$ [kcal/mol]
4-Pregnen-3,20-dion	18.5	– 12.55	–
4,6-Pregnadien-3,20-dion	22.0	– 12.66	– 0.1
19-Norpregn-4-en-3,20-dion	29.0	– 12.81	– 0.3
17-Hydroxy-6α-methyl-4-pregnen-3,20-dion	5.9	– 11.88	–
17-Hydroxy-6-methyl-4,6-pregnadien-3,20-dion	10.2	– 12.21	– 0.3
17β-Hydroxy-4-androsten-3-on	2.9	– 11.47	–
17β-Hydroxy-19-norandrost-4-en-3-on	8.5	– 12.10	– 0.6

[a] Obwohl die 2. Dezimale nicht signifikant ist, wird sie zur Berechnung von $\delta\Delta G^0$ angegeben.
[b] Beitrag der Strukturänderung zur freien Energie der Bindung.

von Hydroxylgruppen, oder durch Konformationsänderungen mit Abbiegung oder Verflachung des Steroidgerüstes, eine allgemeine Erscheinung sind (s. Tab. 9 in Ref.[45]). In Tab. 13 sind Durchschnittswerte für einige charakteristische Beispiele aufgeführt. Die weitgehende Ähnlichkeit der Bindungsspezifitäten dieser verschiedenartigen Progesteron-bindenden Proteine macht die Annahme wahrscheinlich, daß die Natur ähnliche Bauprinzipien bei der Synthese der Bindungsstellen in verschiedenartigen Proteinen anwendet.

Tab. 13. Durchschnittliche relative Steroidaffinitäten für 7 Wirbeltierproteine[a].
Nach Ref.[45].

Steroid	Durchschnittliche relative Bindungsaffinität
Progesteron	100
21-Hydroxyprogesteron	38
17-Hydroxyprogesteron	3.5
11β-Hydroxyprogesteron	4.3
5α-Pregnan-3,20-dion	78
5β-Pregnan-3,20-dion	12
1,4-Pregnadien-3,20-dion	34
19-Norprogesteron	137
Testosteron	4.7
19-Nortestosteron	19

[a] PBG[45]; Rezeptor des Meerschweinchenuterus[48]; Uterusrezeptor der Frau[49]; Rezeptor des Kaninchenuterus[48]; Rezeptor des Schafuterus[48]; Eileiterrezeptor des Huhns[49]; Uteroglobin des Kaninchens[50].

Wenn man diese Befunde und weitere Einzelheiten[45] über den Einfluß der Steroidstruktur auf die Affinität zu PBG zusammenfassend betrachtet, kann man sich eine plausible Vorstellung von den allgemeinen Eigenschaften der Steroidbindungsstelle in PBG machen. Der Versuch eines solchen Bildes ist in Abb. 10 gegeben, in der die Art der Bindung in den verschiedenen Bereichen durch eingeklammerte Nummern angedeutet ist. Die Notwendigkeit der 3-Ketogruppe läßt die Anwesenheit von Wasserstoffdonatoren vermuten (**1**). Die Erscheinung der Fluoreszenzlöschung macht es wahrscheinlich, daß ein Tryptophanrest nahe dem Ring A mit seinem Chromophorsystem gelegen ist. Die hydrophoben Bindungen bei (**2**) und (**3**) sind eng und werden durch Entfernung der C(19)-Methylgruppe gestärkt. An C-11 und C-12 (**4**) finden wir hydrophobe und hydrophile Bindungskräfte, die letzteren sind durch Wasserstoffakzeptoren gegeben. Wasserstoffdonatoren und hydrophobe Affinitäten vermitteln Assoziation an der Seitenkette (**5**). Die enge hydrophobe Bindung an C-16 (**6**) ähnelt der an C-2. An C-6 dominiert hydrophobe Bindung (**7**); die äquatoriale α-Stellung hat einen größeren Effekt als die axiale β-Stellung. Die Konformation mit Streckung oder Abbiegung des A-Ringes (**8**) hat einen starken positiven oder negativen Einfluß auf die Bindungsaffinität.

Dies auf Grund der Bindungsanalyse verschiedenartiger Steroid-PBG-Komplexe gewonnene Bild der Bindungsstelle ist indirekter Natur. Es wird unsere zukünftige Aufgabe sein, durch chemische

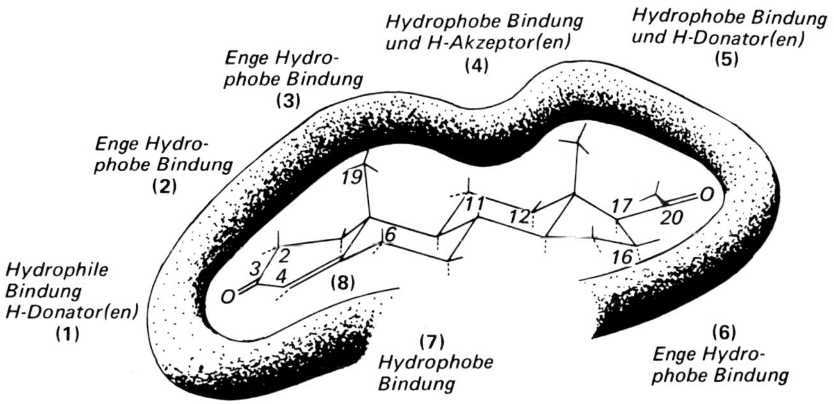

Abb. 10. Schematisches Bild der Steroidbindungsstelle in PBG. Nach Ref.[45].

und physikochemische Analyse dieses Bild zu verifizieren.

Mehr als 40 Jahre sind vergangen seit das Progesteron in Butenandts Laboratorium isoliert[51], in seiner Konstitution aufgeklärt und aus anderen Steroiden im Reagenzglas bereitet wurde[52,53]. Die Arbeit am Progesteron und anderen Steroidhormonen ist bis heute nicht beendet. Der wissenschaftliche Grundstein, den unser Jubilar vor so vielen Jahren gelegt hat, ist inzwischen zu einem großen, weitverzweigten Gebäude gewachsen, zu dem von Jahr zu Jahr mehr Forscher angezogen werden; die Ideen, die er damals auf diesem – wie auch auf manch anderem – Gebiet der biologischen Chemie geformt und an jüngere Generationen weitergegeben hat, sind bis heute lebendig und fruchtbar, und werden für viele zukünftige Jahre die Biochemiker fesseln. Diese Entwicklung ist wohl die schönste Ehrung, die einem Wissenschaftler an seinem 75. Geburtstag zuteil werden kann. In diesem Sinne lege ich meinem Doktorvater diesen Arbeitsbericht als meinen Beitrag vor. Ich verbinde damit meine herzlichsten Wünsche für viele Jahre in Gesundheit und Frische im Kreise seiner Familie, zusammen mit seiner lieben Frau, die so großen Anteil hat am Gelingen des wissenschaftlichen Werkes von Adolf Butenandt.

Literatur

1 Butenandt, A. (1931) Untersuchungen über das weibliche Sexualhormon (Follikel- oder Brunsthormon), Weidmann, Berlin.
2 Butenandt, A. (1934) Mitteil. Universitätsbund Göttingen **15**, 17 - 27.
3 Butenandt, A. (1935) *Dtsch. Med. Wochenschr.* **61**, 781 - 786, 823 - 825.
4 Bennhold, H. (1932) *Ergeb. Inn. Med. Kinderheilkd.* **42**, 273 - 375.
5 Bennhold, H. (1938) in Die Eiweißkörper des Blutplasmas (Bennhold, H., Kylin, E. & Rusznyak, S., Hrsg.) S. 220 - 303, Steinkopf, Dresden.
6 Oppenheimer, E. (1913) *Biochem. Z.* **55**, 134 - 152.
7 Brunelli, B. (1934) *Arch. Int. Pharmacodyn.* **49**, 262 - 271.
8 Landsteiner, K. (1933) Die Spezifität der Serologischen Reaktionen, Springer, Berlin.
9 Meinzinger, E. (1949) Dissertation, Universität Tübingen.
10 Westphal, U. (1971) Steroid-Protein Interactions, Springer, New York–Heidelberg.
11 Jensen, E.V. & Jacobson, H.I. (1962) *Recent Progr. Horm. Res.* **18**, 387 - 414.
12 Greene, G.L., Closs, L.E., Fleming, H., DeSombre, E.R. & Jensen, E.V. (1977) *Proc. Natl. Acad. Sci. USA* **74**, 3681 - 3685.
13 Westphal, U. (1970) *J. Reprod. Fertil., Suppl.* **10**, 15 - 38.
14 Westphal, U. (1975) in Normal and Abnormal Growth of the Prostate (Goland, M., Hrsg.), S. 616 - 631, Thomas, Springfield, Ill.
15 Ewing, L.L., Chubb, C.E. & Robaire, B. (1976) *Nature (London)* **264**, 84 - 86.
16 Westphal, U. & Harding, G.B. (1973) *Biochim. Biophys. Acta* **310**, 518 - 527.
17 Westphal, U., Burton, R.M. & Harding, G.B. (1975) *Methods Enzymol.* **36**, 91 - 104.
18 Stroupe, S.D., Harding, G.B., Forsthoefel, M.W. & Westphal, U. (1978) *Biochemistry* **17**, 177 - 182.
19 Rosner, W. & Smith, R.N. (1975) *Biochemistry* **14**, 4813 - 4820.
20 Stroupe, S.D. & Westphal, U. (1975) *J. Biol. Chem.* **250**, 8735 - 8739.
21 Spector, A.A. (1975) *J. Lipid Res.* **16**, 165 - 179.
22 Kute, T. & Westphal, U. (1976) *Biochim. Biophys. Acta* **420**, 195 - 213.
23 Rosner, W. & Bradlow, H.L. (1975) *Methods Enzymol.* **36**, 104 - 109.
24 Le Gaillard, F., Aubert, J.P., Dautrevaux, M. & Loucheux-Lefebvre, M.H. (1976) *FEBS Lett.* **64**, 278 - 284.
25 Khan, M.S. & Rosner, W. (1977) *J. Biol. Chem.* **252**, 1895 - 1900.
26 Van Baelen, H., Beck, M. & DeMoor, P. (1972) *J. Biol. Chem.* **247**, 2699 - 2703.
27 Le Gaillard, F. & Dautrevaux, M. (1977) *Biochim. Biophys. Acta* **495**, 312 - 323.
28 Mickelson, K.E. & Pétra, P.H. (1975) *Biochemistry* **14**, 957 - 963.
29 Diamond, M., Rust, N. & Westphal, U. (1969) *Endocrinology* **84**, 1143 - 1151.
30 Westphal, U., Stroupe, S.D., Kute, T. & Cheng, S.L. (1977) *J. Steroid Biochem.* **8**, 367 - 374.
31 Stroupe, S.D., Cheng, S.L. & Westphal, U. (1975) *Arch. Biochem. Biophys.* **168**, 473 - 482.
32 Rosner, W., Darmstadt, R. & Tauster, S.J. (1973) *J. Steroid Biochem.* **4**, 249 - 255.
33 Koblinsky, M., Beato, M., Kalimi, M. & Feigelson, P. (1972) *J. Biol. Chem.* **247**, 7897 - 7904.
34 Heyns, W. & DeMoor, P. (1971) *J. Clin. Endocrinol. Metab.* **32**, 147 - 154.
35 Hansson, V., Ritzen, E.M., Wessington, S.C., McLean, W.S., Tindall, D.J., Nayfeh, S.N. & French, F.S. (1974) *Endocrinology* **95**, 690 - 700.
36 Schrader, W.T. & O'Malley, B.W. (1972) *J. Biol. Chem.* **247**, 51 - 59.
37 Kuhn, R.W., Schrader, W.T., Smith, R.G. & O'Malley, B.W. (1975) *J. Biol. Chem.* **250**, 4220 - 4228.
38 Katzenellenbogen, J.A., Johnson, H.J., Jr. & Carlson, K.E. (1973) *Biochemistry* **12**, 4092 - 4099.
39 Westphal, U. (1957) *Arch. Biochem. Biophys.* **66**, 71 - 90.
40 Stroupe, S.D. & Westphal, U. (1978) *Biochemistry*, in press.

41 Tinoco, I., Jr. (1960) *J. Am. Chem. Soc.* **82**, 4785 - 4790.
42 Stroupe, S.D. & Westphal, U. (1975) *Biochemistry* **14**, 3296 - 3300.
43 Kute, T. (1975) Ph.D. Dissertation, University of Louisville.
44 Cheng, S. L. & Westphal, U. (1977) *Fed. Proc.* **36**, 782.
45 Blanford, A.T., Wittman, W., Stroupe, S.D. & Westphal, U. (1978) *J. Steroid Biochem.*, in press.
46 Fieser, L.F. & Fieser, M. (1959) Steroids, S.11, Reinhold, New York.
47 Duax, W.L. & Norton, D.A. (1975) Atlas of Steroid Structure, Bd.I, IFI-Plenum, New York.
48 Kontula, K., Jänne, O., Vihko, R., deJager, E., deVisser, J. & Zeelen, F. (1975) *Acta Endocrinol.* **78**, 574 - 592.
49 Smith, H.E., Smith, R.G., Toft, D.O., Neergard, J.R., Burrows, E.P. & O'Malley, B.W. (1974) *J. Biol. Chem.* **249**, 5924 - 5932.
50 Beato, M. (1976) *J. Steroid Biochem.* **7**, 327 - 334.
51 Butenandt, A., Westphal, U. & Hohlweg, W. (1934) *Hoppe-Seyler's Z. Physiol. Chem.* **227**, 84 - 98.
52 Butenandt, A. & Schmidt, J. (1934) *Ber. Dtsch. Chem. Ges.* **67**, 1901 - 1904.
53 Butenandt, A. & Westphal, U. (1934) *Ber. Dtsch. Chem. Ges.* **67**, 2085 - 2087.

Dem *U.S. Public Health Service, National Institutes of Health*, danke ich für die langjährige Unterstützung meiner Arbeiten durch die Bereitstellung von Research Grant AM-06369, und durch die Verleihung von Research Career Award GM-K6-14,138.

Dr. Ulrich Westphal, Department of Biochemistry, University of Louisville School of Medicine, Health Sciences Center,
P.O. Box 35260, Louisville, Ky. 40232 U.S.A.

Hoppe-Seyler's Z. Physiol. Chem.
Bd. 359, S. 449 - 457, April 1978

29. Mosbacher Kolloquium der Gesellschaft für Biologische Chemie

Energy Conservation in Biological Membranes

April 6th - 8th, 1978

Abstracts

H. Baltscheffsky

Evolutionary Aspects of Energy Metabolism

Currently, an increasingly detailed picture of the molecular evolution of biological energy metabolism is developing. Recent progress in the determination of protein primary, secondary and tertiary structures has initiated consideration of how energy conservation processes such as fermentation, photophosphorylation and oxidative phosphorylation originated and evolved. The tentative evolutionary schemes are based on the strongly supported assumptions that gradually divergent trends in protein structure are evidence for a single common ancestor and that a protein with a certain structure and function may give rise to a protein with different structure and either retained or different function through gene duplication, mutation, etc.

Numerous observations indicate that in protein molecular evolution over billions of years three-dimensional structural features are more conserved than amino acid sequences. The β-pleated sheet structure appears to be of particular evolutionary significance in general[1], as well as in proteins involved in redox reactions[2,3] and in phosphate transfer[4] in the various cellular energy conversion systems. Evidence that β-structures may have already been important when life on earth began[5,6] points to the possibility of a continuity in protein β-structure evolution since that time.

For both electron transport chains[2,3] and fermentation reaction sequences[4] it may be assumed that stepwise evolution from protein to neighbour protein has occurred. The recent solubilization[7] of membrane-bound inorganic pyrophosphatase, which catalyzes the light-induced formation of energy-rich pyrophosphate from orthophosphate[8] from *Rhodospirillum rubrum* chromatophores, may open the possibility of investigating whether stepwise protein molecular evolution has also occurred along the pathway of electron transport-coupled phosphorylation.

1 Rossmann, M. G., Liljas, A., Brändén, C.-I. & Banaszak, L. J. (1975) in The Enzymes (Boyer, P. D., ed.) vol. 11, p. 62 - 102, Academic, New York.
2 Baltscheffsky, H. (1974) *Origins Life* **5**, 387 - 395.
3 Baltscheffsky, H. (1974) in Dynamics of Energy-Transducing Membranes (Ernster, L., Estabrook, R. W. & Slater, E. C., eds.) pp. 21 - 27, Elsevier, Amsterdam.
4 Baltscheffsky, H. (1977) in Living Systems as Energy Converters (Buvet, R., Allen, M. J. & Massué, J.-P., eds.) pp. 81 - 88, North-Holland, Amsterdam.
5 Brack, A. & Orgel, L. E. (1975) *Nature (London)* **256**, 383 - 387.
6 Carter, Jr., C. W. & Kraut, J. (1974) *Proc. Natl. Acad. Sci. U.S.A.* **71**, 283 - 287.
7 Baltscheffsky, M. & Randahl, H., personal communication.
8 Baltscheffsky, H., von Stedingk, L.-V., Heldt, H.-W. & Klingenberg, M. (1966) *Science* **153**, 1120 - 1121.

Herrick Baltscheffsky, Arrheniuslaboratoriet, Avdelningen för Biokemi, S-10691 Stockholm.

Y. Hatefi, Y. M. Galante and R. Kiehl

Organization of the Mitochondrial Respiratory Chain

In mitochondria, the machinery for oxidative phosphorylation is located in the inner membrane in the form of five enzyme complexes. Complexes I, II, III and IV (plus cytochrome *c* and ubiquinone) make up the respiratory chain, while Complex V is responsible for ATP synthesis and hydrolysis. Complex I catalyzes electron transfer from NAD(P)H to ubiquinone. It is composed of 16 - 18 polypeptides, FMN, and 5 iron-sulfur centers. A soluble iron-sulfur flavoprotein with M_r = 75 000 ± 6% (three subunits) is the primary NAD(P)H dehydrogenase. Complex II catalyzes electron transfer from succinate to ubiquinone. It is composed of 4 polypeptides. Two polypeptides belong to succinate dehydrogenase, which is an iron-sulfur flavoprotein with M_r = 97 000 ± 4%. In addition to succinate dehydrogenase, Complex II contains a low potential cytochrome *b* whose reduced form at 77 K exhibits absorption maxima at 557.5, 550, 531, 523 and 422 nm. Complex III catalyzes electron transfer from reduced ubiquinone to cytochrome *c*. It is composed of 7 - 8 polypeptides. The identified electron carriers of Complex III are two *b*-type cytochromes, cytochrome c_1 an iron-sulfur protein, and a component with cytochrome *b*-like electron transfer properties and an absorption peak when reduced at 77 K at 558 nm. Complex IV catalyzes electron transfer from ferrocytochrome *c* to molecular oxygen. It is composed of 7 polypeptides, and contains hemes a, a_3 and two atoms of copper per mole. In addition to the above, each enzyme complex also contains 20 - 30% phospholipids by dry weight, and Complexes I and III contain bound ubiquinone. Complexes I, II, III and IV have been physically and functionally recombined in the presence of cytochrome *c* (and added ubiquinone where necessary) to reconstitute the entire electron transport system ($^{\text{I}}_{\text{II}}$ > III-*c*-IV) or segments thereof (I-III, II-III, $^{\text{I}}_{\text{II}}$ > III, I-III-*c*-IV, II-III-*c*-IV) with the expected overall activities of the participating complexes. In mitochondria, Complexes I, III and IV contain the energy coupling sites 1, 2 and 3, respectively. The oxidation energy captured by these complexes is transferred to Complex V for ATP synthesis. Recently, the latter complex has been obtained in a highly purified form in our laboratory. It is composed of 11 polypeptides, of which 10 have been identified as follows: the five F_1 subunits, the oligomycin-sensitivity-conferring protein, the dicyclohexylcarbodiimide-binding protein, the uncoupler-binding protein, coupling factor F_6, and provisionally coupling factor B. Complexes I to V are present in the mitochondrial inner membrane in the approximate ratio of 1 I : 2 II : 3 III : 7 IV : 3 V, and together they make up about 50% of the inner membrane protein.

Youssef Hatefi, Department of Biochemistry, Scripps Clinic and Research Foundation, 10666 North Torrey Pines Road, La Jolla, California 92037 U.S.A.

H. Weiss

Mitochondrial Cytochrome b

Mitochondrial cytochrome *b* is an approximately 28 000 M_r polypeptide chain containing a single non-convalently bound ferroprotoporphyrin IX prosthetic group. As a dimer (or perhaps as a tetramer) it forms part of the ubiquinone: cytochrome *c* oxidoreductase. This is a 250 000 M_r (or perhaps 500 000 M_r) membraneous multiprotein complex which contains additionally cytochrome c_1 (M_r 31 000), an iron-sulfur protein (M_r 25 000) and five subunits without known prosthetic groups (M_r 8000, 12000, 14000, 45000, 50000). The complex catalyzes electron transport from reduced ubiquinone to ferricytochrome *c* and is a site for the transformation of electrogenic energy into a form suitable for ATP synthesis. Apparently, cytochrome *b* is the membranous section of the multiprotein complex which spans the inner mitochondrial membrane, whereas cytochrome c_1 is located at the cytoplasmic surface of the membrane and the 45 000 and 50 000 M_r subunits at the matrix surface.

Cytochrome *b* is the only polypeptide of the multiprotein complex which is a product of the mitochondrial genetic system. It is coded for by a gene on the mitochondrial DNA and translated on mitochondrial ribosomes. The other subunits of the multiprotein complex are synthesized in the cytoplasm.

A synopsis of our knowledge on the genetics, biogenesis, structure and function of cytochrome *b* will be given.

Wilkström, M. R. F. (1973) *Biochim. Biophys. Acta* **301**, 155 - 193.
Rieske, J. S. (1976) *Biochim. Biophys. Acta* **456**, 195 - 247.
Weiss, H. (1976) *Biochim. Biophys. Acta* **456**, 291 - 313.
Slonimski, P. P. & Tzagoloff, A. (1976) *Eur. J. Biochem.* **61**, 27 - 41.

Hanns Weiss, European Molecular Biology Laboratory, Meyerhofstraße 1, P.O.B. 10.2209, D-6900 Heidelberg.

G. von Jagow, H. Schägger, W. D. Engel, H. Hackenberg and H. Kolb

Beef Heart Cytochrome bc_1 Complex (Complex III): Isolation and Characterization of the Polypeptide Subunits Carrying Heme b

When the bc_1 complex is isolated from antimycin-loaded beef heart mitochondria in Triton X-100 by hydroxyapatite chromatography[1], it consists of six polypeptide subunits with molecular weights of 12 000, 29 000 (cytochrome c_1), 2 × 30 000 (2 × cytochrome *b*) and finally

45 000 and 47 000 (the so-called "core proteins"). This complex lacks the iron sulfur protein (M_r 25 000). It is enzymatically inactive, i.e. it has no ubiquinone cytochrome *c* reductase activity.

From this bc_1 complex the cytochromes *b* were isolated. In the sodium dodecylsulfate gel electrophoresis pattern, they show one band with an M_r of 31 000. Their minimum M_r, as calculated from the heme *b* content, amounts to 30 000. But, in the sedimentation equilibrium runs, the preparation has a corrected M_r of 62 000, i.e. the cytochromes *b* are isolated as a dimer. In redox titrations, the dimer reveals two different midpoint potentials, which are both pH-dependent. Binding experiments with radioactive antimycin show that the isolated cytochrome *b* dimer has no bound antimycin, whereas the isolated bc_1 complex has.

As published recently[1], the isolated bc_1 complex is also present in a dimeric form. Considering that each bc_1 monomer contains two heme *b* centers, it may be speculated that cytochromes *b* function as a tetrameric enzyme complex in the mitochondrial membrane, and may thereon exert a vectorial Bohr effect. It is yet unknown whether the *b* dimer is an α_2 or an $\alpha\beta$ dimer.

1 v. Jagow, G., Schägger, H., Riccio, P., Klingenberg, M. & Kolb, H. J. (1977) *Biochim. Biophys. Acta* **462**, 549 - 558.

Gebhard von Jagow, Institut für Physikalische Biochemie und Physiologische Chemie, Goethestr. 33, D-8000 München 2.

B. Chance

Low Temperature Electron Transport in the Cytochrome c-Cytochrome Oxidase Reaction Evidenced by Tunneling of Cytochrome c to Heme a or Heme a_3

A precise specification of the mechanism of electron transfer seems essential to an understanding of the energy conservation process whether in a chemical or chemiosmotic mechanism. Recent developments in the structure of cytochrome oxidase with respect to membrane bilayer now reveals the cytochrome oxidase as a protein with much of its essential parts on the cytochrome *c* side. Resonance energy transfer using fluorescent forms of cytochrome *c* are capable of serving as donor to the iron and copper components of cytochrome oxidase[1]. Generally, these distance measurements give values between 25 and 35 Å between the heme of cytochrome *c* and the acceptor, either cytochrome *a* or cytochrome a_3. The distance over which nuclear and electron tunneling mechanisms could operate is even less than the distance for resonance energy transfer. Electron transfer to the peroxy compound of cytochrome oxidase may be involved in the region − 50 to − 80 °C. More recent tests (in collaboration with Dr. A. Waring) now indicate that the copper component associated with cytochrome *a* is not a catalyst of this low temperature electron transfer which further limits the distance of the heme of cytochrome a_3 and its associated copper from the cytochrome *c* side of the membrane. In short, one may now visualize cytochrome oxidase as functionally unsymmetrical, with respect to the membrane, with both hemes within electron tunneling distances from cytochrome *c*.

1 Vanderkooi, J., Landesberg, R., Hayden, G. W. & Owen, C. S. (1977) *Eur. J. Biochem.* **81**, 339 - 347.

Britton Chance, Johnson Research Foundation, Dept. of Biophysics, University of Pennsylvania, Philadelphia, Pa. 19104, U.S.A.

L. H. Jensen

Iron-Sulfur Proteins: Structure and Function

Iron-sulfur proteins have been extensively studied in recent years by a variety of physical techniques including X-ray diffraction. Three kinds of iron-sulfur coordinating units are now recognized: $Fe(Cys)_4$ in rubredoxins, $Fe_2S_2(Cys)_4$ in chloroplast ferredoxins, and Fe_4S_4-$(Cys)_4$ in bacterial ferredoxins. A single coordinating unit of the latter occurs in some bacterial ferredoxins and in the so-called high potential iron protein, while two occur in other bacterial ferredoxins. The 4-Fe complex functions in electron transfer between states designated C and C^+ at a midpoint potential of ~ + 350 mV in high potential iron protein and between states C^- and C at a midpoint potential of ~ − 400 mV in what have been termed bacterial ferredoxins. An iron-sulfur protein from *Azotobacter* has been reported with two 4-Fe complexes, one with a midpoint potential of + 340 mV, the other with a midpoint potential of − 420 mV.

A detailed mechanism has been proposed for the electron transfer in high potential iron proteins, and more general suggestions have been made for the relationship between the structure and function of other iron-sulfur proteins. Nevertheless, mechanistic proposals for the function of iron-sulfur proteins have not been definitive.

L. H. Jensen, Department of Biological Structure and the Department of Biochemistry, University of Washington, Seattle, Wash. 98195 U.S.A.

A. Trebst

The Electron Transport System of Chloroplasts

Photosynthetic NADP reduction and oxygen evolution, coupled to ATP formation, is localized in the inner thylakoid membrane system of the chloroplast. The functional sequence of the various electron carriers participating and their structural arrangement in the menbrane support the concept of vectorial electron flow across the membrane. Two energy-conserving sites – leading to coupled ATP formation – have been identified: one connected with photosystem I and plastoquinone, the other with photosystem II and the water-splitting system.

Methodology and evidence for the present schemes of photosynthetic electron transport as well as open questions will be reviewed.

A. Trebst, Abteilung für Biologie der Ruhr-Universität Bochum, Postfach 102148, D-4630 Bochum 1.

A. Kröger

Electron Transport Phosphorylation in Anaerobic Bacteria

The mechanism of coupling between electron transport and phosphorylation is still a matter of hypothesis. This is in part due to the complexity of the mitochondrial respiratory chain, which has been the main object of studies on oxidative phosphorylation. It is, therefore, appropriate to look for simpler systems which nevertheless function according to the same basic principle of oxidative phosphorylation.

The anaerobic growth of some bacteria is sustained by the reduction of certain hydrogen acceptors instead of oxygen. There is evidence that the reduction of the following acceptors is coupled to electron transport phosphorylation:

$$\text{fumarate} + H_2 \rightarrow \text{succinate}$$
$$SO_4^{2\ominus} + 4\,H_2 + 2\,H^{\oplus} \rightarrow H_2S + 4\,H_2O$$
$$HCO_3^{\ominus} + 4\,H_2 + H^{\oplus} \rightarrow CH_4 + 3\,H_2O$$

The composition and function of these systems will be discussed and compared to those of oxidative phosphorylation of mitochondria and bacteria.

A. Kröger, Institut für Physiologische Chemie, Physikalische Biochemie und Zellbiologie der Universität München, Goethestr. 33, D-8000 München 2.

H. T. Witt

Primary Energy Conservation in Chloroplasts

The vectorial pathways of $e^{\ominus}$, $H^{\oplus}$ and H within the functional membrane of photosynthesis are briefly discussed. New results are presented

1) on the plastoquinone pool as a possible hydrogen pump,
2) on molecular organization within the membrane probed by electrochromism,
3) on functional properties of the ATPase obtained by energetization with an external electric field.

H. T. Witt, Max-Volmer-Institut für Physikalische Chemie und Molekularbiologie, Technische Universität Berlin, Straße des 17. Juni 135, D-1000 Berlin 12.

W. Junge

Proton Pumping across the Thylakoid Membrane Resolved in Time and Space

It is generally accepted that vectorial electron transport across the thylakoid membrane is linked to proton uptake from the outer phase of thylakoids and proton release inside. The stoichiometry at each of the two sites, of uptake outside and of release inside, is one proton per electron. We have previously shown (Ausländer & Junge, 1974) that proton uptake from outside is delayed relative to the respective redox reaction (60 ms vs. 1 ms). This led us to postulate that the protolytic reaction sites at the outer side of the thylakoid membrane are shielded from the outer aqueous phase by a diffusion barrier for protons (probably proteinaceous, and removable by detergents or trypsin).

In consequence, electron transport generates a proton gradient not only between the two aqueous phases separated by the thylakoid membrane, but also between each of these phases and a subcompartment at the outer side of the membrane. It is generally accepted that the proton gradient across the thylakoid membrane is used for the synthesis of ATP. However, it is a long-standing challenge to the original chemiosmotic concept of Mitchell (1961, 1966) that there might be a direct flux of protons from specific release sites within the membrane into the ATP-synthesizing system (Williams, e.g. 1975). The possibility that the internal pH "seen" by the enzyme is not always in equilibrium with the pH in the internal osmolar volume has been awarded considerable attention both in mitochondria (van Dam, 1977 and Azzone, 1977) and chloroplasts (Ort et al., 1976). It is evident that this question is a kinetic one: how rapid is the release of protons e.g. by the water-oxidizing system into the internal osmolar volume of thylakoids and how rapidly do protons from the internal osmolar volume enter the ATP synthesizing system? We have previously introduced the dye neutral red as a sensitive and rapid indicator of the small flash-induced pH changes in some internal space of thylakoids (Ausländer & Junge, 1975).

In this communication we will demonstrate that:
1) Neutral red reports pH changes from a compartment inside thylakoids which exchanges protons within less than 100 μs with the internal osmolar space. 2) Protons from the water-oxidizing enzyme system are released into this space within less than 100 μs. 3) Protons enter the ATP-synthesizing system from the internal osmolar volume within some 10 ms.

In conclusion: There is no kinetic evidence for a "micro-chemiosmotic" mechanism of photophosphorylation. Instead it is shown that there is ample time for protons to equilibrate with the internal phase before entry into the enzyme. There is, at the inner side of the thylakoid membrane, no equivalent to the diffusion barrier for protons which exists at the outer side of the membrane.

Wolfgang Junge, Max-Volmer-Institut für Physikalische Chemie und Molekularbiologie, Technische Universität Berlin, Straße des 17. Juni 135, D-1000 Berlin 12.

M. Wikström and K. Krab

Redox-Linked Proton Pumps in Mitochondria

It is presently agreed that the redox reactions of energy-transducing biological membranes are coupled to electrogenic proton transport. According to the chemiosmotic theory[1], this proton transport is an intermediate energy-transducing step prior to oxidative synthesis of ATP.

Two hypotheses have been advanced for the molecular mechanism of redox-linked proton translocation. The redox loop concept[1] is characterised by vectorial electron flow and a direct covalent linkage between the $H^{\oplus}$ and $e^{\ominus}$ (as an H atom) during translocation of the former. The conformational proton pump concept[2,3] does not require vectorial $e^{\ominus}$ transfer, but implies indirect (conformational) coupling, and energy transfer, between the redox reaction and proton transport, through interactions between (and within) polypeptide chains of the protein.

Studies in our laboratory[4-6] now strongly indicate that mitochondrial cytochrome *c* oxidase (EC 1.9.3.1) functions as a redox-linked proton pump, and not as a mere transmembrane electron-transferring device[1-3]. However, since our experiments with mitochondrial membranes[4,5] and reconstituted cytochrome oxidase vesicles[6] suggest a strict $H^{\oplus}/e^{\ominus}$ stoichiometry of 1.0 for the proton translocation process coupled to cytochrome oxidase, proton transport may well be directly coupled to the haem centres.

We tentatively suggest that redox-linked proton transport in biological membranes may be catalysed by a *directly coupled proton pump mechanism*, which (i) does not require vectorial $e^{\ominus}$ transfer, in which (ii) the $H^{\oplus}$ ion is directly linked to the redox centre during transport, and which (iii) does not use conformational changes for energy transfer, but for catalysis only.

1 Mitchell, P. (1976) *Biochem. Soc. Trans.* **4**, 399-430.
2 Skulachev, V.P. (1974) *Ann. N. Y. Acad. Sci.* **227**, 188-202.
3 Papa, S. (1976) *Biochim. Biophys. Acta* **456**, 39-84.
4 Wikström, M.K.F. (1977) *Nature (London)* **266**, 271-273.
5 Wikström, M.K.F. & Saari, H.T. (1977) *Biochim. Biophys. Acta* **462**, 347-361.
6 Krab, K. & Wikström, M. (1978) *Biochim. Biophys. Acta*, submitted.

Mårten Wikström, Department of Medical Chemistry, University of Helsinki, Siltavuorenpenger 10A, SF-00170 Helsinki 17.

D. Oesterhelt, R. Hartmann and G. Wagner

Light-driven Proton Translocation and Energy Conservation in Halobacteria

Halobacteria grow in strong NaCl solutions and are dependent on aerobic metabolism. Under conditions of limiting aeration during growth, these cells synthesize the retinal protein complex bacteriorhodopsin which is found in the cell membrane and serves an energy transducing function. Light energy absorbed by bacteriorhodopsin is converted into that of an electrochemical proton gradient by the coupling of the thermoreversible photochemical reaction in bacteriorhodopsin with the electrogenic extrusion of protons from the cell.

As predicted by the chemiosmotic hypothesis of Peter Mitchell, the enhanced chemical potential of the proton drives various bioenergetic processes in the cell. These include photophosphorylation, amino acid uptake and ion exchange. Phosphorylation can be driven alternatively by the membrane potential ($\Delta\psi$) or by the pH gradient (Δ pH) which together represent the change in the electrochemical proton gradient called proton motive force. In the dark, artificially induced enhancement of $\Delta\psi$ ($K^{\oplus}$ gradient plus ionophore) or Δ pH (addition of HCl) induces a transient increase in the cellular ATP level.

All the reactions driven by light also occur aerobically via the cell's respiratory chain, which acts as a proton pump. Thus light absorption and oxygen consumption are competitive processes for the formation of the electrochemical proton gradient. There is a quantitative relationship between photon absorption, oxygen consumption, proton extrusion and ATP synthesis.

In strong daylight, light energy conversion by bacteriorhodopsin is sufficient to saturate the phosphorylation system of the cell. The capacity for energy storage by the proton gradient and consequent ATP synthesis is small. However, conversion of the proton gradient energy into that of a potassium ion gradient enhances the capacity for energy storage by a factor of at least 10^3. The selective and high permeability of the cell membrane for potassium means that its diffusion potential can be used to maintain an elevated ATP level under anaerobic conditions in the dark for periods of hours.

Dieter Oesterhelt, Institut für Biochemie der Universität Würzburg, Röntgenring 11, D-8700 Würzburg.

P. L. Pedersen, L. M. Amzel, J. W. Soper and N. Cintrón

Structure and Function of the Mammalian Mitochondrial ATPase

The mitochondrial F_1-ATPase of rat liver was first purified from this laboratory[1]. Subsequent work from this laboratory has also led to the isolation from rat liver mitochondria of a small peptide inhibitor of F_1[2] and the isolation of what we so far regard to be a "partially purified" preparation of the oligomycin-sensitive ATPase complex[3].

This lecture will focus on several aspects of the mitochondrial ATPase problem pertinent to the structure and function of this enzyme complex. First, an up-to-date assessment of the subunit structure of the F_1-part of the rat liver enzyme will be presented. This assessment will consider X-ray crystallographic data recently obtained by one of us (L. M. Amzel) on the F_1-portion of mitochondrial ATPase[4]. Secondly, experiments will be presented which will show that the purified F_1-ATPase of rat liver is a highly competent coupling factor which can restore ATP synthesis in F_1-depleted membranes without the addition of other proteins[5]. Thirdly, the role of nucleotide binding sites on the enzyme will be considered as they pertain to ATP synthesis, ATP hydrolysis and ATP-dependent functions[6,7]. Finally, the recent work on an ATPase inhibitor peptide of rat liver mitochondrial ATPase will be presented, and discussed within the context of the role and origin of the inhibitor (i.e., is the inhibitor a true physiological regulator or simply an artifact of isolation procedures).

The take-home message of this lecture will emphasize that the mechanisms underlying the synthesis and hydrolysis of ATP (and the regulation of these processes) are far from being resolved at the chemical and molecular level.

1 Catterall, W. A. & Pedersen, P. L. (1971) *J. Biol. Chem.* **246**, 4987 - 4994.

2 Pedersen, P. L., Levine, H. & Cintrón, N. (1974) in Membrane Proteins in Transport and Phosphorylation (Azzone, G. F. et al., eds.) pp. 43 - 54, North-Holland Publishing Co., Amsterdam.

3 Soper, J. W. & Pedersen, P. L. (1976) *Biochemistry* **15**, 2682 - 2689.

4 Amzel, L. M. & Pedersen, P. L. (1978) *Methods Enzymol.*, in press.

5 Pedersen, P. L. & Hullihen, J. (1978) *J. Biol. Chem.*, in press.

6 Pedersen, P. L. (1975) *Biochem. Biophys. Res. Commun.* **64**, 610 - 616.

7 Pedersen, P. L. (1976) *J. Biol. Chem.* **251**, 934 - 940.

Peter L. Pedersen, Laboratory for Molecular and Cellular Bioenergetics. Dept. of Physiological Chemistry, Johns Hopkins University School of Medicine, 725 N. Wolfe Street, Baltimore, Md. 21205, U.S.A.

G. Schäfer and H. Strotmann

Functional Sites of the Chloroplast Coupling Factor

In comparative studies with a variety of ADP-analogs it has been shown that modification of the ribose moiety affects the properties of the nucleotide molecule with regard to photophosphorylation and binding to the tight nucleotide binding site of CF_1 much less than modification of the nucleobase or the phosphate part[1,2].

Thus, for localization and differentiation of ADP-binding sites by photolabeling, 3′-*O*-arylazidoesters of ADP[3] were expected to be suitable substrates. This expectation has been examined under conditions which exclude photoactivation of the analogs. Independent of the initial approach the results may allow conclusions on the molecular events related to phosphate transfer by the ATP-synthesizing system. In fact, 3′-*O*-[4-(4-azido-2-nitrophenylamino)butyryl]-ADP effectively interacts with the tight nucleotide binding site and the catalytic ADP-site of chloroplast ATP-synthesizing system[4]. This analog, however, is not phosphorylated. On the contrary, it is a powerful inhibitor of photophosphorylation (K_i = 5μM) and shows all characteristics of a typical energy transfer inhibitor. Inhibition is partly abolished by increasing the concentrations of ADP and inorganic phosphate. The corresponding ATP-analog displays a much lower inhibitory effect (K_i > 100μM), and it is not hydrolized by CF_1-ATPase. The AMP-analog does not at all affect phosphorylation.

The results suggest that the ADP-analog with its base- and phosphate-moiety precisely fits into the catalytic ADP-site. The large 3'-*O*-substituent is thought to cover in part the inorganic phosphate site and to prevent by steric hinderance a conformational transition of the ATP-synthesizing system which is essential in phosphate transfer. Molecular models show, that the molecule can easily assume a conformation where access to β-phosphate from one side is completely abolished by the 3'-*O*-substituent. – For evaluation of the essential structural features of the analog in energy-transfer inhibition, a series of related compounds was studied. It turned out that neither the azido- nor the nitro-group are necessary to obtain energy transfer inhibition.

It should be noted that in all essential points corresponding results were obtained in oxidative phosphorylation with beef heart submitochondrial particles, indicating unique structural and functional properties of the catalytic centers of CF_1 and F_1.

1 Strotmann, H., Bickel-Sandkötter, S., Edelmann, K., Schlimme, E., Boos, K. S. & Lüstorf, J. (1977) in Structure and Function of Energy Transducing Membranes (VanDam, K., van Gelder, B.F., eds.) pp. 307 - 317, Elsevier, Amsterdam.
2 Boos, K. S., Lüstorf, J., Schlimme, E., Hesse, H. & Strotmann, H. (1976) *FEBS-Lett.* **71**, 124 - 129.
3 Jeng, S. J. & Guillory, R. G. (1975) *J. Supramol. Struct.* **3**, 448 - 468.
4 Schäfer, G., Onur, G., Edelmann, K., Bickel-Sandkötter, S. & Strotmann, H. (1978) *FEBS-Lett.*, in press.

Günter Schäfer, Institut f. Klinische Biochemie u. Physiologische Chemie der Medizinischen Hochschule Hannover, Karl-Wiechert-Allee 9, D-3000 Hannover-Kleefeld.
Heinrich Strotmann, Botanisches Institut der Tierärztlichen Hochschule Hannover, Bünteweg 17, D-3000 Hannover.

Y. Kagawa

Protonophoric Action of Coupling Factor ATPase

ATPases (EC 3.6.1.3) of mitochondria, chloroplasts, chromatophores and bacterial plasma membranes transport $H^{\oplus}$ and convert chemiosmotic energy of $H^{\oplus}$ into chemical energy of ATP. These ATPases have common subunit structures, but are unstable, except for ATPase from thermophilic bacteria[1]. Using this stable ATPase, the following points became clear.

The catalytic portion of this $H^{\oplus}$-ATPase (F_0F_1) is called F_1. The crystalline F_1 was analyzed by computerized image reconstruction, and its hexagonal molecular structure with a central hole was visualized[2].

F_1 was completely reconstituted from its 5 subunits ($\alpha\beta\gamma\delta\epsilon$)[3]. Of 31 possible combinations of the 5 subunits, only assemblies containing β were active, though β was inactive by itself. The βγ-complex had properties similar to those of F_1, but the αβγ-complex showed azide insensitive, thermolabile ATPase activity. For $H^{\oplus}$-pump activity, α was also required. The γδε complex was a gate for $H^{\oplus}$, and also a connecting bridge between αβ and the $H^{\oplus}$-channel[4].

The molecular weight of F_1 was 390000 by equilibrium centrifugation. Only the α subunit contained one SH group and 3 SH groups were detected in F_1. The subunit stoichiometry is controversial, but should be $\alpha:\beta:\gamma:\delta:\epsilon$ = 3:3:1:1:1. It is a general rule in thermophilic proteins that they contain very few SH groups and large amounts of Arg and Glu to form stable far-reaching salt bridges.

The membrane portion of F_0F_1 was shown to be an $H^{\oplus}$ channel [6 $H^{\oplus}$/(mole × s × 100 mV)][5]. Proteoliposomes containing F_0F_1 and saturated PS3 phospholipids were capable of accumulating protons by ATP hydrolysis (253 mV max.), and also of synthesizing ATP driven by an artificially imposed $H^{\oplus}$ gradient (160 nmol ATP/mg)[6,7].

1 Kagawa, Y., Sone, N., Yoshida, M., Hirata, H. & Okamoto, H. (1976) *J. Biochem.* **80**, 141 - 151.
2 Wakabayashi, T., Kubota, M., Yoshida, M. & Kagawa, Y. (1977) *J. Mol. Biol.* **117**, 515 - 519.
3 Yoshida, M., Sone, N., Hirata, H. & Kagawa, Y. (1977) *J. Biol. Chem.* **252**, 3480 - 3485.
4 Yoshida, M., Okamoto, H., Sone, N. & Kagawa, Y. (1977) *Proc. Natl. Acad. Sci. U.S.A.* **74**, 936 - 940.
5 Okamoto, H., Sone, N., Hirata, H., Yoshida, M. & Kagawa, Y. (1977) *J. Biol. Chem.* **252**, 6125 - 6131.
6 Sone, N., Yoshida, M., Hirata, H. & Kagawa, Y. (1977) *J. Biol. Chem.* **252**, 2956 - 2960.
7 Kagawa, Y., Ohno, K., Yoshida, M., Takeuchi, Y. & Sone, N. (1977) *Fed. Proc.* **36**, 1815 - 1818.

Yasuo Kagawa, Department of Biochemistry, Jichi Medical School, Minamikawachi-machi, Tochigi-ken, Japan 329-04.

W. Sebald and E. Wachter

Amino Acid Sequence of the Major Polypeptide Subunit of the $H^{\oplus}$-Translocating Membrane Factor of the Mitochondrial ATPase Complex

The membrane-integrated part of the mitochondrial ATPase complex is involved in the translocation of $H^{\oplus}$

across the mitochondrial inner membrane. This hydrophobic component provides binding sites for specific inhibitors of the ATPase such as oligomycin and DCCD (dicyclohexylcarbodiimide), and it consists of at least three different subunit polypeptides.

The major subunit of the membrane factor is an extremely hydrophobic polypeptide of 8000 daltons. It is usually referred to as proteolipid due to its solubility in chloroform/methanol. At least six molecules of the proteolipid are present in one ATPase complex. The polypeptide binds specifically and covalently the inhibitor DCCD. It is furthermore involved in the binding of oligomycin as demonstrated by alterations of the amino acid sequence in oligomycin-resistant mutants from *Neurospora crassa* and *Saccharomyces cerevisiae*.

The amino acid sequences of the proteolipid from *N. crassa* (81 residues) and *S. cerevisiae* (76 residues) show extensive homology (50%). The polypeptide contains two hydrophobic segments of 25 residues each, which are separated by a polar segment of 12 residues. The extent of homology is particularly high in the hydrophobic regions. Three basic, one acidic and two polar positions are conserved in the polypeptides from both organisms.

The inhibitor DCCD is bound to a glutamyl residue occurring in the middle of one of the hydrophobic segments. Remarkably, amino acid exchanges identified in four different oligomycin-resistant mutants occur in close vicinity to the DCCD-binding residue.

The structural data obtained from the sequence studies will be discussed in connection with a possible role of the proteolipid in $H^{\oplus}$ translocation.

Walter Sebald, Institut für Physiologische Chemie, Physikalische Biochemie und Zellbiologie der Universität, Goethestraße 33, D-8000 München 2.

D. F. Wilson and M. Erecinska

Control of Energy Flux in Biological Systems

The cellular energy level ([ATP]/([ADP][P_i])) is maintained in a dynamic steady state by the ATP synthesis pathways, glycolysis and mitochondrial oxidative phosphorylation, and the ATP-utilizing pathways, primarily biosynthesis and transport processes. Oxidative phosphorylation supplies more than 90% of ATP synthesis under aerobic conditions. Measurements in vitro and in vivo show that mitochondrial phosphorylation (and respiration) is regulated by the cytosolic [ATP]/([ADP] × [P_i]) and the intramitochondrial [$NAD^{\oplus}$]/[NADH]. A mathematical model has been developed which accurately describes this regulation[1]. Studies on the inhibition of glycolysis by respiration (Pasteur effect) and inhibition of respiration by glucose addition (Crabtree effect) show that glycolysis is also regulated by [ATP]/([ADP][P_i]). In addition to its participation in [ATP]/([ADP][P_i]), low intracellular [P_i] inhibits the ATP-utilizing reactions. The Crabtree effect arises primarily from the decreased intracellular [P_i] following glucose addition and consequent inhibition of ATP utilization rather than from inhibition of ATP synthesis. It is generally true that cellular energy metabolism regulates toward an [ATP]/([ADP][P_i]) characteristic of the cell type.

1 Wilson, D. F., Owen, C. S. & Holian, A. (1977) *Arch. Biochem. Biophys.* **182**, 749 - 762.

David F. Wilson, Dept. Biochem. & Biophys., Univ. of Pennsylvania, Medical School, Philadelphia, Pa. 19104, U.S.A.

J. W. Stucki

Efficiency of Mitochondrial Energy Conversions

A theory of the efficiency of mitochondrial energy conversions, notably oxidative phosphorylation, has been developed within the framework of linear non-equilibrium thermodynamics. The efficiency of oxidative phosphorylation is critically dependent on the degree of coupling (q) of oxidative phosphorylation and on the force ratio, i.e. the ratio of the clamped phosphate potential and the redox potential of the substrate. Efficiency is zero at both maximal (static head) and zero (level flow) force ratios and passes through a maximum (optimal efficiency) in between these values. Experimental observations indicate that mitochondria in vivo operate at optimal efficiency. Under the condition of optimal efficiency there are two distinct degrees of coupling: 1) q_{max} where the *rate* of ATP synthesis is maximal (maximal case), and 2) q_{ideal} where the *output power* of oxidative phosphorylation, i.e. the developed phosphate potential times the rate of ATP production, is maximal (ideal case). The experimentally observed value of 0.9 for the adenylate energy charge in vivo corresponds exactly to the ideal case. This indicates that oxidative phosphorylation has evolved to a state where the output power rather than the rate of ATP production is maximal. On the basis of the principle of minimal entropy production at steady state the following theorem has been proven: Mitochondria reach a state of optimal efficiency whenever the resistance of phosphorylation (R_{phos}) matches the overall resistance (R_{load}) of the cellular ATP-utilizing reactions, according to the relation $R_{phos}/R_{load} = \sqrt{1-q^2}$.

J. W. Stucki, Pharmakologisches Institut der Universität Bern, Friedbühlstr. 49, CH-3010 Bern.

K. van Dam and H. V. Westerhoff

The Energy Balance of Oxidative Phosphorylation

Oxidative and photosynthetic phosphorylation play a key role in the energy metabolism of living matter. According to the chemi-osmotic model, proposed by Mitchell[1], there is a strong analogy in the mechanisms by which the two processes proceed. The essential element of this model is that a gradient of $H^{\oplus}$ ions across the energy-conserving membrane functions as an intermediate between electron transport and ATP synthesis. An energy balance of oxidative and photosynthetic phosphorylation, then, should consider the energetic contributions of all three components: the input (affinity of the redox reaction; ΔG_{ox}), the intermediate (electro-chemical potential difference of $H^{\oplus}$ across the membrane; $\Delta\tilde{\mu}_{H^{\oplus}}$) and the output (affinity of the ATPase reaction; ΔG_p).

In practice, the process of oxidative phosphorylation is to a large extent reversible, i.e. it is possible to drive reversed electron transport by ATP hydrolysis. On the basis of this simple consideration, the difference in energy content (multiplied by the appropriate stoichiometry) of the three components under the usual conditions of the cell cannot be very large. A precise determination of ΔG_{ox}, $\Delta\tilde{\mu}_{H^{\oplus}}$ and ΔG_p under different conditions allows us to draw conclusions concerning the stoichiometry and efficiency of coupling between the different steps of the overall process of oxidative phosphorylation. It appears even to be possible to predict the rate of flow through the different steps from the prevailing affinities.

1 Mitchell, P. (1968) Chemi-osmotic coupling and energy transduction, Glynn Research, Bodmin.

K. van Dam, Laboratorium voor Biochemie, B.C.P. Jansen Instituut, Plantage Muidergracht 12, NL-1018 TV, Amsterdam.

Hoppe-Seyler's Z. Physiol. Chem.
Bd. 359, S. 459 - 464, April 1978

24. Konferenz der Gesellschaft für Biologische Chemie Zugleich 1. Konferenz der Studiengruppe Neurochemie

Adenine Nucleotides and Neuronal Function

Held in München, February 3rd, 1978

Organized by G.W. Kreutzberg and P. Schubert

Sponsored by Stiftung Volkswagenwerk

Abstracts

H. Winkler and W. Aberer

Uptake of Nucleotides by Catecholamine-Storing Organelles

We have recently shown that isolated chromaffin granules take up [^{3}H]ATP by an atractyloside-sensitive carrier[1]. It has been proposed[1,2] that the driving force for this uptake is a proton gradient built up by the Mg-ATPase localized in the granule membranes. This proton gradient also drives catecholamine uptake[3]. In the presence of nigericin and KCl in the incubation medium, catecholamine uptake is inhibited but that of nucleotides is not. This indicates that the electrical component of the proton gradient, i.e. the positive charge inside the granules, drives nucleotide uptake, whereas the chemical part of this gradient is responsible for catecholamine uptake.

Further studies revealed that [^{3}H]ADP and [^{3}H]AMP are transported by the same mechanisms. The affinity of the carrier for ATP and the other nucleotides is rather low ($K_m > 1.4$mM) compared to the mitochondrial carrier for ADP/ATP exchange ($K_m \sim 1\ \mu$M). This difference makes good functional sense. In chromaffin cells ADP is first taken up by the high-affinity system of mitochondria and converted to ATP, which is then released to the cytosol. When the ATP concentration in the cytosol becomes high enough, the low-affinity system of the chromaffin granules can accumulate it. Thus the difference in affinities assures mitochondrial ATP synthesis without interference from the nucleotide uptake by chromaffin granules.

Sympathetic nerves contain, in addition to large dense core vesicles, which are very similar to chromaffin granules, small dense core vesicles. When superfused rat vasa deferentia were pulse-labelled with [^{3}H]-adenosine and [^{14}C]noradrenalin, the small dense core vesicles accumulated both. This indicates that the small dense core vesicles possess the same uptake system as chromaffin granules.

1 Kostron, H., Winkler, H., Peer, L. J. & König, P. (1977) *Neuroscience* **2**, 159 - 166.
2 Winkler, H. (1977) *Neuroscience* **2**, 657 - 683.
3 Bashford, C. L., Casey, R. P., Radda, G. K. & Ritchie, G. A. (1976) *Neuroscience* **1**, 399 - 412.

Hans Winkler, Pharmakologisches Institut der Universität Innsbruck, Peter-Mayr-Straße 1, A-6020 Innsbruck.

H. Zimmermann

Adenine Nucleotide Turnover in Cholinergic Vesicles of the Torpedo Electric Organ

Synaptic vesicles were isolated from perfused blocks of *Torpedo* electric organ on sucrose density gradients in a zonal rotor. Unstimulated tissue contained 83% ATP, 15% ADP and 2% AMP. After stimulation (0.1 Hz, 1260 - 1800 impulses) there is a loss of 50% of total vesicular nucleotides. The corresponding figures for the relative contents in adenine nucleotides after stimulation were 69%, 22% and 6%, respectively. Since vesicle number does not fall under the conditions of the experiments, the results suggest that adenine nucleotides are lost from vesicles on stimulation.

If [2-^{3}H]adenosine is applied to perfused tissue blocks, 85% of the radiolabel recovered in the isolated vesicle fraction is in the form of ATP. On stimulation of tissue blocks at 0.1 Hz, two populations of synaptic vesicles can be isolated, the denser of which (fraction VP_2, 1) contains the bulk (70%) of newly synthesized vesicular

ATP as well as acetylcholine (ACh). The specific radioactivity of ATP of the denser vesicles was about four times higher than that of vesicles isolated at the original sucrose density (fraction VP_1).

The data suggest that adenosine is an effective precursor of vesicular adenine nucleotides. On stimulation nucleotides are lost from synaptic vesicles together with the neurotransmitter. The new population of synaptic vesicles appearing on stimulation has a high turnover rate for both ATP and ACh.

1 Zimmermann, H. & Denston, C. R. (1977) *Neuroscience* **2**, 715 - 730.

H. Zimmermann, Max-Planck-Institut für Biophysikalische Chemie, Abteilung Neurochemie, Am Faßberg, D-3400 Göttingen.

P. Schubert

Intraneuronal Transport and Functionally Related Release of Adenosine Derivatives from Central Axon Terminals

Adenosine is reported to have a remarkable biological activity in the brain. It leads to vasodilatation, stimulates cAMP synthesis, depresses nerve cell activity, and may interfere with Ca fluxes and transmitter release.

Intraneuronal transport and an activity-related release from neurons provide a controlled channeling of adenosine in the brain. Adenosine derivatives are released from dendrites, taken up by glial cells, and reach adjacent vessels. They are also released from axon terminals and taken up by the postsynaptic neurons, as studied in the entorhino-hippocampal system and in several other systems, including cholinergic pathways of the central nervous system. This release could be increased in vivo by synaptic stimulation. More than 94% of the transported and released material was trichloroacetic-acid soluble. In the extracellular space, nucleotides are rapidly broken down to adenosine. The action of adenosine can be terminated by a further enzymatic breakdown and by cellular reabsorption. These findings agree with the concept that adenosine acts as a neuronal modulator which is released on nerve cell activation in addition to the principal transmitter.

In a hippocampal slice preparation, the potent adenosine analogue, chloroadenosine (3 - 10μM), completely inhibited the evoked potentials of CA_1 neurons elicited by synaptic stimulation. This effect was reversible and could be antagonized by theophylline. Adenosine (10 - 100μM) failed to produce a complete inhibition. But in the presence of theophylline (0.1 - 0.3mM), a disinhibition of neuronal activity was observed and considerably lower current strengths were needed to elicit nerve cell responses. This indicates that adenosine has a 'stabilizing' effect on synaptic transmission. Also a longer lasting modulation of nerve cell function based on cAMP-mediated metabolic changes should be considered. The adenosine-induced cAMP increase is reported to be potentiated by monoamines. This may be a mechanism by which the inputs from adenosine-releasing specific afferents and from monaminergic central core neurons could be integrated, resulting in a defined and metabolically-linked effect.

Peter Schubert, Max-Planck-Institut für Psychiatrie, Kraepelinstr. 2, D-8000 München 40.

G. W. Kreutzberg

Cytochemical Localization of 5'-Nucleotidase in the Central Nervous System

With increasing knowledge of the possible function of adenosine as an important intercellular communication molecule the enzymes involved in its metabolism assume increasing interest. 5'-Nucleotidase (EC 3.1.3.5) produces adenosine by hydrolysing AMP. This enzyme is considered a rather general marker for plasma membranes. By means of a cytochemical technique activity of this enzyme has now been demonstrated on the electron microscopical level in the rat central nervous system. The enzyme is present in plasma membranes of astrocytes, oligodendroglial cells and microglial cells but not in neuronal plasma membranes. Activity of myelinated fibers seems to be derived from the oligoglial cells. In several areas of the brain i.e. cingulate cortex, dentate gyrus, caudate nucleus, or cerebellum the astrocytic processes show strong activity as they form juxtasynaptic lamellae or footplates on the local vaculature.

The enzyme seems to be an ectoenzyme in these locations since reaction product is always seen on the outer surface of the glial plasma membranes. This arrangement might be instrumental in order to reach nucleotides in the extracellular spaces after their release, e.g. during synaptic activity. Thus the glial cell lamellae bordering synapses or surrounding synaptic complexes could provide the mechanism for adenosine formation. As one conceivable function adenosine could then be chanelled to the local capillaries where it can act as a vasodilator.

Kreutzberg, G. W., Barron, K. D. & Schubert, P. (1978) *Brain Res.*, in press.

Georg W. Kreutzberg, Max-Planck-Institut für Psychiatrie, Kraepelinstr. 2, D-8000 München 40.

M. Reddington and P. Schubert

Biochemical and Electrophysiological Effects of Adenosine and 2-Chloroadenosine in Hippocampus Slices

2-Chloroadenosine (10μM), a potent analogue of adenosine, completely supresses evoked potentials in hippocampus slices in vitro, probably by interacting directly with adenosine receptors. Since adenosine and 2-chloroadenosine strongly stimulate the synthesis of cAMP in the brain[1] the observed inhibitory effect of 2-chloroadenosine on nerve cell activity might be related to the synthesis of cAMP and the subsequent phosphorylation of neuronal proteins.

In apparent contradiction to this hypothesis, however, adenosine itself failed to mimic the electrophysiological effect of 2-chloroadenosine in hippocampus at concentrations between 10μM and 100μM. This discrepancy could not be explained by differences in adenylate cyclase (EC 4.6.1.1) activation since no significant differences were observed in the abilities of adenosine and 2-chloroadenosine to stimulate cAMP synthesis in this preparation; at concentrations of 50μM and 100μM both substances increased hippocampal cAMP content by ca. 50% and 150%, respectively. Concentrations of 10μM had no measurable effect on cAMP content.

It has been reported that 10μM adenosine inhibits biogenic amine-stimulated protein phosphorylation in slices of brain tissue[2]. Furthermore, we have been able to show that similar concentrations of adenosine significantly inhibit cAMP-stimulated protein kinase activity (EC 2.7.1.37) in synaptosomal plasma membranes and in a soluble fraction from whole rat brain. In contrast, 2-chloroadenosine inhibited only at concentrations above 100μM.

It is therefore possible that the effect of 2-chloroadenosine on nerve cell activity is mediated through stimulation of cAMP synthesis and a subsequent increase in protein kinase activity, whereas adenosine, although stimulating the adenylate cyclase, has no electrophysiological effect due to inhibition of protein phosphorylation.

1 Daly, J. (1977) Cyclic Nucleotides in the Nervous System, Plenum Press, New York.

2 Reddington, M., Rodnight, R. & Williams, M. (1973) *Biochem. J.* **132**, 475 - 482.

M. Reddington, Max-Planck-Institut für Psychiatrie, Kraepelinstr. 2, D-8000 München 40.

M. Wahl, J. Schrader and W. Kuschinsky

The Dilatory Effect of Adenosine on Pial Arteries and its Possible Involvement in the Regulation of Cerebral Blood Flow

Local factors such as extracellular $H^{\oplus}$, $K^{\oplus}$ and $Ca^{2\oplus}$ have been postulated as mediators in the metabolic regulation of cerebral blood flow (CBF). Recently evidence has been found that adenosine also participates in the adjustment of cerebral vascular resistance. Concerning the dilatory effect of adenosine on pial arteries, it has been demonstrated that this nucleoside does not penetrate the blood-brain barrier, but only becomes effective when applied from the extravascular side[1]. By use of a perivascular microapplication technique which permits to mimic the route of penetration of a metabolically released substance, adenosine was shown to induce a dose-dependent vasodilation of pial arteries (50 - 200 μ diameter) in chloralose anesthetized cats[2]. The threshold concentration for adenosine was 10^{-9}M; maximal dilation (30% change in vessel diameter) was observed at 10^{-3}M. As was reported for the coronary vasculature[3], theophylline inhibited the dilatory effect of adenosine on pial arteries, also in a competitive manner[2].

Increases in adenosine content in brain tissue have been demonstrated to occur during ischemia[1,4,5], hypoxia, hyperventilation, hypotension and during electrical stimulation of the cortex[6]. In the case of hypoxia, however, the results could not be confirmed by others, even after prolonged exposure to hypoxia[7]. Furthermore, no changes in tissue content of adenosine were observed during bicuculline-induced seizure, when tissue samples were analyzed 3 - 4 s and 20 min after the onset of the seizure electroencephalogram (EEG)[7].

In order to clarify whether adenosine is formed in the early phase of functional hyperemia, the cortical concentrations of adenosine were determined by an enzymatic spectrophotometric method, 15 and 30 s after the onset of the bicuculline (10 mg i.v.) -induced seizure. Under control conditions, the cortical contents of adenosine and of lactate were 1.4 nmol/g and 0.97 μmol/g, respectively. 15 s after beginning of the typical EEG changes, the adenosine content increased by 190%, that of lactate by 135%. After 30 s, the values for adenosine and lactate were still elevated by 127% and 75%, respectively. From these results it appears likely that adenosine formed at an enhanced rate during the early phase of seizure might participate in the induction of the CBF increase during this condition.

The dilatory action of adenosine on cerebral vessels was shown to become attenuated when the extracellular concentrations of $H^{\oplus}$ or $K^{\oplus}$ are increased[8]. Since rapid changes in $H^{\oplus}$ and $K^{\oplus}$ concentrations are also known to occur during seizure, these ions are likely to modulate the dilatory effect of adenosine on pial vessels under these circumstances. Thus, in order to define the extent to which adenosine might participate in the regulation of CBF, the interplay of adenosine with $H^{\oplus}$ and $K^{\oplus}$ has to be considered in addition.

1 Berne, R. M., Rubio, R. & Curnish, R. R. (1974) *Circ. Res.* **35**, 262 - 271.
2 Wahl, M. & Kuschinsky, W. (1976) *Pflügers Arch.* **362**, 55 - 59.
3 Bünger, R., Haddy, F. J. & Gerlach, E. (1975) *Pflügers Arch.* **358**, 213 - 224.
4 Deuticke, B. & Gerlach, G. (1966) *Pflügers Arch.* **292**, 239 - 254.
5 Nordström, C.-H., Rehncrona, S., Seisjö, B. K. & Westerberg, E. (1977) *Acta Physiol. Scand.* **101**, 63 - 71.
6 Rubio, R., Berne, R. M., Bockman, E. L. & Curnish, R. R. (1975) *Am J. Physiol.* **228**, 1896 - 1902.
7 Rehncrona, S., Nordström, C.-H., Siesjö, B. K. & Westerberg, E. (1977) *Acta Neurol. Scand.* **56**, *Suppl.* **64**, 220 - 221.
8 Wahl, M. & Kuschinsky, W. (1977) *Blood Vessels* **14**, 285 - 293.

Jürgen Schrader, Physiologisches Institut der Universität München, Pettenkoferstraße 14 a, D-8000 München 2.

J. Schrader

Effects of Adenosine on Transmembranal $Ca^{2\oplus}$-Influx and Adenylate Cyclase of the Heart

Adenosine is known to inhibit contractile force development in the atrial myocardium[1]. In experiments carried out in partially depolarized (22mM KCl) guinea pig atrium, the $Ca^{2\oplus}$-mediated action potential elicited by noradrenalin was depressed by adenosine[2]. When applied at micromolar concentrations, adenosine reduced within seconds the rate of rise and amplitude of the slow action potentials. These effects could be reversed by enzymatic degradation of adenosine to inosine, or by raising the extracellular $Ca^{2\oplus}$ concentration.

In order to determine whether adenosine acts via intracellular mechanisms, or by interaction with receptors for adenosine localized at the cell surface, we applied an adenosine derivative of high molecular weight (protein-AMP conjugate) which is confined to the extracellular space[3]. The protein-AMP conjugate was synthesized by covalently coupling AMP by P-N linkage to lysine residues of a carbonate dehydratase (EC 4.2.1.1) preparation (M_r 30 000). Like adenosine, this high-molecular weight compound, which proved to be stable under physiological conditions, also depressed the slow atrial action potential. Thus, the inhibitory effects of adenosine on transmembranal $Ca^{2\oplus}$-influx occurred from the extracellular side and are most likely mediated via a receptor site for adenosine at the surface of atrial muscle cells.

In ventricular myocardium of isolated perfused guinea pig hearts, adenosine induced a dose-dependent inhibition of the isoproterenol-elicited enhancement of cardiac contractile force. Threshold concentrations of adenosine was 10^{-7}M, and maximal inhibition (90%) occurred at 10^{-5}M. In addition to its influence on cardiac performance, adenosine effectively inhibited the isoproterenol-induced rise in myocardial levels of cyclic 3′,5′-AMP, glucose 1-phosphate and glucose 6-phosphate. Adenosine also exhibited a pronounced influence on cardiac adenylate cyclase activity (EC 4.6.1.1). In a crude membrane preparation from guinea pig ventricles, adenosine antagonized the effects of isoproterenol dose-dependently. This nucleoside also inhibited the adenylate cyclase stimulated by $F^{\ominus}$ as well as by GppNHp in a lubrol-PX-solubilized membrane preparation, known to be insensitive to catecholamines. Thus, the effects on the adenylate cyclase are most likely caused by a specific interaction of adenosine with the catalytic unit of the enzyme.

Adenosine constitutes an endogenous metabolite of the heart, which is constantly released into the extracellular space[5]. Upon application of isoproterenol to the isolated beating guinea pig heart, the adenosine content in the effluent perfusate increased within 45 s from 10^{-8}M to 10^{-6}M, while at the same time cardiac contractile force was depressed[4]. It therefore appears conceivable that endogenously formed adenosine may be involved in the limitation of the inotropic and metabolic effects of catecholamines under physiological conditions.

1 DeGubareff, T. & Sleator, W. J. (1965) *Pharmacol. Exp. Ther.* **148**, 202 - 214.
2 Schrader, J., Rubio, R. & Berne, R. M. (1975) *J. Mol. Cell. Cardiol.* **7**, 427 - 433.
3 Schrader, J., Nees, S. & Gerlach, E. (1977) *Pflügers Arch.* **369**, 251 - 257.
4 Schrader, J., Baumann, G. & Gerlach, E. (1977) *Pflügers Arch.* **372**, 29 - 35.
5 Rubio, R., Berne, R. M. & Katori, M. (1969) *Am. J. Physiol.* **216**, 56 - 62.

Jürgen Schrader, Physiologisches Institut der Universität München, Pettenkoferstraße 14 a, D-8000 München 2.

J. Schultz, W. Schocker and G. R. Siggins

Mechanism of Action of Tricyclic Antidepressants: Biochemical and Electrophysiological Studies

Oral treatment of rats with imipramine (30 - 50 mg × kg^{-1} × day^{-1}) for at least 5 days produced a marked subsensitivity (~ 50%) of noradrenergic receptors in the cerebral cortex and striatum, but *not* in the cerebellum, as measured by the noradrenalin-stimulated formation of cyclic AMP in tissue slices. Using cerebral cortical slices, dose-response curves showed a reduced formation of cyclic AMP at all concentrations of noradrenalin (1 - 100 μM). Basal adenylate cyclase activity (EC 4.6.1.1) determined in cell-free homogenates was not altered under these circumstances. Six days after cessation of drug treatment, the cyclic AMP generating system of cerebral cortex and striatum returned to normal sensitivity toward noradrenalin. After 5 days of drug treatment, reuptake of noradrenalin into tissue slices from brain was inhibited by about 50%, although no drug, was added to the incubation medium. Again, this effect faded within 7 days after discontinuation of drug application. Concentrations of noradrenalin in cerebral cortex are decreased by almost two thirds during 5 days of imipramine treatment. The noradrenalin levels obtained were about the same as those seen after treatment of rats with several doses of reserpin.

For electrophysiological testing, animals were treated
1) with desipramine-HCl for 6 - 10 days,
2) with a single dose only, immediately prior to the start of the experiment and
3) with desipramine-HCl for 10 days followed by withdrawal from the drug for 5 days.

Controls received water only. The spontaneous discharge of Purkinje cells was recorded and firing rates and patterns of each neuron were quantified by construction of interspike interval histograms. Six rats chronically treated with desipramine-HCl showed a mean Purkinje firing rate less than one fourth of those of the 6 control rats studied. Even a single dose of drug significantly reduced spontaneous firing by 25%. Purkinje cell firing rates returned to control rates within 5 days after cessation of drug treatment. Responses of Purkinje cells to iontophoretically applied noradrenalin were also tested. The mean threshold noradrenalin current (43 nA) required to depress neurons of chronically drug-treated rats was significantly higher than that for threshold inhibition of control cells (24 nA). Five days after discontinuation of drug treatment, there was no significant difference in threshold currents.

These latter findings are suggestive of a reduced responsiveness of chronically desipramine-HCl-treated cells to noradrenalin. Additional studies will be necessary to assess, whether the observed changes in biochemical and electrophysiological parameters are indeed involved in the therapeutic action of chronic treatment with tricyclic antidepressant drugs.

J. Schultz, Pharmazeutisches Institut der Universität Tübingen, Auf der Morgenstelle 8, D-7400 Tübingen.

W. Seifert, P. Clapshaw, J. Morgan and J. Oey

2',3'-Cyclic Nucleotides and 2',3'-Cyclic-Nucleotide 2'-Phosphodiesterase in Nerve Cell Cultures

We are interested in the biological function of 2',3'-cyclic-nucleotide 2'-phosphodiesterase (EC 3.1.4.16) and its substrates, the 2',3'-cyclic nucleotides. The enzyme serves as a marker enzyme for myelin and is found in relatively high activity in oligodendroglial cells, while neuroblastoma cells or neuronal cells contain only rather low activities. Using tritium-labelled 2',3'-cyclic nucleotides as substrates and separating the products on thin-layer chromatography we find 2',3'-cyclic-nucleotide 2'-phosphodiesterase activity in all cells tested which seems to indicate a rather general function for this membrane-bound enzyme. We purified the enzyme from bovine white matter and found a molecular weight of about 40000. Enzymatic studies with intact cells and with cell homogenates suggest that this enzyme is located as an "ecto-enzyme" on the outside of the plasma membrane. We are now trying to raise antibodies against this enzyme in order to localize it on the plasma membrane by immunofluorescence. It is also interesting to compare the enzymatic properties of the purified enzyme and of the membrane-bound form in glial and neuronal cell lines.

In order to understand the possible role of 2',3'-cyclic-nucleotide 2'-phosphodiesterase in the extra- and intracellular breakdown and the transport of 2',3'-cyclic nucleotides in nerve cell cultures, we have studied these problems using again radioactively labelled 2',3'-cyclic nucleotides. While the uptake of radioactivity was temperature- and serum-dependent and inhibited by jodoacetate and *p*-chloromercuribenzoate, the analysis of extra- and intracellular products showed that more than 90% of this uptake was due to extracellular breakdown to 2'-GMP as well as 3'-GMP and further to guanosine (or resp. adenosine). Less than 1% of the intracellular radioactivity was found in the 2',3'-cyclic nucleotide spot. Although it is possible to separate 2',3'- from 3',5'-cyclic nucleotides and to distinguish both by using 2',3'-cyclic-nucleotide 2'-phosphodiesterase or phosphodiesterase I (EC 3.1.4.1) for enzymatic digestion, it seems important to measure the concentration in vivo of 2',3'-cyclic nucleotides in a more direct way. We therefore developed a radioimmunoassay for 2',3'-cyclic nucleotides and are now in the process of determining the actual concentrations both in brain extracts and in nerve cell cultures.

Clapshaw, P. & Seifert, W. (1977) *Hoppe-Seyler's Z. Physiol. Chem.* **358**, 1189.
Oey, J. & Seifert, W. (1977) *Hoppe-Seyler's Z. Physiol. Chem.* **358**, 1254.

W. Seifert, Friedrich-Miescher-Laboratorium der Max-Planck-Gesellschaft, Spemannstr. 37 - 39, D-7400 Tübingen.

H. Cramer und R. Hammers

Extracelluläre zyklische Nucleotide im Nervensystem: Bedeutung und Regulation

Astrozytenkulturen sezernieren cAMP nach Stimulation von Adenylat-Cyclase (EC 4.6.1.1) gegen Konzentrationsgradienten[1]. Auch in intaktem Nervengewebe wird eine Freisetzung von cAMP in das Medium in Korrelation zur Akkumulation im Gewebe beobachtet[2]. Im intakten Organismus gelangen zyklische Nucleotide aus dem Extracellularraum umgebender Hirnstrukturen in den Ventrikel- und Subarachnoidalraum. Pathologische und pharmakologische Einwirkungen können die Konzentrationen von zyklischen Nucleotiden im Liquor cerebrospinalis verändern, jedoch bleiben die Liquorspiegel bei intaktem Clearance-Mechanismus relativ konstant[3,4]. Liquorraum-begrenzende Strukturen, insbesondere die Plexus chorioidei, nehmen cAMP und cGMP aktiv auf, der Aufnahmemechanismus ist ähnlich wie der für andere schwache Anionen durch Probenezid hemmbar[5]. Im humanen Liquor cerebrospinalis sind die cGMP-Spiegel in der Regel erheblich niedriger als die cAMP-Spiegel. Die Durchsatzrate von cGMP, geschätzt nach Probenezid-Applikation, kann bei neurologischen Patienten ebenso hoch wie die für cAMP sein[4]. Der Plexus chorioideus des Kaninchens enthält zusätzlich zur Fähigkeit, zyklische Nucleotide aufzunehmen, eine Prostaglandin-E_1- und Histamin-sensitive Adenylat-Cyclase und eine durch Cholinergica stimulierbare Guanylat-Cyclase (EC 4.6.1.2) sowie starke Aktivitäten von 3',5'-zykl.-Nucleotid-Phosphodiesterase (EC 3.1.4.17), so daß eine Rolle zyklischer Nucleotide bei den Transport- und Sekretionsfunktionen des Plexus chorioideus zu vermuten ist.

1 Clark, R. B., Su, Y. F., Ortmann, R., Cubeddu, X., Johnson, G. L. & Perkins, J. P. (1975) *Metabolism* **24**, 343 - 358.

2 Cramer, H. & Lindl, T. (1974) *Nature (London)* **249**, 380 - 382.

3 Cramer, H., Ng, L.K.Y. & Chase, T. (1973) *Arch. Neurol. (Chicago)* **29**, 197 - 199.

4 Cramer, H., Hammers, R., Clarenbach, P. & Horstmann, R. (1978) *Int. J. Neurol.*, im Druck.

5 Hammers, R., Clarenbach, P., Lindl, T. & Cramer, H. (1977) *Neuropharmacology* **16**, 135 - 141.

Prof. Dr. H. Cramer, Abt. Klinische Neurologie u. Neurophysiologie der Universität Freiburg, Hansastr. 9, D-7800 Freiburg i.Br.

Hoppe-Seyler's Z. Physiol. Chem.
Bd. 359, S. 465 - 472, April 1978

The Binding of Lysolecithin to Human Serum High Density Apoprotein A-I A ^{13}C-NMR Study

Wilhelm STOFFEL, Peter METZ and Budi TUNGGAL

Institut für Physiologische Chemie der Universität Köln

(Received 16 January 1978)

Summary: 1-[14,15-$^3H_2$;16-^{13}C]Palmitoyl-, 1-[9,10-$^3H_2$;11-^{13}C]- and 1-[9,10,12,13-$^3H_4$;-14-^{13}C]stearoyl-, 1-[9,10-$^3H_2$;11-^{13}C]oleoyl-, 1-[9,10,12,13-$^3H_4$;14-^{13}C]linoleoyl-*sn*-glycero-3-phosphocholine and 1-palmitoyl-*sn*-glycero-3-phospho-[*N*-$^{13}CH_3$,$^{14}CH_3$]choline have been prepared by chemical synthesis. They recombine with apolipoprotein A-I of human serum high density lipoproteins. This binding leads only to minor changes in circular dichroism with an increase of α-helicity (f_H) from 0.44 to 0.54 on the average. The lipid-apoprotein A-I-complexes remain stable during agarose chromatography. 60 - 70 lysolecithin molecules were bound to the apolipoprotein, approximately the number associating with the apolipoprotein A-I during equilibrium dialysis (70 lysolecithins per apolipoprotein A-I molecule).

Spin-lattice relaxation time (T_1) experiments indicate that only the hydrophobic fatty acyl chains of the saturated as well as the unsaturated lysolecithins bind to the polypeptide chains, whereas the mobility of the choline moiety of the hydrophilic, polar group remains completely unaltered. The T_1-times were compared with those of the respective lysolecithin in aqueous micellar solution. The function of lysolecithin in the synthesis of the high density lipoprotein particle during biosynthesis is discussed.

Über die Bindung von Lysolecithinen an Apoprotein A-I des menschlichen Serums: Eine ^{13}C-NMR-Untersuchung

Zusammenfassung: 1-[14,15-$^3H_2$;16-^{13}C]Palmitoyl-, 1-[9,10-$^3H_2$;11-^{13}C]- und 1-[9,10,12,13-$^3H_4$;14-^{13}C]Stearoyl-, 1-[9,10-$^3H_2$;11-^{13}C]-Oleoyl-, 1-[9,10,12,13-$^3H_4$;14-^{13}C]Linoleoyl-*sn*-glycero-3-phosphocholin und Palmitoyl-*sn*-glycero-3-phospho-[*N*-$^{13}CH_3$,$^{14}CH_3$]cholin wurden synthetisiert. Sie rekombinieren mit Apolipoprotein A-I des menschlichen Serum-High-Density-Lipoproteins. Diese Bindung bewirkt nur geringe Änderungen des Circulardichroismus mit einem Anstieg des α-Helix-Gehaltes (f_H) von 0.44 auf 0.54 im Mittel. Die Lysolecithin-Apolipoprotein-Komplexe sind während der Agarose-Chromatographie stabil und binden ungefähr 60 - 70 Lysolecithinmoleküle. Ähnlich viele Lysolecithinmoleküle assoziieren mit Apolipoprotein A-I in der Gleichgewichtsdialyse (70 Lysolecithine pro Apolipoprotein A-I Molekül).

Spin-Gitter-Relaxationszeit-Messungen zeigen, daß nur die hydrophoben Fettsäureketten der gesättigten und ungesättigten Lysolecithine an die Polypeptidketten binden. Die Beweglichkeit der Cholingruppe in der polaren hydrophilen Gruppe ändert sich hingegen nicht. Die T_1-Zeiten wurden jeweils mit denen der entsprechenden Lysolecithine in wäßrigen micellaren Lösungen verglichen. Die Funktion des Lysolecithins in der Synthese des High-Density-Lipoproteins wird diskutiert.

Key words: ^{13}C-labelled lysolecithin species, binding to apolipoprotein A-I, spin-lattice relaxation time studies.

Serum high density lipoproteins (HDL) represent a group of lipoproteins very suitable to study the interactions of their apoproteins A-I and A-II and lipids by physical and chemical methods and finally to construct from these data a model of these particles. This should explain the lipid binding, the lipid transport function and the substrate function of the lipids and the activator function of the apolipoprotein A-I for lecithin-cholesterol acyltransferase[1].

Phosphatidylcholine, cholesterolester and cholesterol are the main classes among the lipid components of HDL, but about 2% lysolecithin is consistently found, which corresponds to about 4 mol lysolecithin per mol apoprotein[2].

In previous studies we have analysed the influence of lysolecithin on the reaggregation of HDL-apolipoproteins, particularly apolipoprotein A-I, with phosphatidylcholine and sphingomyelin species[3,4]. In this recombination procedure, lysolecithin apparently occupies sites of the partially unfolded apoprotein from which it is substituted by the diacylphospholipid or sphingomyelin with two long alkyl chains, one from the fatty acid and the other from the long chain base.

This paper reports experiments which on physical grounds prove that lysolecithin molecules bind to the apoprotein not by interactions with polar side chains of the polypeptide, but by hydrophobic interactions of the alkane chains with hydrophobic sites of the apolipoprotein. This information was derived from ^{13}C-NMR spin-lattice relaxation time experiments with lysolecithins substituted with palmitic, stearic, oleic and linoleic acid, enriched with carbon-13 in specific positions along the fatty acyl chain and in the choline moiety of the polar head groups.

The secondary structure of these lysolecithin-apolipoprotein A-I complexes was scarcely different from the apolipoprotein, as shown by circular dichroism spectra.

The stable complexes had densities between 1.11 - 1.19 g/cm^3.

Methods and Materials

^3H-labelled lysolecithin substituted with ^{13}C-labelled fatty acids ([16-^{13}C]palmitic acid, spec. radioact. 5.8 mCi/mmol, [11-^{13}C]oleic acid, spec. radioact. 22.6 mCi/mmol, [14-^{13}C]linoleic acid, spec. radioact. 3.75 mCi/mmol, [11- and 14-^{13}C]stearic acids) and 1-acylglycerophospho-[*N*-$^{13}CH_3$,$^{14}CH_3$]choline, spec. radioact. 3.52 mCi/mmol, were synthesized by acylation of *sn*-glycero-3-phosphocholine (W. Stoffel and K.P. Salm, to be published). The syntheses of the radioactive ^{13}C-enriched fatty acids have been described before[5]. The fatty acid residue in the 2-position of the diacylglycerophosphocholine molecules was removed by phospholipase A_2 *(Crotalus adamanteus)* hydrolysis and separated from the lysolecithin by silicic acid column chromatography. The fatty acid was eluted with chloroform and lysolecithins with methanol/chloroform 4 : 1 and methanol. All compounds were chemically and radiochemically pure.

Preparation of high density lipoprotein and its purified apolipoproteins: Human serum high density lipoproteins (density 1.063 - 1.21 g/cm^3) were isolated by the sedimentation-flotation ultracentrifugation method[6]. They were delipidated with chloroform/methanol[7] and the apoproteins separated by the combination of DEAE-cellulose[8] and Sephadex G-150 or G-200 chromatography in 8M urea[7]. The main apolipoproteins A-I and A-II were obtained in homogenous form as demonstrated by sodium dodecylsulfate polyacrylamide, disc and slab gel electrophoresis[9-12], in the immunodiffusion assay[13] and amino acid stoichiometry.

Purification of apolipoprotein A-I-lysolecithin complexes: Apolipoprotein A-I was stirred together with different molar amounts of the lysolecithins in aqueous solution. The complexes were separated from unbound lysolecithin

a) by ultracentrifugation in a caesium chloride gradient 10 to 20.5% in a SW 41 Beckman rotor at 38000 rpm for 72 h,

b) by agarose chromatography on a Bio-Gel A-5 m 100 - 200 mesh 90 × 2.5 cm column. The stoichiometry was calculated from the bound radioactive lysolecithin per apolipoprotein.

The radioactivity was assayed in a Tricarb liquid scintillation spectrometer, Packard, model 3380.

Protein was determined according to Lowry et al.[14] with serum albumin as standard.

Proton noise decoupled ^{13}C-nuclear magnetic resonance spectra of lipids (^{13}C-enriched fatty acids and lysolecithins) dissolved in deuterochloroform and apolipoprotein A-I-lysolecithin complexes dissolved in 3 m*l* water-20% D_2O were recorded in the Fourier transform mode in a Bruker WH 90 NMR spectrometer at 22.63 MHz as described before[4]. A temperature control unit Bruker ST 100/700 controlled the temperature within 1 °C.

For T_1-measurements, partially relaxed Fourier transform (PRFT) spectra were recorded, applying the inversion-recovery method[15]. T_1 was calculated from the slope of the computerized semilog plot of ln $(M_0 - M_t)$ against τ, the delay time. The magnetization *(M)* for the $(180° - \tau - 90°)$ pulse sequence is $M_{(t)} = M_0 (1 - 2 \exp - \tau/T_1)$[15]. The experimental error ranges between 5 and 10%.

Circular dichroism in the UV-range was measured with a Jasco J-41A spectropolarimeter. The molar ellipticity ([Θ]) was calculated and corrected according to Chen and Yang[16]. A molecular residue weight (MRW) of 116.5 for apolipoprotein A-I was used.

Table 1. ^{13}C-NMR spin-lattice relaxation times T_1 of ^{13}C-labelled lysolecithins in $CDCl_3$ and water.

Lipid	T_1-times [ms]	
	in $CDCl_3$	in H_2O
Fatty acid		
[16-^{13}C]16 : 0	4479 ± 181	
[11-^{13}C]18 : 1	2060 ± 42	
[14-^{13}C]18 : 2	3560 ± 184	
Lysolecithin containing		
[16-^{13}C]16 : 0	4319 ± 143	1942 ± 15
[14-^{13}C]18 : 0	1536 ± 17	292 ± 10
[11-^{13}C]18 : 0	921 ± 39	221 ± 6
[11-^{13}C]18 : 1	1457 ± 68	344 ± 13
[14-^{13}C]18 : 2	2446 ± 131	728 ± 42
[*N*-$^{13}CH_3$]Choline	95 ± 4	633 ± 32

Results

The NMR spectroscopic properties of six lysolecithins have been studied in 1) organic, 2) aqueous micellar solutions and 3) associated with apolipoprotein A-I. These were 1-[16-^{13}C]-palmitoyl-, 1-[11-^{13}C]- and 1-[14-^{13}C]stearoyl-, 1-[11-^{13}C]oleoyl-, 1-[14-^{13}C]linoleoylglycerophosphocholine and a hydrogenated [*N*-$^{13}CH_3$]-lysolecithin from soy lecithin*. Palmitoyl- and stearoylglycerophosphocholine have phase transition temperatures of 18.5 °C and 31 °C, respectively, as determined by light scattering at 400 nm; the others are liquid in the temperature range of the experiments (4 - 37 °C). All recombinations and spin-lattice relaxation time studies were carried out at 37 °C.

1) Spin-lattice relaxation times in organic solvents. Phospholipid molecules and fatty acids form inverted micelles in apolar solvents with their hydrophilic ionic groups closely packed due to electrostatic interactions. Their fatty acyl chains are directed into the organic solvent. The T_1-times of the aforementioned lysolecithins and fatty acids in $CDCl_3$ underline this suggestion (Table 1).

The choline methyl groups of lysolecithins in $CDCl_3$ are highly immobilized, as indicated by the very short T_1-time. The T_1-times of the ^{13}C-enriched fatty acyl chains in both the free fatty acids and the lysolecithins increased towards the terminal CH_3-group. This is due to the fixation of the carboxy terminal end at the glycerophosphate-backbone, but also due to reduced van der Waals' forces toward the terminal CH_3-group. *Cis*-double bonds in oleic acid and linoleic acid lead to a considerable elevation of the mobility of the carbon segment beyond the double bond most distant from the carboxy group. The *cis*-double bonds cause a disordered arrangement of the acyl chains, with facilitated *trans-gauche* isomerisations and kink formations thereby reducing the rotational correlation time, but prolonging the T_1-time.

2) Lysolecithins form micelles in water with an average molecular weight of 92000[17], which corresponds to about 170 to 190 molecules. The lysolecithin concentrations used in these experiments were about 10mM. This is far above the critical micellar concentration, which has been suggested to be below 120μM[17,18] and above 7μM[19]. The T_1-values of the ^{13}C-enriched lysolecithins in aqueous solution (Table 1) therefore stand for their micellar arrangement. The fatty acyl chains are drastically immobilized due to strong hydrophobic inter-chain interactions, whereas the choline group reaches a high T_1-time, which excluded electrostatic interactions between polar head groups of lipid molecules.

* Fatty acid composition in the 1-position: 16:0 24%; 18:0 76%.

3a) Apolipoprotein A-I-lysolecithin complexes. Lysolecithin easily associates with apolipoprotein A-I at temperatures above their phase transition point. The aforementioned six lysolecithins formed stable complexes in a ratio of 1 apolipoprotein A-I and 55 - 60 lysolecithin molecules. They eluted from an agarose column (2.5 x 90 cm) calibrated with test proteins as single bands with no lysolecithin in the void volume. Lysolecithin produced from highly unsaturated soy lecithin had a retention volume identical with that of the apolipoprotein A-I-lysolecithin complex. The apparent molecular weight determined by this agarose gel chromatography was around 90000 to 100000, very close to that of unsaturated lysolecithin micelles, Fig. 1. Therefore the complexes were centrifuged in a CsCl gradient, Fig. 2. The analysis of the gradients clearly proved that lysolecithin was completely bound to the apolipoprotein with no free apolipoprotein left. Whereas 1-palmitoyl-, 1-oleoyl- and 1-linoleoyl-glycerophosphocholine-A-I complexes banded at $\rho = 1.16 - 1.19$ g/cm^3, 1-stearoylglycerophosphocholine had a density of 1.11 g/cm^3. These results are in good agreement with earlier studies on apolipoprotein A-I-lysolecithin complexes of molar ratios 1 : 73 to 1 : 15, which had densities between 1.17 and 1.23 g/cm^3[3].

b) Circular dichroism. We measured the molar ellipticity ([Θ]) of the different apolipoprotein A-I-lysolecithin complexes in order to follow structural perturbation, particularly in α-helicity (f_H) by the binding of the lipid. Fig. 3 and Ta-

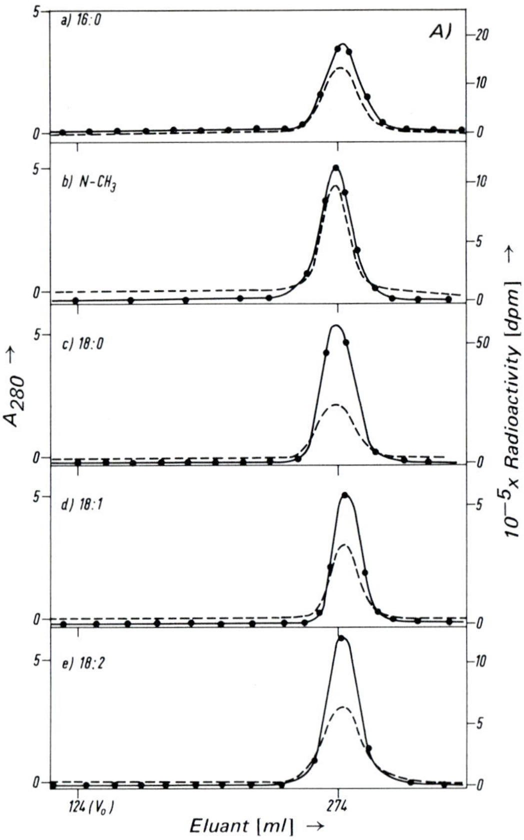

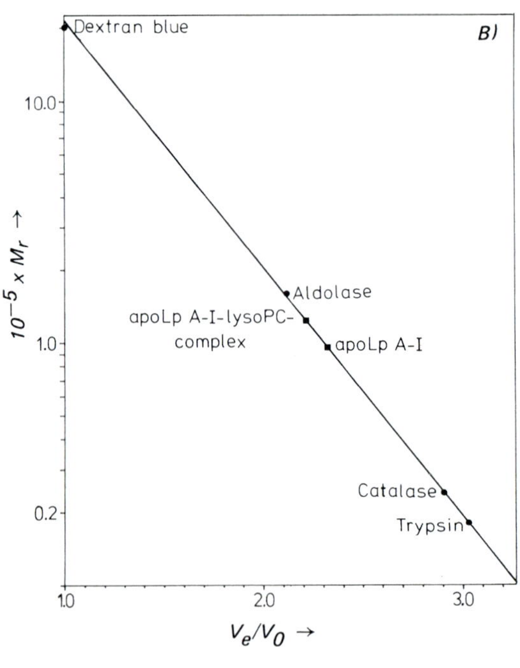

▷

Fig. 1. Determination of molecular weight of apolipoprotein-A-I-lysolecithin complexes by agarose gel chromatography (Bio-Gel A-5 m 200 - 400 mesh (2.5 x 90 cm)) on a calibrated column.

Elution buffer: 0.1M Tris/HCl, pH 8.2, 0.15M NaCl, 0.01% NaN_3, 0.01% EDTA in water.

A) Elution pattern of complexes consisting of apolipoprotein A-I and ^{3}H-labelled a) 1-[16-^{13}C]palmitoylglycerophosphocholine; b) 1-stearoylglycerophospho-[N-$^{13}CH_3$]choline; c) 1-[11-^{13}C]stearoylglycerophosphocholine; d) 1-[11-^{13}C]oleoylglycerophosphocholine; e) 1-[14-^{13}C]linoleoylglycerophosphocholine. ––– A_{280}; —— radioactivity.

B) Semilogarithmic plot of molecular weight against the relative elution volume V_e/V_0 (V_e = elution volume, V_0 = void volume).

ble 2 summarize these results. α-Helicity increases by 14 - 27% due to the binding of lysolecithins. This change in the secondary structure of apolipoprotein A-I is small in comparison to that after recombination of apolipoprotein A-I with lecithin or sphingomyelin. These particles had an f_H of approximately 0.8 - 0.9. An increase in the lysolecithin content of the complexes did not further elevate the α-helical structure.

c) Spin-lattice relaxation studies. The ^{13}C-enriched lysolecithins bound to the apoprotein A-I made spin-lattice relaxation time studies feasible.

Table 2. Molar ellipticity ([Θ]) and α-helicity (f_H) of apolipoprotein A-I free and associated with different lysolecithin species.

Lysolecithin containing	10^{-3} × [Θ] [degree × cm^2 × dmol^{-1}]		f_H
	uncorrected	corrected[16]	
16 : 0	– 14.7	– 18.7	0.54
18 : 0	– 14.3	– 18.2	0.52
18 : 1	– 15.5	– 19.2	0.56
18 : 2	– 13.7	– 17.4	0.50
apoA-I	– 12.3	– 15.6	0.44

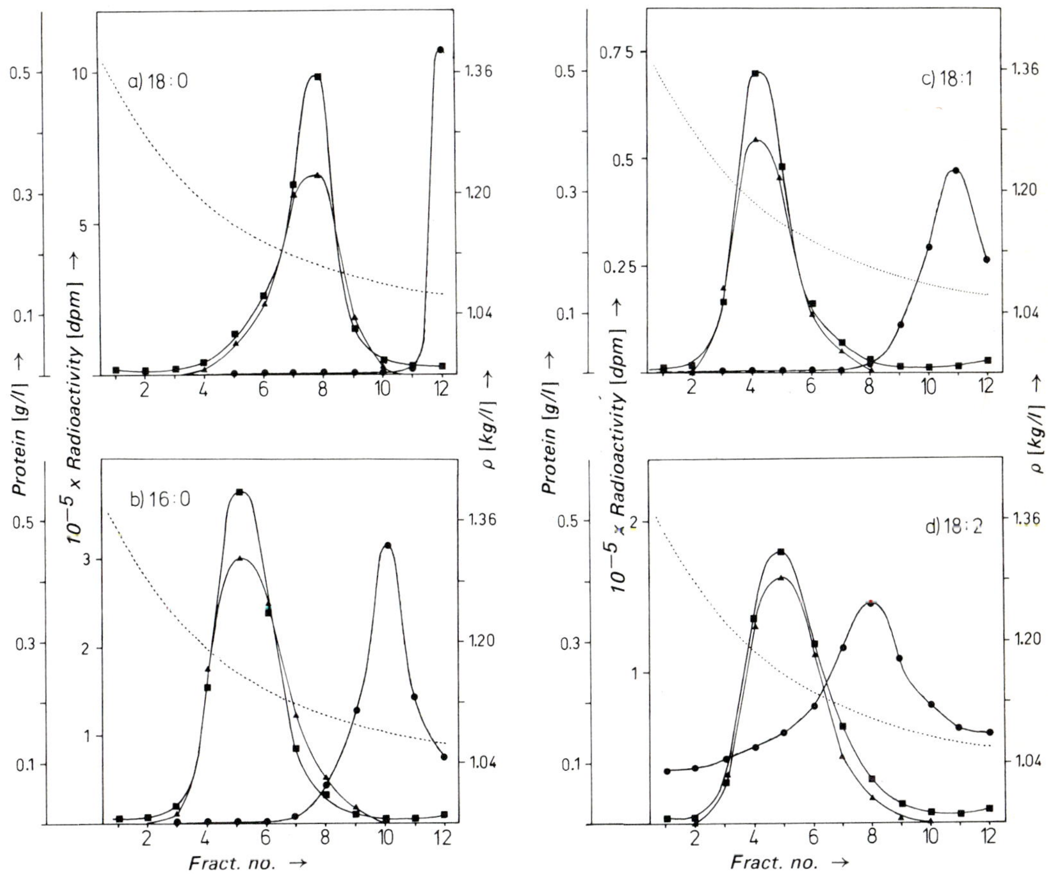

Fig. 2. Patterns of CsCl gradient centrifugation of apolipoprotein A-I complexes with a) 1-stearoyl (18 : 0)-; b) 1-palmitoyl (16 : 0)-; c) 1-oleoyl (18 : 1)-; d) 1-linoleoyl (18 : 2) glycerophosphocholine.
■——■ protein; ▲——▲ radioactivity in complex; ●——● lysolecithin micelles; ······· density.

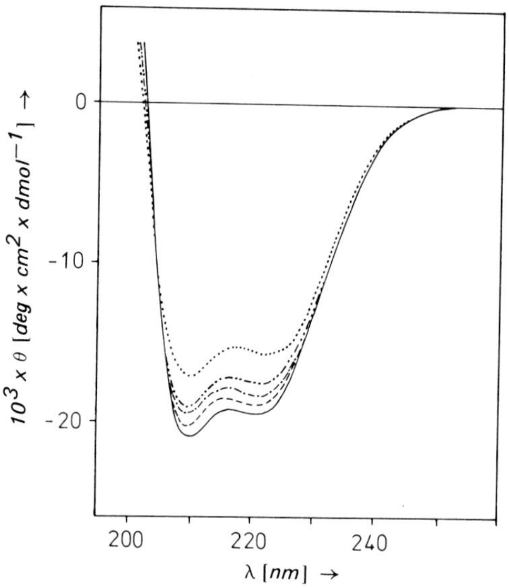

Fig. 3. Ultraviolet circular dichroic spectra of apolipoprotein A-I ········ and associated with 1-oleoylglycerophosphocholine ——; 1-palmitoylglycerophosphocholine – – –; 1-stearoylglycerophosphocholine –.–.–; 1-linoleoylglycerophosphocholine –..–..–.

An example is given in Fig. 4. The T_1-times of the acyl chain carbon of the saturated and unsaturated lysolecithins complexed with the apolipoprotein are considerably reduced by 20 to 25% and 40 to 50% as compared with those of the lysolecithins in aqueous micellar solution (Table 3). The strong reduction of the mobility of the ^{13}C-enriched CH_2-segments of the fatty acyl chains together with the unaltered mobility of the polar head group can only be interpreted by hydrophobic interactions between the lipid and apolar regions of the apoprotein A-I.

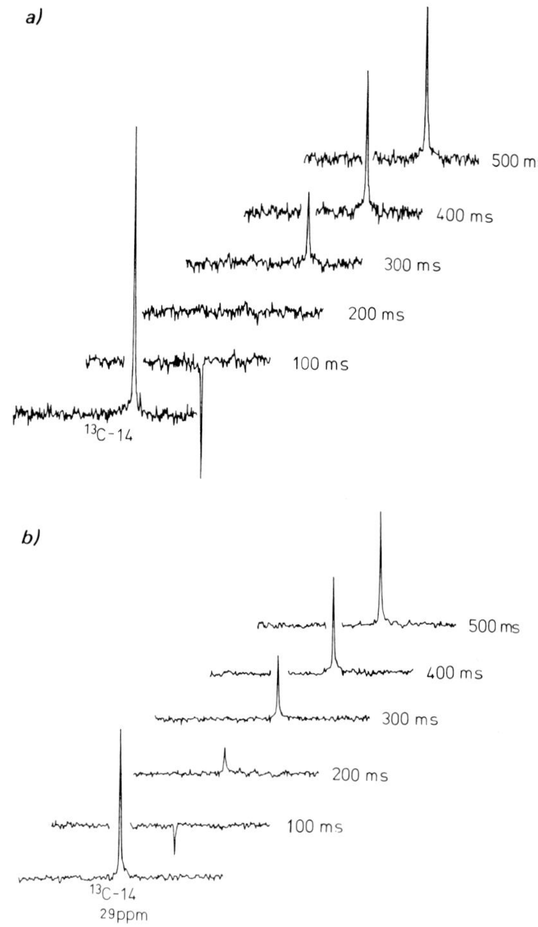

Fig. 4. ^{13}C-NMR spin-lattice relaxation time measurement of
a) micellar aqueous solution of 1-[14-^{13}C]stearoylglycerophosphocholine in 20% D_2O/80% H_2O, (T_1 = 292 ms); b) complex of apolipoprotein A-I and 1-[14-^{13}C]-stearoylglycerophosphocholine (1 : 57) in 20% D_2O/80% H_2O (T_1 = 216 ms).
Numbers at the right side of the set of spectra represent the delay time τ (ms) in the $180° - \tau - 90°$ pulse sequence.

Discussion

Lipid analysis of human serum high density lipoprotein consistently indicates the presence of about 2% lysolecithin[2]. The surface activity of these monoacylphospholipid species led us to use this natural detergent in elaborating recombination methods for HDL apolipoproteins with bisacylphospholipids or sphingomyelin with two hydrophobic alkyl chains[3,4]. In fact the polypeptide A-I reassociates quantitatively with the phospholipid molecules mostly enriched with carbon 13. Our interpretation is that the acyl chains of lysolecithin molecules occupy the hydrophobic sites of apoLp A-I, but then they are substituted by lecithin or sphingomyelin mole-

Table 3. ^{13}C-spin-lattice relaxation times T_1 of apolipoprotein A-I complexed with ^{13}C-labelled lysolecithin after agarose chromatography.

Lysolecithin containing	Apo A-I/lysoPC	T_1-times [ms]	
		ApoA-I-lysoPC complex	Lysolecithin micellar solution
[16-^{13}C]16 : 0	1 : 57	1580 ± 115	1942 ± 15
[14-^{13}C]18 : 0	1 : 57	216 ± 14	292 ± 10
[11-^{13}C]18 : 0	1 : 57	176 ± 6	221 ± 6
[11-^{13}C]18 : 1	1 : 57	190 ± 12	344 ± 13
[14-^{13}C]18 : 2	1 : 57	360 ± 27	728 ± 42
[*N*-$^{13}CH_3$]choline	1 : 57	650 ± 25	633 ± 30

cules monomerized by lysolecithins from the liposomal arrangement.

We have studied in more detail the binding of lysolecithins with apolipoprotein A-I, the main component of HDL. The lysolecithins were labelled with carbon 13 in specific positions along the acyl chain and in the choline group in order to probe the mobility of its hydrophobic and hydrophilic segments in the aqueous micelle and after association with the polypeptide A-I.

The following observations during the physical characterization need interpretation: CD-measurements indicated that the binding of lysolecithin to polypeptide A-I enhances the α-helicity of the free apoprotein (f_H = 0.44) in aqueous solution only slightly (to f_H = 0.56) independent of the lysolecithin concentration in excess of that bound to apolipoprotein A-I by equilibrium dialysis (70 mol lysolecithin per mol apolipoprotein A-I). If we compare the α-helicities of the stearoyl-, oleoyl- and linoleoylsphingomyelin-apolipoprotein A-I complexes[4] or those of 1-stearoyl-2-oleoyl-, 1-stearoyl-2-linoleoyl- and 1,2-dilinoleoylglycerophosphocholine (unpublished results) which reach 80 - 90% (f_H = 0.8 - 0.9) with the lysolecithin-apolipoprotein A-I complexes, we may interpret this large perturbation of secondary structure (induction of α-helix formation) by the lipid binding in the following way: the polypeptide A-I, which in aqueous solution already has a considerable α-helix content, loads all its hydrophobic sites with monoacylphospholipids and changes its surface with the dipolar phosphocholine groups. The lyso-compounds are exclusively bound with their fatty acyl tails to the apoprotein, the immobilisation of which is apparent from the strongly reduced NMR-spin-lattice relaxation times. The zwitterionic group remains freely mobile. Due to the measurement of the T_1-times of the ^{13}C-enriched segments, ^{13}C-NMR spectroscopy permits a direct physical discrimination between hydrophobic and polar lipid-polypeptide interactions.

These studies support our concept of the interactions of bisacylphospholipids and apoproteins of native HDL and in recombined complexes[4,20]. The results of Reynolds et al.[21] and Makino et al.[22] on comparable associations of apolipoprotein A-I from HDL with the detergent anion dodecylsulfate and the cationic tetradecyltrimethylammonium ion allow the same interpretation regarding polar interactions between these detergents and the polypeptide chains.

Vitello and Scanu[23] reported that apoLp A-I can only bind lipids readily in its monomeric form, because of missing protein-protein interactions. The latter have been demonstrated by fluorescence polarisation studies by Jonas[24,25] for bovine apoLp A-I. Obviously the occupation of the hydrophobic sites of apoLp A-I by amphiphilic molecules such as lysolecithins above their phase transition temperature described in this and previous papers[3,4] prohibit the multimer formation (tetramer, octamer[23]) but apparently allow the dimer formation, as concluded from the calibrated gel filtration chromatography.

The in-vitro binding process of lysolecithin described in this paper combined with the displacement of the monoacyl- by the diacylphospholipids of the previous publication[4] could be transposed into the situation in vivo. Lysolecithin present in most subcellular membranes and also in those of the endoplasmic reticulum could well prepare the nascent polypeptide chain for the recombination with the diacylphospholipids which induce the high α-helicity of the final particle structure. Further studies will be concerned with the degree of lipid-polypeptide and lipid-lipid interactions of the diacylphospholipids within the HDL particle.

This study was supported by the *Deutsche Forschungsgemeinschaft*.

Literature

1 Fielding, C. J., Shore, V. G. & Fielding, P. E. (1972) *Biochem. Biophys. Res. Commun.* **46**, 1493 - 1498.
2 Scanu, A. M. (1972) *Biochim. Biophys. Acta* **265**, 471 - 508.
3 Stoffel, W. & Därr, W. (1976) *Hoppe-Seyler's Z. Physiol. Chem.* **357**, 127 - 137.
4 Stoffel, W., Därr, W. & Salm, K.P. (1977) *Hoppe-Seyler's Z. Physiol. Chem.* **358**, 1 - 11.
5 Stoffel, W., Zierenberg, O. & Tunggal, B. (1972) *Hoppe-Seyler's Z. Physiol. Chem.* **354**, 1962 - 1969.
6 Scanu, A. M. (1966) *J. Lipid Res.* **7**, 295 - 306.
7 Scanu, A. M. & Edelstein, C. (1971) *Anal. Biochem.* **44**, 576 - 588.
8 Shore, V. G. & Shore, B. (1969) *Biochemistry* **8**, 4510 - 4516.
9 Weber, K. & Osborn, M. (1969) *J. Biol. Chem.* **244**, 4406 - 4412.
10 Reisfeld, R. A. & Small, P. A. (1966) *Science* **152**, 1253 - 1255.
11 Laemmli, U. K. (1970) *Nature (London)* **227**, 680 - 685.
12 Studier, F. W. (1973) *J. Mol. Biol.* **79**, 237 - 248.
13 Ouchterlony, O. (1953) *Acta Pathol. Microbiol. Scand.* **32**, 231 - 240.
14 Lowry, O. H., Rosebrough, N. J., Farr, A. L. & Randall, R. J. (1951) *J.Biol. Chem.* **193**, 265 - 275.
15 Freeman, R. & Hill, D. D. W. (1971) *J. Chem. Phys.* **54**, 3367 - 3377.
16 Chen, Y. H. & Yang, J. T. (1971) *Biochem. Biophys. Res. Commun.* **44**, 1285 - 1291.
17 Lewis, M. S. & Gottlieb, M. H. (1971) Abstracts of the 62nd Annual Meeting, Federation of American Societies for Experimental Biology, San Francisco, No. 1459.
18 Gwynne, J., Palumbo, G., Brewer, H. B., Jr. & Edelhoch, H. (1975) *J. Biol. Chem.* **250**, 7300 - 7306.
19 Haberland, M. E. & Reynolds, J. A. (1975) *J. Biol. Chem.* **250**, 6636 - 6639.
20 Stoffel, W., Zierenberg, O., Tunggal, B. & Schreiber, E. (1972) *Proc. Natl. Acad. Sci. U.S.A.* **71**, 3696 - 3700.
21 Reynolds, J. A. & Simon, R. H. (1974) *J. Biol. Chem.* **249**, 3937 - 3940.
22 Makino, S., Tanford, C. & Reynolds, J. A. (1974) *J. Biol. Chem.* **249**, 7379 - 7382.
23 Vitello, L. B. & Scanu, A. M. (1976) *J. Biol. Chem.* **251**, 1131 - 1136.
24 Jonas, A. (1975) *Biochim. Biophys. Acta* **394**, 460 - 470.
25 Jonas, A. (1975) *Biochim. Biophys. Acta* **394**, 471 - 482.

Prof. Dr. Dr. W. Stoffel, Dipl.-Chem. P. Metz and Dr. B. Tunggal, Institut für Physiologische Chemie der Universität Köln,
Joseph-Stelzmann-Str. 52, D-5000 Köln 41.

Hoppe-Seyler's Z. Physiol. Chem.
Bd. 359, S. 473 - 480, April 1978

Comparison of tRNA Nucleotidyltransferase from *Escherichia coli* and *Lactobacillus acidophilus*

Michael LEINEWEBER and Georg R. PHILIPPS

Physiologisch-Chemisches Institut der Universität Bonn

(Received 10 October 1977/3 January 1978)

Summary: Purification of tRNA nucleotidyltransferase from *Lactobacillus acidophilus* ATCC 4963 and *Escherichia coli* MRE 600 by preparative polyacrylamide gel electrophoresis is described. Both enzymes gave a single band on analytical polyacrylamide-gel electrophoresis and sodiumdodecylsulfate gels. Chromatography of the high speed supernatant from *Lactobacillus* at low salt concentrations gave three enzyme fractions of molecular weights about 45 000, 90 000, and 120 000. At 1M NaCl only the first enzyme fraction was found. Kinetic data for both enzymes are given.

Vergleich der tRNA-Nucleotidyltransferasen aus Escherichia coli und Lactobacillus acidophilus

Zusammenfassung: Reinigung der tRNA-Nucleotidyltransferase von *Lactobacillus acidophilus* und *Escherichia coli* mittels präparativer Polyacrylamidelektrophorese wird beschrieben. Beide Enzyme zeigten eine einzige Bande in analytischen und Natriumdodecylsulfat-Gelen. Chromatographie des Überstandes nach Ultrazentrifugierung von *Lactobacillus* ergab drei Enzymfraktionen mit Molekulargewichten von 45 000, 90 000 und 120 000. In Gegenwart von 1M NaCl fand sich nur die erste Fraktion. Kinetische Daten für beide Enzyme werden mitgeteilt.

Transfer-RNA nucleotidyltransferases synthesize the 3′-OH terminal trinucleotide sequence of tRNA in the absence of any template. The enzyme has been purified previously from a variety of species by rather elaborate methods (for review see ref.[1]). We report here the purification to electrophoretic homogeneity by preparative polyacrylamide gel electrophoresis of the enzyme from two bacteria, and compare some physical and kinetic characteristics of the enzymes.

Key words: tRNA nucleotidyltransferase, preparative polyacrylamide gel electrophoresis, purification.

Enzyme:
tRNA nucleotidyltransferase [identical with tRNA adenylyltransferase, ATP:tRNA adenylyltransferase (EC 2.7.7.25) and tRNA cytidylyltransferase, CTP:tRNA cytidylytransferase (EC 2.7.7.21).

Abbreviations:
$\overline{\text{tRNA}}$-CC, $\overline{\text{tRNA}}$-C, and $\overline{\text{tRNA}}$ = tRNA from which one, two, or three nucleotides have been removed.

Material and Methods

[8-^{3}H]ATP and [8-^{3}H]CTP (1 mCi/m*l*) were purchased from Amersham Buchler, Braunschweig; they were diluted to a spec. act. of 15 mCi/mmol with the corresponding unlabelled trinucleotides from Boehringer Mannheim GmbH. All other reagents used have been described[2]. *E. coli* MRE 600 tRNA was from Boehringer Mannheim GmbH and purified and digested as described[3]. DEAE-cellulose was from Whatman (DE 52). Marker proteins were all from the MS II collection of Serva Feinbiochemica, Heidelberg.

Determination of protein and tRNA

Protein concentrations were determined according to Lowry et al.[4]. For the estimation of tRNA concentration, an average molecular weight of 26 000 and an absorbance in 0.2M NaCl of $A_{260}^{1\%} = 220$ was used[4a].

Determination of enzyme activity

The standard assay for tRNA nucleotidyltransferase contained in a total volume of 0.075 m*l*: 8 μmol glycine/NaOH (pH 9.2), 0.5 μmol magnesium acetate, 1 μmol reduced glutathione, 0.5 nmol digested tRNA, and varying amounts of enzyme.
For AMP incorporation, the assay contained 14 nmol [^{3}H]ATP and $\overline{\text{tRNA}}$-CC; for CMP incorporation 12 nmol [^{3}H]CTP and a mixture of 20% $\overline{\text{tRNA}}$-C and 80% $\overline{\text{tRNA}}$. Incubations were made for 10 and 20 min at 37 °C. The samples were then processed as in ref.[5].

Preparative polyacrylamide gel electrophoresis

Preparative polyacrylamide gel electrophoresis was performed in an Ultraphor apparatus (Colora Meßtechnik, Lorch, Germany) similar to that described in ref.[2]; the same buffer systems were used. The separating gel contained 0.14% *N,N'*-methylenebisacrylamide and 5.5% acrylamide for the *E. coli* enzyme or 6% acrylamide for the *Lactobacillus* enzyme.

Analytical polyacrylamide gel electrophoresis

The apparent molecular weight of the denatured enzyme was determined using the analytical polyacrylamide gel electrophoresis in the presence of 1% sodium dodecylsulfate. The procedure was performed according to ref.[5]. The marker proteins used are given in Fig. 4.

Gel filtration on Sephadex G-200

The apparent molecular weight estimation of the native enzyme was done on Sephadex G-200. For this purpose, the homogenate was centrifuged twice, at first at low speed and then for 2 h at 105 000 × *g*. This material was then applied to the Sephadex column which was operated by reverse flow. For further details see legend to Fig. 5. In all experiments, NaCl was added to the sample prior to chromatography to give 0.1M, 0.3M or 1.0M, respectively. The columns were equilibrated with buffers containing the same NaCl concentration. Marker proteins were: cytochrome *c*, horse myoglobin, chymotrypsinogen, egg albumin, bovine serum albumin, and aldolase.

Enzyme purification

Both *Escherichia coli* MRE 600 and *Lactobacillus acidophilus* ATCC 4963 were grown to late lograrithmic phase in enriched media.

All operations were performed at 4 °C. Five g *E. coli* cells were suspended in 15 m*l* *buffer A* (10mM Tris/HCl (pH 7.6), 10mM magnesium acetate, 3mM EDTA, 5mM β-mercaptoethanol, 10% glycerol) and broken in a French pressure cell at 60 kp/cm^2. The viscous solution was stirred 20 min with 20 μg deoxyribonuclease and centrifuged in a Sorvall SS 34 rotor at 18 000 rpm for 20 min.

The supernatant was dialysed against buffer A and applied to a DEAE-cellulose column equilibrated in buffer A. The column was washed until absorbance at 280 nm was below 0.5. Fractions of 5 m*l* each were collected and tRNA nucleotidyltransferase was eluted with 60mM NaCl (Fig. 1a). Ten fractions were pooled and concentrated by dialysis against *buffer B* (10mM Tris/HCl (pH 7.5), 10mM magnesium acetate, 5mM β-mercaptoethanol, 10% glycerol) containing 30% polyethylene glycol to about 25 m*l*.

For preparative electrophoresis portions of 6 m*l* containing 12 to 24 mg protein were made 15% in sucrose and 0.01% in bromphenol blue and layered on top of the stacking gel under a protective buffer. The sample was concentrated in the gel at 90 mA and the current was switched then to 150 mA. During the whole electrophoresis (6 to 8 h) the chamber was cooled to 4 °C within the gel. Flow rate of the elution buffer was 24 m*l*/h and 1.75-m*l* fractions were collected. The enzyme was eluted in a double peak of 12 and 14 m*l* (Fig. 1b). Only peak fractions were pooled and bovine serum albumin (0.2 mg/m*l*) was added. The enzyme could be stored without loss of activity for several months at −18 °C.

The purification of tRNA nucleotidyltransferase from *Lactobacillus* was along the same general lines. However, cells (5 g) were broken with 10 g of alumina and the homogenate extracted with 15 m*l* *buffer C* (the same as buffer A but pH 8.4). After centrifugation for 20 min at 4000 rpm the supernatant was dialysed and applied to a DEAE-cellulose column equilibrated and operated in buffer C.

Results

The purification is summarized in Table 1. The stated enzymatic activity of the homogenate is only an approximation because this fraction contained also other RNA-specific enzymes. Thus, the extent of the purification is also approximated.

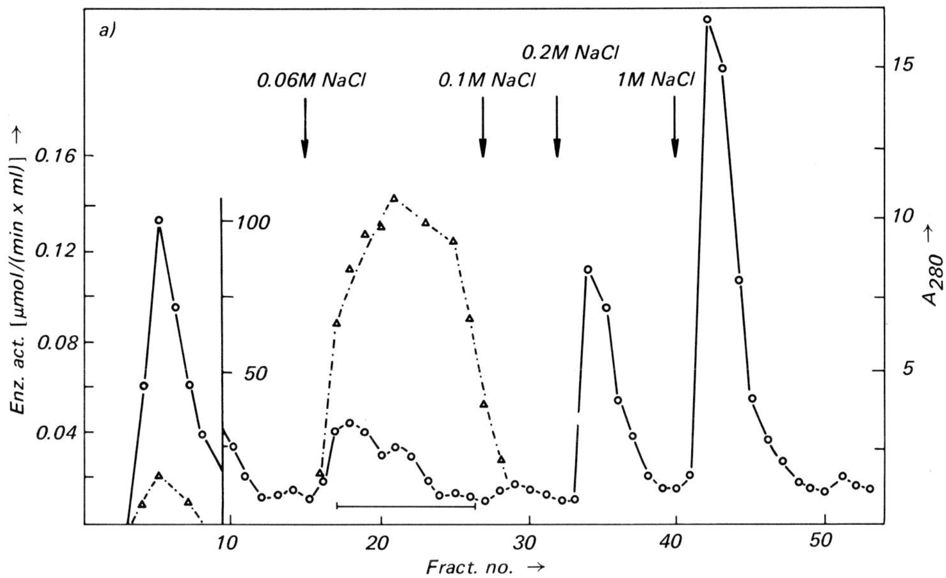

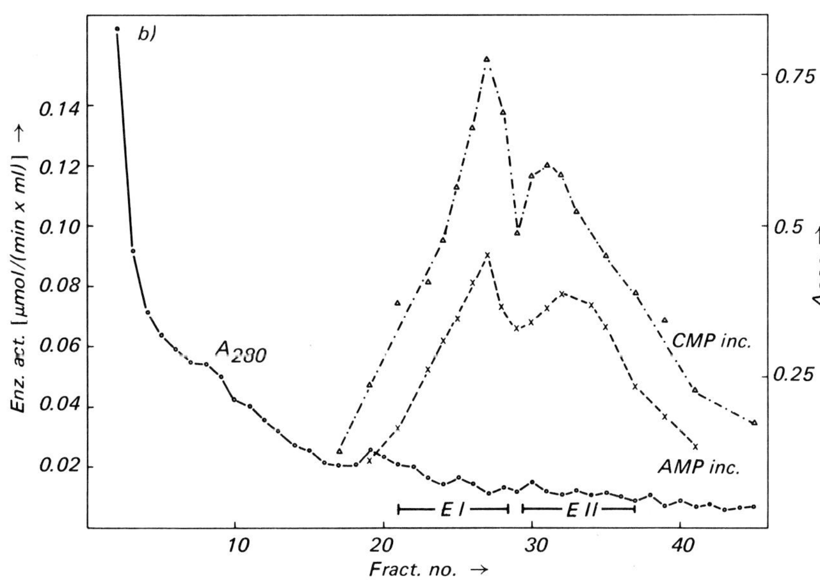

Fig. 1. Purification of *E. coli* tRNA nucleotidyltransferase.

a) Elution profile of a 5 x 2-cm *DEAE-cellulose* column. The low speed supernatant of 5 g cells was applied and the enzymatic activity eluted with 60mM NaCl. Fractions indicated by the horizontal bar were pooled.

b) Elution profile of the *preparative polyacrylamide gel electrophoresis*. 1/4 of the concentrated fraction of the DEAE cellulose column was used. The two peaks were pooled separately as indicated by the bars.

Absorbance at 280 nm (○——○) using a different scale in (a) for the first peak. Enzymatic activity of 50-μl portions: AMP incorporation into $\overline{\text{tRNA}}$-CC (x– – –x), CMP incorporation into ($\overline{\text{tRNA}}$-C/$\overline{\text{tRNA}}$) (△–.–△).

Table 1. Purification of tRNA nucleotidyltransferase.

The standard assay was used and incubations were for 5 and 10 min in the presence of 20 μg enzyme (for step 2 and 3) or 40 μg (step 1).

Fraction	Total protein [mg]	Total activity [U]		Specific activity [U/mg]		Purification *n*-fold		Yield [%]	
		AMP	CMP	AMP	CMP	AMP	CMP	AMP	CMP
I. Escherichia coli MRE 600									
1. Homogenate	800	19	20	0.024	0.025	1	1	–	–
2. DEAE-cellulose	15	69	75	4.6	5	190	200	100	100
3. Electrophoresis	0.3	40	47	133	156	5540	6240	57	62
II. Lactobacillus acidophilus ATCC 4963									
1. Homogenate	500	9	10	0.018	0.02	1	1	–	–
2. DEAE-cellulose	9	20	24	2.2	2.6	120	130	100	100
3. Electrophoresis	0.2	12	13	60	65	3330	3250	60	54

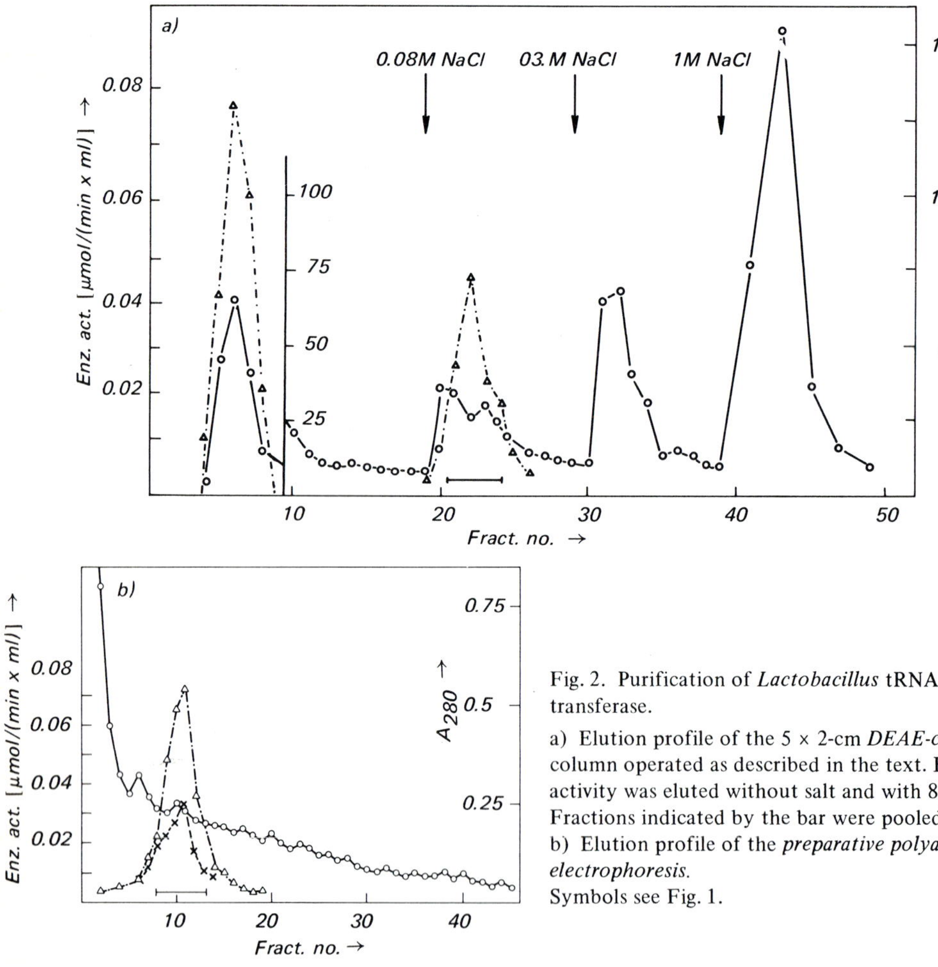

Fig. 2. Purification of *Lactobacillus* tRNA nucleotidyl-transferase.

a) Elution profile of the 5 × 2-cm *DEAE-cellulose* column operated as described in the text. Enzymatic activity was eluted without salt and with 80mM NaCl. Fractions indicated by the bar were pooled.

b) Elution profile of the *preparative polyacrylamide gel electrophoresis.*

Symbols see Fig. 1.

Fig. 2a shows the elution profile of tRNA nucleotidyltransferase from *Lactobacillus* after chromatography on DEAE-cellulose. Compared to the *E. coli* enzyme which was eluted mainly with 60mM NaCl, about half of the enzymatic activity from *Lactobacillus* was not retained by the column, and was collected before the application of NaCl. Upon ultracentrifugation of the first enzymatic peak, varying amounts (40 to 70%) were found in the soluble phase. We assume that the activity in the first peak is associated with ribosomes for unknown reasons; it was discarded, because it was easier to start with more cells then to separate it from ribosomes.

The elution profile of the preparative polyacrylamide gel electrophoresis is shown in Fig. 2b. The *Lactobacillus* enzyme was eluted as a sharp peak in about 10 m*l*. The ratio of AMP:CMP incorporation in this experiment differed throughout the peak probably because of the greater instability of the latter (ref.[6] and unpublished).

Fig. 3. shows the analytical polyacrylamide gel of the two enzymes. Both migrate as a single band but the *Lactobacillus* enzyme migrated faster. After denaturation, both enzymes migrated as a single band about the same distance in sodium-dodecylsulfate gels. Therefore, the different migration of the native enzymes must be due to different charges but not different sizes.

By comparison with markers (Fig. 4) an apparent molecular weight of 45 000 ± 3 000 was determined for the *E. coli* enzyme, and 43 000 ± 3 000 for the *Lactobacillus* enzyme. When the two peaks of Fig. 1b were analysed separately on sodium-dodecylsulfate gels, no significant difference was observed. Thus, it appears that both enzymes consist of a single polypeptide chain with about the same number of amino acids but different charges.

Chromatography on Sephadex G-200 was used to estimate the molecular weight of native tRNA nucleotidyltransferase from *Lactobacillus* (Fig. 5). At 0.1M NaCl, enzymatic activity was found in three peaks. When the NaCl concentration was raised to 0.3M, the first peak disappeared and most of the activity was found in peak II. At 1.0M NaCl. all the enzymatic activity was found in the third peak. The same was true, if the enzyme was kept at 4 °C for about 2 h in 1.0M NaCl and then chromatographed on a column with 0.1M NaCl.

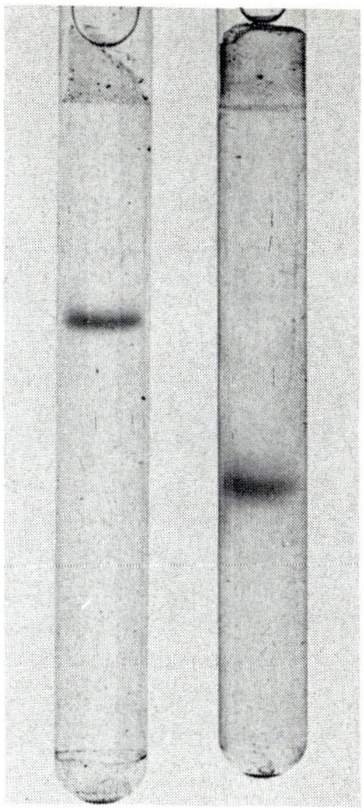

Fig. 3. Analytical polyacrylamide gel electrophoresis of the two purified tRNA nucleotidyltransferases.

About 20 µg of the enzyme from *E. coli* (left) and from *Lactobacillus* (right) were used for the disc electrophoresis on 6.25% gels. For the *E. coli* enzyme, the two peaks of Fig. 1b were combined. Migration from top to bottom.

Thus it appears that tRNA nucleotidyltransferase is predominantely found in aggregates which upon handling (raising the ionic strength of the buffer or probably also chromatography on DEAE-cellulose) dissociate into smaller units. Marker proteins allowed the estimation of the molecular weight. Using the procedure described in ref.[7], values of 45 000, 90000, and about 120000 were determined; the last value is inaccurate due to the lack of proper markers, and to evident limitations in the separation of proteins of this size. The first estimation is in good agreement with the molecular weight after denaturation. A molecular weight of 90000 would then correspond to a dimer molecule, while the third structure could be composed of a dimer enzyme having one or two bound tRNA molecules.

With both enzymes, the time course of the readdition of AMP or CMP showed linearity for 20 min under standard conditions using 20 µg en-

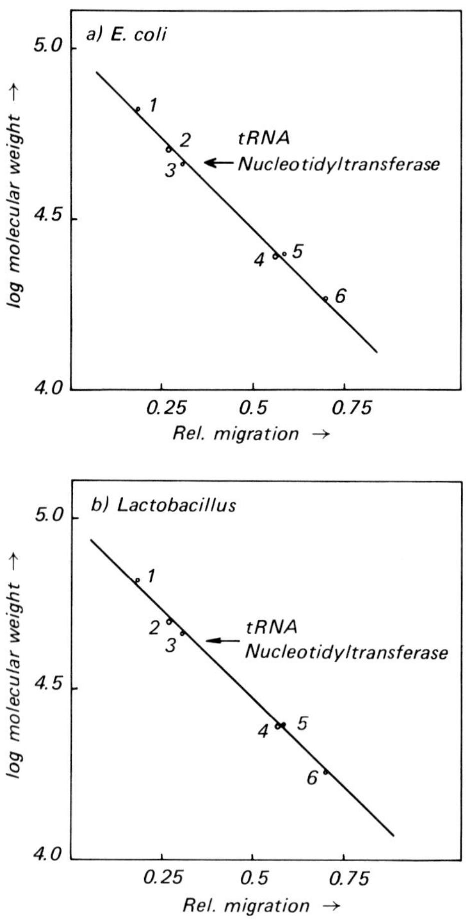

Fig. 4. Estimation of the apparent molecular weight of tRNA nucleotidyltransferase by analytical polyacrylamide gel electrophoresis in the presence of dodecylsulfate.

Plot of the distance of migration versus the log molecular weight. a) *E. coli* enzyme; b) the *Lactobacillus* enzyme. Standard proteins: 1) bovine serum albumin (67000), 2) γ-globulin, H-chain (50000), 3) egg albumin (45000), 4) chymotrypsinogen (24500), 5) γ-globulin, L-chain (23500), 6) myoglobin (17000).

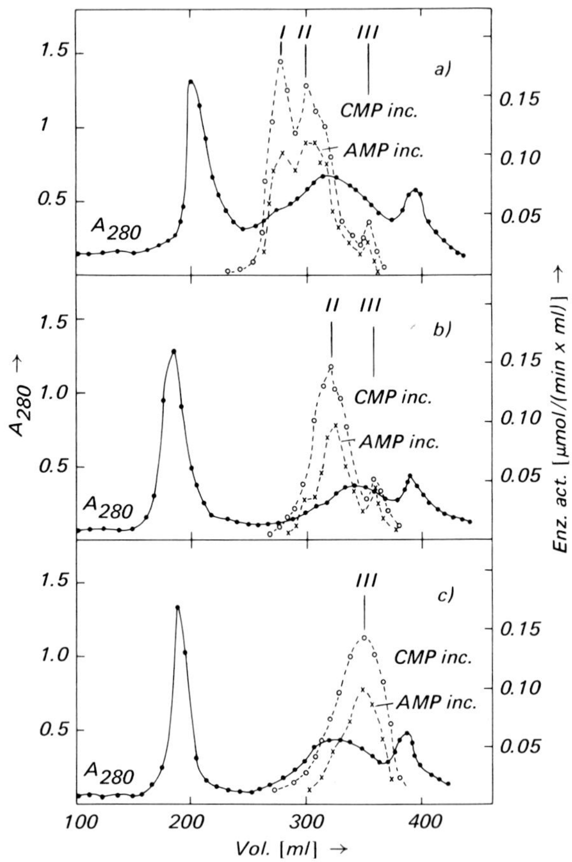

Fig. 5. Chromatography on Sephadex G-200 of tRNA nucleotidyltransferase of *Lactobacillus*.

The 100000 × *g* supernatant of 1 g bacteria was applied to a 90 × 2.5-cm column equilibrated with 0.1M potassium phosphate (pH 6.9)/5mM magnesium acetate/5mM β-mercaptoethanol and a) 0.1M NaCl, b) 0.3M NaCl, c) 1.0M NaCl. Eight m*l* of the supernatant was adjusted to the appropriate NaCl concentration. The flow rate was 0.5 m*l*/min and 6.5-m*l* fractions were collected. Absorbance at 280 nm (●——●), enzymatic activity: AMP incorporation (x– – –x), CMP incorporation (○– – –○).

zyme. At this point, about 90% of all tRNA-CC had been converted to tRNA-CCA and about 50% of (tRNA/tRNA-C) had become tRNA-CC.

The kinetic analysis of the two enzymes is summarized in Table 2 and compared with data from other laboratories.

Preparative polyacrylamide gel electrophoresis of tRNA nucleotidyltransferase from *E. coli* gave two peaks, which we have designated enzyme I and II. In contrast to nearly identical CTP-substrate concentration curves the two enzymes showed different rates of AMP incorporation; in

Table 2. Catalytic properties of tRNA nucleotidyltransferase from different organisms.

Source	Apparent K_m (triphosphates) [mM]		Apparent K_m (tRNA) [μM]			Ref.
	ATP	CTP	t$\overline{\text{RNA}}$-CC	t$\overline{\text{RNA}}$-C	t$\overline{\text{RNA}}$	
L. acidophilus ATCC 4963	0.71	0.029	0.4	1[a]		this paper
E. coli MRE 600	0.33	0.017	1.5	n.d.		8,9
E. coli MRE 600	0.19	0.072	9.6	18[a]		10
E. coli MRE 600						this paper
Enzyme I	3.6	0.017	1.7	n.d.		
Enzyme II	(0.6)	0.017	1.7	n.d.		
E. coli B	0.095	0.015	0.21	0.18	0.20	11
Rabit liver	2	0.004	12	6	4	12

[a] In these determinations t$\overline{\text{RNA}}$-C and t$\overline{\text{RNA}}$ were not distinguished. n.d. = not determined.

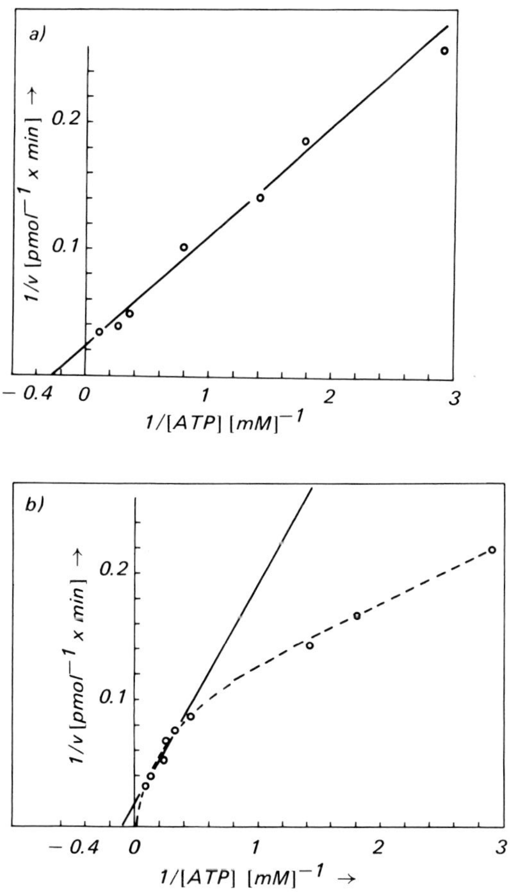

Fig. 6. Lineweaver-Burk plot for AMP incorporation by the *E. coli* enzyme.
a) enzyme I; b) enzyme II.

particular, no real maximum velocity (V) was obtained with enzyme II. The double reciprocal plots for the two reactions are shown in Fig. 6. Enzyme I gave a V value of 45 pmol/min and an apparent K_m of 36×10^{-4}M. No straight line for increasing ATP concentrations could be obtained with enzyme II. A V value could be obtained only for low substrate concentrations, e. g. 12.5 pmol/min; the K_m under these conditions was 6×10^{-4}M. Substrate concentration curves for increasing [tRNA], however, showed no difference.

Discussion

The method employed here for the preparation of electrophoretically homogeneous tRNA nucleotidyltransferase is suitable for the purification of proteins in small quantities. The specific activity of the enzyme described here was about 270-fold higher than that previously obtained by a different method[5]. On that occasion we observed a rapid loss of activity; the preparation took one or two days longer and the enzyme stood at 4 °C in solutions of low protein concentrations. Compared to the *E. coli* enzyme prepared by others (for ref, see[11]) the difference of the specific activities is even larger. Here, tRNA nucleotidyltransferase from two species of bacteria, which are not too closely related, were prepared under nearly identical conditions. This allows comparison of the enzyme directly. The enzymes have the same size but evidently different charges. It remains questionable if the

45 000 mol. weight unit is the active one found in vivo. At low ionic strength, most of the enzymatic activity in *Lactobacillus* appears in structures at least twice the size of the monomer. Similar results had been obtained earlier for the *E. coli* enzyme (unpublished and ref.[9]).

Unquestionably, the same enzyme catalyses the incorporation of AMP and CMP into tRNA, yet the apparent K_m values for the two reactions differ. Most authors reported a K_m for CTP of about 20μM for the bacterial enzyme and somewhat smaller values for the mammalian enzyme (Table 2). However, K_m values for AMP incorporation differed considerably; they range for the *E. coli* enzyme from 0.095 to 3.6mM. This enzyme was not eluted from preparative polyacrylmide gels in a single peak. This might suggest that the *E. coli* enzyme can undergo changes of the tertiary structure at or near the binding site for ATP.

This research was supported by the *Deutsche Forschungsgemeinschaft*. The studies were part of the dissertation of M.L., submitted in partial fulfillment of the requirements for the degree of Dr. rer. nat. to the Naturwissenschaftliche Fakultät der Universität Bonn.

Literature

1 Deutscher, M. P. (1973) *Prog. Nucleic Acid Res. Mol. Biol.* **13**, 51 - 92.
2 Leineweber, M. & Philipps, G.R. (1978) *Biochim. Biophys. Acta*, **517**, 419 - 428.
3 Philipps, G. R. & Chiemprasert, T. (1975) *Hoppe-Seyler's Z. Physiol. Chem.* **356**, 1097 - 1104.
4 Lowry, O.H., Rosebrough, N.J., Farr, A.L. & Randall, R.J. (1951) *J. Biol. Chem.* **193**, 265 - 275.
4a v. Ehrenstein, G. & Lipmann, F. (1961) *Proc. Natl. Acad. Sci. U.S.A.* **47**, 941 - 950.
5 Miller, J. P. & Philipps, G. R. (1971) *J. Biol. Chem.* **246**, 1274 - 1279.
6 Miller, J. P. & Philipps, G. R. (1971) *Biochemistry* **10**, 1001 - 1007.
7 Whitaker, J. R. (1963) *Anal. Chem.* **35**, 1950 - 1953.
8 Carre, D. S., Litvak, S. & Chapeville, F. (1970) *Biochim. Biophys. Acta* **224**, 371 - 381.
9 Carre, D. S., Litvak, S. & Chapeville, F. (1974) *Biochim. Biophys. Acta* **361**, 185 - 197.
10 Gross, H. J., Duerinck, F. R. & Fiers, W. C. (1970) *Eur. J. Biochem.* **17**, 116 - 123.
11 Miller, J. P. & Philipps, G. R. (1971) *J. Biol. Chem.* **246**, 1280 - 1284.
12 Deutscher, M. P. (1972) *J. Biol. Chem.* **247**, 459 - 468.

Dr. M. Leineweber, Department of Biology, MIT, Cambridge, Mass., U.S.A.
PD. Dr. G.R. Philipps*, Neurologische Klinik der Universität Würzburg,
Josef-Schneider-Str. 11, D-8700 Würzburg.

* To whom correspondence should be addressed.

Hoppe-Seyler's Z. Physiol. Chem.
Bd. 359, S. 481 - 488, April 1978

Purification and Characterization of Aromatic-Amino-Acid-Glyoxylate Aminotransferase from Monkey and Rat Liver

Izumi HARADA, Tomoo NOGUCHI and Ryo KIDO

Department of Biochemistry, Wakayama Medical College, Wakayama 640, Japan.

(Received 13 December 1977)

Summary: Aromatic-amino-acid-glyoxylate aminotransferase was highly purified from the mitochondrial fraction of livers from monkey and glucagon-injected rats. The two enzyme preparations showed physical and enzymic properties different from a kynurenine aminotransferase previously described. The two enzymes had nearly identical molecular weights (approximate 80000), isoelectric points (pH 8.0) and pH optima (pH 8.0 - 8.5). However, a difference in substrate specificity was observed between the two enzymes. Both enzymes utilized glyoxylate, pyruvate, hydroxypyruvate and 2-oxo-4-methylthiobutyrate as effective amino acceptors. 2-Oxoglutarate was active for rat enzyme but not for monkey enzyme. With glyoxylate, amino donors were effective in the following order of activity; phenylalanine > histidine > tyrosine > tryptophan > 5-hydroxytryptophan ≫ kynurenine for the rat enzyme, and phenylalanine > kynurenine > tyrosine > histidine > tryptophan > 5-hydroxytryptophan for the monkey enzyme.

Reinigung und Charakterisierung von Aromat.-Aminosäure-Glyoxylat-Aminotransferase aus Affen- und Rattenleber

Zusammenfassung: Aromat.-Aminosäure-Glyoxylat-Aminotransferase wurde nach Glucagoninjektion aus Lebermitochondrien von Affen und Ratten angereichert. Die beiden Präparate unterschieden sich in ihren physikalischen und enzymatischen Eigenschaften von der bisher in der Literatur beschriebenen Kynurenin-Aminotransferase. Beide Enzyme stimmen in Molekulargewicht (ungefähr 80000), isoelektrischem Punkt (pH 8.0) und pH-Optimum (pH 8.0 - 8.5) überein. Sie unterscheiden sich jedoch in ihrer Substratspezifität. Beide Enzyme benutzten Glyoxylat, Pyruvat, Hydroxypyruvat und 2-Oxo-4-methylthiobutyrat als Aminoakzeptoren. 2-Oxoglutarat ist aktiv mit Rattenenzym, aber nicht mit Affenenzym. Mit Glyoxylat als Aminoakzeptor, nimmt die Wirk-

Enzymes:
Alanine-glyoxylate aminotransferase, L-alanine:glyoxylate aminotransferase (EC 2.6.1.44);
Aminoadipate aminotransferase, L-2-aminoadipate:2-oxoglutarate aminotransferase (EC 2.6.1.?);
Aspartate aminotransferase, L-aspartate:2-oxoglutarate aminotransferase (EC 2.6.1.1);
Glutamine-oxo acid aminotransferase, L-glutamine:2-oxo acid aminotransferase (EC 2.6.1.15);
Kynurenine aminotransferase, L-kynurenine:2-oxoglutarate aminotransferase (cyclizing) (EC 2.6.1.7);
L-Kynurenine-glyoxylate aminotransferase (EC 2.6.1.?);
L-Phenylalanine (L-histidine)-pyruvate aminotransferase (EC 2.6.1.?);
Serine-pyruvate aminotransferase, L-serine:pyruvate aminotransferase (EC 2.6.1.51).

samkeit von L-Aminosäuren in nachstehender Reihenfolge ab: Phenylalanin > Histidin > Tyrosin > Tryptophan > 5-Hydroxytryptophan ≫ Kynurenin für Rattenenzym, und Phenylalanin > Kynurenin > Tyrosin > Histidin > Tryptophan > 5-Hydroxytryptophan für Affenenzym.

Key words: Kynurenine, phenylalanine, aminotransferase, rat liver, monkey liver.

Rat liver contains at least three forms of kynurenine aminotransferase, designated isoenzymes 1, 2 and 3, respectively. Isoenzymes 1 and 2 are active with 2-oxoglutarate but not with pyruvate as amino acceptor[1], whereas isoenzyme 3 is active with pyruvate but not with 2-oxoglutarate[2]. Isoenzyme 1 is identical with mitochondrial aspartate-2-oxoglutarate aminotransferase[1] and also with mitochondrial tyrosine-2-oxoglutarate aminotransferase[3]. Isoenzyme 3 is identical with glutamine-2-oxo acid (phenylpyruvate) aminotransferase[2,4]. On the other hand, the purification and characterization of kynurenine aminotransferase has previously not been reported. The present report describes the purification and characterization of aromatic-amino-acid-glyoxylate aminotransferase from rat and monkey liver, which differs from isoenzymes 1, 2 and 3.

Materials and Methods

Chemicals

L-Kynurenine was prepared as previously described[1]. The following compounds were obtained from commercial sources: lithium hydroxypyruvate, sodium pyruvate and glyoxylate from Nakarai Chemicals, Kyoto, Japan; monosodium 2-oxoglutarate and 2-oxoadipate from Sigma Chemical Co., St. Louis, Mo., U.S.A.; 2-oxo-4-methylthiobutyrate from Calbiochem, San Diego, Calif., U.S.A.. Other materials used were obtained as stated previously[1,2,4].

Enzyme assays

Kynurenine aminotransferase activity was in principle assayed at 37 °C by the method of Mason[5]. The reaction mixture (0.2 m*l*) contained, unless specified otherwise, 3mM L-kynurenine, 40μM pyridoxal 5′-phosphate, 3mM glyoxylate, enzyme preparation and 0.1M potassium phosphate buffer. Kynurenine-2-oxoglutarate aminotransferase was assayed at pH 7.5 and aromatic-amino-acid-glyoxylate aminotransferase at pH 8.0. After incubation for 1 h, the reaction was stopped by the addition of 10 μ*l* of 50% (w/v) trichloroacetic acid. The mixture was neutralized with 12 μ*l* of 4N KOH, and then centrifuged. 125 μ*l* of clear supernatant fluid was mixed with 1.0 m*l* of 0.5M potassium phosphate buffer, pH 7.5. The solution was read at 330 and 365 nm against 0.5M potassium phosphate buffer, pH 7.5. These values were used to calculate the amount of kynurenic acid formed[6].

Aminotransferase activities towards other aromatic L-amino acids were assayed as previously described[7].

Transamination between L-glutamate and 2-oxoadipate was assayed by the method of Nakatani et al.[8].

A unit of enzyme activity (U) is defined as the amount of enzyme that catalyses product formation or substrate disappearance at the rate of 1 μmol/min at 37 °C.

Other methods

Mitochondrial fractions (heavy and light mitochondria) of rat and monkey liver were obtained by the method of de Duve et al.[9]. Determination of protein, polyacrylamide disc gel electrophoresis in the presence and absence of sodium dodecylsulfate, isoelectric focusing and determination of approximate molecular weight by sucrose density gradient centrifugation and Sephadex G-150 gel filtration were carried out as previously described[2,7].

Results

Isoelectric focusing of hepatic mitochondrial extracts

Fig. 1 (a, b) shows isoelectric focusing profiles of the mitochondrial extracts from livers of monkey and glucagon-injected rats. A single activity peak of aromatic-amino-acid-glyoxylate aminotransferase with pI 8.0 was obtained with monkey liver (Fig. 1*a*). Kynurenine-glyoxylate aminotransferase activity was not detected in rat liver, presumably owing to its relatively low activity or loss of its activity during the focusing. However, the focusing of the mitochondrial from glucagon-injected rat liver resulted in the appearance of a single activity peak of aromatic-amino-acid-glyoxylate aminotransferase with pI 8.0 (Fig. 1b).

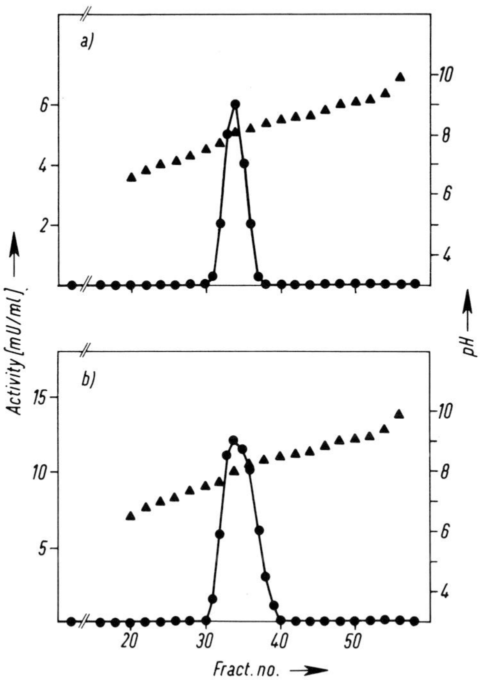

Fig. 1. Isoelectric focusing of hepatic mitochondrial extracts.

A monkey and rats were housed in wire-bottomed cages and maintained at about 20 °C in a room with a 12 h light/12 h dark cycle. Rats were injected subcutaneously with glucagon suspended in 0.15% KCl (0.35 mg, every 8 h for 3 days). The mitochondrial pellet was prepared from each liver as described in the text and homogenized with 5 vol. of ice-cold potassium phosphate buffer, pH 7.5 in a Potter-Elvehjem tissue grinder with a Teflon pestle. After sonication at 10 kHz for 4 min, each homogenate was centrifugated at 105 000 x *g* for 30 min. Each resulting supernatant was separately subjected to isoelectric focusing on a pH 3.5 - 10 Ampholine gradient in amounts corresponding to 4 g of liver. Fractions (2 m*l*) were collected. pH values (▲) and kynurenine-glyoxylate aminotransferase activity (●) were determined as described in the text. The figure shows focusing profiles of the mitochondrial extracts of monkey (a) and glucagon-injected rat (b) liver.

Purification of aromatic-amino-acid-glyoxylate aminotransferase

Aromatic-amino-acid-glyoxylate aminotransferase with pI 8.0 was purified from the mitochondrial extracts of monkey and glucagon-injected rat liver. All operations were carried out at 0-4 °C. Potassium phosphate buffer, containing 200 μM pyridoxal 5′-phosphate, 5mM mercaptoethanol and 20% glycerol were, unless specified otherwise, used throughout.

Purification of the enzyme from rat liver

Male rats (200-250 g) of the Donryu strain were injected subcutaneously with glucagon suspended in 0.15% KCl (0.2 mg, every 8 h for 3 days) and killed by decapitation. The mitochondrial fractions (heavy and light mitochondria) were prepared from seven livers (about 77 g) by the method of de Duve et al.[9], and suspended in 385 m*l* of 5mM buffer. The suspension was sonicated for 4 min at 10 kHz with a Kubota Sonicator (Tokyo, Japan), and then centrifuged at 105 000 x *g* for 30 min.

The supernatant solution was warmed rapidly to 70 °C and maintained at this temperature with constant stirring, fcr 1 min, after which it was quickly chilled in a 0.9% NaCl/ice bath at about − 7 °C.

To the heat-treated solution $(NH_4)_2SO_4$ was added, with gentle stirring, to 30% saturation. After 30 min, the precipitate was removed by centrifugation at 5 000 x *g* for 20 min and discarded. $(NH_4)_2SO_4$ was added to the supernatant to 90% saturation, and the precipitate was collected by centrifugation at 5 000 x *g* for 20 min. The precipitate was disolved in a small volume of 5mM buffer and desalted by dialysis against 5mM buffer overnight.

The enzyme solution was subjected to isoelectric focusing on a pH 3.5 - 10 Ampholine gradient. The focusing resulted in the detection of a single activity peak of aromatic-amino-acid-glyoxylate aminotransferase with pI 8.0. Active fractions were pooled and concentrated by Diaflo ultrafiltration (Amicon far east limited, Tokyo, Japan) by using a pM-10 filter.

The enzyme solution was added to a Sephadex G-150 column (2.5 cm x 100 cm), which had been equilibrated with 50mM buffer. The column was eluted with the same buffer at a flow rate of 25 m*l*/h. The effluent was collected in 3.0-m*l* fractions. The active fractions were pooled, concentrated by ultrafiltration and diluted with water to adjust the buffer concentration to 5mM.

The enzyme solution was applied to a hydroxylapatite column (2.5 cm x 3 cm) which had been equilibrated with 5mM buffer. After washing with 50 m*l* amounts of 15mM and then 50mM buffer, the enzyme was eluted with 100mM buffer. The active fractions were pooled, concentrated by ultrafiltration and diluted with water containing 20% glycerol and 200μM pyridoxal 5′-phosphate to adjust the buffer concentration to 5mM.

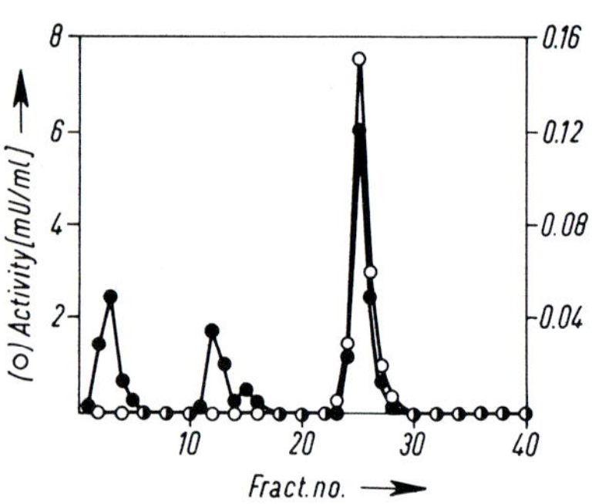

Fig. 2. DEAE-cellulose chromatography of rat liver enzyme preparation from a hydroxylapatite column.

The rat liver enzyme preparation from a hydroxylapatite column was applied to a column (2.5 cm x 5 cm) of DEAE-cellulose and eluted as described in the text. Potassium phosphate buffer, pH 7.5 containing 5mM mercaptoethanol and 20% glycerol, was used throughout. Fractions 1 - 10 were eluted with 5mM buffer and fractions 11 - 20 with 15mM buffer at the fraction size of 20 m*l*. The enzyme was eluted with 25mM buffer in the fraction size of 5 m*l*. ○, Kynurenine-glyoxylate aminotransferase activity; ●, protein (optical density at 280 nm).

The enzyme solution was applied to a DEAE-cellulose column (2.5 cm x 5 cm) equilibrated with 5mM buffer. After washing with 200 m*l* of 5mM and 10mM buffer, the enzyme activity was eluted by 25mM buffer as a single peak coinciding with protein (Fig. 2). In this purification step, potassium phosphate buffer, pH 7.5 containing 5mM mercaptoethanol and 20% glycerol was used throughout. The active fractions were pooled and concentrated by ultrafiltration.

Table 1 shows the result of the purification of aromatic-amino-acid-glyoxylate aminotransferase from liver homogenates of glucagon-injected rats. A 305-fold purification was obtained with a recovery of about 2.2%.

The enzyme preparation (in 50mM potassium phosphate buffer, pH 7.5 containing 200μM pyridoxal 5′-phosphate, 5mM mercaptoethanol and 20% glycerol) may be stored for at least 7 weeks at − 20 °C without loss of activity. Little or no activity was lost when the enzyme was stored at 0 - 4 °C for 3 weeks.

Purification of aromatic-amino-acid-glyoxylate aminotransferase from monkey liver

A male monkey (*Macaca fuscata*, Blyth, about 13 kg) was anesthetized by the muscular injection of pentobarbital and bled by cutting the carotid. The mitochondrial extract was prepared from 36 g of liver as described for rat liver. The extract was warmed rapidly to 65 °C and maintained at this temperature with constant stirring, for 2 min, after which it was quickly chilled in a 0.9% NaCl/ice bath at about − 7 °C. The precipitate was removed by centrifugation at 5000 x *g* for 20 min.

Table 1. Purification of aromatic-amino-acid-glyoxylate aminotransferase from glucagon-injected rat liver. Details of the purification and assay methods are given in the text.

	Prot. [mg]	Spec. act. [mU/mg prot.]	Total act. [U]	Recovery [%]	Purif. [fold]
Homogenate	8723	0.73	6.40	100	1
Mitochondrial extract	1506	2.5	3.77	58.9	3.4
Heat treatment	1500	1.8	2.70	42.2	2.5
$(NH_4)_2SO_4$	145	10.8	1.57	24.5	14.8
Isoelectric focusing and Sephadex G-150	25.8	30.8	0.79	12.3	42.2
Hydroxylapatite	2.6	86.0	0.22	3.4	118
DEAE-cellulose	0.64	223	0.14	2.2	305

Table 2. Purification of aromatic-amino-acid-glyoxylate aminotransferase from monkey liver. Details of the purification and assay methods are given in the text.

	Total prot. [mg]	Spec. act. [mU/mg prot.]	Total act. [U]	Yield [%]	Purif. [fold]
Mitochondrial extract	877	0.43	0.38	100	1
Heat treatment	578	0.67	0.39	10.3	1.6
Isoelectric focusing and Sephadex G-150	33.4	10.2	0.34	89.5	23.7
Hydroxylapatite	3.9	33.2	0.13	34.2	77.1
DEAE-cellulose	0.57	105	0.06	15.8	244

The supernatant solution was then subjected to a sequence of purification procedures as described for rat liver enzyme; isoelectric focusing on a pH 3.5 - 10 Ampholine gradient, Sephadex G-150 gel filtration, $(NH_4)_2SO_4$ fractionation, hydroxylapatite chromatography and DEAE-cellulose chromatography.

The enzyme activity was mainly recovered in the 35-75% saturated fraction of $(NH_4)_2SO_4$. Isoelectric focusing resulted in the detection of a single activity peak with pI 8.0. The enzyme activity was eluted by 50mM buffer from a hydroxylapatite column, and it was eluted by 25mM buffer from a DEAE-cellulose column as a single peak coincided with protein (Fig. 3).

Table 2 shows the results of the purification of aromatic-amino-acid-glyoxylate aminotransferase from monkey liver. About 240-fold purification was obtained with a recovery of about 16%.

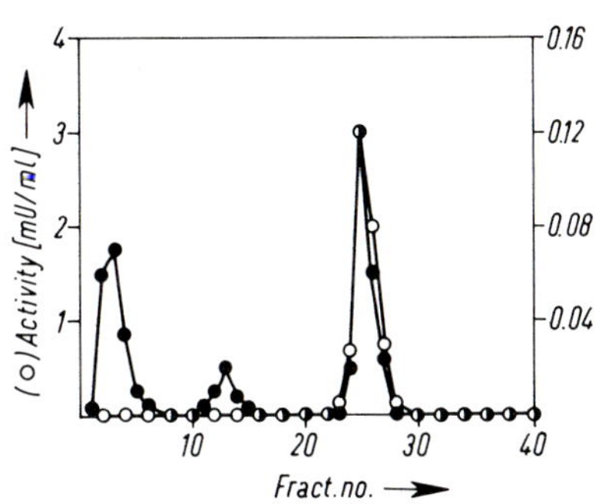

Fig. 3. DEAE-cellulose chromatography of monkey liver enzyme preparation from a hydroxylapatite column. Chromatographic conditions are as described in Fig. 2. ○, kynurenine-glyoxylate aminotransferase; ●, protein (optical density at 280 nm).

The enzyme preparation (in 50mM potassium phosphate buffer, pH 7.5 containing 200 μM pyridoxal 5′-phosphate, 5mM mercaptoethanol and 20% glycerol) may be stored at – 20 °C for at least 4 weeks without loss of activity. Little or no activity was lost when the enzyme was stored at 0 - 4 °C for 2 weeks.

Criteria of purity

Upon polyacrylamide disc gel eletrophoresis at pH 8.9 in a 7% gel, a single protein band with a mobility of 0.16 (relative to that of a Bromophenol Blue marker) was obtained with the rat enzyme and a diffuse single protein band with a mobility from 0.14 to 0.19 with the monkey enzyme.

Polyacrylamide disc gel electrophoresis in the presence of sodium dodecylsulfate resulted in the detection of a single protein band in the case of the rat enzyme. A major protein band which represented about 95% of the total protein and two minor protein bands were obtained with the monkey enzyme.

Sephadex G-150 gel filtration, sucrose density gradient centrifugation and isoelectric focusing on a pH 3.5 - 10 Ampholine gradient gave a single symmetrical peak with protein and aromatic-amino-acid-glyoxylate aminotransferase activity coincident for both enzyme preparations.

pH optimum

The relationship between pH and the enzymic activity was studied. Potassium phosphate buffer (0.2M) was used between pH 7.0 and 8.0, and Tris/HCl (0.2M) between pH 8.0 and 9.6. Both rat and monkey enzyme preparations had pH optima between 8.0 and 8.5.

Determination of approximate molecular weight

Similar molecular weights were estimated for both rat and monkey enzymes by sucrose density gradient centrifugation and Sephadex G-150 gel filtration. Molecular weights of both enzymes were 80 000 ± 6 000. Polyacrylamide disc gel electrophoresis in the presence of sodium dodecylsulfate gave a molecular weight of 40 000 ± 4 500 for both enzymes, showing that they probably consist of two identical subunits.

Substrate specificity

The relative activities of the two enzyme preparations with various 2-oxo acids were determined by using L-kynurenine or L-phenylalanine as amino donor (Table 3). Both enzymes utilized glyoxylate, pyruvate, hydroxypyruvate and 2-oxo-4-methylthiobutyrate with each amino donor. The activity ratio for these four 2-oxo acids with L-kynurenine was similar to that with L-phenylalanine in each case of the two enzymes. This ratio of each enzyme with L-kynurenine or L-phenylalanine was not changed after Sephadex G-150 gel filtration and isoelectric focusing on a pH 3.5 - 10 Ampholine gradient, suggesting that these four activities are associated with the same protein. 2-Oxoglutarate could not serve as amino acceptor with L-kynurenine for the monkey enzyme, and with L-phenylalanine for both monkey and rat enzymes. However, rat enzyme could utilize 2-oxoglutarate with L-kynurenine, suggesting that rat liver enzyme preparation might be contaminated by an additional enzyme, kynurenine-2-oxoglutarate aminotransferase. However, heating the rat enzyme at 70 °C for different lengths of time produced equivalent losses of both kynurenine-2-oxoglutarate and glyoxylate aminotransferase activities (Fig. 4).

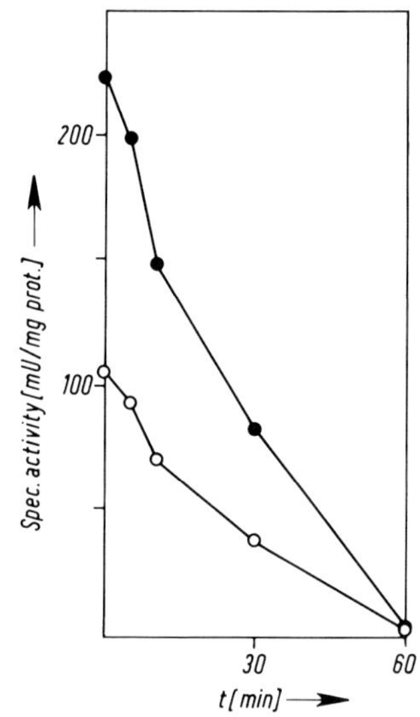

Fig. 4. Effects of heat denaturation of rat liver enzyme.

The purified enzyme preparation from rat liver was maintained at 70 °C for different lengths of time before assays of aromatic-amino-acid-glyoxylate aminotransferase (•) and kynurenine-2-oxoglutarate aminotransferase (○). Details of assays are given in the text.

Table 3. Relative activities of aromatic-amino-acid-glyoxylate aminotransferase from monkey and glucagon-injected rat liver for various 2-oxo acids.

Details of assays are as described in the text. The preparation from the last step of the purification was used as enzyme source. The final concentration of each 2-oxo acid and L-amino acid was 3mM. Relative activity values are given, signifying transamination rates compared with that for glyoxylate. (a) L-Kynurenine was used as amino donor; (b) L-phenylalanine was used as amino donor.

2-Oxo acid (3mM)	Rel. act.			
	Monkey		Rat	
	(a)	(b)	(a)	(b)
Glyoxylate	1	1	1	1
2-Oxoglutarate	0	0	0.61	0
Pyruvate	0.87	0.79	0.95	0.90
Hydroxypyruvate	0.82	0.78	0.90	0.88
2-Oxo-4-methylthiobutyrate	0.55	0.49	1.2	1.1

Furthermore, the activity ratio, kynurenine-glyoxylate aminotransferase/kynurenine-2-oxo-glutarate aminotransferase, was not changed after isoelectric focusing on a pH 3.5 - 10 Ampholine gradient, Sephadex G-150 gel filtration and polyacrylamide disc gel electrophoresis at pH 8.9 in a 7% gel of the purified rat liver enzyme. These results suggest that the two activities were properties of the same protein. Table 4 shows the relative activities of rat and monkey enzyme preparations with various aromatic L-amino acids by using glyoxylate as amino acceptor. Both enzyme preparations showed a broad specifity for aromatic L-amino acids. The most effective amino donors were L-phenylalanine and L-kynurenine in the case of the monkey enzyme. Some activity was detected with L-histidine, L-tyrosine and L-tryptophan. Much higher activity was detected with all aromatic L-amino acids used than with L-kynurenine in the case of the rat enzyme. Amino donors were effective in the following order; phenylalanine > histidine > tyrosine > tryptophan > 5-hydroxytryptophan > kynurenine. The relative activity ratio of each enzyme preparation for L-kynurenine, L-phenylalanine, L-histidine and L-tyrosine was not changed after isoelectric focusing on a pH 3.5 - 10 Ampholine gradient and Sephadex G-150 gel filtration, suggesting that these four activities are associated with the same protein. Neither enzyme had glutamate-2-oxoadipate aminotransferase activity. The apparent K_m values were determined from double-reciprocal plots of initial velocity and substrate concentration. The apparent K_m values for L-kynurenine with glyoxylate (3mM) as amino acceptor and for glyoxylate with L-kynurenine (30mM) are as follows: rat, 1.2 and 0.5mM; monkey, 1.5 and 0.4mM.

Table 4. Relative activities of aromatic-amino-acid-glyoxylate aminotransferase from monkey and glucagon-injected rat liver for various L-amino acids.

Assay conditions are as described in the text. The preparation from the last step of the purification was used as enzyme source. The final concentration of each L-amino acid and glyoxylate was 3mM. Relative activity values are given, signifying transamination rates compared with that for L-kynurenine.

L-Amino acid (3mM)	Rel. act. Monkey	Rel. act. Rat
Kynurenine	1	1
Phenylalanine	2.1	610
Histidine	0.30	259
Tyrosine	0.45	102
Tryptophan	0.25	21
5-Hydroxytryptophan	0.15	15

Inhibition by carbonyl reagents

Rat and monkey enzyme preparations were subjected to Sephadex G-25 gel filtration to remove added pyridoxal 5′-phosphate, glycerol and mercaptoethanol. The two enzyme preparations were inhibited, each to the same degree, by the addition of carbonyl reagents, namely semicarbazide, hydroxylamine, isonicotinic hydrazide and potassium cyanide: this inhibition is probably due to the binding of the inhibitor by the aldehyde group of the coenzyme, pyridoxal phosphate (Table 5). Dialysis of each enzyme preparation against 20mM of L-kynurenine, L-phenylalanine or L-histidine in 50mM potassium phosphate buffer, pH 7.5, did not lead to the resolution of the cofactor from each enzyme. These results suggest that the enzyme contains tightly bound pyridoxal phosphate.

Table 5. Effect of carbonyl reagents on aromatic-amino-acid-glyoxylate aminotransferase from monkey and glucagon-injected rat liver.

Details of assays are given in the text. The preparation from the last step of the purification was used as enzyme source.

Reagent added (1M)	Rel. act. [%] Monkey	Rel. act. [%] Rat
Control	100	100
Semicarbazide	8.6	7.4
Hydroxylamine	60.0	45.7
KCN	67.1	56.5
Isonicotinic hydrazide	71.4	58.7

Discussion

Aromatic-amino-acid-glyoxylate aminotransferase was purified and characterized from glucagon-injected rat liver. The enzyme preparation differed from isoenzymes 1, 2 and 3 of rat liver kynurenine aminotransferase with respect to isoelectric points (isoenzyme 1, 9.4; isoenzyme 3, 5.2) or pH optima (isoenzymes 1 and 3, 8.0 - 8.5; isoenzyme 2, 6.0 - 6.5)[1,2]. The 2-oxo acid specificity of the present enzyme differed from that of these three isoenzymes: the enzyme is active for both pyruvate and 2-oxoglutarate as amino acceptors; isoenzymes 1 and 2 are active with 2-oxoglutarate but not with pyruvate; isoenzyme 3 is active with pyruvate but not with 2-oxoglutarate[1,2]. These data show that rat liver aromatic-amino-acid-glyoxylate aminotransferase differs from isoenzymes 1, 2 and 3. However, substrate specificity, isoelectric point, molecular weight and response to glucagon of the enzyme were similar to those of phenylalanine-pyruvate aminotransferase previously described[1,10], suggesting the identity of the two enzymes. Further studies on the identity of the two enzymes are required.

It has been reported that kynurenine-2-oxoglutarate aminotransferase of rat kidney cytosol is identical with 2-aminoadipate aminotransferase[11,12]. However, rat liver aromatic-amino-acid-glyoxylate aminotransferase showed no such activity. Furthermore, isoenzymes 1 and 3 have been reported to be identical with aspartate-2-oxoglutarate aminotransferase and glutamine-2-oxo acid aminotransferase, respectively. These data suggest the identity of rat liver isoenzyme 2 with 2-aminoadipate aminotransferase. Studies on the purification, characterization and identification of isoenzyme 2 from rat liver are required.

On the other hand, the purification and characterization of kynurenine aminotransferase from monkey liver has previously not been reported. In the present study, aromatic-amino-acid-glyoxylate aminotransferase was purified and characterized from monkey liver; it was found to be active with glyoxylate, pyruvate, 2-oxomethylthiobutyrate and hydroxypyruvate as amino acceptors, but not with 2-oxoglutarate. Isoelectric focusing of the mitochondrial extracts from monkey liver did not result in the detection of kynurenine-2-oxoglutarate aminotransferase activity, presumably owing to its relatively low activity or loss of its activity during focusing. We have found that the activity of kynurenine-glyoxylate aminotransferase was higher than that of kynurenine-2-oxoglutarate aminotransferase in homogenates of human and monkey liver (unpublished data). Therefore it seems likely that aromatic-amino-acid-glyoxylate aminotransferase plays an important role in tryptophan metabolism in human and monkey liver.

Literature

1 Noguchi, T., Minatogawa, Y., Okuno, E., Nakatani, M., Morimoto, M. & Kido, R. (1975) *Biochem. J.* **151**, 399 - 406.
2 Noguchi, T. & Kido, R. (1976) *Hoppe-Seyler's Z. Physiol. Chem.* **357**, 649 - 656.
3 Miller, J.E. & Litwack, G. (1971) *J. Biol. Chem.* **246**, 3234 - 3240.
4 Noguchi, T., Okuno, E. & Kido, R. (1977) *Biochem. J.* **161**, 177 - 179.
5 Mason, M. (1954) *J. Biol. Chem.* **211**, 839 - 844.
6 Knox, W.E. (1953) *Biochem. J.* **53**, 379 - 383.
7 Noguchi, T., Okuno, E., Minatogawa, Y. & Kido, R. (1975) *Biochem. J.* **155**, 107 - 115.
8 Nakatani, Y., Fujioka, M. & Higashino, K. (1970) *Biochim. Biophys. Acta* **198**, 219 - 228.
9 de Duve, C., Pressman, B.C., Gianetto, R., Wattiaux, R. & Appelmans, F. (1955) *Biochem. J.* **60**, 604 - 617.
10 Minatogawa, Y., Noguchi, T. & Kido, R. (1977) *Hoppe Seyler's Z. Physiol. Chem.* **358**, 59 - 67.
11 Tobes, M.C. & Mason, M. (1975) *Biochem. Biophys. Res. Commun.* **62**, 390 - 397.
12 Tobes, M.C. & Mason, M. (1977) *J. Biol. Chem.* **252**, 4591 - 4599.

Izumi Harada, Dr. Tomoo Noguchi and Prof. Dr. Ryo Kido, Department of Biochemistry, Wakayama Medical College, Wakayama 640, Japan.

Hoppe-Seyler's Z. Physiol. Chem.
Bd. 359, S. 489 - 497, April 1978

Immunohistological Studies on the Distribution of Cytochrome b_5 Reductase in Rat Liver Microsomes

Hans-Ulrich SCHULZE, Bernhard FLEISCHER and Liliane POP

Zentrum für Biochemie der Universität Gießen

(Received 2 February 1978)

Summary: The distribution of cytochrome b_5 reductase in rat liver microsomes purified by density gradient centrifugation was studied using anti-cytochrome b_5 reductase/anti-ferritin hybrid antibodies labeled with ferritin. Electron micrographs show that the outer surface and the polar cap of tangential sections of essentially all vesicles derived from endoplasmic reticulum are specifically marked with a few ferritin grains which are not localized in large patches but arranged randomly. This finding was correlated with morphometrical and biochemical measurements. The results, supporting our earlier investigations, suggest that cytochrome b_5 reductase may be clustered at the most in very small assemblies consisting of a few molecules which, in turn, are statistically uniformly distributed over the total membrane system.

Immunhistologische Untersuchungen zur Verteilung der NADH:Ferricytochrom-b_5-Oxidoreduktase in Rattenlebermikrosomen

Zusammenfassung: Wir haben die Verteilung der NADH:Ferricytochrom-b_5-Oxidoreduktase in Rattenlebermikrosomen, die durch Dichtegradientenzentrifugation gereinigt worden sind, mit Ferritin markierten Anti-Cytochrom-b_5-Reduktase/Anti-Ferritin-Hybrid-Antikörpern untersucht. Die elektronenmikroskopischen Bilder zeigen, daß nahezu alle Vesikeln der Mirosomenpräparation mit einigen Ferritingranula spezifisch markiert sind. Die Ferritinkörnchen sind einzeln angeordnet und verteilen sich zufällig über die äußere Oberfläche der Mikrosomen. Eine Korrelation dieser Ergebnisse mit morphometrischen und biochemischen Meßwerten macht deutlich, daß die NADH:Ferricytochrom-b_5-Oxidoreduktase nicht in großen, spezialisierten Abschnitten des endoplasmatischen Retikulums der Leberzelle vorkommt und bestätigt unsere früheren Untersu-

Enzymes:
Acid phosphatase, orthophosphoric-monoester phosphohydrolase (acid optimum) (EC 3.1.3.2);
Cytochrome b_5 reductase, NADH:ferricytochrome b_5 oxidoreductase (EC 1.6.2.2);
Cytochrome c reductase, NADPH:ferricytochrome c oxidoreductase (EC 1.6.99.?);
Glucose-6-phosphatase, D-glucose-6-phosphate phosphohydrolase (EC 3.1.3.9);
Monodehydroascorbate reductase, NADH:monodehydroascorbate oxidoreductase (EC 1.6.5.4);
5′-Nucleotidase, 5′-ribonucleotide phosphohydrolase (EC 3.1.3.5);
Succinate dehydrogenase, succinate:(acceptor) oxidoreductase (EC 1.3.99.1).
Abbreviations:
Cb_5R, cytochrome b_5 reductase; aCb_5R/aF, anti-Cb_5R/anti-ferritin hybrid antibody.

chungen, wonach sich dieses Enzym, wie andere Enzyme auch, statistisch gleichmäßig, möglicherweise in kleinen Ansammlungen von wenigen Enzymmolekülen, über das gesamte Membransystem verteilt.

Key words: Microsomes, enzyme distribution, hybrid antibodies, ferritin labeling.

The endoplasmic reticulum consists of a continuous tubular and cisternal membrane system containing a large number of enzymes. The distribution of the enzymes on the membrane surface is of primary importance to our understanding of the molecular architecture and function of this intracellular membrane system. Therefore, extensive subfractionation experiments on rat liver microsomes have been performed with the aim of estimating the horizontal topology of some structure-bound enzymes in the endoplasmic reticulum[1-12].

The microsomes were found to be heterogeneous, both in density and in their enzyme content [1-5,7-9,12], which was explained in terms of enzymatic macroheterogenity of the endoplasmic reticulum itself[1-5,13,14]. However, the tendency to separate NADH:ferricytochrome b_5 oxidoreductase, NADPH:ferricytochrome *c* oxidoreductase, glucose-6-phosphatase, cytochrome b_5 and cytochrome P-450 by zonal centrifugation of microsomes and to allocate these enzymes to distinct particle populations depends on the size of the microsomal vesicle and is only observed after the microsomes ($\emptyset \approx 200$ nm) have been broken down by sonication[7]. Furthermore, a correlation of morphometrical and biochemical measurements has shown that effective separation of particle populations which were heterogeneous in their enzymatic composition was not achieved until the number of microsomal pieces ($\emptyset \approx 30$ nm) obtained by fragmentation became much greater than the number of molecules of the individual enzymes[7,8]. These investigations seem to indicate that the enzymes studied do not occur in large specialized areas of the membrane, for example near the nucleus or near the plasma membrane exclusively, but rather that, as far as these enzymes are concerned, there is only a single endoplasmic reticulum system in hepatocytes.

Subfractionation of microsomes by immunoadsorption or density gradient centrifugation in the presence of specific antibodies leads to the same conclusion[8,12]. In addition, previous reports from this and other laboratories have suggested that functionally related enzymes are located in the direct neighbourhood in the membrane [4,7,8,11,12,15,16] and, furthermore, it has been supposed that clusters of single enzymes (cytochrome b_5 reductase, cytochrome *c* reductase) [8,12,17] as well as enzyme systems (cytochrome b_5 reductase and cytochrome b_5; cytochrome *c* reductase and cytochrome P-450)[11,12,18,19] may form large[11,18] or small[8,12,17,19] specific patches which in turn are uniformly distributed over the total membrane system. This concept of enzyme clustering, however, was not in accord with the random distribution of cytochrome b_5 reductase, cytochrome *c* reductase and cytochrome b_5 on the membrane concluded from kinetic studies[15,20].

In the present paper we describe immunohistological experiments carried out on rat liver microsomes by means of hybrid antibodies labeled with ferritin which, at first, permit the direct electron microscopic proof that cytochrome b_5 reductase is indeed present in essentially all vesicles ($\emptyset \approx$ 100 - 200 nm) derived from the endoplasmic reticulum. Cytochrome b_5 reductase was chosen as a model enzyme for our topological studies, since cytochrome b_5 reductase, in contrast to cytochrome b_5, which was used by Remacle et al.[10], is a less abundant component of the endoplasmic reticulum and it therefore was expected that, secondly, these investigations should reveal further details concerning the microheterogeneity of the endoplasmic reticulum in its lateral plane.

Materials and Methods

Materials

Cytochrome *c*, 5'-adenosinemonophosphate, glucose 6-phosphate, glycerate 2-phosphate, reduced pyridine nucleotides, and pepsin (from pig; 3600 U/mg) were obtained from Boehringer Mannheim GmbH; donkey anti-rabbit plasma protein serum from Behringwerke, Marburg; and Sephadex and Sepharose from Pharmacia, Uppsala. All the other reagents were of analytical grade. Dehydro-L(+)-ascorbate was produced and purified as described before[21]. Trypsin-solubilized cytochrome b_5 was isolated as an electrophoretically pure protein[22] from pig liver microsomes according to the technique of Strittmatter[23].

Preparation of microsomes

Female Wistar rats (AF/Han.) weighing about 180 g and fasted for 14 h were used. Microsomes were prepared by differential centrifugation in the presence of 1mM EDTA, washed in 1.18% KCl and resuspended in 0.1M Tris/HCl, pH 7.4[7], ribosome-free purified microsomes were isolated as microsomes, but washed and finally resuspended in 0.65M sucrose, 0.1M Tris/HCl, pH 7.5.

Immunization and purification of antibodies

Anti-Cb$_5$R- and anti-ferritin sera were raised in rabbits (yellow-silver strain) as described[22] by injection of highly purified[22] pig liver cytochrome b_5 reductase, prepared according to Takesué and Omura[24] or horse-spleen ferritin (twice cristallized, Cd-free; Serva, Heidelberg).

Specific antibodies were isolated by affinity chromatography at 4 °C on an immunoadsorbent containing cytochrome b_5 reductase or ferritin covalently bound to Sepharose 4B (3.3 mg protein/m*l* packed agarose at pH 9) by means of BrCN (100 mg/m*l* packed agarose, pH 11, 10 min[8,25]. Before use the adsorbents were washed with 0.1M $NaHCO_3$, then with 0.1M glycine/HCl, 1M NaCl, pH 2 (glycine/salt buffer) and finally equilibrated with 0.1M sodium phosphate, pH 7.5 (phosphate buffer). Heat-inactivated (30 min, 56 °C) antisera were slowly passed several times through the columns and unbound protein was removed by washing with phosphate buffer until the absorbance at 280 nm of the effluent was less than 0.02.

Specific antibodies were eluted with glycine/salt buffer, immediately neutralized, concentrated and dialysed (Amicon PM 30) against 0.1M glycine/HCl, pH 7.0.

Preparation of hybrid antibodies

The purified antibodies were subsequently hybridized following in general the procedure of Hämmerling et al.[26] with certain modifications found to be important for our system. In order to reduce aggregation, which is particularly marked with anti-Cb$_5$R immune globulin, antibodies at a concentration of 1 - 1.5 mg protein/m*l* were incubated with pepsin (2 mg/100 mg antibody protein, pH 4.5, 16 h, 37 °C). After digestion, the preparations could be concentrated (Amicon UM 10) for gel filtration without significant loss of protein by aggregation. Reduction of antibody fragments was carried out essentially as described by Hämmerling et al.[26]. The neutralized eluate from the cation exchange column was reoxidized in an atmosphere of O_2, with gentle stirring, for at least 8 h and finally concentrated by ultrafiltration (Amicon PM 30) to 10 mg protein/m*l*.

Formation of hybrid antibody-ferritin complexes

1 m*l* of the concentrated solution containing the anti-Cb$_5$R/anti-ferritin hybrid antibodies (aCb$_5$R/aF) was added dropwise to 5 m*l* (40mg) of ferritin in 0.05M Na phosphate, pH 7.5, while the mixture was rapidly stirred. Before use, the commercial horse spleen ferritin was further purified by crystallization[27] (6 times; Cd-free) and – in order to reduce the nonspecific adsorption on microsomes – by passing the recrystallized ferritin twice through a column (1.2 × 15 cm) containing microsomes (2 mg protein/m*l* packed agarose) coupled at pH 7.0 to CNBr-activated Sepharose 4B and subsequently through a cytochrome b_5 reductase immunoadsorbent column (4 times).

After 1 h of stirring at 0 °C, bivalent and univalent anti-Cb$_5$R antibodies were removed by gel filtration of the mixture on a Sephadex G-150 column (1.8 × 45 cm) which was equilibrated with 0.05M Na phosphate, pH 7.5.

Immunohistological labeling and isolation of labeled microsomes

The fractions of the first peak (fraction A) of the filtration profile eluting in the excluded volume were pooled, concentrated (Amicon PM 30) and incubated (45 min, 30 °C, gentle shaking) with purified and ribosome-free microsomes in a medium (1 - 2 m*l*) containing 0.65M sucrose, 0.1M Tris/HCl, pH 7.5, 10^{-3}M *p*-chloromercuribenzoate. (Preliminary experiments had shown that *p*-chloromercuribenzoate further decreases nonspecific adsorption of ferritin). 10 mg microsomal protein containing about 0.5 nmol cytochrome b_5 reductase[7] was labeled with hybrid antibody-ferritin complex produced from 1.6 mg reoxidized antibodies which have been reported to contain about 4 nmol aCb$_5$R/aF fragments[26], i.e. an eight-fold excess of cytochrome b_5 reductase binding capactiy. 1 m*l* of the incubated suspension was layered onto the following discontinuous sucrose gradient: 1 m*l* 50%, 2 m*l* 40%, and 1.2 m*l* 30% sucrose (w/w) buffered with 0.01M Tris/HCl, pH 7.5 and adjusted by means of an Abbe refractometer. The centrifugation was performed at g_{max} = 248000 × g (Spinco rotor SW65) for 4 h and a significant purification of both microsomes and antibody-ferritin mixture was achieved. Plasma membranes, NADH: monodehydroascorbate oxidoreductase-rich membranes, and presumably Golgi membranes as well (not tested), remained in the upper third of the gradient [21], bivalent and univalent anti-ferritin antibody-ferritin complexes and unreacted hybrid antibody-ferritin complexes were sedimented in the 50% and the lower end of the 40% sucrose; excess ferritin was pelleted. The material at the 30%/40% sucrose interphase was carefully collected with a Pasteur pipette, diluted with 0.01M Na phosphate, pH 7.5, at a final concentration of about 2 mg protein/m*l*, cooled in ice and subsequently fixed overnight at 4 °C in 1% OsO_4. In control experiments, concentrated fraction A containing about 1.6 mg (= about 4 nmol) of hybrid antibody protein was preincubated with 1.1 mg (= about 40 nmol) of purified cytochrome b_5 reductase, shaken at 0 °C for 1 h, then added to the microsome preparation and treated identically as the test. The excess ferritin as well as the anti-ferritin antibody-ferritin complex, both present in this incubation suspension, served as additional control for nonspecific adsorption.

Immunological procedures

Double immunodiffusion analysis of Ouchterlony and immunoelectrophoresis were performed in 1% agarose gels in Michaelis-buffer pH 8.2, $I = 0.05$, containing 0.05% azide. Gels were run for 90 min at 5 mA/plate at 22 °C in a moist chamber (electrode buffer: Michaelis, pH 8.2, $I = 0.1$).

Biochemical and morphological methods

The activities of the enzymes were measured and protein was assayed as described previously[21]. Electron micrographs were obtained with a Phillips 300 electron microscope; experimental details are given in[8].

Results

Immunological analyses

The immunological purity of antigen was demonstrated by reacting anti-Cb_5R serum from rabbit with cytochrome b_5 reductase subjected to electrophoresis. One immunoprecipitin line was seen (Fig. 1).

Fig. 1. Immunoelectrophoretic examination of the purity of the cytochrome b_5 reductase preparation.

Electrophoresis was run for 90 min at 22 °C followed by double immunodiffusion for 24 h at 22 °C.
Well: Purified pig liver enzyme. Trough: Rabbit anti-Cb_5R serum.

Reimmunoadsorption was performed in order to examine the purity of isolated antibodies. All the protein applied to the columns was adsorbed. After washing with 0.1M Na phosphate, pH 7.5, it could be eluted completely with 0.1M glycine/HCl, 1M NaCl, pH 2.0. (Serum proteins of non-immunized rabbits were not adsorbed.) Immunoelectrophoretic analysis revealed that a donkey anti-rabbit plasma protein serum developed a single immunoprecipitin arc with both affinity-purified rabbit anti-ferritin antibody and affinity-purified rabbit anti-Cb_5R antibody, respectively (Fig. 2).

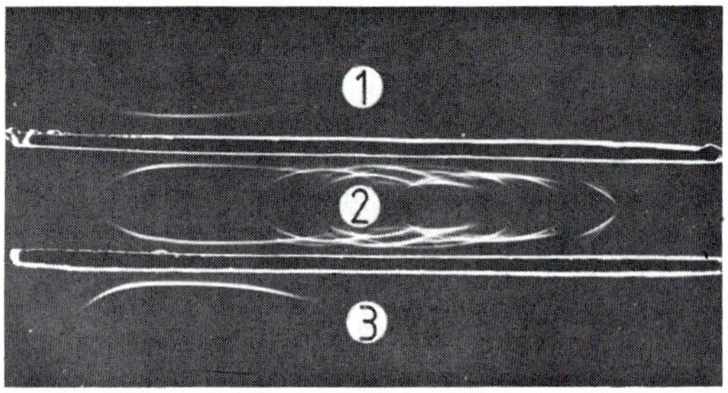

Fig. 2. Immunoelectrophoretic examination of the purity of anti-cytochrome b_5 reductase and anti-ferritin antibodies.

Electrophoresis was run for 90 min at 22 °C followed by double immunodiffusion for 24 h at 22 °C. Wells: 1, rabbit affinity-purified anti-Cb_5R antibody; 2, rabbit nonimmune serum; 3, rabbit affinity-purified anti-ferritin antibody. Troughs: Donkey anti-rabbit plasma protein serum.

The efficacy of each step of the hybridization process was checked by Ouchterlony immunoprecipitation (Fig. 3). The results obtained are in good agreement with those interpreted in detail by Hämmerling et al.[26] and furnish, together with the inhibition of cytochrome b_5 reductase by fraction A (Fig. 4), the evidence for the formation of aCb_5R/aF hybrid molecules in our system.

Biochemical measurements

Fig. 4 shows the results of typical experiments in which rat liver microsomes were preincubated with varying amounts of antiserum and the different preparations of antibodies to the purified

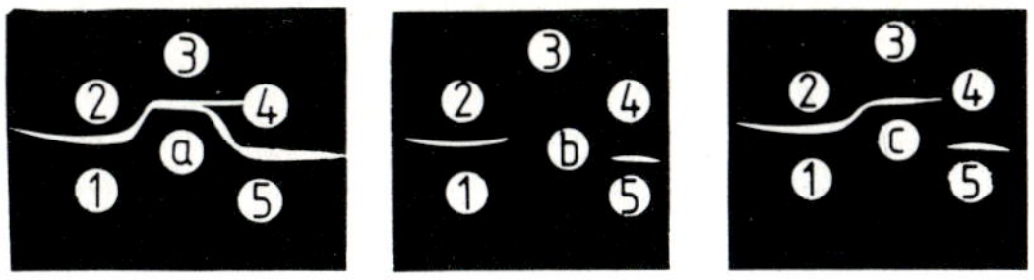

Fig. 3. Immunodiffusion analysis of the hybridization process.

Photographed after 24 h of diffusion at 22 °C. Center wells:

a) Mixture of pepsin-treated affinity-purified anti-Cb_5R antibodies and pepsin-treated affinity-purified anti-ferritin antibodies.

b) Mixture after reduction by cysteamine.

c) Mixture after removal of cysteamine and reoxidation.

Peripheral wells: 1, anti-ferritin serum; 2, ferritin; 3, mixture of ferritin and purified cytochrome b_5 reductase; 4, purified cytochrome b_5 reductase; 5, anti-Cb_5R serum. After reduction of the mixture, bivalent antibody fragments are no longer detectable; after subsequent reoxidation, a single precipitin line between c and 3 reappears, indicating the formation of aCb_5R/aF hybrid antibodies.

cytochrome b_5 reductase prepared from pig liver microsomes. The data clearly show the inhibition of cytochrome b_5 reductase in rat liver microsomes by the heterologous antibodies and demonstrate that pepsin-digested antibodies as well as aCb_5R/aF hybrid antibodies have sufficient antigen-binding activity. This is in agreement with the immunodiffusion analysis.

The table summarizes a typical purification scheme for rat liver microsomal vesicles derived from endoplasmic reticulum. It is seen that material enriched with respect to glucose-6-phosphatase, cytochrome b_5 reductase and cytochrome *c* reductase can be obtained with satisfactory yield. The purification procedure reduced the contaminations with lysosomes, mitochondria, plasma membranes, and NADH: monodehydroascorbate oxidoreductase-rich membranes, which are very low in the microsome fraction isolated by density gradient centrifugation, as estimated by the specific activity of generally accepted marker enzymes.

Morphological observations

Fig. 5a shows the electron microscopic appearance of microsomal vesicles after incubation with aCb_5R/aF hybrid antibody-ferritin complexes (fraction A) and subsequent purification by density gradient centrifugation. Fig. 5b shows the corresponding control. In both pictures, ribosome-free vesicles are visible. The difference between control and test is striking. The outer surface and

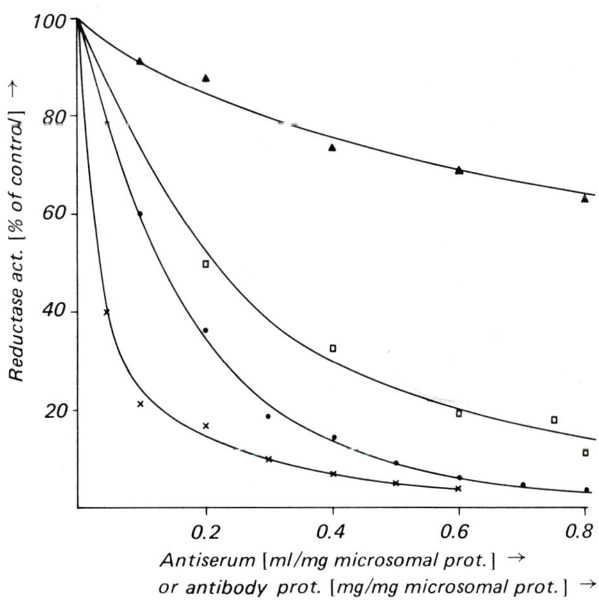

Fig. 4. Inhibition of cytochrome b_5 reductase in rat liver microsomes by antibodies against pig liver reductase.

Microsomes (1 mg protein) were preincubated (1 h at 37 °C and 12 h at 0 °C) in 0.1 M Tris/acetate pH 8.0 in 2 m*l* with varying amounts of anti-Cb_5R serum (x); affinity-purified anti-Cb_5R antibody (•); pepsin-treated anti-Cb_5R antibody (□); aCb_5R/aF hybrid antibody-ferritin complex (fraction A) (▲). The control mixtures contained the same amount of anti-ferritin serum, anti-ferritin antibody; pepsin-treated anti-ferritin antibody, and ferritin, respectively.

Table. Enzyme profiles of different rat liver microsomal fractions.

The results presented were obtained from a single homogenate, and are representative of 6 experiments. The specific activities (mU = nmol × min^{-1} × mg $protein^{-1}$) are expressed as NADH oxidized for monodehydroascorbate reductase, cytochrome reduced for cytochrome reductases, inorganic phosphorus released for phosphatases, and 2,6-dichlorophenolindophenol reduced for succinate dehydrogenase.

Enzyme	Microsomes		Purified microsomes		Microsomal fraction DGC[b]	
	[mU/mg prot.]	[U/g liver]	[mU/mg prot.]	[U/g liver]	[mU/mg prot.]	[U/g liver]
Glucose-6-phosphatase[a]	553	5.70	651	5.42	725	5.05
Cytochrome b_5 reductase	181	1.86	225	1.84	258	1.78
Cytochrome *c* reductase	347	3.57	400	3.28	465	3.21
5'-Nucleotidase[a]	68	0.70	61	0.50	43	0.30
Acid phosphatase[a]	6	0.06	6	0.05	4	0.03
Monodehydro-ascorbate reductase	44	0.45	36	0.30	29	0.20
Succinate dehydrogenase	7	0.07	7	0.06	5	0.03

[a]Assayed at 37 °C, all others assayed at 30 °C.
[b]Microsomes isolated by density gradient centrifugation of purified microsomes.

the polar cap of tangential sections of practically all profiles in the test are characteristically labeled with a few ferritin grains which are randomly arranged; no large patches are seen. Some free aggregates of ferritin are occasionally present in both test and control. The vesicles in the control, however, are nearly completely free of ferritin labeling.

Discussion

Various types of subfractionation techniques employed in recent years have revealed a certain enzyme heterogeneity of microsomes[1-12] which is increased by decreasing the size of the microsomal particles[4,7,8,12,18]. However, those results have been given contradictory explanations in terms of the horizontal topology of enzymes on the endoplasmic reticulum, especially because biochemical and morphometric analyses were not always correlated sufficiently, as discussed in detail earlier[7,8] and summarized recently[28]. Thus, the partial separation of microsomal subfractions having different enzyme compositions was interpreted as demonstrating enzymatic differences between rough and smooth microsomes[1-3], as suggesting the existence of relatively large areas of specialized membrane sections independent of ribosome loading[4,5,11,13,14,18], and finally, as reflecting only small heterogeneities[16] in the endoplasmic reticulum, which were assumed partly to represent very small clusters of a few enzyme molecules[8,12,17] and partly to be based on the different molecular concentration of the several enzyme proteins localized in the membrane[7].

From the electron micrographs presented here, it is evident that all the microsomal vesicles obtained by density gradient centrifugation contain cytochrome b_5 reductase in their membranes. Since,

▷

Fig. 5. Electron microscopic appearance of microsomal vesicles incubated with aCb_5R/aF hybrid antibody-ferritin complex and subsequent purification by density gradient centrifugation.

a) Test: Nearly all the vesicle profiles are labeled with single ferritin grains.

b) Control: Only occassional ferritin grains are seen on the outer surface of the vesicles. Arrow: Free aggregates of ferritin. Pictures are representative for many others obtained in order to find the optimal experimental conditions. Final magnification × 48000.

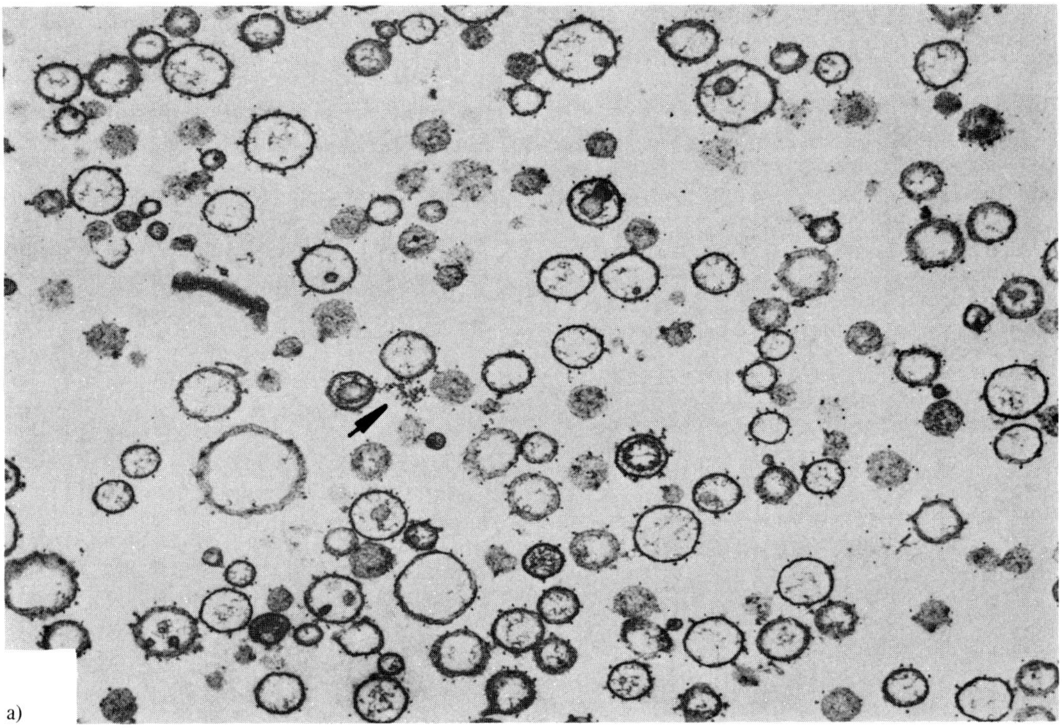

a)

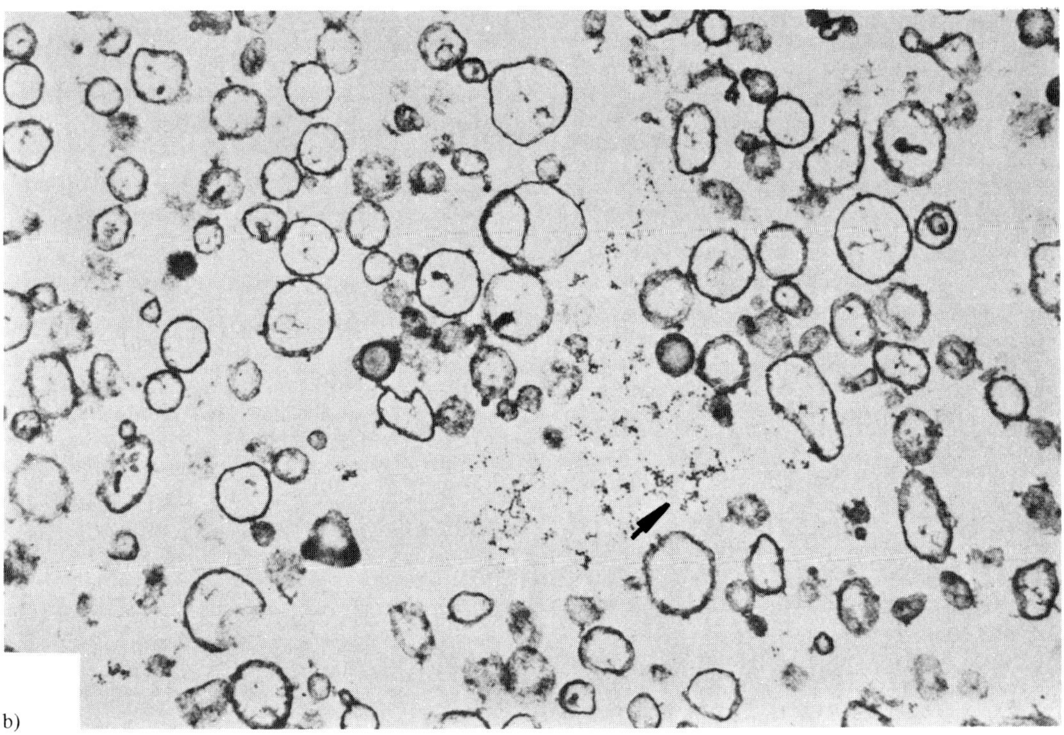

b)

furthermore, almost all profiles are specifically labeled with the aCb_5R/aF-ferritin marker in a similar random manner, first of all, our results provide evidence that cytochrome b_5 reductase is statistically uniformly distributed over the total membrane system (corresponding to the heterogeneity hypothesis of Remacle et al.[10]), confirming our first explanation for biochemical findings on the distribution of cytochrome b_5 reductase in the endoplasmic reticulum obtained by zonal centrifugation[7] and immunoadsorption[8] of several microsomal preparations differing in their vesicle size. Together with results concerning the localization of glucose-6-phosphatase[6,29] and cytochrome b_5[10], it has now been morphologically demonstrated, too, that specialized regions possibly present in the endoplasmic reticulum must be significantly smaller than the entire surface area ($\approx 10^{-6}$ A^2) of a microsome (∅ ≈ 200 nm).

In view of the enlargement of such specialized regions, patches of ferritin grains on the membrane surface cannot be recognized in our electron micrographs. On the contrary, the labeling of profiles shows, as mentioned above, a rather random distribution of single ferritin images around the vesicle surface. It could be possible that the labeling of several cytochrome b_5 reductase molecules within one patch is hindered by steric interference by a neighbouring hybrid ferritin complex attached to the membrane, but the following considerations speak against extensive clusters of many cytochrome b_5 reductase molecules[18] in the endoplasmic membrane. Based on various assumptions[7], the surface area of the endoplasmic reticulum of rat liver cells has previously been estimated to be between 0.44 and 0.63 m^2/mg microsomal protein[30,31]. Biochemical measurements have revealed that 1 mg of microsomal protein contains about 0.05 nmol of cytochrome b_5 reductase (0.2% of the total microsomal protein)[7,22]. Using two different methods[7,19] to estimate the protein content per microsomal vesicle (2×10^{-13} and 8×10^{-13} mg protein/vesicle; ∅ ≈ 200 nm) we calculated between 5 and 25 cytochrome b_5 reductase molecules per microsome. In the present report, direct analysis by electron microscopy revealed about 2 - 6 specific cytochrome b_5 reductase/ferritin hybrid antibody-ferritin complex binding sites on the outer surface of the microsomal profiles, which is consistent with our approximate calculations, assuming a section thickness of about 50 nm.

In view of these results, the clustering of numerous molecules to large complexes[18] of cytochrome b_5 reductase on the endoplasmic membrane is improbable, because in this case cytochrome b_5 reductase, unlike cytochrome b_5 which occurs in microsomes at a 10-15-fold molar excess over the reductase[7], should be present only in a part of the vesicles. Since, however, all the microsomal profiles are labeled with the ferritin marker, we finally conclude from our study that if clustering occurs at all, it is only in assemblies of a few (2 - 4) molecules or that the enzyme is completely randomly (statistically uniformly) distributed in the endoplasmic reticulum membrane as proposed by Rogers and Strittmatter[20].

We thank Miss *R. Kannler* and Mrs. *M. Neu* for excellent technical assistance. The electron microscopy was performed in the *Anatomisches Institut* and the *Strahlenzentrum* of the University of Gießen. This investigation was supported by the *Deutsche Forschungsgemeinschaft* (Schu 54/5) and the *Stiftung Volkswagenwerk* (AZ 11 2454).

Literature

1 Fouts, J. R. (1961) *Biochem. Biophys. Res. Commun.* **6**, 373 - 378.
2 Dallner, G. (1963) *Acta Pathol. Microbiol. Scand., Suppl.* **166**.
3 Gram, T. E., Hansen, A. R. & Fouts, J. R. (1968) *Biochem. J.* **106**, 587 - 591.
4 Dallman, P. R., Dallner, G., Bergstrand, A. & Ernster, L. (1969) *J. Cell Biol.* **41**, 357 - 377.
5 Glaumann, H. & Dallner, G. (1970) *J. Cell Biol.* **47**, 34 - 48.
6 Leskes, A., Siekevitz, P. & Palade, G. E. (1971) *J. Cell Biol.* **49**, 288 - 302.
7 Schulze, H.-U. & Staudinger, Hj. (1971) *Hoppe-Seyler's Z. Physiol. Chem.* **352**, 1659 - 1674.
8 Schulze, H.-U., Pönnighaus, J. M. & Staudinger, Hj. (1972) *Hoppe-Seyler's Z. Physiol. Chem.* **353**, 1195 - 1203.
9 Beaufay, H., Amar-Costesec, A., Thinès-Sempoux, D., Wibo, M., Robbi, M. & Berthet, J. (1974) *J. Cell Biol.* **61**, 213 - 231.
10 Remacle, J., Fowler, S., Beaufay, H., Amar-Costesec, A. & Berthet, J. (1976) *J. Cell Biol.* **71**, 551 - 564.
11 Winquvist, L. & Dallner, G. (1976) *Biochim. Biophys. Acta* **436**, 399 - 412.
12 Kawajiri, K., Ito, A. & Omura, T. (1977) *J. Biochem.* **81**, 779 - 789.

13 Dallner, G. (1969) in Proc. IVth Int. Congr. Pharmacol. 14.-18. July 1969, Basel, Switzerland, vol. 4., pp. 70 - 78, Schwabe & Co., Basel/Stuttgart.
14 Dallner, G., Svensson, H. & Ernster, L. (1972) in Oxidation Reduction Enzymes (Akeson, A. & Ehrenberg, A., eds.) pp. 567 - 574, Pergamon Press, Oxford, New York.
15 Peterson, J. A., Ebel, R. E., O'Keeffe, D. H., Matsubara, T. & Estabrook, R. W. (1976) *J. Biol. Chem.* **251**, 4010 - 4016.
16 Eriksson, L. C., Bergstrand, A. & Dallner, G. (1977) *Cell Tissue Res.* **179**, 17 - 43.
17 Morimoto, T., Matsuura, S., Sasaki, S., Tashiro, Y. & Omura, T. (1976) *J. Cell Biol.* **68**, 189 - 201.
18 Depierre, J. W. & Dallner, G. (1975) *Biochim. Biophys. Acta* **415**, 411 - 472.
19 Ito, A. (1974) *J. Biochem.* **75**, 787 - 793.
20 Rogers, M. J. & Strittmatter, P. (1974) *J. Biol. Chem.* **249**, 895 - 900.
21 Geiß, D. & Schulze, H.-U. (1975) *FEBS Lett.* **60**, 374 - 379.
22 Pönnighaus, J. M., Schulze, H.-U. & Staudinger, Hj. (1972) *Hoppe-Seyler's Z. Physiol. Chem.* **353**, 815 - 824.
23 Strittmatter, P. (1967) *Methods Enzymol.* **10**, 553 - 556.
24 Takesué, S. & Omura, T. (1970) *J. Biochem.* **67**, 267 - 276.
25 Axén, R. & Ernback, S. (1971) *Eur. J. Biochem.* **18**, 351 - 360.
26 Hämmerling, U., Aoki, T., De Harven, E., Boyse, E. A. & Old, L. J. (1968) *J. Exp. Med.* **128**, 1461 - 1473.
27 Andres, G. A., Hsu, K. C. & Seegal, B. C. (1967) in Handbook of Experimental Immunology (Weir, D. M., ed.) p. 528, Blackwell Scientific Publications, Oxford.
28 Schulze, H.-U. & Staudinger, Hj. (1975) *Naturwissenschaften* **62**, 331 - 340.
29 Leskes, A., Siekevitz, P. & Palade, G. E. (1971) *J. Cell Biol.* **49**, 264 - 287.
30 Weibel, E. R., Stäubli, W., Gnägi, H. R. & Hess, F. A. (1969) *J. Cell Biol.* **42**, 68 - 91.
31 Wibo, M., Amar-Costesec, A., Berthet, J. & Beaufay, H. (1971) *J. Cell Biol.* **51**, 52 - 71.

Prof. Dr. H.-U. Schulze, Dr. B. Fleischer, and Dipl.-Biol. L. Pop, Zentrum für Biochemie der Universität Gießen, Friedrichstraße 24, D-6300 Lahn-Gießen.

Hoppe-Seyler's Z. Physiol. Chem.
Bd. 359, S. 499 - 505, April 1978

Localization of Kallikrein in Porcine Pancreas and Submandibular Gland as Revealed by the Indirect Immunofluorescence Technique

Thomas DIETL, Johanna KRUCK and Hans FRITZ

Abteilung für Klinische Chemie und Klinische Biochemie in der Chirurgischen Klinik der Universität München

(Received 16 January 1978)

Dedicated to Professor Dr. Dr. Butenandt on the occasion of his 75th birthday

Summary: The glandular kininogenase kallikrein is known to occur in many mammalian organs and glands but direct histochemical localization has been achieved in only a few cases. We have now been able to localize porcine kallikrein in the acinar cells of the pancreas and in the striated and collecting duct cells of the submandibular gland. Incubation of frozen and fixed sections with one of the crossreacting antibodies, anti-pancreatic, anti-submandibular or anti-urinary kallikrein IgG resulted in the same immunofluorescence pattern. There was evidence of a specific fluorescence neither in the acinar cells, nor in the interstitial tissue or blood cells of the submandibular gland nor in the islets of Langerhans, the interlobular ducts or blood vessels of the pancreas. From all data now available about glandular kallikreins, it seems that the kallikreins in these organs are very similar.

Lokalisierung von Kallikrein in Pankreas und Submandibularis-Drüse des Schweines durch die indirekte Immunfluoreszenz-Methode

Zusammenfassung: Die drüsenständige Kininogenase Kallikrein kommt zwar in vielen Säugetierorganen und -drüsen vor, die direkte histochemische Lokalisation gelang jedoch nur in wenigen Fällen. Wir konnten nun Schweine-Kallikrein in den exokrinen Drüsenendstücken des Pankreas und in den Zellen des Streifenstückes und der Sammelröhren der Submandibularis-Drüse lokalisieren. Die Inkubation der fixierten Gefrierschnitte mit einem der kreuzreagierenden Antikörper, mit Anti-Pankreas-, Anti-Submandibularis- oder Anti-Harnkallikrein-IgG, ergab die gleiche Immunfluoreszenzverteilung. Spezifische Fluoreszenz konnte weder in Drüsenendstücken, im Bindegewebe oder in Blutgefäßen der Submandibularis-Drüse, noch in den Langerhans'schen Inseln, den Ausführungsgängen oder Blutgefäßen des Pankreas beobachtet werden. Aus allen bisher gewonnenen Daten über Drüsenkallikreine wird ersichtlich, daß die Kallikreine in diesen Organen sehr ähnlich sind.

Key words: Kallikrein, localization, pancreas, submandibular gland, immunofluorescence technique.

Enzymes:
Kallikrein (recommended name: kininogenin) (EC 3.4.21.8); trypsin (EC 3.4.21.4).
Abbreviation:
PBS: Phosphate-buffered saline [0.01M phosphate, pH 7.2, 0.9% (w/v) saline].

The glandular kallikreins occur in several organs and body fluids[1], especially in urine[2,3], kidney[4], pancreas[5,6], salivary glands[7,8] and in cells of the intestinal tract[9]. To elucidate the physiological function of the glandular kallikreins, attempts have been made to localize these enzymes at the cellular and subcellular level. Ørstavik et al.[10] and Brandtzaeg et al.[11] localized kallikrein in rat submandibular and sublingual salivary glands by the immunofluorescence technique. In the submandibular gland, most of the kallikrein was found in the cytoplasmic granules of the granular tubules. Cells of the striated ducts showed a faint staining of the cytoplasm with a bright luminal rim that was occasionally also seen in the excretory ducts. In the sublingual gland, kallikrein-specific immunofluorescence was observed in the cytoplasm of the striated duct cells and as a luminal rim in the main ducts.

Hojima et al.[12] showed the location of kallikrein in the submandibular gland of the cat by an indirect immunofluorescence technique. Specific fluorescence was seen only in the apical portion of the striated duct cells. In some sections the collecting ducts also showed a distinct but very fine fluorescent luminal rim.

The three porcine glandular kallikreins from pancreas, submandibular gland and urine purified recently to apparent homogeneity turned out to be very similar or even identical with respect to the following criteria[13,14]: amino acid composition, available sequence data, enzymic behaviour against substrates and inhibitors, and immunological cross-reactivity. The differences observed so far in the carbohydrate portions of the three kallikreins and in their structures – submandibular and urinary kallikrein consists of one polypeptide chain, which seems to be cleaved by limited proteolysis in the pancreatic extracts so that a two-chain "pancreatic" kallikrein results – obviously do not influence significantly the immunoreactivity of the kallikrein molecules. On the other hand, pancreatic kallikrein is known to occur as a zymogen (prekallikrein) in fresh organ extracts[15] or pancreatic juice[16], whereas similar inactive precursor forms have not been found so far for submandibular[17] or urinary[18,19] kallikrein (which should originate from the kidney[1]).

The given results stimulated us to investigate the localization of the kallikreins in pancreas, submandibular gland and kidney with antibodies directed against each of these kallikreins in order to see whether the immunological simularity of the three kallikreins also exists in the native tissue-bound forms.

Materials and Methods

Kallikreins

Porcine pancreatic kallikrein B[6] and porcine submandibular kallikrein[20] were isolated in our department by Dr. Fiedler and Dr. Lemon, respectively. Porcine urinary kallikrein was obtained from Dr. Mair[3].

Anti-kallikrein antibodies

Rabbits were immunized intracutaneously into the back with 0.7 mg each of porcine pancreatic, submandibular or urinary kallikrein dissolved in 1 m*l* physiological saline and emulsified in complete Freund's adjuvant (1/1, v/v). After 3 weeks, every rabbit was boostered intravenously once a week 3 - 5 times with 0.1 mg kallikrein dissolved in 1 m*l* physiological saline until the titer remained constant in an Ouchterlony test (see Table)[21]. The IgG fractions were isolated by precipitation with caprylic acid[22] and purity was examined by immunoelectrophoresis[21] with anti-rabbit serum (from Behringwerke AG). In the double immunodiffusion[21], each of the antikallikrein antibodies crossreacted with each of the kallikreins in a single precipitin line, all of them showing the pattern of identity[13]. Another anti-por-

Table. Determination of anti-kallikrein titer in sera and 1% IgG solutions from immunized rabbits.

Maximal dilution of a pancreatic kallikrein solution (1 μg/5 μ*l* PBS) which still precipitates with 5 μ*l* of serum or 1% (w/v PBS) IgG solution in an Ouchterlony test[21].

Kallikrein from	Antibody fraction from rabbit	Max. dilution of kallikrein solution
Pancreas	serum	1: 20
	1% IgG	1:160
Urine	serum	1: 4
	1% IgG	1: 3
Submandibularis gland		
rabbit no. 1	serum	1: 64
rabbit no. 2	serum	1: 32
rabbit no. 3	serum	1: 32
rabbit no. 3	1% IgG	1: 8

cine pancreatic kallikrein antibody fraction (IgG)[13] isolated by ammonium sulfate precipitation was a gift of Dr. Truscheit (Bayer AG).

Tissue sections and fixation

Small pieces of fresh organs were quickly frozen in cold isopentane[23]. The frozen organ pieces were cut into 6 μm thick sections and air dried. Best results were obtained by fixation of the dry sections with phosphate-buffered (20 mg Na_2HPO_4, 100 mg KH_2PO_4, 30 m*l* H_2O) acetone containing 9% formaldehyde (30 s, 25 °C)[24]. The concentration of acetone was 45% (v/v) in the case of pancreas sections and 65% (v/v) when sections of the submandibular gland were applied. The wet sections were immediately washed in PBS (3 × 10 min) prior to incubation with antibody solutions.

Indirect immunofluorescence technique[25]

The sections were incubated (15 min, 25 °C) with a solution of 1% (w/v PBS) bovine serum albumin[26] (Sigma, RIA grade) to reduce non-specific adsorption of IgG. Excess solution was removed before the sections were covered with 0.1% (w/v PBS) IgG containing one of the anti-kallikrein antibodies (30 min, 37 °C). Thereafter excess IgG was washed off and the sections were incubated (30 min, 37 °C) with fluorescein-isothiocyanate-labeled goat antirabbit IgG (from Behringwerke AG; 1% (w/v) IgG containing 10% (w/v) specific IgG; 4% human albumin; molar ratio fluorescein/protein = 4.0). After final washings with PBS, the sections were embedded in 0.1M phosphate buffer (pH 7.2)/glycerol (1/9, v/v). Microphotographs were taken using a Leitz fluorescence microscope (Ploem system) and a daylight Kodak high-speed Ektachrome film (23 DIN, exposed and developed according to 27 DIN).

Three types of controls were run. The specific anti-kallikrein IgG fraction was substituted by

1) anti-kallikrein IgG previously saturated with porcine pancreatic kallikrein,
2) unspecific IgG from rabbit and
3) PBS.

Results and Discussions

Porcine submandibular gland

Best fixation of kallikrein was achieved using a mixture of 3.5 vol. of phosphate buffer containing 9.25% (w/v) formaldehyde and 6.5 vol. of acetone. An acetone concentration of 45% (v/v) was too low to make kallikrein insoluble in submandibular gland sections during fixation with formaldehyde. In this connection it is remarkable that submandibular kallikrein is reversibly precipitated at an acetone concentration between 35 and 65% during the first purification step of the isolation procedure. On the other hand, fixation of cryo-sections with 100% acetone or 95% ethanol did not yield satisfactory results.

The strong and specific fluorescence obtained with the developed procedure indicates location of kallikrein in striated and also in collecting ducts of the submandibular gland (Fig. 1). Kallikrein is present there in the apical portion of these duct cells. This region contains the zymogen granules. Kallikrein could not be found in acinar, interstitial or other cells of the tissue. The same immunofluorescence pattern was observed if one of the other antibody-IgG preparations directed against each of the kallikreins was applied. These results are in agreement with the location of kallikrein described by other authors in submandibular glands of rat[10,11] and cat[12] as discussed in the introduction. The only difference is that in our experiments the fluorescence in the collecting duct cells was as strong as the fluorescence in the striated duct cells.

Intracellular kallikrein seems to be located in the excretory granules. Barton et al.[27] showed that prolonged intermittent sympathetic nerve stimulation completely depleted membrane-bound electron-dense organelles present in the apical region of the striated duct cells. After depletion, 90-95% of the kallikrein content was lost. These observations are in accordance with the assumption of a local physiological role of submandibular kallikrein, namely the increase in local blood circulation by vasodilation and regulation of the ionic composition of the secretory fluid due to local kallikrein-induced kinin release[28].

Porcine pancreas

Best results in the fixation of pancreatic kallikrein in the cryo-sections were achieved with a mixture of 5.5 vol. of phosphate buffer containing 9.25% (w/v) formaldehyde and 4.5 vol. of acetone. The following fixatives or procedures turned out to be unsuitable to show proper kallikrein localization: acetone at room remperature or at − 20 °C; 95% (v/v) ethanol; 95% ethanol/acetone (4:1, v/v); absolute ethanol/acetic acid (3:1 or 20:1, v/v); subsequent treatment with 95% (v/v) ethanol, absolute ethanol and xylol; treatment with 95% (v/v) ethanol followed by incubation with trypsin; methanol; 4% (w/v) formaldehyde in buffer; paraformaldehyde vapor or *p*-benzoquinone vapor. This shows clearly that the most suitable fixation conditions have to be worked out for each kallikrein-organ system. Obviously, kallikrein is relatively easily dissolved and thus detached from the

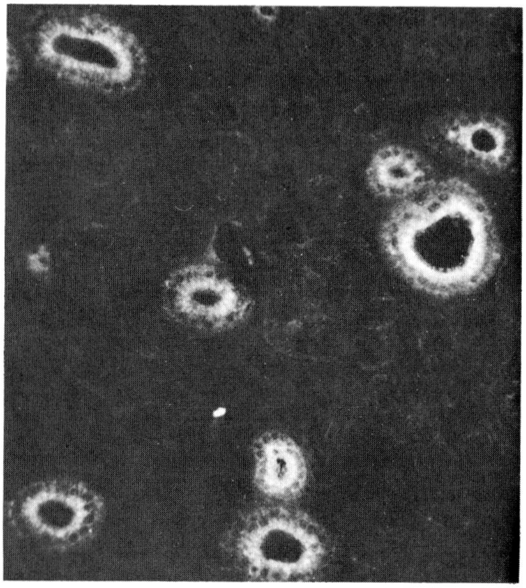

a)

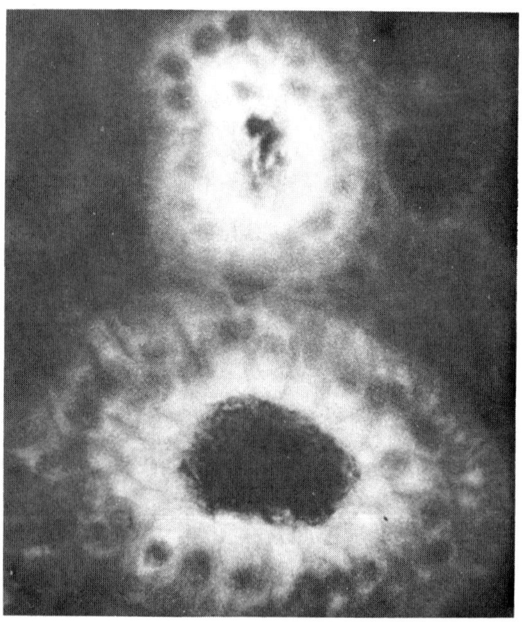

b)

◁

Fig. 1. Localization of kallikrein in porcine submandibular gland as revealed by immunofluorescence.

Anti-submandibular kallikrein IgG from rabbit and fluorescein-labeled goat and anti-rabbit IgG were employed (see Materials and Methods).

a) Location of kallikrein in the striated ducts of porcine submandibular gland (x 107). The same pattern of immunofluorescence could be seen in the larger collecting ducts.

b) At higher magnification the location of kallikrein mainly in the apical portion of the duct cells is clearly to be seen (× 427).

In the control sections no unspecific fluorescence was observed.

tissue by too mild a treatment (perhaps followed by precipitation at the section surface, resulting in an unspecific distribution), whereas too harsh a treatment may cause denaturation of the antigenic sites of the kallikrein molecules, thus preventing binding of the specific antibodies.

Using the developed fixation procedure, a strong and specific fluorescence was to be seen in the acinar cells, and there mainly in the apical portions (Fig. 2). In contrast, specific fluorescence could not be observed in the interstitial tissue, the interlobular ducts or in the islets of Langerhans. The same immunofluorescence pattern was observed, if one of the other antibody-IgG preparations directed against each of the kallikreins was applied. That means that pancreatic kallikrein is located in the exocrine cells, as are the proteinases chymotrypsinogen[29,30], trypsinogen[30], elastase[31] and carboxypeptidase[29]. We suppose, therefore, that the zymogen granules in the apical region of the excretory cells contain the kallikrein[32,33]. This assumption is in agreement with previously published results demonstrating parallel secretion (in response to the same stimuli) of kallikrein and other secretory proteinases with the pancreatic juice[34].

The similar intensity of light in the immunofluorescence pattern observed in pancreas and submandibular gland sections implies that the antibodies produced against the three active kallikreins also cross-react with prekallikrein. This is not unusual, in view of the immunological relationship between other zymogen-enzyme systems of the pancreas, e.g. trypsinogen/trypsin and chymotrypsinogen/chymotrypsin, etc.[35]. On the other hand, activation of pancreatic prekallikrein during handling of the tissue sections cannot be excluded.

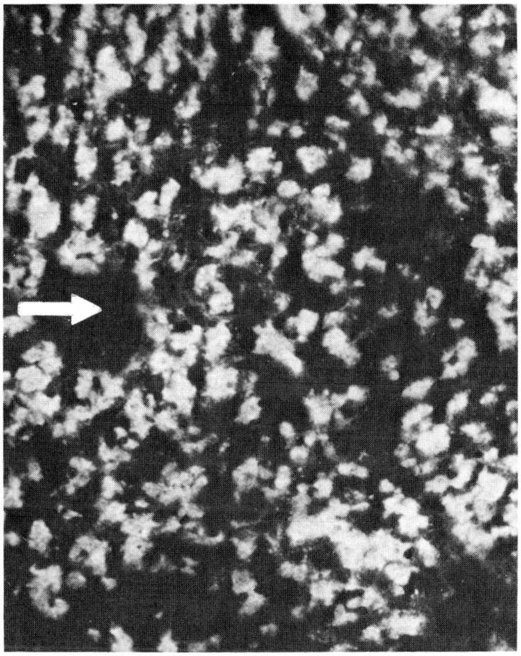

a)

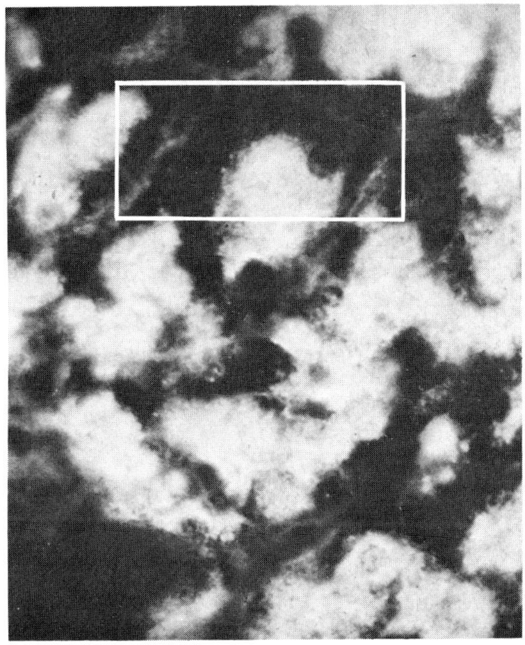

b)

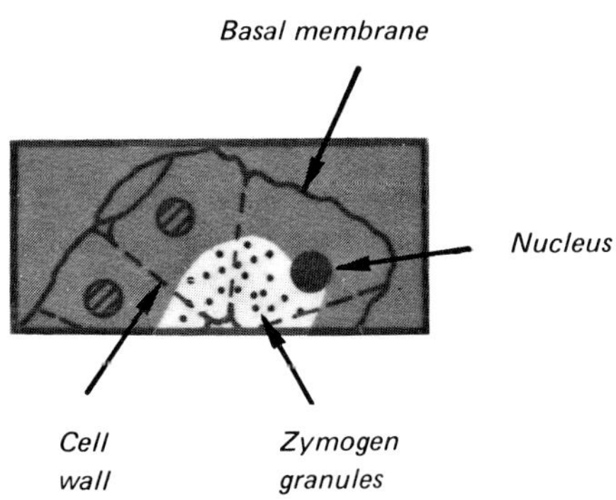

c)

Fig. 2. Localization of kallikrein in porcine pancreas as revealed by immunofluorescence.

Anti-pancreatic kallikrein IgG (Bayer AG) from rabbit and fluorescein-labeled goat anti-rabbit IgG were employed (see Materials and Methods).
a) Location of kallikrein in acinar cells of porcine pancreas. The small islet of Langerhans in the middle of the left side (arrow) is dark (× 107).
b) At higher magnification strong kallikrein-specific fluorescence can be seen in the apical region of the acinar cells (× 427).
In the control sections no unspecific fluorescence was observed. For better orientation, acinar cells (window) are drawn schematically in Fig. 2c.
c) Schematic drawing of acinar cells. Nucleus and basal membrane, but no cell walls, are visible in Fig. 2b. Zymogen granules are located in the region of fluorescence.

The high substrate specificity of kallikrein and its low concentration in the pancreatic juice[1] is not in favour of a simple protein-digestive function of the kallikrein. It is much more likely that pancreatic kallikrein has a special function, perhaps concerned with the absorption of water, ions[36] and digested foodstuffs such as amino acids[37] or carbohydrates[38] in the intestine as well as in the stimulation of gut motility[1] and cell proliferation[39,40]. Recently kininogen was found in gastric mucus[41]. This completes the digestive tract kallikrein-kinin system, in so far as the kallikrein-specific substrate would thus be available directly in the intestine, allowing local kinin generation. However, the absorption of kallikrein via the gut and subsequent systemic kinin generation in the circulation also has to be considered[42,43].

Porcine kidney

Many experiments have been performed to localize kallikrein in the kidney by the immunofluorescence technique. In addition to the procedures used to get suitable tissue sections from pancreas and submandibular glands, paraffine and cryosections combined with various fixation methods were applied. However, proper localization of renal kallikrein could not be achieved. This might be due either to the much lower amount of kallikrein present in the kidneys compared to the other two glands[1] or to a special mode of binding of kallikrein to kidney organelles so that its antigenic sites are buried. This problem has to be investigated further, therefore, by ultrahistochemical studies on tissue sections or radioimmunological or biological detection of kallikrein in isolated organelles[44,45]. It is to be hoped that these techniques may help to answer the question whether urinary kallikrein[13,46] is simply filtered pancreatic and/or submandibular kallikrein (and thus kidney kallikrein only absorbed kallikrein originating from the other glands) or the excretion product of a kidney-specific kallikrein[44,47].

These investigations have been supported by the *Deutsche Forschungsgemeinschaft*, Sonderforschungsbereich 51, München (grant B/3). We are indebted to *Dr. G. Mair* for samples of porcine urinary kallikrein and to *Dr. E. Truscheit* (Bayer AG) for anti-porcine pancreatic kallikrein immunoglobulin G. We are also indebted to *Dr. M. Warwas* (grant of *Alexander von Humboldt-Stiftung*) for immunizations of rabbits and for isolation of anti-kallikrein IgG fractions.

Literature

1 Frey, E., Kraut, H. & Werle, E. (1968) Das Kallikrein-Kinin System und seine Inhibitoren. Ferdinand Enke Verlag, Stuttgart.
2 Frey, E. & Kraut, H. (1928) *Naunyn-Schmiedebergs Arch. Exp. Pathol. Pharmakol.* **133**, 1 - 56.
3 Tschesche, H., Mair, G., Förg-Brey, B. & Fritz, H. (1976) in Kinins (Sicuteri, F., Back, N. & Haberland, G.L., eds.) *Adv. Exp. Med. Biol.* **70**, 119 - 122, Plenum Press, New York.
4 Werle, E. & Vogel, R. (1961) *Archs Int. Pharmacodyn. Thér.* **131**, 257 - 261.
5 Frey, E., Kraut, H. & Schultz, F. (1930) *Naunyn-Schmiedebergs Arch. Exp. Pathol. Pharmakol.* **158**, 334 - 347.
6 Fiedler, F. (1976) *Methods Enzymol.* **45**, 289 - 303.
7 Werle, E. & von Roden, P. (1939) *Biochem. Z.* **301**, 328 - 337.
8 Erdös, E. (ed.) (1970) Bradykinin, Kallidin and Kallikrein, Handbook of Experimental Pharmacology, vol. 25, Springer Verlag, Berlin.
9 Seki, T., Nakajima, T. & Erdös, E. (1972) *Biochem. Pharmacol.* **21**, 1227 - 1235.
10 Berg Ørstavik, T., Brandtzaeg, P., Nustad, K. & Halvorsen, K. (1975) *Acta Histochem.* **54**, 183 - 192.
11 Brandtzaeg, P., Gautvik, K., Nustad, K. & Pierce, J. (1976) *Brit. J. Pharmacol.* **56**, 155 - 167.
12 Hojima, Y., Maranda, B., Moriwaki, C. & Schachter, M. (1977) *J. Physiol.* **268**, 793 - 801.
13 Fritz, H., Fiedler, F., Dietl, T., Warwas, M., Truscheit, E., Kolb, H.J., Mair, G. & Tschesche, H. (1977) in Kininogenases 4 (Haberland, G.L., Rohen, J.W. & Suzuki, T., eds.) pp. 15 - 28, Schattauer Verlag, Stuttgart.
14 Mair, G. (1977) Ph. D. thesis at the Technische Universität München.
15 Fiedler, F. & Werle, E. (1967) *Hoppe-Seyler's Z. Physiol. Chem.* **348**, 1087 - 1089.
16 Werle, E. & Urhahn, K. (1940) *Biochem. Z.* **304**, 387 - 396.
17 Werle, E. & von Roden, P. (1936) *Biochem. Z.* **286**, 213 - 219.
18 Kraut, H., Frey, E., Werle, E. & Schultz, F. (1934) *Hoppe-Seyler's Z. Physiol. Chem.* **230**, 259 - 277.
19 Rüdel, G. (1961) Ph. D. thesis at the Universität München.
20 Lemon, M.J.C., Förg-Brey, B. & Fritz, H. (1976) in Kinins (Sicuteri, F., Back, N. & Haberland, G.L., eds.) *Adv. Exp. Med. Biol.* **70**, 209 - 216. Plenum Press, New York.
21 Ouchterlony, Ö. (1968) Handbook of Immunodiffusion and Immunoelectrophoresis, Ann. Arbor Science Publishers, Inc. Ann Arbor, MI.
22 Steinbuch, M. & Audran, R. (1969) *Arch. Biochem. Biophys.* **134**, 279 - 284.
23 Nei, T. (1976) Recent Progress in Electron Microscopy of Cells and Tissues (Yamada, E., Mizuhira, V., Kurosumi, K. & Nagano, T., eds.) pp. 213 - 243, Georg Thieme Publishers, Stuttgart, Igaku Shoin Ltd., Tokyo.
24 Mason, D.Y., Farrell, C. & Taylor, C.R. (1975) *Brit. J. Haematol.* **31**, 361 - 370.
25 Coons, A.H. & Kaplan, M.H. (1950) *J. Exp. Med.* **91**, 1 - 13.
26 Brandtzaeg, P. (1973) *Scand. J. Immunol.* **2**, 333 - 348.
27 Barton, S., Saunders, E., Schachter, M. & Uddin, M. (1975) *J. Physiol.* **251**, 363 - 369.
28 Hilton, S.M. (1970) in Bradykinin, Kallidin, Kallikrein. Handbook of Experimental Pharmacology

(Erdös, E., ed.) vol. 25, pp. 389 - 399, Springer Verlag, Berlin.
29 Marshall, J. (1954) *Exp. Cell Res.* **6**, 240 - 242.
30 Yasuda, K. & Coons, A. (1966) *J. Histochem. Cytochem.* **14**, 303 - 313.
31 Moon, H. & McIvor, B. (1960) *J. Immunol.* **85**, 78 - 80.
32 Siebert, G., Werle, E., Jung, G. & Maier, L. (1955) *Biochem. Z.* **326**, 420 - 423.
33 Bhoola, K. (1970) *Adv. Exp. Med. Biol.* **8**, 615 - 619.
34 Forell, M. (1960) in Polypeptides which Affect Smooth Muscles and Blood Vessels (Schachter, M., ed.) pp. 247 - 252, Pergamon Press, Oxford.
35 Arnon, R. (1973) in The Antigens (Sela, M., ed.) vol. I, pp. 88 - 159, Academic Press, New York.
36 Dennhardt, R. & Haberich, F.J. (1973) Kininogenases (Haberland, G.L. & Rohen, J.W., eds.) pp. 81 - 88, Schattauer Verlag, Stuttgart.
37 Caspary, W. & Creutzfeldt, W. (1973) in Kininogenases (Haberland, G.L. & Rohen, J.W., eds.) pp. 67 - 73, Schattauer Verlag, Stuttgart.
38 Meng, K. & Haberland, G.L. (1973) in Kininogenases (Haberland, G.L. & Rohen, J.W., eds.) pp. 75 - 80, Schattauer Verlag, Stuttgart.
39 Rixon, R. & Whitfield, J. (1973) in Kininogenases (Haberland, G.L. & Rohen, J.W., eds.) pp. 131 - 145, Schattauer Verlag, Stuttgart.
40 Rohen, J.W. & Peterhoff, I. (1973) in Kininogenases (Haberland, G.L. & Rohen, J.W., eds.) pp. 147 - 157, Schattauer Verlag, Stuttgart.
41 Geiger, R., Feifel, G. & Haberland, G.L. (1977) *Hoppe-Seyler's Z. Physiol. Chem.* **358**, 931 - 933.
42 Fink, E., Seifert, J. & Güttel, C. (1978) *Fresenius' Z. Anal. Chem.*, in press.
43 Moriwaki, C., Moriya, H., Yamaguchi, Y., Kizuki, K. & Fujimori, H. (1973) in Kininogenases (Haberland, G.L. & Rohen, J.W., eds.) pp. 57 - 66, Schattauer Verlag, Stuttgart.
44 Mann, K., Geiger, R. & Werle, E. (1976) in Kinins (Sicuteri, F., Back, N. & Haberland, G.L., eds.) *Adv. Exp. Med. Biol.* **70**, 65 - 73, Plenum Press, New York.
45 Geiger, R., Heidrich, H. & Mann, K. (1977) *Kidney Int.* **11**, 214.
46 Frey, E., Kraut, H. & Werle, E. (1950) Kallikrein – Padutin, Ferdinand Enke Verlag, Stuttgart.
47 Ward, P., Gedney, C., Dowben, R. & Erdös, E. (1975) *Biochem. J.* **151**, 755 - 758.

Dr. T. Dietl, J. Kruck, Prof. Dr. H. Fritz, Abteilung für Klinische Chemie und Klinische Biochemie in der Chirurgischen Klinik der Universität München,
Nußbaumstraße 20, D-8000 München 2.

Hoppe-Seyler's Z. Physiol. Chem.
Bd. 359, S. 507 - 515, April 1978

A New Method for the Preparation of Band 3, the Main Integral Protein of the Human Erythrocyte Membrane

Dieter SCHUBERT and Brigitte DOMNING

Max-Planck-Institut für Biophysik, Frankfurt a. M.

(Received 21 November 1977)

Summary: Band-3 protein from human erythrocyte membranes was isolated, without using detergents, by a two-step procedure:

(1) The peripheral proteins were removed from the membrane by treatment with 10% acetic acid.
(2) The remaining lipoprotein complex was solubilized in approximately 92% (v/v) acetic acid and then separated into its components by preparative zonal electrophoresis in a gradient made up of acetic acid, water and sucrose.

Band 3 was recovered from the gradient at a yield of 60 - 70% and a purity of about 95%. Approximately 25 mg of band 3 could be prepared in one run. The protein is soluble in aqueous solutions, even in the absence of organic solvents or detergents. In addition to band 3, the proteins stained by periodic acid/Schiff's reagent (the sialoglycoproteins) are also separated from the other proteins.

Eine neue Methode zur Darstellung von „Band 3", dem häufigsten integralen Protein der menschlichen Erythrozytenmembran

Zusammenfassung: Das „Band 3"-Protein aus menschlichen Erythrozytenmembranen wurde ohne Verwendung von Detergenzien in einem Zwei-Stufen-Verfahren isoliert:

(1) Die peripheren Membranproteine wurden durch Behandlung der Membran mit 10proz. Essigsäure aus ihr herausgelöst.
(2) Der zurückbleibende Lipoprotein-Komplex wurde in ca. 92proz. Essigsäure solubilisiert und dann durch präparative Zonenelektrophorese in einem Gradienten aus Essigsäure, Wasser und Saccharose in seine Komponenten getrennt.

Die Ausbeute an „Band 3"-Protein betrug 60 - 70%, die Reinheit etwa 95%. Pro Lauf wurden etwa 25 mg des Proteins erhalten. Die Präparationsprozedur erlaubt zusätzlich eine Isolierung der mit Perjodsäure/Schiffs-Reagenz anfärbbaren Proteine (Sialoglykoproteine).

Key words: Band-3 protein from erythrocyte membranes, isolation of integral membrane proteins, preparative electrophoresis in organic solvents.

Abbreviation: PAS 1 - 3: bands stained by periodic acid/Schiff's reagent.

"Band 3", a polypeptide (or a class of closely related polypeptides) of molecular weight 100000, is the predominant integral protein of the human erythrocyte membrane and constitutes approximately 25% of the total membrane protein[1,2]. It is very interesting from a structural point of view: it spans the membrane[1,2] (traversing the membrane core probably at least twice[3,4]), and it is the main (or perhaps the only) constituent of the "membrane particles" of the erythrocyte membrane seen in freeze-fracture electronmicrographs[1,5]. In addition, it is highly interesting also from a functional point of view: it is involved in anion transport, the most prominent transport function of the erythrocyte membrane [6-8] and, possibly, in the transport of water[9] and sugar[10,11] (but see ref.[12]). As a consequence, band 3 is now among the most widely studied of all membrane proteins.

Efforts to isolate band-3 protein have been successful only during the last few years. All described methods of purification require the use of detergents, either ionic[8,13-18] or non-ionic[19-21] or both[4,22], for the solubilization of the protein. Many chemical as well as functional studies can be performed with the protein samples thus isolated. For certain studies, however, protein which is entirely free of detergent, but still soluble in aqueous buffers, would be desirable or even necessary. Important examples are studies on the interactions of the protein with lipid monolayers at the air-water interface, on lipid binding below the critical micelle concentration of the lipid, and on the influence of protein-lipid interactions on the electrical properties of planar lipid bilayers.

In earlier publications, we have described a procedure for the preparation of a mixture of the integral proteins from human erythrocyte membranes ("strongly bound" protein fraction)[23,24]. This procedure used 10% acetic acid for the extraction of the peripheral ("loosely bound") membrane proteins and 90% acetic acid for the solubilization of the residual lipoprotein pellet, followed by separation of the lipid by gel filtration on Sephadex LH 20 (in 90% acetic acid) and subsequent removal of the organic solvent by dialysis against water. Using preparative zonal electrophoresis in acetic acid/water/sucrose mixtures instead of gel filtration, we are now able to achieve large scale preparation of detergent-free band-3 protein. A brief description of the new procedure has already been given[25].

Materials and Methods

Protein preparation

1) Preparation of the "strongly bound" proteins[23,24]

25 m*l* of a suspension of human erythrocyte membranes (prepared according to ref.[23], free of hemoglobin and containing approximately 80 mg of protein) were mixed with 5 vol. of 12% (v/v) acetic acid and incubated for 10 min at 0 °C. The sample was then centrifuged for 45 min at 28000 x *g*, and the supernatant containing the peripheral membrane proteins was aspirated. The lipoprotein pellet (approx. 45 mg of protein) was resuspended in 10% (v/v) acetic acid and centrifuged for 180 min at 28000 x *g*. The pellet was then dissolved in glacial acetic acid, and 100% and (if necessary) 92.5% (v/v) acetic acid were added to obtain an acetic acid concentration of 92.0 - 92.5% (v/v) and a final volume of 40 m*l* (or, if possible, 30 m*l*) ("solution A").

2) Preparative electrophoresis in a density gradient

Stock solutions: A, see above; B, 80 g sucrose dissolved (under gentle warming) in 240 g 75% (v/v) acetic acid; C, 87% (v/v) acetic acid; D, 0.1% methylene blue B in glacial acetic acid.

The electrofocussing column LKB 8100-2 (LKB Producter AB, Bromma, Sweden) was filled, from bottom to top, with the following solutions (Fig. 1): (a) 70 m*l* 1M sodium acetate containing 0.6 g sucrose per m*l* (cathode solution); (b) linear gradient made up of 150 m*l* B/55 m*l* B plus 95 m*l* C; (c) linear gradient made up of 18.3 m*l* A plus 7.7 m*l* B plus 10 μl D/21.6 m*l* A plus 4.4 m*l* B plus 10 μl D (or, if the volume of A was 30 m*l*: 13.7 m*l* A plus 5.8 m*l* B plus 10 μl D/16.2 m*l* A plus 3.3 m*l* B plus 10 μl D); (d) 4.0 m*l* B plus 32.5 m*l* C; (e) 20 - 30 m*l* 0.1M H_2SO_4 (anode solution). The column was filled and the density gradients formed by means of a peristaltic pump[26] (model 131900, Desaga (Heidelberg, Germany), silicon tubing). Electrophoresis (with anode at top) was carried out at a voltage of 120 V overnight and at 350 V for an additional 48 h (current: approximately 1.5 mA and 4.8 mA, respectively). The temperature of the column was kept at approximately 5.0 °C. Afterwards, the column was emptied from the bottom, by collecting fractions of 150 drops each. We found it convenient to fill and run two columns in parallel (with the same peristaltic pump and power supply), thus handling twice the amount of protein at one time.

For some preliminary experiments, the smaller column LKB 8100-1 was used. With this column, the maximum amount of protein applied was 10 mg.

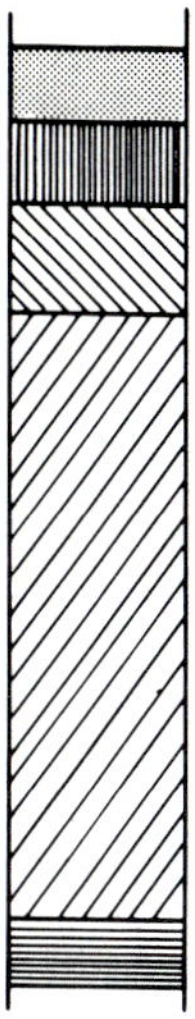

Fig. 1. Schematic diagram of the zonal composition in the electrophoresis column.

3) Subsequent treatment[23]

The fractions from the column containing the band-3 protein were pooled and dialyzed extensively against 0.1mM HCl to remove the organic solvent and sucrose. Afterwards, 0.1M NaOH was added to pH 8 - 10. Transfer into buffer solution of the desired composition was done by dialysis or by addition of concentrated buffer. Protein concentration in these final solutions was up to 1.0 mg/m*l*. Occasionally, the protein was concentrated by isoelectric precipitation (normally in 2mM sodium acetate, pH 5.4).

Chemical analyses

Measurement of protein concentration was done according to the procedure of Lowry et al.[27] using bovine serum albumin as a relative standard (albumin concentration being measured by its absorbance at 280 nm using an extinction $A^{1\%}_{1\,cm} = 6.6$[28]), or from the absorbance difference at 280 nm and 320 nm[29] at alkaline pH or in acetic acid. For the purified band-3 protein, these methods were calibrated in terms of protein weight by measuring the dry weight of protein samples (in 0.1mM NaOH) as described by Kickhöfen and Warth[30]. Sample volume in these determinations was 100 μl, and the measured dry weight was around 400 μg. Phosphorus was determined according to Bartlett[31], and sialic acid content according to Warren[32]. For analysis of samples containing acetic acid or sucrose, the solutions were first dialyzed against water or 0.1% sodium dodecylsulfate. In some cases, the samples (in 0.1% sodium dodecylsulfate) were freeze-dried and then redissolved in a smaller volume of water.

Determinations of cholesterol in samples of band-3 protein were done according to Zak et al.[33] either in isoelectric precipitates (after dissolution in 90% acetic acid) or in extracts prepared from freeze-dried samples with chloroform/methanol (2:1).

Physical methods

Polyacrylamide gel electrophoresis in the presence of 0.5% sodium dodecylsulfate was performed by the method of Fairbanks et al.[34], using Coomassie Brilliant Blue and periodic acid/Schiff's reagent for staining. To prevent incomplete penetration of the protein into the gels, all protein samples were incubated before electrophoresis in 4% sodium dodecylsulfate, 1mM EDTA, 0.5% mercaptoethanol, 10mM glycine/NaOH (pH 12.5) for 30 min at 37 °C, followed by dialysis against 0.5% sodium dodecylsulfate, 1mM EDTA, 10mM Tris/HCl (pH 7.4). Densitometry of the gels was done at 550 nm by means of a Gilford 240 spectrophotometer equipped with scanning attachment.

Analytical ultracentrifugation was performed as described earlier[23]. For measurements of the specific electrical conductivity of acetic acid/H_2O/sucrose mixtures, a Radiometer CDM3 conductivity meter was used. UV absorbance was measured by a Zeiss PMQ II photometer.

Materials

Bovine serum albumin and the reagents for gel electrophoresis were purchased from Serva (Heidelberg, Germany). All other reagents were obtained from Merck (Darmstadt, Germany) and were of analytical grade (if available).

Results

Principles and results of the separation procedure

Extraction of erythrocyte membranes with 10% acetic acid solubilizes most of the peripheral proteins of the membrane[23,24]. It yields a lipoprotein pellet which is highly enriched in band 3, though it also contains the integral proteins stained by periodic acid/Schiff's reagent ("PAS 1 - 3") and traces of other proteins[24] (Fig. 2). The pellet can be solubilized in 90% acetic acid [23,35]. Since, in this solvent, the protein is probably completely disaggregated[35], it seemed useful to employ it not only for the separation of the lipid[23,35] but also for the further purification of band 3. Consequently, we tried to achieve purification by preparative zonal electrophoresis in acetic acid, stabilized by a density gradient.

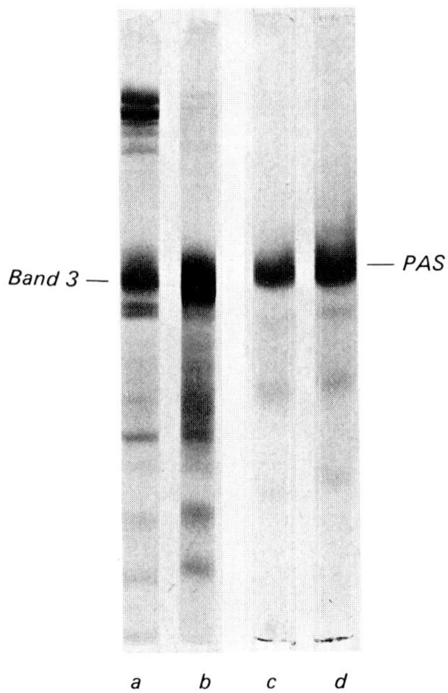

Fig. 2. Sodium dodecylsulfate gel electrophoresis patterns of stroma (a, c) and stroma after extraction by 10% acetic acid (b, d).
Staining was done with Coomassie Brilliant Blue (a, b) and periodic acid/Schiff's reagent (c, d). Amount of protein applied: 30 μg for gels a, b; 300 μg for gels c, d.

However, a difficulty had to be overcome: solutions of sucrose in acetic acid/water mixtures show a decrease in specific electrical conductivity λ with increasing density, whereas, for a good separation during electrophoresis in a density gradient, λ should be independent of density or slightly increase with increasing density*. For similar reasons, the specific conductivity of the sample solution should be smaller than that of the gradient. To meet these conditions, we superimposed on the gradient of sucrose concentration an oppositely directed gradient of acetic acid concentration and added the sample in a solution with acetic acid concentration higher than that in the gradient (see Materials and Methods). The latter step, plus the application of the sample to the column as a sucrose gradient to prevent convection, also allowed the use of large sample volumes: the bands sharpened during the initial phase of electrophoresis to up to one-fourth of the original width; they maintained this reduced width during the subsequent migration (as indicated by the behaviour of the zone of methylene blue applied with the sample).

Distribution of the sample components after electrophoresis

1) Proteins

When, after electrophoresis, the UV absorbance and thus the protein content of the fractions collected from the column are measured, nearly all of the protein is found in a single but broad and non-symmetrical peak which apparently represents the superposition of 2 - 3 bands (Fig. 3). The electrophoretic mobility of this peak is about half that of the tracking dye; under the conditions described it migrates for approximately 170 mm. In addition, a small peak of lower electrophoretic mobility is observed (Fig. 3). Dodecylsulfate gel electrophoresis reveals that this small peak is composed entirely of the proteins stained by periodic acid/Schiff's reagent (Fig. 4). Thus, most or all of these proteins are clearly separated from band-3 protein. The gels show, in addition, that within the main protein peak some separation of the other protein components has also been achieved:

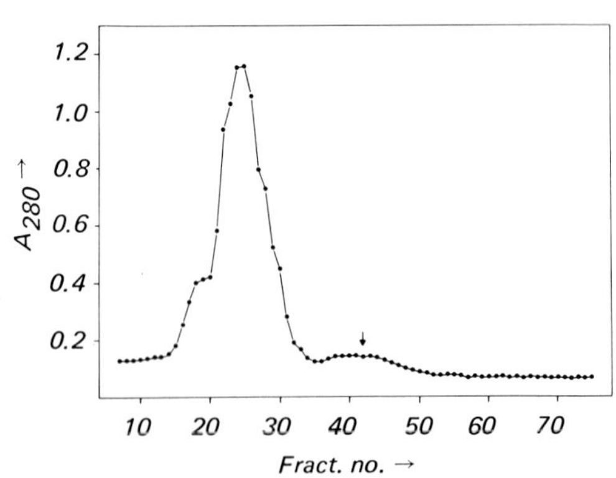

Fig. 3. Absorbance at 280 nm of the fractions (150 drops each) from the electrophoresis column. Original sample position is at fractions 76 - 83. The second (small) protein peak is marked by "↓".

* Otherwise, the leading edge of a protein peak would be in a region of higher electrical field than the trailing edge. This would result in band broadening.

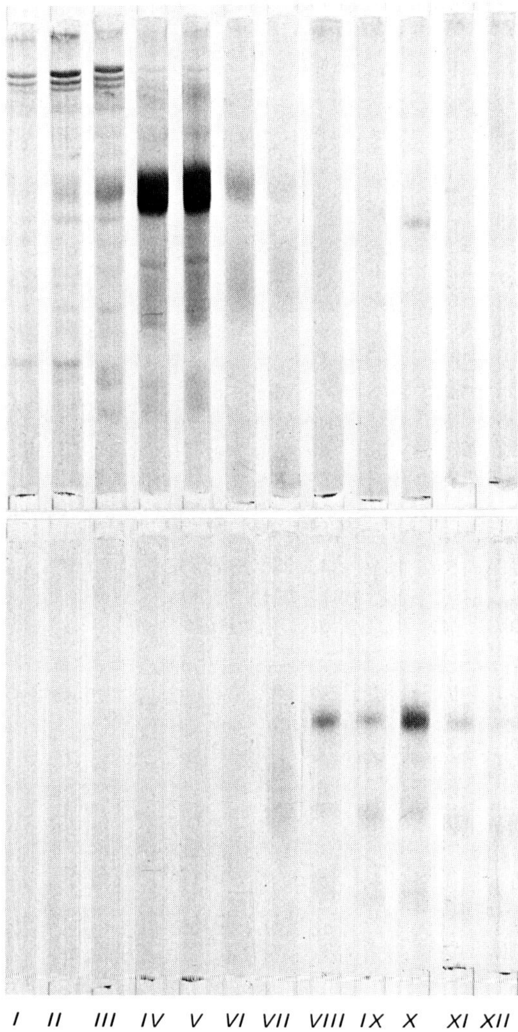

Fig. 4. Sodium dodecylsulfate gel electrophoresis patterns of the fractions from the electrophoresis column corresponding to the regions of the two protein peaks. Fraction size is 3 times that of Fig. 3. Staining: Coomassie Brilliant Blue (upper row), periodic acid/Schiff's reagent (lower row). Sample content: 0.005 of fraction volume for upper row, 0.05 of fraction volume for lower row.

(a) The gels corresponding to the fractions of maximum UV absorbance (fractions 22 - 26 of Fig. 3, fractions IV and V of Fig. 4) are dominated by band-3 protein. Other components are represented only by very weak bands (with respect to these bands, it should be noted that the gels shown in Fig. 4 are heavily overloaded).
(b) The leading part (the shoulder) of the main protein peak (fractions 15 - 21 of Fig. 3 or I - III of Fig. 4) contains, besides some band-3 protein, nearly all of the peripheral proteins left after extraction of the membranes with 10% acetic acid.
(c) Most of the protein in the trailing part of the main peak (fractions 27 - 33 of Fig. 3 or VI and VII of Fig. 4) is band-3 protein. A significant amount is, however, found in a broad band corresponding to molecular weights around 55 000.

Apparently, the fractions from the center of the main protein peak (corresponding to fractions 22 - 26 of Fig. 3) yield a preparation of band 3 which, for many purposes, is of sufficient purity. These fractions normally contain 50 - 55% of the total UV absorbance in the two protein peaks or 25 mg of protein, equivalent to about 60 - 70% of the amount of band-3 protein in the starting material. In addition, the fractions containing the small protein peak represent a preparation consisting entirely of the sialoglycoproteins (PAS 1 - 3). However, since a number of isolation methods exist for these proteins[1,2], including methods which avoid the use of detergents, we did not pursue this aspect.

2) Lipids

In contrast to the proteins, most of the phospholipids and the cholesterol do not enter the gradient during electrophoresis and are finally found at the origin or near the anode. Only a few percent of the phosphorus applied to the column migrates with the main protein peak and practically none with the second peak. Thus, the procedure described is an effective method for the separation of protein and lipid, at least as effective as gel filtration on Sephadex LH 20[23,35].

Properties of the band-3 preparations

1) Protein impurities

To determine the content of protein impurities in the band-3 preparations (the combined fractions described above), densitometry of dodecylsulfate gels stained with Coomassie Brilliant Blue was performed. The scans obtained show one single peak, with no clearly visible other bands (Fig. 5).

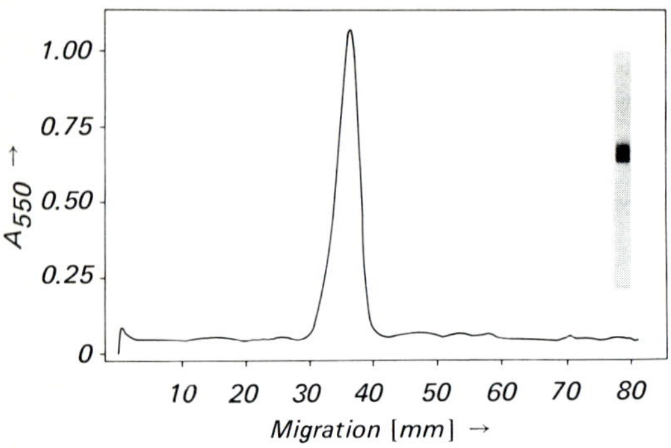

Fig. 5. Dodecylsulfate gel electrophoresis of the combined fractions from the central region of the main protein peak (fractions 22 - 26 of Fig. 3).

Photograph and densitometric scan of a gel after staining with Coomassie Brilliant Blue. Amount of protein applied: 30 μg.

However, a thorough inspection reveals small deviations of the measured absorbance from the baseline, the area under these parts of the curve amounting to approx. 5% of the area under the main peak. More than two-thirds of the additional material is found in a broad, diffuse peak around a molecular weight of 55000 (electrophoretic migration in the gel of Fig. 5: 45 - 60 mm), as does part of the protein in the trailing edge of the main protein peak of Fig. 3. It could possibly be correlated to the region 4.5, a diffuse band of similar molecular weight in dodecylsulfate gels of human erythrocyte membranes[1]. However, since its peak height increases with increasing time of incubation of the protein in high concentrations of acetic acid, and also with increasing time of incubation at pH 12.5 during the treatment of the samples before dodecylsulfate gel electrophoresis*, it seems much more probable that the band, or at least most of it, is a degradation product of band 3.

Since periodic acid/Schiff's reagent staining is of low sensitivity, the content of PAS 1 - 3 in the band-3 preparations was assayed by analytical determination of sialic acid. It was found that the preparations contained, per mol of band 3, 1.7 ± 0.2 mol of sialic acid (n = 10). This compares well with the values of 1 - 2 mol reported for other preparations of band 3[8,15]. Since the main component of the proteins stained by periodic acid/Schiff's reagent, "glycophorin A" (M_r 31000[36], bears approximately 25 mol of sialic acid per polypeptide chain[37-39], this could mean that the band-3 preparations contain about 2% (w/w) glycophorin A. It seems, however, not unlikely that most or all of the carbohydrate retained is bound to band 3. Interestingly, the ratio sialic acid/protein increases towards the trailing edge of the main protein peak. This could indicate that band 3 is heterogeneous with respect to its content of sialic acid, as was already observed with respect to lectin binding[40] and to the susceptibility to labeling of sugars[20].

In summary, the combined fractions from the central region of the main protein peak contain approximately two-thirds of the band-3 protein applied to the column, at a purity of approx. 95%. The only significant protein contaminant is probably a degradation product of band 3. Thus, the purity of the isolated material will be sufficient for many kinds of studies, especially of the physico-chemical type. Moreover, purity as well as yield of the preparation compare well with those of the other preparations of band 3 described[4,8,13-22].

2) *Lipid content*

As described above, the bulk of the phospholipid applied to the electrophoresis column is well separated from the fractions containing band-3 protein. However, the purified protein still contained 5.1 ± 0.4 mol of phosphorus per mol of protein, as determined from 10 different preparations. This value is not changed by repeated electrophoresis or by gel filtration on Sephadex LH 20 (in 90% acetic acid). When the protein (dissolved at pH 10) was lyophilized and then extracted with neutral or acidic mixtures of chloroform and methanol (2:1), only about 20% of the phosphorus could be solubilized. Attempts for complete extraction of the phosphorus or for the

* After sample treatment at neutral pH, a large part of the protein often did not enter the gel during dodecylsulfate electrophoresis. This effect was greatly reduced by alkaline preincubation of the samples. The percentage of protein depolymerized followed a titration curve with an apparent pK of approx. 12. The titration of a basic group or, more probably, a buried tyrosine residue thus seems to play an important role in the depolymerization process.

identification of its nature have not yet been successful. However, it still seems possible that most of the phosphorus represents tightly bound phospholipid.

Determination of cholesterol in samples of isolated band-3 protein by the method of Zak et al.[33] yielded absorbance values at 560 nm corresponding to a cholesterol content of approximately 1.4 mol per mol of protein, whereas 0.3 mol per mol of protein were found in chloroform:methanol (2:1) extracts of the lyophilized protein. In both cases, however, the color developed by the samples was yellow instead of purple. Since similar results were observed with pure, cholesterol-free proteins (e.g. trypsin, chymotrypsin and aldolase), we conclude that the isolated band-3 protein is practically free of cholesterol.

3) Measures of protein concentration

For the determination of absolute protein concentration by the measurement of absorbance in the near ultraviolet and by the assay of Lowry et al.[27], these methods were calibrated against protein dry weight. The following values were found for the extinction coefficient of the protein (mean and standard error obtained with three different protein samples): $\Delta A = A^{1\%}_{1\,cm}$ (280 nm) – $A^{1\%}_{1\,cm}$ (320 nm) = 8.85 ± 0.20 for protein dissolved in 0.1mM NaOH, ΔA = 8.55 ± 0.30 for protein in 0.1mM NaOH plus 1% sodium dodecylsulfate, ΔA = 9.00 ± 0.20 for protein in 80% acetic acid. In the Lowry assay, 1 mg of band-3 protein showed the same absorbance at 750 nm as (0.450 ± 0.025) mg of bovine serum albumin*.

4) Solubility. Binding to lipid layers

After removal of the acetic acid by dialysis against 0.1mM HCl and titration to pH 8 - 10, band-3 protein forms clear or slightly opalescent solutions from which it can only be removed by prolonged ultracentrifugation. However, under these conditions the protein is not monomeric (though monomeric protein can be isolated from the samples according to ref.[41]) but shows a complex aggregational behavior. Addition of salt favours aggregation, as does increasing the protein concentration. Precipitation of the protein is, however, only observed at pH values between 4 and 6, the exact width of this range of isoelectric precipitation depending on protein concentration and ionic strength.

Band-3 protein, as isolated by the method described above, readily binds to spherical and planar lipid bilayers, as does the material from which it had been isolated, the "strongly bound" protein fraction[29,42]. A detailed description of these interactions will be given elsewhere. The protein also binds to lipid monolayers[25,43].

* In these data, the protein weight includes all molecules possibly bound to the protein, especially the phosphorus-containing compound of unknown nature mentioned above. The values for the uncomplexed protein will differ probably only by a few percent.

Discussion

The classical techniques of protein preparation, based on the use of normal aqueous buffers, have turned out to be not directly applicable to the isolation of integral membrane proteins. Various attempts have been made to overcome this difficulty. Undoubtedly, the most successful and most widely applicable of them is the utilization of detergents, both for solubilizing the proteins and for keeping them in solution. This paper describes another attempt which is based on the use of an organic acid for protein solubilization[35] but adds a technique not applied before: zonal electrophoresis in the organic solvent, stabilized by a density gradient. This method may be applicable to other membrane proteins besides the band-3 protein from erythrocyte membranes. Probably it could be used not only with acetic acid, but also with other organic solvents, including aprotic solvents. The main advantage of our method is that the protein isolated is free of detergents but nevertheless soluble in aqueous buffers. Thus, the protein could be suitable for physico-chemical studies which require the absence of detergents.

The most crucial question in deciding on the usefulness of the preparation method described is whether or not the protein is still in its native state. In this respect, the protein may be inferior to proteins isolated by the use of non-ionic detergents (Triton X-100, Ammonyx, etc.). However, several arguments suggest that the protein obtained by our method is, at least, not grossly denatured. Some of these arguments have already been mentioned[23]. Others can be derived from the following properties of the band-3 preparations:

(1) In aqueous solutions at low protein concentrations, the protein shows dimer-tetramer and, probably, monomer-dimer association equilibria (H.-J. Dorst and D. Schubert, in preparation).
(2) When inserted into black phosphatidylserine membranes, the protein shows association equilibria involving at least monomers and tetramers[42].
(3) In experiments on the interactions of the protein with lipid monolayers, a very high affinity to cholesterol, a main lipid component of the erythrocyte membrane, is observed[25,43].
(4) The interaction of the protein with cholesterol is stronger than or at least as strong as with similar sterols which are not components of the erythrocyte membrane. Sterols differing from cholesterol with respect to the polar group or showing larger differences with respect to the aliphatic chain have a considerably lower affinity to the protein than cholesterol (E. Klappauf and D. Schubert, in preparation).

The described properties seem to be much too specific to be artifacts due to denaturation; some of them correspond to known or supposed properties of the band-3 protein in the intact erythrocyte membrane[1,44]. A fully satisfactory answer to the question of the intactness of the protein isolated by the described method will, however, not be possible before a reconstitution in lipid bilayers of the anion transport system of the erythrocyte membrane has been achieved.

We thank *Prof. H. Passow* for stimulating discussions and *Mr. R. Warth* for the dry weight determinations. This work was supported by the *Deutsche Forschungsgemeinschaft* (Fa 48/18 (D) and Fa 48/21 (I)).

Literature

1 Steck, T.L. (1974) *J. Cell Biol.* **62**, 1 - 19.
2 Marchesi, V.T. & Furthmayr, H. (1976) *Annu. Rev. Biochem.* **45**, 667 - 698.
3 Jenkins, R.E. & Tanner, M.J.A. (1975) *Biochem. J.* **147**, 393 - 399.
4 Drickamer, L.K. (1976) *J. Biol. Chem.* **251**, 5115 - 5123.
5 Bächi, T., Whiting, K., Tanner, M.J.A., Metaxas, M.N. & Anstee, D.J. (1977) *Biochem. Biophys. Acta* **464**, 635 - 639.
6 Cabantchik, Z.I. & Rothstein, A. (1974) *J. Membr. Biol.* **15**, 207 - 226.
7 Zaki, L., Fasold, H., Schuhmann, B. & Passow, H. (1975) *J. Cell Physiol.* **86**, 471 - 494.
8 Ho, M.K. & Guidotti, G. (1975) *J. Biol. Chem.* **250**, 675 - 683.
9 Brown, P.A., Feinstein, M.B. & Sha'afi, R.I. (1975) *Nature (London)* **254**, 523 - 525.
10 Lin, S. & Spudich, J.A. (1974) *Biochem. Biophys. Res. Commun.* **61**, 1471 - 1476.
11 Kahlenberg, A. (1975) *J. Biol. Chem.* **251**, 1582 - 1590.
12 Kasahara, M. & Hinkle, P.C. (1977) in Biochemistry of Membrane Transport (Semenza, G. & Carafoli, E., eds.), pp. 346 - 350, Springer Verlag, Berlin.
13 Tanner, M.J.A. & Boxer, D.H. (1972) *Biochem. J.* **129**, 333 - 347.
14 Tanner, M.J.A., Jenkins, R.E., Anstee, D.J. & Clamp, J.R. (1976) *Biochem. J.* **155**, 701 - 703.
15 Furthmayr, H., Kahane, J. & Marchesi, V. (1976) *J. Membr. Biol.* **26**, 173 - 187.
16 Barratt, D.G., Sharom, F.J., Thede, A.E. & Grant, C.W.M. (1977) *Biochim. Biophys. Acta* **465**, 191 - 197.
17 Ross, A.H. & McConnell, H.M. (1977) *Biochem. Biophys. Res. Commun.* **74**, 1318 - 1325.
18 Schiechl, H. (1977) *Z. Naturforsch.* **32c**, 67 - 71.
19 Findlay, J.B. (1974) *J. Biol. Chem.* **249**, 4398 - 4403.
20 Yu, J. & Steck, T.L. (1975) *J. Biol. Chem.* **250**, 9170 - 9175.
21 Kahlenberg, A. (1976) *Anal. Biochem.* **74**, 337 - 342.
22 Bhakdi, S., Bjerrum, O.J. & Knüfermann, H. (1976) *Biochim. Biophys. Acta* **446**, 419 - 431.
23 Schubert, D., Poensgen, J. & Werner, G. (1972) *Hoppe-Seyler's Z. Physiol. Chem.* **353**, 1034 - 1042.
24 Schubert, D. (1973) *Hoppe-Seyler's Z. Physiol. Chem.* **354**, 781 - 790.
25 Schubert, D. & Klappauf, E. (1977) 11th FEBS Meeting, Copenhagen, Abstract A4-13-663.
26 Wallach, D.F.H. (1970) *Anal. Biochem.* **37**, 138 - 141.
27 Lowry, O.H., Rosebrough, N.J., Farr, A.L. & Randall, R.J. (1951) *J. Biol. Chem.* **193**, 265 - 275.
28 Kirchenbaum, D.M. (1970) in Handbook of Biochemistry, Selected Data for Molecular Biology (Sober, H., ed.), 2nd ed., pp. C-71 - C-98, Chemical Rubber Company, Cleveland, Ohio.
29 Lossen, O., Brennecke, R. & Schubert, D. (1973) *Biochim. Biophys. Acta* **330**, 132 - 140.
30 Kickhöfen, B. & Warth, R. (1976) *Hoppe-Seyler's Z. Physiol. Chem.* **357**, 745 - 749.
31 Bartlett, G.R. (1959) *J. Biol. Chem.* **234**, 466 - 468.
32 Warren, L. (1959) *J. Biol. Chem.* **234**, 1971 - 1975.
33 Zak, B., Dickenman, R.C., White, E.G., Burnett, H. & Cherney, P.J. (1954) *Am. J. Clin. Pathol.* **24**, 1307 - 1315.
34 Fairbanks, G., Steck, T.L. & Wallach, D.F.H. (1971) *Biochemistry* **10**, 2606 - 2617.
35 Zahler, P.H., Wallach, D.F.H. & Lüscher, E.F. (1967) *Protides Biol. Fluids, Proc. Colloq. (Bruges)* **15**, 69 - 78.
36 Marchesi, V.T. & Furthmayr, H. (1976) *Annu. Rev. Biochem.* **45**, 667 - 698.

37 Grefrath, S.P. & Reynolds, J.A. (1974) *Proc. Natl. Acad. Sci. U.S.A.* 71, 3913 - 3916.
38 Kahane, J., Furthmayr, H. & Marchesi, V.T. (1976) *Biochim. Biophys. Acta* **426**, 464 - 476.
39 Liljas, L., Lundahl, P. & Hjertén, S. (1976) *Biochim. Biophys. Acta* **426**, 526 - 534.
40 Tanner, M.J.A. & Anstee, D.J. (1976) *Biochim. J.* **153**, 265 - 270.
41 Schubert, D., Bleuel, H., Domning, B. & Wiedner, G. (1977) *FEBS Lett.* **74**, 47 - 49.
42 Bleuel, H., Wiedner, G. & Schubert, D. (1977) *Z. Naturforsch.* **32c**, 375 - 378.
43 Klappauf, E. & Schubert, D. (1977) *FEBS Lett.* **80**, 423 - 425.
44 Margaritis, L.H., Elgsaeter, A. & Branton, D. (1977) *J. Cell. Biol.* **72**, 47 - 56.

Prof. Dr. D. Schubert and B. Domning, Max-Planck-Institut für Biophysik, Abt. Zellphysiologie, Heinrich-Hoffmann-Str. 7, D-6000 Frankfurt/Main 71.

Hoppe-Seyler's Z. Physiol. Chem.
Bd. 359, S. 517 - 528, April 1978

Studies on the Involvement of Lysosomes in Estrogen Action, I

Isolation and Enzymatic Properties of Pig Endometrial Lysosomes

Walter SIERRALTA, Anne J. TRUITT and Peter W. JUNGBLUT

Max-Planck-Institut für Zellbiologie, Wilhelmshaven

(Received 9 January 1978)

Herrn Professor Dr. Adolf Butenandt zum 75. Geburtstag gewidmet.

Summary: Pig endometrium cells, collected by curettage and homogenized in an all-glass Potter Elvehjem homogenizer, gave a considerably higher yield of intact mitochondria and lysosomes than homogenates of whole uterus obtained with the Ultraturrax or the Parr bomb.

After homogenization of the cells and subfractionation in the presence of $Mg^{2\oplus}$, mitochondria and lysosomes equilibrated at the same modal density in isopycnic centrifugation. Homogenization and subfractionation in buffers devoid of divalent cations and containing EDTA resulted in a decrease in the buoyant density of mitochondria, allowing for a separation from lysosomes.

The pH optima and the specific activities of two mitochondrial enzymes and eight hydrolyases used as marker enzymes were determined. The morphological characteristics of fractions were established by electron microscopy.

Preliminary results indicate an involvement of lysosomes in steroid metabolism rather than in steroid and receptor translocation into the nucleus.

Untersuchungen über die Beteiligung von Lysosomen an der Östrogenwirkung, I. Isolierung und enzymatische Eigenschaften von Lysosomen aus Schweineendometrium

Zusammenfassung: Zur Prüfung der Frage, ob und in welcher Weise Lysosomen am Wirkungsmechanismus von Östrogenen beteiligt sind, wurde ein Verfahren zur Isolierung der Organellen aus dem Erfolgsorgan Uterus ausgearbeitet. Gesamtuterus erwies sich auch bei vorsichtiger Homogenisation mit dem Ultraturrax und der Parr-Bombe als ungeeignetes Ausgangsmaterial. Mit Zirkularküretten

---

Enzymes:
Acid phosphatase, orthophosphoric-monoester phosphohydrolase (acid optimum) (EC 3.1.3.2);
Arylsulfatase, aryl-sulfate sulfohydrolase (EC 3.1.6.1);
Cathepsin D (EC 3.4.23.5);
Cytochrome *c* oxidase, ferrocytochrome *c*: oxygen oxidoreductase (EC 1.9.3.1);
β-Galactosidase, β-D-galactoside galactohydrolase (EC 3.2.1.23);
α-Glucosidase, α-D-glucoside glucohydrolase (EC 3.2.1.20);
β-Glucosidase, β-D-glucoside glucohydrolase (EC 3.2.1.21);
β-Glucuronidase, β-D-glucuronide glucuronosohydrolase (EC 3.2.1.31);
Isocitrate dehydrogenase ($NADP^{\oplus}$), *threo*-D_s-isocitrate: $NADP^{\oplus}$ oxidoreductase (decarboxylating) (EC 1.1.1.42);
Ribonuclease I, ribonucleate 3′-pyrimidino-oligonucleotidohydrolase (EC 3.1.4.22).

abgelöste und in Potter-Elvehjem Homogenisatoren aufgeschlossene Endometriumzellen ergaben dagegen Homogenate mit intakten Organellen in guter Ausbeute.

Intrauterine Instillation von Triton WR 1399 – in vivo – verringerte die Schwebedichte von Lysosomen im Vergleich zu der von Mitochondrien nicht.

Die Trennung der Partikeln in Saccharosedichtegradienten gelang nach – in vitro – Verringerung der Schwebedichte von Mitochondrien in EDTA-haltigen Puffern. Dabei erfolgte eine partielle Ablösung der inneren von der äußeren Mitochondrienmembran, die mit einer Kondensierung von Cristae und Matrix einherging. Die veränderten Mitochondrien äquilibrierten sich als scharfer Gipfel im weniger dichten Schenkel des breiten Lysosomengipfels. Ausgangsmaterial und Fraktionen wurden enzymatisch und morphologisch charakterisiert.

Die Ergebnisse von ersten Versuchen mit isolierten Lysosomen sprechen für eine Beteiligung am Östradiolstoffwechsel und gegen eine Beteiligung an der Steroid-Rezeptor-Translokation in den Zellkern.

Key words: Lysosomes, pig endometrium, estrogen action.

The binding of steroids to "receptors" is known to be essential for their hormonal activity. In the case of estradiol, the sequence of events comprises the formation of a hormone-receptor complex and its dimerization to $\genfrac{}{}{0pt}{}{\text{ɹoʇdǝɔǝɹ}}{\text{receptor}}$ in the cytoplasm, followed by a fast translocation of the dimer into the nucleus, where it enhances the rate of transcription, including that of the receptor message[1].

Although the emphasis placed on the importance of the steroid in the basic mechanism has been somewhat lessened by the discovery that receptors can act as transcription-regulators even in the absence of the hormone[2], (cytoplasmic) receptors are still widely considered as steroid "transporters". This concept, seemingly supported by the non-covalent nature of the steroid-receptor-complex and the reversibility of the steroid attachment in vitro, may have helped to sustain the idea of a receptor shuttle, implying a repeated use. Reviewing the classical experiments of Jensen and Jacobson[3] with Sprague-Dawley rats, we favored instead a directional flux – into and out of the nucleus – as a more probable explanation for the physiological events[4] and later, indeed, found no evidence for receptor recycling[1]. If the receptor, on leaving the nucleus (together with the steroid) does not return to a cytoplasmic pool, its degradation is likely to occur in lysosomes. An investigation of the possible involvement of lysosomes requires their isolation in reasonable purity and yield.

Materials and Methods

Uteri of normal, prepuberal pigs of the German Landrace breed were obtained from the local slaughterhouse; uteri of normal and of ovariectomized pigs of the same breed from the Federal Institute for Animal Husbandry and Animal Behaviour at Mariensee.

For detergent treatment, Triton WR 1339 (1 - 2%) was dissolved either in saline or in diluted (20%) pig serum and injected into the uterine lumen of unanesthetized, trained pigs via the cervix.

The uteri were removed immediately after slaughter, sealed in polyethylene bags, chilled in ice and processed within 1/2 to 3 h.

Homogenization

All operations took place at 0 - 2 °C. Whole organs were washed with saline and passed through a chilled meat grinder. For Ultraturrax homogenization (TP 18, Janke & Kunkel), 5 vol. of 0.25M sucrose buffered with 0.05M Tris/HCl (pH 7.3) were added to the mince. The working voltage was 70 V (max. 220 V) and 10 bursts of 10 s each with intermittant pauses of 30 s were applied.

For homogenization in the Parr bomb (Parr Instruments, Moline, Illinois, U.S.A.), the mince was suspended in 10 vol. of 0.25M buffered sucrose and 300-m*l* portions were placed in a glass container. The suspension was magnetically stirred while being exposed to 900 psi of nitrogen for 30 min at 1 °C and then slowly released within 45 min.

Endometrium

Uterine horns were rinsed with ice-cold saline and cut into 4 - 5 cm long pieces. A rod-shaped curette (Fig. 1) was forced 2 - 3 times through each piece and the detached endometrium collected in a glass beaker containing 0.25M sucrose, 0.1% ethanol, 0.05M Tris/HCl (pH 7.3) with either 5mM $MgCl_2$ or 1mM EDTA. The curette was moistened with buffer before each scraping. The final ratio of buffer to tissue was adjusted to 10:1. The suspension was homogenized with 5 up and down strokes in a tightly fitting, motor-driven, all-glass Potter-Elvehjem homogenizer.

Processing of homogenates

All homogenates were first spun for 10 min at 600 x *g* in a Christ Zeta 20 at 2 °C. The sediments from the Ultraturrax and the Potter homogenates were twice resuspended in buffer, rehomogenized and respun. The combined supernatants were filtered through nylon gauze (Ultraturrax and Parr bomb homogenate: 300 μm followed by 100 μm; Potter homogenate 20 μm), and then centrifuged for 30 min at 16000 rpm and 1 °C in an SW 27 rotor, Beckman L-2 65B centrifuge. The sediments were washed once by resuspension in 10 vol. of buffered sucrose and recentrifuged. The final sediments (mitochondrial/lysosomal fraction, ML) were resuspended-ed in 10 vol. of buffered sucrose by homogenization in a loosely fitting teflon/glass Potter Elvehjem homogenizer with 2 strokes at 500 rpm and then passed through 5 μm nylon gauze.

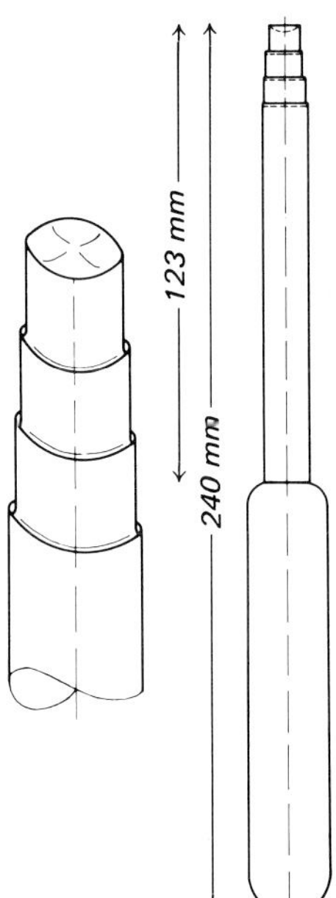

Fig. 1. Stainless steel curette.

Two curettes were used with maximum diameters of 10 and 12 mm. Diameter reduction to the tip was 1 mm/step. Curettes were selected to distend the uterine segments slightly and cell layers were detached by the scraping action of the edges.

Density gradient centrifugation

1.5 m*l* of the mitochondrial/lysosomal suspensions were layered on 11 m*l* sucrose gradients, containing 0.1% ethanol, of either linear (23 - 52% w/w) or convex exponential (25 - 40% w/w) shape in SW 40 nitrocellulose tubes. The gradient maker used was of the Britten and Roberts[5] type. For the preparation of the exponential gradient, the volume in the mixing chamber was kept constant by closing it with a silicone rubber stopper. For theoretical and methodological details see ref.[6].

The gradients were run for 150 - 180 min at 22 500 rpm and 1 °C in a SW 40 rotor, after which time equilibrium was obtained. (No difference in distribution was seen after 12 h at 12 500 rpm, 1 °C, SW 40.)

After centrifugation, the tubes were punctured at the bottom and constant volume fractions (0.42 m*l*) collected by displacement with light mineral oil. The refractive index of each fraction was measured in an Abbé refractometer (Zeiss) at 20 °C and the density computed.

Analysis of enzymatic activities

All substrates and chemicals used were of analytical grade and purchased from Sigma Chemie GmbH, München; Serva, Heidelberg; Boehringer Mannheim GmbH, and Merck, Darmstadt. The conditions of incubation and the analytical procedures are listed in Tables 1 and 2. The linearity of the analyses was previously checked. The pH optima of enzymes in mitochondrial/lysosomal suspensions were measured in citrate (pH 2.7 - 3.8), acetate (pH 3.7 - 5.7) and phosphate buffers (pH 6.7 - 7.5). All incubations contained 0.1% Triton X-100. Suitable blanks were included in each determination. Enzymatic activities are expressed as nmol of substrate converted per min (mU) with the exception of cytochrome oxidase, for which the Cooperstein and Lazarow unit[7] was used, and ribonuclease activity (1 unit = $1.0 A_{260}$/min). For comparing the results from various isopycnic centrifugations, histograms were constructed according to Bowers and De Duve[8], but with density steps of 0.015. Protein was determined with Campbell and Sargent's[9] modification of the Lowry method[10] using bovine serum albumin as standard.

Electron microscopy

Scrapings: samples were washed in 0.5M sodium cacolylate buffer and fixed for 1 h in 2% glutaraldehyde in the same buffer. After washing in buffer they were post-fixed in 1% OsO_4 for 2 h. Dehydration was carried out in a graded acetone series, and the samples were embedded in an epon-araldite mixture[11]. Thin sections were cut with a Reichert Om V3 ultramicrotome, stained

Table 1. Incubation conditions for enzyme assays.
Sample volume: 50 or 100 μl.
Incubation temperatures: cytochrome *c* oxidase 25 °C; isocitrate dehydrogenase 30 °C; others 37 °C.
0.1% (w/v) Triton X-100 present in all incubation mixtures.

Enzyme	Buffer	pH	Substrate	Final vol. [m*l*]	Time [min]
Acid phosphatase	90mM acetate	4.8	12.5mM β-glycerophosphate	1.0	60
Arylsulfatase	90mM acetate	4.5	10mM nitrocatechol sulfate	1.0	30
Cathepsin D	100mM acetate	3.9	1% hemoglobin	2.0	90
Cytochrome *c* oxidase	100mM acetate	5.4	34μM reduced cytochrome *c*	3.0	2 - 5
β-Galactosidase	90mM acetate	3.9	8mM *o*-nitrophenyl-β-galactoside	1.0	30
α-Glucosidase	90mM acetate	3.9	10mM *p*-nitrophenyl-α-glucoside	1.0	30
β-Glucosidase	80mM acetate	3.9	10mM *p*-nitrophenyl-β-glucoside	1.0	30
β-Glucuronidase	90mM acetate	4.1	10mM *p*-nitrophenyl-β-glucuronide	1.0	60
Isocitrate dehydrogenase	35mM phosphate/ 3.3mM $MgCl_2$	7.4	7mM-DL-isocitrate, 0.3mM NADP	3.0	2 - 5
Ribonuclease	33mM phosphate	6.5	3 mg RNA/m*l*	1.0	60

with uranyl acetate and lead citrate[12] and viewed with a Philips 301 electron microscope.
Fractions from density gradients: samples were diluted 1 : 1 with 2% glutaraldehyde in 10mM phosphate buffer containing 0.25M sucrose and after 15 min were centrifuged for 15 min at 13 000 rpm in a Hettich Microrapid and the supernatants discarded. Fresh fixative was added to the pellets and they were fixed for 1 h, washed in buffer and postfixed for 2 h in 1% OsO_4. Dehydration was accomplished in a graded ethanol series, and embedding in Spurr's low viscosity epoxy resin[13]. Sections were cut, stained and viewed as above.

Table 2. Analytical procedures for enzyme assays.

Enzyme	Manipulations	Analysis
Acid phosphatase	+ 2.5 m*l* 12% trichloroacetic acid, filtered	P_i, Fiske and Subbarow[14]
Arylsulfatase	+ 1 m*l* 0.4% hydroquinone 0.2M Na_2SO_3 in 2.5M NaOH, stirred, + 1 m*l* 0.05M deoxycholate	Free nitrocatechol at 510 nm
Cathepsin D	+ 4 m*l* 5% trichloroacetic acid 30 min in cold, filtered	Tyrosine after addition of 0.35M NaOH/ 2% Na_2CO_3 and Folin-Ciocalteu's reagent, spectrophotometry at 750 nm
Cytochrome *c* oxidase		Oxidation of cytochrome *c* at 550 nm
β-Galactosidase	+ 2 m*l* 0.2M Tris/0.3M dodecylsulfate (pH 10.5)	Free *o*-nitrophenol at 405 nm
α-Glucosidase	+ 2 m*l* 0.2M Tris/0.3M dodecylsulfate (pH 10.5)	Free *p*-nitrophenol at 405 nm
β-Glucosidase	+ 2 m*l* 0.2M Tris/0.3M dodecylsulfate (pH 10.5)	Free *p*-nitrophenol at 405 nm
β-Glucuronidase	+ 2 m*l* 0.2M Tris/0.3M dodecylsulfate (pH 10.5)	Free *p*-nitrophenol at 405 nm
Isocitrate dehydrogenase		Reduction of $NADP^{\oplus}$ at 340 nm
Ribonuclease	+ 0.5 m*l* 0.75% uranyl acetate in 25% perchloric acid, filtered after 30 min in cold, diluted 6 ×	Absorbance at 260 nm

Table 3. Distribution (%) of lysosomal enzymes after different homogenization procedures.
It was assumed that the total initial activity (100%) was equal to the sum of the total activities present in each of the three fractions analyzed. ML = mitochondrial/lysosomal.

	Whole uteri				Endometrium	
	Ultraturrax		Parr bomb		All-glass Potter-Elvehjem	
	Cathepsin D	Phosphatase	Cathepsin D	Phosphatase	Cathepsin D	Phosphatase
Low speed sediment	48.6	41.9	42.0	34.5	25.7	16.5
ML fraction	4.3	5.1	8.5	8.5	30.4	36.9
Post-ML supernatant	47.1	53.0	49.5	57.0	43.9	46.6

Results

1) The effects of homogenization procedures on the distribution of lysosomal enzymes in subcellular fractions of the uterus

The quantitative isolation of cellular organelles from a tissue in an intact state requires the disruption of all cells by a thorough homogenization which leaves subcellular structures intact. This is difficult to achieve with an organ like uterus, which is composed of muscle, a layer of connective tissue lined with epithelial cells and interspersed with tubular glands, and additionally contains considerable amounts of extracellular collagen, glycosaminoglycans and proteoglycans.

The most frequently used procedure is the homogenization of minced uterus with instruments of the Ultraturrax (Janke & Kunkel) type. These instruments homogenize by a combination of shearing and cutting action, the ratio of which depends on the speed of rotation of the two inner "teeth" and their distance from the outer, serrated stainless steel tube. In order to reduce the cutting action, we applied 6 000 of the maximal 20 000 rpm, but with unsatisfactory results (Table 3). Very little cathepsin D or acid phosphatase activity was found in the mitochondrial/lysosomal fraction after differential centrifugation. Most of the enzymatic activity of the homogenate, in about equal parts, resided in the low-speed sediment, due to incomplete cell rupture, and in the particle-free supernatant as a consequence of lysosome destruction.

Nitrogen cavitation of uterus mince in a Parr bomb nearly doubled the yield of intact lysosomes and reduced the enzyme retention in the low speed sediment, but also increased the solubilization of the marker enzymes (Table 3). The results obtained with both methods could not be appreciably improved by shorter bursts and continuous flow in the Ultraturrax operation or by applying a variety of pressures in the Parr bomb. We therefore proceeded to use endometrium instead of whole uterus as a starting material. The inner lining was detached by forcing a stainless-steel rod (Fig. 1) of slightly larger diameter through the uterine lumen. The scrapings contained cell clusters, isolated cells and cell fragments in an apparently well preserved morphological state (Fig. 2). Homogenization in all-glass Potter-Elvehjem homogenizers was carried on until differential cell/nuclei counts showed a disruption of more than 98% of the cells. The marker enzyme distribution in the homogenate was markedly improved in favor of the lysosomal/mitochondrial fraction as compared to the two other methods (Table 3). The still close to 30% retention of enzymatic activity in the low-speed sediment might be attributable to the glueing effect of glycosaminoglycans and proteoglycans, by which lysosomes and other cytoplasmic structures are trapped. The enzyme content of the supernatant can be explained by the fragility of lysosomes. A distribution survey for a number of lysosomal and mitochondrial enzymes (Table 4) shows that mitochondria are very resistant to rupture during homogenization. The bulk of cytochrome-oxidase activity was found in the mitochondrial/lysosomal fraction, little in the low-speed sediment and almost none in the supernatant. It is known from other tissues that NADP-dependent isocitrate dehydrogenase is not a purely mitochondrial enzyme, but also occurs in the cytosol[15]. A dual localisation has also been described for the lysosomal marker glucuronidase[16]. Its relatively high concentration in the

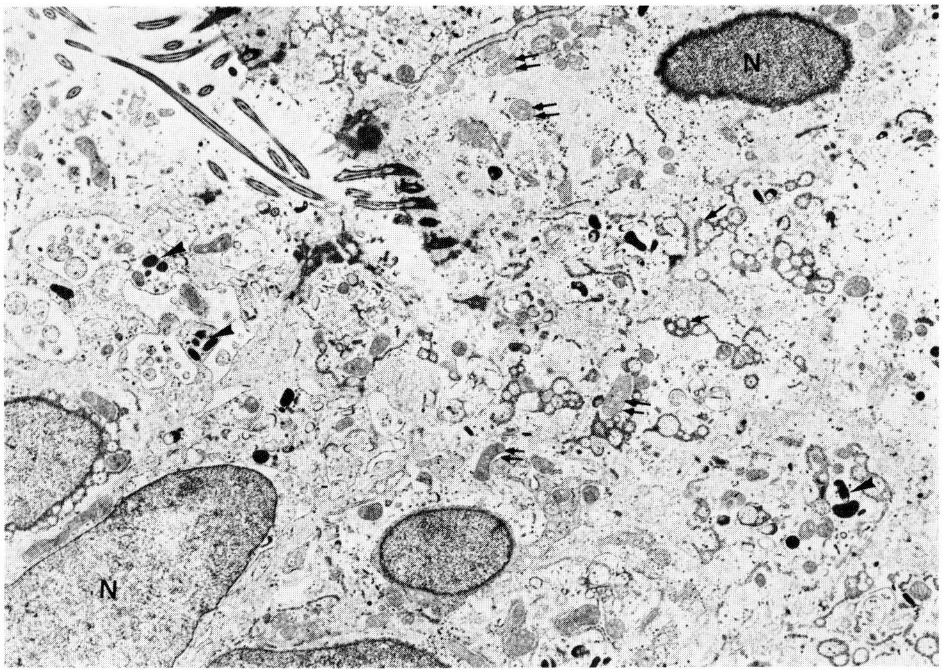

Fig. 2. Electron micrograph of a section of scraping showing ciliated endometrial cells, typical of the pig uterus. Cells are partially intact and show nuclei (N), mitochondria (double arrows), lysosomes (arrow heads), endoplasmic reticulum (arrows), × 5000.

Table 4. Subcellular distribution and specific activities of lysosomal and mitochondrial enzymes in endometrium homogenates.
ML = mitochondrial/lysosomal.

		Percentage of total homogenate			Spec. act. in homogenate [mU/mg prot.]
		Nuclei + cell debris	ML fraction	Post-mitochondrial supernatant	
Protein	(16)[a]	29.6 ± 2.9[b]	8.1 ± 0.9	61.8 ± 2.7	
Arylsulfatase	(7)	28.6 ± 2.1	53.5 ± 6.0	15.4 ± 6.2	4.56 ± 0.88
β-Glucosidase	(7)	38.6 ± 3.1	47.7 ± 4.1	14.4 ± 4.7	1.96 ± 0.31
β-Galactosidase	(11)	29.4 ± 4.5	38.4 ± 2.0	32.3 ± 3.7	9.62 ± 1.76
Ribonuclease I	(11)	31.0 ± 4.5	32.0 ± 2.3	36.4 ± 3.6	474.0 ± 80.4
Cathepsin D	(10)	27.3 ± 5.0	31.5 ± 3.4	38.1 ± 6.3	8.73 ± 0.56
Acid phosphatase	(7)	27.4 ± 6.2	31.4 ± 3.5	42.8 ± 4.1	21.6 ± 2.47
α-Glucosidase	(7)	32.7 ± 4.2	31.1 ± 2.5	35.2 ± 2.8	1.42 ± 0.26
β-Glucuronidase	(14)	41.7 ± 3.9	27.6 ± 3.0	30.9 ± 3.6	0.88 ± 0.07
Cytochrome *c* oxidase	(4)	9.5 ± 6.3	85.1 ± 5.9	1.7 ± 0.8	1488.0 ± 90.0
Isocitrate dehydrogenase	(2)	17.4 ± 2.5	47.8 ± 0.7	34.8 ± 3.3	29.5 ± 0.65

[a] No. of determinations [b] mean ± s.e.m.

low speed sediment could be due to trapped endoplasmic reticulum. .

2) pH optima of lysosomal and mitochondrial enzymes of pig endometrial cells

The pH optima were determined in the mixed mitochondrial/lysosomal fraction (Fig. 3). The acidic values for most of the hydrolases agree with the data reported for lysosomal enzymes from other tissues and species[17]. The pH optimum of 5.4 for the mitochondrial marker cytochrome *c* oxidase is rather low, but close to that of beef heart cytochrome oxidase[18].

3) Attempts to separate mitochondria and lysosomes by Triton WR 1339 accumulation

We tried to adapt Wattiaux's[19] method for the isolation of lysosomes from rat liver to pig uterus by intraluminal administration of Triton WR 1399. Using a variety of treatment schedules and different vehicles, we have not been able to change the density of endometrium lysosomes by detergent ingestion. In most of these experiments the scrapings were collected and homogenized in a $Mg^{2\oplus}$-containing, buffered sucrose solution. Isopycnic centrifugation in continuous gradients did not result in a separation of lysosome- and mitochondrion-linked enzymatic activities (Fig. 4). Electron

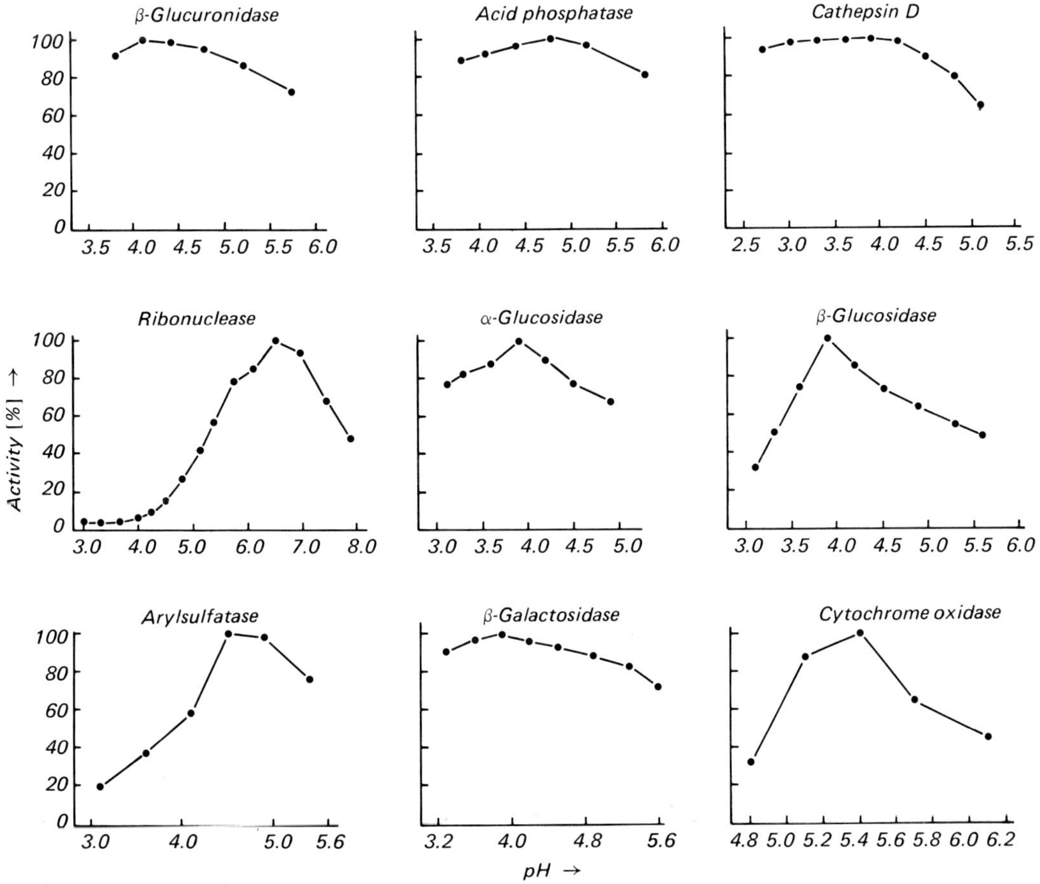

Fig. 3. Effect of pH on the activity of eight hydrolases and cytochrome oxidase from pig endometrium.
0.1M acetate, 0.1M citrate and 0.033M phosphate buffers were used (see Materials and Methods). Other conditions for the assays are listed in Tables 1 and 2. Sources of enzymes were mitochondrial/lysosomal suspensions. Results, from duplicate measurements, are expressed as percentage of the maximal activity, plotted against the pH.

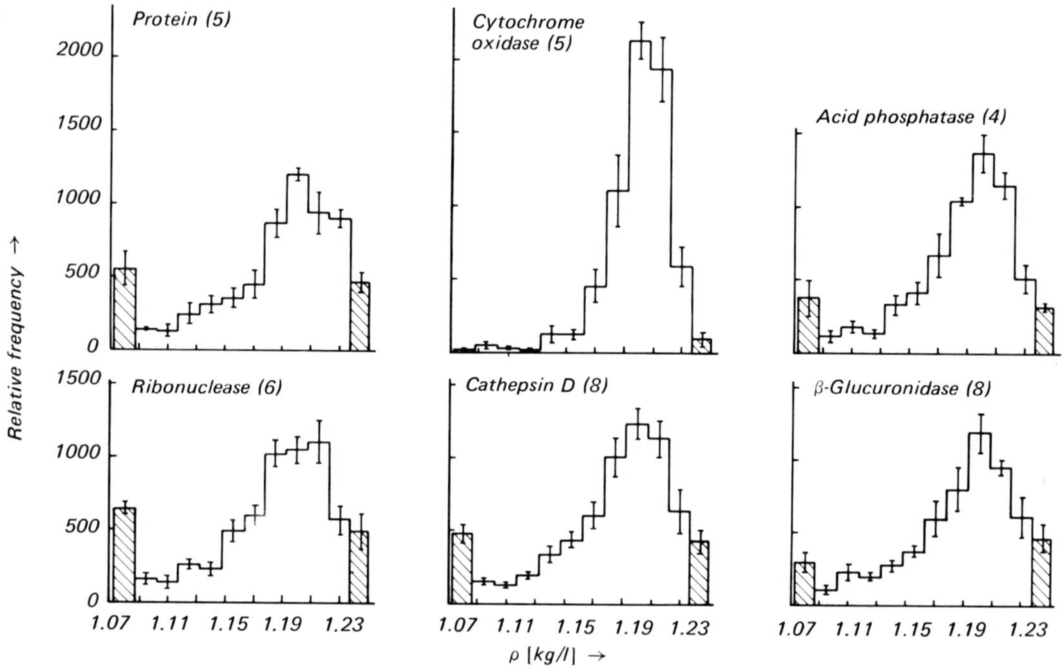

Fig. 4. Isopycnic subfractionation of mitochondrial/lysosomal fractions from pig endometrium homogenized with buffered sucrose containing 5mM $MgCl_2$.

Average density frequency distribution histograms. Vertical lines in the bars show s.e.m. Numbers of experiments are in parenthesis.

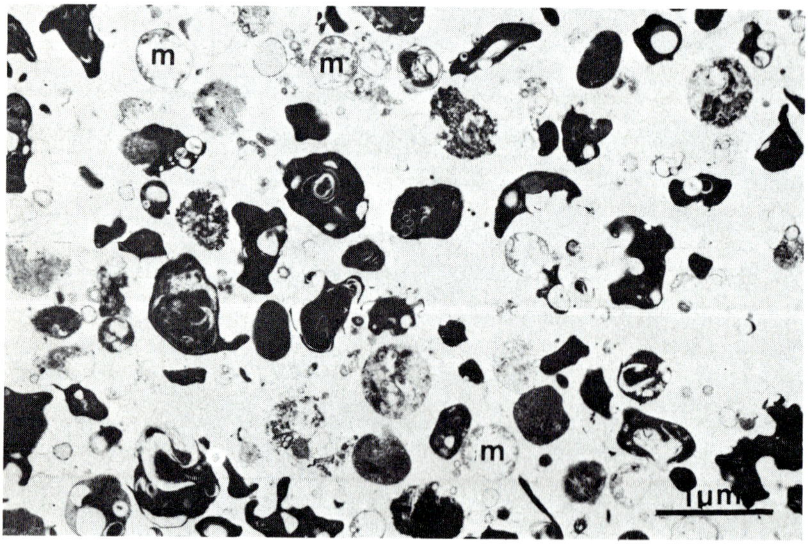

Fig. 5. Electron micrograph of fraction from isopycnic centrifugation containing the highest activities for mitochondrial and lysosomal enzymes (Fig. 4).

Buffered sucrose used during homogenization and differential centrifugation contained $Mg^{2\oplus}$. m: mitochondrion.

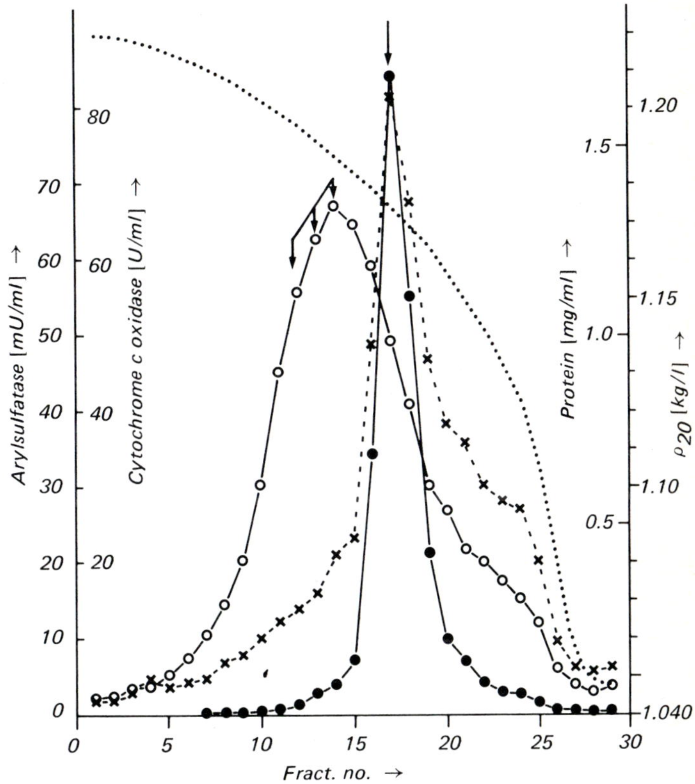

Fig. 6. Isopycnic subfractionation of a mitochondrial/lysosomal fraction from pig endometrium homogenized with buffered sucrose containing 1 mM EDTA.

Arrows show the fractions used for electron microscopic analysis (Fig. 7).

x– – –x protein;
o——o arylsulfatase;
●——● cytochrome oxidase;
······· density.

microscopy of the fraction equilibrating at ρ = 1.200 revealed the presence of lysosomes in a multitude of shapes and of distended mitochondria with poorly discernable structure in a 1 : 1 ratio (Fig. 5). The same results were obtained with endometrium from untreated uteri.

4) The effect of EDTA on the buoyant densities of mitochondria and lysosomes

When endometrium scrapings were collected and homogenized in buffered sucrose devoid of divalent cations and containing EDTA, distributions as shown in Fig. 6 were observed after isopycnic centrifugation of the mitochondrial/lysosomal fraction. While the density pattern of lysosomal enzymes remained virtually unchanged, the peak of mitochondrial enzymatic activity was narrowed and shifted from ρ 1.200 to 1.174. By this shift, also evident in a coinciding, sharp protein peak, the mitochondria were essentially concentrated in the lower density limb of the broad lysosome peak. The maximal enrichment of enzymatic activity/mg of protein from the original homogenate was 30.9-fold for cytochrome *c* oxidase in fraction No. 17 (ρ = 1.174) and 29.4-fold for arylsulfatase in fraction No. 12 (ρ = 1.196).

The withdrawal of divalent cations brought about a drastic change in the morphological appearance of the mitochondria (Fig. 7a). The inner membrane system and matrix were condensed in shell-like structures at the periphery of the organelle. The evacuated area was either empty or contained some less electron-dense, poorly structured material and was enclosed by the outer membrane.

In contrast to the mitochondria, no morphological change was observed in the structure of lysosomes (Fig. 7b). Since three fractions had to be pooled for the analysis, an overlapping with the mitochondrial peak could not be avoided. The electron micrograph, therefore, still shows the presence of some mitochondria.

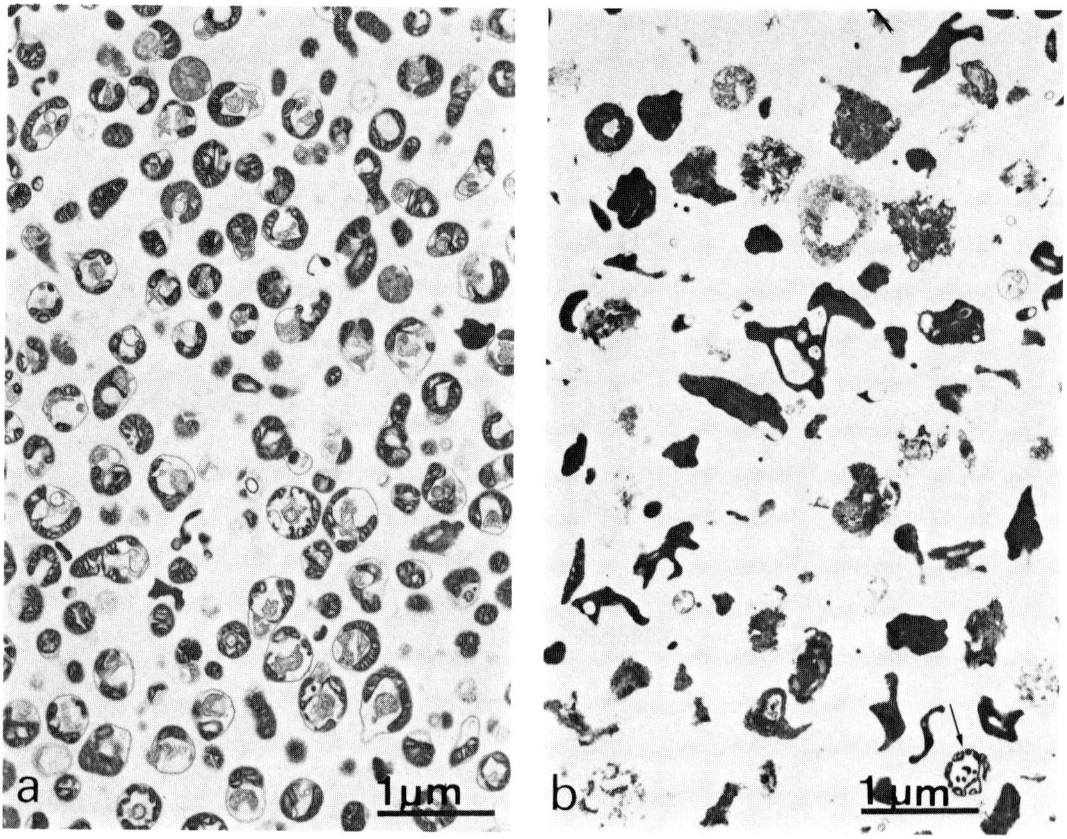

Fig. 7. Electron micrographs of
a) fraction having the highest cytochrome *c* oxidase activity, × 15 000, and
b) three pooled fractions with the highest arylsulfatase activity, shown in Fig. 6. Mitochondria indicated by arrows; × 15 000.

The reproducibility of the separation, both on convex exponential and on linear sucrose gradients can be judged from the histograms of Fig. 8. The heterogeneity of the lysosomal population is indicated by differences in the distribution patterns of the enzymatic activities, for example of arylsulfatase and glycosidases.

Discussion

The reasons for not obtaining tritosomes from endometrium cells after intrauterine administration of Triton WR 1339, although the detergent disappeared from the lumen, could be manifold. The substance might be preferentially ingested by mobile macrophages, which move into the deeper layer of the endometrium or even out of the organ. It is also possible that indigenous endometrial cells are either unable to take up WR 1339 or do so in quantities too small to achieve a noticeable effect on the buoyant density of lysosomes. We might further, in spite of the many administration schedules tried, have failed to pinpoint the time of maximum pinocytotic activity in endometrium cells.

The observed effect of EDTA on the structure of uterus mitochondria is a phenomenon well known for mitochondria from rat liver and ox adren-

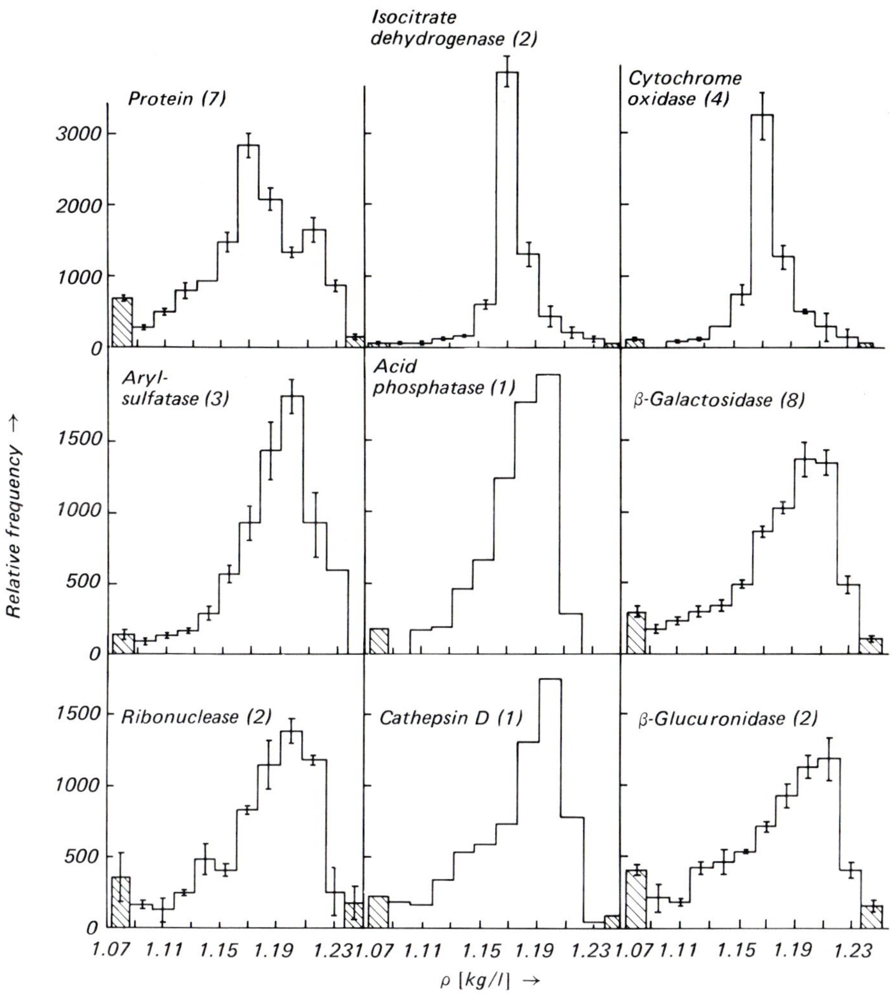

Fig. 8. Isopycnic subfractionation of mitochondrial/lysosomal fractions from pig endometrium homogenized with buffered sucrose containing 1mM EDTA.

Averaged density frequency distribution histograms. Vertical lines in the bars show s.e.m. Numbers of experiments are in parenthesis.

als[20,21]. The contraction of the contents and the decrease in buoyant density of the organelle is most pronounced in the anaerobic state[20] here favored by the experimental conditions chosen. By taking advantage of this change, we were able to obtain endometrial lysosomes comparable in purity to those previously isolated from liver and kidney[22,23].

Although we had to restrict the source of uterine lysosomes to endometrial cells, this has no adverse effect on the intended purpose. Endometrium is a "target" for estradiol and rich in estradiol receptor. Meaningful studies, of the kind mentioned in the introduction, can now be envisaged, where after a well-defined exposure of endometrial cells to estradiol in vivo, lysosomes are harvested and analysed for receptors, receptor degradation products and steroid metabolites.

Experiments with lysosomes in vitro might also be useful, but carry the danger of misinterpretations. After incubations of a highly purified lysosome fraction with tritiated estradiol at 0 - 2 °C, subsequent osmotic rupture of the organelles and charcoal treatment of the extract, we found that some label remained unabsorbed, suggesting the presence of an estradiol-receptor complex. This, however, could not be substantiated by agar gel electrophoresis[24] and density gradient centrifugation, both of which gave no evidence of receptor-bound estradiol. On chromatographic analysis, all radioactivity was found to be associated with a polar derivative of estradiol, the structure of which we have not yet identified.*

Previous reports[25,26], in which the presence of estradiol receptor in lysosomes was inferred solely on the basis of binding studies with the charcoal technique, therefore, should be reconsidered.

Literature

1 Little, M., Szendro, P., Teran, C., Hughes, A. & Jungblut, P.W. (1975) *J. Steroid Biochem.* **6**, 493 - 500.
2 Jungblut, P.W., Gaues, J., Hughes, A., Kallweit, E., Sierralta, W., Szendro, P. & Wagner, R.K. (1976) *J. Steroid Biochem.* 7, 1109 - 1116.
3 Jensen, E.V. & Jacobson, H.I. (1962) *Recent Progr. Horm. Res.* **18**, 387 - 414.
4 Jungblut, P.W. (1968) in Advances in the Biosciences (Raspé, G., ed.) vol. 2, pp. 157 - 181, Pergamon Press, Braunschweig, Germany.
5 Britten, R.J. & Roberts, R.B. (1960) *Science* **131**, 32 - 33.

* Manuscript in preparation.

6 Morris, C.J. & Morris, P. (1970) Separation Methods in Biochemistry, 1st edn., pp. 146 - 149, Pitman, London.
7 Cooperstein, S.J. & Lazarow, A. (1951) *J. Biol. Chem.* **189**, 665 - 670.
8 Bowers, W.E. & De Duve, C. (1967) *J. Cell Biol.* **32**, 339 - 348.
9 Campbell, P.N. & Sargent, J.R. (1967) in Techniques in Protein Biosynthesis (Campbell, P.N. & Sargent, J.R., eds.) vol. 1, pp. 299 - 300, Academic Press, London.
10 Lowry, O., Rosebrough, N., Farr, A. & Randall, R. (1951) *J. Biol. Chem.* **193**, 265 - 275.
11 Mollenhauer, H.H. (1963) *J. Stain Tech.* **39**,111 - 114.
12 Reynolds, E.S. (1963) *J. Cell Biol.* **17**, 208 - 212.
13 Spurr, A.R. (1969) *J. Ultrast. Res.* **26**, 31 - 43.
14 Fiske, C.M. & Subbarow, J. (1925) *J. Biol. Chem.* **66**, 375.
15 Henderson, N.S. (1965) *J. Exp. Zool.* **158**, 263 - 274.
16 Ganschow, R. & Paigen, K. (1967) *Proc. Natl. Acad. Sci. (U.S.A.)* **58**, 938 - 945.
17 Barrett, A.J. (1972) in Lysosomes. A Laboratory Handbook (Dingle, J.T., ed.) pp. 46 - 126, North Holland, Amsterdam.
18 Yonetani, T. (1961) *J. Biol. Chem.* **236**, 1680 - 1688.
19 Wattiaux, R., Wibo, M. & Baudhuin, P. (1963) in Ciba Symposium: Lysosomes (de Renck, A.V.S. & Cameron, M.P., eds.) pp. 176 - 200, J. & A. Churchill Ltd., London.
20 Deamer, D.W., Utsumi, K. & Packer, L. (1967) *Arch. Biochem. Biophys.* **121**, 641 - 651.
21 Allmann, D.W., Wakabayashi, T., Korman, E.F. & Green, D.E. (1970) *J. Bioenerg.* **1**, 73 - 86.
22 Leighton, F., Poole, B., Beaufay, H., Baudhuin, P., Coffey, J., Fowler, S. & De Duve, C. (1968) *J. Cell. Biol.* **37**, 482 - 513.
23 Shibko, S. & Tappel, A.L. (1965) *Biochem. J.* **95**, 731 - 741.
24 Wagner, R.K. (1972) *Hoppe-Seyler's Z. Physiol. Chem.* **353**, 1235 - 1245.
25 Hirsch, P. & Szego, C. (1974) *J. Steroid Biochem.* **5**, 533 - 542.
26 Szego, C. (1976) *Gynocol. Invest.* **7**, 257 - 279.

Dr. W. Sierralta*, Dr. A.J. Truitt and Prof. Dr. P.W. Jungblut, Max-Planck-Institut für Zellbiologie, Postfach 1009, D-2940 Wilhelmshaven

* Present address: Department of Biochemistry, I.N.T.A., Universidad de Chile, Casilla 15138, Santiago II, Chile

Hoppe-Seyler's Z. Physiol. Chem.
Bd. 359, S. 529 - 537, April 1978

Aggregation of Sponge Cells, XVIII*

Glycosyltransferases Associated with the Aggregation Factor

Werner E. G. MÜLLER[a,b], Rudolf K. ZAHN[a,b],
Branko KURELEC[b], Gerhard UHLENBRUCK[c], Peter VAITH[c] and Isabel MÜLLER[a,b]

[a] Institut für Physiologische Chemie Universität Mainz,
[b] Laboratory for Marine Molecular Biology, Institute Rudjer Boskovic, Rovinj, Yugoslavia, and
[c] Medizinische Universitätsklinik Köln, Abteilung Immunbiologie

(Received 16 January 1978)

Summary: This paper presents the experimental evidence that one function of the extracellular high molecular weight particles from the siliceous sponge *Geodia cydonium*, is to carry a series of glycosyltransferases. These particles have the further function of promoting aggregation.

In a previous study (J. Biol. Chem. **252**, 3836; 1977) it was demonstrated that the particles contain a sialyltransferase as one subunit. Here we proved the existence of two further glycosyltransferases: β-D-galactosyltransferase and glucuronosyltransferase. We were not able to demonstrate the existence of an α-mannosyltransferase, or an acetylgalactosaminyltransferase.

The galactosyltransferase uses both the degalactosylated anti-aggregation receptor and the degalactosylated aggregation receptor as acceptors for the sugar transfer reaction. Kinetic studies revealed a Michaelis constant for UDPgalactose of 0.1mM and 0.4mM, depending on the acceptor used. The Michaelis constant for degalactosylated anti-aggregation receptor was found to be lower (0.28nM) than that for degalactosylated aggregation receptor (660nM). The enzyme has a pH optimum of approximately 8.2. For full activity the presence of $Mg^{2\oplus}$ and 300mM NaCl is required. The temperature optimum lies around 20 °C.

The glucuronosyltransferase uses deglucuronylated aggregation receptor as an acceptor. This enzyme is characterized by a Michaelis constant for UDP-glucuronic acid of 1.1mM, while that for the acceptor is 216nM. The other properties of the reaction are almost identical with those described for the galactosyltransferase.

Enzymes:
CMP-*N*-acetylneuraminate-galactosyl-glycoprotein sialyltransferase, CMP-*N*-acetylneuraminate:D-galactosyl-glycoprotein *N*-acetylneuraminyltransferase (= sialyltransferase) (EC 2.4.99.1);
GDPmannose α-mannosyltransferase, GDPmannose:heteroglycan 2,3-α-mannosyltransferase (EC 2.4.1.48);
β-Galactosidase, β-D-galactoside galactohydrolase (EC 3.2.1.23);
β-Glucuronidase, β-D-glucuronide glucuronosohydrolase (EC 3.2.1.31);
UDPacetylgalactosamine-protein acetylgalactosaminyltransferase, UDP-2-acetamido-2-deoxy-D-galactose:protein acetamidodeoxygalactosyltransferase (= acetylgalactosaminyltransferase) (EC 2.4.1.41);
UDPgalactose-glucosaminyl-galactosyl-glucosylceramide β-D-galactosyltransferase, UDPgalactose:2-acetamido-2-deoxy-D-glucosyl-(1,3)-D-galactosyl-(1,4)-D-glucosylceramide β-D-galactosyltransferase (= β-D-galactosyltransferase) (EC 2.4.1.86);
UDPglucuronosyltransferase, UDPglucuronate β-glucuronosyltransferase (acceptor-unspecific) (= glucuronosyltransferase) (EC 2.4.1.17).

* XVII. Communication: Müller, W. E. G., Zahn, R. K., Kurelec, B. & Müller, I. (1978) *Biochem. System.*, submitted.

Aggregation von Schwammzellen, XVIII: Glykosyltransferasen, assoziiert mit dem Aggregationsfaktor

Zusammenfassung: In der vorliegenden Arbeit werden die experimentellen Ergebnisse vorgelegt, die die hochmolekularen, extrazellulären Partikeln aus dem Kieselschwamm *Geodia cydonium* als Trägerkomponente für eine Reihe von Glykosyltransferasen ausweisen. Diese Partikeln enthalten auch die Aggregations-auslösende Aktivität.

In einer früheren Arbeit (J. Biol. Chem. **252**, 3836; 1977) wurde nachgewiesen, daß die Partikeln eine Sialinsäure-Transferase als Untereinheit besitzen. In der vorliegenden Untersuchung wird die Existenz von zwei weiteren Glykosyltransferasen gezeigt: β-D-Galaktosyltransferase und Glucuronosyltransferase. Es gelang jedoch nicht, α-Mannosyltransferase- oder Acetylgalaktosaminyltransferase-Aktivitäten nachzuweisen.

Die Galaktosyltransferase benutzt sowohl den degalaktosylierten Anti-Aggregations-Rezeptor als auch den degalaktosylierten Aggregationsrezeptor als Akzeptor für die Zuckerübertragung. Kinetische Untersuchungen ergaben eine Michaelis-Konstante für UDPGalactose von 0.1mM bzw. 0.4mM in Abhängigkeit von dem eingesetzten Akzeptor. Die Michaelis-Konstante war für den degalaktosylierten Anti-Aggregations-Rezeptor bedeutend geringer (0.28nM) als für den degalaktosylierten Aggregationsrezeptor (660nM). Das Enzym hat ein pH-Optimum von etwa 8.2. Für die volle Enzymaktivität sind $Mg^{2\oplus}$-Ionen und 300mM NaCl notwendig. Das Temperaturoptimum liegt bei 20 °C.

Die Glucuronosyltransferase benutzt den deglucuronylierten Aggregationsrezeptor als Akzeptor. Dieses Enzym ist durch eine Michaelis-Konstante von 1.1mM für UDPGlucuronsäure und eine von 216nM für den Akzeptor ausgezeichnet. Die anderen Reaktionseigenschaften für dieses Enzym entsprechen weitgehend denen, die für die Galaktosyltransferase aufgeführt wurden.

Key words: Marine sponge, aggregation factor, glycosyl transferases, cell-cell interaction.

Marine sponges were the first material from which a soluble aggregation factor was obtained[1]. Ten years later this factor was purified nearly to homogeneity from the siliceous sponge *Geodia cydonium*; it has been characterized with regard to its physical and biological properties[2,3]. The aggregation factor from *Geodia* is associated with large particles, which are characterized by a sedimentation coefficient ($s^0_{20,w}$) of 3000 and a molecular weight of 2 to 5 × 10^9[3]. Later we succeeded in the isolation and purification of that macromolecule from the cell surface which specifically binds with the aggregation factor; it was termed aggregation receptor[4]. The binding of the factor to the aggregation receptor provokes cell-cell aggregation. The factor-mediated reaggregation of cells occurs initially at random; in a second phase, there occurs a reconstitution of a functional system of water-containing channels by rearrangement of the cells in the aggregates[2]. Hence a controlled segregation and subsequent reaggregation of the cells in the aggregates must be postulated. The molecular mechanism of the process of specific cell movement is unknown.

Concerning the sponge model, evidence has been presented, that a glycosyltransferase (sialyltransferase), which is associated with the aggregation factor[5], is involved in a controlled cell-cell interaction. Very recently a component, which inactivates reversibly the aggregation factor, from the same source, was isolated and purified from cell membranes of the sponge *Geodia cydonium*; this molecule was termed anti-aggregation receptor[6]. In this study it was determined that one biologically active site of the anti-aggregation receptor is the D-galactose moiety. Removal of this moiety by enzymic treatment with β-galactosidase inactivates the biological activity of the anti-aggregation receptor.

It is the aim of this paper to determine, whether the sponge system contains an extracellular enzymic mechanism which might be able to reactivate the β-galactosidase-treated anti-aggregation receptor. Thus, it was necessary to search for a β-D-galactosyltransferase. In addition the possible presence of other glycosyltransferases was investigated.

Materials and Methods

MATERIALS

β-Galactosidase (bovine liver; 5 μU/mg protein) and UDPglucuronic acid were from Sigma Chemicals, St. Louis, Mo. (U.S.A.); β-glucuronidase (*E. coli*; 10 μU/mg) from Boehringer Mannheim GmbH (Germany); UDP-D-[U-^{14}C]galactose (spec. activity 337 mCi/mmol), UDP-D-[U-^{14}C]glucuronic acid (spec. activity 321 mCi/mmol), UDP-*N*-acetyl-D-[1-^{14}C]galactosamine (spec. activity 61 mCi/mmol) and GDP-[U-^{14}]mannose (spec. activity 179 mCi/mmol) from the Radiochemical Centre, Amersham (England); UDPgalactose from Calbiochem, San Diego, Calif. (U.S.A); aquasol from NEN, Dreieichenhain (Germany); Whatman No. 3 MM filter paper discs from Hormuth & Vetter, Heidelberg (Germany); Sepharose 6B and dextran blue from Deutsche Pharmacia, Freiburg (Germany).

The siliceous sponge (*Tetractinellida*) *Geodia cydonium* Jameson was collected in the vicinity of Rovinj (Yugoslavia).

METHODS

Isolation of the aggregation factor

The purification procedure of the aggregation factor from *Geodia cydonium* was performed as described[7]; briefly: extraction of sponge cubes, stepwise ammonium sulfate fractionation of the crude extract and Sepharose 4B gel filtration. The partially purified fraction contained 0.74 mg protein/m*l* and had a specific activity of 24 × 10^5 AU/mg protein. The aggregation activity is expressed in aggregation units (AU) as described[2].

Isolation of the aggregation receptor

The aggregation receptor was isolated as previously described[4]. The fraction used (Step III) contained 0.28 mg protein/m*l* and had as specific activity of 650 RU/mg carbohydrate. The receptor activity is expressed in receptor units (RU) as described[4]

Isolation of the anti-aggregation receptor

The procedure used has been described previously[6]. The fraction used (Step IV) had a neutral carbohydrate content of 69 μg/m*l* and a specific activity of 250 aAR units/μg carbohydrate. The anti-aggregation receptor activity is expressed in units (aAR units) as described[6].

Acceptor preparations

In order to convert the aggregation receptor as well as the anti-aggregation receptor into a from lacking D-galactose or D-glucuronic acid the following incubations were performed:

Assay A; 0.5 m*l* containing the aggregation receptor (or the anti-aggregation receptor preparation) was added to 0.5 m*l* of a solution containing 40mM Tris/HCl, pH 7.2, 10mM $MgCl_2$, 100mM NaCl and 0.83 μU β-galactosidase. *Assay B;* 0.5 m*l* of the aggregation receptor preparation was added to 0.5 m*l* of 0.04M Na-phosphate, pH 6.8, 5mM EDTA and 33.3 μU β-glucuronidase.
Assay mixtures A and B were incubated for 2 h at 32 °C and subsequently heated for 5 min at 100 °C to inactivate the enzymes. These preparations (degalactosylated aggregation receptor, degalactosylated anti-aggregation receptor, deglucuronylated aggregation receptor) served as acceptors for the glycosyltransferases.

Enzyme assays

The *galactosyltransferase* was measured in a reaction mixture containing 10mM $MgCl_2$, 300mM NaCl, 50mM Tris/HCl (pH 8.2), 0.1mM UDP-[U-^{14}C]galactose (5 × 10^6 dpm/μmol), 20 μ*l* of degalactosylated aggregation receptor solution (= 2.8 μg protein) or 20 μ*l* of degalactosylated anti-aggregation receptor solution (= 0.4 μg carbohydrate) and 30 μ*l* of the aggregation factor (= 22 μg of protein) in a final volume of 80 μ*l*. The incubation was usually carried out for 60 min at 20 °C. After termination of the incubation 10 μ*l* of bovine serum albumin solution (3 mg/m*l*) was added to the assays. Portions of 60 μ*l* were pipetted onto Whatman 3 MM (2.3 cm diameter) filter paper discs. The discs were dried and transfered into 10% trichloroacetic acid (for 30 min), into 5% trichloroacetic acid (for 30 min), into ethanol (for 15 min) and finally into ether (for 15 min). Radioactivity was determined by counting the dried discs in 20 m*l* aquasol.

Glucuronosyltransferase assay was composed and performed as described for the galactosyltransferase assay with the exception that 0.1mM UDP-[U-^{14}C]glucuronic acid (4 × 10^6 dpm/μmol) was used instead of UDPgalactose deglucuronylated aggregation receptor (= 2.8 μg protein/assay) served as the acceptor.

The assay conditions for the *mannosyltransferase* and the *acetylgalactosaminyltransferase* were identical with those described for glucuronosyltransferase; either 0.1mM GDP-[U-^{14}C]mannose (6 × 10^6 dpm/μmol) or 0.1mM UDP-*N*-acetyl-D-[1-^{14}C]galactosamine (2.5 × 10^6 dpm/μmol) were used as substrates.

1 U of enzyme is defined as the amount which catalyzes the transfer of 1 μmol of sugar moiety/min at 20 °C under the conditions described above. The specific activity is expressed as U per mg protein (refered to the protein content of the aggregation factor).

For protein determination the method of Lowry et al.[8] was used. Neutral carbohydrates were determined according to Dubois et al.[9].

Results

1) β-D-Galactosyltransferase

This section reports the characteristics of the enzymes, UDPgalactose–anti-aggregation receptor β-D-galactosyltransferase, and UDPgalactose–aggregation receptor β-D-galactosyltransferase, which is linked with the aggregation factor. It was not the purpose of this study to determine, whether this enzyme is an entire subunit of the extracellular large particles that carry the aggregation factor. Therefore it is not possible to answer the question as to whether the observed enzyme activity, using the two acceptor molecules (degalactosylated anti-aggregation receptor and degalactosylated aggregation receptor), is due to the same enzyme molecule. The purified large particles, referred to hereafter as aggregation factor were used as the source of enzyme.

Effect of time and enzyme concentration

The galactosyltransferase standard assay was performed under conditions such that product formation was proportional to time of incubation and protein concentration. Fig. 1 shows that for 60 min incorporation is proportional to the time of incubation; a prolonged incubation results in a decreased incorporation rate. In addition it was shown that the reaction was proportional to the amount of enzyme protein up to 67 μg protein per assay (Fig. 2).

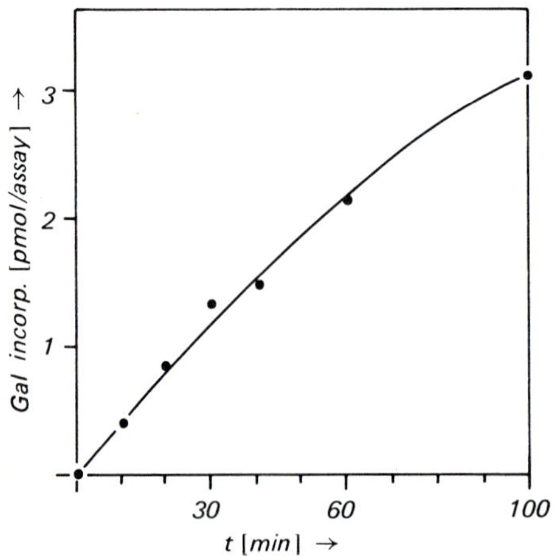

Fig. 1. The effect of the incubation time on the incorporation of galactose into degalactosylated anti-aggregation receptor.

Reactions were performed as described under the standard conditions described in "Methods", using 22 μg of enzyme protein (= aggregation factor).

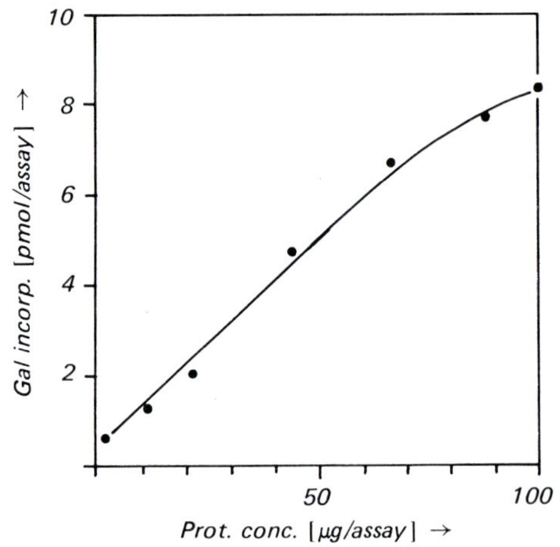

Fig. 2. The effect of galactosyltransferase concentration (= aggregation-factor preparation) on the incorporation of galactose into degalactosylated anti-aggregation receptor.

Incubations were carried out as described for the standard assay procedure (see "Methods").

The proof that galactose is indeed incorporated into the anti-aggregation receptor moiety is documented in Fig. 3. Using the method of gel filtration (on Sepharose 6B), the authentic anti-aggregation receptor elutes between the V_e/V_o values of 2.1 and 2.5. After incubation at 0 °C (Fig. 3a) no radioactivity was found in this range; the non-incorporated UDP-[^{14}C]galactose elutes at a V_e/V_o value of greater than 3.5. In contrast to the findings at 0 °C, incubation at 20 °C leads to a high incorporation of radioactivity, which elutes from the column at the same position as the anti-aggregation receptor (Fig. 3b).

Variation of substrate concentration

Fig. 4 shows the dependence of the enzyme reaction upon the concentration of increasing amounts of UDPgalactose. From the Lineweaver-Burk plot,

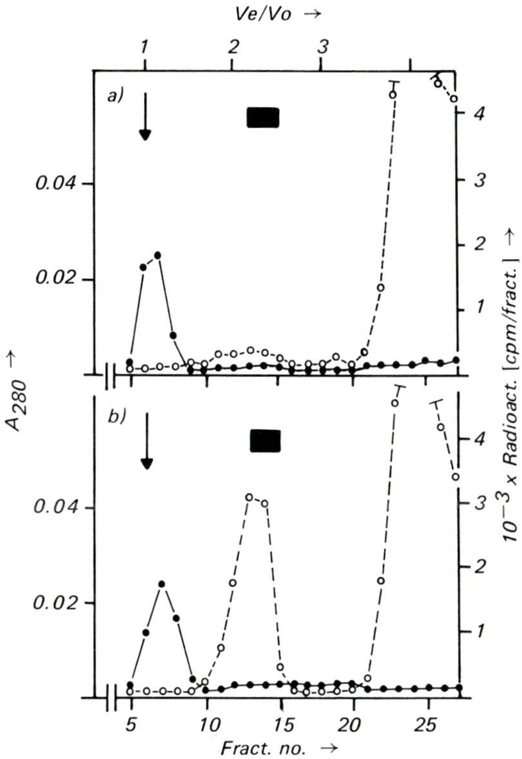

Fig. 3. Galactosylation of anti-aggregation receptor.

The degalactosylated anti-aggregation receptor was regalactosylated by the preparation of the aggregation factor using UDP-[^{14}C]galactose as substrate.

a) Incubation of the standard reaction mixture at 0 °C;

b) Incubation at 20 °C.

After incubation, the mixture was separated on Sepharose 6B. The column (1 x 18 cm) was loaded with the total reaction mixture. which was supplemented with 2M NaCl at the end of the incubation period. Elution was performed at 0 °C with 100M Tris/HCl (pH 8.2), containing 2M NaCl. Fractions of 0.8 m*l* were collected and both the radioactivity of the fractions, counted in 20 m*l* aquasol (○— —○) and the A_{280} values of the fractions (●——●) were determined. The arrow marks the position of V_o (= dextran blue marker). Abscissa (upper scale): V_e/V_o values[10]. The bar gives the position of the purified anti-aggregation receptor.

a K_m value of 0.1mM was calculated when the degalactosylated anti-aggregation receptor was used as acceptor. The K_m value with the degalactosylated aggregation receptor as the acceptor was 0.4mM. The maximum reaction velocity calculated from

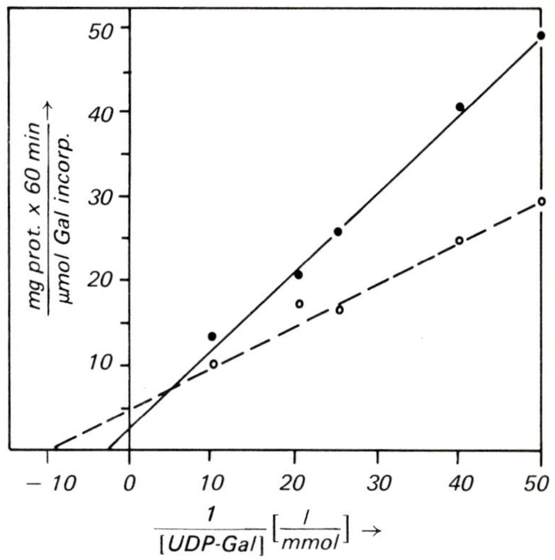

Fig. 4. Effect of varying the UDP-[^{14}C]galactose concentration on the galactosyltransferase activity using both degalactosylated aggregation receptor (●——●) and degalactosylated anti-aggregation receptor (○— —○) as acceptors.

The incubation period was 60 min. The plot was performed according to the method of Lineweaver and Burk[11].

the same plot, was 0.2 μmol of galactose incorporated/(mg prot. x 60 min) for the degalactosylated anti-aggregation receptor, and 0.4 μmol of galactose incorporated/(mg prot. x 60 min) using the degalactosylated aggregation receptor as the acceptor molecule.

Variation of acceptor concentration

The enzyme reaction is dependent upon both the concentration of the degalactosylated acceptor and the nature of the acceptor used in the assay (Fig. 5). A K_m value of 0.045 μg for the degalactosylated anti-aggregation receptor per assay was determined from the Lineweaver-Burk plot and a K_m value of 1.1 μg for the degalactosylated aggregation receptor per assay was obtained by the same method. A rough estimation of these K_m values on the basis of molarity can be given using the physical characteristics as well as the purification state of the acceptor preparations added to the reaction mixture. The anti-aggregation receptor has a molecular weight of around 210000 and was found to be to at least 80% pure[6]; the aggregation receptor is characterized

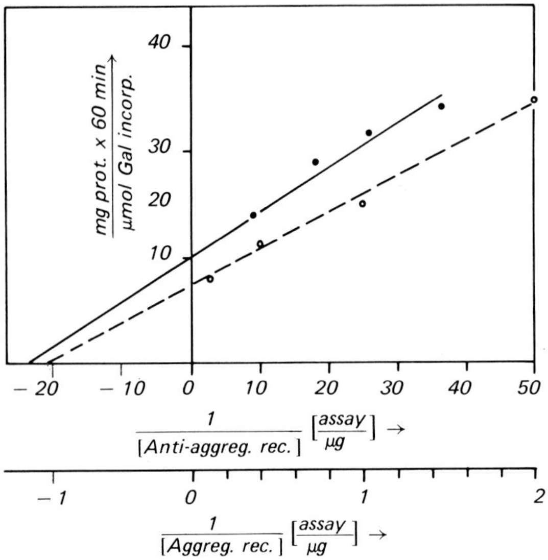

Fig. 5. The effect of acceptor concentration on galactose incorporation into degalactosylated anti-aggregation receptor (○– – –○) or into degalactosylated aggregation receptor (●——●).

The incubation was carried out under standard assay conditions. The concentration of the degalactosylated anti-aggregation receptor is expressed in µg carbohydrate per assay, that of the degalactosylated aggregation receptor in µg protein per assay. The data are plotted by the method of Lineweaver and Burk[11].

by a molecular weight of 18 000 and a purity of around 64%[4] (and P. Vaith, W.E.G. Müller and G. Uhlenbruck, unpublished result). From these figures the K_m value in the case of the degalactosylated anti-aggregation receptor was calculated to be 0.28nM and the K_m value for the degalactosylated aggregation receptor was 660nM.

The total number of sugar moieties incorporated per acceptor molecule can be calculated from the K_m values for the two enzyme acceptor molecules and the incorporation rates of the enzymes under standard incubation conditions. It was found that 1 molecule of galactose is incorporated per 4.6×10^5 molecules of degalactosylated anti-aggregation receptor or per 7.1×10^6 degalactosylated aggregation receptor molecules.

Effect of pH

The galactosyltransferase shows a single fairly broad pH optimum in the range of 7.5 to 8.6. At pH 6.5 the enzyme activity amounts to 27% of the optimal activity, and at pH 9.5 to only 35%. The pH effects on galactosyltransferase are identical, irrespective of the nature of acceptor used in the assay.

Cation and NaCl requirement

The galactosyltransferase is completely inhibited if 20mM EDTA is added to the standard reaction mixture. Only the metal ion $Mg^{2\oplus}$ was tested for its activity on the enzyme activity. Varying its concentration between 1 and 20mM had only little effect on the incorporation rate. Above and below this value the activity decreases.

The enzyme activity is strongly dependent on the ionic strength of the reaction mixture. Using NaCl, the optimal concentration was found to be 300mM. At 0mM NaCl and 600mM NaCl the incorporation rate was nearly zero.

These characteristics are observed in assays, that contain either degalactosylated anti-aggregation receptor or degalactosylated aggregation receptor.

Temperature optimum

The temperature optimum of the enzyme lies around 20 °C. The temperature coefficient Q_{10}[12] 30 °C/20 °C is 2.1, and that for 20 °C/10 °C is 2.6. These values are independent of the nature of the acceptor used in the assay.

2) Glucuronosyltransferase

Here we describe the characteristics of the UDP-glucuronic acid–aggregation receptor glucuronosyltransferase. As described for the galactosyltransferase, the purified large particles, carrying the aggregation factor are used as the enzyme source. Deglucuronylated aggregation receptor is added to the assay as the acceptor for the transferred sugar moiety.

Effect of time and enzyme concentration

The glucuronosyltransferase assay was designed so that the incorporation rate was proportional to time of incubation and protein concentration. The limits necessary to maintain these conditions were found to be 90 min and 90 µg protein/assay.

Variation of substrate concentration

Kinetic studies revealed a Michaelis constant of 1.1mM for UDPglucuronic acid (Fig. 6). The maximal reaction velocity was found to be 0.2 μmol of glucuronic acid incorporated/(mg prot. x 60 min).

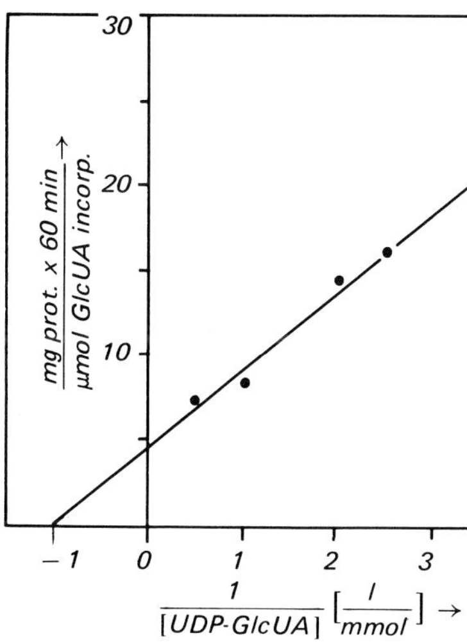

Fig. 6. Determination of the K_m value of glucuronosyltransferase.

The plot was performed according to Lineweaver and Burk[11]. Deglucuronylated aggregation receptor was used as acceptor. GlcUA = glucuronic acid.

Variation of acceptor concentration

Fig. 7 shows the dependence of the enzyme reaction upon the concentration of deglucuronylated aggregation receptor. A K_m value of 0.36 μg deglucuronylated aggregation receptor per assay was determined. Applying the same conversion as used for the galactosyltransferase, the K_m value on the basis of molarity is 216nM.

Effect of pH

The enzyme has a broad pH optimum between 7.4 and 8.4; at pH 6.5 the enzyme activity is 41% and at pH 9.5 only 21% of the optimal activity.

Cation and NaCl requirement

Like the galactosyltransferase, the glucuronsyltransferase requires $Mg^{2\oplus}$ for its activity. Addition

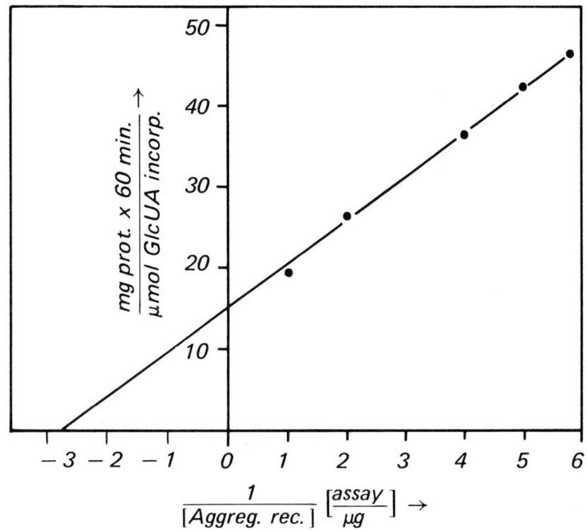

Fig. 7. Influence of the acceptor concentration (deglucuronylated aggregation receptor) on glucuronic acid incorporation.

The incubation was carried out under standard assay conditions. The plot was performed according to the method of Lineweaver and Burk[11].

of 20mM EDTA to the reaction mixture results in an incorporation rate of only 15% of the optimal values.

For optimal activity, NaCl in a concentration of 250 - 300mM is required. At a lower ionic strength (0mM NaCl) the activity decreases by 80%; in the presence of 600mM the incorporation rate was reduced by 65%.

Temperature optimum

The temperature optimum is 20 °C. The temperature coefficient Q_{10}[12] for 30 °C/20 °C is 2.6, and that for 20 °C/10 °C is 1.8.

3) Mannosyltransferase and acetylgalactosaminyltransferase

Some attempts were made to detect mannosyltransferase- and acetylgalactosaminyltansferase activity. Under the conditions described under "Methods", using GDPmannose or UDP-*N*-acetyl-D-galactosamine as substrate, deglucuronylated aggregation receptor as acceptor and the extracellular large particle as enzyme source, no activity of either transferase could be detected.

Discussion

It was Roseman[13] who proposed that a "multiglycosyltransferase system" is involved in the function of intercellular adhesion. This concept was an excellent working hypothesis for a series of experimental studies (review[14]). However, till now, only the sponge cell system was suitable for the purpose of approaching this problem on a molecular basis. This is because the sponge cell is the only source from which a soluble aggregation factor[1,2], a soluble aggregation receptor[4] and a soluble anti-aggregation receptor[6] could be isolated in a highly purified state. All these molecules are involved in the reaggregation process. As mentioned above, the aggregation factor and the aggregation receptor (linked with the cell membrane) cause reaggregation of single cells to cell clumps at random[2], while the anti-aggregation receptor is probably involved in the controlled "sorting out" mechanism[6]. To do justice to these specific functions, the aggregation receptor as well as the anti-aggregation receptor must be accessible for modifying enzyme systems, e.g. an inactivating – and a reactivating enzyme system. From the previous data[6], we know that the isolated anti-aggregation receptor can be inactivated by the action of the enzyme β-galactosidase, an enzyme which is also present on the cell surface[15]. The involvement of another carbohydrate group, glucuronic acid, in the aggregation factor-guided reassociation of sponge cells was shown[16].

In the present study, it is shown, that the sponge organism contains extracellularly enzyme systems (β-D-galactosyltransferase and glucuronosyltransferase), which catalyze the transfer of a monosaccharide (either galactose or glucuronic acid) to a glycoprotein acceptor (aggregation receptor and/or anti-aggregation receptor). These enzymes are linked with the aggregation factor; the molecules are "subunits" of the high molecular weight particles, first described by us[2]. The galactosyltransferase and the glucuronosyltransferase are enzymes, which have been known for more than 10 years[17,18], and have been isolated primarily from mammalian tissue[14]. The extraordinary properties of these two enzymes from the sponge material are:

a) they are particle-bound and not present in a cell membrane in bound or soluble form[14];

b) they have a low temperature optimum of around 20 °C, instead of one around 37 °C as usually determined for the isoenzymes from other sources (e.g.[19]); and

c) they require a high ionic strength (250 - 300mM NaCl) for full activity; in contrast, the enzymes from mammalian tissue exert their highest activity in a low ionic strength medium (e.g.[20]).

The main outcome of this paper is the finding, that a "multiglycosyltransferase system"[13] does exist, and its most likely function is as an "activator" of the aggregation receptor and the anti-aggregation receptor from the sponge *Geodia cydonium*. In subsequent work, the "inactivating" enzymes (β-galactosidase and β-glucuronidase) on the cell membranes of the same biological system will be studied.

This work was supported by grants from the *Bundesministerium für Forschung und Technologie*, Bonn, and by the *Academy of Science and Letters*, Mainz, Germany. We are indebted to Ms. *R. Steffen* and Mr. *R. Beyer* for excellent technical assistance.

Literature

1 Moscona, A. A. (1963) *Proc. Natl. Acad. Sci. U.S.A* **49**, 742 - 747.
2 Müller, W. E. G. & Zahn, R. K. (1973) *Exp. Cell Res.* **80**, 95 - 104.
3 Zahn, R. K., Müller, W. E. G., Geisert, M., Reinmüller, J., Michaelis, M., Pondeljak, V. & Beyer, R. (1976) *Cell Diff.* **5**, 129 - 137.
4 Müller, W. E. G., Müller, I., Zahn, R. K. & Kurelec, B. (1976) *J. Cell Sci.* **21**, 227 - 241.
5 Müller, W. E. G., Arendes, J., Kurelec, B., Zahn, R. K. & Müller, I (1977) *J. Biol. Chem.* **252**, 3836 - 3842.
6 Müller, W. E. G., Zahn, R. K., Kurelec, B. & Müller, I. (1978) *Proc. Natl. Acad. Sci. U.S.A* **75**, in press.
7 Müller, W. E. G., Müller, I. & Zahn, R. K. (1974) *Experientia* **30**, 899 - 902.
8 Lowry, O. H., Rosebrough, N. J., Farr, A. L. & Randall, R. J. (1951) *J. Biol. Chem.* **193**, 265 - 275.
9 Dubois, M., Gilles, K. A., Hamilton, J. K., Rebers, P. A. & Smith, F. (1956) *Anal. Chem.* **28**, 350 - 356.
10 Determann, H. (1967) Gelchromatographie. Springer Verlag Berlin.
11 Lineweaver, H. & Burk, D. (1934) *J. Am. Chem. Soc.* **56**, 408 - 412.
12 Netter, H. (1959) Theoretische Biochemie – Physikalisch-chemische Grundlagen der Lebensvorgänge. p. 557. Springer Verlag Berlin.
13 Roseman, S. (1970) *Chem. Phys. Lipids* **5**, 270 - 297.
14 Shur, B. D. & Roth, S. (1975) *Biochim. Biophys. Acta* **415**, 473 - 512.

15 Aoyaki, T., Suda, H., Nagai, M., Ogawa, K., Suzuki, J., Takeuchi, T. & Umezawa, H. (1976) *Biochim. Biophys. Acta* **452**, 131 - 143.
16 Turner, R. S. & Burger, M. M. (1973) *Nature New Biol.* **244**, 509 - 510.
17 Bosman, H. B. & Eylar, E. H. (1968) *Biochem. Biophys. Res. Commun.* **30**, 89 - 94.
18 Dutton, G. J. (1966) Glucuronic Acid Free and Combined, pp. 185 - 299, Academic Press, New York.
19 McLean, R. J. & Bosmann, H. B. (1975) *Proc. Natl. Acad. Sci. U.S.A* **72**, 310 - 313.
20 Hodson, S. & Brenchley, G. (1976) *J. Cell Sci.* **20**, 167 - 182.

Prof. Dr. W.E.G. Müller*, Prof. Dr. R.K. Zahn und I. Müller, Institut für Physiologische Chemie der Universität Mainz,
Duesbergweg, D-6500 Mainz.

Dr. B. Kurelec, Laboratory for Marine Molecular Biology, Institute Rudjer Boskovic,
Y-5 22 10 Rovinj.

Prof. Dr. G. Uhlenbruck und Dr. P. Vaith, Medizinische Universitätsklinik Mainz, Abteilung Immunbiologie, Kerpenerstr. 15, D-5000 Köln 41.

* To whom correspondence should be sent.

Hoppe-Seyler's Z. Physiol. Chem.
Bd. 359, S. 539 - 542, April 1978

SHORT COMMUNICATION

Physiological Ligands of Putative LSD-Serotonin Receptors: Heterogeneity of LSD-Displacing Factors in Human Body Fluids and Nervous Tissue

Ehrenfried MEHL and Lisa GUIARD

Max-Planck-Institut für Psychiatrie, Neurochemische Abteilung, München

(Received 3 February 1978)

Herrn Prof. Dr. Horst Jatzkewitz zum 65. Geburtstag

Physiologische Liganden eines vermutlichen LSD-Serotonin-Rezeptors: Heterogenität von LSD-Verdrängungsfaktoren aus Nervengewebe und Körperflüssigkeiten des Menschen

Zusammenfassung: Bei [2-^{3}H]LSD-Bindungsstudien mit Mikrosomenfraktionen des Gehirns wurden von uns Faktoren erfaßt, die schon in physiologischen Konzentrationen bewirken, daß die LSD-Bindung an die hochaffinen Bindungsstellen der vermutlichen LSD-Serotonin-Rezeptoren gehemmt wird. Diese LSD-Verdrängungsfaktoren wurden in Körperflüssigkeiten wie der Zerebrospinalflüssigkeit, Blutplasma und Urin gefunden, ebenso in Extrakten von Gehirn und Milz. Aus Gehirnextrakten wurden mindestens vier Typen dieser Faktoren chromatographisch abgetrennt. Neben einer höhermolekularen wurden die α-, β-Fraktion und die Serotonin-haltige γ-Fraktion als niedermolekulare Anteile erhalten. Es bestehen Hinweise für eine organspezifische Verteilung von bestimmten Typen dieser Faktoren im Organismus: Die β-Fraktion wurde im Nervensystem gefunden und war in Urin und Milzextrakt nicht nachweisbar. Die ausgeprägte Bindungsspezifität der hochaffinen LSD-Serotonin-Bindungsstellen gegenüber verschiedenen Pharmaka und physiologischen Substanzen spricht ebenfalls dafür, daß es sich bei der β-Fraktion um einen spezifischen Faktor handeln könnte.

Key words: Nervous tissue, lysergic acid diethylamide, serotonin, receptor, LSD-displacing factors.

Competitive binding assays such as radioligand-receptor or immunoassays have been widely and successfully employed for the determination of hormone levels in crude subfractions of body fluids. As a highly sensitive method for detecting endogenous psychedelic ligands, we have used a radioligand receptor assay that measures the binding of the hallucinogenic serotonin antagonist, [^{3}H](+)-LSD, to high-affinity LSD-serotonin binding sites in neuronal membrane preparations. These binding sites in rat brain were found to be stereospecific[1-4] and have been purified[5] from pig brain. The equilibrium dissociation constant was 2×10^{-9} mol/*l* with [^{3}H](+)-LSD[6]. The pharmacological specificity was very pronounced[1-4,6,7]. Within a wide group of physiological ligands tested, only indolealkylamines such as serotonin, 5-methoxytryptamine and tryptamine were found to inhibit [^{3}H]LSD binding at concentrations below 1μM. Therefore, these binding sites are regarded as putative LSD-serotonin receptors. When comparing unprocessed cerebrospinal fluid from nonpsychotic and psychotic patients and from controls, we detected the presence of unidentified substances[8] which were capable of reversibly interfering with [^{3}H]LSD binding in a radioligand receptor assay. The substances were heat-stable, of low molecular weight (500) and anionic at pH 7.4. They were provisionally named LSD-displacing factors. The experiments reported here concern the questions of the existence of subtypes of these factors, their distribution in body fluids for diagnostic evaluation, and the existence of a brain-specific type of the LSD-displacing factor.

Abbreviation: LSD = (+)-lysergic acid diethylamide.

Materials and Methods

For chromatographic separation, 100-m*l* samples were prepared by pooling cerebrospinal fluid (100 neurological patients), urine (18 neurotic patients), and a low-molecular weight fraction of blood plasma (4 healthy men). The latter fraction was obtained by gel filtration (Sephadex G-50, 10mM sodium phosphate buffer pH 7.0) and was supplemented with the γ-fraction (method B) which was prepared by mixing 160 m*l* of plasma with 400 m*l* of methanol and by evaporating the filtrate. Cerebral cortex and spleen samples were collected within 6 h after death from a man who died of cardioclasia. The tissues (20 g) were extracted with 180 m*l* of distilled water at 80 °C by homogenizing for 10 min (Ultraturrax homogenizer). The homogenate was centrifuged (100 000 × *g*, 1 h at 20 °C). After lyophilization of body fluids and tissue extracts, the residues were taken up with 20 m*l* of distilled water and were adjusted to pH 7.0. The solutions were clarified by centrifugation (as above). Chromatography of 19-m*l* samples was performed on a 450-m*l* column (91 × 2.5 cm) of Sephadex G-10 at 20 °C using 0.1M ammonium acetate buffer, pH 7.0 (containing 0.002% NaN_3) and collecting one 15-m*l* fraction per h. When necessary, the eluted fractions were adjusted to pH 7.0 and were diluted with water to give the same electrical conductivity as the elution buffer. In general, 0.3-m*l* samples were tested in the [^{3}H]LSD-receptor assay. The LSD-displacing capacity of the eluted fractions was analyzed by a radioligand receptor assay using 2nM [2-^{3}H](+)-LSD (21 Ci/mmol, purity at time of use > 95%) and subcellular fractions from pig brain as described[8].

Results and Discussion

LSD-displacing factors of body fluids and extracts from brain and spleen were subfractionated by gel chromatography with Sephadex G-10. The eluted fractions were assayed for their inhibitory capacity. Their interference with [^{3}H]LSD-binding to high affinity sites was tested with a microsomal (P_3) fraction from pig cerebral cortex. Typical chromatographic separations are shown in the figure. Various subfractions could be clearly separated from each other as well as from macromolecules in the void volume (fraction 12), from the bulk of UV-absorbing substances (fraction 12), from inorganic salts (fractions 20 and 21) and from endogenous serotonin (fractions 47 - 53). As observed earlier[9] for indolealkylamines, the elution of serotonin was retarded due to adsorption on Sephadex (K_d = 8.7). In addition to the macromolecular fraction E in the void volume, brain extract contained the low-molecular weight fractions α (fractions 14 and 15; K_d = 0.40), β (fractions 17, 18; K_d = 0.65) and the serotonin-containing γ fraction. The broad peak of the γ fraction of urine and spleen may be indicative of

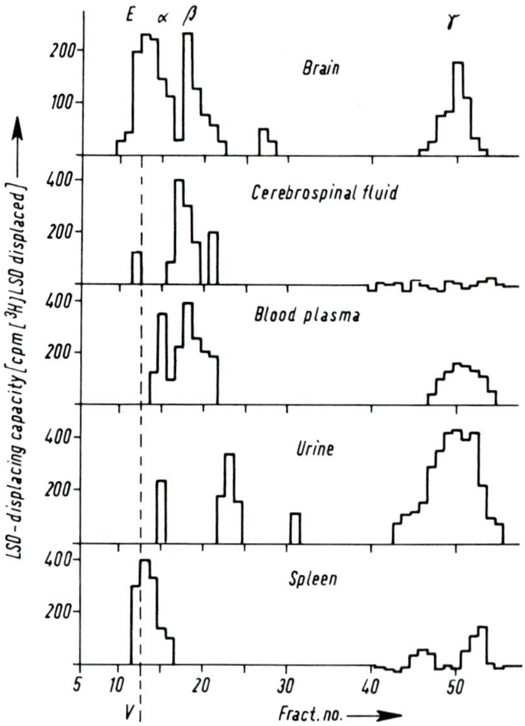

Figure. Chromatographic separation of LSD-displacing factors in crude extracts from brain, body fluids and spleen.

The crude extracts from 100 m*l* of body fluid, 25 g of brain or 70 g of spleen were separated on 450 m*l*-columns of Sephadex G-10 (nominal exclusion limit at 700 daltons). Fractions 1 - 30 consisted of 15 m*l* each, and fractions 31 - 60 of 45 m*l* each. Samples of 300 μl of the eluted LSD-displacing fraction were incubated with 2nM [^{3}H]LSD plus P_3 fraction and compared to [^{3}H]LSD binding of standard samples containing 0.1M ammonium acetate buffer, pH 7.4 instead of the LSD-displacing fraction. The amount of LSD displaced was calculated as the difference between the cpm of [^{3}H]LSD bound in the absence and presence of the eluted fractions. The standard binding was 900 cpm, residual binding of samples inhibited by the LSD-displacing factor was at least 500 cpm. The standard error was 50 cpm. When [^{14}C]serotonin was added as an internal standard to brain extract, it was eluted with the γ fraction. E = macromolecular fraction. V = void volume (broken line at fraction 12). Sodium chloride was used as a marker for the internal volume and was eluted with the fractions 20 and 21.

inhomogeneity. The recovery of the LSD-displacing factors was between 70 and 90%.

Substances isographic with the cerebral α fraction (Figure, fractions 14 and 15) were detectable in spleen extract, in blood plasma and urine but not in the cerebrospinal fluid, whereas substances having the elution characteristics of the cerebral β fraction (fractions 17 and 18) were detectable in the cerebrospinal fluid and plasma, but not in urine or in spleen extract. When the chromatographic fractions 12 and 13 from brain or spleen extracts were dialysed (Visking cellulose tubing, exclusion limit 10000 - 20000 daltons) a macromolecular LSD-displacing fraction was found to be retained. This macromolecular fraction of blood plasma comprised about 85 % of the total LSD-displacing capacity. It was eluted in the void volume on Sephadex G-50 and thereby separated from the low-molecular weight fraction of plasma. The latter was fractionated further in this study (Figure). In urine, an additional LSD-displacing fraction (fractions 22 - 24) was detectable.

It was excluded that the factors might influence LSD-binding by degradation of LSD itself. In that hypothetical case, they could, for instance, function as a cofactor of an LSD-degrading enzyme in the P_3 membrane fraction. By thin-layer chromatography in three different systems, however, it was found that the radioactivity isographic with LSD was as high in samples incubated with LSD-displacing factors as in the control samples. In addition, the inhibition by LSD-displacing factors was found to be fully reversible (Table 1). The original LSD-binding capacity of the microsomal (P_3) fraction was restored by washing out the reversibly

Table 1. Restoration of original LSD-binding capacity by washing-out of reversibly bound factors.

Samples of 250 μl of cerebrospinal fluid, 50 μl of urine and 20 μl of the other fractions were adjusted to 300 μl with buffer (25mM sodium phosphate, pH 7.4 containing 50mM NaCl) and were exposed to 80 °C for 10 min. After cooling and addition of 150 μl of microsomal fraction to the LSD displacing fractions, the samples were incubated for 1 h at 25 °C. For washing-out of the factors, one set of quadruplicate samples was layered over 0.8 ml of a 0.5M sucrose solution (containing 25mM buffer) and centrifuged for 1 h at 150 000 x g at 25 °C. After removal of the supernatant, the pellets were resuspended by mild ultrasonication. [^{3}H]LSD binding of the washed microsomal fraction was compared to [^{3}H]-LSD binding in the presence of the displacing factors.

Source of the factors	LSD-displacing fractions Present	Washed out
	[^{3}H]LSD bound [cpm]	
Brain	575	1040
Blood plasma	890	1070
Urine	420	1020
Spleen	615	1020
Standard binding without the factors	1025	1045

Table 2. Selectivity of high-affinity [^{3}H]LSD binding sites of synaptosomal plasma membranes in the presence of various substances and 2nM [^{3}H]LSD.

The IC_{50} values represent the concentrations of a substance necessary for 50% inhibition of high-affinity LSD binding. Mean ± SD with the number of experiments in brackets.

Compound added	IC_{50} value [μM]
[^{3}H](+)-LSD*	0.002 ± 0.001 (6)
Methysergide	0.090 ± 0.014 (2)
5-Methoxytryptamine	0.095
Cyproheptadine	0.22
Serotonin (5-hydroxytryptamine)	0.23 ± 0.1 (2)
5-Hydroxy-N^ω-methyltryptamine	0.24
5-Methyltryptamine	0.41 ± 0.1 (2)
Bufotenine (N^ω-dimethyl-5-hydroxytryptamine)	0.50
N^ω-Methyltryptamine	0.62
7-Methyltryptamine	2.3
Tryptamine	2.7 ± 0.6 (2)
(+)-Lysergic acid	3.7
6-Hydroxytryptamine	12.0
Tyramine	34.0

No effect at 10μM:

Corticotropin (ACTH), bradykinin, delta sleep-inducing factor, glucagon, insulin, Leu-enkephalin, gonadoliberin (LH/FSH-RF), α-melanotropin (α-MSH), melanostatin (MIF), neurotensin, oxytocin, somatotropin, substance P, thyroliberin (TRH), vasopressin; adrenochrome and (−)-noradrenalin.

No effect at 100μM:

(+)-Amphetamine, dopamine, harmaline, harmine, histamine, γ-aminobutyric acid, glycine, 5-hydroxyindole acetic acid, 5-hydroxytryptophol, DL-muscarine, naloxone, N-acetyl-5-hydroxytryptamine, spermidine and L-tryptophan.

* by Scatchard[10] plot, equilibrium dissociation constant.

bound factors. This reversibility and the heat stability also exclude that it might be a receptor-destroying enzyme activity.

We have tested whether the high-affinity LSD-binding sites of pig brain have the same high degree of binding specificity and affinity as the binding sites of rat brain[1-4]. After simultaneous addition of [^{3}H]LSD and various drugs in at least five different concentrations, residual LSD binding was assayed in triplicate and the concentrations for half-maximal inhibition (IC_{50} values) were obtained (Table 2). Aside from cyproheptadine and methysergide, both antagonistic drugs, only physiologically occurring indolealkylamines and some of their derivatives had high affinity (IC_{50} value below 1μM). Affinity was only slightly changed by methylation of serotonin at the amino and hydroxyl groups, provided that the hydroxyl group was in the 5-position. With other physiological ligands, such as putative neurotransmitters and hormones, no influence on LSD-binding was detectable, even at concentrations of 10 or 100μM.

Further studies are necessary to test whether LSD-displacing factors resembles such a functional ligand of the LSD receptors, as was established for the endogenous opioid peptides of the morphine receptors[11].

The technical assistance of Mrs. *Jutta Redemann* and Mr. *C. Suchanek* are gratefully acknowledged. We also thank *Dr. M. Reddington* for correction of the English text and the *Deutsche Forschungsgemeinschaft* for generous support.

Literature

1 Farrow, J.T. & van Vunakis, H. (1973) *Biochem. Pharmacol.* **22**, 1103 - 1113.
2 Bennett, J.L. & Aghajanian, G.K. (1974) *Life Sci.* **15**, 1935 - 1944.
3 Bennett, J.P., Jr. & Snyder, S.H. (1975) *Brain Res.* **94**, 523 - 544.
4 Lovell, R.A. & Freedman, D.X. (1976) *Mol. Pharmacol.* **12**, 620 - 630.
5 Mehl, E. & Weber, L. (1974) in Advances in Biochemical Psychopharmacology (Costa, E., Gessa, G.L. & Sandler, M., eds.) vol. 11, pp. 105 - 108, Raven, Press, New York.
6 Mehl, E. (1975) *Exp. Brain Res. Suppl.* **23**, 139.
7 Mehl, E. & Weber, L. (1974) in Biochemistry of Sensory Functions, 25. Mosbacher Colloquium (Jaenicke, L., ed.) pp. 593 - 595. Springer-Verlag, Berlin, Heidelberg, New York.
8 Mehl, E., Rüther, E. & Redemann, J. (1977) *Psychopharmacologica (Berlin)* **54**, 9 - 16.
9 Schloßberger, H.G. & Buhrow, I. (1963) *Hoppe-Seyler's Z. Physiol. Chem.* **333**, 152 - 153.
10 Scatchard, G. (1949) *Ann. N.Y. Acad. Sci.* **51**, 660 - 672.
11 Goldstein, A. (1976) *Science* **193**, 1081 - 1086.

Dr. E. Mehl and L. Guiard, Max-Planck-Institut für Psychiatrie, Kraepelinstr. 2, D-8000 München 40.

Hoppe-Seyler's Z. Physiol. Chem.
Bd. 359, S. 543 - 545, April 1978

KURZMITTEILUNG

Anreicherung und Charakterisierung eines Leberzell-Proliferationsfaktors aus teilhepatektomierten Rattenlebern

Gerhard RUHENSTROTH-BAUER, Michel GOLDBERG, Siegfried SILZ und Wolf STRECKER

Max-Planck-Institut für Biochemie, Abteilung für Experimentelle Medizin, Martinsried bei München

(Der Schriftleitung zugegangen am 16. Februar 1978)

Adolf Butenandt zum 75. Geburtstag in herzlicher Verbundenheit gewidmet.

Concentration and Characterization of a Liver Cell Proliferation Factor from Partially Hepatectomized Rat Livers

Summary: An extract of rat liver, made 12 h after partial hepatectomy, was injected into untreated rats. The synthesis of DNA in the liver cells of the injected animals showed a 3.3 fold increase in comparison with controls. A similar increase was observed in the mitosis index. Injection of liver extracts from normal rats showed no effect when compared with the injection of NaCl.

The effect of the active extract was not altered by neuraminidase treatment. Trypsin-chymotrypsin treatment, however, removes the activity. The active factor must therefore be either a protein or a protein-like substance. The molecular weight of the factor was shown to be between about 30000 and 50000 by molecular filtration. The specific activity after molecular filtration was raised by a factor of 100.

The factor is inert with respect to the proliferation of kidney and spleen. It is apparently not species-specific, however, since it also causes increased proliferation of liver cells in normal mice.

Key words: Rat, partial hepatectomy, liver cell proliferating factor, concentration, characterization.

Die Leber von Säugetieren ist u.a. durch zwei Regelkreise gekennzeichnet. Der eine betrifft das mittlere Volumen der Leberzellen; dieses ist mit dem Körpergewicht korreliert. Der zweite Regelkreis betrifft die Zahl der Leberzellen: Diese Größe ist bei jugendlichen und erwachsenen Tieren ungefähr gleich[1]. Die beiden Regelkreise sind voneinander unabhängig[2]. Nach einer Teilhepatektomie erhöht sich die normal sehr niedrige Proliferationsrate der Leberzellen außerordentlich, bis die Ausgangszahl der Leberzellen ungefähr wieder erreicht ist; dann fällt sie auf das niedrige Ausgangsmaß zurück. Im folgenden wird ein Faktor beschrieben, der ein Glied dieses Regelkreises darzustellen scheint.

Es gibt mehrere Hinweise dafür, daß die Proliferationserhöhung von Leberzellen nach einer Teilhepatektomie durch einen proliferationsaktiven Faktor verursacht wird[1,3]. Ein Teil dieser Hinweise konnte allerdings bei Nachuntersuchungen nicht bestätigt werden. Dies gilt z.B. für die Angabe von Levi und Zeppa[4], daß das In-vitro-Perfusat von teilhepatektomierten Rattenlebern einen Faktor enthält, der bei gleichzeitiger Perfusion in normale Lebern kurzfristig zu einer Proliferationserhöhung führt. Dies konnten wir jedoch bei unserer Nachprüfung nicht finden (unveröffentlicht).

Auf der anderen Seite konnte bisher auch noch nicht völlig ausgeschlossen werden, daß die Proliferationsstei-

Enzym:
Neuraminidase, Acylneuraminyl-Hydrolase (EC 3.2.1.18).

Abkürzungen:
NL-Extrakt = normaler Leberextrakt;
TL-Extrakt = Extrakt aus den Lebern teilhepatektomierter Tiere.
SPF = spezifisch pathogenfrei.

gerung nach Teilhepatektomie durch das Verschwinden eines im Normalzustand vorhandenen Proliferationsinhibitors zustande kommt.

Im folgenden beschreiben wir die Anreicherung eines Proliferations-stimulierenden Faktors aus teilhepatektomierten Lebern sowie seine erste biochemische und biologische Charakterisierung.

Material und Methoden

Tiere

Für alle Versuche wurden weibliche 95 - 105 g schwere, SPF-Wistar-Ratten (Institut für Strahlen- und Umweltforschung, Neuherberg/München) verwendet. Sie bekamen vor und nach der Behandlung als Nahrung Wasser und Altromin ad libitum. Sämtliche Tiere wurden zwischen 19 und 21 Uhr operiert, der Zeitpunkt der Tötung lag stets 12 h danach. Die etwa 68proz. Hepatektomie wurde nach Higgins und Anderson[5] unter leichter Äthernarkose durchgeführt.

Präparation des proliferationsaktiven Faktors

Zur Gewinnung des Extrakts wurden die Tiere unter leichter Narkose entblutet und die Restleber vor Entfernung von der V. portae aus mit 0.9proz. NaCl-Lösung durchspült. Zur selben Zeit wurden unbehandelte Ratten als Kontrolle getötet und wie beschrieben verfahren. Jeweils drei der Restlebern wurden mit der 4fachen Menge (w/w) bidest. Wasser in einem Elvehjem-Potter homogenisiert; mit einer äquivalenten Lebermenge der Kontrolltiere wurde ebenso verfahren. Die Leberhomogenate wurden mit 0.1N HCl auf pH 5.5 gebracht und hitzedenaturiert (20 min, 95 °C). Anschließend wurden die Homogenate zentrifugiert (15 min, 4000 × *g*, Minifuge Christ, Osterode/Harz). Die Niederschläge wurden mit bidest. Wasser auf das ursprüngliche Volumen aufgefüllt und die Hitzedenaturierung und Zentrifugation 2mal wiederholt. Die vereinigten drei Überstände wurden lyophilisiert und bis zur Weiterverwendung bei − 20 °C aufbewahrt.

Meßmethode des Proliferations-stimulierenden Faktors

Normalen Ratten wurden die in 2 m*l* 0.9proz. NaCl-Lösung gelösten lyophilisierten Überstände i.p. injiziert; Kontrolltiere erhielten 2 m*l* der von Normaltieren gewonnenen Extrakte bzw. 2 m*l* 0.9proz. NaCl-Lösung. 19 h nach der Injektion bekamen die Tiere i.p. 50 µCi [^{3}H]Methylthymidin (spez. Aktivität 25 Ci/mmol; Radiochem. Center Amersham) und wurden 1 h später getötet. Die Leber wurde entfernt und bei − 20 °C aufbewahrt. Die DNA der Leber wurde nach Weinbren und Woodward[6] extrahiert. Mit einem Teil der Lösung wurde die DNA-Konzentration nach Burton[7] bestimmt, der andere Teil wurde zur Radioaktivitätsmessung verwendet. Dazu wurde 1.5 m*l* der Perchlorsäurehaltigen Lösung mit 0.5 m*l* 1N NaOH neutralisiert und im Szintilationsgläschen mit 5 m*l* Triton X-100 und 10 m*l* Toluol (0.6% PPO) versetzt. Gemessen wurde die Radioaktivität in einem Liquid-Szintilationszähler (Intertechnique, Paris). Die spezifische Aktivität wurde in Zpm/(µg DNA × g Leber) ausgedrückt.

Enzymatische Behandlung des Faktors

1. Trypsin-Chymotrypsin-Behandlung

Die lyophilisierten Überstände wurden in 20 m*l* bidest. Wasser gelöst, auf pH 7.6 gebracht und bei 30 °C für 2 h mit jeweils 80 U Trypsin, reinst und 90 U α-Chymotrypsin, reinst (Serva, Heidelberg) inkubiert. Zur Inaktivierung der restlichen Enzymaktivitäten wurde dann 30 min bei 95 °C inkubiert und zentrifugiert. Danach wurde die Lösung lyophilisiert, in 2 m*l* 0.9proz. NaCl-Lösung aufgenommen und i.p. injiziert.

2. Neuraminidase-Behandlung

Die lyophilisierten Überstände wurden jeweils in 20 m*l* bidest. Wasser gelöst, auf pH 5.5 gebracht und 1 h bei 37 °C mit 250 U Neuraminidase/Ansatz (Behringwerke, Marburg) inkubiert. Anschließend wurde die überschüssige Neuraminidase bei 95 °C während 30 min inaktiviert, zentrifugiert (15 min, 3000 × *g*) und wieder lyophilisiert. Das Lyophilisat wurde in 2 m*l* 0.9proz. NaCl-Lösung aufgenommen und i.p. injiziert.

Mitosezählung

24 h nach Injektion von normalem Leber-(NL-)Extrakt oder dem Leber-Extrakt aus teilhepatektomierten Ratten (TL) wurden die Ratten entblutet und die Lebern vor Entfernung mit 0.9proz. NaCl-Lösung gespült. Zur histologischen Präparation nach Pera[8] und Silz et al.[9] wurde ein Teil des Lobus caudatus verwendet. Die 50-µm-Schnitte wurden mit Hämatoxylin-Eosin angefärbt und die Mitoserate durch Auszählen von 10000 Zellen/Schnitt bestimmt.

Ergebnisse und Besprechung

Die Injektion von NL-Extrakt in normale Ratten ergab nach der Injektion von [H^3]Methylthymidin und bei den übrigen genannten Versuchsbedingungen in den Lebern eine mittlere spezifische Aktivität von 3350 ± 1320 (n = 8). Bei der Injektion von physiologischer Kochsalzlösung (n = 8) wurde 3470 ± 740 gefunden. Die Extrakte aus normalen Lebern entsprechen demnach hinsichtlich ihrer Einbauaktivität den NaCl-Kontrollen. TL-Extrakt dagegen ergab eine mittlere spezifische Aktivität von 11240 ± 4750. Dies ist gegenüber den beiden Kontrollwerten eine Steigerung der Einbauaktivität um den Faktor 3.3. Der Unterschied ist statistisch hoch gesichert ($p \ll 0.01$).

Die Zählung der Mitoserate ergab, daß die Injektion von TL-Extrakt nicht nur zu einer Steigerung der DNA-Synthese, sondern zu einer wirklichen Erhöhung der Proliferationsrate der Leberzellen führt: der Mitoseindex betrug bei einer Doppelbestimmung 7 bzw. 4 Mitosen/10^4 Zellen gegenüber 0 - 2 Mitosen/10^4 Zellen bei normalen Ratten[8].

Die Behandlung aktiver Extrakte mit einem Trypsin-Chymotrypsingemisch führt zur Zerstörung der Aktivität (3230 ± 1340 (n = 5)), dagegen vermindert die Neuraminidaseeinwirkung die Einbauaktivität nicht (13900 ± 5650 (n = 3). Demnach ist die aktive Substanz entweder ein Protein oder ein Proteid, das keine für die Aktivität bedeutsame Neuraminsäure besitzt. Der TL-Extrakt wurde durch Molekularfilter von Amicon (Lexington, USA) verschiedener Porenweite druckfiltriert. Bei der Filtration durch die Filter PM 30 wurde die gesamte Aktivität zurückgehalten. Das Molekulargewicht liegt demnach oberhalb 30000. Bei Druckfiltrationen durch XM-50-Filter blieb etwa die Hälfte der Aktivitäten im Filtrat, die anderen Hälften wurden zurückgehalten. Dieses Verhalten[10] spricht dafür, daß das Molekulargewicht etwa zwischen 30000 und 50000 liegt.

Die spezifische Aktivität des Extrakts nach Druckfiltration durch ein PM-30-Filter zeigte gegenüber dem Ausgangsmaterial eine etwa 100fache Anreicherung. Die Aktivität von TL-Extrakt ist (mindestens zum Teil) organspezifisch: der radioaktive Einbau in Milz oder Nierengewebe war bei Injektionen von TL-Extrakt, NL-Extrakt und NaCl nicht unterscheidbar. Dagegen scheint der Extrakt nicht speziesspezifisch zu sein, da die Injektion eines aktiven Rattenleberextrakts in NMRI-Mäuse (21 - 24 g, SPF, Institut für Strahlen- und Umweltforschung, Neuherberg/München) zu dem gleichen Ergebnis wie in Ratten führte (TL-Extrakt: 12050 ± 6510; NL-Extrakt: 4350 ± 530; NaCl: 3240 ± 1800; n = 5).

Herrn *D.-A. Feeny* sei für seine geschickte und exakte technische Assistenz sehr gedankt. Für die Anfertigung der histologischen Präparate danken wir Frau *Friederike Ring*.

Literatur

1 Bucher, N. L. R. & Malt, R. A. (1971) Regeneration of Liver and Kidney, Little, Brown & Company, Boston.
2 Tongendorff, J., Trebin, R. & Ruhenstroth-Bauer, G. (1975) *Am. J. Pathol.* **80**, 519 - 524.
3 LaBrecque, D. R. & Pesch, L. A. (1975) *J. Physiol.* **248**, 273 - 284.
4 Levi, J. U. & Zeppa, R. (1972) *J. Surg. Res.* **12**, 114 - 119.
5 Higgins, G. M. & Anderson, R. M. (1931) *Arch. Pathol. (Chicago)* **12**, 186 - 202.
6 Weinbren, K. & Woodward, E. (1964) *Br. J. Exp. Pathol.* **45**, 442 - 449.
7 Burton, K. (1956) *Biochem. J.* **62**, 315 - 323.
8 Pera, F. (1972) *Histochemie* **30**, 82.
9 Silz, S., Valet, G., Tongendorff, J., Strecker, W. & Ruhenstroth-Bauer, G. (1976) *Acta Hepato-Gastroenterol.* **23**, 255 - 261.
10 Amicon Ultrafiltration, Anleitung und Katolog (1977).

Prof. Dr. G. Ruhenstroth-Bauer, M. Goldberg, Dr. S. Silz und W. Strecker, Max-Planck-Institut für Biochemie, Abt. für Experimentelle Medizin,
D-8033 Martinsried bei München.

Titel Nr. 3 109 002 691

Anzeigenverwaltung: Merkur Werbung GmbH, Postfach 12 45, 5210 Troisdorf 1, Tel. (0 22 41) 4 20 51.
Für den Anzeigenteil verantwortlich: Wolfgang Frieben, Troisdorf. – Printed in Germany – Satz: R. & J. Blank, München; Druck: Grafik + Druck, München; Bindearbeiten: F. Steinberger & Sohn, München.

Redaktionelle Hinweise

(Wichtige Änderungen ab Januar 1976 sind am Rand markiert)

Die Zeitschrift bringt nur Experimantalarbeiten aus den Gebieten der Biochemie und Physiologischen Chemie. Die Manuskripte werden in deutscher oder in englischer Sprache veröffentlicht. Es wird vorausgesetzt, daß alle Koautoren mit der eingereichten Fassung einverstanden sind.

Aktuelle Kurzmitteilungen, deren Umfang 4 Seiten in Kleindruck (ca. 8 Manuskriptseiten) nicht übersteigen darf, werden bevorzugt bearbeitet und in das nächsterreichbare Heft aufgenommen. Die Dringlichkeit zur Prioritätssicherung ist vom Autor in einem Begleitschreiben kurz zu begründen.

Manuskripte sind in zweifacher Ausfertigung an eine der folgenden Anschriften zu senden:

Prof. Dr. A. Butenandt oder Prof. Dr. F. Lynen, Max-Planck-Institut für Biochemie, D-8033 Martinsried b. München;

Prof. Dr. Dr. G. Weitzel, Physiologisch-Chemisches Institut der Universität, Hoppe-Seyler-Str. 1, D-7400 Tübingen;

Dr. A. Dillmann, Hoppe-Seyler-Redaktion, Am Klopferspitz, D-8033 Martinsried b. München, Tel. (089) 85 85-823.

Bei der Abfassung der Manuskripte bitten wir dringend, folgende Regeln zu beachten, die im wesentlichen mit den „Instructions to Authors" von *Biochimica et Biophysica Acta* und *European Journal of Biochemistry* übereinstimmen:

Es soll in Maschinenschrift (2zeilig, auch vorgesehener Kleindruck!) und mit 4 cm breitem Rand geschrieben werden. Vorgesehene Absätze sind deutlich zu markieren.

Seite 1 des Manuskripts soll enthalten (in normaler Schrift, nicht in Großbuchstaben):

a) Titel der Arbeit
b) Autorennamen (jeweils ein Vorname ausgeschrieben)
c) Name des Instituts oder Fachbereichs, in dem die Arbeit ausgeführt wurde (möglichst in der Sprache des Herkunftlandes), vorausgesetzt, daß die Leitung des Instituts oder Fachbereichs der Veröffentlichung zugestimmt hat.
d) Kurztitel, maximal 70 Satzzeichen
e) Adresse für den Versand der Korrekturen

Auf Seite 2 folgen

a) Eventuelle Widmung
b) Vollständige Postadresse der Autoren in der Sprache des Herkunftslandes
c) Ad-hoc-Abkürzungen (nur in seltenen Fällen zugelassen)
d) Enzyme in alphabetischer Reihenfolge der empfohlenen Trivialnamen mit systematischen Namen und EC-Nummern

Auf Seite 3 folgt die *Zusammenfassung* sowie eine Übersetzung des Titels und der (eventuell erweiterten) Zusammenfassung ins Englische (bzw. Deutsche). Diese Übersetzung oder deren sprachliche Überprüfung kann auch von der Redaktion übernommen werden.

Eventuelle *Danksagungen* sind an den Schluß des Manuskriptes zu stellen.

Literaturzitate im Text sind fortlaufend zu numerieren (nicht alphabetisch) und am Schluß des Manuskripts zu bringen. Die in Chemical/Biological Abstracts üblichen Zeitschriftenabkürzungen sind zu verwenden. Zur Anordnung s. IUB Commission of Editors of Biochemical Journals (1973) *Eur. J. Biochem.* **37**, 201 - 202; (1974) *diese Z.* **355**, 1 - 3.

Beispiel: Spackman, D.H., Stein, W.H. & Moore. S. (1958) *Anal. Chem.* **30**, 1190 - 1206.

Zeichnungen sollen einmal in reproduzierbarer Form mit schwarzer Tusche gezeichnet eingereicht werden; Zahlenskalen, gemessene Größen, Einheiten und sonstige Beschriftungen sind jedoch nur leicht mit weichem Bleistift einzutragen. Die Abbildungen sollen so groß sein, daß sie auf 1/2 oder 1/3 verkleinert werden können. Deutlich beschriftete Kopien für die Referenten sind beizufügen.

Alle *Tabellen* und *Abbildungen* sind mit Legenden zu versehen, so daß die Ergebnisse auch ohne Zuhilfenahme des Textes ersichtlich werden. Es wird gebeten, die Legenden nicht zu unterstreichen und die Abbildungslegenden nicht auf die Abbildungen selbst, sondern auf ein eigenes Blatt zu schreiben.

Symbole für die am häufigsten vorkommenden Größen und ihre Einheiten sind in Tab. 1 zusammengestellt. Wir bitten streng zu unterscheiden, ob bei Ansätzen molare Konzentrationen (mol/l = M ; z.B. 0.25M) oder zugesetzte Mengen (z. B. in mol, mmol, μmol oder nmol) gemeint sind.

Hinsichtlich der *Nomenklatur* richte man sich nach der IUPAC-Nomenklatur sowie nach den Empfehlungen der "IUPAC-IUB Commission on Biochemical Nomenclature": Kohlenhydrate, *Eur. J. Biochem.* **21**, 455 - 477 (1971); **25**, 4 (1972); α-Aminosäuren, *Eur. J. Biochem.* **53**, 1 - 14 (1975);

Tab. 1. Symbole für Größen und ihre Einheiten

Größe		Einheit[a]
l	Länge	m, mm, μm, nm, etc.
V	Volumen	m^3, l (dm^3), ml, etc.
λ	Wellenlänge	nm (nicht mμ oder Å)
t	Zeit	h, min, s, ms, μs, etc.
ν	Frequenz	Hz
m	Masse	kg, g, mg, μg, etc.
ρ	Dichte	g/cm^3
s	Sedimentationskoeffizient	$S = 10^{-13}$ s
n	Substanzmenge	mol, mmol, etc.
M	molare Masse (m/n)	g/mol
–	Masse eines Moleküls	Dalton[b]
M_r	Molekulargewicht	(dimensionslos)
c_B, [B]	molare Konzentration einer Substanz B	M, mM, μM, etc. (besser als 10^{-3} M, etc.)
T	Temperatur (absolut)	K (nicht °K)
t, θ	Temperatur	°C
q, Q	Wärmemenge	J (besser als cal)[c]
I	Ionenstärke	M (mol/l), mM, etc.
K	Gleichgewichtskonstante	
K_m	Michaeliskonstante	
K_i	Inhibitorkonstante	
k	Geschwindigkeitskonstante	
v	Reaktionsgeschwindigkeit	mol/s (oder mmol, μmol oder nmol/s, etc.), besser als μmol/min
V (nicht V_{max})	Geschwindigkeit der enzymkatalysierten Reaktion bei unendlich großer Substratkonzentration	
I	Lichtintensität	cd
T	Durchlässigkeit (I/I_0)	(dimensionslos)
A	Absorption[d] ($-\lg T$)	(dimensionslos)
ϵ	molarer Extinktionskoeffizient	$M^{-1} \times cm^{-1}$

[a] Präfixe für Einheiten: T, Tera (10^{12}); G, Giga (10^9); M, Mega (10^6), k, Kilo (10^3); m, Milli (10^{-3}); μ, Mikro (10^{-6}); n, Nano (10^{-9}); p, Piko (10^{-12}); f, Femto (10^{-15}); a, Atto (10^{-18}).

[b] Das Dalton ist 1/12 der Masse eines Atoms des Nuklids ^{12}C.

[c] Das Joule ist die SI-Einheit der Energie. Die Kalorie (1 cal = 4.184 J; 1 J = 0.239 cal) ist für eine Übergangszeit noch zugelassen; in jedem Fall sollten aber die Angaben auch in Joule ausgedrückt werden.

[d] Die Ausdrücke optische Dichte, Extinktion oder Absorbancy sollen nicht mehr verwendet werden.

Cyclite, *diese Z.* **350**, 523 - 535 (1969), *Eur. J. Biochem.* **57**, 1 - 7 (1975); Lipide, *diese Z.* **350**, 279 - 285 (1969); Steroide, *diese Z.* **351**, 663 - 689 (1970); Chinone mit isoprenoiden Seitenketten, Folsäure u. verschiedene andere Verbindungen, *diese Z.* **348**, 266 - 276 (1967); Corrinoide, *Eur. J. Biochem.* **45**, 7 - 12 (1974); Vitamine B_6, *Eur. J. Biochem.* **40**, 325 - 327 (1973); Carotinoide, *Eur. J. Biochem.* **25**, 397 - 408 (1972), **57**, 317 - 318 (1975); synthetische Analoga natürlicher Peptide, *diese Z.* **348**, 262 - 265 (1967); Enzyme Nomenclature, Recommendations (1972), Elsevier Publ. Comp., Amsterdam; multiple Enzymformen, *diese Z.* **353**, 852 - 854 (1972); Eisen-Schwefel-Proteine, *Eur. J. Biochem.* **35**, 1 - 2 (1973); Human-Immunglobuline, *Eur. J. Biochem.* **45**, 5 - 6 (1974); Peptidhormone, *Eur. J. Biochem.* **55**, 485 - 486 (1975). Für die Rechtschreibung ist „Jansen/Mackensen, Rechtschreibung der technischen und chemischen Fremdwörter" maßgebend.

Tab. 2 Empfohlene Abkürzungen.

Atome, isotope	=	z. B. [16-^{14}C]Östron
Atome, numeriert	=	z. B. C-5
Curie	=	Ci
Impulse pro Minute	=	Ipm (engl. cpm)
Prozent der Theorie	=	% d. Th.
prozentig	=	proz. (nicht %-ig)
Relative Wanderungsgeschwindigkeit	=	R_F
Schmelzpunkt	=	Schmp. (engl. m.p.)
Siedepunkt	=	Sdp. (engl. b.p.)
Umdrehungen pro Minute	=	U./min (engl. rpm)
Zerfälle pro Minute	=	Zpm (engl. dpm)
Zersetzung	=	Zers. (engl. decomp.)

Bezüglich der *Abkürzungen und Symbole für chemische Namen,* die in der biologischen Chemie von besonderem Interesse sind, wird auf die Empfehlungen der „IUPAC-IUB Commission on Biochemical Nomenclature" verwiesen (siehe z.B.: Allgemeines, Aminosäuren, Zucker, Phosphate, Coenzyme, *diese Z.* **348**, 245 - 265 (1967); Aminosäurederivate und Peptide, *Eur. J. Biochem.* **27**, 201 - 207 (1972); synthet. Polypeptide, *diese Z.* **349**, 1013 - 1016 (1968); Nucleinsäuren, Polynucleotide und deren Bestandteile, *diese Z.* **351**, 1055 - 1063 (1970). Andere als die dort empfohlenen Abkürzungen werden nur in seltenen Ausnahmefällen zugelassen. Sie müssen am Beginn jeder Arbeit in einer Fußnote erklärt werden.

Sonderdrucke: Von jeder Arbeit werden dem Verfasser 75 Sonderdrucke kostenlos geliefert. Die Bestellung weiterer Sonderdrucke gegen Berechnung ist möglich, muß jedoch spätestens bei der Rücksendung der Korrekturen erfolgen.

Der Verlag erwirbt mit der Annahme des Manuskriptes das Verlagsrecht und das ausschließliche Recht der Vervielfältigung, Verbreitung und Übersetzung. Der Verlag behält sich vor, den Inhalt dieser Zeitschrift für fremdsprachige Ausgaben zu verwenden.

Bezugspreis: Der Jahresbezugspreis (1 Jahrg. zu 12 Heften) beträgt DM 590,–, das Einzelheft kostet DM 56,–. Bestellungen nimmt jede Buchhandlung entgegen.

Verlag: WALTER DE GRUYTER & Co.
Genthiner Str. 13, D-1 Berlin 30. Tel. (030) 261 1341
Telegrammadresse: Wissenschaft Berlin
Postscheck: Berlin-West 10307-108

WALTER DE GRUYTER Inc.
3 Westchester Plaza
Elmsford, N.Y. 10523, U.S.A.
Tel. (9 14) 592-5890

Für die nächsten Hefte sind folgende Beiträge vorgesehen:

Synthesis of protected decapeptide (B_{21-30}) of human insulin, *V. K. Naithani and J. Föhles*

Human Plasma Kallikrein: Reinigung, enzymatische Charakterisierung und quantitative Bestimmung in Plasma, *H. Heber, R. Geiger und N. Heimburger*

Multiple protein kinases from *Trypanosoma gambiense*, *R. D. Walter*

Adenosine 3',5'-cyclic monophosphate binding proteins from *Trypanosoma gambiense*, *R. D. Walter*

Investigations on the functional role of rat liver carboxylesterase isoenzymes, *R. Arndt, H.-E. Schlaak, D. Uschtrin, D. Südi, K. Michelssen and W. Junge*

Vitamin-B_6-Biosynthese in *Bacillus subtilis*, *W. Pflug and F. Lingens*

Immunoreactive trypsin in serum and peritoneal fluid in acute pancreatitis, *A. Borgström and K. Ohlsson*

Release of granulocyte eleastase in lethal canine endotoxin shock, *A. O. Aasen and K. Ohlsson*

Studies on the influence of Trasylol on the partition of trypsin between the human plasma protease inhibitors in vitro, *G. Balldin, K. Ohlsson and A.-S. Ohlsson*

Uptake of thymidine into isolated rat hepatocytes. Evidences for two transport systems, *F. R. Ungemach and D. Hegner*

Changes in the microheterogeneity of histone H1 after mengovirus infection of Ehrlich ascites tumor cells, *U. Traub and P. Traub*

Enhanced phosphorylation of ribosomal protein S6 and other cytoplasmic proteins after mengovirus infection of Ehrlich ascites tumor cells, *I. Rosnitschek, U. Traub and P. Traub*

Studies on the biosynthesis of cyclitols, XXXVI Purification of *myo*-Inositol-1-phosphate synthase of the duckweed, *Lemna gibba*, to homogeneity by affinity chromatography on NAD-sepharose. Molecular and catalytic properties of the enzyme, *E. Olugbenga Ogunyemi, F. Pittner and O. Hoffmann-Ostenhof*

Hämoglobine, XXII: Phosphat/Protein-Wechselwirkung und die Atmung des Lamas, des menschlichen Foetus und die des Pferdes, *G. Braunitzer, B. Schrank, A. Stangl und C. Bauer*

Snake venoms. The amino-acid sequence of protein S_5C_4 from *Dendroaspis jamesoni kaimosae* (Jameson's mamba) venom, *F. J. Joubert, A. J. C. Strydom and N. Taljaard*

d 359 Heft 5 Mai 1978

N 0018-4888 A 20679 E

Hoppe-Seyler's Zeitschrift für Physiologische Chemie

Herausgegeben von

A. Butenandt, F. Lynen, G. Weitzel

WALTER DE GRUYTER · BERLIN · NEW YORK

Hoppe-Seyler's Z. Physiol. Chem. Bd. 359 S. 547 - 658 Mai 1978
HSZPAZ 359 (5) 547 - 658 (1978)

ACTA BIOCHIMICA ET BIOPHYSICA
ACADEMIAE SCIENTIARUM HUNGARICAE

VOLUME 12 1977 NUMBER 3

Contents

ACTA BIOCHIMICA ET BIOPHYSICA, a journal of the Hungarian Academy of Sciences, appeares four times a year, making up a volume of some 400 pages. Subscription rate per volume: $32.00; DM 90,—. Size: 17 × 25 cm.

AKADÉMIAI KIADÓ, Budapest
Distributors: KULTURA, H-1389 Budapest, P. O. B. 149

Hoppe-Seyler's Zeitschrift für Physiologische Chemie
Begründet 1877 · Fortgeführt von A. Kossel, F. Knoop und K. Thomas

Bd. 359 Mai 1978 Heft 5

Inhaltsverzeichnis / Contents

Fortsetzung nächste Seite *Continuous*

Redaktionelle Hinweise siehe gegenüber Seite 658

Indexed in Current Contents

Autoren-Verzeichnis

Hoppe-Seyler's Z. Physiol. Chem.
Bd. 359, S. 547 - 558, Mai 1978

Hämoglobine, XXII[1]
Phosphat/Protein-Wechselwirkung und die Atmung des Lamas, des menschlichen Fötus und die des Pferdes

Gerhard BRAUNITZER, Barbara SCHRANK, Anton STANGL und Christian BAUER

Max-Planck-Institut für Biochemie, Abteilung Proteinchemie, Martinsried bei München, und
Institut für Physiologie, Universität Regensburg, Regensburg

(Der Schriftleitung zugegangen am 16. Februar 1978)

Prof. A. Butenandt zum 75. Geburtstag gewidmet

Zusammenfassung: Die Sequenzanalyse des Hämoglobins von Lama *(Lama glama, Camelidae)* wird beschrieben. Die Ketten wurden getrennt, tryptisch gespalten, quantitativ charakterisiert und nach der Filmmethode sequenziert. Die α- bzw. β-Ketten unterscheiden sich in 25 bzw. 24 Aminosäuren gegenüber dem Hämoglobin des Menschen.

Wir diskutieren die Wechselwirkung zwischen Protein und Phosphat. Der früher erhobene Befund, daß die O_2-Affinität des Lamahämoglobins vom Gehalt an 2,3-Bisphosphoglycerat abhängig ist, wird gedeutet als Mutation der 2,3-Bisphosphoglycerat-Kontaktstelle β2 His beim Menschen zum β2 Asn beim Lama, wodurch eine der vier Kontaktstellen außer Funktion gesetzt wird. Diese Änderung bewirkt eine verminderte Reduktion der intrinsischen Sauerstoffaffinität der Hämoglobinpartikel und deutet molekular die erhöhte Sauerstoffaffinität des Hämoglobins und somit die Höhenatmung des Lamas.

Analog läßt sich die erhöhte O_2-Affinität des fötalen Hämoglobins beim Menschen aufgrund früherer physiologischer Untersuchungen am Blut und fötalen Hämoglobin durch die Inaktivierung der 2,3-Bisphosphoglycerat-Kontaktstelle β143 His des adulten Hämoglobins durch Mutation zu γ143 Ser des fötalen Hämoglobins deuten.

Es wird weiterhin die Atmung beim Pferdehämoglobin (2,3-Bisphosphoglycerat-Kontaktstelle β2 Gln) diskutiert und durch Vermessung der atomaren Parameter gezeigt, daß die Amidgruppe des Glutamins nahe genug am 2,3-Bisphosphoglycerat-Sauerstoff liegt, um eine Wasserstoffbrücke zum Phosphat zu bilden. Somit ist die Niederland-Atmung des Pferdes erklärt: Glutamin und Histidin erfüllen stereochemisch dieselbe Funktion.

The Interaction Between Phosphate and Protein, and the Respiration of the Llama, the Human Fetus and the Horse

Summary: The sequence analysis of llama *(Lama glama, Camelidae)* hemoglobin is described. The chains were separated, cleaved by trypsin as previously described, quantitatively characterized and sequenced in the sequenator. The llama hemoglobin differs from the human hemoglobin in that it has 25 different amino acids in the α chain and 24 different amino acids in the β chain.

Abkürzungen:
Quadrol = *N,N,N',N'*-Tetrakis(2-hydroxypropyl)äthylendiamin; I = 4-(Isothiocyanato)benzol-sulfonsäure, Natriumsalz. IV = 7-(Isothiocyanato)naphthalin-1,3,5-trisulfonsäure, Trinatriumsalz. HbA = adultes Hämoglobin; HbF = fötales Hämoglobin. TPCK = *N*-Tosyl-L-phenylalanylchlormethan.

The interaction between protein and phosphate is discussed. The earlier finding that the O_2 affinity of the llama hemoglobin is dependent on its content of 2,3-bisphosphoglycerate is interpreted here as a mutation of the 2,3-bisphosphoglycerate contact position β2 His in human hemoglobin to β2 Asn in llama hemoglobin, whereby one of the four 2,3-bisphosphoglycerate contact points is interrupted. This interrruption gives rise to a diminished reduction of intrinsic oxygen affinity in the hemoglobin molecule and explains, on a molecular basis, the increased oxygen affinity of the llama hemoglobin, and consequently, the high-altitude respiration of the llama.

By analogy, the increased O_2 affinity of human fetal hemoglobin is interpreted according to previous physiological investigations on blood and fetal hemoglobin by the inactivation of the phosphoglycerate contact point β143 His in the adult hemoglobin by mutation to γ143 Ser in the fetal hemoglobin.

With respect to respiration in horses (2,3-bisphosphoglycerate contact β2 Gln), measurement of atomic parameters show that the amido group of the glutamine is situated close enough to the 2,3-bisphosphoglycerate oxygen to build a hydrogen bond with the phosphate. Consequently, the explanation of the low-altitude respiration of the horse lies in the fact that glutamine and histidine fulfill sterochemically an identical function.

Key words: Respiration, high altitude, mammals, hemoglobin, 2,3-bisphosphoglycerate.

Wir berichten über die Sequenz der α- und β-Ketten des Hämoglobins des Lamas *(Lama glama, Camelidae)*, nachdem wir früher die tryptischen Peptide[2] und in zwei Kurzmitteilungen die Sequenz der α-[3] und β-Ketten[4] sowie weiterhin die 2,3-Bisphosphoglycerat-Abhängigkeit der intrinsischen Sauerstoffaffinität des Hämoglobins des Lamas[5] beschrieben haben.

In dieser Arbeit bringen wir die Belege, die zur Ableitung der Konstitution geführt haben. Weiterhin erklären wir die Aspekte der Primärstruktur und interpretieren – in Zusammenhang mit früheren Versuchen – die Höhenatmung und die fötale Atmung des Menschen als Folge der Inaktivierung einer der 2,3-Bisphosphoglycerat-Kontaktstellen. Weiterhin berichten wir über Glutamin als 2,3-Bisphosphoglycerat-Kontaktstelle β2 beim Pferd.

Experimenteller Teil

1) Gewinnung des Lamahämoglobins und seiner tryptischen Peptide

Blut des Lamas wurde vom Zoo durch dankenswertes Entgegenkommen von Direktor Dr. Lutz Heck, Tierpark Hellabrunn, erhalten. Mit diesem Präparat, das in der Elektrophorese nur eine Bande zeigte, wurden sämtliche Ergebnisse erarbeitet. Die Gewinnung der Erythrozyten, die Lyse und die Gewinnung des Rohhämoglobins erfolgte nach Standardmethoden. Die Hämoglobinlösung wurde bei – 30 °C aufbewahrt. Die Darstellung des Globins erfolgte nach Rossi-Fanelli und Antonini[6,3].

Die Gewinnung der α- bzw. β-Ketten erfolgte durch Gegenstromverteilung nach Hill und Craig[7,3] im System sec.-Butanol/1 proz. Dichloressigsäure (Glasgerät der Fa.: O. Post, Inc., New York, N.Y., USA).

Die tryptische Spaltung wurde am oxidierten und am nativen Präparat durchgeführt, die Tryptophanreste der α- und β-Ketten haben wir durch Sequenzanalyse der intakten Ketten zusätzlich identifiziert: Die Oxidation der Ketten erfolgte mit Perameisensäure, die tryptische Spaltung mit TPCK-Trypsin (Firma Worthington).

Fraktionierung der Peptide: Die Peptide wurden über eine Sephadex-Säule G-50 (2.5 × 110 cm) in 0.1N Essigsäure vorfraktioniert. Die einzelnen Gipfel wurden lyophilisiert und über Dowex 1X2[8,9] oder Dowex 50X4 rechromatographiert. So erhielten wir reine Peptide, die anschließend der Sequenzanalyse unterworfen wurden.

2) Sequenzanalyse

Es wurde nach der Filmtechnik[10] gearbeitet. Programme: A) Quadrolprogramm, B) 3-Dimethyl-[11] oder 3-Diäthylaminopropin-Programm. Programm A) verwendeten wir an den intakten Ketten. Es wurden meist 45 Abbauschritte durchgeführt. Es diente weiterhin zur Sequenzierung der tryptischen lysinhaltigen Peptide mit den hydrophilen Senfölen I[12] oder IV[13].

A) Quadrolprogramm

Das Quadrolprogramm wurde in starker Anlehnung an die Originalarbeit[10] und mit zwei Cyclisierungen angewendet. Die Chemikalien wurden selbst gereinigt, lediglich das 1N Quadrol wurde von der Firma Pierce (Rockfort, Ill., USA) bezogen.

a) Sequenzierung der tryptischen lysinhaltigen Peptide: 0.5 - 1.0 μmol eines Peptides und 2 - 3 mg I oder IV (8 - 10facher molarer Überschuß) wurden in 200 μl Wasser gelöst und in den Sequenator eingebracht und getrocknet. Anschließend wurden in den *stehenden* Becher ein Gemisch aus 100 μl 3-Dimethylaminopropin und 100 μl Wasser eingebracht, die Zelle verschlossen, mit N_2 gespült und (niedrige Geschwindigkeit) mit den Reagenzien I oder IV 15 min bei 50 °C inkubiert. Das überschüssige Amin wurde durch Ausblasen mit Stickstoff entfernt, der Film getrocknet und der Abbau mit Programmpunkt 1 begonnen (Kupplung).

b) Reinigung der Peptide bei automatischer Sequenzanalyse: Diese Peptide wurden unabhängig von der Art der Isolierung über einer kleinen Säule (0.9 × 30 cm Sephadex G-10) rechromatographiert[1].

B) Propinprogramm

Wir verwendeten 3-Dimethyl- oder 3-Diäthylaminopropin. Beide Präparate wurden von der Firma Fluka (Buchs, Schweiz) erhalten. 3-Dimethylaminopropin wurde, wie früher beschrieben[1], gereinigt, 3-Diäthylaminopropin verwendeten wir in jüngster Zeit, da dieses (Siedepunkt 118 °C) konstantere Filme liefert. Die Reinigung muß jedoch modifiziert werden[1]. Die Reinheit des Präparats wurde spektroskopisch, insbesondere durch NMR getestet. Gute Präparate zeigen auch nach 30 Abbauschritten oder mehr einen farblosen Film, schlechte Präparate hingegen liefern gelbe Filme.

a) Einstellung des Puffers: 11.2 g 3-Diäthylaminopropin + 35 g n-Propanol + Wasser ad. 100 ml mit 1N Essigsäure auf pH 9.2 eingestellt.

b) Abbau von lysinhaltigen Peptiden: 0.5 - 1 μmol Peptid wurde in Wasser suspendiert, kurz kräftig gerührt und mit 1.5 - 2.0 mg I versetzt, geschüttelt und die Suspension bei niedriger Geschwindigkeit in den Becher gebracht und getrocknet. In den stehenden Becher wurde ein Gemisch aus 125 μl 3-Dimethylaminopropin und 150 μl Wasser gebracht, die Kammer geschlossen und einige Minuten mit N_2 die Luft gespült. Die Kupplung von I wurde durch Einschalten der niederen Geschwindigkeit in Gang gesetzt. Nach 15 min wurde mit N_2 und anschließend im reduzierten Vakuum getrocknet und der Abbau mit Programmpunkt 1 gestartet (2 cleavages). So wird erreicht, daß der erste Schritt, falls die Kupplung mit I nicht quantitativ verläuft, die Reaktion mit Phenylsenföl komplett ist. Als Sequenator stand der Sequencer 89 B der Firma Beckman Instruments, Palo Alto, Calif., USA, zur Verfügung. Die Analysen wurden mit dem Gerät 121 C der Firma Beckman durchgeführt, dem ein Rechner angeschlossen war. Einwaage 600 μg Protein, 300 - 400 μg Peptid.

Konversion und Identifizierung: Wir konvertierten meist mit 3N Trifluoressigsäure oder Heptafluorbuttersäure 14 min bei 80 °C. Die Proben wurden mit Essigester extrahiert[1]. Die Esterphasen, in denen sich auch die Phenylthiohydantoine des Histidins und des Arginins befinden, wurden eingeengt, in 50 μl Essigester Methanol (1:1) gelöst und in einem aliquoten Teil (5 μl) die Ausbeute durch Messung im UV bei 259 nm bestimmt. Die Proben wurden automatisch in einem Gerät der Firma Woelm Pharma GmbH, Eschwege, strichförmig getüpfelt. Es werden besonders scharfe Flecken erzielt. Die Chromatographie erfolgt an derselben Platte hintereinander in den Systemen

a) Propionsäure-tert.-butylester,
b) Äthylenchlorid/Essigsäure (60:15) und
c) Äthylenchlorid/Methanol (80:20).

Das letzte System dient zur Identifizierung der Phenylthiohydantoine des Histidins und des Arginins.

Ergebnisse

Abbau des intakten Proteins

Die Analyse im Sequenator wurde mit einem oxidierten Produkt oder mit einem Carboxymethylderivat durchgeführt und beide Ketten wurden über 45 Schritte analysiert. Hiermit konnte auch die Sequenz β39 - 40 (-Arg-Arg-) gesichert werden. In beiden Ketten wurden sämtliche Aminosäuren durch Dünnschicht allein identifiziert.

Sequenzanalyse der tryptischen Peptide

Diese erfolgte ohne Probleme, da die Anzahl der Aminosäuren eines tryptischen Peptides im Vergleich zur Leistung des Sequenators klein ist.

α-Ketten (Tab. 1)

Tp1a (Pos. 1 - 5): Methode: A; Reagenz IV:
Sequenz:
Val-Leu-Ser-Ser-Lys.

Tp1b (Pos. 6 - 7): Methode: B; Sequenz:
Asp-Lys.

Tp2 (Pos. 8 - 11): Methode: A; Reagenz: IV;
Sequenz:
Ala-Asn-Ile-Lys.

Tp3 (Pos. 12 - 16): Methode: A; Reagenz: IV;
Sequenz:
Thr-Ala-Phe-Gly-Lys.

Tp4 (Pos. 17 - 31): Methode: A; Reagenz: IV;
Sequenz:
Ile-Gly-Gly-His-Ala-Ala-Asp-Tyr-Gly-Ala-Glu-
-Ala-Leu-Glu-Arg.

Tab. 1. Tryptische Peptide der α-Ketten des Hämoglobins des Lamas.
Hydrolyse 6N HCl, 110°C, 20 h. Lediglich die Serin- und Threoninwerte sind korrigiert. Tryptophan ist nicht vorhanden.

Lösl. Peptide	Tp1a	Tp1b	Tp2	Tp3	Tp4	Tp5	Tp6	Tp7	Tp8	Tp9a	Tp9b	Tp10	Tp11	Tp12	Tp13	Tp14	Tp12+13	Summe
Lys	1.14	1.03	1.50	1.08		1.37	1.03	1.02	1	1.04	1.13		1.00	1.05	1.05		2.27	11.82 (12)
His					0.93		1.89	0.91			2.80			2.95			3.35	10.07 (10)
Arg					1.00							0.98				1.11		3.10 (3)
Asp		0.96	1.20		1.29		1.09			0.94	4.17		2.01	2.60	1.10		3.69	15.06 (15)
Thr				1.01		1.90	0.97			1.14				1.80	1.96		2.78	8.62 (9)
Ser	1.92						1.95				3.11			2.02	1.91		2.69	10.90 (11)
Glu					1.94		1.03											3.94 (3)
Pro						0.89	1.03				0.83		0.88	2.07			1.98	6.02 (7)
Gly				1.01	2.95	1.07	1.03	1.01		1.00				1.30			1.59	9.86 (9)
Ala			1.00	0.97	4.18		1.09	1.06		0.97	4.85			4.20	1.26		4.82	18.80 (19)
Cys(O_3H)														1.00			1.00	(1)
Val	1.02						1.03			0.83			2.01	2.86	1.91		3.97	10.49 (10)
Met(O_2)						0.95												0.93 (1)
Ile			0.80		0.79													1.55 (2)
Leu	0.98				1.08	0.95	0.97			1.04	5.00	1.01		5.10	1.86		6.02	18.49 (18)
Tyr					0.72		0.85									0.88		2.49 (3)
Phe				0.91		1.84	1.89						1.07	1.17	0.90		2.38	7.79 (8)
Summe	5	2	4	5	15	9	16	4	1	7	22	2	7	28	12	2	40	141

Tp5 (Pos. 32 - 40): Methode: A; Reagenz: IV;
Sequenz:
Met-Phe-Leu-Gly-Phe-Pro-Thr-Thr-Lys.

Tp6 (Pos. 41 - 56): Methode: A; Reagenz: IV;
Sequenz:
Thr-Tyr-Phe-Pro-His-Phe-Asp-Leu-Ser-His-Gly-
-Ser-Ala-Gln-Val-Lys.

Tp7 (Pos. 57 - 60): Methode: B; Sequenz:
Ala-His-Gly-Lys.

Tp8 (Pos. 61): Lys.

Tp9a (Pos. 62 - 68): Methode: A; Reagenz: IV;
Sequenz:
Val-Gly-Asp-Ala-Leu-Thr-Lys.

Tp9b (Pos. 69 - 90): Methode: A; Reagenz: IV;
Sequenz:
Ala-Ala-Asp-His-Leu-Asp-Asp-Leu-Pro-Ser-Ala-
-Leu-Ser-Ala-Leu-Ser-Asp-Leu-His-Ala-His-Lys.

Tp10 (Pos. 91 - 92): Methode: B; Sequenz:
Leu-Arg.

Tp11 (Pos. 93 - 99): Methode: A; Reagenz: IV;
Sequenz:
Val-Asp-Pro-Val-Asn-Phe-Lys.

Tp12 (Pos. 100 - 127): Methode: A; Reagenz: IV;
Sequenz:
Leu-Leu-Ser-His-Cys-Leu-Leu-Val-Thr-Val-Ala-
-Ala-His-His-Pro-Gly-Asp-Phe-Thr-Pro-Ala-Val-
-Asp-Ala-Ser-Leu-Asp-Lys.

Tp13 (Pos. 128 - 139): Methode: A; Reagenz: IV;
Sequenz:
Phe-Leu-Ala-Asn-Val-Ser-Thr-Val-Leu-Thr-Ser-
-Lys.

Tp14 (Pos. 140 - 141): Methode: B; Sequenz:
Tyr-Arg.

β-Ketten (Tab. 2)

Tp1 (Pos. 1 - 8): Methode: A; Reagenz: IV;
Sequenz:
Val-Asn-Leu-Ser-Gly-Asp-Glu-Lys.

Tp2 (Pos. 9 - 17): Methode: A; Reagenz: IV;
Sequenz:
Asn-Ala-Val-His-Gly-Leu-Trp-Ser-Lys.

Tp3a (Pos. 18 - 19): Methode: B; Sequenz:
Val-Lys.

Tp3b (Pos. 20 - 30): Methode: B; Sequenz:
Val-Asp-Glu-Val-Gly-Gly-Glu-Ala-Leu-Gly-Arg.

Tab. 2. Tryptische Peptide der β-Ketten des Hämoglobins des Lamas. Hydrolyse 6N HCl, 110 °C, 20 h. Lediglich die Serin- und Threoninwerte sind korrigiert. Die Peptide Tp3a und Tp6 sind identisch. Isoleucin ist nicht vorhanden.

Lösl. Peptide	Tp1	Tp2	Tp3a	Tp3b	Tp4a	Tp4b	Tp5	Tp7+8	Tp9	Tp10a	Tp10b	Tp11	Tp12a	Tp12b	Tp13	Tp14	Tp15	Summe
Lys	0.95	0.97	0.96				0.92	0.92	1.10	1.02	1.03			1.10	0.98			11.13 (11)
His		0.75						0.99	1.02		0.83	1.03		0.91		1.11	1.05	8.08 (8)
Arg				0.89	1.08	1						1.01	0.97			1.14		6.28 (6)
Asp	1.98	0.90		1.07			4.10		3.86		1.29	2.09	1.36		1.05	1.00		18.49 (18)
Thr					0.87		0.92			0.89					1.18			4.31 (4)
Ser	1.06	1.05					2.08	0.99	2.12		1.03							8.26 (8)
Glu	1.08			1.97			1.10				1.03	0.96			2.98			9.15 (9)
Pro					1.08		0.83					0.84			0.81			4.13 (4)
Gly	0.95	1.25		2.84			1.10	1.03	2.04	1.06			1.08	1.02		1.26		13.29 (13)
Ala		1.05		1.07			2.10	1.03		1.13			1.08		2.03	3.87		13.29 (13)
Cys(O_3H)											0.78							1.07 (1)
Val	1.03	0.90	1.03	2.05	1.80*		1.18		1.18			1.02	3.85*			2.70		15.80* (18)
Met(O_2)							0.83											
Leu	1.06	1.12		1.07	2.06		1.18		3.78		1.97	1.04	3.75		1.18	1.20		19.21 (19)
Tyr					0.89					0.85					0.84		0.94	3.95 (4)
Phe							2.73		0.94			0.95		0.95	0.94			7.36 (7)
Trp		1.00			1.00													1.80 (2)
Summe	8	9	2	11	9	1	19	5	16	5	8	9	12	4	12	12	2	146

* 200 h Hydrolyse

Tp4a (Pos. 31 - 39): Methode: B; Sequenz: Leu-Leu-Val-Val-Tyr-Pro-Trp-Thr-Arg.

Tp4b (Pos. 40): Arg.

Tp5 (Pos. 41 - 59): Methode: A; Reagenz: IV; Sequenz: Phe-Phe-Glu-Ser-Phe-Gly-Asp-Leu-Ser-Thr-Ala-Asp-Ala-Val-Met-Asn-Asn-Pro-Lys.

Tp6 (Pos. 60 - 61): Methode: B; Sequenz: Val-Lys.

Tp7+8 (Pos. 62 - 66): Methode: A; Reagenz: IV; Sequenz: Ala-His-Gly-Ser-Lys.

Tp9 (Pos. 67 - 82): Methode: A; Reagenz: IV; Sequenz: Val-Leu-Asn-Ser-Phe-Gly-Asp-Gly-Leu-Ser-His-Leu-Asp-Asn-Leu-Lys.

Tp10a (Pos. 83 - 87): Methode: A; Reagenz: IV; Sequenz: Gly-Thr-Tyr-Ala-Lys.

Tp10b (Pos. 88 - 95): Methode: A; Reagenz: IV; Sequenz: Leu-Ser-Glu-Leu-His-Cys-Asp-Lys.

Tp11 (Pos. 96 - 104): Methode: B; Sequenz: Leu-His-Val-Asp-Pro-Glu-Asn-Phe-Arg.

Tp12a (Pos. 105 - 116): Methode: B; Sequenz: Leu-Leu-Gly-Asn-Val-Leu-Val-Val-Val-Leu-Ala-Arg.

Tp12b (Pos. 117 - 120): Methode: A; Sequenz: His-Phe-Gly-Lys.

Tp13 (Pos. 121 - 132): Methode: A; Reagenz: IV; Sequenz: Glu-Phe-Thr-Pro-Asp-Leu-Gln-Ala-Ala-Tyr-Gln-Lys.

Tp14 (Pos. 133 - 144): Methode: B; Sequenz: Val-Val-Ala-Gly-Val-Ala-Asn-Ala-Leu-Ala-His-Arg.

T15 (Pos. 145 - 146): Methode: B; Sequenz: Tyr-His.

Diskussion

Die Sequenz

Die Konstitution des Lamahämoglobins ist in Abb. 1 wiedergegeben. α- und β-Kette sind in homologer Weise[14] einander gegenübergestellt

```
H                 Pro-Ala         Thr      Val  11 Ala     Trp          Val     Ala  20     Gly-Glu
• L α  Val-   -Leu-Ser-Ser-Lys-Asp-Lys-Ala-Asn-Ile-Lys-Thr-Ala-Phe-Gly-Lys-Ile-Gly-Gly-His-Ala-Ala-Asp-Tyr-Gly-Ala-Glu-Ala-Leu-
L β    Val-Asn-Leu-Ser-Gly-Asp-Glu-Lys-Asn-Ala-Val-His-Gly-Leu-Trp-Ser-Lys-Val-Lys-        -Val-Asp-Glu-Val-Gly-Gly-Glu-Ala-Leu-
H          His     Thr-Pro-Glu         Ser  10     Thr-Ala        Gly         Asn          20

        30               Ser                   40                                50
  α  -Glu-Arg-Met-Phe-Leu-Gly-Phe-Pro-Thr-Thr-Lys-Thr-Tyr-Phe-Pro-His-Phe-   -Asp-Leu-Ser-His-                    -Gly-Ser-Ala-
  β  -Gly-Arg-Leu-Leu-Val-Val-Tyr-Pro-Trp-Thr-Arg-Arg-Phe-Phe-Glu-Ser-Phe-Gly-Asp-Leu-Ser-Thr-Ala-Asp-Ala-Val-Met-Asn-Asn-Pro-
          30                                  Gln  40                            50 Pro                 Gly

                Gly          60          Ala                Asn  69 Val-Ala      Val         Met     Asp       80
  α  -Gln-Val-Lys-Ala-His-Gly-Lys-Lys-Val-Gly-Asp-Ala-Leu-Thr-Lys-Ala-Ala-Asp-His-Leu-Asp-Asp-Leu-Pro-Ser-Ala-Leu-Ser-Ala-Leu-
  β  -Lys-Val-Lys-Ala-His-Gly-Ser-Lys-Val-Leu-Asn-Ser-Phe-Gly-Asp-Gly-Leu-Ser-His-Leu-Asp-Asn-Leu-Lys-Gly-Thr-Tyr-Ala-Lys-Leu-
          60                Lys              Gly-Ala     Ser             Ala              80                 Phe      Thr

                             90                               100                                  Leu 110        Leu
  α  -Ser-Asp-Leu-His-Ala-His-Lys-Leu-Arg-Val-Asp-Pro-Val-Asn-Phe-Lys-Leu-Leu-Ser-His-Cys-Leu-Leu-Val-Thr-Val-Ala-Ala-His-His-
  β  -Ser-Glu-Leu-His-Cys-Asp-Lys-Leu-His-Val-Asp-Pro-Glu-Asn-Phe-Arg-Leu-Leu-Gly-Asn-Val-Leu-Val-Val-Val-Leu-Ala-Arg-His-Phe-
          90                                  100                                  110     Cys                His

        Ala-Glu             120      His                             130 Ser                          140
  α  -Pro-Gly-Asp-Phe-Thr-Pro-Ala-Val-Asp-Ala-Ser-Leu-Asp-Lys-Phe-Leu-Ala-Asn-Val-Ser-Thr-Val-Leu-Thr-Ser-Lys-Tyr-Arg.
  β  -Gly-Lys-Glu-Phe-Thr-Pro-Asp-Leu-Gln-Ala-Ala-Tyr-Gln-Lys-Val-Val-Ala-Gly-Val-Ala-Asn-Ala-Leu-Ala-His-Arg-Tyr-His.
        120                 Pro-Val          130                               140             Lys
```

Abb. 1. Die Sequenz der α- und β-Ketten des Hämoglobins des adulten Lamas, in homologer[14] Weise einander gegenübergestellt.*
Die 25 (α-Ketten) bzw. 24 Austausche (β-Ketten) gegenüber dem adulten Hämoglobin des Menschen wurden dadurch gekennzeichnet, daß die im Humanhämoglobin vorkommenden unterschiedlichen Reste der Sequenz übergestellt werden.

* In der vorläufigen Mitteilung über die Sequenz der β-Ketten des Lamahämoglobins war versehentlich in Pos. 19 Asn statt Lys angegeben; dadurch ergeben sich nun 24 statt 23 Austausche im Vergleich zu den β-Ketten des HbA.

und mit den α- und β-Ketten des Menschen verglichen. Wie ersichtlich unterscheiden sich die α-Ketten in 25 und die β-Ketten in 24 Aminosäuren gegenüber dem Humanhämoglobin. Die Austausche sind relativ gleichmäßig über den Ketten verteilt, an einigen Stellen kann jedoch eine gewisse Anhäufung gefunden werden, z.B. in den *N*-terminalen Bereichen der α-Ketten, wo 10 der Austausche bereits in den ersten 24 Aminosäuren vorkommen. Ein ähnliches Bild ergibt sich für die β-Ketten. Zu erwähnen ist, daß die α-Ketten des Lamas kein Tryptophan enthalten, da α15 Trp des Humanhämoglobins hier durch Phenylalanin ersetzt ist; nur ganz selten wird in dieser Position das Tryptophan substituiert.

Helices, Proteinkontakte und Hämkontakte

Die helicalen Bereiche sind bei Lama und Mensch identisch, auch die Kontaktstellen der Ketten miteinander und zum Häm[15] zeigen keine Besonderheiten, mit einer Ausnahme: In Pos. β85 (F1) steht Tyrosin anstelle des Phenylalanins. Der Phenylalaninrest markiert beim Menschen- und Pferdehämoglobin einen Hämkontakt. Untersuchungen am Modell, die freundlicherweise Dr. Perutz durchführte, zeigten, daß diese Seitenkette interessant ist: Es besteht nämlich die Möglichkeit, daß die Phenolgruppe des Tyrosins mit der Carboxyl- oder mit der Hydroxylgruppe von Serin in Pos. β70 eine Wasserstoffbrücke bildet, also die Helices E und F verbindet. Diese Bindung ist beim Human- und Pferdehämoglobin nicht möglich, es kann aber zur Zeit noch nicht abgeschätzt werden, was diese Substitution für die Ligandenaffinität des Moleküls bedeutet. Die restlichen Hämkontakte sind identisch.

Die Atmungsfunktion des Blutes beim Lama

Schon vor vielen Jahren haben Physiologen die besondere Fähigkeit des Lamas registriert, sich auch im Hochgebirge der Anden normal zu bewegen und Sauerstoff aufzunehmen. Zur Erklärung dieser Hypoxietoleranz hat man die hohe Sauerstoffaffinität des Lamablutes mit der anderer Säuger verglichen. Eine genaue Analyse des Gaswechsels[16-19] ergab, daß unter den reduzierten atmosphärischen Bedingungen beim Lama einerseits mehr Sauerstoff im arteriellen Blut gebunden werden kann und andererseits die Abgabe des Sauerstoffs im Gewebe dadurch erleichtert ist, daß sie auch bei niedrigen Sauerstoffpartialdrucken im Gewebe stattfindet. Diese Atmung wurde Höhenatmung genannt und die Frage nach dem molekularen Mechanismus war Gegenstand zahlreicher Untersuchungen.

Aber erst heute und mit dem Modell des Pferde-[20] und des Humanhämoglobins[15], mit den neueren physiologischen Daten[5] und der vorliegenden Sequenz kann eine molekulare Interpretation der höheren Sauerstoffaffinität des Hämoglobins und der Höhenatmung des Lamas gegeben werden.

Bereits zu Beginn des Jahrhunderts berichtete Adair[21], daß die Sauerstoffaffinität einer Hämoglobinlösung höher ist als diejenige des entsprechenden Blutes. Diese Daten wurden zunächst kaum berücksichtigt und erst auf Umwegen einer Klärung zugeführt. Der Ausgangspunkt war der hohe Phosphatgehalt der Erythrozyten. Bald konnte festgestellt werden, daß es sich hierbei um einen organischen Phosphorsäureester, das 2,3-Bisphosphoglycerat[22,23], handelt.

Spätere Untersuchungen zeigten, daß Vögel statt dessen Inosinpentaphosphat[23,24] in ihren Erythrozyten haben. Bei Fischen werden Adenosintriphosphat[23] und Guanosintriphosphat als organische Phosphate gefunden. 2,3-Bisphosphoglycerat entsteht in einem Seitenzyklus der Erythrozyten-Glykolyse aus 3-Phosphoglyceroylphosphat, das durch Bisphosphoglyceromutase (EC 2.7.5.3) in 2,3-Bisphosphoglycerat übergeht und nach Abspaltung eines Phosphatrestes als 3-Phosphoglycerat wieder in den glykolytischen Abbau eingeschleust wird (Rapoport-Luebering-Zyklus)[25].

Die physiologische Bedeutung des Phosphats für die Regulation der O_2-Affinität selbst wurde vor knapp 10 Jahren geklärt, indem gezeigt wurde, daß 1 mol Phosphat stöchiometrisch an 1 mol Hämoglobintetramer gebunden wird[26], wobei die Affinität zum Desoxyhämoglobin um etwa 2 Zehnerpotenzen höher ist als zur oxygenierten Struktur. Weiterhin wurde durch Titration des Hämoglobins mit Phosphat gezeigt, daß 2,3-Bisphosphoglycerat die Sauerstoffaffinität des Moleküls *herabsetzt*[27]. Die intrinsische O_2-Affinität von 2,3-Bisphosphoglycerat-freiem Hämoglobin („stripped hemoglobin“) ist also höher als die des Hämoglobins in Anwesenheit von Phosphat, wie es in den Erythrozyten vorkommt. Der Sauerstofftransport ist daher nicht ein Zwei- sondern ein Drei-Komponenten-System[28] (Sauerstoff-Kohlendioxid, Hämoglobin, Phosphat).

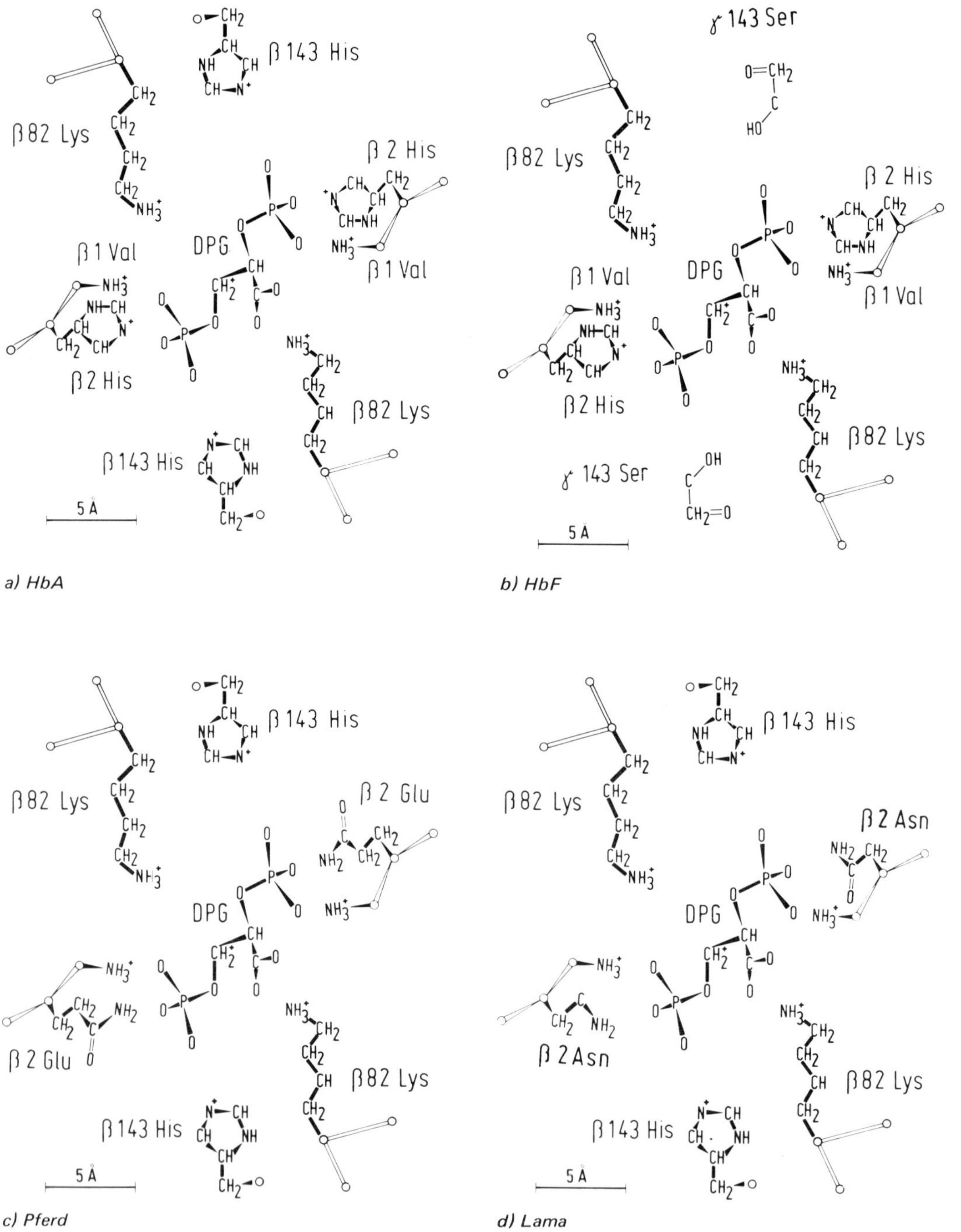

Abb. 2. a) Die Position des 2,3-Bisphosphoglycerats in der Struktur der Desoxyform des menschlichen Hämoglobins (HbA) (in Anlehnung an Arnone, Perutz[29,30] und Kilmartin[30a]).

b-d) Die Positionen der restlichen 2,3-Bisphosphoglycerat-Kontaktstellen mit unterschiedlichen Seitenketten beim Hämoglobin des menschlichen Fötus (HbF; b) des Pferdes (c) und des Lamas (d) wurden analog eingezeichnet.

Diese Arbeiten haben vor einigen Jahren insofern ihre Abrundung gefunden, als die Kontaktstellen, sowohl für das 2,3-Bisphosphoglycerat[29], wie auch für das Inosinhexaphosphat[30] durch Röntgenanalysen an dem Desoxyhämoglobin festgelegt wurden. Die Kontaktstellen sind die *N*-terminale Aminogruppe der β-Ketten, sowie β2 His, β82 Lys und β143 His (Abb. 2a). Die Kontaktstelle β82 wird nur von der einen der beiden β-Ketten besetzt und bildet eine Wasserstoffbrücke zur Carboxylgruppe des 2,3-Bisphosphoglycerats, während die Phosphat-Kontaktstellen jeweils an *beiden* β-Ketten, also 6 Kontakte bilden. Es bestehen somit insgesamt 7 Wasserstoffbrücken[29]. Für das Inosinpentaphosphat ergaben sich 4 weitere Phosphat-Wasserstoffbrücken zum Hämoglobin[30]. Diese Daten wurden am Hämoglobin des Menschen und an der Desoxyform erarbeitet.

Hohe Sauerstoffaffinität als Folge einer Mutation an der 2,3-Bisphosphoglycerat-Kontaktstelle (β2 His→Asn)

Unsere früheren Untersuchungen haben dieses Problem wieder aufgerollt und den Einfluß des Phosphats auf die Reduktion der Sauerstoffaffinität im Vergleich zum Humanhämoglobin untersucht. Die Ergebnisse waren insofern überraschend, als das phosphatfreie Hämoglobin des Menschen sogar eine *höhere* (intrinsische) Sauerstoffaffinität besitzt als das Hämoglobin des Lamas[5]. Wird jedoch die O_2-Affinität in Gegenwart des Phosphats titriert, so zeigt sich, daß ihre Verminderung bei äquimolarer Phosphatmenge beim Lama wesentlich geringer ist. Hiermit wurde die Basis zum Verständnis der durch die hohe O_2-Affinität bedingten Hypoxietoleranz gefunden, denn aus diesen Daten geht eindeutig hervor, daß die Wechselwirkung zwischen Phosphat und Protein beim Lama geringer ist: der Unterschied muß in einem definierten Sequenzunterschied liegen. Diesen zu deuten erlaubt die vorliegende Sequenz des Lamahämoglobins. In Tab. 3 sind die Kontaktstellen vom Humanhämoglobin, vom Pferdehämoglobin, vom fötalen Hämoglobin und vom Hämoglobin des Lamas zum Phosphat wiedergeben. Wie ersichtlich, ist beim Lama die Kontaktstelle β2 His→Asn mutiert; sie kann daher keinen Kontakt mehr mit dem Phosphat halten. Dadurch erklärt sich die verminderte Wechselwirkung des Lamahämoglobins mit 2,3-Bisphosphoglycerat aus der veränderten Sequenz in Pos. β2 und daraus ergibt sich die höhere Sauerstoffaffinität. *So erhält die Höhenatmung eine molekulare Deutung.*

Die hohe Sauerstoffaffinität des fötalen menschlichen Blutes: Die 2,3-Bisphosphoglycerat-Kontaktstelle in Pos. β143

Wir beschäftigten uns mit dem fötalen Hämoglobin des Menschen. Es ist die Frage nach der Ursache der hohen Sauerstoffaffinität menschlichen fötalen Blutes, die die Aufnahme von Sauerstoff in der Placenta erleichtert [31,32]. Das fötale Hämoglobin wurde 1866 durch Körber beim Menschen[33] entdeckt, das Problem erst in den Dreißigerjahren[34,35] weiter bearbeitet und durch den Nachweis des fötalen Hämoglobins beim Menschen, dessen Isolierung[36], Sequenzanalysen[37], Charakterisierung seiner physiologischen Eigenschaften[38,39] und durch die vor einigen Monaten veröffentlichte Röntgenanalyse[40] zu einem gewissen Abschluß geführt.

Bereits vor 10 Jahren wurde gezeigt, daß 2,3-Bisphosphoglycerat die Sauerstoffaffinitat von Nabelschnurhämolysaten[41,42] und von chromatographisch gereinigtem fötalem menschlichem Hämoglobin[43,44] viel weniger stark beeinflußt als die von adultem Hämoglobin. In analoger Weise zu den Daten, die am Lama erarbeitet wurden, wird diese verminderte Wechselwirkung durch den

Tab. 3. Die vier 2,3-Bisphosphoglycerat-Kontaktstellen des Hämoglobins des adulten Menschen (HbA), des menschlichen Foetus (HbF), des Hämoglobins des Pferdes und des Lamas. Die Positionen der β- bzw. der γ-Ketten.

Position	1	2	82	143	O_2-Affinität
Mensch adult	α-NH_2	His	Lys	His	Niederlandatmung
Pferd	α-NH_2	Gln	Lys	His	Niederlandatmung
Lama	α-NH_2	Asn	Lys	His	Höhenatmung
Fötus (Mensch)	α-NH_2	His	Lys	Ser	hochaffines Hb

Austausch einer 2,3-Bisphosphoglycerat-Bindungsstelle erreicht: Das Histidin in Pos. β143 des HbA ist im fötalen Hämoglobin durch γ143 Ser ersetzt (Abb. 2b). Damit erklärt sich die hohe Sauerstoffaffinität menschlichen fötalen Blutes. Sie erleichtert die Sauerstoffaufnahme des Fötus in der Placenta. *Der hohen O_2-Affinität des Blutes vom menschlichen Fötus und vom Lama liegt also ein analoger Mechanismus zugrunde.* Es ist bemerkenswert, daß die Natur an ganz verschiedenen Stellen in der Evolution sich desselben Mechanismus zur Anpassung an Hypoxie bediente.

Abgesehen von den Aminosäure-Austauschen, die zu einer Verminderung der 2,3-Bisphosphoglycerat-Kontakte des Moleküls führen, könnten auch noch Änderungen der Tertiärstruktur eine Rolle spielen. So wurde vor einigen Monaten die Raumstruktur des menschlichen fötalen Hämoglobins publiziert[40]. Die röntgenographischen Daten zeigen insofern eine Verfeinerung, als die Substitutionen β3 Leu→γ3 Phe und β7 Glu→γ7 Asp eine schwache Drehung der A-Helix bewirken, so daß die Kontaktstellen γ1 NH_2(Gly) und γ2 His eine größere Distanz zum 2,3-Bisphosphoglycerat als beim adulten Hämoglobin haben. Die Vergrößerung des Abstandes zwischen dem Phosphat und diesen Kontakten hat vermutlich eine weitere Verminderung der 2,3-Bisphosphoglycerat-Bindung zur Folge.

Um die reduzierten Atmungsbedingungen des Fötus anschaulich zu erklären, war früher der Ausspruch[45] geprägt worden: „Der Fötus atmet am Mount Everest". Der hier erhobene Befund am Lama und der analoge Mechanismus des Fötus zeigen, wie anschaulich dieser Vergleich ist; er kann heute erweitert und formuliert werden: „Der Fötus am Mount Everest und das Lama in den Anden atmen unter ähnlichen Bedingungen mit einem analogen Mechanismus".

Die Regulation der fötalen Atmung beim Schwein erfolgt ebenfalls durch 2,3-Bisphosphoglycerat, jedoch auf anderem Wege. Es kann auf diese Situation nur hingewiesen werden[1,46-50].

Die O_2-Affinität des Pferde-Hämoglobins: Die 2,3-Bisphosphoglycerat-Kontaktstelle in Pos. β2

Vor vielen Jahren berichteten wir über die *N*-terminale Sequenz des Hämoglobins des Pferdes. Diese unterscheidet sich u.a. von der beim Menschen dadurch, daß in Position β2 (also einer 2,3-Bisphosphoglycerat-Kontaktstelle) anstelle von Histidin Glutamin[51] gefunden wird. Bemerkenswerterweise zeigt aber das Pferdehämoglobin dieselbe starke Wechselwirkung mit 2,3-Bisphosphoglycerat wie adultes Humanhämoglobin[52]. Bei der Niederschrift der Publikation zur Höhenatmung des Lamas erinnerten wir uns dieses Befundes und glaubten erst nicht, daß Asparagin und Glutamin so grundlegend verschiedene Funktion ausüben könnten. Diese Frage wurde durch Vermessung der atomaren Distanzen am Modell geklärt, die freundlicherweise Herr Dr. Perutz durchgeführt hat. Für den Abstand zwischen dem Amidstickstoff des Glutamins und dem Phosphatsauerstoff des 2,3-Bisphosphoglycerats wurde beim Pferdehämoglobin 0.35 bis 0.40 nm gefunden (Abb. 2c), also eine Distanz, die ausreicht, um eine Wasserstoffbrücke zwischen Phosphat und Seitenkette des Glutamins zu bilden. Für die γ-Amidgruppe des Asparagins im Lamahämoglobin beträgt dagegen der entsprechende Abstand in Pos. β2 Asn 0.45 - 0.50 nm (Abb. 2d), wodurch an dieser Stelle der Kontakt unterbrochen wird. Natürlich ist der pK-Wert der Amidgruppe sehr tief und zwar außerhalb des physiologischen Bereiches, so daß eine Ionisierung des Säureamids nicht in Frage kommt: Es ist aber bekannt, daß eine Ionisierung zur Ausbildung der Wasserstoffbindung nicht notwendig ist. Damit wäre die niedere O_2-Affinität und die hohe 2,3-Bisphosphoglycerat-Affinität des Pferdehämoglobins erklärt. Dieser Befund zeigt, wie wichtig die Bestimmung und korrekte Wiedergabe der Säureamide ist: Andererseits ist überraschend, daß Glutamin und Histidin stereochemisch, in diesem Falle die gleiche Wasserstoffbrücke zum Phosphat bilden und die gleiche Funktion ausüben. Hiermit ist der oben erwähnte scheinbare Widerspruch geklärt und die Atmung und die maximale Reduktion der intrinsischen Sauerstoffaffinität des Hämoglobins des Pferdes auch molekular erklärt.

Hypoxietoleranz des Menschen

Daß eine hohe O_2-Affinität des Blutes bei Anpassung an schwere akute Hypoxie von entscheidender physiologischer Bedeutung sein kann, wurde von Eaton[53] und Mitarbeitern gezeigt. Diese Autoren erhöhten die O_2-Affinität des Blutes von Ratten durch Fütterung von Natriumcyanat. Diese Substanz carbamoyliert die α-NH_2-Gruppen des Hämoglobins und erhöht die O_2-Affinität des Blutes durch eine Verminderung der Wechselwirkung mit 2,3-Bisphosphoglycerat und durch Erhöhung der intrinsischen Ligandenaffinität des Hämoglobins[53]. Eaton und Mitarbeiter fanden, daß die Tiere mit hoher O_2-Affinität eine akute Hypoxie zu 100% überlebten, während die Kontrollgruppe nur eine Überlebensquote von 20% aufwiesen. Diese Befunde legen den Gedanken nahe, ob nicht auch eine Hypoxie-Toleranz des Menschen durch eine pharmakologische Erhöhung der O_2-Affinität des Blutes, z.B. eine spezifische chemische Modifizierung einer Seitenkette der 2,3-Bisphosphoglycerat-Kontakte verbessert werden könnte. Diese „symptomatische" Behandlung würde praktikabel sein; da die Lebensdauer der Erythrozyten 120 Tage beträgt, würde das modifizierte Hämoglobin nach dieser Zeit aus dem Kreislauf wieder entfernt.

Wir danken Herrn *Dr. Lutz Heck*, Direktor am Tierpark Hellabrunn, für die großzügige Unterstützung dieser Arbeiten. Herr *Dr. Max Ferdinand Perutz* hat freundlicherweise unsere physiologischen und chemischen Daten mit uns diskutiert: wir danken ihm weiterhin für die Vermessung der Seitenketten in Pos. β2 an seinem Modell und Überlassung der Daten.

Literatur

1 XXI: Braunitzer, G., Schrank, B., Stangl, A. & Scheithauer, U. (1978) *Hoppe-Seyler's Z. Physiol. Chem.* **359**, 137 - 146.
2 Braunitzer, G., Hilse, K., Rudloff, V. & Hilschmann, N. (1964) *Adv. Protein Chem.* **19**, 1 - 71.
3 Braunitzer, G., Schrank, B. & Stangl, A. (1977) *Hoppe-Seyler's Z. Physiol. Chem.* **358**, 409 - 412.
4 Braunitzer, G., Schrank, B., Stangl, A. & Bauer, C. (1977) *Hoppe-Seyler's Z. Physiol. Chem.* **358**, 921 - 925.
5 Petschow, D., Würdinger, I., Baumann, R., Duhm, J., Braunitzer, G. & Bauer, C. (1977) *J. Appl. Physiol.* **42**, 139 - 143.
6 Rossi-Fanelli, A. & Antonini, E. (1958) *Biochim. Biophys. Acta* **30**, 608 - 615.
7 Hill, R. J. & Craig, L. C. (1959) *J. Am. Chem. Soc.* **81**, 2272 - 2273.
8 Hilse, K. & Braunitzer, G. (1959) *Z. Naturforsch.* **14b**, 604 - 606.
9 Rudloff, V. & Braunitzer, G. (1961) *Hoppe-Seyler's Z. Physiol. Chem.* **323**, 129 - 144.
10 Edman, P. & Begg, G. (1967) *Eur. J. Biochem.* **1**, 80 - 91.
11 Braunitzer, G. & Schrank, B. (1970) *Hoppe-Seyler's Z. Physiol. Chem.* **351**, 417.
12 Braunitzer, G., Schrank, B. & Ruhfus, A. (1970) *Hoppe-Seyler's Z. Physiol. Chem.* **351**, 1589 - 1590.
13 Braunitzer, G., Schrank, B., Ruhfus, A., Petersen, S. & Petersen, U. (1971) *Hoppe-Seyler's Z. Physiol. Chem.* **352**, 1730 - 1732.
14 Braunitzer, G., Gehring-Müller, R., Hilschmann, N., Hilse, K., Hobom, G., Rudloff, V. & Wittmann-Liebold, B. (1961) *Hoppe-Seyler's Z. Physiol. Chem.* **325**, 283 - 286.
15 Perutz, M. F., Muirhead, H., Cox, J. M. & Goaman, L. C. G. (1968) *Nature (London)* **219**, 131 - 139.
16 Hall, F. G., Dill, D. B. & Guzman Barron, E. S. (1936) *J. Cell. Comp. Physiol.* **8**, 301 - 313.
17 Bartels, H., Hilpert, P., Barbey, K., Betke, K., Riegel, K., Lang, E. M. & Metcalfe, J. (1963) *Am. J. Physiol.* **205**, 331 - 336.
18 Banchero, N., Grover, R. & Will, J. A. (1971) *Respir. Physiol.* **13**, 102 - 115.
19 Reynafarje, C., Faura, J., Villavicencio, D., Curaca, A., Reynafarje, B., Oyola, L., Contreras, L., Vallenas, E. & Faura, A. (1975) *J. Appl. Physiol.* **38**, 806 - 810.
20 Perutz, M. F. (1969) *Proc. R. Soc. London, Ser.B* **173**, 113 - 140.
21 Adair, G. S., Barcroft, J. & Bock, A. V. (1921) *J. Physiol. (London)* **55**, 332.
22 Greenwald, I. (1925) *J. Biol. Chem.* **63**, 339.
23 Rapoport, S. & Guest, G. M. (1941) *J. Biol. Chem.* **138**, 269 - 282.
24 Johnson, L. & Tate, M. E. (1969) *Can. J. Chem.* **47**, 63 - 73.
25 Rapoport, S. & Luebering, J. (1950) *J. Biol. Chem.* **183**, 507 - 516.
26 Chanutin, A. & Curnish, R. R. (1965) *Proc. Soc. Exp. Biol. Med.* **120**, 291 - 292.
27 Benesch, R. & Benesch, R. (1967) *Biochem. Biophys. Res. Commun.* **26**, 162 - 167.
28 Benesch, R. E. & Benesch, R. (1974) *Adv. Protein. Chem.* **28**, 211 - 237.
29 Arnone, A. (1972) *Nature (London)* **237**, 146 - 149.
30 Arnone, A. & Perutz, M. F. (1974) *Nature (London)* **249**, 34 - 36.
30a Kilmartin, J. V. (1976) *Brit. Med. Bull.* **32**, 209 - 212.
31 Künzer, W. (1951) Über den Blutfarbstoffwechsel gesunder Säuglinge und Kinder. S. Karger, Basel, Bibliotheka Paediatrika, fasc. 51.
32 Betke, K. (1954) Der menschliche rote Blutfarbstoff bei Fötus und reifem Organismus. Springer-Verlag, Heidelberg.
33 Körber, E. (1866) Über Differenzen des Blutfarbstoffes. Diss. Dorpat, *Zbl. Med. Wiss.* **1867**, 117.
34 Jonxis, I. H. P. (1937) *Monatsschr. Kindergeneeskd.* **6**, 356 - 373. Vgl. *Zbl. Kinderheilkd.* **34**, 108 (1938).
35 Haurowitz, F. (1929) *Hoppe-Seyler's Z. Physiol. Chem.* **183**, 78 - 87.
36 Rhinesmith, H. S., Schroeder, W. & Martin, N. (1958) *J. Am. Chem. Soc.* **80**, 3358 - 3361.
37 Schroeder, W. A., Shelton, J. R., Shelton, J. B., Cormick, J. & Jones, R. T. (1963) *Biochemistry* **2**, 992 - 1008.

38 Haselhorst, G. (1929) *Z. Geburtshilfe* **95**, 400.
39 Haselhorst, G. & Stromberger, K. (1931) *Z. Geburtshilfe* **100**, 48 - 69.
40 Frier, J. H. & Perutz, M. F. (1977) *J. Mol. Biol.* **112**, 97 - 112.
41 Bauer, C., Ludwig, M. & Ludwig, J. (1968) *Life Sci.* **7**, 1379 - 1343.
42 Bauer, C., Ludwig, M., Ludwig, J. & Bartels, H. (1969) *Respir. Physiol.* **7**, 271 - 277.
43 Tyuma, I. & Shimizu, K. (1969) *Arch. Biochem. Biophys.* **129**, 404 - 405.
44 Tyuma, I. & Shimizu, K. (1970) *Fed. Proc. Fed. Am. Soc. Exp. Biol.* **29**, 1112 - 1114.
45 Dawes, G. W. (1973) Foetal and Neonatal Physiology. Year Book Medical Publishers, Chicago. (Betke, K., priv. Mitteilung.)
46 Kleihauer, E. & Stöffler, G. (1968) *Mol. Gen. Genet.* **101**, 59 - 69.
47 Tautz, C. H. & Kleihauer, E. (1972) *Res. Exp. Med.* **159**, 44 - 49.
48 Duhm, K. & Kim, H. D. (1973) in Erythrocytes, Thrombocytes, Leukocytes (Gerlach, E., Moser, K., Deutsch, E. & Willmanns, W., Hrsg.) S. 164 - 168, Georg Thieme Publishers, Stuttgart.
49 Baumann, R., Teischel, F., Koch, R. & Bartels, H. (1973) *Respir. Physiol.* **19**, 153 - 161.
50 Kim, H. D. & Duhm, K. (1974) *Am. J. Physiol.* **226**, 1001 - 1006.
51 Braunitzer, G. (1958) *Hoppe-Seyler's Z. Physiol. Chem.* **312**, 72 - 84.
52 Bunn, H.F. & Kitchen, H. (1973) *Blood* **42**, 471 - 479.
53 Eaton, J. W., Skelton, T. D. & Berger, E. (1974) *Science* **183**, 743 - 744.

Prof. Dr. Gerhard Braunitzer, Barbara Schrank und Anton Stangl, Max-Planck-Institut für Biochemie, Am Klopferspitz, D-8033 Martinsried.

Prof. Dr. Christian Bauer, Institut für Physiologie, Universität Regensburg, Universitätsstr. 31, D-8400 Regensburg.

Hoppe-Seyler's Z. Physiol. Chem.
Bd. 359, S. 559 - 570, Mai 1978

Vitamin-B_6-Biosynthese in *Bacillus subtilis*

Werner PFLUG und Franz LINGENS

Institut für Mikrobiologie der Universität Hohenheim, Stuttgart-Hohenheim

(Der Schriftleitung zugegangen am 25. Januar 1978)

Herrn Professor Dr. A. Butenandt zum 75. Geburtstag gewidmet

Zusammenfassung:

1) Durch Mutation und Selektion wurde eine Vitamin-B_6-Ausscheidermutante von *Bacillus subtilis* gewonnen. Von dieser Mutante BA 1 konnten auf Gradientenplatten mit dem Vitamin-B_6-Antagonisten Isonicotinohydrazid 3 Stämme isoliert werden, die 2 - 5 mg Vitamin B_6/l ins Medium abgeben.

2) Von den Vitamin-B_6-Überproduzenten wurden nach Mutagenbehandlung 49 Vitamin-B_6-Mangelmutanten isoliert. Alle Mutanten wachsen mit Pyridoxin, Pyridoxal, Pyridoxamin und sogar mit 4'-Desoxypyridoxin. Mit Glykolaldehyd oder Nicotinsäure war kein Wachstum nachzuweisen. Ein Teil der Mutanten wächst auch mit Isoleucin. Die übrigen Mutanten werden in Gegenwart einer Vitamin-B_6-Komponente durch Isoleucin zusätzlich stimuliert. Isoleucin kann durch 3-Methyl-2-oxovaleriansäure ersetzt werden. Die Mutanten ließen sich durch Fütterungsversuche in 2 Gruppen einteilen. Wachstumswirksame Produkte konnten chromatographisch angereichert, aber nicht identifiziert werden. Bei Zusatz von 4'-Desoxypyridoxin läßt sich aus dem Medium 4'-Desoxypyridoxin-5'-phosphat isolieren.

3) Enzymatische Untersuchungen ergaben, daß die Threonin-Dehydratase und die Transaminasen B und C von Vitamin-B_6-Überproduzenten und -auxotrophen Mutanten eine höhere spezifische Aktivität im Vergleich zum Ausgangsstamm zeigen. Die multivalente Repression durch Isoleucin, Leucin und Valin ist bei der Threonin-Dehydratase und der Transaminase B nicht mehr vorhanden. Die Leucin-Dehydrogenase, das erste Enzym des Isoleucinabbaues, wird von den Vitamin-B_6-Überproduzenten und -auxotrophen Mutanten konstitutiv gebildet. Bei den Vitamin-B_6-Mangelmutanten sind die spezifische Aktivität der Leucin-Dehydrogenase und die Wachstumsausbeute miteinander korreliert. In *Bacillus subtilis* konnte keine Pyridoxaminphosphat-Oxidase nachge-

---

Enzyme:
Leucin-Dehydrogenase, L-Leucin:$NAD^{\oplus}$-Oxidoreduktase, desaminierend (EC 1.4.1.9);
Pyridoxal-Kinase, ATP:pyridoxal-5'-phosphotransferase (EC 2.7.1.35);
Pyridoxaminphosphat-Oxidase, Pyridoxaminphosphat:O_2-Oxidoreduktase, desaminierend (EC 1.4.3.5); (in dieser Arbeit wird Pyridoxin-5'-phosphat als Substrat benützt);
Threonin-Dehydratase, L-Threonin-Hydro-Lyase, desaminierend (EC 4.2.1.16);
Als Transaminase B wird eine Isoleucin:2-Oxoglutarat-transaminierende und als Transaminase C eine Isoleucin:Pyruvat-transaminierende Aktivität bezeichnet.

Trivialnamen:
Entsprechend den letzten Empfehlungen der IUPAC-IUB Commission on Biochemical Nomenclatur [(1973) *Eur. J. Biochem.* **40**, 325 - 327] wird hier der Ausdruck Pyridoxin nur für die Verbindung 4,5-Bis(hydroxymethyl)-2-methylpyridin-3-ol verwendet; ihr Oxidationsprodukt 3-Hydroxy-5-(hydroxymethyl)-2-methylpyridin-4-carbonsäure trägt den Trivialnamen 4-Pyridoxsäure.

wiesen werden. Pyridoxal-Kinase-Aktivität war im Rohextrakt vorhanden.

4) Für *Bacillus subtilis* wird ein Vitamin-B_6-Biosyntheseschema postuliert.

Biosynthesis of Vitamin B_6 in Bacillus subtilis

Summary:

1) A vitamin-B_6-producing mutant, BA 1, was selected by treatment of *Bacillus subtilis* with *N*-methyl-*N'*-nitro-*N*-nitrosoguanidine. Using gradient plates supplemented with the vitamin B_6 antagonist isonicotinohydrazide, three mutants of BA 1 were isolated, which excrete 2 - 5 mg of vitamin B_6/l of growth medium.

2) Mutation of the three vitamin-B_6-producing strains BA 1, BA 11 and L 71 led to the isolation of 49 vitamin-B_6deficient mutants. All mutants are able to grow with pyridoxine, pyridoxal, pyridoxamine, and even with 4'-deoxypyridoxine. Glycolaldehyde or nicotinic acid do not support growth of the mutants. Some of these vitamin-B_6-deficient mutants can also grow in the absence of vitamin B_6, providing isoleucine is present. Others show a growth stimulation, when isoleucine is added to a medium containing a vitamin B_6 compound. Isoleucine can be replaced by 3-methyl-2-oxovalerate. Cross-feeding experiments indicated a division of the mutants into two groups. Using chromatographic methods, substances which support growth of the mutants were purified, but have not yet been identified. Following the addition of 4'-deoxypyridoxine, 4'-deoxypyridoxine 5'-phosphate was isolated from the growth medium of a vitamin B_6-deficient mutant.

3) Threonine dehydratase, transaminase B and transaminase C from wild-type *Bacillus subtilis* were compared with the enzymes from vitamin-B_6-producing strains and with the enzymes from vitamin-B_6-deficient mutants. Both the vitamin-B_6-producing and the vitamin B_6-deficient mutants show higher specific activities than wild-type. In the mutant strains no multivalent repression of the threonine dehydratase and transaminase B by isoleucine, leucine and valine could be demonstrated. Leucine dehydrogenase, the first enzyme of the isoleucine catabolic pathway, is constitutively produced in the vitamin-B_6-producing and in the vitamin-B_6-deficient mutants. In the vitamin-B_6-deficient mutants, there is a correlation between growth yield in the presence of isoleucine and the specific activity of leucine dehydrogenase. In the crude extract of *Bacillus subtilis* no pyridoxamine-phosphate oxidase activity could be demonstrated, whereas pyridoxal kinase was readily detectable.

4) A hypothetical pathway for the biosynthesis of vitamin B_6 in *Bacillus subtilis* is proposed.

Key words: Bacillus subtilis, biosynthesis of vitamin B_6, vitamin-B_6-producing mutants.

Die Biosynthese des Vitamins B_6 ist von Dempsey et al.[1-8], Hill et al.[9-12] bei *Escherichia coli* und von Ishida et al.[13-15] bei *Achromobacter cycloclastes* untersucht worden, es konnte aber noch kein gesichertes Biosyntheseschema aufgestellt werden. Potentielle Zwischenprodukte der Biosynthese wurden ebenfalls noch nicht isoliert. Hill et al.[9,10] formulierten ein hypothetisches Formelschema, das von drei Triosephosphat-Einheiten ausgehend zum Kohlenstoffskelett des B_6-Moleküls führen soll. Wir haben unsere Untersuchungen an *Bacillus subtilis* ausgeführt, über die im folgenden berichtet wird.

Material und Methoden

Geräte: Für die Extinktionsmessungen der enzymatischen Tests standen die Geräte PMQ II und PMQ III der Firma Zeiss zur Verfügung. Trübungsmessungen wurden mit dem Photometer Eppendorf der Firma Netheler & Hinz durchgeführt. Schüttelkulturen wurden mit Rundschüttelmaschinen der Firma E. Bühler & Infors gezüchtet. Der Zellaufschluß erfolgte mit dem Ultraschallgerät Sonifier S 125 der Firma Branson Inc., die Enzymextrakte wurden mit Hilfe der Kühlzentrifuge J 21 von Beckman Inc. gewonnen. Für die Radioaktivitätsmessungen kam ein Liquid-Scintillation-Spektrometer ABAC SL 40 der Firma Intertechnique zur Anwendung.

Chemikalien: Folgende Reagenzien wurden verwendet: *N*-Methyl-*N'*-nitro-*N*-nitrosoguanidin, synthetisiert von R. Süßmuth[16], Pyridoxin, 4'-Desoxypyridoxin und Aminosäuren von E. Merck AG, Pyridoxamin und Isonicotinohydrazid von Fluka AG, Pyridoxal und Pyridoxal-5'-phosphat von Serva, ATP, $NAD^{\oplus}$ und NADH von Boehringer Mannheim GmbH, 4'-Desoxypyridoxin-5'-phosphat[17] und Pyridoxin-5'-phosphat[17,18] wurden synthetisiert, [^{3}H]Isoleucin und [^{3}H]Valin von Buchler, Dowex 50 WX8 und 3-Methyl-2-oxovaleriansäure von Serva.

Stämme: Bacillus subtilis B 23 und *Saccharomyces carlsbergensis* wurden der Stammsammlung des Instituts für Mikrobiologie der Universität Hohenheim entnommen.

Medien: Für *B. subtilis* wurde das Mineralsalzmedium nach Spizizen[19] verwendet. Das Komplettmedium enthielt: 4 g Nutrient Broth, 3 g Hefe Extrakt, 2 g Ammoniumsulfat, 14 g K_2HPO_4, 6 g KH_2PO_4, 5 g Glucose und 1 *l* deionisiertes Wasser. Das Testmedium für *S. carlsbergensis* setzte sich zusammen aus: 5 g Trinatriumcitrat × 2 H_2O, 5 g „Casamino acids" (vitaminfrei), 2.25 g KH_2PO_4, 6.35 g K_2HPO_4, 2 g $(NH_4)H_2PO_4$, 0.222 g $CaCl_2 \cdot 6\,H_2O$, 1.35 mg $FeCl_3 \cdot 6\,H_2O$, 25 mg $MgSO_4 \cdot 7\,H_2O$, 0.5 mg $MnSO_4 \cdot 1\,H_2O$, 0.5 mg Thiamin · HCl, 5 mg Nicotinsäure, 50 mg Inosit, 5 mg (+)-Calciumpantothenat, 0.016 mg Biotin, 10 g Glucose und 1 *l* deionisiertem Wasser.

Isolierung der Mutanten: Der Wildstamm *B. subtilis* B 23 wurde mit *N*-Methyl-*N'*-nitro-*N*-nitrosoguanidin bei pH 6 behandelt[20]. Die Selektion der Vitamin B_6-Ausscheidermutanten erfolgte auf Isonicotinohydrazid-Gradientenplatten[21,22]. Vitamin B_6-Mangelmutanten wurden mittels der Lederberg-Stempeltechnik[23] isoliert und die Mutanten im Strich- und Blättchentest weiter untersucht[24].

Die *Züchtung* der Bakterien erfolgte bei 37 °C in Schüttelkulturen. Für die Akkumulierung von Vitamin B_6 in Flüssigmedium wurde bei 37 °C angezüchtet und dann 4 Tage bei 30 °C im Dunkeln geschüttelt.

Die *Vitamin-B_6-Bestimmung* wurde mikrobiologisch mit *S. carlsbergensis* durchgeführt[25,26]. Die Testansätze in Kapsenberg-Röhrchen wurden liegend bei 30 °C inkubiert. Die Identifizierung der verschiedenen B_6-Komponenten erfolgte mittels Bioautographie[27]. Dazu wurden die Vitamin-B_6-Komponenten aufsteigend in 1-Propanol/10proz. Ameisensäure[28] auf Papier MN 261 von Macherey, Nagel & Co. chromatographiert. Das Chromatogramm wurde dann nach Abdampfen des Laufmittels mit *S. carlsbergensis* in Glaspetrischalen (Ø 19 cm) getestet. Das Wachstum der Vitamin-B_6-Mangelmutanten wurde in Schüttelkulturen bei 37 °C untersucht. Den Testansätzen mit den zu prüfenden Substanzen wurden 0.05 m*l* der gewaschenen Mutantensuspension zupipettiert.

Reversionsversuche: 0.05 m*l* einer gewaschenen Mutantensuspension (ca. 10^8 Zellen/m*l*) wurden auf Minimal-Agarmedium ausplattiert und 2 - 3 Tage bei 37 °C inkubiert.

Zellaufschluß: Die Zellen wurden mit dem Ultraschallgerät 8 × 45 s beschallt und die Zellbruchstücke in der Kühlzentrifuge bei 18 000 U./min abzentrifugiert.

Die *Proteinbestimmung* erfolgte nach der Biuret-Methode[29].

Enzymtests

Die *Threonin-Dehydratase* wurde nach Hatfield und Umbarger[30], die *Transaminase B* mit [^{3}H]Isoleucin und 2-Oxoglutarat als Substraten nach Szentirmai und Umbarger[31] bestimmt. Der Test auf *Transaminase C* entspricht dem der Transaminase B mit dem Unterschied, daß anstelle des 2-Oxoglutarats Pyruvat als Aminogruppenacceptor verwendet wird und in einigen Versuchen [^{3}H]Valin als Aminodonator. Die *Leucin-Dehydrogenase* wurde nach Sanwal und Zink[32] getestet.

Die Aktivität der *Pyridoxaminphosphat-Oxidase* wurde nach Wada und Snell[33] in leicht modifizierter Form bestimmt. Die Versuchsansätze enthielten in 1 m*l*: 50 µmol Phosphatpuffer, pH 8, 0.1 µmol Pyridoxin-5'-phosphat (oder Pyridoxamin-5'-phosphat), 0.05 - 0.4 m*l* Rohextrakt (30 mg Protein/m*l*). Inkubiert wurde in 25-m*l*-Erlenmeyerkölbchen bei 37 °C unter Schütteln. Nach 30 min wurden 0.2 m*l* 50proz. Trichloressigsäure zugegeben und das ausgefällte Protein abzentrifugiert. Der Überstand wurde mit 0.1 m*l* Phenylhydrazin-Reagenz (2 g Phenylhydrazin in 100 m*l* 10N H_2SO_4) versetzt und nach 10 min das gebildete Phenylhydrazon bei 410 nm gemessen.

Die Testansätze für die *Pyridoxal-Kinase* enthielten in 1 m*l*: 50 µmol Phosphatpuffer, pH 6, 0.5 µmol Pyridoxal, 1 µmol ATP, 5 µmol $MgCl_2$, 0.1 - 0.4 m*l* Rohextrakt (30 mg Protein/m*l*). Nach 30 min bei 37 °C wurden 0.2 m*l* 50proz. Trichloressigsäure zupipettiert, das Protein abzentrifugiert und der Überstand auf 0 °C abgekühlt. Dann wurden 0.1 m*l* 18N H_2SO_4 und 0.1 m*l* Phenylhydrazin-Reagenz zugegeben und nach 30 min bei 0 °C die Extinktion bei 410 nm bestimmt.

Ergebnisse

1) Isolierung von Vitamin-B_6-Ausscheidermutanten von Bacillus subtilis B 23

Ausgehend vom Wildstamm B 23 von *B. subtilis* konnten nach Mutation und Selektion Mutanten isoliert werden, die erhöhte Mengen von Vitamin B_6 ins Medium ausscheiden. Zur Mutationsauslösung wurde *N*-Methyl-*N'*-nitro-*N*-nitrosoguanidin verwendet, und die Selektion erfolgte mit Hil-

Tab. 1. Nomenklatur und Eigenschaften der isolierten Vitamin-B_6-Ausscheidermutanten von *B. subtilis*.

Ausgangsstamm	Anzahl	Bezeichnung	Charakteristik
Wildstamm B 23	1	BA 1	scheidet 1 - 2 μg/m*l* aus
BA 1	3	BA 11, 12, 13	scheiden 3 - 5 μg/m*l* aus
BA 1	1	L 71	Lysin-Mangelmutante (wächst mit Vitamin B_6) scheidet 4 - 5 μg B_6/m*l* aus.

fe eines Testmediums, in dem die Vitamin-B_6-bedürftige Hefe *S. carlsbergensis* suspendiert war.

Eine Ausscheidermutante, BA 1, synthetisiert unter optimalen Bedingungen ca. 1 - 2 μg Vitamin B_6/m*l*.

Für die weitere Optimierung der Ausscheidereigenschaften wurden Vitamin-B_6-Antagonisten eingesetzt. Dabei wurde erfolgreich eine Gradientenplattentechnik mit dem Vitamin-B_6-Antagonisten Isonicotinohydrazid nach Pardini und Argoudelis[22] angewandt. Andere B_6-Antagonisten wie z.B. 4'-Desoxypyridoxin oder 4-Amino-5-(hydroxymethyl)-2-methylpyrimidin (Toxopyrimidin) führten nicht zum Erfolg. Zur Erhöhung der Mutantenausbeute wurde der *Bacillus*-Stamm BA 1 mit dem oben genannten Mutagen behandelt. Mit dieser Mutations- und Selektionsmethode wurden 3 Mutanten erhalten, die maximal 3 - 5 μg Vitamin B_6/m*l* synthetisieren. Die Bestimmung der Vitamin-B_6-Konzentration im Akkumulationsmedium erfolgte im Flüssigtest mit *S. carlsbergensis* als Testorganismus[25,26]. In Tab. 1 sind die verschiedenen B_6-Ausscheidermutanten und die jeweiligen Ausgangsstämme aufgeführt.

Bei der Suche nach potentiellen Vorläufern und Verbindungen, die stimulierend auf die Vitamin-B_6-Biosynthese einwirken, wurden verschiedene Stickstoff- und Kohlenstoffquellen getestet. Es war bekannt, daß Glycerin[9,10] bei *E. coli* und 4-Aminobuttersäure oder β-Alanin in Kombination mit Glycerin[13] bei *A. cycloclastes* zu einer Steigerung der Vitamin-B_6-Produktion führen. Überraschenderweise hatte die Zugabe von Glycerin bei den *Bacillus*-Mutanten aber eine 80proz. Hemmung der B_6-Synthese zur Folge. Dabei wachsen die Bakterien mit Glycerin als C-Quelle. Aufgrund dieser Ergebnisse wurde bei weiteren Akkumulationsversuchen immer Glucose eingesetzt. Die optimale Glucosekonzentration wurde für BA 1 bestimmt und liegt bei 0.5 - 2%. Ober- und unterhalb dieses Bereichs sinkt die B_6-Konzentration im Akkumulationsmedium stark ab. Die Zugabe bestimmter Aminosäuren zum Testmedium brachte eine Steigerung der Vitamin-B_6-Produktion. Mit Isoleucin, Asparaginsäure, Glutamin und Alanin wurde eine Verdoppelung der Vitamin-B_6-Menge im Minimalmedium erreicht. Alle anderen Aminosäuren hatten keinen Einfluß. Um eine maximale Stimulierung der B_6-Synthese zu erhalten, müssen die aktiven Aminosäuren in einer Konzentration von 50 μmol/m*l* eingesetzt werden, wie aus Abb. 1 hervorgeht.

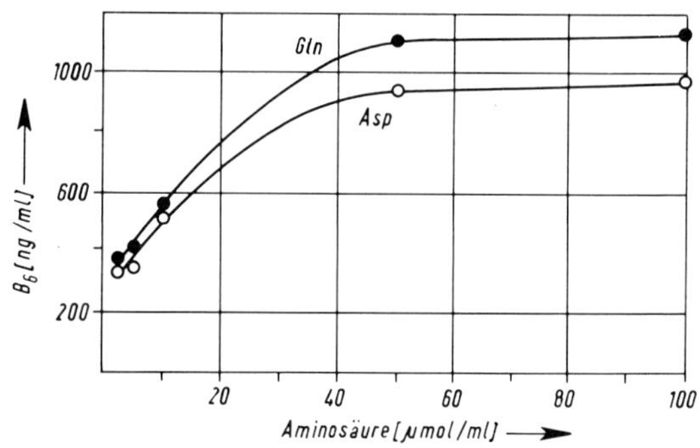

Abb. 1. Abhängigkeit der Vitamin-B_6-Produktion von der Aminosäurekonzentration (Glutamin, Asparaginsäure) bei BA 1.

Bei Kombination der wirksamen Aminosäuren verhalten sich die Stimulierungseffekte bei den Paaren Isoleucin + Asparaginsäure und Glutamin + Asparaginsäure bei BA1 additiv (Tab. 2). Dies könnte bedeuten, daß diese Aminosäuren unabhängig voneinander in die B_6-Biosynthese eingreifen. Bei BA 11 ist dieser Effekt weniger ausgeprägt (Tab. 2).

Tab. 2. Stimulierung der Vitamin-B_6-Produktion bei den Ausscheidermutanten BA 1 und BA 11 durch verschiedene Aminosäuren.

Aminosäure [50 μmol/ml]	Vitamin B_6 [ng/ml Medium]		
	BA 1	BA 11	Wildstamm B 23
Kontrolle (Minimalmedium)	450	1300	90
+ Ile	1100	2700	
+ Gln	1050	3000	130
+ Asp	1100	2800	
+ Ala	1000	2400	
+ Ile + Asp	1900	–	
+ Gln + Asp	1900	4000	

Die Identifizierung der verschiedenen B_6-Komponenten bei BA 1 und BA 11 wurde mittels Bioautographie durchgeführt. Da sich Pyridoxal und Pyridoxin nur sehr schwer trennen lassen, wurde das Chromatographiepapier vor dem Lauf mit Borsäure (2.5% in Äthanol) besprüht. Dadurch wird Pyridoxin durch die Bildung eines Borsäurekomplexes in seinem Lauf gegenüber Pyridoxal verzögert. Aus den R_F-Werten (Tab. 3) und aus der Intensität der Wachstumshöfe kann geschlossen werden, daß hauptsächlich Pyridoxamin und wenig Pyridoxal im Akkumulat vorliegen. Pyridoxin ließ sich *nicht* nachweisen.

Tab. 3. Identifizierung der verschiedenen Vitamin-B_6-Komponenten mittels Bioautographie auf Boratpapier. Als Laufmittel wurde ein Gemisch von 1-Propanol/10proz. Ameisensäure 80 : 20 (v/v) verwendet.

	R_F-Wert
Vergleichssubstanzen:	
Pyridoxal	0.55
Pyridoxin	0.45
Pyridoxamin	0.28
wachstumswirksame Komponenten im Akkumulat	0.56 0.27

2) *Isolierung von Vitamin-B_6-Mangelmutanten*

Zur weiteren Untersuchung der Vitamin-B_6-Biosynthese in *B. subtilis* wurden von den drei Vitamin-B_6-Ausscheidermutanten BA 1, BA 11 und L 71 Vitamin-B_6-Mangelmutanten selektioniert. Um das ganze Spektrum potentieller B_6-Mutanten zu erfassen, wurden den Selektionsmedien außer Pyridoxin und Pyridoxal noch verschiedene Aminosäuren zugesetzt. Auch das Komplettmedium für die Zwischenzüchtung wurde mit Pyridoxal supplementiert, da diese leicht oxidierbare Verbindung in vollsynthetischen Nährmedien höchstens in Spuren vorhanden ist. Insgesamt wurden 49 Vitamin-B_6-Mangelmutanten isoliert, die in Tab. 4 aufgeführt und näher charakterisiert sind.

Tab. 4. Charakterisierung und Nomenklatur der isolierten Vitamin-B_6-Mangelmutanten.

Ausgangsstamm	Anzahl der davon induzierten Vitamin-B_6-Mangelmutanten	Bezeichnung	Nährstoffbedarf (alternatives Wachstum)
BA 1	21	PM 1, 4, 6, 7, 8, 10 - 15 17 - 23, 26 - 28	Pyridoxin, Pyridoxal, Pyridoxamin
BA 1	6	PM 3, 5, 9, 16, 24, 25	Pyridoxin, Pyridoxal, Pyridoxamin, Isoleucin
L 71	6	PM 101, 102, 104, 105 107, 109	Pyridoxin, Pyridoxal, Pyridoxamin
L 71	3	PM 103, 106, 108	Pyridoxin, Pyridoxal, Pyridoxamin, Isoleucin
BA 11	11	PM 201, 202, 205 - 209 213, 215 - 217	Pyridoxin, Pyridoxal, Pyridoxamin
BA 11	2	PM 210, 211	Pyridoxin, Pyridoxal, Pyridoxamin, Isoleucin

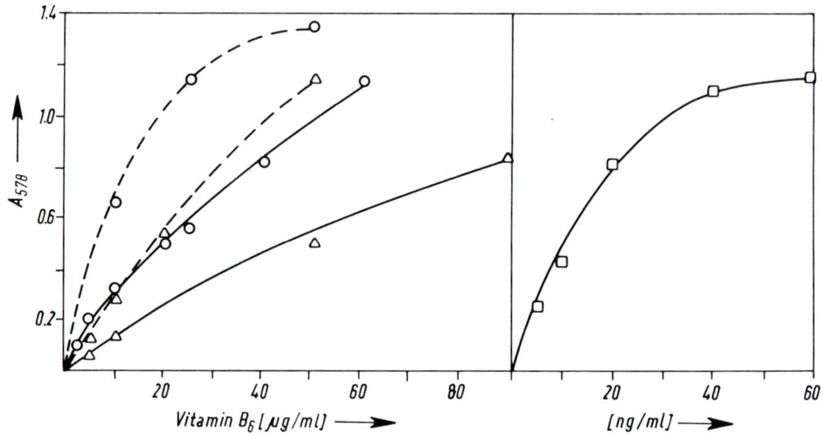

Abb. 2. Wachstum der Vitamin-B_6-Mangelmutanten PM 4 (– – –) und PM 17 (——) bei verschiedenen Konzentrationen an Vitamin B_6.
○ = Pyridoxin; △ = 4′-Desoxypyridoxin; □ = Pyridoxal.

Die Untersuchung der Nährstoffansprüche ergab, daß alle B_6-Mangelmutanten sowohl mit Pyridoxal, als auch mit Pyridoxamin oder Pyridoxin wachsen können. Die verschiedenen Vitamin-B_6-Formen sind in unterschiedlicher Weise aktiv. So induziert Pyridoxal schon in Konzentrationen von 60 - 80 ng/m*l* maximales Bakterienwachstum. Pyridoxin und Pyridoxamin dagegen müssen in sehr viel höherer Konzentration zugesetzt werden (60 - 100 µg/m*l*).

Besonders interessant war die Beobachtung, daß der wohl bekannteste Vitamin-B_6-Antagonist 4′-Desoxypyridoxin, bei allen B_6-Mangelmutanten von *B. subtilis* fast so wirksam wie Pyridoxin oder Pyridoxamin ist. Für maximales Wachstum müssen von dieser Verbindung 80 - 120 µg/m*l* Medium zugegeben werden. In Abb. 2 ist das durch die verschiedenen B_6-Formen ermöglichte Wachstum am Beispiel der Mutanten PM 4 und PM 17 dargestellt. Andere B_6-Analoge wie z.B. 4-Pyridoxsäure oder [3-Hydroxy-4-(hydroxymethyl)-2-methyl-5-pyridyl]methansulfonsäure sind nicht wachstumswirksam.

Außer den verschiedenen B_6-Formen wurden auch noch andere Wuchsstoffe wie z.B. Aminosäuren getestet. Aus den Wachstumsversuchen ging hervor, daß von den eingesetzten Aminosäuren nur Isoleucin bei einigen B_6-Mangelmutanten das Vitamin ersetzen kann (s. Tab. 4). Allerdings ist die Lag-Phase beim Wachstum in Flüssigmedium mit Isoleucin verlängert (5 - 8 h). Bietet man Isoleucin in Verbindung mit einer der verschiedenen Vitamin-B_6-Formen im Medium an, so erhält man eine deutliche Stimulierung des Wachstums bei *allen* B_6-Mangelmutanten. In

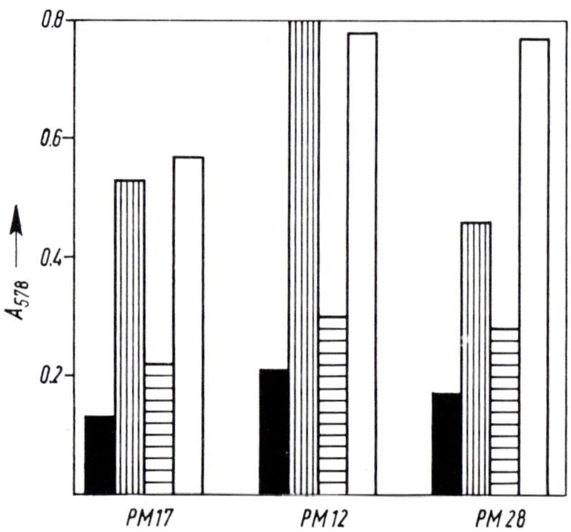

Abb. 3. Wachstum der Vitamin-B_6-Mangelmutanten in Minimalmedium supplementiert mit Pyridoxin (5 µg/m*l*), 4′-Desoxypyridoxin (10 µg/m*l*) und Isoleucin (500 µg/m*l*).
■ 4′-Desoxypyridoxin; ▥ 4′-Desoxypyridoxin + Isoleucin; ▤ Pyridoxin; □ Pyridoxin + Isoleucin.

Abb. 3 ist dies für die Mutanten PM 12, 17 und 28 graphisch dargestellt. Andere Aminosäuren zeigen diesen Effekt nicht. Valin bewirkt sogar eine leichte Hemmung.

Versuche mit verschiedenen Isoleucinkonzentrationen deuten darauf hin, daß eine optimale Wachstumsstimulierung erst bei 0.5 - 1 mg/m*l* erreicht wird. Ober- und unterhalb dieses Bereichs tritt der Effekt deutlich zurück. 3-Methyl-2-oxovaleriansäure, das Desaminierungsprodukt von Isoleucin, verhält sich im Wachstumstest wie die Aminosäure. Dagegen wirkt 2-Oxobuttersäure in Konzentrationen von 0.5 - 1 mg/m*l* toxisch und hemmt das Wachstum der *Bacillus*-Mutanten. Daß diese Wachstumsstimulierung durch Isoleucin für die B_6-Mangelmutanten spezifisch ist, geht aus unseren Untersuchungen mit Thiaminmangelmutanten von *B. subtilis* hervor; diese werden in ihrem Wachstum durch Isoleucin nicht beeinflußt[34].

Mit Glykolaldehyd, wie für *E. coli* beschrieben[7], und mit Nicotinsäure[35] konnte bei den B_6-Mangelmutanten von *B. subtilis* kein Wachstum nachgewiesen werden. Zur weiteren Charakterisierung der B_6-Mangelmutanten wurde versucht, eine Gruppierung durch Kreuzfütterungsversuche zu erzielen. Dabei ließ sich eine Einteilung in 2 Gruppen vornehmen, wobei die eine die andere füttert. Versuche, den fütterungswirksamen Stoff anzureichern, führten bislang nicht zum Erfolg, da die Verbindung labil zu sein scheint.

3) Enzymatische Untersuchungen

Zur Deutung der Befunde aus den Akkumulations- und Wachstumsversuchen der Vitamin-B_6-Ausscheider- und -Mangelmutanten wurden vergleichende Untersuchungen bestimmter enzymatischer Umsetzungen bei den verschiedenen *Bacillus*-Mutanten unternommen. In diesem Zusammenhang waren vor allem die Enzyme des Isoleucinstoffwechsels von Bedeutung, wie z.B. die Threonin-Dehydratase, die Transaminase B und die Leucin-Dehydrogenase. Außerdem sollten die Enzyme der letzten beiden Schritte der Vitamin-B_6-Biosynthese, wie sie von Dempsey[2] bei *E. coli* postuliert werden, untersucht und eine enzymatische Umsetzung von Pyridoxin bzw. 4'-Desoxypyridoxin bei *B. subtilis* überprüft werden.

a) Threonin-Dehydratase

Aus eigenen Untersuchungen an *E. coli*[36] und von Guirard et al. an *Salmonella typhimurium*[37] war bekannt, daß ein alternatives Wachstum mit Isoleucin und Vitamin B_6 gegebenenfalls durch eine Mutation der Threonin-Dehydratase zu deuten ist. Es handelt sich also in diesem Falle um echte Isoleucin-Mangelmutanten. In vitro ist bei diesen Mutanten eine geringe Aktivität der Threonin-Dehydratase nachzuweisen, die durch Pyridoxalphosphat etwas erhöht werden kann. Diese Mutanten wachsen aber mit B_6 schlechter als mit der Aminosäure.

Aufgrund dieser Kenntnisse war es naheliegend, die Threonin-Dehydratase bei den *Bacillus*-Mutanten zu untersuchen. Die Versuche bei *B. subtilis* zeigten, daß dieses Enzym bei den Mutanten immer vorhanden ist und daß es im Vergleich zum Wildstamm sogar mit doppelter spezifischer Aktivität im Rohextrakt vorliegt (Tab. 5). Eine echte Isoleucin-Mangelmutante (Ile 134) wies im Kontrollversuch nur ca. 4% Restaktivität auf. Eine Bedürftigkeit für Isoleucin kann auch in einer Hypersensitivität der Threonin-Dehydratase gegenüber dem allosterischen Effektor Isoleucin begründet sein. Eine derartige Hypersensitivität, wie sie von *Salmonella-typhimurium*-Mutanten berichtet wurde[38], liegt ebenfalls nicht vor, da die Inhibitorkonstanten dieses Enzyms bei allen untersuchten Stämmen unverändert sind.

Auch die Regulation des Enzyms ist bei den Mutanten im Vergleich zum Wildstamm verändert. Beim Wildstamm läßt sich die bekannte multivalente Repression der Enzyme der Isoleucin- und Valinbiosynthese, die eine Überproduktion von Isoleucin, Valin und Leucin verhindert, nachweisen[39]. Bei den *Bacillus*-Mutanten ist diese multivalente Repression aufgehoben oder zumindest stark vermindert (Tab. 5).

b) Transaminase B

In der Literatur sind auch Mutanten beschrieben, die im letzten Enzym der Isoleucin- und Valinbiosynthese, der Transaminase B, genetisch blockiert sind und nur Isoleucinbedarf zeigen[40].

Da die Threonin-Dehydratase bei unseren *Bacillus*-Mutanten die volle Aktivität aufwies, bestand die Möglichkeit, daß die Wachstumsstimulierung durch Isoleucin bzw. das alternative Wachstum

Tab. 5. Aktivität der Threonin-Dehydratase und der Transaminase B unter Repressions- und Derepressionsbedingungen.

Stamm	Spezifische Aktivität [mU/mg]					
	Threonin-Dehydratase			Transaminase B		
	Derepression	Repression	Repr. Fakt.	Derepression	Repression	Repr. Fakt.
Wildstamm B 23	67	36	1.8	3.6	2.2	1.65
BA 1	125	133	1	6.1	5.8	1.05
BA 11	115	116	1	–	–	
PM 16	145	–		6.9		
PM 17	130	–		5.4		
Ile 134	4	–				

einiger B_6-Mangelmutanten mit Isoleucin oder B_6 auf einem Defekt der Transaminase B beruhen könnte. Aus Tabelle 5 geht hervor, daß dies bei den *Bacillus*-Mutanten nicht der Fall ist. Auch bei diesem Enzym war die spezifische Aktivität im Rohextrakt fast doppelt so hoch wie beim Wildstamm. Außerdem ist die multivalente Repression aufgehoben, ähnlich wie bei der Threonin-Dehydratase. Dies war auch zu erwarten, da die Kontrolle das gesamte Operon betrifft.

c) Transaminase C

Die Untersuchungen wurden auch auf die Transaminase C ausgedehnt, da kürzlich über *E.-coli*-Mutanten berichtet worden ist, die mit Isoleucin bzw. Leucin oder Vitamin B_6 wachsen und keine Transaminase-C-Aktivität besitzen[41]. Die Transaminase C katalysiert den Übergang Valin ↔ Alanin (2-Aminobuttersäure). In diesen Mutanten sind Threonin-Dehydratase und Transaminase B normal vorhanden.

In unseren *Bacillus*-Mutanten ließ sich Transaminase C nachweisen. Die spezifische Aktivität im Rohextrakt war dreimal so hoch wie beim Wildstamm (Tab. 6). Im übrigen konnte gezeigt werden, daß der Rohextrakt nicht nur den Übergang Valin ↔ Alanin wie bei *E. coli*, sondern auch den Übergang Isoleucin ↔ Alanin katalysiert.

Aus den enzymatischen Versuchen ergibt sich, daß die B_6-Mangelmutanten von *B. subtilis* nicht identisch mit bisher beschriebenen Mutantentypen sind. Ein Defekt in der Isoleucin-Biosynthese ist demnach *nicht* für das Wachstumsverhalten verantwortlich.

Tab. 6. Aktivität der Transaminase C im Rohextrakt verschiedener *Bacillus*-Stämme.

Stamm	Spezifische Aktivität [mU/mg]	
	Ile ↔ Ala	Val ↔ Ala
Wildstamm B 23	1.7	1.2
BA 1	5.2	4.1
PM 16	4.8	3.2
PM 17	3.8	3

d) Leucin-Dehydrogenase

Ein weiteres Enzym, das in den Isoleucinstoffwechsel eingreift, ist die Leucin-Dehydrogenase. Sie katalysiert im ersten Schritt des Isoleucin-Abbaus die Umwandlung in 3-Methyl-2-oxovalerat. Es war bekannt, daß das Enzym in *B. subtilis* durch Isoleucin und Valin, nicht aber durch Leucin induziert wird[42].

Unsere Untersuchungen ergaben, daß bei den *Bacillus*-Mutanten die Leucin-Dehydrogenase vorhanden ist, aber *konstitutiv* gebildet wird (Tab. 7).

Tab. 7. Aktivität der Leucin-Dehydrogenase unter induktiven und nichtinduktiven Bedingungen mit 3-Methyl-2-oxovalerat als Substrat.

Stamm	Spezifische Aktivität [U/mg]	
	Minimalmedium	Minimalmedium + Isoleucin
Wildstamm B 23	0.05	0.40
BA 1	0.43	0.40
BA 11	0.37	0.42
PM 17	0.40	0.37

Tab. 8. Bestimmung der K_m-Werte für die Substrate und Cofaktoren der Hin- und Rückreaktion bei der Leucin-Dehydrogenase.
Die Werte wurden beim Wildstamm B 23 und den Mutanten BA 1, BA 11 und PM 17 bestimmt.

Stamm	K_m [mmol/l] für				
	Ile	$NAD^{\oplus}$	3-Methyl-2-oxovalerat	NADH	$NH_4^{\oplus}$
B 23	2.7	0.2	0.77	0.035	33
BA 1	2.7	0.2	0.7	0.03	–
BA 11	–	–	0.8	–	33
PM 17	2.7	–	0.77	–	40

Das Enzym selbst scheint unverändert zu sein, zumindest ergab die Bestimmung der verschiedenen K_m-Werte keinen Unterschied im Vergleich zum Wildstamm (Tab. 8).

Eine Untersuchung der Enzymaktivität in Abhängigkeit von der Isoleucinkonzentration zeigt, daß bei hohen Aminosäuremengen eine Substrathemmung auftritt. Die optimale Isoleucinkonzentration liegt in dem Bereich, der auch im Wachstumstest bei den B_6-Mangelmutanten eine optimale Stimulierung bewirkt (Abb. 4).

e) Pyridoxaminphosphat-Oxidase und Pyridoxal-Kinase

Die Umsetzung des Pyridoxins zum aktiven Coenzym Pyridoxal-5'-phosphat konnte von Dempsey[2] an *E. coli* aufgeklärt werden. Dabei wird zuerst Pyridoxin von der Pyridoxal-Kinase zu Pyridoxin-5'-phosphat phosphoryliert. Dann erfolgt die Oxidation zum aktiven Coenzym durch die Pyridoxaminphosphat-Oxidase. Diese Oxidationsreaktion ist auch bei Säugetieren beschrieben[33].

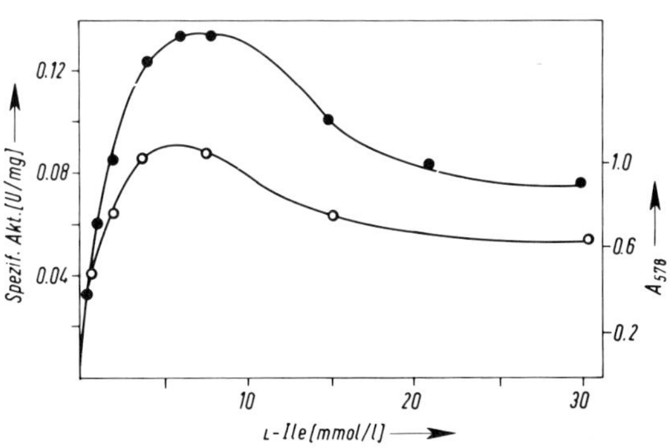

Abb. 4. Bestimmung der optimalen Isoleucinkonzentration für die Leucin-Dehydrogenase (●——●) im Vergleich zur optimalen Isoleucinkonzentration im Wachstumstest (○——○) bei PM 17.

In unseren Untersuchungen konnte bei *B. subtilis* und auch bei den *Bacillus*-Mutanten keine Pyridoxaminphosphat-Oxidase-Aktivität nachgewiesen werden. Die Pyridoxal-Kinase ist dagegen im Rohextrakt des Wildstammes und der Mutanten vorhanden. Weitere Einzelheiten zu diesem Thema, das zu einer vergleichenden Untersuchung beider Enzyme in grampositiven und gramnegativen Bakterien führte, sind einer nachfolgenden Veröffentlichung zu entnehmen.

Diskussion

Vergleicht man die hier beschriebenen B_6-Ausscheidermutanten von *B. subtilis*, die maximal 5 µg B_6/m*l* ausscheiden, mit anderen B_6-Überproduzenten, so findet man bei *E. coli* 0.5 µg/m*l*[2] und bei *A. cycloclastes* 2 - 3 µg B_6/m*l* im Medium[13]. Eine Ausnahme bildet *Pichia guilliermondi*, eine Alkanhefe, die 25 µg B_6/m*l* produziert[43].

Bei der Suche nach potentiellen Vorläufern oder Substanzen, die die B_6-Biosynthese stimulieren, erwiesen sich Isoleucin, Asparaginsäure, Glutamin und Alanin bei unseren Ausscheidermutanten als wirksam. Mit ihnen erhält man eine Verdoppelung der B_6-Menge im Akkumulationsmedium. Von den verschiedenen Vitamin-B_6-Komponenten ließ sich hauptsächlich Pyridoxamin und wenig Pyridoxal bioautographisch nachweisen. Zu einem ähnlichen Ergebnis gelangen auch Ishida und Shimura bei *A. cycloclastes.*[13]

Pyridoxin, das hauptsächlich bei der *E.-coli*-Mutante von Dempsey[2] gefunden wird, läßt sich bei *B. subtilis* nicht nachweisen. Dies könnte bedeuten, daß es als freies Produkt in der Biosynthese nicht vorkommt.

Aus den Wachstumsversuchen mit B_6-Mangelmutanten von *B. subtilis* war ersichtlich, daß von allen getesteten Aminosäuren nur Isoleucin einen stimulierenden Einfluß ausübt. Einige B_6-Mutanten wachsen auch alternativ mit Isoleucin oder Vitamin B_6 (Tab. 4), wobei das Wachstum mit Isoleucin stark verzögert einsetzt (eventuell Leaky-Mutanten). Dieses alternative Wachstum der *Bacillus*-Mutanten mit Isoleucin oder Vitamin B_6 kann nicht auf einem genetischen Defekt bestimmter Enzyme der Isoleucinbiosynthese beruhen, wie es von anderen Autoren[36,37] beschrieben wurde. Die enzymatischen Untersuchungen der Threonin-Dehydratase und der Transaminase B bei *B. subtilis* schließen dies aus. Eine Doppelmutation der B_6-Mangelmutanten von *B. subtilis* (Ile- und B_6-) kommt aufgrund der Reversionsraten (10^{-5} - 10^{-8}) und des Wachstumsverhaltens ebenfalls nicht in Betracht. Dies würde bedeuten, daß Isoleucin oder eine analoge Verbindung in die B_6-Biosynthese eingreift. Als Hinweis dafür können auch die Untersuchungen der Leucin-Dehydrogenase gelten. Bei diesem Enzym, das Isoleucin in die entsprechende 2-Oxosäure umsetzt, wird das Optimum der Enzymaktivität bei einer Isoleucinkonzentration erreicht, die auch im Wachstumstest bei den B_6-Mangelmutanten eine optimale Stimulierung bewirkt (Abb. 4).

Überraschend war, daß der Vitamin-B_6-Antagonist 4'-Desoxypyridoxin das Vitamin bei den *Bacillus*-Mutanten ersetzen kann. Dies legt den Schluß nahe, diesen Vitamin-B_6-Antagonisten bei *B. subtilis* als Zwischenprodukt in der B_6-Biosynthese zu postulieren. Für die Funktion von Vitamin B_6 als Coenzym im Aminosäurestoffwechsel ist eine Aldehydgruppe in 4-Stellung erforderlich. Zur Einschleusung des Antagonisten in den B_6-Biosyntheseweg muß also die Oxidation der 4-Methyl- zur Aldehydgruppe gefordert werden.

Bei Akkumulationsversuchen mit 4'-Desoxypyridoxin ließ sich aus dem Medium 4'-Desoxypyridoxin-5'-phosphat isolieren. Dies steht im Einklang mit dem Nachweis der Pyridoxal-Kinase bei *B. subtilis*.

Eine Umwandlung der Methyl- zur Hydroxymethyl- bzw. Aldehydgruppe konnte bislang enzymatisch nicht nachgewiesen werden. Auch die Oxidation des Pyridoxins zum aktiven Coenzym läßt sich bei *B. subtilis* nicht nachweisen. Pyridoxaminphosphat-Oxidase, wie sie bei *E. coli*[44,2] und bei Säugetieren[33] als freies Enzym nachgewiesen wurde, liegt bei *B. subtilis* in der Form offenbar nicht vor.

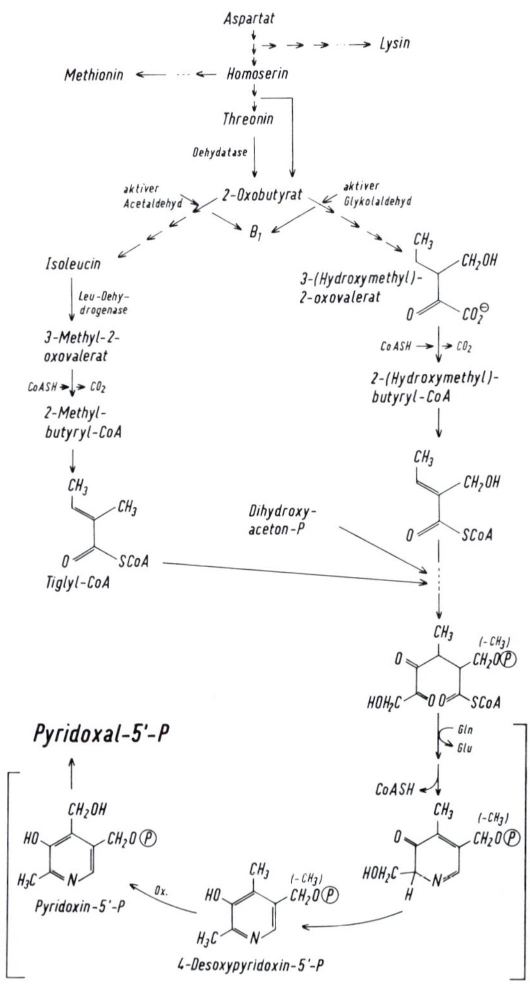

Abb. 5. Hypothetisches Vitamin B_6-Biosyntheseschema für *Bacillus subtilis*.

Die verschiedenen Ergebnisse aus den Akkumulations- und Wachstumsversuchen bzw. den enzymatischen Untersuchungen bei *B. subtilis*, sollen in einem hypothetischen Formelschema dargestellt werden (Abb. 5). Es bleibt dabei ungeklärt, ob dieses Schema auch auf andere Bakterien übertragbar ist. Danach wird ein Teil des C-Skeletts im B_6-Molekül von einer Verbindung geliefert, die das C-Gerüst der Aminosäure Isoleucin besitzt. Dafür kommen zwei Möglichkeiten in Betracht:

1) Der Weg über Isoleucin und dessen Abbauprodukte 3-Methyl-2-oxovalerat bzw. Tiglyl-CoA. Dies wäre denkbar, wenn man annimmt, daß es sich bei den Vitamin-B_6-Mangelmutanten, die ja durch Isoleucin stimuliert werden, um „Leaky-Mutanten" handelt. So würde die Synthese durch die Vorprodukte Isoleucin bzw. 3-Methyl-2-oxovalerat stimuliert, obwohl die genetische Blokkierung dahinter liegen muß. Damit steht in Übereinstimmung die verlängerte Lag-Phase bei der Inkubation mit Isoleucin (PM 16).

2) Die Trennung der Synthesewege für Isoleucin und Vitamin B_6 auf der Stufe des Homoserins bzw. Homoserinphosphats. Homoserin bzw. dessen Phosphorsäureester kann direkt in 2-Oxobutyrat überführt werden[45]. Wenn nun aktiver Glykolaldehyd anstelle des aktiven Acetaldehyds wie im ersten Schritt der Isoleucin-Biosynthese mit 2-Oxobutyrat reagiert, gelangt man letztlich zum 3-(Hydroxymethyl)-2-oxovalerat.

In weiteren hypothetischen Reaktionsschritten würde die Biosynthese dann über 4'-Desoxypyridoxin, das ja bei allen Vitamin-B_6-Mangelmutanten von *B. subtilis* wachstumswirksam ist, zum aktiven Coenzym Pyridoxalphosphat führen.

Hill et al. konnten bei B_6-Mangelmutanten von *E. coli* zeigen, daß markierter Glykolaldehyd als C-5 und 5' in das C-Skelett des Pyridoxins eingebaut wird.[12] An derselben Stelle würde man auch die in Form von aktivem Glykolaldehyd eingeschleuste C_2-Einheit im hypothetischen Formelschema wiederfinden. Für eine Trennung der Synthesewege könnte auch sprechen, daß Homoserin-Mangelmutanten von BA 11 in unseren Untersuchungen doppelt so viel Vitamin B_6 synthetisieren, wenn dem Medium außer Homoserin (optimale Wachstumskonzentration) noch Methionin, Threonin und Isoleucin (50 μg/m*l*) zugesetzt werden. Wenn der Biosyntheseweg über 3-(Hydroxymethyl-2-oxovalerat und 4'-Desoxypyridoxin geht, müßte man fordern, daß Isoleucin bzw. dessen Abbauprodukte in die Synthesekette eingeschleust werden, da die B_6-Mangelmutanten mit Isoleucin wachsen können.

Die Tatsache, daß Pyridoxaminphosphat-Oxidase-Aktivität und freies Pyridoxin nicht nachzuweisen sind und daß bei den B_6-Mangelmutanten Pyridoxal in sehr viel geringerer Konzentration wachstumswirksam ist, könnte mit folgender Annahme erklärt werden: Wenn die letzten Schritte der Biosynthese bei *B. subtilis* durch einen Enzymkomplex katalysiert werden, ließe sich freie Oxidase-Aktivität wie bei *E. coli* und Säugern und freies Pyridoxin (bzw. 4'-Desoxypyridoxin) gar nicht nachweisen. Die Reaktionskette würde dann intramolekular am Enzymkomplex ablaufen. Exogen zugesetztes Pyridoxal dagegen kann von der Pyridoxal-Kinase, die bei *B. subtilis* im Rohextrakt vorhanden war, in Pyridoxalphosphat überführt werden. Ähnliche Beispiele sind aus der Literatur bekannt. So wird die Nicotinsäurebiosynthese, die bei Pflanzen und Bakterien von Asparaginsäure und Dihydroxyacetonphosphat ausgeht, zum großen Teil von einem Enzymkomplex katalysiert[46,47].

Im hypothetischen Biosyntheseschema wurde als mögliche C_3-Verbindung, welche die C-Atome 2', 2 und 3 des C-Skeletts bereitstellt, Dihydroxyacetonphosphat ausgewählt. Dadurch ließe sich auch die Funktion des Glutamins (Stimulierung der B_6-Synthese bei den B_6-Überproduzenten) als N-Donator erklären. Genauso könnte jedoch auch 3-Phosphoserin, das nach Dempsey als potentieller B_6-Vorläufer in Frage kommt[5] oder Alanin bzw. Pyruvat eingezeichnet werden.

Wir danken dem *Fonds der Chemischen Industrie* für Unterstützung dieser Arbeit und Herrn *Dr. J. Eberspächer* für die Durchsicht des Manuskripts.

Literatur

1 Dempsey, W.B. & Pachler, P.F. (1966) *J. Bacteriol.* **91**, 642 - 645.
2 Dempsey, W.B. (1966) *J. Bacteriol.* **92**, 333 - 337.
3 Dempsey, W.B. (1969) *J. Bacteriol.* **97**, 1403 - 1410.
4 Dempsey, W.B. (1969) *J. Bacteriol.* **100**, 295 - 300.
5 Dempsey, W.B. (1969) *Biochem. Biophys. Res. Commun.* **37**, 89 - 93.

6 Dempsey, W.B. & Itoh, H. (1970) *J. Bacteriol.* **104**, 658 - 667.

7 Dempsey, W.B. (1971) *J. Bacteriol.* **108**, 1001 - 1007.

8 Tani, Y. & Dempsey, W.B. (1973) *J. Bacteriol.* **116**, 341 - 345.

9 Hill, R.E. & Spenser, I.D. (1970) *Science* **169**, 773 - 775.

10 Hill, R.E., Rowell, F.J., Gupta, R.N. & Spenser, I.D. (1972) *J. Biol. Chem.* **247**, 1869 - 1882.

11 Hill, R.E. & Spenser, I.D. (1973) *Can. J. Biochem.* **51** 1412 - 1416.

12 Hill, R.E., Horsewood, P. & Spenser, I.D. (1975) *J. Chem. Soc. (Perkin I)* 1622 - 1627.

13 Ishida, M. & Shimura, K. (1970) *Agric. Biol. Chem.* **34**, 327 - 334.

14 Ishida, M., Nagayama, H. & Shimura, K. (1973) *Agric. Biol. Chem.* **37**, 1277 - 1283.

15 Ishida, M., Nagayama, H. & Shimura, K. (1973) *Agric. Biol. Chem.* **37**, 1881 - 1892.

16 McKay, A.F. & Wright, G.F. (1947) *J. Am. Chem. Soc.* **69**, 3028 - 3030.

17 Peterson, E.A. & Sober, H.A. (1954) *J. Am. Chem. Soc.* **76**, 169 - 175.

18 Heyl, D., Luz, E., Harris, S.A. & Folkers, K. (1951) *J. Am. Chem. Soc.* **73**, 3430 - 3433.

19 Spizizen, J. (1958) *Proc. Natl. Acad. Sci. U.S.A.* **44**, 1072 - 1078.

20 Süssmuth, R., Chandra, P., Wacker, A. & Lingens, F. (1969) *Z. Naturforsch.* **24b**, 707 - 712.

21 Scherr, G.H. & Rafelson, M.E., Jr. (1962) *J. Appl. Bacteriol.* **25**, 187 - 194.

22 Pardini, R.S. & Argoudelis, C.J. (1968) *J. Bacteriol.* **96**, 672 - 677.

23 Lederberg, J. & Lederberg, E.M. (1952) *J. Bacteriol.* **63**, 399 - 406.

24 Lingens, F. & Oltmanns, O. (1964) *Z. Naturforsch.* **19b**, 1058 - 1065.

25 Atkin, L., Schultz, A.S. & Frey, C.N. (1943) *Arch. Biochem.* **1**, 9 - 16.

26 Atkin, L., Schultz, A.S., Williams, W.L. & Frey, C.N. (1943) *Ind. Eng. Chem., Anal. Ed.* **15**, 141 - 144.

27 Winsten, W.A. & Eigen, E. (1948) *Proc. Soc. Expt. Biol. Med.* **67**, 513 - 517.

28 Gstirner, F. (1965) Chemisch-physikalische Vitaminbestimmungsmethoden, 5. Auflage, S. 156, Ferdinand-Enke-Verlag, Stuttgart.

29 Layne, E. (1957) *Methods Enzymol.* **3**, 447 - 454.

30 Hatfield, G.W. & Umbarger, H.E. (1971) *Methods Enzymol.* **17B**, 561 - 566.

31 Szentirmai, A. & Umbarger, H.E. (1968) *J. Bacteriol.* **95**, 1666 - 1671.

32 Sanwal, B.D. & Zink, M.W. (1961) *Arch. Biochem. Biophys.* **94**, 430 - 435.

33 Wada, H. & Snell, E.E. (1961) *J. Biol. Chem.* **236**, 2089 - 2095.

34 Pflug, W. (1977) Dissertat. Univ. Hohenheim.

35 Vatsala, T.M., Shanmugasundaram, K.R. & Shanmugasundaram, E.R.B. (1976) *Biochem. Biophys. Res. Commun.* **72**, 1570 - 1575.

36 Grimminger, H. & Lingens, F. (1969) *FEBS Lett.* **5**, 225 - 226.

37 Guirard, B.M., Ames, B.N. & Snell, E.E. (1971) *J. Bacteriol.* **108**, 359 - 363.

38 Springer, W., Grimminger, H. & Lingens, F. (1972) *J. Bacteriol.* **112**, 259 - 263.

39 Freundlich, M., Burns, R.O. & Umbarger, H.E. (1962) *Proc. Natl. Acad. Sci. U.S.A.* **48**, 1804 - 1808.

40 Rudman, D. & Meister, A. (1953) *J. Biol. Chem.* **200**, 591 - 604.

41 Falkinham III. J.O. (1977) *J. Bacteriol.* **130**, 566 - 568.

42 Poralla, K. (1971) *Arch. Mikrobiol.* **77**, 339 - 343.

43 Nishio, N., Sakai, K., Fujii, K. & Kamikubo, T. (1973) *Agric. Biol. Chem.* **37**, 553 - 559.

44 Henderson, H.M. (1965) *Biochem. J.* **95**, 775 - 779.

45 Schildkraut, I. & Greer, S. (1973) *J. Bacteriol.* **115**, 777 - 785.

46 Ogasawara, N., Chandler, J.L.R., Gholson, R.K., Rosser, R.J. & Andreoli, A.J. (1967) *Biochim. Biophys. Acta.* **141**, 199 - 201.

47 Griffith, G.R., Chandler, J.L.R. & Gholson, R.K. (1975) *Eur. J. Biochem.* **54**, 239 - 245.

Prof. Dr. F. Lingens, Institut für Mikrobiologie der Universität Hohenheim, Garbenstr. 30, D-7000 Stuttgart 70.

Hoppe-Seyler's Z. Physiol. Chem.
Bd. 359, S. 571 - 579, Mai 1978

High-Resolution Polyacrylamide Gradient Slab Gel Electrophoresis of Histones H1, H3 and H4

Peter TRAUB and Gunda BOECKMANN

Max-Planck-Institut für Zellbiologie, Wilhelmshaven

(Received 2 February 1978)

Dedicated to Prof. Dr. Adolf Butenandt on the occasion of his 75th birthday

Summary: The resolving power of the histone electrophoresis system described by Panyim and Chalkley [(1969) *Arch. Biochem. Biophys.* **130**, 337] has been improved by the introduction of polyacrylamide gradient slab gel electrophoresis. The histone classes H1, H3 and H4 could be clearly separated into multiple subspecies on a 10 to 20% polyacrylamide gradient gel using 6M urea, 6% acetic acid as a buffer system. Depending on the cell source and the growth conditions of the cells, histone H1 was found to contain 3 to 6 subspecies, histone H_3 also 3 to 6 subspecies, and histone H4, 2 components. In various tissues of the adult mouse, the microheterogeneity of histone H3 with up to 6 components was especially striking in comparison with that in various mouse cell lines propagated in vitro which normally contained only 3 subspecies of histone H3. Mouse tissues with a high extent of histone H3 heterogeneity showed a low degree of histone H1 heterogeneity. The equidistance of the bands in the histone H1 and H3 fractions points at a regularly progressing modification of parental histone molecules. Under the above electrophoretic conditions, the histones H2A and H2B were only poorly resolved; only one subspecies could be separated from the bulk of histone H2A and H2B. It is phosphorylated and co-migrates with histone H2B on polyacrylamide gradient gels containing sodium dodecylsulfate. A better electrophoretic resolution of the histones H2A and H2B was achieved, under otherwise identical electrophoretic conditions, at a urea concentration of 2.5M. At lower urea concentration, however, the histone bands turned out to be less sharp. On the basis of the high resolving power of the present electrophoresis system, a simple method for the quantitation of histone H1, H3 and H4 subspecies could be established.

Hochauflösende Polyacrylamid-Gradienten-Flachgel-Elektrophorese der Histone H1, H3 und H4

Zusammenfassung: Das Auflösungsvermögen des von Panyim und Chalkley [(1969) *Arch. Biochem. Biophys.* **130**, 337] zur Auftrennung von Histonen entwickelten Elektrophoresesystems wurde durch Einführung der Polyacrylamid-Gradienten-Flachgel-Elektrophorese verbessert. Die Histonklassen H1, H3 und H4 konnten unter Verwendung eines 10 bis 20proz. Polyacrylamid-Gradientengels und 6M Harnstoff, 6proz. Essigsäure als Puffersystem deutlich in mehrere Untereinheiten aufgetrennt werden. In Abhängigkeit vom Ursprung der Zellen und von deren Wachstumsbedingungen wurde gezeigt, daß Histon H1 aus 3 bis 6, Histon H3 ebenfalls aus 3 bis 6 und Histon H4 aus 2 Komponenten besteht. In verschiedenen Geweben der erwachsenen Maus war die Mikroheterogenität von Histon H3 mit bis zu 6 Untereinheiten besonders auffallend. Im Vergleich dazu enthielten verschiedene in vitro fortgeführte Mäusezelllinien nur 3 Histon-H3-Untereinheiten. Mäusegewebe mit hoher Mikroheterogenität im Histon-H3-Bereich zeichneten sich durch geringe Mikroheterogenität der Histone H1 aus. Die Äquidistanz der einzelnen Banden der Histon-H1- und -H3-Fraktionen deutet auf eine gleichmäßig fortschreitende Modifikation von Elternmolekülen hin. Unter den oben angeführten Elektrophoresebedingungen ließen sich die Histonklassen H2A und H2B nur geringfügig auflösen, indem nur eine Untereinheit von den mit gleicher Geschwindigkeit wandernden Histonen H2A und H2B abgetrennt werden konnte. Diese Untereinheit ist phosphoryliert und zeigt im Natriumdodecylsulfat-Polyacrylamid-Gradientengel dieselbe elektrophoretische Beweglichkeit wie Histon H2B. Eine bessere elektrophoretische Auftrennung der Histone H2A und H2B wurde, unter sonst identischen Bedingungen, bei einer Harnstoffkonzentration von 2.5M erzielt. In Gradientengelen niedriger Harnstoffkonzentration erwies sich die Histontrennung jedoch als weniger scharf. Das hohe Auflösungsvermögen des beschriebenen Elektrophoresesystems erlaubte die Ausarbeitung einer einfachen Methode zur Quantifizierung der Untereinheiten der Histonklassen H1, H3 und H4.

Key words: Histones, polyacrylamide gradient gel electrophoresis.

In 1969, Panyim and Chalkley described a high-resolution gel electrophoresis system for the separation of histones which is characterized by extraordinary simplicity of composition[1]. It has been sucessfully employed in the characterization of the histone complements of many tissues and cell species grown under various environmental conditions and it was especially useful for the characterization of the microheterogeneity in the histone H1 fraction[2-4]. In order to clearly separate the histone H1 subspecies, however, long electrophoretic runs lasting up to 80 h had to be accepted. It was desirable, therefore, to search for conditions which allow considerably shorter electrophoresis times without loss of resolving power. In the present communication we show that this goal can be partially achieved by introducing polyacrylamide gradient slab gel electrophoresis under essentially the same conditions as described by Panyim and Chalkley[1].

This modified electrophoresis system is applicable to the separation of all basic cellular proteins. We have used it for the electrophoretic characterization of the basic proteins synthesized in a cell-free system from Ehrlich ascites tumor cells in response to poly(A)$^{\oplus}$ mRNA from normal and mengovirus-infected ascites tumor cells[5]. Moreover, in the course of our studies on the picornavirus-induced cessation of host-specific macromolecular synthesis, we were able to demonstrate a characteristic change in the microheterogeneity of histone H1 in mengovirus-infected cells as a function of time post-infection, employing the modified Panyim and Chalkley system[6].

Materials and Methods

Growth of cells

All cells used in this study were propagated in minimum essential medium supplemented with 5 to 10% fetal calf serum as described previously[7]. While Ehrlich ascites tumor cells, L-929 cells and HeLa cells were grown in suspension culture, mouse embryo cells were cultured on monolayers and scraped off the glass wall before confluency, using saline, 1mM phenylmethanesulfonyl fluoride[8] as suspension buffer. Ascites tumor cells were harvested at cell densities of 3×10^5 cells/m*l* (logarithmic growth phase) and 2×10^6 cells/m*l* (stationary growth phase), L-929 cells at 4×10^5 cells/m*l* and HeLa cells at 3.7×10^5 cells/m*l*.

Excision of mouse tissues and calf thymus

Adult Swiss mice were killed by cervical luxation and lung, kidney, liver and spleen were excised. Immediately after excision, the organs were rinsed with cold saline and frozen on dry ice to prevent enzymatic degradation of histones. Calf thymus was frozen in liquid N_2 right after excision and kept at – 70 °C.

Preparation of nuclei

Mouse organs and calf thymus were chopped in buffer A (10mM Tris/HCl (pH 7.5), 2.5mM $Mg(OAc)_2$, 10mM KCl, 6mM 2-mercaptoethanol, 1mM phenylmethanesulfonyl fluoride, 1% Triton X-100) and briefly sheared in a Potter-Elvejhem homogenizer at 0 °C. The homogenates were filtered through three layers of cheese cloth and then centrifuged at 9000 × *g* for 5 min. The supernatants were discarded, the pellets suspended in a large volume of buffer A and extracted three times with buffer A in a tightly fitting all-glass Dounce homogenizer. The nuclear residues which were sedimented by centrifugation at 9000 × *g* for 5 min were used for histone extraction. The nuclei of cells grown in vitro were prepared in the same way by repeated extraction of the cells with buffer A. Under these conditions, the nuclei were virtually free of cytoplasmic tags.

Extraction of histones from nuclei

The nuclear material was suspended in 10 vol. dist. water and the homogeneous suspension was mixed with 2.5N HCl to achieve a final concentration of 0.25N. The mixture was shaken at 0 °C for 2 h and then centrifuged at 37000 × *g* for 10 min. Solubilized protein was precipitated from the supernatant by the addition of 10 vol. cold acetone. After standing at 0 °C for 1 h, the precipitate was collected by centrifugation at 3000 × *g* for 5 min, washed once with cold acetone and finally dried in vacuo at 0 °C. For gel electrophoretic analysis, 4 mg of each histone preparation was dissolved in 1 m*l* 6M urea, 0.6% acetic acid, 10% 2-mercaptoethanol and 15 μ*l* of each solution was applied to the gel.

Polyacrylamide gradient slab gel electrophoretic separation of histones

The electrophoretic separation of crude histones was performed in a Bio-Rad Dual Vertical Slab Gel Electrophoresis Cell, Modell 221 (Bio-Rad Laboratories, Richmond, Calif.). Acrylamide and *N,N'*-methylenebis-(acrylamide) were purchased from Eastman Kodak Co., Rochester, N.Y. and used without further purification.

Linear polyacrylamide gradient slab gels were poured at 0 °C from degassed solutions of the following compositions: (A) 20% acrylamide, 0.5% *N,N'*-methylenebis-(acrylamide), 6M urea, 6% acetic acid, 0.5% *N,N,N',N'*-tetramethylethylenediamine, 0.05% ammonium persulfate, 10% sucrose; (B) 10% acrylamide, 0.25% *N,N'*-methylenebis(acrylamide), 6M urea, 6% acetic acid, 0.5% *N,N,N',N'*-tetramethylethylenediamine, 0.05% ammonium persulfate. A second set of acrylamide solutions contained 2.5M urea instead of 6M urea. The gradients, 27 cm long, 16 cm wide and 0.15 cm thick, were overlayered with 6% acetic acid and allowed to polymerize at room temperature which normally took 2 h. The 1 cm long spacer gel consisted of 4% acrylamide, 0.25% *N,N'*-methylenebis(acrylamide), 6M urea, 0.6% acetic acid, 0.25% *N,N,N',N'*-tetramethylethylenediamine and was polymerized in the presence of 0.05% ammonium persulfate and 0.0002% riboflavin in the fluorescent light at room temperature. 0.8-cm wells were formed by inserting a 10-well comb into the spacer gel solution before polymerization. After gentle removal of the comb, the sample slots were rinsed twice with 0.6% acetic acid.

For better temperature control during electrophoresis, the slab gel plates were fully surrounded by 0.6% acetic acid which was used as lower and upper electrode buffer. Water of 20 °C was circulated through the inner core of the apparatus which served as a support for the gel slab plates. In general, 15 μ*l* of a histone solution (4 mg/m*l*) in 6M urea, 0.6% acetic acid, 10% 2-mercaptoethanol was applied to the sample slots by underlayering and the electrophoresis was started at a constant current of 25 mA/slab gel. Since acetic acid of low concentration served as electrophoresis buffer and *N,N,N',N'*-tetramethylethylenediamine and ammonium persulfate as the only completely dissociated ionic substances in the slab gels were replaced by protons and acetate anions during electrophoresis, the current dropped to 12.5 mA/gel plate during a 24 h run when the protein separation was performed in the presence of 6M urea and to 20 mA/gel plate when the urea concentration in the gels was 2.5M. A Beckman Duostat with a maximum voltage of 500 V was used as a power supply. The protein separation can also be carried out at 25 mA/gel plate constant current throughout electrophoresis, but then more effective cooling of the gels is necessary since the voltage rises from 280 V to 1200 V. Under these conditions, the advantage of saving time is neutralized by blurring of the histone bands.

The slab gels were stained in 0.25% Coomassie Brilliant Blue, 25% 2-propanol-2, 10% acetic acid for 1 h followed by 0.025% Coomassie Brilliant Blue, 10% 2-propanol, 10% acetic acid for a second hour. Finally, the gels were destained in 10% acetic acid which was circulated through Amberlite IRA-400 for adsorption of free dye. They were scanned in a Gilford 2400 S spectrophotometer at a wave length of 550 nm.

The assignment of the different histone bands seen in Fig. 1 and 2 to the various histone classes was achieved in the following way: After semipreparative electrophoresis of crude histones from logarithmically growing Ehrlich ascites tumor cells, the slab gel was superficially stained with Coomassie Brilliant Blue for 5 min. The bands were cut out, the gel pieces ground in a mortar and the proteins partially extracted from the gel with 0.25N HCl at 0 °C overnight. After their precipitation with acetone, they were subjected to sodium dodecylsulfate polyacrylamide gradient gel electrophoresis[7] using unfractionated histone from Ehrlich ascites tumor cells as a reference.

Results and Discussion

Fig. 1a) illustrates the separation of crude histones from logarithmically growing as well as stationary Ehrlich ascites tumor cells, L-929 cells, mouse embryo cells, mouse lung, mouse kidney, mouse liver, mouse spleen, HeLa cells and calf thymus

a)

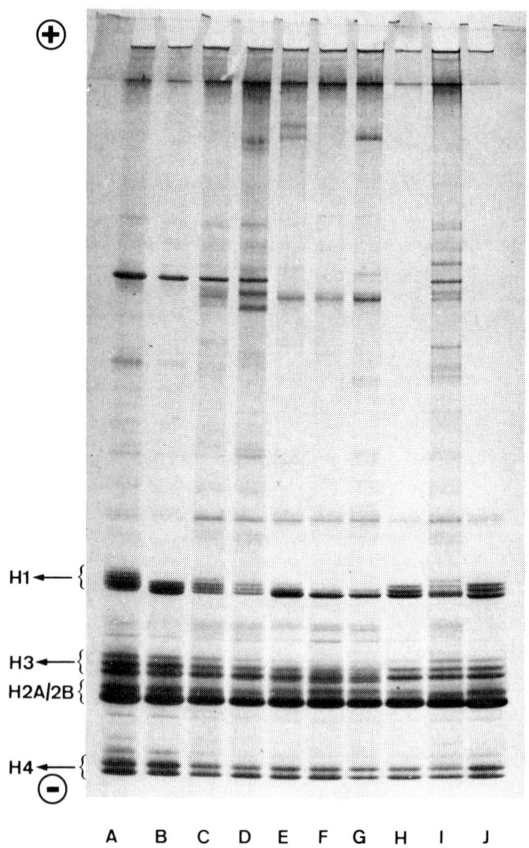

Fig. 1. a) Polyacrylamide gradient slab gel electrophoresis of crude histones from (A) logarithmically growing Ehrlich ascites tumor cells, (B) stationary Ehrlich ascites tumor cells, (C) logarithmically growing L-929 cells, (D) logarithmically growing mouse embryo cells, (E) mouse lung, (F) mouse kidney, (G) mouse liver, (H) mouse spleen, (I) logarithmically growing HeLa cells, (J) calf thymus.

The electrophoretic separation was performed on linear 10 to 20% polyacrylamide gradients in 6M urea, 6% acetic acid as a buffer system. The lower part of the stained gel (ca. 5 cm at a total length of 31 cm) was cut off since no bands with a higher electrophoretic mobility than that of histone $H4_a$ were visible. The conditions of electrophoresis and staining are described in Materials and Methods.

b) Enlargement of the histone regions of the gel patterns presented in a).

The individual bands were assigned to the various histone classes by extraction of the proteins from the gel with 0.25N HCl and polyacrylamide gradient slab gel electrophoresis in the presence of sodium dodecylsulfate with unfractionated histone from ascites tumor cells as a reference[7].

b)

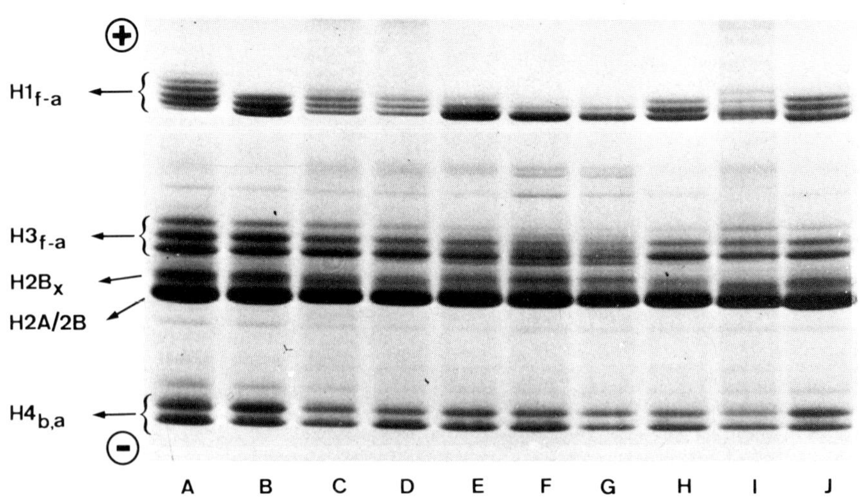

by polyacrylamide gradient slab gel electrophoresis. In Fig. 1b), an enlargement of the histone regions of the various gel patterns is presented. In all cases, a clear separation of the histone classes H1, H3 and H4 into several subspecies is visible.

Whereas the histone H1 fraction from logarithmically growing ascites tumor cells contained 6 components of different electrophoretic mobility, designated as histone $H1_a$, $H1_b$, $H1_c$, $H1_d$, $H1_e$ and $H1_f$ in the order of decreasing mobility, in the case of histone H1 from stationary ascites tumor cells only 3 subspecies were detectable, namely histone $H1_a$, $H1_b$ and $H1_c$. The same reduction of the microheterogeneity of histone H1 was observed in mouse embryo cells, mouse spleen and calf thymus. It was even more diminished in mouse lung, mouse kidney and mouse liver, whereas in L-929 and HeLa cells 4 subspecies of histone H1 were demonstrable. Fig. 1 also shows that the histone H1 pattern of HeLa cells was different from that of mouse cells in that the individual bands were not equidistant.

It has been shown mainly by Chalkley and coworkers[2,3], but also by others, that the microheterogeneity of histone H1 is due, at least in part, to increasing phosphorylation of one or few parental molecules. The covalent attachment of phosphate groups to these parent molecules effects a reduction of their positive net charges and consequently a diminution of their electrophoretic mobilities. If, therefore, highly heterogeneous, ^{32}P-labeled histone H1 is treated with alkaline

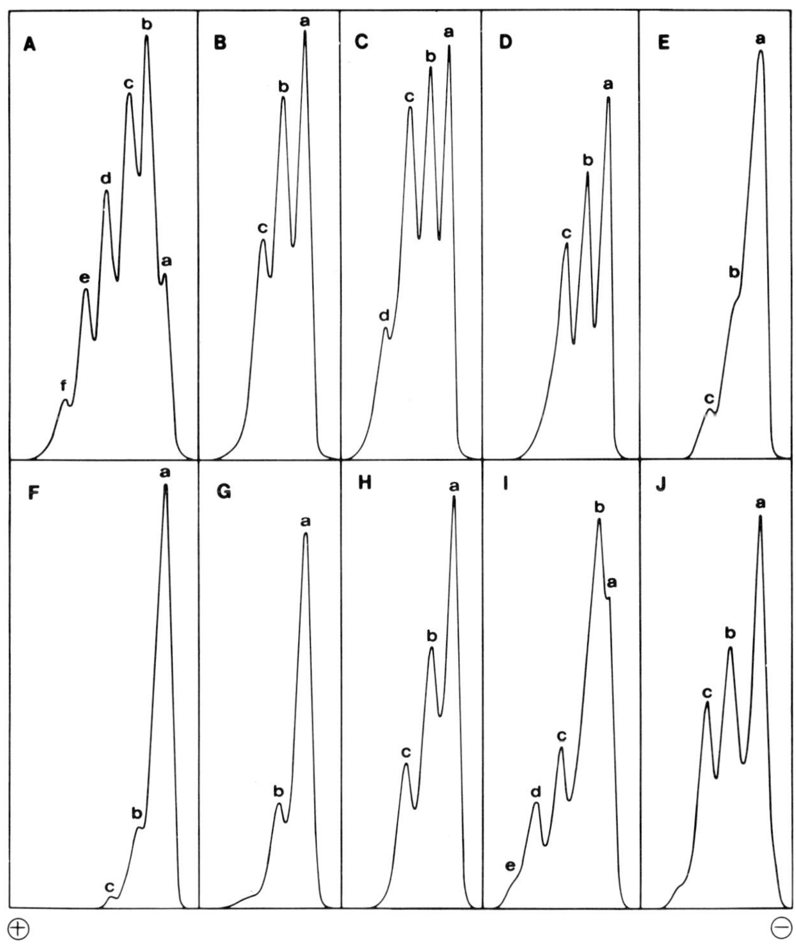

Fig. 2. Densitometer tracings at 550 nm of the histone H1 regions of the gel patterns presented in Fig. 1.

The profiles are lettered in the same order as their corresponding gel patterns in Fig. 1.

phosphatase, radioactive phosphate is released with a concomitant reduction of the band multiplicity[2,3].

A similar situation was observed when exponentially growing ascites tumor cells entered the stationary growth phase. As depicted in Fig. 1 and as is particularly obvious from the densitometer tracings of the various gel patterns (Fig. 2), histones $H1_d$, $H1_e$ and $H1_f$ disappeared from the gel pattern and a pronounced increase in the relative amount of histone $H1_a$ was observed. These changes in the relative intensities of the multiple histone H1 bands could either be the result of a transformation of slowly migrating histone H1 components into molecules of high electrophoretic mobility through dephosphorylation or to a specific proteolytic degradation of histones $H1_d$, $H1_e$ and $H1_f$ combined with a preferential synthesis of histone $H1_a$.

Furthermore, there seems to be a positive correlation between the degree of microheterogeneity of histone H1 and the rate of cell division. As already pointed out above, in vitro the extreme case was verified when ascites tumor cells changed from logarithmic into stationary growth phase. Yet, we also encountered intermediate situations, e.g. in the case of L-929 or mouse embryo cells. Although these cells were harvested in their logarithmic growth phases, the microheterogeneity of their histone H1 complements was less well expressed than that of logarithmically multiplying ascites tumor cells. This difference might well be due to the longer generation times of the two cell species which were approximately 24 h in comparison to 11 to 12 h in the case of logarithmically growing ascites cells. In this respect, it is noteworthy that Ehrlich ascites tumor cells grown in the presence of heated calf serum had a less heterogeneous histone H1 population than those propagated in the presence of fetal calf serum (data not shown). And finally, with regard to essentially nonreplicative tissues, as e.g. mouse lung, mouse liver or mouse kidney, the band multiplicities in the histone H1 fractions were reduced to one major band, histone $H1_a$ (Fig. 1 and 2). Thus our preliminary results are in accordance with observations made by Chalkley and coworkers who have found a positive correlation between the replicative activities of cell cultures or tissues and the extent of histone H1 heterogeneity[9-11]. The dynamic interrelationships between the multiple components of histone H1 from Ehrlich ascites tumor cells and their influenceability by changes in the growth conditions of these cells are currently under investigation.

As can be seen from Fig. 1 and 3, the polyacrylamide gradient had also a high resolving power for histone H3. In general, 3 major subspecies, $H3_b$, $H3_d$ and $H3_f$, were detectable which in the case of mouse kidney and mouse liver were joined by 3 additional components, designated as $H3_a$, $H3_c$ and $H3_e$ in the order of decreasing electrophoretic mobility. Again the equidistance between the multiple bands suggests a regularly proceeding reduction of the positive net charge of a basic parent molecule, presumably through modification by acetylation[12,13]. It is striking that the 6-band multiplicity of histone H3 was observed only in those mouse tissues which showed little microheterogeneity in the histone H1 region. It should be pointed out, that in such cases 3 additional basic proteins were detected which migrated between histone H1 and histone H3 and which were expressed faintly in the remaining

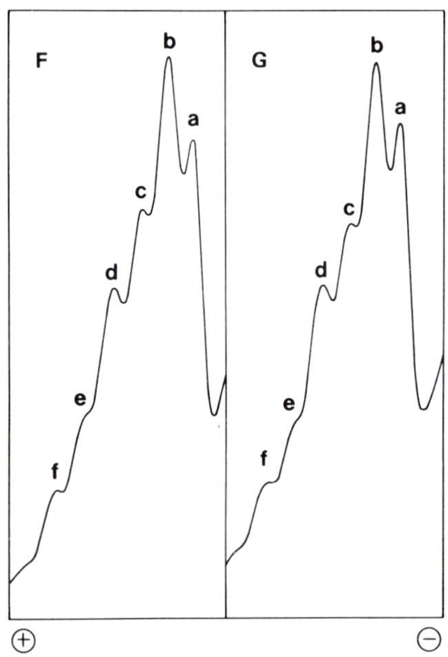

Fig. 3. Densitometer tracings at 550 nm of the histone H3 regions of the gel patterns F and G presented in Fig. 1. (F) Mouse kidney; (G) mouse liver.

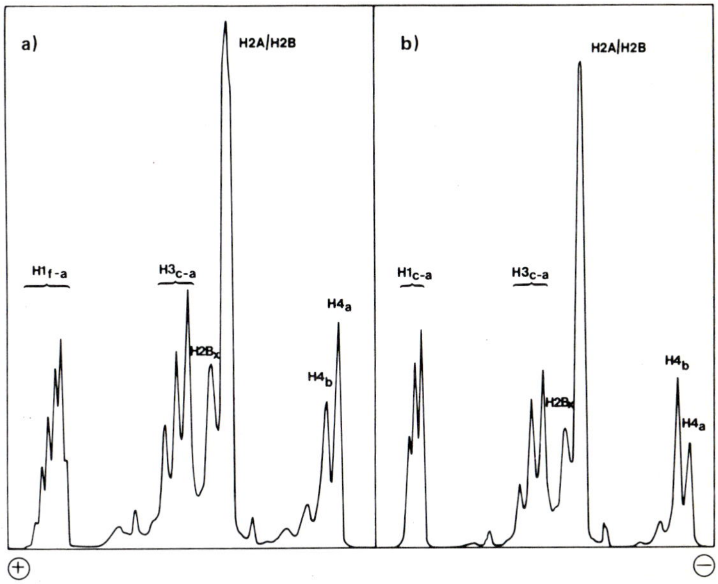

Fig. 4. Densitometer tracings at 550 nm of the total histone regions of the gel patterns A and B presented in Fig. 1.

(A) Logarithmically growing and (B) stationary Ehrlich ascites tumor cells.

8 band patterns of Fig. 1. Fig. 3 shows the densitometer scans of histone H3 from mouse kidney and mouse liver.

Under the conditions of polyacrylamide gradient slab gel electrophoresis, as they are delineated in Materials and Methods, histone H4 of all cell lines and tissues investigated was split into a doublet. Densitometric scanning of the electrophoresis profiles revealed that only in stationary ascites tumor cells the component with the lower electrophoretic mobility ($H4_b$) was predominant while in all other instances the faster migrating band, $H4_a$, displayed a higher intensity (Fig. 4).

In contrast to histones H1, H3 and H4, the histones H2A and H2B were only poorly resolved on a 10 to 20% polyacrylamide gradient gel with 6M urea, 6% acetic acid as a buffer system. Only one subspecies could be separated from the bulk of histone H2A/H2B under these conditions. It turned out to be strongly phosphorylated (data not shown) and to comigrate with histone H2B on polyacrylamide gradient gels containing sodium dodecylsulfate. This finding is not in accord with the experimental result published by Balhorn et al.[14] who documented phosphorylation of histone H2A in Morris hepatoma cells.

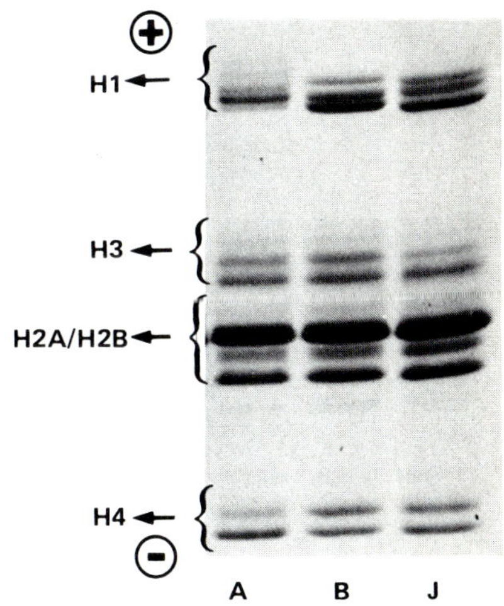

Fig. 5. Polyacrylamide gradient slab gel electrophoresis of crude histones from (A) logarithmically growing and (B) stationary Ehrlich ascites tumor cells as well as (J) calf thymus on a 10 to 20% polyacrylamide gradient in 2.5M urea, 6% acetic acid as a buffer system.

The patterns represent only the histone regions of the total gel patterns which are not shown. For experimental details see Materials and Methods.

Panyim and Chalkley[11] reported that in the presence of 6.25M urea and 0.9M acetic acid histones H2A and H2B are not separable from each other on normal polyacrylamide gels, in contrast to their clear separation in an acidic buffer system containing 2.5M urea. The gel patterns of Fig. 1 demonstrate that, with respect to histones H2A and H2B, the resolving power of the 6M urea-electrophoresis system could not be improved by employing a 10 to 20% polyacrylamide gradient gel. However, when in such gradient gels the urea concentration was reduced to 2.5M, at least a partial splitting of the dense histone H2A/H2B band into subspecies was recorded (Fig. 5); though at the expense of sharpness of the histone bands and an almost complete disappearance of the phosphorylated H2B subspecies from its previous position in the electrophoresis profile.

In summary, the resolution of the histone classes H1, H3 and H4 into multiple subspecies and their separation from the histone classes H2A and H2B seems to be optimal in 6M urea, 6% acetic acid if the electrophoresis is performed on 10 to 20% polyacrylamide gradients. The separation of the histone classes H2A and H2B requires different conditions among which the reduction of the urea concentration appears to be most effective. The influence of preelectrophoresis of the gradient gels on the separation of the individual histones has not been investigated yet.

Finally, employing the modified Panyim and Chalkley system described above, we were able to establish a linear relationship between the peak heights of the individual histone bands and the amount of total histone applied to the gel (Fig. 6). Due to the high resolving power of the polyacrylamide gradient gel, corrections of the peak heights because of overlapping of adjacent bands were not necessary. Only when the amount of histone $H1_a$ was low compared to that of histone $H1_b$ (Fig. 2), the most prominent band in most histone

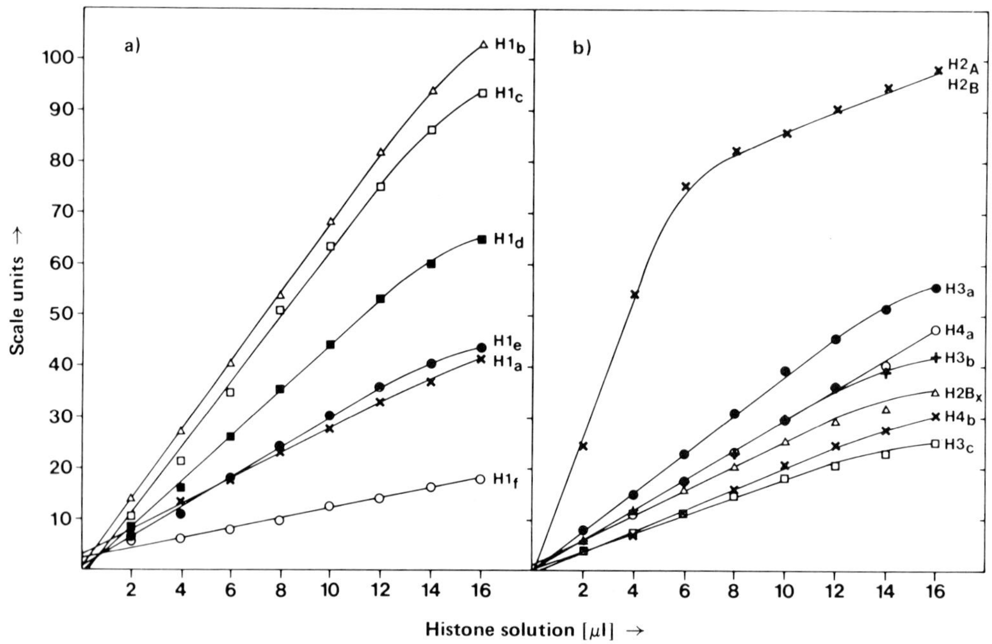

Fig. 6. Peak heights of (a) histones H1 and (b) histones H3, $H2B_x$, H2A/H2B, H_4 from logarithmically growing Ehrlich ascites tumor cells as a function of the amount of total histone applied to the gel.

The electrophoretic separation was carried out as described in the legend to Fig. 1 and in Materials and Methods. Histone solution: 4 mg/m*l*. The peak heights were taken from densitometer tracings (550 nm) of Coomassie Brilliant Blue-stained gels similar to those presented in Fig. 2 and 4. The profiles of the histones H1 (a) and the remaining histones (b) were scanned at two different calibrated absorbancies.

H1 profiles, the peak height of histone $H1_a$ had to be corrected. With this quantitation method at our disposal, we were able to follow the kinetics of changes in the microheterogeneity of histone H1 after mengovirus infection of Ehrlich ascites tumor cells and upon alteration of their growth conditions (papers in preparation).

We are greatly indebted to *Mrs. Margot Bialdiga* and *Mrs. Ulrike Traub* for providing us with mammalian cells and to *Mrs. Ingrid Buchwald* for expert technical assistance.

Literature

1 Panyim, S. & Chalkley, R. (1969) *Arch. Biochem. Biophys.* **130**, 337 - 346.
2 Sherod, D., Johnson, G. & Chalkley, R. (1970) *Biochemistry* **9**, 4611 - 4615.
3 Balhorn, R., Rieke, W.O. & Chalkley, R. (1971) *Biochemistry* **10**, 3952 - 3959.
4 Tanphaichitr, N., Balhorn, R., Granner, D. & Chalkley, R. (1974) *Biochemistry* **13**, 4249 - 4254.
5 Hackett, P.B., Egberts, E. & Traub, P. (1977) *Mol. Biol. Rep.* **3**, 387 - 396.
6 Traub, U. & Traub, P. (1978) *Hoppe-Seyler's Z. Physiol. Chem.* **359**, 581 - 592, following paper.
7 Egberts, E., Hackett, P.B. & Traub, P. (1976) *Hoppe-Seyler's Z. Physiol. Chem.* **357**, 1779 - 1792.
8 Carter, D.B. & Chae, C.-B. (1976) *Biochemistry* **15**, 180 - 185.
9 Balhorn, R., Balhorn, M. & Chalkley, R. (1972) *Dev. Biol.* **29**, 199 - 203.
10 Balhorn, R., Balhorn, M., Morris, H.P. & Chalkley, R. (1972) *Cancer Res.* **32**, 1775 - 1784.
11 Balhorn, R., Chalkley, R. & Granner, D. (1972) *Biochemistry* **11**, 1094 - 1098.
12 Dixon, G.H., Candido, E.P.M., Honda, B.M., Louie, A.J., MacLeod, A.R. & Sung, M.T. (1975) in The Structure and Function of Chromatin (Ciba Foundation Symposium 28 (new series)) pp. 229 - 250, Elsevier Excerpta Medica, North Holland, Amsterdam.
13 Wangh, L., Ruiz-Carrillo, A. & Allfrey, V.G. (1972) *Arch. Biochem. Biophys.* **150**, 44 - 56.
14 Balhorn, R., Bordwell, J., Sellers, L., Granner, D. & Chalkley, R. (1972) *Biochem. Biophys. Res. Commun.* **46**, 1326 - 1333.

Prof. Dr. Peter Traub* and Gunda Boeckmann, Max-Planck-Institut für Zellbiologie, Wilhelmshaven, Anton-Dohrn-Weg 59, Postfach 10 09, D-2940 Wilhelmshaven.

* To whom correspondence should be addressed.

Hoppe-Seyler's Z. Physiol. Chem.
Bd. 359, S. 581 - 592, Mai 1978

Changes in the Microheterogeneity of Histone H1 after Mengovirus Infection of Ehrlich Ascites Tumor Cells

Ulrike TRAUB and Peter TRAUB

Max-Planck-Institut für Zellbiologie, Wilhelmshaven

(Received 8 February 1978)

Dedicated to Prof. Dr. Adolf Butenandt on the occasion of his 75th birthday

Summary: Employing high-resolution polyacrylamide gradient slab-gel electrophoresis in 6M urea and 6% acetic acid, a distinct change in the microheterogeneity of histone H1 was detected after mengovirus infection of Ehrlich ascites tumor cells. Whereas in uninfected cells the band multiplicity was found to be 6, it was reduced to 4 in the course of the infectious cycle (8 to 9 h). It could be demonstrated that, while the electrophoretically slowly moving subspecies $H1_e$ and $H1_f$ disappeared from the band profile of histone H1, the faster migrating bands $H1_a$ and $H1_b$ increased in relative intensity. The relative intensity of band $H1_d$ was also drastically reduced, that of band $H1_c$ stayed practically constant throughout infection. When Ehrlich ascites tumor cells were labeled with a mixture of [^{14}C]amino acids prior to mengovirus infection, the radioactivity incorporated into histone H1 subspecies of low electrophoretic mobility was chased into histone H1 components of high mobility in the course of infection, suggesting a virus-induced, unidirectional interconversion of the multiple histone H1 subunits. With the exception of a histone H2B subspecies, which also decreased in amount, the relative quantities of the other histones stayed constant throughout infection.

Änderung der Mikroheterogenität von Histon H1 nach Mengovirus-Infektion von Ehrlich-Ascites-Tumorzellen

Zusammenfassung: Mit Hilfe der hochauflösenden Polyacrylamid-Gradienten-Flachgel-Elektrophorese in 6M Harnstoff/6proz. Essigsäure als Puffersystem wurde eine deutliche Veränderung der Mikroheterogenität von Histon H1 nach Mengovirus-Infektion von Ehrlich-Ascites-Tumorzellen nachgewiesen. Während in nichtinfizierten Ehrlich-Ascites-Tumorzellen die Bandenmultiplizität 6 war, verringerte sie sich im Verlaufe des Infektionszyklus (8 bis 9 h) auf 4. Es konnte gezeigt werden, daß, während die Untereinheiten niedriger elektrophoretischer Beweglichkeit ($H1_e$ und $H1_f$) in Abhängigkeit von der Zeit nach Virusinfektion aus dem Bandenmuster der Histon-H1-Fraktion verschwanden, die relative Intensität der elektrophoretisch schneller wandernden Banden $H1_a$ und $H1_b$ zunahm. Die relative Intensität der Bande $H1_d$ nahm ebenfalls drastisch ab, die der Untereinheit $H1_c$ blieb über den gesamten Infektionszyklus hinweg praktisch konstant. Wurden Ehrlich-Ascites-Tumorzellen vor Mengovirus-Infektion mit einem Gemisch aus [^{14}C]Aminosäuren vormarkiert, ging im Verlaufe des Infektionszyklus die in Histon-H1-Untereinheiten niedriger elektrophoretischer Beweglichkeit eingebaute Radioaktivität in schnell-wandernde Histon-H1-Komponenten über. Dieser Befund läßt auf eine durch Mengovirus induzierte, einseitig verlaufende Überführung der Histon-H1-Komponenten ineinander schließen. Mit Ausnahme einer Histon-H2B-Untereinheit, deren Menge ebenfalls abnahm, blieben die relativen Mengen der übrigen Histone im Verlaufe der Virusinfektion konstant.

Key words: Histone H1, subspecies, microheterogeneity, mengovirus infection, Ehrlich ascites tumor cells.

Though the cessation of host-specific protein and nucleic acid synthesis after picornavirus infection of mammalian cells has been a well established fact for many years[1,2], its molecular mechanism[3] is still widely unknown. Only on the translational level has some progress been made during the last years insofar as the initiation of host protein synthesis has been found to be inhibited after virus infection[4-13]. With respect to transcription and DNA replication, however, almost no experimental evidence has been obtained for virus-induced cellular events leading to cessation of host-specific nucleic acid synthesis. But it should be mentioned that, in the past, research on the virus-induced inhibition of host macromolecular synthesis has concentrated on protein synthesis and that only few notable efforts have been made to unravel the mechanism of the virus-induced cessation of host-specific nucleic acid synthesis.

So it has been shown for example that the inhibition of host cell RNA synthesis following picornavirus infection is dependent on viral protein synthesis[14-18], but that gross changes in the DNA template cannot be implicated in the loss of RNA synthesizing activity in mengovirus-infected L cells[19]. It has also been reported that DNA-dependent RNA polymerases can be recovered at control levels from mengovirus-infected L cells[20]. Furthermore, evidence has been presented for an inhibitor of cellular RNA synthesis in poliovirus-infected HeLa cells which, however, does not inhibit either nuclear DNA-dependent RNA polymerase or partially purified RNA polymerase from *E. coli* in vitro[21]. Since it is insensitive to ribonuclease or proteolytic enzymes, but is destroyed by phospholipases[21], it seems to be different from another, trypsin-sensitive inhibitor found in mengovirus-infected L cells[22,23]. Finally, the possibility was diminished that suppression of RNA synthesis after infection of Ehrlich ascites tumor cells with Maus-Elberfeld virus is caused by the synthesis of a new histone fraction, although it was not definitely excluded that minor differences in the histone patterns of normal and virus-infected cells are present[24].

On the basis of the experimental facts accumulated so far concerning the picornavirus-induced cessation of host-specific nucleic acid synthesis, it must be assumed that the inhibition is either due to alterations of the chromatin structure or to changes in the intranuclear concentrations of low-molecular weight compounds essential for DNA and RNA synthesis, comparable e.g. to the increase of the $Na^{\oplus}/K^{\oplus}$ ratio which, according to a hypothesis put forward by Carrasco and Smith[12], is responsible for the picornavirus-induced cessation of host protein synthesis. In the present communication we demonstrate a modification of the chromatin structure in mengovirus-infected Ehrlich ascites tumor cells consisting of a distinct reduction of the microheterogeneity of histone H1 as a function of time post-infection.

Materials and Methods

Growth of Ehrlich ascites tumor cells and mengovirus; mengovirus infection of Ehrlich ascites tumor cells

Ehrlich ascites tumor cells were grown in suspension culture using minimum essential medium supplemented with 5% fetal calf serum. Uninfected cells were harvested at a density of 5.5×10^5 cells/m*l*. Mengovirus was multiplied on monolayers of L-929 cells and ascites tumor cells were infected with ca. 50 plaque-forming units/cell when the cell density had reached 5.5×10^5 cells/m*l*. All methods employed have been described previously[25].

Preparation of nuclei and extraction of histones

Nuclei were prepared from uninfected and mengovirus-infected Ehrlich ascites tumor cells. At zero time (uninfected cells) and at various times post-infection (see figures in the Results section), 500-m*l* portions of suspension culture were centrifuged at $600 \times g$ for 5 min. The pelleted cells were transferred with 10 m*l* cold 10mM Tris/HCl (pH 7.5), 0.15M NaCl into 15-m*l* tightly-fitting, all-glass Dounce homogenizers, sedimented by centrifugation at 600 × *g* for 5 min, and extracted 3 times with 10-m*l* portions of buffer A (10mM Tris/HCl (pH 7.5), 2.5mM $Mg(OAc)_2$, 10mM KCl, 1mM phenylmethanesulfonyl fluoride[26], 6mM 2-mercaptoethanol and 1% Triton X-100) by 30 up and down strokes at 0 °C. Phenylmethanesulfonyl fluoride was used in order to prevent proteolytic degradation of histones. Immediately after homogenization, 1.1 m*l* buffer B (0.23M Tris/HCl (pH 7.5), 4mM $Mg(OAc)_2$, 1.22M KCl, 6mM 2-mercaptoethanol) was mixed into the homogenates and the nuclear material was sedimented by centrifugation at 9000 × *g* for 5 min. The supernatants were discarded and the final nuclear pellets were suspended in 4 m*l* dist. water containing 1mM phenylmethanesulfonyl fluoride by brief homogenization. 0.45 m*l* 2.5N HCl was added to each suspension and the mixtures were shaken for 2 h at 0 °C. Insoluble material was removed by centrifugation at 37 000 × *g* for 10 min. The solubilized crude histones were precipitated from the supernatants by the addition of 50 m*l* cold acetone. After standing at 0 °C for 1 h, the precipitated protein was collected by centrifugation at 3 000 × *g* for 5 min. The pellets were washed once with 20 m*l* cold acetone, dried in vacuo at 0 °C and dissolved in 2 m*l* 6M urea, 0.6% acetic acid and 10% 2-mercaptoethanol. The protein solutions were kept at – 20 °C.

Residual synthesis of histone H1 after mengovirus infection of Ehrlich ascites tumor cells

Ehrlich ascites tumor cells were grown in suspension culture to a density of 5.5 × 10^5 cells/m*l* and infected with mengovirus at a multiplicity of ca. 50 plaque-forming units/cell[25]. 100-m*l* portions of uninfected (0 h) or mengovirus-infected cell culture were taken at the times indicated in Fig. 5 and centrifuged at 450 × *g* for 5 min at 37 °C. The supernatants were aspirated and the cells washed with 50 m*l* prewarmed minimum essential medium lacking amino acids (except glutamine, tryptophan, cysteine and methionine) and containing 5% fetal calf serum previously dialyzed against 0.15M NaCl. After centrifugation at 450 × *g* for 5 min at 37 °C, the cell pellets were suspended in 5 m*l* warm washing medium to which 50 μCi [^{14}C]protein hydrolysate (55 mCi/mmol carbon, Amersham-Buchler) had been added. The radioactively labeled amino acids were allowed to be incorporated during a 10 min pulse at 37 °C. Then 20 m*l* prewarmed, normal minimum essential medium supplemented with 5% fetal calf serum was added and the incubation was continued for 5 min at 37 °C. Following cooling in ice water, the cells were harvested by centrifugation at 450 × *g* for 5 min at 0 °C, transferred with 7 m*l* 10mM Tris/HCl (pH 7.5), 0.15M NaCl into 10-m*l* centrifuge tubes and pelleted by centrifugation at 450 × *g* for 5 min. After aspiration of the supernatants, the cells were suspended in 2 m*l* buffer A and vortexed vigorously at 0 °C. The homogenates were then diluted with an additional 2 m*l* buffer A and centrifuged at 3000 × *g* for 5 min. The extraction of the nuclear material was repeated twice under the same conditions. Finally, it was suspended in 0.9 m*l* dist. water, 0.1 m*l* 2.5N HCl was added and the suspensions were shaken for 2 h. Insoluble material was removed by centrifugation at 9000 × *g* for 10 min and the solubilized crude histones were precipitated from the supernatants with 10 m*l* cold acetone. The precipitates were allowed to stand at 0 °C for 1 h. Thereafter, they were collected by centrifugation at 3000 × *g* for 5 min, washed with 5 m*l* cold acetone, dried in vacuo at 0 °C and dissolved in 0.5 m*l* 6M urea, 0.6% acetic acid and 10% 2-mercaptoethanol. 10 μ*l* of each sample was used for polyacrylamide gradient slab gel electrophoretic analysis[27]. After staining[25,27], the gels were prepared for fluorography as described by Bonner and Laskey[28]. The fluorographs were densitometrically scanned at 540 nm.

Prelabeling of Ehrlich ascites tumor cells with [^{14}C] amino acids followed by mengovirus infection

Ehrlich ascites tumor cells were grown in 0.5 *l* suspension culture to a density of 6.6 × 10^5 cells/m*l* as described previously[25]. They were harvested by centrifugation at 600 × *g* for 5 min at 37 °C and suspended in 25 m*l* minimum essential medium lacking amino acids (except glutamine, tryptophan, cysteine and methionine) and containing 5% fetal calf serum previously dialyzed versus 0.15M NaCl. 1 mCi [^{14}C]protein hydrolysate (55 mCi/mmol carbon) was added to the culture that, after incubation at 37 °C for 1 h, was diluted with 500 m*l* of warm, original medium (supernatant) in which the cells had been propagated prior to radioactive labeling, plus 500 m*l* prewarmed normal medium. After a chase period of 14 h at 37 °C and at a cell density of 6 × 10^5 cells/m*l*, the cells were infected with mengovirus at a multiplicity of 50 plaque-forming units/cell in a final volume of 50 m*l* according to the protocol described previously[25]. Immediately after virus infection (0 h) and at 0.5 h post-infection, 2.5-m*l* portions of suspension culture were withdrawn and diluted with 7.5 m*l* cold, normal medium. The cells were pelleted by centrifugation at 450 × *g* for 5 min at 0 °C and kept at – 70 °C. The major part of the concentrated, infected suspension culture (45 m*l*) was diluted with 855 m*l* of the warm supernatant of a normal cell culture whose density had been 6 × 10^5 cells/m*l*. 50-m*l* portions were taken at the times indicated in Fig. 6, the cells harvested by centrifugation at 450 × *g* for 5 min and kept at – 70 °C. For histone preparation, the cell pellets were thawed at 0 °C and processed as described above. The crude histones were finally dissolved in 0.5 m*l* 6M urea, 0.6% acetic acid and 10% 2-mercaptoethanol and

15 μl of each protein sample was used for polyacrylamide gradient slab gel electrophoretic analysis[27]. After electrophoresis, the gels were stained and prepared for fluorography[28]. The fluorographs were densitometrically scanned at 540 nm.

Polyacrylamide gradient slab gel electrophoresis of histones

10- to 15-μl samples of the histone solutions prepared as described above were applied to 10 to 20% polyacrylamide gradient slab gels. The composition of the gels and the conditions of electrophoresis, staining and densitometric scanning were as outlined previously[27].

Results

When crude histones were isolated from Ehrlich ascites tumor cells at various times after mengovirus infection and subjected to polyacrylamide gradient slab gel electrophoresis in 6M urea plus 6% acetic acid[27], the gel patterns shown in Fig. 1 a were obtained. Fig. 1 a illustrates only the histone regions of the original, Coomassie Brilliant Blue-stained gels. Fig. 1 b presents an enlargement of the histone H1 patterns of Fig. 1 a. Under the electrophoretic conditions employed, the histone classes H1, H3 and H4 were resolved into multiple subspecies. As can be seen from Fig. 1 a and b, histone H1 from uninfected cells (time 0) contained 6 subspecies ($H1_{f-a}$), histone H3 3 components ($H3_{c-a}$) and histone H4 2 subunits ($H4_{b,a}$). The individual subspecies are ordered alphabetically with decreasing electrophoretic mobility. Only one component, $H2B_x$, could be separated from the bulk of histone H2A/H2B.

For comparison of the amounts of the different histones present in mengovirus-infected cells at various times post-infection, the gel patterns of Fig. 1 a were densitometrically scanned at 550 nm. The densitometer tracings of the total histone regions and of the histone H1 patterns of uninfected and 4.5 and 8.5 h mengovirus-infected cells are depicted in Fig. 2 and Fig. 3, respectively. Whereas mengovirus infection had no significant effect on the relative amounts of histones H3, H2A/H2B and H4, the extent of microheterogeneity in the histone H1 region decreased from a 6 band to a 4 band multiplicity during the course of the infectious cycle. Moreover, Fig. 1 b and 3 show that, while histone $H1_f$ and $H1_e$ completely disappeared from the nuclei of mengovirus-infected Ehrlich ascites tumor cells, the relative amount of histone $H1_a$ dramatically increased during the same time period. The amount of histone $H2B_x$ diminished slightly with the time post-infection.

In order to follow the kinetics of the virus-induced changes in the histone H1 microheterogeneity more closely, the densitometer tracings of all histone H1 patterns, determined at 0.5 h intervals throughout the infectious cycle, were evaluated quantitatively. We have shown previously that under our conditions of electrophoresis and staining[27], firstly, the intensity (peak height) of each individual histone band seen on the Coomassie Brilliant Blue-stained gel is linearly related to the amount of total histone applied to the gel, and, secondly, due to the high resolving power of the gradient gel, there is practically no overlapping of adjacent histone bands in their peak positions. At low histone $H1_a$ concentrations, however, the peak height of the histone $H1_a$ band, which adjoins the prominent histone $H1_b$ band, needs correction. On the basis of these characteristics, the peak heights of the histone H1 bands can be taken as a measure of the quantities of the different H1 subspecies present in mengovirus-infected Ehrlich ascites tumor cells. Fig. 4 shows the relative amount of each one of the 6 histone

Fig. 1. a) Polyacrylamide gradient slab gel electrophoretic separation of histones isolated from mengovirus-infected. Ehrlich ascites tumor cells at various times post-infection. 10 μl of each histone solution was applied to a 10 to 20% polyacrylamide gradient slab gel containing 6M urea and 6% acetic acid and electrophoresed as described in Materials and Methods. Only the histone regions of the original, Coomassie Brilliant Blue-stained gels are shown. The numbers at the bottoms of the columns indicate the times [h] at which the mengovirus-infected Ehrlich ascites tumor cells were harvested for histone isolation.

b) Enlargement of the histone H1 patterns of the electrophoresis profiles presented in a.

The individual subspecies are ordered alphabetically with decreasing electrophoretic mobility. The numbers at the bottoms of the profiles indicate the times at which the infected cells were harvested for histone isolation.

Fig. 2. Densitometer tracings of the total histone regions of the gel patterns presented in Fig. 1 a.

Only the 0-h (control), 4.5-h and 8.5-h gels of Fig. 1 a were densitometrically scanned.

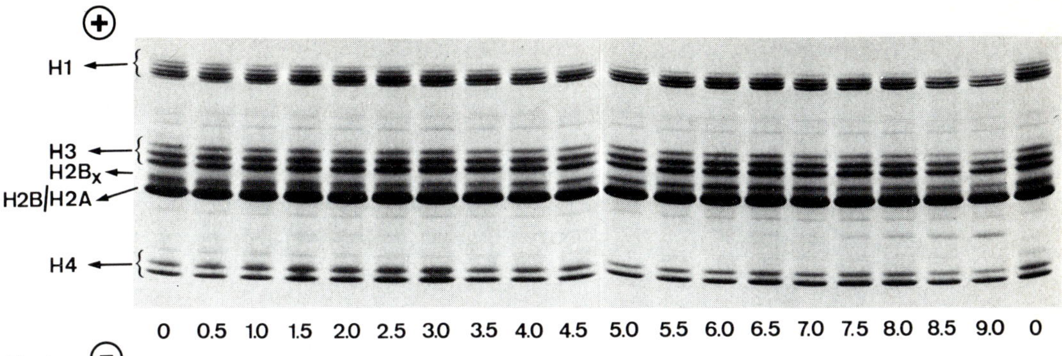

Fig. 1a

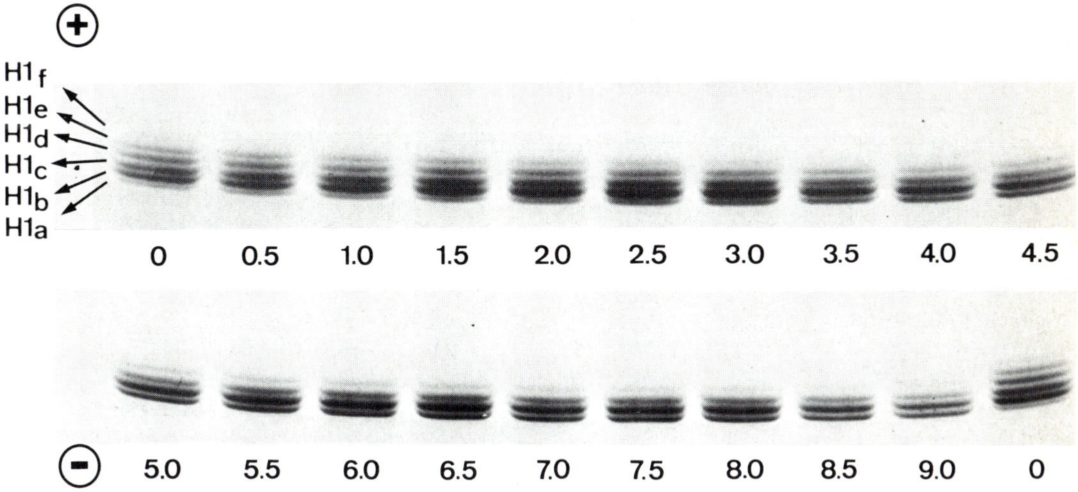

Fig. 1b

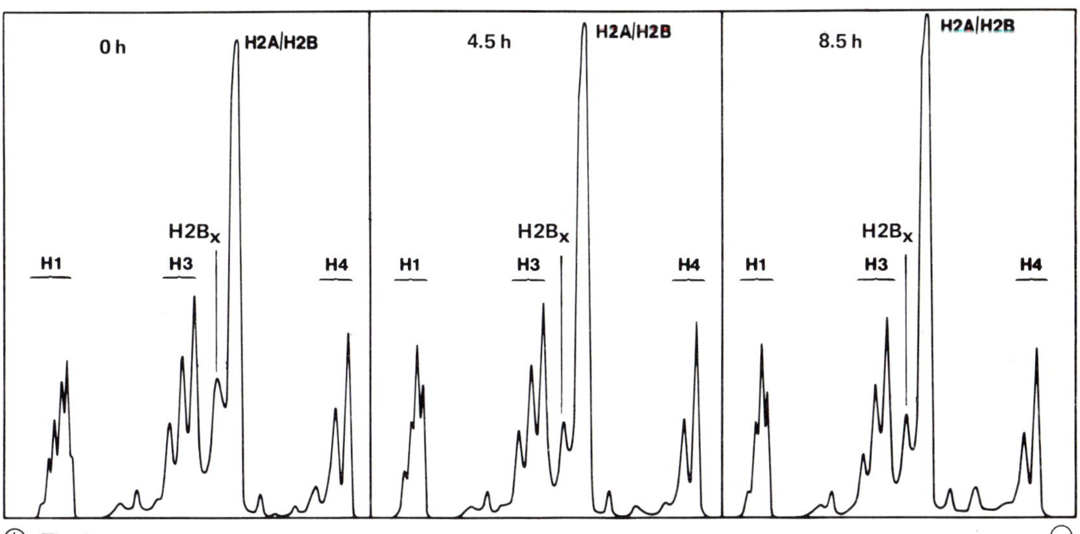

⊕ Fig. 2 ⊖

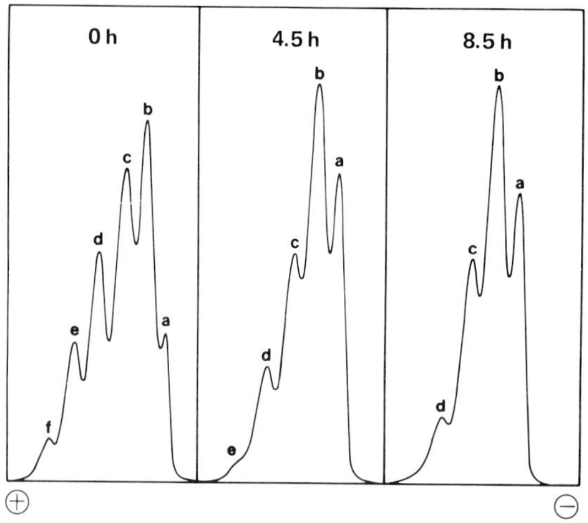

Fig. 3. Extended densitometer tracings of the histone H1 regions of the patterns presented in Fig. 1.

Only the 0-h (control), 4.5-h and 8.5-h gels of Fig. 1 were densitometrically scanned.

H1 subspecies as a function of time post-infection. Whereas histones $H1_f$ and $H1_e$ disappeared entirely from the gel patterns as a consequence of mengovirus infection, only an incomplete though substantial reduction of the relative amount of histone $H1_d$ was observed. By way of contrast, the relative amount of histone $H1_a$ increased dramatically and also that of histone $H1_b$ was augmented considerably. Throughout the infectious cycle, the relative quantity of histone $H1_c$ stayed almost constant. Approximately 5 h post-infection, the relative amounts of the different histone H1 subunits had essentially reached their final values.

The enhancement of the relative quantities of the faster moving histone H1 subspecies concomitant with the disappearance of the electrophoretically slowly migrating H1 histones could be the result of the interconversion of the various protein species. On the other hand, it is also possible that during the course of mengovirus infection the histone H1 components of low electrophoretic mobility are selectively degraded by specific proteases and, since host protein synthesis is diminished progressively, the increase in intensity of histone $H1_b$ and particularly of histone $H1_a$ is due to residual synthesis of the faster migrating subunits.

In order to distinguish between the two possibilities, we first followed the cessation of histone H1 synthesis after mengovirus infection of Ehrlich ascites tumor cells. In a pulse-chase experiment, ^{14}C-labeled amino acids were allowed to be incorporated into the total protein of mengovirus-infected ascites tumor cells at different times post-infection. The histones were isolated and

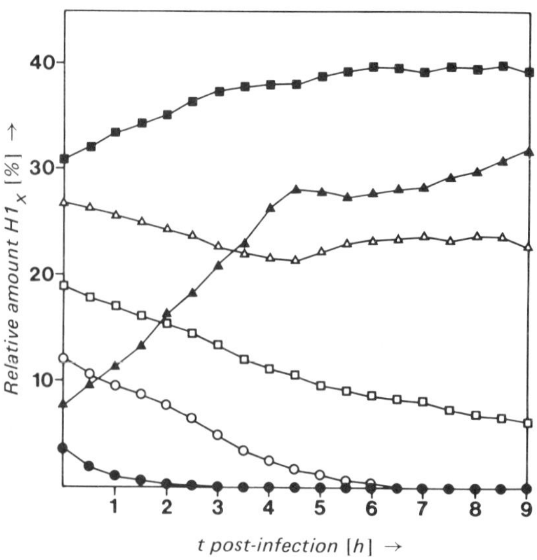

Fig. 4. Time-dependent changes in the relative quantities of individual histone H1 subspecies after mengovirus infection of Ehrlich ascites tumor cells.

The densitometer tracings of the type presented in Fig. 3 were evaluated by measuring the peak heights of the individual histone H1 bands. Since there is only very little overlapping of adjacent bands in the peak positions and because of the linear relationship between peak heights and total amount of histone applied to the gel [27], the relative amount of each histone H1 subspecies was determined by summing up the peak heights of all bands present and calculating the percentage each peak height contributes to the sum of the peak heights. This relative peak height is proportional to the relative quantity of a certain histone H1 subspecies present in mengovirus-infected cells at various times post-infection. All points are averages of 4 separate experiments. Histone $H1_a$, ▲——▲; histone $H1_b$, ■——■; histone $H1_c$, △——△; histone $H1_d$, □——□; histone $H1_e$, ○——○; histone $H1_f$, ●——●.

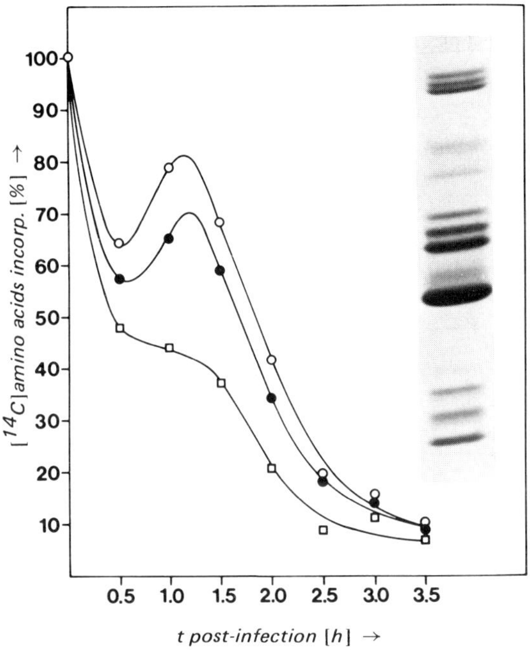

Fig. 5. Mengovirus-induced cessation of histone H1 synthesis.

At various times after mengovirus infection, Ehrlich ascites tumor cells were pulse-labeled with [^{14}C]amino acids for 10 min. After a chase period of 5 min, the histones were isolated and subjected to polyacrylamide gradient slab gel electrophoresis. Following fluorography of the Coomassie Brilliant Blue-stained gels, the fluorographs were densitometrically evaluated at 540 nm. The intensities of the $H1_a$ (○——○), $H1_b$ (●——●) and $H1_c$ (□——□) bands are expressed in percent of control (uninfected cells, 0 h). The inset represents the fluorograph of the total histone region of the control.

subjected to polyacrylamide gradient slab gel electrophoresis and fluorography. The fluorographs were densitometrically scanned and the peak heights of the histone H1 bands were plotted against the time post-infection (Fig. 5). Under the conditions employed, i.e. during a 10-min pulse and a chase period of 5 min, only the 3 fastest moving histone H1 subspecies, histones $H1_{a-c}$, were radioactively labeled (see inset of Fig. 5). In later experiments, we found that the labeling of the remaining 3 histone H1 components, histones $H1_{d-f}$, requires a substantially longer chase period (paper in preparation). According to Fig. 5, the synthesis of total histone H1 was stopped with similar kinetics as the production of the other histone classes[29]. After an initial decrease during the first half-hour post-infection, histone H1 synthesis recovered slightly, passed through a maximum between 1 and 1.5 h post-infection and then fell off steeply to very low activities in the further course of virus infection. The cessation of the synthesis of histones $H1_a$, $H1_b$ and $H1_c$ followed similar kinetics, with the exception that in the case of histone $H1_b$ the maximum of histone production between 1 and 1.5 h post-infection was less well expressed and that for histone $H1_c$ it was reduced to a shoulder. In general, the kinetics of cessation of the production of histone $H1_c$ were faster than those of the synthesis of histone $H1_b$ and $H1_a$ in that order. Hence it is quite possible that, after mengovirus infection of Ehrlich ascites tumor cells, the observed enhancement of the relative quantities of histones $H1_a$ and $H1_b$ is the consequence of the preferential, residual synthesis of these two subspecies.

Because, however, under the labeling conditions described no radioactivity was incorporated into the histone H1 subspecies $H1_{d-f}$, the above experimental data were not conclusive. A less ambiguous result should be obtained by radioactive labeling of histone H1 prior to mengovirus infection and by following the fate of the radioactivity as a function of time post-infection. In order to label all histone H1 components, a mixture of [^{14}C]amino acids was offered to logarithmically growing Ehrlich ascites tumor cells for 1 h and the chase period was extended to 14 h. Free label was removed by centrifugation, the prelabeled cells were infected with mengovirus and, at different times post-infection, their histone complements were analyzed by polyacrylamide gradient slab gel electrophoresis followed by fluorography. The histone H1 regions of the fluorographs, which are presented in Fig. 6, were densitometrically evaluated by measuring the relative peak heights of the multiple histone H1 bands and plotting them versus the time post-infection. The densitometer tracings of the histone H1 patterns of control and 4.5 and 8.5 h mengovirus-infected cells are depicted in Fig. 7, whereas the results of the quantitative evaluation of the densitometer scans are summarized in Fig. 8. As in the case of

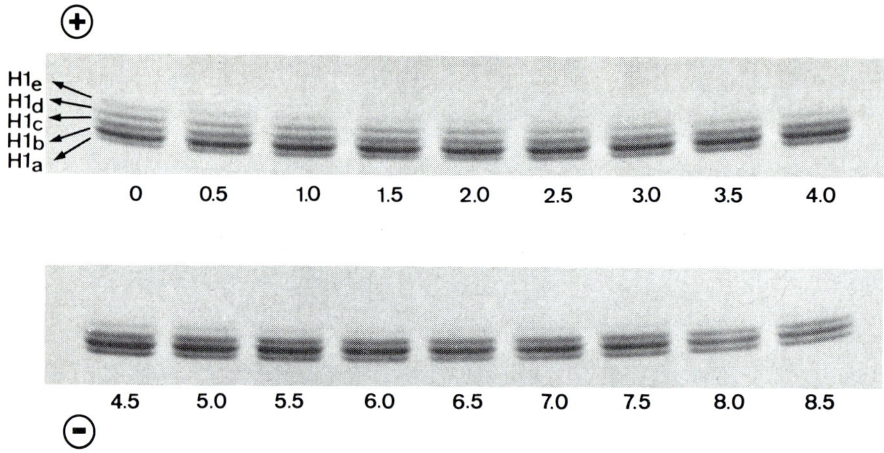

Fig. 6. Fluorographs of the electrophoresis profiles of histone H1 which had been prelabeled with [^{14}C]amino acids prior to mengovirus infection of Ehrlich ascites tumor cells.

Logarithmically growing ascites tumor cells were prelabeled with [^{14}C]amino acids during a 1-h period followed by a chase period of 14 h. Thereafter the cells were infected with mengovirus as described in Materials and Methods. At various times post-infection, histones were isolated and separated by polyacrylamide gradient slab gel electrophoresis. The Coomassie Brilliant Blue-stained gels were fluorographed. The numbers at the bottoms of the individual profiles indicate the times [h] post-infection at which the prelabeled, mengovirus-infected cells were harvested for histone isolation. The individual subspecies are ordered alphabetically with decreasing electrophoretic mobility.

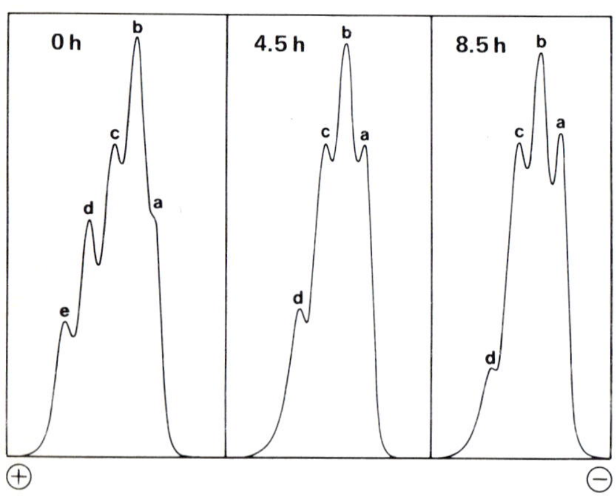

Fig. 7. Densitometer tracings of the fluorographs of histones H1 presented in Fig. 6.

The densitometric scans of the 0-h (control), 4.5-h and 8.5-h patterns of Fig. 6 are presented.

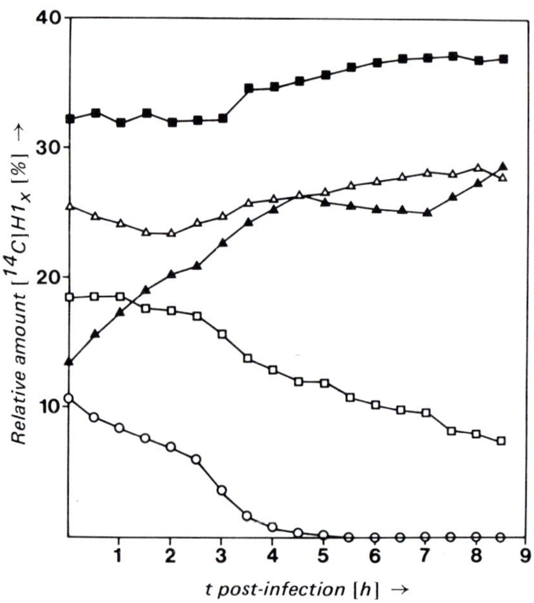

Fig. 8. Fate of ^{14}C radioactivity incorporated into histones H1 prior to mengovirus infection of Ehrlich ascites tumor cells.

The densitometer tracings of the gel patterns depicted in Fig. 6 and of the type presented in Fig. 7 were quantitatively evaluated as described in the legend to Fig. 4. Histone H1$_a$, ▲——▲; histone H1$_b$, ■——■; histone H1$_c$, △——△; histone H1$_d$, □——□; histone H1$_e$, ○——○.

Coomassie Brilliant Blue staining, the relative intensities of the histone $H1_a$ and $H1_b$ bands increased while those of the bands $H1_d$ and $H1_e$ diminished in the course of the infectious cycle. In this experiment, the histone H1 subspecies $H1_f$ could not be detected, either by Coomassie Brilliant Blue staining or by fluorography. Compared with the data presented in Fig. 4, at the beginning of the infectious cycle the relative quantity of radioactively labeled histone $H1_a$ was rather high, despite the long chase period of 14 h. The reason for the poor convertability of pulse-labeled histone $H1_a$ into histone H1 subspecies of lower electrophoretic mobility is unknown.

Discussion

The experimental data presented in the Results section clearly demonstrate pronounced changes in the histone H1 region of the electrophoresis profile of total histone after mengovirus infection of Ehrlich ascites tumor cells. Mengovirus-infected ascites tumor cells actually approach the histone H1 profile of stationary ascites tumor cells which is characterized by the predominance of the histone bands $H1_a$ and $H1_b$ and the lack of the subspecies $H1_{d-f}$[27]. However, they cannot fully reach this profile because the cells lyse 8 to 9 h post-infection, thus causing the interruption of the transformation. Both processes, mengovirus infection as well as the transition from logarithmic to stationary growth (paper in preparation), have in common the transition of electrophoretically slowly migrating histone H1 subspecies into faster moving components.

That we are indeed dealing with an interconversion of histone H1 subspecies could be convincingly demonstrated by the experimental fact that the radioactivity incorporated into slowly moving histone H1 subunits prior to mengovirus infection was chased into histone H1 components of higher electrophoretic mobility. Therefore, the second possibility, that the changes in the microheterogeneity of histone H1 are solely the consequence of the selective, proteolytic degradation of the slowly migrating histone H1 subspecies concomitant with the preferential, residual synthesis of the faster moving histone H1 components, could be excluded from consideration. This eventuality was suggested by the observation that the virus-induced cessation of the production of the fast moving histone H1 subunits proceeds more slowly than that of the synthesis of the slower histone H1 components, thus enabling the mengovirus-infected cells to accumulate preferentially histones $H1_a$ and $H1_b$ in the phase of progressively declining, residual host protein synthesis. Presumably, both possibilities are verified in the mengovirus-infected ascites tumor cell, with a preponderance, however, of the interconversion of the histone H1 subunits.

It has been shown recently, mainly by Chalkley and coworkers[30,31], that the microheterogeneity of histone H1 is the consequence, at least in part, of the progressive phosphorylation of one or a few parent histone H1 molecules. On the basis of this observation, we must conclude therefore that the transformation of slowly migrating histone H1 subspecies into components of higher electrophoretic mobility in the course of mengovirus infection of these ascites tumor cells is caused by dephosphorylation of the slowly moving histone H1 molecules. That such a dephosphorylation of histone H1 actually takes place in the mengovirus-infected cells, we have demonstrated in separate experiments which are not described here (paper in preparation).

In actively growing mammalian cells, protein kinases and phosphoprotein phosphatases operate concurrently, resulting to a certain extent in phosphorylation of cellular proteins. When the balance between the two activities is disturbed, e.g. by changes in the environmental conditions, increased phosphorylation or dephosphorylation of cellular proteins takes place. In this respect, mengovirus infection constitutes a severe interference with cellular processes as well, leading among other effects to a preponderance of the phosphatase activity specific for phosphorylated histone H1. There are two possibilities which might explain the presumptive enhancement of the phosphatase activity in nuclei of mengovirus-infected Ehrlich ascites tumor cells. Either the protein-kinase activity is reduced and the phosphatase activity stays constant or the protein-kinase activity stays constant and the phosphatase activity is stimulated. Since mengovirus infection effectuates a general inhibition of host protein synthesis, it is unlikely that the augmentation of the phosphatase activity is due to newly synthesized enzyme; at the most, an activation of preexisting enzyme may occur, although the production of a virus-specific phosphatase cannot be disregarded. The detailed investigation of this problem is in progress in our laboratory.

On the assumption that mengovirus infection of Ehrlich ascites tumor cells causes a predominance of phosphatase over protein-kinase activity, the differential cessation of the synthesis of individual histone H1 subspecies could also be explained (Fig. 6). Since all histone H1 components, except histone $H1_a$, are phosphorylated (paper in preparation), an enhancement of the ratio phosphatase activity/protein-kinase activity must necessarily result in a relatively higher synthesis of the parent histone H1 molecules, as long as histone H1 is still produced in the mengovirus-infected Ehrlich ascites tumor cells.

The alteration of the histone H1 microheterogeneity after mengovirus infection of Ehrlich ascites tumor cells of course implies a modification of the chromatin structure. Especially in the light of the presumed function of histone H1 in chromatin condensation[32-36], such an alteration certainly has some influence on the synthetic activities of chromatin. In Fig. 9, the mengovirus-induced inhibitions of host nucleic acid synthesis are compared with the changes in the amounts of the various histone H1 subspecies. It is striking that the host DNA- and RNA-synthesizing activities diminish as the alteration of the microheterogeneity of histone H1 proceeds. Approximately 5 h post-infection, the various processes are essentially terminated. From this coincidence we deduce that the cessation of host-specific DNA and RNA synthesis might well be the consequence, at least in part, of the structural modifications of the histone H1 subspecies. In this context it is particularly noteworthy that Chalkley and coworkers[30,31,38-40] detected a positive correlation between histone H1 phosphorylation and the replicative activities of various tissues and cell lines, and Watson and Langan[41] an effect of histone H1 phosphorylation on the template activity of chromatin. These findings are the more comprehensible as unphosphorylated histone H1 is more effective in producing conformational changes in DNA upon

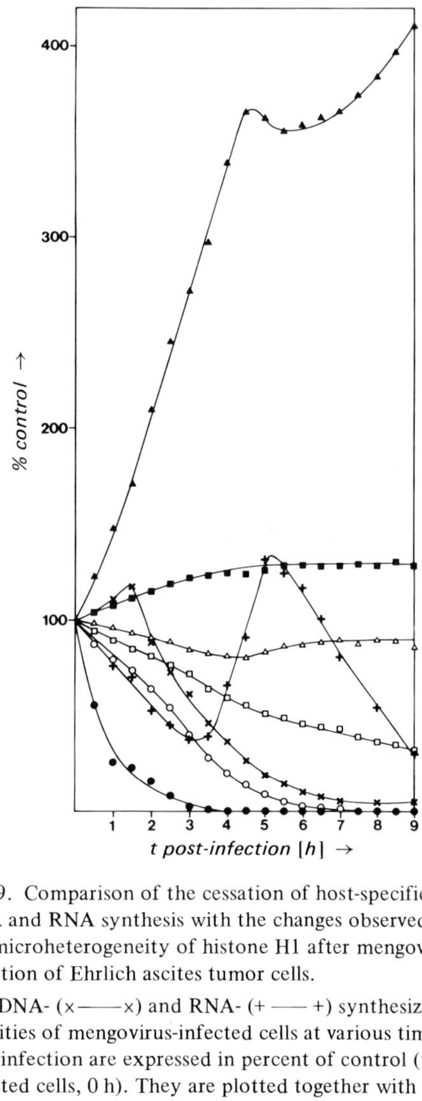

Fig. 9. Comparison of the cessation of host-specific DNA and RNA synthesis with the changes observed in the microheterogeneity of histone H1 after mengovirus infection of Ehrlich ascites tumor cells.

The DNA- (x——x) and RNA- (+——+) synthesizing activities of mengovirus-infected cells at various times post-infection are expressed in percent of control (uninfected cells, 0 h). They are plotted together with the relative amounts of the individual histone H1 subspecies which are depicted in Fig. 4 and which are also expressed in percent of control (▲——▲, histone $H1_a$; ■——■, histone $H1_b$; △——△, histone $H1_c$; □——□, histone $H1_d$; ○——○, histone $H1_e$; ●——●, histone $H1_f$). The DNA-synthesizing activities of mengovirus-infected cells were determined as described previously[37] and the corresponding RNA-synthesizing activities by incorporation of [^{3}H]uridine during a 10-min pulse (conventional method). The maximum of RNA synthesis at 5 h post-infection is due to the production of viral RNA.

complex formation than is the phosphorylated one[42].

However, virus-induced changes in the populations of other nuclear constituents, especially of non-histone chromosomal proteins, might exert a strong influence on the synthetic activities of chromatin as well. There is suggestive evidence for specific binding between histone H1 and non-histone chromosomal proteins[43-45]; it might be affected by the modulation of the phosphorylation state of histone H1. Since non-histone chromosomal proteins fulfill an important regulatory function in gene expression[46-52], it might well be that, on the basis of this interaction, the mengovirus-induced structural changes of histone H1 have some influence on the transcriptional activity of the virus-infected cell.

We thank Mrs. *Margot Bialdiga* for the production of cells and virus and Mrs. *Ingrid Buchwald* and Miss *Gunda Boeckmann* for excellent technical assistance.

Literature

1 Martin, E.M. & Kerr, J.M. (1968) in The Molecular Biology of Viruses (Crawford, L.V. & Stoker, M.G.P., eds.) pp. 15 - 46, University Press, Cambridge.
2 Baltimore, D. (1969) in The Biochemistry of Viruses (Levy, H.B., ed.) pp. 101 - 176, Marcel Dekker, New York.
3 Leibowitz, R. & Penman, S. (1971) *J. Virol.* **8**, 661 - 668.
4 Lawrence, C. & Thach, R.E. (1974) *J. Virol.* **14**, 598 - 610.
5 Wigle, D.T. & Smith, A.E. (1973) *Nature (London), New Biol.* **242**, 136 - 140.
6 Nudel, U., Lebleu, B. & Revel, M. (1973) *Proc. Natl. Acad. Sci. U.S.A.* **70**, 2139 - 2144.
7 Abreu, S.L. & Lucas-Lenard, J. (1976) *J. Virol.* **18**, 182 - 194.
8 Golini, F., Thach, S.S., Birge, C.H., Safer, B., Merrick, W.C. & Thach, R.E. (1976) *Proc. Natl. Acad. Sci. U.S.A.* **73**, 3040 - 3044.
9 Kaufmann, Y., Goldstein, E. & Penman, S. (1976) *Proc. Natl. Acad. Sci. U.S.A.* **73**, 1834 - 1838.
10 Jenkins, N., Taylor, M.W. & Raff, R.A. (1973) *Proc. Natl. Acad. Sci. U.S.A.* **70**, 3287 - 3291.
11 Nuss, D.L., Oppermann, H. & Koch, G. (1975) *Proc. Natl. Acad. Sci. U.S.A.* **72**, 1258 - 1262.
12 Carrasco, L. & Smith, A.E. (1976) *Nature (London)* **264**, 807 - 809.
13 Hackett, P.B., Egberts, E. & Traub, P. (1978) *Eur. J. Biochem.* **83**, 353 - 361.
14 Baltimore, D., Franklin, R.M. & Callender, J. (1963) *Biochim. Biophys. Acta* **76**, 425 - 430.
15 Verwoerd, D.W. & Hausen, P. (1963) *Virology* **21**, 628 - 635.
16 Penman, S. & Summers, D. (1965) *Virology* **27**, 614 - 620.
17 Collings, F.D. & Roberts, W.K. (1972) *J. Virol.* **10**, 969 - 978.
18 Holland, J.J. (1964) *J. Mol. Biol.* **8**, 574 - 581.
19 Baltimore, D. & Franklin, R.M. (1962) *Proc. Natl. Acad. Sci. U.S.A.* **48**, 1383 - 1390.
20 Apriletti, J.W. & Penhoet, E. (1974) *Virology* **61**, 597 - 601.
21 Ho, P.P.K. & Washington, A.L. (1971) *Biochemistry* **10**, 3646 - 3651.
22 Balandin, I.G. & Franklin, R.M. (1964) *Biochem. Biophys. Res. Commun.* **15**, 27 - 32.
23 Balandin, I.G. & Kastrikina, L.N. (1967) *Acta Virol.* **11**, 403 - 406.
24 Holoubek, V. & Crocker, T.T. (1971) *Virology* **43**, 527 - 530.
25 Egberts, E., Hackett, P.B. & Traub, P. (1976) *Hoppe Seyler's Z. Physiol. Chem.* **357**, 1779 - 1792.
26 Carter, D.B. & Chae, C.-B. (1976) *Biochemistry* **15**, 180 - 185.
27 Traub, P. & Boeckmann, G. (1978) *Hoppe Seyler's Z. Physiol. Chem.* **358**, 571 - 579, preceding paper.
28 Bonner, W.M. & Laskey, R.A. (1974) *Eur. J. Biochem.* **46**, 83 - 88.
29 Egberts, E., Hackett, P.B. & Traub, P. (1977) *Hoppe Seyler's Z. Physiol. Chem.* **358**, 463 - 474.
30 Sherod, D., Johnson, G. & Chalkley, R. (1970) *Biochemistry* **9**, 4611 - 4615.
31 Balhorn, R., Rieke, W.O. & Chalkley, R. (1971) *Biochemistry* **10**, 3952 - 3959.
32 Renz, M., Nehls, P. & Hozier, J. (1977) *Proc. Natl. Acad. Sci. U.S.A.* **74**, 1879 - 1883.
33 Lindigkeit, R., Bellmann, K., Fenske, H., Böttger, M., Holtzhauer, M. & Eichhorn, I. (1974) *FEBS Lett.* **44**, 146 - 152.
34 Thoma, F. & Koller, T. (1977) *Cell* **12**, 101 - 107.
35 Gurley, L.B., Walters, R.A. & Tobey, R.A. (1974) *J. Cell Biol.* **60**, 356 - 364.
36 Bradbury, E.M., Inglis, R.J. & Matthews, H.R. (1974) *Nature (London)* **247**, 257 - 261.
37 Gallwitz, D., Traub, U. & Traub, P. (1977) *Eur. J. Biochem.* **81**, 387 - 393.
38 Balhorn, R., Balhorn, M. & Chalkley, R. (1972) *Dev. Biol.* **29**, 199 - 203.
39 Balhorn, R., Balhorn, M., Morris, H.P. & Chalkley, R. (1972) *Cancer Res.* **32**, 1775 - 1784.
40 Balhorn, R., Chalkley, R. & Granner, D. (1972) *Biochemistry* **11**, 1094 - 1098.
41 Watson, G. & Langan, T.A. (1973) *Fed. Proc.* **32**, 588.
42 Adler, A.J., Langan, T.A. & Fasman, G.D. (1972) *Arch. Biochem. Biophys.* **153**, 769 - 777.
43 Goodwin, G.H. & Johns, E.W. (1973) *Eur. J. Biochem.* **40**, 215 - 219.
44 Alfageme, C.R., Zweidler, A., Mahowald, A. & Cohen, L.H. (1974) *J. Biol. Chem.* **249**, 3729 - 3736.
45 Smith, J.A. & Stocken, L.A. (1973) *Biochem. J.* **13**, 859 - 861.
46 Spelsberg, T.C., Wilhelm, J.A. & Hnilica, L.S. (1972) *Sub-Cell. Biochem.* **1**, 107 - 145.

47 Stein, G.S., Spelsberg, T.C. & Kleinsmith, L.J. (1974) *Science* **183**, 817 - 824.
48 Stein, J.L., Reed, K. & Stein, G.S. (1976) *Biochemistry* **15**, 3291 - 3295.
49 Park, W., Jansing, R., Stein, J. & Stein, G. (1977) *Biochemistry* **16**, 3713 - 3721.
50 Stein, G.S., Hunter, G. & Lavie, L. (1974) *Biochem. J.* **139**, 71 - 76.
51 Barrett, T., Maryanka, D., Hamlyn, P.H. & Gould, H.J. (1974) *Proc. Natl. Acad. Sci. U.S.A.* **71**, 5057 - 5061.
52 Blüthmann, H., Mroszek, S. & Gierer, A. (1975) *Eur. J. Biochem.* **58**, 315 - 326.

Ulrike Traub and Prof. Dr. Peter Traub*, Max-Planck-Institut für Zellbiologie, Anton-Dohrn-Weg 59, Postfach 10 09, D.2940 Wilhelmshaven.

* To whom correspondence should be addressed.

Hoppe-Seyler's Z. Physiol. Chem.
Bd. 359, S. 593 - 600, Mai 1978

Enhanced Phosphorylation of Ribosomal Protein S6 and Other Cytoplasmic Proteins after Mengovirus Infection of Ehrlich Ascites Tumor Cells

Ingeborg ROSNITSCHEK, Ulrike TRAUB and Peter TRAUB

Max-Planck-Institut für Zellbiologie, Wilhelmshaven

(Received 13 February 1978)

Dedicated to Prof. Dr. Adolf Butenandt on the occasion of his 75th birthday

Summary: After mengovirus infection of Ehrlich ascites tumor cells, an increased incorporation of radioactive phosphate into ribosomal protein S6 was detected. As judged from the shape of the radioautography spot and the location of the modified protein on the two-dimensional polyacrylamide gel, it must be concluded that more than one phosphate group was incorporated into ribosomal protein S6. Concomitantly, multiple cytoplasmic proteins, particularly those with high affinity for ribosomes, were phosphorylated. Compared with the controls (mock infection), mengovirus infection induced only a slight additional phosphorylation of the proteins of the ribosome-free supernatant. Employing the radio isotope dilution test with 3′,5′-cyclic AMP-binding protein, no enhancement of the intracellular concentration of 3′,5′-cyclic AMP in response to mengovirus infection could be demonstrated.

Erhöhte Phosphorylierung des ribosomalen Proteins S6 und anderer cytoplasmatischer Proteine nach Mengovirus-Infektion von Ehrlich-Ascites-Tumorzellen

Zusammenfassung: Nach Mengovirus-Infektion von Ehrlich-Ascites-Tumorzellen wurde ein erhöhter Einbau radioaktiven Phosphats in ribosomales Protein S6 festgestellt. Der Form des Radioautographieflecken und der Lage des modifizierten Proteins auf dem zweidimensionalen Polyacrylamidgel nach zu urteilen, muß mehr als eine Phosphatgruppe in ribosomales Protein S6 eingebaut worden sein. Gleichzeitig wurde eine Vielzahl cytoplasmatischer Proteine, hauptsächlich solche hoher Ribosomen-Affinität, phosphoryliert. Im Vergleich zur Kontrolle wurde die Phosphorylierung der Proteine des ribosomenfreien Überstandes durch Mengovirus-Infektion nur gerinfügig stimuliert. Unter Verwendung des Radioisotopen-Verdünnungstests mit 3′,5′-cyclisches AMP bindendem Protein konnte keine Erhöhung der intrazellulären Konzentration des 3′,5′-cyclischen AMP als Antwort auf Mengovirus-Infektion nachgewiesen werden.

Key words: Mengovirus infection, Ehrlich ascites tumor cells, phosphorylation, ribosomal protein S6, cytoplasmic proteins.

Enzymes:
Guanyloribonuclease, ribonucleate 3′-guanylo-oligonucleotidohydrolase (EC 3.1.4.8) (= RNase T1);
Ribonuclease I, ribonucleate 3′-pyrimidino-oligonucleotidohydrolase (EC 3.1.4.22) (= RNase A).

It is well documented that picornavirus infection of mammalian cells leads to the inhibition of host-specific protein synthesis[1,2]. During recent years, it has been shown that this shutoff is very likely due to the interference of viral infection with the initiation step of protein synthesis[3-5] and it was hypothesized that one or more of the constituents of the host's protein synthesizing machinery were structurally modified. Since host template RNA could be excluded as a candidate for modification[6-8], the further elucidation of the shutoff mechanism must concentrate on the characterization of the remaining components of the translational apparatus. That such changes, particularly the phosphorylation of initiation factors and ribosomal proteins, may indeed play a regulatory role on the translational level, has been demonstrated in the case of protein synthesis in rabbit reticulocytes[9]. In the present communication, we show that mengovirus-infected Ehrlich ascites tumor cells also show an enhancement of the phosphorylation of the ribosomal protein S6 and of proteins included in the salt wash fraction of ribosomes. This phosphorylation study was initiated by our recent finding, that in cell-free systems from mengovirus-infected Ehrlich ascites tumor cells, the preferential translation of mengovirus RNA over host template RNA is largely due to the protein fraction that can be dissociated from ribosomes with buffers of high ionic strength[5].

Materials and Methods

Growth of Ehrlich ascites tumor cells and mengovirus

Ehrlich ascites tumor cells and mengovirus were grown as described previously[10].

Mengovirus infection and [^{32}P]orthophosphate-labeling of Ehrlich ascites tumor cells

Cells from a 2-*l* suspension culture at 5.9×10^5 cells/m*l* were harvested by centrifugation at 600 × *g* for 5 min at 37 °C; they were then washed with 700 m*l* of prewarmed, phosphate-free minimum essential medium which was supplemented with 5% fetal calf serum previously dialyzed against 0.15M NaCl (washing medium), and finally suspended in prewarmed washing medium to give a final volume of 80 m*l*. The suspension culture was divided into halves. One half was mixed with 50 m*l* prewarmed washing medium containing 5 mCi [^{32}P]orthophosphoric acid (New England Nuclear) (mock-infected cells), whereas the other portion received 10 m*l* of a mengovirus solution (titre: 5×10^9 plaque-forming units/m*l*) also containing 5 mCi [^{32}P]orthophosphoric acid (mengovirus-infected cells). After shaking in closed vessels at 37 °C for 75 min, 50 m*l* prewarmed washing medium was added to each culture and the incubation was continued for another 75 min. The cells were harvested by centrifugation at 600 × *g* for 5 min, washed twice with cold 10mM sodium phosphate (pH 7.5), 0.15M NaCl and suspended in 2.7 m*l* buffer A (10mM Tris/HCl (pH 7.5), 2.5mM $Mg(OAc)_2$, 10mM KCl, 6mM 2-mercaptoethanol) in 5-m*l* tightly-fitting all-glass Dounce homogenizers.

Preparation and fractionation of postnuclear supernatants

The cells were allowed to swell for 10 min at 0 °C. 0.3 m*l* 10% Triton X-100 in buffer A and 30 µ*l* 0.1M phenylmethanesulfonylfluoride[11] in dimethylsulfoxide were added and the cells were disrupted by 30 up and down strokes. Immediately after cell breakage, 0.3 m*l* buffer B (0.23M Tris/HCl (pH 7.5), 4mM $Mg(OAc)_2$, 1.22M KCl, 6mM 2-mercaptoethanol) was mixed into the homogenates which were centrifuged at 3000 × *g* for 10 min. The extraction of the sedimented material was repeated twice under the above conditions and the supernatants were combined and processed as follows:

The combined supernatants (10.5 m*l*) were divided into 3 equal portions, layered on top of 0.3-m*l* cushions of 1M sucrose in buffer A and centrifuged at 370 000 × *g* for 3 h at 2 °C in the SW 60 Ti rotor of the Spinco L2 65B centrifuge. The supernatants (11 m*l*) were recovered quantitatively from the centrifuge tubes, adjusted to 4M urea, 10mM EDTA, 2M LiCl (final volume: 18 m*l*), layered on top of 7 m*l* 70% CsCl in 3 centrifuge tubes and centrifuged at 150 000 × *g* for 40 h at 2 °C in the SW 40 Ti rotor of the Spinco L2 65B centrifuge[12]. The gradients were processed as described below.

The ribosomal pellets of the first high-speed centrifugation (see above) were resuspended in 2.5 m*l* buffer A by thorough douncing. After the addition of 0.6 m*l* 4M KCl (final concentration: 0.75M KCl), the ribosome solutions were centrifuged at 370 000 × *g* for 3 h at 2 °C in the SW 60 Ti rotor of the Spinco L2 65B centrifuge on top of 0.3 m*l* 1M sucrose in buffer A. The supernatants (3.8 m*l*) were adjusted to 4M urea, 10mM EDTA, 2M LiCl (final volume: 6.5 m*l*), layered on top of 6.5 m*l* 70% CsCl and centrifuged at 150 000 × *g* for 40 h at 2 °C in the SW 40 Ti rotor of the Spinco L2 65B centrifuge. The gradients were processed as described below.

The pelleted, salt-washed ribosomes were suspended in 4 m*l* Tris/HCl (pH 7.5), 7M urea and the solutions were again adjusted to 4M urea, 10mM EDTA, 2M LiCl (final volume: 6 m*l*). They were layered on top of 7 m*l* 70% CsCl and centrifuged as described above.

Following centrifugation, the urea-containing protein solutions plus the upper 1.5 m*l* of the CsCl layers were removed and, after the addition of ribonucleases A and T1 to final concentrations of 10 μg/m*l*, dialyzed excessively against 10mM Tris/HCl (pH 7.5), 5mM EDTA, 7M urea at 6 °C. For polyacrylamide gel electrophoretic analysis of the total protein of each solution, a small part was mixed with an equal volume 20% trichloroacetic acid. The precipitates were collected by centrifugation at 3000 x *g* for 10 min, washed with acetone and dissolved in a solution containing 10% glycerol, 5% 2-mercaptoethanol, 2% sodium dodecylsulfate, and 0.002% bromphenol blue at room remperature.

The major part of each dialyzed protein solution was passed through carboxymethyl cellulose (0.7 mequ./g) previously equilibrated with 10mM Tris/HCl (pH 7.5, 0 °C), 7M urea. Depending on the amount of protein contained in each solution, 0.2 to 0.5 *g* of the ion exchanger was used. The columns were washed with equilibration buffer, then the basic protein fractions adsorbed to the ion exchanger were eluted with a solution containing 4M urea and 2M LiCl[12]. The eluates were divided into two equal portions; in the case of ribosomal proteins, each eluate was divided into 3 fractions and the largest fraction (85% of total basic protein) was used for two-dimensional polyacrylamide gel electrophoresis. The proteins of the various acidic and basic-protein solutions were precipitated by the addition of an equal volume 20% trichloroacetic acid and washed with acetone. Whereas the acidic flow through proteins and the basic proteins of one portion of the eluates were dissolved in a solution containing 10% glycerol, 5% 2-mercaptoethanol, 2% sodium dodecylsulfate and 0.002% bromphenolblue at room temperature[10], the basic proteins of the second half of the eluates were dissolved in a solution containing 6M urea, 0.6% acetic acid and 10% 2-mercaptoethanol at 0 °C[13].

For two-dimensional polyacrylamide gel electrophoretic analysis, the major part of the basic ribosomal protein was dissolved in 500 μ*l* of sample buffer of the composition described by Kaltschmidt and Wittmann[14]; however, the concentrations of EDTA and boric acid were reduced to one half. The ribosomal protein solutions were supplemented with 40 μ*l* 2-mercaptoethanol and 2 μ*l* 1% pyronin red B. 220 μ*l* protein solution was used for gel electrophoretic analysis.

Polyacrylamide gel electrophoresis

The one-dimensional polyacrylamide gradient slab gel electrophoresis of the various protein samples in the presence of sodium dodecylsulfate or 6M urea and 6% acetic acid was performed as described previously[10,13].

Two-dimensional polyacrylamide gel electrophoresis was carried out essentially as described by Kaltschmidt and Wittmann[14]. In order to shorten the electrophoresis time in the first dimension, the concentrations of EDTA, boric acid and Tris were reduced to one half; the proteins were separated on 19 x 0.6 cm cylindrical separation gels at a current of 4 mA/gel within 14 h at room temperature. The second dimension gel plates, which measured 20 x 20 x 0.5 cm, consisted of 18% acrylamide, 0.5% *N,N'*-methylenebis(acrylamide) (both from Eastman Kodak), 7M urea, 6% acetic acid and 0.5% *N,N,N',N'*-tetramethylethylenediamine. After electrophoresis in the first dimension, the gel rods were not dialyzed against equilibration buffer but directly polymerized on top of the prepolymerized gel plates using a small volume of second dimension gel solution with ammonium persulfate and riboflavin as catalysts. The electrophoresis time was 48 h at a constant voltage of 150 V and at room temperature. Following staining in 0.5% amido black, 5% acetic acid for 2 h and destaining in tap water and 1% acetic acid, the gel plates were photographed, wrapped in a very thin plastic sheet and radioautographed without drying.

Results and Discussion

For a quantitative comparison of cytoplasmic protein phosphorylation in mock- and mengovirus-infected Ehrlich ascites tumor cells, equal amounts of [^{32}P]orthophosphate-labeled cells of both types of cell culture were processed identically. Postnuclear supernatants were prepared in the presence of 1% Triton X-100 and fractionated into postribosomal supernatants, 0.75M KCl-washed ribosomes and the corresponding ribosomal wash fractions. After the nucleic acids contained in the various cytoplasmic fractions had been removed by CsCl density gradient centrifugation in the presence of EDTA and urea[12], the total proteins were separated into acidic and basic fractions by carboxymethyl cellulose chromatography[12] and analyzed by polyacrylamide gradient slab gel electrophoresis in the presence of sodium dodecylsulfate[10] or in 6M urea, 6% acetic acid[13]. All polyacrylamide gels were subjected to radioautography.

Fig. 1 shows the radioautographs of the sodium dodecylsulfate electrophoresis profiles of total protein from ribosome-free supernatant (A), 0.75M KCl-washed ribosomes (B) and 0.75M KCl-ribosomal wash fraction (C), all derived from mock- as well as mengovirus-infected cells. It can be clearly seen that in mengovirus-infected cells the extent of protein phosphorylation is much higher than in mock-infected cells and that the

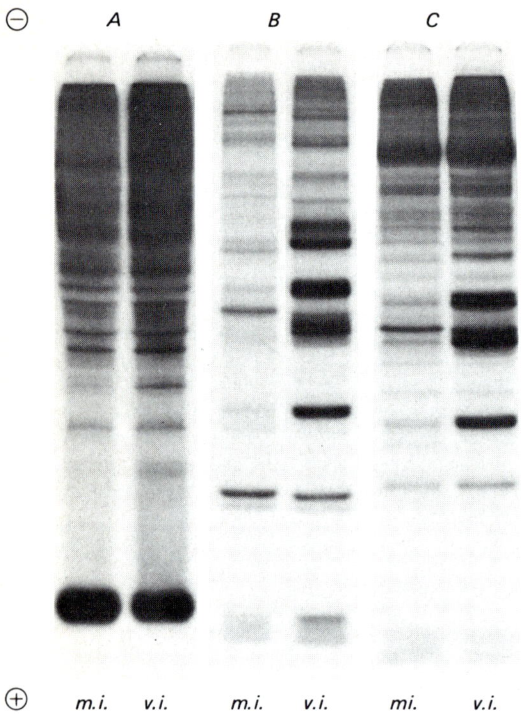

Fig. 1. Radioautographs of the sodium dodecylsulfate polyacrylamide gel electrophoretic separation of total [^{32}P]orthophosphate-labeled protein contained in the (A) ribosome-free supernatant, (B) 0.75M KCl-washed ribosomes, and (C) 0.75M KCl ribosomal wash fraction from mock- and mengovirus-infected Ehrlich ascites tumor cells.

m.i. indicates that the proteins were derived from mock-infected cells (control), v.i. that they were isolated from virus-infected cells. For experimental details see Materials and Methods.

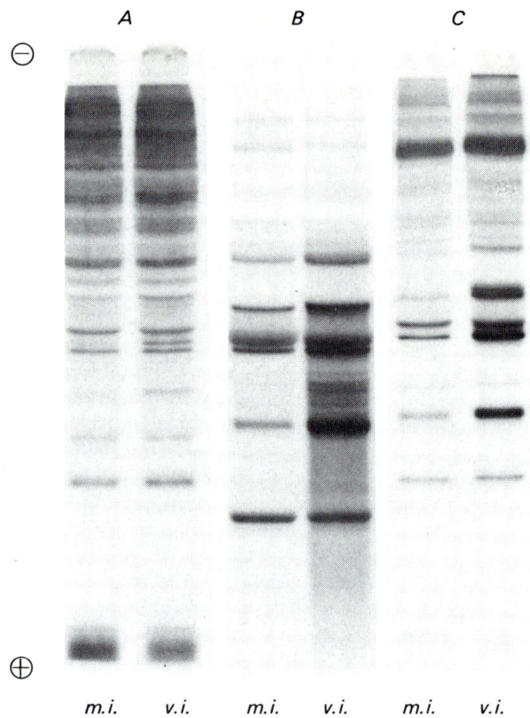

Fig. 2. Radioautographs of the sodium dodecylsulfate polyacrylamide gel electrophoretic separation of the acidic, [^{32}P]orthophosphate-labeled protein contained in the (A) ribosome-free supernatant, (B) 0.75M KCl-washed ribosomes, and (C) 0.75M KCl ribosomal wash fraction from mock- and mengovirus-infected Ehrlich ascites tumor cells.

The acidic proteins were obtained by carboxymethyl cellulose chromatography of the total protein of the 3 cytoplasmic fractions. m.i. indicates that the proteins originated from mock-infected cells, v.i. that they were prepared from virus-infected cells. For experimental details see Materials and Methods.

differences in [^{32}P]orthophosphate labeling are almost exclusively restricted to the ribosomal protein and ribosomal wash fractions. There is a close similarity in the phosphorylation patterns of these two protein fractions, which might be explained by assuming that a substantial part of the phosphorylated, non-ribosomal protein with high affinity for ribosomes is still attached to the ribonucleoprotein particles even after washing with 0.75M KCl. The differences between the phosphorylation patterns of ribosomal protein and the constituents of the ribosomal wash fractions, however, became more pronounced when the total protein of the two fractions was separated into acidic and basic parts by carboxymethyl cellulose chromatography (Fig. 2 and 3).

The phosphorylation patterns of the acidic proteins of all 6 cytoplasmic protein fractions are depicted in Fig. 2, those of the corresponding basic proteins in Fig. 3. Whereas, in response to mengovirus infection, only minor changes were detected in the profiles of the acidic and basic ribosome-free supernatant proteins (Fig. 2A and 3A) and in the patterns of the acidic ribosomal proteins

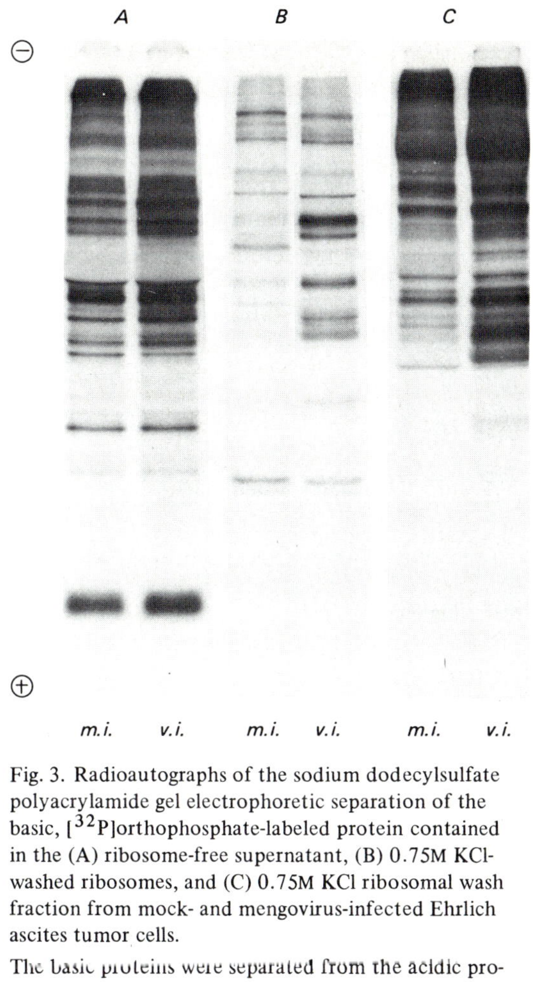

Fig. 3. Radioautographs of the sodium dodecylsulfate polyacrylamide gel electrophoretic separation of the basic, [^{32}P]orthophosphate-labeled protein contained in the (A) ribosome-free supernatant, (B) 0.75M KCl-washed ribosomes, and (C) 0.75M KCl ribosomal wash fraction from mock- and mengovirus-infected Ehrlich ascites tumor cells.

The basic proteins were separated from the acidic proteins by carboxymethyl cellulose chromatography of the total protein of the 3 cytoplasmic fractions. m.i. indicates that the proteins were prepared from mock-infected cells, v.i. that they were isolated from virus-infected cells. For experimental details see Materials and Methods.

(Fig. 2B), major differences could be demonstrated between the corresponding patterns of the basic ribosomal proteins (Fig. 3B) and those of the acidic (Fig. 2C) and the basic (Fig. 3C) proteins of the ribosomal wash fractions.

One might argue that some of the bands seen on the radioautographs of Fig. 1 to 3 are due to the presence of ^{32}P-labeled nucleic acids in the various protein mixtures. However, this is unlikely since the different cytoplasmic fractions were subjected, under denaturing conditions, to long duration CsCl density gradient centrifugation which efficiently separates the dense nucleic acids from the lighter proteins[12,15]. Furthermore, it was notably the basic protein fractions from mock- and mengovirus-infected ascites tumor cells that showed considerable diversity in their phosphorylation patterns. In the course of carboxymethyl cellulose chromatography, however, the negatively charged nucleic acids are rejected by the ion exchanger and should appear therefore in the acidic flow through fractions. In addition, after CsCl density gradient centrifugation, all protein fractions were treated with ribonucleases A and T1, though only at 6 °C; but to compensate for this, a long incubation period was chosen. It should be noted that the incubation of the various protein samples at higher temperatures was avoided, in order to keep the spontaneous, hydrolytic cleavage of the protein-phosphoric acid ester bonds as low as possible.

The notion that radioactively labeled nucleic acids might contribute to the radioautographic band patterns of Fig. 1 to 3 could be further refuted by the result of polyacrylamide gradient slab gel electrophoretic analysis of the proteins in 6M urea, 6% acetic acid[8] (data not shown). Under the electrophoretic conditions employed, the negatively charged nucleic acids do not enter the gel, in contrast to the positively charged basic proteins which migrate toward the cathode at the bottom of the gel. With all 3 cytoplasmic, basic-protein fractions, distinct radioautography profiles were obtained showing clear differences in the extent of phosphorylation between components derived from mock- and mengovirus-infected Ehrlich ascites tumor cells.

Finally, the possibility must be considered that the bands seen on the radioautographs of Fig. 1 to 3 are due, at least in part, to radioactively-labeled phospholipids which certainly contaminate the various protein fractions. However, this eventuality can also be excluded since, firstly, sodium dodecylsulfate dissociates complexes between protein and phospholipids and, secondly, because the low molecular weight phospholipids migrate much faster in sodium dodecylsulfate polyacrylamide gel electrophoresis than the larger cellular proteins.

From the radioautographs presented in Fig. 1 to 3 it is evident that the major differences between the phosphorylation patterns of proteins from mock- and mengovirus-infected Ehrlich ascites tumor cells reside in the crude ribosome fraction, and that after mengovirus infection the non-ribosomal proteins with high affinity for ribosomes are preferentially phosphorylated. Since many of the non-ribosomal constituents of the protein-synthesizing machinery interact with the ribosome in some way in the course of the initiation, elongation and termination of protein synthesis, their virus-induced phosphorylation might well play a regulatory role on the translational level and lead to the cessation of host-specific protein synthesis. Our further investigation of this problem will concentrate on the characterization of initiation factors as possible candidates of phosphorylation. This is because we have found recently that in cell-free systems from mengovirus-infected Ehrlich ascites tumor cells the preferential translation of mengovirus RNA over that of host template RNA is largely due to the initiation factor fraction[5], and because in reticulocytes the phosphorylation of initiation factor eIF-2 has already been shown to play an important role in the regulation of protein synthesis[9].

As indicated in Fig. 3 B, not only the non-ribosomal proteins associated with ribosomes are modified after mengovirus infection of Ehrlich ascites tumor cells but also one or more of the basic structural proteins of the ribosomal particle itself. To investigate the virus-induced phosphorylation of the ribosome more closely, the basic ribosomal proteins from mock- and mengovirus-infected Ehrlich ascites tumor cells were subjected to two-dimensional polyacrylamide gel electrophoretic analysis and radioautography. In Fig. 4 A, the amido black-stained gels showing the electrophoresis patterns of the ribosomal proteins from mock- and mengovirus-infected Ehrlich ascites tumor cells are compared with each other, while in Fig. 4 B the corresponding radioautographs are depicted. The radioautographs show that one ribosomal protein which is only slightly phosphorylated in mock-infected Ehrlich ascites tumor cells becomes strongly phosphorylated after mengovirus infection. The position of this protein relative to the other ribosomal proteins is indicated by a dotted circle on the stained gels. Judging from the structured appearance and the inclined position of the elongated spot on the radioautograph, the ribosomal protein seems to have more than one phosphorylation site. Since under the electrophoretic conditions employed, in both dimensions the basic ribosomal proteins are separated mainly according to charge, the reduction of the positive net charge must result in an even but discontinuous decrease of the protein's electrophoretic mobility, i.e. in equidistant spots arranged on a line which points toward the origin.

On a somewhat closer look at the stained gels of Fig. 4 A, the original protein spot seems to lose intensity after mengovirus infection and to assume a streaked appearance with decreasing density toward the anodic site. From a comparison of the staining density and the radioactivity of different sections of the elongated spot, we could deduce that the specific radioactivity increases with diminishing electrophoretic mobility of the ribosomal protein, consistent with the notion that it has more than one phosphorylation site.

The designation of the radioactively labeled protein as one of the structural proteins of the 80S ribosome was established by comparing our gel electrophoresis patterns with previously published patterns obtained under similar conditions[16-25], and which clearly show that mainly ribosomal protein S6 is strongly phosphorylated. In the majority of the cases referred to, the enhanced phosphorylation of protein S6 was paralleled by relocation of proteinaceous material from the original site of protein S6 toward the anode, resulting in elongated spots with inclined positions on the two-dimensional gels. Multiple phosphorylation sites on ribosomal protein S6 have also been detected by others[21,25].

Since washing with 0.75M KCl obviously did not remove all non-ribosomal protein from the ribosomes (see Fig. 1), it was not surprising to detect additional spots on the radioautographs of the two-dimensional gels. They also increased in density as a result of mengovirus infection, but could not be assigned to the ribosomal protein spots seen in Fig. 4 A.

As demonstrated by Datta et al.[26], the activation of the eIF-2 kinase, the inhibitor of polypeptide chain initiation in hemin-deficient reticulo-

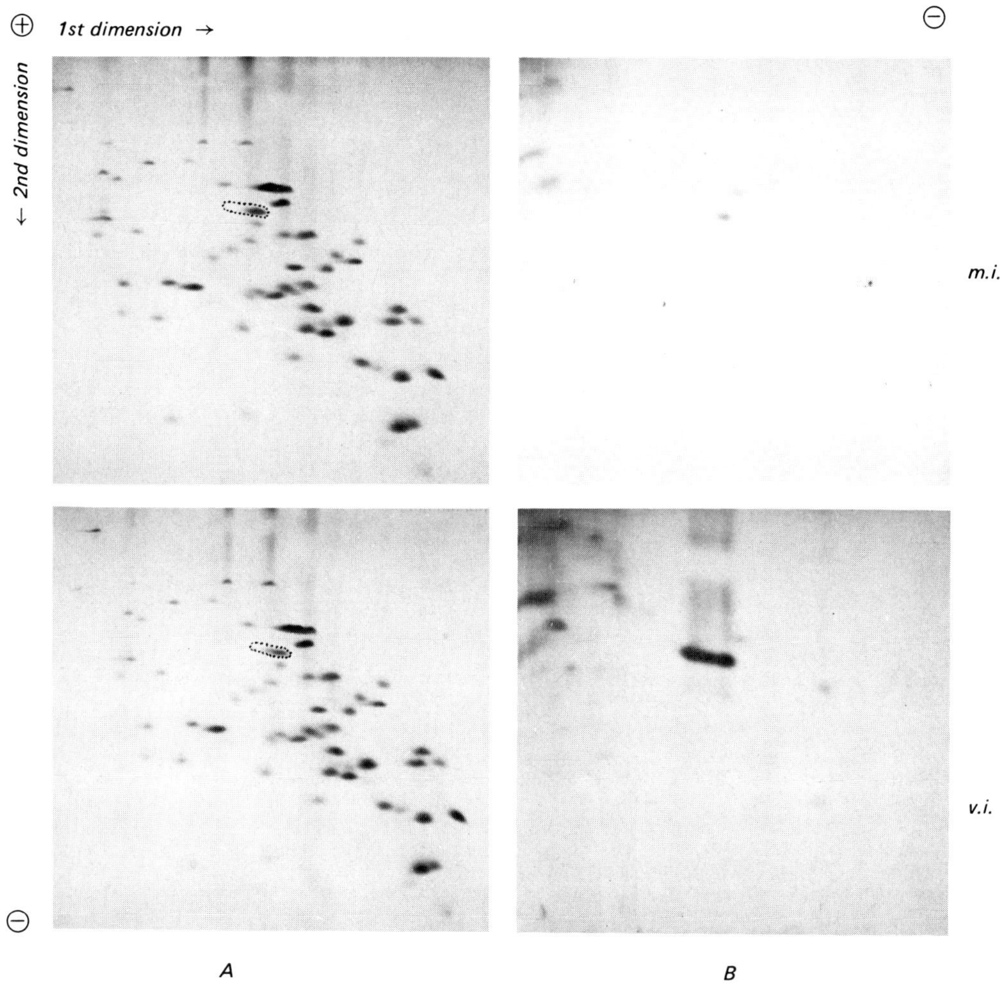

Fig. 4. Two-dimensional polyacrylamide gel electrophoresis of the basic ribosomal proteins from [^{32}P]orthophosphate-labeled mock- and mengovirus-infected Ehrlich ascites tumor cells.

(A) Amido black-stained gels, (B) radioautographs of the amido black-stained gels. The basic ribosomal protein was isolated from 0.75M KCl-washed ribosomes. m.i. indicates that the proteins originated from mock-infected cells, v.i. that they were derived from virus-infected cells. The dotted circle in (A) localizes the radioactivity of ribosomal protein S6 as visualized on the radioautograph of (B). For experimental details see Materials and Methods.

cyte lysates, appears to involve a phosphorylation catalyzed by 3′,5′-cyclic AMP-dependent protein kinase. Moreover, in several instances, the phosphorylation of ribosomal protein S6 was found to be dependent on an enhancement of the intracellular 3′,5′-cyclic AMP level[22,27], though this correlation is not well established. By analogy with these observations, we tried to detect a mengovirus-induced augmentation of the intracellular concentration of 3′,5′-cyclic AMP. Employing the radio isotope dilution test with 3′,5′-cyclic AMP-binding protein[28], we were unable to observe any change of the 3′,5′-cyclic AMP concentration in response to mengovirus infection. As far as the phosphorylation of ribosomal protein S6 is concerned, this result seems to be in accordance with the finding that in pre-confluent baby hamster kidney fibroblasts the enhanced phosphorylation of protein S6 is not paralleled by the augmentation of the 3',5'-cyclic AMP concentration[17]. In this context, however, it should be pointed out that sera of several species including fetal calf serum, which was present in culture medium of the Ehrlich ascites tumor cells, contain factors that block the rise of 3',5'-cyclic AMP in cultured mammalian cells[29].

We express our gratitude to Mrs. *Margot Bialdiga* for the production of Ehrlich ascites tumor cells and mengovirus and to Mrs. *Ingrid Buchwald* and Miss *Gunda Boeckmann* for expert technical assistance.

Literature

1 Martin, E.M. & Kerr, I.M. (1968) in The Molecular Biology of Viruses (Crawford, L.V. & Stoker, M.G.P., eds.) pp. 15 - 46, University Press, Cambridge.
2 Baltimore, D. (1969) in The Biochemistry of Viruses (Levy, H.B., ed.) pp. 101 - 176, Marcel Dekker, New York.
3 Golini, F., Thach, S.S., Birge, C.H., Safer, B., Merrick, W.C. & Thach, R.E. (1976) *Proc. Natl. Acad. Sci. U.S.A.* **73**, 3040 - 3044.
4 Kaufmann, Y., Goldstein, E. & Penman, S. (1976) *Proc. Natl. Acad. Sci. U.S.A.* **73**, 1834 - 1838.
5 Hackett, P.B., Egberts, E. & Traub, P. (1978) *Eur. J. Biochem.* **83**, 353 - 361.
6 Lawrence, C. & Thach, R.E. (1974) *J. Virol.* **14**, 598 - 610.
7 Golini, F., Thach, S., Lawrence, C. & Thach, R. (1976) in ICN-UCLA Symposia on Molecular and Cellular Biology. Animal Virology (Baltimore, D., Huang, A.S. & Fox, C.F., eds.) vol. 4, pp. 717 - 734, Academic Press, New York.
8 Hackett, P.B., Egberts, E., Buchwald, I. & Traub, P. (1977) *Mol. Biol. Rep.* **3**, 387 - 396.
9 Kramer, G., Henderson, A.B., Pinphanichakarn, P., Wallis, M.H. & Hardesty, B. (1977) *Proc. Natl. Acad. Sci. U.S.A.* **74**, 1445 - 1449.
10 Egberts, E., Hackett, P.B. & Traub, P. (1976) *Hoppe Seyler's Z. Physiol. Chem.* **357**, 1779 - 1792.
11 Carter, D.B. & Chae, C.-B. (1976) *Biochemistry* **15**, 180 - 185.
12 Egberts, E., Hackett, P.B. & Traub, P. (1977) *Hoppe Seyler's Z. Physiol. Chem.* **358**, 463 - 474.
13 Traub, P. & Boeckmann, G. (1978) *Hoppe-Seyler's Z. Physiol. Chem.* **359**, 571 - 579.
14 Kaltschmidt, E. & Wittmann, H.G. (1970) *Anal. Biochem.* **36**, 401 - 412.
15 Sonnenbichler, J., Reichhart, F. & Tobien, M. (1977) *Hoppe Seyler's Z. Physiol. Chem.* **358**, 1647 - 1648.
16 Rankine, A.D. & Leader, D.P. (1975) *FEBS Lett.* **52**, 284 - 287.
17 Leader, D.P., Rankine, A.D. & Coia, A.A. (1976) *Biochem. Biophys. Res. Commun.* **71**, 966 - 974.
18 Rankine, A.D., Leader, D.P. & Coia, A.A. (1976) *Biochim. Biophys. Acta* **474**, 293 - 307.
19 Kaerlein, M. & Horak, I. (1976) *Nature (London)* **259**, 150 - 151.
20 Lastick, S.M., Nielsen, P.J. & McConkey, E.H. (1977) *Mol. Gen. Genet.* **152**, 223 - 230.
21 Gressner, A.M. & Wool, I.G. (1974) *J. Biol. Chem.* **249**, 6917 - 6925.
22 Gressner, A.M. & Wool, I.G. (1976) *J. Biol. Chem.* **251**, 1500 - 1504.
23 Gressner, A.M. & Wool, I.G. (1976) *Nature (London)* **259**, 148 - 150.
24 Gressner, A.M. & Greiling, H. (1977) *FEBS Lett.* **74**, 77 - 81.
25 Treloar, M.A., Treloar, M.E. & Kisilevsky, R. (1977) *J. Biol. Chem.* **252**, 6217 - 6221.
26 Datta, A., De Haro, C., Sierra, J.M. & Ochoa, S. (1977) *Proc. Natl. Acad. Sci. U.S.A.* **74**, 3326 - 3329.
27 Cawthon, M.L., Bitte, L.F., Krystosek, A. & Kabat, D. (1974) *J. Biol. Chem.* **249**, 275 - 278.
28 Gilman, A.G. (1970) *Proc. Natl. Acad. Sci. U.S.A.* **67**, 305 - 312.
29 Van Calker, D., Müller, M. & Hamprecht, B. (1977) *Hoppe Seyler's Z. Physiol. Chem.* **358**, 1188.

Dr. Ingeborg Rosnitschek, Ulrike Traub and Prof. Dr. Peter Traub*, Max-Planck-Institut für Zellbiologie, Anton-Dohrn-Weg 59, Postfach 1009, D-2940 Wilhelmshaven.

* To whom correspondence should be addressed.

Hoppe-Seyler's Z. Physiol. Chem.
Bd. 359, S. 601 - 606, Mai 1978

Multiple Protein Kinases from *Trypanosoma gambiense*

Rolf D. WALTER

Bernhard-Nocht-Institut für Schiffs- und Tropenkrankheiten, Abt. Biochemie

(Received 5 January 1978)

Summary: Three protein kinases were distinguished in *Trypanosoma gambiense* extract. The enzymes preferred phosvitin, histone, and protamine as acceptor proteins, respectively. The amino acid residues of the acceptor proteins which were phosphorylated by these protein-kinase activities were serine and – to less extent – threonine. The protein kinase activities were neither affected by cyclic nucleotides nor by cyclic AMP receptors. The molecular weights of these protein kinases were determined to be > 200 000, 95 000 and 37 000, respectively. The activities of all three protein kinases were affected to varying degrees by nucleotides and nucleosides.

Protein-Kinasen von Trypanosoma gambiense

Zusammenfassung: Im Extrakt von *Trypanosoma gambiense* konnten drei Protein-Kinasen bestimmt werden. Sie bevorzugten Phosvitin, Histon bzw. Protamin als Substrate. Es wurde überwiegend Serylphosphat gebildet. Die Aktivität aller drei Protein-Kinasen wurde weder durch zykl. Nukleotide stimuliert noch durch zykl. AMP-Rezeptoren gehemmt. Die Aktivität dieser Enzyme wurde aber im unterschiedlichen Maße durch Nucleotide und Nucleoside beeinflußt. Ihre Molekulargewichte wurden mit > 200 000, 95 000 bzw. 37 000 bestimmt.

Key words: Trypanosoma, protein kinases, cyclic nucleotide-independent, nucleosides.

Protein kinases, dependent on or independent of cyclic nucleotides, are capable of phosphorylating a wide range of proteins. Whereas the role of cyclic AMP-dependent protein kinases as mediators of hormonal actions is well documented, the role of cyclic nucleotide-independent protein kinases has been less investigated. Cyclic nucleotide-independent protein kinases are reported from animal sources[1-10] as well as from *Tetrahymena*[11,12], *Neurospora*[13], yeast[14-16], *Euglena*[17] and plants[18-21]. Recently the occurrence of nucleoside-stimulated protein-kinase activity in *Trypanosoma gambiense* and *T. cruzi*[22,23] has been described. This paper presents a description and characteristics of protein kinases from *T. gambiense*, the pathogenic agent of sleeping sickness.

Enzyme: Protein kinase, ATP:protein phosphotransferase, (EC 2.7.1.37).

Materials and Methods

Materials

[8-^{3}H]Adenosine 3′,5′-monophosphate and [^{32}P]adenosine triphosphate were purchased from the Radiochemical Centre, Amersham. Nucleosides and nucleotides were obtained from Boehringer Mannheim GmbH. Calf thymus histones (mixture, lysine-rich and arginine-rich), protamine (salmon sperm) and phosvitin (egg vitellin) were purchased from Sigma, St. Louis. Casein was obtained from Merck, Darmstadt. Bovine serum albumin was purchased from Behringwerke, Marburg and phosphothreonine and phosphoserine from Sigma, St. Louis. Silica gel-coated thin-layer plates and all other chemicals were products of Merck, Darmstadt. Sephadex G-200 superfine was obtained from Pharmacia, Uppsala. Ultrogel AcA 34 was purchased from LKB Instruments and HAWP 25 filters from Millipore, Neu-Isenburg.

Protein-kinase assay

Protein kinase activity was assayed by measuring ^{32}P incorporation from [^{32}P]ATP into acceptor proteins[22]. The standard assay contained in a total volume of 0.1 m*l*: 0.1M Tris maleate (pH 7.0), 0.01M $MgCl_2$, 0.1 mg bovine serum albumin, 0.1mM [^{32}P]ATP (5×10^5 - 8×10^5 cpm), 0.1 mg mixed histones, phosvitin or protamine and enzyme preparation. Incubation was carried out at 37 °C for 30 min. The reaction rate was linear within this period. The reaction was terminated by addition of 1 m*l* 10% trichloroacetic acid. In the case of protamine as acceptor protein 1 m*l* 20% trichloroacetic acid solution was used. The enzyme activity is expressed in units (1 U = 1 μmol of ^{32}P transferred to acceptor protein per min).

Acid hydrolysis of ^{32}P-labelled acceptor proteins

The enzymatically phosphorylated radioactive phosvitin and protamine preparations were hydrolysed for 4 h in 6N HCl at 100 °C in ampules sealed under nitrogen. The hydrolysates were dried in vacuo over KOH. Thin-layer electrophoresis was performed on silica gel (2 h, 40 V/cm) in formic/acetic acid buffer, pH 1.9[22,24]. The radioactivity was determined by using a thin-layer scanner (Berthold, Wildbad). The positions of authentic phosphoserine and phosphothreonine added as carriers were located by staining with ninhydrin.

Cyclic AMP-binding assay

Binding of cyclic AMP to protein was determined by a modification of the method of Gilman[25]. The reaction mixture contained in a total volume of 0.1 m*l*: 0.02M potassium phosphate, pH 6.0 and 1 μM cyclic [8-^{3}H] AMP. Incubation was started by addition of protein solution and carried out at 0 °C for 60 min.

Protein determination

Protein was determined by the method of Lowry et al.[26] with bovine serum albumin as reference standard.

Separation of protein-kinase activities from Trypanosoma extract

Bloodstream forms of *T. gambiense* were obtained from NMRI mice. The preparation of *Trypanosoma* extract was performed as described in a preceding paper[22]. The homogenate (118 mg protein) was centrifuged at $100000 \times g$ for 60 min. To achieve further purification ammonium sulfate was added slowly to the supernate (54 mg) to give a final concentration of 65% saturation. The suspension was stirred for 1 h and the precipitate was collected by centrifugation at $10000 \times g$ for 30 min. No protein-kinase activity remained in the supernate. The precipitate (32 mg) was dissolved and dialysed against 2 *l* of 50mM potassium phosphate buffer (pH 7.0) for 12 h. The protein solution was then placed on a Sephadex G-200 column (1.5 × 85.0 cm), which had been previously equilibrated with 50mM potassium phosphate buffer (pH 7.0). The column was eluted with a flow rate of 3 m*l*/h and fractions of 3 m*l* were collected. The protein-kinases activities are referred to as PK I, II and III in the order of elution from the column (Fig. 1). The protein kinases were pooled (Fract. 20 - 22, PK I; 29 - 32, PK II; 38 - 42, PK III) and stored frozen at − 76 °C until used.

Preparation of cyclic AMP receptors from T. gambiense

Cyclic AMP-binding proteins were prepared from *Trypanosoma* extract[27]. The homogenate was extracted by 0.5M NaCl in 0.1M potassium phosphate buffer (pH 6.0). Partial purification and separation of cyclic AMP receptor I and II were performed by ammonium sulfate precipitation followed by gel chromatography on an Ultrogel AcA 34 column.

Results

Separation of protein kinases

The filtration of *T. gambiense* extract, concentrated by ammonium sulfate precipitation, on a Sephadex G-200 column resulted in the separation of three protein kinase activities, differing in their preference for acceptor proteins. The elution pattern is shown in Fig. 1. The fractions containing protein-kinase activity were reassayed in the presence of 10^{-6}M cyclic AMP or 10^{-4}M inosine, respectively. The addition of cyclic AMP was without effect on the protein-kinase activities, whereas inosine stimulated the activity of Pk II

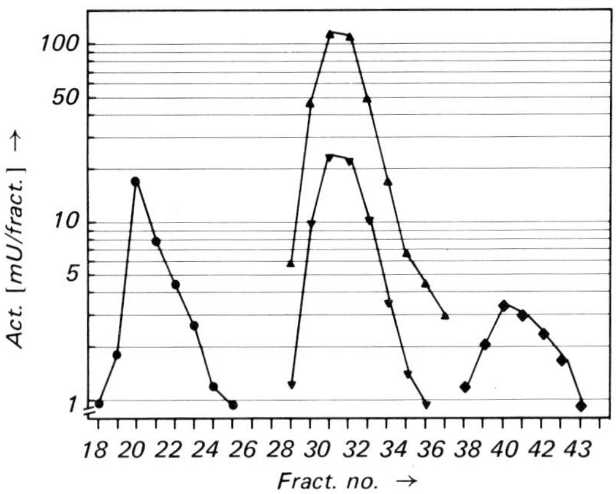

Fig. 1. Separation of protein kinase I, II, and III by Sephadex G-200 chromatography.

Ammonium sulfate fraction was applied to Sephadex G-200 column (1.5 × 85.0 cm) and eluted with 50mM potassium phosphate buffer (pH 7.0) at a flow rate of 3 ml/h. Fractions of 3 ml were collected. Protein-kinase activity was measured as described in "Methods". The following acceptor proteins were used: phosvitin (●——●), mixed histones (▼——▼), and protamine (◆——◆). Stimulation of protein-kinase activity by nucleosides was determined in the presence of 0.1mM inosine with mixed histones as acceptor protein (▲——▲).

about 5-fold (Fig. 1). A large part of the stimulatory effect of inosine on the enzyme activity was lost when PK II was stored at 0 °C or at − 76 °C. This loss could not be prevented by addition of 10^{-4}M inosine. Properties of this nucleoside stimulated protein kinase have been described in detail previously[22,23].

Molecular weight

The molecular weights of these protein kinases were estimated according to Whitaker[28]. Chromatography on Sephadex G-200 superfine column was performed as described in "Methods". The column was calibrated with xanthine oxidase, pyruvate kinase, aldolase, bovine serum albumin, ovalbumin, and chymotrypsinogen. Crude extract as well as partially purified protein kinases from *T. gambiense* were subjected to chromatography. The molecular weights were determined to be > 200 000 for protein kinase I, 95 000 for protein kinase II, and 37 000 for protein kinase III.

Effect of cyclic nucleotides and cyclic AMP receptors

The activities of these protein kinases from *T. gambiense* were found to be not dependent on cyclic nucleotides. Furthermore, none of the partially purified protein kinases bound cyclic AMP. The activities of these protein kinases were not influenced by 10^{-7}M to 10^{-4}M cyclic AMP or cyclic GMP when phosvitin, mixed histones or protamine were used as acceptor proteins. In addition, protein kinases I and III were reassayed with mixed histones as substrate in the presence of 10^{-6}M cyclic AMP and cyclic GMP. None of these enzyme activities was affected by cyclic nucleotides. Furthermore, when crude extract, dialysed against 50mM potassium phosphate buffer (pH 7.0) for 18h, was used as source of enzyme, the phosphorylation of phosvitin, protamine or mixed histones was not influenced by 10^{-6}M cyclic AMP or cyclic GMP, respectively.

The activities of protein kinases I, II, and III were not inhibited by addition of cyclic AMP receptors prepared from *T. gambiense* as demonstrated in Table 1. The increase of enzyme activity when receptor I was added to the reaction mixture, was due to contamination of this preparation with protein kinase I activity.

Effect of ions and variation of pH

The protein kinases demonstrated an absolute requirement for divalent metal ions. In the case

Table 1. Effect of cyclic AMP receptors on protein kinase activities.

Enzyme activity was tested in the presence of cyclic AMP receptors. 5 μg of receptor I and 3.5 μg of receptor II were added to the standard reaction mixture. For details of isolation of cyclic AMP binding proteins from *T. gambiense* see under "Methods". The activities of protein kinase I, II, and III were measured with phosvitin, mixed histones, and protamine as acceptor proteins, respectively. Enzyme activity is expressed in μU.

Protein kinase	Acceptor protein	Control	Addition of	
			receptor I	receptor II
		[μU]	[μU]	[μU]
PK I	Phosvitin	12.7	16.5	12.5
PK II	Histones	18.2	19.7	18.6
PK III	Protamine	8.5	9.4	8.8

of protein kinases I and III maximal activity was achieved at a concentration of 10mM $Mg^{2\oplus}$, whereas the optimal concentrations for $Mn^{2\oplus}$ and $Co^{2\oplus}$ were found to be 0.5mM and 1.0mM, respectively. In the presence of $Mn^{2\oplus}$ or $Co^{2\oplus}$ at optimal concentration the activities of protein kinases I and III were found to be about 50% with respect to 10mM $Mg^{2\oplus}$. $Ca^{2\oplus}$, $Ba^{2\oplus}$, $Ni^{2\oplus}$, $Zn^{2\oplus}$, and $Cu^{2\oplus}$ at concentrations of 0.5mM and 5mM were without effect on the protein-kinases activities.

Protein kinases I and III exhibited maximum activity around pH 7.5. The pH optimum of protein kinase II activity has been reported to be about 7.0[22].

Acceptor proteins and site of phosphorylation

The relative degree of phosphorylation of various protein substrates by protein kinases from *T. gambiense* is demonstrated in Table 2. Protein kinase I showed preference for phosvitin, compared to protamine, lysine-rich histone and casein. Protein kinases II and III preferred histones and protamine, respectively. Both were inactive towards phosvitin and casein. The Michaelis constants of protein kinases I and III for phosvitin and protamine were estimated to be 0.2 mg/m*l* and 0.08 mg/m*l*, respectively. In the case of PK III the K_m value for mixed histones was determined to be 0.25 mg/m*l*. This value is in agreement with that reported for protein kinase II[22]. Characterization of phosphorylated histones by protein kinase II has been reported[22]. After partial acid hydrolysis and thin-layer electrophoresis the main amount of radioactivity was associated with serine phosphate and only a small amount with threonine phosphate. Protein kinases I and III were assayed under the same conditions except that phosvitin and protamine were used as acceptor proteins, respectively. The ratio of radioactivity associated with phosphoserine and phosphothreonine was found to be 95% to 5%, and 98% to 2% for protein kinase I and III, respectively.

Table 2. Substrate specificity of protein kinases.

The standard assay was used except for the variation of acceptor protein. The concentration of acceptor protein was 1 mg/m*l*. Results represent means of triplicate determinations.

Acceptor protein	Relative reaction rate [%]		
	PK I	PK II	PK III
Mixed histones	42	100	64
Arginine-rich histone	22	77	43
Lysine-rich histone	40	4	20
Phosvitin	100	< 1	< 1
Protamine	38	59	100
Casein	36	< 1	< 1

Effect of adenine derivatives

ATP was used as phosphoryl donor. The Michaelis constants for ATP were calculated to be 0.08mM and 0.1mM for protein kinases I and III, respectively (Fig. 2 and 3). In the case of protein kinase II the K_m value for ATP was reported to be 0.2mM[22]. The inhibitory effects of adenine derivatives on the activities of protein kinases are shown in Tab. 3. ADP proved to be an especially potent inhibitor of the activities of all three protein kinases from *T. gambiense*. The type of inhibition by ADP was found to be competitive with respect to ATP (Fig. 2 and 3). The inhibition constants were calculated to be 0.14mM and 0.13mM for protein kinases I and III, respectively. The inhibition constant of protein kinase II for

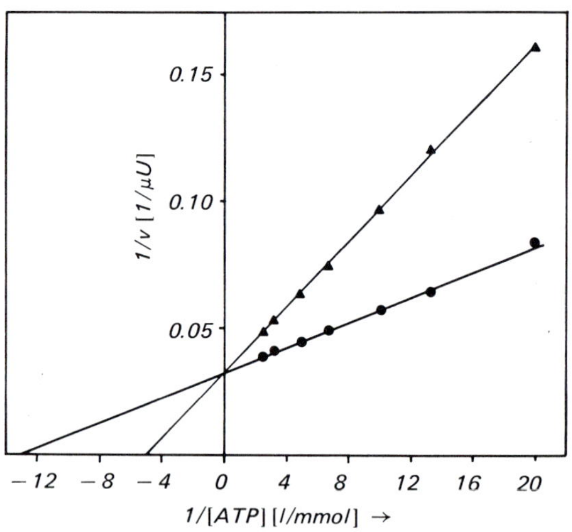

Fig. 2. Effect of ATP concentration on the activity of protein kinase I.

Double reciprocal plots of phosphorylation of phosvitin as a function of ATP in the absence (●——●) and in the presence of 0.2mM ADP (▲——▲).

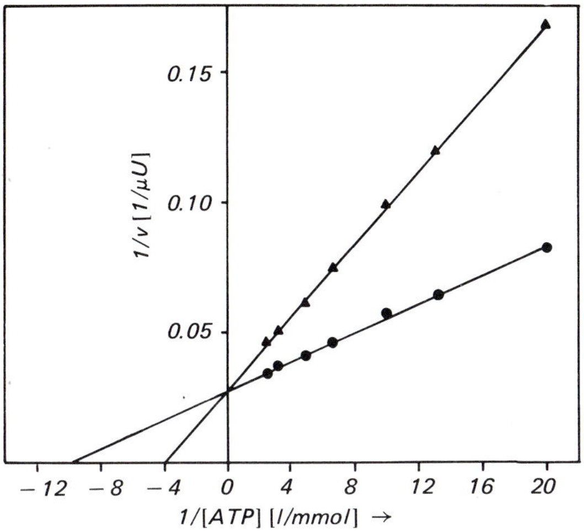

Fig. 3. Effect of ATP concentration on the activity of protein kinase III.

Double reciprocal plots of phosphorylation of protamine as a function of ATP in the absence (●——●) and in the presence of 0.2mM ADP (▲——▲).

ADP has been reported to be 0.25mM[22]. Deoxyadenosines as well as adenosine are competitive inhibitors of protein kinases activities from *T. gambiense.* 3′-Deoxyadenosine was found to be a more potent inhibitor of protein kinase III activity compared to 2′-deoxyadenosine. The inhibition constants were calculated to be 0.08mM and 0.2mM, respectively. In so far the data on protein kinase III resemble those on protein kinase II[23].

Table 3. Effect of adenine derivatives.

The standard assay was used. The concentration of ATP was 0.1mM. Results represent means of triplicate determinations.

Additions (0.5mM)	Relative reaction rate [%]		
	PK I	PK II	PK III
Adenosine	58	146	43
2'-Deoxyadenosine	55	68	44
3'-Deoxyadenosine	55	26	20
Adenine	71	75	56
AMP	84	88	81
ADP	34	47	34
ITP	72	28	37
GTP	77	34	46

In the case of protein kinase I the inhibitory effects of 2' and 3'-deoxyadenosines were equal. The K_i values were determined to be 0.4mM. The inhibition constants for adenosine were calculated to be 0.5mM and 0.2mM for protein kinases I and III, respectively.

Discussion

Three protein kinases could be separated on the basis of their molecular weight from *T. gambiense* extract. None of these protein kinases was dependent on cyclic AMP or cyclic GMP. Cyclic nucleotides neither stimulated the protein-kinase activity of the *T. gambiense* extract nor the activities of the partially purified protein kinases I, II and III. Furthermore, these protein-kinase activities were not influenced when cyclic AMP-binding factors, partially purified from *T. gambiense*, were added to the assay.

At present the function of these protein kinases in *T. gambiense* remains unknown. The protein kinases were tested with phosvitin, histones and protamine as acceptor proteins, but nothing is still known about the endogenous substrates of these kinases. The isolation and characterization of endogenous protein substrates should inform us about the biological function of these protein kinases. Thus far it remains open to further investigation, whether the phosphorylation of endogenous proteins is affected by cyclic nucleotides or not.

The possible role of purine derivatives on the activities of protein kinases of *T. gambiense* is speculative, because no data about its concentration within these cells are available as yet. An outstanding result is the stimulatory effect of nucleosides on the activity of protein kinase II[22,23]. The inhibitory effect of adenosine and other purine derivatives on protein-kinase activity is common to kinases from many sources. The higher potency of cordycepin compared to 2′-deoxyadenosine as inhibitor of protein kinase III has been reported from protein kinase II from *T. gambiense* as well as *T. cruzi* and from protein kinases from mammalian sources, too[23,29,30].

The excellent technical assistance of Miss *H. Jahn* is gratefully acknowledged. This work was supported by the *Deutsche Forschungsgemeinschaft*.

Literature

1 Farago, A., Antoni, F., Takats, A. & Fabian, F. (1973) *Biochim. Biophys. Acta* **297**, 517 - 526.
2 Farago, A., Antoni, F. & Fabian, F. (1974) *Biochim. Biophys. Acta* **370**, 459 - 467.
3 Traugh, J. A. & Traut, R. R. (1974) *J. Biol. Chem.* **249**, 1207 - 1212.
4 Langan, T. A. (1971) *Ann. N. Y. Acad. Sci.* **185**, 166 - 179.
5 Kemp, B. E., Froscio, M., Rogers, A. & Murray, A.W. (1975) *Biochem. J.* **145**, 241 - 249.
6 Murofushi, H. (1974) *Biochim. Biophys. Acta* **364**, 260 - 271.
7 Clari, G., Pinna, L. A. & Moret, V. (1976) *Biochim. Biophys. Acta* **451**, 484 - 490.
8 Sankaran, K., Pawse, A. R. & Nadkarni, G. B. (1976) *Biochim. Biophys. Acta.* **438**, 412 - 423.
9 Maragoudakis, M. E. & Hankin, H. (1977) *Biochim. Biophys. Acta* **480**, 122 - 136.
10 Hosey, M. M. & Tao, M. (1977) *Biochim. Biophys. Acta* **482**, 348 - 357.
11 Majumder, G. C., Shrago, E. & Elson, C. E. (1975) *Biochim. Biophys. Acta* **384**, 399 - 412.
12 Murofushi, H. (1973) *Biochim. Biophys. Acta* **327**, 354 - 364.
13 Gold, M. H. & Segel, I. H. (1974) *J. Biol. Chem.* **249**, 2417 - 2423.
14 Takai, Y., Yamamura, H. & Nishizuka, Y. (1974) *J. Biol. Chem.* **249**, 530 - 535.
15 Lerch, K., Muir, L. W. & Fisher, E. (1975) *Biochemistry* **14**, 2015 - 2023.
16 Nakashima, M., Takai, Y., Yamamura, H. & Nishizuka, Y. (1975) *Biochim. Biophys. Acta* **397**, 117 - 123.
17 Keirns, J. J., Carrit, B., Freeman, J., Eisenstadt, J.M & Bitensky, M. W. (1973) *Life Sci.* **13**, 287 - 302.
18 Giannattasio, M., Carratu, G. & Tucci, G. F. (1974) *FEBS Lett.* **49**, 249 - 253.
19 Keates, R. A. B. (1973) *Biochem. Biophys. Res. Commun.* **54**, 655 - 661.
20 Nakaya, N., Sugano, N., Nishi, A. & Tsukada, K. (1975) *Biochim. Biophys. Acta* **410**, 273 - 278.
21 Pai, H. S., Dehm, P., Schweiger, M., Rahmsdorf, H.J., Ponta, H., Hirsch-Kauffmann, M. & Schweiger, H.G. (1975) *Protoplasma* **85**, 209 - 218.
22 Walter, R. D. (1976) *Biochim. Biophys. Acta* **429**, 137 - 146.
23 Walter, R. D. & Ebert, F. (1977) *Hoppe-Seyler's Z. Physiol. Chem.* **358**, 23 - 26.
24 Allerton, S. E. & Perlmann, G. E. (1965) *J. Biol. Chem.* **240**, 3892 - 3898.
25 Gilman, A. G. (1970) *Proc. Natl. Acad. Sci. U.S.A* **67**, 305 - 312.
26 Lowry, O. H., Rosebrough, N. J., Farr, A. L. & Randall, R. J. (1951) *J. Biol. Chem.* **193**, 265 - 275.
27 Walter, R. D. (1978) *Hoppe-Seyler's Z. Physiol. Chem.* **359**, 607 - 612, succeeding paper.
28 Whitaker, J. R. (1963) *Anal. Chem.* **35**, 1950 - 1953.
29 Hirsch, J. & Martelo, O. J. (1976) *Life Sci.* **19**, 85 - 90.
30 Glazer, R. I. & Kuo, J. F. (1977) *Biochem. Pharmacol.* **26**, 1287 - 1290.

Dr. Rolf D. Walter, Bernhard-Nocht-Institut für Schiffs-und Tropenkrankheiten, Abt. Biochemie, Bernhard-Nocht-Straße 74, D-2000 Hamburg 4.

Hoppe-Seyler's Z. Physiol. Chem.
Bd. 359, S. 607 - 612, Mai 1978

Adenosine 3′,5′-Cyclic Monophosphate-Binding Proteins from *Trypanosoma gambiense*

Rolf D. WALTER

Bernhard-Nocht-Institut für Schiffs- und Tropenkrankheiten, Abt. Biochemie

(Received 5 January 1978)

Summary: Two cyclic AMP-binding proteins, not identical with regulatory subunits of protein kinases, have been isolated from *Trypanosoma gambiense*. The cyclic AMP receptors were separated by gel chromatography on the basis of their molecular weights. The binding constants of the high and the low molecular weight receptors for cyclic AMP were determined to be 0.4 μM and 0.6 μM, respectively. Cyclic IMP and cyclic GMP compete with cyclic AMP for the binding sites of both receptors. The cyclic AMP binding of the low molecular weight receptor was competitively inhibited by adenine derivatives. The binding capacity of the high molecular weight receptor was enhanced about two-fold by proteolytic modification with trypsin.

Trypanosoma gambiense: spezifische Bindungsproteine für zyklisches AMP

Zusammenfassung: Aus *Trypanosoma gambiense* wurden zwei für zyklisches AMP spezifische Bindungsproteine isoliert. Diese Rezeptoren waren nicht identisch mit Untereinheiten von für zykl. AMP abhängigen Protein-Kinasen. Die Bindungskonstanten für Rezeptor I und II wurden mit 0.4 μM und 0.6 μM bestimmt. Zykl. IMP und zykl. GMP hemmten die Bindung von zykl. AMP kompetitiv im Falle beider Rezeptoren. Die Bindungskapazität von Rezeptor I konnte durch Inkubation mit Trypsin verdoppelt werden. Im Falle von Rezeptor II wurde die Bindung von zykl. AMP durch Adeninderivate kompetitiv gehemmt.

Key words: Trypanosoma, cyclic AMP receptors, protein kinases, trypsin, adenine derivatives

Binding proteins for cyclic AMP, besides those acting as regulatory subunits of cyclic AMP-dependant protein kinases, have been reported from yeast[1], *Drosophila*[2] and various mammalian sources[3-8]. The function of these cyclic AMP receptors remains still unknown. Recently the

Enzymes:
Adenylate cyclase, ATP pyrophosphate-lyase (cyclizing) (EC 4.6.1.1);
3′,5′-Cyclic-nucleotide phosphodiesterase, 3′,5′-cyclic-nucleotide 5′-nucleotidohydrolase, also named 3′,5′-cyclic-AMP phosphodiesterase (EC 3.1.4.17);
Protein kinase, ATP:protein phosphotransferase (EC 2.7.1.37).

existence of such cyclic AMP-binding proteins has been reported from *Trypanosoma gambiense* and *T. cruzi*[9]. In this paper the partial purification and some properties of two cyclic AMP receptors, not related to protein kinases, from *T. gambiense* are described.

Materials and Methods

Materials

[8-^{3}H]Adenosine 3',5'-monophosphate (30 Ci/mmol) and [2-^{3}H]adenosine (20 Ci/mmol) were obtained from the Radiochemical Centre, Amersham. Nucleosides and nucleotides, as well as trypsin, subtilisin and trypsin inhibitor were purchased from Boehringer Mannheim GmbH. Ultrogel AcA 34 was obtained from LKB Instruments, Stockholm. Membrane filters (HAWP 25) were purchased from Millipore, Neu-Isenburg. All other chemicals used were products of Merck, Darmstadt.

Enzyme assays

The activities of adenylate cyclase, cyclic-nucleotide phosphodiesterase, and protein kinases were determined as described before[10-12].

Protein determination

Protein was determined according to Lowry et al.[13] with bovine serum albumin as reference standard.

Cyclic AMP- and adenosine-binding assay

Binding of cyclic AMP and separation of bound and unbound cyclic [^{3}H]AMP were performed according to Gilman[14] and Ueland and Døskeland[7]. The standard assay contained in a final volume of 100 μl: 0.02M potassium phosphate buffer, pH 6.0, 1 μM cyclic [^{3}H]-AMP, and receptor preparation. Alternatively 1 μM [^{3}H]adenosine was used instead of cyclic AMP. Incubation was started by the addition of receptor preparation and carried out at 0 °C for 60 min. The reaction was terminated by dilution with 1 m*l* ice-cold potassium phosphate buffer (20mM, pH 6.0), followed by filtration through wet membrane filters. The filters were washed with 15 m*l* of cold 20mM potassium phosphate buffer, pH 6.0. Alternatively the separation of bound and unbound cyclic [^{3}H]AMP or [^{3}H]adenosine was performed by ammonium sulfate precipitation followed by filtration.

Preparation of T. gambiense extract

NMRI mice infected with *T. gambiense* were killed by decapitation and the blood was collected in heparin and phosphate/saline/glucose buffer (PSG buffer, pH 8.0, ionic strength 0.22)[15]. The blood was centrifuged at 4000 rpm in a Sorvall-HB 4 rotor for 20 min. Blood cells and *Trypanosoma* were suspended in PSG buffer and applied to a DEAE-cellulose column (4.0 x 10.0 cm), previously equilibrated with PSG buffer. The *Trypanosoma* were eluted with the same buffer according to Lanham[15], while the other cells remained on the column. All further steps were carried out at 2 °C. The *Trypanosoma* were washed four times in PSG buffer and 0.9% NaCl in the last step (4000 rpm for 20 min). The cells were suspended in 50mM potassium phosphate buffer, pH 6.0, and disrupted by sonication or centrifugation for 90 min at 100 000 x *g* (to test the particulate fraction for cyclic AMP-binding activity). The homogenate was stored at − 76 °C until used.

Molecular weight estimation

To estimate the molecular weight range of both cyclic AMP binding proteins the procedure of Whitaker[16] was used. Preparations of receptor I and II were applied to a Sephadex G-200 superfine column (1.5 x 85.0 cm), previously equilibrated with 50mM potassium phosphate buffer (pH 6.0) and standardized with chymotrypsinogen, ovalbumin, bovine serum albumin, aldolase, pyruvate kinase, and blue dextran. The column was eluted with a flow rate of 3 m*l*. Fractions of 1 m*l* were collected.

Results

Distribution and partial purification of cyclic AMP receptors

Cyclic AMP-binding activity was found in the supernatant as well as in the particulate fractions of *T. gambiense* homogenate as demonstrated in Table 1. The particle-bound cyclic AMP-binding activity could not be solubilized by treatment with 0.3% Triton X-100. About 60% of this particle bound activity could be solubilized by adding 0.5M NaCl to the homogenisation buffer. Therefore the *Trypanosoma* homogenate was extracted in 50mM potassium phosphate (pH 6.0) containing 0.5M NaCl. The homogenate was centrifuged for 90 min at 100 000 x *g*. The supernatant was fractionated by the addition of solid ammonium sulfate. The precipitate between 30 and 60% saturation was collected by centrifugation for 30 min at 10 000 x *g*, dissolved in 2 m*l* potassium phosphate buffer (20mM, pH 6.0) and directly applied to an Ultrogel AcA 34 column (2.5 x 35.0 cm), previously equilibrated with the same buffer. Elution was performed at a flow rate of 15 m*l*/h, and fractions of 4 m*l* were collected.

Table 1. Distribution of cyclic AMP binding activity in *T. gambiense*.

Binding capacities are expressed as pmol cyclic [^{3}H]-AMP bound. The standard assay was used. *Trypanosoma* cells were disrupted in 50mM potassium phosphate buffer (pH 6.0). The homogenate was centrifuged in a Sorvall HB4 rotor for 30 min at 3000 rpm. The supernatant was recentrifuged in a MSE 59587 rotor for 90 min at 35000 rpm.

Fraction	Protein [mg]	Activity [pmol]	Spec. act. [pmol/mg]
Homogenate	172	620	3.6
Pellet 3000 rpm	70	225	3.2
Pellet 35000 rpm	28	65	2.3
Supernatant	74	288	3.9

The fractionation by gel chromatography yielded an elution pattern which showed two peaks of cyclic AMP-binding activity. These cyclic AMP-binding activities were eluted in Fract. 14 - 17 and 26 - 30, respectively. The cyclic AMP receptors are referred to as receptor I and II in the order of elution from the column. Those fractions of the two peaks, containing the main activity, were pooled, concentrated by ultrafiltration (Diaflo PM10) and stored at – 76 °C until used. A summary of the partial purification is demonstrated in Table 2.

Properties of cyclic AMP receptors

To demonstrate the specificity of both receptors, they were incubated at 0 °C with 1 μM cyclic [^{3}H]AMP and alternatively with 1 μM [^{3}H]adenosine. It was impossible to bind adenosine instead of cyclic AMP under the conditions described, even when the pH of the binding assay was changed to pH 7.5. There was no change in the binding actitivities of both receptors if 10^{-3}M and 10^{-2}M $Mg^{2\oplus}$ or $Mn^{2\oplus}$ were added to the incubation mixture.

Fig. 1 and 2 show double reciprocal plots of cyclic AMP-binding to receptor I and II with binding constants of 0.4 μM and 0.6 μM, respectively. The specificity of the binding sites was studied by adding purine derivatives simultaneously with cyclic [^{3}H]AMP. As demonstrated in Tab. 3 the binding activities of both receptors are reduced by cyclic IMP and cyclic GMP, whereas adenosine 2′,3′-monophosphate, cyclic CMP, and cyclic UMP did not affect the binding. Adenine nucleoside and nucleotides were found to be potent inhibitors of cyclic [^{3}H]AMP-binding to receptor II, but they did not inhibit the binding to receptor I. The potency of these adenine de-

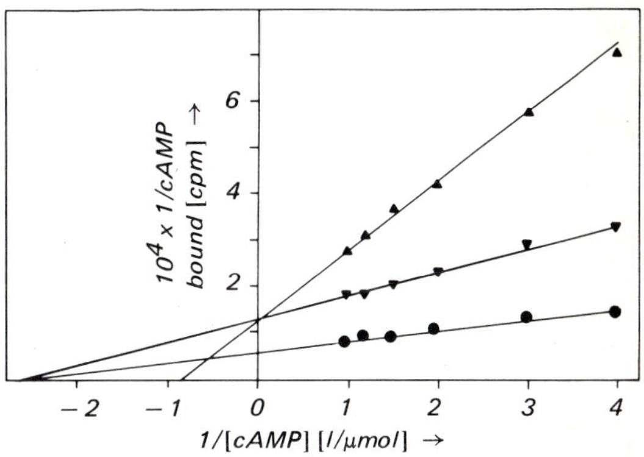

Fig. 1. Double reciprocal plots for the binding of cyclic [^{3}H]AMP to receptor I as a function of cyclic AMP in the absence (▼——▼) and in the presence of 50 μM cyclic GMP (▲——▲).

In addition receptor I preparation was preincubated at 0 °C for 30 min with 1 mg trypsin/m*l*, terminated by addition of trypsin inhibitor (●——●). The binding was measured by the standard assay.

Table 2. Partial purification of cyclic AMP receptors.

Fraction	Volume [m*l*]	Protein [mg]	Activity [pmol]	Spec. act. [pmol/mg]
Extract	14	68	315.0	4.6
Ammonium sulfate	2	44	304.0	6.9
Ultrogel (14 - 17)	16	7.8	118.5	15.2
Ultrogel (26 - 30)	20	6.5	76.0	11.7

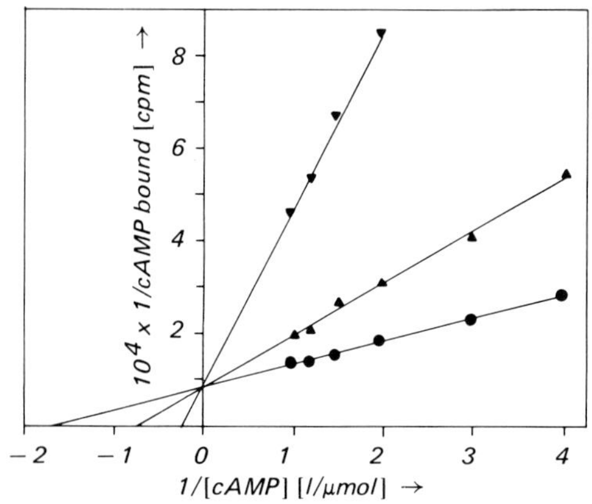

Fig. 2. Double reciprocal plots for the binding of cyclic [^{3}H]AMP to receptor II as a function of cyclic AMP in the absence (●——●) and in the presence of 50 μM cyclic GMP (▲——▲) or 50 μM ADP (▼——▼).

The standard assay was used.

Table 4. Effect of purine derivatives on cyclic AMP binding of receptor II.

Purine derivatives were added at a concentration of 50 μM. The standard assay with cyclic [^{3}H]AMP at a concentration of 1 μM was used. Binding capacity is expressed as pmol cyclic [^{3}H]AMP bound. Receptor II preparation was not contaminated by protein-kinase activity[16].

Addition	Binding activity	
	[pmol]	[%]
None	0.204	100
ATP	0.067	33
ADP	0.062	30
AMP	0.073	36
Adenosine	0.152	75
Adenine	0.172	84
Deoxyadenosine	0.177	87
Guanosine	0.205	100

rivatives towards the cyclic AMP-binding of receptor II is shown in Tab. 4. These inhibitors compete with cyclic AMP for the binding site of the receptor as demonstrated for cyclic GMP and ADP in Fig. 1 and 2.

To exclude the possibility that cyclic AMP-binding proteins are related to possible enzyme activities both receptors were checked for adenylate cyclase, cyclic-nucleotide phosphodiesterase and protein kinase activities. The receptors did not correlate with any of these enzyme activities.

Table 3. Effect of cyclic nucleotides on cyclic AMP binding.

The standard assay was used. Binding capacities are expressed as pmol cyclic [^{3}H]AMP bound. Cyclic nucleotides were added at a concentration of 50μM. The concentration of cyclic [^{3}H]AMP was 1 μM.

Addition	Receptor I		Receptor II	
	[pmol]	[%]	[pmol]	[%]
None	0.247	100	0.204	100
2',3'-cyclic AMP	0.246	100	0.205	100
Cyclic CMP	0.234	95	0.192	94
Cyclic UMP	0.231	94	0.197	97
Cyclic GMP	0.088	36	0.128	63
Cyclic IMP	0.094	38	0.124	61

The cyclic AMP-binding activities of both receptors were stable for weeks if stored at − 76 °C or − 30 °C. Heat treatment by preincubating the receptors for 5 min at 95 °C completely destroyed the binding activities of both receptors.

An outstanding result was the stimulatory effect of trypsin (1 mg/ml) on the cyclic AMP-binding of receptor I. The preincubation was performed at 0 °C for 5 or 30 min and was terminated by addition of trypsin inhibitor. As demonstrated in Fig. 2 the proteolytic modification of receptor I promoted a two-fold increase in the cyclic AMP-binding capacity. No change in the binding constant for cyclic AMP was found. In the case of receptor II about 15% of the binding capacity was lost after 30 min preincubation with trypsin at 0 °C. When receptor I was preincubated with trypsin for 5 or 30 min there was no difference in the increase of binding capacity. After 60 min preincubation the rate of stimulation gradually decreased. After 20 h of preincubation with trypsin about 50% of the binding capacity of receptor I as well as of receptor II was destroyed. If preincubation was performed with subtilisin (0 °C, 1 mg/ml) the binding capacity of both receptors was totally degraded after 20 h.

The cyclic AMP-binding activity of the particulate fraction (Table 1) was shown to be identical with receptor I. The cyclic AMP-binding activity of the particulate fraction was determined in the presence of ADP and trypsin. The addition of 0.5mM ADP did not inhibit the binding activity, whereas the preincubation with 1 mg trypsin/m*l* for 30 min increased the binding capacity of the particulate preparation about two-fold.

By gel filtration on Sephadex G-200 V_e/V_0 ratios were obtained indicating a molecular weight of about 150 000 for receptor II, whereas receptor I was eluted in the void volume.

Discussion

Cyclic AMP is reported to bind to regulatory subunits of protein kinases, resulting in dissociation of holoenzymes into catalytic and regulatory subunits. A second group of cyclic AMP-binding proteins with unknown function has been found in eukaryotic cells[1-9]. Previously the existence of protein kinases independent of cyclic nucleotides was demonstrated in *T. gambiense*. It was shown that these cyclic AMP-binding factors are not identical with regulatory subunits or protein kinases[17]. These cyclic AMP receptors did not inhibit the activities of protein kinases prepared from *T. gambiense*. Both receptors were found to be proteinaceous. They differed with respect to molecular weight, binding constant and effect of adenine derivatives on cyclic AMP-binding. The finding of two cyclic AMP-binding activities may reflect alterations as a result of solubilization procedures. The low molecular weight receptor was found to be soluble, whereas the high molecular weight receptor was strongly bound to the particulate fraction and could only be solubilized by high ionic concentration in the homogenisation buffer.

The affinity constants for cyclic AMP of both receptors were found to be in the range of 10^{-7}M. Thus they agree well with data reported on cyclic AMP-binding proteins from *Drosophila*[2] and various mammalian sources[4,7,8].

The binding capacity of receptor I was enhanced by preincubation with trypsin. This increase was independent of the affinity of receptor I to cyclic AMP. The proteolytic activation might be due to an unmasking of new binding sites for cyclic AMP. An increased capacity of cyclic AMP-binding was recently reported for mouse liver[7]. This increase of binding sites resulted from treatment of the receptor with $Mg^{2\oplus}$-ATP.

The competitive inhibition of cyclic AMP binding by adenine derivatives has been reported for *Drosophila*[2] and from mammalian sources[4-8]. In the case of *T. gambiense* neither receptor was able to bind [^{3}H]adenosine under the experimental conditions described. But it could be demonstrated that the binding of cyclic AMP to receptor II was competitively inhibited by adenosine and especially by AMP, ADP, and ATP.

The function of these two cyclic AMP-binding proteins has yet to be identified. Both receptors were found to be not identical with adenylate cyclase, cyclic-nucleotide phosphodiesterase or protein-kinase activities. It is concluded that these receptors might be involved in regulating the cyclic AMP concentration within *Trypanosoma* cells by arresting excessive amounts of cyclic AMP. Besides synthesis and degradation of cyclic AMP by adenylate cyclase and cyclic-nucleotide phosphodiesterase there might be another system regulating the cyclic AMP level. Cyclic AMP-binding proteins might be involved in this regulation by binding cyclic AMP and thereby determining the concentration of free, available cyclic AMP within the cell. Adenine derivatives are expected to interfere with the cyclic AMP metabolism. Recently the inhibitory effect of adenine derivatives on the activity of cyclic-nucleotide phosphodiesterase was reported from *T. gambiense*[18]. The present result on the inhibition of cyclic AMP binding to receptor II indicate that there might be another interaction of adenine derivatives and cyclic AMP metabolism.

The excellent technical assistance of Miss *H. Jahn* is gratefully acknowledged. This work was supported by the *Deutsche Forschungsgemeinschaft*.

Literature

1 Sy, J. & Richter, D. (1972) *Biochemistry* **11**, 2784 - 2787.

2 Tsuzuki, J. & Kiger, J. A. (1975) *Biochim. Biophys. Acta* **393**, 225 - 235.

3 Chambaut, A. M., Leray, F. & Hanoune, J. (1971) *FEBS Lett.* **15**, 328 - 334.

4 Yuh, K. M. & Tao, M. (1974) *Biochemistry* **13**, 5220 - 5226.

5 Døskeland, S. O. & Ueland, P. M. (1975) *Biochem. Biophys. Res. Commun.* **66**, 606 - 613.

ugden, P.H. & Corbin, J.D. (1976) *Biochem. J.* **159**, 423 - 437.
Ueland, P.M. & Døskeland, S.O. (1977) *J. Biol. Chem.* **252**, 677 - 686.
8 Døskeland, S.O. & Ueland, P.M. (1977) *Biochem. J.* **165**, 561 - 573.
9 Walter, R.D. (1977) Vth Intern. Congr. of Protozoology, New York, Abstracts p. 219.
10 Walter, R.D., Nordmeyer, J.P. & Königk, E. (1974) *Hoppe-Seyler's Z. Physiol. Chem.* **355**, 427 – 430.
11 Walter, R.D. (1974) *Hoppe-Seyler's Z. Physiol. Chem.* **355**, 1443 - 1450.
12 Walter, R.D. (1976) *Biochim. Biophys. Acta* **429**, 137 - 146.
13 Lowry, O.H., Rosebrough, N.J., Farr, A.L. & Randall, R.J. (1951) *J. Biol. Chem.* **193**, 265 - 275.
14 Gilman, A.G. (1970) *Proc. Natl. Acad. Sci. U.S.A.* **67**, 305 - 312.
15 Lanham, S.M. (1968) *Nature (London)* **218**, 1273 - 1274.
16 Whitaker, J.R. (1963) *Anal. Chem.* **35**, 1950 - 1953.
17 Walter, R.D. (1978) *Hoppe-Seyler's Z. Physiol. Chem.* **359**, 601 - 606, preceding paper.
18 Walter, R.D. (1975) *Hoppe-Seyler's Z. Physiol. Chem.* **356**, 43 - 45.

Dr. Rolf D. Walter, Bernhard-Nocht-Institut für Schiffs- und Tropenkrankheiten, Abt. Biochemie, Bernhard-Nocht-Straße 74, D-2000 Hamburg 4.

Hoppe-Seyler's Z. Physiol. Chem.
Bd. 359, S. 613 - 616, Mai 1978

Studies on the Biosynthesis of Cyclitols, XXXVI[1]

Purification of *myo*-Inositol-1-phosphate Synthase of the Duckweed, *Lemna gibba*, to Homogeneity by Affinity Chromatography on NAD-Sepharose Molecular and Catalytic Properties of the Enzyme

Ebenezer Olugbenga OGUNYEMI, Fritz PITTNER and Otto HOFFMANN-OSTENHOF

Institut für Allgemeine Biochemie der Universität Wien, and Ludwig-Boltzmann-Forschungsstelle für Biochemie, Wien

(Received 14 February 1978)

Summary: Using the technique of affinity chromatography on NAD-Sepharose the *myo*-inositol-1-phosphate synthase of *Lemna gibba* was purified to homogeneity. The molecular and catalytic properties of this enzyme differ very much from those of *myo*-inositol-1-phosphate synthase from animal sources. Thus the specific activity of the duckweed enzyme is more than two orders of magnitude higher than that of the enzyme from rat testes. It is inhibited by EDTA and can be reactivated by $Mn^{2\oplus}$. Its molecular weight (135 000 ± 5 000), its subunit composition (3 subunits with identical electrophoretic behaviour) and its isoelectric point (pH 7.7) are also very different from the corresponding parameters for the animal enzyme.

Untersuchungen über die Biosynthese der Cyclite, XXXVI
Reinigung der myo-Inosit-1-phosphat-Synthase aus der Wasserlinse Lemna gibba bis zur Homogenität mit Hilfe der Affinitätschromatographie auf NAD-Sepharose. Katalytische und molekulare Eigenschaften des Enzyms

Zusammenfassung: Mit Hilfe der Affinitätschro matographie an NAD-Sepharose war es möglich, aus der Wasserlinse die *myo*-Inosit-1-phosphat-Synthase in homogener Form zu erhalten. Die katalytischen und molekularen Eigenschaften dieses Enzyms unterscheiden sich sehr stark von denjenigen der *myo*-Inosit-1-phosphat-Synthase tierischer Herkunft. So zeigt das Enzym aus der Wasserlinse eine um mehr als zwei Größenordnungen höhere Spezifität als die *myo*-Inosit-1-phosphat-Synthase aus Rattenhoden; es wird durch EDTA gehemmt und kann durch $Mn^{2\oplus}$ reaktiviert werden. Sein Molekulargewicht (135 000 ± 5 000), seine Zusammensetzung aus Untereinheiten (3 Untereinheiten mit gleichem elektrophoretischen Verhalten) und sein isoelektrischer Punkt (pH 7.7) sind ebenfalls sehr verschieden von den entsprechenden Daten für das tierische Enzym.

Enzymes
Alcohol dehydrogenase, alcohol: $NAD^{\oplus}$ oxidoreductase (EC 1.1.1.1);
L-Amino-acid oxidase, L-amino-acid:oxygen oxidoreductase (deaminating) (EC 1.4.3.2);
Catalase, hydrogen-peroxide:hydrogen-peroxide oxidoreductase (EC 1.11.1.6);
Fumarate hydratase, L-malate hydro-lyase (EC 4.2.1.2);
Glutamate dehydrogenase ($NADP(P)^{\oplus}$), L-glutamate:$NADP(P)^{\oplus}$ oxidoreductase (deaminating) (EC 1.4.1.3);
myo-Inositol-1-phosphate synthase, 1L-*myo*-inositol-1-phosphate lyase (isomerizing) (EC 5.5.1.4);
Peroxidase, donor: hydrogen-peroxide oxidoreductase (EC 1.11.1.7).

Key words: 1L-*myo*-Inositol-1-phosphate, D-glucose 6-phosphate, *myo*-inositol-1-phosphate lyase (isomerizing), metalloenzyme.

myo-Inositol-1-phosphate synthase from the duckweed *Lemna gibba* was previously partly purified by means of affinity chromatography on NAD-Sepharose[2]. In the present communication its purification to homogeneity and some of its chemical and catalytic properties are described. The enzyme differs considerably from *myo*-inositol-1-phosphate synthase from animal sources[3]. In contrast to the usual situation that enzymes from higher plants are less active than animal enzymes, the enzyme from duckweed exhibits a considerably higher activity than all animal enzymes of the same specificity so far known. The duckweed enzyme can also be inhibited by metal chelating agents, whereas the animal enzymes of this specificity are not sensitive to such compounds.

Materials and Methods

NAD$^{\oplus}$, D-glucose 6-phosphate, serum albumin: Sigma, St. Louis; Lactate dehydrogenase, fumarate hydratase, glutamate dehydrogenase, catalase, L-amino-acid oxidase, cytochrome *c*, myoglobin, horse radish peroxidase, yeast alcohol dehydrogenase, liver catalase: Boehringer Mannheim GmbH; Sephadex G-200, dextran blue: Pharmacia, Uppsala; all other chemical reagents were purchased from E. Merck, Darmstadt.

NAD-Sepharose was prepared according to Mosbach et al.[4]. The enzyme activity at its optimum pH 6.4 was determined according to Barnett et al.[5]. Protein concentration was measured following the method of Bradford[6].

The purity of the enzyme was determined by electrophoresis on polyacrylamide gels containing 7.5% acrylamide. For the staining of the bands Coomassie Brilliant Blue B was used[7].

The apparent molecular weight of the enzyme protein was determined by the gel filtration method on Sephadex G-200[8]; various pure proteins with known molecular weight were used as markers.

The dissociation of the enzyme was achieved by treatment with sodium dodecylsulfate in the presence of mercaptoethanol[9]. The molecular weight of the enzyme subunits was determined with the help of standard marker proteins.

The isoelectric point was determined by electrofocusing with the help of LKB Ampholine Carrier Ampholyte (LKB, Stockholm).

Results

Purification of the enzyme

Lemna gibba, grown in a sterile medium according to Pierson and Seidl[10] under long day conditions (20 h light, 4 h darkness; the plant flowered under these conditions), was harvested on a Buchner funnel 2 weeks after planting, and then washed several times with tap water and finally with distilled water. Excess water was removed by sucking air through the funnel. About 50 g wet-weight of the plants were then homogenized in the cold room in a glass homogenizer in a buffer solution 0.1M with respect to Tris/acetate and 5mM with respect to mercaptoethanol (this buffer solution will hereafter be referred to as Tris/acetate buffer, pH 6.4).

The homogenate was centrifuged at 4 °C at 20000 rpm for 45 min in order to remove the undissolved solid material. The supernatant was progressively made 20%, 40%, 60%, 80% and 100% saturated, with respect to ammonium sulfate.

The precipitates obtained by this method were tested for protein concentration and for activity of *myo*-inositol-1-phosphate synthase. It was found that the highest *myo*-inositol-1-phosphate synthase activity was located in the fraction between 60 and 80% ammonium sulfate saturation, whereas the highest protein concentration was found in the fraction between 40 and 60%, where also the highest phosphatase activity could be observed.

The homogeneous *myo*-inositol-1-phosphate synthase was prepared by affinity chromatography on NAD-Sepharose.

About 5 m*l* of the NAD-Sepharose was put on a column of about 1 cm diameter and washed with 10 m*l* of 0.1M Tris/acetate buffer, pH 6.4. The fraction obtained with 60 - 90% ammonium sulfate saturation of the homogenate from *Lemna gibba* described before was put on the column and allowed to pass through it slowly, i.e. 4 drops/min, in order to equilibrate between the NAD$^{\oplus}$ on the column and the NAD$^{\oplus}$ attached to the enzyme.

The column was well washed with Tris/acetate buffer, pH 6.4, and then eluted slowly (2 - 3 drops/min) first with 10mM NAD$^{\oplus}$ solution in the same buffer and later – after the 10th fraction – with an analogous 20mM NAD$^{\oplus}$ solution. Eighteen fractions of 3 m*l* each were collected. All fractions were twice dialyzed against 0.1M Tris/acetate buffer, pH 6.4.

All 18 samples were tested for *myo*-inositol-1-phosphate synthase activity. Those six fractions where the activity was highest were subjected to electrophoresis on 7.5% polyacrylamide gel, to test the purity of the enzyme protein. In all cases only one band was observed, which indicates that the enzyme protein is homogeneous.

Properties of the enzyme: The *apparent molecular weight* of the enzyme was determined by the gel filtration method on Sephadex G-200 and was found to be 135 000 ± 5 000.

The enzyme, dissociated by treatment with sodium dodecylsulfate in the presence of mercaptoethanol[9], yielded only one band on electrophoresis on polyacrylamide gel. By comparison with protein markers with known molecular weight the molecular weight of the *subunits* was calculated as 45 000 ± 2 000. This means that the enzyme consists of 3 subunits, which behave identically on electrophoresis on polyacrylamide, but does, however, not necessarily mean that they are in fact identical.

By the above described method the *isoelectric point* of the enzyme was determined to be pH 7.7.

The *specific activity* of the enzyme was found to be 842 mkat/kg protein; the corresponding value for the enzyme of rat testes with the same specificity is 1.62 mkat/kg protein[3].

The *Michaelis constant* for the substrate was found to be 35 μM. A determination of the Michaelis constant for $NAD^{\oplus}$ was not possible, because removal of $NAD^{\oplus}$ from the enzyme protein by means of norite, a method successfully applied to the enzyme of the same specificity from rat testes, apparently denatures the enzyme protein.

Inhibition by EDTA: 1mM EDTA inhibits by about 30%, whereas 10mM EDTA inhibits completely. This inhibition can be reversed by addition of a 5mM solution of $MnCl_2 \cdot 4\,H_2O$; addition of the same molar quantity of $MgCl_2 \cdot 6\,H_2O$ does not reverse the inhibition caused by EDTA. $Mn^{2\oplus}$ also shows a considerable activating action (up to 36%) on solutions of the homogeneous enzyme.

Discussion

The key enzyme in the biosynthesis of *myo*-inositol, the other inositols, and the derivatives of *myo*-inositol like phytic acid and the inositol phosphatides, the *myo*-inositol-1-phosphate synthase, seems to exist in two different forms. All animal enzymes of this specificity so far investigated seem to act by means of a Schiff base mechanism[1,11] (cf. however the paper of Sherman et al.[12]). The *myo*-inositol-1-phosphate synthases from higher plants and microbes are obviously metalloproteins and differ considerably in all properties from the animal enzymes. This situation shows a certain analogy to that of the aldolases.

In the present paper we describe the *myo*-inositol-1-phosphate synthase from *Lemna gibba*, which to our knowledge is the first enzyme of the metalloprotein group of these enzymes obtained in a homogeneous state.

The differences between this enzyme and that from animal sources are really striking, although there is hardly any doubt that in both cases the general mechanism is the same. This means that $NAD^{\oplus}$ bound to the enzyme proteins is used for the dehydrogenation of the hydroxyl group at the 5-position in the substrate to a keto group, thus inducing ring closure. The formed NADH is then used for reduction of the cyclic keto compound to 1L-*myo*-inositol 1-phosphate. The detailed mechanisms of the plant and animal enzyme might be, however, very different.

Perhaps the most surprising difference is that the enzyme from duckweed shows a specific activity about 500 times higher than those of the animal enzymes so far investigated. The affinity of the plant enzyme to the substrate is much higher than that of the animal enzyme, as shown by the Michaelis constants.

The molecular properties of the two kinds of enzymes also differ considerably. Whereas the animal enzyme has a molecular weight around 215 000 and consists of four subunits (two different pairs in the case of the enzyme from rat testes), the plant enzyme is somewhat smaller having a molecular weight of about 135 000, and consists of three subunits with identical electrophoretic behaviour.

Whereas the animal enzymes are not inhibited by EDTA, our plant enzyme is, and this inhibition can be reversed by $Mn^{2\oplus}$. This makes it probable that manganese is the metal part of this metalloprotein.

It may be added that *myo*-inositol-1-phosphate synthases from other plant and microbial sources, which have been studied in this laboratory, also show considerable similarities to the enzyme from *Lemna gibba*.

This investigation was supported by grants of the *Fonds zur Förderung der wissenschaftlichen Forschung in Österreich*, projects 2639 and 3389, and of the *Hochschuljubiläumsstiftung der Stadt Wien*.

Literature

1 XXXVth Commun.: Pittner, F. & Hoffmann-Ostenhof, O. (1976) *Hoppe-Seyler's Z. Physiol. Chem.* **357**, 1667 - 1671.

2 Schwarcz, R., Fried, W., Pittner, F. & Hoffmann-Ostenhof, O. (1974) *Monatsh. Chem.* **105**, 445 - 451.

3 Pittner, F. & Hoffmann-Ostenhof, O. (1976) *Monatsh. Chem.* **107**, 793 - 797.

4 Mosbach, K., Guilford, H., Ohlsson, R. & Scott, M. (1972) *Biochem. J.* **127**, 625 - 631.

5 Barnett, J. E. G., Brice, R. E. & Corina, D. L. (1970) *Biochem. J.* **119**, 183 - 186.

6 Bradford, M. M. (1976) *Anal. Biochem.* **72**, 248 - 254.

7 Maurer, H.R. (1968) Diskelektrophorese, W. de Gruyter, Berlin

8 Andrews, P. (1965) *Biochem. J.* **96**, 595 - 606.

9 Weber, K., Pringle, J. R. & Osborn, M. (1973) *Methods Enzymol.* **26c**, 3 - 27.

10 Pirson, A. & Seidel, F. (1950) *Planta* **38**, 431 - 473.

11 Naccarato, W. F., Ray, R. E. & Wells, W. W. (1974) *Arch. Biochem. Biophys.* **164**, 194 - 201.

12 Sherman, W. R., Rasheed, A., Mauck, L. A. & Wiecko, J. (1977) *J. Biol. Chem.* **262**, 5672 - 5676.

E. O. Ogunyemi, Dr. F. Pittner and Prof. Dr. Hoffmann-Ostenhof, Institut für Allgemeine Biochemie der Universität Wien, and Ludwig-Boltzmann-Forschungsstelle für Biochemie, Währingerstr. 38, A-1090 Wien.

Hoppe-Seyler's Z. Physiol. Chem.
Bd. 359, S. 617 - 628, Mai 1978

Präparative Isolierung von Di-, Tri- und Tetrapyrimidinnucleotiden aus depurinierter Heringsspermen-DNA

Herbert SCHOTT und Margarete SCHWARZ

Institut für Organische Chemie der Universität Tübingen

(Der Schriftleitung zugegangen am 15. März 1978)

Zusammenfassung: Die Pyrimidinnucleotide $p(dT)_3$, $p(dT)_2$ p, pdCp, $p(dC)_2$ p und $p(dC)_3$ sowie die Mischungen der Sequenzisomeren p(dC, dT), p(dC, dT) p, $p(dT_2, dU)$, (dT_2, dU) p, $p(dC, dT_2)$, $p(dC, dT_2)$ p, $p(dC, dT_3)$, $p(dC_2, dT)$, $p(dC_2,dT)$p und $p(dC_2,dT_2)$ werden aus depurinierten Hydrolysaten der Heringsspermen-DNA nach folgendem Trennungsgang im präparativen Maßstab isoliert. Das DNA-Hydrolysat wird zunächst an DEAE-Cellulose säulenchromatographisch in eine nieder- und eine hochmolekulare Pyrimidinnucleotidmischung vorgetrennt. Aus der Mischung der niedermolekularen Pyrimidinnucleotide werden die Nucleosidbisphosphate pdCp, pdTp, die Dinucleotide $p(dC)_2$, $p(dT)_2$ und die Sequenzisomeren p(dC, dT) weitgehend abgetrennt. Das verbleibende Gemisch wird bei pH 3.5 an QAE-Sephadex rechromatographiert. Hierbei werden Pyrimidinnucleotide mit überwiegend Cytidyl- von denen mit überwiegend Thymidylsäurebausteinen getrennt, die anschließend bei pH 7.5 nach der Anzahl ihrer Kettenglieder fraktioniert werden. Die säulenchromatographisch, in 83 - 99proz. Reinheit isolierten Pyrimidinnucleotide und Mischungen der Sequenzisomeren werden papierchromatographisch nachgetrennt und hierbei in chromatographisch reiner Form erhalten. Die Struktur der isolierten DNA-Fragmente wird aus chromatographischen Daten, Absorptionsverhältnissen und durch enzymatischen Abbau ermittelt.

Preparative Isolation of Di-, Tri- und Tetrapyrimidine Nucleotides from Depurinated Herring Sperm DNA

Summary: The pyrimidine nucleotides $p(dT)_3$, $p(dT)_2$ p, pdCp, $p(dC)_2$ p and $p(dC)_3$ and the mixtures of sequence isomers p(dC, dT), p(dC, dT) p, $p(dT_2, dU)$, (dT_2, dU) p, $p(dC, dT_2)$, $p(dC,dT_2)$ p, $p(dC,dT_3)$, p(dC,dT), $p(dC_2,dT)$ p, and $p(dC_2, dT_2)$ are isolated on a preparative scale from depurinated hydrolysates of herring sperm DNA by the following procedure. The DNA hydrolysate is first separated into a low and a high molecular weight mixture of pyrimidine nucleotides by column chromatography on DEAE-cellulose. The nucleoside bisphosphates pdCp and pdTp, the dinucleotides $p(dC)_2$ and $p(dT)_2$ and the sequence isomers p(dC, dT) are largely separated out of the mixture of the low molecular weight pyrimidine nucleotides. The

---

Enzyme:
Alkalische Phosphatase, Orthophosphorsäure-Monoester-Phosphohydrolase (alkalisches Optimum) (EC 3.1.3.1); Phosphodiesterase I, Oligonucleat-5-Nucleotidohydrolase (EC 3.1.4.1).

Abkürzung:
1 A_{260}-Einheit ist die Menge bei 260 nm absorbierenden Materials, die, gelöst in 1 m*l*, eine Extinktion von 1 hat.

remaining mixture is rechromatographed at pH 3.5 on QAE-Sephadex. This separates the pyrimidine nucleotides containing a majority of cytidylic acid units from those containing more thymidylic acid units, which are then fractionated at pH 7.5 according to the number of bases in the chain. The pyrimidine nucleotides and mixtures of sequence isomers separated to 83 - 99% purity by column chromatography are further separated by paper chromatography and are obtained in chromatographically pure form after this step. The structures of the isolated DNA fragments are determined from chromatographic data, absorption and enzymatic degradation.

Key words: DNA, DNA hydrolysates, ion exchange chromatography pyrimidine nucleotides.

In anfänglichen Versuchen zur Sequenzierung von Desoxyribonucleinsäuren (DNA) wurde DNA bspw. mit Ameisensäure in Gegenwart von Diphenylamin chemisch hydrolysiert. Unter diesen Bedingungen werden alle Purinnucleobasen der DNA zerstört, während die Pyrimidinnucleotidsegmente intakt bleiben und aus der depurinierten DNA freigesetzt werden. Aus dem komplex zusammengesetzten DNA-Hydrolysat können verschiedene Pyrimidinnucleotidfragmente isoliert und identifiziert werden[1-5].

Wir haben die ursprünglich für analytische Ansätze konzipierte chemische DNA-Hydrolyse in den präparativen Maßstab übertragen und einen Trennungsgang gefunden, nach dem die Pyrimidinnucleotide $p(dC)_2$, $p(dT)_2$, pdCp, pdTp, p(dC, dT) chromatographisch rein im präparativen Maßstab aus dem DNA-Hydrolysat erhalten werden[6]. Hierbei gehen wir so vor, daß käufliche DNA aus Heringsspermen im 100-g-Ansatz zunächst chemisch hydrolysiert wird. Das Hydrolysat wird durch Ultrafiltration entsalzt, lyophilisiert und anschließend im 10-g-Ansatz an DEAE-Cellulose säulenchromatographisch im zweistufigen NaCl-Gradienten in eine nieder- und eine hochmolekulare Pyrimidinnucleotidmischung vorgetrennt. Die Mischung der niedermolekularen Pyrimidinnucleotide wird an QAE-Sephadex wiederum im 10-g-Ansatz im dreistufigen NaCl-Gradienten rechromatographiert und hierbei in drei Fraktionen aufgetrennt. Aus den beiden ersten Fraktionen, die ca. 50 - 60% der niedermolekularen Pyrimidinnucleotide enthalten, werden die oben aufgeführten Nucleotide isoliert. Aus der 3. Fraktion, in der die restlichen 40% der niedermolekularen Pyrimidinnucleotide enthalten sind, werden nach dem in der vorliegenden Arbeit beschriebenen Trennungsgang längerkettige Pyrimidinnucleotide und definierte Mischungen von Sequenzisomeren im präparativen Maßstab chromatographisch rein isoliert.

Beschreibung der Versuche

Reagenzien

Chemikalien werden in „chemisch reiner" Form verwendet. QAE-Sephadex A-25 (Pharmacia, Uppsala, Schweden); Membranen (Amicon, Lexington, USA); Chromatographiepapier 2316 (Schleicher & Schüll, Dassel); Enzyme (Boehringer Mannheim GmbH); Nucleoside dT, dC; Nucleotide pdT, pdC; DNA (Waldhof, Mannheim); die übrigen Referenznucleotide pdC-dT; dC-dT; $(dC)_3$; $(dC)_2$; $(dT)_2$; dC_2-dT wurden in unserem Labor synthetisiert.

Pufferlösungen

a) 1M Natriumacetat-Puffer, pH 3.5:
57 m*l* (1 mol) Eisessig werden zu 800 m*l* Wasser gegeben. Der pH-Wert wird mit 40 m*l* 2M NaOH auf pH 3.5 eingestellt und die Lösung mit Wasser auf 1000 m*l* aufgefüllt.
b) 1M Tris/HCl-Puffer, pH 7.5:
121 g (1 mol) Tris werden in 900 m*l* H_2O gelöst, mit 50 m*l* konz. HCl versetzt, dann mit HCl auf pH 7.5 eingestellt und schließlich mit H_2O auf 1 *l* aufgefüllt.

Laufmittelsysteme

A: Äthanol/1M NH_4Ac (pH 7.5) 7:3; (v/v);
B: 1-Propanol/konz. NH_4OH/H_2O 55:10:35; (v/v/v)

Im folgenden Teil wird ausschließlich die Fraktionierung der niedermolekularen Pyrimidinnucleotide beschrieben, die in Peak III (Fig. 1) der vorangegangenen Arbeit[6], in der sogenannten „3. Fraktion" eluiert wurden.

1) Säulenchromatographische Fraktionierung an QAE-Sephadex A-25 bei pH 3.5 (vgl. Fig. 1)

5 g lyophilisierter Pyrimidinnucleotide der „3. Fraktion" werden in 20 m*l* 25mM NaCl, 25mM NaAc (pH 3.5) gelöst und auf eine mit 50mM NaCl, 50mM NaAc äquilibrierte QAE-Sephadex-A-25-Säule aufgetragen. Die Säulenfüllung beträgt 70 x 3 cm. Die anschließende Elution (vgl. Fig. 1) der Säule erfolgt bei 20 °C mit einer Durchflußgeschwindigkeit von ca. 400 m*l*/h im dreistufigen NaCl-Gradienten, der mit 50mM NaAc auf pH 3.5 ge-

puffert ist. Fraktionen zu ca. 20 m*l* werden gesammelt. Die Absorption jeder 10. Fraktion wird bei dieser und den folgenden Trennungen spektralphotometrisch bei 260 nm gemessen und graphisch gegen das Elutionsvolumen aufgetragen (vgl. Fig. 1). Nach dem Auftragen der Probelösung wird die Säule in der 1. Stufe mit 4 *l* 50mM NaCl, 50mM NaAc (pH 3.5) eluiert. In der 2. Stufe folgt ein linear steigender Gradient: 5 *l* 50mM NaCl, 50mM NaAc im Mischgefäß; 5 *l* 100mM NaCl, 50mM NaAc im Vorratsgefäß. In der 3. Stufe wird die Säule nochmals mit einem linear steigenden NaCl-Gradienten eluiert: 4 *l* 100mM NaCl, 50mM NaAc im Mischgefäß, 4 *l* 200mM NaCl, 50mM NaAc im Vorratsgefäß. Zum Schluß werden die noch adsorbierten Substanzen mit ca 1 *l* 1M NaCl von der QAE-Sephadex-Säule eluiert. Die Peakfraktionen innerhalb der senkrechten Strichelung werden vereinigt, mit NaOH neutralisiert und am Rotationsverdampfer soweit konzentriert, bis Salz ausfällt. Vom Salz wird abgesaugt und mit Eiswasser nachgewaschen. Die mit dem Waschwasser vereinigte gelbe Lösung wird durch Ultrafiltration an einer UM-2-Membran bei ca. 500 - 600 kPa* entsalzt und anschließend lyophilisiert. Die Fraktionen zwischen den Peaks und die mit 1M NaCl eluierten Fraktionen werden verworfen.

2) Rechromatographie der Pyrimidinnucleotide aus Peak I - V an QAE-Sephadex A-25 bei pH 7.5 (vgl. Fig. 2)

In den folgenden 5 Trennungen werden jeweils 1 g der lyophilisierten Peakfraktionen aus der 1. Trennung (vgl. Fig. 1) in 10 m*l* 60mM NaCl, 25mM Tris/HCl gelöst und auf eine mit 120mM NaCl, 50mM Tris/HCl äquilibrierte Säule aufgetragen. Die Säulenfüllung beträgt 45 x 2 cm. Die Säule wird bei 20 °C mit einer Durchflußgeschwindigkeit von ca. 600 m*l*/h, die mit einer Schlauchpumpe eingestellt wird, im steigenden NaCl-Gradienten, der mit 50mM Tris/HCl auf pH 7.5 gepuffert ist, eluiert.

Die Gradienten (vgl. Fig. 2) weisen folgende molaren NaCl Konzentrationen auf:

Fig.	1. Stufe	2. Stufe	
		Mischgefäß	Vorratsgefäß
2a	4 *l* 0.12	2 *l* 0.12	2 *l* 0.15
2b	5 *l* 0.12	2 *l* 0.12	2 *l* 0.15
2c	3 *l* 0.12	1) 2 *l* 0.12	2 *l* 0.15
		2) 2 *l* 0.15	2 *l* 0.19
2d	4 *l* 0.12	1) 2 *l* 0.12	2 *l* 0.15
		2) 2 *l* 0.15	2 *l* 0.19
		3) 1 *l* 0.19	
2e	4 *l* 0.12	1) 2 *l* 0.12	2 *l* 0.15
		2) 2 *l* 0.15	2 *l* 0.19
		3) 1 *l* 0.19	

* = 4 - 5 atü.

Nach Durchlauf der Gradienten werden die noch adsorbierten Substanzen jeweils mit ca. 500 m*l* 1M NaCl von der QAE-Sephadex-Säule eluiert. Die Isolierung der in den Peakfraktionen enthaltenen Pyrimidinnucleotide erfolgt nach dem unter 1) beschriebenen Weg.

Ergebnisse

Das niedermolekulare Pyrimidinnucleotidgemisch, das in Peak III, Fig. 1 der vorangegangenen Arbeit[6] enthalten ist und im folgenden als „3. Fraktion" bezeichnet wird, läßt sich wie folgt auftrennen. Zur Isolierung chromatographisch reiner Pyrimidinnucleotide aus dem sehr komplex zusammengesetzten Lyophilisat der „3. Fraktion" nutzen wir wiederum die pH-abhängigen Unterschiede in der Gesamtladung, die für Pyrimidinnucleotide unterschiedlicher Sequenz berechnet ist[7]. Aufgrund der Ladungsunterschiede läßt sich die Reihenfolge, in der Pyrimidinnucleotide einen Anionenaustauscher passieren, etwa vorhersagen. So werden bspw. im Sauren Pyrimidinnucleotide mit überwiegend Cytidyl- vor denen mit überwiegend Thymidylsäureeinheiten eluiert. Im Alkalischen werden Pyrimidinnucleotide dagegen nicht nach der Sequenz, sondern nach der Anzahl ihrer Kettenglieder fraktioniert.

Etwa 5 g des Lyophilisats der „3. Fraktion" werden zunächst bei pH 3.5 an QAE-Sephadex A-25 säulenchromatographisch im dreistufigen NaCl-Gradienten fraktioniert. Das Elutionsprofil ist in Fig. 1 abgebildet. Es weist 6 deutlich voneinander getrennte Peaks auf. Peak I - V werden nacheinander bei steigender NaCl-Konzentration eluiert, während der VI. Peak mit 1M NaCl die Säule verläßt. Mit 50mM NaCl verlassen in Peak I nur 2.7% des aufgetragenen Lyophilisats die Säule (Tab. 1). In der 2. Gradientenstufe, in der die NaCl-Konzentration des Elutionspuffers von 50 - 100mM ansteigt, werden mit 60 - 75mM NaCl in Peak II 12.2% und mit 90 - 100mM NaCl in Peak III 29.2% der Ausgangsmischung eluiert. Im Verlauf der 3. und letzten Gradientenstufe, die von 100 - 200mM NaCl steigt, verlassen in Peak IV mit 120 - 130mM NaCl 3.2% und in Peak V mit 155 - 190mM NaCl 31.9% der aufgetragenen Pyrimidinnucleotidmischung die Säule. Nach dem dreistufigen Gradienten werden die restlichen 10% der Mischung mit 1M NaCl-Lösung eluiert. Die Fraktionen des 1M Peaks sowie die Zwischen-

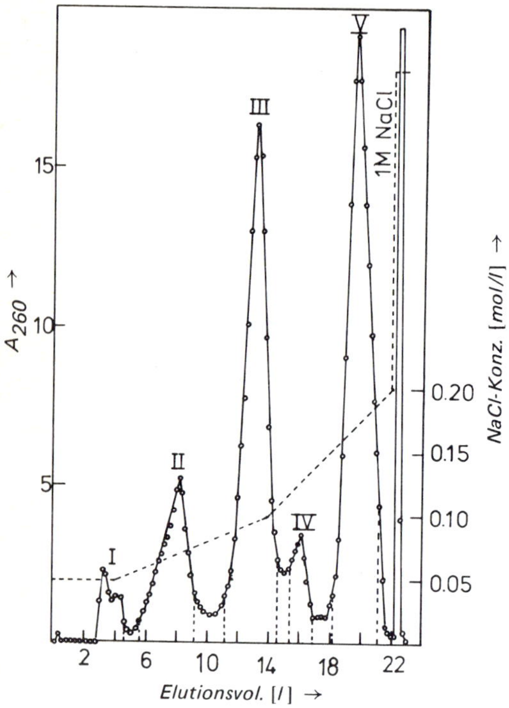

Fig. 1. Elutionsprofil der säulenchromatographischen Fraktionierung von 5 g niedermolekularer Pyrimidinnucleotide an QAE-Sephadex A-25.

Experimentelle Einzelheiten s. Beschreibung der Versuche. Die Peakfraktionen innerhalb der senkrechten Strichelung werden vereinigt, entsalzt und rechromatographiert (vgl. Fig. 2).

fraktionen, die zusammen 20.9% der Ausgangsmischung enthalten, werden verworfen.

Zur Anreicherung der in Peak I und IV eluierten Pyrimidinnucleotide, die in der „3. Fraktion" als Nebenprodukte enthalten sind, wird die Trennung zehnmal wiederholt, so daß insgesamt 50 g der „3. Fraktion" fraktioniert werden. Bei allen 10 Trennungen erhält man im wesentlichen das gleiche, in Fig. 1 abgebildete Elutionsprofil, wobei aber Peak IV von Trennung zu Trennung unterschiedlich gut von Peak V getrennt eluiert wird. Die Folge ist, wie wir später sehen werden, daß das Hauptprodukt aus Peak V, $p(dT)_2$ p, als Nebenprodukt in Peak IV gefunden wird. Die Ergebnisse des 1. Trennschritts sind in Tab. 1 zusammengefaßt.

Aus den Absorptionsverhältnissen 250/260 und 280/260 nm der Peakfraktionen folgt, daß bei allen Trennungen, wie erwartet, in Peak I und II Pyrimidinnucleotide mit vorwiegend Cytidylsäureresten eluiert werden, während die Pyrimidinnucleotide in Peak III - V überwiegend aus Thymidylsäureeinheiten aufgebaut sind. In sich entsprechenden Peaks werden grundsätzlich die gleichen Pyrimidinnucleotide eluiert, so daß wir die Lyophilisate von Peaks gleicher Nummer aus allen Trennungen vereinigen.

Im folgenden Trennschritt werden die vereinigten Lyophilisate von Peak I - V der Fig. 1 an QAE-Sephadex A-25 im NaCl-Gradienten bei pH 7.5 rechromatographiert. Hierbei werden die Komponenten der in den verschiedenen Peaks eluierten Pyrimidinnucleotidgemische nach der Anzahl ihrer Ladungen fraktioniert. Bei der Rechromatographie (vgl. Tab. 2) werden ca. 70 - 90% einer Mischung im steigenden NaCl-Gradienten eluiert, während etwa 5 - 10% der aufgetragenen Mischung erst in 1 M NaCl die Säule verlassen. Die Elutionsprofile der fünf Fraktionen sind in Fig. 2 abgebildet, ihre weitere Charakterisierung zeigen Tab. 3 - 5. Die Zwischenfraktionen und die mit 1 M NaCl eluierten Fraktionen werden nicht näher untersucht.

Die Fraktionen innerhalb der senkrechten Strichelung werden vereinigt, durch Ultrafiltration entsalzt und lyophilisiert. Bei Ultrafiltration treten, bedingt durch die unterschiedliche Güte der Membran, schwankende Verluste auf, so daß nur etwa 80 - 90% der in den Peaks eluierten Pyrimidinnucleotide isoliert werden. Die lyophilisierten Peakfraktionen werden zunächst papierchromatographisch untersucht und mit Hilfe von Vergleichssubstanzen charakterisiert. Hierbei zeigt sich, daß jeder Peak nur ein Hauptprodukt enthält, das im Papierchromatogramm einen Hauptfleck bildet, in dem ca. 83 - 99% des aufgetragenen Lyophilisats wandern. Die restlichen 1 - 17% verteilen sich meistens auf 1 - 2 schwache Nebenflecken. Alle im Papierchromatogramm auftretenden Flecken werden ausgeschnitten und eluiert. Im Eluat der einzelnen Flecken bestimmt man spektralphotometrisch die A_{260}-Einheiten und errechnet daraus die prozentuale Zusammensetzung des aufgetragenen Lyophilisats. Die Eluate der Hauptflecken werden konzentriert und in einem zweiten Laufmittelsystem rechromatogra-

Tab. 1. Ergebnisse der säulenchromatographischen Fraktionierung (vgl. Fig. 1) von 5 g (93900 A_{260}-Einheiten) eines niedermolekularen Pyrimidinnucleotidgemisches aus depurinierten DNA-Hydrolysaten an QAE-Sephadex A-25. Alle Prozentangaben sind auf die A_{260}-Einheiten der aufgetragenen Mischung bezogen und aus 10 Trennungen gemittelt.

Peak	NaCl-Konz. [mM]	Pyrimidinnucleotidgemische: Bezeichnung	Eluierte Mengen [A_{260}-Einh.]	Eluierte Mengen [%]	Isolierte Mengen [g]	Isolierte Mengen [%]	Isolierte Mengen bez. auf DNA [%]
I	50	p(dC, dT), pdCp, $p(dC)_3$	2500	2.7	0.13	2.6	0.21
II	60 - 75	$p(dC_2, dT)$, $p(dC)_2p$	11400	12.2	0.60	12.0	0.96
III	90 - 100	$p(dC, dT_2)$, p(dC, dT) p	27500	29.2	1.40	28.0	2.24
IV	120 - 130	$(dT_2,dU)p$, $p(dT_2,dU)$, $p(dT)_2$ p, $p(dC_2,dT_2)$, $p(dC_2, dT)$ p	3000	3.2	0.14	2.8	0.22
V	155 - 190	$p(dT)_3$, $p(dT)_2p$, $p(dC, dT_3)$,[a] $p(dC, dT_2)$ p	29900	31.9	1.40	28.0	2.24
Zwischenfraktionen + 1M-NaCl-Peak			19600	20.9	–	–	–

[a] Enthält vermutlich ca. 20% $p(dT)_4$.

Tab. 2. Ergebnisse der säulenchromatographischen Nachtrennung von Pyrimidinnucleotidgemischen (vgl. Fig. 2), die in den Peakfraktionen I-V (vgl. Fig. 1 und Tab. 1) eluiert werden.

Aufgetragene Pyrimidinnucleotidmischung: Aus Peak	Menge [A_{260}-Einh.]	Getrennte Pyrimidinnucleotide: In Peak	Bei NaCl-Konz. [M]	Bezeichnung	Reinheitsgrad [%][a]	Eluierte Mengen [A_{260}-Einh.]	Eluierte Mengen [%]	Isolierte Mengen [mg]	Isolierte Mengen [%]
I	15600	Ia	0.12	p(dC, dT)	83	1800	11.5	110	11
		Ib	0.12	pdCp	91	5800	37.2	360	36
		Ic	0.12-0.13	$p(dC)_3$	92	2800	18.0	150	15
II	17000	IIa	0.12	$p(dC_2, dT)$	98	6400	37.6	350	35
		IIb	0.12-0.15	$p(dC)_2p$	96	6800	40.0	390	39
III	19500	IIIa	0.12	$p(dC, dT_2)$	94	3000	15.4	140	14
		IIIb	0.12-0.15	p(dC, dT) p	96	14700	75.4	680	68
IV	19900	IVa	0.12	(dT_2, dU) p, $p(dT_2, dU)$	83	2300	11.6	100	10
		IVb	0.12-0.13	$p(dT)_2p$	84	2400	12.0	110	11
		IVc	0.14-0.16	$p(dC_2, dT_2)$	95	4300	21.6	210	21
		IVd	0.17-0.19	$p(dC_2, dT)$ p	95	8000	40.2	400	40
V	19200	Va	0.12	$p(dT)_3$	90	1700	8.9	70	7
		Vb	0.12-0.14	$p(dT_2)$ p	99	10800	56.3	540	54
		Vc	0.15-0.17	$p(dC, dT_3)$[b]	94	1300	6.8	60	6
		Vd	0.17-0.19	$p(dC, dT_2)$ p	98	4000	20.8	190	19

[a] Aus den A_{260}-Einheiten der papierchromatographisch getrennten Substanzen ermittelt.
[b] Enthält ca. 20% pdT_4.

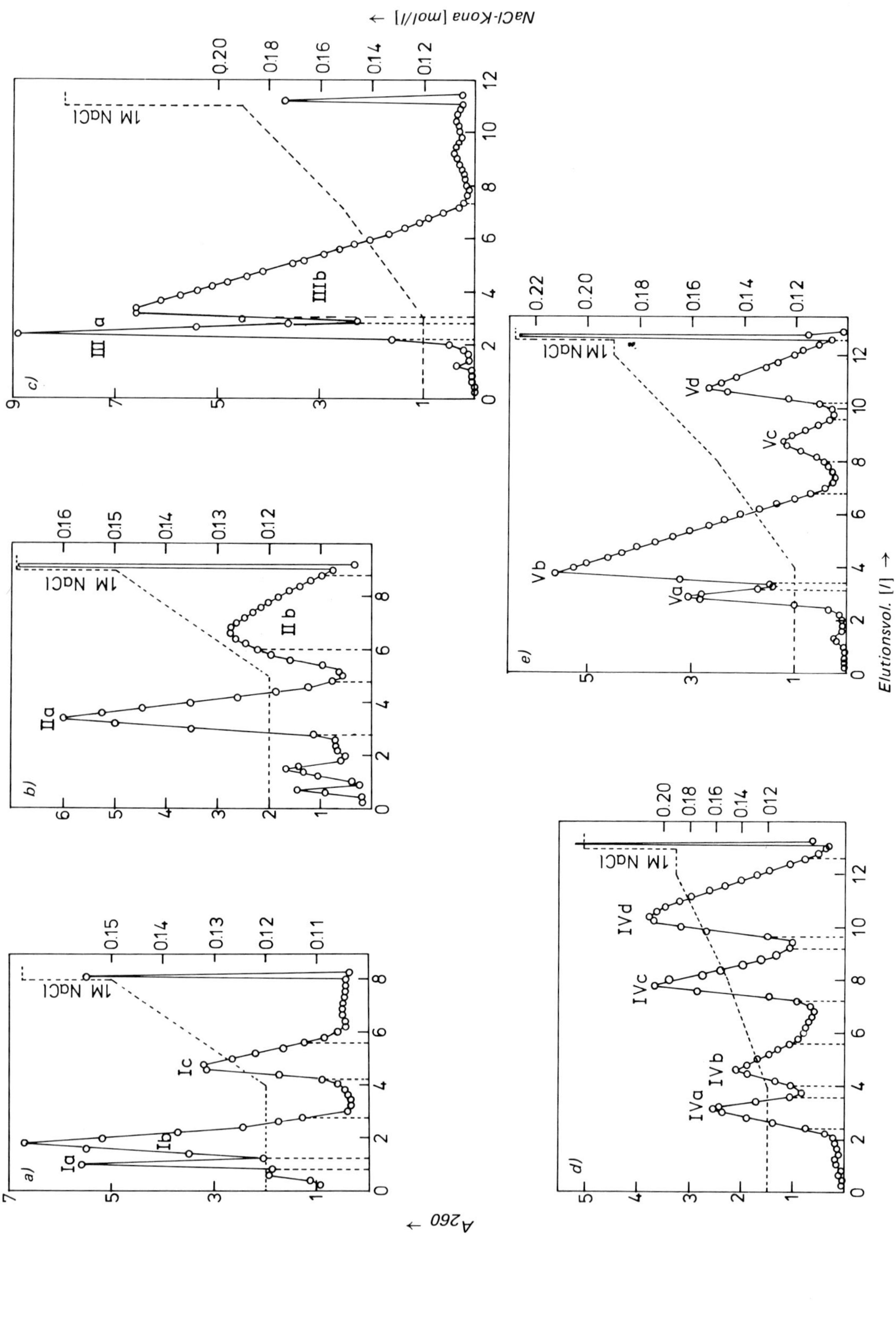
NaCl-Konz [mol/l] →
A260 →
Elutionsvol. [l] →
a)
b)
c)
d)
e)
1M NaCl
Ia
Ib
Ic
IIa
IIb
III
a
IIIb
IVa
IVb
IVc
IVd
Va
Vb
Vc
Vd

◁

Fig. 2. Elutionsprofil der Rechromatographie von jeweils 1 g der Pyrimidinnucleotide aus Peak I - V der Fig. 1 an QAE-Sephadex A-25.

Experimentelle Einzelheiten s. Beschreibung der Versuche. Die Peakfraktionen innerhalb der senkrechten Strichelung werden vereinigt, entsalzt und lyophilisiert.

phiert. Wandert das DNA-Fragment hierbei wiederum als einheitlicher Fleck, so wird die Substanz überlicherweise mit „chromatographisch rein" bezeichnet. Da auf einem Papierbogen (60 x 58 cm) bis zu 600 - 800 A_{260}-Einheiten chromatographiert werden können, sind auf diesem Weg praktisch alle Komponenten eines Lyophilisats in präparativen Mengen chromatographisch rein zu isolieren.

Die Struktur der isolierten, chromatographisch reinen Nucleotide wird aus R_F-Werten, Absorptionsverhältnissen und den Ergebnissen des enzymatischen Abbaus ermittelt. Auf diesem Weg, der im folgenden näher beschrieben wird, läßt

Tab. 3. R_F-Werte der aus depurinierten DNA-Hydrolysaten isolierten Pyrimidinnucleotide vor und nach dem Abbau mit alkalischer Phosphatase.

Absteigende Papierchromatographie, Laufmittel s. Beschreibung der Versuche.

Pyrimidin-nucleotide (Bezeichnung)	R_F-Werte relativ zu denen von pdT		
	Laufmittel A		Laufmittel B
	ohne Phosphatase	nach Phosphatase-Behandlung	ohne Phosphatase
p(dC, dT)	0.40	1.74	0.83
$p(dT)_3$	0.34	1.42	0.80
$p(dC)_3$	0.09	0.69	0.65
$p(dC, dT_2)$	0.22	1.07	0.76
$p(dC_2, dT)$	0.16	0.84	0.73
$p(dT_2, dU)$, (dT_2, dU) p	0.22	1.07	0.76
$p(dC, dT_3)$	0.12	0.74	0.68
$p(dC_2, dT_2)$	0.09	0.69	0.63
pdCp	0.17	2.40	0.70
$p(dT)_2$p	0.14	2.06	0.71
$p(dC)_2$p	0.09	1.34	0.51
p(dC, dT) p	0.10	1.74	0.56
$p(dC, dT_2)$ p	–	1.07	0.53
$p(dC_2, dT)$ p	0.02	0.84	0.50

sich bekanntlich die Struktur synthetischer Oligonucleotide eindeutig beweisen.

Aus dem säulenchromatographischen Verhalten, den R_F-Werten und dem Vergleich mit gekauften und, soweit vorhanden, synthetischen Vergleichssubstanzen wird zunächst die Kettenlänge der isolierten Pyrimidinnucleotide ermittelt.

Aus dem Vergleich der gemessenen Absorptionsverhältnisse 250/260 and 280/260 nm mit den berechneten Werten, läßt sich dann der Anteil an Thymidyl- und Cytidylsäurebausteinen im DNA-Fragment bestimmen (vgl. Tab. 5). Die Absorptionsverhältnisse bei pH 7 werden additiv aus den molaren Extinktionskoeffizienten errechnet, die bekanntlich bei 250, 260 und 280 nm folgende Werte annehmen: pdT (5 700, 8 800, 6 400); pdC (6 100, 7 300, 7 300); pdU (7 400, 10 000, 3 800).

Die aus den chromatographischen Daten und Absorptionsverhältnissen ermittelte Struktur wird abschließend durch enzymatischen Abbau mit alkalischer Phosphatase und Phosphodiesterase aus Schlangengift überprüft und bestätigt.

Phosphatase entfernt aus Oligonucleotiden endständige Phosphatgruppen ohne die Polymerketten anzugreifen. Beim Abbau mit Phosphatase verlieren Nucleotide eine, Nucleotidphosphate zwei terminale Phosphatendgruppen. Der Verlust der Phosphatendgruppen hat zur Folge, daß sich der R_F-Wert deutlich erhöht.

Nach dem Phosphataseabbau wird papierchromatographisch einerseits die Vollständigkeit des Abbaus und andererseits nochmals die Reinheit der Produkte überprüft. Aus dem Vergleich der R_F-Werte vor und nach dem Abbau läßt sich mit Sicherheit entscheiden, ob das ursprüngliche Nucleotid eine, zwei oder keine terminalen Phosphatgruppen aufweist.

Die dephosphorylierten Oligonucleotide werden vom Papier eluiert und zur Bestimmung der Kettenlänge mit Phosphodiesterase aus Schlangengift hydrolysiert. Phosphodiesterase aus Schlangengift baut bekanntlich als Exonuclease lineare Oligonucleotide vom 3'-Ende schrittweise ab und setzt dabei aus dephosphorylierten Oligonucleotiden 5'-Mononucleotide und den 5'-terminalen Baustein als Nucleosid frei. Bei der enzymatischen Hydrolyse von 5'-Oligodesoxyribopyrimi-

Tab. 4. Enzymatische Hydrolyse der aus depurinierten DNA-Hydrolysaten isolierten Pyrimidinnucleotide mit Phosphodiesterase aus Schlangengift.

Pyrimidinnucleotide (Bezeichnung)	Spaltprodukte	Verhältnis der Spaltprodukte	
		erwartet	beobachtet
$p(dT)_3$	pdT	–	–
$(dT)_3$	pdT, dT	2:1	1.99:1.00
$p(dT)_2p$	pdT, pdTp	1:1	1.00:1.08
$(dT)_2$	pdT, dT	1:1	1.00:1.03
$p(dC)_3$	pdC	–	–
$(dC)_3$	pdC, dC	2:1	2.16:1.00
pdCp	keine Spaltung	–	–
(dT_2, dU)	pdT, pdU, dT, dU	(dU + pdU) : (dT + pdT) 1:2	1.00:1.82
$p(dC, dT_2)$	pdC, pdT	1:2	1.00:1.96
$p(dC_2, dT)$	pdC, pdT	2:1	2.09:1.00
$p(dC, dT_3)$[a]	pdC, pdT	1:3	1.00:4.00
$p(dC_2, dT_2)$	pdC, pdT	1:1	1.09:1.00
p(dC, dT)	pdC, pdT	1:1	1.00:1.02
(dC, dT)	pdC, pdT, dT, dC	(dC + pdC) : (dT + pdT) 1:1	0.99:1.00
(dC_2, dT)	pdC, pdT, dT, dC	(dC + pdC) : (dT + pdT) 2:1	2.16:1.00
$(dC)_2$	pdC, dC	1:1	0.89:1.00
(dC, dT_2)	pdC, pdT, dC, dT	(dT + pdT) : (dC + pdC) 2:1	2.21:1.00

[a] Enthält vermutlich ca. 20% $p(dT)_4$.

Tab. 5. Berechnete und beobachtete Absorptionsverhältnisse der aus depurinierten DNA-Hydrolysaten isolierten Pyrimidinnucleotide.

Pyrimidinnucleotide (Bezeichnung)	Absorptionsverhältnisse bei pH 7.5			
	250/260 nm		280/260 nm	
	ber.	beob.	ber.	beob.
p(dC, dT)	0.73	0.76	0.84	0.84
$p(dT)_3$	0.65	0.65	0.73	0.69
$p(dC)_3$	0.82	0.82	0.98	0.95
$p(dC, dT_2)$	0.70	0.71	0.80	0.79
$p(dC_2, dT)$	0.76	0.75	0.88	0.88
$p(dT_2, dU)$ + (dT_2, dU) p	0.68	0.71	0.61	0.57
$p(dC, dT_3)$	0.69	0.69	0.78	0.77
$p(dC_2, dT_2)$	0.73	0.74	0.84	0.85
pdCp	0.82	0.84	0.98	1.00
$p(dT)_2p$	0.65	0.65	0.73	0.70
$p(dC)_2p$	0.82	0.84	0.98	0.97
p(dC, dT) p	0.73	0.75	0.84	0.82
$p(dC, dT_2)$ p	0.70	0.71	0.80	0.77
$p(dC_2, dT)$ p	0.76	0.76	0.88	0.88

dinnucleotiden mit Phosphodiesterase bilden sich ausschließlich 5′-Mononucleotide. 3′-Pyrimidindinucleotide werden von Phosphodiesterase zu Nucleosiden und Nucleosiddiphosphaten abgebaut, während längere 3′-Oligodesoxyribopyrimidinnucleotide zu Nucleosiden, Nucleotiden und Nucleosiddiphosphaten gespalten werden. Der Abbau von Oligopyrimidinnucleotidphosphaten führt auf diesem Weg zu Nucleotiden und Nucleosiddiphosphaten. Aus der Art der Spaltprodukte kann man eindeutig feststellen, ob die Oligopyrimidinnucleotide am 5′- oder 3′- bzw. an beiden Enden Phosphatgruppen tragen.

Eine 3′-Phosphatendgruppe erschwert den enzymatischen Abbau mit Phosphodiesterase, so daß kurze Oligonucleotide nur sehr langsam und längere nicht mehr abgebaut werden. Aus diesem Grund sind zur enzymatischen Hydrolyse der 3′-Nucleotide bzw. Nucleotidphosphate höhere Enzymkonzentrationen und längere Reaktionszeiten als bei 5′-Nucleotiden bzw. Oligomeren ohne terminale Phosphatgruppen erforderlich.

Nach dem enzymatischen Abbau werden die Spaltprodukte papierchromatographisch getrennt und mit Referenzsubstanzen identifiziert. Auftretende Nucleosiddiphosphate werden vom Papier eluiert, mit Phosphatase abgebaut und dann als Nucleoside nachgewiesen. Zur Ermittlung der Kettenlänge eines DNA-Fragments werden die einzelnen Spaltprodukte quantitativ vom Papier eluiert. Aus den im Eluat gemessenen A_{260}-Einheiten wird der molare Anteil für jedes Spaltprodukt errechnet. Aus dem Verhältnis der molaren Anteile ergibt sich die Kettenlänge des ursprünglichen Pyrimidinnucleotids.

Mischungen sequenzisomer Pyrimidinnucleotide, die im Papier als einheitlicher Fleck wandern, werden auf folgendem Weg nachgewiesen. Die isolierte, chromatographisch reine Mischung wird zunächst mit Phosphatase behandelt und auf Papier chromatographiert. Die dephosphorylierten Sequenzisomeren, die im Papierchromatogramm einen Fleck bilden, werden vom Papier eluiert und mit Phosphodiesterase aus Schlangengift hydrolysiert. Treten neben den 5'-Mononucleotiden gleichzeitig die Nucleoside Cytidin und Thymidin auf, so liegt eine Mischung von Sequenzisomeren vor.

Die Charakterisierung und Identifizierung der lyophilisierten Peakfraktionen, die in Tab. 3 - 5 zusammengefaßt sind, führen im einzelnen zu folgenden Ergebnissen.

Bei der Rechromatographie (vgl. Fig. 2a) wird das Lyophilisat von Peak I der Fig. 1 in Peak I a - c aufgetrennt. Peak I a, der mit 0.12M NaCl eluiert wird, enthält zu 83% die Sequenzisomeren p(dC, dT), die zusammen 11.5% der Mischung ausmachen. Die Mischung der Sequenzisomeren wandert im Papierchromatogramm in beiden Laufmittelsystemen als ein einheitlicher Fleck, dessen R_F-Wert mit dem von synthetischem pdC-dT identisch ist. Der Abbau, der vom Papier eluierten Sequenzisomeren, mit Phosphodiesterase führt zu pdC und pdT im Verhältnis 1 : 1.02. Nach der Behandlung mit Phosphatase wandert die Mischung der Sequenzisomeren wiederum als einheitlicher Fleck, dessen R_F-Wert mit dem von synthetischem dC-dT übereinstimmt. Beim Abbau der dephosphorylierten Verbindungen mit Phosphodiesterase erhält man etwa zu gleichen Teilen dT, dC, pdT und pdC, die mit käuflichen Referenzsubstanzen identifiziert werden. Die durch enzymatischen Abbau ermittelte Struktur wird zusätzlich durch die gute Übereinstimmung zwischen gefundenen und berechneten Absorptionsverhältnissen bestätigt.

Peak I b, der ebenfalls wie Peak I a mit 0.12M NaCl eluiert wird, enthält zu 91% pdCp, das mit 37.2% das Hauptprodukt der Mischung aus Peak I bildet. Beim Abbau mit Phosphatase erhält man einen einheitlich wandernden Fleck, der im R_F-Wert und in den Absorptionsverhältnissen mit käuflichem dC identisch ist. Mit Phosphodiesterase erhält man keine Abbauprodukte, so daß die verwendete Phosphodiesterase frei von Phosphatase ist.

In Peak I c (18% der Mischung), der mit 0.12 - 0.13M NaCl die Säule verläßt, ist zu 92% das Trinucleotid $p(dC)_3$ enthalten. Beim Abbau mit Phosphodiesterase bildet sich ausschließlich pdC, das papierchromatographisch mit käuflichem pdC identisch ist. Nach der Behandlung mit Phosphatase wandert das dephosphorylierte Produkt als einheitlicher Fleck, dessen R_F-Wert mit dem von synthetischem $(dC)_3$ übereinstimmt und der mit Phosphodiesterase zu pdC und dC im Verhältnis 2.16 : 1.00 gespalten wird. Die gemessenen Absorptionsverhältnisse stimmen mit den berechneten Werten überein und bestätigen zusätzlich die ermittelte Struktur.

Die Pyridinnucleotidmischung aus Peak II der Fig. 1 wird bei der Rechromatographie in die Hauptpeaks II a, b der Fig. 2b getrennt, die jeweils ca. 40% der aufgetragenen Mischung enthalten. In Peak II a verläßt mit 0.12M NaCl die Mischung der sequenzisomeren Trinucleotide $p(dC_2, dT)$ in 98proz. Reinheit die Säule. Die Absorptionsverhältnisse der chromatographisch reinen, in einem Fleck wandernden Sequenzisomeren stimmen sehr gut mit den berechneten Werten überein (vgl. Tab. 5). Bei der hydrolytischen Spaltung der Sequenzisomeren mit Phosphodiesterase erhält man pdC und pdT im Verhältnis 2.09 : 1.00 (vgl. Tab. 4). Nach Phosphatasebehandlung findet man wiederum einen einheitlich wandernden Fleck, der mit Phosphodiesterase zu dC, dT, pdT und pdC abgebaut wird.

Mit 0.12 - 0.15M NaCl wird in Peak II b das Dinucleotidphosphat $p(dC)_2$ p in 96proz. Reinheit eluiert. Die gemessenen Absorptionsverhältnisse des chromatographisch reinen Dinucleotidphosphats stimmen mit den für $p(dC)_2$ p berechneten Werten überein. Nach Phosphatasebehandlung erhält man ein einheitliches Produkt, das im R_F-Wert mit dem von synthetischem $(dC)_2$ übereinstimmt und mit Phosphodiesterase zu pdC und dC im Verhältnis 0.89 : 1.00 gespalten wird.

Das Lyophilisat von Peak III der Fig. 1 wird bei der Rechromatographie in Peak III a, b aufgetrennt (vgl. Fig. 2c). Peak III a, in dem 15.4% der Ausgangsmischung eluiert werden, die mit 0.12M NaCl die Säule verlassen, enthält zu 94% eine Mischung der sequenzisomeren Trinucleotide p(dC, dT_2). Vor und nach Behandlung mit Phosphatase wandert die Mischung der Sequenzisomeren in beiden Laufmittelsystemen als einheitlicher Fleck, der in den Absorptionsverhältnissen mit den für p(dC, dT_2) berechneten Werten gut übereinstimmt. Phosphodiesterase baut die Mischung der Sequenzisomeren zu pdC und pdT im Verhältnis 1.00 : 1.96 ab. Aus den dephosphorylierten Sequenzisomeren werden mit Phosphodiesterase dC, dT, pdT und pdC freigesetzt.

Peak III b, der mit 0.12 - 0.15M NaCl eluiert wird, enthält zu 96% die sequenzisomeren Dinucleotidphosphate p(dC, dT) p, die zusammen 75.4% der aufgetragenen Mischung ausmachen. Die Sequenzisomeren wandern in beiden Laufmittelsystemen als ein einheitlicher Fleck, dessen Absorptionsverhältnisse mit den für p(dC, dT) berechneten Werten gut übereinstimmen. Nach Phosphatasebehandlung erhält man wiederum nur einen Fleck, der im R_F-Wert mit dem von synthetischem dC-dT identisch ist. Das dephosphorylierte Produkt wird von Phosphodiesterase zu dC, dT, pdT und pdC im Verhältnis 1.00 : 1.09 : 1.00 : 1.07 abgebaut.

Die Rechromatographie des Lyophilisats aus Peak IV der Fig. 1 führt zu 4 getrennten Peaks (vgl. Fig. 2d). Mit 0.12M NaCl wird in Peak IV a zu 83% eine Mischung der sequenzisomeren Trinucleotide p(dT_2, dU), (dT_2, dU) p eluiert, die ca. 11.6% des aufgetragenen Lyophilisats ausmacht. Die Mischung wandert in beiden Laufmittelsystemen als einheitlicher Fleck, dessen Absorptionsverhältnisse den für p(dT_2, dU) berechneten Werten entsprechen. Der Uridylsäureanteil der Mischung äußert sich deutlich im 280/260 Verhältnis (0.57), das unter den Werten für pdT (0.73) und pdC (0.98) liegt. Durch Phosphodiesterase wird die chromatographisch reine Mischung der Trinucleotide in pdT, pdU, pdTp, pdUp, dT und dU hydrolysiert. Die Spaltprodukte pdUp, pdTp und pdU werden vom Papier eluiert und anschließend mit Phosphatase zu dU, dT, dU abgebaut, die mit käuflichen Referenzsubstanzen papierchromatographisch identifiziert werden. Nach Behandlung mit Phosphatase wandert die Mischung der Sequenzisomeren im Laufmittelsystem A wiederum als einheitlicher Fleck. Das dephosphorylierte Gemisch wird vom Papier eluiert und mit Phosphodiesterase hydrolysiert. Im Hydrolysat findet man papierchromatographisch dU, pdU, dT, pdT, wobei die Summen aus (dU + pdU) : (dT + pdT) im Verhältnis 1:1.82 auftreten.

In Peak IV b wird mit 0.12 - 0.13M NaCl das Dinucleotidphosphat $p(dT)_2$ p in 84proz. Reinheit eluiert, das zu 12% im Lyophilisat von Peak IV enthalten ist. Das Dinucleotidphosphat wandert in beiden Laufmittelsystemen als einheitlicher Fleck, dessen Absorptionsverhältnisse den für $p(dT)_2$ berechneten Werten entsprechen. Beim Abbau mit Phosphodiesterase bilden sich pdT und pdTp im Verhältnis 1 : 1.08. Durch Behandlung mit Phosphatase verliert das Dinucleotidphosphat seine beiden terminalen Phosphatgruppen, da sich die R_F-Werte vor und nach der Phosphatasebehandlung sehr stark unterscheiden (0.14 - 2.06, vgl. Tab. 3). Das Abbauprodukt ist im R_F-Wert mit dem von synthetischem $(dT)_2$ identisch und wird mit Phosphodiesterase zu pdT und dT im Verhältnis 1.00 : 1.03 gespalten.

Peak IV c enthält zu 95% die sequenzisomeren Tetranucleotide p(dC_2, dT_2), die mit 0.14 - 0.16M NaCl eluiert werden und zu 21.6% in der ursprünglichen Mischung vorhanden sind. Die Mischung der Sequenzisomeren wandert in beiden Laufmittelsystemen als ein Fleck, dessen Absorptionsverhältnisse sehr gut mit den für p(dC_2, dT_2) berechneten Werten übereinstimmen. Durch Phosphodiesterasebehandlung wird die Mischung in pdC und pdT im Verhältnis 1.09 : 1.00 gespalten. Nach Phosphataseabbau wandern die Sequenzisomeren im Laufmittelsystem A als ein einheitlicher Fleck, der mit Phosphodiesterase zu dT, dC, pdC und pdT abgebaut wird.

Peak IVd, der mit 0.17 - 0.19M NaCl die Säule verläßt, enthält zu 95% die sequenzisomeren Trinucleotidphosphate $p(dC_2, dT)p$, die mit 40.2% den Hauptbestandteil der Pyrimidinnucleotidmischung bilden. Die Mischung der Sequenzisomeren wandert im Laufmittelsystem B als einheitlicher Fleck, dessen Absorptionsverhältnisse mit den für $p(dC_2, dT)$ berechneten Werten identisch sind. Nach Phosphatasebehandlung steigt der R_F-Wert der Sequenzisomeren von 0.02 auf 0.8, so daß durch Dephosphorylierung die Verbindung zwei terminale Phosphatgruppen verliert. Die dephosphorylierten Sequenzisomeren wandern als ein Fleck, dessen R_F-Wert mit dem von synthetischen $(dC_2\text{-}dT)$ identisch ist und der mit Phosphodiesterase in dC, dT, pdT und pdC hydrolysiert wird, wobei die Summen von dC + pdC und dT + pdT im Verhältnis 2.16 : 1.00 auftreten.

Das Lyophilisat von Peak V der Fig. 1 enthält zu 56.3% $p(dT)_2$ p, das als Hauptprodukt in 99proz. Reinheit mit 0.12 - 0.14M NaCl in Peak Vb der Fig. 2e eluiert wird. Das Pyrimidinnucleotid aus Peak Vb ist mit dem in Peak IVb der Fig. 2d eluierten $p(dT)_2$ p in den Absorptionsverhältnissen, R_F-Werten und enzymatischen Abbauprodukten identisch.

Vor dem Hauptprodukt verläßt in Peak Va das Trinucleotid $p(dT)_3$, das zu 8.9% in der Mischung enthalten ist, in 90proz. Reinheit mit 0.12M NaCl die Säule. Die Absorptionsverhältnisse des chromatographisch reinen Produkts entsprechen den für $p(dT)_3$ berechneten Werten. Vor und nach der Behandlung mit Phosphatase verhält sich das Trinucleotid im Papierchromatogramm wie synthetisches $p(dT)_3$ bzw. $(dT)_3$.

Mit Phosphodiesterase wird das Produkt zu pdT hydrolysiert, während die dephosphorylierte Verbindung in pdT und dT im Verhältnis 1.99 : 1.00 gespalten wird.

In Peak Vc werden vermutlich zu 94% die sequenzisomeren Tetranucleotide $p(dC, dT_3)$ mit 0.15 - 0.17M NaCl eluiert, die 6.8% der Pyrimidinnucleotidmischung ausmachen. Die chromatographisch reine Mischung der Sequenzisomeren stimmt in den Absorptionsverhältnissen gut mit den berechneten Werten überein. Nach Phosphatasebehandlung findet man aber einen zusätzlichen Flecken, der im R_F-Wert mit synthetischem $(dT)_4$ übereinstimmt. Beim Abbau der Sequenzisomeren mit Phosphodiesterase werden die Spaltprodukte dpC und dpT im Verhältnis 1.00 : 4.00 gebildet. Der zu hohe pdT Anteil rührt vermutlich aus $p(dT)_4$, das wir in der Mischung der Sequenzisomeren vermuten.

Peak Vd enthält in 98proz. Reinheit die sequenzisomeren Trinucleotidphosphate $p(dC, dT_2)p$, die mit 0.17 - 0.19M NaCl gemeinsam desorbiert werden und zu 20.8% in der Ausgangsmischung vorhanden sind. Nach Phosphatasebehandlung erhält man ein Abbauprodukt, das im R_F-Wert und in den Absorptionsverhältnissen dem dephosphorylierten Produkt aus Peak IIIa, Fig. 2c entspricht und mit Phosphodiesterase in dT, dC, pdT, pdC gespalten wird, wobei die Summen von dT + pdT und dC + pdC im Verhältnis 2.21 : 1.00 auftreten.

Diskussion

In Tab. 6 sind die Pyrimidinnucleotide und Mischungen der Sequenzisomeren aufgeführt, die nach dem bisherigen Trennungsgang aus dem Hydrolysat der Heringsspermen-DNA chromatographisch rein, im präparativen Maßstab erhalten werden. Die Oligodesoxythymidyl- und Cytidylsäuren sind bereits mit wesentlich geringerem Aufwand aus dem DNA-Hydrolysat als durch chemische Synthese zugänglich.

Der Reinheitsgrad aller isolierten Pyrimidinnucleotide erfüllt vollständig die bei synthetischen Oligonucleotiden angewendeten Kriterien. So stimmen die gefundenen Absorptionsverhältnisse im Rahmen der Meßgenauigkeit bei allen Pyrimidinnucleotiden mit den berechneten Werten überein. In 2 unterschiedlichen Laufmittelsystemen wandern die DNA-Fragmente vor und nach Behandlung mit Phosphatase einheitlich. Die Oligonucleotide können mit Phosphodiesterase vollständig abgebaut werden, wobei die gefundenen molaren Anteile der auftretenden Spaltprodukte die erwarteten Werte bestätigen.

Alle in Tab. 6 aufgeführten Pyrimidinnucleotide sind auch in dephosphorylierter Form zugänglich, da die terminalen Phosphatgruppen der isolierten Nucleotide mit Phosphatase quantitativ entfernt werden können.

Die Mengenangaben in Tab. 6 sind aus den A_{260}-Einheiten der papierchromatographisch isolierten

Tab. 6. Berechnete Gewichtsanteile von Pyrimidinnucleotiden, die aus 100 g Heringsspermen-DNA chromatographisch rein isoliert werden können.

Pyrimidinnucleotide (Bezeichnung)	Gewichtsanteile [mg]
$p(dT)_2$	720
$p(dC)_2$	220
p(dC, dT)	1460
$p(dT)_3$	160
$p(dC)_3$	30
$p(dC, dT_2)$	310
$p(dC_2, dT)$	340
$p(dT_2, dU), (dT_2, dU)\,p$	20
$p(dC, dT_3)$	130
$p(dC_2, dT_2)$	50
pdTp	2520
pdCp	990
$p(dT)_2 p$	1230
$p(dC)_2 p$	370
p(dC, dT) p	1520
$p(dC, dT_2)\,p$	430
$p(dC_2, dT)\,p$	90
Nach Phosphatase-Behandlung	
$(dT)_2$	1950
$(dC)_2$	590
(dC, dT)	2980
$(dT)_3$	160
$(dC)_3$	30
(dC, dT_2)	740
(dC_2, dT)	430
(dT_2, dU)	20
(dC, dT_3)	130
(dC_2, dT_2)	50

Nucleotide hochgerechnet und geben nur bedingt die Größenordnung wieder, in der die einzelnen Pyrimidinnucleotide im Hydrolysat auftreten. Die für Pyrimidinnucleotide mit 3 und 4 Monomereinheiten angegebenen Gewichtsanteile sind sicherlich zu niedrig angesetzt, da diese Fragmente vermutlich noch in der höhermolekularen, bisher nicht aufgearbeiteten Fraktion erhalten sind.

Bemerkenswert ist, daß im DNA-Hydrolysat außer den erwarteten Pyrimidinnucleotidphosphaten auch ein erheblicher Anteil an 5′-Pyrimidinnucleotiden enthalten ist, während dagegen 3′-Pyrimidinnucleotide außer in der Mischung der Sequenzisomeren $p(dT_2, dU)$, $(dT_2, dU)\,p$ nicht gefunden werden. In analytischen Trennungen werden 5′-Pyrimidinnucleotide nicht nachgewiesen. Wir nehmen an, daß bei diesen Trennungen die 5′-Pyrimidinnucleotide als Nebenprodukte im sogenannten Untergrund eluiert und daher nicht gefunden werden. Die 5′-Pyrimidinnucleotide werden, wie W.E. Cohn vermutet[8], nicht direkt bei der chemischen Hydrolyse, sondern erst in einer anschließenden Sekundärreaktion gebildet, bei der aus den Nucleotidphosphaten bevorzugt die 3′-Phosphatgruppe entfernt wird.

Zur Zeit versuchen wir, die längerkettigen Fragmente des DNA-Hydrolysats im präparativen Maßstab chromatographisch rein zu isolieren und die Mischung der Sequenzisomeren in ihre Komponenten aufzutrennen.

Der *Deutschen Forschungsgemeinschaft* danken wir für die großzügige Unterstützung dieser Arbeit.

Literatur

1 Burton, K. & Peterson, G.B. (1960) *Biochem. J.* **75**, 17 - 27.
2 Shapiro, H.S. & Chargaff, E. (1964) *Biochim. Biophys. Acta* **91**, 262 - 270.
3 Peterson, G.B. & Reeves, J.M. (1969) *Biochim. Biophys. Acta* **179**, 510 - 512.
4 Ling, V. (1972) *J. Mol. Biol.* **64**, 87 - 102.
5 Burton, K. (1967) *Methods Enzymol.* **12a**, 222 - 224.
6 Schott, H. & Schwarz, M. (1977) *Hoppe-Seyler's Z. Physiol. Chem.* **358**, 949 - 957.
7 Mazin, A.L. & Vanyushin, B.F. (1967) *Biokhimiya* **32**, 377 - 387.
8 Cohn, W.E. & Volkin, E. (1957) *Biochim. Biophys. Acta* **24**, 359 - 364.

Priv.-Doz. Dr. H. Schott und Dr. Margarete Schwarz, Institut für Organische Chemie der Universität Tübingen, Auf der Morgenstelle 18, D-7400 Tübingen 1.

Hoppe-Seyler's Z. Physiol. Chem.
Bd. 359, S. 629 - 640, Mai 1978

The Amino Acid Sequences of the Tryptic, Chymotryptic and Peptic Peptides from the L-2 Light Chain of Rabbit Skeletal Muscle Myosin

Genji MATSUDA, Tetsuo MAITA, Yoji SUZUYAMA, Masayuki SETOGUCHI and Toshiyo UMEGANE

Department of Biochemistry, Nagasaki University School of Medicine, Nagasaki, Japan

(Received 13 February 1978)

Summary: The amino acid sequences of the tryptic peptides from the aminoethylated L-2 light chain of rabbit skeletal muscle myosin were determined by various enzymatic hydrolyses, partial hydrolysis with dilute acetic acid and Edman degradation. The amino acid sequences of the chymotryptic and peptic peptides from the carboxymethylated L-2 light chain were partially analysed in the same manner as the tryptic peptides. The primary structure of the L-2 light chain of rabbit skeletal muscle myosin was deduced from the above results.

Die Aminosäuresequenzen der tryptischen, chymotryptischen und peptischen Peptide der leichten Kette L-2 des Myosins aus Kaninchenskelettmuskel

Zusammenfassung: Die leichte L-2-Kette des Kaninchenskelettmuskel-Myosins wurde nach Aminoethylierung mit Pepsin gespalten; die Aminosäuresequenzen der dabei erhaltenen Peptide wurden durch Nachspaltung mit verschiedenen Enzymen, durch partielle Hydrolyse mit Essigsäure und durch Edman-Abbau bestimmt. Ebenso wurden die chymotryptischen und die peptischen Peptide der carboxymethylierten L-2-Kette analysiert. Aus den Ergebnissen wird die Primärstruktur der L-2-Kette abgeleitet.

Key words: Skeletal muscle myosin, L-2 light chain, primary structure, sequence analysis.

We isolated the L-2 light chain of rabbit skeletal muscle myosin as previously reported[1]. After aminoethylation, we digested it with trypsin, purified the resulting tryptic peptides, and determined their amino acid compositions. We also carboxymethylated the L-2 light chain, digested

Enzymes:
Carboxypeptidase A, peptidyl-L-amino-acid hydrolase (EC 3.4.12.2);
Carboxypeptidase B, peptidyl-L-lysine(-L-arginine) hydrolase (EC 3.4.12.3);
Chymotrypsin (EC 3.4.21.1);
Pepsin (EC 3.4.23.1);
Subtilisin (EC 3.4.21.14);
Thermolysin (one of the microbial metalloenzymes) (EC 3.4.24.4);
Trypsin (EC 3.4.21.4).

it with chymotrypsin and pepsin, purified the chymotryptic and peptic peptides obtained, and determined their amino acid compositions.

Subsequently, we completed the amino acid sequences of the tryptic peptides mentioned above and further partially analysed the amino acid sequences of the chymotryptic and peptic peptides. We deduced the primary structure of the L-2 light chain of rabbit skeletal muscle myosin from the results of these experiments. It is composed of 168 amino acids as reported in a preliminary communication[2]. We report the details of the results of the sequence analyses in this paper.

Materials and Methods

All adult rabbits used in this experiment were the Japanese white native strain. The isolation and purification of tryptic, chymotryptic and peptic peptides from the L-2 light chain of rabbit skeletal muscle myosin and the analyses of their amino acid compositions were carried out in the same manner as previously reported[1]. The methods used in the further analyses of these peptides are given in the following.

Subtractive Edman degradation[3]

Peptides were lyophilized and dissolved in 1 m*l* of 66% pyridine. 0.1 m*l* of 5% phenylisothiocyanate was added to them and they were reacted at pH 8.0, 40 °C for 1.5 h and lyophilized. After addition of 0.5 m*l* of trifluoroacetic acid, they were sealed and reacted at 40 °C for 0.5 h and lyophilized. They were then dissolved in 1 m*l* of water and extracted with three 2-m*l* of portions of a mixture of n-butyl acetate and water (2:1). A part of the water layer in the reacting solution was used for the amino acid analysis and the remainder for the next step of the reaction.

Digestion with carboxypeptidase A and B

The digestion with carboxypeptidase A and B (Worthington Co.) was carried out according to Hayashi et al.[4] using 0.01M borate buffer (pH 8.1) at 30 °C for 20 - 30 min. Later the isolated amino acids were analysed using an amino acid analyser.

Identification of the location of amides

The locations of amides in the peptides were deduced from their mobilities on paper electrophoresis at pH 6.4 and 2 kV. The electric charges of peptides were determined according to Offord's method[5].

Fragmentation of larger peptides

Larger peptides were enzymatically redigested into fragments with chymotrypsin (Sigma Co.), pepsin (Sigma Co.), subtilisin (Sigma Co.), thermolysin (Nakarai Chemical Co.) and trypsin (Worthington Co.). That is, chymotryptic digestion was carried out at pH 8.5, 37 °C for 2 - 6 h, peptic digestion, at pH 2.0, 37 °C for 4 h, subtilitic digestion, at pH 8.5, 37 °C for 5 h, thermolitic digestion, at pH 8.5, 45 °C for 4 h and tryptic digestion, at pH 8.5, 37 °C for 2 - 16 h. Larger peptides, also, were partially hydrolysed into fragments with 0.25M acetic acid at 105 °C for 16 h.

Purification of peptide fragments

Column chromatography, paper chromatography and paper electrophoresis were used to purify the peptide fragments obtained by partial hydrolysis of larger peptides. The column chromatography was performed using a Chromo Beads P (Technicon Co.) column (0.9 x 20 cm) as adsorbent, and pyridine acetate buffer as eluent in the same manner as in the previous report[1]. Paper chromatography was performed using Toyo filter paper No. 50 (40 x 40 cm) and the upper phase of the mixture of 1-butanol, acetic acid and water (4:1:5) as developer in the same manner as in the previous report[1]. Electrophoresis was performed using Toyo filter paper No. 50 (30 x 60 cm) and the mixture (pH 6.4) of pyridine, acetic acid and water (100:4:900) at 2 kV for 1.5 h.

Results and Discussion

Sequence analyses of the tryptic peptides from the L-2 light chain

The main tryptic peptides from the aminoethyl-L-2 light chain of rabbit skeletal muscle myosin were designated as Try-S-1 - Try-I-2 (for details see Fig. 1 and Ref.[1]). Fig. 1 shows the amino acid sequences of the 26 tryptic peptides.

The amino acid sequence of Try-S-1 containing 3 amino acids was directly determined by subtractive Edman degradation. The *N*- and *C*-terminal sequences of Try-S-2 (13 amino acids) were firstly analysed and then the peptide was digested with thermolysin. The seven peptide fragments obtained were purified by column and paper chromatography as shown in Fig. 2 and their amino acid compositions and partial amino acid sequences were analysed. From these experimental results, the amino acid sequence of Try-S-2 was deduced.

Try-S-3 (13 amino acids) was digested with chymotrypsin and 6 peptide fragments obtained were purified by column and paper chromatography as shown in Fig. 3. Try-S-4 (9 amino acids) was digested with thermolysin and the 3 peptide fragments obtained were purified by column chromatography using Chromo Beads P. Tyr-S-5a (10 amino acids) was digested with chymotrypsin and the 2 peptide fragments obtained were purified by column chromatography using Chromo Beads P.

Try-S-5b (17 amino acids) was digested with chymotrypsin and the 3 peptide fragments obtained were purified by column and paper chromatography as shown in Fig. 4 (A, B) and then designated Try-S-5b-CH-1, Try-S-5b-CH-2a and Try-S-5b-CH-2b, respectively. Try-S-5b-CH-1 was further digested with subtilisin and the 2 peptide fragments obtained were purified by column and paper chromatography as shown in Fig. 4 (C, D). It was also partially hydrolysed with 0.25M acetic acid and the four kinds of peptide fragments obtained were purified by paper electrophoresis and paper chromatography.

Try-S-6, Try-S-7a and Try-S-7b contain 7, 7 and 8 amino acids, respectively. Their amino acid sequences were directly determined by subtractive Edman degradation and carboxypeptidase A and B digestion. Try-S-8a was identified to be free lysine by amino acid analysis and subtractive Edman degradation. It was determined by amino acid analysis that Try-S-8b contains Pro (1) and Lys (1), but its *N*-terminal amino acid was not identified by subtractive Edman degradation. Therefore Try-S-8b was deduced to be X-Pro-Lys, supposing that the *N*-terminal is blocked.

Try-S-8c (10 amino acids) was digested with thermolysin and the 4 peptide fragments obtained were purified by column chromatography using Chromo Beads P. The *C*-terminal amino acid of Try-S-9a (12 amino acids) was not identified by carboxypeptidase A and B digestion. This tryptic peptide was further redigested with trypsin at pH 8.5, 37 °C for 6 h. The 2 peptide fragments obtained were purified by paper electrophoresis and designated Try-S-9a-TR-1 and Try-S-9a-TR-2. Try-S-9a-TR-2 was digested with chymotrypsin and the 6 peptide fragments obtained were purified by column and paper chromatography as shown in Fig. 5.

Try-S-9b, Try-S-9c and Try-S-9d contain 3, 4 and 4 amino acids, respectively. Their amino acid sequences were directly determined by subtractive Edman degradation.

The *N*-terminal sequence of Try-S-9e (12 amino acids) was analysed and then the amino acid sequence was deduced from the homology of Try-S-3 with this peptide. Peptides Try-S-10a, Try-S-10b, Try-S-11, Try-S-12a, Try-S-12b, Try-S-13 and Try-S-14 contain 2, 1, 2, 9, 2, 5 and 3 amino acids, respectively. These amino acid sequences were directly determined by amino acid analyses, subtractive Edman degradation and carboxypeptidase A and B digestion.

Try-I-1 (18 amino acids) was digested with chymotrypsin. The five peptide fragments obtained were purified by column and paper chromatography as shown in Fig. 6. Try-I-2 (23 amino acids) was digested with chymotrypsin. The 3 peptide fragments obtained were purified by paper electrophoresis and paper chromatography and designated Try-I-2-CH-1, Try-I-2-CH-2 and Try-I-2-CH-3, respectively. Try-I-2-CH-1 was further digested with pepsin and the 3 peptide fragments obtained were purified by paper electrophoresis and paper chromatography and designated Try-I-2-CH-1-PE-1, Try-I-2-CH-1-PE-2a and Try-I-2-CH-1-PE-2b, respectively. Moreover, Try-I-2-CH-1-PE-1 and Try-I-2-CH-2 were digested with thermolysin and the 4 and 5 peptide fragments obtained were purified by paper electrophoresis and paper chromatography, respectively.

Sequence analyses of the chymotryptic peptides from the L-2 light chain

The main chymotryptic peptides from the carboxymethylated L-2 light chain of rabbit skeletal muscle myosin were designated as Chy-I-1 - Chy-III-10 (for details see Fig. 7 and Ref.[1]).

The amino acid sequences of these 21 chymotryptic peptides were determined as shown in Fig. 7 by their partial sequence analyses and by the comparison of the amino acid sequences of the corresponding tryptic peptides mentioned in the previous paragraph.

Try-S-1 (C-terminus)

Asp-Gln-Glu

Try-S-2 (T-15)

Gly-Ala-Asp-Pro-Glu-Asp-Val-Ile-Thr-Gly-Ala-Phe-Lys

←— S-2 —→

←— TH-1a —→ ←TH-5→

←— TH-2a —→←— TH-3b —→

←— TH-2b —→←— TH-3a —→

←— TH-4 —→

Try-S-3 (T-19ab)

Gln-Phe-Leu-Glu-Glu-Leu-Leu-Thr-Thr-Gln-Cys-Asp-Arg

←— S-3 —→

←— CH-1 —→ CH-2b ←— CH-5 —→

←CH-2a→←— CH-3 —→←— CH-4 —→

Try-S-4 (T-12)

Asn-Glu-Glu-Leu-Asp-Ala-Met-Met-Lys

←— S-4 —→

← TH-1' —→←— TH-2 —→←TH-3→

Try-S-5a (T-7)

Glu-Ala-Phe-Thr-Val-Ile-Asp-Gln-Asn-Arg

←— S-5a —→

← CH-1 →←— CH-2 —→

Try-S-5b (T-21)

Asn-Met-Trp-Ala-Ala-Phe-Pro-Pro-Asp-Val-Gly-Gly-Asn-Val-Asp-Tyr-Lys

←— S-5b —→

← CH-2b →←— CH-1 —→ CH-2a

←— SU-1 —→← SU-2 →

←— AC-2c —→ AC-1 ←— AC-2b —→ AC-1 AC-2a

Try-S-6 (T-16)

Val-Leu-Asp-Pro-Glu-Gly-Lys

Try-S-7a (T-20)

Phe-Ser-Gln-Glu-Glu-Ile-Lys

Try-S-7b (T-10)

Asp-Thr-Phe-Ala-Ala-Met-Gly-Arg

Try-S-8a (T-2, T-18)

Lys

Try-S-8b (T-1)

X-Pro-Lys

Try-S-8c (T-8, 9)

Asp-Gly-Ile-Ile-Asp-Lys-Glu-Asp-Leu-Arg

←— S-8c —→

←TH-1→←— TH-3 —→←TH-4→

←— TH-2 —→

Try-S-9a (T-22b)

Tyr-Val-Ile-Tyr-His-Gly-Asp-Ala-Lys-Asp-Gln-Glu

S-9a

TR-2 TR-1

CH-1 CH-4

CH-2b CH-2a

CH-3b CH-3a

Try-S-9b (T-22a)

Asn-Ile-Cys

Try-S-9c (T-17)

Gly-Thr-Ile-Lys

Try-S-9d (T-11)

Leu-Asn-Val-Lys

Try-S-9e (T-18, 19a)

Lys-Gln-Phe-Leu-Glu-Glu-Leu-Leu-Thr-Thr-Gln-Cys

Try-S-10a (T-19b)

Asp-Arg

Try-S-10b (T-4, T-5)

Arg

Try-S-11 (T-3)

Ala-Lys

Try-S-12a (pos. 157~165)

Tyr-Val-Ile-Thr-His-Gly-Asp-Ala-Lys

Try-S-12b (T-14)

Leu-Lys

Try-S-13 (T-17, 18)

Gly-Thr-Ile-Lys-Lys

Try-S-14 (T-2, 3)

Lys-Ala-Lys

Try-I-1 (T-13)

Glu-Ala-Ser-Gly-Pro-Ile-Asn-Phe-Thr-Val-Phe-Leu-Thr-Met-Phe-Gly-Glu-Lys

I-1

CH-1 CH-3 CH-2 CH-5

CH-4

Try-I-2 (T-6)

Ala-Ala-Ala-Glu-Gly-Gly-Ser-Ser-Asn-Val-Phe-Ser-Met-Phe-Asp-Gln-Thr-Gln-Ile-Gln-Glu-Phe-Lys

I-2

CH-2 CH-1 CH-3

TH-3b TH-2 TH-3c PE-2b PE-1 PE-2a

TH-1 TH-3a TH-2a TH-2b

TH-1 TH-3

Fig. 1. Amino acid sequences of the tryptic peptides from the aminoethylated L-2 light chain of rabbit skeletal muscle myosin.

TH, CH, SU and TR are peptide fragments obtained from the digestions with thermolysin, chymotrypsin, subtilisin and trypsin, respectively, and AC, peptide fragments obtained from the partial hydrolyses with 0.25M acetic acid. Sequences marked ⟶ were determined by subtractive Edman degradation and those marked ⟵, by carboxypeptidase A and B digestion. Parenthesized T-1, T-2, ..., T-22 show the final designations of tryptic peptides.

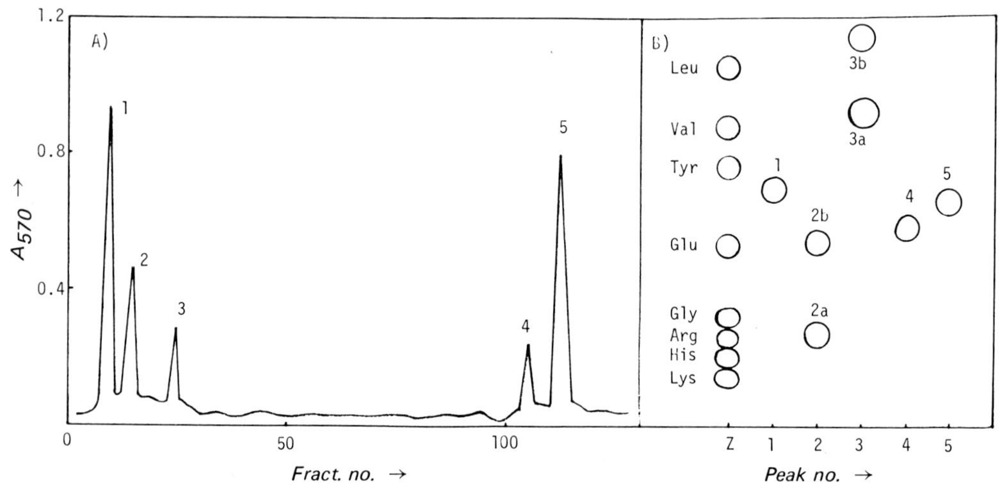

Fig. 2. Column and paper chromatograms of the thermolytic peptides from Try-S-2.
A) Column chromatogram. B) Paper chromatogram. The numbers 1 - 5 refer to the peaks in (A) and the spots (Z) to the left in the figure are the authentic amino acids.

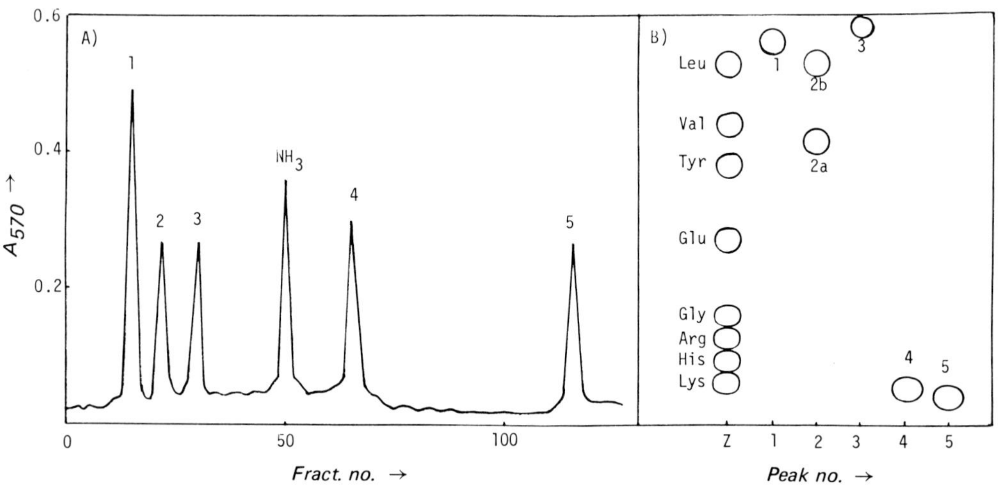

Fig. 3. Column and paper chromatograms of the chymotryptic peptides from Try-S-3.

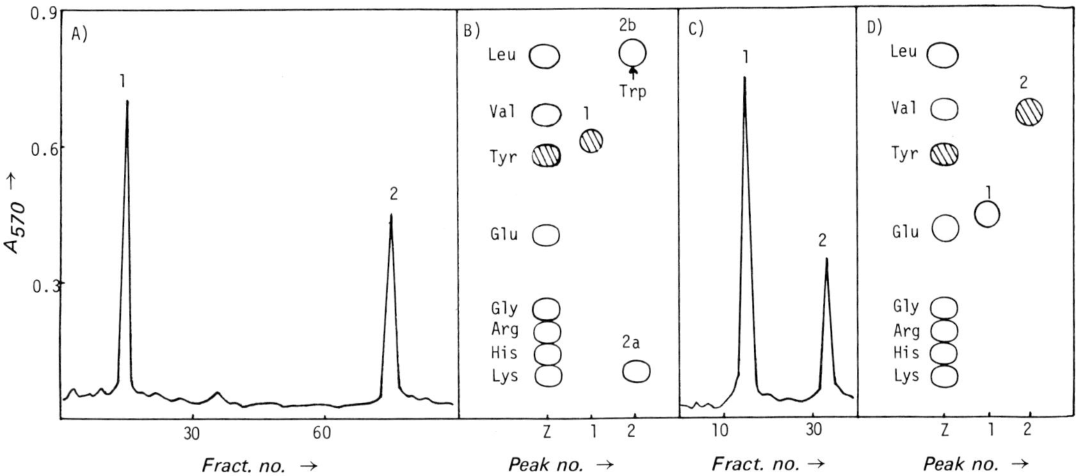

Fig. 4. Column and paper chromatograms of the chymotryptic and subtilitic peptides from Try-S-5b. A) and B) chymotryptic peptides. C) and D) subtilitic peptides.

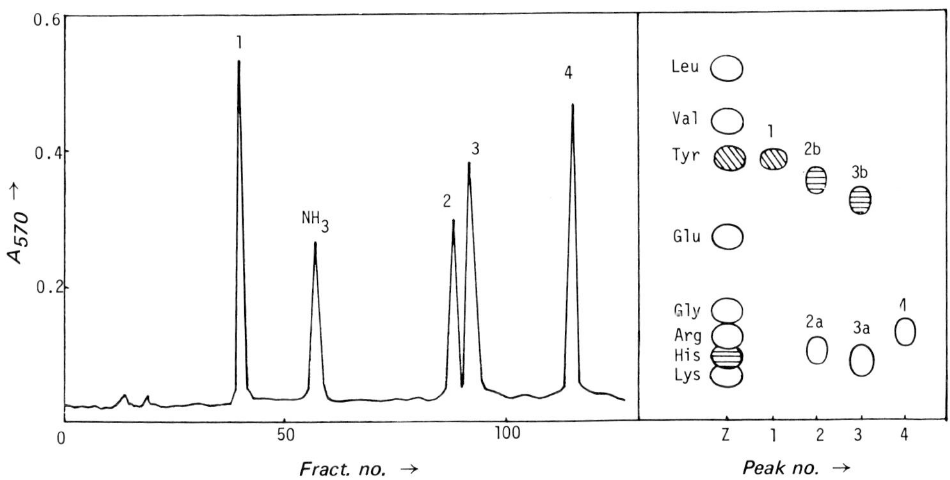

Fig. 5. Column and paper chromatograms of the chymotryptic peptides from Try-S-9a.

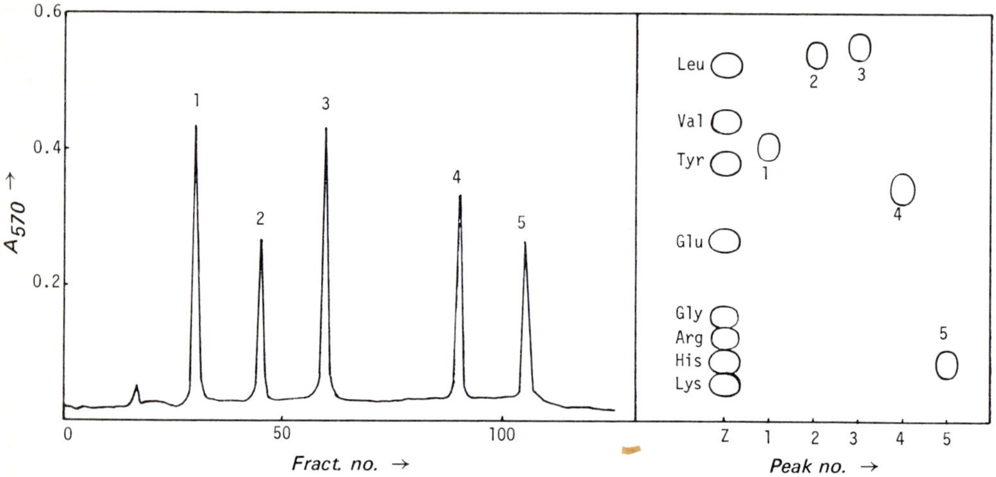

Fig. 6. Column and paper chromatograms of the chymotryptic peptides from Try-I-1.

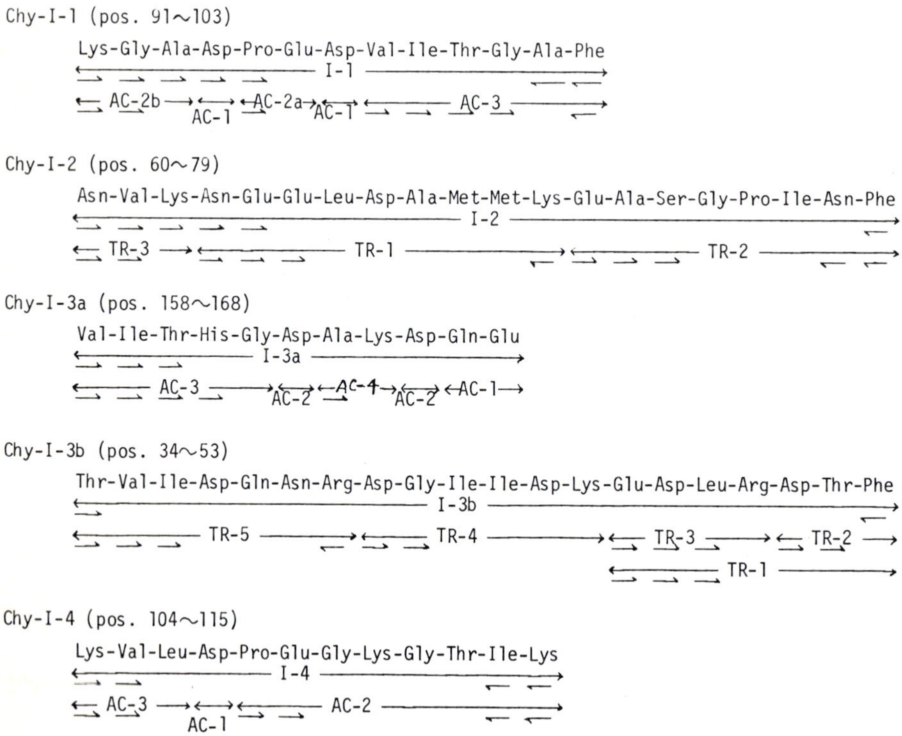

Chy-II-S-1 (pos. 140∼152)

Ala-Ala-Phe-Pro-Pro-Asp-Val-Gly-Gly-Asn-Val-Asp-Tyr

II-S-1

AC-3 AC-1 AC-2 AC-1 AC-4

Chy-II-S-2 (pos. 131∼139)

Ser-Gln-Glu-Glu-Ile-Lys-Asn-Met-Trp

II-S-2

TR-1 TR-2

Chy-II-S-3 (pos. 124∼130)

Thr-Thr-Gln-Cys-Asp-Arg-Phe

Chy-II-S-4 (pos. 1∼5)

X-Pro-Lys-Lys-Ala-Lys

II-S-4

TR-4 TR-1 TR-3

TR-2

Chy-II-I-1 (pos. 19∼29)

Ser-Met-Phe-Asp-Gln-Thr-Gln-Ile-Gln-Glu-Phe

II-I-1

TH-4a TH-4b TH-1a TH-1b

TH-2

TH-3

Chy-III-1 (pos. 8∼18)

Ala-Ala-Ala-Glu-Gly-Gly-Ser-Ser-Asn-Val-Phe

Chy-III-2 (pos. 54∼56)

Ala-Ala-Met

Chy-III-3 (pos. 119∼123)

Leu-Glu-Glu-Leu-Leu

Chy-III-4 (pos. 80∼82)

Thr-Val-Phe

Chy-III-5a (pos. 7∼18)

Arg-Ala-Ala-Ala-Glu-Gly-Gly-Ser-Ser-Asn-Val-Phe

III-5a

TH-1a

TH-4 TH-3b TH-1b TH-3a TH-3c

TH-2

Chy-III-5b (pos. 83∼90)

Leu-Thr-Met-Phe-Gly-Glu-Lys-Leu

Chy-III-6 (pos. 6)

Arg

Chy-III-7 (pos. 30∼33)

Lys-Glu-Ala-Phe

Chy-III-8 (pos. 116∼118)

Lys-Gln-Phe

Chy-III-9 (pos. 57∼59)

Gly-Arg-Leu

Chy-III-10 (pos. 153∼157)

Lys-Asn-Ile-Cys-Tyr

Fig. 7. Amino acid sequences of the chymotryptic peptides from the carboxymethylated L-2 light chain of rabbit skeletal muscle myosin.
The symbols are the same as in Fig. 1.

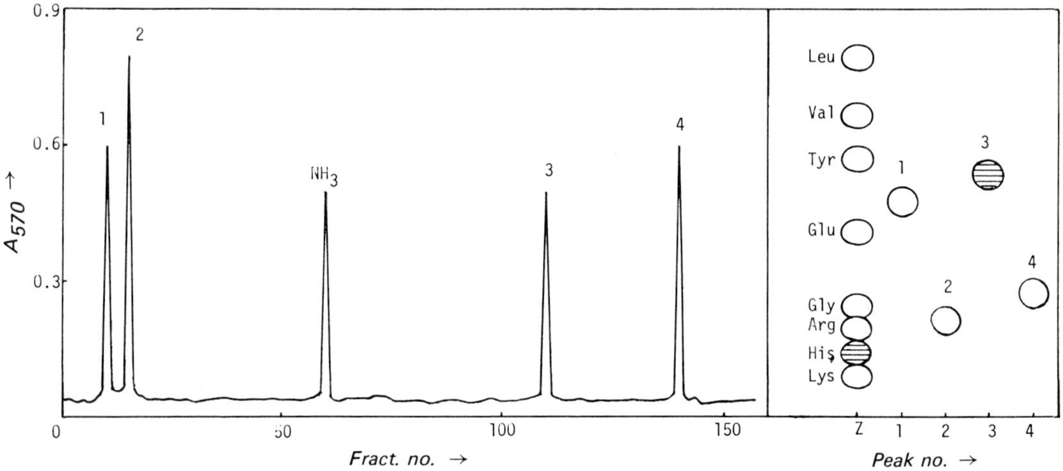

Fig. 8. Column and paper chromatograms of the peptide fragments obtained by partial hydrolysis of Chy-I-3a with 0.25M acetic acid.

The *N*- and *C*-terminal sequences of Chy-I-1 containing 13 amino acids were firstly analysed, and then the peptide was partially hydrolysed with 0.25M acetic acid. The four kinds of peptide fragments obtained were purified by paper electrophoresis and paper chromatography and their amino acid compositions and partial amino acid sequences were analysed.

Chy-I-2 (20 amino acids) was digested with trypsin and the 3 peptide fragments obtained were purified by paper electrophoresis and paper chromatography. The *N*-terminal sequence of Chy-I3a (11 amino acids) was analysed, but its *C*-terminal amino acid was not identified by carboxypeptidase A and B digestion. This peptide was partially further hydrolysed with 0.25M acetic acid and

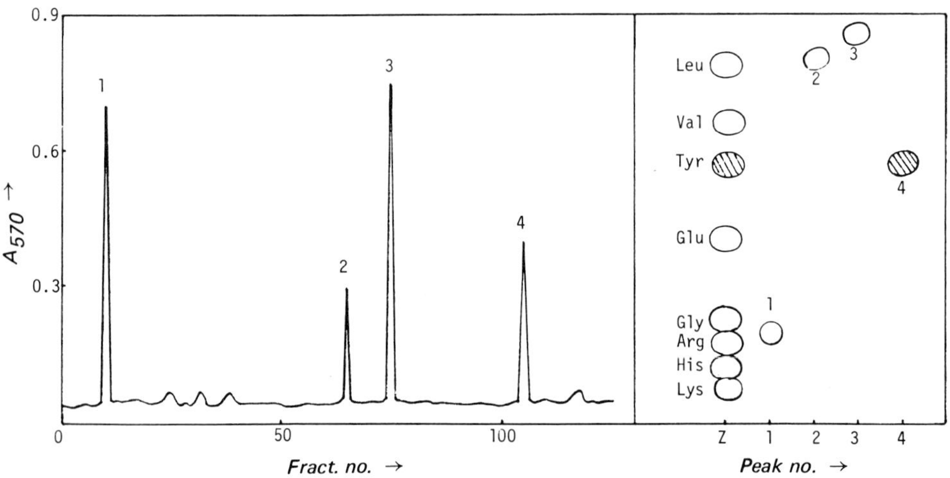

Fig. 9. Column and paper chromatograms of the peptide fragments obtained by partial hydrolysis of Chy-II-S-1 with 0.25M acetic acid.

the 4 kinds of peptide fragments obtained were purified by column and paper chromatography as shown in Fig. 8.

Chy-I-3b (20 amino acids) was digested with trypsin and the 5 peptide fragments obtained were purified by paper electrophoresis and paper chromatography. Chy-I-4 (12 amino acids) was partially hydrolysed with 0.25M acetic acid and the 3 peptide fragments obtained were purified by paper electrophoresis and paper chromatography. Chy-II-S-1 (13 amino acids) was partially hydrolysed with 0.25M acetic acid and the 4 kinds of peptide fragments obtained were purified by column and paper chromatography as shown in Fig. 9.

Chy-II-S-2 (9 amino acids) was digested with trypsin and the 2 peptide fragments obtained were purified by paper chromatography. The amino acid composition of Chy-II-S-3 (7 amino acids) was analysed and after that, its amino acid sequence was analysed directly by subtractive Edman degradation and carboxypeptidase A and B digestion. It was determined by the amino acid analysis and carboxypeptidase A and B digestion that Chy-II-S-4 contains 5 amino acids and the *C*-terminal sequence is Ala-Lys. From this fact,

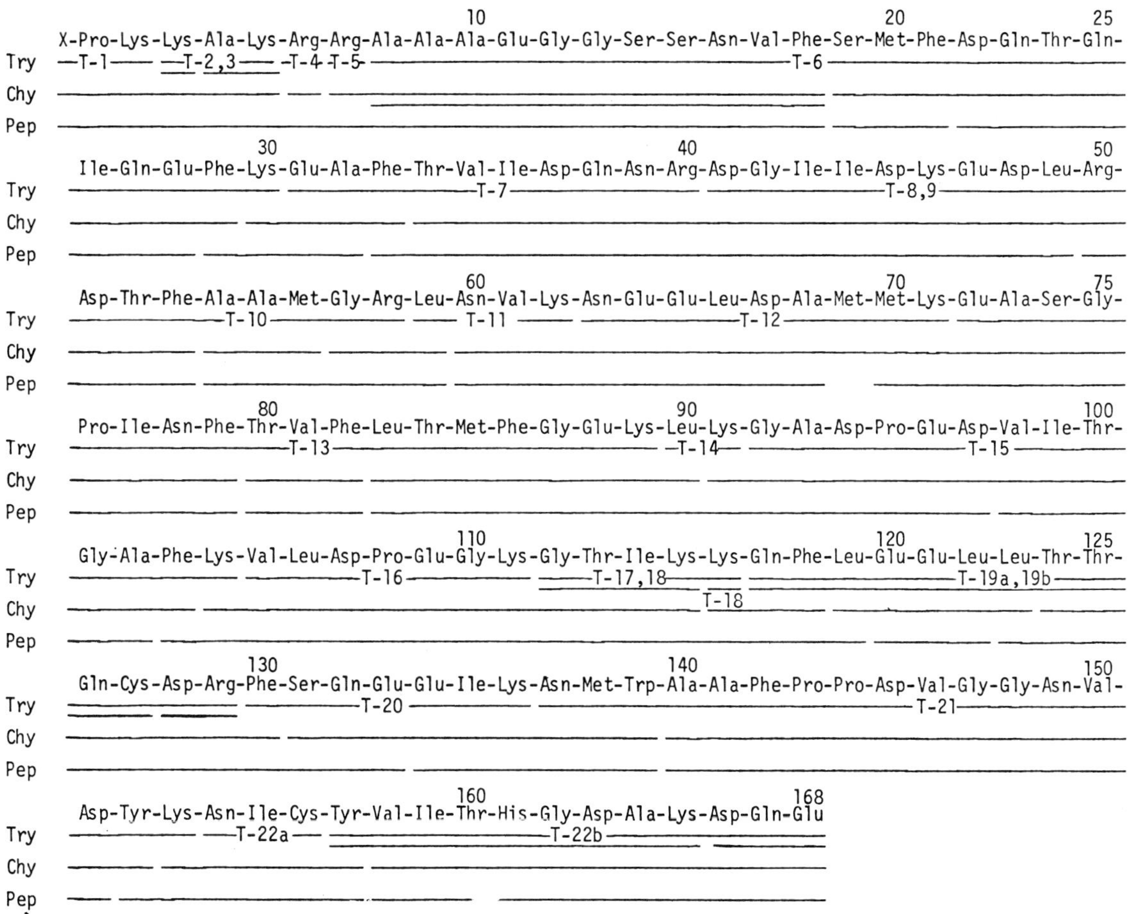

Fig. 10. The primary structure of the L-2 light chain of rabbit skeletal muscle myosin.

In this, tryptic peptides (Try), chymotryptic peptides (Chy) and peptic peptides (Pep) are shown. T-1, T-2, ..., T-22 show the final designations of tryptic peptides.

it was assumed that enzyme material used in this experiment had some trypsin-like activity. As the *N*-terminal amino acid was not identified by subtractive Edman degradation, we assumed that the *N*-terminal was blocked. This peptide was further digested with trypsin, and the peptide fragments obtained were purified by paper electrophoresis.

Chy-II-I-1 (11 amino acids) was digested with thermolysin and the 6 peptide fragments obtained were purified by paper chromatography and paper electrophoresis. The amino acid sequence of Chy-III-1 (11 amino acids) was deduced from the amino acid composition, the *N*- and *C*-terminal sequence and by the comparison of the amino acid sequence of the corresponding tryptic peptide with it. Peptides Chy-III-2, Chy-III-3 and Chy-III-4 contain 3, 5 and 3 amino acids, respectively. After their amino acid compositions were analysed, their amino acid sequences were directly analysed.

Chy-III-5a (12 amino acids) was digested with thermolysin. The 7 peptide fragments obtained were purified by paper chromatography and paper electrophoresis. Peptide Chy-III-5b, Chy-III-6, Chy-III-7, Chy-III-8, Chy-III-9 and Chy-III-10 contain 8, 1, 4, 3, 3 and 5 amino acids, respectively. After their amino acid compositions were analysed, their amino acid sequences were directly analysed by subtractive Edman degradation and carboxypeptidase A and B digestion.

Sequence analyses of the peptic peptides from the L-2 light chain

The main peptic peptides from the carboxymethylated L-2 light chain of rabbit skeletal muscle myosin were designated as Pep-I-1 - Pep-II-4 (for details see Ref.[1]). The amino acid compositions and the *N*- and *C*-terminal sequences of the 19 peptic peptides were analysed. The amino acid sequences of these peptides were deduced from these experimental results and by the comparison of the amino acid sequences of the corresponding tryptic and chymotryptic peptides mentioned in the previous paragraph with them.

The primary structure of the L-2 light chain

The primary structure of the L-2 light chain of rabbit skeletal muscle myosin was deduced from all the results of the experiments described above, as shown in Fig. 10. This protein is composed of 168 amino acids and its *N*-terminal proline might be blocked by some means. These results have been reported as a preliminary communication[2].

On the other hand, Collins[6] also reported the primary structure of the L-2 light chain (DTNB light chain) of rabbit skeletal muscle myosin independently. He reported that this protein is composed of 169 amino acids, that is one more than ours. Besides, there are 8 differences between the two as described in the preliminary communication[2]. These differences might be due to the difference in the type of rabbits used.

We wish to thank Miss Michiko Ishibashi for her help in preparing this paper.

Literature

1 Matsuda, G., Suzuyama, Y., Setoguchi, M., Maita, T., Watanabe, B., Maekawa, T. & Umegane, T. (1978) *Hoppe-Seyler's Z. Physiol. Chem.* **359**, 357 - 368.

2 Matsuda, G., Maita, T., Suzuyama, Y., Setoguchi, M. & Umegane, T. (1977) *J. Biochem. (Tokyo)* **81**, 809 - 811.

3 Konigsberg, W. (1967) *Methods Enzymol.* **XI**, 461 - 469.

4 Hayashi, R., Moore, S. & Stein, W.H. (1972) *J. Biol. Chem.* **248**, 2296 - 2302.

5 Offord, R.E. (1966) *Nature (London)* **211**, 591 - 593.

6 Collins, J.H. (1976) *Nature (London)* **259**, 699 - 700.

Prof. Dr. G. Matsuda*, Dr. T. Maita, Dr. Y. Suzuyama, Dr. M. Setoguchi and Mrs. T. Umegane: Department of Biochemistry, Nagasaki University School of Medicine, Nagasaki 852, Japan.

* To whom correspondence should be addressed.

Hoppe-Seyler's Z. Physiol. Chem.
Bd. 359, S. 641 - 651, Mai 1978

Investigations on the Functional Role of Rat Liver Carboxylesterase Isoenzymes

Rüdiger ARNDT*, Heinz-Eberhard SCHLAAK**, Dieter USCHTRIN**, Dörte SÜDI**, Konrad MICHELSSEN** and Wolfgang JUNGE***

* Abteilung für Immunologie, I. Medizinische Universitätsklinik Hamburg-Eppendorf,
** Biochemisches Institut im Fachbereich Medizin der Universität Kiel
and *** Zentrallabor des Städtischen Krankenhauses Kiel

(Received 19 January 1978)

This paper is dedicated to the late Prof. Dr. Klaus Krisch. His work exerted a great influence on the study of carboxylesterases

Summary: We have compared the substrate specificities of two isoenzymes of rat liver esterase. Isoenzyme E1 which we isolated earlier, and isoenzyme EA, the isolation of which is reported here, have very similar molecular weights, subunit weights, and isoelectric points. In contrast, the two enzymes have distinctly different substrate optima when tested with a series of esterase substrates. Our finding poses the question, as to whether the rather unspecific in vivo behaviour of liver carboxylesterase could be explained by an ensemble of isoenzymes with overlapping substrate specificities and mutually complementary substrate optima. The biological function of esterase isoenzymes appears to represent an extension of the range of hydrolytic competence of a single enzyme.

Untersuchungen zur biologischen Bedeutung der Rattenlebercarboxylesterase-Isoenzyme

Zusammenfassung: Die Substratspezifität von zwei Isoenzymen der Rattenlebercarboxylesterase wird verglichen. Das kürzlich von uns isolierte Isoenzym E1 und das Isoenzym EA, über dessen Isolierung in dieser Arbeit berichtet wird, sind hinsichtlich des Molekulargewichtes, der Untereinheitenstruktur und des isoelektrischen Punktes sehr ähnlich. Sie unterscheiden sich jedoch durch charakteristische Substratoptima, die mit verschiedenen Serien von Esterasesubstraten gefunden werden. Es wird diskutiert, ob die große Unspezifität der Lebercarboxylesterasen in vivo durch ein Zusammenwirken verschiedener Isoenzyme mit übergreifender und sich ergänzender Substratspezifität erklärt werden kann. Die biologische Bedeutung der Esterase-Isoenzyme liegt offensichtlich in der Erweiterung des Bereichs der hydrolytischen Kompetenz eines einzelnen Enzymes.

Key words: Carboxylesterases, isoenzymes, rat liver, substrate specificity, gas-liquid chromatography.

The broad substrate specificities of carboxylesterases for aliphatic and aromatic esters as well as for aromatic amides, is characteristic for this subclass of enzymes. The so called unspecificity is responsible for the difficulties in differentiating the carboxylesterases with respect to their

Enzyme: Carboxylesterase, carboxylic-ester hydrolase (EC 3.1.1.1).

catalytic function. In recent communications we reported on the chromatographic differentiation of five rat liver carboxylesterases[1] and the isolation of the rat liver esterase E1[2,3]. The aim of this investigation is to elucidate the functional role of rat liver carboxylesterase isoenzymes by isolating esterase EA and comparing its catalytic properties with that of esterase E1.

The experiments presented here are an approach to the understanding of the physiological function of the carboxylesterase isoenzymes which may lie in their complementary and overlapping substrate specificity.

Methods

1. *Animals*

Male Wistar rats, weighing approximately 200 g were kept on a standard diet (Altromin) and fasted 24 h prior to sacrifice.

2. *Protein determination*

The protein concentration of the enzyme samples was determined by a microbiuret procedure[4].

3. *Enzyme assay*

If not stated otherwise, all enzyme assays were performed at pH 8.0 and at a temperature of 30 °C.

a) Anilides

The enzymatically formed aniline was determined by a modification[5] of the diazotization procedure of Brodie and Axelrod[6].

b) Aliphatic esters

The hydrolysis of these substrates was followed by the pH-stat method using the autotitrator TTT 11 with automatic burette (ABU 13; Radiometer Copenhagen/Denmark) and continuous registration (titrigraph SBR 2 C).

c) Gas chromatographic analyses of the enzymatic hydrolysis of triglycerides

Triglycerides from triacetin to tricaproin (Serva, BRD) were emulsified in water by sonification. The enzymatic hydrolysis of 5 m*l* substrate emulsion was performed by the pH-stat method. The amount of triglycerides in the assay was 1.71×10^{-5} mol. After different incubation times a sample of 1 m*l* of the assay was lyophilized at exactly 10^{-1} Torr. The residue was dissolved in 96 μ*l* hexane or ether and 4 μ*l* *N*-methyl-*N*-trimethylsilyl-trifluoracetamide (MSTFA, Macherey & Nagel, BRD). For gaschromatography a GC 990 (Perkin Elmer) and a calculator (Minigrator Spectra-Physics) was used. Glass columns (length 140 cm or 200 cm, internal diameter 0.2 cm) were filled with Chromosorb G-AW-DMCS 60 - 80 mesh (Macherey & Nagel) or Gas Chrom Q 80 - 100 mesh (Applied Sience Lab. U.S.A.) layered with 5.0% OV 17 (Macherey & Nagel)/0.5% Dexsil 300 GC (Riedel - de Haen, BRD)[7].

4. *Isolation of rat liver carboxylesterase (EA)*

Unless otherwise mentioned all operations were carried out at a temperature of 0 - 4 °C.

a) Preparation of the microsomal fraction, solubilization of the esterase EA from the microsomes, ammonium sulfate fractionation and gel-filtration on Sephadex G-100 were carried out as described recently in detail for the esterase E1[2].

b) Chromatography on Sephadex A-50

The acetanilide-active fractions of the preceding chromatography on Sephadex G-100 or G-200 were combined and immediately applied to a Sephadex A-50 (Pharmacia, Uppsala, Sweden) ion exchange chromatography column (dimensions 2.5 x 25 cm) and eluted with a phosphate buffer gradient. The gradient was prepared by a 4-chamber elution system; the vessels each contained 150 m*l* of phosphate buffer (pH 8.0) in the following concentrations; (1) 0.01M; (2) 0.03M; (3) 0.06M and (4) 0.12M. The elution pattern was characterized by two esterase-active peaks. The first peak eluting in the range of 0.035 - 0.05M phosphate buffer had a high activity against methyl butyrate corresponding to the recently described esterase E_1 while the second peak eluting at 0.08 - 0.11M phosphate buffer showed only small activity against methyl butyrate but high activity against acetanilide. The fractions of the second peak, containing the acetanilide hydrolyzing isoenzyme EA were pooled and concentrated by lyophilisation.

c) Gelfiltration on Sephadex G-200

The combined and concentrated eluates from the previous step were applied to a Sephadex G-200 column and eluted with 0.005M phosphate buffer pH 6.0. The elution diagram essentially consisted of one acetanilide active peak, recently designated as EA_2. The combined fractions of this peak were taken for further purification.

d) Isoelectric focusing

After dialysis of the combined fractions for 12 h against distilled water the sample was then submitted to isoelectric focusing on a 110 m*l* column (LKB 8101) with pH 5 - 7 ampholyte (ampholine 8153) according to Vesterberg and Svensson[8]. The elution pattern yielded one peak where the protein curve and the enzyme activity ran in parallel (Fig. 1).

5) Analytical polyacrylamide gel electrophoresis

a) Polyacrylamide gel electrophoresis for the estimation of the subunit weight was performed in the presence of 0.1% sodium dodecylsulfate and 2-mercaptoethanol, as outlined by Weber and Osborn[9]. For calibration, the following proteins were used: Cytochrome *c*, β-lactoglobulin, ovalbumin, γ-globulin and bovine serum albumin (all obtained from Serva, BRD).

b) Disc electrophoresis was performed according to Rodbard and Crambach[10]. The buffer system "B" of these authors, consisting of glycine/phosphate chloride, was employed. In addition, the molecular weight was estimated by disc electrophoresis in polyacrylamide gels of different concentrations[11].

6) Analytical gel filtration

Analytical gel filtration experiments were performed at 4 °C using columns (1 cm x 100 cm) under the constant hydrostatic pressure of a 1-m water column. Standard proteins were dissolved in elution buffer at a concentration of 2 mg/m*l*. Blue dextran 2000 and dinitrophenylalanine were added as markers for the void and internal volume of the gel. Experimental data were expressed in terms of the distribution coefficient σ[12] which is defined as

$$\sigma = \frac{V_e - V_o}{V_i}$$

where V_e is the elution volume of the protein, V_o the void volume (determined by dextran blue 2000) and V_i the internal volume of the gel (determined by dinitrophenyl-alanine). σ values were plotted versus the logarithm of the Stokes radius of the reference proteins and a linear equation was calculated by regression analysis*. The logarithm of the Stokes radius of the esterase EA was determined from the linear equation using the experimentally determined distribution coefficient σ. The molecular weight M_r was calculated according to the equation

$$R_S = \frac{f}{f_0} \times (\bar{V} \times M_r)^{\frac{1}{3}} \times \left(\frac{3}{4 \times \pi \times N}\right)^{\frac{1}{3}}$$

where R_S is the Stokes radius, N Avogadro's number, f/f_0 the frictional coefficient (a value of 1.12 was used, as calculated for solvated proteins with normal hydratation of 0.3 g water/g protein[13], $\bar{V}$ the partial specific volume (values between 0.7 and 0.75 g cm^3 g^{-1} were used for the various proteins[14]. For associated proteins a frictional coefficient of 1.17 was used, to account for deviation from the proposed approximately spherical shape[15].

* Schlaak, H.E. & Gersondo, K. (1978) *Biochim. Biophys. Acta,* submitted.

Results

1) Purification of the esterase EA

The solubilization procedure solubilized more than 90% of the total acetanilide hydrolyzing activity. The purification steps were similar to those reported previously[2] for the isolation of the esterase E_1. The enzyme described here corresponds to the esterases previously described as EA_1 and EA_2[1]. The results of the purification procedure are shown in Table 1. The esterase EA was purified more than 1000-fold and the overall yield was 6.3%. The specific activity of the purified enzyme for acetanilide was about 6 - 7 μmol x min^{-1} x mg^{-1} in all preparations, although the degree of purification differed slightly depending on the specific activity of the microsomes. The elution profile of the enzyme after isoelectric focusing on ampholine of the pH range 5 - 7 shows one single protein peak, eluted between pH 5.3 and 5.6 running coincidentally with the

Table 1. Purification of carboxylesterase (EA) from rat-liver microsomes.

Purification step	Spec. act.[b] [U/mg]	Activity [U]	Purification [-fold]	Yield [%]
I. 100 000 x *g* sediment	0.0061	319	1	100
II. Crude extract	0.26	382	42.5	119
III. Ammonium sulfate fraction (45 - 65%)	0.38	290	62.3	91
IV. Gel filtration (Sephadex G-200 pH 8.0)[a]	0.54	234	88.5	73
V. DEAE-Sephadex A-50 chromatography[a]	0.90	173	147.5	54
VI. Gel filtration (Sephadex G-200 pH 6.0)[a]	1.58	78.5	259	24.6
VII. Isoelectric focusing[a]	6.92	20.25	1134	6.3

[a] Combined fractions, [b] Substrate: acetanilide.

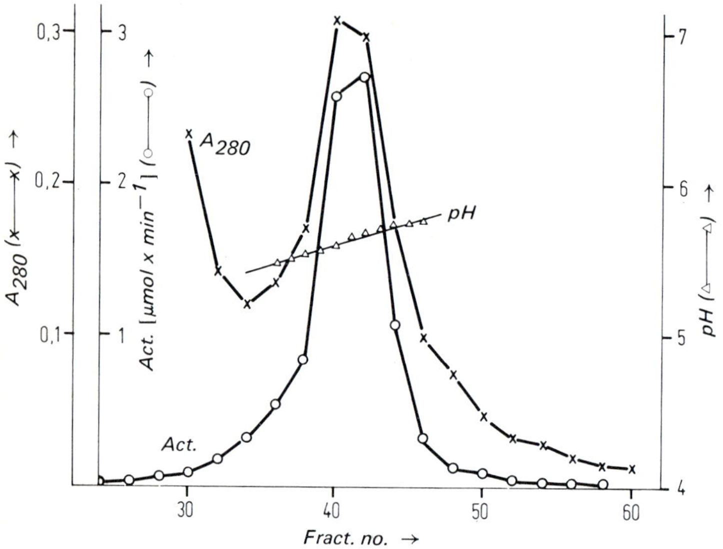

Fig. 1. Isoelectric focusing of rat-liver esterase (EA) with ampholine pH 5 - 7.

acetanilide hydrolyzing activity (Fig. 1). The purified enzyme proved to be homogeneous on disc electrophoresis in polyacrylamide gels (Fig. 2a).

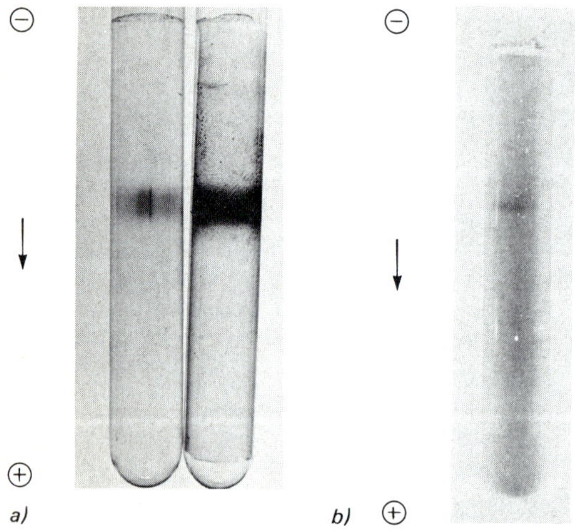

Fig. 2 a). Disc electrophoresis of native rat-liver esterase (EA) in 5% polyacrylamide gel.

Left column protein stained with coomassie blue; 20 µg enzyme applied. Right column substrate stained with α-naphthylacetate; 50 µg enzyme applied.

b). Dodecylsulfate electrophoresis of rat-liver esterase (EA) in 6% polyacrylamide gel.

Protein stained with coomassie blue; 20 µg enzyme was applied.

2) Interconversion of two molecular forms of the esterase EA

In gel filtration experiments the esterase EA appeared as two molecular forms of different size. The association and dissociation of the enzyme was dependent on pH and ionic strength. Gel filtration at pH 8.0 and high salt concentration (0.05M phosphate buffer; 1M KCl) yielded two acetanilide active peaks designated as peak I and II in Fig. 3. A change of the conditions to higher pH and lower ionic strength (pH 9.5; 0.05M glycine/NaOH) yielded a similar pattern. After rechromatography of either of the separated peaks I or II under identical experimental conditions two active peaks were observed again in the same positions as shown in Fig. 3. If the enzyme was first chromatographed at pH 6.0, using 0.05M phosphate buffer and subsequently rechromatographed under the conditions stated above, no shift of enzyme activity from peak II to peak I position was observed. This molecular form of the enzyme was subjected to dodecylsulfate electrophoresis and a single protein band was obtained (Fig. 2b). The electrophoretic mobility of this protein band corresponded to a molecular weight of 61 500 (Table 2), which was in good agreement with the results obtained by gel filtration (60 000) and by disc electrophoresis in polyacrylamide gels of different concentrations (58 000) (Table 2). For the larger form of the enzyme a Stokes radius of

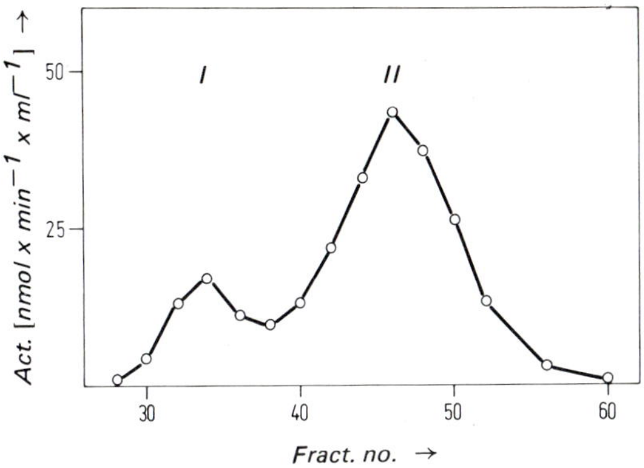

Fig. 3. Rechromatography of peak II of the rat-liver esterase EA on Sephadex G-100 in 0.05M phosphate buffer, pH 8.0, 1M KCl.

0.1 mg enzyme was applied to the column.

43 Å corresponding to a molecular weight of 180 000 was estimated by gel filtration, which indicated that it was a trimeric form of the 60 000 molecular weight unit[16,17].

3) pH-optimum

The pH dependence of the esterase EA activity was characterized by a sharp peak between pH 9 and 10 (Fig. 4).

Table 2. Some physical properties of rat-liver carboxylesterase (EA).

Property	Method	Result
Molecular weight	Gel filtration (Sephadex G-200)	180 000
Subunit weight	Gel filtration (Sephadex G-100)	60 000
	Disc electrophoresis[11]	58 000
	Dodecylsulfate polyacrylamide gel electrophoresis[9]	61 500
Stokes radius	Gel filtration	
monomer	(Sephadex G-100)	29 Å
trimer	(Sephadex G-200)	43 Å
Isoelectric point	Isoelectric focusing	pH 5.3 - 5.6

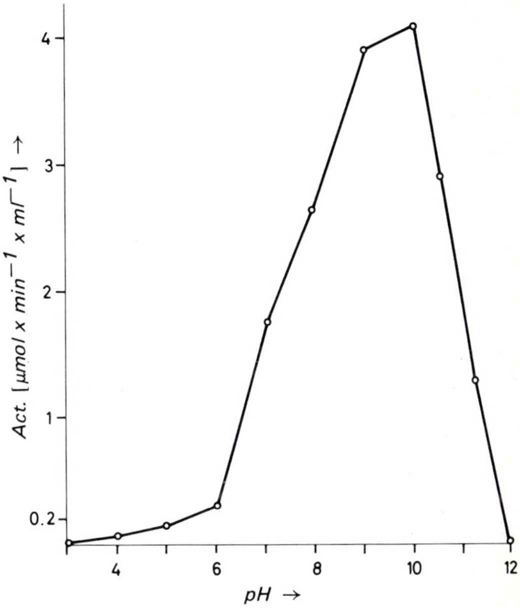

Fig. 4. pH optimum of rat-liver esterase EA.

Substrate: 10^{-3}M acetanilide.

Buffer system: 0.04M phosphoric acid, 0.04M acetic acid and 0.04M boric acid. The desired pH was achieved by addition of 0.2M NaOH.

4) Inhibition of the hydrolytic activity of the esterase EA

Three inhibitors specific for the serine function and two SH-reagents were used in inhibition experiments (Table 3). Of these only *O,O*-diethyl-*O*-(*p*-nitrophenyl)phosphate inhibited the acetanilide activity completely in both its membrane-bound state and in its solubilized form. In contrast, bis(*p*-nitrophenyl)phosphate inhibited the membrane-bound and solubilized enzyme to different extents. The inhibition was complete only with the membrane-bound enzyme, while the solubilized enzyme retained a residual activity of 45%. Pretreatment of the isoenzyme EA with *p*-chloromercuribenzoate, $HgCl_2$ and phenylmethanesulfonyl fluoride led to a loss of only 30 to 50 percent of the original acetanilide activity, irrespective of whether the enzyme was membrane-bound or not. To investigate whether or not this pattern of inhibition was specific for the substrate acetanilide, we performed a further set of experiments with tributyrin as substrate. In sharp contrast to the inhibition of the acetanilide hydrolyzing activity, the hydrolysis of tributyrin was completely inhibited by all five reagents when the enzyme was in its purified form. Obviously the different isoenzymes EA and E1 of rat liver esterases differ with respect to their sensitivity to SH reagents. This is apparent from comparison of the last two columns of Table 3, showing that *p*-chloromercuribenzoate had no influence on the catalytic activities of the isoenzyme E1 whereas EA was completely inhibited with tributyrin as substrate.

5) Comparison of substrate specificity of esterases EA and E1

Of all vertebrate esterases so far characterized, the isoenzyme EA had the highest activity for acetanilide, but only very low activity for the aliphatic ester methylbutyrate. The isoenzyme E1 preferentially cleaved methylbutyrate and showed almost no activity for acetanilide. The question as to which of the different substrate properties, long chain or short chain acyl residue, amide or ester bond, aromatic or aliphatic residue, are responsible for the substrate specificities of the isoenzymes remains unresolved.

5.1) Acyl residues of aromatic amides

The influence of the chain length of unbranched acyl residues of anilides on the rate of hydrolysis is demonstrated in Fig. 5. The plot of the maximal velocity *V* against the number of acyl chain carbon atoms shows the highest activity with acetanilide as the substrate, followed by a steep decline with substrates containing more than two carbon atoms. In contrast to this, the isoenzyme E1 did not significantly hydrolyze the C_2 or C_3 derivatives and the activity curve is characterized by a sharp increase from C_4 to C_5 (Fig. 5).

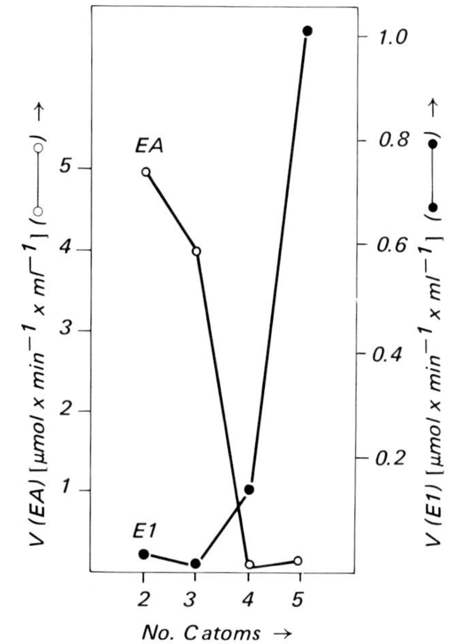

Fig. 5. Influence of the chain length of the acyl groups of anilides on the maximal velocity *V* of rat-liver esterases EA and E1.

Table 3. Influence of inhibitors on rat-liver carboxylesterase EA and E1.

Inhibitor	Conc. [mol/*l*]	% residual activity after 30 min preincubation			
		Acetanilide		Tributyrin	
		Microsomes	Purified enzyme EA	Purified enzyme EA	Purified enzyme E1
Phenylmethanesulfonyl fluoride (PMSF)	10^{-4}	54	48	0	0
Bis(*p*-nitrophenyl)phosphate (BNPP)	10^{-3}	0	45	0	0
O,O-Diethyl-*O*-(*p*-nitrophenyl)phosphate (Paraoxon, E 600)	10^{-3}	0	0	0	0
p-Chloromercuribenzoate (PCMB)	10^{-3}	78	63	0	100
$HgCl_2$	10^{-3}	66	69	0	41

5.2) Alkyl and acyl residues of aliphatic esters

Fig. 6 and 7 show that the activities of the isoenzymes EA and E1 differ with respect to their dependence on the chain length of the alkyl group of acetate esters and the chain length of the acyl group of methyl esters. The esterase E1 preferred acetate esters with alkyl groups of C_1 - C_3, whereas the esterase EA has a substrate optimum at C_3 - C_5 (Fig. 6). A similar pattern was observed when acyl derivatives of methyl esters were used as substrates although E1 showed higher activity against these substrates. While E1 had a sharp optimum at C_4 the esterase EA preferred C_5 - C_6 derivatives of methyl esters (Fig. 7).

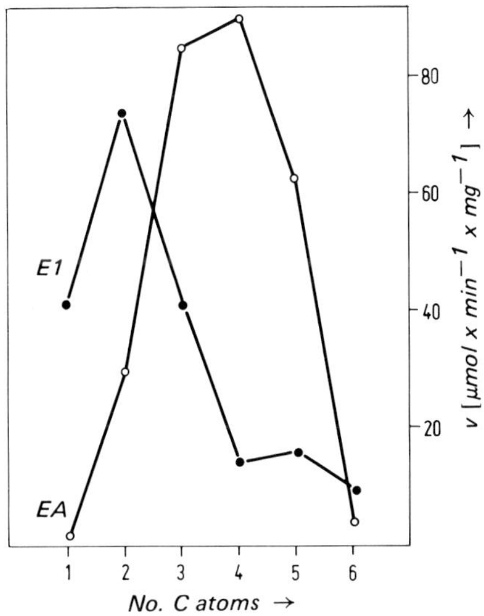

Fig. 6. Influence of the chain length of the alkyl group of acetate esters on the reaction velocity at 10^{-3}M substrate concentration of rat-liver esterases EA and E1.

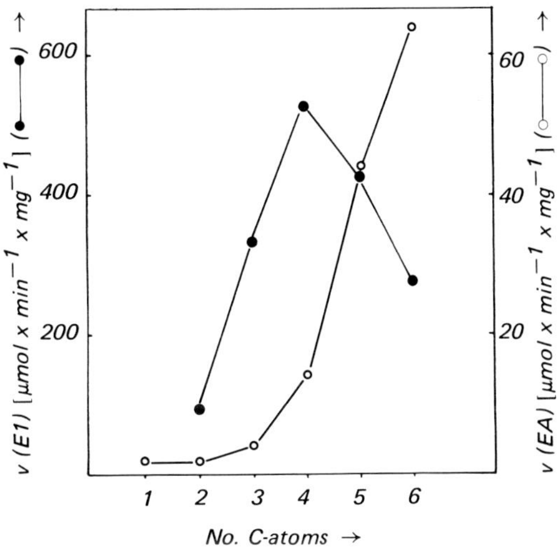

Fig. 7. Influence of the chain length of the acyl group of methyl esters on the reaction velocity at 10^{-3}M substrate concentration of rat-liver esterases EA and E1.

Table 4 Enzymatic hydrolysis of triglycerides by rat liver carboxylesterase EA and E1.
One assay mixture contained 5 *ml* substrate emulsion (1.71×10^{-5} mol of triglycerides) and 60 or 80 μg of E1 or EA, respectively.

Carboxylesterase isoenzyme	Chain length of acyl residues of triglycerides	Relative rate of hydrolysis of		
		triglycerides to diglycerides	diglycerides to monoglycerides	monoglycerides to glycerol and free fatty acids
E1	C_2	+++	++	+
EA		+++	+	–
E1	C_3	++++	+++	+
EA		++++	++	–
E1	C_4	++++	+++	+++
EA		++++	++	(+)
E1	C_5	+++	++++	+++
EA		+	++	+
E1	C_6	+++	++++	+++
EA		–	–	–

▷

Fig. 8. Gas chromatograms of tributyrin hydrolysis catalyzed by isoenzyme EA.
(Details of hydrolysis given in Materials and Methods.) A = without enzyme; B, C, and D = incubation for 4, 10, and 20 min., resp. with 80 μg isoenzyme EA; E = final result obtained in 40 min. 1 = tributyrin; 2 = TMS-dibutyrin; 3 = $(TMS)_2$-monobutyrin; 4 = $(TMS)_3$-glycerol. Column length 140 cm, column diameter 2 mm; injector: 220 °C; program: 2 min; isotherm at 80 °C, then 4 °C per min; nitrogen: 30 m*l* per min. Five per cent OV 17/ 0.5% Dexsil 300GC with Gas Chrom Q 80 - 100 mesh.

5.3) Glycerides of aliphatic acids

The foregoing experiments demonstrated the differential dependence of the two isoenzymes on the chain length of acyl residues of monoesters. The analysis of the substrate specificity was therefore extended to aliphatic esters of glycerol to see whether, in addition to the chain length, the degree of esterification of polyalcohols was responsible for the specificity of the isoenzymes. The mono-, di- and triesters of glycerol with short chain fatty acids from C_2 - C_6 were used. The enzymatic hydrolysis of esters emulsified in water by the esterases EA and E1 is shown in Table 4. In the experiments shown in Table 4 we did not generally succeed in resolving the consecutive reaction steps of tri-, di- and monoglyceride hydrolysis. Accordingly the table only contains rough estimates of relative rates, expressed by plus or minus signs. However, we could always decide by gas chromatographic analysis, whether a particular intermediate was enriched and to what extent a polyester was hydrolyzed by the enzyme. For instance triacetin, tripropionin and tributyrin were cleaved by EA to monoglycerides with the transient accumulation of diglyceride intermediates (Fig. 8). The final products of the reaction were monoglycerides with only traces of glycerol present. The rate of enzymatic hydrolysis of monoglycerides was too small to be detectable by the pH-stat method. In contrast to EA the isoenzyme E1 efficiently cleaved the monoesters of the substrates mentioned above to glycerol and free fatty acids (Fig. 9). When the monoesters generated by EA were presented as substrates to E1, this enzyme (Fig. 10) completed the degradation. Further interesting differences between the two isoenzymes could be observed in the ratios

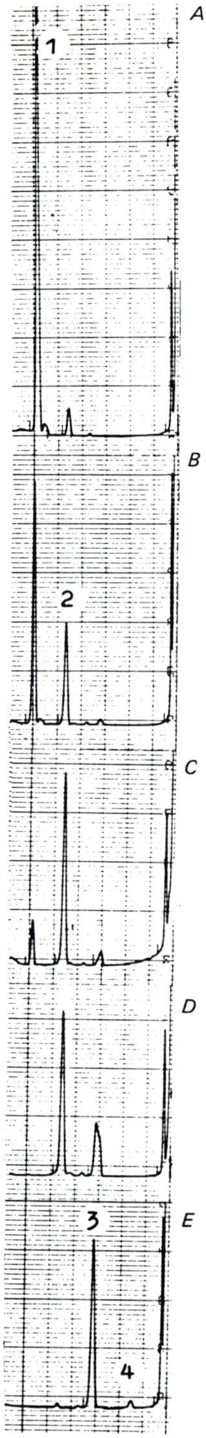

Fig. 8.

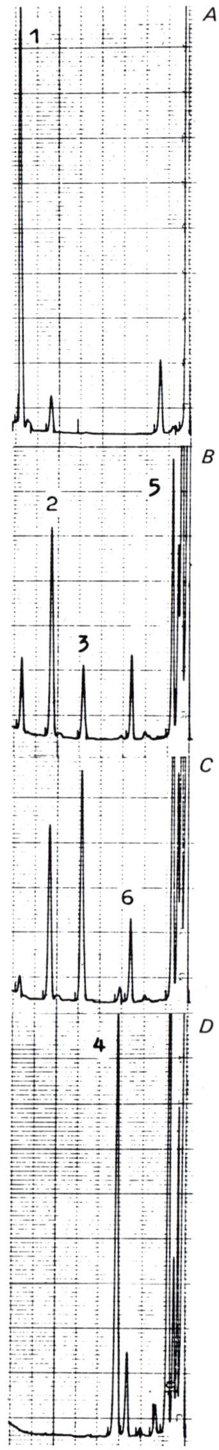

Fig. 9.

◁

Fig. 9. Gas chromatograms of tributyrin hydrolysis catalyzed by isoenzyme E1.

Details of hydrolysis are given in Materials and Methods. A = without enzyme; B and C = incubation for 4 and 8 min, resp., with 60 μg isoenzyme E1; D = final result obtained with isoenzyme E1 in 70 min. 1 = tributyrin; 2 = TMS-dibutyrin; 3 = $(TMS)_2$-monobutyrin; 4 = $(TMS)_3$-glycerol; 5 = TMS-butyric acid; 6 = dodecane standard[7]. Column length 200 cm, column diameter 2 mm; injector: 220 °C; program: 1 min isotherm at 70 °C, then 4 °C per min; nitrogen: 30 m*l* per min. Five per cent OV 17/0.5 % Dexsil 300 GC with Chromosorb G-AW-DMCS 60 - 80 mesh.

of the rates at which the substrate and its first intermediate were hydrolyzed. The ratio, rate of hydrolysis of triglyceride to rate of hydrolysis of diglyceride, was high with EA and low with E1. In addition tri- and diglycerides with acyl residues of C_5 and C_6 were more efficiently hydrolyzed by E1 than by EA. While C_6 di- and triglycerides were not cleaved by EA, the enzyme showed some activity for C_5 monoglycerides. No accumulation of intermediates was observed during enzymatic hydrolysis of C_5 and C_6-triglycerides. Interestingly neither enzyme hydrolyzed triglycerides of saturated fatty acids with acyl residues longer than C_6 under these conditions.

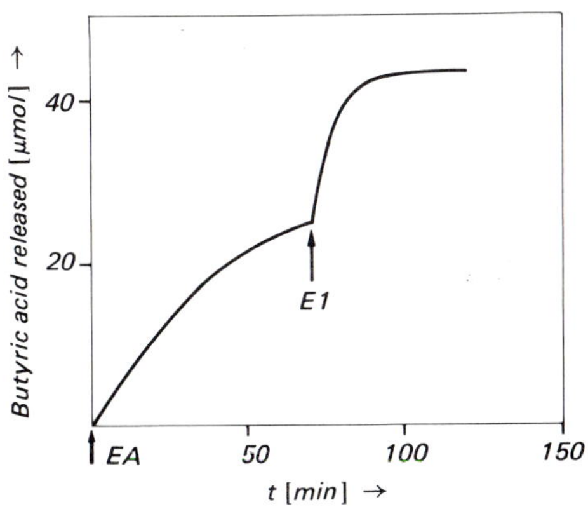

Fig. 10. Enzymatic hydrolysis of tributyrin.

The liberation of butyric acid is determined by pH-stat method. Volume of the assay: 5 m*l* tributyrin (1.71 × 10^{-5} mol); EA: addition of 100 μg EA; E1: addition of 60 μg E1.

Discussion

The broad substrate specificity of carboxylesterases has prevented their exact classification amongst the hydrolases, and also hindered the understanding of their physiological function. In our previous studies[1,2] we have already obtained evidence for multiple enzyme forms of rat liver carboxylesterases. In this paper we report the characteristic preferences of the isolated isoenzymes E1 and EA with regard to different substrates. These isoenzymes were highly sensitive to *O,O*-diethyl-*O*-(*p*-nitrophenyl)phosphate (E 600) inhibition and therefore could not be identical to the genetically defined rat esterase isoenzymes described by Womack and Sharp[18]. Various attempts have been made to classify unspecific esterases on the basis of their hydrolytic specificities[19], but each of these classifications is established on the rate of hydrolysis of a few arbitrarily chosen substrates. Although several unspecific rat liver esterases are described[20-25] some of these enzymes may represent identical esterases. It should be noted that such attempts at characterization of particular isoenzymes can hardly contribute to a clarification of the functional significance of the heterogeneity in carboxylesterases. Some positive progress in this direction is made in the present paper by the isolation of two isoenzymes with similar physical chemical properties, and the comparison of their catalytic efficiencies toward a series of structurally related substrates. The prerequisite of our experimental design is the isolation of the two isoenzymes, so that they are essentially free of cross contamination. This was achieved , as evidenced by the fact that isoenzyme EA has very high activity towards acetanilide, and as its dissociation behaviour has a characteristic dependence on ionic strength. Each of the two isoenzymes used in our studies is electrophoretically homogeneous, and essentially free of cross contamination.

The results presented clearly indicate that isoenzyme E1 and EA do not markedly differ in their catalytic properties. However, it is quite easy to define the difference between the two isoenzymes in terms of substrate optima within a homologous series. For instance, especially in comparison to isoenzyme EA, isoenzyme E1 preferentially cleaves short aliphatic chains both in the acyl, and in the alkyl moiety. A different rule is obeyed in the hydrolysis of aromatic amides: Isoenzyme EA is more active than isoenzyme E1 toward short chain aromatics. These observations also indicate that the mechanisms of hydrolysis of aliphatic esters and aromatic amides are not identical. Obviously, there is no fundamental difference between the hydrolysis of ester and amide bonds. Therefore, this last conclusion apparently agrees with the suggestion of Greenzaid and Jencks[26], according to which the catalytic mechanism of the hydrolysis of aromatic and aliphatic esters is distinctly different.

Comparison of the hydrolysis of aliphatic esters and aromatic amides by isoenzymes E1 and EA can also be used to illustrate the suggestion that the physiological role of these isoenzymes is not limited to the hydrolysis of a single substrate. Instead, the two isoenzymes complement each other and in cooperation they hydrolyse a wide range of ester and amide substrates. This interpretation of the physiological significance of esterase isoenzymes is particularly obvious with regard to the complete hydrolysis of triglycerides. Since in our experimental arrangement the hydrolysis of tri-, di-, and monoglycerides partially overlaps, we cannot assign rate constants to the individual hydrolytic reactions. However, gas chromatographic analysis of the hydrolysis products of a series of triglycerides by isoenzyme E1 and EA convincingly demonstrates that these two isoenzymes differ markedly in their substrate specificities. The hydrolytic activity of the two enzymes is also different at intermediate stages of triglyceride hydrolysis. For example, the hydrolysis of tributyrate by isoenzyme E1 and EA leads to the transient accumulation of different intermediates, the further hydrolysis of which obviously limits the overall rate. It is tempting to explain the physiological function of esterase isoenzymes on the basis of the specificities reported here for isoenzymes E1 and EA, even if the tested mono and polyesters are not necessarily physiological substrates, since our studies involve several series of structurally related substances. It appears that a gradual shift of substrate optima among members of an ensemble of esterase isoenzymes, especially with regard to physiologically relevant substrates, would lead to the competent hydrolysis of a wide spectrum of compounds by "unspecific liver esterase". The physiological significance of this suggestion is especially obvious with regard to the complete hydrolysis of polyesters in consecutive reaction steps, where the preference of a particular isoenzyme for each esterification level would result in a very favourable rate of the overall hydrolysis reaction.

Gratitude is expressed to *Professor H.-J. Staudinger* for stimulating this paper. The authors thank *Drs. A. Corfield, O.H.W. Martini* and *J. Südi for help in editing the manuscript.*

Literature

1 Arndt, R. & Krisch, K. (1972) *Hoppe-Seyler's Z. Physiol. Chem.* **353**, 589 - 598.
2 Arndt, R. & Krisch, K. (1973) *Eur. J. Biochem.* **36**, 120 - 128.
3 Arndt, R., Heymann, E., Junge, W. & Krisch, K. (1973) *Eur. J. Biochem.* **36**, 129 - 134.
4 Beisenhertz, G., Boltze, H.J., Bücher, Th., Czok, R., Garbade, K.H., Meyer-Arendt, E. & Pfleiderer, G. (1953) *Z. Naturforsch.* **8b**, 555 - 577.
5 Krisch, K. (1963) *Biochem. Z.* **337**, 531 - 545; 546 - 573.
6 Brodie, B.B. & Axelrod, I. (1948) *J. Pharmacol. Exp. Ther.* **94**, 22 - 28.
7 Donike, M. (1973) *Chromatographia* **6**, 190 - 195.
8 Vesterberg, O. & Svensson, H. (1966) *Acta Chem. Scand.* **20**, 820 - 834.
9 Weber, K. & Osborn, M. (1969) *J. Biol. Chem.* **176**, 259 - 266.
10 Rodbard, D. & Crambach, A. (1971) *Anal. Biochem.* **40**, 95 - 134.
11 Hedrick, J.L. & Smith, A.J. (1968) *Arch. Biochem. Biophys.* **126**, 155 - 164.
12 Ackers, G.U. (1970) *Adv. Protein Chem.* **24**, 343 - 446.
13 Squire, Ph. G. (1964) *Arch. Biochem. Biophys.* **107**, 471 - 478.
14 Schachmann, K.H. (1957) *Methods Enzymol.* **4**, 32 - 103.
15 Oncley, J.L. (1941) *Ann. N.Y. Acad. Sci.* **41**, 121 - 150.
16 Junge, W., Krisch, K. & Hollandt, H. (1974) *Eur. J. Biochem.* **43**, 379 - 389.
17 Junge, W., Heymann, E., Krisch, K. & Hollandt, H. (1974) *Arch. Biochem. Biophys.* **165**, 749 - 763.
18 Womack, J.E. & Sharp, M. (1976) *Genetics* **82**, 665 - 675.
19 Junge, W. & Krisch, K. (1975) *Crit. Rev. Toxicol.* **3**, 371 - 434.
20 Hayase, K. & Tappel, A.L. (1969) *J. Biol. Chem.* **244**, 2269 - 2274.
21 Ljungquist, A. & Augustinsson, K.-B. (1971) *Eur. J. Biochem.* **23**, 303 - 313.
22 Akao, T. & Omura, R. (1972) *J. Biochem.* **72**, 245 - 256.

23 Haugen, D.A. & Suttie, J.W. (1974) *J. Biol. Chem.* **249**, 2717 - 2722.
24 Ikeda, Y., Okamura, K., Arima, R. & Fujii, S. (1977) *Biochem. Biophys. Acta* **487**, 189 - 203.
25 Ikeda, Y., Okamura, K. & Fujii, S. (1977) *Biochem. Biophys. Acta* **488**, 128 - 139.
26 Greenzaid, P. & Jencks, W.P. (1971) *Biochemistry* **10**, 1210 - 1222.

Dr. R. Arndt, Abteilung für Immunologie, I. Medizinische Universitätsklinik
Martinstr. 52, D-2000 Hamburg 20;

Dr. H.-E. Schlaak, D. Uschtrin, D. Südi und K. Michelssen, Biochemisches Institut im Fachbereich Medizin der Universität Kiel,
Ohlshausenstr. 40 - 60, D-2300 Kiel;

Dr. W. Junge, Zentrallabor des Städtischen Krankenhauses Kiel,
Metzstr. 53, D-2300 Kiel.

Hoppe-Seyler's Z. Physiol. Chem.
Bd. 359, S. 653 - 656, Mai 1978

SHORT COMMUNICATION

Sex-Dependent Composition of Cuticular Lipids from the Beetle *Rhagonycha fulva*

Jürgen Jacob

Biochemisches Institut für Umweltcarcinogene, Ahrensburg/Holstein

(Received 24 February 1978)

Geschlechtsabhängige Zusammensetzung der Cuticularlipide des Weichkäfers, Rhagonycha fulva

Zusammenfassung: Mit Hilfe der Gaschromatographie/Massenspektrometrie-Kombination wurden die Integumentlipide verschiedener Weichkäfer identifiziert und bei *Rhagonycha fulva* eine signifikante Geschlechtsabhängigkeit derselben nachgewiesen. Bei dieser Art wurden Tricosan, Pentacosan, *cis*-11-Pentacosen und 10,14-Pentacosadien identifiziert. Das Verhältnis $C_{25:2}/C_{25:1}$ beträgt bei ♂♂ 2.1, bei den ♀♀ dagegen 0.8. Innerhalb der Käfer scheint ein hoher Gehalt an ungesättigten, aliphatischen Kohlenwasserstoffen in den Cuticularlipiden für Weichkäfer charakteristisch zu sein.

Key words: Cuticular lipids, beetles, alkanes, alkenes, chemotaxonomy.

Although only limited data on cuticular lipid patterns from insects are available[1], it seems possible to use them as a chemotaxonomic parameter. Unbranched and 3-methylalkanes predominate in cockroaches[2-4], whereas mixtures of n-, 2-methyl-, and mono- as well as dimethyl branched saturated alkanes with substituents in the middle of the carbon chain occur in the integumental lipids of beetles of the family *Curculionidae*[5] and related families (Jacob, J., unpublished). Though not generally observed, in some cockroaches a sex-dependent composition of the cuticular lipid pattern has been reported[2].

This paper presents for the first time evidence of a sex dependence of integumental lipid composition in beetles from the family *Cantharidae*.

Materials and Methods

The beetles [in all: *Cantharis livida* (10), *Cantharis fusca* (20), and *Rhagonycha fulva* (209)] were washed with chloroform for 5 s each, and the extracts filtered, evaporated and purified by column chromatography on SiO_2 (14.0% water content), eluting with cyclohexane (50 m*l*); 0.4 mg, 0.6 mg and 4.7 mg lipids were obtained, respectively. In a second experiment, 50 couples of *Rhagonycha fulva* were separated during copulation in order to identify the males and females. The integumental lipids were separately analysed by gas chromatography. Thin-layer chromatography on SiO_2 plates (E. Merck) with CCl_4 developement indicated saturated and unsaturated hydrocarbons to be present.

In the case of *Rhagonycha*, the hydrocarbons were treated with OsO_4 in pyridine/dioxane according to ref.[6] and the resulting diols and tetraols subsequently silylated with Tri-Syl (Pierce) and *N,O-bis*(trimethylsilyl)-trifluoroacetamide (Pierce) according to ref.[7].

Gas-liquid chromatography was performed with a Perkin-Elmer F 20 instrument using 10 m glass columns with 3% OV 101 impregnation on GasChrom Q at 250 °C (column, injection, and detection). Mass spectra were recorded on a MAT 111 (GNOM) mass spectrometer at 80 eV using the same gas chromatography column. Details are published elsewhere[8].

Results

The quantitative composition of the integumental lipids from three *Cantharidae* species (♂ and ♀ mixed) are listed

Table 1. Composition of the cuticular hydrocarbons from several Cantharidae species (♂, ♀ mixed). (ECL = equivalent chain length). The data are given as % of the total hydrocarbons.

ECL	Hydrocarbon	*Cantharis fusca*	*Cantharis livida*	*Rhagonycha fulva*
	Unbranched (total)	(19.4)	(25.7)	(11.7)
21.00	n-C_{21}	10.0	15.2	–
23.00	n-C_{23}	5.2	10.5	8.1
25.00	n-C_{25}	3.5	–	3.6
27.00	n-C_{27}	0.7	–	–
	Mono-unsaturated (total)	(79.1)	(73.4)	(32.5)
20.76	$C_{21:1}$	0.3	–	–
21.75	$C_{22:1}$	–	1.4	–
22.75	$C_{23:1}$	32.8	72.0	–
23.75	$C_{24:1}$	–	–	0.2
24.75	$C_{25:1}$	20.9	–	32.3[a]
26.74	$C_{27:1}$	25.1	–	–
	Di-unsaturated (total)	(–)	(–)	(55.8)
24.65	$C_{25:2}$	–	–	55.8[b]
	unidentified	1.5	0.9	–

[a] double bond in Δ^{11}, [b] double bond in $\Delta^{10,14}$. Double bonds of other unsaturated hydrocarbons were not determined.

in Table 1. The individual hydrocarbons were identified by their retention times and mass spectrometry. Increments of – 0.25 and – 0.35 for the equivalent chain length of mono- and di-unsaturated hydrocarbons were found, in accordance with earlier published data[9].

From a second population, couples of *Rhagonycha fulva* were separated during copulation and the integumental lipids from two groups of males and females analysed. The results are given in Table 2. There were significant differences in the percentages of mono- and di-unsaturated hydrocarbons, yielding a relation of $C_{25:2}/C_{25:1}$ of 2.1 for males and of 0.8 for females.

To locate the double bond of the unsaturated hydrocarbons, the latter were converted by OsO_4 treatment and subsequent silylation to the polyol-trimethylsilane derivatives, which yielded excellent mass spectra. As can be seen from Fig. 1 and 2, the double bond of $C_{25:1}$ is located in Δ^{11}, whereas the double bonds of $C_{25:2}$ are in $\Delta^{10,14}$ according to the fragmentation patterns:

(a)

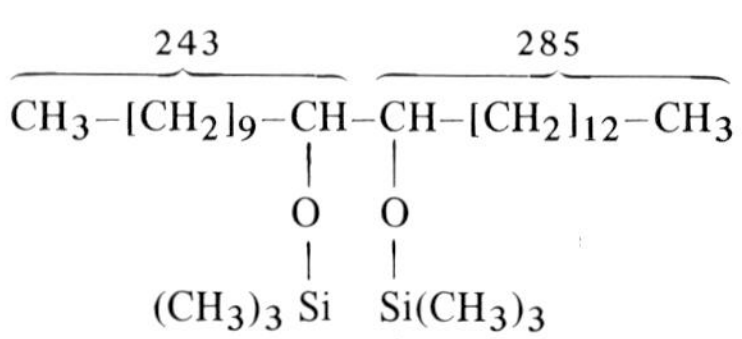

Very small fragments at *m/e* 215, 229, 299 and 313 seem to indicate that minor amounts of positional isomers with double bonds in Δ^{10} and Δ^{9} are present.

(b)

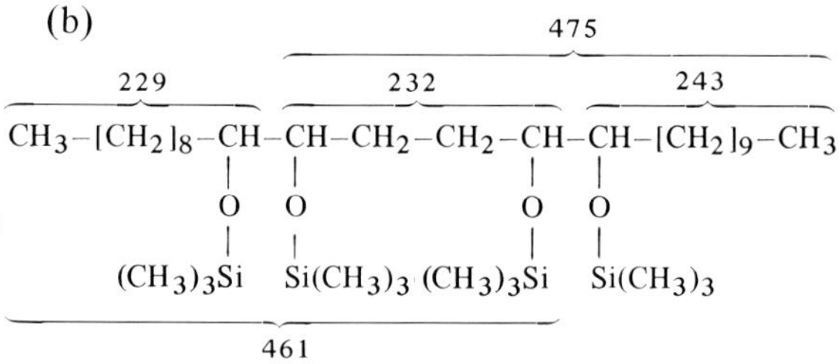

According to ref.[6], intense signals of fragments containing one $(CH_3)_3SiO$ group are observed (*m/e* 229 and 243), whereas the corresponding fragments lose 1 or 2 equivalents of trimethylsilanol, yielding intense signals at *m/e* 385 (= 475 – 90) and *m/e* 295 (= 475 – 2 × 90), indicating the original double bond at C-10, and signals at *m/e* 371 (= 461 – 90) and *m/e* 281 (461 – 2 × 90), indicating an original double bond at C-14.

Table 2. Composition of cuticular hydrocarbons from males and females of *Rhagonycha fulva*.
(Figures in parentheses from a second group.

Hydrocarbon	♂	♀
n-C_{23}	6.7 (4.3)	9.6 (6.1)
n-C_{25}	3.4 (4.2)	3.9 (4.2)
$C_{24:1}$	0.3 (0.2)	0.3 (0.3)
$C_{25:1}$	28.3 (30.8)	48.3 (49.7)
$C_{25:2}$	61.3 (60.5)	37.9 (39.7)

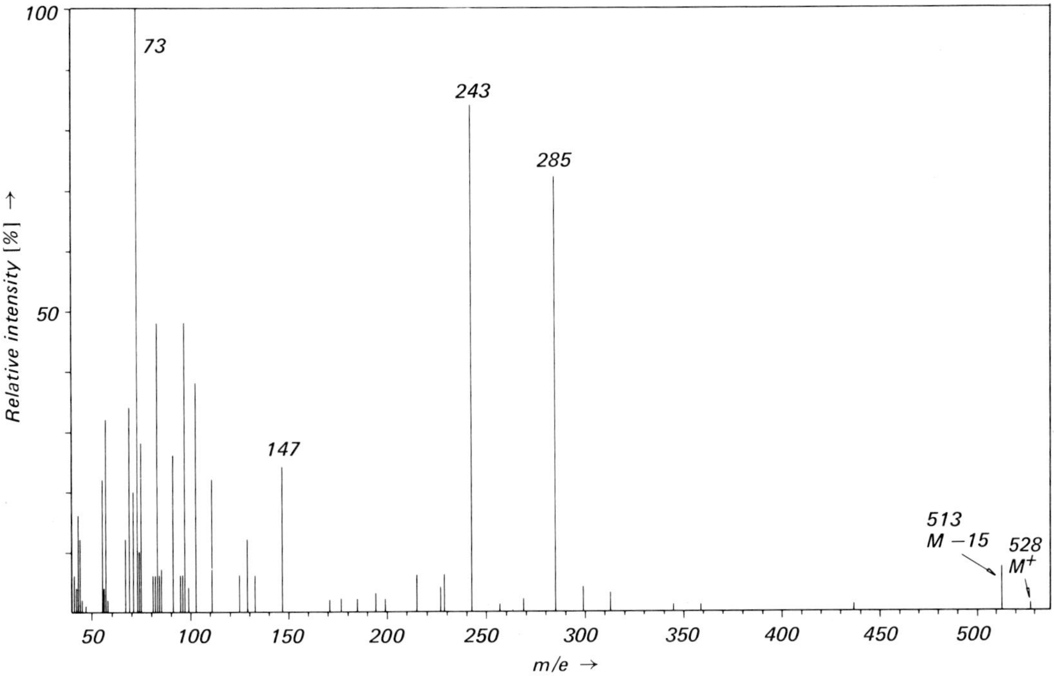

Fig. 1. Mass spectrum of 11,12-dihydroxypentacosyl trimethylsilyl ether.

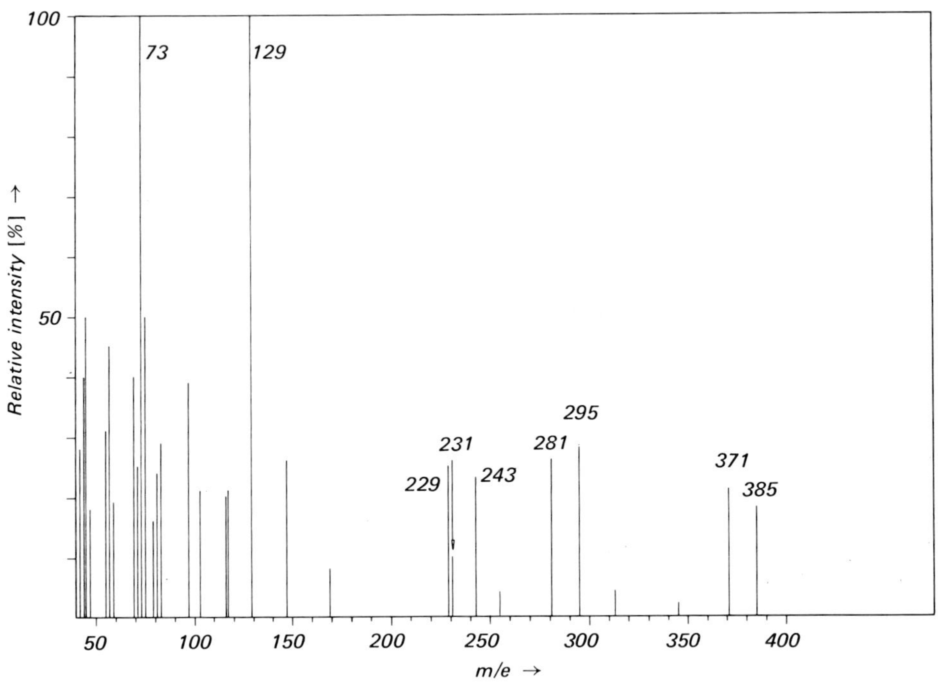

Fig. 2. Mass spectrum of 10,11,14,15-tetrahydroxypentacosyl trimethylsilyl ether.

The structure, moreover, is reconfirmed by the intense signal at *m/e* 129, which has been correlated with the cyclic structure

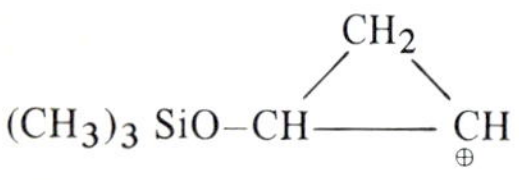

resulting from the ethylene-interrupted double bonds[6]. The middle part

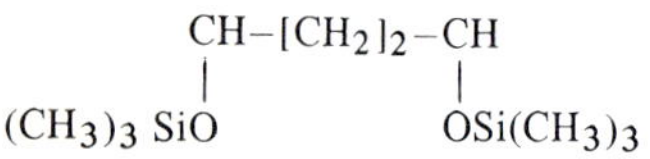

which is supposed to cyclize after loss of hydrogen[6] is represented by *m/e* 231 in Fig. 2.

IR absorption at 700 cm^{-1} and the absence of a *trans* band in the range of 960 cm^{-1} indicate that all hydrocarbons possess the *cis*-configuration.

Discussion

In contrast to the family *Curculionidae* (snout beetles) beeltes of the family *Cantharidae* (soldier beetles) possess high contents of unbranched unsaturated hydrocarbons (73 - 88%). Branched alkanes which are common in other insect integuments could not be detected, not even in traces. This pattern seems to be suitable as a first chemotaxonomic parameter for *Cantharidae*.

Both *cis*-11-pentacosene and *cis*-10,14-pentacosadiene, occurring in *Rhagonycha*, could derive by decarboxylation from either a C_{26}-acid with one double bond in Δ^{15} (Δ^{15}-C_{26} and $\Delta^{11,15}$-C_{26}) or with one double bond in Δ^{12} (Δ^{12}-C_{26} and $\Delta^{12,16}$-C_{26}), but no experimental evidence for this can be offered at present.

The reason for sexual differences in the cuticular lipid composition in *Rhagonycha*, which has also been observed in some cockroaches, is hitherto unknown. We suppose that it is involved in the sex recognition of these insects.

The author thanks *Prof. Dr. H. Strümpel*, University of Hamburg, for the exact determination of the species investigated.

Literature

1 Jackson, L.L. & Blomquist, G.J. (1976) in Chemistry and Biochemistry of Natural Waxes (Kolattukudy, P.E., ed.) pp. 201 - 233, Elsevier Sci. Publ. Amsterdam.
2 Tartivita, K. & Jackson, L.L. (1970) *Lipids* **5**, 35 - 37.
3 Jackson, L.L. (1970) *Lipids* **5**, 38 - 41.
4 Jackson, L.L. (1972) *Comp. Biochem. Physiol.* **41B**, 331 - 336.
5 Jacob, J. (1977) *Hoppe-Seyler's Z. Physiol. Chem.* **358**, 1375 - 1377.
6 Murawski, U., Egge, H. & Zilliken, F. (1971) *Z. Naturforsch.* **26b**, 1241 - 1252.
7 Stoming, T.A. & Bresnick, E. (1973) *Science* **181**, 951 - 952.
8 Jacob, J. (1975) *J. Chromatogr. Sci.* **13**, 415 - 422.
9 Jacob, J. (1978) *J. Lipid Res.* **19**, 148 - 153.

Priv.-Doz. Dr. Jürgen Jacob, Biochemisches Institut für Umweltcarcinogene, Sieker Landstr. 19, D-2070 Ahrensburg/Holstein.

Hoppe-Seyler's Z. Physiol. Chem.
Bd. 359, S. 657, Mai 1978

Corrections to

Nomenclature of Phosphorus-Containing Compounds of Biochemical Importance

(Recommendations 1976)

IUPAC-IUB Commission on Biochemical Nomenclature

Hoppe-Seyler's Z. Physiol. Chem. **358**, 599 - 616 (1977)

In Table IV, item 7 - 9 (pp. 611 - 612), the stereochemical prefixes should be D instead of L. The structural formulae show the naturally occuring 1-(3-*sn*-phosphatidyl)-D-*myo*-inositol and its derivatives. The names in column 1 and 2 are to be corrected accordingly. The correct names are given below.

Moreover, it should be pointed out that the naturally occuring bisphosphate is the 1-(3-*sn*-phosphatidyl)-D-*myo*-inositol-4,5-bis(phosphate), not the 3,4-bis(phosphate) shown in Table IV, item 9.

Correct names for Table IV, 7 to 9:

Recommended names	*Other names*
7) 1-(3-*sn*-Phosphatidyl)-D-*myo*-inositol	1,2-Diacyl-*sn*-glycero(3)-phospho(1)--D-*myo*-inositol; Phosphoinositide[d]
8) 1-(3-*sn*-Phosphatidyl)-D-*myo*-inositol 4-phosphate	1,2-Diacyl-*sn*-glycero(3)phospho(1)--D-*myo*-inositol 4-phosphate; 1,2-Diacyl-*sn*-glycero(3)phospho(1)--4-phospho-D-*myo*-inositol; Phosphoinositide 4-phosphate[d]; Diphosphoinositide[d]
9) 1-(3-*sn*-Phosphatidyl)-D-*myo*-inositol 3,4-bis(phosphate)	1,2-Diacyl-*sn*-glycero(3)phospho(1)--D-*myo*-inositol 3,4-bis(phosphate); 1,2-Diacyl-*sn*-glycero(3)phospho(1)--3,4-bis(phospho)-D-*myo*-inositol; Phosphoinositide 3,4-bis(phosphate)[d]; Triphosphoinositide[d]

[d] Trivial names occasionally used in the past; not recommended. Included here only for reference.

Titel Nr. 3 109 002 692

 – Printed in Germany – Satz: R. & J. Blank, München; Druck: Grafik + Druck, München; Bindearbeiten: F. Steinberger & Sohn, München.

Redaktionelle Hinweise

(Wichtige Änderungen ab Januar 1976 sind am Rand markiert)

Die Zeitschrift bringt nur Experimantalarbeiten aus den Gebieten der Biochemie und Physiologischen Chemie. Die Manuskripte werden in deutscher oder in englischer Sprache veröffentlicht. Es wird vorausgesetzt, daß alle Koautoren mit der eingereichten Fassung einverstanden sind.

Aktuelle Kurzmitteilungen, deren Umfang 4 Seiten in Kleindruck (ca. 8 Manuskriptseiten) nicht übersteigen darf, werden bevorzugt bearbeitet und in das nächsterreichbare Heft aufgenommen. Die Dringlichkeit zur Prioritätssicherung ist vom Autor in einem Begleitschreiben kurz zu begründen.

Manuskripte sind in zweifacher Ausfertigung an eine der folgenden Anschriften zu senden:

Prof. Dr. A. Butenandt oder Prof. Dr. F. Lynen, Max-Planck-Institut für Biochemie, D-8033 Martinsried b. München;
Prof. Dr. Dr. G. Weitzel, Physiologisch-Chemisches Institut der Universität, Hoppe-Seyler-Str. 1, D-7400 Tübingen;
Dr. A. Dillmann, Hoppe-Seyler-Redaktion, Am Klopferspitz, D-8033 Martinsried b. München, Tel. (089) 85 85-823.

Bei der Abfassung der Manuskripte bitten wir dringend, folgende Regeln zu beachten, die im wesentlichen mit den „Instructions to Authors" von *Biochimica et Biophysica Acta* und *European Journal of Biochemistry* übereinstimmen:

Es soll in Maschinenschrift (2zeilig, auch vorgesehener Kleindruck!) und mit 4 cm breitem Rand geschrieben werden. Vorgesehene Absätze sind deutlich zu markieren.

Seite 1 des Manuskripts soll enthalten (in normaler Schrift, nicht in Großbuchstaben):

a) Titel der Arbeit
b) Autorennamen (jeweils ein Vorname ausgeschrieben)
c) Name des Instituts oder Fachbereichs, in dem die Arbeit ausgeführt wurde (möglichst in der Sprache des Herkunftlandes), vorausgesetzt, daß die Leitung des Instituts oder Fachbereichs der Veröffentlichung zugestimmt hat.
d) Kurztitel, maximal 70 Satzzeichen
e) Adresse für den Versand der Korrekturen

Auf Seite 2 folgen

a) Eventuelle Widmung
b) Vollständige Postadresse der Autoren in der Sprache des Herkunftslandes
c) Ad-hoc-Abkürzungen (nur in seltenen Fällen zugelassen)
d) Enzyme in alphabetischer Reihenfolge der empfohlenen Trivialnamen mit systematischen Namen und EC-Nummern

Auf Seite 3 folgt die *Zusammenfassung* sowie eine Übersetzung des Titels und der (eventuell erweiterten) Zusammenfassung ins Englische (bzw. Deutsche). Diese Übersetzung oder deren sprachliche Überprüfung kann auch von der Redaktion übernommen werden.

Eventuelle *Danksagungen* sind an den Schluß des Manuskriptes zu stellen.

Literaturzitate im Text sind fortlaufend zu numerieren (nicht alphabetisch) und am Schluß des Manuskripts zu bringen. Die in Chemical/Biological Abstracts üblichen Zeitschriftenabkürzungen sind zu verwenden. Zur Anordnung s. IUB Commission of Editors of Biochemical Journals (1973) *Eur. J. Biochem.* **37**, 201 - 202; (1974) *diese Z.* **355**, 1 - 3.

Beispiel: Spackman, D.H., Stein, W.H. & Moore. S. (1958) *Anal. Chem.* **30**, 1190 - 1206.

Zeichnungen sollen einmal in reproduzierbarer Form mit schwarzer Tusche gezeichnet eingereicht werden; Zahlenskalen, gemessene Größen, Einheiten und sonstige Beschriftungen sind jedoch nur leicht mit weichem Bleistift einzutragen. Die Abbildungen sollen so groß sein, daß sie auf 1/2 oder 1/3 verkleinert werden können. Deutlich beschriftete Kopien für die Referenten sind beizufügen.

Alle *Tabellen* und *Abbildungen* sind mit Legenden zu versehen, so daß die Ergebnisse auch ohne Zuhilfenahme des Textes ersichtlich werden. Es wird gebeten, die Legenden nicht zu unterstreichen und die Abbildungslegenden nicht auf die Abbildungen selbst, sondern auf ein eigenes Blatt zu schreiben.

Symbole für die am häufigsten vorkommenden Größen und ihre Einheiten sind in Tab. 1 zusammengestellt. Wir bitten streng zu unterscheiden, ob bei Ansätzen molare Konzentrationen (mol/l = M; z.B. 0.25M) oder zugesetzte Mengen (z. B. in mol, mmol, μmol oder nmol) gemeint sind.

Hinsichtlich der *Nomenklatur* richte man sich nach der IUPAC-Nomenklatur sowie nach den Empfehlungen der "IUPAC-IUB Commission on Biochemical Nomenclature": Kohlenhydrate, *Eur. J. Biochem.* **21**, 455 - 477 (1971); **25**, 4 (1972); α-Aminosäuren, *Eur. J. Biochem.* **53**, 1 - 14 (1975);

Tab. 1. Symbole für Größen und ihre Einheiten

Größe		Einheit[a]
l	Länge	m, mm, μm, nm, etc.
V	Volumen	m^3, l (dm^3), ml, etc.
λ	Wellenlänge	nm (nicht mμ oder Å)
t	Zeit	h, min, s, ms, μs, etc.
ν	Frequenz	Hz
m	Masse	kg, g, mg, μg, etc.
ρ	Dichte	g/cm^3
s	Sedimentationskoeffizient	S = 10^{-13} s
n	Substanzmenge	mol, mmol, etc.
M	molare Masse (m/n)	g/mol
—	Masse eines Moleküls	Dalton[b]
M_r	Molekulargewicht	(dimensionslos)
c_B, [B]	molare Konzentration einer Substanz B	M, mM, μM, etc. (besser als 10^{-3} M, etc.)
T	Temperatur (absolut)	K (nicht °K)
t, θ	Temperatur	°C
q, Q	Wärmemenge	J (besser als cal)[c]
I	Ionenstärke	M (mol/l), mM, etc.
K	Gleichgewichtskonstante	
K_m	Michaeliskonstante	
K_i	Inhibitorkonstante	
k	Geschwindigkeitskonstante	
v	Reaktionsgeschwindigkeit	mol/s (oder mmol, μmol oder nmol/s, etc.), besser als μmol/min
V (nicht V_{max})	Geschwindigkeit der enzymkatalysierten Reaktion bei unendlich großer Substratkonzentration	
I	Lichtintensität	cd
T	Durchlässigkeit (I/I_0)	(dimensionslos)
A	Absorption[d] ($-\lg T$)	(dimensionslos)
ϵ	molarer Extinktionskoeffizient	$M^{-1} \times cm^{-1}$

[a] Präfixe für Einheiten: T, Tera (10^{12}); G, Giga (10^9); M, Mega (10^6), k, Kilo (10^3); m, Milli (10^{-3}); μ, Mikro (10^{-6}); n, Nano (10^{-9}); p, Piko (10^{-12}); f, Femto (10^{-15}); a, Atto (10^{-18}).

[b] Das Dalton ist 1/12 der Masse eines Atoms des Nuklids ^{12}C.

[c] Das Joule ist die SI-Einheit der Energie. Die Kalorie (1 cal = 4.184 J; 1 J = 0.239 cal) ist für eine Übergangszeit noch zugelassen; in jedem Fall sollten aber die Angaben auch in Joule ausgedrückt werden.

[d] Die Ausdrücke optische Dichte, Extinktion oder Absorbancy sollen nicht mehr verwendet werden.

Cyclite, *diese Z.* **350**, 523 - 535 (1969), *Eur. J. Biochem.* **57**, 1 - 7 (1975); Lipide, *diese Z.* **350**, 279 - 285 (1969); Steroide, *diese Z.* **351**, 663 - 689 (1970); Chinone mit isoprenoiden Seitenketten, Folsäure u. verschiedene andere Verbindungen, *diese Z.* **348**, 266 - 276 (1967); Corrinoide, *Eur. J. Biochem.* **45**, 7 - 12 (1974); Vitamine B_6, *Eur. J. Biochem.* **40**, 325 - 327 (1973); Carotinoide, *Eur. J. Biochem.* **25**, 397 - 408 (1972), **57**, 317 - 318 (1975); synthetische Analoga natürlicher Peptide, *diese Z.* **348**, 262 - 265 (1967); Enzyme Nomenclature, Recommendations (1972), Elsevier Publ. Comp., Amsterdam; multiple Enzymformen, *diese Z.* **353**, 852 - 854 (1972); Eisen-Schwefel-Proteine, *Eur. J. Biochem.* **35**, 1 - 2 (1973); Human-Immunglobuline, *Eur. J. Biochem.* **45**, 5 - 6 (1974); Peptidhormone, *Eur. J. Biochem.* **55**, 485 - 486 (1975). Für die Rechtschreibung ist „Jansen/Mackensen, Rechtschreibung der technischen und chemischen Fremdwörter" maßgebend.

Tab. 2 Empfohlene Abkürzungen.

Atome, isotope	=	z.B. [16-^{14}C]Östron
Atome, numeriert	=	z.B. C-5
Curie	=	Ci
Impulse pro Minute	=	Ipm (engl. cpm)
Prozent der Theorie	=	% d. Th.
prozentig	=	proz. (nicht %-ig)
Relative Wanderungsgeschwindigkeit	=	R_F
Schmelzpunkt	=	Schmp. (engl. m.p.)
Siedepunkt	=	Sdp. (engl. b.p.)
Umdrehungen pro Minute	=	U./min (engl. rpm)
Zerfälle pro Minute	=	Zpm (engl. dpm)
Zersetzung	=	Zers. (engl. decomp.)

Bezüglich der *Abkürzungen und Symbole für chemische Namen,* die in der biologischen Chemie von besonderem Interesse sind, wird auf die Empfehlungen der „IUPAC-IUB Commission on Biochemical Nomenclature" verwiesen (siehe z.B.: Allgemeines, Aminosäuren, Zucker, Phosphate, Coenzyme, *diese Z.* **348**, 245 - 265 (1967); Aminosäurederivate und Peptide, *Eur. J. Biochem.* **27**, 201 - 207 (1972); synthet. Polypeptide, *diese Z.* **349**, 1013 - 1016 (1968); Nucleinsäuren, Polynucleotide und deren Bestandteile, *diese Z.* **351**, 1055 - 1063 (1970). Andere als die dort empfohlenen Abkürzungen werden nur in seltenen Ausnahmefällen zugelassen. Sie müssen am Beginn jeder Arbeit in einer Fußnote erklärt werden.

Sonderdrucke: Von jeder Arbeit werden dem Verfasser 75 Sonderdrucke kostenlos geliefert. Die Bestellung weiterer Sonderdrucke gegen Berechnung ist möglich, muß jedoch spätestens bei der Rücksendung der Korrekturen erfolgen.

Der Verlag erwirbt mit der Annahme des Manuskriptes das Verlagsrecht und das ausschließliche Recht der Vervielfältigung, Verbreitung und Übersetzung. Der Verlag behält sich vor, den Inhalt dieser Zeitschrift für fremdsprachige Ausgaben zu verwenden.

Bezugspreis: Der Jahresbezugspreis (1 Jahrg. zu 12 Heften) beträgt DM 590,–, das Einzelheft kostet DM 56,–. Bestellungen nimmt jede Buchhandlung entgegen.

Verlag: WALTER DE GRUYTER & Co.
Genthiner Str. 13, D-1 Berlin 30. Tel. (030) 261 1341
Telegrammadresse: Wissenschaft Berlin
Postscheck: Berlin-West 10307-108

WALTER DE GRUYTER Inc.
3 Westchester Plaza
Elmsford, N.Y. 10523, U.S.A.
Tel. (914) 592-5890

Für die nächsten Hefte sind folgende Beiträge vorgesehen:

Synthesis of protected decapeptide (B_{21-30}) of human insulin, *V.K. Naithani and J. Föhles*

Human Plasma Kallikrein: Reinigung, enzymatische Charakterisierung und quantitative Bestimmung in Plasma, *H. Heber, R. Geiger und N. Heimburger*

Studies of the ligand binding to cholera toxin, III: Cooperativity of cholera toxin binding, *J. Sattler, G. Schwarzmann, I. Knack, K.H. Röhm and H. Wiegandt*

Determination of superoxide dismutase activity of D-penicillamine copper, *E. Lengfelder and E.F. Elstner*

A radioimmunoassay for measurement of human pancreatic secretory trypsin inhibitor in different body fluids, *A. Eddeland and K. Ohlsson*

Immunoreactive trypsin in serum and peritoneal fluid in acute pancreatitis, *A. Borgström and K. Ohlsson*

Release of granulocyte eleastase in lethal canine endotoxin shock, *A.O. Aasen and K. Ohlsson*

Studies on the influence of Trasylol on the partition of trypsin between the human plasma protease inhibitors in vitro, *G. Balldin, K. Ohlsson and A.-S. Ohlsson*

Increased catabolism of α-Macroglobulins after intravenous infusion of trypsin-α_1-antitrypsin complexes in dog, *G. Balldin, C.-B. Laurell and K. Ohlsson*

Uptake of thymidine into isolated rat hepatocytes. Evidences for two transport systems, *F.R. Ungemach and D. Hegner*

The modification of the lone tryptophan residue in human serum albumin by 2-hydroxy-5-nitrobenzyl bromide. Characterization of the modified protein and the binding of L-tryptophan and benzodiazepines to the tryptophan-modified albumin, *K.J. Fehske, W.E. Müller and U. Wollert*

A study of microsomal oxidative functions with the aid of a series of *N*-colchiceyl derivatives, *B. Hasenmüller, M. Schönharting and G. Siebert*

Fructose, a substrate for energy yielding metabolism in the heart?, *R. Bünger, O. Sommer, G. Walter and R. Scholz*

Acid precipitation of fibrinogen: a technique for measurement and isolation of fibrinogen and its derivatives fromh highly diluted solutions, *R. Hafter, R. von Hugo and H. Graeff*

Snake venoms. The amino-acid sequence of protein S_5C_4 from *Dendroaspis jamesoni kaimosae* (Jameson's mamba) venom, *F.J. Joubert, A.J.C. Strydom and N. Taljaard*

Subcellular distribution of [^{14}C]5-hydroxytryptamine in the blood platelets of rabbits: effects of imipramine and haloperidol, *L. Hussein, H.W. Goedde, M. Rosenfeld and E. Freerksen*

Nebenreaktionen bei Peptidsynthesen, III. Synthese und Charakterisierung von N^{in}-tert.-butylierten Tryptophan-Derivaten, *E. Jaeger, P. Thamm, S. Knof, E. Wünsch, M. Löw und L. Kisfaludy*

Nebenreaktionen bei Peptidsynthesen, IV. Charakterisierung von C- und C,N-tert.-butylierten Tryptophan-Derivaten, *E. Jaeger, P. Thamm, S. Knof und E. Wünsch*

Direkte tert.-Butylierung des Tryptophans: Herstellung von 2',5',7'-Tri-tert-butyl-tryptophan, *M. Löw, L. Kisfaludy, E. Jaeger, P. Thamm, S. Knof und E. Wünsch*

tert.-Butylierung des Tryptophan-Indolringes während der Abspaltung der tert.-Butyloxycarbonyl-Gruppe bei Peptidsynthesen, *M. Löw, L. Kisfaludy und P. Sohár*

Maintenance and regeneration in vitro of bulk isolated cortical neurones from adult rats, *H.H. Althaus, V. Neuhoff, W.B. Huttner, R. Monzain and A. Shahar*

Determination of the superoxide dismutating activity of D-penicillamine copper, *E. Lengfelder and E.F. Elstner*

d 359 Heft 6 Juni 1978

N 0018-4888 A 20679 E

.oppe-Seyler's Zeitschrift für 'hysiologische Chemie

Mitteilungen der Gesellschaft für Biologische Chemie

WALTER DE GRUYTER · BERLIN · NEW YORK

Hoppe-Seyler's Z. Physiol. Chem. Bd. 359 S. 659 - 776 Juni 1978
HSZPAZ 359 (6) 659 - 776 (1978)

Mitteilungen der Gesellschaft für Biologische Chemie

35. Rundschreiben (Juni 1978)

Niederschrift über die ordentliche Mitgliederversammlung

am Donnerstag, dem 6. April 1978 in Mosbach/Baden, Stadthalle

Vorsitz:	Prof. Dr. K. Decker
Teilnehmer:	60
Beginn:	17.30 Uhr
Ende:	18.45 Uhr

Der Vorsitzende eröffnet die Sitzung und stellt fest, daß die Mitgliederversammlung ordnungsgemäß einberufen ist.

Er gedenkt der seit der letzten Mitgliederversammlung verstorbenen Mitglieder

Dr. Gesche Brand, Heidelberg
Prof. Dr. Hans Netter, Kiel
Prof. Dr. Gerhard Orzechowski, Köln
Dr. Wolfgang Philipp-Dormston, Marburg
Prof. Dr. Willi Riechert, Offenburg

1. Bericht des Vorstandes

1.1 Mitgliederstand am 15.4.1977: 1 934 plus 12 korporative
Mitgliederstand am 1.4.1978: 2 153 plus 12 korporative

Die erfreuliche Steigerung der Mitgliederzahl kann mindestens teilweise auf die inzwischen gelaufene Mitgliederwerbung zurückgeführt werden.

Allerdings wird bei dieser Mitgliederzahl in Zukunft eine Kassenführung ohne Einstellung zusätzlichen bezahlten Personals nur noch mit Hilfe einer EDV möglich sein. **Die Mitglieder werden deshalb dringend gebeten, zur Erstellung der Grunddatei die bereits zugesandten Fragebogen ausgefüllt zurückzusenden.** Der Umfang der Datei ist so ausgelegt, daß damit auch – einem allgemeinen Wunsche entsprechend – die Decker/Hartmann-Studie zur Berufssituation der Biochemiker weitergeführt werden kann. Dadurch können den Mitgliedern nützliche Informationen geliefert werden, die auch als Grundlage für Aktivitäten gegenüber Ministerien und Politikern dienen.

1.2 An allen Hochschulorten sind inzwischen *Kontaktpersonen* zur Mitarbeit gewonnen worden (Auflistung siehe Anhang). Auf diesem Wege wird eine vollständige Erfassung der biochemisch arbeitenden Institute und anderer Ein-richtungen erhofft, was einen raschen Informationsfluß von dort (z.B. Adressen- und Organisationsänderungen, Wechsel in der Leitung), andererseits eine gezielte Informationsmöglichkeit dorthin (z.B. für Versendung von Plakaten) sichern sollte.

1.3 *Unterrichtsfragen*

Die von Fasold für den naturwissenschaftlichen Bereich schon im vorigen Jahr erwähnten Besprechungen und Erhebungen wurden und werden fortgesetzt. Ergebnisse konnten in den letzten Mitteilungen (34. Rundschreiben), wie angekündigt, veröffentlicht werden. Folgerungen hieraus werden vom Vorstand den zuständigen Ministerien bzw. Unterrichtskommissionen offiziell zugeleitet werden.

Im medizinischen Bereich verfolgt Buddecke weiterhin die notwendigen Aktivitäten im Zusammenhang mit den Lernziel- und Prüfungsfragenkatalogen.

1.4 *Tagungen der GBCh*

im vergangenen Jahr

Herbsttagung in Freiburg, 3.–5.10.1977 (Decker), zusammen mit der schweizerischen und französischen biochemischen Gesellschaft, einschließlich der 3. Fritz-Lipmann-Vorlesung (Haenni). Teilnehmer: über 500.

Frühjahrstagung in Frankfurt/M., 12.–15.3.1978 (Fasold), einschließlich der 3. Adolf-Butenandt-Vorlesung (U. Westphal) und Verleihung der Otto-Warburg-Medaille an Stoffel. Teilnehmer: über 400.

zukünftige

Herbsttagung in Konstanz, 10.-13.9.1978 (Sund), zusammen mit der British Biochemical Society, einschließlich der 4. Fritz-Lipmann-Vorlesung (P. Mitchell).

Frühjahrstagung in Berlin, 12.-14.3.1979 (Kleinkauf), in Gemeinschaft mit der Gesellschaft für Genetik und der Gesellschaft für Zellbiologie, einschließlich der 4. Adolf-Butenandt-Vorlesung (Gerisch).

Herbsttagung in Tübingen, 26.–28.9.1979 (Mecke/Weitzel).

Frühjahrstagung in Münster, 1980 (Buddecke/Barz).

Herbsttagung in Innsbruck, 1980, zusammen mit der Österreichischen Biochemischen Gesellschaft.

1.5 *Kleinkonferenzen*

bisherige

Nr. 22 „Charakterisierung von Proteinen energieübertragender Prozesse", Freiburg, 6.10.1977; Organisatoren: G. Schäfer und E. Bäuerlein.

Nr. 23 „Biochemical Aspects of Kidney Function", Reisensburg, 23.–26.10.1977; Organisatoren: W.G. Guder and U. Schmidt, Basel.

Nr. 24 „Adeninnukleotide und neuronale Funktion", MPI für Psychiatrie München, 3.2.1978; Organisatoren: G.W. Kreutzberg und P. Schubert.

bevorstehende Kleinkonferenzen bzw. Konferenzen von Studiengruppen

„Functions of Glutathion in Hepatocytes", Reisensburg, 8.–11.7.1978; Organisatoren: H. Sies und A. Wendel.

ferner in zeitlichem Zusammenhang mit der Herbsttagung Konstanz 1978:

a) Konferenz der Studiengruppe „Physikalische Chemie biologischer Makromoleküle" zum Thema „Thermodynamik, Kinetik und Struktur von Biopolymeren"; Organisatoren: R. Jaenicke und R. Huber.

b) Konferenz der Studiengruppe „Membranstruktur und Membrantransport"; Organisatoren: H. Passow und W. Boos.

1.6 *Reisefonds der GBCh*

Es wird daran erinnert, daß der Reisefonds der GBCh von jüngeren Mitgliedern in Anspruch genommen werden kann. In der Diskussion wird die starre Altersgrenze von 30 Jahren kritisiert. Der Vorstandsrat wird sich bei seiner nächsten Sitzung damit befassen.

Für das FEBS Meeting in Dresden (2.–8.7.1978) stehen uns öffentliche Mittel zur Verfügung. Wegen der Kürze der verfügbaren Zeit (Termin für die Einreichung von Anträgen: 5.6.1978) wird eine Sonderinformationsaktion über die Kontaktleute gestartet.

1.7 *DFG-Fachgutachter*

Im Frühjahr 1979 sind durch unsere Gesellschaft der DFG Kandidaten für die Wahl der Fachgutachter zu benennen. Es wird nur noch einen Fachausschuß „Biochemie und Biophysik" mit 12 Gutachtern geben, so daß 24 Kandidaten nominiert werden sollten, und zwar für den Unterausschuß „Biochemie" 12 Kandidaten sowie für die Unterausschüsse „Klinische Chemie und Pathobiochemie", „Molekulare Biologie" und „Biophysikalische Chemie und Biophysik" je 4 Kandidaten. **Es werden Vorschläge hierfür aus Mitgliederkreisen bis zum 15. Oktober 1978 an Decker erbeten.** Daraus wird der Vorstandsrat eine Auswahl von insgesamt 24 Kandidaten treffen und einreichen.

1.8 Eine *European Society of Biotechnology* befindet sich in der Entstehung. Als nationale Vorstufe dazu ist eine *Arbeitsgemeinschaft Biotechnologie* unter Federführung der DECHEMA gegründet worden, an der außer unserer Gesellschaft noch das Institut für Gärungsgewerbe in Berlin, die Gesellschaft für Hygiene und Mikrobiologie und der Verein der Ingenieure für Chemie beteiligt sind. An der Gründungsversammlung hat von unserer Seite Fasold teilgenommen.

1.9 *Kassenbericht*

Decker referiert den Bericht des erkrankten Schatzmeisters. Die Einnahmen der Gesellschaft betrugen 156 044,90 DM, die Ausgaben 134 080,59 DM. Der Überschuß von 21 964,31 DM besteht zum Teil aus nicht realisierten Kursgewinnen der Gesellschaftsvermögensanlage.

2. Bericht der Kassenprüfer

Frau Debuch berichtet auch namens von Truscheit, daß eine eingehende Prüfung keinerlei Beanstandungen ergeben hat.

3. Auf Antrag der Kassenprüfer wird bei drei Enthaltungen der anwesenden Vorstandsmitglieder ohne Gegenstimme dem Vorstand Entlastung erteilt.

4. Festsetzung des Beitrages für 1979

Auf Antrag des Vorstandes unverändert DM 30,–, Studenten DM 5,– pro Jahr.
Ein Antrag, den studentischen Beitrag auf DM 10,– zu erhöhen, verfiel der Ablehnung bei einer Enthaltung und zwei Ja-Stimmen.

5. Wahl der Kassenprüfer

Für 1978 bei zwei Enthaltungen, ohne Gegenstimme Wiederwahl von Frau Debuch und Truscheit.

6. Beiratsnachwahl

Für die ausscheidenden Beiratsmitglieder Fasold, Staudinger und Breuer wurden ordnungsgemäß Fasold (Wiederwahl), V. Ullrich, Homburg und R. Jaenicke, Regensburg nominiert. Leider wurden keine weiteren Kandidaten benannt, so daß es sich diesmal nicht um eine echte Wahl handeln wird.

7. Themen der nächsten Mosbacher Kolloquien

Unter Berücksichtigung früher vorgelegter und in der Zwischenzeit überarbeiteter Themen wurde für 1979 (30. Kolloquium, 26. - 28.4.1979) der Arbeitstitel „Proteolytische Aktivität und biologische Kontrolle" unter Leitung von Holzer und Tschesche einstimmig gebilligt. Tschesche hat eine ausführliche Erläuterung der Planungen gegeben*.

Für 1980 wird von Bücher, der sich auch bezüglich möglicher Organisatoren bemühen wird, als Thema vorgeschlagen: „Assembly and Biogenesis of Cellular Organelles".

Für 1981 oder 1982 ist an ein Thema „Glykokonjugate" (Buddecke) zusammen mit „sekretorischen Proteinen" (Winkler) gedacht. Die Vorplanungen finden allgemeine Zustimmung.

Ein Antrag, die wissenschaftlichen Veranstalter des Mosbacher Kolloquiums im Programm zu zitieren, wird vom Vorstandsrat behandelt werden.

---

* Endgültiges Thema: „Biologische Funktionen der Proteinasen" – Information: Prof. Dr. H. Holzer, Biochemisches Institut der Universität Freiburg, Hermann-Herder-Str. 7, D-7800 Freiburg/Br.

8. Genehmigung einer Studiengruppe „Glykokonjugate"

Die endgültige Etablierung einer Studiengruppe „Glykokonjugate", mit Buddecke und Schauer als Sprechern, wird ohne Enthaltungen und Gegenstimmen genehmigt.

Eine Studiengruppe für molekulare Virologie wird sich bis auf weiteres *nicht* etablieren.

9. Verschiedenes

Anträge und Anfragen hierzu wurden nicht vorgebracht.

20. April 1978	Für die Richtigkeit	Prof. Dr. K. Decker (Präsident, Leiter der Versammlung)	Prof. Dr. H. Gibian (Schriftführer, Protokollant)

Tagungsankündigungen

(neu aufgenommene Ankündigungen sind mit • gekennzeichnet)

Informationen über die zahlreichen *EMBO-Kurse und -Symposien* über molekularbiologische, zellbiologische und virologische Themen gibt das Sekretariat der EMBO, Postfach 10 22 40, D-6900 Heidelberg 1. (Vorläufiges Programm 1978 s. unser 34. Rundschreiben vom Februar 1978.) Auch mehrere *„NATO Advanced Study Institutes"* betreffen biochemische und molekularbiologische Themen. Information: NATO Scientific Affairs Division, B-1110 Brüssel. Im 34. Rundschreiben sind auch die *GDCH-Fortbildungskurse 1978* aus dem Fachbereich Biochemie zusammengestellt.

3rd International Congress of Plant Pathology
16.-31. August 1978, München
Information: Dr. E. Fuchs, Biologische Bundesanstalt, Messeweg 11/12, D-3300 Braunschweig.

XIV. Internationaler Kongreß für Genetik
21.-30. August 1978, Moskau
Grundidee: Genetics and Human Welfare
Information: Nationales Organisationskomitee des XIV. Internationalen Genetik-Kongresses, Fersmana Street 11, n. 4, BOTuC um. N. I. Wawilowa, Moscow 117312, U. S. S. R.

European Society for Neurochemistry, 2nd Meeting
28. August - 1. September 1978, Göttingen
Information: Prof. Dr. Volker Neuhoff, Max-Planck-Institut für Experimentelle Medizin, Hermann-Rein-Str. 3, D-3400 Göttingen.

4th International Congress for Virology
30. August - 6. September 1978, The Hague
Information: 4th International Congress for Virology, Netherlands Congress Centre, P.O. Box 9000, The Hague, The Netherlands.

12th International Congress for Microbiology
3. - 8. September 1978, München
Information: Prof. Dr. K.-H. Schleifer, Arcisstraße 21, D-8000 München 2.

IUPAB, 6th International Biophysics Congress
3. - 9. September 1978, Kyoto, Japan

Information: Prof. Fumio Oosawa, Secretary General, 6th International Biophysics Congress, Dept. of Biophysical Engineering, Osaka University, Toyonaka, Osaka 560, Japan.

International Congress of Bacteriology and Mycology
4. - 12. September 1978, München

Information: Prof. G. Pulverer, Fürst-Pückler-Str. 56, D-5000 Köln 41.

Gemeinsame Tagung der Biochemical Society und der Gesellschaft für Biologische Chemie
10. - 13. September 1978, Konstanz

Themen:

1) Biochemie der Nukleinsäuren
2) Membranen, Lipide und Transportprozesse
3) Biochemische Oxidoreduktion, Enzymologie des Redox-Transports

Neben diesen Symposien sind Poster-Präsentationen vorgesehen. Abstracts hierfür müssen bis zum 1. Juli 1978 der Tagungsleitung vorliegen. Anweisungen zur Abfassung der Kurzreferate wurden im 33. Rundschreiben abgedruckt.
Anträge auf Reisekostenzuschuß für junge Biochemiker (siehe 31. und 32. Rundschreiben) sind bis zum 10. Juli 1978 beim Schriftführer, Prof. Gibian, einzureichen.

Information: Prof. Dr. Horst Sund, Universität Konstanz, Postfach 7733, D-7750 Konstanz.

Anläßlich dieser Tagung plant die *Studiengruppe Physikalische Chemie biologischer Makromoleküle* ein erstes Treffen unter dem Thema
‚Thermodynamik, Kinetik und Struktur von Biopolymeren'.
In Form von Kurzvorträgen und Posters sollen im Anschluß an die Tagung am 14.9.78 neue Ergebnisse diskutiert werden. Abstracts vorgesehener Beiträge sind bis zum 15. Juli 1978 zu senden an Prof. Dr. Robert Huber, Max-Planck-Institut für Biochemie, Am Klopferspitz, D-8033 Martinsried bei München.

● Ferner findet eine Kleinkonferenz der *Studiengruppe Membranstruktur und Membrantransport* statt.

Themen:

1) Biosynthese von Membranproteinen und Membranassembly
2) Rekonstitution von Membranen und Transportsystemen
3) Struktur von Transportproteinen
4) Regulation von Transportsystemen

Information: Prof. Dr. W. Boos; Universität Konstanz, FB Biologie, Postfach 7733, D-7750 Konstanz

12th Harden Conference
Mechanism of action of the reproductive hormones
10. - 15. September 1978, Ashford, Kent

13th Harden Conference
Specificity and plasticity in brain development
17. - 22. September 1978, Ashford, Kent

Information: Meetings Officer, The Biochemical Society, 7 Warwick Court, High Holborn, London WC1R 5DP, U.K.

Workshop Course: Modern Techniques in Centrifugation
11. - 16. September 1978, University of Essex

Information: Dr. D. Rickwood, Dept. of Biology, University of Essex, Wivenhoe Park, Colchester, CO4 3SQ, U.K.

110. Versammlung der Gesellschaft Deutscher Naturforscher und Ärzte
17.-21. September 1978, Innsbruck
Generalthema: Das naturwissenschaftliche Weltbild heute – Grundlagen und Bausteine
Information: Prof. Dr. E. Auhagen, Gesellschaft Deutscher Naturforscher und Ärzte, Friedrich-Ebert-Str. 217, D-5600 Wuppertal 1.

6th Meeting, International Research Group for Carcino-Embryonic Proteins (I.R.G.C.P.)
17.-21. September 1978, Marburg/Lahn
Themen: Carcino-embryonale Proteine: 1. Klinische Anwendung; 2. Biologie, Immunchemie und Bedeutung in der Embryo- und Carcinogenese; 3. Methoden, Standardisierung.
Information: Prof. Dr. F.-G. Lehmann, Med. Univ.-Klinik, E. Mannkopfstraße 1, D-3550 Marburg/Lahn.

FEBS Advanced Course No. 53
Hormonal Regulation of Specific Gene Expression
19.-22. September 1978, London
Organized by P.N. Campbell and R.K. Craig.
Information: Prof. P. N. Campbell, Courtauld Institute, The Middlesex Hospital Medical School, London, W1P 7PN, England.

XX. Symposion der Gesellschaft für Histochemie
20.-23. September 1978, Gargellen/Montafon, Österreich
Hauptthemen:
1. Cytochemische Merkmale als Ordnungsprinzip in Histologie und Pathohistologie
2. Pathohistochemie von Magen-Darm-Biopsien

Information: Prof. Dr. Ch. Pilgrim, Abt. Klinische Morphologie der Universität Ulm, Postfach 4066, D-7900 Ulm/Donau.

Tagung der Gesellschaft für Genetik
21.-23. September 1978, Innsbruck
Hauptthema: Molekularbiologische Grundlagen der Humangenetik: 1. Techniken zur Charakterisierung des Genoms (Genomstruktur, Kartierung); 2. Läsion des Genoms; 3. Biochemische Manifestation von Genläsionen.
Information: Institut für Medizinische und Klinische Chemie der Universität Innsbruck, Fritz-Pegl-Straße 3/VI, A-6020 Innsbruck.

● *XIII. Internationale Konferenz für Embryologie*
25. - 29. September 1978, Berlin.
Themen: Gene expression in early embryonic development; control of gene expression in terminal differentiated tissues; cell interaction and pattern formation; eye development; etc.
Information: Prof. Dr. Dr. H. Tiedemann, FB1, Institut für Molekularbiologie und Biochemie (WE 03) der Freien Universität Berlin, Arnimallee 22, D-1000 Berlin 33.

● Die in Gründung befindliche Europäische Föderation für Biotechnologie organisiert den
1. Europäischen Kongreß für Biotechnologie
25.–29. September 1978 in Interlaken, Schweiz
Hauptthemen: 1) Bioreaktoren; 2) Immobilisierte Biokatalysatoren; 3) Regulation und Kontrolle in der Biotechnologie.
Information: DECHEMA Deutsche Gesellschaft für Chemisches Apparatewesen e.V., Postfach 97 01 46, D-6000 Frankfurt/M. 97.

- *50. Tagung der Deutschen Physiologischen Gesellschaft*
  3.–6. Oktober 1978 in Göttingen
  Information: Prof. Dr. H.-D. Henatsch, Physiologisches Institut der Universität Göttingen, Lehrstuhl II, Humboldtallee 7, D-3400 Göttingen.

3rd European Symposium on Hormones and Cell Regulation
4. - 7. October 1978, Bischenberg, near Strasbourg, France

Topics: Organization and expression of the genome; growth factors; kinases and multisite phosphorylation; insulin secretion; hormones and adenylate cyclase activities; steroid effects on gene expression.

Information: H. J. van der Molen, Dept. of Biochemistry II, Medical Faculty, Erasmus University, P.O. Box 1738, Rotterdam, The Netherlands.

- *Diskussionstagung „Elektrophorese Forum"*
  25. - 27. Oktober 1978, München
  Information: Dr. B. J. Radola, Technische Universität München, 8050 Freising-Weihenstephan.

- **Frühjahrstagung der Gesellschaft für Biologische Chemie gemeinsam mit den Gesellschaften für Genetik und Zellbiologie**
  12. - 14. März 1979 in Berlin

Vorläufiges Programm:

Replikation, Transkription, Translation, Genetic Engineering, Chromosomen, Immunologie, Differenzierung.

Als Redner haben zugesagt: P. Berg, Stanford; J.E. Darnell, New York; H. Harris, Oxford; H. Hoffmann-Berling, Heidelberg; N.K. Jerne, Basel; S. Ochoa, Nutley; J. Schell, Gent; W. Szybalski, Madison; Ch. Weissmann, Zürich.

Geplant sind 9 Plenarvorträge an den Vormittagen, die Nachmittage sind für Diskussionen, Posterpräsentationen und Kurzvorträge offen.

Die *4. Adolf-Butenandt-Vorlesung* wird von Prof. Dr. G. Gerisch, Biozentrum der Universität Basel, gehalten.
Geplant ist ferner ein Treffen der *Studiengruppe Neurochemie.*

Für Anmeldung wird ein eigenes Rundschreiben versandt werden. Abstracts für Kurzvorträge und Poster müssen bis zum 19. Januar 1979 eingereicht werden. Anweisungen zur Abfassung der Kurzreferate sind im 33. Rundschreiben abgedruckt.

Anträge auf Reisekostenzuschuß für junge Biochemiker (siehe 31. und 32. Rundschreiben) sind bis zum 15. Januar 1979 beim Schriftführer, Prof. Gibian, einzureichen.

Information: Prof. Dr. H. Kleinkauf, Technische Universität Berlin, Max-Volmer-Institut für Physikalische Chemie und Molekularbiologie, Abt. Biochemie, Franklinstraße 29, D-1000 Berlin 10, Telefon: (030) 314-4205.

- *3rd European Symposium on Vitamin B_{12} and Intrinsic Factor*
  5. - 8. March 1979, Zürich
  Information: Doz. Dr. habil. B. Zagalak, Chemisches Laboratorium, KISPI, Universität Zürich, Steinweißstraße 75, CH-8032 Zürich.

FEBS Special Meeting on
Enzymes
17. - 21. April 1979, Dubrovnik-Cavtat, Jugoslavia

Information: Prof. Dr. P. Mildner, Laboratory of Biochemistry, Technological Faculty, Pierottijeva 6/VI, YU-41000 Zagreb.

XII. Acta Endocrinologica Congress
26.-30. Juni 1979, München

Information: Prof. Dr. Peter C. Scriba, Secretary General, Medizinische Klinik Innenstadt der Universität München, Ziemssenstraße 1, D-8000 München 2.

XIth International Congress of Biochemistry
8.-14. July 1979, Toronto, Ontario, Canada

Topics:

1) Structure, organization and replication of genetic elements
2) Genetic expression and its regulation
3) Structure and function of proteins
4) Mechanism and regulation of enzyme action
5) Membrane structure and function
6) Bioenergetics and transport
7) Biochemistry of growth and differentiation
8) Biochemical aspects of immunology
9) Neurochemistry
10) Biochemistry of sensory, contractile and motile systems
11) Hormones: biochemistry and molecular mechanisms of action
12) Molecular mechanisms of disease
13) Environmental biochemistry

The Congress is combined with an *International Exhibition.*

Information: R. W. Dolan, Executive Secretary, XIth International Congress of Biochemistry, c/o National Research Council of Canada, Ottawa, Ontario, Canada K1A OR6.

13th FEBS Meeting
24.-29. August 1980, Jerusalem, Israel

Information: Alisa Tietz-Devir (President of the Israel Biochemical Society), Dept. of Biochemistry, Tel-Aviv University, Ramat Aviv, Tel Aviv.

Nathan Sharon (Chairman of the Organizing Committee), Dept. of Biophysics, The Weizmann Institute of Science, P. O. Box 26, Rehovot.

The Secretariat, 13th FEBS Meeting, P. O. Box 16271, Tel Aviv, Israel.

● *21st International Conference on the Biochemistry of Lipids*
30. August - 1. September 1979, Köln

Topics:

1) Analysis by physicochemical and chemical methods.
2) Impact of hydrophilic and hydrophobic groups on phospholipid metabolism.
3) Impact of lipid structures on model membranes, lipoprotein structures, membrane asymmetry, lipid protein interactions.

Information: Prof. Dr. Dr. W. Stoffel, Institut für Physiologische Chemie der Universität Köln, Joseph-Stelzmann-Straße 52, D-5000 Köln 41.

● *VIth International Symposium on Glycoconjugates*
3.-7. September 1979, Kiel

Topics:

1) Alfred Gottschalk Memorial Lecture.
2) Chemical and physical properties of glycoconjugates.
3) Metabolism of glycoconjugates.
4) Role of glycoconjugates in recognition.
5) Glycoconjugates in cells, tissues and organes.

Information: Prof. Dr. R. Schauer, Biochemisches Institut der Universität Kiel, Olshausenstraße 40 - 60, D-2300 Kiel.

ANHANG

Kontaktpersonen der Gesellschaft für Biologische Chemie

(Bitte aufbewahren)

Aachen	Prof. Dr. Helmut Zahn Deutsches Wollforschungsinstitut an der RWTH Aachen, Veltmanplatz, 5100 Aachen T. (02 41) 3 64 54 und 2 07 21
Augsburg	Prof. Dr. Rainer Jaenicke, s. Regensburg
Bayreuth	Prof. Dr. Walter Kersten, s. Erlangen
Berlin	Prof. Dr. Horst Kleinkauf Technische Universität Berlin, Max-Volmer-Institut, Abt. Biochemie Franklinstr. 29, 1000 Berlin 10 T. (0 30) 314 - 42 05
Bielefeld	Prof. Dr. Peter von Sengbusch Universität Bielefeld, Fakultät für Biologie Postfach 8640, Universitätsstr., 4800 Bielefeld T. (05 21) 1 06 - 56 23
Bochum	Prof. Dr. Achim Trebst Ruhr-Universität Bochum, Abt. für Biologie Postfach 21 48, 4630 Bochum-Querenburg T. (02 34) 7 00 - 36 33
Bonn	Prof. Dr. Heinz Breuer Institut für Klinische Biochemie der Universität 5300 Bonn-Venusberg T. (0 22 21) 19 - 21 11
Braunschweig-Stöckheim	Prof. Dr. Fritz Wagner Technische Universität Braunschweig, Lehrstuhl für Biochemie und Biotechnologie Mascheroder Weg 1, 3300 Braunschweig T. (05 31) 70 08 - 3 89
Bremen	Prof. Dr. Volker Kasche Studienbereich 3 der Universität Bremen, Physikalische Biologie Postfach 33 04 40, 2800 Bremen 33 T. (04 21) 2 18 - 29 06
Darmstadt	Prof. Dr. Hans-Günter Gassen Technische Hochschule Darmstadt, Institut für Organische Chemie, Fachgebiet Biochemie Petersenstr. 22, 6100 Darmstadt T. (0 61 51) 16 36 57
Clausthal	Prof. Dr. Walther Lamprecht, s. Hannover
Dortmund	Prof. Dr. Benno Hess Max-Planck-Institut für Ernährungsphysiologie Rheinlanddamm 201, 4600 Dortmund T. (02 31) 12 06 - 3 80

Düsseldorf	Prof. Dr. Wolfgang Staib Physiologisch-Chemisches Institut der Universität Witzelstr. 111, 4000 Düsseldorf T. (02 11) 3 11 - 2 72
Erlangen	Prof. Dr. Walter Kersten Institut für Physiologische Chemie der Universität Erlangen–Nürnberg Wasserturmstr. 5, 8520 Erlangen T. (0 91 31) 85 23 05
Essen	Prof. Dr. Herbert Schriefers Institut für Physiologische Chemie des Universitäts-Klinikums Hufelandstr. 55, 4300 Essen T. (02 01) 7 23 - 21 00
Frankfurt/Main	Prof. Dr. Hugo Fasold Institut für Biochemie der Universität Frankfurt, Mehrzweckgebäude der Chemischen Institute, Gebäude 75A, Sandhofstraße, 6000 Frankfurt/M. 71 T. (06 11) 63 01 - 60 61
Freiburg	Prof. Dr. Helmut Holzer Biochemisches Institut der Universität Freiburg Hermann-Herder-Str. 7, 7800 Freiburg i. Br. T. (07 61) 2 03 - 31 92
Giessen	Prof. Dr. Wilhelm Schoner Universität Giessen, Institut für Biochemie und Endokrinologie, Fachbereich Veterinärmedizin Frankfurter Str. 112, 6300 Giessen T. (06 41) 7 02 - 48 40
Göttingen	Prof. Dr. Hans-Dieter Söling Medizinische Klinik und Poliklinik der Universität Göttingen, Abt. Klinische Biochemie Humboldtallee 1, 3400 Göttingen T. (05 51) 39 - 29 20
Hamburg	Prof. Dr. Helmuth Hilz Physiologisch-Chemisches Institut der Universität Hamburg Martinistr. 52, 2000 Hamburg 20 T. (0 40) 4 68 - 23 90
Hannover	Prof. Dr. Walther Lamprecht Med. Hochschule Hannover, Institut für Klinische Biochemie und Physiologische Chemie, Karl-Wiechert-Allee 9, Postfach 61 01 80, 3000 Hannover 61 T. (05 11) 5 32 - 28 25
Heidelberg	Prof. Dr. Joachim Knappe Institut für Biologische Chemie der Universität Heidelberg Im Neuenheimer Feld 501, 6900 Heidelberg T. (0 62 21) 56 24 21
Kaiserslautern	Prof. Dr. Volker Ullrich, s. Saarland
Karlsruhe	Prof. Dr. Janos Rétey Universität Karlsruhe, Lehrstuhl für Biochemie im Institut für Organische Chemie Richard-Willstätter-Allee, Postfach 63 80, 7500 Karlsruhe T. (07 21) 6 08 20 92
Kiel	Prof. Dr. Roland Schauer Biochemisches Institut im Fachbereich Medizin, Christian-Albrechts-Universität Olshausenstr. 40/60, Haus N 11, Otto-Meyerhof-Haus, 2300 Kiel T. (04 31) 8 80 - 22 10

Kassel	Prof. Dr. Wilhelm Schoner, s. Gießen
Köln	Prof. Dr. Lothar Jaenicke Universität Köln, Institut für Biochemie An der Bottmühl 2, 5000 Köln 1 T. (02 21) 31 13 31
Konstanz	Prof. Dr. Dirk Pette Universität Konstanz, Fachbereich Biologie Postfach 733, 7750 Konstanz T. (0 75 31) 88 - 22 62
Lübeck	Prof. Dr. E. Harbers Institut für Medizin, Molekularbiologie der Medizinischen Hochschule Lübeck Ratzeburger Allee 160, 2400 Lübeck T. (04 51) 50 01 - 24 06
Mainz	Prof. Dr. Rudolf K. Zahn Physiologisch-Chemisches Institut der Johannes-Gutenberg-Universität Duesbergweg, Postfach 39 80, 6500 Mainz T. (0 61 31) 39 - 59 11
Mannheim	Prof. Dr. Joachim Knappe, s. Heidelberg
Marburg	Prof. Dr. Peter Karlson Universität Marburg, Physiologisch-Chemisches Institut, Lahnberge Deutschhausstr. 1 - 2, 3550 Marburg T. (0 64 21) 28 - 50 11
München	Prof. Dr. Guido Hartmann Chemisches Laboratorium der Universität München, Institut für Biochemie Karlstr. 23, 8000 München 2 T. (0 89) 59 02 - 2 52
Münster	Prof. Dr. Eckhart Buddecke Westfälische Wilhelms-Universität, Physiologisch-Chemisches Institut Waldeyer Str. 15, 4400 Münster T. (02 51) 4 90 - 55 71
Oldenburg	Prof. Dr. Thomas Höpner Universität Oldenburg, Fachbereich IV Ammerländer Heerstr. 67 - 99, Postfach 943, 2900 Oldenburg T. (04 41) 5 10 61
Osnabrück	Prof. Dr. Peter von Sengbusch, s. Bielefeld
Paderborn	Prof. Dr. Dr. Eckhard Schlimme Gesamthochschule Paderborn, Fachbereich 13, Naturwissenschaften II, Warburger Str. 100, Gebäude J, Postfach 16 21, 4790 Paderborn T. (0 52 51) 60 23 26
Passau	Prof. Dr. R. Jaenicke, s. Regensburg
Regensburg	Prof. Dr. R. Jaenicke Biochemie II, Fachbereich Biologie der Universität Regensburg Universitätsstr. 31, 8400 Regensburg T. (09 41) 9 43 - 30 15
Saarland	Prof. Dr. Volker Ullrich Universität des Saarlandes, Physiologisch-Chemisches Institut 6650 Homburg/Saar T. (0 68 41) 16 - 22 59

Stuttgart	Prof. Dr. Gerhard Pfleiderer Institut für Organische Chemie der Universität Stuttgart Pfaffenwaldring 55, 7000 Stuttgart 80 T. (07 11) 7 84 - 43 90
Tübingen	Prof. Dr. Ulf Henning Max-Planck-Institut für Biologie Corrensstr. 38, 7400 Tübingen T. (0 70 71) 60 12 30
Ulm	Prof. Dr. Helmut Thomas Universität Ulm (Med.-Naturwiss. Hochschule), Abteilung Biochemie Oberer Eselsberg N 26, 7900 Ulm T. (07 31) 1 76 - 22 71
Würzburg	Prof. Dr. Ernst Helmreich Universität Würzburg, Physiologisch-Chemisches Institut Koellikerstr. 2, 8700 Würzburg T. (09 31) 3 17 46
Wuppertal	Prof. Dr. A. Trebst, s. Bochum

Hoppe-Seyler's Zeitschrift für Physiologische Chemie
Begründet 1877 · Fortgeführt von A. Kossel, F. Knoop und K. Thomas

Bd. 359 Juni 1978 Heft 6

Inhaltsverzeichnis

Fortsetzung nächste Seite

Contents

Continued

Redaktionelle Hinweise siehe gegenüber Seite 776

Indexed in Current Contents

Autoren-Verzeichnis

Hoppe-Seyler's Z. Physiol. Chem.
Bd. 359, S. 659 - 669, Juni 1978

Human Plasma Kallikrein: Purification, Enzyme Characterization and Quantitative Determination in Plasma

Helmut HEBER, Reinhard GEIGER and Norbert HEIMBURGER

Forschungslaboratorien der Behringwerke AG, Marburg/Lahn,
Institut für Klinische Chemie und Klinische Biochemie der Universität München und
Klinikum Großhadern, Medizinische Klinik II

(Received 23 December 1977)

Summary: Human plasma kallikrein was isolated from a plasma fraction related to Cohn fraction IV_4 by affinity- and Sephadex G-150 chromatography yielding a material with 17.3 TAME-U/A_{280} unit. The preparation was characterized by immunological and enzymatic methods. Complex formation with α_2-macroglobulin, $\overline{C1}$-inactivator and aprotinin was demonstrated by immunoelectrophoresis. The bradykinin release from high-molecular weight kininogen and its inhibition by antibodies to kallikrein, AT III and AT III-heparin complex were measured using a biological test system (rat uterus). Time dependent inactivation of kallikrein by AT III, and AT III-heparin complex was shown by means of a synthetic kallikrein substrate: Bz-Pro-Phe-Arg--pNan. The same substrate was used to measure the activation of prekallikrein in plasma by kaolin and F XIIa. Antibodies raised against kallikrein were shown to inhibit the reaction specifically. A quantitative determination of plasma prekallikrein by electroimmunodiffusion according to Laurell was developed: the plasma concentration in normal individuals was found to be 1.8 - 2.2 TAME-U/m*l* related to kallikrein activity; this corresponds approximately to 9 - 11 mg antigen/100 m*l* plasma.

Enzymes:
Kallikrein (recommended name: kininogenin (EC 3.4.21.8); Plasmin (EC 3.4.21.7); Thrombin (EC 3.4.21.5);
F XII = blood coagulation factor XII; FXI = blood coagulation factor XI.

Abbreviations:
Bz-Arg-OEt = N^{α}-benzoyl-L-arginine ethyl ester (= BAEE);
Tos-Arg-OMe = tosyl-L-arginine methyl ester (= TAME);
Bz-Pro-Phe-Arg-pNan·HCl = *N*-benzoyl-L-prolyl-L-phenylalanyl-L-arginine-*p*-nitranilide (= CPK; = Chromozym plasma kallikrein);
AT III = antithrombin III;
IgG fraction = immunoglobulin fraction;
SBTI-Sepharose 4B = soybean trypsin inhibitor-Sepharose 4B;
Polybrene = polymeric 1,1,5,5-tetramethyl-1,5-diazoniaundecamethylene dihydroxide (recommended name: hexadimethrine bromide);
1 U = amount of enzyme that converts 1 μmol of the substrate/min under the conditions described;
BAEE-U = enzyme activity in U using Bz-Arg-OEt as substrate;
TAME-U = enzyme activity in U using Tos-Arg-OMe as substrate;
CPK-U = enzyme activity in U using Bz-Pro-Phe-Arg-pNan·HCl as substrate;
(1 CPK-U = 3.2 - 3.4 BAEE-U; 1 BAEE-U = 1.3 - 1.6 TAME-U).
USP unit = unit for heparin activity corresponding to the United States Pharmacopeia;
1 Unit of factor XI and XII, respectively, is related to the amount of these factors in 1 m*l* normal pooled citrated plasma = 100% of the norm;
A_{280} unit = amount of absorbing material in 1 m*l* that gives an A_{280} value of 1.

Human-Plasma-Kallikrein: Reinigung, enzymatische Charakterisierung und quantitative Bestimmung in Plasma

Zusammenfassung: Human-Plasma-Kallikrein wird durch Affinitäts- und Sephadex-G-150-Chromatographie aus einer Cohn IV_4 entsprechenden Plasmafraktion isoliert. Die Präparation, die eine Aktivität von 17.3 TAME-U/A_{280}-Einheit hatte, wird immunologisch und enzymatisch charakterisiert. Mit der Technik der Immunelektrophorese werden Komplexe von Plasma-Kallikrein mit α_2-Makroglobulin, $\overline{C1}$-Inaktivator und Aprotinin nachgewiesen. Die Bradykinin-Freisetzung aus hochmolekularem Kininogen; deren Hemmung durch Antikörper gegen Kallikrein, durch Antithrombin III und den Antithrombin-III-Heparin-Komplex werden in einem biologischen Testsystem (Rattenuterus) gemessen. Die Kinetik der Kallikrein-Hemmung durch AT III und den AT III-Heparin-Komplex wird mit dem synthetischen Substrat Bz-Pro-Phe-Arg-pNan gemessen. An diesem Substrat wird auch die Aktivierung von Präkallikrein in Plasma durch Kaolin und deren Inhibition durch Antikörper gegen Kallikrein untersucht. Die quantitative Präkallikreinbestimmung in Plasma erfolgt nach dem Prinzip der Elektroimmundiffusion nach Laurell: Bei Normalpersonen beträgt der Gehalt, gemessen als Kallikrein-Aktivität, 1.8 - 2.2 TAME-U/m*l* oder annähernd 9 - 11 mg/100 m*l* Plasma.

Key words: Human plasma kallikrein, affinity chromatography, soybean-trypsin inhibitor-Sepharose 4B, bradykinin release, Laurell electrophoresis.

Human plasma kallikrein is a serine protease circulating in plasma as an inactive precursor like other plasma proteases. Originally its main physiological function was suggested to be the release of bradykinin from kininogen[1], a vasoactive nonapeptide which lowers the blood pressure, increases the permeability of blood vessels and mimics some features of the inflammatory response[2,3]. However, according to more recent results, both human plasma kallikrein (identical with Fletcher-factor) and high-molecular weight kininogen (identical with Fitzgerald-, Williams- and Flaujeac-factor) play a central role in the contact phase of haemostasis[4-7]. High molecular weight kininogen stimulates the activation of surface-adsorbed factor XII and prekallikrein. The generated kallikrein reciprocally catalyses the activation of surface-adsorbed factor XII[8,9]. In this scheme both factors are necessary for a potent activation of the factor XII-dependent pathway of coagulation and fibrinolysis. The lack of some of these factors is associated with a defect of the contact phase of the intrinsic pathway, indicated by a prolonged partial thromboplastin time.

This communication concerns the purification of plasma kallikrein by affinity (SBTI-Sepharose 4B)- and Sephadex G-150 chromatography. The preparation is characterized by immunological and enzymatic methods. A quantitative determination of prekallikrein in plasma using electroimmunodiffusion according to Laurell is described.

Materials and Methods

Chemicals

Sepharose 4B and Sephadex G-150 were products from Pharmacia Fine Chemicals, Frankfurt; Cyanogen bromide and benzamidine were obtained from Merck, Darmstadt, Soybean Trypsin Inhibitor from Serva, Heidelberg, Tosyl-L-arginine methyl ester and N^{α}-benzoyl-L-arginine ethyl ester from Merck, Darmstadt. Chromozym PK: the test kit from Pentapharm Ltd., Basel containing N^{α}-benzoyl-L-prolyl-L-phenylalanyl-L-arginine-*p*-nitroanilide was used. Heparin (liquemin 5 000 USP units/m*l*) and protamine sulfate (Protamin 1 000) were purchased from Hoffmann-La Roche AG, Grenzach. Antisera and plasma proteins were products of the Behringwerke AG, Marburg/Lahn. Human high-molecular weight kininogen was partially purified according to Mann and Geiger[10]. Excess porcine pancreatic kallikrein liberates 250 ng bradykinin equivalent/mg protein from this kininogen preparation. Aprotinin (Trasylol, ampoule 100 000 KIE/5 m*l* from Bayer AG, Leverkusen and Antagosan, ampoule 100 000 KIE/5 m*l* from Behringwerke GA) were used, Polybrene from EGA-Chemie, Steinheim.

Preparation of SBTI-Sepharose 4B was performed according to Cuatrecasas[11]. The synthesis of the benzamidine spacer and the coupling on Sepharose 4B was performed as described by Shaw et al.[12].

Starting material

The kallikrein containing material corresponded to a Cohn IV_4 fraction, a side fraction from the Behringwerke plasma fractionation procedure. In a typical preparation, 1 kg of the Cohn IV_4 fraction was suspended in 2 *l* of cold 0.9% sodium chloride. After an incubation time of 5 h, the pH of the suspension was adjusted to 7.4 and insoluble material was removed by centrifugation; the supernatant was diluted with saline to an optical density of 25 - 30 at 280 nm, yielding 3 - 4 *l* starting material with an esterolytic activity of 0.5 - 0.7 TAME-U/m*l*.

Purification methods

Affinity adsorption was performed essentially according to Shaw et al.[12] on SBTI-Sepharose 4B. The benzamidine eluate was concentrated and dialysed against the equilibration buffer using the "Hollow-Fiber Dialyzer Concentrator DC 2" from Amicon. This material was readsorbed to a Sepharose-gel containing spacer-bound benzamidine[12] or purified further at room temperature by Sephadex G-150 chromatography on a column (4.5 x 100 cm) eluted with 0.1M Tris/HCl buffer, pH 8.0, containing 0.5M NaCl.

Polyacrylamide-gel electrophoresis was run in a 7.5% polyacrylamide gel, pH 8.6, 5 - 6 V/cm. Immunological identification of the bands was performed after covering the gel with an agarose layer[14].

Immunoelectrophoresis was carried out according to the method of Schwick et al.[15].

Immunoelectrophoresis according to Laurell[16] was run in a 1% agarose gel with 0.02M Barbital buffer, pH 8.6, at 2 V/cm for 8 h. The concentration of the antiserum against human plasma kallikrein in the gel was 1%. In order to establish calibration curves, either 5 μl pooled plasma from normal donors in the concentrations 100%; 50%; 25% or 12.5% of the norm, or purified kallikrein containing 2 TAME-U/m*l* in 5% human albumin, diluted with 5% albumin corresponding to 2; 1.5; 1; 0.5 or 0.25 TAME-U/m*l* were used.

Determination methods

Kallikrein activity was measured using the synthetic substrate Tos-Arg-OMe[12], Bz-Arg-OEt[13] and Bz-Pro-Phe--Arg-pNan, using the test kit Chromozym PK of Pentapharm Ltd., Basel. The activity values for the different substrates differed somewhat; the relations are given under Abbreviations.

The molecular weight of human plasma kallikrein was determined by sodium dodecylsulfate-polyacrylamide-gel electrophoresis according to Helting et al.[17].

Kallikrein/kallikrein-inhibitor complex formation

To study the complex formation with some inhibitors, 0.2 m*l* of 1% solutions of α_2-macroglobulin, $\overline{C1}$-inactivator and aprotinin were separately incubated with 0.2 m*l* kallikrein solution (8.7 TAME-U/m*l*) for 3 h at room temperature. Immunoelectrophoresis was carried out according to Schwick and Störiko[15] with doubled running time.

Human plasma kallikrein-induced bradykinin liberation

Incubation buffer: 0.15M Tris/HCl 4mM EDTA, 10mM 8-hydroxyquinoline, pH 8.6; substrate solution: 1 mg high-molecular weight kininogen in 0.1 m*l* incubation buffer. Assay system: 0.1 m*l* incubation buffer and 0.1 m*l* high-molecular weight kininogen were preincubated for 30 min at 37 °C, the enzyme reaction was started by addition of 50 μl kallikrein solution (0.2 BAEE-mU/m*l*). Bradykinin liberation was inhibited after 10, 20, 40, 60, and 100 min by addition of 50 μl aprotinin (20 000 KIE/m*l*) and heating of the samples to 100 °C for 10 min. Blanks contained 50 μl buffer instead of kallikrein solution. The samples were stored at – 80 °C until testing.

Inhibition of bradykinin liberation

Inhibition of bradykinin liberation *by the IgG fraction from anti-kallikrein serum:* 5 μl human plasma kallikrein solution (2 BAEE-mU/m*l*) was preincubated for 30 min at 37 °C with 5 μl of the IgG fraction (1.3 mg/m*l*) and 40 μl 0.9% sodium chloride solution. The following steps were identical to those described above.

Inhibition of bradykinin liberation *by AT III and AT III-heparin complex:* samples containing 50 μl AT III (0.15% in 0.9% saline) or 50 μl AT III and 5 μl heparin (2.5 USP units/m*l*) were filled up to 110 μl with 0.9% sodium chloride and incubated for 5 min at 37 °C. From such incubation mixtures aliquots (0, 5, 10 and 20 μl) were added to 5 μl kallikrein solution (2 BAEE-mU/m*l*) and filled up to 50 μl with 0.9% sodium chloride. After 10 min of preincubation at 37 °C, 200 μl of high-molecular weight kininogen was added to each sample and incubated for 4 h at 37 °C. The enzyme activity was stopped as described above.

Inhibition of human plasma kallikrein by AT III and AT III-heparin complex using the substrate Bz-Pro-Phe--Arg-pNan.

Samples with 3 m*l* kallikrein solution (200 CPK-mU/m*l* were incubated at 37 °C with (1): 10 μl heparin (50 USP units), 75 μl 0.9% saline. After 90 min, 10 μl AT III solution (3% in saline) was added; (2): 10 μl AT III solution, 75 μl 0.9% saline. After 90 min 10 μl of heparin was added; (3): 10 μl AT III solution and 10 μl heparin solution, after 6 min 75 μl of protamine sulfate solution (75 USP units) was added. After various times of incubation samples (0.25 ml) were taken to measure the kallikrein activity. Test buffer: (a): 3.03 g Tris, 1.70 g imidazole and 500 ml 0.1N HCl, diluted to 1 l with dist. water; (b): 4.04 g Tris, 2.27 g imidazole and 1.95 g sodium chloride diluted to 1 l with dist. water; (c): 11.69 g sodium chloride, diluted in 1 l of dist. water. Solutions (a) and (b) were mixed to give pH 7.9 and this solution was mixed with an equal volume of solution (c) to give a final buffer solution with an ionic strength of 0.15M and pH 7.9. Test system: Samples (0.125 ml) were mixed with 1 ml of the test buffer and incubated for 2 min at 37 °C. Then, 0.125 ml substrate solution (1.5mM Bz-Pro-Phe-Arg-pNan) was added, thoroughly mixed and the change in optical density at 405 nm was recorded.

Human plasma kallikrein activation in plasma by kaolin

Citrated plasma (0.5 ml) was mixed with 0.5 ml of kaolin suspension (10 mg/ml) in a plastic centrifuge tube and left for 10 min at 0 °C. After centrifugation, the supernatant was used to measure kallikrein activity as described above.

Celite-eluate: Citrated plasma was adsorbed with 30 mg Hyflow/ml plasma, washed with 0.38% citrate, pH 8.0 containing 0.135M NaCl. The resin was eluted with 0.1M Tris, pH 10 containing 1M NaCl. The eluate contained 13 units of factor XIa and 3.7 units of factor XIIa. The activities were measured according to Landell et al.[18]. 1 unit is related to the amount of these factors in 1 ml normal pooled citrated plasma corresponding to 100% activity

Results

During the first step of purification of human plasma kallikrein by affinity adsorption with SBTI-Sepharose 4B gel, the esterase activity was found enriched about 260-fold in the concentrated and dialyzed SBTI-Sepharose 4B eluate. The total yield of this step was about 70%. As main contaminations IgG and IgA were detected by means of immunoelectrophoresis and Ouchterlony technique (Fig. 1, 2). Further purification was performed by two different procedures: one aliquot of the eluate was dialyzed and readsorbed to an affinity gel containing spacer-bound benzamidine[12]; the other part was chromatographed on Sephadex G-150. A typical elution profile obtained by the Sephadex gel chromatography is given in Fig. 3. The esterolytic activity is eluted in the descending shoulder of the main peak and in the following small peak. The top fractions had

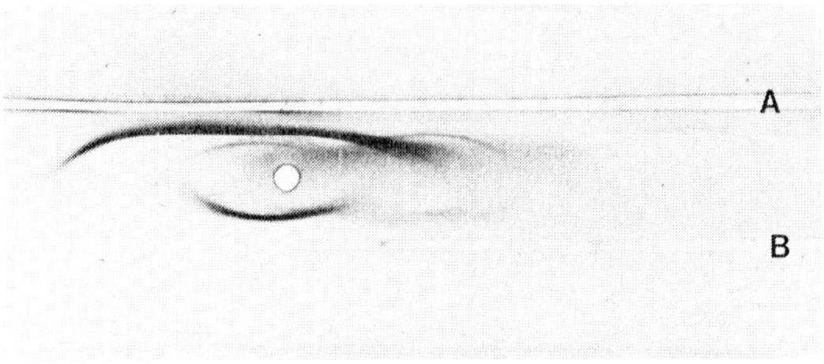

Fig. 1. Immunoelectrophoresis of the SBTI-Sepharose eluate.

The plate was developed with antiserum from rabbits against human plasma proteins (A) and with our nonadsorbed anti-kallikrein-serum (B).

an activity of 25 TAME-U/A_{280} unit. However, they still contained traces of IgG. All fractions of peak 2 with specific activity higher than 10 TAME-U/A_{280} unit were combined, giving a pool with 17.6 TAME-U/A_{280} unit. In comparison, the material obtained with the benzamidine-Sepharose 4B showed 11 TAME-U/A_{280} unit. A flow-sheet for the preparation is given in Table 1. The purest fractions were used for immunisation. The antiserum raised in rabbits showed two unspecific precipitation lines against plasma; they could be removed by adsorption with immunoglobulins and small amounts of plasma.

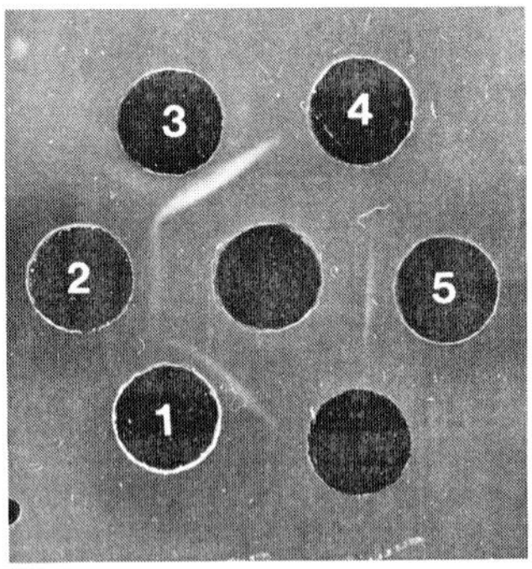

Fig. 2. Immunodiffusion of the SBTI-Sepharose eluate with the Ouchterlony technique.

Center well: SBTI-Sepharose eluate, outside wells: 1. anti-kallikrein-serum, 2. anti-human γ-globulin serum, 3. anti-IgG-serum, 4. anti-IgM-serum, 5. anti-IgA-serum. The diffusion was carried out at room temperature for 24 h.

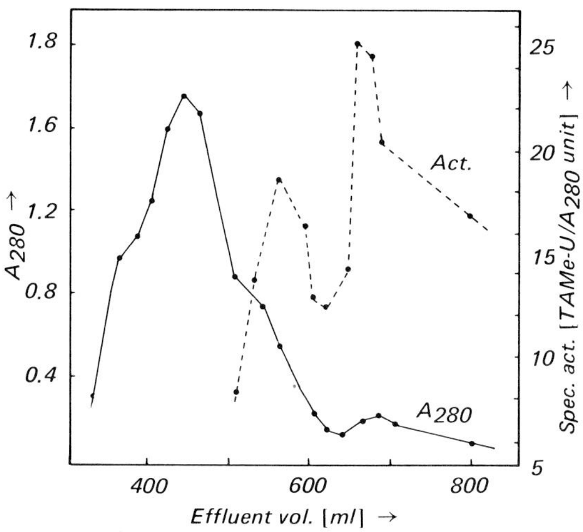

Fig. 3. Sephadex G-150 chromatography of the benzamidine eluate from SBTI-Sepharose.
For details see Materials and Methods.

The preparation was investigated for homogeneity in polyacrylamide gel electrophoresis (Fig. 4). Samples run in duplicate were stained with amido black or developed with antisera after they had been covered with a thin layer of agarose. The broad band in the region of transferrin represents kallikrein. The two minor bands in the region of albumin are due to contaminating materials. On immunoelectrophoresis, our preparation migrates like a β_1-protein (Fig. 1). Using the Ouchterlony technique (Fig. 5), the preparation can be shown to be free from several proteinases of human

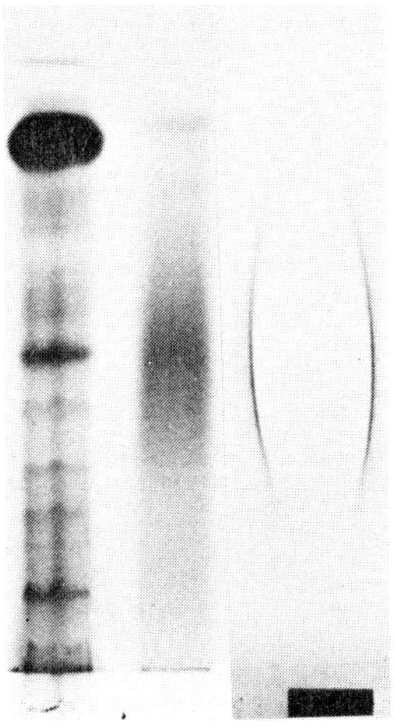

Fig. 4. Polyacrylamide gel electrophoresis of the Sephadex G-150 eluate.
1. Serum, 2. Sephadex G-150 eluate, A_{280} = 4.75,
3. duplicate gel developed with anti-kallikrein-serum.
For details see Materials and Methods.

Table 1. Purification of human plasma kallikrein from Cohn fraction IV_1.
The assay was performed as described[12].

Step	Vol. [ml]	Total prot. [A_{280} units]	Total esterase act. [TAME-U]	Spec. act. [TAME-U/ A_{280} unit]	Esterase act. recovered [%]
Starting material	900	16 800	380	2.25×10^{-2}	100
Benzamidine eluate from SBTI-Sepharose column after dialysis and concentration	50	41.6	265	5.9	69
Benzamidine eluate from Benzamidine Sepharose after dialysis and concentration	70	15.5	170	11	45
Sephadex G-150 chromatography	175	12.3	210	17.6	55

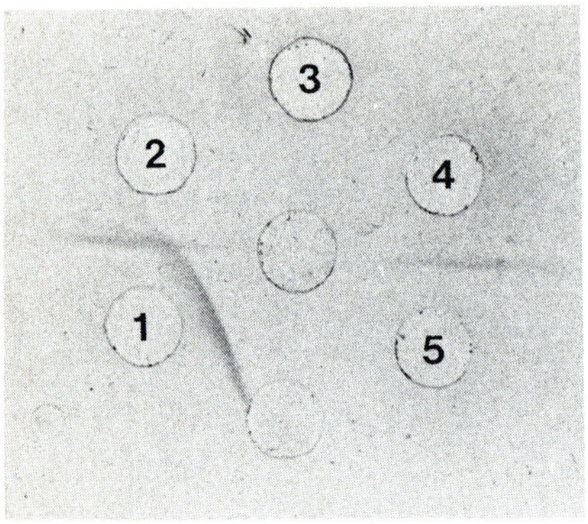

Fig. 5. Immunodiffusion of the Sephadex G-150 eluate with the Ouchterlony technique.

Center well: Sephadex G-150 eluate, 16 BAEE-U/m*l*. Outside wells: 1. anti-kallikrein-serum, 2. anti-plasminogen-serum, 3. anti-prothrombin-serum, 4. anti-$\overline{C1}$-esterase-serum, 5. anti-$\overline{C3}$-activator-serum. The diffusion was carried out at room temperature for 24 h.

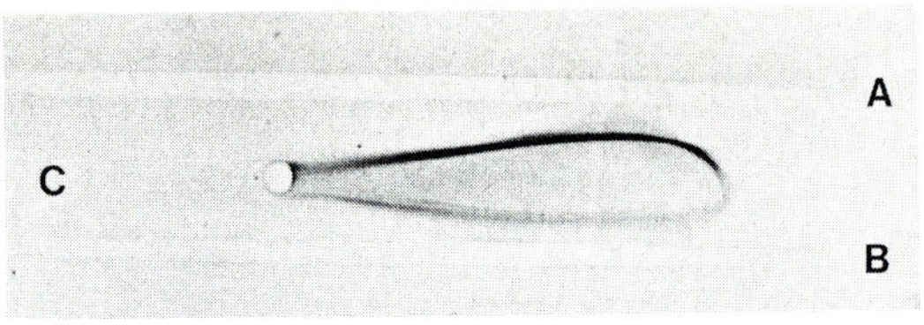

Fig. 6. Immunoelectrophoresis of kallikrein-α_2-macroglobulin complex.

Center well: kallikrein-α_2-macroglobulin complex (C); developed with anti-α_2-macroglobulin-serum (A) and anti-kallikrein-serum (B). For details see Materials and Methods.

Fig. 7. Immunoelectrophoresis of kallikrein-$\overline{C1}$-inactivator complex.

Center well: kallikrein-$\overline{C1}$-inactivator complex (C); developed with anti-$\overline{C1}$-inactivator (A) and anti-kallikrein serum (B). For details see Materials and Methods.

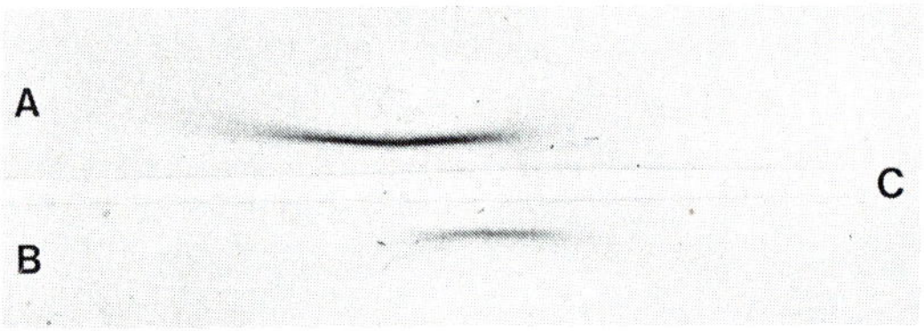

Fig. 8. Immunoelectrophoresis of kallikrein-aprotinin complex (well A) and kallikrein (well B).

Developed with anti-kallikrein serum (C). For details see Materials and Methods.

plasma: plasmin, thrombin, $\overline{C1}$-esterase and $\overline{C3}$-activator.

Furthermore, the preparation showed neither a caseinolytic, nor a clotting activity against fibrinogen. On immunoelectrophoresis, the interaction between kallikrein and α_2-macroglobulin (Fig. 6), $\overline{C1}$-inactivator (Fig. 7) and aprotinin (Fig. 8) were characterized by formation of new precipitation arcs.

Molecular weight estimation of the material corresponding to the second peak by dodecylsulfate-polyacrylamide-gel electrophoresis gave an apparent M_r of 98 000 ± 5 000.

Activation of prekallikrein in plasma

These studies were performed using the synthetic substrate Bz-Pro-Phe-Arg-pNan. The addition of kaolin or of activated contact factors XI and XII to plasma generates kallikrein activity[19]. The celite eluate used in this study contained 13 units of factor XIa/m*l* and 3.7 units factor XIIa/m*l*. The requirement of factor XIIa for the activation of prekallikrein with kaolin was demonstrated by the addition of Polybrene prior to kaolin into the activation mixture (Table 2). Furthermore, kallikrein mixed with its corresponding antibody was shown to be inactive when tested with the synthetic substrate Bz-Pro-Phe-Arg-pNan (Fig. 9). Consequently preincubation of plasma with the IgG fraction of anti-human plasma kallikrein serum and subsequent kaolin activation results in a diminution or total decrease of esterase activity. In contrast IgG fraction from antisera against plasminogen and prothrombin did not influence the esterase activity (Fig. 10). Since antibodies against factor XII inhibit the surface activation of this clotting factor, the IgG fraction from anti-human plasma kallikrein-serum was tested for antibodies to F XII in a one-stage F XI/XII neutralization test, in principal according to Landell et al.[18]. The results are summarized in Table 3.

Table 2. Activation of prekallikrein in plasma, 10 min, 0 °C.
Total incubation volume 500 *μl*, the substrate used was Bz-Pro-Phe-Arg-pNan. The celite eluate used contained 13 units factor XI a and 3.7 units factor XII a/m*l*. For further details see Materials and Methods.

Citrated Plasma [*μl*]	Activator	Polybrene [ng]	Activity [CPK-mU/m*l*]
250	–	–	< 1
–	150 *μl* kaolin suspension	–	0
–	20 *μl* celite eluate	–	0
250	250 *μl* kaolin suspension	–	41
250	20 *μl* celite eluate	–	80
250	250 *μl* kaolin suspension	8	28
250	250 *μl* kaolin suspension	80	< 1

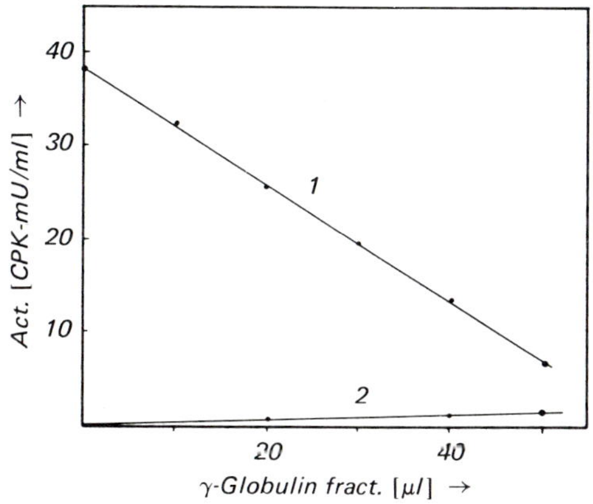

Fig. 9. Inhibition of kallikrein activity by γ-globulin fraction (1.3 mg/m*l*) of anti-kallikrein serum using the synthetic substrate Bz-Pro-Phe-Arg-pNan.
Curve 1: samples of 0.2 m*l* of kallikrein solution (37 CPK-mU/m*l*) were incubated with increasing volumes of γ-globulin fraction at 37 °C for 10 min prior to testing; curve 2: amidolytic activity of the IgG fraction (1.3 mg/m*l*) used.

Table 3. Examination of the anti-kallikrein γ-globulin fraction for antibodies against factor XI and XII using the one-stage activity test according to Landell[18].
The anti-kallikrein γ-globulin solution contained 1.3 mg/m*l*.

Plasma [m*l*]	Anti-kallikrein γ-globulin [m*l*]	F XI act. [% of the norm]	F XII act. [% of the norm]
0.5	–	72	98
0.5	0.05	70	97
0.5	0.10	71	98
0.5	0.15	71	97

Fig. 10. Inhibition of the kaolin-induced esterase activity in plasma by IgG fraction of different sera.
Curve 1: anti-kallikrein serum; curve 2 and 3: samples containing IgG fraction of anti-plasminogen (1.0 mg/m*l*) or anti-prothrombin serum (1.3 mg/m*l*) from rabbit.
Substrate: Bz-Pro-Phe-Arg-pNan. ▷

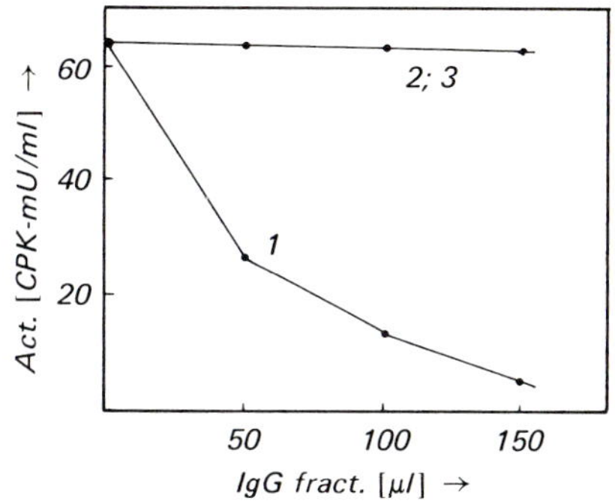

Bradykinin liberation from high-molecular weight kininogen

The bradykinin liberation from high-molecular weight kininogen was determined on the rat uterus as described by K. Mann, R. Geiger and E. Werle[10]. As can be seen in Fig. 11, curve 1, the bradykinin liberation is very fast during the first 10 min. After 60 min, only small amounts of bradykinin are liberated although the maximal amount that can be liberated in the test sample was 46.7 ng. Curve 2 shows the diminished bradykinin liberation in the presence of the IgG fraction from anti-kallikrein serum. In the same test system, the influence of antithrombin III (AT III), heparin and the preformed AT III-heparin complex was measured (Fig. 12). AT III shows only a weak inhibitory effect whereas the preformed AT III-heparin complex strongly inhibits the bradykinin release by kallikrein. Heparin alone had no effect on the enzyme activity. Using the synthetic substrate Bz--Pro-Phe-Arg-pNan, the inhibition studies with kallikrein revealed a progressive inhibition by AT III[20] which could be accelerated by subsequent addition of heparin (Fig. 13, curve 2[21,22]. The AT III-heparin complex caused rapid rate of inactivation of kallikrein (curve 3) which could be slowed down to the normal rate of AT III by adding protamine sulfate. Heparin itself had no effect on kallikrein activity (curve 1) but subsequent addition of AT III caused a rapid inactivation of the enzyme activity.

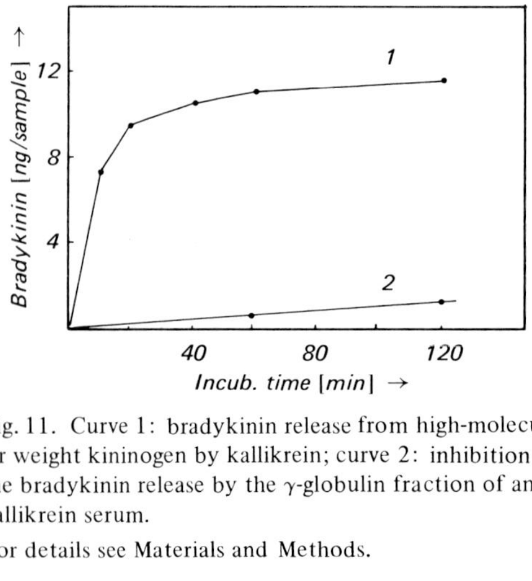

Fig. 11. Curve 1: bradykinin release from high-molecular weight kininogen by kallikrein; curve 2: inhibition of the bradykinin release by the γ-globulin fraction of anti-kallikrein serum.

For details see Materials and Methods.

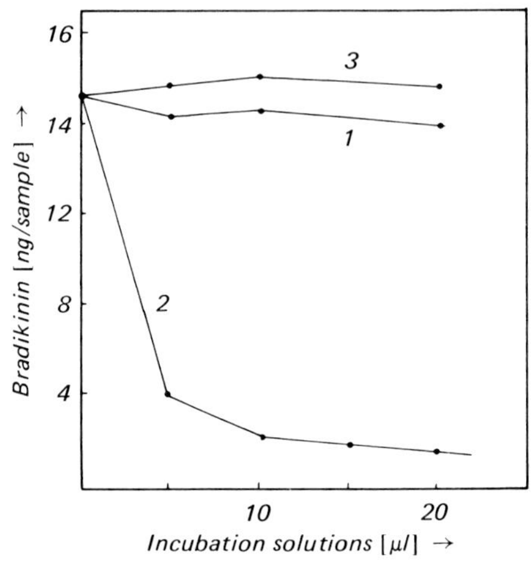

Fig. 12. Curve 1: bradykinin release from high-molecular weight kininogen by kallikrein in the presence of AT III; curve 2: AT III-heparin complex and curve 3: heparin alone.

Ordinate: ng bradykinin released; abscissa: μl of AT III, AT III-heparin complex and heparin solutions added to the samples. For details see Materials and Methods.

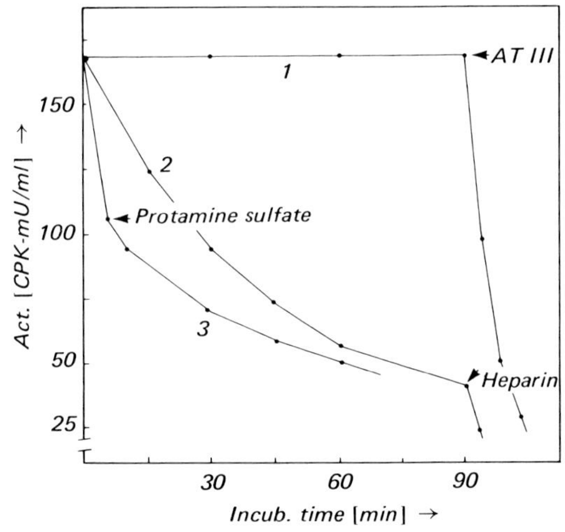

Fig. 13. Inhibition of kallikrein by AT III using the synthetic substrate Bz-Pro-Phe-Arg-pNan.

Curve 1: preincubation of kallikrein with heparin for 90 min prior to the addition of AT III; curve 2: incubation of kallikrein with AT III for 90 min prior to the addition of heparin; curve 3: inhibition of kallikrein by AT III-heparin complex for 6 min prior to the addition of protamine sulfate. For details see Materials and Methods.

Quantitative determination of human plasma kallikrein in plasma

Since anti-kallikrein serum from rabbits also precipitates prekallikrein, it is possible to estimate the concentration of prekallikrein in human plasma by means of usual immunological techniques[23]. The electroimmunodiffusion technique according to Laurell[16] was applied to quantitate this antigen. Calibration curves (Fig. 14, 15) were set up with purified preparations containing 2; 1.5; 1; 0.5 or 0.25 TAME-U of human plasma kallikrein per m*l* diluted with a solution of human albumin, or with pooled human plasma in the concentrations 100%; 50%; 25% or 12.5% of the norm. By this method the prekallikrein content of plasma from eight healthy donors was found to vary between 1.8 and 2.2 TAME-U/m*l*. Assuming a specific activity of about 20 TAME-U/mg kallikrein the level of kallikrein in normal individuals is in the range of 9 - 11 mg/100 m*l* plasma.

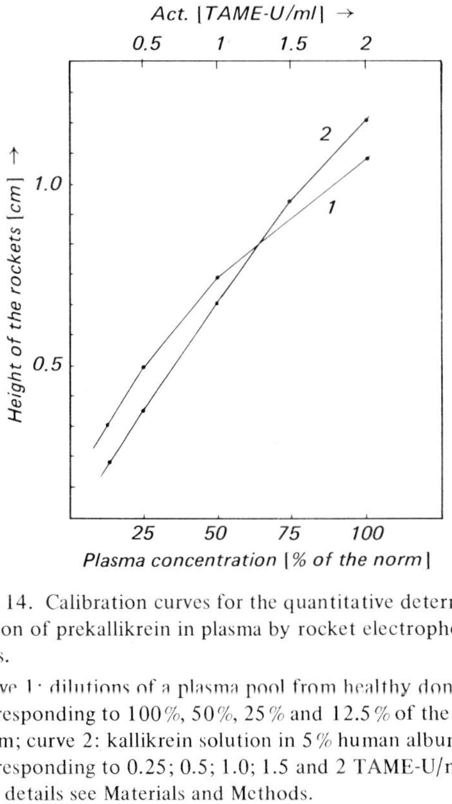

Fig. 14. Calibration curves for the quantitative determination of prekallikrein in plasma by rocket electrophoresis.

Curve 1: dilutions of a plasma pool from healthy donors corresponding to 100%, 50%, 25% and 12.5% of the norm; curve 2: kallikrein solution in 5% human albumin corresponding to 0.25; 0.5; 1.0; 1.5 and 2 TAME-U/m*l*. For details see Materials and Methods.

Discussion

As starting material the Cohn fraction IV_4 was chosen since this material contains a lot of kallikrein and is nearly free from other plasma proteases which could also be adsorbed to the resin coated with soybean trypsin inhibitor. Isolation of plasma kallikrein was achieved by adsorption to SBTI-Sepharose 4B and further purification by a second affinity chromatography step using spacer-bound benzamidine-Sepharose 4B[12]. Instead of the benzamidine resin, gel filtration on a Sephadex G-150 column was performed as an alternative procedure for purifying the enzyme. In principle both methods worked well and proved to be suitable. However, gel filtration led to a material with higher specific activity; in addition it had the advantage that two dialysis steps which diminished the total yield could be omitted.

The kallikrein preparation was characterized by polyacrylamide gel electrophoresis, immunoelec-

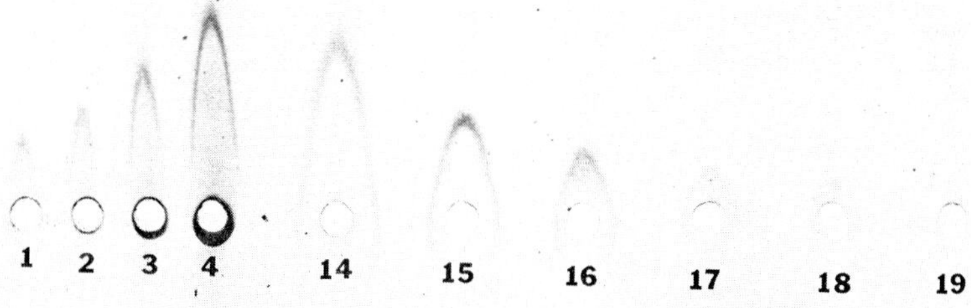

Fig. 15. Rocket electrophoresis of pooled human plasmas.
No. 1 - 4 are plasma dilutions prepared from a pool of healthy donors corresponding to 100%, 50%, 25% or 12.5% of the norm; no. 14 - 18 are purified kallikrein samples in 5% albumin containing 2; 1.5; 1.0; 0.5 and 0.25 TAME-U/m*l*. No. 19 is a blank value containing only 5% albumin. For details see Materials and Methods.

trophoresis, molecular weight determination and susceptibility to some proteinase inhibitors of human plasma. The results obtained were in good agreement with those given by other authors [12,19,23]. Colman[19] separated three forms of plasma kallikrein and classified them according to their molecular weights as kallikrein I (99 800), kallikrein II (163 000) and kallikrein III (124 000). Our material might be identical with the lower molecular weight derivative I. This is in good agreement with the molecular weight determined by means of dodecylsulfate gel electrophoresis M_r = 98 000 ± 5 000) and the elution position as the last protein and activity peak during filtration on Sephadex G-150 (Fig. 3). It is further characterized by the complex formation with aprotinin (Fig. 8) and inhibition by soybean trypsin inhibitor. The kallikrein activity in the descending shoulder of the first peak (Fig. 3) might correspond to Colman's kallikrein II; it was not further characterized but it could be inhibited by soybean trypsin inhibitor. The fact that we did not observe kallikrein III may be due to the inability of form III to combine strongly with soybean trypsin inhibitor[19]. Consequently, it could not be expected to be adsorbed to the SBTI-Sepharose gel.

The highest specific activity detected in the last elution peak rose to 25 TAME-U/A_{280} unit. This material was only contaminated with traces of IgG detected by means of the double immunodiffusion technique according to Ouchterlony with antisera against the whole plasma proteins and IgG. Shaw et al.[12] reported specific activities for human plasma kallikrein isolated according to a similar procedure which varied from batch to batch between 45 - 50 und 60 TAME-U/A_{280} unit. The lower specific activity of the material prepared according to the procedure described might be due to differences in the starting material. In contrast to the material used by Shaw[12], this contained kallikrein in form of the activated enzyme: acid treatment did not cause an increase in the TAME-cleaving activity; it was not stable and decreased during storage. However, after elution from SBTI-Sepharose and after Sephadex G-150 chromatography there was no considerable further loss of activity. It thus can not be excluded that we isolated a relatively stable but not so active form of kallikrein. During the course of our work we used three synthetic substrates for the determination of kallikrein activity, Tos-Arg-OMe, Bz-Arg-OEt and the tripeptide Bz-Pro-Phe-Arg-pNan.

The isolated enzyme could be shown to be identical with the proteinase activated when kaolin is added to human plasma. These studies were performed with the synthetic substrate Bz-Pro-Phe--Arg-pNan which is not attacked by the activated clotting factors XI and XII. However, Bz-Pro-Phe--Arg-pNan is also cleaved by plasmin and thrombin. Therefore, it was necessary to exclude that the esterolytic activity induced by contact activation was due to the presence of one of these enzymes. This could be achieved by adding antibodies against plasminogen and prothrombin to plasma prior to the addition of kaolin.

The results showed clearly that both enzymes are not involved or identical with the arginine esterase activity which is generated during contact activation.

The action of antithrombin III on kallikrein was measured using two substrates: (I) the natural one, high-molecular weight kininogen and (II) the synthetic one, Bz-Pro-Phe-Arg-pNan. In the presence of kininogen, AT III shows only a weak inhibitory effect; however, using Bz-Pro-Phe-Arg-pNan, there is a typical progressive inhibition. These results may suggest that kallikrein has a higher affinity to high-molecular weight kininogen than to AT III. Consequently, AT III itself seems to play only a secondary role in the regulation of kallikrein activity in plasma. However, in the presence of heparin AT III is a very potent kallikrein inhibitor probably controlling the fibrinolysis via the F XIIa pathway[8,24] and kinin formation. In connection with these biological functions the determination of kallikrein in plasma may be of diagnostic importance. Colman et al.[23] found a good correlation between antigen concentration and the esterase activity generated by kaolin addition to plasma. The protein concentration was determined by radial immunodiffusion according to Mancini, the esterase activity by the hydrolysis rate of Tos-Arg-OMe. The prekallikrein concentration of plasma found by electroimmunodiffusion according to Laurell[16] is of the same magnitude 9 - 10 mg/100 m*l* plasma, as described by Colman et al. (10 - 13 mg/100 m*l* plasma). However, there was no good correlation with the Bz-Pro-Phe-Arg-pNan activity: using kaolin or a factor XIa/XIIa mixture for activating prekallikrein in plasma, varying amounts of kallikrein activity were generated, probably due to an incomplete activation. However, our activation conditions were different from those of Colman[23] and this might explain the differences observed. The possibility to determine prekallikrein in plasma by its protein concentration in addition to the functional assay offers a new and interesting field for diagnosis.

Literature

1 Webster, M.E. (1970) Handbook of Experimental Pharmacology, vol. 25 (Erdös, E.G., ed.) pp. 659 - 665, Springer Verlag, New York.
2 Margolis, J. (1958) *J. Physiol. (London)* **144**, 1 - 22.
3 Werle, E (1970) Handbook of Experimental Pharmacology, vol. 25 (Erdös, E.G., ed.) pp. 1 - 6, Springer Verlag, New York.
4 Webster, M.E. & Pierce, J.V. (1973) *Fed. Proc.* **32**, 854.
5 Griffin, J.H. & Cochraine, C.G. (1976) *Proc. Natl. Acad. Sci. U.S.A.* **73**, 2554 - 2558.
6 Chung, Y.L., Bagdasarian, A., Meier, H., Scott, C.F., Pierce, J., Kaplan, A.P. & Colman, R.W. (1976) *Fed. Proc.* **35**, 692.
7 Colman, R.W., Bagdasarian, A., Talamo, R.C., Scott, C.F., Seavey, M., Guimaraes, J.A., Pierce, J.V. & Kaplan, A.P, (1975) *J. Clin. Invest.* **56**, 1650 - 1662.
8 Meier, H.L., Pierce, J.V., Colman, R.W., Kaplan, A.P. (1977) *J. Clin. Invest.* **60**, 18 - 31.
9 Liu, C.Y., Scott, C.F., Bagdasarian, A., Pierce, J.V., Kaplan, A.P. & Colman, R.W. (1977) *J. Clin. Invest.* **60**, 7 - 17.
10 Mann, K., Geiger, R. & Werle, E (1976) *Adv. Exp. Med. Biol.* **70**, 65 - 73.
11 Cuatrecasas, P. (1970) *J. Biol. Chem.* **245**, 3059 - 3065.
12 Sampaio, C., Wong, S.-C., & Shaw, E. (1974) *Arch. Biochem.* **165/1**, 133 - 139.
13 Trautschold, J., Werle, E. & Schweitzer, G. (1974) in Methoden d. Enzymat. Analyse (Bergmeyer, H.U., ed.), 3 rd. edn., pp. 1071 - 1080, Verlag Chemie, Weinheim/Bergstraße.
14 Zwisler, O. & Biel, H. (1966) *Z. Klin. Chem.* **4**, 58 and *Behringwerke-Mitt.* **46**, 129.
15 Schwick, H.-G., Störiko, K. & Becker, W., (1969) *Laboratory notes (Behringwerke)* **3**, 1 - 50.
16 Laurell, C.-B. (1966) *Anal. Biochem.* **15**, 45 - 52.
17 Helting, T.B. & Zwisler, O. (1977) *J. Biol. Chem.* **252**, 187 - 193.
18 Landell, R.D., Wagner, R.H. & Brinkhous, K.M. (1953) *J. Lab. Clin. Med.* **41**, 637 - 647
19 Colman, R.W., Mattler, L. & Sherry, S. (1969) *J. Clin. Invest.* **48**, 23 - 32; 11 - 22.
20 Fritz, H., Wunderer, G., Kummer, K., Heimburger, N. & Werle, E. (1972) *Hoppe-Seyler's Z. Physiol. Chem.* **353**, 906 - 910.
21 Veneröd, A.M. & Laake, K. (1975) *Thromb. Res.* **7**, 223 - 226.
22 Veneröd, A.M., Laake, K., Solberg, A.K. & Strömland, S. (1976) *Thromb. Res.* **9**, 457 - 466.
23 Bagdasarian, A., Lahiri, B., Talamo, R.C., Wong, P. & Colman, R.W. (1974) *J. Clin. Invest.* **54**, 1444 - 1454.
24 Laake, K. & Veneröd, A.M. (1974) *Thromb. Res.* **4**, 285 - 302.

Dr. H. Heber und Dr. N. Heimburger, Behringwerke AG, Postfach 1140, D-3550 Marburg 1;

Dr. R. Geiger, Abteilung für Klinische Chemie und Klinische Biochemie in der Chirurgischen Klinik der Universität München,
D-8000 München 2, Nußbaumstr. 20.

Hoppe-Seyler's Z. Physiol. Chem.
Bd. 359, S. 671 - 675, Juni 1978

A Radioimmunoassay for Measurement of Human Pancreatic Secretory Trypsin Inhibitor in Different Body Fluids

Allan EDDELAND and Kjell OHLSSON

Departments of Surgery and Clinical Chemistry, University of Lund, Malmö General Hospital, Malmö, Sweden

(Received 16 December 1977)

Summary: A radioimmunoassay for measurement of human pancreatic secretory trypsin inhibitor in nanogram quantities has been developed. The sensitivity of the assay now permits examination of the inhibitor content of various body fluids, wherein other methods exhibit serious shortcomings.

In healthy blood donors the serum level was 8.1 μg/*l*. In patients with acute pancreatitis levels as high as 320 μg/*l* have been measured, and patients who underwent endoscopic retrograde cholangiopancreatography showed an elevated inhibitor level in serum immediately after the examination without any clinical signs of disease, the highest registered value being 128 μg/*l*. In peritoneal lavage fluid from patients with severe acute pancreatitis levels of 5–304 μg/*l* have been measured. In urine the inhibitor level is about 14 μg/*l* in healthy persons. The urine from one patient with proteinuria of glomerulo-tubular type contained 380 μg/*l*.

Ein Radioimmunoassay zur Bestimmung des sekretorischen Trypsininhibitors aus menschlichem Pankreas in verschiedenen Körperflüssigkeiten

Es wurde ein Radioimmunoassay zur Bestimmung von Nanogramm-Mengen des sekretorischen Trypsininhibitors aus mesnchlichem Pankreas entwickelt. Seine Empfindlichkeit erlaubt die Messung des Inhibitorgehalts verschiedener Körperflüssigkeiten, bei denen andere Methoden versagen.
Bei gesunden Blutspendern beträgt der Blutspiegel 8.1 μg/*l*. Bei Patienten mit akuter Pankreatitis wurden Werte bis 320 μg/*l* gemessen und bei Patienten nach endoskopischer retrograder Cholangio-Pankreatographie ohne klinische Symptome Werte bis 128 μg/*l*. In peritonealer Spülflüssigkeit von Patienten mit akuter Pankreatitis betrugen die Werte 5–304 μg/*l*. Der Urin gesunder Personen enthielt 14 μg des Inhibitors/*l*, der Urin eines Patienten mit Proteinurien vom glomerulo-tubularen Typ 380 μg/*l*.

Key words: Pancreatitis, radioimmunoassay, trypsin, trypsin inhibitor.

Abbreviations: ERCP = endoscopic retrograde cholangiopancreatography;
PSTI = pancreatic secretory trypsin inhibitor.

The acini of the pancreas secrete a trypsin inhibitor known as the Kazal inhibitor[1] or the pancreatic secretory trypsin inhibitor (PSTI)[2], because it is secreted into the juice together with the exocrine secretory proteins[3]. The presence of the inhibitor in the pancreatic juice is thought to prevent an autocatalytic digestion of the pancreatic gland, should a premature activation of trypsinogen to trypsin occur within the gland[4-6]. A reduction or total lack of the PSTI might be part of the etiology of pancreatic disease, especially pancreatitis[7]. It is thus of interest to measure the PSTI in various body fluids. Previously the PSTI in pancreatic juice was measured by enzymological methods. Recently we described an immunochemical method[8] that allowed quantitation of PSTI in human pancreatic juice to a lowest level of 0.2 μg/m*l*. This method is not sensitive enough to detect PSTI in serum, peritoneal fluid or urine. Bovine pancreatic trypsin inhibitors have been measured by radioimmunoassay by Fink and Greene[9,10] but no such method has to our knowledge been presented for the human inhibitor. We therefore developed a radioimmunoassay sensitive enough to measure the PSTI in these body fluids.

Methods and Results

Special materials

Sephadex G-50, G-75 and G-200 were obtained from AB Pharmacia Fine Chemicals, Uppsala, Sweden. Human serum albumin was purchased from AB KABI, Stockholm, Sweden, as 20% Albumin KABI. Goat antirabbit IgG serum and normal rabbit serum were available in our laboratory. The same purified human PSTI and antiserum against it were used as in a previous investigation[8].

Human biological material

Serum samples were obtained from 67 healthy blood donors, from 18 patients immediately before and after they had undergone endoscopic retrograde cholangiopancreatography (ERCP) and also from 16 patients with acute pancreatitis. The serum samples and pool were first kept at + 4 °C and later frozen and kept at − 20 °C. Urine from 10 healthy individuals was sampled and used immediately and urine from one patient with glomerulotubular proteinuria was stored frozen at − 20 °C and thawed and used as needed. The first fluid return from 12 patients who were treated for severe acute pancreatitis with peritoneal lavage was collected and stored at − 20 °C until used.

Radioactive labelling of pancreatic secretory trypsin inhibitor

5 μg of the purified human PSTI was labelled with 1 mCi ^{125}I, using a lactoperoxidase method[11], to a specific activity measuring about 10 mCi/mg on different labellings. Unbound ^{125}I was separated by gel filtration on Sephadex G-50 (column 1.6 × 45 cm; buffer: 10mM Tris/HCl, pH 7.4, containing 0.12M NaCl and 5mM EDTA). 1 m*l* of 20% serum albumin was run through the column previously in order to coat the Sephadex with protein before the labelled PSTI was applied to the column. Tubes containing 75 μ*l* of 2% serum albumin were used for collecting 1-m*l* fractions from the column.

Properties of the labelled tracer

Precipitation experiments were performed with the tracer and antibodies against PSTI. A solution of the tracer containing 2.5 ng/m*l* was prepared and 200 μ*l* of this tracer solution was incubated with 200 μ*l* antiserum against PSTI in dilutions between 1 : 10 and 1 : 50000. 100 μ*l* buffer (10mM Tris/HCl, pH 7.4, containing 0.12M NaCl, 5mM EDTA, 23μM Trasylol and 0.2% serum albumin) was added and the solutions were agitated in a test-tube mixer and left at + 4 °C overnight. 100 μ*l* normal rabbit serum (diluted 1 : 200) and 100 μ*l* goat antirabbit IgG serum (1 : 3), both diluted in the same buffer as above, were added and incubated again at + 4 °C overnight. Finally, 500 μ*l* of this buffer was added and the reaction mixture was centrifuged at 2000 × *g* at + 4 °C for 15 min. The supernatant was decanted and the radioactivity of the precipitate was measured. The labelled tracer was easily precipitated by antibodies against PSTI. 90% of the radioactivity could be precipitated with an antiserum dilution of 1 : 1000. With the antiserum in a dilution of 1 : 25000, about 2/3 of the radioactivity could be precipitated. When the labelled tracer was mixed with pancreatic juice and subjected to gel filtration on Sephadex G-75 (column 1.6 × 36 cm; buffer: 10mM Tris/HCl, pH 7.4, containing 0.12M NaCl and 5mM EDTA) it was eluted in the same volume as unlabelled PSTI (Fig. 1).

Radioimmunological assay procedure

For all dilutions the following buffer was used: 10mM Tris/HCl, pH 7.4, containing 0.12M NaCl, 5mM EDTA, 0.2% serum albumin and 23μM Trasylol. Standard solutions in concentrations from 0 to 16 μg/*l* were prepared with this buffer and PSTI. Small portions of these solutions were frozen and used as needed after having been thawed only once. The labelled PSTI at a concentration of about 2.5 ng/m*l* was used as tracer. This corresponded to about 30000 cpm/200 μ*l*. The antiserum was used in a dilution of 1 : 25000, permitting about 2/3 precipitation of the radioactivity in the assay when no additional unlabelled PSTI was present. In the assay each tube contained 100 μ*l* standard solution or 100 μ*l* of a sample (or

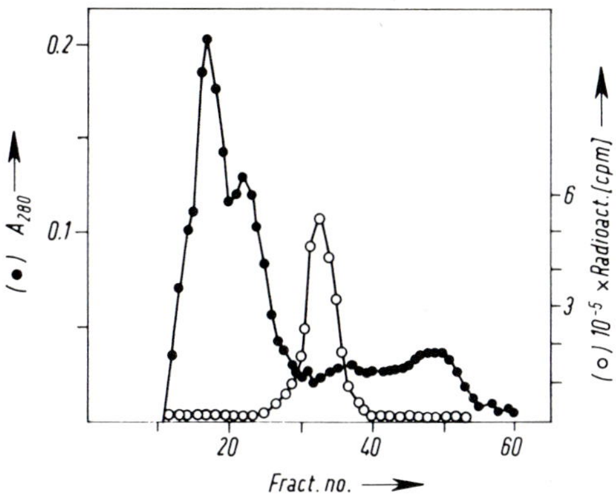

Fig. 1. Gel filtration on Sephadex G-75 of 1 m*l* of normal pancreatic juice mixed with 0.03 nmol ^{125}I-labelled PSTI.

Column: 1.6 × 36 cm; buffer: 10mM Tris/HCl, pH 7.4, containing 0.12M NaCl and 5mM EDTA; fraction volume: 1.4 m*l*. •——• = protein; ○——○ = ^{125}I-labelled PSTI.

a dilution of the sample appropriate for the standard curve) and 200 μl antibody solution. The tube was thoroughly agitated for several seconds and incubated overnight at + 4 °C, after which 200 μl tracer solution was added. After thorough agitation and overnight incubation, 100 μl normal rabbit serum (1:200), 100 μl goat antirabbit IgG serum (1:3) and 500 μl of the buffer were added and the contents were thoroughly mixed for a few seconds. After an additional overnight incubation at + 4 °C, the tubes were centrifuged at 2000 × *g* at + 4 °C for 15 min, the supernatant was decanted, and the radioactivity of the pellet was measured. In an assay with samples of serum, plasma, peritoneal fluid or urine a series of standard dilutions was always measured along with the samples, and both standards and samples were run in duplicate assay tubes.

Specificity

In Fig. 2 the standard dilution curve for PSTI is shown together with a dilution curve which is the mean of the dilution curves for serum samples from six healthy blood donors. The curves appear parallel, indicating that the same immunoreactive substance is measured in the different dilution series.

*Sensitivity**

The sensitivity of the assay was 0.4 $\mu g/l$ which was adequate to measure the PSTI concentration in serum, plasma, peritoneal lavage fluid and urine.

* The least amount of ligand distinguishable from zero.

Accuracy

The measurements of duplicate samples had a coefficient of variation (in the interval 1 to 8 $\mu g/l$) of ± 1.5% within-batch and ± 6.3% between-batch, even with different lots of labelled tracer.

The pancreatic secretory trypsin inhibitor in normal serum and plasma

Freshly drawn serum from healthy blood donors contains 8.1 ± 0.3 $\mu g/l$ PSTI (mean ± SEM, n = 67). A serum pool from 10 other blood donors contained 8.4 $\mu g/l$. The concentration is the same in plasma and also in serum and plasma that has been thawed after freezing.

The pancreatic secretory trypsin inhibitor in serum from patients who had undergone endoscopic retrograde cholangiopancreatography (ERCP)

Serum samples from 18 patients who had undergone this treatment were drawn immediately after the examination was completed. Three of these serum samples were subjected to gel filtration on Sephadex G-200 (column 1.3 × 53 cm; buffer: 10mM Tris/HCl, pH 7.4, containing 0.12M NaCl and 5mM EDTA). Samples from the fractions were measured with the radioimmunoassay and compared to a control pooled serum filtered in the same way. The immunoreactive PSTI in normal serum as well as after ERCP was eluted in the same volume as free inhibitor. A rise in PSTI in serum after ERCP was demonstrated (Fig. 3). In the three patients the levels were 16, 29 and 80 $\mu g/l$, respectively. In the 18 patients who underwent ERCP, the serum level immediately before the examination was 13 ± 1 $\mu g/l$ and immediately after 44 ± 8 $\mu g/l$ (mean ± SEM), the highest registered value being 128 $\mu g/l$.

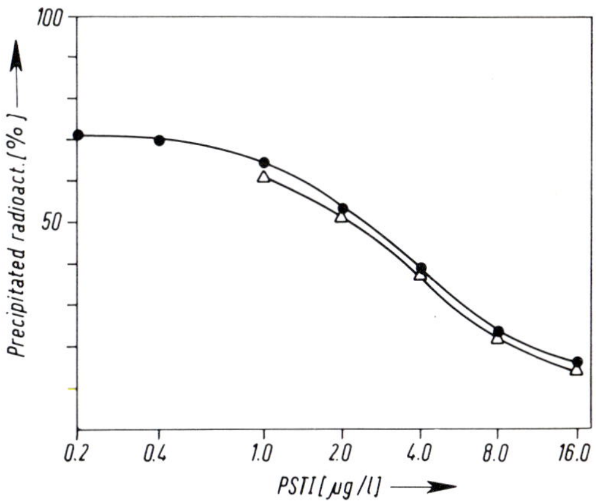

Fig. 2. Comparison between standard dilution curves of PSTI in the dilution buffer (•) and mean of six normal sera (△).

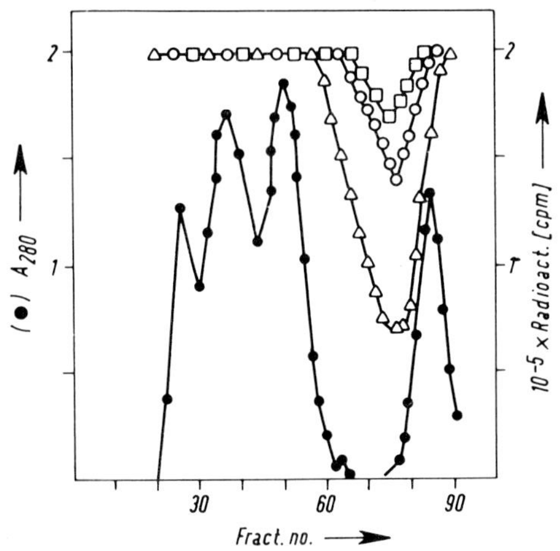

Fig. 3. Partition of immunoreactive PSTI expressed as cpm precipitated (right ordinate) on gel filtration on a Sephadex G-200 column.

Column, 1.3 x 53 cm; buffer 10mM Tris/HCl, pH 7.4, containing 0.12M NaCl and 5mM EDTA; fraction volume 1.0 m*l*. □——□ = a serum pool from 10 healthy individuals; ○——○ = mean of sera from three patients immediately after ERCP; △——△ = mean of sera from three patients with acute pancreatitis. ●——● = protein (left ordinate).

The pancreatic secretory trypsin inhibitor in serum from patients with acute pancreatitis

Serum samples from three patients with severe acute pancreatitis were gel filtered (same column and buffer as above) and samples of the fractions were measured in the same way as mentioned above. In Fig. 3 the mean of these radioimmunological determinations is shown. The elution pattern of the immunoreactive PSTI is the same as in normal serum but the amount of PSTI was markedly increased in the three patients: 41, 99 and 128 μg/*l*, respectively. In 16 patients with acute pancreatitis the serum level was 147 ± 28 μg/*l* (mean ± SEM), the highest registered value being 320 μg/*l*.

The pancreatic secretory trypsin inhibitor in urine

The radioimmunoassay was used to detect the inhibitor in urine. In healthy individuals the concentration was 14 ± 3 μg/*l* (mean ± SEM, n = 10). The urine from a patient who had proteinuria of the glomerulotubular type had a concentration of 380 μg/*l*. The peak of measurable PSTI in urine corresponded to the same molecular weight on gel filtration as that measured in serum.

The pancreatic secretory trypsin inhibitor in peritoneal lavage fluid from patients with acute pancreatitis

In peritoneal lavage fluid from 12 patients with severe acute pancreatitis, the concentration of the inhibitor was 66 ± 24 μg/*l* with a highest value of 304 μg/*l*.

Discussion

In his review on immunoassay, Walker predicted that enzyme assay by mass rather than activity will undergo exponential growth[12]. Working with an enzyme inhibitor, the PSTI, we were in search of a suitable method to measure this inhibitor in different body fluids such as serum, plasma, urine and peritoneal lavage fluid wherein other methods exhibit serious shortcomings. Although a radioimmunoassay has more experimental variables than most other analytical techniques[12], it was possible to develop an assay which proved to be quite reproducible both within-batch (coefficient of variation ± 1.5%) and between-batch (coefficient of variation ± 6.3%).

The standard values were chosen so that the test standard curve fell in the range of 1 - 10 μg/*l* (Fig. 2) which is the range of the serum level of PSTI from normal individuals, with the mean for healthy blood donors being 8.1 μg/*l*. Gel filtration of serum both from healthy individuals and from patients with acute pancreatitis show that the PSTI is eluted in a homogeneous peak in the same fractions as free PSTI (Fig. 3). In most patients, serum levels were elevated immediately after an ERCP, in some cases 8-fold, but without any clinical signs of disease. Skude et al. have shown that hyperamylasemia also occurs after duodenoscopy and ERCP[13].

Patients with acute pancreatitis also exhibited greatly elevated serum levels of PSTI, of which the highest registered was 320 μg/*l*. Patients with severe acute pancreatitis who were treated by peritoneal lavage showed high values of the inhibitor in the lavage fluid. It would appear that it is released into the peri-pancreatic tissues and into the circulation when pancreatitis develops. We have recently shown that PSTI does not form complexes with trypsin to any appreciable extent in the presence of excess α_2-macroglobulin or α_1-antitrypsin*. Furthermore, PSTI-trypsin

* Eddeland, A., to be published.

complexes already formed are rapidly cleaved on admixture of serum *in vitro* and *in vivo*, the trypsin being taken over mainly by α_2-macroglobulin[14]. The PSTI thus released is only bound to a limited extent by the trypsin-α_2-macroglobulin[14]. Native PSTI, however, readily forms complexes with trypsin-α_2-macroglobulin* and thus inhibits the hydrolytic activity of these complexes against low molecular weight substrates[15]. This means that PSTI exists in serum either free or bound by trypsin-α_2-macroglobulin complexes. The PSTI of these complexes would probably not be recognized by our antibodies against it[16] and would thus not be measured by the assay. Such complexes are, however, rapidly eliminated by the cells of the reticuloendothelial system[17]. Hence the amount of PSTI present in serum bound by such complexes is probably small compared to the amount of free inhibitor, which is readily quantitated by our assay. This is further suggested by the parallel relation between the dilution curves for the tracer and serum. Low molecular weight substances are known to be eliminated into the urine, the kidneys being considered the primary organ of catabolism[6,14,18]. In agreement with this PSTI could be measured in normal urine and the level was considerably raised in a patient with proteinuria of the glomerulo-tubular type.

It is obvious that the PSTI is involved in a complex relationship with the plasma protease inhibitors. In further studies on this interplay and the endogenous turnover of the PSTI in health and disease, it will be of importance to measure human PSTI at the low levels which the radioimmunoassay presented here now permits.

The skilful technical assistance of Mrs *Ulla Åkesson* is greatly appreciated.

* Eddeland, A., to be published.

This investigation was supported by the *Swedish Medical Research Council* (project no. B78-17X-03910-06C), the *Medical Faculty, University of Lund*, the *Swedish Association for Medical Research* and *Bayer AG.*

Literature

1 Kazal, L. A., Spicer, D. S. & Brahinsky, R. A. (1948) *J. Am. Chem. Soc.* **70**, 3034 - 3040.
2 Greene, L. J. & Giordano Jr, J. S. (1969) *J. Biol. Chem.* **244**, 285 - 298.
3 Greene, L. J. (1968) *Ann. N. Y. Acad. Sci.* **146**, 386 - 387.
4 Fritz, H., Hüller, I., Wiedemann, M. & Werle, E. (1967) *Hoppe-Seyler's Z. Physiol. Chem.* **348**, 405 - 418.
5 Pubols, M. H., Bartelt, D. C. & Greene, L. J. (1974) *J. Biol. Chem.* **249**, 7, 2235 - 2242.
6 Fritz, H., Oppitz, K.-H., Meckl, D., Kemkes, B., Haendle, H., Schult, H. & Werle, E. (1969) *Hoppe-Seyler's Z. Physiol. Chem.* **350**, 1541 - 1550.
7 Dimagno, E. P. & Go, V. L. (1976) *Dis. Mon.* **22**, 3 - 36.
8 Eddeland, A. & Ohlsson, K. (1978) *Scand. J. Clin. Lab. Invest.*, in press.
9 Fink, E. & Greene, L. J. (1974) in Bayer-Symposium V "Proteinase Inhibitors" (Fritz, H., Tschesche, H., Greene, L. J. & Truscheit, E., eds.) pp. 243 - 249, Springer-Verlag, Berlin, Heidelberg, New York.
10 Fink, E. & Greene, L. J. (1976) in Protides of the Biological Fluids, Proc. of XXIII Colloquium, Brugge, 1975 (Peeters, H., ed.) Pergamon Press, Oxford and New York.
11 Thorell, J. I. & Johansson, B. G. (1971) *Biochim. Biophys. Acta* **251**, 363 - 369.
12 Walker, W. H. C. (1977) *Clin. Chem.* **23**, 384 - 402.
13 Skude, G., Wehlin, L., Moruyama, T. & Ariyama, J. (1976) *Gut* **17**, 127 - 132.
14 Eddeland, A. & Ohlsson, K. (1978) *Scand. J. Clin. Lab. Invest.*, in press.
15 Eddeland, A. & Ohlsson, K. (1976) *Scand. J. Clin. Lab. Invest.* **36**, 815 - 820.
16 Ohlsson, K. (1974) in Bayer-Symposium V "Proteinase Inhibitors (Fritz, H., Tschesche, H., Greene, L. J. & Truscheit, E., eds.) pp. 96 - 105, Springer-Verlag, Berlin, Heidelberg, New York.
17 Eddeland, A. & Ohlsson, K. (1978) *Hoppe-Seyler's Z. Physiol. Chem.* **359**, 379 - 384.
18 Strober, W. & Waldmann, T. A. (1974) *Nephron.* **13**, 35 - 66.

Dr. Allan Eddeland, Kirurgiska kliniken, Malmö allmänna sjukhus, S-214 01 Malmö.

Doc. Kjell Ohlsson, Kliniskt kemiska centrallaboratoriet, Malmö allmänna sjukhus, S-214 01 Malmö.

Hoppe-Seyler's Z. Physiol. Chem.
Bd. 359, S. 677 - 681, Juni 1978

Immunoreactive Trypsin in Serum and Peritoneal Fluid in Acute Pancreatitis

Anders BORGSTRÖM and Kjell OHLSSON

Departments of Clinical Chemistry and Surgery, University of Lund, Malmö General Hospital, Malmö, Sweden

(Received 23 December 1977/7 March 1978)

Summary: High levels of immunoreactive cathodal trypsin are shown to be present in serum and peritoneal fluid in acute pancreatitis. The immunoreactive material is contained in two fractions with different molecular weights: free trypsinogen and trypsin in complex with protease inhibitors, mainly α_1-antitrypsin. The presence of trypsin in complex with protease inhibitors indicates the formation of active trypsin during acute pancreatitis.

Immunreaktives Trypsin in Serum und Peritonealflüssigkeit bei akuter Pankreatitis

Zusammenfassung: Bei akuter Pankreatitis wurden in Serum und in der Peritonealflüssigkeit hohe Konzentrationen an immunreaktivem kationischem Trypsin gefunden. Das immunreaktive Material besteht aus zwei Fraktionen mit unterschiedlichen Molekulargewichten: freies Trypsinogen und Trypsin im Komplex mit Proteaseinhibitoren, überwiegend mit α_1-Antitrypsin.

Das Vorhandensein von Trypsin-Proteaseinhibitor-Komplexen ist ein Hinweis auf die Bildung von aktivem Trypsin bei akuter Pankreatitis.

Key words: Acute pancreatitis, human cathodal trypsin, protease inhibitors, radioimmunoassay.

Despite much experimental work, the pathophysiological mechanisms in the development of acute pancreatitis are still obscure. For a long time it has been thought that the fundamental cause of acute pancreatitis is autodigestion of the gland by pancreatic enzymes[1]. Most pancreatic enzymes are produced and secreted as enzymatically inactive proenzymes. Trypsin has a key role in activating these proenzymes. One hypothesis is that small amounts of trypsinogen may be autocatalytically activated and subsequently activate other proenzymes.

The significance of the proteolytic enzymes has, however, been questioned, since most investigators have been unable to demonstrate significant amounts of active trypsin or any other protease in the inflamed pancreas, serum or peritoneal fluid during acute pancreatitis[2]. This may, however, be explained by the presence of large amounts of endogenous protease inhibitors in

Enzymes: Enteropeptidase (EC 3.4.21.9); trypsin (EC 3.4.21.4).
Abbreviation: iPr_2PF, diisopropylfluorophosphate.

serum and tissue fluids. Active proteolytic enzymes released from the pancreatic gland into peritoneal fluid and serum are rapidly inactivated by these inhibitors. They are then not detectable by enzymatic methods.

Specific antisera against the inhibitors and the pancreatic proteases, however, permit detection and characterization of such enzyme-inhibitor complexes. With the aid of such techniques a large release of active trypsin, chymotrypsin and elastase has been demonstrated in experimentally bile-induced pancreatitis in dogs[3]. Similar studies concerning human pancreatitis have not been reported.

We have recently developed a radioimmunological method for determination of human cathodal trypsin[4]. Approximately 25 μg immunoreactive trypsin/m*l* was shown to be present normally in the circulation. This immunoreactivity was found to be due only to trypsinogen. We have found an increase in immunoreactive trypsin in serum and peritoneal fluid during acute pancreatitis, which is in accordance with other recent reports[5,6]. The immunoreactivity has, however, not yet been characterized.

The purpose of the present investigation was to characterize the immunoreactive trypsin in acute pancreatitis.

Experimental

Materials

Peritoneal fluid was obtained through a catheter inserted into the abdominal cavity through a small incision just below the umbilicus. Blood samples were drawn at the same time. Samples from two patients with a typical clinical and laboratory picture of acute pancreatitis were studied. All samples were kept at -20 °C up to 1 month before analysis. Human pancreatic juice was obtained from patients having their main pancreatic duct drained after pancreatic surgery as described earlier[4]. Agarose (Batch 520 A) was purchased from Miles-Seravac, Maidenhead, England. CNBr-activated Sepharose 4-B, Sephadex G-100 and DEAE-Sephadex A-50 were obtained from AB Pharmacia Fine Chemicals, Uppsala, Sweden. Human serum albumin was purchased as 20% Albumin Kabi from AB Kabi, Stockholm, Sweden. Diisopropylfluorophosphate (iPr_2PF) was obtained from Merck, Darmstadt, West Germany, and porcine enteropeptidase (Batch 11 A) from Miles Laboratories (PTY) Ltd U.K. Human α_1-antitrypsin purified according to Laurell[7] was available at the laboratory as well as monospecific rabbit antisera against human albumin, α_1-antitrypsin and α_2-macroglobulin.

Methods

Albumin, α_1-antitrypsin and α_2-macroglobulin concentrations were determined by electroimmunoassay[8]. Crossed immunoelectrophoresis was performed according to Ganrot[9]. Radioimmunoassay for human cathodal trypsin was done as described earlier[4]. Human cathodal trypsin was purified according to Ohlsson[10]. Protein concentration was measured according to Lowry[11] using human albumin as a standard.

Gel filtration was done on a Sephadex G-100 column (0.9 × 50 cm) at a flow rate of 2.4 m*l*/h. 0.4 m*l* of each sample was applied and eluted with 0.01M Tris/HCl buffer, containing 0.12M NaCl and 5mM EDTA at pH 7.4; 0.8-m*l* fractions were collected. The immunoglobulin fraction of rabbit anti-human-α_1-antitrypsin was purified using a DEAE-Sephadex A-50 column in 0.05M Tris/HCl buffer at pH 7.0. These antibodies were coupled to CNBr-activated Sepharose 4B according to the manufacturer's instructions. Samples containing the high molecular weight form of immunoreactive trypsin, found after gel filtration of peritoneal fluid, were passed through a 0.9 × 10 cm column packed with Sepharose-conjugated anti-human-α_1-antitrypsin. Immunoreactive trypsin, α_1-antitrypsin and albumin were determined before and after column passage. The albumin concentrations were used to evaluate the degree of dilution of the samples after passage through the column.

0.4 m*l* peritoneal fluid from a patient with acute pancreatitis was incubated with 0.1 m*l* 0.2M Tris/HCl buffer, 10mM $CaCl_2$, containing 1 mg enteropeptidase/m*l* at pH 7.0 for 24 h at room temperature. The incubation mixture was then filtered through a Sephadex G-100 column as described above. The effluent was analyzed for immunoreactive cathodal trypsin, α_2-macroglobulin and α_1-antitrypsin. The same amount of the same peritoneal fluid sample was incubated with 0.1 m*l* of the same buffer without enteropeptidase for 3 days at room temperature and gel filtered as a control.

Standards

0.1 m*l* of fresh pancreatic juice was diluted with 5 m*l* of fresh normal human serum and a dilution series of this stock solution was used to produce the standard dilution curve for trypsinogen. The results are expressed in per cent of the value obtained for a pool of serum from 10 healthy men run together with the standard curve in each assay.

For the formation of trypsin-α_1-antitrypsin complexes 0.3 mg of active human cathodal trypsin (10^{-8} mol) was incubated for 10 min at room temperature with

5 mg of pure α_1-antitrypsin (9×10^{-8} mol) in 1 ml 10mM Tris buffer at pH 7.4, containing 50mM $CaCl_2$. This reaction mixture was then diluted 100 times with the same buffer, containing 2 g human serum albumin/l, divided into small aliquots and stored at −20 °C. This stock solution was then thawed and diluted further to produce the standard dilution curve for trypsin in complex with α_1-antitrypsin.

If not otherwise mentioned, all dilutions were made with a 10mM Tris/HCl buffer, pH 7.4, containing 0.12M NaCl, 5mM EDTA and 2 g human serum albumin/l.

For determination of α_1-antitrypsin concentrations a dilution series of blood donor pool was used as standard. The α_1-antitrypsin concentration of this pool was 1.35 g/l[12].

Results

The standard curves for radioimmunoassay of trypsinogen and of the trypsin-α_1-antitrypsin complexes prepared in vitro using our antiserum to trypsin are shown in Fig. 1. It is evident that they give non-parallel curves, indicating different reactions with the antibody used. The results of the assay on the effluents from gel filtration of the standard stock solutions of trypsinogen and trypsin-α_1-antitrypsin complexes, respectively, are shown in Fig. 2 (A and B). The trypsin-α_1-antitrypsin complex is eluted earlier than trypsinogen in accordance with its larger size. No immunoreactivity can be detected corresponding to the size of free trypsin in Fig. 2B.

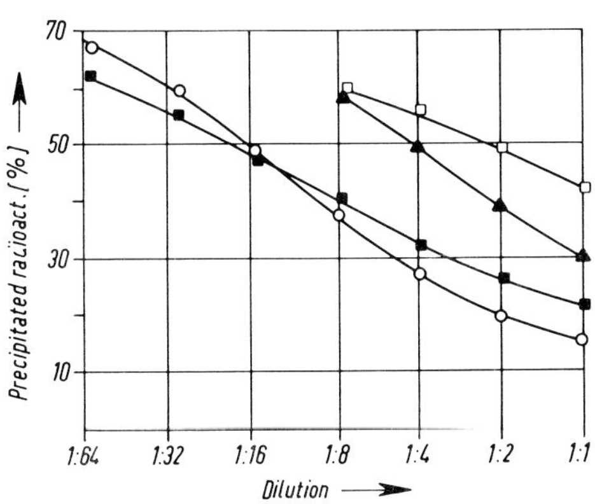

Fig. 1. Dilution curves for
○——○ inactive panreatic juice (trypsinogen),
■——■ trypsin-α_1-antitrypsin complexes,
□——□ 1st peak on gel filtration peritoneal fluid,
▲——▲ 2nd peak on gel filtration peritoneal fluid.

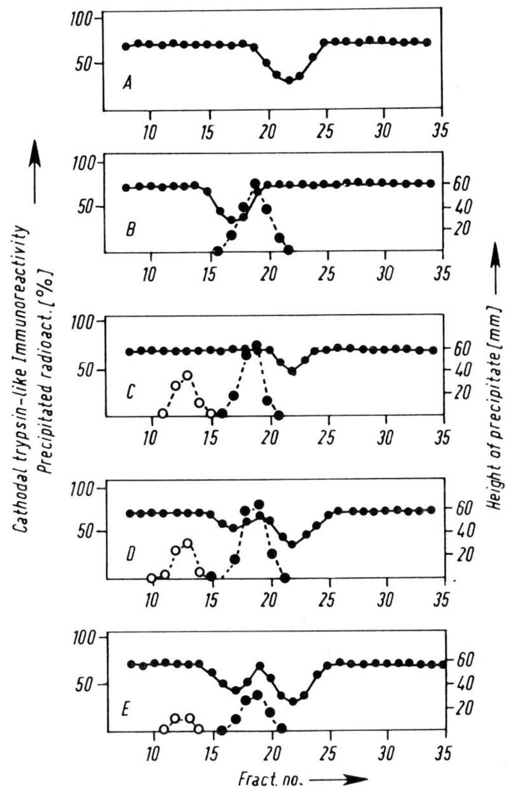

Fig. 2. Partition of cathodal trypsin-like immunoreactivity ●——● (left abscissa) and α_1-antitrypsin (α_1-AT) ●- - -● and α_2-macroglobulin (α_2-M) ○- - -○ (right abscissa) on gel filtration on a Sephadex G-100 column, 0.9 x 50 cm, of:

A) Inactive pancreatic juice (trypsinogen)
B) Trypsin-α_1-antitrypsin standard
C) Normal serum
D) Serum during acute pancreatitis
E) Peritoneal fluid during acute pancreatitis.

Dilution curves of a solution of iPr_2PF-treated pure cathodal trypsin and a solution of the same amount of active trypsin incubated with an excess of α_1-antitrypsin are shown in Fig. 3 (for experimental design see legend). The immunoreactivity of trypsin is clearly altered and diminished after complex formation with α_1-antitrypsin. The immunoreactivity of the standard serum pool is composed only of trypsinogen according to Fig. 2C.

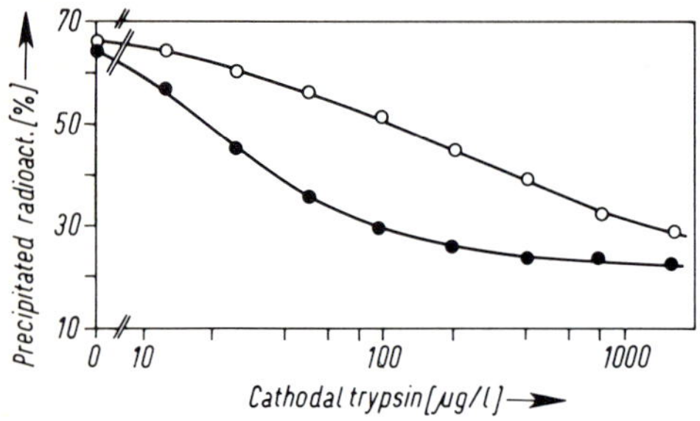

Fig. 3. Dilution curves for active cathodal trypsin after addition of buffer containing 1 mM i Pr_2 *PF* •——• and the same amount of trypsin after addition of the same volume of buffer containing α_1-antitrypsin (α_1-antitrypsin in molar excess) ○——○.

Immunoreactive cathodal trypsin in peritoneal fluid in two patients with acute hemorrhagic pancreatitis was 26 and 37 times the value of the standard serum pool. α_1-Antitrypsin concentrations were 1.2 and 1.1 g/*l*. Crossed immunoelectrophoresis with antibodies against α_1-antitrypsin showed that 10 - 15% of the α_1-antitrypsin in both peritoneal fluid samples had a slower electrophoretic mobility, indicating complex formation.

Immunoreactive cathodal trypsin concentrations in the serum samples taken at the same time were 7 and 11 times the standard serum pool value. After gel filtration the immunoreactivity in both serum and peritoneal fluid separated into two fractions (Fig. 2D and E). Dilution curves for these different fractions from peritoneal fluid are shown in Fig. 1. All immunoreactive α_1-antitrypsin disappeared after one passage of the peak fraction containing the high molecular weight trypsin immunoreactivity from peritoneal fluid through the column containing Sepharose-conjugated anti-human-α_1-antitrypsin antibodies. Immunoreactive trypsin, however, only decreased to about 30% of the original level. When the trypsin-α_1-antitrypsin standard prepared in vitro was run through the same column, all α_1-antitrypsin disappeared and less than 5% of the original immunoreactive trypsin remained. Using the same trypsin-α_1-antitrypsin complexes prepared in vitro as standard, the cathodal trypsin concentrations in the first peak from peritoneal fluid were 2.3 and 5.2 mg/*l*, respectively, for the two patients. The corresponding values for peaks from serum samples were 0.08 and 0.13 mg/*l*, respectively.

Gel filtration profiles before and after enteropeptidase activation of peritoneal fluid from a patient with acute hemorrhagic pancreatitis is shown in Fig. 4. There is a very drastic decrease in the second peak accompanied by a slight increase in the first peak after enteropeptidase activation. Incubating the sample with buffer without enteropeptidase at room temperature for 3 days did not change the gel filtration profile.

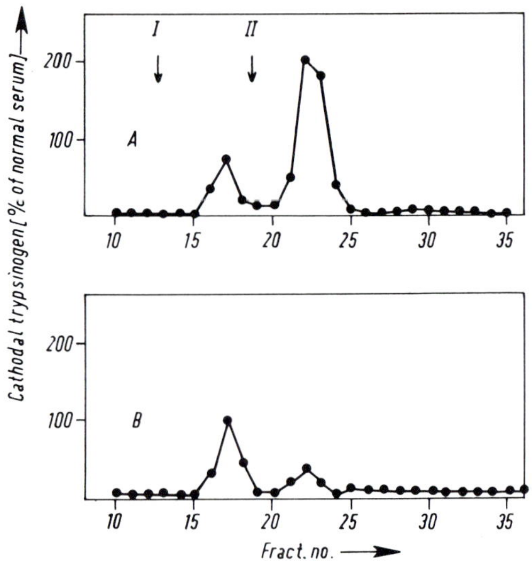

Fig. 4. Partition of cathodal trypsin-like immunoreactivity in peritoneal fluid from a patient with acute hemorrhagic pancreatitis •——• using cathodal trypsinogen as standard on a Sephadex G-100 column 0.9 x 50 cm.

A) Before incubation with enteropeptidase
B) After 24 h incubation with enteropeptidase
I) Elution volume for α_2-macroglobulin
II) Elution volume for α_1-antitrypsin.

Discussion

In acute pancreatitis immunoreactive cathodal trypsin rises in both serum and peritoneal fluid. In serum from healthy persons the immunoreactivity consists of only trypsinogen[4], but in sera and peritoneal fluid from subjects with acute pancreatitis, an immunoreactive fraction appears with a size corresponding to that of trypsin-α_1-antitrypsin complexes in addition to the trypsinogen fraction. Two-thirds of this high molecular weight immunoreactive fraction was bound by antibodies specific for α_1-antitrypsin and is thus composed of trypsin in complex with α_1-antitrypsin. The remaining one-third of the immunoreactivity towards trypsin in this high molecular fraction might be explained by dissociation of the trypsin-α_1-antitrypsin complexes during handling or by a decrease in immunoreactivity of the α_1-antitrypsin when in complex with trypsin. This is, however, a less likely explanation, since the trypsin-α_1-antitrypsin complexes prepared in vitro, when run through the immunosorbent column, lost 95% of the immunoreactive trypsin together with the α_1-antitrypsin. It seems more probable that part of the high molecular weight immunoreactive trypsin is composed of trypsin in complex with other protease inhibitors.

Collen et al. recently showed that the fast-acting plasmin inhibitor with a molecular weight of 60 000 was a fast-acting trypsin inhibitor equal to α_1-antitrypsin in vitro[13]. The molar concentration ratio antiplasmin/α_1-antitrypsin in serum is about 1/50[13]. Thus, trypsin-antiplasmin complexes can only account for part of the unidentified high molecular trypsin immunoreactivity.

The second peak on gel filtration eluted in the same volume as trypsinogen in inactive pancreatic juice. Furthermore, the enteropeptidase activation experiment clearly indicates that this peak contains mainly trypsinogen.

When trace amounts of trypsin are added to serum in vitro, a large fraction is complexed with α_2-macroglobulin, and this trypsin is not immunoreactive[14] and therefore not measured in this assay. Only a minor part of the trypsin is complexed with α_1-antitrypsin, and the immunoreactivity of this trypsin is clearly reduced, as shown in Fig. 3. These facts suggest that the trypsin found in complex with protease inhibitors in this study only represent a part of the active trypsin formed during acute pancreatitis.

The inhibitor-bound trypsin may represent trypsinogen activated in and around the pancreas during acute pancreatitis. It can also be due to activation of the trypsinogen after its appearance in the circulation. The reproducibility of the gel filtration patterns suggests that there is no activation of the trypsinogen after the sampling. The absence of α_1-antitrypsin complexes in normal sera argues against any substantial activation of trypsinogen in the circulation. The finding of trypsin in complex with protease inhibitors is most probably a specific finding in acute pancreatitis, and then reflects the amount of activated trypsinogen in and around the pancreas. Further studies on the possible mechanisms for activation of trypsinogen in the presence of protease inhibitors are needed for more elucidation of the significance of the present findings.

The skilful technical assistance of Mrs. *Eva Ljungberg* is greatly appreciated.
This investigation was supported by the *Swedish Medical Research Council* (project no. B78-17X-03910-06C), The *Medical Faculty, University of Lund*, the *Swedish Society for Medical Research* and *Kungliga Fysiografiska Sällskapet i Lund*.

Literature

1 Creutzfeldt, W. & Schmidt, H. (1970) *Scand. J. Gastroenterol. Suppl.* **6**, **5**, 47 - 62.
2 Haverback, B.J., Swanson, V.L., Dyce, B., Rindernecht, H., Wick, M. & Silverman, P. (1970) Symposium of the exocrine pancreas, pp. 121 - 130, J.A. Churchill Ltd.
3 Ohlsson, K. & Eddeland, A. (1975) *Gastroenterology* **69**, 668 - 675.
4 Borgström, A. & Ohlsson, K. (1976) *Scand. J. Clin. Lab. Invest.* **36**, 809 - 814.
5 Temler, R. & Felber, I.-P. (1976) *Biochim. Biophys. Acta* **445**, 720 - 728.
6 Elias, E., Wood, T. & Redshaw, M. (1977) *Lancet* **ii**, 66 - 68.
7 Laurell, C.-B., Pierce, J., Persson, U. & Thulin, E. (1975) *Eur. J. Biochem.* **57**, 107 - 113.
8 Laurell, C.-B. (1972) *Scand. J. Clin. Lab. Invest.* **29**, Suppl. 124.
9 Ganrot, P.O. (1972) *Scand. J. Clin. Lab. Invest.* **29**, 124; 39 - 47.
10 Ohlsson, K. & Skude, G. (1976) *Clin. Chim. Acta* **66**, 1 - 7.
11 Lowry, O.M. (1951) *J. Biol. Chem.* 265 - 275.
12 Jeppsson, J.-O., Laurell, C.-B. & Fagerhol, M. (1978) *Eur. J. Biochem.* **83**, 143 - 153.
13 Edy, J. & Collen, D. (1977) *Biochim. Biophys. Acta*, **484**, 423 - 432.
14 Ohlsson, K. (1974) in Bayer-Symposium V "Proteinase Inhibitors" (Fritz, H., Tschesche, H., Greene, L.J. & Truscheit, E., eds.) pp. 96 - 105, Springer-Verlag, Berlin, Heidelberg, New York.

Dr. A. Borgström and Doc. K. Ohlsson, Kliniskt kemiska centrallaboratoriet, Malmo allmänna sjukhus, S-21401 Malmö.

Hoppe-Seyler's Z. Physiol. Chem.
Bd. 359, S. 683 - 690, Juni 1978

Release of Granulocyte Elastase in Lethal Canine Endotoxin Shock

Ansgar O. AASEN and Kjell OHLSSON

Institute for Surgical Research, Rikshospitalet, Oslo, Norway, and
Department of Surgery, Malmö General Hospital, Malmö, Sweden

(Received 6 February 1978)

Summary: The release of granulocyte elastase and its interaction with plasma protease inhibitors was studied in dogs receiving a slow infusion of a lethal dose of *Escherichia coli* endotoxin. During endotoxin infusion a marked decline in leucocyte counts was parallelled by a rapid increase in plasma granulocyte elastase concentrations. Maximal values were reached after 3 h, when the infusion was ended.

Crossed immunoelectrophoresis with antiserum against granulocyte elastase did not reveal the presence of elastase components with the electrophoretic mobility of free elastase, but elastase-α_1-antitrypsin complexes were detected.

A gradually decreasing plasma concentration of α_2-macroglobulin was noted during the experiments. Crossed immunoelectrophoresis, however, did not reveal any electrophoretic heterogeneity. It is concluded that the release of granulocyte proteases might be of significance for several pathophysiological changes seen in endotoxin shock.

Freisetzung von Granulozyten-Elastase bei tödlich verlaufendem Endotoxin-Schock beim Hund

Zusammenfassung. Die Freisetzung von Granulozyten-Elastase sowie deren Reaktion mit Protease-Inhibitoren des Plasmas wurde anhand von Tierexperimenten verfolgt, bei denen Hunden eine letale Dosis von *Escherichia-coli*-Endotoxin langsam infundiert wurde. Während der Endotoxin-Infusion stieg die Konzentration der Granulozyten-Elastase rasch und parallel zur Abnahme der Zahl der Leukozyten im Plasma an. Maximale Werte wurden nach 3 h, also am Ende der Infusion gemessen.

Mit Hilfe der „Crossed" Immunelektrophorese (unter Verwendung von Antiserum gegen Granulozyten-Elastase) konnten Elastase-α_1-Antitrypsin-Komplexe nachgewiesen werden, nicht jedoch freie Elastase. Die Konzentration des α_2-Macroglobulins fiel im Verlaufe des Experiments ebenfalls deutlich ab. Die Freisetzung von Granulozyten-Proteasen während des Endotoxin-Schocks dürfte die Ursache mehrerer dabei beobachteter pathophysiologischer Veränderungen sein.

Key words: Canine endotoxin shock, granulocyte elastase, α_1-antitrypsin, α_2-macroglobulin, complement factor 3 concentrations.

Enzyme: Elastase (EC 3.4.21.11).

Canine endotoxin shock is characterized by a rapid fall in blood leucocyte counts[1]. These cells have therefore been included in several studies on endotoxemia, and found to participate in several of the phenomena observed during this ailment[2-4]. These include the increased proteolytic activity of plasma observed during endotoxemia both in man[5] and in several experimental animals[6-8].

Components of granulocytes have been suggested as responsible for triggering of disseminated intravascular coagulation[9,10]. In the presence of endotoxin, granulocytes release fibrin-digesting enzymes in vitro[11]. The presence of fibrinolytic activity with neutral pH optimum in human granulocytes was an early finding[12]. Only recently this activity was further characterized, and it was demonstrated that human polymorphonuclear granulocytes contain at least three neutral proteases which degrade fibrinogen. These proteases, collagenase, elastase and chymotrypsin-like proteins, have been purified from human granulocytes by several groups during the last few years [13-22]. The proteases which are released during phagocytosis[23-25] degrade a variety of substances, and include among their substrates, besides fibrin(ogen), components of the complement system: C_3 is cleaved by all of these enzymes and C_5 by elastase and chymotrypsin-like proteases[26,27]. Granulocyte collagenase, elastase and the chymotrypsin-like proteins are inactivated by α_1-antitrypsin (α_1-AT) and α_2-macroglobulin (α_2-M)[14,19,28-31]. Both elastase and chymotrypsin-like proteins are inactivated by a low molecular weight inhibitor produced by the mucosa of the respiratory tract[32-34]. In addition, the chymotrypsin-like proteins are inactivated by α_1-antichymotrypsin[19,20,35]. Human leucocytes are known to release kinins from kininogen[36,37]. The addition of endotoxin to the incubated leucocytes doubles the amount of kinins generated[37].

Dog granulocytes, like the human granulocytes, have been shown to contain large amounts of neutral proteases[38]. Elastase was recently purified from dog granulocytes[39], and a specific antiserum was raised against this enzyme. The purpose of the present investigation was to study any in vivo release of granulocyte elastase and its interaction with plasma protease inhibitors after intravenous infusion in dogs of endotoxin in sufficient amounts to cause lethal shock. Changes in plasma C_3 concentration and electrophoretic homogeneity were also studied during various stages of the shock.

Materials and Methods

Animals

Ten Labrador retriever dogs (17 - 20 kg) of either sex, from a veterinary controlled kennel, were studied. Four of the animals were used for control experiments.

Experimental procedure

Anesthesia was induced by intravenous application of 30 mg sodium pentobarbitone (Mebumal)/kg body weight. Additional small doses of this drug were given if necessary during the experimental period. A cuffed tube was placed in the trachea. Nylon catheters were inserted into both femoral veins and advanced to the inferior vena cava. A catheter was also passed through the left femoral artery and advanced to the descending aorta. Electrocardiogram and heart rate were continuously recorded as well as mean aortic blood pressure (p_a) using a strain gauge transducer EMT 34 (Elema Schønander AB, Solna, Sweden). All data were recorded directly by a multichannel recorder (Mingograph 81, Elema Schønander AB).

Endotoxin

Endotoxin shock was induced by infusing 6 mg of *E. coli* endotoxin/kg body weight (Lipopolysaccaride B. *E. coli* 026:B6, Difco Laboratories, Detroit, Mich.) dissolved in isotonic saline, into the inferior vena cava over a 3-h-period.

Control animals underwent the same experimental procedure as animals receiving endotoxin, with sterile isotonic saline being infused instead of endotoxin.

Sampling procedure

At hourly intervals 4.5 m*l* samples of blood were withdrawn from the inferior vena cava into plastic syringes containing 0.5 m*l* of saturated EDTA solution. In dogs receiving endotoxin the sampling procedure was performed at these intervals until the animals died. The last sample in each experiment was obtained at circulatory collapse when p_a was below 4 kPa and marked bradycardia had developed. Sampled blood volumes were restored by equal amounts of isotonic saline. During the whole experiment the animals also received a constant infusion of isotonic saline (40 m*l*/h). The control experiments lasted for 8 to 10 h. All dogs in the control group made a good recovery.

Leucocyte counts

Leucocyte counts were performed using a cell counter (Celloscope 401, AB Lars Ljungberg & Co., Sweden). Differential counts were obtained on 100 leucocytes in a blood film stained with May-Grünwald Giemsa stain.

Antisera

Rabbit anti-dog α_1-antitrypsin, α_2-macroglobulin and C_3 were available at the laboratory and rabbit anti-dog granulocyte elastase was prepared as recently described[39].

Immunochemical techniques

Agarose A-37 was obtained from Miles-Seravac, Maidenhead, England. α_1-Antitrypsin, α_2-macroglobulin and C_3 were quantitated with the electroimmuno assay method of Laurell[40] and granulocyte elastase by the single radial immunodiffusion technique according to Mancini et al.[41]. The electrophoretic homogeneity of α_1-antitrypsin and C_3 was analyzed with crossed immunoelectrophoresis[42]. The α_1-antitrypsin complexes thus found were analyzed for granulocyte elastase with intermediary gel containing antiserum against this enzyme as described earlier[43].

Statistical methods

The mean and one standard error (SE) of the mean was calculated for each of the parameters measured at the scheduled intervals. The Wilcoxon rank sum test was used to test statistical significance. p values less than 0.05 were considered significant.

Results

Circulatory collapse occurred within 7.5 to 15 h after the start of endotoxin infusion. In order to compare data from the same stages of shock in the different experiments, values obtained from the first six and the last three samples of each experiment were evaluated.

Mean aortic blood pressure

Before endotoxin was infused (zero time) p_a was 17.7 ± 0.7 kPa. At 3 h,,when the endotoxin infusion was ended, p_a was reduced by an average of 44% of preinfusion values. Thereafter a 2 - 3 h period of slight increases in p_a occurred, followed by a gradual decrease which was maintained until the animals died. In control animals p_a was not significantly altered during the whole observation period.

Leucocyte counts

Endotoxin infusion was parallelled by a marked decline of the leucocyte counts ($p < 0.001$ at 1 h) (Fig. 1). Reduced values were maintained for at least the succeeding 4 h of the experiments. During the last hours of the shock, however, a leucocytosis was noted (Fig. 1). In this phase of shock, leucocyte values exceeded the preendotoxin infusion values and were kept rather stable during the last hours of the experiments. Control animals showed a significantly increasing pattern of leucocyte counts during the experiments. This increase was most marked during the first hours (Fig. 1). Differential cell counts revealed a most marked reduction of neutrophil granulocytes. These changes, however, were also parallelled by significantly reduced numbers of circulating lymphocytes.

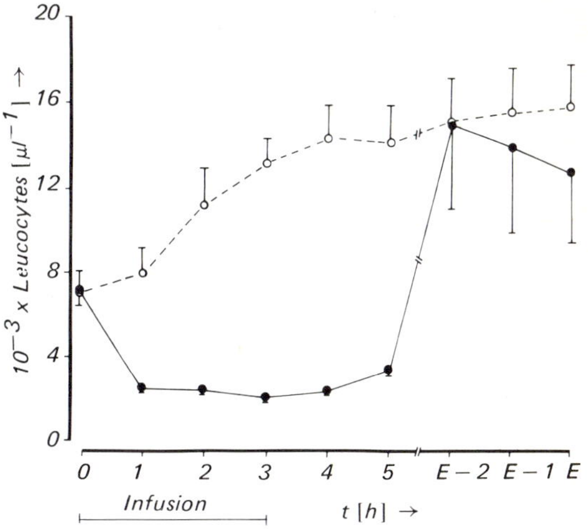

Fig. 1. Changes in leucocyte counts during various stages of lethal canine endotoxin shock, (●——●); control experiments, (○---○).

(Mean values ± SE). The last three samples of each experiment are designated $E - 2$, $E - 1$ and E (E = time of collaps).

Granulocyte elastase concentrations

The concentration of granulocyte elastase was below the detection limit of the immunoassay utilized in the zero time blood specimens from dogs receiving endotoxin, and in all blood specimens obtained from the control animals. During endotoxin infusion the granulocyte-elastase concentration rapidly increased, as shown in Fig. 2, reaching a maximal value at 3 h. Analyses by crossed immunoelectrophoresis with antiserum against granulocyte elastase did not reveal the presence of elastase components with the electrophoretic mobility of free elastase, but only elastase α_1-antitrypsin complexes (Fig. 3).

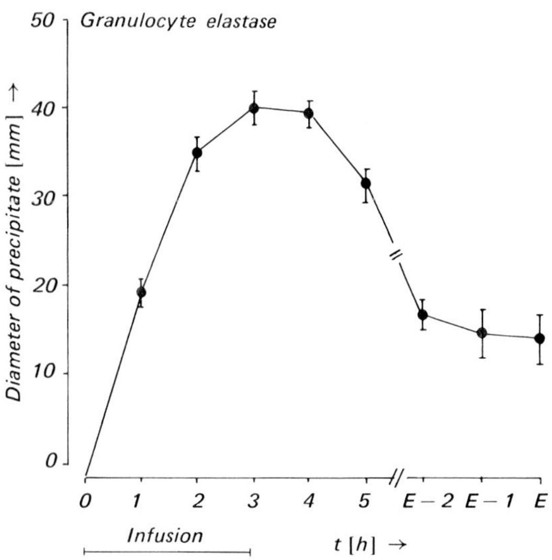

Fig. 2. Changes in serum concentration of granulocyte elastase during various stages of lethal canine endotoxin shock.

(Mean value ± SE). The last three samples of each experiment are designated $E-2$, $E-1$ and E.

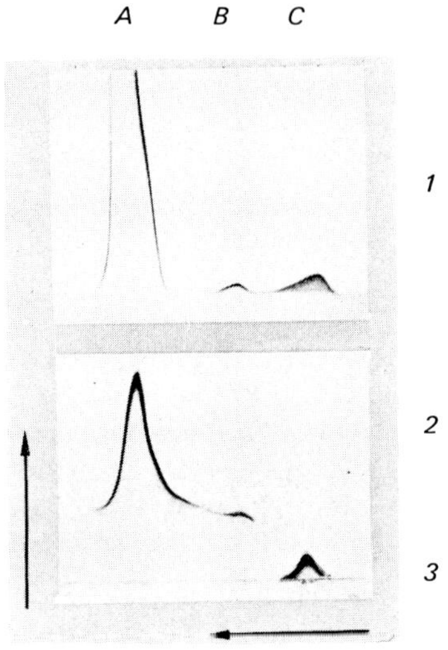

Fig. 3. Precipitate patterns obtained on crossed immunoelectrophoresis with antiserum against α_1-antitrypsin of a plasma specimen prepared from venous blood drawn at the termination of a slow infusion of a lethal dose of endotoxin (1).

The same specimen was also analyzed with the intermediary gel technique with antiserum against dog granulocyte elastase in the lower compartment (3) and against α_1-antitrypsin in the upper (2). Peak A has the electrophoretic mobility of free α_1-antitrypsin. The α_1-antitrypsin component C is precipitated by antibodies against granulocyte elastase and thus consists of elastase-α_1-antitrypsin complexes.

α_1-Antitrypsin concentrations

The α_1-antitrypsin concentration decreased only slightly in dogs receiving endotoxin. The lowest mean values versus zero time were 85 ± 3% ($p < 0.005$) at 4 h and 88 ± 5% ($p < 0.05$) at circulatory collapse (Table). The control animals showed no significant changes in α_1-antitrypsin concentration. Crossed electrophoresis between anti-dog α_1-antitrypsin serum and plasma specimens prepared from blood drawn after the start of endotoxin infusion revealed at least three α_1-antitrypsin components with different electrophoretic mobilities. One was located in the α_1-region and consisted of native α_1-antitrypsin. Another showed α_2-mobility, and the third migrated as a β-globulin. This last component was precipitated by antibodies against granulocyte elastase, and thus consisted of complexes between α_1-antitrypsin and granulocyte elastase (Fig. 3).

α_2-Macroglobin concentrations

The α_2-macroglobin levels decreased gradually during the experiments in dogs receiving endotoxin, reaching 69 + 6% of preendotoxin infusion values at circulatory collapse (Table). Analyses with crossed immunoelectrophoresis did not reveal any electrophoretic heterogeneity. In control animals no significant changes in α_2-macroglobin concentration occurred during the observation period.

Table. α_1-Antitrypsin (α_1-AT), α_2-macroglobulin (α_2-M) and complement factor 3 (C_3) levels during various stages of lethal canine endotoxin shock (Mean ± SE).
The designations E, $E-1$ and $E-2$ refer to samples taken at the end of the experiment (collapse), and 1 and 2 h before, respectively.

	Time [h]	Concentration [%]								
		0	1	2	3	4	5	$E-2$	$E-1$	E
α_1-AT	Shock	100	96 ± 5	91 ± 3	90 ± 3	85 ± 3	93 ± 3	93 ± 4	90 ± 3	88 ± 5
	Control	100	100 ± 1	98 ± 1	100 ± 1	99 ± 1	98 ± 1	95 ± 2	98 ± 1	100 ± 2
α_2-M	Shock	100	89 ± 4	89 ± 3	82 ± 4	83 ± 4	85 ± 5	81 ± 8	79 ± 7	69 ± 6
	Control	100	100 ± 1	100 ± 1	99 ± 1	100 ± 1	98 ± 1	99 ± 1	100 ± 1	99 ± 1
C_3	Shock	100	84 ± 4	81 ± 5	78 ± 5	69 ± 6	64 ± 6	41 ± 6	38 ± 6	38 ± 8
	Control	100	96 ± 1	98 ± 1	99 ± 1	96 ± 1	97 ± 1	96 ± 2	100 ± 1	100 ± 1

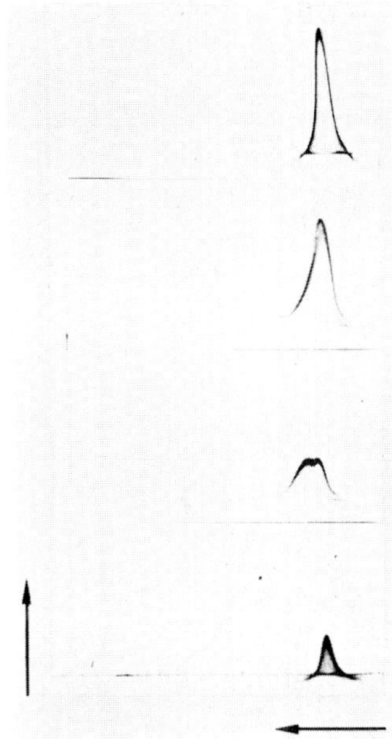

A

B

C

D

Fig. 4. Precipitate patterns obtained on crossed immunoelectrophoresis with antiserum against C_3 of plasma specimens prepared from venous blood drawn at different time intervals after the start of a slow infusion of a lethal dose of endotoxin.

(A) before, (B) 2 h, (C) 5 h and (D) at circulatory collapse.

C_3 *concentrations*

C_3 levels determined with the immunochemical assay were gradually reduced to about 50% of initial values within 5 h after the start of endotoxin infusion (Table). At circulatory collapse, C_3 levels were 38 ± 8% of pre-infusion values ($p < 0.005$). In control animals only insignificant changes in C_3 levels occurred during the observation period. Crossed immunoelectrophoresis with anitserum against C_3 revealed a C_3 component with slightly increased electrophoretic mobility in samples obtained after the start of endotoxin infusion, thus indicating a cleavage of C_3 and formation of cleavage products (Fig. 4).

Discussion

The present study demonstrates release of granulocyte elastase in canine endotoxin shock. This phenomenon was regularly accompanied by pronounced falls in the concentration of factor C_3 of the complement system, which also is known to be a substrate for granulocyte elastase[26]. This indicates an activation of the complement system during this ailment, and is in accordance with previous studies by others[44].

The α_2-macroglobin concentration decreased continuously during the experiments. Analyses by crossed immunoelectrophoresis, however, did not reveal any heterogeneity, indicating complexation. This was not surprising and does not exclude protease binding. The half-life in the circulation of complexes between α_2-macroglobin and proteases in dogs is only about 5 min. These complexes thus are very rapidly cleared by the cells of the reticuloendothelial system and degraded[45]. Any accumulation of such complexes in the circulation in amounts recognizable with the techniques utilized in this study was therefore hardly likely. The drop in α_2-macroglobin concentration observed is, however, nevertheless most likely due to binding of proteases. This is suggested by the demonstration of several electrophoretic α_1-antitrypsin fractions indicating protease α_1-antitrypsin complexes. One of these components was shown to consist of α_1-antitrypsin-elastase complexes. As dog granulocyte elastase shows a partition between α_1-antitrypsin and the α_2-macroglobin on release into plasma[46], α_2-macroglobin complexes must have been formed too, but cleared too rapidly to allow identification. The half-life of α_1-antitrypsin protease complexes in the circulation of the dog is about 45 min[46], which probably facilitates identification. The drop in concentration of α_1-antitrypsin was less pronounced than for α_2-macroglobin. This may be explained by a transfer of elastase to α_2-macroglobin before elimination. The α_1-antitrypsin thus released has a half-life in the circulation of about 4 to 6 h[47] and does not bind any proteases again[48]. This relatively long half-life, at least when compared with α_2-macroglobin complexes, may also account for the patterns of α_1-antitrypsin changes observed. Furthermore, α_1-antitrypsin is an "acute phase reactant"[49]. This mechanism also may counteract a drop in concentration of this inhibitor.

In the present studies, no free granulocyte elastase was demonstrable in the circulation. This finding is in accordance with the study of Egbring et al. on patients with septicemia, where no free elastase-like protease, but complexes between serum α_1-antitrypsin and this proteolytic enzyme could be demonstrated[4]. These findings do not, however, exclude a local saturation of the inhibitors in areas with massive protease release. This might occur in the lungs, where sequestration of large amounts of granulocytes has been noted in the pulmonary capillaries[50] and in the liver, whereas a trapping of blood occurs within the portal system during canine endotoxin shock[51].

The proteolytic activity of granulocyte elastase is known to cause degradation of human lung elastin, arterial walls and basement membrane [52,53]. Furthermore, this enzyme was recently shown to destroy rapidly the mucociliary activity of respiratory epithelium in vitro[54]. It is also reported that homogenates of dog polymorphonuclear leucocytes instilled into the tracheobronchial tree of dogs produced emphysema-like lesions[55].

Extracellular release of elastase from granulocytes has also been found in several studies to be parallelled by release of collagenase and chymotrypsin-like proteins from these cells[24,25]. It thus appears likely that release of granulocyte proteases may participate to cause the organ damage seen in experimental endotoxin shock[51,56] and in humans suffering from septicemia[57].

The exact mechanism for the release of proteolytic enzymes from granulocytes during endotoxemia is unknown. It has, however, been reported that human granulocytes and mononuclear leucocytes possess endotoxin binding structures on their surfaces as well as intracellularly[58]. Binding of endotoxin to cell membranes has been found to cause morphological changes in these structures[59]. The binding of endotoxin to granulocyte membranes might therefore directly facilitate "leakage" of proteolytic enzymes from these cells.

As reported elsewhere, we have also noted consumption of coagulation factors*, reduced levels of components of the plasma kallikrein-kinin sys-

* Aasen, A.O., Dale, J., Ohlsson, K. & Gallimore, M.J., submitted for publication.

tem[60] and activation of the fibrinolytic system[61] during these experiments. It is obvious that release of granulocyte proteases might significantly have influenced the occurrence of these proteolytic processes.

This work was financially supported by the *Dr. Alexander Malthes Fund* and the *Swedish Medical Research Council* (project no. B77-17X-03910-05B) and the *Swedish Association against Heart and Chest Diseases.*

Literature

1 Bennet, I.L., Jr. & Beeson, P.B. (1950) *Medicine* **29**, 365 - 400.
2 Müller-Berghaus, G., Bohn, E. & Høbel, W. (1976) *Br. J. Haemotol.* **33**, 213 - 220.
3 Golde, D.W. & Cline, M.J. (1975) *Proc. Soc. Exp. Biol. Med.* **149**, 845 - 848.
4 Egbring, R., Schmidt, W., Fuchs, G. & Havemann, K. (1977) *Blood* **49**, 219 - 231.
5 Mason, J.W., Kleeberg, U., Dolan, P. & Coiman, R. (1970) *Ann. Intern. Med.* **73**, 545 - 551.
6 Gans, H. & Krivit, W. (1961) *Ann. Surg.* **153**, 453 - 458.
7 Hardaway, R.M. (1962) *Ann. Surg.* **155**, 325 - 338.
8 Sardesai, V.M. & Rosenberg, J.C. (1974) *J. Trauma* **14**, 945 - 949.
9 Bohn, E. & Müller-Berghaus, G. (1976) *Am. J. Pathol.* **84**, 239 - 258.
10 Schmidt, W., Egbring, R. & Havemann, K. (1975) *Thromb. Res.* **6**, 315 - 326.
11 Havemann, K. & Schmidt, W. (1974) in Proc. 8th Leuc. Cult. Conf. (Lindahl-Kiessling, K., ed.) pp. 181 - 185. Academic Press, New York.
12 Opie, E.L. (1905) *J. Exp. Med.* **7**, 316 - 334.
13 Lazarus, G.S., Daniels, J.R., Brown, R.S., Bladen, H.A. & Fullmer, H.M. (1968) *J. Clin. Invest.* **47**, 2622 - 2629.
14 Ohlsson, K. (1971) *Scand. J. Clin. Lab. Invest.* **28**, 251 - 253.
15 Janoff, A. (1973) *Lab. Invest.* **29**, 458 - 464.
16 Ohlsson, K. & Olsson, I. (1973) *Eur. J. Biochem.* **36**, 473 - 481.
17 Ohlsson, K. & Olsson, I. (1974) *Eur. J. Biochem.* **42**, 519 - 527.
18 Olsson, I. & Venge, P. (1972) *Scand. J. Haematol.* **9**, 204 - 214.
19 Schmidt, W. & Havemann, K. (1974) *Hoppe-Seyler's Z. Physiol. Chem.* **355**, 1077 - 1082.
20 Gerber, A.C., Carson, J.H. & Hadorn, B. (1974) *Biochim. Biophys. Acta* **364**, 103 - 112.
21 Rindler-Ludwig, R. & Braunsteiner, H. (1975) *Biochim. Biophys. Acta* **379**, 606 - 617.
22 Baugh, R.J. & Travis, J. (1976) *Biochemistry* **15**, 836 - 841.
23 Henson, P.M. (1971) *J. Immunol.* **107**, 1535 - 1546.
24 Ohlsson, K. & Olsson, I. (1977) *Scand. J. Haemotol.* **19**, 145 - 152.
25 Venge, P. (1975) Thesis. Acta Univ. Uppsala.
26 Johnson, U., Ohlsson, K. & Olsson, I. (1976) *Scand. J. Immunol.* **5**, 421 - 426.
27 Venge, P. & Olsson, I. (1975) *J. Immunol.* **115**, 1505 - 1508.
28 Janoff, A. (1972) *Am. Rev. Respir. Dis.* **105**, 121 - 122.
29 Ohlsson, K. & Olsson, I. (1974) *J. Clin. Lab. Invest.* **34**, 349 - 355.
30 Ohlsson, K. & Olsson, I. (1977) *J. Lab. Clin. Med.* **89**, 269 - 277.
31 Venge, P., Olsson, I. & Odeberg, H. (1975) *Scand. J. Clin. Lab. Invest.* **35**, 737 - 744.
32 Hochstrasser, K., Reichert, R., Werle, E. & Haendle, H. (1973) Protides of the biological fluids – 20th Coloquium 20, p. 417, Pergamon Press, Oxford.
33 Ohlsson, K. & Tegner, H. (1976) *Scand. J. Clin. Lab. Invest.* **36**, 437 - 445.
34 Tegner, H., Ohlsson, K. & Olsson, I. (1977) *Hoppe-Seyler's Z. Physiol. Chem.* **358**, 431 - 433.
35 Ohlsson, K. & Åkesson, U. (1976) *Clin. Chim. Acta* **73**, 285 - 291.
36 Melmon, K.L. & Cline, M.J. (1967) *Nature (London)* **213**, 90 - 92.
37 Miller, R.L., Webster, M.E. & Melmon, K.L. (1975) *Eur. J. Pharmacol.* **33**, 53 - 60.
38 Ohlsson, K. (1971) *Scand. J. Clin. Lab. Invest.* **28**, 225 - 230.
39 Delshammar, M. & Ohlsson, K. (1976) *Eur. J. Biochem.* **69**, 125 - 131.
40 Laurell, C.-B. (1972) *Scand. J. Clin. Lab. Invest.* **29**, *Suppl.* **124**, 21 - 37.
41 Mancini, G., Carbonara, A.O. & Heremans, J.F. (1965) *Immunochemistry* **2**, 235 - 254.
42 Ganrot, P.O. (1972) *Scand. J. Clin. Lab. Invest.* **29**, Suppl. 124, 39 - 47.
43 Ohlsson, K. & Eddeland, A. (1975) *Gastroenterology* **69**, 668 - 675.
44 Mergenhagen, S.E., Snyderman, R., Gewurz, H. & Shin, H.S. (1969) *Curr. Top. Microbiol. Immunol.* **50**, 37 - 77.
45 Ohlsson, K. (1971) *Acta Physiol. Scand.* **81**, 269 - 272.
46 Ohlsson, K. & Delshammar, M. (1975) in Dynamics of Connective Tissue Macromolecules (Burleigh, P.M.C. & Poole, A.R., eds.) pp. 259 - 275, North Holland Publishing Co.
47 Balldin, G., Laurell, C.-B. & Olsson, K. (1978) *Hoppe-Seyler's Z. Physiol. Chem.* **359**, 699 - 708.
48 Johnson, D.A. & Travis, J. (1976) *Biochem. Biophys. Res. Commun.* **72**, 33 - 39.
49 Ohlsson, K. (1971) *Biochim. Biophys. Acta* **236**, 84 - 91.
50 Coalson, J.J., Hinshaw, C.B. & Guenter, C.A. (1970) *J. Exp. Mol. Pathol.* **12**, 84 - 103.
51 Lillehei, R.C., Longerbeam, J.K., Bloch, J.H. & Manax, W.G. (1964) *Clin. Pharmacol. Ther.* **5**, 63 - 101.
52 Janoff, A. & Zelig, J.D. (1968) *Science* **161**, 702 - 704.
53 Janoff, A., Sandhaus, R.A., Hospelhorn, V.D. & Rosenberg, R. (1972) *Proc. Soc. Exp. Biol. Med.* **140**, 516 - 519.

54 Tegner, H. (1977) Thesis. Univ. of Lund.
55 Kimbel, G.W. (1975) in Lung Metabolism (Junod & de Haller, eds.) pp. 25 - 41, Academic Press Inc., New York, San Francisco, London.
56 De Palma, R.G., Coil, J., Davis, J.H. & Holden, W.D. (1967) *Surgery* **62**, 505 - 515.
57 Caridis, D.T., Reinhold, R.B., Woodruff, P.W.H. & Fine, J. (1972) *Lancet* **1** 1381 - 1386.
58 Springer, G.F. & Adye, J.C. (1975) *Infect. Immun.* **12**, 978 - 986.
59 Shands, J.W., Jr. (1973) *J. Infect. Dis.* **128**, 197 - 201.
60 Aasen, A.O. & Frøhlich, W. (1976) *Microvasc. Res.* **11**, 422.
61 Aasen, A.O. & Larsbråten, M. (1976) Third Int. Conf. on Synthetic Fibrinolytic Thrombolytic Agents, Glasgow.

Ansgar O. Aasen, Institute for Surgical Research, Rikshospitalet, Oslo 1, Norway.
Kjell Ohlsson, Department of Surgery, Malmö General Hospital, S-21401 Malmö.

Hoppe-Seyler's Z. Physiol. Chem.
Bd. 359, S. 691 - 697, Juni 1978

Studies on the Influence of Trasylol on the Partition of Trypsin between the Human Plasma Protease Inhibitors in vitro

Göran BALLDIN, Kjell OHLSSON and Ann-Sofie OLSSON

Departments of Clinical Chemistry and Surgery, Malmö General Hospital, University of Lund, Sweden

(Received 10 February 1978)

Summary: The distribution of trypsin between the protease inhibitors of human serum with and without Trasylol was studied in vitro.

1) Trypsin was preferentially bound by α_2-macroglobulin on addition of small amounts of the enzyme to normal serum in both the presence and absence of Trasylol in a molar concentration equal to that of α_2-macroglobulin.

2) On saturation of α_2-macroglobulin, a considerable amount of trypsin was bound by Trasylol even when most of the serum α_1-antitrypsin was in a free form.

3) In reaction mixtures containing small amounts of trypsin, Trasylol was identified in a free form as well as in complex with trypsin-α_2-macroglobulin complex and to a limited extent with trypsin.

4) With larger amounts of trypsin, sufficient to saturate α_2-macroglobulin, increasing amounts of Trasylol were bound to trypsin. The relative amount of Trasylol bound to trypsin-α_2-macroglobulin complexes was now smaller. This was explained by a higher affinity (or binding rate) of Trasylol for trypsin than for trypsin-α_2-macroglobulin complexes.

5) Trypsin-Trasylol complexes showed no signs of dissociation after 5 h incubation at 37 °C in serum.

In-vitro-Untersuchungen über den Einfluß des Trasylols auf die Verteilung von Trypsin auf die Protease-Inhibitoren des menschlichen Plasmas

Zusammenfassung: Es wurde die Verteilung von Trypsin auf die Protease-Inhibitoren des menschlichen Serums mit und ohne Trasylolzusatz untersucht.

1) Bei Zusatz kleiner Enzymmengen zu Normalseren in Abwesenheit und in Gegenwart von Trasylol wurde Trypsin bevorzugt an α_2-Makroglobulin gebunden. Die zugegebene Trasylolmenge entsprach dabei der molaren α_2-Makroglobulinkonzentration.

2) Nach Sättigung des α_2-Makroglobulins wurde eine beträchtliche Menge des Trypsins an Trasylol gebunden, obwohl α_1-Antitrypsin überwiegend in freier Form vorlag.

3) Bei Zugabe kleiner Trypsinmengen wurde Trasylol sowohl in freier Form, als auch gebunden an den Trypsin-α_2-Makroglobulin-Komplex, gefunden, in kleiner Menge auch im Komplex mit Trypsin allein.

4) Nach Sättigung des α_2-Makroglobulins wurden bei weiterer Trypsinzugabe steigende Mengen an Trasylol gebunden. Die relative an den Trypsin-α_2-Makroglobulin-Komplex gebundene Trasylolmenge war dann geringer. Dies wird erklärt durch eine höhere Affinität des Trasylol für

Trypsin als für den Trypsin-α_2-Makroglobulin-Komplex bzw. durch eine raschere Bildung des entsprechenden Komplexes.

5) Trypsin-Trasylol-Komplexe zeigten keine Anzeichen einer Dissoziation nach 5-stündiger Inkubation bei 37 °C im Serum.

Key-words: α_1-Antitrypsin, α_2-macroglobulin, trypsin, trypsin inhibitors, kallikrein-trypsin inactivator.

In 1936 Kunitz and Northrop[1] isolated and characterized a trypsin inhibitor from bovine pancreas. This inhibitor proved identical[2,3] to the kallikrein inhibitor found by Kraut, Frey and Werle[4] in various bovine organs. It inhibited not only trypsin and kallikrein, but also chymotrypsin and plasmin. Species differences exist, however, as human chymotrypsin is not inactivated by the inhibitor[5]. Trasylol is a commercially available preparation of this inhibitor, purified from bovine lung and parotid gland[6].

Trasylol has been used for the treatment of acute pancreatitis for almost two decades, mainly based on the findings that it is a strong trypsin inhibitor and an effective remedy for experimental pancreatitis if given before or shortly after the induction of pancreatitis[7]. The efficacy in human pancreatitis is, however, still a matter of conjecture. Most reports, with a few exceptions[8], have failed to show any beneficial effects of the inhibitor, at least as regards mortality[9-12].

However, these results should not exclude a pathophysiological role of the pancreatic proteases in acute pancreatitis. Recent results with experimental[13] and human pancreatitis[14] indicate, on the contrary, that a pronounced activation of all the pancreatic proteases occurs with local saturation of the inhibitors. Trypsin thus probably plays a key role. Not much is known, however, about the interaction of Trasylol with the endogenous protease inhibitors. In vitro studies in pure systems have shown that Trasylol is a very strong inhibitor of trypsin[6], but no direct comparisons in serum have been made with the endogenous trypsin inhibitors such as α_1-antitrypsin and α_2-macroglobulin. The results of such studies may provide a firmer basis for the judgment of appropriate dosage, often mentioned as a major problem[8], and route of administration of Trasylol in acute pancreatitis.

The purpose of the present investigation was to study the influence of Trasylol on the interactions between human trypsin and the plasma protease inhibitors in vitro.

Materials and Methods

Agarose was a product of Miles-Seravac, Maidenhead, England, Sephadex G-200 and cyanogen-bromide-activated Sepharose 4-B were purchased from AB Pharmacia, Uppsala, Sweden. Trasylol was a gift of *Bayer AG*, Wuppertal, West-Germany.

Cationic human trypsin was purified as described previously[15]. Trasylol and trypsin were coupled to cyanogen-bromide-activated Sepharose 4-B according to the instructions provided by AB Pharmacia. ^{125}I free of reducing agents was obtained from AB Studsvik, Sweden. Trypsin and Trasylol were labelled according to a lactoperoxidase method[16] to a spec. act. of about 0.5 mCi/mg protein. After the trypsin was labelled it was applied to a Trasylol-Sepharose column and the column was washed with 0.05M Tris/HCl buffer, pH 8.0, containing 0.01M $CaCl_2$ and 0.5M NaCl until the absorbance had reached the baseline. The column was then washed with 0.05M acetate buffer, pH 3.6, 0.5M in NaCl and finally the labelled trypsin was eluted with 0.1M formic acid, pH 2.5. The labelled trypsin was concentrated in a Diaflo Ultrafiltration cell with a UM 10 filter and was then dialysed against 1mM HCl, 0.05M in $CaCl_2$. After labelling the Trasylol, low molecular weight radioactive material was separated by gel filtration on a Sephadex G-25 column (0.9 × 20 cm) equilibrated with 0.05M Tris/HCl buffer, pH 7.4, 0.15M in NaCl. Specific antisera against human cationic trypsin, α_2-macroglobulin, α_1-antitrypsin and inter-α-trypsin-inhibitor were produced in the laboratory (Fig. 1).

Trasylol, purified by affinity chromatography on trypsin-Sepharose 4B, as described above, was used for antiserum production. 1 mg of Trasylol in 1 m*l* 0.15M NaCl was emulsified with 1 m*l* of Freund's complete adjuvant and then injected into the footpads and subcutaneously in the neck of a rabbit. This was repeated after four weeks and harvesting of antiserum was started after another two to three weeks. Immunoelectrophoresis[17] of crude Trasylol with this antiserum showed one homogeneous precipitate cathodal to the site of origin (Fig. 1).

α_2-Macroglobulin, inter-α-trypsin inhibitor and α_1-antitrypsin were quantitated with the electroimmunoassay

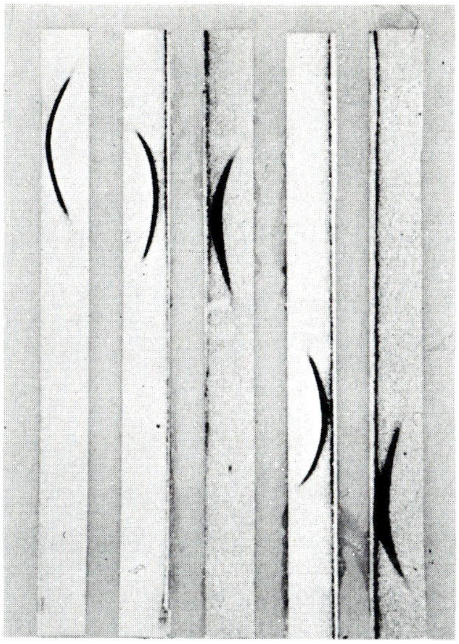

1 2 3 4 5

Fig. 1. Precipitate pattern obtained on immunoelectrophoresis of human serum with rabbit anti-α_1-antitrypsin (1), anti-inter-α-trypsin inhibitor (2) and anti-α_2-macroglobulin (3). Human pancreatic juice with anti-cathodal trypsin (4). Crude Trasylol with rabbit anti-Trasylol (5).

of Laurell[18] and trypsin with the single radial immunodiffusion method of Mancini[19]. Precision of the methods was ± 5%. The amount of radioactivity bound to the different proteins was measured as follows. The gels with the immunoprecipitates after electroimmunoassay or single radial immunodiffusion were washed in three changes of 0.15M NaCl, 2 h in each bath. The gel containing a single precipitate was then cut out and transferred to a test tube containing 250 μl of 1M NaOH for measurement of the radioactivity in a well-type scintillation detector.

Gel filtration was done on Sephadex G-200 or G-75 at + 4 °C. 1-ml samples were applied to a column (2.6 × 40 cm) equilibrated with 0.04M Tris/HCl buffer, pH 7.6, containing 0.2M NaCl. A constant flow of 13.8 ml/h was maintained by a technicon pump and fractions of 4.6 ml were collected.

The protein concentration was determined according to Lowry et al.[20] using human serum albumin as standard or monitored as absorbance at 280 nm on gel filtration.

Human serum pools from 20 healthy blood donors were stored at + 4 °C, after addition of a small amount of sodium azide, for 2 to 5 days. The concentration of α_2-macroglobulin and α_1-antitrypsin was 1.8 ± 0.2 mg/ml and 1.3 ± 0.3 mg/ml, respectively, and the concentration of inter-α-trypsin inhibitor 0.4 ± 0.1 mg/ml. The binding of two trypsin molecules per molecule of α_2-macroglobulin[21] was considered when calculating molarity. In such calculations a molecular weight of 725 000 was used for α_2-macroglobulin and 50 000 and 160 000 for α_1-antitrypsin and inter-α-trypsin inhibitor[22], respectively.

Reaction mixtures of human serum, Trasylol and ^{125}I-trypsin

To three series of test tubes were added 100 μl of human serum and 0, 1.0 and 3.0 μg of Trasylol in 0.15M NaCl. To these reaction mixtures were added varying amounts of ^{125}I-trypsin (0 to 50 μg in 1mM HCl, containing 0.01M $CaCl_2$). 0.05M Tris/HCl, pH 7.4, 0.15M in NaCl was added to give a final volume of 300 μl for the reaction mixtures.

Reaction mixtures of human serum ^{125}I-Trasylol and human cationic trypsin

To two series of test tubes were added 100 μl of human serum and 1.0 and 3.0 μg of ^{125}I-Trasylol, respectively. Varying amounts of human cationic trypsin (0 to 50 μg in 1mM HCl and 0.01M in $CaCl_2$) were added to the test tubes and the final reaction volume was brought to 300 μl by the addition of 0.05M Tris/HCl buffer, pH 7.4, containing 0.15M NaCl.

The reaction mixtures were analysed by electroimmunoassay and single radial immunodiffusion and gel filtration on Sephadex G-200 after 30 min incubation at room temperature. Each experiment was performed four different times.

Trypsin-Trasylol complexes were purified by gel filtration on Sephadex G-75 from reaction mixtures of 100 μg of ^{125}I-Trasylol in 0.05M Tris/HCl buffer, pH 7.4, 0.15M in NaCl and 200 μg of cationic human trypsin in 1mM HCl, 0.01M $CaCl_2$. A final volume of 1 ml was obtained by the addition of the Tris/HCl buffer. The reaction mixture was incubated at room temperature for 10 min before gel filtration. The ^{125}I-Trasylol eluted in two peaks. The first peak was precipitable by antibodies against either Trasylol or trypsin, and thus contained trypsin-Trasylol complexes. The fractions of this peak were pooled. The stability of the complexes in vitro in serum was tested by incubation at 37 °C of reaction mixtures containing 100 μl of the pool of trypsin-Trasylol complexes and 1 ml of the serum pool for 1 h and 5 h. The mixtures were then subjected to gel filtration on Sephadex G-200 analysis of the partition of the radioactivity between the fractions. The amount of added trypsin bound to ^{125}I-Trasylol corresponded to about 5% of the binding capacity of the α_2-macroglobulin of the serum.

Results

Partition of ^{125}I-trypsin between the major plasma protease inhibitors before and after the addition of varying amounts of Trasylol

On addition of increasing amounts of ^{125}I-trypsin to a constant amount of human serum, trypsin is preferentially bound by α_2-macroglobulin, as is shown in Fig. 2. On saturation of α_2-macroglobulin, increasing amounts of enzyme are bound by α_1-antitrypsin, while only slight amounts are bound by inter-α-trypsin inhibitor. Correspondingly, on gel filtration on Sephadex G-200 of reaction mixtures of serum and ^{125}I-trypsin, trypsin is eluted in two peaks, the macroglobulin (I) and the albumin (II) (Fig. 3). More than 90% of the radioactivity of the macroglo-

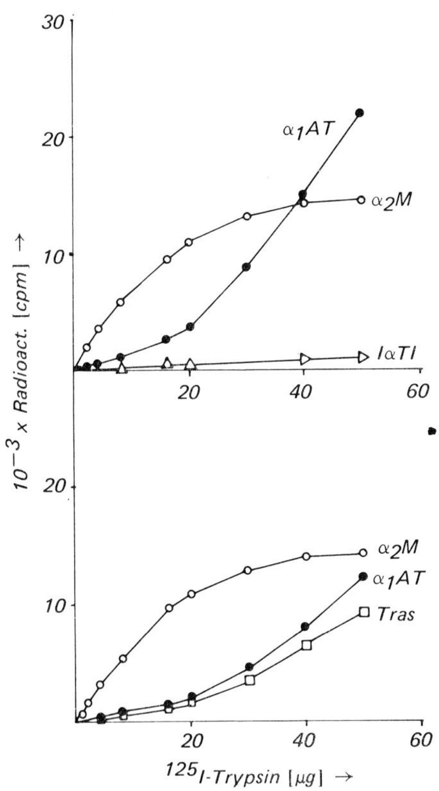

Fig. 2. Analysis by electroimmunoassay of reaction mixtures of 100 μl of serum and increasing amounts of ^{125}I-trypsin (above) and (below) 100 μl of serum and 3 μg of Trasylol followed by increasing amounts of ^{125}I-trypsin.

Amount of radioactivity precipitated by antibodies against α_2-macroglobulin (○——○), α_1-antitrypsin (●——●), inter-α-trypsin inhibitor (△——△) and Trasylol (□——□).

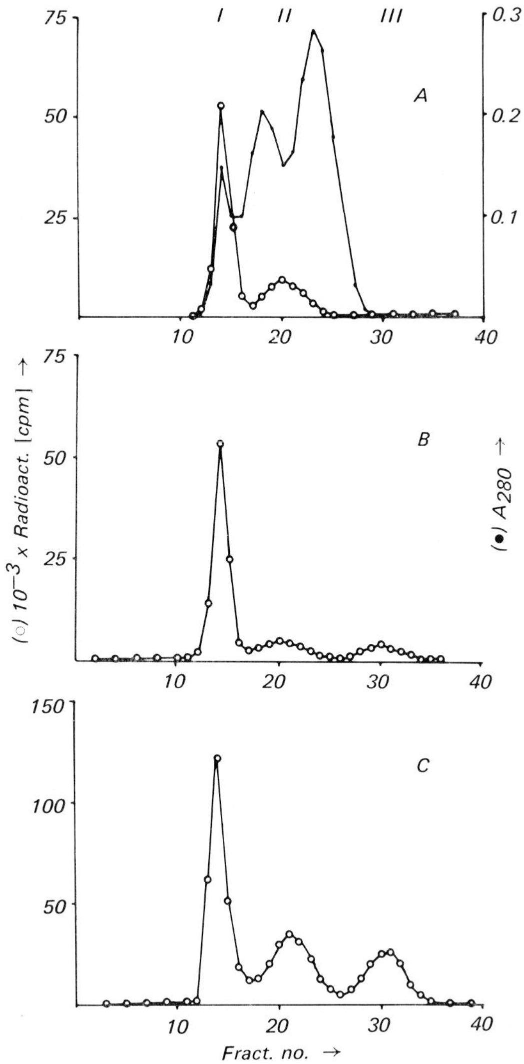

Fig. 3. Distribution of protein (●——●) and radioactivity (○——○) among the fractions obtained on gel filtration (Sephadex G-200) of

(A) a mixture of 200 μl of serum and 20 μg of ^{125}I-trypsin,

(B) 200 μl of serum and 6 μg of Trasylol followed by 20 μg of ^{125}I-trypsin and

(C) 200 μl of serum, 6 μg of Trasylol and 80 μg of ^{125}I-trypsin.

bulin and the albumin peaks was precipitated by antibodies against α_2-macroglobulin and α_1-antitrypsin, respectively, thus proving that ^{125}I-trypsin was bound in complex by these two inhibitors.

Analysis of reaction mixtures of serum containing Trasylol in amounts corresponding to the molar concentration of α_2-macroglobulin and increasing amounts of ^{125}I-trypsin showed that the ^{125}I-trypsin is bound mainly to α_2-macroglobulin initially (Fig. 2). With increasing amounts of trypsin a gradual binding of trypsin to Trasylol and α_1-antitrypsin is seen. Thus, less trypsin is bound to α_1-antitrypsin than in the situation without Trasylol. Similarly, on analysis of such reaction mixtures by gel filtration on Sephadex G-200, the labelled enzyme is eluted in three protein peaks (Fig. 3), the dominating macroglobulin peak, the albumin peak and an equally small third peak. Specific antibodies against α_2-macroglobulin and α_1-antitrypsin precipitate the radioactivity of the macroglobulin and the albumin peaks, respectively, proving that trypsin is bound by α_2-macroglobulin and α_1-antitrypsin. The third trypsin-containing peak is eluted after the trypsin-α_1-antitrypsin-containing fraction in an elution volume corresponding to the ^{125}I-Trasylol complex. Furthermore, about 90% of this radioactivity is precipitated by antibodies against Trasylol. Thus, this fraction contained ^{125}I-trypsin-Trasylol complexes. With increasing amounts of ^{125}I-trypsin added, this peak shows an increase in size parallel to that of the α_1-antitrypsin-containing peak.

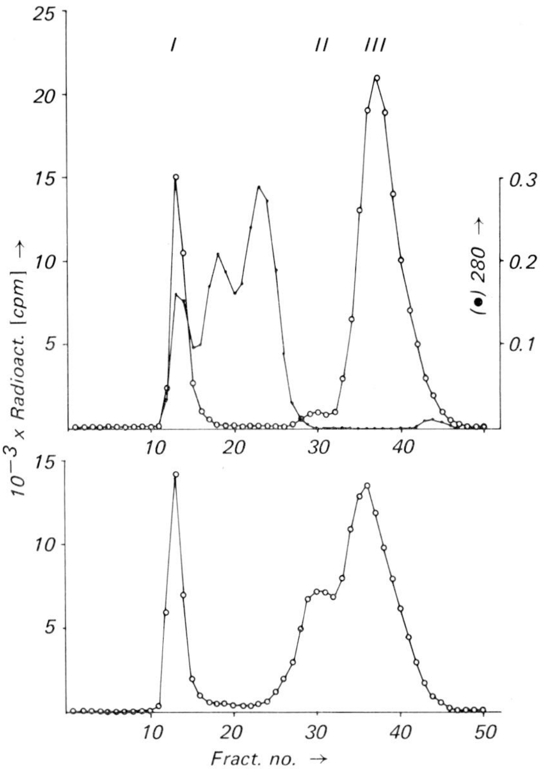

Fig. 4. Distribution of protein (●——●) and radioactivity (○——○) among the fractions obtained on gel filtration (Sephadex G-200) of 200 μl of serum and 6 μg of ^{125}I-Trasylol, followed by 20 μg of trypsin (above), and 200 μl of serum, 6 μg of ^{125}I-Trasylol and 80 μg of trypsin (below).

Identification of Trasylol in reaction mixtures of serum ^{125}I-Trasylol and varying amounts of trypsin

Analysis by gel filtration on Sephadex G-200 of reaction mixtures of serum containing ^{125}I-Trasylol and α_2-macroglobulin in equal molar concentrations and small amounts of trypsin (100 μg/ml) showed three Trasylol-containing peaks (Fig. 4). One was eluted with the void volume (I) and contained some ^{125}I-Trasylol which was bound by α_2-macroglobulin, as the radioactivity was precipitated by antibodies against this inhibitor. The radioactivity of the second (II) minute peak was eluted in a lesser volume than free Trasylol and was precipitated by antibodies against Trasylol as well as against trypsin, and thus consisted of trypsin-^{125}I-Trasylol complexes. The third peak showed an elution volume corresponding to free Trasylol. With larger amounts of trypsin (400 μg/ml), the second ^{125}I-Trasylol-containing peak showed a considerable increase and the peak of free ^{125}I-Trasylol a corresponding decrease in size. The amount of ^{125}I-Trasylol bound by trypsin-α_2-macroglobulin complexes decreased slightly (Fig. 4). As shown in Fig. 5 the amount of ^{125}I-Trasylol bound by trypsin-α_2-macroglobulin complex on addition of trypsin reaches a maximum and then shows a tendency to decrease. Fig. 4 indicates the same phenomenon. With small amounts of trypsin relatively more Trasylol is recovered in the macroglobu-

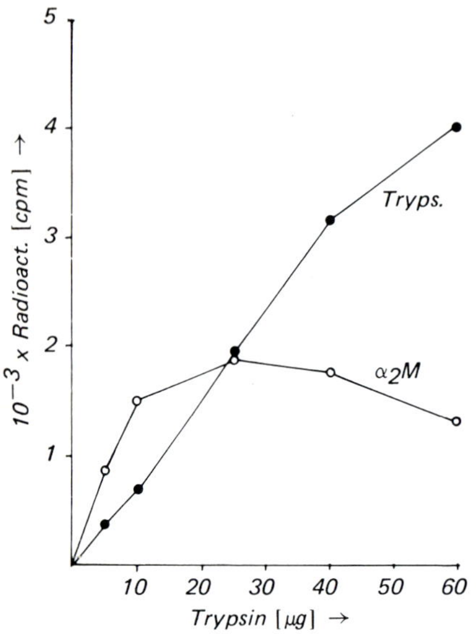

Fig. 5. Amount of radioactivity precipitated by antibodies against α_2-macroglobulin (○——○) and against trypsin (●——●) from reaction mixtures of 100 μl of serum and 3 μg of ^{125}I-Trasylol followed by increasing amounts of trypsin.

lin peak than after the addition of trypsin in amounts saturating α_2-macroglobulin.

Trypsin-^{125}I-Trasylol complexes mixed with serum showed no signs of dissociation after 1 h and 5 h incubation at 37 °C as judged from the gel filtration studies (Fig. 6).

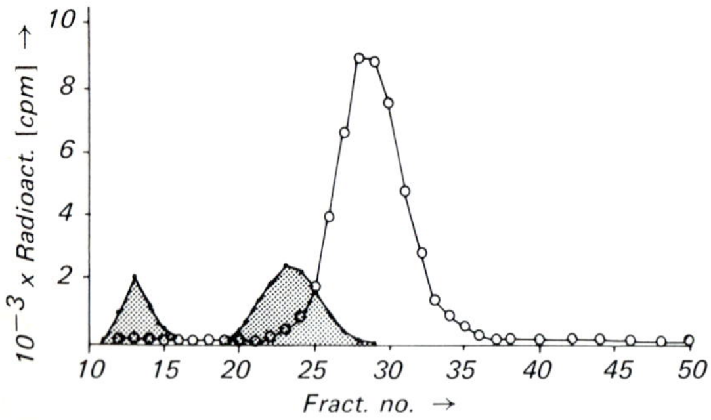

Fig. 6. Distribution of radioactivity between the fractions obtained on gel filtration (Sephadex G-200) of a reaction mixture of serum and trypsin-^{125}I-Trasylol complexes after 5 h incubation at 37 °C.

Stippled areas denote fractions containing α_2-macroglobulin and α_1-antitrypsin, respectively, as demonstrated by electroimmunoassay.

Discussion

Trasylol was recently shown to inhibit the hydrolytic activity against low molecular weight substrates of the trypsin-α_2-macroglobulin complexes[23]. We found that such Trasylol-trypsin-α_2-macroglobulin complexes had the same short half-life in the circulation of dogs[24] (about 5 min) as did the trypsin-α_2-macroglobulin complex itself[25]. Thus, binding of Trasylol to the trypsin-α_2-macroglobulin complexes with subsequent inhibition of the activity against low molecular weight substrates did not impair the natural elimination mechanism of such complexes in the reticuloendothelial system. The present study increases our knowledge about the complicated interplay between Trasylol and the endogenous protease inhibitors in the defence against active trypsin. Trasylol-trypsin complexes appear rather stable in serum, as no transfer of trypsin is seen to any of the plasma protease inhibitors after 5 h incubation at 37 °C.

Trypsin is, however, preferentially bound by α_2-macroglobulin in the presence of Trasylol in equal molar concentration to α_2-macroglobulin. Trasylol is identified in such reaction mixtures in a free form, but also to considerable extent bound to the trypsin-α_2-macroglobulin complex. When α_2-macroglobulin approaches saturation with trypsin, however, significant amounts of trypsin are bound by Trasylol. This also means that less α_1-antitrypsin is complexed by trypsin in the presence of Trasylol than in its absence. After the addition of large amounts of trypsin, relatively less Trasylol is seen bound by the trypsin-α_2-macroglobulin complex than at lower trypsin concentrations. This may be explained by higher affinity (or binding rate) of Trasylol for trypsin than for trypsin-α_2-macroglobulin complexes. This means that on further addition of trypsin to serum beyond the saturation of α_2-macroglobulin, increasing amounts of Trasylol are bound by trypsin directly and less by the trypsin-α_2-macroglobulin complexes.

The interplay between Trasylol and the plasma protease inhibitors on release of trypsin seems, however, to be even more complicated, as judged from preliminary results. Trypsin initially bound to α_1-antitrypsin is to some extent transferred to

Trasylol, at least in pure systems containing only trypsin, α_1-antitrypsin and Trasylol.

α_2-Macroglobulin is thus consumed on release of trypsin, even in the presence of large amounts of Trasylol. On saturation of α_2-macroglobulin, considerable amounts of trypsin are bound to Trasylol, even in the presence of much higher concentrations of α_1-antitrypsin. A pertinent question is then, if Trasylol can prevent the trypsin shock regularly seen on saturation of α_2-macroglobulin in the circulation of dogs[26]. Preliminary results from studies in our laboratory indicate that this is the case, and this finding is currently the subject of detailed investigation.

This investigation was supported by the *Swedish Medical Research Council* (project no B78-17X-03910-06C) and *Bayer AG*.

Literature

1 Kunitz, M. & Northrop, I. (1936) *J. Gen. Physiol.* **19**, 991 - 1007.
2 Anderer, F.A. (1965) *Z. Naturforsch.* **206**, 499 - 501.
3 Anderer, F.A. & Hörnle, S.Z. (1965) *Z. Naturforsch.* **206**, 457 - 462.
4 Kraut, H., Frey, E.K. & Werle, E. (1930) *Z. Physiol. Chem.* **192**, 1 - 21.
5 Coan, M.H. & Travis, J. (1971) in Proceedings of the International Research Conference on Proteinase Inhibitors, pp. 294 - 298, Walter de Gruyter, Berlin, New York.
6 Trautschold, I., Werle, E. & Fritz, H. (1966) in Neue Aspekte der Trasylol-Therapie (New Aspects of Trasylol Therapy) vol. I, pp. 3 - 19, F.K. Schattauer, Stuttgart.
7 Thomson, A.G. (1968) *Ann. N. Y. Acad. Sci.* **146**, 540 - 546,
8 Trapnell, J.E., Rigby, C.C., Talbot, C.H. & Duncan, E. (1974) *Brit. J. Surg.* **61**, 177 - 182.
9 Hansen, H.Th., Drube, H.G., Brüning, W. & Borm, D. (1966) *Med. Klin.* **61**, 1254 - 1257.
10 Baden, H., Jordal, K., Lund, F. & Zachariae, F. (1969) *Scand. J. Gastroenterol.* **4**, 291 - 295.
11 Trapnell, J.E., Talbot, C.H. & Capper, W.M. (1967) *Am. J. Dig. Dis.* **12**, 409 - 412.
12 Welbourn, R.B., Armitage, P., Gilmore, O.J.A., McKay, C., Trapnell, J.E., Williamson, R.C.N. & Cox, A.G. (1977) *Lancet* **24**, 632 - 635.
13 Ohlsson, K. & Eddeland, A. (1975) *Gastroenterology* **69**, 668 - 675.
14 Ohlsson, K. (1974) in Bayer-Symposium V "Proteinase Inhibitors", pp. 96 - 105, Springer-Verlag.
15 Ohlsson, K. & Skude, G. (1976) *Clin. Chim. Acta* **66**, 1 - 7.
16 Johansson, B.G. & Thorell, J.I. (1971) *Biochim. Biophys. Acta* **251**, 363 - 369.
17 Scheidegger, J.J. (1955) *Int. Arch. Allergy Appl. Immunol.* **7**, 103 - 114.
18 Laurell, C.B. (1972) *Scand. J. clin. Lab. Invest.* **29**, *Suppl.* **124**, 21 - 37.
19 Mancini, G., Carbonara, A.O. & Heremans, J. (1965) *Immunochemistry* **2**, 235 - 254.
20 Lowry, O.H., Rosebrough, N.J., Farr, A.L. & Randall, R.J. (1951) *J. Biol. Chem.* **193**, 265 - 257.
21 Ganrot, P.O. (1966) *Acta Chem. Scand.* **20**, 2299 - 2300.
22 Jeppsson, J.O. & Laurell, C.B. (1975) in Plasma Proteins, Protease Inhibitors, pp. 229 - 264, Academic Press.
23 Ganrot, P.O. (1976) *Ark. Kemi* **26**, 583 - 591.
24 Eddeland, A. & Ohlsson, K. (1978) *Hoppe-Seyler's Z. Physiol. Chem.* **359**, 379 - 384.
25 Ohlsson, K. (1971) *Acta Physiol. Scand.* **81**, 269 - 272.
26 Ohlsson, K., Ganrot, P.O. & Laurell, C.-B. (1971) *Acta Chir. Scand.* **137**, 113 - 121.

Dr. Göran Balldin, Kirurgiska kliniken, Malmö allmänna sjukhus,
S-214 01 Malmö;

Doc. Kjell Ohlsson and Laboratorieassistent Ann-Sofie Olsson, Kliniskt kemiska centrallaboratoriet, Malmö allmänna sjukhus,
S-214 01 Malmö;

crease in enzymatic activity was observed after labelling. ^{131}I-Trypsin was added to unlabelled trypsin as a marker before admixture of α_1-AT.

The trypsin-binding capacity. Increasing amounts of trypsin were added to α_1-AT-containing fractions and analysed by agarose gel electrophoresis, starting within 5 min after mixing to find the saturation limit. The trypsin showed two major cathodal electrophoretic components. The trypsin-α_1-AT complexes migrated electrophoretically in the β-zone as a double band. These complexes were sensitive to digestion by free trypsin. For plasma and impure α_1-AT-fractions, the complex formation was analysed by crossed immunoelectrophoresis. On slight supersaturation the electrophoretic mobility of α_1-AT and complexes was slightly altered, but the immunoreactivity was mainly preserved. Approximately 10% of α_1-AT in the plasma and prepared fractions did not link with trypsin.

Canine α_1-antitrypsin preparations

Individual canine sera were collected for preparation of α_1-AT. The purified protein was later infused as a T-α_1-AT complex into the same dog which had served as donor as described below.

Enrichment of α_1-AT by DEAE cellulose chromatography (A). The α_1-AT was precipitated with ammonium sulfate (50 - 72% saturation) followed by DEAE-cellulose chromatography starting with a 0.02M phosphate buffer, pH 6.8, producing a linear gradient of decreasing pH and increasing ionic strength by admixture of 0.2M NaH_2PO_4. The fractions with highest relative α_1-AT content were pooled, dialysed against 0.05M Tris/HCl buffer, pH 7.4, 0.15M in NaCl and concentrated by ultrafiltration. Selection of fractions was made according to the results of electroimmunoassay with anti-α_1-AT. About one fourth of the protein in the pools prepared was α_1-AT. Predominant contaminating proteins in the selected fractions were albumin, α-lipoproteins and haptoglobins. These concentrates of α_1-AT are called A in the following discussion.

Partially purified α_1-AT (B). An ammonium sulfate precipitate (50 - 72% saturation) of serum was dialysed against 0.1M Tris buffer, pH 8.1, and fractionated by thiol-disulfide interchange chromatography on Thiol-Sepharose (Pharmacia Fine Chemicals, Uppsala, Sweden) as described for human α_1-AT[12]. The α_1-AT-containing fractions eluted from the Thiol-Sepharose by mild reduction were filtered through an immunosorbent column packed with Sepharose-conjugated anti-albumin, anti-haptoglobins and "anti-prealbumin" to eliminate major contaminants. About 85% of the protein content remaining after passage through this column was α_1-AT. These partially purified fractions are denoted B.

Almost pure α_1-AT (> 95%) (C). This was obtained by preparative agarose gel electrophoresis[6] of fraction B, and is denoted C. This α_1-AT was used for labelling with ^{125}I or ^{131}I. The specific activity varied between 1 and 0.6 mCi/mg protein. ^{125}I-α_1-AT was added to the other α_1-AT preparations as a marker before admixture of trypsin. Unless stated otherwise, α_1-AT was dissolved in 0.05M Tris/HCl buffer, pH 7.4, 0.15M in NaCl, and trypsin in 2mM HCl.

Formation and administration of trypsin-α_1-antitrypsin complexes.

For experiment 1 T-α_1-AT-complexes were formed and purified in the following way to eliminate α_1-AT (free and degraded). An amount of trypsin calculated to give 80% saturation of α_1-AT in 0.02M phosphate buffer (pH 6.5, 0.02M NaCl) was added to an α_1-AT preparation of type B enriched with ^{125}I-α_1-AT. 0.38 g moist DEAE-cellulose was added per mg α_1-AT used and stirred for 2 min. The cellulose had been equilibrated with the same buffer. The suspension was filtered through a glass filter and the cellulose was resuspended in 1 *ml* buffer per mg α_1-AT initially used. The filtration was repeated, and the two filtrates were pooled. Agarose gel electrophoresis showed the two electrophoretically adjacent T-α-AT-complexes in the α-β-zone but no free α_1-AT. The recovery of α_1-AT as pure complex was 40%. To minimize the consumption of α_1-AT, the complexes were not isolated after mixing T and α_1-AT in the following experiments.

In experiments 2 to 6, the α_1-AT fraction (type A) was divided into four equal parts. 5 min before the infusion, trypsin was added to each part in an amount giving 80% saturation of the trypsin-binding capacity. The ^{131}I-labelled trypsin and ^{125}I-labelled-α_1-AT were only added to the first of the four T-α_1-AT mixtures in each experiment.

In experiments 7, 8 and 9, serum was prepared from freshly drawn venous blood and reinfused into the same dog as follows. Serum and trypsin were divided into four equal parts. One part of serum and trypsin in an amount giving about 80% saturation of the trypsin-binding capacity of the serum were mixed 5 min before infusion.

The mixtures were given by continuous infusion at a rate of 4 mg of trypsin per min in all experiments.

For the studies in vitro, the ^{131}I-T-^{125}I-α_1-AT-complexes were prepared from ^{131}I-T (in 2mM HCl) and ^{125}I-α_1-AT (in 0.05M Tris/HCl buffer, pH 7.4, 0.15M in NaCl) to give about 80% saturation of the trypsin-binding capacity of the ^{125}I-α_1-AT. The reaction mixtures were incubated 10 min at room temperature before the addition of serum.

PROCEDURES

9 mongrel dogs weighing 7 to 17 kg were anesthetized with pentobarbital (30 mg per kg body weight). The dogs were intubated during the experiments. One teflon catheter was placed in a femoral artery for registration of the blood pressure, one in a femoral vein and one in a brachial vein for injection and for sampling. Pulse and blood pressure were registered continuously. The plasma volume of the dog was estimated[13]. In all experiments 4 m*l* of venous blood was sampled every fifth minute during the first 30 min and then at longer intervals. In experiments 7 to 9, sampling was continued for 21 days, but the animals were anesthetized only during the initial 3 h.

Comparison of plasma disappearance of ^{131}I-α_1-AT and T-^{125}I-α_1-AT with α_2-M level and its heterogeneity

Experiment 1: Infusion of about 1 mg ^{131}I-α_1-AT (purity C) was followed by 131 m*l* isolated T-^{125}I-α_1-AT complexes to which 700 mg canine albumin had been added as a protection for autodigestion. The α_1-AT and trypsin content of the infused complexes was 41 mg α_1-AT and 17 mg trypsin. Infusion of the complexes started about 20 min after mixing α_1-AT and trypsin. The infused trypsin corresponded to 32% of the trypsin-binding capacity of αM of the dog (Table 1).

Experiment 2: This was in principle performed as experiment 1, but the complexes obtained on mixing of a B-fraction and trypsin were infused without further purification. The saturation degree of the infused α_1-AT was about 100% and the amount of complex injected was larger (67 mg T and 165 mg α_1-AT) than in experiment 1. 250 mg albumin was added to each 40 mg batch of α_1-AT. The infused trypsin corresponded to 125% of the trypsin binding capacity of αM of the plasma pool of the dogs.

In experiments 3 to 6 four different A-preparations of α_1-AT were used. The amount of T infused as α_1-AT complexes was chosen to approach the trypsin-binding capacity of αM in the plasma of the dogs, except in experiment 5 where a larger amount was used (Table 1). Trypsin was added to about 80% saturation of the α_1-AT to avoid any risk of autodigestion of the complexes before the infusion.

Disappearance and reappearance of αM after infusion of plasma enriched with T

Experiments 7 to 9. The αM were saturated and the α_1-AT about 70% saturated in the reaction mixtures of serum and trypsin that were infused. The amount of α_1-AT-bound trypsin infused was chosen to approach 75% of the trypsin-binding capacity of αM in plasma of the dogs (Table 1).

Studies in vitro on exchange of ^{131}I-T between ^{125}I-α_1-AT and αM

In addition to the in vivo experiments, 2-m*l* samples of serum from the dog were mixed with ^{131}I-T-^{125}I-α_1-AT

Table 1. Comparison between major trypsin inhibitors in plasma and the amount of trypsin infused as α_1-AT complexes.

For experiments 1 - 6, α_1-AT was prepared from individual canine sera. The purified protein was later infused as trypsin-α_1-AT complex into the same dog which had served as donor. In experiments 7 - 9, serum was prepared from freshly drawn venous blood and reinfused into the same dog after the addition of appropriate amounts of trypsin (for details: see Experimental). For calculations of molar concentrations the following molecular weights were used: α_1-AT = 55 000 and α-M = 725 000. A molar combining ratio of 2:1 was assumed for trypsin/α-M[14,15].

Exp. no.	Plasma vol. [m*l*]	α_1-AT [g/*l*]	α_1-M + α_2-M [g/*l*]	Trypsin-binding capacity of intravascular αM [mg]	Amount infused of trypsin bound to α_1-AT [mg]
1	406	1.5	1.88	52.6	17
2	522	1.7	1.46	52.6	67
3	1102	1.4	1.16	88.2	67
4	638	1.4	1.82	80.1	50
5	522	1.3	1.88	87.7	103
6	986	1.6	1.28	87.0	70
7	1160	1.3	1.56	125.0	105
8	528	1.4	1.70	61.0	56
9	800	1.2	1.67	92.1	86

complexes in a series of test tubes. The amount of trypsin in the complexes corresponded to about 50% of the trypsin binding capacity of the αM of the serum sample. The reaction mixtures were incubated at 37 °C and analysed by gel filtration on Sephadex G-200 at 4 °C after 0, 10 min, 1 h and 4 h incubation. The column (40 × 2.5 cm) was equilibrated with 0.04M Tris/HCl buffer, pH 7.4, 0.2M NaCl and the flow rate was 20 m*l*/h. The fraction volume was 3.6 m*l*.

Results

Preliminary experiments

The α_1-AT complexes with β-mobility were recognized on agarose gel electrophoresis on addition of increasing amounts of trypsin to α_1-AT, started within 1 min after addition of trypsin, and showed unchanged electrophoretic mobility (pH 5 to 8.6) at least during the first 30 min if the trypsin-binding capacity of α_1-AT was not surpassed. On addition of excess trypsin, the β-complexes were partially converted to faster compounds within minutes. Addition of albumin in 10 molar excess to α_1-AT inhibited the mobility change of T-α_1-AT on addition of minute amounts of excess T.

Preliminary experiments showed that injection of freshly prepared T-α_1-AT complexes caused a drop in blood pressure if more than 5 to 6 mg T was infused per min, in spite of the unsaturated trypsin-binding capacity of α_1-AT. Therefore, in the experiments reported, the motor-driven injection syring was adjusted to deliver about 4 mg trypsin per minute.

Plasma volumes and concentrations of α_1-AT and αM in the dogs at the beginning of the experiments are summarized in Table 1 together with amounts of trypsin administered as T-α_1-AT complexes.

Disappearance rate of ^{131}I-α_1-AT and of ^{131}I-T-^{125}I-α_1-AT complexes after intravenous infusion into dogs

The initial disappearance rate of ^{131}I-α_1-AT is considerably slower than that of ^{125}I-α_1-AT simultaneously infused as trypsin complex (Fig. 1). The ^{131}I-T of the complexes is, however, eliminated much faster than the ^{125}I-α_1-AT, with a half-life of only about 1.5 h in the circulation.

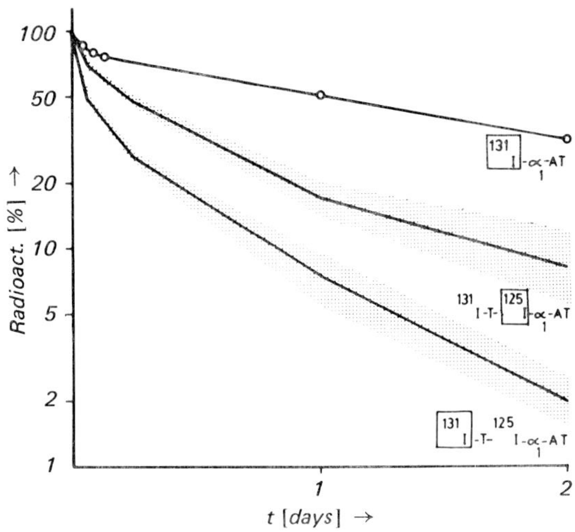

Fig. 1. The elimination of the radioactivity after infusion of ^{131}I-α_1-AT (one experiment for control) and ^{131}I-trypsin-^{125}I-α_1-AT (mean ± SE, n = 7).
In the middle curve the ^{125}I (boxed) of the complex was measured and in the upper and lower curves, ^{131}I (boxed). Ordinate: radioactivity expressed as percentage of the plasma radioactivity at the end of infusion. Abscissa: time in days. For details see Experimental.

Variation in concentration and electrophoretic homogeneity of the α-macroglobulins in the circulation on intravenous infusion of trypsin-α_1-AT complexes

The concentration of αM in plasma decreased rapidly during the first 2 h after infusion of T-α_1-AT (Fig. 2 and 4). Analyses with crossed immunoelectrophoresis showed that the concentration of free αM decreased even more rapidly (Fig. 3). In dogs 2 and 5, circulating αM was saturated after 1 and 1.5 h, respectively. Both these dogs showed a drop of blood pressure within 1/2 h and they died about 2 h later. The other dogs had a maximum percentage of their αM in complexed form after about 45 min to 1 h (Fig. 3) and the blood pressure remained within the normal range. The amount of αM lost from plasma was equivalent to about 80% of the trypsin infused as T-α_1-AT-complexes (Table 2), provided that αM binds 2 mol T/mol. The two dogs that died had at death about 30 and 11% αM left in circulation as T-αM-complexes.

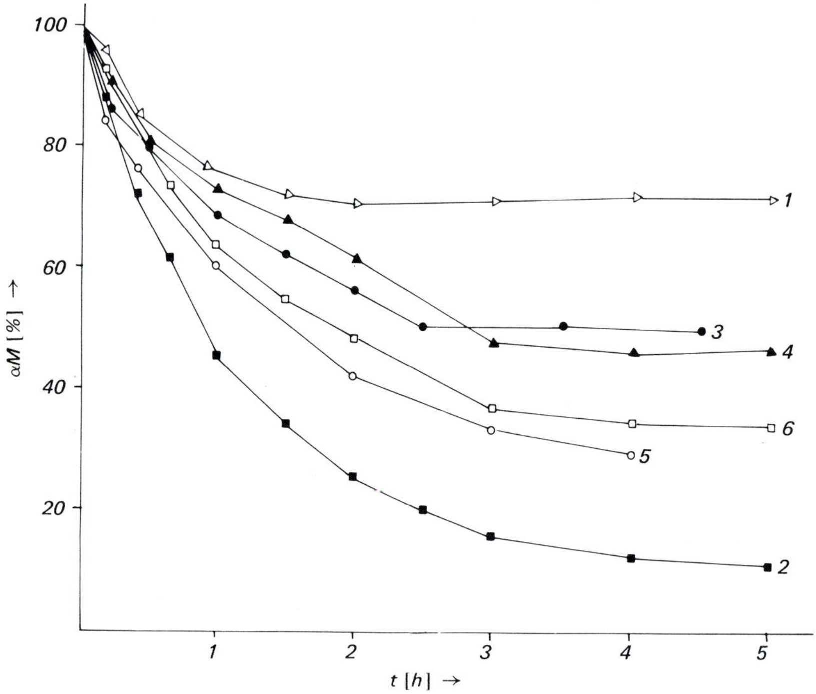
100
80
60
40
20
αM [%] →
1
3
4
6
5
2
1
2
3
4
5
t [h] →

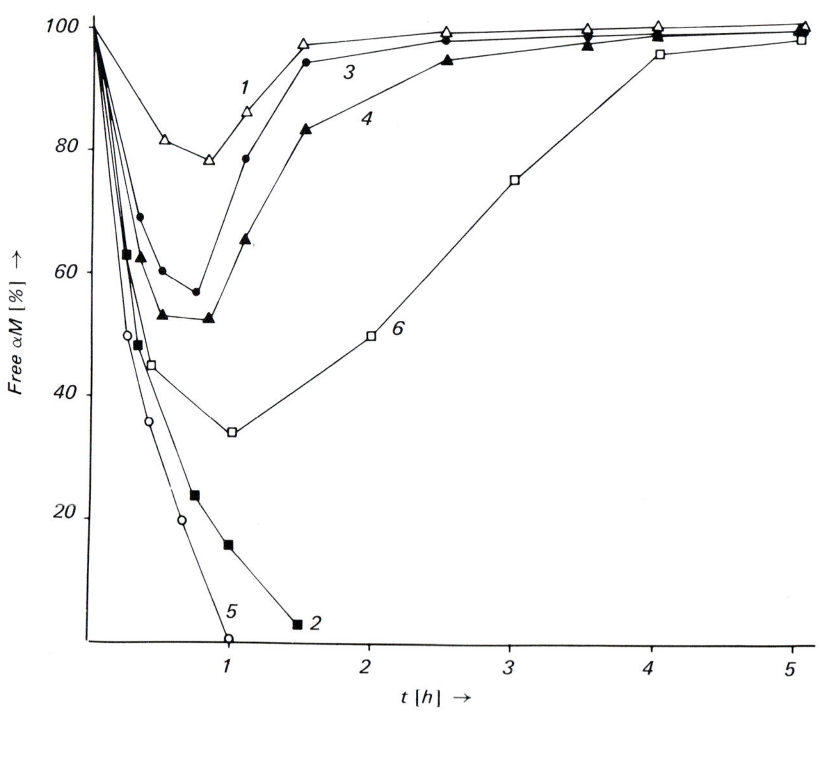
100
80
60
40
20
Free αM [%] →
1
3
4
6
5
2
1
2
3
4
5
t [h] →

◁ Fig. 2. The changes in concentration of the αM (expressed as percentage of the initial values of plasma) during and after the infusion of trypsin-α_1-AT complexes.
Experiments 1 - 6; the number of each experiment is indicated beside the respective curve.

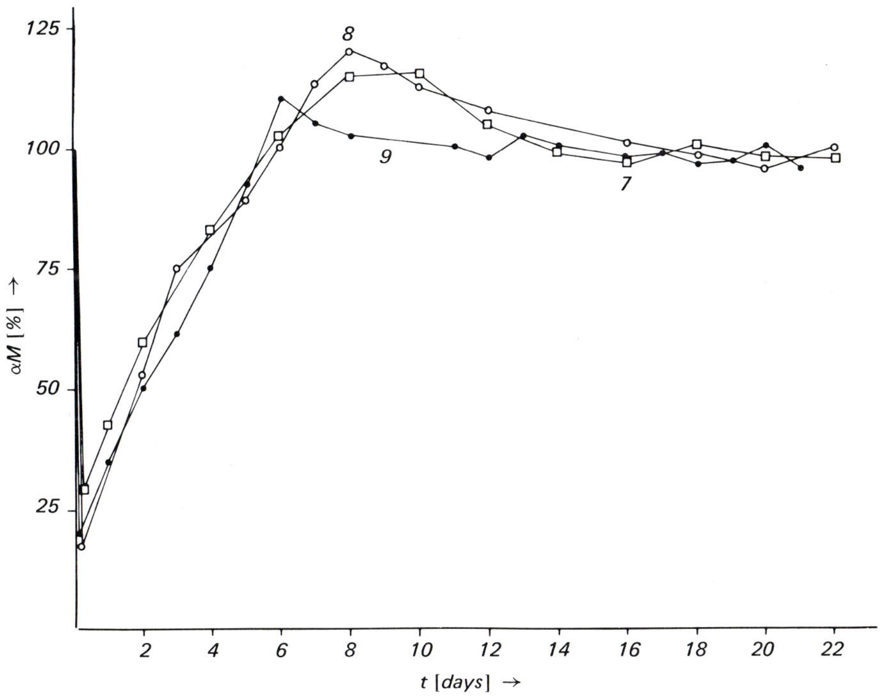

Fig. 4. The changes in concentration of αM (expressed as percentage of the initial values of plasma) after the infusion of trypsin-α_1-AT complexes.
Experiments 7 - 9; the number of the experiment is indicated beside the curve.

The concentration was followed for 3 weeks in experiments 7 to 9 to study the kinetics of αM recovery. The α_1-AT bound trypsin infused (Table 2) was equivalent to about 80% of the initial intravascular αM pool. About 75% of the αM pool disappeared the first day and the αM-concentration normalized during the following 5 days (Fig. 4). An overshoot 10 to 20% above the initial αM level was found after a week and a slow decline to the initial αM level occurred in the second week.

◁ Fig. 3. Free α-macroglobulins in per cent of total plasma αM during and after the infusion of trypsin-α_1-AT complexes.
Experiments 1 - 6; the number of each experiment is indicated beside the respective curve.

Table 2. α-Macroglobulin changes in plasma after infusion of T-α_1-AT complexes.
The experiments are the same as in Table 1, where the plasma volume for each dog is given. The following molecular weights were used for the calculations: trypsin, 25 000 and α-M, 725 000. A molar combining ratio of 2:1 was assumed for trypsin/αM[14,15].

Exp. no.	Initial αM conc. [g/l]	Lowest αM conc. [g/l]	αM loss [g]	Amount of α_1-AT-bound trypsin eliminated as αM complexes [mg]	[% of total]
1	1.88	1.35	0.22	14.8	87
2	1.46	0.43	0.53	37	55
3	1.16	0.58	0.65	44	66
4	1.82	0.85	0.62	42	84
5	1.88	0.21	0.87	60	58
6	1.28	0.43	0.83	58	83
7	1.56	0.44	1.30	89	84
8	1.70	0.34	0.72	49	87
0	1.67	0.30	1.10	76	88

In vitro exchange of ^{125}I-α_1-AT bound ^{131}I-trypsin mixed with α-macroglobulin

50 to 60% of T added to serum as a T-α_1-AT complex was recovered in the macroglobulin containing fractions on gel filtration after 1/2 to 1 h incubation (Fig. 5). No immunoreactive α_1-antitrypsin was found in the macroglobulin-containing fractions. More than 90% of the radioactivity of the macroglobulin and the α_1-antitrypsin-containing fractions were precipitable with the corresponding specific antibodies.

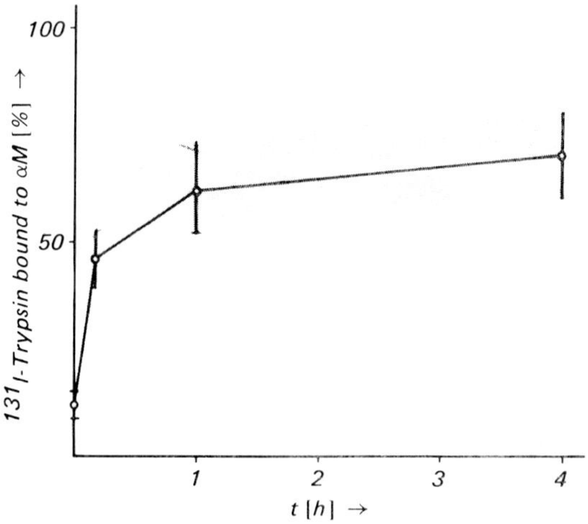

Fig. 5. α_2-M uptake of ^{131}I-trypsin after addition of ^{131}I-trypsin-^{125}I-α_1-AT to serum in vitro.

The amount of ^{131}I-trypsin added corresponded to half the trypsin-binding capacity of αM of the serum samples. (Mean ± SE, n = 6).

Discussion

Double labelled T-α_1-AT complexes are formed when trypsin and α_1-AT labelled with different iodine isotopes are mixed. The two isotopes were used as markers to compare the metabolism of the two proteins. The commercial trypsin contained some 30% proteolytically inactive enzyme, which was also labelled. Furthermore, α_1-AT is claimed to lose a peptide of 8000 daltons on interaction with trypsin[4]. Therefore, the T-α_1-AT complex should preferably be separated from labelled inactive trypsin and any peptides before the infusion for metabolic studies. No addition of trypsin above the trypsin-binding capacity of α_1-AT is acceptable because this rapidly results in proteolytic attack on the complex, which is apparent on electrophoretic analysis[16]. Some 80% saturation of α_1-AT with trypsin and fast isolation of the T-α_1-AT complexes the hour before the infusion started was strictly observed in experiment 1, but involved large preparative losses and is the reason why less pure T-α_1-AT complexes were accepted in the later experiments. No principle differences in results were observed when the purity of the α_1-AT was varied. Any released peptides and/or proteolytically inactive trypsin probably had little influence on the results because of a reasonably fast extravascular distribution and turnover with $t_{1/2}$ around 30 min, as for trypsinogen (unpublished finding) and other proteins of similar or smaller size. The inactive trypsin fraction is partly responsible for the early disappearance of ^{131}I-trypsin from plasma, but does not affect the αM metabolism. Because α_1-AT was not saturated to 100% with trypsin, on complex formation some 5 to 10% of the ^{125}I-α_1-AT injected was still in free native form. Free native α_1-AT has a slower disappearance rate from plasma than the complexes and α_1-AT released from trypsin (Fig. 1). This implies that the small native α_1-AT fraction will mainly influence the final slope of the elimination curve of the injected preparation of the iodine-labelled T-α_1-AT complex. In all experiments, the infusion of 200 to 300 mg of trypsin as T-α_1-AT complexes was followed by a marked, continuous decrease in concentration of αM in plasma during the first few hours (Fig. 2 and 4). Furthermore, analyses of plasma with crossed immunoelectrophoresis showed that the electrophoretic mobility pattern characteristic of T-αM complexes appeared rapidly (Fig. 3), indicating a fast transfer of trypsin from α_1-AT complexes to αM in vivo.

In plasma of dogs receiving large amounts of T-α_1-AT complexes containing some trypsin in excess of the binding capacity of the plasma αM, 50% of the plasma αM was complexed within 20 to 30 min. A transfer rate of trypsin of similar order was found in experiments in vitro (Fig. 5), and indicates that the α-AT linkage of bovine trypsin is relatively weak in the dog. The decrease in αM concentration (Fig. 2 and 4, Table 2) gives a minimum value of the trypsin transfer. The experiments do not show whether the formation rate of T-αM complexes or the uptake of the T-αM complexes by reticuloendothelial cells[3] is rate limiting for the αM decrease in plasma. Theoretically, uptake of one trypsin per αM molecule could result in elimination of the complex. The fall in plasma αM has been used to calculate the percentage of trypsin in T-α_1-AT complexes that was eliminated by secondary complexation with αM (Table 2). Figures between 55 and 87% were obtained. This indicates that two trypsin molecules were bound by most of the disappearing αM. 50% of the trypsin in the infused T-α_1-AT complexes disappeared from the circulation in about 1.5 h, while the corresponding figure for the labelled α_1-AT of the complex was 4 to 6 h (Fig. 1). The dissociation of the complex results in release of α_1-AT molecules with higher fractional catabolic rate than of native α_1-AT. This is probably explained by a cleavage of α_1-AT effected by trypsin on the initial interaction[5,17]. The fast disappearance of α_1-AT injected as T-α_1-AT is in agreement with the results of studies in vitro showing that α_1-AT released from its trypsin complex by interaction with benzamidine[5] or on incubation of a 1:1 complex of α_1-AT and trypsin at pH 8.0[17] is not able to bind or inhibit trypsin again.

The amount of αM which disappeared from the circulation during the first hour was roughly equivalent to 80% of the trypsin infused as T-α_1-AT complex, assuming a molar ratio of αM to T of 1:2.

Two dogs (2 and 5) received trypsin bound to α_1-AT in amounts exceeding the binding capacity of the circulating αM. Both showed total complexation of all circulating αM and they both died within 3 to 5 h. The same end result has earlier been obtained on slow intravenous infusion of free, active trypsin[2], which was preferentially bound to αM. When the circulating αM was saturated, about 80% of the α_1-AT was still free. A further injection of only minute amounts of trypsin was regularly followed by irreversible shock. This proves that α_1-AT forms too labile a trypsin complex or binds trypsin too slowly to protect the animal against the disastrous effect of trypsin if released intravenously. One pathogenic factor explaining the irreversible shock is probably activation of the kinogens of plasma when the saturation of the αM with trypsin is surpassed[18].

No pressor or depressor effect was observed on injection of solutions with newly formed T-α_1-AT complexes. This suggests that any peptide released during the α_1-AT interaction are without circulatory effects.

The faster diffusion and higher concentration of α_1-AT than of αM in the viscous intercellular matrix makes α_1-AT here more easily available than αM as a first defence against released proteases. Local intercellular release of kinins via released proteases may induce increased permeability and thus faster renewal of the inhibitor by infusion from plasma. Further possible biological advantages inherent in the two-step inactivations of trypsin prior to its elimination remain to be demonstrated.

This investigation was supported by the *Swedish Medical Research Council* (project no. B78-17X-03910-05B and B78-13X-0581-14A), the *Medical Faculty, University of Lund* and *Bayer AG.*

Literature

1 Ohlsson, K. (1970) *Biochim. Biophys. Acta* **236**, 84 - 91.
2 Ganrot, P.O., Laurell, C.-B. & Ohlsson, K. (1970) *Acta Physiol. Scand.* **79**, 280 - 286.
3 Ohlsson, K. (1971) *Acta Physiol. Scand.* **81**, 269 - 272.
4 Ohlsson, K. & Laurell, C.-B. (1976) *Clin. Sci. Mol. Med.* **51**, 87 - 92.
5 Johnsson, D.A. & Travis, J. (1976) *Biochem. Biophys. Res. Commun.* **72**, 33 - 39.
6 Johansson, B.G. (1972) *Scand. J. Clin. Lab. Invest.* **29** *Suppl.* **124**, 7 - 19.
7 Laurell, C.-B. (1972) *Scand. J. Clin. Lab. Invest.* **29**, *Suppl.* **124**, 21 - 37.
8 Ganrot, P.O. (1972) *Scand. J. Clin. Lab. Invest.* **29**, *Suppl.* **124**, 39 - 47.
9 Thorell, J.I. & Johansson, B.G. (1971) *Biochim. Biophys. Acta* **251**, 363 - 370.
10 Ohlsson, K. (1971) *Scand. J. Clin. Lab. Invest.* **28**, 219 - 223.
11 Chase, T., Jr. & Shaw, E. (1967) *Biochem. Biophys. Res. Commun.* **29**, 508 - 514.
12 Laurell, C.-B., Pierce, J., Persson, U. & Thulin, E. (1975) *Eur. J. Biochem.* **57**, 107 - 113.
13 Woodward, K.T., Berman, A.R., Michaelsson, S.M. & Odland, L. (1968) *Am. J. Vet. Res.* **29**, 1935 - 1944.
14 Ganrot, P.O. (1966) *Acta Chem. Scand.* **20**, 2299 - 2300.
15 Harpel, P.C. (1973) *J. Exp. Med.* **138**, 508 - 518.
16 Moroi, M. & Yamasaki, M. (1974) *Biochim. Biophys. Acta* **359**, 130 - 141.
17 Oda, K., Laskowski, SR, M., Kress, L. & Kowalski, D. (1977) *Biochim. Biophys. Res. Commun.* **76**, 1062 - 1070.
18 Ferreira, S.H. & Rocha E Silva, M. (1969) *Br. J. Pharmacol.* **36**, 611 - 622.

Dr. G. Balldin, Kirurgiska kliniken, Malmö allmänna sjukhus, S-214 01 Malmö;

Prof. C.-B. Laurell and Doc. K. Ohlsson, Kliniskt kemiska centrallaboratoriet, Malmö allmänna sjukhus, S-214 01 Malmö.

Hoppe-Seyler's Z. Physiol. Chem.
Bd. 359, S. 709 - 717, Juni 1978

The Modification of the Lone Tryptophan Residue in Human Serum Albumin by 2-Hydroxy-5-nitrobenzyl Bromide

Characterization of the Modified Protein and the Binding of L-Tryptophan and Benzodiazepines to the Tryptophan-Modified Albumin

Klaus J. FEHSKE*, Walter E. MÜLLER and Uwe WOLLERT

Pharmakologisches Institut der Universität Mainz

(Received 10 February 1978)

Summary: The possible function of the lone tryptophan residue of human serum albumin in the stereospecific binding site for indole and benzodiazepine compounds was investigated by chemical modification. This residue can be selectively modified with 2-hydroxy-5-nitrobenzyl bromide. The modification alters the conformation of the albumin only slightly, as revealed by circular dichroism, fluorescence, and ultraviolet absorption measurements.

A decrease in the association constants of L-tryptophan and diazepam of about 30 - 50% and a decrease in the extrinsic Cotton effects of four benzodiazepine derivatives of about 10 - 15% were found as specific effects of the tryptophan modification. The tryptophan modification itself did not change the number of binding sites of diazepam and L-tryptophan.

It is suggested that the lone tryptophan residue of human serum albumin is not directly involved in the specific binding site for indole and benzodiazepine compounds. However, the modification alters the properties of the binding site either by an incomplete refolding of the albumin after urea treatment, or a more selective allosteric effect of the modified tryptophan residue.

Die Modifikation des Tryptophanrestes in Humanserumalbumin durch 2-Hydroxy-5-nitrobenzylbromid. Charakterisierung des modifizierten Albumins und die Bindung von L-Tryptophan und Benzodiazepinen an das modifizierte Albumin

Zusammenfassung: Die mögliche Beteiligung des einzigen Tryptophanrestes des Humanserumalbumin-Moleküls an der stereospezifischen Bindungsstelle für Indol- und Benzodiazepinverbindungen wurde über eine chemische Modifikation untersucht. Der Tryptophanrest wird selektiv durch 2-Hydroxy-5-nitrobenzylbromid modifiziert. Die Modifikation hat nur einen sehr geringen Einfluß auf die Sekundärstruktur des Proteins, was durch Zirkulardichroismus-, Fluoreszenz- und UV-Absorptionsmessungen festgestellt wurde.
Als spezifische Effekte der Tryptophanmodifikation wurden eine Abnahme der Bindungskonstanten von L-Tryptophan und Diazepam um ca. 30 - 50% und eine Abnahme der induzierten Cotton-Effekte von vier Benzodiazepinderivaten

* The results are part of a thesis of K.J.F.

Abbreviations: I, 2-hydroxy-5-nitrobenzyl; I-Br, 2-hydroxy-5-nitrobenzyl bromide;
Chlordiazepoxide, 7-chloro-2-methylamino-5-phenyl-3*H*-1,4-benzodiazepine 4-oxide;
Diazepam, 7-chloro-1-methyl-5-phenyl-1*H*-1,4-benzodiazepin-2(3*H*-one;
Oxazepam, 7-chloro-3-hydroxy-5-phenyl-1*H*-1,4-benzodiazepin-2(3*H*)-one;
Tetrazepam, 7-chloro-5-(1-cyclohexen-1-yl)-1-methyl-1*H*-1,4-benzodiazepin-2(3*H*)-one.

um ca. 10 - 15% gefunden. Die Anzahl der Bindungsstellen von L-Tryptophan und Diazepam wurde durch die Modifikation nicht geändert.

Es wird geschlossen, daß der einzige Tryptophanrest von Humanserumalbumin nicht direkt an der spezifischen Bindungsstelle für Indolkörper und Benzodiazepine beteiligt ist. Die beobachteten Effekte der Tryptophanmodifikation auf die Bindung von L-Tryptophan und Benzodiazepinen könnten auf eine nicht vollständige Rückfaltung des Albumins nach Harnstoffdenaturierung oder auf einen allosterischen Effekt des modifizierten Tryptophans zurückzuführen sein.

Key words: Albumin binding, benzodiazepines, L-tryptophan, chemical modification, tryptophan residue.

The carrier function of human serum albumin for endogenous substances, e.g. fatty acids, bilirubin, L-tryptophan, and for many drugs is well documented[1,2]. In general, the binding to serum albumin is a rather unspecific mechanism, but for some substances very specific binding sites on the protein molecule have been demonstrated[3-12]. The most striking example is the indole binding site, which binds several indole compounds with a high degree of structural specificity[5-7]. The binding of L-tryptophan to this binding site is highly stereospecific[5]. A binding site with quite similar properties, including a high stereospecificity, has been found for some benzodiazepine tranquillizers[9-12]. Interestingly, it has been shown by competition experiments that the specific high affinity binding sites of the benzodiazepines and the indoles are identical[13].

The serum albumin molecule contains only one tryptophan residue[14,15]. It has been suggested that the apolar sequence surrounding the lone tryptophan[16] may be an important binding site for several small molecules, e.g. detergents[17], steroids[18], dicoumarol[19], and 8-anilino-1-naphthalenesulfonic acid[20]. Contradictory evidence exists for a possible localisation of the specific binding site for benzodiazepines and L-tryptophan near or at this lone tryptophan residue[21-24]. Both indole and benzodiazepine binding can be found in peptic fragments of serum albumin containing the tryptophan[22,24]. Modification of arginine residues in serum albumin strongly inhibits the binding of benzodiazepines[23]. One arginine residue adjoins the apolar sequence surrounding the tryptophan[16]. Thyroxine[4], which competes with L-tryptophan for the specific binding site[25], and benzodiazepines (Müller and Wollert, unpublished observations) are potent quenchers of the intrinsic tryptophan fluorescence when bound to serum albumin. Positively charged amino acid residues have been suggested to be located near or at the binding site for indoles and benzodiazepines[5,10]. On the other hand, Gambhir et al.[24] have found that a histidine residue (located at position 23) and a lysine residue (located at position 67) are involved in the binding of indoles to a peptic fragment of serum albumin, which seemed to contain the intact indole binding site. The lone tryptophan was located at position 86 of this fragment[24].

We have tried by a different approach, i.e. the modification with a specific reagent, to evaluate the possible role of the tryptophan residue of serum albumin in the specific binding site for benzodiazepine and indole compounds.

Materials and Methods

Materials

Human serum albumin was obtained from Behringwerke, Marburg (trocken, reinst, electrophoretic purity (100%). The benzodiazepine derivatives were gifts of the manufacturers: Chlordiazepoxide, diazepam and [5-^{14}C]-diazepam from Hoffmann-La Roche AG, Grenzach; oxazepam from Boehringer, Ingelheim; tetrazepam from Clin Byla, Paris. L-[3'-^{14}C]Tryptophan was obtained from NEN, Dreieichenhain; 2-hydroxy-5-nitrobenzyl bromide (I-Br) from Merck. All other chemicals were of reagent grade.

Titration of serum albumin with I

The modification of serum albumin was performed at room temperature according to the methods of Koshland et al.[26,27] and Karkhanis et al.[28]. 1.0 g serum albumin was dissolved in 200 m*l* 10M urea adjusted to pH 4.4 by acetic acid. For each experiment 1.0 m*l* methanol was added to 8.0 m*l* of this solution. Increasing amounts of I-Br were added and the reaction mixture was shaken occasionally. After 2 h the insoluble hydrolysed reagent was separated by centrifugation. The supernatant was dialyzed against water for 60 h and then lyophylised. Preliminary experiments showed that lower urea concentrations and less acidic pH values resulted in a more unspecific modification of serum albumin. The secondary structure of the protein seemed to be less affected by removing the excess of I-Br by dialysis than by gel filtration.

Preparation of I-serum albumin

Serum albumin was modified with a 1 100-fold molar excess of I-Br in the reaction mixture as described above. Another amount of serum albumin was treated the same way but without adding I-Br. It was used as control and is called "urea-serum albumin".

Determination of the degree of modification

The degree of modification was determined by an ultraviolet absorption quotient according to the method of Karkhanis et al.[28], as shown below. The measurements were performed at pH 7.4, instead of at alkaline pH values[28].

Number of residues of I bound to one molecule serum

$$\text{albumin} = \frac{A_{410} \times 69\,000 \times 0.498}{13\,800 \times (A_{280} - 0.167 \times A_{410})}.$$

Measurements of fluorescence emission

Fluorescence measurements were performed with a Zeiss PMQ 2 spectrofluorimeter. The wavelength of excitation was 291.5 nm. Emission was measured at 350 nm.

UV absorption spectra

Ultraviolet absorption was measured with a Gilford spectrophotometer, model 2 400.

Circular dichroism spectra

CD measurements were recorded with a Cary 61 CD spectral polarimeter calibrated with *d*-camphorsulfonic acid. All solutions were prepared in 67mM phosphate buffer adjusted to pH 7.4. Results are expressed as molar ellipticities $[\theta]$ (deg. x cm^2 x $dmol^{-1}$), calculated with reference to the serum albumin concentration using a molecular weight of 69 000. In the range 350 - 250 nm the optical pathlength was 1 cm and the protein concentration 13.1μM. In the range 250 - 200 nm the optical pathlength was 1 mm and the protein concentration was 2.62μM.

Amino acid analysis

Amino acid analysis of native and modified serum albumin were performed on a Biotronic LC 4010 amino acid analyser. HCl and methansulfonic acid hydrolysates were used according to Moore[29].

Equilibrium dialysis

Binding of L-tryptophan and diazepam was determined by equilibrium dialysis. Concentrations ranged from molar ratios of 0.2 to 1.5 and 0.2 to 1.3, respectively. Protein concentration was 36.2μM. 0.9 m*l* of the solution was dialyzed for 16 h at 25 °C against 0.9 m*l* buffer using cellophane dialysis membranes (Union Carbide). The drug concentration was determined by tracer amounts of ^{14}C-labeled drugs. The activity was measured in a Tri-Carb liquid scintillation spectrometer (Packard model 3380).

Results

Titration of serum albumin by I-Br

Human serum albumin was titrated with I-Br following the methods of Barman and Koshland[27] and Karkhanis et al.[28] as described in Materials and Methods. The reaction of the reagent with serum albumin is a saturable process (Fig. 1). About 1.1 mol of I is incorporated when about 1 000 mol of I-Br has been added to the reaction mixture (Fig. 1). Increasing the molar excess of the reagent to more than 3 000 increases the mol I incorporated per mol serum albumin only slightly (Fig. 1). The maximal degree of modification was about 1.3, suggesting only negligible modification of other amino acid residues like cysteine, even under saturating conditions.

The modification of serum albumin by I strongly quenches the native tryptophan fluorescence[30] of the protein (Fig. 1). When about 1.1 molecule I has been incorporated per serum albumin molecule (corresponding to a 1 000-fold excess of I-Br in the reaction mixture), less than 10% of the intrinsic tryptophan fluorescence of the albumin is evident (Fig. 1). Because of the possibility of an inner filter effect of the bound I residues, due to an unspecific absorption of the emission light, the fluorescence quenching curve was corrected by the effects of equal amounts of hydrolysed I-Br on the fluorescence of a serum albumin solution with a similar concentration. The fluorescence quenching curve corrected for the inner filter effect (closed circles in Fig. 1) indicates no further decrease in the tryptophan fluorescence at a degree of modification of higher than 1.15 (corresponding to about 1 100 molecules I-Br per molecule serum albumin in the reaction mixture).

These data suggest that the specific quenching of the serum albumin fluorescence by I-Br is only associated with the modification of the lone tryptophan.

The titration curve (Fig. 1) shows that by using a molar excess of I-Br of about 1 100-fold, a modified serum albumin can be obtained, in which the lone tryptophan residue is specifically and nearly quantitatively modified. All further experiments were performed with serum albumin which had been modified using a 1 100-fold excess of I-Br. For this tryptophan-modified serum albumin, in the following called I-serum albumin, a degree of modification of 1.15 was found. Since the modification by I includes an unfolding of the albumin in 10M urea solution at acidic pH values (see Materials and Methods), another part of the albumin was prepared under the same conditions but without I-Br (= "urea-serum albumin"). Both albumins, the I-serum albumin and the urea-serum albumin, were further characterized before they were used in the binding studies.

Characterization of I-serum albumin and urea-serum albumin

The UV absorbance of I-serum albumin is only slightly increased between 250 - 350 nm, while that of urea-serum albumin is unchanged (Fig. 2). The increase in the absorbance of I-serum albumin corresponds to the theoretical value, calculated with the molar extinction coefficient of I and the amount of the reagent incorporated. In I-serum albumin a large maximum is obvious around 410 nm (Fig. 2), corresponding to the maximum of the free reagent[26].

Tryptophan modification and urea treatment have only small effects on the intrinsic circular dichroism spectra of serum albumin in the near and far UV range (Fig. 3), suggesting only small effects on the ordered secondary structure of the protein. The molar ellipticities of the albumins at 208, 222, and 260 nm are decreased by about 5 or 10% for urea-serum albumin and I-serum albumin, respectively (Fig. 3). A positive ellipticity band of I-serum albumin at 310 nm could be due

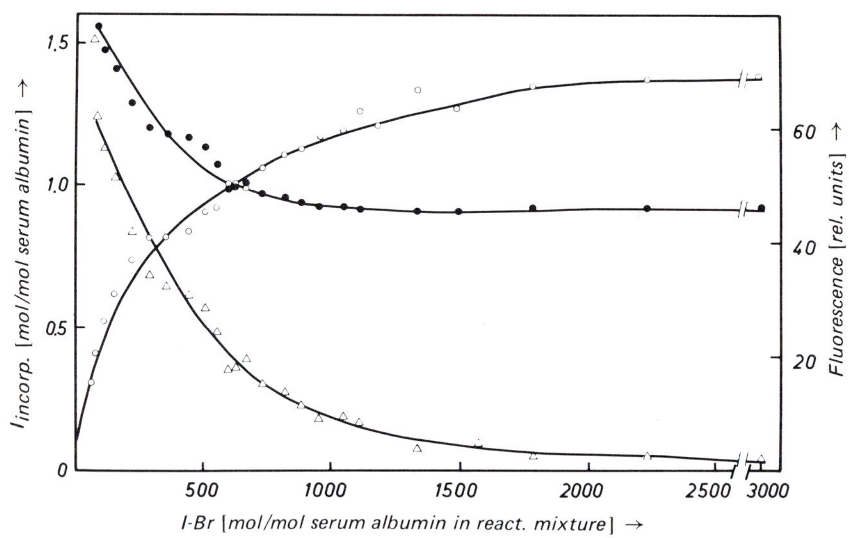

Fig. 1. The effect of increasing amounts of I-Br on the I residues incorporated per mol albumin and on the native tryptophan fluorescence of the albumin.

Abscissa: molar excess of I-Br in the reaction mixture. Left ordinate: mol I incorporated per mol of albumin. Right ordinate: fluorescence at 350 nm in relative fluorescence units. Activation wavelength was 291.5 nm. ○–○–○ = mol I incorporated per mol of serum albumin. △–△–△ = tryptophan fluorescence, uncorrected data. ●–●–● = tryptophan fluorescence, data corrected for the inner filter effect of the reagent. Each point represents the mean of two or three experiments.

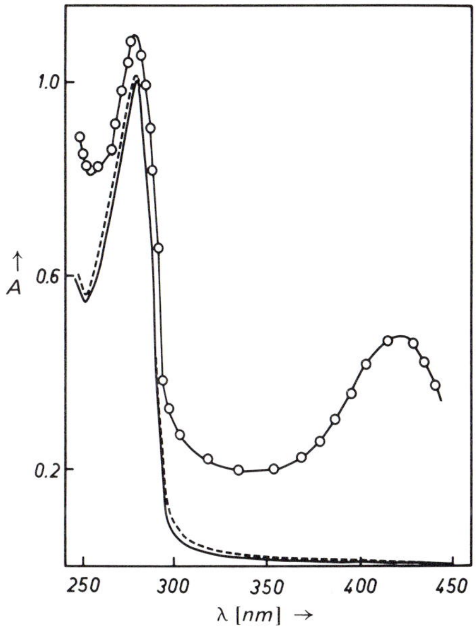

Fig. 2. UV-absorption spectra of serum albumin (——), urea-serum albumin (– – –) and I-serum albumin (o–o–o).
The spectra were recorded in 67mM phosphate buffer, pH 7.4, using an albumin concentration of 29μM. Each spectrum represents the mean spectrum of three experiments.

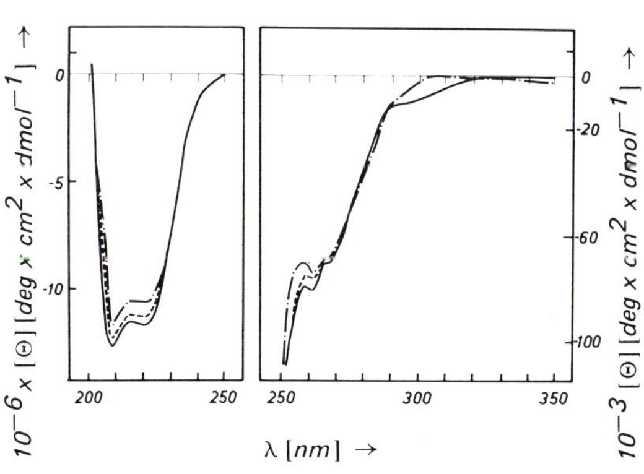

Fig. 3. CD-spectra of serum albumin (——), urea-serum albumin (– – –) and I-serum albumin (–·–·–).
Ordinate: molar ellipticity calculated with reference to the albumin concentration. Abscissa: wavelength in nm. The near UV-spectra were recorded in 67mM phosphate buffer, pH 7.4, using an albumin concentration of 13.1μM. The far UV spectra were recorded in 6.7μM phosphate buffer, pH 7.4, using an albumin concentration of 2.62μM. Each spectrum represents the mean spectrum of three experiments.

to an extrinsic Cotton-effect of the bound I molecule, since I has an absorption band about 315 nm[26]. Using the method of Greenfield and Fasman[31], which is based on the molar ellipticity at 208 nm, the observed decrease in the molar ellipticities may suggest a reduction of the α-helix content of the albumins from 58.3% (native serum albumin) to 55.9% (urea-serum albumin) to 52.6% (I-serum albumin).

Compared to native serum albumin the fluorescence intensity of the lone tryptophan residue is decreased by about 10% in urea-serum albumin (Fig. 4). By contrast, the tryptophan fluorescence of I-serum albumin is decreased by more than 90% (Fig. 4). For I-serum albumin and urea-serum albumin no changes of the wavelength positions of the fluorescence excitation and emission maxima, compared to those of native serum albumin, can be detected (Fig. 4).

The modification of serum albumin by I does not change the amino acid composition, with the exception that no tryptophan can be detected in I-serum albumin (Table 1). The data found by the amino acid analysis are in fairly good agreement with data of Behrens et al.[15] and Meloun[14], based on the total sequence of the serum albumin molecule (Table 1).

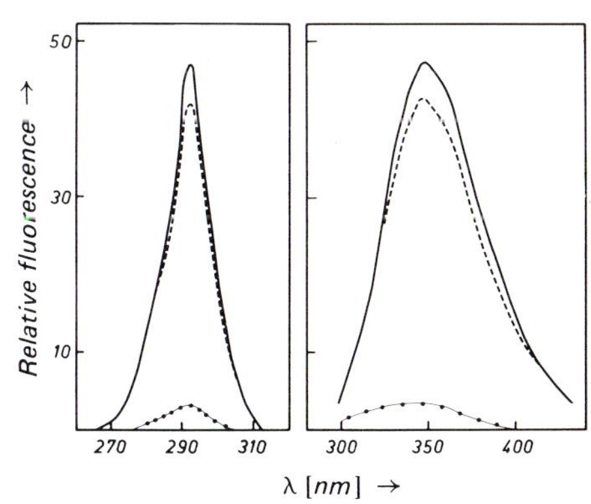

Fig. 4. Fluorescence spectra of serum albumin (——), urea-serum albumin (– – –) and I-serum albumin (–·–·–).
Left side: excitation spectra, emission was measured at 350 nm. Right side: emission spectra, activation wavelength was 291.5 nm. The spectra were recorded in 67mM phosphate buffer, pH 7.4, using an albumin concentration of 29μM. Each spectrum represents the mean spectrum of three experiments.

Table 1. Amino acid composition (mol/mol protein) of serum albumin and I-serum albumin compared to the amino acid composition of serum albumin found by Behrens et al.[15] and by Meloun et al.[14].

Amino acid	Behrens	Meloun	Serum albumin	I-Serum albumin
Asp	54	53	53.9	52.7
Thr	30	28	27.4	27.0
Ser	22	24	21.9	21.6
Glu	83	82	81.8	80.2
Pro	25	24	23.5	20.9
Gly	12	12	11.8	11.7
Ala	63	62	62.8	61.0
Cys	35	35	29.6	29.5
Val	39	41	37.3	35.8
Met	6	6	7.0	6.7
Ile	8	8	6.7	6.4
Leu	61	61	63.0	61.2
Tyr	18	18	17.3	16.8
Phe	30	31	29.9	28.9
His	16	16	17.7	17.3
Lys	58	59	59.2	58.4
Arg	23	24	23.8	23.2
Trp	1	1	1.0	0.0
Total	584	585	575.6	559.3

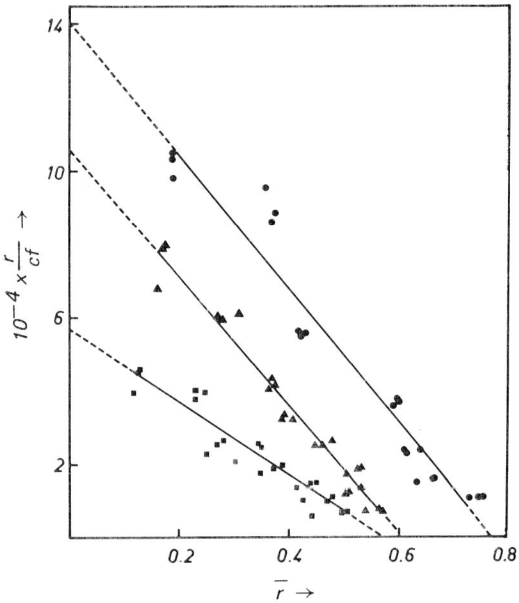

Fig. 5. Scatchard plot of the binding of L-tryptophan to serum albumin (●), urea-serum albumin (▲), and I-serum albumin (■).

Ordinate: $\bar{r}/c_f$, $\bar{r}$ = mol drug bound/mol albumin, c_f = concentration of the free drug [M]. The experiments were performed in 6.7mM phosphate buffer, pH 8.00, using an albumin concentration of 36.2μM. The straight lines represent the regression lines, calculated for 21 experiments.

The binding of L-tryptophan and benzodiazepines to I-serum albumin and urea-serum albumin

The binding of L-[^{14}C]tryptophan and of [^{14}C]-diazepam to native serum albumin, I-serum albumin, and urea-serum albumin was investigated by equilibrium dialysis. Both compounds bind to one primary binding site of the serum albumin molecule with high affinity and at least L-tryptophan binds to some secondary binding sites with much lower affinity[5-7,9-12]. The results of the dialysis experiments were plotted following the method of Scatchard[32]. The graphical representations, as shown for L-tryptophan in Fig. 5, revealed straight lines, suggesting only one set of binding sites. Obviously, the binding of L-tryptophan and diazepam to the secondary binding sites is negligible within the concentration range investigated. The number of the primary binding sites of L-tryptophan and diazepam is decreased by about 10 - 15% in I-serum albumin and in urea-serum albumin (Fig. 5, Table 2). The reason may be an incomplete refolding of the protein after the unfolding in acidic urea solution. The decrease in the ordered secondary structure, indicated by the circular dichroism measurements, and the quantitatively similar decrease in the intrinsic tryptophan fluorescence in urea-serum albumin (Fig. 3, 4) may be further evidence of incomplete refolding. Unfolding of serum albumin decreases the fluorescence of the lone tryptophan residue, as shown by Chen[33]. The decrease in the specific binding sites of L-tryptophan and diazepam does not lower the affinity of either compound to the remaining part of the specific binding sites (Fig. 5), as shown by the association constants (Table 2) of both compounds to urea-serum albumin. While the tryptophan modification in I-serum albumin does not have any specific effect on the number of binding sites available for L-tryptophan and diazepam (Fig. 5, Table 2), the association constants of both compounds

Table 2. Binding parameters for the interaction of diazepam and L-tryptophan with serum albumin, I-serum albumin and urea-serum albumin.

The total binding constant, *nk*, the association constant, *k*, the number of binding sites on the serum albumin molecule, *n*, and the correlation coefficients of the regression lines of the Scatchard plots, *r*, calculated for 21 experiments.

	Diazepam				L-Tryptophan			
	$10^{-4} \times nk$ [M^{-1}]	$10^{-4} \times k$ [M^{-1}]	n	r	$10^{-4} \times nk$ [M^{-1}]	$10^{-4} \times k$ [M^{-1}]	n	r
Serum albumin	102.16	102.36	0.998	– 0.978	14.031	18.050	0.777	– 0.965
I-serum albumin	63.92	72.96	0.876	– 0.972	5.658	9.957	0.571	– 0.941
Urea-serum albumin	103.17	120.09	0.859	– 0.978	10.600	17.402	0.609	– 0.979

are decreased by about 30 - 50% for I-serum albumin compared to native serum albumin and urea-serum albumin (Table 2).

In Table 3, the effects of the tryptophan modification on the extrinsic Cotton effects of four benzodiazepines bound to serum albumin are summarized. For all four benzodiazepines at both concentrations a decrease in the molar ellipticities of about 10 - 15% was found for urea-serum albumin (Table 3), which can be explained by the decrease in the number of binding sites by the same percentage (Table 2). In I-serum albumin the molar ellipticities are decreased by a further 10 - 15% at both benzodiazepine concentrations (Table 3), which may be an effect of the tryptophan modification.

Using the method of Rosen[34], which is based on a concentration dependence of the extrinsic Cotton effects, quite similar effects of urea treatment and tryptophan modification on the associa-

Table 3. Extrinsic Cotton effects of four benzodiazepines bound to native serum albumin, I-serum albumin and urea-serum albumin.

The molar ellipticities, [Θ], calculated with reference to the albumin concentration (13.1μM), of four benzodiazepines bound to serum albumin at two different drug/albumin (D/P) ratios at the wavelength of the extrinsic CD maxima λ. The data are difference values, using the molar ellipticities of the albumin at the same wavelength as blank. The values in parentheses are percent of the corresponding value for native serum albumin.

Drug	λ [nm]	D/P [mol/mol]	[Θ] [deg. × cm^2 × $dmol^{-1}$] Serum albumin x ± S.E.M. (n = 6)	I-serum albumin x ± S.E.M. (n = 6) [%]	Urea-serum albumin $\bar{x}$ ± S.E.M. (n = 3) [%]
Diazepam	317	1	– 21.84 ± 0.60	– 17.87 ± 0.24 (81.8)	– 18.05 ± 0.72 (82.7)
	317	3	– 27.08 ± 0.36	– 19.80 ± 0.24 (73.1)	– 23.59 ± 0.24 (87.1)
	258	1	+ 111.50 ± 1.14	+ 90.74 ± 1.02 (81.4)	+ 102.71 ± 1.44 (92.1)
	258	3	+ 141.16 ± 0.24	+ 106.02 ± 0.66 (75.1)	+ 124.37 ± 0.18 (88.1)
Chlordiazepoxide	267	1	+ 143.99 ± 0.84	+ 110.29 ± 0.54 (76.6)	+ 135.20 ± 0.90 (93.9)
	267	3	+ 190.62 ± 1.20	+ 151.51 ± 0.54 (79.5)	+ 174.67 ± 0.54 (91.6)
Tetrazepam	305	1	– 19.13 ± 0.54	– 14.20 ± 0.60 (74.2)	– 17.15 ± 0.18 (89.6)
	305	3	– 25.27 ± 0.30	– 17.69 ± 0.18 (70.0)	– 21.06 ± 0.30 (83.3)
	258	1	+ 108.79 ± 1.62	+ 83.51 ± 1.20 (76.8)	+ 99.46 ± 0.90 (91.4)
	258	3	+ 144.59 ± 1.62	+ 102.41 ± 1.32 (70.8)	+ 127.38 ± 1.44 (88.1)
Oxazepam	317	1	– 17.69 ± 0.18	– 14.26 ± 0.54 (80.6)	– 16.73 ± 0.24 (94.6)
	317	3	– 25.39 ± 0.42	– 19.25 ± 0.42 (75.8)	– 22.26 ± 0.18 (87.6)
	258	1	+ 57.64 ± 0.30	+ 46.27 ± 0.66 (80.3)	+ 53.73 ± 0.54 (93.2)
	258	3	+ 85.68 ± 0.72	+ 68.77 ± 0.78 (80.3)	+ 77.50 ± 0.06 (90.4)

tion constants and the number of binding sites of diazepam were found.

Discussion

Using *o*-nitrophenylsulfenyl chloride, Jacobsen has shown that the lone tryptophan residue of serum albumin can be modified and that the modification does not influence the binding of bilirubin[35,36]. A possible effect of the tryptophan modification by the same reagent on the binding of a homopyrimidazole derivative to serum albumin was reported by Szekerke and Horváth[37]. I is another reagent known to react specifically with tryptophan residues in proteins under suitable conditions[26-28]. In preliminary experiments it was found that the intrinsic Cotton effects of serum albumin in the near UV range (250 - 350 nm) are influenced by the modification with I to a much smaller extent than with *o*-nitrophenylsulfenyl chloride. Because the extrinsic Cotton effects of the benzodiazepines bound to serum albumin are located within this wavelength range, and because *o*-nitrophenylsulfenyl chloride leads to higher degrees of modification[35-37], the modification with I-Br was preferred.

The data reported in this study strongly suggest that the lone tryptophan residue of serum albumin is not directly involved in the specific binding site for benzodiazepines and L-tryptophan. A loss of about 15% of the binding sites available for diazepam and L-tryptophan can be found in I-serum albumin as well as in urea-serum albumin, so that the reason seems to be the acidic unfolding in urea solution rather than the modification. The alternative possibility, that the tryptophan residue is involved in the specific binding site for benzodiazepines and L-tryptophan, but its modification does not influence the number of binding sites available, seems not to be very likely. The binding site for the indole and benzodiazepine compounds was shown to possess "receptor-like" specificity[5,10]. Its specific modification would result in a much higher decrease of the association constants. Furthermore, a much larger change in the extrinsic Cotton effects of the benzodiazepines must be obtained, if the binding site is directly modified. Extrinsic Cotton effects of ligands bound to serum albumin have been shown to be extremely sensitive even to small changes of the environment of optically active binding sites[9,38-42].

On the other hand, the decrease in the association constants of L-tryptophan and diazepam in I-serum albumin seems to be a specific effect of the tryptophan modification and cannot be found in urea-serum albumin. Furthermore, the extrinsic Cotton effects of four benzodiazepines are decreased in I-serum albumin to a higher extent than in urea-serum albumin. A smaller concentration of the ligands at the binding site cannot be the reason, since this additional decrease in the extrinsic Cotton effects is evident at molar ratios of one and three, and the higher concentration corresponds to saturating conditions of the optically active binding site[10,12]. Interestingly, the affinities to serum albumin and the intensities of the extrinsic Cotton effects of the benzodiazepines bound to serum albumin seem to depend on similar factors[9], so that the same mechanism may be the cause of the decreased affinities of L-tryptophan and diazepam and of the decreased extrinsic Cotton effects in I-serum albumin. Possibly the modification of the tryptophan residue prevents the serum albumin molecule from a complete refolding in the neighbourhood of the specific binding site. This is supported by the further decrease in the ordered secondary structure in I-serum albumin. Alternatively, the tryptophan modification might alter the environment of the common binding site of the benzodiazepines and of L-tryptophan by an allosteric mechanism, resulting in a decreased association constant and a decreased optical perturbation of the benzodiazepines bound to this site.

Many drugs can bind to the common binding site of benzodiazepines and indoles, as was shown by competition studies with L-tryptophan[43,44]. Based on the data reported, the important role of the lone tryptophan residue of human serum albumin for the binding of many drugs, as suggested several times[16,19,24,37], seems to be questionable.

This work was supported by the *Deutsche Forschungsgemeinschaft.*

Literature

1 Meyer, M.C. & Guttman, D.E. (1968) *J. Pharm. Sci.* 57, 895 - 918.

2 Jusko, W.J. & Gretch, M. (1976) *Drug Metab. Rev.* **5**, 43 - 140.
3 O'Reilly, R.A. (1973) *Ann. N.Y. Acad. Sci.* **226**, 293 - 308.
4 Steiner, R.F., Roth, J. & Robbins, J. (1966) *J. Biol. Chem.* **241**, 560 - 567.
5 McMenamy, R.H. & Oncley, J.L. (1958) *J. Biol. Chem.* **233**, 1436 - 1447.
6 McMenamy, R.H. & Seder, R.H. (1963) *J. Biol. Chem.* **238**, 3241 - 3248.
7 McMenamy, R.H. (1963) *Arch. Biochem. Biophys.* **103**, 409 - 417.
8 Brodersen, R. (1974) *J. Clin. Invest.* **54**, 1353 - 1364.
9 Müller, W.E. & Wollert, U. (1973) *Naunyn-Schmiedeberg's Arch. Pharmacol.* **280**, 229 - 237.
10 Müller, W.E. & Wollert, U. (1974) *Naunyn-Schmiedeberg's Arch. Pharmacol.* **283**, 67 - 82.
11 Müller, W.E. & Wollert, U. (1975) *Mol. Pharmacol.* **11**, 52 - 60.
12 Sjödin, T., Roosdorp, N. & Sjöholm, I. (1976) *Biochem. Pharmacol.* **25**, 2131 - 2140.
13 Müller, W.E. & Wollert, U. (1975) *Naunyn-Schmiedeberg's Arch. Pharmacol.* **288**, 17 - 27.
14 Meloun, B., Moravek, L. & Kostka, V. (1975) *FEBS Lett.* **58**, 134 - 137.
15 Behrens, P.Q., Spiekerman, A.M. & Brown, J.R. (1975) *Fed. Proc.* **34**, 591.
16 Swaney, J.B. & Klotz, I.M. (1970) *Biochemistry* **9**, 2570 - 2574.
17 Williams, E.J., Herskovits, T.T. & Laskowski, M., Jr. (1965) *J. Biol. Chem.* **240**, 3574 - 3579.
18 Ryan, M.T. & Gibbs, G. (1970) *Arch. Biochem. Biophys.* **136**, 65 - 72.
19 Perrin, J.H., Vallner, J.J. & Nelson, D.A. (1975) *Biochem. Pharmacol.* **24**, 769 - 774.
20 Santos, E.C. & Spector, A.A. (1974) *Mol. Pharmacol.* **10**, 519 - 528.
21 Sjöholm, I. & Grahnén, A. (1972) *FEBS Lett.* **22**. 109 - 112.
22 Sjöholm, I. & Ljungstedt, I. (1973) *J. Biol. Chem.* **248**, 8434 - 8441.
23 Roosdorp, N., Wänn, B. & Sjöholm, I. (1977) *J. Biol. Chem.* **252**, 3876 - 3880.
24 Gambhir, K.K., McMenamy, R.H. & Watson, F. (1975) *J. Biol. Chem.* **250**, 6711 - 6719.
25 Tritsch, G.L. & Tritsch, N.E. (1963) *J. Biol. Chem.* **238**, 138 - 142.
26 Koshland, D.E., Jr., Karkhanis, Y.D. & Latham, H.G. (1964) *J. Am. Chem. Soc.* **86**, 1448 - 1450.
27 Barman, T.E. & Koshland, D.E., Jr., (1967) *J. Biol. Chem.* **242**, 5771 - 5776.
28 Karkhanis, Y.D., Carlo, D.J. & Zeltner, J. (1975) *Anal. Biochem.* **69**, 55 - 60.
29 Moore, S. (1972) in Chemistry and Biology of Peptides (Meienhofer, J., ed.) p. 629, Ann Arbor Science Publishers Inc., Ann Arbor.
30 Theale, F.W.J. (1960) *Biochem. J.* **76**, 381 - 388.
31 Greenfield, N. & Fasmann, G.D. (1969) *Biochemistry* **8**, 4108 - 4116.
32 Scatchard, G. (1949) *Ann. N.Y. Acad. Sci.* **51**, 660 - 672.
33 Chen, R.F. (1966) *Biochim. Biophys. Acta* **120**, 169 - 171.
34 Rosen, A. (1970) *Biochem. Pharmacol.* **19**, 2075 - 2081.
35 Jacobsen, C. (1972) *Eur. J. Biochem.* **27**, 513 - 519.
36 Jacobsen, C. Funding, L., Møller, N.P.H. & Steensgaard, J. (1972) *Eur. J. Biochem.* **30**, 329 - 402.
37 Szekerke, M. & Horváth, M. (1976) *Arzneim. Forsch.* **26**, 478 - 482.
38 Blauer, G., Harmatz, D. & Snir, J. (1972) *Biochim. Biophys. Acta* **278**, 68 - 88.
39 Devaux, P., Viennet, R. & Legrand, M. (1974) *FEBS Lett.* **40**, 18 - 24.
40 Chignell, C.F. & Starkweather, D.K. (1971) *Mol. Pharmacol.* **7**, 229 - 237.
41 Perrin, J.H. & Vallner, J.J. (1975) *J. Pharm. Sci.* **64**, 1508 - 1513.
42 Müller, W.E. (1976) *Hoppe-Seyler's Z. Physiol. Chem.* **357**, 1487 - 1494.
43 McArthur, J.N., Dawkins, P.D. & Smith, M.J.H. (1971) *J. Pharm. Pharmacol.* **23**, 393 - 398.
44 Müller, W.E. & Wollert, U. (1975) *Res. Commun. Chem. Pathol. Pharmacol.* **10**, 565 - 568.

K. J. Fehske, Dr. W. E. Müller and Prof. Dr. U. Wollert, Pharmakologisches Institut der Universität Mainz, Obere Zahlbacher Str. 67, D-6500 Mainz.

Hoppe-Seyler's Z. Physiol. Chem.
Bd. 359, S. 719 - 723, Juni 1978

Studies of Ligand Binding to Cholera Toxin, III

Cooperativity of Oligosaccharide Binding

Jörg SATTLER, Günter SCHWARZMANN, Ingo KNACK, Klaus-H. RÖHM and Herbert WIEGANDT

Physiologisch-Chemische Institute I und II der Philipps-Universität Marburg

(Received 4 January / 20 February 1978)

Summary: Binding parameters for the interaction of cholera toxin and choleragenoid with monosialo-gangliotetraose, the oligosaccharide moiety of ganglioside II^3NeuAc-GgOse$_4$-Cer, have been measured by equilibrium displacement dialysis. The experimental data were evaluated by a curve fitting computer program.

The binding curves obtained at 6 °C reflected positive cooperativity with average Hill coefficients in the range of 1.16 to 1.25.

The cholera toxin as well as its ganglioside-binding protomer B-protein bind 4 mol of monosialo-gangliotetraitol per mol of protein.

The presence of disulfide-cleaving agents like 2-mercaptoethanol did not eliminate the cooperativity of the binding characteristics, but increased the Hill coefficient.

Untersuchungen zur Bindung von Liganden an Choleratoxin, III:
Die Kooperativität der Wechselwirkung mit Oligosaccharid

Zusammenfassung: Die bei der Bindung zwischen Choleratoxin bzw. Choleragenoid und Monosialogangliotetraose, dem Kohlenhydrat des Gangliosids II^3NeuAc-GgOse$_4$-Cer, vorliegenden kinetischen Parameter wurden mit Hilfe der Gleichgewichtsdialyse bestimmt und die hierbei erhaltenen experimentellen Werte mit Hilfe eines speziellen Computerprogramms ausgewertet.

Die für 6 °C errechneten Isothermen weisen eindeutig kooperativen Bindungscharakter mit durchschnittlichen Hill-Koeffizienten von 1.16 bis 1.25 auf. Sowohl das intakte Choleratoxin als auch das Choleragenoid binden 4 mol Monosialogangliotetrait/mol Protein. Die Anwesenheit von disulfidspaltenden Agentien wie 2-Mercaptoäthanol hat keinerlei negativen Einfluß auf den kooperativen Bindungscharakter. Vielmehr scheint dadurch die Kooperativität noch gesteigert zu werden.

Key words: Cholera toxin, ganglioside.

Abbreviation: (as recommended by IUPAC-IUB Lipid Nomenclature Document 1976)
Monosialo-gangliotetraose = Galβ1→3 GalNAcβ1→4 Gal[3←2αNeuAc]β1→4 Glc.

Cholera toxin exerts its biological activity by stimulating cellular adenylate cyclase, which results in elevated levels of cAMP[1]. The toxin consists of two non-covalent protomers, an A-protomer (M_r = 29 000) and a B-protomer (M_r = 58 000).

The latter is composed of five small non-covalent subunits[2].

The A-protomer enzymatically catalyses the transfer of ADP-ribose from NAD to an intracellular adenylate-cyclase-linked GTPase, that regulates the nucleotide cyclase. ADP-ribosylation of this GTPase appears irreversibly to unlock a negative GTP feedback regulation, leaving the adenylate cyclase in a permanently highly active state[3-6].

For the A-protomer to become biologically active with intact cells, it must first be fixed to the outer cell membrane. This is achieved by the tight binding of the B-protomer to a specific cell surface receptor believed to be ganglioside II^3NeuAc-GgOse$_4$-Cer[7,8]. This ganglioside readily precipitates the toxin from aqueous solutions. Although the specificity of this interaction had to reside in the hydrophilic portion of the ganglioside, its isolated free oligosaccharide moiety by itself was incapable of precipitating the toxin and eliminating its in vivo toxicity[9].

However, monosialo-gangliotetraose interfered to some extent with the precipitation of cholera toxin-ganglioside II^3NeuAc-GgOse$_4$-Cer[10] in gel double diffusion experiments. Equilibrium displacement dialysis did demonstrate a strong interaction between monosialo-gangliotetraose and cholera toxin[11]. In these latter experiments, binding isotherms were obtained only by graphical evaluation of the experimental data, whereby the very low and the very high concentrations of bound ligand could not easily be determined. Therefore no correct and unequivocal information, e.g. on the number of binding sites and possible cooperativity effects of ligand binding to the protein, could be obtained. In the meantime a computer program for calculating the parameters of the Hill equation became available. We therefore performed new binding measurements and now report more valid data on ligand interaction with cholera toxin or its B-protomer that were obtained by non-linear regression analysis[12].

Materials and Methods

Cholera toxin and *choleragenoid* were donated by *Dr. J. Staerk*, Behring Werke AG, Marburg. *^{3}H-Monosialo-gangliotetraitol* was prepared as described earlier[11].

Equilibrium displacement dialysis: Small acrylic glass chambers with total volumes of 0.4 m*l* and dialysis membranes (Visking tubing from Union Carbide, Chicago, Ill.) were used as in the previous experiments[11]. Equilibrium between the sugar ligand (range of concentration 10 - 100nM) and solutions of cholera toxin or B-protein (concentration 25nM) was reached at 6 °C after 12 to 15 h. Shaking was continued for an additional 12 h without denaturation of the protein.

In experiments with 2-mercaptoethanol all solutions were brought to 0.1% 2-mercaptoethanol.

Binding experiments were evaluated with the aid of a computer program[12], which fits the Hill equation to the data:

$$\bar{v} = \frac{[LB]}{[LB]_{max}} = \frac{[LF]^{n_H}}{K + [LF]^{n_H}} \qquad (I)$$

where

$\bar{v}$ = degree of saturation,
$[LB]$ = concentration of bound ligand,
$[LB]_{max}$ = maximal concentration of bound ligand,
$[LF]$ = concentration of free ligand,
K = empirical parameter defined by (II),
n_H = Hill coefficient

If n_H deviates from unity, K is not a true binding constant but is correlated to the ligand concentration at half saturation by

$$[LF]_{0.5} = \sqrt[n_H]{K} \qquad (II)$$

The Hill coefficient n_H indicates deviations from normal hyperbolic binding ($n_H > 1$, positive cooperativity; $n_H = 1$, hyperbolic curve; $n_H < 1$, negative cooperativity). Since the value of maximal saturation, $[LB]_{max}$, is not known a priori, a straightforward determination of the coefficients n_H and K is impossible. The computer program therefore proceeds by alternately varying all three parameters until the sum of squares of the residuals i.e. the differences between experimental and calculated values $R = \sum(Y_{exp.} - Y_{calcd.})^2$ reaches a minimum. Since the fit is based on the original set of data rather than a linearized form of the Hill equation, the problem of statistical weighting of the data is considerably simplified. The following procedure was used: In a first fit based on the assumption of constant relative error for all points, a preliminary set of Hill parameters, $[LB]_{max}$, K and n_H was established. Then for each point the actual deviation from the resulting curve was calculated. Assuming a Gaussian distribution of the errors around their average, statistical weights were computed for each data pair. The weights then were introduced into a second fitting cycle to give the final values of $[LB]_{max}$, K and n_H.

Since the Hill coefficients emerging from the fitting procedures proved to be rather close to unity, it had to be ascertained whether the observed cooperativity is statistically significant under the conditions of our experiments or not. This was performed according to Hurst[13] by fitting both the Hill equation and the Michaelis-Menten function ($n_H = 1$) to each set of data (Fig. 1 d). From the resulting residual errors R_1 and R_2, partial F-values F_p were calculated according to

$$F_p = \frac{(R_1 - R_2)\, m_2}{R_2\,(m_1 - m_2)} \qquad \text{(III)}$$

Values of m are obtained by subtracting the number of parameters (3 for the Hill equation, 2 for the Michaelis-Menten case) from the number of data points n. The resulting F_p values are then compared with the tabulated values of F for a given significance level P and the degrees of freedom $f_1 = 1$ and $f_2 = m_2$. If $F_p > F$, the assumption that the difference between R_1 and R_2, and thus the observed cooperativity, is significant, may be accepted.

Results

The binding of cholera toxin and B-protein to [^{3}H]monosialo-gangliotetraitol was studied quantitatively by equilibrium displacement dialysis. The sugar dialysed freely and reached full equilibrium after 12 to 15 h at 6 °C. Because of the very low concentration of toxin (25nM), all solutions had to contain 0.2% egg albumin. This served to prevent any unspecific binding to the acrylic material of the dialysis cell. Egg albumin itself had no binding affinity to the sugar ligand.

Fig. 1 a, b, c shows the typical binding curves of monosialo-gangliotetraitol to cholera toxin and B-protein, respectively. The curves are clearly of sigmoidal shape, thus indicating positive cooperativity of ligand binding. Some of the calculated

▷

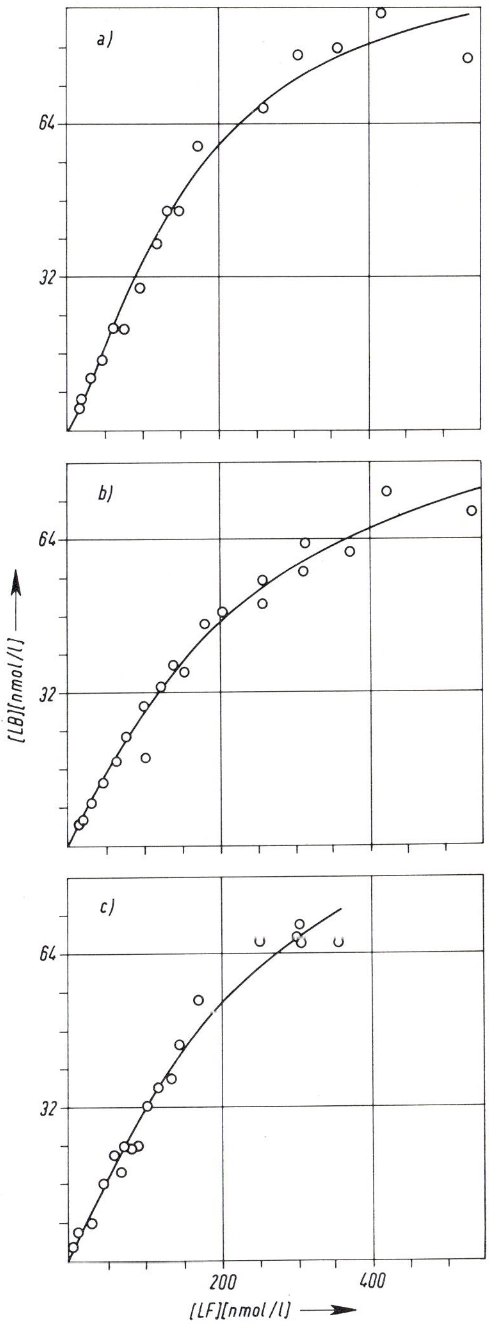

Fig. 1. Binding curves of monosialo-gangliotetraitol to cholera toxin (a) in the presence and (b) in absence of 2-mercaptoethanol and to cholera toxin B-protomer (c) in absence of 2-mercaptoethanol.

See text for experimental conditions. The solid lines represent the best fit by computer analysis. Circles are experimental values.

[LB] = concentration of bound ligand
[LF] = concentration of free ligand

Protein concentration in all experiments 25nM.
The binding parameters arrived at are given in the table.

Table. Binding parameters.
The protein concentration in each case was 25nM.

Protein	Mercapto-ethanol	Hill coefficient	LB_{max} [nmol/l]	K [nmol/l]	$LF_{0.5}$ [nmol/l]	Statistical parameters (see Methods)
Cholera toxin (Fig. 1 a)	+	1.44	102	0.16	158	$F_p = 11.3; n = 16$ F at $P = 99\%: 9.1$
Choleratoxin (Fig. 1 b)	–	1.16	104	20.5	236	$F_p = 3.8; n = 20$ F at $P = 90\%: 3.03$
B-Protein (Fig. 1 c)	–	1.25	109	4.1	195	$F_p = 3.8; n = 19$ F at $P = 90\%: 3.05$

values of the binding parameters are listed in the table.

From the values of $[LB]_{max}$ the molar ratio of the ligand-toxin complexes could be estimated. At maximal saturation, 4 molecules oligosaccharide bind to 1 molecule toxin or B-protein.

This result is not changed by the presence of disulfide-cleaving agents. Only an increase in the value of the Hill-coefficient was obtained when 2-mercaptoethanol was included in the dialysis experiments.

Discussion

A number of qualitative and quantitative conclusions may be drawn from the results of the experiments described above.

First, a positive cooperativity has been observed for sugar binding to cholera toxin or its B-protomer. The statistical treatment of the data showed rather high levels of significance (see Table). Even with a Hill-coefficient of 1.16, the fit of the Hill equation is significantly superior to the hyperbolic model above the 90% level. Second, this cooperativity of binding is not decreased by cleavage of disulfide bridges of the protein with 2-mercaptoethanol. Instead, the average Hill-coefficient increased, indicating an even stronger positive cooperatively with a value of $n_H = 1.44$.

This change in shape of the curve possibly is accompanied by structural alteration of the monomers within the B-protomer that may lead to a better conformational fit of the B-protein towards the specific sugar ligand. Third, the presence of the A-protomer obviously does not influence the binding of the B-protomer to the sialo-oligosaccharide. It is surprising to see that the number of ligand binding sites within the toxin molecule is 4 instead of the expected 5, corresponding to the 5 B-protomer subunits. The calculation of 4 ligand binding sites per molecule confirms our earlier observations with a value of 3.8 extrapolated from the curvilinear Scatchard plot[11]. This finding may in fact influence hypotheses on the structural architecture of the toxin molecule with its pentameric B-protomer. Thus the one of the 5 very similar B-subunit monomers ($M_r = 11600$) that does not bind to ganglioside possibly is located in a different position than the other 4.

It is, however, as yet unclear, whether the inability of the one B-monomer to bind to ganglioside is due to structural accessibility or only to an alteration of the net charge. Studies from this laboratory have indicated the correlation between the degree of deamination of B-protein and concommitant decrease of ligand binding capacity as well as cooperativity*. Thus from our data it may be argued that the binding affinity of cholera toxin cannot be expressed by a single value of K_D or K_A, respectively. Therefore all such data reported so far[14-17] must be interpreted with caution and not be taken as true association constants. For information concerning the strength of the ligand binding it is therefore only possible to compare the ligand concentration at the point of half saturation, which is obtained by correcting the calculated K values according to (II).

* Sattler, J., Knack, I., Schwarzmann, G., Wiegandt, H., Staerk, J. & Röhm, K.-H. (1978) in preparation.

The assumption of a positive cooperativity of ligand binding to the choleragenoid not only leads to a closer fit, but also yields the lower value for $[LB]_{max}$, i.e. a number of 4 ligand molecules bound per pentamer (Fig. 1 a, b, c). Fitting of the Michaelis-Menten equation to the data always led to value of $[LB]_{max}$ corresponding to a number of 5 - 6 ligand molecules bound (Fig. 2).

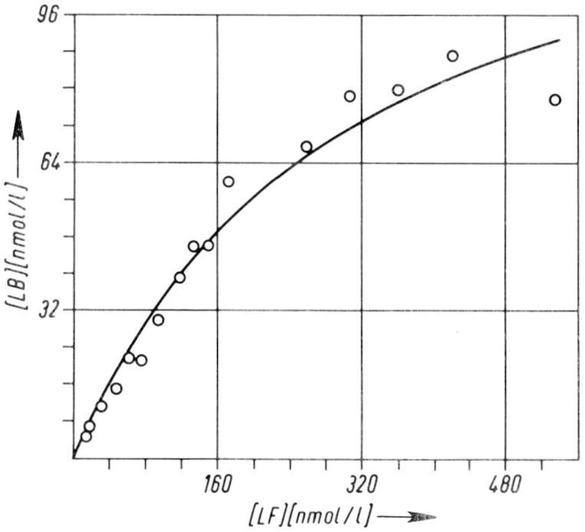

Fig. 2. Fit of the Michaelis-Menten equation to the data shown in Fig. 1a.

The fit is obviously much less satisfactory than the one given in Fig. 1a. The resulting parameters were: $[LB]_{max}$ = 142 nmol/l; K = 300 nmol/l.

Literature

1 Gill, D.M. (1977) *Adv. Cyclic Nucleotide Res.* **8**, 85 - 118.
2 Sattler, J., Wiegandt, H., Staerk, J., Kranz, Th., Ronneberger, H.J., Schmidtberger, R. & Zilg, H. (1975) *Eur. J. Biochem.* **57**, 309 - 316.
3 Moss, J., Manganiello, V.C. & Vaughan, M. (1976) *Proc. Natl. Acad. Sci.* **73**, 4424 - 4427.
4 Helmreich, E.J.M., Zenner, H.P., Pfeuffer, T. & Cori, C.F. (1976) *Curr. Top. Cell. Regul.* **10**, 41 - 87.
5 Cassel, D. & Zelinger, Z. (1977) *Proc. Natl. Acad. Sci.* **74**, 3307 - 3311.
6 Moss, J. & Vaughan, M. (1977) *Proc. Natl. Acad. Sci.* **74**, 4396 - 4440.
7 Van Heyningen, W.E., Carpenter, C.C.J., Pierce, N.F. & Greenough, W.B. III, (1971) *J. Infect. Dis.* **124**, 415 - 418.
8 Staerk, J., Ronneberger, H.J., Wiegandt, H. & Ziegler, W. (1974) *Eur. J. Biochem.* **48**, 103 - 110.
9 Wiegandt, H., Ziegler, W., Staerk, J., Kranz, Th. Ronneberger, H.J., Zilg, H., Karlsson, K.A. & Samuelson, B.E. (1976) *Hoppe-Seyler's Z. Physiol. Chem.* **357**, 1637 - 1646.
10 Holmgren, J., Manson, J.E. & Svennerholm, L. (1974) *Med. Biol.* **52**, 229 - 233.
11 Sattler, J., Schwarzmann, G., Staerk, J., Ziegler, W. & Wiegandt, H. (1977) *Hoppe-Seyler's Z. Physiol. Chem.* **358**, 159 - 163.
12 Herrmann, V., Knack, I. & Röhm, K.-H. (1978) *Hoppe-Seyler's Z. Physiol. Chem.* **359**, 47 - 61.
13 Hurst, R.O. (1974) *Can. J. Biochem.* **52**, 1137 - 1142.
14 Holmgren, J., Lindholm, L. & Lönnroth, J. (1974) *J. Exp. Med.* **139**, 801 - 819.
15 O'Keefe, E. & Cuatrecasas, P. (1974) *J. Biol. Chem.* **71**, 2502 - 2504.
16 Holmgren, J., Lönnroth, J., Mannson, J.E. & Svennerholm, L. (1975) *Proc. Natl. Acad. Sci.* **72**, 2520 - 2524.

Dr. J. Sattler, Dr. G. Schwarzmann and Prof. Dr. H. Wiegandt, Physiologisch-Chemisches Institut I der Universität Marburg,
Lahnberge, D-3550 Marburg/Lahn.
Dipl.-Chem. I. Knack and Dr. K.-H. Röhm, Physiologisch-Chemisches Institut II der Universität Marburg, Lahnberge, D-3550 Marburg/Lahn.

Hoppe-Seyler's Z. Physiol. Chem.
Bd. 359, S. 725 - 731, Juni 1978

A Study of Microsomal Oxidative Functions with the Aid of a Series of *N*-Colchiceyl Derivatives*

Bernd HASENMÜLLER, Martin SCHÖNHARTING and Günther SIEBERT

Institut für Biologische Chemie der Universität Hohenheim, Stuttgart

(Received 21 February 1978)

Summary: *N*-Colchiceyl derivatives are presented as a class of substances which, provided certain structural requirements are met, are oxidatively degraded by microsomal suspensions both at the nitrogen-linked side chain at ring C and at the oxygen-linked side chain(s) at Ring A. The chemical structures of unknown metabolites have been elucidated by comparison with authentic samples whose syntheses are described. Comparative studies indicate structure-dependent preferred sites for the introduction of oxygen, whose extent correlates rather closely with the partition coefficient in the system n-octanol/ethanol/water. The results demonstrate, with the inclusion of some inhibition data, the suitability of *N*-colchiceyl derivatives for detailed investigations of microsomal degradative processes.

N-Colchiceylderivate als geeignete Substanzklasse zur Untersuchung oxidativer mikrosomaler Prozesse

Zusammenfassung: *N*-Colchiceylderivate werden als Substanzklasse präsentiert, die unter bestimmten strukturellen Voraussetzungen sowohl an der über Stickstoff verknüpften Seitenkette am Ring C als einer der über Sauerstoff verknüpften Seitenketten am Ring A durch eine Mikrosomensuspension oxidativ angegriffen werden. Es wurde eine ganze Reihe bisher unbekannter Substanzen als Metabolite gefunden, deren Struktur mit Hilfe von auf synthetischem Wege hergestellten Vergleichssubstanzen ermittelt werden konnte. Vergleichende Metabolisierungsstudien weisen auf strukturell bedingte, bevorzugte Reaktionsorte bei der Einführung des Sauerstoffs hin. Das Metabolisierungsausmaß ist in guter Näherung mit dem Wert des jeweiligen Verteilungskoeffizienten in dem System n-Octanol/Äthanol/Wasser korreliert. Zusammen mit vorläufigen Daten aus Hemmversuchen lassen die Ergebnisse erkennen, daß *N*-Colchiceylverbindungen eine geeignete Substanzklasse zur Untersuchung mikrosomaler Abbauvorgänge darstellen.

Key words: colchicine derivatives, cytochrome P-450, microsomal degradation, liver.

Enzyme:
Isocitrate dehydrogenase (NADP$^{\oplus}$), *threo*-D_S-isocitrate: NADP$^{\oplus}$ oxidoreductase (decarboxylating) (EC 1.1.1.42).

* Taken in part from the Ph.D. Thesis of B. Hasenmüller, University of Hohenheim (1975).

Derivatives of colchicine which are formed by substitution of the methoxy group at C-10 (ring C) by a nitrogenous residue $-NR_2$ have been known for many years[1-10]; the parent compound colchiceinamide, where $-OCH_3$ is replaced by $-NH_2$, was synthesized by Zeisel[11] in 1888. Nearly all substances with primary or secondary amino groups may be coupled with the alkaloid, unless steric or electronic conditions are too unfavourable. Some of these colchicine derivatives have been supposed to be useful in the chemotherapeutic treatment of benign or malignant tumors or in the treatment of plants to effect polyploidy[12]. All these derivatives are characterized by the amide-like linkage of the nitrogen atom of the introduced amino compound with C-10 of the alkaloid moiety, whose free acid O^{10}-demethylcolchicine[13] is known as colchiceine. Therefore, these derivatives will be named, in the light of presently confusing designations[4,6-10,14,15], as *N*-colchiceyl amino compounds, which is in accordance with international agreements on nomenclature.

In the course of our studies on a representative series of *N*-colchiceyl derivatives, which are concerned with detailed information on rates of synthesis, sensitivity to hydrolysis, and biological activities, we have also studied the biodegradation of these compounds. We have found that *N*-colchiceyl derivatives, depending on their structures, may be oxidatively degraded by a series of different microsomal attacks; it became apparent that this class of substances provides suitable structures for a closer investigation of the cytochrome P-450 enzyme system. Therefore, the data obtained with hamster liver microsomes will be presented in the following because they permit a deeper insight into the complexity of function of the various microsomal oxidative processes.

Materials and Methods

Chemicals

Colchicine DAB 7 grade was purchased from E. Merck AG, Darmstadt, and O^3-demethylcolchicine and O^2-demethylcolchicine were isolated after incubation of colchicine in a Udenfriend system* as described[16] and characterized by chromatographic and spectral procedures[16]. Amino acids ("chromatographically pure") were obtained from Mann Lab., New York; amines from Schuchardt GmbH, München; Sephadex LH-20 and DEAE-Sephadex from Pharmacia, Uppsala; DL-isocitrate trisodium salt · 2 H_2O from Serva Feinbiochemica, Heidelberg; $NADP^{\oplus}$ from Boehringer Mannheim GmbH; all other chemicals from E. Merck AG, Darmstadt.

Preparation of N-colchiceyl derivatives

Colchicine at a final concentration of 20mM is kept in the dark in a closed vessel with a 25-fold excess of the respective amino compound at pH 10.5 and 37 °C for 48 h. The yellow solution is then brought to pH 7 with 5% acetic acid and mixed with 4 vol. of acetone. The precipitate is removed on a filter and washed with acetone, then the combined filtrates are concentrated to a small volume in a rotatory evaporator and chromatographed on Sephadex LH-20 with 1% (v/v) acetic acid[16]**. The resulting pure fractions are lyophilized to give an amorphous yellow powder.

This procedure is generally applicable for a wide range of derivatives which, depending on steric factors, are obtained in 65 to 98% final yield. The rate of the substitution reaction correlates in most cases with the p*K* values of the nucleophilic attacking amino groups.

In the experiments described below the following amino compounds were used: glycine, L-alanine, L-lysine, L-arginine, β-alanine, γ-aminobutyrate, taurine, putrescine, ethylenediamine, methylamine, dimethylamine, ethylamine, 1-propylamine, ethanolamine, 1-aminopropan-2-ol, and 1-aminopropan-3-ol. The synthesis of colchiceinamide was performed as described elsewhere[17]. Some derivatives were also synthesized, as authentic samples for comparative purposes, following the procedure described above with O^3-demethylcolchicine or O^2-demethylcolchicine replacing colchicine.

Identification and characterization reactions of N-colchiceyl compounds

Whereas some *N*-colchiceyl derivatives were already characterized by a limited set of physico-chemical parameters[4,5,6-10,14], the following data were determined for the purpose of identification and characterization.

a) Special studies concerning the chemical structure: Irradiation with UV light (intact ring C), deacetylation

* This system consists of phosphate buffered (pH 6.8) $FeSO_4$, EDTA and ascorbate and is usually regarded as simulating oxidative reactions of liver microsomes in drug metabolism.

** If aromatic amino acids are used to produce *N*-colchiceyl derivatives, a column of DEAE-Sephadex with 5 g NaCl, 0.5 g Na_2CO_3, and 0.5 g $NaHCO_3$ per liter as eluant is used for separation and purification.

and reacetylation (intact side-chain at ring B) and formation of [^{14}C]methanol from [*methoxy-C*^{10}-^{14}C] colchicine, together with literature data[1,6-10,14,15] leave no doubt as to the structures of the *N*-colchiceyl derivatives which are given in the table, and are therefore not presented here in detail. In the case of diamino compounds, however, it is necessary to determine which and how many amino groups have reacted with the alkaloid.

The reaction product of colchicine with L-lysine is chromatographed on Sephadex LH-20 as a double peak whose components were identified by specific blocking of α- or ε-amino groups by acetylation or benzyloxycarbonylation, respectively, before synthesis, and cleavage of the protective group afterwards. Thus a clear distinction between the N^{α}- and N^{ϵ}-colchiceyl lysines which constitute the double peak could be achieved. In the experiments to be described below, N^{ϵ}-colchiceyl-L-lysine was always used.

However, the reaction of colchicine with arginine, putrescine, or ethylenediamine always yielded, under the conditions described above, one product only. Experiments with arginine, specifically blocked in analogy to the studies with lysine just described, demonstrated that N^{α}-colchiceyl arginine was the compound actually formed. This agrees well with the observation that the guanidino group does not react with colchicine unless extreme alkalinity is maintained (e.g. sodium alcoholate)[18].

If the conditions of synthesis are changed so that the alkaloid concentration exceeds that of the amino compound, putrescine and ethylene diamine lead in addition to the formation of dumb-bell-shaped *N,N'*-dicolchiceyl derivatives whose spectral and chromatographic properties are profoundly changed from mono-substituted compounds. Obviously, syntheses with putrescine or ethylenediamine in high excess over colchicine, as described above, permit only monosubstitution. In the experiments to be described below, *N*-(mono)colchiceyl ethylenediamine and *N*-(mono)colchiceyl putrescine were used.

b) Nitrogen determination with the Kjeldahl procedure[19]:
40 - 50 mg lyophilized derivative was digested in 30 m*l* conc. H_2SO_4 containing 10 g K_2SO_4 and 2 g selenium mixture for 1.5 h. After steam distillation, NH_3 was titrated with 10mM HCl. Observed values coincided within 3% with calculated data.

c) Spectral analysis from 210 to 500 nm in the Acta V photometer (Beckman Instr., München) at acid (0.1M HCl), neutral (0.1M phosphate buffer, pH 7.0) and alkaline (0.1M NaOH) conditions. Molar absorption coefficients were determined at 355 nm in 0.1M phosphate buffer, pH 7.0.

d) Thin-layer chromatography on silica gel plates 60 F_{254} (E. Merck AG, Darmstadt):
With the solvent mixtures

A) acetone/25% NH_3 82:18,
B) chloroform/acetic acid/H_2O 60:20:1,
C) acetone/chloroform/25% NH_3 15:75:8

it was found that all *N*-colchiceyl derivatives behaved as a single compound.

e) Electrophoresis in the Elphor apparatus (Desaga, Heidelberg) on paper no. 2043 a (Schleicher & Schüll, Dassel):
With the buffer solutions

i) 0.2M ammonium acetate, pH 7.0,
ii) 0.5M sodium citrate, pH 4.0,
iii) 15% acetic acid/5% formic acid pH 1.9,

migration in the electric field coincided with the data expected from theoretical considerations.

f) Determination of the partition coefficient:
A biphasic mixture of n-octanol/ethanol/H_2O 50:45:50 was equilibrated at 21 °C and then 3 m*l* of each phase was mixed with 100 nmol of colchicine or *N*-colchiceyl derivatives. With the aid of molar absorption coefficients (see above), the distribution of the compound was examined in the two phases by measuring the absorption at 355 nm.

δ values represent the ratio of concentrations in the organic versus aqueous phase.

Experiments with hamster liver microsomes

Preparation and incubation: Livers were taken from female Syrian golden hamsters (Bäumler, Wolfratshausen) of 70 g body weight which were pretreated with 80 mg sodium phenobarbital/kg intraperitoneally for 4 consecutive days prior to sacrifice. The post-mitochondrial supernatant from 33% (w/v) liver homogenates in Krebs-Ringer phosphate solution was prepared as described previously[20]. The protein concentration of the microsomal suspension was brought to 20 mg/m*l* by dilution with Krebs-Ringer-phosphate solution.

Except where stated otherwise, incubation mixtures contained in a final volume of 10 m*l* 7.5 μmol substrate (colchicine or *N*-colchiceyl derivatives), 100 mg protein, 60 μmol $MgCl_2$, 5 μmol $NADP^{\oplus}$, 50 μmol DL-isocitrate and 1 U isocitrate dehydrogenase in Krebs-Ringer phosphate solution. The reaction was started by the addition of substrate and continued for 90 min at 37 °C under air in a metabolic shaker.

Analytical methods

The reaction was stopped by accition of 1 vol. of ice-cold acetone, and after standing for 30 min at 0 °C the mixture was centrifuged. After washing of the precipitate with 50% acetone, the combined supernatants were reduced to a small volume in a rotatory evaporator, and the slightly turbid solution chromatographed on Sephadex LH-20 (50 x 2.8 cm) with 1% acetic acid as the eluant. The separated peaks were identified, with the aid of authentic substances, by their chromatographic behaviour on Sephadex LH-20 columns and on silica gel thin-layer plates with solvent C as well as by their spectral behaviour, and were determined quantitatively by the molar absorption coefficients at 355 nm and pH 7.0 (see above). The recovery of total alkaloids always exceeded 94% of the substrate used for incubations.

All experimental data are the average of at least three independent incubations, with livers of 3 to 4 hamsters per experiment.

Results

The table shows the results obtained with a number of *N*-colchiceyl derivatives. If the residue R, covalently bound through the nitrogen to the alkaloid, consists of an aliphatic alkyl chain, oxidative attack is observed at this side chain at ring C as well as at the methoxy group at C-3 of ring A. Oxidative demethylations at other positions of ring A are extremely rare in these experiments and do not exceed, in almost all cases, the level of detectability (< 1% of total alkaloid); this agrees well with previous data on the metabolism of colchicine itself by microsomes from hamster liver[20].

Depending on the length of the side chain attached to C-10, aliphatic hydroxylation or *N*-dealkylation is observed, with a pronounced preference for the introduction of oxygen in the $\omega - 1$ position. If, however, no choice exists in this side chain for the entry of oxygen, as in compound II of the table (*N*-colchiceyl methylamine), ω-oxidation is observed to a considerable extent instead,

Table. Physico-chemical and biological data of some *N*-colchiceyl derivatives.

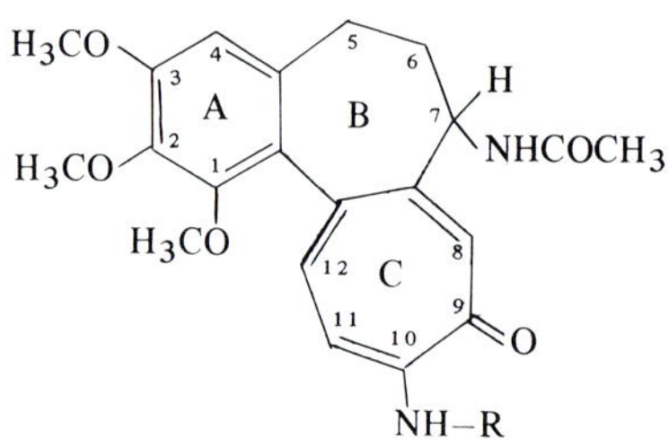

No.	Derivatives R	$\log \epsilon_{355nm}$ [a]	δ [b]	Metabolic rate [c] Total [%]	Oxidation of side chains [%] –NHR ω	ω-1	ω-2	–OCH_3
I	H	4.136	2.7	35	x	x	x	30 (86)[d]
II	CH_3	4.336	3.5	60	39 (65)	x	x	10 (17)
III	CH_2CH_3	4.358	5.2	51	2 (4)	25 (49)	x	15 (29)
IV	$CH_2CH_2CH_3$	4.356	11.5	69	–	33 (48)	3 (4)	22 (32)
V	CH_2CH_2OH	4.352	2.0	24	–	–	x	24 (100)
VI	$CH_2CH_2CH_2OH$	4.340	2.4	30	–	–	–	30 (100)
VII	$CH_2CH(OH)CH_3$	4.339	2.5	29	–	–	–	29 (100)
VIII	$CH_2CH_2NH_2$	4.302	0.2	0				
IX	$[CH_2]_4NH_2$	4.294	1.5	0				

[a] Logarithm of molar absorption coefficient ϵ determined at 355 nm in 0.1M phosphate buffer pH 7.0.
[b] Partition coefficient δ determined as described under Methods.
[c] Formation of metabolites after incubation with hamster liver microsomes (see Methods), given as percent of total alkaloid recovered after incubation (in brackets as percent of total metabolites); x = no oxidative attack possible; – = no metabolic degradation observed.
[d] Another 5% of the substrate (14% of total metabolites) is converted into an as yet unidentified product.

followed in this case by *N*-demethylation (see below).

It is interesting to note how some of the various reactions with such substances seem to compete with each other. If an oxidative attack may occur at –OR as well as at –NHR, as in compounds II to IV, the latter site is definitely favoured, as can be demonstrated best with *N*-colchiceyl methylamine containing identical alkyl substituents at all positions. In all derivatives which were investigated, the side chain at ring C undergoes a single attack only: if directed towards the C-atom next to the nitrogen, dealkylation takes place, presumably through a semi-acetal as intermediate[21], whereas attack at more remote C-atoms leads to hydroxylated products which cannot be degraded further in the ring C region (see compounds V to VII). However, if a substrate contains two side chains at the ring C-linked nitrogen, both side chains may be attacked, as in the case of *N*-colchiceyl dimethylamine, where various processes can be observed. Whereas oxidative *N*-demethylation first produces *N*-colchiceyl (mono)methylamine, whose properties as a suitable substrate are already described in the table (compound II), oxidative *O*-demethylation at ring A occurs in addition, resulting in the following complete metabolite pattern consisting of *N*-colchiceyl methylamine (32%), *N*-(O^3-demethylcolchiceyl) dimethylamine (8%), colchiceinamide (12%), *N*-(O^3-demethylcolchiceyl) methylamine (4%), and O^3-demethylcolchiceinamide (1%), with 43% *N*-colchiceyl dimethylamine remaining undegraded after 90 min incubation.

This metabolite pattern demonstrates that ring A may be attacked, though to a smaller extent, independently of the degradative steps taking place at ring C; even metabolites – whether hydroxylated or dealkylated at –NHR of ring C (e.g. compounds I, V to VII) – still may undergo further metabolization at the rather distant site of ring A. On the other hand, once ring A has undergone oxidative demethylation, no further attack at the side chain of ring C has been found, which can be demonstrated by the inability of microsomal suspensions to form any metabolite from *N*-(O^3-demethylcolchiceyl) methylamine (data not shown). The formation of a phenolic OH-group on ring A apparently terminates the oxidative degradability of these alkaloid derivatives and of colchicine itself, since a second methoxy group on ring A has also never been observed to be attacked[20,22].

The extent to which degradation occurs apparently depends on the partition coefficient δ in the n-octanol/ethanol/water system (see the table), since there is a close correspondence between the hydrophobicity and degradability with the compounds I, III to VII. If the partition coefficient, under the experimental conditions given, falls below 2, no metabolic degradation is observed at all (compounds VIII and IX of the table). In these latter derivatives, free amino groups in the side chain are positively charged at pH 7, which obviously suffices to suppress degradation. Similarly, the *N*-colchiceyl amino acid derivatives of glycine, alanine, lysine, arginine, β-alanine, γ-aminobutyrate, and taurine, which have partition coefficients between 0.1 and 0.7 and are negatively charged at pH 7, are not degradable by microsomes from hamster liver.

Nevertheless, some interaction is observed between non-oxidizable derivatives and the microsomal enzyme system; whereas the uncharged metabolite O^3-demethylcolchicine[20] and the positively charged *N*-colchiceyl ethylene diamine, both not degradable by themselves, inhibit by about 50% the oxidative *O*-demethylation of colchiceinamide in equimolar concentrations, the negatively charged N^ϵ-colchiceyl-L-lysine is totally inactive.

Discussion

In this paper the microsomal degradation of *N*-colchiceyl derivatives is described. With one exception, all the various metabolites formed in our studies have been elucidated in their chemical structure, their relative amounts indicating certain preferences in the microsomal oxidative pathways. *N*-Dealkylation exceeds *O*-dealkylation (table) with secondary carbon atoms being more easily oxidizable than primary ones, as has already been demonstrated by Frommer et al.[23]. If various possibilities exist for side chain attack, $\omega - 1$ oxidation is preferred. In contrast to *O*-dealkylation, *N*-dealkylation does not prevent further metabolic degradation, which may to some extent be influenced by the different electronic states of the hetero atoms involved. Lack of metabolic degradation, as shown for compounds VIII and IX, does not necessarily mean lack of interaction with the microsomal enzyme system, because inhibitory effects may be observed. However, determinations of type of inhibition and binding site of the enzyme system were not performed in this study. They would require separate measurements for each individual inhibitory derivative, since minor structural differences have already been shown, e.g. with quinoline versus isoquinoline, to result in changing the type I binding spectrum into type II binding spectrum[24].

A decisive factor for oxidative metabolism of a compound is given by its lipophilic character[25], which can easily be assayed with the partition coefficient δ as a parameter. It is demonstrated in the table that partition in the n-octanol/ethanol/water system corresponds well with the degree of metabolic transformation. One may assume that a first approach to a general prediction of the biotransformation of foreign compounds should be possible when partition coefficient δ and chemical structure are known.

The microsomal degradation of *N*-colchiceyl derivatives, described in this paper, may stimulate further investigations of the enzyme system containing cytochrome P-450. The wide range of amino compounds which are available for syntheses permits the selection of suitable derivatives for comparison of the role of steric and electronic conditions in microsomal oxidative attack. Aromatic hydroxylation could also be included in such comparative studies by making derivatives of colchicine with anilines or phenylenediamines, respectively.

It has become increasingly apparent that the substrate specificity for these metabolic pathways resides in the cytochrome P-450 or P-448 fraction of a monoxygenase enzyme system[26,27], and that there are multiple forms of these heme proteins, depending on species differences[28] and on the pretreatment with different inducers[29]. This enzyme system is of particular interest to environmental research because of its ability to hydroxylate (or otherwise modify) not only steroids and fatty acids, but also drugs, insecticides, carcinogens, and a variety of other foreign compounds[30,31]. The class of *N*-colchiceyl derivatives, with its ease of preparation and multiplicity of possible structures, is apparently well suited to contribute to a deeper understanding of a metabolic pathway whose importance has been stressed in recent times because of its role in environmental biology.

This work was financially supported by *Fonds der Chemie* and *Universitätsbund Hohenheim.*

Literature

1 Wildman, W.C. (1960) in The Alkaloids (Manske, R.H.F., ed.), vol. VI, pp. 247 - 288, Academic Press, New York/London.

2 May & Baker Ltd. (1946) (Ewins, A.J., Ashley, J.N. & Harris, J.O., inventors) Brit. Patent 577606; C.A. **41**: 1716.

3 Shchukina, M.N., Borodina, G.M. & Scheĭnker, Y.N. (1951) *J. Gen. Chem. USSR* **21**, 735 - 739.

4 Fourneau, J.P. (1955) *Bull. Soc. Chim. Fr.* 1569 - 1571.

5 Biniecki, S. & Kolodynska, Z. (1964) *Acta Pol. Pharm.* **21**, 417 - 420.

6 Kiselev, V.V. (1961) *J. Gen. Chem. USSR* **31**, *305 - 306.*

7 Kiselev, V.V. (1964) *J. Gen. Chem. USSR* **34**, 620 - 622.

8 Kiselev, V.V. & Serova, N.A. (1966) *J. Gen. Chem. USSR* **36**, 36 - 39.

9 Kiselev, V.V. & Men'shikov, G.P. (1970) *J. Gen. Chem. USSR* **40**, 892 - 893.

10 Kiselev, V.V. (1971) *J. Gen. Chem. USSR* **41**, 456 - 458.

11 Zeisel, S. (1888) *Monatsh. Chem.* **9**, 1 - 30.

12 E. Merck AG (1953) (Lettré, H. & Küssner, W., inventors) Ger. Patent 867093; C.A. **50**: 4241.

13 Huber, G. (1951) *Angew. Chemie* **63**, 401 - 508.

14 Uffer, A (1952) *Helv. Chim. Acta* **35**, 2135 - 2139.

15 Rapoport, H., Williams, A.R., Campion, J.E. & Pack, D.E. (1954) *J. Am. Chem. Soc.* **76**, 3693 - 3698.

16 Schönharting, M., Pfaender, P., Rieker, A. & Siebert, G. (1973) *Hoppe-Seyler's Z. Physiol. Chem.* **354**, 421 - 436.

17 Horowitz, R.M. & Ullyot, G.E. (1952) *J. Am. Chem. Soc.* **74**, 587 - 592.

18 Nozoe, T., Ikemi, T. & Ito, S. (1954) *Proc. Jpn Acad.* **30**, 609 - 613.

19 Hadorn, H. & Obrist, Ch. (1973) *Dtsch. Lebensm. Rundsch.* **69**, 109 - 115.

20 Schönharting, M., Mende, G. & Siebert, G. (1974) *Hoppe-Seyler's Z. Physiol. Chem.* **355**, 1391 - 1399.

21 Keberle, H., Riess, W., Schmid, K. & Hoffmann, K. (1963) *Arch. Int. Pharmacodyn. Thér.* **142**, 125 - 140.

22 Schönharting, M., Mende, G. & Siebert, G. (1973) *Hoppe-Seyler's Z. Physiol. Chem.* **354**, 1242 - 1243.

23 Frommer, U., Ullrich, V. & Staudinger, H.J. (1970) *Hoppe-Seyler's Z. Physiol. Chem.* **351**, 903 - 917.

24 Gorrod, J.W., Disley, L.G. & Temple, D.J. (1971) *Xenobiotica* **1**, 521 - 522.

25 Gaudette, L.E. & Brodie, B.B. (1959) *Biochem. Pharmacol.* **2**, 89 - 96.
26 Lu, A.Y.H., Levin, W., West, S.B., Jacobson, M., Ryan, D., Kuntzman, R. & Conney, A.H. (1973) *J. Biol. Chem.* **248**, 456 - 460.
27 Van der Hoeven, T.A. & Coon, M.J. (1974) *J. Biol. Chem.* **249**, 6302 - 6310.
28 Williams, R.T. (1974) *Biochem. Soc. Trans.* **2**, 359 - 377.
29 Cooper, D.Y., Schleyer, H., Thomas, J.H., Vars, H.M. & Rosenthal, O. (1975) *Adv. Exp. Med. Biol.* **58**, 81 - 102.
30 Conney, A.H. (1967) *Pharmacol. Rev.* **19**, 317 - 366.
31 Schuster, I. (1977) in Progress in Molecular and Subcellular Biology (Hahn, F.E., Kersten, H., Kersten, W. & Szybalski, W., eds.) vol. 5, pp. 31 - 71, Springer Verlag Berlin/Heidelberg/New York.

Prof. Dr. G. Siebert, Institut für Biologische Chemie, Universität Hohenheim 04100, Garbenstr. 30, D-7000, Stuttgart 70.

Hoppe-Seyler's Z. Physiol. Chem.
Bd. 359, S. 733 - 740, Juni 1978

Transfer of C_2-Units Across the Mitochondrial Membrane Direct Recording of Citrate, Acetate and Acetoacetate Production Rates

Georg E. HOFFMANN and Ludwig WEISS

Forschergruppe Diabetes in München und Institut für Klin. Chemie, Städt. Krankenhaus Harlaching, München.

(Received 27 December 1977)

Summary: Methods are described for the direct optical assay of citrate, acetate, and acetoacetate production by isolated, incubated rat liver mitochondria. Each metabolite is converted into acetyl-CoA, using ATP:citrate lyase or acetyl-CoA synthetase or acetoacetyl-CoA synthetase and acetyl-CoA acetyltransferase, respectively. Arylamine acetyltransferase acts as auxiliary enzyme. It was shown that isolated rat liver mitochondria produce citrate, acetate and acetoacetate, and that production rates are stimulated by pyruvate and hexanoate. It was concluded that these three products might contribute to the transport of acetyl units across the mitochondrial membrane and thus serve as precursors in fatty acid synthesis. The rate of acetyl transfer does not seem to be rate-limiting with regard to the overall-process of fatty acid synthesis from carbohydrates.

Transport von C_2-Einheiten durch die Mitochondrienmembran. Direkte Registrierung der Transportgeschwindigkeiten von Citrat, Acetat und Acetacetat

Zusammenfassung: Es werden Methoden zur direkten optischen Bestimmung der Citrat-, Acetat- und Acetacetatproduktion durch isolierte, inkubierte Rattenlebermitochondrien angegeben. Die drei Metabolite werden durch ATP:Citrat-Lyase bzw. Acetyl-CoA-Synthetase bzw. Acetacetyl-CoA-Synthetase und Acetyl-CoA-Acetyltransferase zu Acetyl-CoA umgewandelt. Durch Arylamin Acetyltransferase wird der Acetylrest anschließend auf *p*-Nitroanilin übertragen, was zu einer Abnahme der Extinktion bei 405 nm Wellenlänge führt. Pyruvat und Hexanoat, zwei Substrate der mitochondrialen Acetyl-CoA-Bildung, stimulieren die Produktionsgeschwindigkeiten aller drei Metabolite. Es wird deshalb geschlossen, daß Citrat, Acetat und Acetacetat den Transport

---

Enzymes:
Acetoacetyl-CoA synthetase, acetoacetate:CoA ligase (not yet registered);
Acetyl-CoA-acetyltransferase, acetyl-CoA:acetyl-CoA *C*-acetyltransferase (= thiolase) (EC 2.3.1.9);
Acetyl-CoA synthetase, acetate:CoA ligase (AMP-forming) (EC 6.2.1.1);
Arylamine acetyltransferase, acetyl:CoA:arylamine *N*-acetyltransferase, (EC 2.3.1.5);
ATP:citrate (*pro-3S*)-lyase, ATP:citrate oxaloacetate-lyase (*pro-3S*-$CH_2CO_2^{\ominus}$→acetyl-CoA; ATP-dephosphorylating) (= citrate cleavage enzyme) (EC 4.1.3.8);
Glutamate dehydrogenase (NAD(P)$^{\oplus}$), L-glutamate:NAD(P)$^{\oplus}$ oxidoreductase (deaminating) (EC 1.4.1.3);
3-Hydroxybutyrate dehydrogenase, D-3-hydroxybutyrate:NAD$^{\oplus}$ oxidoreductase (EC 1.1.1.30).

Abbreviation:
CCP = carbonyl cyanide *p*-(trifluoromethoxy)phenylhydrazone = [4-(trifluoromethoxy)phenylhydrazono]propanedinitrile.

von C_2-Einheiten durch die Mitochondrienmembran vermitteln und so als Vorstufen der Fettsäurebiosynthese dienen. Die Transportgeschwindigkeiten scheinen für den Gesamtprozeß der Fettsäurebiosynthese aus Kohlenhydraten nicht geschwindigkeitsbestimmend zu sein.

Key words: Transport of acetyl units, citrate, acetate, acetoacetate, rat liver mitochondria.

The transfer of acetyl units across the inner mitochondrial membrane plays an important role in the de novo synthesis of fatty acids. In rat liver and adipose tissue acetyl-CoA generated from carbohydrates within the mitochondria is transported to the cytosol predominantly as citrate[1,2]. This "citrate pathway" involves intramitochondrial citrate synthase, the mitochondrial tricarboxylate carrier, and cytosolic ATP:citrate lyase. Recent observations, however, indicate that metabolites other than citrate contribute to the transfer of acetyl units[3]. Acetate[4] and acetoacetate[5] have been discussed in this connection. In this paper it is shown that isolated rat liver mitochondria produce citrate as well as acetate and acetoacetate and that production rates are stimulated by pyruvate and hexanoate, two substrates from which acetyl-CoA is generated intramitochondrially. The citrate, acetate, and acetoacetate output is, however, not likely to be rate-limiting with respect to lipogenesis, when compared to the activities of key enzymes reported in the literature.

Materials and Methods

Acetyl-CoA synthetase, β-hydroxybutyrate dehydrogenase, ATP, ADP, CoASH, acetyl-CoA, NADH, and CCP were purchased from Boehringer Mannheim GmbH. Mercaptoethanol and dithioerythritol were from Serva, Heidelberg, bovine serum albumin (low in fatty acids) from Behring-Werke, Marburg, the detergent Lubrol from ICI GmbH, Frankfurt, and 1,2,3-benzenetricarboxylic acid from ICI pharmaceutical Inc. Plain View, N.Y. All other reagents were from Merck, Darmstadt. Published procedures were used for the preparation of arylamine acetyltransferase[6], ATP:citrate lyase from pigeon liver[7], Li-acetoacetate[8] as well as for the determination of acetoacetate[9], protein[7] and for measurements of the following enzyme activities: acetyl-CoA-acetyltransferase[10], ATP:citrate lyase[7], acetyl-CoA-synthetase[7], and glutamate dehydrogenase[11].

Male Sprague-Dawley rats (160 - 220 g) from W. Gassner, Sulzfeld were fed ad libitum normal laboratory chow (Herilan RM 204, obtained from H. Eggersmann, Rinteln, Germany). They were killed at 9 a.m. by neck blow and liver mitochondria were prepared immediately according to Portenhauser and Wieland[12] except that the media contained 0.5% albumin to stabilize the mitochondria as suggested by Löffler et al.[13].

The lysosome-poor mitochondrial pellet, obtained by differential centrifugation between 1000 and 4000 × *g* was used for the experiments. One rat was used for each experiment.

Acetate production was followed optically using a double-beam spectrophotometer (Perkin-Elmer model 156). Mitochondria were incubated at 25 °C in an isotonic buffered medium (40mM Tris pH 7.2, 40mM KCl, 10mM KH_2PO_4, 10mM mercaptoethanol, 4mM $MgCl_2$, 4mM ADP, 0.04% albumin, 0.1mM EDTA) containing a trapping system for acetate (acetyl-CoA synthetase, 0.05mM CoASH, 4mM ATP), an optical indicator system (arylamine acetyltransferase, 0.15mM *p*-nitroaniline), and substrates as indicated in the experimental section. 20 mU of each of the enzymes and 10 - 50 μg of mitochondrial protein were present in a total volume of 1 m*l*. In the reference cuvette acetyl-CoA synthetase was omitted, so that acetyl-CoA synthetase-independent reactions (see below) were subtracted.

Reaction scheme:

$$\text{Acetate} \xrightarrow[\text{ATP, CoASH}]{1} \text{acetyl-CoA} \xrightarrow[p\text{-nitroaniline}]{2} p\text{-nitroacetanilide.}$$

1 = acetyl-CoA synthetase; 2 = arylamine acetyltransferase.

The acetylation of *p*-nitroaniline produces a decrease of the optical density at 405 nm, the molar extinction coefficient being 10×10^6 cm^2/mol[7]. The signal of the photometer was amplified up to 100 times.

It was shown that the test system reacted with acetate and propionate but not with hexanoate, pyruvate, citrate, acetylcarnitine, and acetylphosphate.

Traces of acetate present in the incubation medium were allowed to react completely before the reaction was started by addition of the mitochondrial suspension. Only neglegible amounts of acetate were measurable in the supernatant of the mitochondrial suspensions after storage in an icebath for up to 4 h.

Citrate production was measured in a similar way, where ATP:citrate lyase was used instead of acetyl-CoA synthetase. Malate (1mM) was added as counter anion for the citrate transport and as a source of mitochondrial oxaloacetate.

A somewhat different method was used for the measurement of acetoacetate production. The one-cuvette-dual-wavelength procedure was applied (405/480 nm wavelength), and 60 mU of arylamine acetyltransferase per m*l* incubation medium was used. Other auxiliary enzymes were acetoacetyl-CoA synthetase and acetyl-CoA acetyltransferase, both present at sufficient amount in the arylamine acetyltransferase preparation used (5 and 24%, respectively, when referred to arylamine acetyltransferase activity).

Each single experiment of acetate, citrate, or acetoacetate production was performed 2 - 4 times with different amounts of mitochondria, to ascertain linearity of the test system.

For the measurement of glutamate-dehydrogenase activity mitochondria were broken by freezing and thawing in 15mM Tris buffer, pH 7 with addition of 0.2% Lubrol. Production of metabolites was calculated as nmol C_2-units × min^{-1} × U^{-1} glutamate dehydrogenase. Glutamate dehydrogenase was chosen as a reference value, because of the presence of albumin in the incubation medium. As a basis for comparison with the values reported by other authors, specific activities of glutamate dehydrogenase of 5.7 ± 0.7 U/mg mitochondrial protein (n = 5) or 1.8 ± 0.1 U/mg total rat liver protein (n = 5) were used.

The oxygen consumption of incubated mitochondria was measured polarographically (Clark electrode) in the complete incubation medium described already for acetate determination, but with the omission of ADP. By repeated administration of small amounts of ADP it was proven that the mitochondria remained in a coupled state during the incubation.

Results

Acetate production rates of isolated rat liver mitochondria

Exogenous acetate added to the test medium was recovered quantitatively. Acetate production was directly proportional to the amount of mitochondrial protein, up to about 70 μg of mitochondrial protein/m*l*. The rate of acetate production was linear during the first 5 - 10 min (Fig. 1) and declined thereafter gradually. In some cases lag phases of up to 30 min occurred.

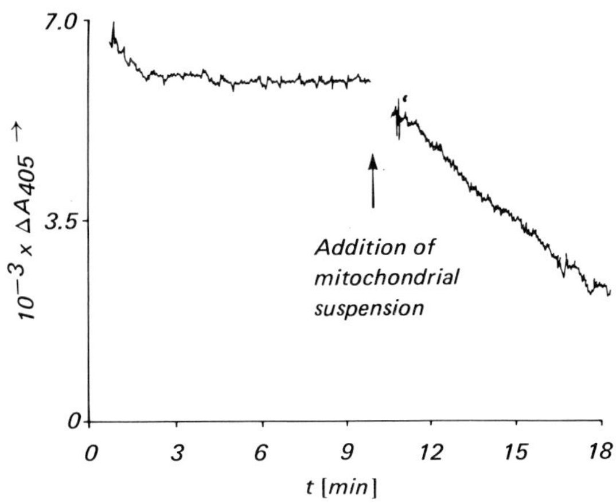

Fig. 1. Time course of acetate production by isolated rat liver mitochondria.

Reproduction of an original registration curve. For experimental details see methods. Ordinate: absorption difference at 405 nm.

Table 1. Maximal rates of acetate production by isolated rat liver mitochondria in the presence and absence of hexanoate or pyruvate, respectively.

For experimental details see text. p-Values are calculated according to student's test for paired analyses. Glu deh. = glutamate dehydrogenase.

	Experiment 1 (n = 7)		Experiment 2 (n = 7)	
	Control	Hexanoate [2mM]	Control	Pyruvate [2mM]
Mean [nmol × min^{-1} × $U_{Glu\ deh.}^{-1}$]	0.37	0.54	0.28	0.47
SEM	0.03	0.05	0.03	0.05
p	–	< 0.01	–	< 0.05

Values listed in Table 1 are calculated from the linear parts of the curves (3rd to 4th minute).

In two sets of experiments pyruvate or hexanoate were added as substrates (2mM each). Controls from the same mitochondrial preparation but without substrate were run at each time (Tab. 1).

Acetate production in the presence of ADP, omitting ATP, decreased by not more than 30% indicating that mitochondria produced sufficient ATP to serve as substrate for the acetyl-CoA synthetase reaction. Addition of CCP (0.005mM) abolished this reaction completely while with uncoupler and ATP added, 40% of the production rate remained measurable.

Citrate production rates of isolated incubated rat liver mitochondria

The rate of citrate production increased gradually within the first 5 - 20 min of incubation, then became linear for at least 15 min. Maximal production rates were proportional to the amount of mitochondrial suspension added, as depicted in Fig. 2. Rates calculated from the linear parts of the registration curves are shown in Table 2. Measurements were performed in the presence of hexanoate (2mM), or pyruvate (2mM), or pyruvate (2mM) plus hydrogencarbonate (10mM). Controls were without substrate.

Malate was present as a counter-anion in all experiments. No citrate production was detectable in the absence of a counter-anion or in the presence of an inhibitor of the tricarboxylate carrier (15mM 1,2,3-benzenetricarboxylate). Hydrogencarbonate alone was without effect on the basal citrate production.

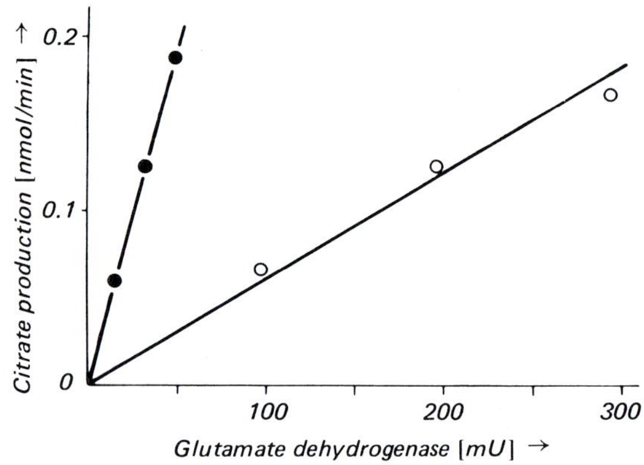

Fig. 2. Citrate production by isolated, incubated rat liver mitochondria as a function of the concentration of mitochondrial protein.

5.7 U glutamate dehydrogenase are equivalent to 1 mg mitochondrial protein. ○——○: no substrate added. ●——●: pyruvate and hydrogencarbonate present as substrates. Further experimentel details are given in the methods section.

Acetoacetate production rates of isolated incubated rat liver mitochondria

The production of acetoacetate, like that of citrate, showed a "lag-phase" of 10 - 20 min. In Table 3 production rates of acetyl units in the form of acetoacetate are given. Rates were calculated from the linear parts of the registration curves between the 20th and 30th minute.

To evaluate the arylamine acetyltransferase method the following experiments were performed. It was shown that the arylamine acetyltransferase preparation used in the experiments was able to

Table 2. Maximal rates of citrate production by isolated rat liver mitochondria in the presence and absence of hexanoate, pyruvate, and pyruvate plus hydrogencarbonate.

For experimental details see text. p-values refer to controls. Glu deh. = glutamate dehydrogenase. n = 9.

	Control	Hexanoate [2mM]	Pyruvate [2mM]	Pyruvate + hydrogen-carbonate [2mM + 10mM]
Mean [nmol × min^{-1} × $U_{Glu\ deh.}{}^{-1}$]	0.94	1.25	1.43	1.96
SEM	0.18	0.22	0.19	0.27
p	–	< 0.10	< 0.01	< 0.001

Hoppe-Seyler's Z. Physiol. Chem.
Bd. 359, S. 741 - 749, Juni 1978

Snake Venoms

The Amino-Acid Sequence of Protein S_5C_4 from *Dendroaspis jamesoni kaimosae* (Jameson's mamba) Venom

Francois J. JOUBERT, Andries J.C. STRYDOM and Nico TALJAARD

National Chemical Research Laboratory, Pretoria, South Africa

(Received 14 March 1978)

Summary: A major component (S_5C_4) was purified from Jameson's mamba by gel filtration on Sephadex G-50 and by ion-exchange chromatography on CM-cellulose. Protein S_5C_4 contains 60 amino acid residues and is cross-linked by four intrachain disulphide bridges. The complete primary structure of the protein has been elucidated. The toxicities, the immunochemical properties, the sequence and the invariant amino acid residues of protein S_5C_4 resemble subgroup II of the *angusticeps*-type proteins.

Schlangengift: Die Aminosäuresequenz von Protein S_5C_4 aus dem Gift von Dendroaspis jamesoni kaimosae (Jameson-Mamba)

Zusammenfassung: Protein S_5C_4 enthält 60 Aminosäuren und vier Disulfidbrücken. Die vollständige Aminosäuresequenz von S_5C_4 wurde aufgeklärt. Die Toxizitäten, die serologischen Eigenschaften, die Sequenz und die invarianten Aminosäurereste des Proteins S_5C_4 zeigen einen hohen Grad von Homologie mit der Gruppe II der Proteine vom *Angusticeps*-Typ.

Key words: Snake venom, protein S_5C_4, properties, amino acid sequence.

A number of papers have appeared in the last decade on the toxic and less-toxic components of mamba venoms; viz. several of the species of the genus *Dendroaspis*. The venom from *Dendroaspis angusticeps* (green mamba) does not appear to contain either long or short neurotoxins[1]. However, from this venom Viljoen and Botes[2,3] purified and sequenced major polypeptides F_7 and F_8 (the so-called *angusticeps*-type proteins[4]) of low toxicity. They are structurally homologous to the short neurotoxin and the cytotoxin groups but of completely distinct immunochemical properties. The venom of *Dendroaspis polylepis polylepis* (black mamba) contains, apart from the neurotoxins, a numer of basic polypeptides of low toxicity[4]. From the latter, Strydom[5] sequenced a major polypeptide FS_2. The structure of FS_2 is related to the short neurotoxin group and also to the *angusticeps*-type proteins. In another paper Joubert and Taljaard (unpublished results) elucidated the primary structure of toxin C, also from black mamba venom. The structure of toxin C shows a very high degree of homology to the *angusticeps*-type protein F_7. Besides the

Enzymes: Trypsin EC (3.4.21.4); Chymotrypsin (EC 3.4.21.1).

neurotoxins, *Dendroaspis viridis* venom contains several low-toxicity polypeptides[6]. Among the latter, Shipolini and Banks[7] purified and sequenced polypeptide 4.9.6. This molecule resembles the *angusticeps*-type protein F_8.

Concerning the components of *Dendroaspis jamesoni kaimosae* venom, Strydom[8] purified and established the sequence of a short neurotoxin and a long neurotoxin. The present communication describes the purification and the complete amino acid sequence of protein S_5C_4, another *angusticeps*-type protein, also from this venom.

Experimental Procedure

Desiccated *Dendroaspis jamesoni kaimosae* venom was supplied by D. Muller, Professional Snake Catcher (Pty) Ltd., 215 Brakston Drive, Blairgowrie, Johannesburg, 2001. The source of the trypsin, α-chymotrypsin and chemical reagents have been described previously[9,10]. The physicochemical methods, the toxicity determinations by intravenous injections; the immunochemical examination by Ouchterlony technique, the reduction and *S*-carboxymethylation; the digestion with trypsin and chymotrypsin; the fractionation of enzyme digests by chromatography on DEAE-cellulose and paper chromatography and/or electrophoresis; the amino acid analyses of the protein and peptides; the sequence determination of reduced and *S*-carboxymethylated protein and peptides by Edman degradation with the Beckman sequenator or manually, and the nomenclature of the peptides have also been detailed in previous communications[9,10].

Results

Purification and some of the properties of protein S_5C_4

Fig. 1 shows the elution profile obtained for the separation of Jameson's mamba venom on Sephadex G-50. Three major and three minor peaks were evident. Peak S_5 was further fractionated on CM-cellulose using a linear ammonium acetate solution gradient of pH 6.3 and the elution pattern is depicted in Fig. 2. This revealed seven or six major and a number of minor peaks. Peak S_5C_4 was further purified by rechromatography on a CM-cellulose column at pH 5.0 as above and the chromatogram showed a major and several minor peaks. The major peak afforded protein S_5C_4.

Protein S_5C_4 appeared to be homogeneous by chromatography, amino acid analysis, *N*-terminal end group determination and immunodiffusion. The amino acid composition of S_5C_4 is given in Table 1. Examination of protein S_5C_4 with Ellman's reagent, both in the presence and absence of 6M guaninium chloride, revealed that the protein was devoid of free sulphydryl groups.

Amino acid sequence of reduced and S-carboxymethylated protein S_5C_4

The tryptic and chymotryptic digests of reduced and *S*-carboxymethylated protein S_5C_4 were fractionated on columns of DEAE-cellulose as depicted in Fig. 3 and 4. Peptides were further purified, where necessary, by methods indicated

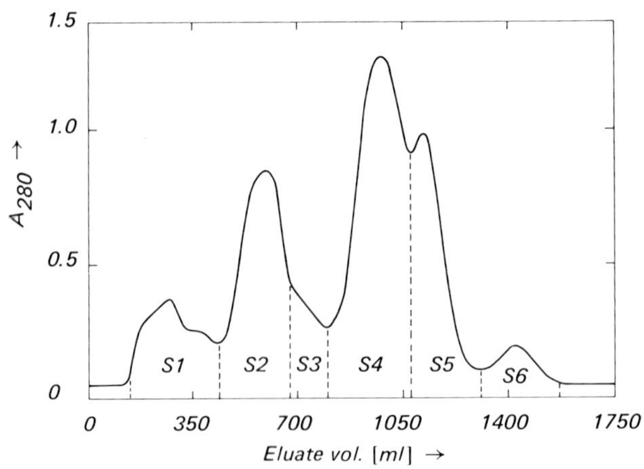

Fig. 1. Gel filtration of *Dendroaspis jamesoni kaimosae* venom.

Crude venom (5 g) was loaded on Sephadex G-50 column (3.8 × 150 cm) and elution effected with 0.2M ammonium acetate solution (pH 6.3) at a flow rate of 60 ml/h. The column temperature was 20 °C and the eluate was monitored at 280 nm.

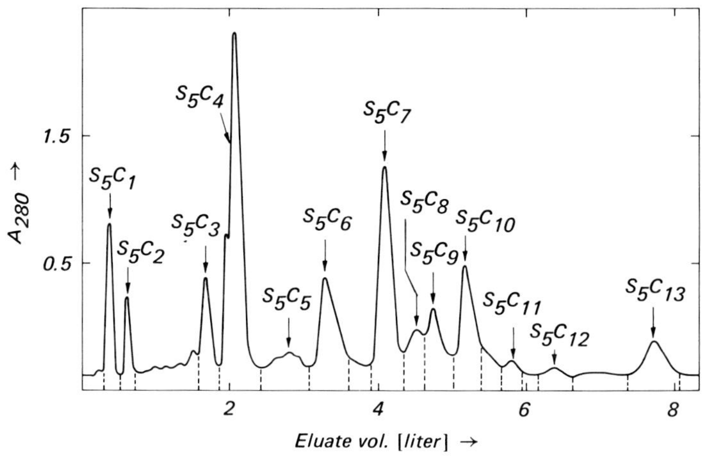

Fig. 2. Chromatography of fraction S_5 on CM-cellulose.

Fraction S_5 (4.0 g) was loaded on the column (3.8 × 50 cm) and elution effected by a linear gradient of 0.05 - 0.65M ammonium acetate solutions (pH 6.3) over 12 *l* at a flow rate of 200 m*l*/h. The column temperature was 5 °C and the eluate was monitored at 280 nm.

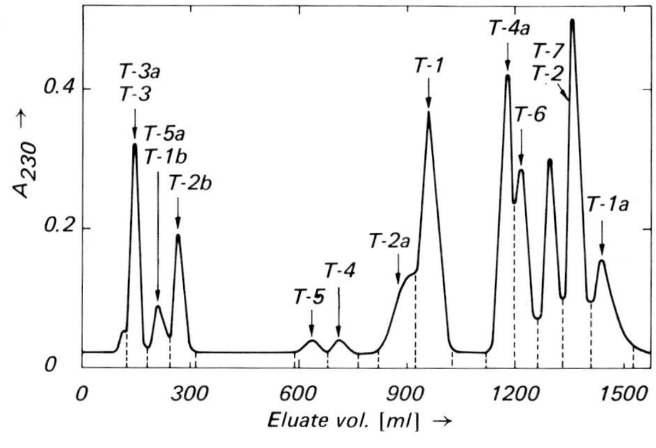

Fig. 3. Chromatography of the tryptic digest of reduced and *S*-carboxymethylated protein S_5C_4 on DEAE-cellulose.

The digest (10 µmol) was applied to the column (0.9 × 150 cm) and eluted with a linear gradient of 0.025 to 0.6M ammonium hydrogencarbonate solution over 2 *l* at a flow rate of 50 m*l*/h. The column temperature was 20 °C and the eluate was monitored at 230 nm.

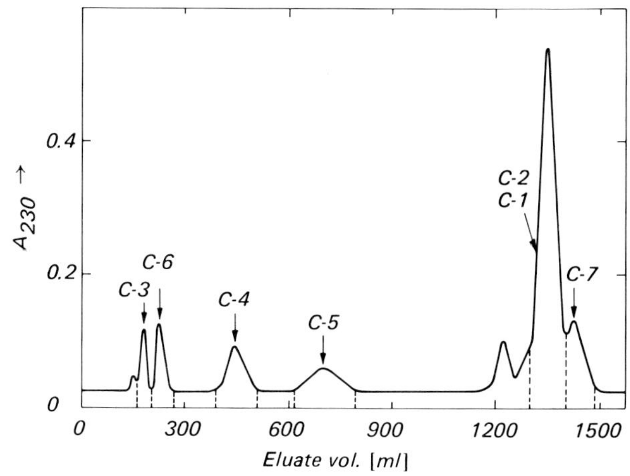

Fig. 4. Chromatography of the chymotryptic digest (10 µmol) of reduced and *S*-carboxymethylated protein S_5C_4 on DEAE-cellulose.

The experimental conditions were the same as in Fig. 3.

Table 1. Amino acid composition of reduced and *S*-carboxymethylated protein S_5C_4 and its tryptic peptides. Values are given as mol of residue per mol.

Amino acid	Protein S_5C_4	T-1	T-1a	T-1b	T-2	T-2a	T-2b
CH_2CO_2H Cys	7.6 (8)	1.0 (1)	0.6 (1)		1.7 (2)	0.8 (1)	0.9 (1)
Asp	4.2 (4)				1.3 (1)		
Thr	4.8 (5)				3.6 (4)	0.9 (1)	1.0 (1)
Ser	3.6 (4)	1.0 (1)		0.9 (1)	1.2 (1)		
Glu	5.0 (5)				3.1 (3)		
Pro	3.2 (3)				0.7 (1)		
Gly	3.1 (3)						
Ala	2.0 (2)				0.9 (1)		
Val	2.5 (3)						
Met	0.9 (1)	1.0 (1)	0.7 (1)				
Ile	4.4 (5)	0.9 (1)	1.0 (1)		1.0 (1)		
Leu	1.1 (1)						
Tyr	1.9 (2)	1.0 (1)	1.0 (1)		0.7 (1)	0.9 (1)	1.0 (1)
Phe	1.8 (2)						
Lys	7.0 (7)	0.9 (1)		1.0 (1)	1.8 (2)	1.1 (1)	2.1 (2)
His	1.9 (2)	1.0 (1)		1.0 (1)			
Arg	3.1 (3)						
Trp	0						
Total	60	7	4	3	17	4	5
Yield [%]		16.9	4.0	10.3	8.4	7.1	15.0
Purif.		BPAW[a]	BPAW[a]	BPAW[a]	pH 1.9[b]	pH 4.5[c]	

Amino acid	T-3	T-3a	T-4	T-4a	T-5	T-5a	T-6	T-7
CH_2CO_2H Cys			2.0 (2)	2.1 (2)			2.1 (2)	1.0 (1)
Asp	1.0 (1)	1.0 (1)						2.0 (2)
Thr	0.9 (1)	0.7 (1)						
Ser							1.0 (1)	1.0 (1)
Glu					1.0 (1)	1.0 (1)		1.1 (1)
Pro	0.9 (1)	1.0 (1)	1.0 (1)	1.0 (1)				
Gly	1.0 (1)	1.0 (1)	2.1 (2)	1.8 (2)				
Ala	1.0 (1)	1.0 (1)						
Val	2.4 (3)	2.8 (3)						
Met								
Ile	0.6 (1)	0.6 (1)			1.0 (1)	1.0 (1)	0.8 (1)	
Leu	0.9 (1)	1.0 (1)						
Tyr								
Phe	0.9 (1)	0.8 (1)			1.0 (1)	0.9 (1)		
Lys	1.1 (1)		2.0 (2)	1.0 (1)		1.0 (1)		1.0 (1)
His							0.8 (1)	
Arg	1.0 (1)	1.0 (1)			1.0 (1)	1.0 (1)	0.8 (1)	
Trp								
Total	13	12	7	6	4	5	6	6
Yield [%]	66.7	12.0	6.9	25.9	21.3	19.3	25.6	39.2
Purif.		BPAW[a]	BPAW[a]	BPAW[a]	BPAW[a]	BPAW[a]	BPAW[a]	pH 1.9[b]

[a] Paper chromatography with 1-butanol/pyridine/acetic acid/water. [b] Paper electrophoresis at pH 1.9.
[c] Paper electrophoresis at pH 4.5.

Table 2. Amino acid composition of the chymotryptic peptides of reduced and *S*-carboxymethylated protein S_5C_4. Values are given as mol of residue per mol.

Amino acid	C-1	C-2	C-3	C-4	C-5	C-6	C-7
CH_2CO_2H-Cys	0.9 (1)	2.3 (2)			2.0 (2)		3.2 (4)
Asp		1.0 (1)		1.0 (1)			1.9 (2)
Thr		4.0 (4)		1.0 (1)			
Ser		1.6 (2)				0.8 (1)	1.0 (1)
Glu		3.1 (3)			0.9 (1)		1.0 (1)
Pro		0.9 (1)		1.1 (1)	0.9 (1)		
Gly				1.1 (1)	1.9 (2)		
Ala		1.0 (1)			0.9 (1)		
Val				2.6 (3)			
Met	0.8 (1)						
Ile	1.1 (1)	1.0 (1)		0.6 (1)	0.9 (1)	0.9 (1)	
Leu				1.0 (1)			
Tyr	1.0 (1)	0.9 (1)					
Phe			1.0 (1)		0.9 (1)		
Lys		2.0 (2)	2.0 (2)		1.8 (2)		0.8 (1)
His		1.0 (1)				1.0 (1)	
Arg					0.9 (1)	1.0 (1)	0.8 (1)
Trp							
Total	4	19	3	9	12	4	9
Yield [%]	23.5	38.4	49.6	54.0	25.4	37.2	27.0
Purif.	BPAW[a]	BPAW[a]	BPAW[a]		pH 1.9[b]	BPAW[a]	BPAW[a]

[a] Paper chromatography with 1-butanol/pyridine/acetic acid/water.
[b] Paper electrophoresis at pH 1.9.

in Tables 1 and 2, which also list the amino acid composition of the peptides. The aminoterminal sequence of reduced and *S*-carboxymethylated protein S_5C_4, determined by using the Beckman sequenator, is given in Table 3. Additional sequence studies on some of the tryptic peptides are summarised in Table 4. Consequently, the sequence of S_5C_4 was completely elucidated.

Fig. 5 reveals the complete primary structure of reduced and *S*-carboxymethylated protein S_5C_4. The known amino-terminal sequence of S_5C_4 established directly the alignment of peptides T-1 and T-2 and also C-1 and C-2. Tryptic peptide T-7 contained no *C*-terminal lysine or arginine, and must be derived from the carboxyterminal peptide of protein S_5C_4. The known amino acid composition of chymotryptic peptides C-3, C-5 and C-6 provided, respectively, the necessary overlaps between peptides T-2 and T-3, T-3 and T-4 and T-5, and T-5 and T-6.

Discussion

A major component, viz. protein S_5C_4, was purified from Jameson's mamba venom utilizing gel filtration and ion-exchange chromatography. Concerning the properties of S_5C_4, the protein contains 60 amino acid residues and comprises 8 half-cystine residues. Since no free sulphydryl groups could be demonstrated, it is cross-linked by four intra-molecular disulphide bridges. Protein S_5C_4 is particularly rich in lysine (7 residues), threonine (5 residues), glutamic acid and its corresponding amide (5 residues) and also isoleucine (5 residues). In contrast the protein contains 2 residues each of alanine, tyrosine, phenylalanine

Table 3. *N*-terminal sequence of reduced and *S*-carboxymethylated protein S_5C_4.
Determined with the automatic Beckman sequenator; 850 nmol was loaded. The Quadrol programme (Beckman no. 122974 Mod.) was used. The numbers signify yields of phenylthiohydantoin-amino acids in nmol.

Step	Normal	Silylated	TLC[a]	Identification
1	Met 336	Met 560	Met	Met
2	Leu/Ile 144	Ile 184	Ile	Ile
3	Cys/Ser 104	Cys 64	Cys	Cys[b]
4	Tyr 176	Tyr 416	Tyr	Tyr
5	Cys/Ser 26	Ser 22	Ser	Ser
6	–	–	–	His[c]
7	Lys 24	Lys 40	Lys	Lys
8	Thr 36	Thr 28	–	Thr
9	Pro 142	Pro 204	Pro	Pro
10	Gln 36	Gln 24	Gln	Gln
11	–	Asn 28	–	Asn
12	Cys/Ser 36	Ser 24	–	Ser
13	Ala 180	Ala 304	Ala	Ala
14	Thr 18	Thr 28	–	Thr
15	Leu/Ile 130	Ile 180	Ile	Ile
16	Thr 64	Thr 54	–	Thr
17	Cys/Ser 42	Cys 34	–	Cys[b]
18	–	Glu 110	Glu	Glu
19	–	Glu 84	Glu	Glu
20	Lys 16	Lys 24	Lys	Lys

[a] Thin-layer chromatography. [b] Identified as *S*-carboxymethylcysteine.
[c] Identified by colour reaction with Pauly reagent.

Table 4. Amino acid sequences of tryptic peptides of reduced and *S*-carboxymethylated protein S_5C_4.
The sequence of peptide T-1 (residues 1 - 7) and the partial sequence of T-2 (residues 8 - 24) follow from the sequence of the amino-terminal segment of protein S_5C_4 (Table 3). The lower half-arrows indicate the residues identified by the manual Edman procedure.

Peptide	Residues	Sequences
T-2b	21 - 25	Thr-Cys-Tyr-Lys-Lys
T-3a	26 - 37	Phe-Val-Thr-Asn-Val-Pro-Gly-Val-Ile-Leu Ala-Arg
T-4a	38 - 43	Gly-Cys-Gly-Cys-Pro-Lys
T-5a	44 - 48	Lys-Glu-Ile-Phe-Arg
T-6	49 - 54	Ser-Ile-His-Cys-Cys-Arg
T-7	55 - 60	Ser-Asp-Lys-Cys-Asn-Glu

```
                                    10                                      20
H2N-Met-Ile-Cys-Tyr-Ser-His-Lys-Thr-Pro-Gln-Asn-Ser-Ala-Thr-Ile-Thr-Cys-Glu-Glu-Lys-
<---------T-1---------> <-------------------------T-2-----------------------------
<---T-1a---> <---T-1b--->
<---C-1---> <--------------------------C-2----------------------------------------
<-----------------------------Sequenator----------------------------------------->

                                    30                                      40
Thr-Cys-Tyr-Lys-Lys-Phe-Val-Thr-Asn-Val-Pro-Gly-Val-Ile-Leu-Ala-Arg-Gly-Cys-Gly-
-----T-2-----> <---------------------T-3-------------------------> <----T-4------
<----T-2a----->  <--------------------T-3a------------------------> <----T-4a-----
<------T-2b------->
---C-2--> <--C-3--> <------------C-4--------------> <--------C-5-----------------

                                    50                                      60
Cys-Pro-Lys-Lys-Glu-Ile-Phe-Arg-Ser-Ile-His-Cys-Cys-Arg-Ser-Asp-Lys-Cys-Asn-Glu-OH
------T-4-----> <-----T-5-----> <--------T-6--------> <----------T-7------------->
---T-4a--> <------T-5a------>
                    •
----------C-5-----------> <----C-6----> <--------------C-7----------------------->
```

Fig. 5. The complete amino acid sequence of reduced and *S*-carboxymethylated protein S_5C_4 from *Dendroaspis jamesoni kaimosae.*

and histidine, single residues of methionine and leucine, and is devoid of tryptophan. The amino acid composition of protein S_5C_4 from Jameson's mamba venom and polypeptide 4.9.6 from *Dendroaspis viridis*[7] differ only in their contents of aspartic acid, valine, leucine and lysine.

The complete primary structure of protein S_5C_4 was readily elucidated by using trypsin and chymotrypsin digestions. Apart from the cleavage of the Tyr^4 - Ser^5 bond by trypsin, the proteolytic enzymes in general exhibited their anticipated specificities.

Concerning the toxicities of the *angusticeps*-type proteins. Shipolini and Banks[7] obtained a LD_{50} value > 3.3 µg/g for polypeptide 4.9.6 from *Dendroaspis viridis* venom. For protein F_8 from *Dendroaspis angusticeps* venom, Viljoen and Botes[3] reported a subcutaneous LD_{100} value > 100 µg/g mouse. Recently, Joubert (unpublished results) determined a LD_{100} value > 250 µg/g mouse for protein F_8. The toxicity of protein S_5C_4 from Jameson's mamba venom was determined and an intravenous LD_{100} value of > 250 µg/g mouse was found. These values confirm the non-toxic nature of this group of *angusticeps*-type proteins.

The immunochemical properties of protein S_5C_4, investigated by the Ouchterlony technique, showed a single precipitin band for the protein. Further, protein S_5C_4 was compared to a long neurotoxin (α), a short neurotoxin (β), a cytotoxin ($V^{II}1$), from *Naja nivea*[11,12] and the *angusticeps*-type protein (F_7 and F_8) from *Dendroaspis angusticeps*[2,3]. Complete crossing of the precipitin bands was observed for the long neurotoxin, the short neurotoxin, the cytotoxin and the *angusticeps*-type protein F_7, but the *angusticeps*-type protein F_8 revealed coalescence with protein S_5C_4.

To date, the complete primary structure of five *angusticeps*-type proteins have been reported from several species of the genus *Dendroaspis*. In Fig. 6 the amino acid sequence of protein S_5C_4 from *Dendroaspis jamesoni kaimosae* venom is compared to the known sequences of the *angusticeps*-type protein and also to those of the short neurotoxin and the cytotoxin groups. The similarity between the sequences is apparent. However, the higher degree of homology of protein S_5C_4 to polypeptide 4.9.6 is quite obvious. The sequence of protein S_5C_4 differs from that of polypeptide 4.9.6 in only two positions; lysine (position 31) and leucine (position 32) replace, respectively, asparagine and valine in protein S_5C_4.

Strydom[5] compared the known sequences of the *angusticeps*-type proteins from mamba venoms among themselves and also to those of mamba short neurotoxins and concluded, that (i) the only characteristic of the *angusticeps*-type proteins which binds them together as a group is the invariable seryl residue in position 5, (ii) the few other invariable amino acids are all shared with the mamba short neurotoxins, and (iii) based on chemical evidence, the *angusticeps*-type proteins should be divided in three subgroups.

On the basis of the degree of homology between the sequences of the proteins, the three subgroups

```
     Sub-
     groups         10                  20                  30                  40                  50                  60
(a)  I     T M C Y S H T T T S R A I L T N C G - - E N S C Y R K S R R H P P K M V L G R G C - - G C P P G D D N L E V K C C T S P D K C N Y
(b)  I     T I C Y S H T T T S R A I L K D C G - - E N S C Y R K S R R H P P K M V L G R G C - - G C P P G D D Y L E V K C C T S P D K C N Y
(c)  II    M I C Y S H K T P Q P S A T I T C E - - E K T C Y K K S V R K L P A V V A G R G C - - G C P S K E M L V A I H C C R S - D K C N E
(d)  II    M I C Y S H K T P Q N S A T I T C E - - E K T C Y K K F V T K L P G V I L A R G C - - G C P K K E I F R S I H C C R S - D K C N E
(e)  II    M I C Y S H K T P Q N S A T I T C E - - E K T C Y K K F V T N V P G V I L A R G C - - G C P K K E I F R S I H C C R S - D K C N E
(f)  III   R I C Y S H K A S L P R A T K T C V - - E N T C Y K M F I K T H R Q Y I S E R G C - - G C P T A M W P Y Q T E C C K G - D R C N K
(g)        X X C X N X X S X X X X T X X X C X X X X X X C Y X K X W X D X R G X X X E R G C - - G C P X V K X G X X X X C C X X - X X C N X
(h)        X X C X X - - X L X P X X X K T C P X G K N L C Y X M X X X X X X X X P V K R G C I X V C P K X S X X X K Y X C C X T - X X C N -
```

Fig. 6. Comparison of the amino acid sequences of various *angusticeps*-type proteins. (a) *Dendroaspis angusticeps*, F_7[2]; (b) *Dendroaspis polylepis polylepis*, C (Joubert and Taljaard, unpublished results); (c) *Dendroaspis angusticeps*, F_8[3]; (d) *Dendroaspis viridis*, 4.9.6[7]; (e) *Dendroaspis jamesoni kaimosae*, S_5C_4; (f) *Dendroaspis polylepis polylepis*, FS_2[5]; (g) the short neurotoxin group (Joubert[3]) and (h) the cytotoxin group (Joubert[10]). ▷

In the sequence of *Dendroaspis viridis* 4.9.6, a lysine residue is inserted at position 27 and a lysine deleted between position 52 and 53, for reasons outlined by Strydom[5]. The X shows the variant amino acids residues. The positions of the invariant amino acids are boxed. The IUPAC one-letter notation for amino acids is used [(1969) *Hoppe-Seyler's Z. Physiol. Chem.* **350**, 793 - 797].

are indicated in Fig. 6. The immunochemical properties of the subgroups are summarised in Table 5. It is evident that the three subgroups are immunochemically unrelated, but within a subgroup the immunochemical properties of the proteins are similar.

Table 5. Immunochemical properties of the subgroups of the *angusticeps*-type proteins.

Angusticeps-type proteins	Precipitin bands
D. angusticeps, F_7(I)[a] – *D. polylepis polylepis*, C(I)	Coalescence
D. angusticeps, F_8(II) – *D. jamesoni kaimosae*, S_5C_4(II)	Coalescence
D. angusticeps, F_7(I) – *D. angusticeps*, F_8(II)	Cross[3]
D. angusticeps, F_7(I) – *D. jamesoni kaimosae*, S_5C_4(II)	Cross
D. polylepis polylepis, C(I) – *D. jamesoni kaimosae*, S_5C_4(II)	Cross
D. polylepis polylepis, FS_2(III) – *D. angusticeps*, F_7(I)	Cross
D. polylepis polylepis, FS_2(III) – *D. angusticeps*, F_8(II)	Cross

[a] Subgroup of the *angusticeps*-type proteins.

In conclusion, the toxicity, the immunochemical properties, the primary structure and the invariant amino acid residues of protein S_5C_4 from *Dendroaspis jamesoni kaimosae* venom reveal a high degree of homology with the corresponding properties of peptide 4.9.6 from *Dendroaspis viridis* venom.

The authors are indebted to *Mrs Ruelle* for her assistance with the Beckman sequencer.

Literature

1 Strydom, D. J. & Botes, D. P. (1970) *Toxicon* **8**, 203 - 209.
2 Viljoen, C. C. & Botes, D. P. (1973) *J. Biol. Chem.* **248**, 4915 - 4919.
3 Viljoen, C. C. & Botes, D. P. (1974) *J. Biol. Chem.* **249**, 366 - 372.
4 Strydom, D. J. (1976) *Eur. J. Biochem.* **69**, 169 - 176.
5 Strydom, D. J. (1977) *Eur. J. Biochem.* **76**, 99 - 106.
6 Shipolini, R. A., Bailey, G. S., Edwardson, J. E. & Banks, B. E. C. (1973) *Eur. J. Biochem.* **40**, 337 - 344.
7 Shipolini, R. A. & Banks, B. E. C. (1974) *Eur. J. Biochem.* **49**, 399 - 405.
8 Strydom, A. J. C (1973) *Biochim. Biophys. Acta* **328**, 491 - 509.
9 Joubert, F. J. (1975) *Hoppe-Seyler's Z. Physiol. Chem.* **356**, 53 - 72.
10 Joubert, F. J. (1976) *Eur. J. Biochem.* **64**, 219 - 232.
11 Botes, D. P., Strydom, D. J., Anderson, C. G. & Christensen, P. A. (1971) *J. Biol. Chem.* **246**, 3132 - 3139.
12 Botes, D. P. & Viljoen, C. C. (1976) *Biochim. Biophys. Acta* **446**, 1 - 9.

Dr F. J. Joubert, c/o The Director, National Chemical Research Laboratory, P. O. Box 395, Pretoria 0001, Republic of South Africa.

Hoppe-Seyler's Z. Physiol. Chem.
Bd. 359, S. 751 - 757, Juni 1978

Determination of the Superoxide Dismutating Activity of D-Penicillamine Copper

Edmund LENGFELDER[a] and Erich F. ELSTNER[b]

[a] Strahlenbiologisches Institut der Universität München und
[b] Institut für Botanik und Mikrobiologie der Technischen Universität, München.

(Received 31 March 1978)

Summary: The superoxide dismutating activity of the D-penicillamine copper complex was determined and compared with the activities of Cu-Zn and Mn superoxide dismutase in four $O_2^{\ominus}$ generating systems.

I. Nitrite formation from hydroxylamine.
II. Crocin destruction by xanthine/xanthine oxidase.
III. Ethylene production by isolated chloroplasts.
IV. Nitrite formation from hydroxylamine by chloroplasts in the presence of diquat (1,1′-dimethylene-2,2′-bipyridylium dibromide).

In all four test systems a high dismutative activity of the complex was found, which is not sensitive to KCN as demonstrated with test system III. The results are discussed with regard to the antiinflammatory activity of D-penicillamine.

Bestimmung der Superoxid-dismutierenden Aktivität von Penicillamin-Kupfer

Zusammenfassung: Die Superoxid dismutierende Aktivität des D-Penicillamin-Kupferkomplexes wurde bestimmt und verglichen mit den Aktivitäten von Cu-Zn- und Mn-Superoxid-Dismutase in vier $O_2^{\ominus}$ generierenden Testsystemen:

I. Bildung von Nitrit aus Hydroxylamin;
II. Abbau von Crocin durch Xanthin/Xanthin-Oxidase;
III. Bildung von Äthylen durch isolierte Chloroplasten;
IV. Bildung von Nitrit aus Hydroxylamin in Anwesenheit von Diquat (1,1′-Dimethylen-2,-2′-bipyridylium-dibromid).

In allen vier Testsystemen wurde eine hohe dismutative Aktivität des Komplexes gefunden, der, wie mit dem Testsystem III gezeigt wurde, gegen KCN unempfindlich ist. Die Ergebnisse werden im Hinblick auf die entzündungshemmende Wirkung von D-Penicillamin diskutiert.

Key words: Penicillamine, superoxide dismutase, copper complex, inflammation.

Enzymes:
Superoxide dismutase, superoxide:superoxide oxidoreductase (EC 1.15.1.1);
Xanthine oxidase, xanthine:oxygen oxidoreductase (EC 1.2.3.2).

D-Penicillamine is a drug of considerable pharmacological importance. According to various biochemical properties it is successfully used in the clinical treatment of numerous diseases. In the case of Wilson's disease (hepatolenticular degeneration) and in other cases of abnormally high concentrations of copper and other heavy metal ions, D-penicillamine is the drug of choice due to its high chelating activity[1].

Because of the high stability constant of the penicillamine copper complex ($\log K = 16.5$)[2] D-penicillamine chelates not only copper ions bound to proteins, which are normally free of copper, but also normal serum copper from albumin or ceruloplasmin. Due to its antiinflammatory activity, D-penicillamine is valued in the treatment of rheumatoid arthritis. Sorenson[3] investigated the antiinflammatory activities of numerous drugs using the cotton-wad granuloma, the carrageenan foot edema and the therapeutic polyarthritis as model inflammations. He observed that the copper chelates of acetylsalicylic acid, D-penicillamine and many other tested drugs had much more antiinflammatory activity than the parent compounds or cupric acetate, which had low activity in the carrageenan foot edema, but none in the other inflammation models. On the other hand, superoxide dismutase shows significant efficacy in the treatment of inflammatory processes[4,5], indicating that superoxide radicals are involved in inflammation[6]. A pulse radiolysis study of the reactions of D-penicillamine copper complex with $O_2^{\ominus}$ revealed the superoxide-dismutase activity of this complex (K = approx. $10^9 M^{-1} s^{-1}$)[7].

In order to corroborate these results, the activity of D-penicillamine copper complex was determined in four different biochemical systems.

Materials and Methods

The red-violet D-penicillamine copper complex was a gift of *M. Younes* (University Tübingen). It was prepared as described in ref.[8], a molecular weight of about 2 200 being determined by gel chromatography. The complex formed by the reaction of Cu(II) and D-penicillamine is a mixed-valence Cu(I)Cu(II) cluster. Its structure has been studied intensively[9,10] and was finally determined by Birker and Freeman[11] by x-ray diffraction: $(Cu^{II}{}_6Cu^{I}{}_8(\text{D-penicillamine})_{12}Cl)^{5\ominus}$. The measurements of the absorption spectra were carried out on Perkin Elmer 575 and Zeiss PMQ III spectrophotometers. The cyanide-sensitive Cu-Zn superoxide dismutase was isolated from dry green peas[12,13] and the cyanide-insensitive Mn superoxide dismutase from *Euglena gracilis* *. Crocin[14] was prepared from saffron. Crocin destruction was measured photometrically at 440 nm[15] The superoxide-dismutase activity (expressed in units) was determined by measuring nitrite formation from hydroxylamine[16], correlated to the superoxide dismutase test with cytochrome *c*[17]. Chloroplasts were isolated from spinach[18]. Ethylene production by chloroplast lamellae was measured as previously described[19]. Xanthine oxidase was obtained from Boehringer, Mannheim. The comparison of activities of superoxide dismutase with D-penicillamine copper complex was based on statistical means of the values presented in the figures.

Results and Discussion

1. Nitrite formation from hydroxylamine

The systems for testing superoxide-dismutase activities are normally based on the inhibition of an $O_2^{\ominus}$-dependent reaction. In the test described by Elstner and Heupel[16] hydroxylammonium chloride is oxidized with xanthine oxidase as the $O_2^{\ominus}$ generator. The reaction product is nitrite, which is determined photometrically. Addition of superoxide dismutase inhibits nitrite formation. The amount of superoxide dismutase that inhibits 50% of nitrite formation, is defined as 1 unit[17]. The inhibitory effects of various concentrations of Cu-Zn and Mn superoxide dismutase and D-penicillamine copper complex on nitrite formation from NH_2OH are shown in Fig. 1. According to the above definition[16,17] 0.9 μl of Cu-Zn superoxide dismutase and 5 μl of Mn superoxide dismutase represent 1 unit, respectively. In this test 21 nmol D-penicillamine copper complex exhibited the activity of 1 superoxide dismutase unit in 3 ml test volume. At very low concentrations of D-penicillamine copper complex (0.6 - 6 nmol/3 ml) nitrite formation from hydroxylamine was slightly stimulated. This effect was possibly due to free Cu(II) ions from the complex preparation, since the addition of 5μM $CuSO_4$ to this system very effectively stimulated nitrite formation. The Cu(II)-dependent stimula-

* Elstner, E.F. & Frommeyer, D., unpublished results.

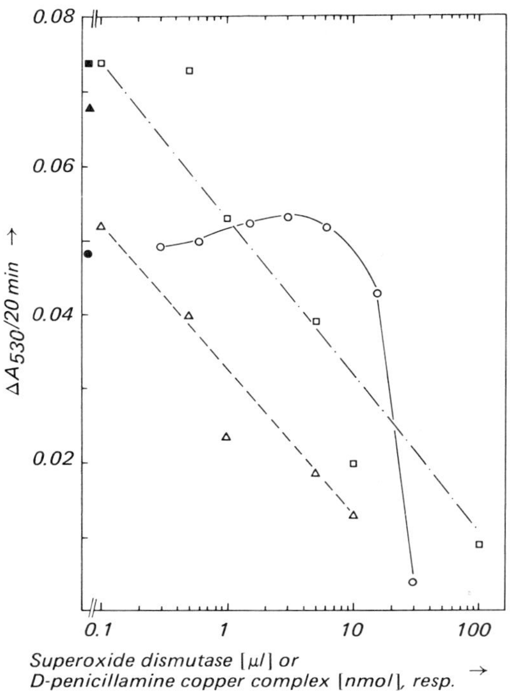

Fig. 1. Inhibitory effects of Cu-Zn superoxide dismutase △---△, Mn superoxide dismutase □-.-.-□, and D-penicillamine copper complex ○——○ on nitrite formation from hydroxylamine by xanthine/xanthine oxidase.

Closed symbols: nitrite formation without superoxide dismutase and D-penicillamine copper complex. Nitrite formation was monitored photometrically by increase in absorption at 530 nm after 20 min reaction time. The superoxide-dismutase activity of 1 unit is defined as the amount (µl of the enzyme preparation) that inhibits 50% of nitrite formation[16,17]. The reaction mixture contained in 3 ml: 180 µmol phosphate buffer pH 7.8; 5 µmol xanthine; 40 µg xanthine oxidase; 1 µmol hydroxylammonium chloride; the reaction was conducted at 25 °C.

tion of nitrite formation, however, was clearly diminished in the presence of 0.1µM EDTA, while the superoxide-dismutase activity of the complex remained unaltered (data not shown).

2. *Crocin destruction*

Crocin, the digentiobiose ester of the carotenoid acid crocetin, was used as a model compound for studies of carotene bleaching[14,15,24]. At pH 5 the rate of crocin destruction in the presence of xanthine/xanthine oxidase was monitored by the decrease in absorption at 440 nm. The inhibitory effects of superoxide dismutase and D-penicillamine copper complex are shown in Fig. 2. The rate of crocin destruction was decreased with increasing amounts of either superoxide dismutase or D-penicillamine copper complex. Over the range of a factor of 100 in concentration, the

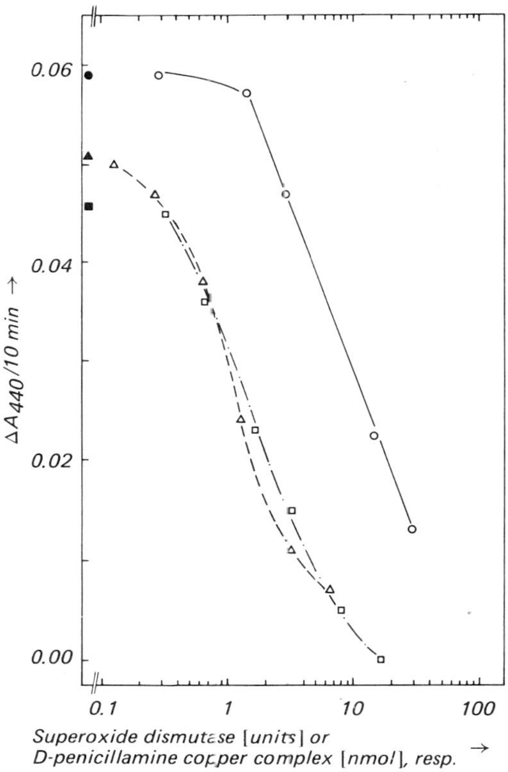

Fig. 2. Inhibitory effects of Cu-Zn superoxide dismutase △---△, Mn superoxide dismutase □-.-.-□ and D-penicillamine copper complex ○——○ on crocin destruction by xanthine/xanthine oxidase, measured by the decrease of absorption at 440 nm after 10 min reaction time.

Closed symbols: rates without superoxide dismutase or D-penicillamine copper complex. The reaction mixture contained in 3 ml: 80 µmol acetate buffer pH 5; 6 nmol crocin; 5 µmol xanthine; 40 µg xanthine oxidase. The reaction was conducted at 22 °C.

inhibitory effect of D-penicillamine copper complex on crocin destruction was compared with the inhibition by either Cu-Zn or Mn superoxide dismutase. In this test 11 nmol D-penicillamine copper complex had the equivalent activity of 1 unit Cu-Zn superoxide dismutase and 8.9 nmol had the equivalent activity of 1 unit Mn superoxide dismutase (determined according to ref.[16],

Table 1. Comparison of the quantities of D-penicillamine copper complex (nmol) with equivalent dismutating activities of 1 unit Mn superoxide dismutase and 1 unit Cu-Zn superoxide dismutase, respectively.

The values were determined from the curve presented in the figures of the particular test systems. I. Nitrite formation from hydroxylamine; II. Crocin destruction by xanthine/xanthine oxidase; III. Ethylene production by isolated chloroplasts; IV. Nitrite formation from hydroxylamine by chloroplasts in the presence of diquat.

	nmol D-penicillamine copper complex, test system I	II	III	IV
Equival. to 1 unit Mn superoxide dismutase	21	8.9	0.18	1
Equival. to 1 unit Cu-Zn superoxide dismutase	21	11	–	–

Table 1). When 10^{-4}M KCN was added to this test system, the inhibitory effect of Cu-Zn superoxide dismutase was considerably reduced, as was to be expected, whereas the activities of both Mn superoxide dismutase and D-penicillamine copper complex remained nearly unchanged (Table 2). From the cyanide-insensitive action of the D-penicillamine copper complex we suggest that its $O_2^{\ominus}$-dismutating activity cannot be explained satisfactorily by a change of valence of copper alone, as is assumed for the action of Cu-Zn superoxide dismutase[25] and for copper complexes of peptides with dismutating activity, which are known to be Cu(II) chelates[26,27,28]. This assumption is in agreement with recent findings of Younes and Weser[29]. Their x-ray photoelectron spectrometric studies of the D-penicillamine copper complex revealed that the actual oxidation state of copper before and after the reaction with $O_2^{\ominus}$ was + I. They suggest therefore a possible stabilized sulfur radical, present as Cu(I)·SR, which undergoes the redox cycle.

3. *Ethylene production by isolated chloroplasts*

The production of ethylene from methionine by chloroplasts in the presence of pyridoxal phosphate and ferredoxin has been reported[19]. In this reaction both $O_2^{\ominus}$[13] and H_2O_2[20,23] are formed. The production of ethylene was strongly increased by the addition of superoxide dismutase as well as by D-penicillamine copper complex. Fig. 3 shows the correlation between increasing

Table 2. Effect of Cu-Zn superoxide dismutase, Mn superoxide dismutase or D-penicillamine copper complex on crocin destruction in the presence or absence of KCN.

The reaction is essentially as described in Fig. 2. Number of determinations in brackets.

Addition	Crocin bleached [pmol × min^{-1} × ml^{-1}] – KCN	+ KCN (10^{-4}M)
None	80.9 ± 9.0 (3)	63.7 ± 3.5 (4)
Cu-Zn superoxide dismutase (1.3 units)	15.0 ± 4.24 (4)	26.4 ± 3.29 (5)
Mn superoxide dismutase (3.3 units)	9.0 ± 3.0 (3)	10.8 ± 5.02 (4)
D-penicillamine copper complex (1.5 × 10^{-8} mol)	22.5 ± 3.0 (5)	20.1 ± 3.46 (4)

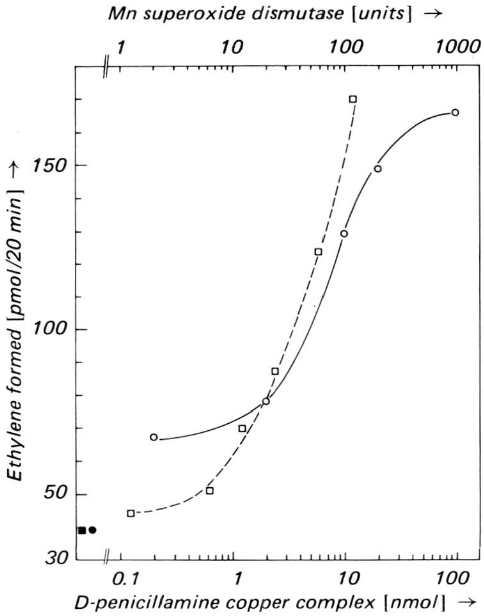

Fig. 3. Rates of ethylene production (pmol/20 min) by isolated chloroplast lamellae as stimulated by either Mn superoxide dismutase □– – –□ or D-penicillamine copper complex ○——○.

Mn superoxide dismutase units were determined as described in Fig. 1. The reaction mixture contained in 2 m*l*: 80 μmol phosphate buffer pH 7.0; 5 μmol NH_4Cl; 5 μmol $MgCl_2$; 0.5 mmol NaN_3; isolated chloroplast lamellae with 0.1 mg chlorophyll; 40 nmol ferredoxin; 20 μmol methionine and 0.5 mmol pyridoxal phosphate. The light reaction was conducted for 20 min at 18 °C in white light (30 000 lux). Closed symbols: rates without superoxide dismutase or D-penicillamine copper complex.

concentrations of superoxide dismutase and D-penicillamine copper complex and the concomitant increase of ethylene formation. 9×10^{-8} M D-penicillamine copper complex exhibited an $O_2^{\ominus}$-dismutating activity equivalent to 1 unit Mn superoxide dismutase. Referring to the test volume of 2 m*l*, 1.8×10^{-10} mol D-penicillamine copper complex had the same stimulating effect on ethylene formation as 1 unit Mn superoxide dismutase. It should be pointed out that in this test the superoxide dismutase activity was determined by a positive effect on the reaction.

IV. Nitrite formation from hydroxylamine by chloroplasts in the presence of diquat (1,1'-dimethylene-2,2'-bipyridylium dibromide)

Isolated chloroplast lamellae produce superoxide radicals[13,30], which in turn form nitrite from hydroxylamine. In the presence of an artificial low potential electron acceptor like diquat, the generation of $O_2^{\ominus}$ by chloroplasts is strongly stimulated[13,20,21]. In the presence of diquat

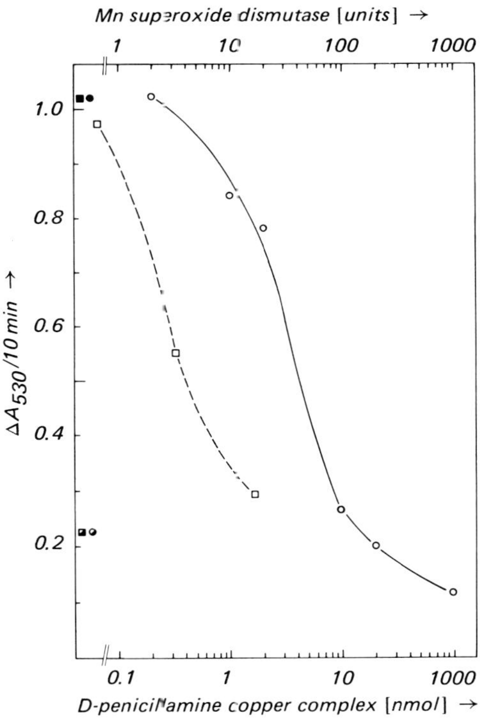

Fig. 4. Rates of nitrite formation from hydroxylamine by chloroplast lamellae in the presence of diquat as electron acceptor, measured by difference in absorption at 530 nm after 10 min reaction time.

Nitrite formation is stimulated from ΔA_{530} = 0.22 in the absence of diquat (□○) to ΔA_{530} = 1.02 in the presence of diquat (■●). This stimulating effect is inhibited by Mn superoxide dismutase □– – –□ or D-penicillamine copper complex ○——○. The reaction mixture contained in 2 m*l*: 100 μmol phosphate buffer pH 7.8; 2.5 μmol NH_4Cl; 2.5 μmol $MgCl_2$; 1 μmol NH_2OH; chloroplast lamellae with 100 μg chlorophyll and 40 nmol diquat (see also ref.[31]).

the formation of nitrite by illuminated chloroplast lamellae decreased with increasing concentrations of Mn superoxide dismutase or D-penicillamine copper complex. In this test 1 nmol D-penicillamine copper complex had the same effect as 1 unit Mn superoxide dismutase (Fig. 4).

Conclusions

The superoxide-dismutating activity of D-penicillamine copper complex was determined and compared with the activities of Cu-Zn and Mn superoxide dismutases. Different $O_2^{\ominus}$-generating systems were used in order to detect possible interferences and influences on activity. D-Penicillamine copper complex turned out to be an effective "superoxide dismutase" at low concentrations. Enzymes which catalyze the reaction

$$O_2^{\ominus} + O_2^{\ominus} + 2\,H^{\oplus} \rightarrow H_2O_2 + O_2$$

are metalloproteins and have been called superoxide dismutases[17]. Apart from enzymes, Cu(II) and low molecular weight copper chelates[27,28] display marked superoxide dismutase activity in "in vitro" systems. However, in the presence of different chelating agents such as bovine serum albumin[28] or EDTA*, this effect is considerably diminished. The fact that the D-penicillamine copper complex is stable and of high $O_2^{\ominus}$-dismutating activity even in the complex systems of the chloroplast experiments, and its insensitivity to cyanide and EDTA permits the supposition that a copper complex of D-penicillamine, formed *in vivo*[9], should also show superoxide dismutating activity. Sorenson[3] has suggested that in vivo formed copper chelates of antiinflammatory drugs are the active species. In this respect D-penicillamine alone has no superoxide dismutating activity[7]. It is therefore entirely possible that D-penicillamine owes its efficacy as an antiinflammatory drug to its ability to form in vivo a copper complex which dismutates the $O_2^{\ominus}$ radicals thought to be associated with some inflammatory processes.

We wish to thank Mrs. *Christine Fuchs* and Mr. *Horst Zeller* for expert technical assistance and the *Deutsche Forschungsgemeinschaft* for financial support.

* Halliwell, B., private communication.

Literature

1 Walshe, J.M. (1966) in The Biochemistry of Copper (Peisach, J., Aisen, P. & Blumberg, W.E., eds.) pp. 475 - 498, Academic Press, New York.

2 Henschler, D. (1975) in Pharmakologie und Toxikologie (Forth, W., Henschler, D. & Rummel, W., eds.) pp. 542 - 546, Bibliographisches Institut, Mannheim, Wien, Zürich.

3 Sorenson, J.R. (1976) *J. Med. Chem.* **19**, 135 - 148.

4 Lund-Olesen, K. & Menander, K. (1974) *Curr. Ther. Res.* **16**, 706 - 717.

5 Marberger, H., Huber, W., Bartsch, G., Schulte, T. & Swoboda, P. (1974) *Int. Urol. Nephrol.* **6**, 61 - 74.

6 Salin, M.L. & McCord, J.M. (1975) *J. Clin. Invest.* **56**, 1319 - 1323.

7 Lengfelder, E., Saran, M. & Bors, W. (1977) International Conference on Singlet Oxygen and Related Species in Chemistry and Biology, August 21–26, 1977, Pinawa, Canada, Abstract Nr. L-3.

8 Rupp, H. & Weser, U. (1976) *Biochim. Biophys. Acta* **446**, 151 - 165.

9 Wright, J.R. & Frieden, E. (1975) *Bioinorg. Chem.* **4**, 163 - 175.

10 Birker, P.J. & Freeman, H.C. (1976) *J. Chem. Soc. D* **164**, 312 - 313.

11 Birker, P.J. & Freeman, H.C. (1977) *J. Am. Chem. Soc.* **99**, 1247 - 1253.

12 Sawada, J., Ohyama, T. & Yamazaki, I. (1972) *Biochim. Biophys. Acta* **268**, 305 - 312.

13 Elstner, E.F. & Kramer, R. (1973) *Biochim. Biophys. Acta* **314**, 340 - 353.

14 Friend, J. & Mayer, A.M. (1960) *Biochim. Biophys. Acta* **41**, 422 - 429.

15 Weber, F. & Grosch, W. (1976) *Z. Lebensm. Unters. Forsch.* **161**, 223 - 230.

16 Elstner, E.F. & Heupel, A. (1976) *Anal. Biochem.* **70**, 616 - 620.

17 McCord, J.M. & Fridovich, I. (1969) *J. Biol. Chem.* **244**, 6049 - 6055.

18 Jensen, R.G. & Bassham, J.A. (1966) *Proc. Natl. Acad. Sci. U.S.A.* **56**, 1095 - 1098.

19 Konze, J.R. & Elstner, E.F. (1976) *FEBS Lett.* **66**, 8 - 11.

20 Allen, J.F. & Hall, D.O. (1973) *Biochem. Biophys. Res. Commun.* **52**, 856 - 862.

21 Epel, B.L. & Neumann, J. (1973) *Biochim. Biophys. Acta* **325**, 520 - 529.

22 Elstner, E.F. & Heupel, A. (1973) *Biochim. Biophys. Acta* **325**, 182 - 188.

23 Elstner, E.F. & Konze, J.R. (1974) *FEBS Lett.* **45**, 18 - 21.

24 Dicks, J.W. (1970) *Phytochemistry* 9, 1433 - 1441.

25 Fielden, E.M., Roberts, P.B., Bray, R.C., Lowe, D.J., Mautner, G.N., Rotilio, G. & Calabrese, L. (1974) *Bioch. J.* **139**, 49 - 60.

26 Younes, M. & Weser, U. (1976) *FEBS Lett.* **61**, 209 - 212.

27 Brigelius, R., Spöttli, R., Bors, W., Lengfelder, E., Saran, M. & Weser, U. (1974) *FEBS Lett.* **47**, 72 - 75.

28 Brigelius, R., Hartmann, H. J., Bors, W., Saran, M., Lengfelder, E. & Weser, U. (1975) *Hoppe-Seyler's Z. Physiol. Chem.* **356**, 739 - 745.
29 Younes, M. & Weser, U. (1977) *Biochem. Biophys. Res. Commun.* 78, 1247 - 1253.
30 Elstner, E. F. & Heupel, A. (1974) *Z. Naturforsch.* **29**, 564 - 571.
31 Elstner, E. F., Stoffer, C. & Heupel, A. (1974) *Z. Naturforsch.* **30**, 53 - 56.

Dr. E. Lengfelder*, Strahlenbiologisches Institut der Universität, Bavariaring 19, D-8000 München 2 und

Dr. E. F. Elstner, Institut für Botanik und Mikrobiologie der Technischen Universität, Arcisstr. 21, D-8000 München 2.

* To whom correspondence should be sent.

Hoppe-Seyler's Z. Physiol. Chem.
Bd. 359, S. 759 - 763, Juni 1978

SHORT COMMUNICATION

Acid Precipitation of Fibrinogen
A Technique for Isolation of Fibrinogen and its Derivatives from Highly Diluted Solutions

Reimar HAFTER, René VON HUGO and Henner GRAEFF

I. Frauenklinik der Universität München

(Received 6 March 1978)

Säurefällung von Fibrinogen: Eine Methode zur Isolierung von Fibrinogen und seinen Derivaten

Zusammenfassung: Eine Fällungsmethode wird beschrieben, durch die Fibrinogen und seine hochmolekularen Derivate aus hochverdünnten Lösungen durch Ansäuern auf pH 3.7 abgeschieden werden können. Das Verfahren ist geeignet, um Fibrinogenderivate aus chromatographischen Fraktionen oder anderen verdünnten Lösungen in praktisch reiner Form zu isolieren, wodurch eine weitere Charakterisierung mittels Gelelektrophorese möglich ist. Wegen ihrer hohen Empfindlichkeit ist die Methode anderen Verfahren überlegen. Auch die Fibrinogen- und Fibrin-Abbauprodukte können zum Teil erfaßt werden. So sind Fragment X vollständig und die Fragmente Y, D und D-dimer teilweise fällbar. Fragment E ist nur geringfügig fällbar.

Key words: Acid precipitation, fibrinogen, high molecular weight fibrinogen derivatives, fibrinogen degradation products, agarose gel filtration.

It is often necessary to increase the concentration of a chromatographic fraction before the protein can be characterized. Either precipitation procedures or dialysis techniques can be applied for this purpose. The methods commonly used to precipitate fibrinogen and fibrin(ogen) derivatives, which involve the salting out with non-specific protein precipitants such as sodium sulfate[1], phosphate[2] or others[3,4] did not prove to be sensitive enough. The same holds true for the more specific precipitants such as glycine[5], β-alanine[6] or 12% ammonium sulfate[7]. Sodium sulfite[8,9], which is more sensitive than other precipitants, proved unsatisfactory for this purpose, in that it splits the molecules into the single polypeptide chains. A change in molecular structure is also involved if clotting enzymes like thrombin or snake venoms are used for fibrinogen precipitation, thus rendering further characterization uncertain.

The alternative, i.e. dialysis, often leads to large losses of protein due to adsorption effects in the dialysis membrane. Furthermore, contaminating proteins cannot be removed by these techniques. A precipitation procedure that combines high specificity for fibrin(ogen) and its derivatives with high sensitivity is presented in this paper. It is achieved simply by adjusting the pH of the protein-containing buffer to a value between 2.8 and 4.4. We have studied the reaction regarding its application to chromatographic fractions and its use for determination of fibrinogen in plasma.

Further, it was investigated under which conditions and to what extent high molecular weight derivatives of fibrinogen and fibrin(ogen) degradation products take part in the precipitation reaction.

Materials and Methods

Fibrinogen: Human fibrinogen (FFBH Deutsche Kabi München) was dissolved in distilled water to give a stock solution of 8 mg/m*l*.

Plasma: Plasma was prepared from fresh human blood, anticoagulated with citrate (1 vol. 3.8% trisodium citrate/9 vol. blood). Defibrinated plasma was prepared by heat precipitation of the fibrinogen at 60 °C for 20 min. The fibrinogen was then removed by highspeed centrifugation and the supernatant collected.

Thrombin-clotting technique: A slightly modified version of the Blombäck and Blombäck method[10-12] was used.

Buffers for gel filtration and for precipitation studies: Buffer: Tris/NaCl/citrate, pH 7.6, ionic strength $\mu = 0.39$; containing in 1 *l*: 6.05 g Tris (0.05 mol), 6.76 g sodium chloride (0.116 mol), 3.8 g trisodium citrate · 2 H_2O (0.0129 mol), and 26.2 g ϵ-aminocaproic acid (0.2 mol), pH adjusted with 25% hydrochloric acid.
For precipitation studies, the buffer was adjusted to pH values between 2.6 and 4.4 with 37% hydrochloric acid.

Preparation of fibrinogen degradation products and D-dimer: The procedure of Gaffney et al.[9,13] was used.

Precipitation studies: In order to find the optimal pH condition of precipitation, 0.1 - 0.2 m*l* samples of the fibrinogen stock solution and of the individual fibrinogen digests, respectively, were diluted to 10 m*l* with buffer of varied pH values between pH 2.8 and 4.4. Turbidities were measured at 325 nm after 1 h at 24 °C. Recoveries were calculated from the precipitates, spun down after 2 h and dissolved in 5 m*l* alkaline urea (40% urea in 0.2N NaOH) according to the formula:
mg fibrinogen/m*l* = $(A_{280} - A_{325}) \times 0.629$[12].
For further studies buffer of pH 3.7 was used. Recoveries from very dilute fibrinogen solutions were calculated from the differences between the absorbance differences $(A_{280} - A_{325})$ in the supernatants and in fibrinogen samples in buffer, pH 7.6.

Kinetic measurements: 0.05-m*l* and 0.1-m*l* samples of fibrinogen stock solution or 0.5-m*l* and 1.0-m*l* plasma samples were diluted with buffer, pH 3.7, to 10 m*l*. The solutions were immediately transferred into 1-cm cuvettes and the absorbances caused by the developing turbidities were measured at short intervals at 325 nm against a buffer blank.

Treatment of gel filtration fractions containing fibrinogen and fibrin(ogen) derivatives: Gel filtration of the β-alanine-precipitated fibrinogen fraction from the plasma of a patient with advanced renal carcinoma was performed as described elsewhere[11] using the above buffer, pH 7.6. The elution profile was established by plotting the absorbance differences $(A_{280} - A_{325})$ vs. effluent volume. The single fractions were combined to fractions of the void volume (*V*o), the shoulder, and the fibrinogen peak, respectively, and the pH adjusted to 3.7 with hydrochloric acid. The precipitates developed on standing were spun down after 2 h, dissolved in 0.2 to 1 m*l* 8M urea and subjected to sodium dodecylsulfate polyacrylamide gel electrophoresis. The yield of precipitated protein was calculated from the remaining absorbances in the supernatants.

Polyacrylamide gel electrophoresis: Fibrinogen, samples of its degradation products, and their constituent polypeptide chains were characterised by electrophoresis on sodium dodecylsulfate polyacrylamide gels by the method of Weber and Osborn[14] modified by Timpl and Furthmayr[15].

Results

Studies on fibrinogen: The pH optimum of fibrinogen precipitation in Tris/NaCl/citrate was found to be 3.7 - 3.8. In this range the recovery amounts to 98% (Fig. 1, curve F).

The course of precipitation follows a sigmoid curve, as shown by turbidity measurements (Fig. 2). In pure solutions the maximal turbidity develops within 20 min, irrespective of the concentration of fibrinogen. In the presence of other plasma proteins, however, precipitation is retarded and dependent on concentration. However, a similar maximal turbidity is reached and the recovery percentage is as high as that from pure solutions.

To evaluate the sensitivity of the reaction, various dilutions of fibrinogen between 4 and 168 μg/m*l* were pre-

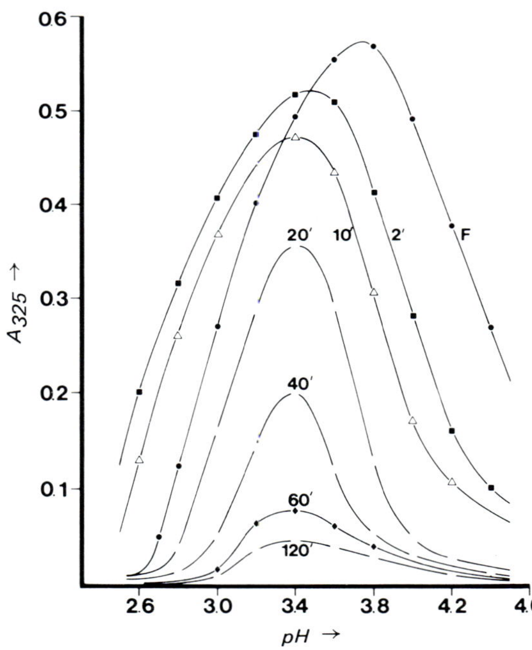

Fig. 1. Precipitation of fibrinogen (F) and plasmic fibrinogen digests after various digestion times (in min) in Tris/NaCl/citrate at varied pH values (100 μg protein/m*l*).

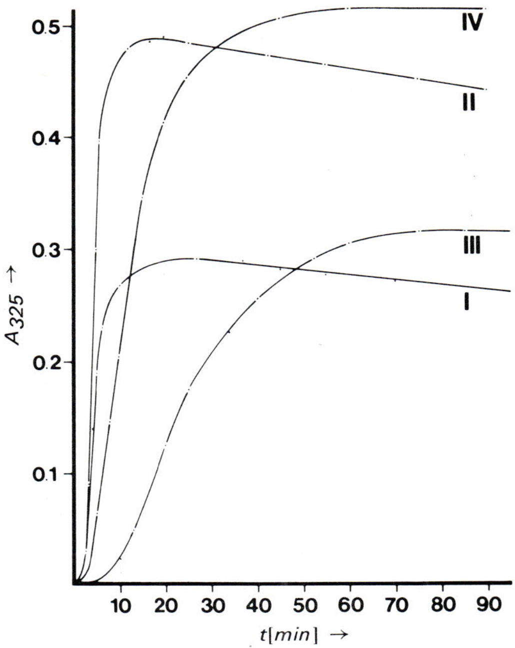

Fig. 2. Progress of fibrinogen precipitation in Tris/NaCl/citrate, pH 3.7 at 24 °C.

I and II: 40 and 80 μg fibrinogen/m*l*.

III and IV: 0.05 and 0.1 m*l* plasma/m*l*, containing 66 and 132 μg fibrinogen/m*l*, respectively.

cipitated at pH 3.7. The turbidities showed a linear relationship to the fibrinogen concentration up to 84 μg fibrinogen/m*l*. At concentrations above 100 μg/m*l*, a curvilinear relationship became evident. The recoveries in all samples were close to 100%. Concentrations as small as 4 μg fibrinogen/m*l* can be estimated by turbidity measurements.

The precipitates dissolve readily in 5M urea. After reduction with 2-mercaptoethanol, sodium dodecylsulfate polyacrylamide gel electrophoresis revealed a pattern of α, β and γ-chains identical to that of non-precipitated fibrinogen. Precipitates obtained at a lower pH than the optimal one showed bands of aggregated fibrinogen in addition. Depolymerization occurs when a drop of aqueous ammonia (10 - 20 μ*l*) is added to the precipitates prior to their dissolution in urea. Fibrinogen from plasma showed less tendency to form urea-stable aggregates than fibrinogen from pure solutions. By adding a drop of ammonia the precipitates can also be made soluble without urea and can be clotted with thrombin after neutralization.

Studies on plasma: The influence of plasma proteins on the measurement of fibrinogen was investigated by adding increasing amounts of fibrinogen to 0.5-m*l* samples of defibrinated plasma and subjecting the mixtures to acid precipitation. In the low concentration range of 16 to 40 μg fibrinogen/m*l*, the recoveries remained below 100%. In the upper concentration range (80 - 400 μg/m*l*), however, the recoveries exceeded 100%. The deviation in percentage was found to be greatest at 160 μg fibrinogen/m*l* (mean 12%) and decreased with increasing fibrinogen concentration.

A comparison between the acid and the thrombin method to measure fibrinogen in plasma was carried out with 16 plasma samples from different donors. (Fibrinogen content: 107 - 483 mg/100 m*l*). The values obtained by the acid technique were higher in all samples by 11 - 33% (mean: 22.6%; standard deviation, 5.89; coefficient of variation 25.9) than the ones found with the thrombin method. Less variation was observed at higher fibrinogen concentrations.

Studies on fractions of gel filtration chromatography: High molecular weight derivatives of fibrin(ogen), which separate from the majority of fibrinogen on gel filtration of a fibrinogen fraction from plasma of diseased persons, are also susceptible to acid precipitation. When a sample from a patient suffering from advanced renal carcinoma was fractionated, the precipitable fibrinogen derivatives migrating with the void volume amounted to 75% of the total protein in these fractions, and those in a shoulder preceding the fibrinogen peak amounted to 69 - 81% of the measured protein content. From the fibrinogen peak, 95% of the protein was precipitable.

Characterization of the precipitates by sodium dodecylsulfate polyacrylamide gel electrophoresis after reduction with 2-mercaptoethanol revealed the presence of $\gamma - \gamma$-dimers and diminished α-chains in the fractions from the void volume and the shoulder, thus indicating the precipitability of slightly degraded cross-linked fibrin.

Studies on fibrin(ogen) degradation products: Precipitation studies on fibrinogen digests produced by action of plasmin on fibrinogen revealed a shift in the pH optimum from 3.7 to 3.4 occurring within the first 2 min of plasmic attack in which fibrinogen is degraded into fragments X and Y (Figs. 1 and 3). The recovery of precipitated degradation products obtained after 2 min of digestion still amounted to 87% of the initial fibrinogen concentration, but dropped rapidly with further digestion (Fig. 1) in which the concentration of X and Y decreased while that of D and E inversely increased. The electrophoretograms of the individual digestion mixtures show that fragment X is fully precipitable whereas fragments Y and D only partially precipitate (Fig. 3). Only traces

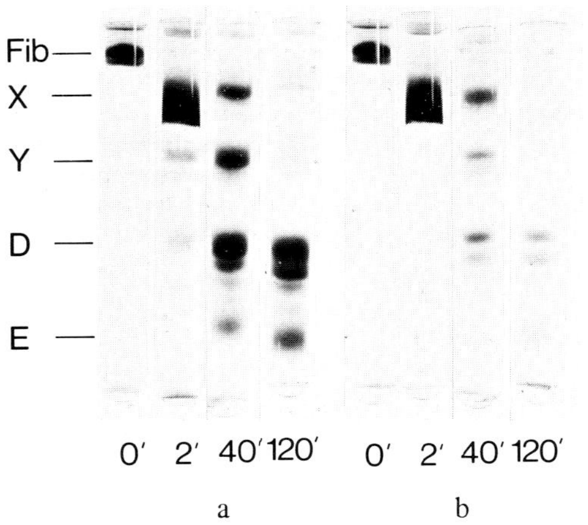

Fig. 3. 5% sodium dodecylsulfate polyacrylamide electrophoretic patterns of fibrinogen and fibrinogen digests: non precipitated (a), and precipitated samples at optimal pH (b).

of fragment E are susceptible to acid precipitation. The precipitable portion of fragment D is comprised of two bands compatible with D_g and D_p according to Gaffney et al.[13].

Fragment D-dimer derived from plasmin digestion of cross-linked fibrin is also susceptible to acid precipitation to an extent of approximately 20%. This was calculated on the assumption that one mol fibrin with the assumed molecular weight of 340 000 will give rise to one mol D-dimer, with a molecular weight of 160 000[13]. The pH optimum of precipitation was found to be at pH 3.9

Discussion

Acid precipitation of fibrinogen as presented in this paper seems to be related to the phenomenon of isoelectric precipitation of proteins. The shift of the optimal pH, which differs from the one usually described for the isoelectric point (pI) of fibrinogen[16,17] could be due to the buffer used, since the pI of proteins can be influenced by composition and concentration of ions[18].

The Tris/NaCl/citrate buffer used in this study is the same which we applied for gel filtration chromatography to separate the higher molecular derivatives of fibrinogen from the parent fibrinogen[11]. Its buffer capacity on the acid side permits the optimal pH to be adjusted precisely and stably enough to facilitate a high yield, close to 100%, even in very dilute solutions. The pH optimum is not changed when the buffer is used with a lower concentration of ϵ-aminocapronic acid (0.025M).

The technique results in an additional purification of the fibrinogen derivatives contained in chromatographic fractions.

The specificity of the reaction was also checked on mixtures of fibrinogen with defibrinated plasma and on plasma samples. The recoveries were found to be higher in the presence of other plasma proteins, indicating that some coprecipitation occurs.

In studies on fibrinogen degradation products the plasmic fragment X seemed fully susceptible to acid precipitation. By contrast, the fragments Y and D were found to be only partially susceptible to acid precipitation and fragment E only in traces. In this respect our method is similar to the sulfite technique described by Rampling and Gaffney[9], although a considerably higher portion of fragments X, Y and D is recovered by acid precipitation.

The pH optimum for precipitation of the fragments is different from that for fibrinogen. The pH optimum is shifted from pH 3.7 to 3.4 as fibrinogen is split into fragment X. A corresponding shift in the pI values is reported by Arnesen[19]: pH 5.95 - 6.10 for fibrinogen and pH 5.65 - 6.10 for product X.

The fragment D-dimer originating from cross-linked fibrin is of special interest in diagnosing intravascular coagulation biochemically, since it reflects the occurrence of both thrombin- and plasmin-mediated catabolic pathways of fibrinogen turnover. It was found that about 20% of the experimentally obtained D-dimer is susceptible to acid precipitation. The pH optimum, at 3.9, seems to differ from that for the fibrinogen-derived fragment D.

The finding that the plasmic fragments are only partially susceptible to acid precipitation, with the exception of fragment X, might be interpreted by their strongly altered molecular structures and their different net charges, which may induce a higher solubility.

The authors wish to thank Miss *B. von Brunn* for excellent technical assistance. The study was supported by the *Deutsche Forschungsgemeinschaft*, Bonn-Bad Godesberg, Germany (Sonderforschungsbereich 51, München, grant no. B/6).

Literature

1 Campbell, W.R. & Hanna, M.J. (1937) *J. Biol. Chem.* **119**, 15 - 22.
2 Martinek, R.G. & Berry, E.R. (1965) *Clin. Chem.* **11**, 10 - 16.
3 Podmore, D.A. (1959) *Clin. Chim. Acta* **4**, 242 - 251.
4 Jacox, R.F. (1954) *J. Lab. Clin. Med.* **44**, 885 - 891.
5 Kazal, L.A., Amsel, S., Miller, O.P. & Tocantins, L.M. (1963) *Proc. Soc. Exp. Biol. Med.* **113**, 989 - 994.
6 Straughn, W. & Wagner, R.H. (1966) *Thromb. Diath. Haemorrh.* **16**, 197 - 206.

7 Partfentjev, I.A., Johnson, M.L. & Cliffton, E.E. (1953) *Arch. Biochem. Biophys.* **46**, 470 - 480.
8 Goodwin, J.F. (1965) *Clin. Chem.* **11**, 63 - 73.
9 Rampling, M.W. & Gaffney, P.J. (1976) *Clin. Chim. Acta* **67**, 43 - 52.
10 Blombäck, B. & Blombäck, M. (1956) *Ark. Kemi* **10**, 415 - 443.
11 Hafter, R. & Graeff, H. (1976) in Progress in chemical fibrinolysis and thrombolysis (Davidson, J.F., Samama, M.M. & Desnoyers, P.C., eds.) vol. 2, pp. 137 - 149, Raven Press, New York.
12 Hörmann, H. & Gollwitzer, R. (1966) *Hoppe Seyler's Z. Physiol. Chem.* **346**, 21 - 41.
13 Gaffney, P.J., Lane, D. & Brasher, M. (1975) *Clin. Sci. Mol. Med.* **49**, 149 - 156.
14 Weber, K. & Osborn, M. (1969) *J. Biol. Chem.* **244**, 4406 - 4412.
15 Furthmayr, H. & Timpl, R. (1971) *Anal. Biochem.* **41**, 510 - 516.
16 Mihalyi, E. (1954) *J. Biol. Chem.* **209**, 723 - 732.
17 Henschen, A. (1964) *Ark. Kemi* **22**, 397 - 415.
18 Alberty, R.A. (1953) in The Proteins (Neurath, H. & Baily, K., eds.) vol. 1 A, pp. 461 - 548, Academic Press, New York.
19 Arnesen, H. (1974) *Thromb. Res.* **4**, 861 - 868.

Dr. R. Hafter, Dr. R. von Hugo and Prof. H. Graeff, I. Frauenklinik der Universität München, Maistraße 11, D-8000 München 2.

Hoppe-Seyler's Z. Physiol. Chem.
Bd. 359, S. 765 - 767, Juni 1978

SHORT COMMUNICATION

Action of Chondroitinases ABC and AC on Chondroitin 4-Sulfate-Oligosaccharides with Terminal *N*-Acetylgalactosaminitol Residues

Ulrich KLINGBEIL, Udo KLEIN and KURT VON FIGURA

Physiologisch-Chemisches Institut der Westfälischen Wilhelms-Universität Münster

(Received 13 April 1978)

Einfluß von terminalen N-Acetylgalaktosaminit-Resten auf den Abbau von Chondroitin-4-sulfat-Oligosacchariden durch die Chondroitinasen ABC und AC

Zusammenfassung: Der Vergleich der Wirkung der Chondroitinasen ABC und AC auf reduzierte und nicht reduzierte Oligosaccharide, die aus Chondroitin-4-sulfat durch Abbau mit Testes Hyaluronidase gewonnen wurden, zeigt, daß Chondroitinasen galaktosaminidische Bindungen dann nicht endoglykosidisch spalten, wenn das benachbarte Disaccharid reduziert ist.

Key words: Chondroitinases, chrondroitin sulfates, reduced oligosaccharides.

Chondroitinases are a group of bacterial lyases that split β-1,4 linkages of *N*-acetyl-D-glucosamine and *N*-acetyl-D-galactosamine to D-glucoronic acid and L-iduronic acid[1]. Chondroitin sulfates, dermatan sulfate, chondroitin and hyaluronic acid are degraded by these lysases into unsaturated disaccharides with a $\Delta^{4,5}$-enuronic acid at the non-reducing terminal. Chondroitinase ABC is able to degrade all these glycosaminoglycans, whereas the action of chondroitinase AC is limited to hexosaminidic residues linked to D-glucuronic acid, and thus it does not degrade dermatan sulfate. The chondroitinases have been widely used for structural analyses of glycosaminoglycans because of their different substrate specificities and their commercial availability in sufficiently pure forms. By comparison of the chondroitinases' action on reduced and unreduced chondroitin 4-sulfate oligosaccharides it is demonstrated in the present study that chondroitinases do not split endogalactosaminidic bonds neighbouring to reduced disaccharide units.

Materials and Methods

[^{3}H]$NaBH_4$ (spec. act. 280 mCi/mmol) was obtained from NEN (Dreieichenhain), Sephadex gels from Pharmacia (Uppsala), chondroitinases ABC and AC from *Proteus vulgaris* from Seikagaku Kogyo Co. (Tokyo). Disaccharides, tetrasaccharides and a hexasaccharide-octasaccharide mixture obtained from nasal cartilage chondroitin 4-sulfate by digestion with testicular hyaluronidase[2] were a gift of *Dr. E. Buddecke* and *R. Niemann*, this institute.
The tetrasaccharide and the hexa-octasaccharide mixtures were reduced with $NaBH_4$ or with [^{3}H]$NaBH_4$, spec. act. 20 mCi/mmol, as described[3].
The ^{3}H-reduced oligosaccharides were chromatographed on a 0.8 cm x 125 cm Sephadex G-25 column in 0.2M pyridine/acetic acid, pH 5.5, in order to remove acid-resistant ^{3}H-labelled low molecular weight material present in the [^{3}H]$NaBH_4$ preparation. The [^{3}H]-oligosaccharides were pooled and dried by high vacuum

---

Enzymes:
Chondroitin ABC lyase (EC 4.2.2.4); Chondroitin AC lyase (EC 4.2.2.5); Testicular hyaluronidase (recommended name: hyaluronoglucosaminidase), hyaluronate 4-glycanohydrolase (EC 3.2.1.35)

evaporation. After acid hydrolysis as described for isolation of hexosaminitol residues[4], [^{3}H]galactosaminitol was shown by high voltage electrophoresis[5] to be the only ^{3}H-compound present in the reduced oligosaccharides. Reduced and unreduced oligosaccharides were digested with chondroitinases ABC and AC as described[6]. The chondroitinases were stored as stock solutions of 2 U/m*l* (activity as determined by the producer) in the Tris/acetate buffer, pH 8.0[6] at 4 °C and were used within four weeks. The chondroitinase activity was recorded continuously, in a heated cuvette of an Acta III spectrophotometer (Beckman Instruments, München) at 37 °C, using the increase of absorbance at 232 nm as indicator for the formation of unsaturated disaccharides. Uronic acids[7] and hexosamines[8] were determined as described. Radioactivity was determined in Unisolve I (Zinsser, Frankfurt/Main) with a 2450 B liquid scintillation counter (Packard, Frankfurt/Main).

Results and Discussion

Chondroitinases ABC and AC split the chondroitin-4-sulfate tetrasaccharide of the structure GlcUA-GalNAc-(4 S)-GlcUA-GalNAc(4 S) into a saturated disaccharide (from the non-reducing terminal) and an unsaturated disaccharide (from the reducing terminal). The appearance of the unsaturated disaccharide can be followed by recording the increase of absorbance at 232 nm (Fig. 1 a). The reduced tetrasaccharide with an *N*-acetylgalactosaminitol 4-sulfate residue at the reducing terminal is not split by chondroitinases ABC and AC (Fig. 1 a). Digestion of an equimolar mixture of reduced and unreduced tetrasaccharide with chondroitinase ABC yields the intermediate absorbance (Fig. 1 a), indicating that the sodium borohydride reduction does not produce an inhibitor for the chondroitinase.

Similar experiments were done with an approximately equimolar mixture of hexa- and octasaccharides from chondroitin 4-sulfate (estimated from the reducing-sugar to uronic-acid ratio of 0.29). The absorbance ratio at 232 nm after complete digestion with chondroitinase ABC for the reduced and non-reduced oligosaccharide mixtures was 0.65 : 1 (Fig. 1 b). This ratio fits well with the ratio of 0.60 : 1 calculated under the assumption that conversion of the terminal *N*-acetylgalactosamine 4-sulfate residue into the corresponding sugar alcohol prevents only the cleavage of the endohexosaminidic bond of the tetrasaccharide at the reducing terminal. This was confirmed by the separation of the chondroitinase ABC digest of a mixture of reduced hexa- and octasaccharides on a Sephadex G-25 column into a tetra- and a disaccharide fraction (Fig. 2). The tetrasaccharide fraction contained all the ^{3}H radioactivity originally present at the reducing terminal, whereas the disaccharide fraction was free of ^{3}H radioactivity. The molar ratio of tetra- to disaccharide was 0.68. Under the assumption that only the endohexosaminidic linkage of the tetrasaccharide at the reducing terminal is not split after sodium borohydride reduction, a theoretical value of 0.67 is expected. The hexosamine/uronic acid ratios of the [^{3}H]tetrasaccharide and the disaccharide were 0.56 (0.50) and 1.05

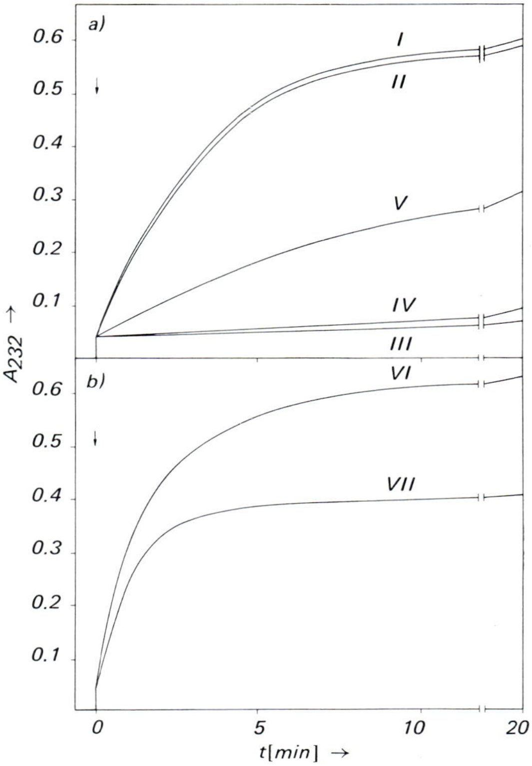

Fig. 1. Chondroitinase ABC and AC digestion of reduced and unreduced chondroitin 4-sulfate oligosaccharides.

a) 90 nmol of reduced or unreduced tetrasaccharide was digested in a final volume of 0.37 m*l* with 40 mU chondroitinase ABC or AC in a heated cuvette at 37 °C. Absorbance was recorded at 232 nm. Curves I - V represent: unreduced tetrasaccharide digested with chondroitinase ABC (I) or AC (II); reduced tetrasaccharide digested with chondroitinase ABC (III) or AC (IV); mixture of 45 nmol reduced tetrasaccharide and 45 nmol unreduced tetrasaccharide digested with chondroitinase ABC (V).

b) A reduced or unreduced hexa/octasaccharide mixture from chondroitin 4-sulfate was digested in a final volume of 0.21 m*l* with 20 mU chondroitinase ABC at 37 °C. Absorbance was recorded at 232 nm. Curves VI and VII represent the digestion of the unreduced and reduced oligosaccharides, respectively.

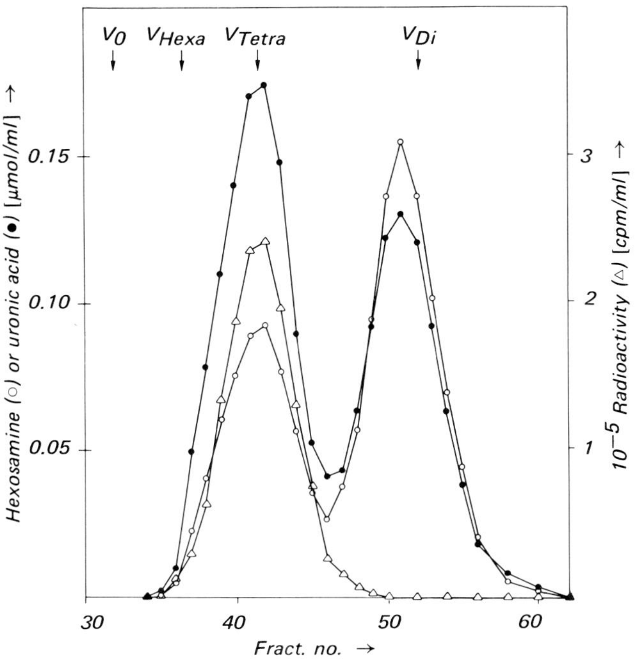

Fig. 2. Elution profile of the [^{3}H]hexasaccharide-[^{3}H]octasaccharide mixture digested with chondroitinase ABC on a Sephadex G-25 column.

A [^{3}H]hexasaccharide-[^{3}H]octasaccharide mixture (1.8 µmol uronic acid) was digested with 180 mU chondroitinase ABC in a final volume of 3.5 m*l* for 5 min at 37 °C. After addition of 1 m*l* 0.2M pyridine/acetic acid (0.2M with regard to pyridine), pH 5.3, the reaction mixture was dried under reduced pressure. The sample was dissolved in 0.15 m*l* of the pyridine buffer and loaded on a 0.8 cm × 125 cm Sephadex G-25 column, equilibrated and eluted with the pyridine buffer at a rate of 5 m*l*/h into fractions of 0.72 m*l*. The fractions were assayed for radioactivity (△), hexosamine (○) and uronic acid (●). The column was calibrated with dextran blue (V_0), hexasaccharide (V_{Hexa}), tetrasaccharide (V_{Tetra}) and disaccharide (V_{Di}) from chondroitin 4-sulfate and with ^3H_2O (V_T at fraction 70).

(1.00), respectively (theoretical values for a reduced tetrasaccharide and an unreduced disaccharide in brackets).

These results clearly indicate that chondroitinases ABC and AC recognize not only the hexosaminidic linkage between a hexosamine residue and an uronic acid, but also the hexosamine residue in the reducing position with regard to the uronic acid. The sodium borohydride reduction does not influence the splitting of endohexosaminidic bonds within polymeric chondroitin sulfate chains as suggested by Hiyama and Okada[9]. The effect of sodium borohydride reduction is restricted to the chondroitinase ABC and AC action on the endohexosaminidic bond between the ultimate and penultimate disaccharide at the reducing terminal. Thus the exhaustive digestion of [^{35}S]dermatan sulfate with a terminal [^{3}H]*N*-acetylgalactosaminitol residue with chondroitinase ABC yielded only [^{35}S]disaccharides and a [^{3}H/^{35}S]-tetrasaccharide fraction. The ratio of tetra- to disaccharides indicated that the [^{3}H]tetrasaccharide fraction represented only the reducing terminal of the [^{35}S]dermatan sulfate chains (U. Klein, U. Klingbeil, K. v. Figura, unpublished results).

This work was supported by a grant of the *Deutsche Forschungsgemeinschaft* (SFB 104).

Literature

1 Yamagata, T., Saito, H., Habuchi, O. & Suzuki, S. (1968) *J. Biol. Chem.* **243**, 1523 - 1535.
2 Flodin, P., Gregory, J.D. & Rodén, L. (1964) *Anal. Biochem.* **8**, 424 - 433.
3 Ögren, S. & Lindahl, U. (1975) *J. Biol. Chem.* **250**, 2690 - 2697.
4 Klein, U., Kresse, H. & von Figura, K. (1976) *Biochem. Biophys. Res. Commun.* **69**, 158 - 166.
5 Takasaki, S. & Kobata, A. (1974) *J. Biochem.* **76**, 783 - 789.
6 Saito, H., Yamagata, T. & Suzuki, S. (1968) *J. Biol. Chem.* **243**, 1536 - 1542.
7 Bitter, T. & Muir, H. (1962) *Anal. Biochem.* **4**, 330 - 334.
8 Boas, N. (1953) *J. Biol. Chem.* **204**, 553 - 563.
9 Hiyama, K. & Okada, S. (1976) *J. Biochem.* **80**, 1201 - 1207.

Prof. Dr. K. von Figura, U. Klingbeil and Dr. U. Klein, Physiologisch-Chemisches Institut der Universität Münster, Waldeyerstraße 15, D-4400 Münster.

Hoppe-Seyler's Z. Physiol. Chem.
Bd. 359, S. 769 - 772, Juni 1978

SHORT COMMUNICATION

Glyceraldehyde-3-Phosphate Dehydrogenase (NADP$^{\oplus}$) from *Sinapis alba* L.

Isolation and Electrophoretic Characterization of Isoenzymes

Rüdiger CERFF and Susan E. CHAMBERS

Biologisches Institut II, Universität Freiburg i. Br.

(Received 20 April 1978)

Die Glycerinaldehyd-3-phosphat-Dehydrogenase (NADP$^{\oplus}$) aus Sinapis alba L. Isolierung und elektrophoretische Charakterisierung von Isoenzymen

Zusammenfassung: Die Glycerinaldehydphosphat-Dehydrogenase (NADP$^{\oplus}$) aus Extrakten Weißlicht-gewachsener *Sinapis-alba*-Keimlinge wurde in die zwei Ammoniumsulfatfraktionen 55 bis 70% und 70 bis 95% Sättigung zerlegt, anschließend für jede Fraktion getrennt bis zur Homogenität gereinigt und elektrophoretisch charakterisiert. Jede Fraktion enthält eines von zwei verschiedenen Isoenzymen (M_r 170000), deren isoelektrische pH-Werte 4.4 und 4.6 betragen und die sich in der Untereinheitenzusammensetzung unterscheiden. Letztere entspricht vermutlich den Typen A_2B_2 (Frakt. 55 - 70%) und A_4 (Frakt. 70 - 95%), wobei die Untereinheiten A (M_r 39000) und B (M_r 41000) durch verschiedene Molekulargewichte charakterisiert sind.

Key words: Chloroplast glyceraldehyde-3-phosphate dehydrogenase (NADP$^{\oplus}$); quaternary structure; *Sinapis alba* L.

In two preceding investigations[1,2] it was demonstrated that glyceraldehyde-phosphate dehydrogenase (NADP$^{\oplus}$) from *Sinapis alba* L. exists in an abnormal conformation in the presence of NAD$^{\oplus}$, which associates with a separate binding fraction and/or interacts with Sepharose 6 B, leading to variable elution profiles during gel filtrations. While most previous reports on enzyme multiplicity[1-8,13] may be interpreted in terms of different conformational and/or aggregational states of one and the same enzyme protomer, no classical isoenzymes have so far been described, that contain a constant number of different subunits arranged in different combinations. The present paper describes the isolation and electrophoretic characterization of such isoenzymes of glyceraldehyde-phosphate dehydrogenase (NADP$^{\oplus}$) from *Sinapis alba* L.

Material and Methods

Seeds of *Sinapis alba* L. were purchased from Asgrow Company (Hamburg, Germany). Seedlings were grown for 7 days under continuous fluorescent white light on soil in a growth chamber.

The following experimental procedures have been described elsewhere[1,2]: a) extraction procedure; b) ammonium sulfate fractionation of the extract; c) gel filtration, affinity chromatography, and column isoelectric focusing of the partially purified enzyme; d) dodecylsulfate/discontinuous electrophoresis of the enzyme on gradient polyacrylamide slab gels; e) enzyme assay. In the present investigation the extraction and gel filtration media contained 25 μg of the protease inhibitor phenyl-

Enzyme:
Glyceraldehyde-phosphate dehydrogenase (NADP$^{\oplus}$) (phosphorylating), D-glyceraldehyde-3-phosphate:NADP$^{\oplus}$ oxidoreductase (phosphorylating) (EC 1.2.1.13).

methanesulfonyl fluoride per m*l* and 0.2% (w/w) sodium azide, in addition to the other substances.

Polyacrylamide slab gel electrophoresis under non-denaturing conditions

Electrophoresis of the native isoenzymes (Fig. 2a) was performed at 100 V for 3 h and 30 min in a cold room (+ 4 °C) using a continuous gel system (5% acrylamide) with 0.1M potassium phosphate buffer (pH 6.8) containing 1mM EDTA.

Carbohydrate determination by gas-liquid chromatography

The purified enzyme was dialysed against bi-distilled water for 48 h and then lyophilized. The lyophilized enzyme (2 mg) was hydrolysed according to Lehnhardt and Winzler[9] and the sugars were determined by the method of Sawardeker et al.[10]. The sugar determination was done by *Dr. I. Fromme*, Max-Planck-Institut für Immunbiologie, Freiburg.

Results

An extract from white light-grown *Sinapis alba* seedlings was separated into three ammonium sulfate fractions of 40 to 55%, 55 to 70%, and 70 to 95% saturation, respectively (table). The 40 to 55% fraction, containing less than 5% of the total recovered activity of glyceraldehyde-3-phosphate dehydrogenase (NADP$^{\oplus}$), was discarded. The other two enzyme fractions were purified separately to homogeneity by exploiting the differential transport behaviour of the enzyme during filtration with NAD$^{\oplus}$ and NADP$^{\oplus}$ on Sephadex G-200, as shown in Fig. 1 and the table, and analysed in detail previously[1,2]. Conventional polyacrylamide gel electrophoresis of the two filtered enzyme fractions (Fig. 2a) demonstrates that the enzyme is pure after this treatment and that each fraction contains one of two distinct isoenzymes: an isoenzyme showing fast migration (fractions 55 to 70%) and an isoenzyme that migrates more slowly (70 to 95%). Both isoenzymes are active with NADP *and* NAD and they have similar molecular weights ($\approx$ 170000) in the presence of NADP$^{\oplus}$, judging from the position of the marker lactate dehydrogenase (M_r 145000) in Fig. 1b and d (see also ref. 1). Discontinuous polyacrylamide gel electrophoresis of the two isoenzymes in the presence of dodecylsulfate shows that the two species differ in subunit composition. Whereas the "fast" species (Fig. 2b, channel 1) contains two separate polypeptide chains with apparent M_r values 39000 and 41000, the "slow" species (Fig. 2b, channel 2) is composed of a single type of subunit with M_r 39000. Affinity chromatography of the "fast" species on NAD-Sepharose (for experimental details see ref.[11]) has no

Table. Purification scheme of glyceraldehyde-3-phosphate dehydrogenase (NADP$^{\oplus}$) from extracts of white light-grown *Sinapis alba* seedlings.

400 g fresh weight of seedlings was used as starting material. The assay mixture contained in 1.05 m*l* total volume: 100 μmol Tris/Cl (pH 7.8), 4.5 μmol 3-phosphoglycerate, 2 μmol ATP, 2 μmol dithioerythritol, 1 μmol EDTA, 8 μmol $MgSO_4$, and 1.8 U (4 μg) 3-phosphoglycerate kinase. After addition of NADP-linked glyceraldehyde-3-phosphate dehydrogenase the reaction was initiated with 0.32 μmol NADPH. The temperature was 25 °C. 1 U activity is defined as the amount which will catalyze the formation of 1 μmol of product per min under these conditions.

	Enzyme act. [U]	Protein [mg]	Spec. act. [U/mg]
Extract (1800 m*l*)	5500	3063	1.8
Ammonium sulfate (40 to 55%)	130	710	0.2
Ammonium sulfate (55 to 70%)	2970	1130	2.6
NAD filtration (Fig. 1a)	1170	114	10.3
NADP filtration (Fig. 1b)	790	6.5	122
Ammonium sulfate (70 to 95%)	1030	230	4.5
NAD filtration (Fig. 1c)	523	5.8	90
NADP filtration (Fig. 1d)	470	5.5	85

influence on the band pattern (Fig. 2b, channel 3), indicating that both polypeptide chains are true enzyme subunits. In another experiment column isoelectric focusing was performed with the two isoenzymes in the presence of NADP$^{\oplus}$ (for details see ref.[2]). The isoelectric pH values were found to be 4.4 and 4.6 for the "fast" and "slow" species respectively.

The migration rate of a protein in dodecylsulfate gels may be influenced by its carbohydrate content[12]. To exclude this possibility for the present enzyme, the "fast" species after affinity chromatography was submitted to a sugar analysis by gas liquid chromatography (see under "Material and Methods"). The result was negative, that is the determined sugar content of the enzyme sample was less than 0.5% (w/w).

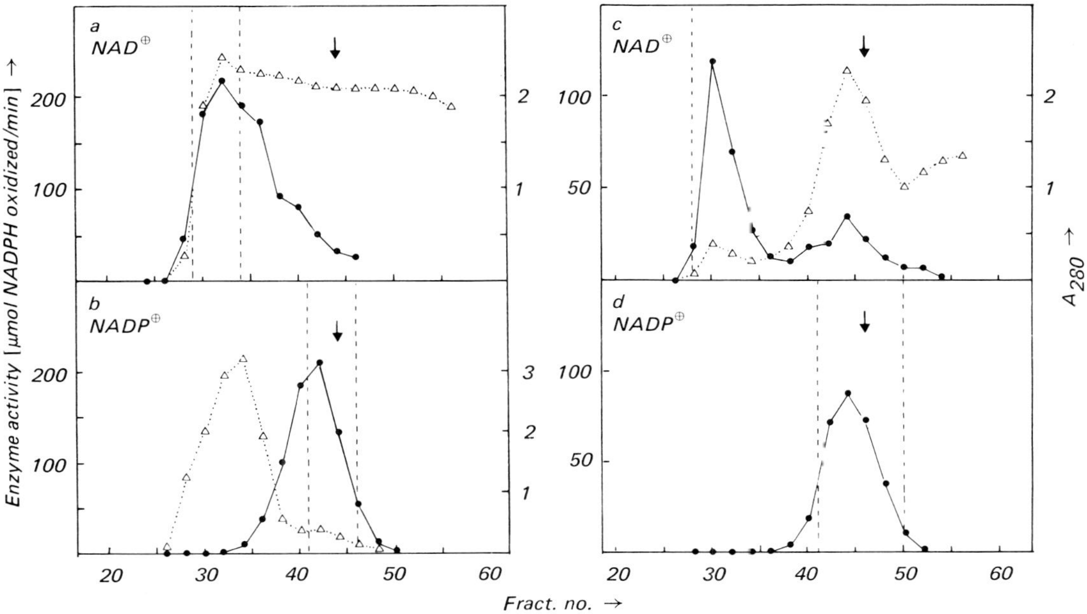

Fig. 1. Isolation of glyceraldehyde-3-phosphate dehydrogenase (NADP$^\oplus$) from ammonium sulfate fractions 55 to 70% (a, b) and 70 to 95% (c, d) by consecutive filtrations with NAD$^\oplus$ and NADP$^\oplus$ on Sephadex G-200.

The ammonium sulfate fractions (Table 1) were first filtered with 0.5mM NAD$^\oplus$ (a and c). The enzyme fractions enclosed by the vertical dashed lines in a and c were pooled, concentrated, and refiltered with 0.05mM NADP$^\oplus$ on the same column (b and d). The pure enzyme (fractions enclosed by the vertical dashed lines in b and d) was recovered and analysed electrophoretically. Column size: 2.6 x 100 cm. Filtration buffer: 0.1M Tris/Cl (pH 7.4), containing 2mM dithioerythritol and 1mM EDTA in addition to NAD$^\oplus$ and NADP$^\oplus$. Fraction volume: 6 m*l*. ●——●, enzyme activity (NADPH oxidation); △.....△, absorbance profile (280 nm). Arrow indicates the position of the marker lactate dehydrogenase (M_r 145 000) which was determined in a separate filtration. Recoveries of glyceraldehyde-3-phosphate dehydrogenase (NADP$^\oplus$) activity (a-d): 73, 93, 70, 102%.

Discussion

The present results demonstrate that glyceraldehyde-3-phosphate dehydrogenase (NADP$^\oplus$) from *Sinapis alba* L. exists as two major isoenzymes, both of which are active with both NADP and NAD and which can be separated by ammonium sulfate precipitation. The molecular weights of the native isoenzymes in the presence of NADP$^\oplus$ ($\approx$ 170000, Fig. 1b and d, ref. 1) and the molecular weights of the subunits (39000 and 41000, Fig. 2b) suggest that both species are composed of 4 subunits. The "slow" species appears to be a homotetramer of the type A_4, as is the NAD-specific enzyme (EC 1.2.1.12). It resembles the NAD-specific enzyme also in its solubility at high ammonium sulfate concentrations (both enzymes precipitate above 70% saturation, see the table and ref.[1]. However it differs from the latter in subunit M_r and in its tendency to associate in the presence of NAD$^\oplus$[1]. The subunit pattern of the "fast" species indicates a heterooligomer. The relative proportion of the 39000 subunit seems somewhat higher than that of the 41000 polypeptide chain. Taking into account a certain contamination of the 55 to 70% ammonium sulfate fraction with the A_4 species, a subunit composition A_2B_2 may be suggested for the "fast" species. Pawlitzki and Latzko[5] separated the spinach enzyme into a high and a low molecular weight form showing different relative proportions of the two subunits (1 : 1 and > 2 : 1). From the present results it may be inferred that this separation coincided with a partial separation of the two isoenzymes. Anderson and Lim[13] obtained two band zymograms of the NADP enzyme from peas after electrophoresis of crude extracts and after staining of the enzyme with tetrazolium salts. It seems likely from the present results that the two bands represent two different isoenzymes rather

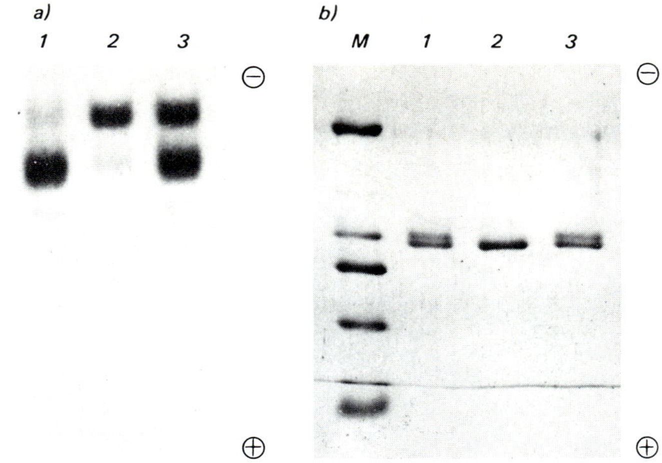

Fig. 2. a) Polyacrylamide gel electrophoresis of the isoenzymes under non-denaturing conditions (see "Material and Methods").
Channel 1: "fast" species (55 to 70% ammonium sulfate fraction); channel 2: "slow" species (70 to 95% fraction); channel 3: mixture of both isoenzymes.
b) Discontinuous polyacrylamide gel electrophoresis of the isoenzymes in the presence of dodecylsulfate.
Channel 1: "fast" species; channel 2: "slow" species; channel 3: "fast" species after affinity chromatography on NAD-Sepharose; channel M: marker proteins (from top to bottom): bovine serum albumin, M_r 68000; aldolase, M_r 40000; glyceraldehyde-3-phosphate dehydrogenase (NAD$^{\oplus}$) from muscle, M_r 36000; chymotrypsinogen, M_r 25000; cytochrome *c*, M_r 12500. Dodecylsulfate electrophoresis was performed on gradient polyacrylamide slab gels (7.5 to 18% acrylamide) as described by Studier[11] (see also ref.[1]). Proteins in a and b were stained with Coomassie Brilliant Blue.

than two different conformations of a single enzyme species as assumed by Anderson and Lim.

In previous studies[1,2] with seedlings of a different batch of *Sinapis alba* seeds a 3 band subunit pattern (apparent M_r values 39000, 42000, 43000) of the enzyme was obtained. This can possibly be explained by polymorphism of the B-subunit.

The authors thank *Professor H. Mohr* for research facilities, *Dr. I. Fromme* (Max-Planck-Institut für Immunbiologie, Freiburg) for the sugar determinations, and *Professor J. W. Bradbeer* (University of London King's College) for his critical comments on the manuscript. This work was supported by the *Deutsche Forschungsgemeinschaft* (Forschungsstipendium and SFB 46).

Literature

1 Cerff, R. (1978) *Eur. J. Biochem.* **82**, 45 - 53.
2 Cerff, R. (1978) *Plant Physiol.* **61**, 369 - 372.
3 Pupillo, P. & Giuliani-Piccari, G. (1973) *Arch. Biochem. Biophys.* **154**, 324 - 331.
4 McGowan, R. E. & Gibbs, M. (1974) *Plant Physiol.* **54**, 312 - 319.
5 Pawlitzki, K. & Latzko, E. (1974) *FEBS Lett.* **42**, 285 - 288.
6 Bradbeer, J. W. (1976) in Perspectives in Experimental Biology (N. Sunderland, ed.,) vol. 2 Botany, pp. 131 - 143, Pergamon Press Oxford and New York.
7 O'Brien, M. J. & Powls, R. (1976) *Eur. J. Biochem.* **63**, 155 - 161.
8 Wolosiuk, R. A. & Buchanan, B. B. (1976) *J. Biol. Chem.* **251**, 6456 - 6461.
9 Lehnhardt, W. F. & Winzler, R. (1968) *J. Chromatogr.* **34**, 471 - 479.
10 Sawardeker, J. S., Sloneker, J. H. & Jeanes, A. (1967) *Anal. Chem.* **37**, 1602 - 1604.
11 Studier, F. W. (1973) *J. Mol. Biol.* **79**, 237 - 248.
12 Segrest, J. P. & Jackson, R. L. (1972) *Methods Enzymol.* **XXVII** *Part B*, 54 - 63.
13 Anderson, L. E. & Lim, T.-C. (1972) *FEBS Lett.* **27**, 189 - 191.

Dr. Rüdiger Cerff* and Susan E. Chambers (MSc), Biologisches Institut II, Universität Freiburg i. Br., Schänzlestr. 1, D-7800 Freiburg.

* To whom correspondence should be addressed.

Hoppe-Seyler's Z. Physiol. Chem.
Bd. 359, S. 773 - 775, Juni 1978

SHORT COMMUNICATION

Maintenance and Regeneration in vitro of Bulk-Isolated Cortical Neurones from Adult Rats

Hans-Hinrich ALTHAUS[a], Volker NEUHOFF[a], Wieland B. HUTTNER[b], Rachel MONZAIN[c] and Abraham SHAHAR[c]

[a] Max-Planck-Institut für Experimentelle Medizin, Forschungsstelle Neurochemie, Göttingen
[b] Yale University School of Medicine, New Haven, Conn.
[c] Israel Institute for Biological Research, Ness-Ziona.

(Received 20 March 1978)

Erhaltung und Regeneration von präparativ gewonnenen Hirnnervenzellen erwachsener Ratten in vitro

Zusammenfassung: Ob Hirnnervenzellen erwachsener Ratten, die nach einer neuen Methode präparativ gewonnen wurden, als lebensfähig bezeichnet werden können, wird in ihrem Verhalten unter Kulturbedingungen untersucht. Polylysin-beschichtete Petrischalen oder Kollagen-beschichtete Maximow-Kammern wurden mit Nervenzellen beschickt. Nach einer „Ruhephase", in der die Nervenzellen morphologisch intakt bleiben, kommt es bei Nervenzellen der Hirn-Zellsuspension zu einem Auswachsen ihrer Fortsätze. Dieses läßt sich kinematographisch dokumentieren. Bei Gradienten-gereinigten Nervenzellen konnte bisher kein Aussprossen beobachtet werden.

Key words: Bulk isolation, neurones, culture, regeneration.

The viability of bulk-prepared nerve cells is difficult to estimate. Metabolic studies performed immediately after the isolation procedure reflect the viability of the nerve cells only to some extent since the measured activity might be a residual one. A statement about their viability might be more meaningful if the cells could be established in culture for several days. An attempt in this direction has recently been made in which neuronal perikarya isolated from 10 - 20 day-old rat brains were maintained for up to 24 h[1].

The bulk preparation method previously described[2] allows the isolation of well preserved cortical nerve cells from adult rats (150 - 180 g), and it was of interest to test their vitality by subjecting them to culture conditions. The culturing procedure followed techniques previously described[3,4].

Portions of the initial brain cell suspension (minced brain tissue after perfusion with a medium containing hyaluronidase/collagenase) or of the purified neuronal fraction[2] were placed either in polylysine-coated petri dishes or on collagen-coated coverslips in Maximow chambers. Different media such as Dulbecco's modified Eagle's medium or modified Ham's F 12* (containing 24mM potassium) were used. All media contained different amounts of horse serum (5-20%) and 0.1% (w/v) gentamycin. The cells were incubated at 36 °C and observed daily. Within the first 12 h most of the nerve cells settled and adhered to the coating material (Fig. 1a - c). Some of the isolated nerve cells (containing a granular cytoplasm) were obviously suffering, although a few of these seemed to recover after this period. A few neurones did not attach, disintegrated and died.

During the following 7 days the nerve cells remained morphologically intact and no fiber regeneration was oberserved. After this period (8 days and on) some of the nerve cells began to regenerate. A few neurones became surrounded by a halo of very fine nerve fibers radiating from the pre-existing apical dendrite or from the cell body (Fig. 2a - c). These processes either grew and branched or retracted (Fig. 3). Concomitantly with nerve fiber regeneration, growth and multiplication of glial elements and fibroblasts was observed.

a) b) c)

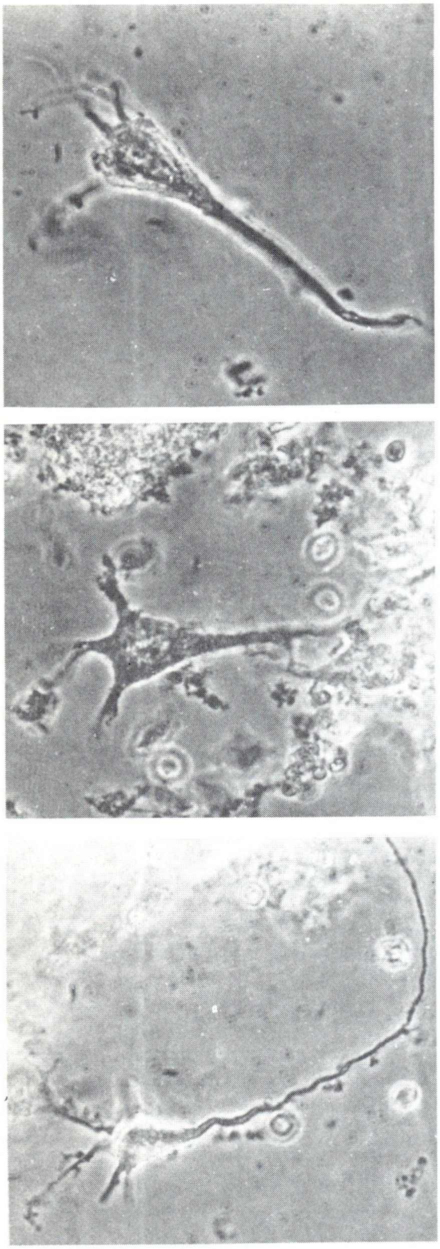

△

Fig. 1a - c. Phase-contrast photomicrographs of nerve cells from the initial cell suspension.
1 h in culture. Magnif.: × 480.

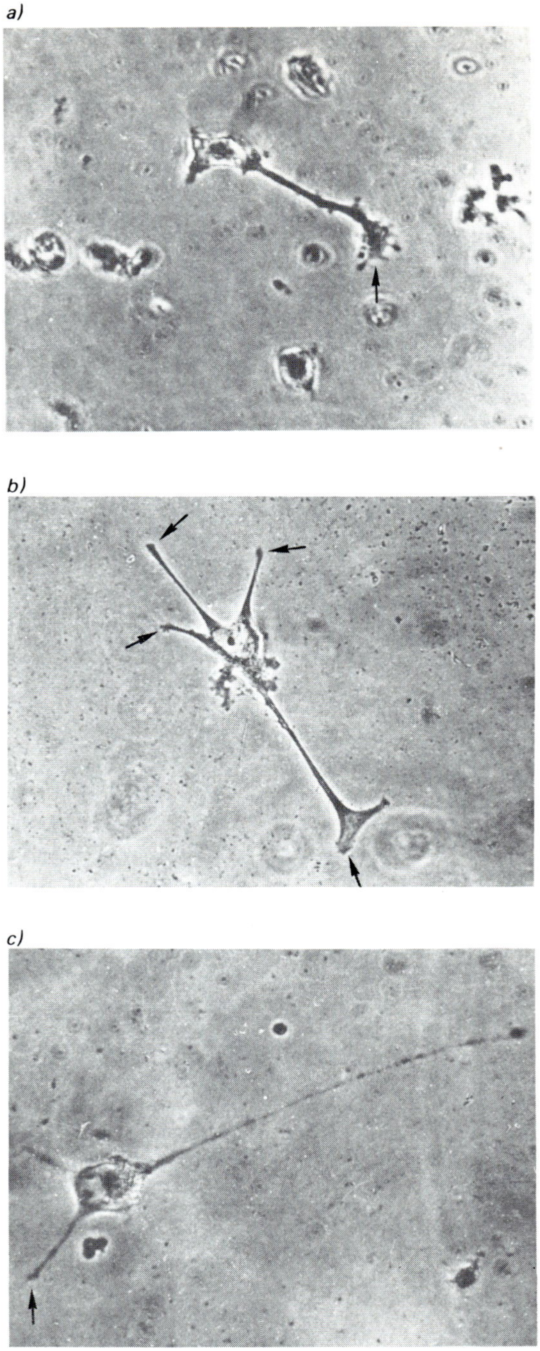

◁

Fig. 2a - c. Phase-contrast photomicrographs showing various types of nerve fiber growth in neurones after one week in vitro.

Neurones were cultivated in Petri dishes on polylysine substrate in Ham's F 12 modified medium. Arrows indicate growth cones. Magnif.: × 255.

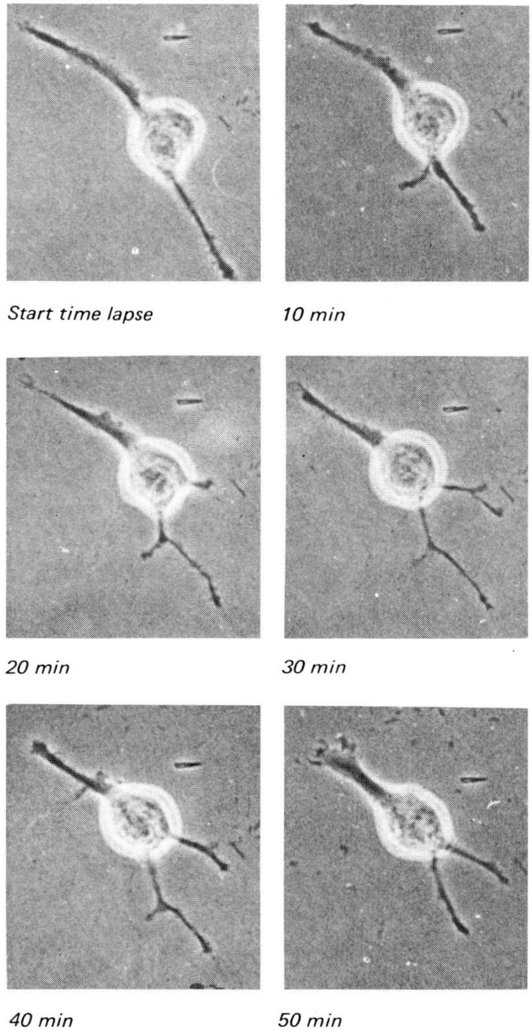

Fig. 3. A cinematographic analysis of fiber regeneration in an isolated neurone from adult rat brain 12 days in vitro.

Magnif.: × 330.

Growing cells were obtained so far from cultures of the initial brain cell suspension. Several attempts to establish cultures of gradient purified neurones were unsuccessful because of bacterial contamination. The problem of how to obtain sterile cultures without the use of high doses of antibiotics is now under investigation.

In summary, these preliminary results demonstrate that differentiated neurones from adult rat brain can be cultured for two weeks.

It should be noted that while explanted embryonic neurones begin to regenerate their fibers within the first day[5] in vitro, the outgrowth of neuronal process in adult differentiated neurones is delayed for more than a week.

Experiments are now in progress to quantify the growth and establishment of regenerating neurones, to determine their metabolic behaviour and to examine the role of various factors necessary for nerve cell growth in vitro.

Literature

1 Poduslo, E.S. & McKhann, G.M. (1976) *Neurosci. Lett.* **2**, 267 - 271.
2 Althaus, H.H., Huttner, W.B. & Neuhoff, V. (1977) *Hoppe-Seyler's Z. Physiol. Chem.* **358**, 1155 - 1159.
3 Shahar, A., Grundfeld, Y., Spiegelstein, M. & Monzain, R. (1975) *Brain Res.* 88, 44 - 51.
4 Lasher, R.S. & Zagon, I.S. (1972) *Brain Res.* **41**, 482 - 488.
5 Shahar, A. & Saar, M. (1970) *Brain Res.* **23**, 315 - 321.

Dr. H.H. Althaus and Prof. Dr. V. Neuhoff, Max-Planck-Institut für Experimentelle Medizin, Hermann-Rein-Str. 3, D-3400 Göttingen;

Dr. W.B. Huttner, Yale University School of Medicine, New Haven, Conn. 06510, U.S.A.;

R. Monzain and Dr. A. Shahar, Israel Institute for Biological Research, Ness-Ziona, Israel.

Einem Teil dieser Auflage liegt ein Prospekt über neue Bücher auf dem Gebiet der Biochemie bei, den wir zu beachten bitten.

Titel Nr. 3 109 002 693

Anzeigenverwaltung: Merkur Werbung GmbH, Postfach 12 45, 5210 Troisdorf 1, Tel. (0 22 41) 4 20 51.
Für den Anzeigenteil verantwortlich: Wolfgang Frieben, Troisdorf. – Printed in Germany – Satz: R. & J. Blank, München; Druck: Grafik + Druck, München; Bindearbeiten: F. Steinberger & Sohn, München.

Redaktionelle Hinweise

(Wichtige Änderungen ab Januar 1976 sind am Rand markiert)

Die Zeitschrift bringt nur Experimantalarbeiten aus den Gebieten der Biochemie und Physiologischen Chemie. Die Manuskripte werden in deutscher oder in englischer Sprache veröffentlicht. Es wird vorausgesetzt, daß alle Koautoren mit der eingereichten Fassung einverstanden sind.

Aktuelle Kurzmitteilungen, deren Umfang 4 Seiten in Kleindruck (ca. 8 Manuskriptseiten) nicht übersteigen darf, werden bevorzugt bearbeitet und in das nächsterreichbare Heft aufgenommen. Die Dringlichkeit zur Prioritätssicherung ist vom Autor in einem Begleitschreiben kurz zu begründen.

Manuskripte sind in zweifacher Ausfertigung an eine der folgenden Anschriften zu senden:

Prof. Dr. A. Butenandt oder Prof. Dr. F. Lynen, Max-Planck-Institut für Biochemie, D-8033 Martinsried b. München;

Prof. Dr. Dr. G. Weitzel, Physiologisch-Chemisches Institut der Universität, Hoppe-Seyler-Str. 1, D-7400 Tübingen;

Dr. A. Dillmann, Hoppe-Seyler-Redaktion, Am Klopferspitz, D-8033 Martinsried b. München, Tel. (089) 85 85-823.

Bei der Abfassung der Manuskripte bitten wir dringend, folgende Regeln zu beachten, die im wesentlichen mit den „Instructions to Authors" von *Biochimica et Biophysica Acta* und *European Journal of Biochemistry* übereinstimmen:

Es soll in Maschinenschrift (2zeilig, auch vorgesehener Kleindruck!) und mit 4 cm breitem Rand geschrieben werden. Vorgesehene Absätze sind deutlich zu markieren.

Seite 1 des Manuskripts soll enthalten (in normaler Schrift, nicht in Großbuchstaben):

a) Titel der Arbeit
b) Autorennamen (jeweils ein Vorname ausgeschrieben)
c) Name des Instituts oder Fachbereichs, in dem die Arbeit ausgeführt wurde (möglichst in der Sprache des Herkunftlandes), vorausgesetzt, daß die Leitung des Instituts oder Fachbereichs der Veröffentlichung zugestimmt hat.
d) Kurztitel, maximal 70 Satzzeichen
e) Adresse für den Versand der Korrekturen

Auf Seite 2 folgen

a) Eventuelle Widmung
b) Vollständige Postadresse der Autoren in der Sprache des Herkunftslandes
c) Ad-hoc-Abkürzungen (nur in seltenen Fällen zugelassen)
d) Enzyme in alphabetischer Reihenfolge der empfohlenen Trivialnamen mit systematischen Namen und EC-Nummern

Auf Seite 3 folgt die *Zusammenfassung* sowie eine Übersetzung des Titels und der (eventuell erweiterten) Zusammenfassung ins Englische (bzw. Deutsche). Diese Übersetzung oder deren sprachliche Überprüfung kann auch von der Redaktion übernommen werden.

Eventuelle *Danksagungen* sind an den Schluß des Manuskriptes zu stellen.

Literaturzitate im Text sind fortlaufend zu numerieren (nicht alphabetisch) und am Schluß des Manuskripts zu bringen. Die in Chemical/Biological Abstracts üblichen Zeitschriftenabkürzungen sind zu verwenden. Zur Anordnung s. IUB Commission of Editors of Biochemical Journals (1973) *Eur. J. Biochem.* **37**, 201 - 202; (1974) *diese Z.* **355**, 1 - 3.

Beispiel: Spackman, D.H., Stein, W.H. & Moore. S. (1958) *Anal. Chem.* **30**, 1190 - 1206.

Zeichnungen sollen einmal in reproduzierbarer Form mit schwarzer Tusche gezeichnet eingereicht werden; Zahlenskalen, gemessene Größen, Einheiten und sonstige Beschriftungen sind jedoch nur leicht mit weichem Bleistift einzutragen.

Die Abbildungen sollen so groß sein, daß sie auf 1/2 oder 1/3 verkleinert werden können. Deutlich beschriftete Kopien für die Referenten sind beizufügen.

Alle *Tabellen* und *Abbildungen* sind mit Legenden zu versehen, so daß die Ergebnisse auch ohne Zuhilfenahme des Textes ersichtlich werden. Es wird gebeten, die Legenden nicht zu unterstreichen und die Abbildungslegenden nicht auf die Abbildungen selbst, sondern auf ein eigenes Blatt zu schreiben.

Symbole für die am häufigsten vorkommenden Größen und ihre Einheiten sind in Tab. 1 zusammengestellt. Wir bitten streng zu unterscheiden, ob bei Ansätzen molare Konzentrationen (mol/l = M; z.B. 0.25M) oder zugesetzte Mengen (z. B. in mol, mmol, µmol oder nmol) gemeint sind.

Hinsichtlich der *Nomenklatur* richte man sich nach der IUPAC-Nomenklatur sowie nach den Empfehlungen der "IUPAC-IUB Commission on Biochemical Nomenclature": Kohlenhydrate, *Eur. J. Biochem.* **21**, 455 - 477 (1971); **25**, 4 (1972); α-Aminosäuren, *Eur. J. Biochem.* **53**, 1 - 14 (1975);

Tab. 1. Symbole für Größen und ihre Einheiten

Größe		Einheit[a]
l	Länge	m, mm, µm, nm, etc.
V	Volumen	m^3, l (dm^3), ml, etc.
λ	Wellenlänge	nm (nicht mµ oder Å)
t	Zeit	h, min, s, ms, µs, etc.
ν	Frequenz	Hz
m	Masse	kg, g, mg, µg, etc.
ρ	Dichte	g/cm^3
s	Sedimentations-koeffizient	$S = 10^{-13}$ s
n	Substanzmenge	mol, mmol, etc.
M	molare Masse (m/n)	g/mol
–	Masse eines Moleküls	Dalton[b]
M_r	Molekulargewicht	(dimensionslos)
c_B, [B]	molare Konzentration einer Substanz B	M, mM, µM, etc. (besser als 10^{-3}M, etc.)
T	Temperatur (absolut)	K (nicht °K)
t, θ	Temperatur	°C
q, Q	Wärmemenge	J (besser als cal)[c]
I	Ionenstärke	M (mol/l), mM, etc.
K	Gleichgewichtskonstante	
K_m	Michaeliskonstante	
K_i	Inhibitorkonstante	
k	Geschwindigkeits-konstante	
v	Reaktionsgeschwindig-keit	mol/s (oder mmol, µmol oder nmol/s, etc.), besser als µmol/min
V (nicht V_{max})	Geschwindigkeit der enzymkatalysierten Reaktion bei unendlich großer Substrat-konzentration	
I	Lichtintensität	cd
T	Durchlässigkeit (I/I_0)	(dimensionslos)
A	Absorption[d] (– lg T)	(dimensionslos)
ϵ	molarer Extinktions-koeffizient	$M^{-1} \times cm^{-1}$

[a] Präfixe für Einheiten: T, Tera (10^{12}); G, Giga (10^9); M, Mega (10^6), k, Kilo (10^3); m, Milli (10^{-3}); µ, Mikro (10^{-6}); n, Nano (10^{-9}); p, Piko (10^{-12}); f, Femto (10^{-15}); a, Atto (10^{-18}).

[b] Das Dalton ist 1/12 der Masse eines Atoms des Nuklids ^{12}C.

[c] Das Joule ist die SI-Einheit der Energie. Die Kalorie (1 cal = 4.184 J; 1 J = 0.239 cal) ist für eine Übergangszeit noch zugelassen; in jedem Fall sollten aber die Angaben auch in Joule ausgedrückt werden.

[d] Die Ausdrücke optische Dichte, Extinktion oder Absorbancy sollen nicht mehr verwendet werden.

Cyclite, *diese Z.* **350**, 523 - 535 (1969), *Eur. J. Biochem.* **57**, 1 - 7 (1975); Lipide, *diese Z.* **350**, 279 - 285 (1969); Steroide, *diese Z.* **351**, 663 - 689 (1970); Chinone mit isoprenoiden Seitenketten, Folsäure u. verschiedene andere Verbindungen, *diese Z.* **348**, 266 - 276 (1967); Corrinoide, *Eur. J. Biochem.* **45**, 7 - 12 (1974); Vitamine B_6, *Eur. J. Biochem.* **40**, 325 - 327 (1973); Carotinoide, *Eur. J. Biochem.* **25**, 397 - 408 (1972), **57**, 317 - 318 (1975); synthetische Analoga natürlicher Peptide, *diese Z.* **348**, 262 - 265 (1967); Enzyme Nomenclature, Recommendations (1972), Elsevier Publ. Comp., Amsterdam; multiple Enzymformen, *diese Z.* **353**, 852 - 854 (1972); Eisen-Schwefel-Proteine, *Eur. J. Biochem.* **35**, 1 - 2 (1973); Human-Immunglobuline, *Eur. J. Biochem.* **45**, 5 - 6 (1974); Peptidhormone, *Eur. J. Biochem.* **55**, 485 - 486 (1975). Für die Rechtschreibung ist „Jansen/Mackensen, Rechtschreibung der technischen und chemischen Fremdwörter" maßgebend.

Tab. 2 Empfohlene Abkürzungen.

Atome, isotope	=	z. B. [16-^{14}C]Östron
Atome, numeriert	=	z. B. C-5
Curie	=	Ci
Impulse pro Minute	=	Ipm (engl. cpm)
Prozent der Theorie	=	% d. Th.
prozentig	=	proz. (nicht %-ig)
Relative Wanderungsgeschwindigkeit	=	R_F
Schmelzpunkt	=	Schmp. (engl. m.p.)
Siedepunkt	=	Sdp. (engl. b.p.)
Umdrehungen pro Minute	=	U./min (engl. rpm)
Zerfälle pro Minute	=	Zpm (engl. dpm)
Zersetzung	=	Zers. (engl. decomp.)

Bezüglich der *Abkürzungen und Symbole für chemische Namen,* die in der biologischen Chemie von besonderem Interesse sind, wird auf die Empfehlungen der „IUPAC-IUB Commission on Biochemical Nomenclature" verwiesen (siehe z. B.: Allgemeines, Aminosäuren, Zucker, Phosphate, Coenzyme, *diese Z.* **348**, 245 - 265 (1967); Aminosäurederivate und Peptide, *Eur. J. Biochem.* **27**, 201 - 207 (1972); synthet. Polypeptide, *diese Z.* **349**, 1013 - 1016 (1968); Nucleinsäuren, Polynucleotide und deren Bestandteile, *diese Z.* **351**, 1055 - 1063 (1970). Andere als die dort empfohlenen Abkürzungen werden nur in seltenen Ausnahmefällen zugelassen. Sie müssen am Beginn jeder Arbeit in einer Fußnote erklärt werden.

Sonderdrucke: Von jeder Arbeit werden dem Verfasser 75 Sonderdrucke kostenlos geliefert. Die Bestellung weiterer Sonderdrucke gegen Berechnung ist möglich, muß jedoch spätestens bei der Rücksendung der Korrekturen erfolgen.

Der Verlag erwirbt mit der Annahme des Manuskriptes das Verlagsrecht und das ausschließliche Recht der Vervielfältigung, Verbreitung und Übersetzung. Der Verlag behält sich vor, den Inhalt dieser Zeitschrift für fremdsprachige Ausgaben zu verwenden.

Bezugspreis: Der Jahresbezugspreis (1 Jahrg. zu 12 Heften) beträgt DM 590,—, das Einzelheft kostet DM 56,—. Bestellungen nimmt jede Buchhandlung entgegen.

Verlag: WALTER DE GRUYTER & Co.
Genthiner Str. 13, D-1 Berlin 30. Tel. (030) 2611341
Telegrammadresse: Wissenschaft Berlin
Postscheck: Berlin-West 10307-108

WALTER DE GRUYTER Inc.
3 Westchester Plaza
Elmsford, N.Y. 10523, U.S.A.
Tel. (914) 592-5890

Für die nächsten Hefte sind folgende Beiträge vorgesehen:

The granulocyte chalone. Partial purification and specificity testing by differentiating assays, *R. Maschler and H. R. Maurer*

Uptake of thymidine into isolated rat hepatocytes. Evidences for two transport systems, *F. R. Ungemach and D. Hegner*

Fructose, a substrate for energy yielding metabolism in the heart?, *R. Bünger, O. Sommer, G. Walter and R. Scholz*

Structural properties of the human MN blood group antigen receptor sites, *W. Dahr and G. Uhlenbruck*

Enzymatic hydrolysis of sulpho-sphingolipids and -glycerolipids by sulphatase A in the presence and absence of the activator protein, *G. Fischer, S. Reiter and H. Jatzkewitz*

Subcellular distribution of [^{14}C]5-hydroxytryptamine in the blood platelets of rabbits: effects of imipramine and haloperidol, *L. Hussein, H. W. Goedde, M. Rosenfeld and E. Freerksen*

Effect of organic anions on bile acid uptake by isolated rat hepatocytes, *M. S. Anwer and D. Hegner*

Distribution of metabolites between the cytosolic and mitochondrial compartments of hepatocytes isolated from fed rats, *E. A. Siess, D. G. Brocks and O. H. Wieland*

Lipids with photosensitive groups as chemical probes for the regio-analysis of biological membranes. On the localization of the G- and M-protein of vesicular stomatitis virus, *W. Stoffel, Ch. Schreiber and H. Scheefers*

Pentagastrin analogs containing α-aminooxy acids, I. Substitution at *N*-terminus, *L. Kisfaludy, I. Schön, J. Náfrádi, L. Varga and V. Varró*

Pentagastrin analogs containing α-aminooxy acids, II. L- and D-phenylglycine analogs, *I. Schön, L. Kisfaludy, J. Náfrádi, L. Varga, V. Varró*

Pentagastrin analogs containing α-aminooxy acids, III. Modifications at the *C*-terminus, *I. Schön, L. Kisfaludy, J. Náfrádi, L. Varga and V. Varró*

Nebenreaktionen bei Peptidsynthesen, III. Synthese und Charakterisierung von N^{in}-tert.-butylierten Tryptophan-Derivaten, *E. Jaeger, P. Thamm, S. Knof, E. Wünsch, M. Löw und L. Kisfaludy*

Nebenreaktionen bei Peptidsynthesen, IV. Charakterisierung von *C*- und *C,N*-tert.-butylierten Tryptophan-Derivaten, *E. Jaeger, P. Thamm, S. Knof und E. Wünsch*

Direkte tert.-Butylierung des Tryptophans: Herstellung von 2,5,7-Tri-tert-butyl-tryptophan, *M. Löw, L. Kisfaludy, E. Jaeger, P. Thamm, S. Knof und E. Wünsch*

tert.-Butylierung des Tryptophan-Indolringes während der Abspaltung der tert.-Butyloxycarbonyl-Gruppe bei Peptidsynthesen, *M. Löw, L. Kisfaludy und P. Sohár*

Zur Synthese von [Trp(1-Bu^t)^9]- und [Trp(2,5,7-Bu$_3^t$)^9]Corticotropin-(1–19)-nonadecapeptidamid, *N. Löw, L. Kisfaludy und M. Sarközi*

Untersuchungen über den enzymatischen Abbau von tert.-Butyl-tryptophan-haltigen Peptiden, *M. Sajgó, M. Löw und L. Kisfaludy*

Untersuchungen über den N^{in}-Formylrest als mögliche Schutzgruppe gegen Tryptophan-tert.-butylierung, *M. Löw und L. Kisfaludy*

A